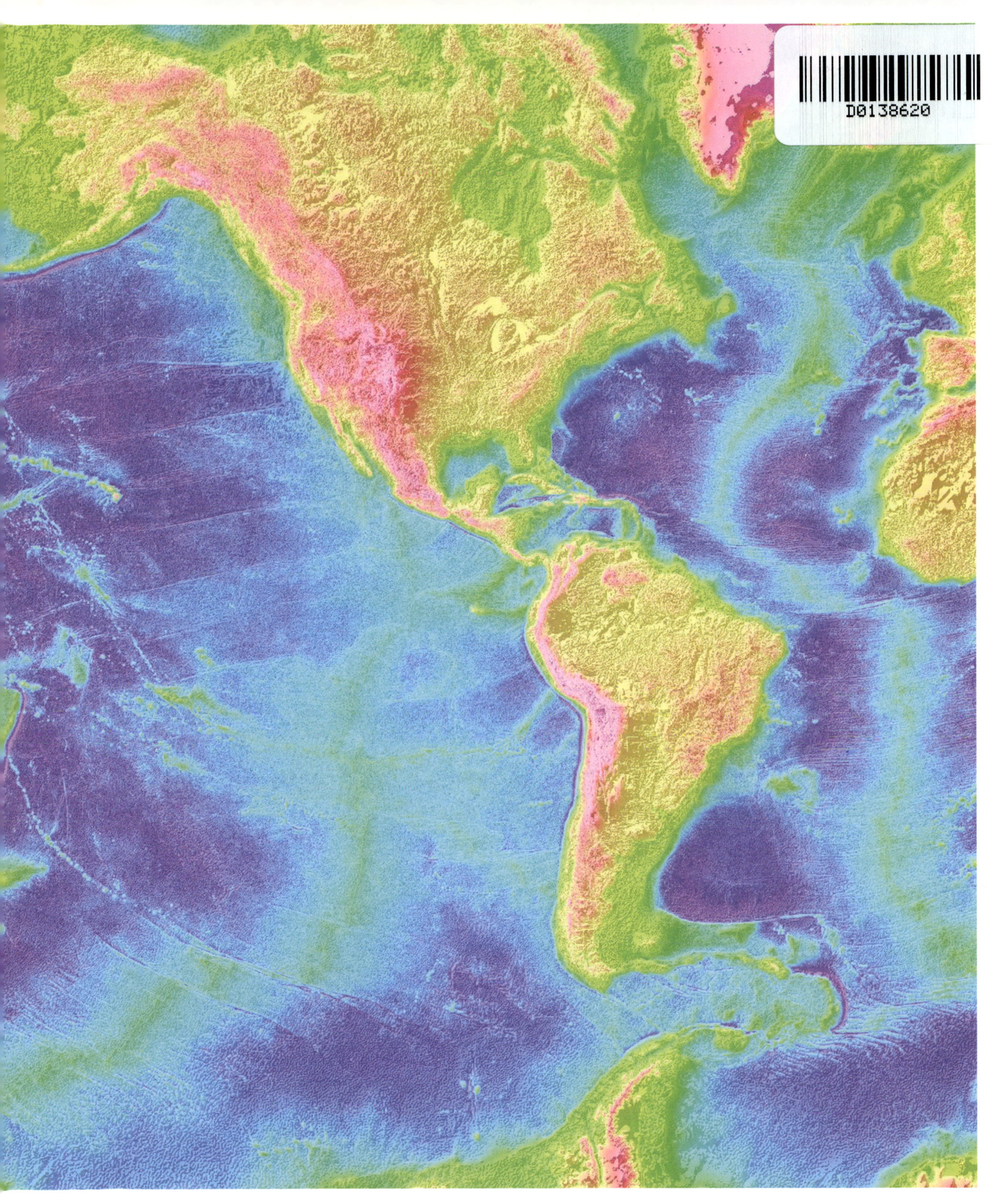

Mercator projection map of seafloor and land surface elevation. Seafloor elevation data were obtained from ship generated precision depth recorder data and satellite measurements of sea surface height data (Data obtained from many sources). The satellite sea surface height data provides resolution sufficient enough for seafloor topography features of a scale greater than about 20-25 km to be identified in the map. Analysis of data and imagery primarily provided by NOAA and Scripps Institute of Oceanography, University of California.

INTRODUCTION TO OCEAN SCIENCES

SECOND EDITION

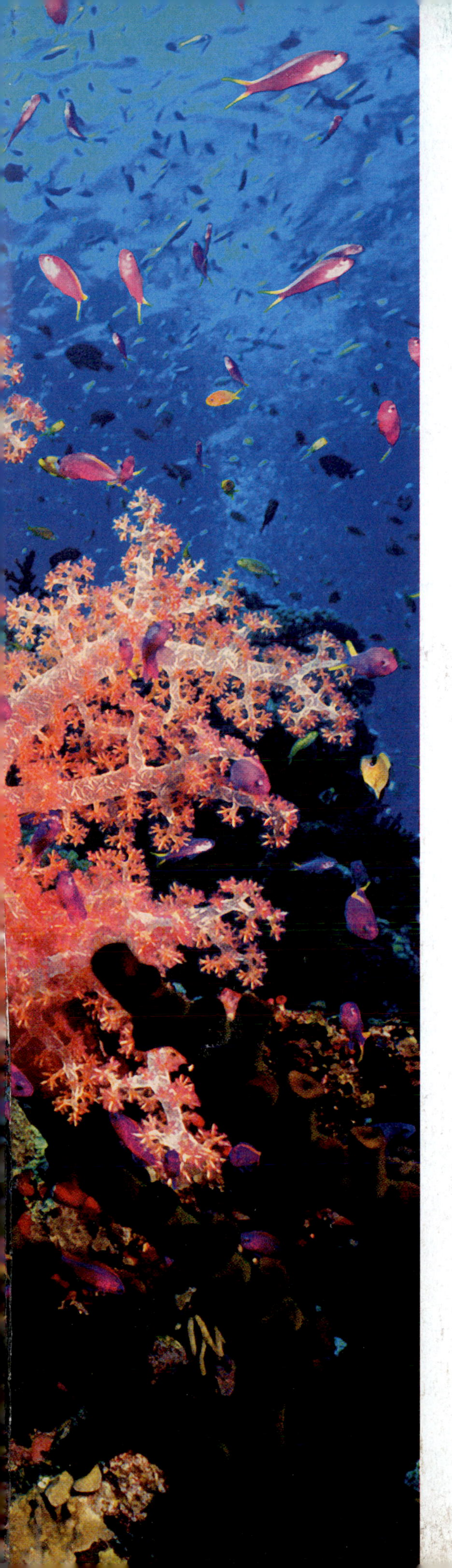

Introduction to Ocean Sciences

Second Edition

DOUGLAS A. SEGAR

Contributing author
Elaine Stamman Segar

W. W. NORTON & COMPANY
NEW YORK LONDON

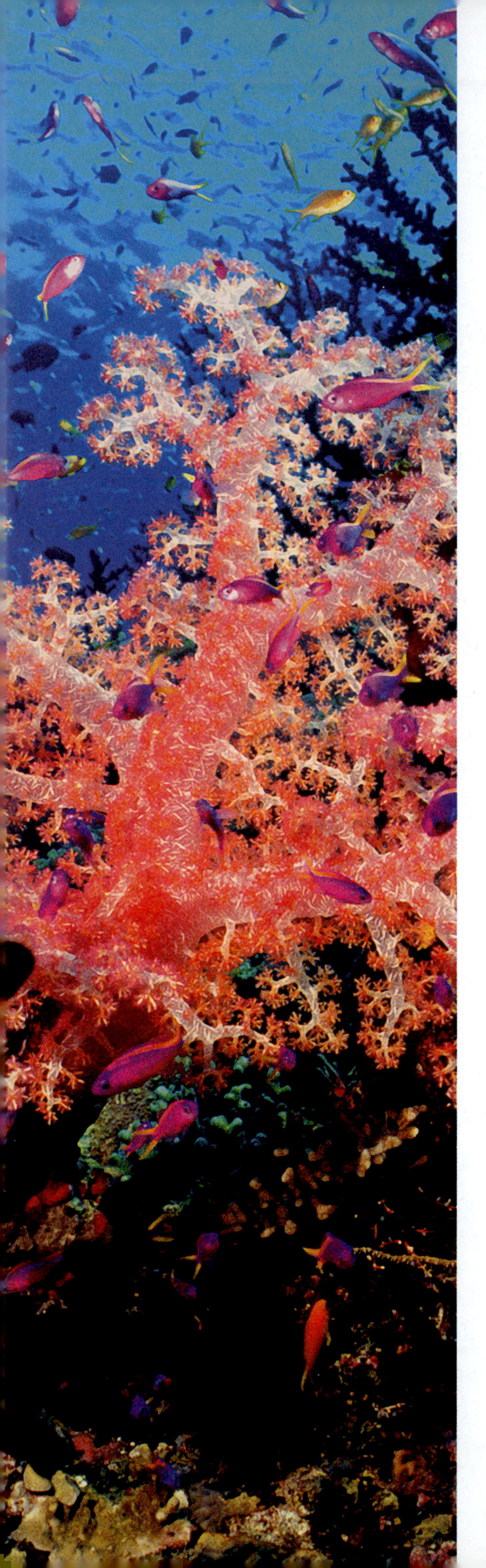

W. W. Norton & Company has been independent since its founding in 1923, when William Warder Norton and Mary D. Herter Norton first published lectures delivered at the People's Institute, the adult education division of New York City's Cooper Union. The Nortons soon expanded their program beyond the Institute, publishing books by celebrated academics from America and abroad. By mid-century, the two major pillars of Norton's publishing program—trade books and college texts—were firmly established. In the 1950s, the Norton family transferred control of the company to its employees, and today—with a staff of four hundred and a comparable number of trade, college, and professional titles published each year—W. W. Norton & Company stands as the largest and oldest publishing house owned wholly by its employees.

Printed in the United States of America.
Second Edition

Composition by Precision Graphics
Manufacturing by Courier
Illustrations and photo manipulation for the Second Edition by Precision Graphics

Editor: Leo A. W. Wiegman
Associate editor: Sarah England
Project editor: Thomas Foley
Copy editor: Stephanie Hiebert
Developmental editor: Spencer Cotkin
Director of manufacturing: Roy Tedoff
Layout artist: Roberta Flechner
Book designer: Joan Greenfield
Photo research: Kelly Mitchell
Photo permissions: Michele Riley
Ancillary project editor: Lory Frenkel
Editorial assistants: Lisa Rand and Sarah Mann
Managing editor, College: Marian Johnson

Library of Congress Cataloging-in-Publication Data

Segar, Douglas A.
Introduction to ocean sciences / Douglas A. Segar ; with contributions from Elaine Stamman Segar.—2nd ed.
p. cm.
Includes index.

ISBN-13: 978-0-393-92629-3
ISBN-10: 0-393-92629-X

1. Oceanography. I. Segar, Elaine Stamman. II. Title

GC11.2.S443 2006
551.46—cd22 2006047235

W. W. Norton & Company, Inc., 500 Fifth Avenue, New York, N.Y. 10110
www.wwnorton.com

W. W. Norton & Company Ltd., Castle House, 75/76 Wells Street, London W1T 3QT

1 2 3 4 5 6 7 8 9 0

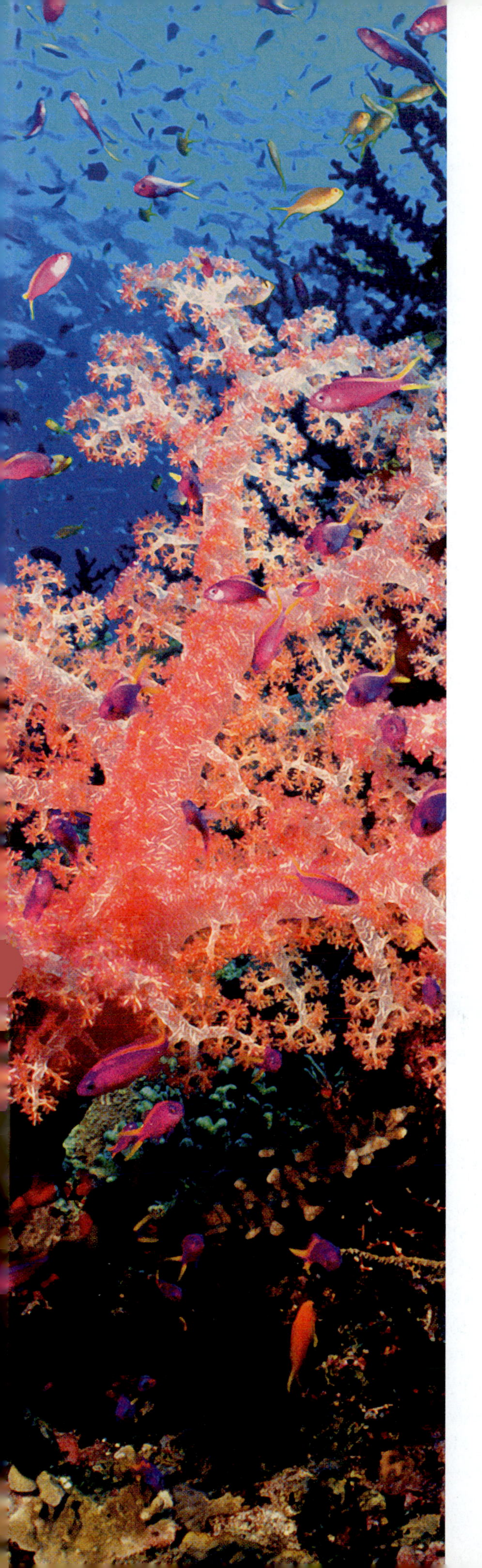

Brief Contents

Contents

CRITICAL CONCEPTS

APPENDIXES

Preface

The title of this text is *Introduction to Ocean Sciences* although most introductory courses in this field use the term "Oceanography" and my own degrees are in "Oceanography." I chose to use the term "Ocean Sciences" because, within the span of my career, this field has grown and matured. Oceanography, an activity whose name parallels the term "geography," implies an emphasis on exploration and mapping, which is still an important part of any study of the oceans. In fact, the vast majority of the seafloor and volumes of water in the oceans, especially below the shallow surface layer where the sun's light never penetrates, are still largely unexplored. We have, however, now progressed beyond oceanography, in the same way that alchemy evolved into chemistry with increased scientific knowledge. So much progress has been made in understanding the nature of the oceans and the physical, chemical, geological, and biological processes that occur in the oceans, that students are now able to learn not just oceanography, but **ocean sciences**. There are still frequent and exciting discoveries made about the oceans, but many of those discoveries are now about the scientific processes that affect the ocean environment and ocean life, or are about ocean processes that affect the atmosphere and our climate. Accordingly, this textbook was written with four main objectives:

- To introduce students to the exciting world of ocean discovery, both geographic and scientific.
- To support and engage the extraordinarily wide range of students who take ocean science courses, from non-science majors with varying science and math skills to science majors who may go on to take advanced courses in different aspects of ocean sciences.
- To impart an understanding of the interdisciplinary nature of ocean sciences, while focusing on ocean processes and how they are connected to our lives.
- To present a flexible framework to meet the needs of the different curricula and course lengths in different schools and departments.

These objectives were met in the previous edition of *Introduction to Ocean Sciences* with the help of organizational and pedagogical features that aimed to make the text easy to navigate. As any good ocean navigator would do, we have retained this successful approach the second time around. However, we have also made some improvements to the journey in this second edition.

For Students: How to Navigate *Ocean Sciences*

Understanding the oceans requires a basic knowledge of geological, chemical, physical, and biological processes and of how they interact in different parts of the ocean environment. If you are new to science, it may sometimes seem like you have an ocean of information to navigate in this course, but look for the following navigation aids in each chapter to help you on your way toward a mastery of ocean sciences.

VISUALIZE OCEAN SCIENCES AROUND YOU

Each chapter opens with a story to inspire you as the oceans have long fascinated and inspired me. Some of these stories are from my own experiences on and in the ocean. As an underwater photographer with over twenty-five years of experience, I want you to see what the world of ocean sciences actually looks like from close range. But it is also important that you see the big picture, and understand the impact and importance of the oceans both in your own region and on a global scale.

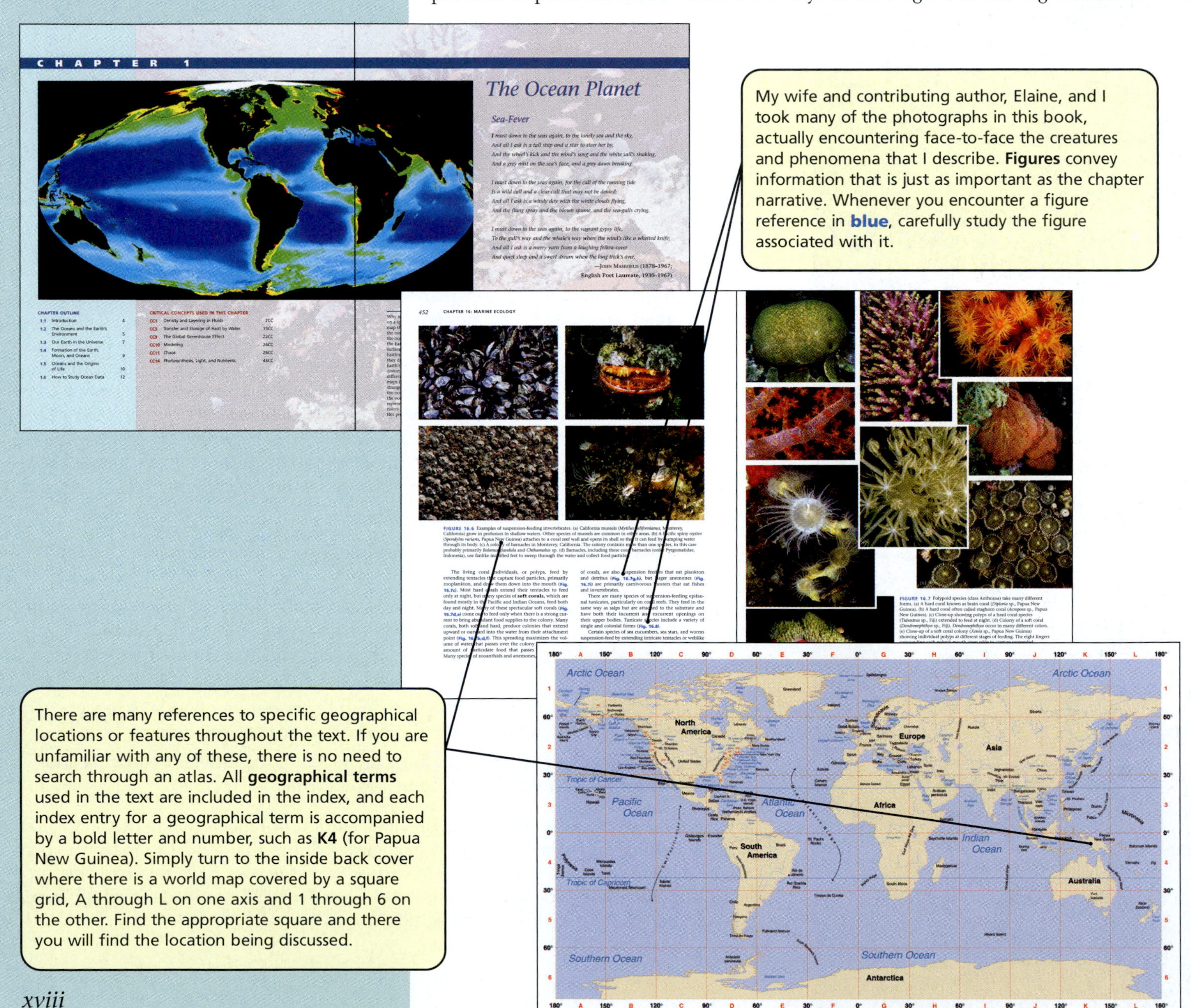

My wife and contributing author, Elaine, and I took many of the photographs in this book, actually encountering face-to-face the creatures and phenomena that I describe. **Figures** convey information that is just as important as the chapter narrative. Whenever you encounter a figure reference in **blue**, carefully study the figure associated with it.

There are many references to specific geographical locations or features throughout the text. If you are unfamiliar with any of these, there is no need to search through an atlas. All **geographical terms** used in the text are included in the index, and each index entry for a geographical term is accompanied by a bold letter and number, such as **K4** (for Papua New Guinea). Simply turn to the inside back cover where there is a world map covered by a square grid, A through L on one axis and 1 through 6 on the other. Find the appropriate square and there you will find the location being discussed.

DEVELOP SCIENTIFIC LITERACY

Each chapter has a variety of tools to increase your familiarity with scientific concepts and the ability to think critically, which will help you excel in this course. But those skills will be of use to you even after completion of the course, whether you are inspired to pursue further scientific studies, are reading about the latest hurricane in the newspaper, or are just visiting the beach.

Glossary terms are in boldface type at their first use in *each* chapter. These terms are important and you should know their meaning.

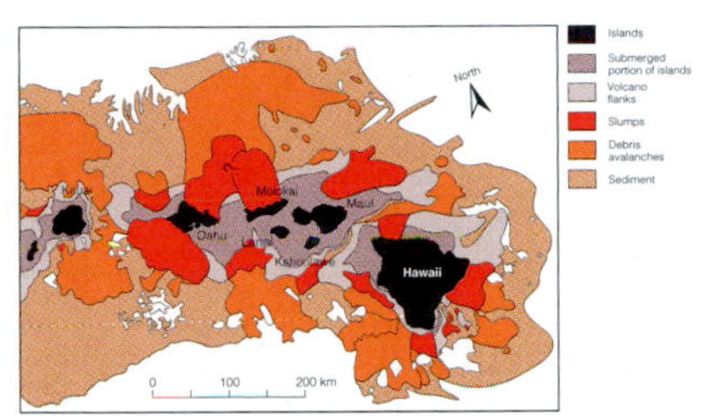

Critical Thinking Questions appear in boxes in each chapter and are designed to exercise your problem-solving ability by asking you to apply concepts you have just learned.

Key Term Lists include terms that you are less likely to be familiar with but are important to know and be comfortable with for classroom discussion and exams.

Each chapter concludes with a **Summary** that reviews the main topics covered in the chapter.

Critical Concept Reminders summarize the Critical Concept topics referenced in the chapter.

Study Questions help you review the main concepts presented in the chapter.

LEARN CRITICAL CONCEPTS

I have identified eighteen **Critical Concepts** that are fundamental to understanding important aspects of ocean sciences. Look for the Critical Concepts section between chapter 3 and chapter 4. The pages of this section are marked with red tabs on the corners for easy reference.

Each Critical Concept is presented in two parts. First, **"Essential to Know"** lists the key information you will need to apply the concept as you read the chapter.

Second, **"Understanding the Concept"** provides a detailed explanation of the Critical Concept, information that is not essential to your understanding of the chapters, but is useful if you want to master the concepts or are just interested in learning more.

Detailed **figures** help you visualize the concept.

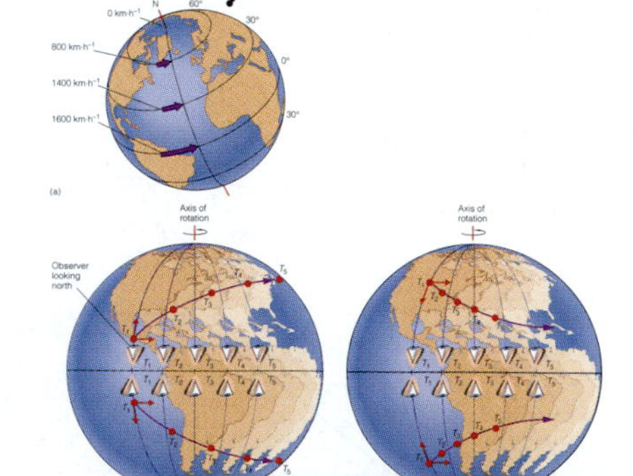

The Critical Concepts are the backbone of this book and of understanding ocean sciences. Though they are gathered in a separate section for easy reference, they are completely integrated into the chapters. Look for the following signposts:

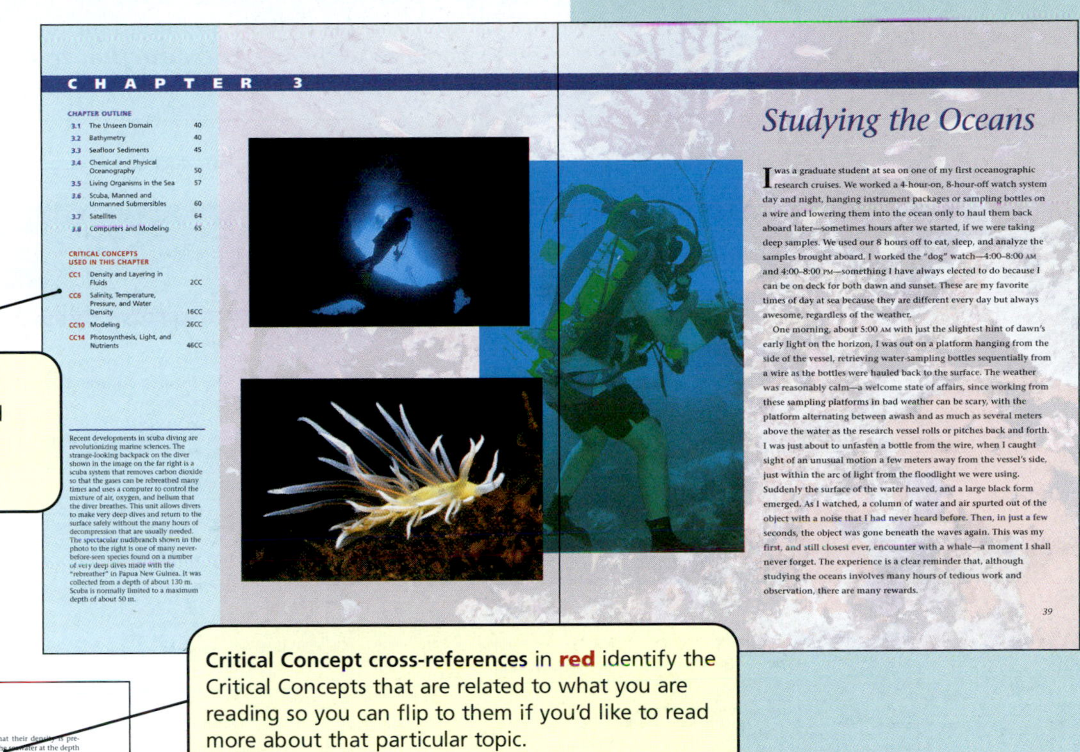

CHAPTER OUTLINE

CRITICAL CONCEPTS USED IN THIS CHAPTER

I was a graduate student at sea on one of my first oceanographic research cruises. We worked a 4-hour-on, 8-hour-off watch system day and night, hanging instrument packages or sampling bottles on a wire and lowering them into the ocean only to haul them back aboard later—sometimes hours after we started, if we were taking deep samples. We used our 8 hours off to eat, sleep, and analyze the samples brought aboard. I worked the "dog" watch—4:00–8:00 AM and 4:00–8:00 PM—something I have always elected to do because I can be on deck for both dawn and sunset. These are my favorite times of day at sea because they are different every day but always awesome, regardless of the weather.

One morning, about 5:00 AM with just the slightest hint of dawn's early light on the horizon, I was out on a platform hanging from the side of the vessel, retrieving water-sampling bottles sequentially from a wire as the bottles were hauled back to the surface. The weather was reasonably calm—a welcome state of affairs, since working from these sampling platforms in bad weather can be scary, with the platform alternating between awash and as much as several meters above the water as the research vessel rolls or pitches back and forth. I was just about to unfasten a bottle from the wire, when I caught sight of an unusual motion a few meters away from the vessel's side, just within the arc of light from the floodlight we were using. Suddenly the surface of the water heaved, and a large black form emerged. As I watched, a column of water and air spurted out of the object with a noise that I had never heard before. Then, in just a few seconds, the object was gone beneath the waves again. This was my first, and still closest ever, encounter with a whale—a moment I shall never forget. The experience is a clear reminder that, although studying the oceans involves many hours of tedious work and observation, there are many rewards.

The opening page lists the **Critical Concepts Used in That Chapter** and their page numbers in the Critical Concept section for easy reference.

Critical Concept cross-references in **red** identify the Critical Concepts that are related to what you are reading so you can flip to them if you'd like to read more about that particular topic.

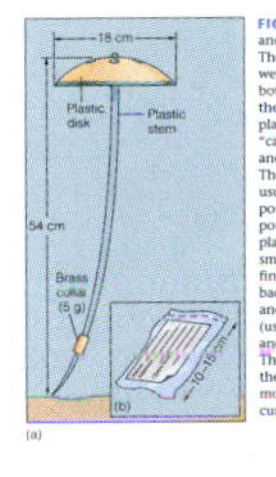

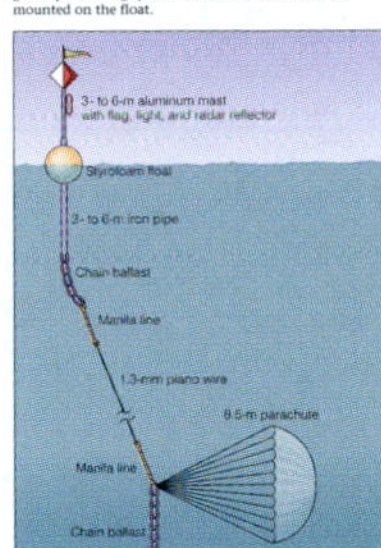

At the end of each chapter is a **Critical Concept Reminder** that summarizes the connection of the Critical Concepts to that particular chapter.

Most of the **Critical Concepts** are of key importance in several chapters. Learning about them when they're first mentioned in the early chapters will make your progress through the rest of the text easier.

STUDY ONLINE

Once you have visualized ocean sciences through the photos and figures in this text, developed your scientific literacy using the study tools in each chapter, and learned the critical concepts that are so important to understanding the processes discussed in this book, go to the free ***Introduction to Oceans Sciences* Student Web Site** at wwnorton.com/web/ocean for additional review and testing resources.

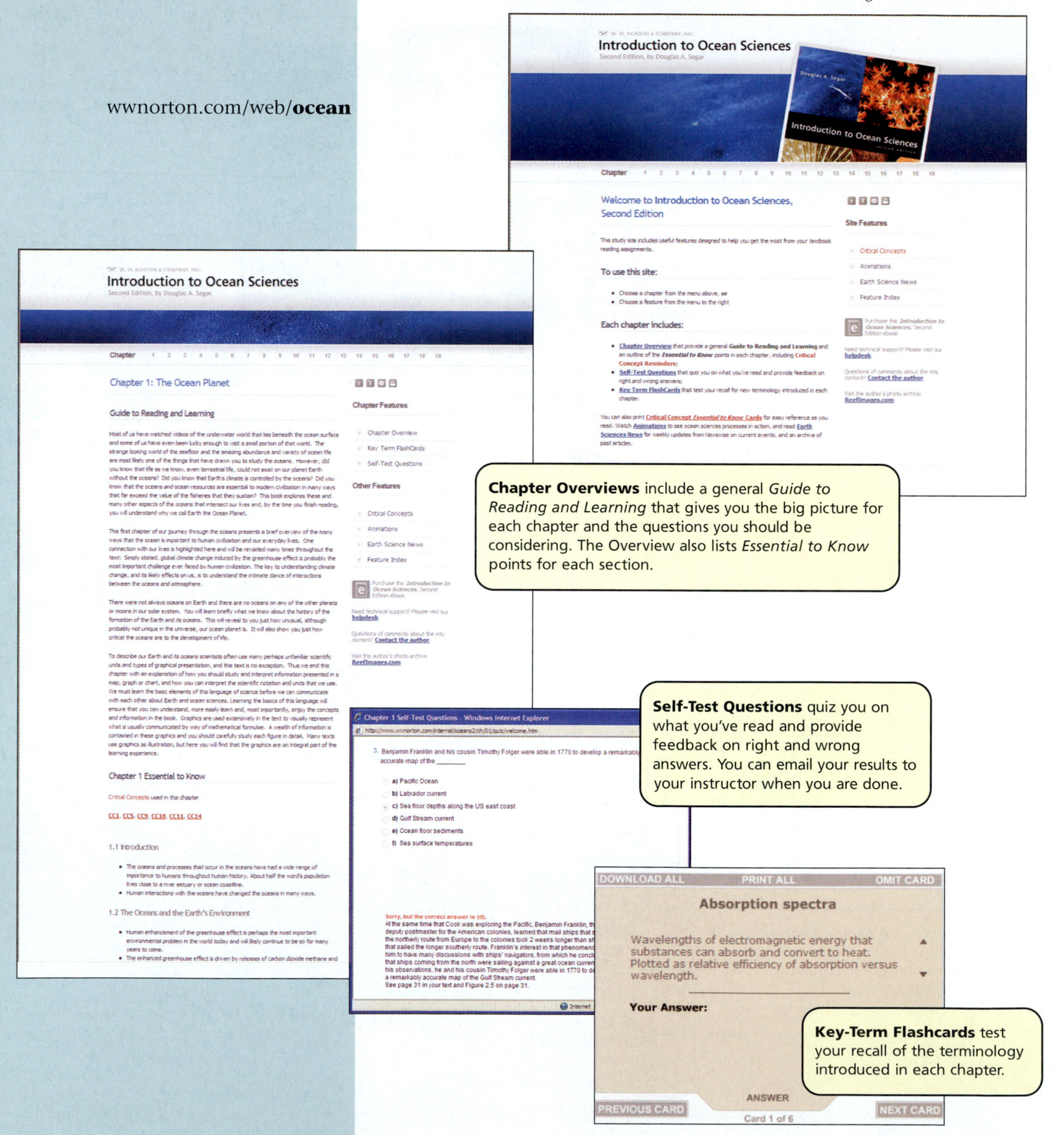

wwnorton.com/web/**ocean**
Introduction to Ocean Sciences
Welcome to Introduction to Ocean Sciences, Second Edition
Chapter 1: The Ocean Planet
Guide to Reading and Learning
Chapter 1 Essential to Know
Chapter 1 Self-Test Questions - Windows Internet Explorer
Chapter Overviews include a general *Guide to Reading and Learning* that gives you the big picture for each chapter and the questions you should be considering. The Overview also lists *Essential to Know* points for each section.
Self-Test Questions quiz you on what you've read and provide feedback on right and wrong answers. You can email your results to your instructor when you are done.
DOWNLOAD ALL
PRINT ALL
OMIT CARD
Absorption spectra
Wavelengths of electromagnetic energy that substances can absorb and convert to heat. Plotted as relative efficiency of absorption versus wavelength.
Your Answer:
ANSWER
PREVIOUS CARD
NEXT CARD
Card 1 of 6
Key-Term Flashcards test your recall of the terminology introduced in each chapter.

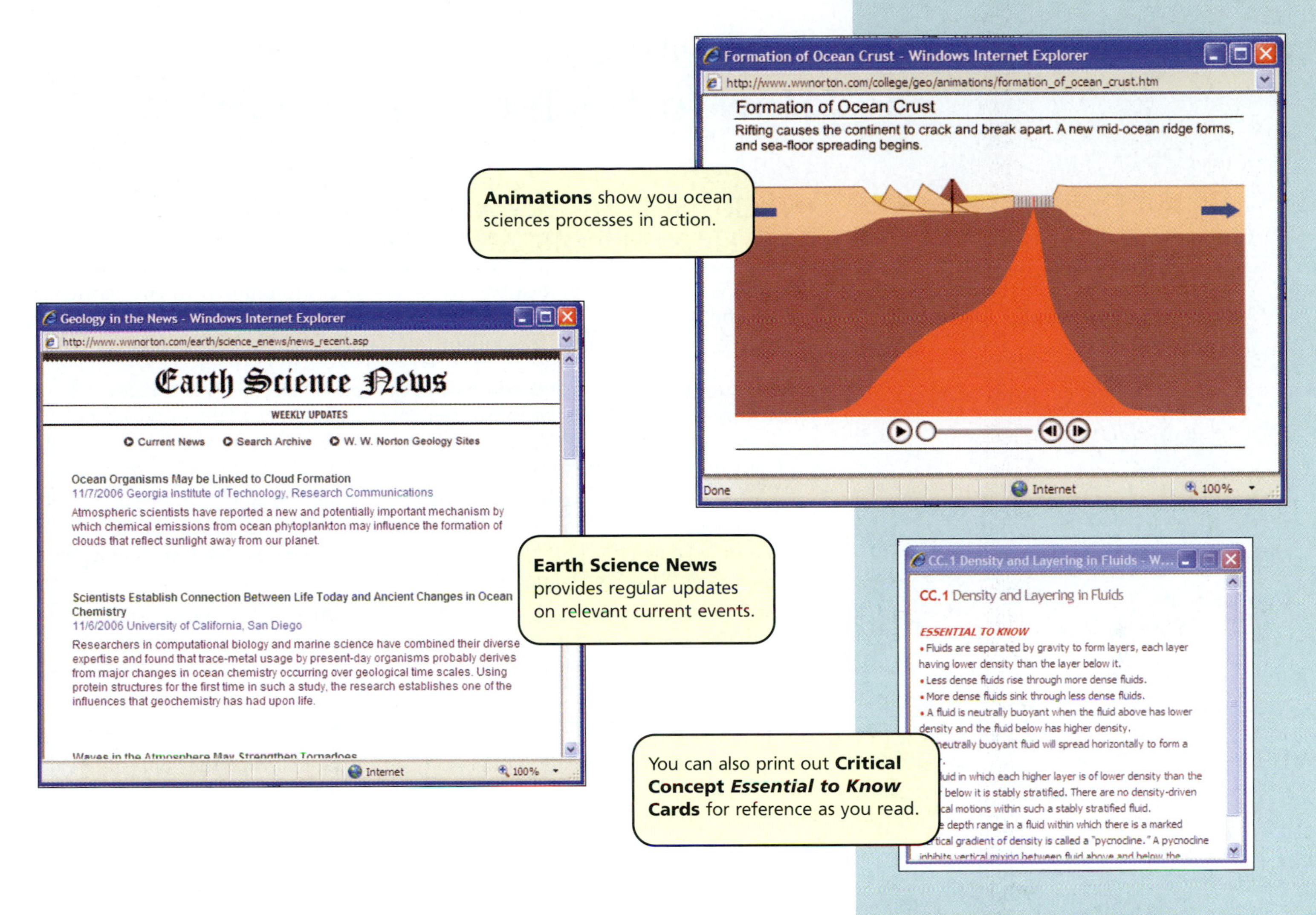

Introduction to Ocean Sciences, Second Edition, is also available as a **Norton ebook.** The ebook features actual book pages for a pleasant reading experience on your computer screen, and allows you to take notes, highlight, print pages as needed, and electronically search the text. Visit *nortonebooks.com* for more information.

norton**ebooks**.com

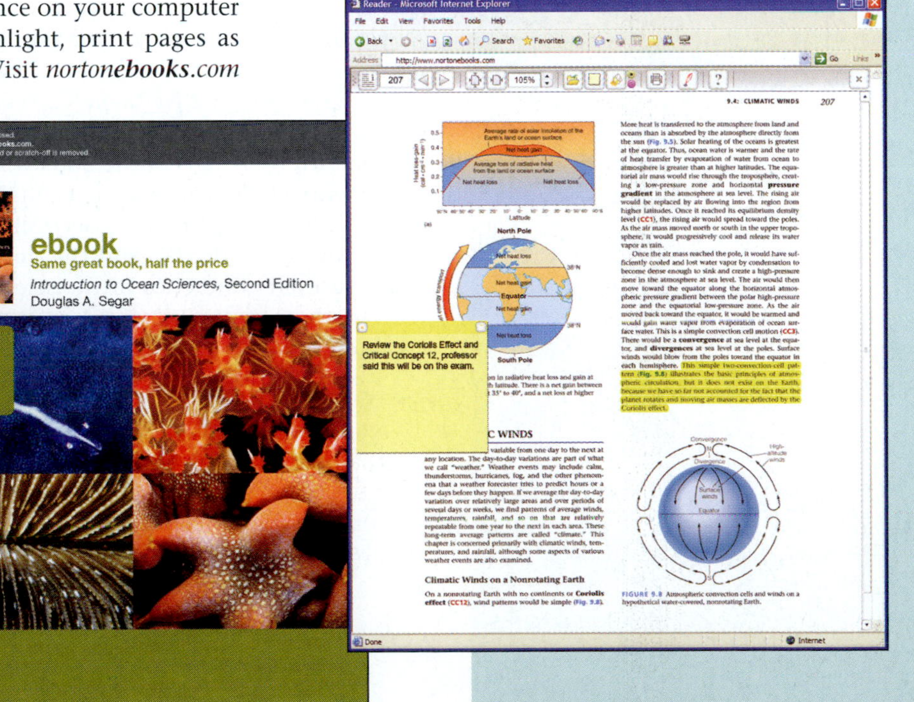

For Instructors: How to Teach *Ocean Sciences*

INTERDISCIPLINARY APPROACH AND FLEXIBLE ORGANIZATION

The text is designed to be flexible, so it can be easily tailored to the particular length and disciplinary emphases of your course. Chapters may be eliminated completely or assigned only for student reading. For example:

- **A biologically oriented course** might omit Chapter 5 ("Plate Tectonics: History and Evidence"); or for shorter courses Chapters 4 ("Plate Tectonics: Evolution of the Ocean Floor"), 5 ("Plate Tectonics: History and Evidence"), 9 ("Ocean-Atmosphere Interactions"), and 12 ("Tides") could all be omitted.
- **A physical science-oriented course** may omit Chapter 17 ("Ocean Ecosystems"); shorter physical science courses could omit Chapters 16 ("Marine Ecology"), 17 ("Ocean Ecosystems"), 18 ("Ocean Resources and the Impacts of Their Use"), and 19 ("Pollution").
- Also, **in shorter courses,** especially those offered to non-science majors, students can focus on just the "Essential to Know" bulleted lists of the Critical Concepts, while only reading or even skipping over the more difficult explanations.

The main features that support the flexibility of this textbook are the Critical Concepts and the identification of glossary terms the first time they appear in each chapter. Each chapter is also written to be as self-contained as possible. More information on how you can use this flexibility to meet your specific needs can be found in the Instructors' Manual.

UPDATED AND EXPANDED SECOND EDITION

As you might expect in a dynamic field like ocean sciences, there was much to update and add to this second edition. Embarking on this task, we were mindful of the responses that we had received from reviewers, as well as the students and instructors who used the first edition. The principal message in these responses was that the strengths of the book were its flexibility; its refusal to "dumb down" the essential scientific concepts that are the underpinning of oceans sciences (while still making them as accessible and easy-to-understand as possible for a wide range of students); and the quality of the figures, both their attractive design and their adherence to the science. As a result, we have fine-tuned and enhanced the essential features of the text, rather than making any wholesale changes to it.

- We established a better connection between the chapter material and the world around us by redesigning the chapter opening pages. In addition to a chapter outline and a list of the Critical Concepts used in that chapter, we have added opening vignettes that highlight at least one aspect of how the contents of the chapter relate to the human experience. Many of these are stories from my own experiences at sea or under the water.

- In the first edition, we included a prologue that reviewed the basic information needed to make sense of maps and charts, including understanding vertical exaggeration. This feature benefited so many students that we have expanded it to include an introduction to scientific notation and to the international system of scientific units. We have elevated this section to become part of a new **Chapter 1** (**The Ocean Planet**). This chapter includes an entirely new section that describes the formation of the universe, stars, and our solar system and its planets to illustrate just how unique our water covered planet is. It also includes a brief history of the Earth to provide the student with a better perspective on the enormous range of time scales on which ocean processes occur.
- **Chapter 2** (**History and Challenges of Ocean Studies**) and **Chapter 3** (**Studying the Oceans**) present the history of ocean exploration, the reasons that the oceans are so difficult to study, and the tools that ocean scientists use to perform these studies. In addition to discusssing exciting new technologies, we provide the historical framework that explains why we still know so little about the oceans.
- **Chapter 4** (**Plate Tectonics: Evolution of the Ocean Floor**) has been updated to include superplumes and other emerging concepts. One particularly important update is discussion of how we learned that the bend in the Hawaiian island-Emperor Seamount chain is caused by a movement of the hot spot, not a change in the direction of motion of the Pacific Plate. This has become my favorite story for illustrating to students how progress in science is made.
- Now that plate tectonics is no longer a theory, the information in **Chapter 5** (**Plate Tectonics: History and Evidence**) is less important to introductory students. But the chapter has been retained largely unchanged from the first edition because it provides perhaps the best possible historical example of how scientific paradigms change.
- The single chapter on water and seawater from the first edition has been broken up into two shorter chapters by separating the discussion of the physical properties of water into its own chapter. The new **Chapter 6** (**Water and Seawater**) and **Chapter 7** (**Physical Properties of Water and Seawater**) will make it easier to introduce several of the most important Critical Concepts at this point in your course and help students to maintain focus and learn this rather dense material.
- Aside from many individual updates and the addition of a few figures and images, **Chapters 8 through 13** are largely as they were in the first edition except that we have expanded the coverage of tsunamis and added a short section on the interactions of marine species with the tides.
- **Chapter 14** (**Foundations of Life in the Oceans**) has a new explanation of taxonomy and the tree of life. It has also been substantially updated to reflect the growing understanding of the importance of microbial and submicrobial species in the oceans, as well as the importance of iron in controlling primary production.
- **Chapter 15** (**Coastal Oceans and Estuaries**) and **Chapter 16** (**Marine Ecology**) have been enhanced by many new images, especially those that illustrate reproductive strategies and species associations. The material in these chapters is often the primary reason that students take a course in ocean sciences so we have tried to provide both maximum visual impact and enjoyment and a clear presentation of the basic scientific concepts.
- **Chapter 17** (**Ocean Ecosystems**) examines several special ocean ecosystems, including hydrothermal vents both on the oceanic ridges and elsewhere.

- We have moved the discussion of the Law of the Sea from the beginning of the text to a new **Chapter 18** (**Ocean Resources and the Impacts of Their Use**). We have added more detailed discussions of ocean resources and of methane hydrates, and of how exploitation of these resources conflicts with other uses. This chapter now provides a good lead-in to **Chapter 19** (**Pollution**).
- Finally, we have added one new **Critical Concept** (CC10 Modeling). Deciding what NOT to include as a critical concept in order to keep the list of concepts manageable and preserve their overarching importance was, and is, difficult. However, we felt that modeling in general, and computer modeling in particular, is of such great importance to modern ocean science that it merited a Critical Concept of its own.

MULTIMEDIA RESOURCES FOR INSTRUCTORS

The support program for *Introduction to Ocean Sciences,* authored by Doug Segar, has been fully revised and updated for the second edition:

- The **Instructor's Manual** offers tips on planning for the course and making the most of the textbook material, including Chapter Goals and Main Concepts, Chapter Outlines, Suggestions for Presentation, and Suggested Readings for each chapter. Also included are alternative outlines for courses with a biological or physical sciences focus.
- The **Test Bank** includes almost 1,600 questions, featuring multiple-choice, true/false, and completion questions for every chapter and Critical Concept. It is available in *Examview® Assessment Suite,* BlackBoard, and WebCT formats.
- **Answers to the Critical Thinking Questions** in the textbook are available separately and are a great resource for instructors who want more opportunities to test their students. They can also be used to generate classroom discussion.
- The **Instructor's Web Site** (wwnorton.com/instructors) provides the Instructor's Manual and Critical Thinking Question Answers in PDF and RTF formats, and the Test Bank in RTF and *Examview® Assessment Suite* formats. It also directs you to the author's website, where you will find commentary on the latest ocean sciences news, and where you and other instructors can post and share your own insights.

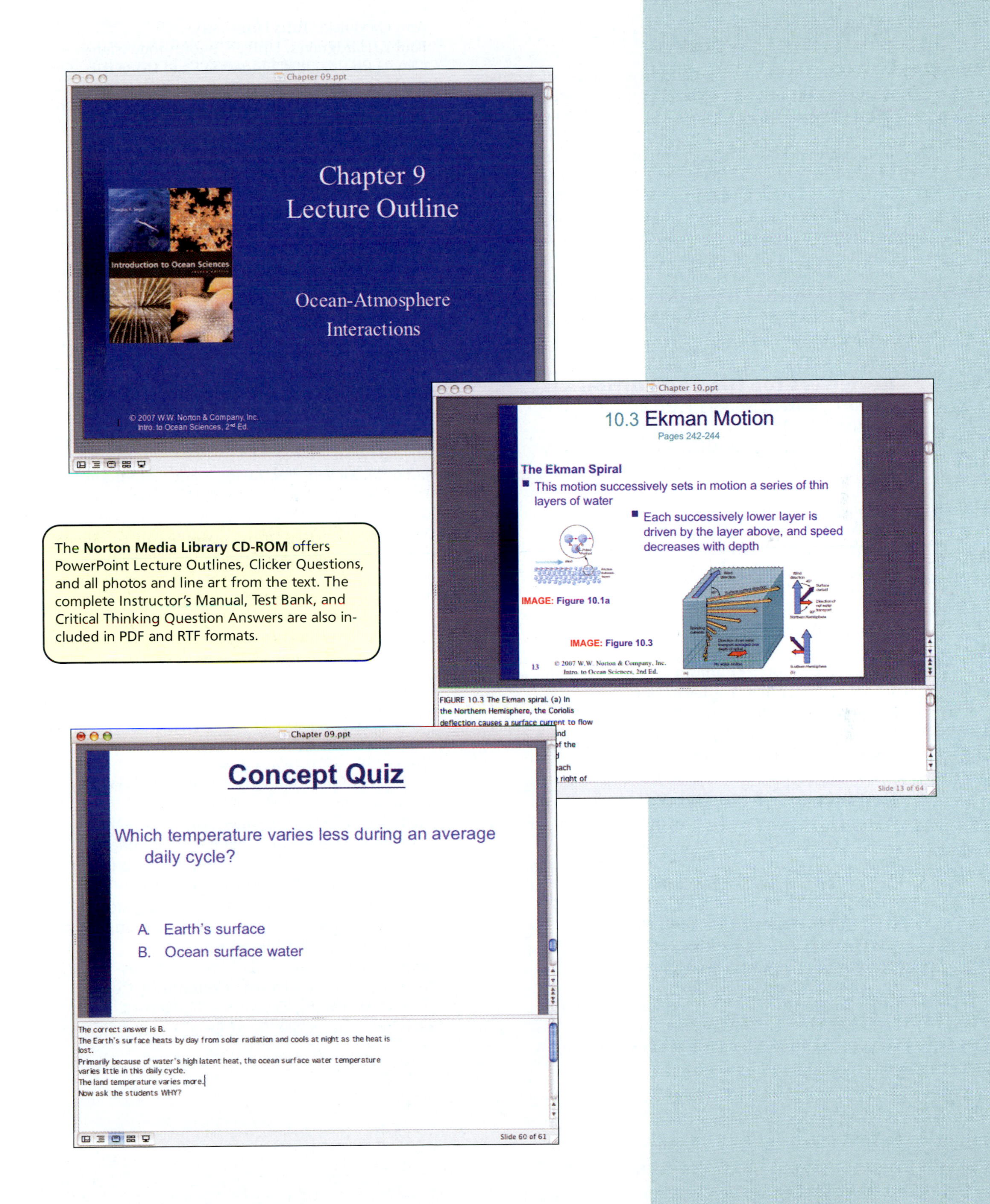

The **Norton Media Library CD-ROM** offers PowerPoint Lecture Outlines, Clicker Questions, and all photos and line art from the text. The complete Instructor's Manual, Test Bank, and Critical Thinking Question Answers are also included in PDF and RTF formats.

Acknowledgments

This book represents the collective wisdom of so many teachers, students, and colleagues that it is impossible to name them all. To all of those not mentioned specifically below, I express my sincere appreciation. I have learned from each and every one of you and I am richer for the experience. Particular thanks are due to one special student, George Welles, who provided me with both the opportunity and the motivation to write this text.

In an interdisciplinary text that covers a rapidly advancing field such as ocean sciences, advice and input from a wide range of reviewers is essential. I feel fortunate to have received such advice from the following individuals, each of whom contributed greatly to the text. I thank you all:

Reviewers for the 2nd Edition

Ronald C. Blakely, Northern Arizona University
Thomas A. Bush, Pierce College
Jim T. Byrd, Armstrong Atlantic State University
Walter J. Conley, St. Petersburgh Junior College, SUNY-Potsdam
William C Cornell, University of Texas, El Paso
Iver W. Duedall, Florida Institute of Technology
Robert Duncan, Oregon State University
Yoram Eckstein, Kent University
Bruce W. Fouke, University of Illinois
Gary Hitchcock, University of Miami
Jackie Huntoon, Michigan Tech. University
Brian McAdoo, Vassar College
T. James Noyes, Jr., El Camino Community College
Suzzanne O'Connell, Wesleyan University
Jonathan Sharp, University of Delaware
Stan L. Ulanski, James Madison University
Michael J. Valentine, University of Puget Sound

Reviewers for the 1st Edition

William B. N. Berry, University of California at Berkeley
Laurie Brown, University of Massachusetts
Theodore Chamberlain, Colorado State University
Karl Chauffe, St. Louis University
Tom M. Dillon, Oregon State University
Larry Doyle, University of South Florida
Iver Duedall, Florida Institute of Technology
W. Crawford Elliot, Case Western University
Ken Finger, Irvine Valley College
James Foxworthy, Loyola Marymount University
Dirk Frankenberg, University of North Carolina
Robert B. Furlong, Wayne State University
Ann Gardulski, Tufts University
Paul E. Hardgraves, University of Rhode Island
Clay Harris, Middle Tennessee State University
Ted Herman, West Valley College
William H. Hoyt, University of Northern Colorado
Gary Jacobsen, Grossmont College
Barbara Javor, San Diego State University
Ronald E. Johnson, Old Dominion University
Phyllis Kingsbury, Drake University
John Klasik, California State Polytechnic University, Pomona
Ernest Knowles, North Carolina State University
Paul LaRock, Florida State University
Paul H. LeBlond, University of British Columbia
Stephen Lebsack, Linn-Benton Community College
Stephen Lee, Los Angeles Pierce College
Larry Leyman, Fullerton College
Michael E. Lyle, Tidewater Community College
James E. Mackin, State University of New York, Stony Brook
Neil Maloney, California State University, Fullerton
Stanley V. Margolis, University of California, Davis
James M. McWhorter, Miami-Dade Community College
Robert Meade, California State University, Los Angeles
A. Lee Meyerson, Kearn College of New Jersey
Maurice A. Meylan, University of Southern Mississippi
John E. Mylroic, Mississippi State University
Charles L. Nelson, St. Cloud State University
Brian Novak, University of Michigan
Bernard Oostdam, Millersville University
John Plansky, Laney College
Mark Plunkett, Bellevue Community College
C. Nicholas Raphael, Eastern Michigan University
Mary Jo Richardson, Texas A&M University
David Roach, California State Polytechnic University, San Luis Obispo
Peter Roth, University of Utah
Fred Schlemmer, Texas A&M University, Galveston
David L. Schwartz, Cabrillo College
Frederick M. Soster, DePauw University
William R. Stephenson, Diablo Valley College
Joan Stover, South Seattle Community College
James F. Stratton, East Illinois University
Edward D. Stroup, University of Hawaii, Manoa
Keith A. Sverdrup, University of Wisconsin, Milwaukee
D. D. Trent, Citrus College
Stan L. Ulanski, James Madison University
Joe Valencic, Santa Barbara Community College
John H. Wormuth, Texas A&M University
Melvin B. Zucker, Skyline College

Authors create manuscripts. These manuscripts are turned into books by a team of editors, artists, designers, proofreaders, and production managers. The team assembled by Norton for this book was indeed an impressive one, and I am grateful for all their hard work and dedication. The team was led by the following individuals:

Editor: Leo Wiegman
Associate Editor: Sarah England
Copyeditor: Stephanie Hiebert
Project Editor: Thom Foley
Managing Editor, College: Marian Johnson
Developmental Editor: Spencer Cotkin
Production Manager: Roy Tedoff
Designer: Joan Greenfield
Layout Artist: Roberta Flechner
Art Studio/Composition: Precision Graphics, Stan Maddock, Dave Atteberry, Andrew Troutt
Photo Researcher: Kelly Mitchell
Photo Permissions: Michele Riley
Photo Manager: Neil Ryder Hoos
Editorial Assistants: Lisa Rand, Sarah Mann

In addition to those on this list, many other professionals contributed their special talents. These others have also earned my gratitude and I thank you all. Those who deserve special mention include Leo Wiegman, without whose continuing commitment this second edition would never have been produced; Sarah England, who guided this project to completion with commendable patience for my endless search for perfection; Stan Maddock, who performed his magic on many of the figures; and Stephanie Hiebert, who copyedited the text with great care and a good nature, even when I complained from time to time that the only true dictionary was the Oxford.

It is extremely gratifying for me and my wife and coauthor, Elaine, to be able to provide well over 270 of the photographs in this text and to share our love of marine life and underwater photography with you. Our thanks are due to the many boat captains and crews who took us to some of the best dive sites in the world to photograph the subjects that we sought for this text.

About the Authors

Douglas A. Segar (Ph.D., Oceanography, University of Liverpool, England) has performed research on diverse aspects of marine sciences, natural resource management, and environmental pollution, spending many months at sea on ocean research vessels of all sizes. He has taught ocean sciences at universities in Florida, California, and Alaska, and has directed several marine and environmental academic research centers. He has also worked in both private industry and government, including as a scientist with the U.S. Congress, where he investigated issues related to marine pollution, ocean dumping, fisheries conservation, habitat protection, marine mammal management, and endangered species. He also acted as a representative in the negotiation of several international marine conservation treaties.

Elaine Stamman Segar (M.S., marine emphasis, Texas A&M University) has pursued a career in research administration, beginning with marine environmental studies for the National Oceanic and Atmospheric Administration. She is currently the Director of Laboratory Programmatic Assessment and Oversight for the Office of the President of the University of California System. Both Doug and Elaine are experienced scuba divers and underwater photographers with at least twenty-five years experience each. Together, they created the marine photo archive and educational web site **www.reefimages.com** which now has over 5,000 of their images online and is used by schools and colleges worldwide.

INTRODUCTION TO OCEAN SCIENCES

CHAPTER 1

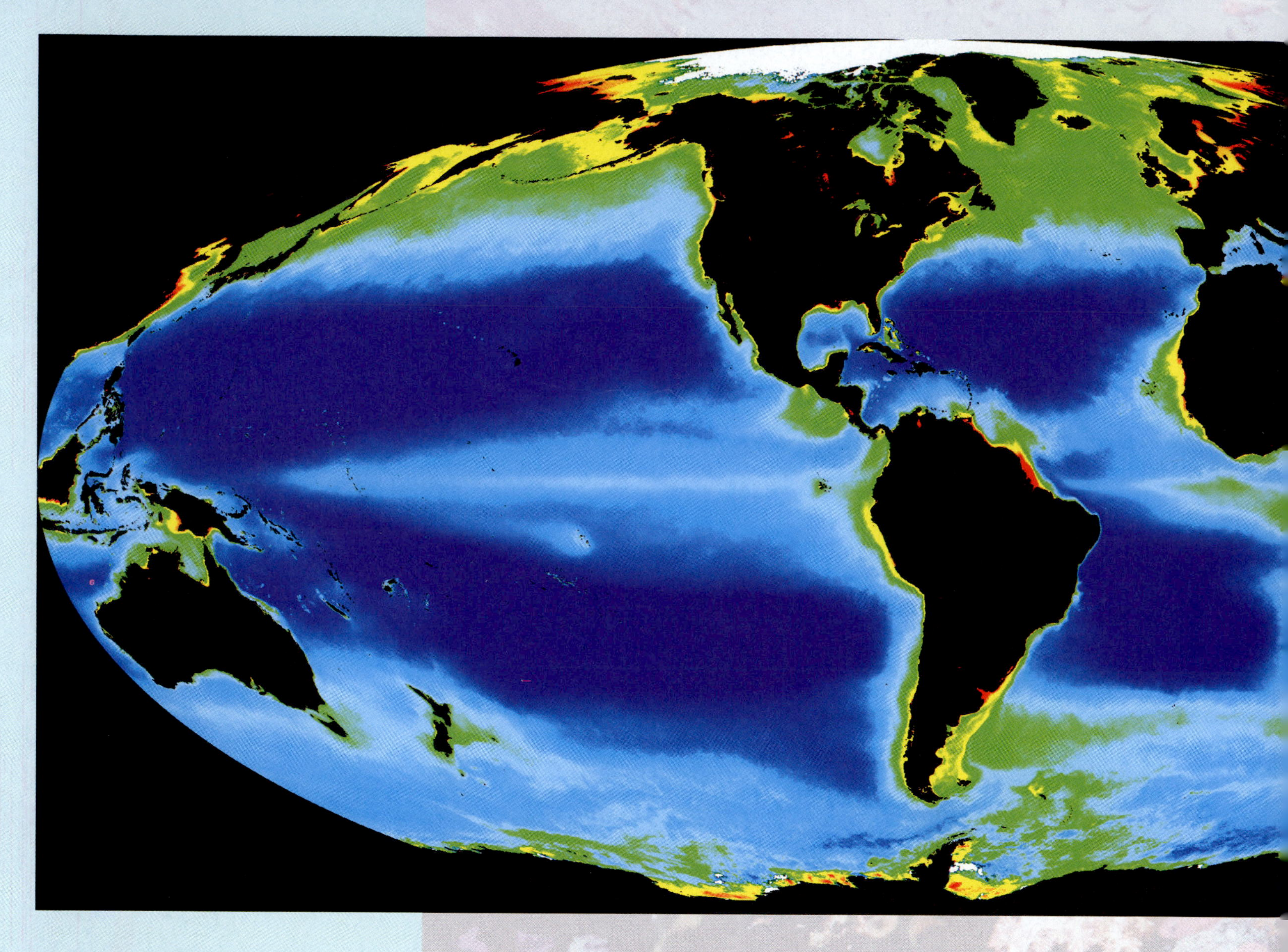

CHAPTER OUTLINE

CRITICAL CONCEPTS USED IN THIS CHAPTER

The Ocean Planet

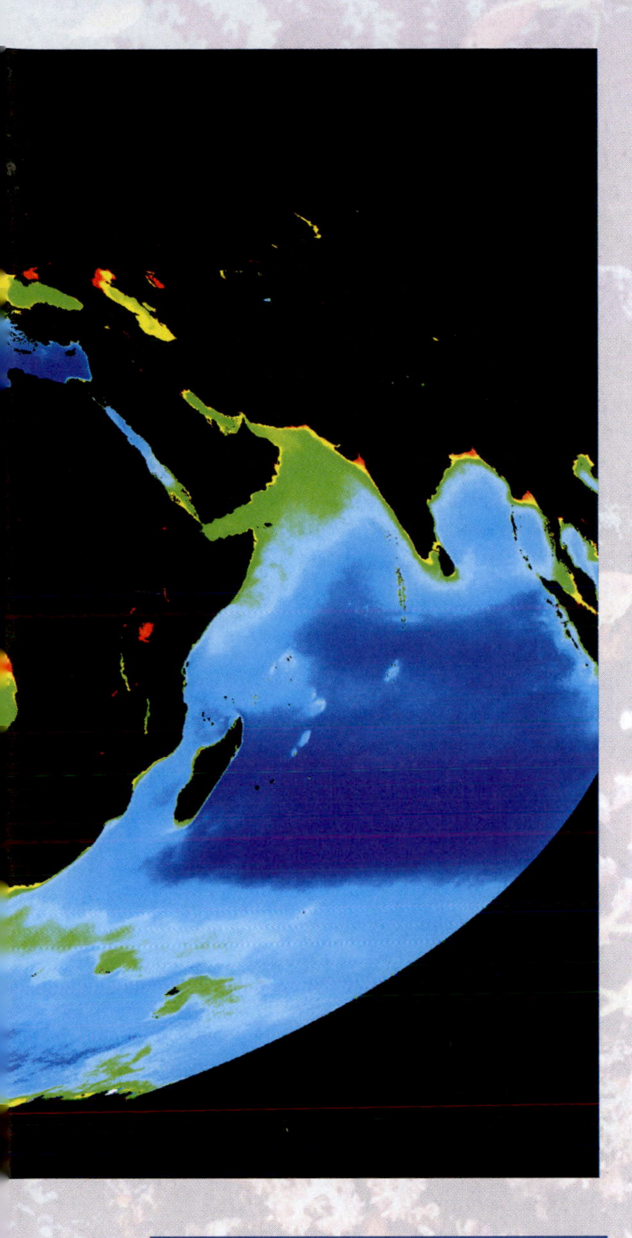

Why is this map of the world not drawn on a sphere or represented in a rectangular map shape such as we are used to? Why are the continents oddly misshapen? Why are the oceans those odd colors? First, maps of the Earth are made on two-dimensional surfaces. Because they must represent the Earth's continents, which lie on a sphere, they distort the shapes and sizes of the Earth's features. This map is drawn to distort the Earth's surface features in a different way than is done in most of the maps that are familiar to us. Second, the strange colors that shade various areas of the oceans do not represent the color of the ocean surface itself, but instead represent biological productivity. The colors reveal details of the distribution of this property in the oceans.

Sea-Fever

I must down to the seas again, to the lonely sea and the sky,
And all I ask is a tall ship and a star to steer her by,
And the wheel's kick and the wind's song and the white sail's shaking,
And a grey mist on the sea's face, and a grey dawn breaking.

I must down to the seas again, for the call of the running tide
Is a wild call and a clear call that may not be denied;
And all I ask is a windy day with the white clouds flying,
And the flung spray and the blown spume, and the sea-gulls crying.

I must down to the seas again, to the vagrant gypsy life,
To the gull's way and the whale's way where the wind's like a whetted knife;
And all I ask is a merry yarn from a laughing fellow-rover
And quiet sleep and a sweet dream when the long trick's over.

—John Masefield (1878–1967;
English Poet Laureate, 1930–1967)

This famous poem expresses very well the irresistible attraction that the oceans hold for many, including me. Perhaps this attraction is also part of the reason you have decided to learn more about the oceans.

1.1 INTRODUCTION

About half of the world's human population lives within a few tens of kilometers of the oceans. A large proportion of the remainder lives within a few tens of kilometers of a river or **estuary** that is connected to the oceans. These facts hint at, but do not fully reveal, the importance of the oceans to human civilizations.

The opening paragraph of a 2004 preliminary report by the U.S. Commission on Ocean Policy states the following:

> The oceans affect and sustain all life on Earth. They drive and moderate weather and climate, provide us with food, transportation corridors, recreational opportunities, pharmaceuticals and other natural products, and serve as a national security buffer. But human beings also influence the oceans. Pollution, depletion of fish and other living marine resources, habitat destruction and degradation, and the introduction of invasive non-native species are just some of the ways people harm the oceans, with serious consequences for the entire planet. (Preliminary Report of the U.S. Commission on Ocean Policy Governors' Draft, Washington, DC, April 2004.)

This paragraph appropriately summarizes and highlights the interdependence of all human civilization and the oceans. Unfortunately, the final report replaces this paragraph with a statement that is more specific to the national interests of the United States in the oceans.

In this text, we will explore the fundamental physical, chemical, geological, and biological features and processes of the oceans and review how humans have explored and studied the oceans. Armed with this knowledge, we will then discuss the range of resources that the oceans provide for us, and the **pollution** and other impacts that result from human use of the oceans and exploitation of their resources. After reading this text, you will have a new appreciation for the intimate and intricate linkages between our lives and the oceans. More importantly, you will be equipped to more fully analyze, understand, and evaluate the ever-increasing stream of scientific and popular reports of findings or theories concerning the oceans and the implications of human activities on the oceans.

The ocean has always held a fascination for humans. Generations of artists have created an entire genre of painting devoted to seascapes. For centuries, poets and authors have written about the romance, beauty, and moods of the oceans, ocean voyages, and marine species. Nautical literature extends as far back as ancient Greece with Homer's *Iliad* and *Odyssey*, and even earlier. There are far too many great tales of the sea to mention here, but consider the range expressed by some classics: *20,000 Leagues under the Sea* by Jules Verne; *Two Years before the Mast: A Personal Narrative of Life at Sea* by Richard Henry Dana, Jr.; *Moby Dick* by Herman Melville; *The Caine Mutiny* by Herman Wouk; *The Old Man and the Sea* by Ernest Hemingway; *Jaws* by Peter Benchley; and *The Hunt for Red October* by Tom Clancy. Perhaps you have your own favorites.

In recent decades, human interest in the oceans has grown in response to our rapidly deepening knowledge and understanding of the creatures that inhabit the underwater world, especially **coral reefs**. **Scuba** diving has become a pastime, avocation, or sport for millions, and millions more are fascinated by video programs that allow them to see the amazing array of ocean life in the comfort of their homes. Ocean aquariums have been built in many cities and attract millions every year. All forms of ocean recreation, not just scuba, have grown explosively. For a large proportion of the planet's population, recreation or vacation pastimes are directed toward the **beaches**, boating, and other water sports, or to sailing the oceans on cruise ships. On a more somber note, humans are also fascinated and terrified by the death and destruction caused by **tsunamis**, **hurricanes**, severe storms, ships sinking or running aground, and sharks.

The explanations in this text of how the oceans and ocean life function provide much that can inspire the poets and artists among us. However, there is also much more to be learned from this text that is of practical value to each of our lives. For example, after studying this text you will have answers to such questions as

- Why do some say that we know less about the oceans than we do about the moon or the planet Mars?
- Why are the world's active volcanoes located where they are?
- Why does the east coast of the United States have many **barrier islands** and tidal **wetlands** (marshes and **mangrove** swamps), whereas the west coast does not?
- What is a tsunami, and what should I do if I am ever on a beach when one is approaching?
- What is a **rip current** (rip tide), and how can I survive one?
- What is unique about water, and how do its unique properties affect daily life and our **environment**?
- What do we need to know about the oceans to understand and predict **climate** changes?
- Why are fisheries concentrated mostly in **coastal zones**, and why do some areas of the coastal ocean produce more fish than others?
- Why do most hurricanes cross onto land along the shores of the Gulf of Mexico and the southeastern United States rather than along the northeastern Atlantic coast or on the west coast of the United States?
- Is pollution of the oceans by oil spills and industrial discharges as bad as newspaper and other reports often portray it?

- What are the main causes of the widely reported destruction of coral reefs and damage to other ocean **ecosystems**?

Answers to some of these questions may be found in a single section or chapter of this text. However, the oceans are a complex environment in which geology, physics, chemistry, and biology are all linked in intricate ways. As a result, finding answers to many other questions requires fitting together information from several different chapters. You will be reminded numerous times of the interdisciplinary nature of ocean sciences as you study this text.

One particular theme that exemplifies the interdisciplinary nature of ocean sciences is revisited many times throughout this book. This theme concerns what is perhaps the most important scientific question facing contemporary human society: Is human activity permanently altering the environment in a manner that will ultimately damage or diminish civilization itself? Specifically, will the release of gases from the combustion of **fossil fuels** alter the oceans and atmosphere in such a way that major changes will occur in the global climate?

1.2 THE OCEANS AND THE EARTH'S ENVIRONMENT

Only very recently have we come to realize that humans have already caused profound changes, not just in local environments, but in the global ocean, atmosphere, and terrestrial environments as a whole. We have become aware that too little is known about the global or regional consequences of environmental changes already caused by human activities. Even less is known about the future environmental consequences if our civilization maintains its current pattern of exponential development and growth. The urgent need to assess the unknowns has been felt throughout the environmental-science community. The oceanographic community has been particularly affected, because most global environmental problems involve the oceans and ocean ecosystems, and our knowledge of the oceans is much poorer than our knowledge of the terrestrial realm.

Changes in the marine environment caused by human activities are many and varied. They include many forms of pollution (Chap. 19) and physical changes in the coastal environment (Chaps. 13, 15). Many pieces of information must be obtained by different oceanographic disciplines before human impacts on the ocean can be identified, assessed, and effectively managed. Throughout this text, reference is made to the application of various oceanographic findings, principles, or studies to the practical problems of ocean management. The intent is to facilitate understanding of similar problems reported almost daily in the media.

Among contemporary environmental problems associated with human activities, one stands out as perhaps the most important and complex: global climate change due to enhancement of the **greenhouse effect** (**CC9**). A greenhouse is effective at maintaining higher internal temperatures compared to external temperatures because its glass windows allow more of the sun's energy to pass through into the greenhouse than it allows to pass out. The Earth's atmosphere acts in a similar way to control temperatures at the Earth's surface.

In the atmosphere, several gases, especially carbon dioxide, function like the glass in a greenhouse. Since the Industrial Revolution, the burning of fossil fuels has steadily increased the concentration of carbon dioxide in the atmosphere (**Fig. 1.1**). The concentrations of other greenhouse gases, such as methane and **chlorofluorocarbons**, also are increasing as a result of human activities. If there were no other changes to compensate for the consequent increase in greenhouse efficiency, the Earth's temperature would rise. Indeed, some scientists have predicted that the average temperature of the Earth's lower atmosphere, in which we live, will rise by several degrees Celsius in the next two to three decades.

If such an increase in the global temperature occurs, it will cause dramatic climate changes throughout the world. These changes could be devastating to agriculture, the environment, and our entire civilization. In addition, if the predicted warming occurs, sea level will rise, primarily as a result of thermal expansion of the warmed ocean water and, to a lesser extent, melting of ice. Some experts predict that the sea level will rise as much as half a meter during the next several decades. If this happens, large areas of low-lying coastal land will be inundated, and some entire low-lying island chains may disappear entirely.

FIGURE 1.1 Change in the concentration of carbon dioxide in the atmosphere since 1880. Data for the smooth part of the curve were obtained from ice cores. Data from 1960 were obtained from direct measurements at Mauna Loa, Hawaii, and show both annual oscillations and a long-term upward trend.

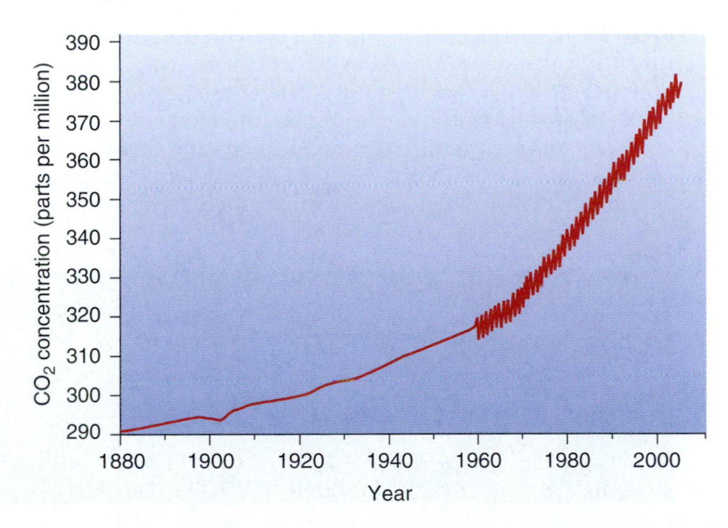

FIGURE 1.2 A telescope image of a typical spiral galaxy. In our galaxy, the solar system lies at the edge of one of the spiral arms.

The oceans and atmosphere act together as a complex system that regulates our climate and the concentrations of carbon dioxide and other gases in the atmosphere. Within the ocean–atmosphere system, numerous complicated changes and **feedbacks** occur as a result of increases in atmospheric concentrations of carbon dioxide and other gases. Some changes would add to the predicted global warming, whereas others would reduce or even negate it. However, our current understanding of the ocean–atmosphere system is poor, especially those segments of the system that are associated with ocean processes. If we are to be able to predict our climatic future with greater certainty, so that we can take appropriate actions, we must improve our understanding of the oceans.

In addition to studying contemporary ocean processes, oceanographers study the oceans to uncover important information about past changes in the world's climate. Such historical information, found primarily in ocean seafloor **sediment** and sedimentary rocks, can

FIGURE 1.3 The birth of the solar system, the Earth, and the moon. (a) A nebula that contains hydrogen and helium, plus the heavier elements formed in earlier supernovae, is disturbed. (b) The nebula begins to condense and spin, forming a flattened disk that is denser at its center. (c) The central ball grows hot, fusion starts, and the ball becomes a star. Dust particles in the disk collect into rings and then into larger objects. (d, e) The forming Earth becomes large enough for gravity to compress and heat it until its interior melts and density stratification occurs. (f, g) Soon after the Earth forms, it is struck by a smaller planet-sized object. (h, i) The debris ejected in the collision collects in a ring and then condenses to form the moon. (j) Eventually the Earth develops an atmosphere from volcanic gases and cools until water condenses to form an ocean.

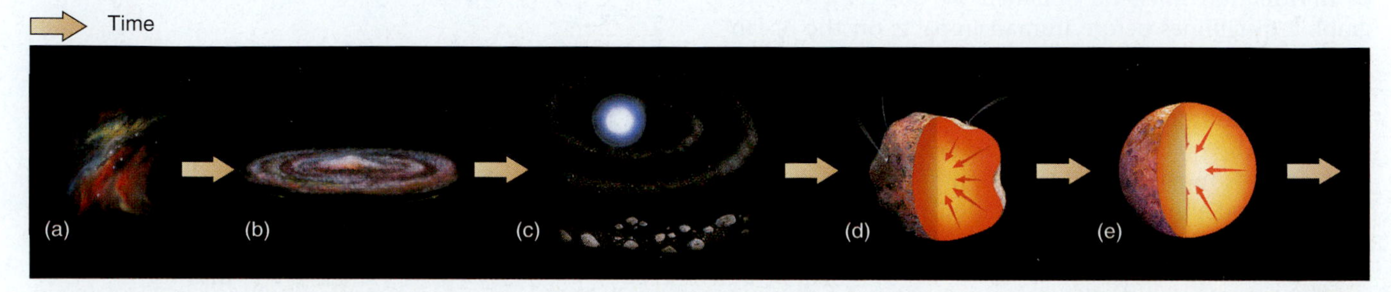

reveal how the Earth's climate has changed over tens of millions of years and help us assess how it might change in the future, either as a result of natural changes or as a result of the enhanced greenhouse effect.

Consideration of the enhanced greenhouse effect and other environmental problems throughout this text is intended to convey an appreciation for the complexity of ocean and atmospheric environmental issues. Many important decisions must be made by current and future generations about human use and the protection of our environment. It is important for all of us to participate in these decisions with an understanding of the uncertainties inherent in all scientific studies and predictions of complex environmental systems (**CC10, CC11**). We must recognize that science cannot provide definitive answers to even the most intensively studied environmental problems.

Before we begin to explore the intricacies of the relationships among humans, the oceans, and the environment, it is important to have an understanding of just how unique the circumstances are that have created and sustained oceans on the Earth, the "Ocean Planet," and of how critical to life on the Earth these oceans are. It is also important to review some of the basic tools that oceanographers and other scientists use to describe what they have learned about the Earth and its oceans. Many of those tools will be used in this text, especially various forms of maps, graphical representations of data that vary geographically or with time, and the standard scientific notation for numbers and units of measurement. The remainder of this chapter is devoted to establishing the groundwork for the rest of the book.

1.3 OUR EARTH IN THE UNIVERSE

The Earth is one of eight planets orbiting an average-sized star located near the outer edge of a spiral **galaxy** (**Fig. 1.2**) called the Milky Way galaxy. The Milky Way galaxy contains an estimated 300 billion stars and is just one of more than 100 billion galaxies that constitute the known universe. The scale of the universe is almost unimaginable. For example, within our own galaxy, the nearest star to our own is a little more than 4 light-years away. To put this distance in perspective, a light-year is the distance that light travels in one year—9,500,000,000,000 km—or the equivalent of more than 200 million trips around the world. The spiral disk of the Milky Way galaxy is about 100,000 light-years in diameter. The known universe is many orders of magnitude larger. The most distant object so far observed in the universe is approximately 13 billion light-years away.

What do we know about the universe and how the Earth and its oceans came to be formed? A detailed review of the answers to these questions is far beyond the scope of this text. What follows is a very brief summary of what current theories suggest is the history of the universe.

The Big Bang

Approximately 15 billion years ago, all matter and energy of the universe was packed into one point that was so small it occupied almost no volume. The universe as we know it began when this point, for unknown reasons, exploded in what is known as the "Big Bang." Little is known about the early history of the universe after the Big Bang. However, it is thought that, for hundreds of thousands of years after the initial explosion, the intense cloud of matter and energy that exploded outward was so hot that no atoms could form. As the cloud continued to cool, hydrogen and helium atoms began to form. By about 1 million years after the Big Bang, the cloud consisted mostly of hydrogen (98%) and helium (2%) atoms.

As this cloud, or **nebula**, expanded further, portions began to clump, eventually forming many individual smaller nebulae separated from each other by space that contained much lower concentrations of matter. Matter in these nebulae was by now relatively cool. Within the nebulae, however, further clumping took place and some of the resulting clumps became large enough that their **gravity** began to attract other matter from around them. As these clumps became larger, gravity compressed this matter, mostly hydrogen, into tighter balls that began to spin ever faster. (This is the same principle that explains why the speed of a skater's spin increases when the skater's extended arms are pulled tight into the body.) The energy released by compression of the gases heated the spinning balls of gas until some were hot enough to initiate nuclear fusion reactions, in which two hydrogen atoms are combined to form a helium atom. These fusion reactions released additional energy as heat and light, and the balls eventually became stars (**Fig. 1.3**). Nebulae are found and stars still form in today's

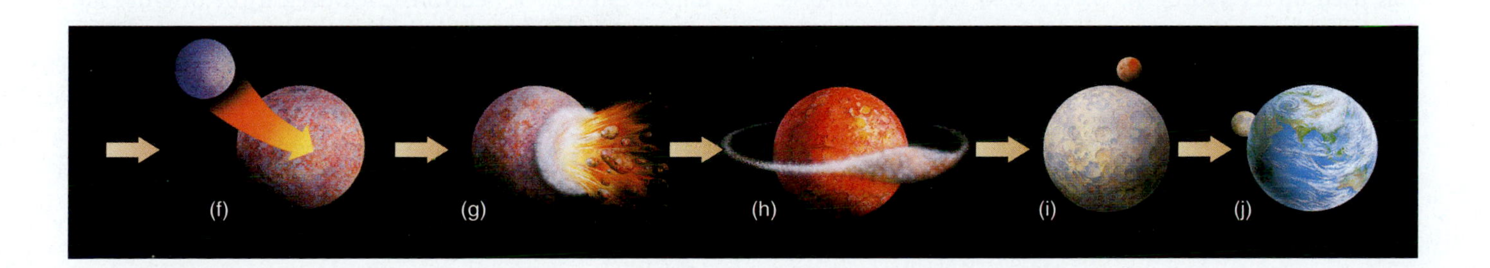

FIGURE 1.4 Taken by the Hubble Space Telescope, this image shows distinct cloudlike nebulae with already formed stars seen between the clouds.

universe (**Fig. 1.4**), but as we shall see, these younger nebulae consist of gas and dust produced by a continuing process of star formation and destruction.

Life of the Stars

Stars do not last forever. Eventually they use up all of their nuclear fuel. As a result of their greater mass, the cores of large stars are more compressed and hotter than smaller stars and, therefore, burn faster. Whereas a smaller star, such as our sun, may burn for about 10 billion years, large stars may burn for only a few million years. Thus, small stars and large stars have different life cycles.

Once the hydrogen in any star is converted to helium, the helium can undergo other fusion reactions to form heavier elements, such as carbon, nitrogen, and oxygen. Even heavier elements, such as iron and nickel, can be formed if the star is hot enough. In smaller stars, the temperature is not high enough to continue fusion reactions to create heavier elements. When the core of a small star has used up its hydrogen, it cools and starts to collapse. Compression of the core causes the star to heat up again, and it then burns the small amount of hydrogen that remains in a shell wrapped around its core. This burning shell provides energy that causes the star to expand again to become a red giant. When this happens to our sun, it will become so large that its surface will extend beyond Mercury. Don't worry—this catastrophe will not happen for another 5 billion years! The core temperature of the red giant progressively increases again until it is hot enough to burn all the helium it created before becoming a red giant. Eventually, the helium is converted into carbon and other heavier elements.

The sun will spend only about a billion years as a red giant. When the red giant has exhausted its helium fuel, it is not hot enough to be able to burn the carbon it created. At this point the core of the star contracts, releasing energy that makes the envelope of the star expand and the star again increases in size. However, the star is not very stable, and it eventually blows its outer layers off into surrounding space. The core of the star remains intact and becomes a white dwarf, surrounded by an expanding shell of gas in a nebula. White dwarfs are very dense, typically about half the mass of our sun, but compressed into a star about the size of the Earth.

Large stars that are more than approximately eight times as massive as our sun end their lives very differently. When the fuel for the fusion process is depleted in the core of the star, the star swells to become a red supergiant. However, the high gravitational force due to its large mass begins to shrink the core inward. As it shrinks, the core grows hotter and denser. At the extremely high temperatures generated, nuclear reactions fuse even carbon and other moderate-sized atoms into heavier elements. This temporarily halts the collapse of the core, but when the core eventually contains essentially just iron, it has nothing left to fuse (because iron's nuclear structure does not permit its atoms to fuse into heavier elements).

When fusion stops, the star begins a catastrophic gravitational collapse that occurs within only about a second, during which the core temperature rises to over 100 billion degrees as the iron atoms are crushed almost instantly together until the repulsive force between the iron nuclei becomes greater than the force of gravity. This collapse happens so fast that the core compresses, but then it rebounds. The energy of the core's rebounding outward is transferred to the outer part of the star, which then explodes violently. As the shock wave from this explosion impacts matter in the star's outer layers, this matter is heated, and atoms are fused to form new elements and radioactive **isotopes**, including elements heavier than iron. The shock propels this material into space. This event is called a **supernova**. It is believed that all the elements in the universe heavier than iron must have been formed in a supernova. After a supernova, all that remains of the original star is often a neutron star, which is a small, superdense star composed almost entirely of neutrons. However, if the original star was extremely massive, even the neutrons cannot survive and the core itself collapses to form a black hole, a region of space from which nothing can escape, not even light.

Formation of the Solar System

Our star, the sun, is believed to have formed less than 5 billion years ago, some 10 billion years after the Big Bang.

Because our solar system contains all elements, including the heavy elements that are thought to have been formed only in a supernova, the solar system must have formed from material at least partially originating in a supernova.

The solar system is thought to have originated when a cloud of gas and dust was disturbed, perhaps by a supernova, which caused it to collapse and spin (**Fig. 1.3**). Eventually the cloud grew hotter and denser in its center, with this central mass surrounded by a disk of gas and dust. The central mass condensed sufficiently to begin fusion and form our sun. The disk became progressively thinner, the rotation rate increased, and particles within the disk began to come together to form clumps. Some clumps grew larger, as particles and other small clumps stuck to them.

In addition to light, all stars produce an outward stream of subatomic particles such as protons and electrons known as the "**solar wind**." The solar wind from our sun had enough energy to blow hydrogen, helium, and other light gaseous elements out of the inner portion of the solar system. Thus, the composition of the planets varies with distance from the sun. Mercury, which is closest to the sun, consists mostly of iron. Venus, Earth, and Mars have less iron, but greater amounts of elements such as silicon, magnesium, and oxygen. The outer planets—Jupiter, Saturn, Uranus, and Neptune—are composed mostly of compounds of the light elements, primarily methane and ammonia. The eight planets all orbit within the same plane. Pluto, once considered to be the ninth planet, has an orbit that is angled to this plane. As a result, Pluto is thought to have formed by a different process and is now considered to be an **asteroid** rather than a planet.

1.4 FORMATION OF THE EARTH, MOON, AND OCEANS

As material from the rotating disk accumulated to form the planets, the material of each planetary mass was compressed by gravity, which caused the planets' temperatures to rise. Additional heat was supplied by the decay of **radioactive** elements. For the inner planets, including the Earth, an important stage in development was reached when they became warmer than the melting point of iron and nickel. These two high-**density** metals migrated toward the center of the forming planets. The lower-density material, rich in silicon, aluminum, carbon, and oxygen, migrated preferentially upward to the cooler surface, where it formed a rocky **crust**. This segregation by density, called "density **stratification**" (**CC1**), led to the layered structure that exists today on the Earth and almost certainly on the other inner planets. The Earth's layered structure is described in Chapter 4.

A similar density layering process occurred in all the planets. However, in the outer planets, which are rich in hydrogen and helium, the layers formed into a deep atmosphere of gaseous hydrogen and helium. It is believed that below that atmosphere is a solid layer of metallic hydrogen (solidified by extreme pressure and low temperature) and probably a small core of heavier elements, perhaps also arranged in density-stratified layers.

When the Earth first formed, it is believed that its original atmosphere was blown away by the solar wind, which was much stronger at that time. However, as iron and nickel migrated toward the core and lighter elements migrated toward the surface, gases, including hydrogen, hydrogen chloride, carbon monoxide, carbon dioxide, nitrogen, and water vapor, migrated to the surface and were released during the intense volcanism that was part of the reorganization process. These gases formed the Earth's early atmosphere. Note that there was virtually no oxygen in the early atmosphere. This is because any released oxygen reacted rapidly with elements, especially iron, that were abundant in the Earth's crust.

The Earth's surface slowly cooled until the water vapor in the atmosphere condensed and accumulated on the surface to form the oceans. For perhaps as long as 25 million years, water that fell as rain evaporated quickly because the Earth's surface was hotter than it is today. During this time, many elements were dissolved from the rocks and provided the salts in the early seawater. In addition, gases from the atmosphere were dissolved in the rain. Carbon dioxide was dissolved to form carbonic acid, much of which reacted with the crustal rocks to form carbonates.

Water has continued to be added to the oceans over time as more of the water vapor that was trapped in the mantle when the Earth was formed continues to be released through volcanism. In the 1980s and 1990s, many scientists supported the hypothesis that, because comets are believed to contain substantial amounts of water (as ice), much of the water in the oceans may have been derived from comets that impacted the Earth's surface. Recent studies of the composition of comets indicate that this hypothesis is unlikely.

Why Oceans on the Earth and Not on Venus or Mars?

In terms of size, composition, and structure, Venus, Earth, and Mars are quite similar, yet only the Earth has oceans. Why is this so? We do not know the complete answer to this question, although we now know that oceans may have existed on Mars at some time in the past. Some scientists believe that Venus, too, may have had oceans early in its history. Indeed, it is now believed that these three planets may have had very similar early histories and similar early atmospheres and oceans. However, the atmosphere of Venus is now extremely dry, and its surface is too hot for liquid water to exist. Mars has a very thin atmosphere with little water vapor and appears to have only a limited amount of water deposited as ice in its polar regions.

Scientists currently believe that, at one time, water may have been abundant both on Mars and on Venus, but that this water has largely been lost to space. On Venus, water in the atmosphere may have decomposed to hydrogen and oxygen by the action of ultraviolet radiation, which is more intense on Venus than on the Earth because Venus is closer to the sun. The hydrogen was then progressively lost to space. Water vapor on Mars may have been lost to space through the very thin atmosphere by sublimation (evaporation of water vapor directly from ice). It is also possible that some water is locked away in ice in the Martian polar regions and below the Martian surface.

The surface of Venus is much hotter (about 480°C) than the Earth's surface. The temperature difference is due partly to the fact that Venus orbits closer to the sun. However, the surface of Mercury is not hotter than the surface of Venus, even though Mercury orbits much closer to the sun than Venus. Thus, the primary reason for the high surface temperature on Venus is that its atmosphere consists mostly of carbon dioxide and, as a result, traps heat much more effectively than the Earth's atmosphere does. Earlier in this chapter and in **CC9**, we discuss the role of carbon dioxide gas in the greenhouse effect on the Earth. Some scientists have suggested that Venus may be an extreme model of what could occur on the Earth if the carbon dioxide concentration continues to rise in our atmosphere.

The Earth, it seems, had a unique set of conditions that allowed it to develop and retain oceans. Several factors may be involved, including the following:

- The Earth and Mars both rotate much faster (one full revolution every 24 and 24.62 hours, respectively) than Venus (one revolution every 243 days). This difference is important for two reasons. First, the faster rotation reduces the temperature variations between day and night. Second, faster rotation of the iron/nickel core produces a stronger planetary magnetic field, and this magnetic field deflects much of the solar wind.
- The Earth's orbit is more nearly circular than the more elliptical orbits of Mars and other planets. This more nearly circular orbital shape minimizes the seasonal range of globally averaged solar radiation and thus surface temperatures on the Earth.
- Venus is closer to the sun than is the Earth. As a result, ultraviolet radiation that reaches its upper atmosphere is more intense, which causes more water vapor to break down to hydrogen and oxygen. The stronger solar wind was also more effective at blowing away this hydrogen.
- Most carbon dioxide in the early atmosphere of the Earth was removed by rain and stored in carbonate rocks. However, sufficient carbon dioxide remained to create a moderate greenhouse effect that maintains most of the Earth's surface within the temperature range at which water is liquid. The oceans themselves contribute to this balance as they retain a large amount of heat and moderate atmospheric temperature changes (**CC5**).

Although we do not fully understand how all these factors combine to make the Earth the only ocean planet in the solar system, or what other factors may have been involved, the essential fact is that oceans exist on the Earth because of many unique characteristics of the Earth and its history.

Formation of the Moon

Very early in the Earth's history, perhaps tens of millions of years after it first formed about 4.6 billion years ago, a large planetary object is thought to have smashed into the Earth (**Fig. 1.3f,g**). Computer modeling suggests that this object was approximately 10% of the Earth's mass and one-half of the Earth's diameter. In the collision, most of the metallic core of the impacting object remained attached to the Earth, where it melted and merged with the Earth's own core by density stratification. However, much of the rocky mantle of the impacting object was ejected in the collision to form a ring of debris around the Earth that eventually was collected to form our moon (**Fig. 1.3h,i**).

1.5 OCEANS AND THE ORIGINS OF LIFE

To investigate the origins of life, a number of laboratory studies have attempted to reproduce the conditions thought to exist in the early oceans and atmosphere, including high ultraviolet radiation levels and frequent lightning discharges. Under these conditions, a wide range of organic compounds are created from carbon dioxide, methane, ammonia, hydrogen, and water. The organic compounds formed in these experiments included **amino acids**, which are considered the most important building blocks of the complex molecules needed for life to exist. However, we do not understand how or know where these building blocks were assembled into the much more complex molecules that were needed to form the first living matter. Several hypotheses have been proposed. The chemical reactions necessary to construct complex molecules may have taken place on the surface of solid particles, or deep within the Earth, or at **hydrothermal vents** (Chap. 17). However, another possible explanation for the origin of these complex compounds is that they first reached the Earth in meteorites.

All life as we know it depends on the transport of chemical elements and compounds within cells, and between cells and the surrounding environment. In all known living organisms, from **archaea** to mammals, chemical substances are transported dissolved in water. Therefore, water is essential to all life as we know it. As a result, it is not surprising that life appears to have been nurtured and developed in the oceans for most of the billions of years that it has existed on the Earth. Evidence from the **fossil** record indicates that life has existed on the Earth for a very long time. The first known life forms were bacteria-like microorganisms that existed about 3.6 billion years ago (**Fig. 1.5**), which is approximately 1 billion years after the Earth was formed. However, it is likely that more primitive forms of life were present much earlier in the Earth's history.

When the oceans were first formed, about 4 billion years ago, the atmosphere consisted mostly of nitrogen, with smaller amounts of carbon dioxide and methane. Because there was no free oxygen in the atmosphere or oceans, it is thought that some of the earliest life forms were similar to the **chemosynthetic** microbes that are found at present in many isolated environments such as hydrothermal vents and deep within the Earth's crust (**CC14**, Chap. 17). These environments contain no free oxygen and are often characterized by conditions of extreme temperature and pressure, similar to the conditions that may have existed in the oceans for many hundreds of millions of years after they were formed.

If indeed the first organisms utilized chemical energy to support their energy needs, then at some unknown time the first species of microorganisms that utilized the sun's light energy developed. These organisms were **photosynthetic** (**CC14**). Early photosynthetic species probably used hydrogen sulfide as their source of hydrogen, which is needed for building chemical compounds. Later, photosynthetic microorganisms developed that were able to split water molecules, using the hydrogen to build their chemical compounds and releasing oxygen in the process. Over a period of time, between about 1 and 2 billion years ago, these photosynthetic organisms changed the composition of the Earth's atmosphere. Oxygen concentrations steadily increased until they reached approximately the 20% now present. This

FIGURE 1.5 The history of the Earth. The Earth formed about 4.6 billion years ago. Animals have existed on the Earth for only an extremely brief period of its history. If the age of the Earth were only 1 year, animals would have existed for only 2 months, humans for about 6 hours, and the science of oceanography for less than a single second. Most actual dates on the figure are expressed as millions of years ago (MYA).

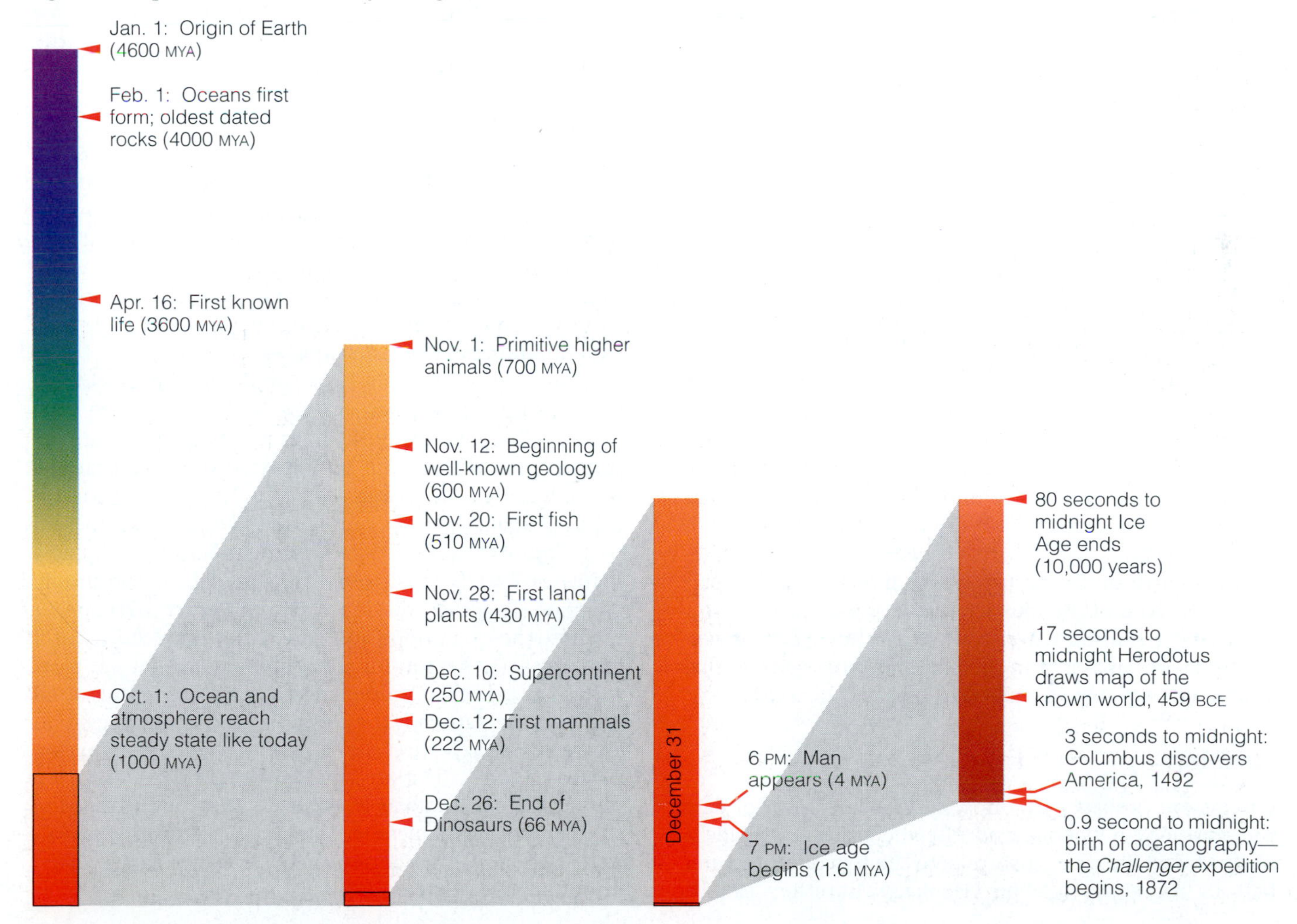
Jan. 1: Origin of Earth (4600 MYA)
Feb. 1: Oceans first form; oldest dated rocks (4000 MYA)
Apr. 16: First known life (3600 MYA)
Oct. 1: Ocean and atmosphere reach steady state like today (1000 MYA)
Nov. 1: Primitive higher animals (700 MYA)
Nov. 12: Beginning of well-known geology (600 MYA)
Nov. 20: First fish (510 MYA)
Nov. 28: First land plants (430 MYA)
Dec. 10: Supercontinent (250 MYA)
Dec. 12: First mammals (222 MYA)
Dec. 26: End of Dinosaurs (66 MYA)
December 31
6 PM: Man appears (4 MYA)
7 PM: Ice age begins (1.6 MYA)
80 seconds to midnight Ice Age ends (10,000 years)
17 seconds to midnight Herodotus draws map of the known world, 459 BCE
3 seconds to midnight: Columbus discovers America, 1492
0.9 second to midnight: birth of oceanography—the *Challenger* expedition begins, 1872

change was fundamental to the development of almost all living species now on the Earth. As the free oxygen accumulated in the atmosphere, it also reacted with sulfides and other chemical compounds in the oceans that support the life cycles of chemosynthetic species. As a result, chemosynthetic species disappeared or became restricted to a few extreme environments, such as those where they are found today. Furthermore, the free oxygen permitted the development of the animal kingdom. All species of animals depend on **respiration** using oxygen obtained from their environment.

The oceans and atmosphere reached a **steady state** (maintaining approximately the same chemical compositions that they have today) approximately 1 billion years ago, and this relative stability permitted the development of more complex life forms. The first primitive higher animals were marine **invertebrates**, perhaps similar to **sponges** and jellyfish, that developed about 700 million years ago, followed by the first fishes about 500 million years ago (**Fig. 1.5**). The first plants appeared on land about 430 million years ago, and the first mammals only about 220 million years ago. Humans are latecomers. Hominids (humanlike species) have existed for only about 4 million years—less than one-tenth of 1% of the history of the Earth.

Although the specific environment in which life first developed is not known, clearly the oceans have played a major role in nurturing the development of life on the Earth.

1.6 HOW TO STUDY OCEAN DATA

As is true for other sciences, describing what we know about the oceans and ocean processes requires the use of graphs and similar diagrams, and sometimes mathematics. For ocean sciences, we must also be able to represent data in a geographic context, which requires the use of maps or charts. Graphs, diagrams, and maps make it easy to understand certain features of even complex data sets without using mathematics (a goal we have set for this text). However, these visual representations can be properly interpreted only if we understand how a particular graph or map presents, and often distorts, the data. To make it possible for you to properly study and understand the rest of this text, the remainder of this chapter discusses some of the characteristics of graphs and maps that will be utilized extensively in this text. Even if you are a science major and are familiar with science diagrams, you are strongly urged to review this material.

Graphs

Graphs are probably the most widely used form of data presentation in science and elsewhere. They provide a means by which relationships between two or more numbers or properties can be visualized, but they can be extremely misleading unless they are read properly. To understand a graph, you must not only look at the general shape of the line or curve connecting points, or the apparent difference in sizes of bars in a bar chart, for example, but also examine the axes. Two simple examples illustrate why.

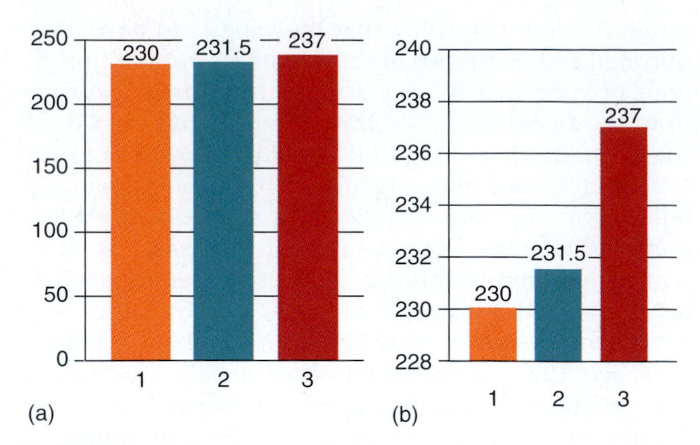

FIGURE 1.6 These two bar charts show the same data but use different y-axis ranges. Notice that they give very different impressions about the differences among the three values.

In **Figure 1.6**, three simple values that could be, for example, the prices of an item at different stores or the depths of the **thermocline** at three locations in the ocean are plotted in a simple bar chart. The same data are plotted in both parts of the figure. Why do the plots look so different? In **Fig. 1.6a**, the y-axis (vertical axis) extends from 0 to 250. In **Fig. 1.6b**, the data are "expanded" by plotting on the y-axis only the range from 228 to 240. Items 1, 2, and 3 have nearly the same values, but one plot seems to show that item 3 has a much higher value than items 1 and 2. Have you ever seen this technique used in advertisements or newspaper reports to present data differences in a misleading way?

The second example is a simple nonlinear (**CC10**) relationship between two variables ($x^2 = y$). These variables could be, for example, light intensity and depth, or primary production rate and **nutrient** concentrations (although the relationships between these pairs of parameters are actually more complicated). Notice that the first plot (**Fig. 1.7a**) shows that y increases more rapidly as x increases, exactly what we would expect from the simple equation. However, the same data in **Figure 1.7b** appear to show the exact opposite (y increases more slowly as x increases). The reason for this difference is that y is plotted as the log of the value. Log plots are commonly used for scientific data, but they may be confusing unless you examine the axis carefully. One of the reasons why log plots are used is illustrated by **Figure 1.7c**. This figure is a plot of the same data as in **Figure 1.7a,b**, but it has a log scale for both variables. The plot is a straight line, revealing a feature of this nonlinear relationship that is often important to scientists.

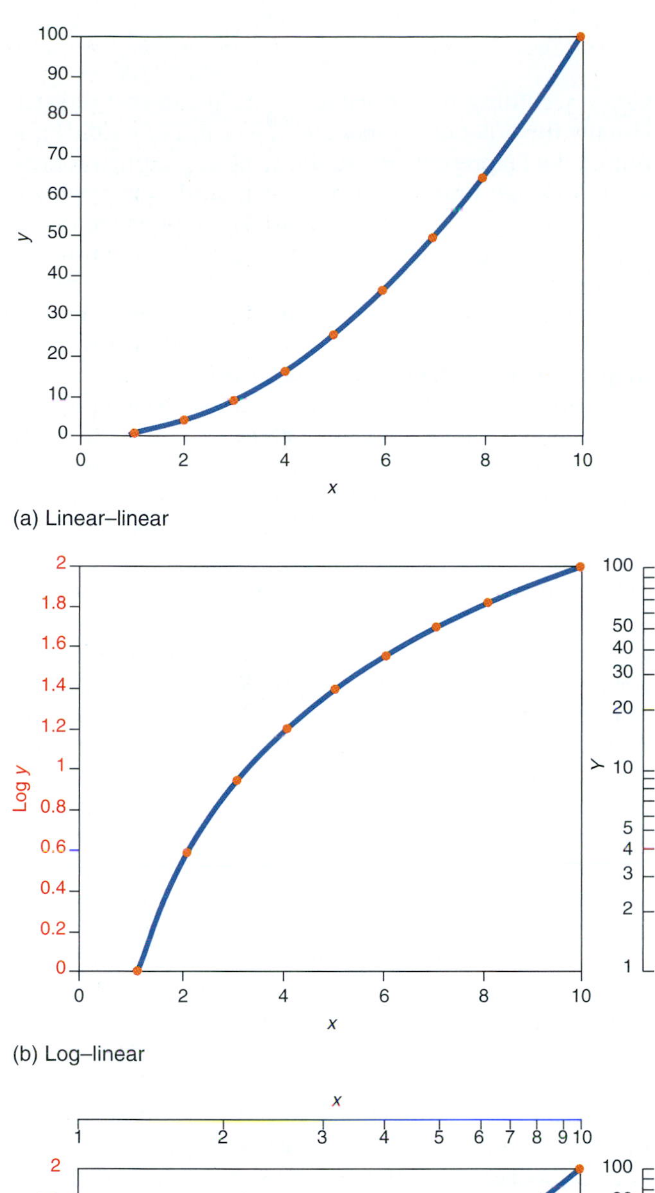

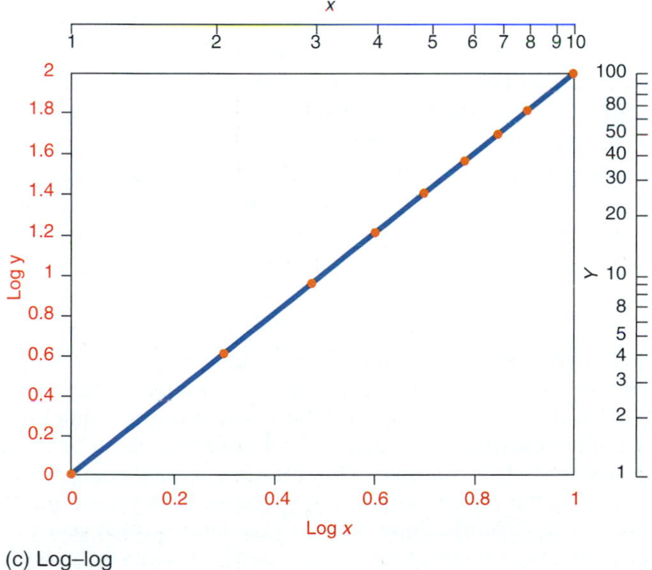

FIGURE 1.7 A simple nonlinear relationship is plotted in three different ways to show how logarithmic scales can change the apparent nature of the relationship between the two variables: (a) a linear–linear plot, (b) a log–linear plot, (c) a log–log plot.

Contour Plots and Profiles

Contour plots are used to display the two-dimensional **spatial** distribution of a variable such as atmospheric pressure on **weather** maps. The variable plotted can be any parameter such as pressure, temperature, soil moisture, vegetation type, or concentration of a substance. Contours of pressure are usually called **isobars**, temperature contours are **isotherms**, and density contours are **isopycnals**.

Contour plots describe the distribution of variables on flat surfaces. In some cases, the plotted variable is actually a measurement of something in the third dimension, perpendicular to the surface on which the plot is drawn. The most common examples are **topographic** maps that depict the height of the land surface above sea level, and **bathymetric** maps that depict the depth of the ocean below sea level. Sea level is the flat surface on which the third dimension variable, the height or depth, is plotted.

Contours are lines that connect points on the surface that have the same value of the variable. A number of such contour lines are drawn to represent different values of the plotted variable. Often, but not always, the values contoured are spaced at equal intervals of the variable, such as 50 m, 100 m, 150 m, and 200 m above sea level for a topographic map. One very important rule of contouring is that contours of different elevations (or values of any other parameter that is contour plotted) cannot cross or merge with each other. This is because each contour line always follows its own elevation (or parameter value). If the plotted parameter is elevated or depressed (forming hills or depressions) in two or more areas, contours around any two such features must not connect with each other. In addition, unless the feature is at the edge of the plot, contours around each such feature must connect with themselves in a closed loop (**Fig. 1.8**).

Contour lines show the distribution and magnitude of highs and lows of the plotted parameter. In any particular contour plot, high and low features (hills and valleys in a topographic map) that have more contours surrounding them are of greater magnitude than those with fewer surrounding contours. For example, higher hills are depicted by more contours between the sea level and the hilltop than would be present for lower hills. Contours can also reveal the strength of gradients at different points on the plotted surface. Where the contours are closely bunched together, the value of the parameter must change quickly with distance across the surface, and hence the gradient is relatively strong (steep terrain on a topographic map). Conversely, where the contours are spread out, the gradient is relatively weak. The term *relatively* is important because the number of contours and the intervals between the contours can be different in different plots, even plots of the same data (compare **Fig. 1.8a and b**).

When you study contour maps, make sure you look at the values on each of the contour lines. They will enable you to see which features are highs and lows and to assess accurately the relative magnitude of gradients at different points on the plot. Many contour plots are now produced by computers, and in some of these, contour lines are not used. Instead, the entire plot is filled with color that varies according to the value of the parameter plotted. Usually the order of colors in the spectrum of visible light is used. Red represents the highest value, grading progressively through orange, yellow, green, and blue to violet, which represents the lowest value. These plots are essentially contour plots that have an infinitely large number of contour lines and in which the range of values between each pair of adjacent contours is represented by a slightly different shade of color. The **Inside front cover map** and **Figures 7.11 and 10.18** are good examples. This convention, using the spectrum from red to violet to represent higher values to lower values, is now universally accepted and makes it easy to visually identify the areas of high and low values. Consequently, with a few exceptions made for specific reasons, we have used this convention to color-code the regions between contours in all of the contour plots in this text. **Figure 1.8** is an example.

Contour plots are often displayed on cross-sectional profiles. Cross-sectional profiles represent the distributions of properties on slices (cross sections) through, for example, the Earth or an ocean. Oceanographers commonly use cross-sectional profiles that represent a vertical slice through the Earth and oceans. Because the oceans and the Earth's crust through which these vertical profiles are drawn are extremely thin in comparison with the widths of the oceans and continents, these profiles almost always have a large vertical exaggeration. For example, the maximum depth of the oceans is approximately 11 km, whereas their widths are several thousand kilometers. If we were to draw a vertical profile with no vertical exaggeration on a page in this book, an ocean depth of 11 km scaled to 11 cm long would necessitate a diagram that had a width of 1000 cm (more than 50 page widths) for each 1000 km of ocean width plotted. Because this is obviously not possible, the scale on which distance across the ocean is plotted is reduced. This decreases the width of the plot but also produces a vertical exaggeration that distorts the data.

Figure 1.9 shows the effects of 10 and 100 times vertical exaggerations of a topographic profile. Most profiles used in ocean sciences have exaggerations greater than 100 to 1. Vertically exaggerated profiles make the gradi-

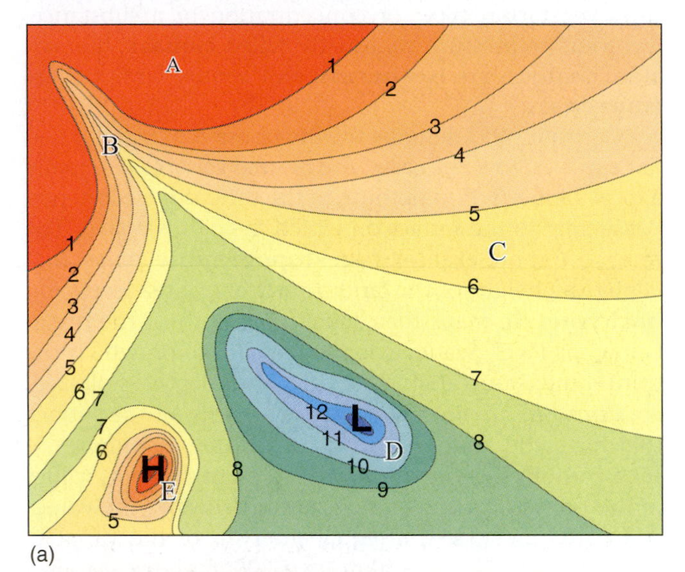

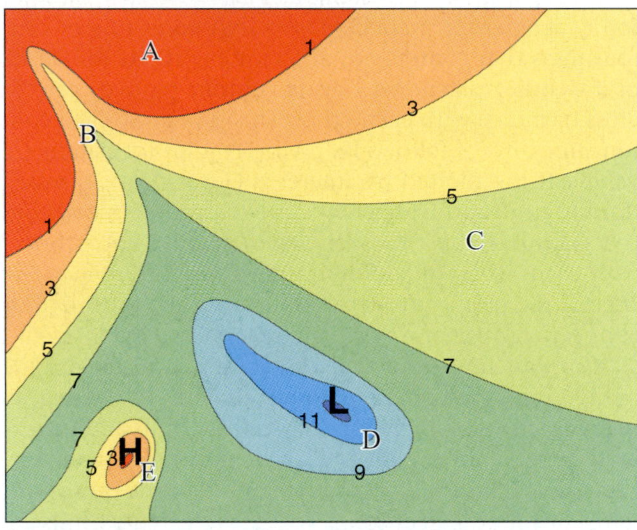

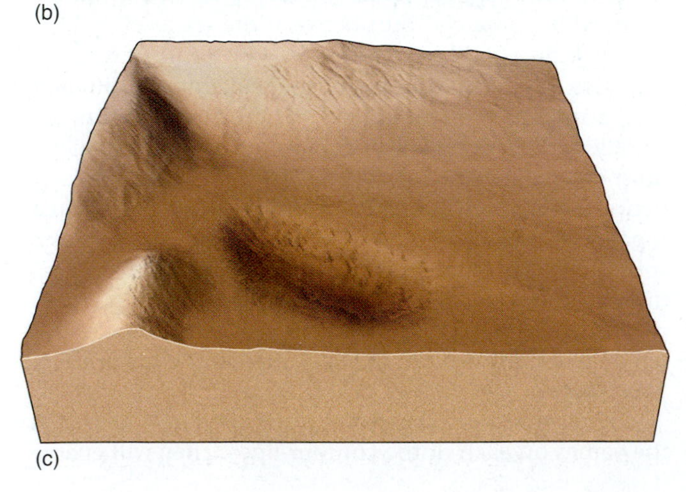

FIGURE 1.8 Contour plots. These plots reveal areas of higher and lower values of the plotted parameter, identified as H and L. They also indicate the strength of the gradient, which is stronger where adjacent contours are closer to each other. (a) In this example, which is a topographic map, there is a flat plateau at point A, a steep-sided valley at point B, an area of gentle slope at point C, a depression at point D, and a hill or mountain at point E, which has a very steep slope, especially on its right side. (b) Here, the same data are plotted with a greater interval between contours. Unless you examine it carefully, this plot can give the appearance that gradients are not as steep as they appear in part (a). (c) A three-dimensional representation of the topography plotted in parts (a) and (b).

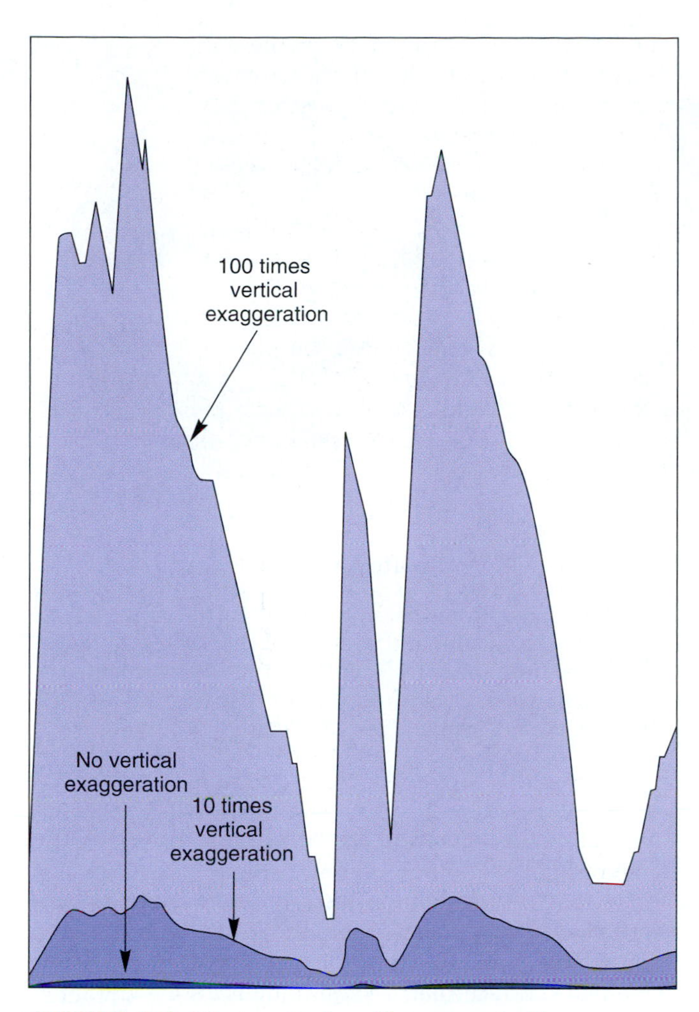

FIGURE 1.9 Most vertical profiles are vertically exaggerated. Topography appears progressively more distorted as the exaggeration increases. The relatively flat topography depicted in this figure looks like extremely rugged mountain chains with very steep slopes when it is plotted at a vertical exaggeration of 100 times. Most oceanographic profiles are plotted with vertical exaggerations greater than 100 times.

ents of topographic features appear much greater than they really are. Two points are important to remember as you examine vertically exaggerated profiles in this textbook. First, the seafloor topography is much smoother and the slopes are much shallower than depicted in the profiles. Second, ocean **water masses** are arranged in a series of layers that are very thin in comparison with their aerial extent, and this characteristic is not adequately conveyed by the profiles.

Maps and Charts

The Earth is spherical, and the ideal way to represent its geographic features would be to depict them on the surface of a globe. Although virtual-reality computer displays may eventually make such representation possible, it is currently impractical for most purposes. Consequently, geographic features of the Earth's surface are almost always represented on two-dimensional maps or charts (*chart* is the name given to maps that are designed and used for navigation). To represent the spherical surface of the Earth in two dimensions on a map, a projection must be used. A projection consists of a set of rules for drawing locations on a flat piece of paper that represent locations on the Earth's surface. All projections distort geographic information, each in a different way. Several different projections are used in this textbook, and you may see other projections elsewhere.

To draw accurate maps of the Earth, we must relate each location on the Earth's surface to a particular location on the flat map surface. Therefore, we must be able to identify each point on the Earth's surface by its own unique "address." **Latitude** and **longitude** are used for this purpose.

In a city, the starting "address" is usually a specific point downtown, and the numbers of streets and of houses on each street increase with distance from that point. On the Earth, there is no center from which to start. However, two points, the North and South Poles, are fixed (or almost so), and the equator can be easily defined as the circle around the Earth equidistant from the two poles. This is the basis for latitude. The equator is at latitude 0°, the North Pole is 90°N, and the South Pole is 90°S. Every location on the Earth other than at the poles or on the equator has a latitude between either 0° and 90°N or 0° and 90°S. Why degrees? **Figure 1.10a** shows that if we draw a circle around the Earth parallel to the equator, the angle between a line from any point on this circle to the Earth's center and a line from the Earth's center to the equator is always the same. If the circle is at the equator, this angle is 0°; if it is at the pole, the circle becomes a point and this angle is 90°. If the circle is not at the equator or one of the poles, this angle is between 0° and 90° and is either south or north of the equator.

Latitude can be measured without modern instruments. Polaris, the Pole Star or North Star, is located exactly over the North Pole, so at 90°N it is directly overhead. At lower latitudes, Polaris appears lower in the sky until, at the equator, it is on the horizon (at an angle of 90° to a vertical line). In the Northern Hemisphere, we can determine latitude by measuring the angle of the Pole Star to the horizon. In the Southern Hemisphere, no star is directly overhead at the South Pole, but other nearby stars can be used and a correction made to determine latitude. Star angles can be measured accurately with very simple equipment, and the best early navigators were able to measure these angles with reasonable accuracy without instruments.

Latitude is a partial address of a location on the Earth. It specifies only the hemisphere in which the location lies, and that the location is somewhere on a circle (line of latitude) drawn around the Earth at a specific distance from the pole (equivalent to "somewhere on 32nd

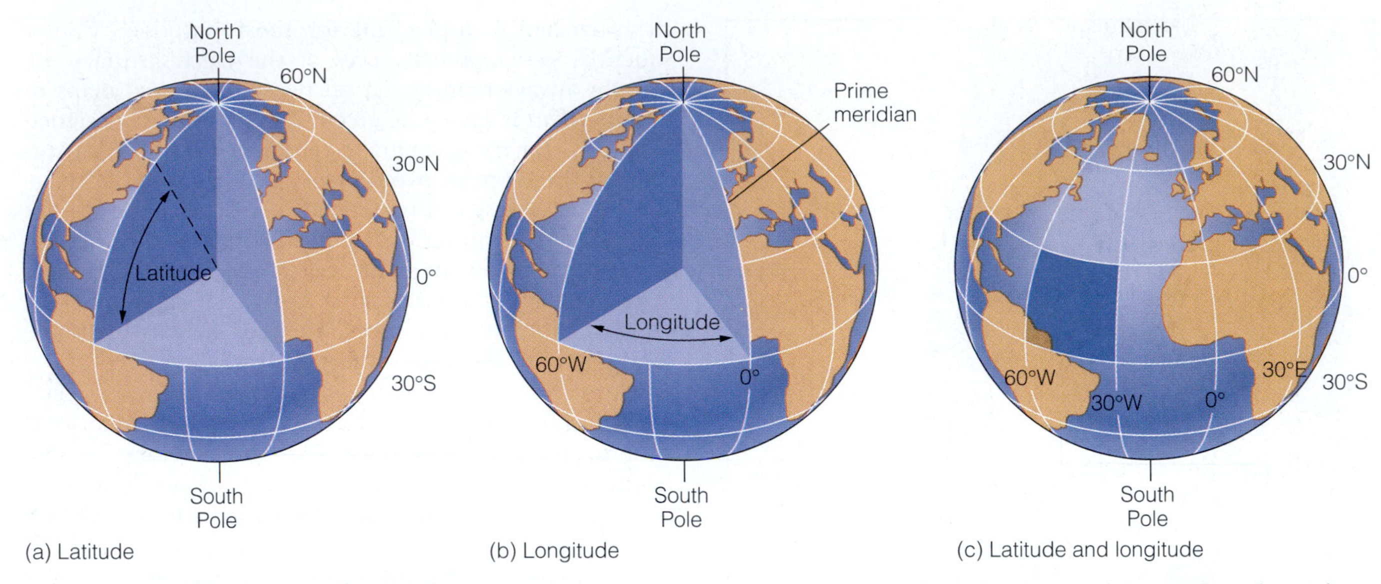

FIGURE 1.10 Latitude and longitude are measured as angles between lines drawn from the center of the Earth to the surface. (a) Latitude is measured as the angle between a line from the Earth's center to the equator and a line from the Earth's center to the measurement point. (b) Longitude is measured as the angle between a line from the Earth's center to the measurement point and a line from the Earth's center to the prime (or Greenwich) meridian, which is a line drawn from the North Pole to the South Pole passing through Greenwich, England. (c) Lines of latitude are always the same distance apart, whereas the distance between two lines of longitude varies with latitude.

Street"). The other part of the address is the longitude. **Figure 1.10b** shows that the relative locations of two points on a line of latitude can be defined by measurement of the angle between lines drawn from the Earth's center to each of the locations. This angle is longitude, but it has no obvious starting location. Consequently, a somewhat arbitrarily chosen starting location, the line of longitude (north–south line between the North and South Poles) through Greenwich, England, has been agreed upon as 0° longitude and is known as the "prime meridian."

Longitude is measured in degrees east or west of the prime meridian (**Fig. 1.10b**). Locations on the side of the Earth exactly opposite the prime meridian can be designated as either 180°E or 180°W. All other locations are between 0° and 180°E or between 0° and 180°W. Longitude is not as easy to measure as latitude, because there is no fixed reference starting point and no star remains overhead at any longitude as the Earth spins. The only way to measure longitude is to accurately fix the time difference between noon (sun directly overhead) at the measurement location and noon at Greenwich. The Earth rotates through 360° in 24 hours, so a 1-hour time difference indicates a 15° difference in longitude. To determine longitude, the exact time (not just the time relative to the sun) must be known both at Greenwich and at the measurement location. Before radio was invented, the only way to determine the time difference was to set the time on an accurate clock at Greenwich and then carry this clock to the measurement location. Consequently, longitude could not be measured accurately until the invention of the **chronometer** in the 1760s.

Latitude and longitude lines provide a grid system that specifies any location on the Earth with its own address and enables us to draw maps. Before we look at these maps, notice that 1° of latitude is always the same length (distance) wherever we are on the Earth, but the distance between lines of longitude is at a maximum at the equator and decreases to zero at the poles (**Fig. 1.10c**).

Figure 1.11a is the familiar map of the world. This representation of the Earth's features is a Mercator projection, which is used for most maps. For the Mercator projection, the lines of latitude and longitude are drawn as a rectangular grid. This grid distorts relative distances and areas on the Earth's surface. The reason is that, on the Earth, the distance between two lines of longitude varies with latitude (**Fig. 1.10c**), but the Mercator projection shows this distance as the same at all latitudes. Furthermore, the Mercator shows the distance between lines of latitude to be greater at high latitudes than near the equator, even though on the Earth's surface they are the same.

Why, then, has the Mercator projection been used for so long? The answer is that it preserves one characteristic that is important to travelers: on this projection, the angle between any two points and a north–south line can be used as a constant compass heading to travel between the points. However, the constant-compass-heading path is not the shortest distance between the two points. The shortest distance is a great-circle route. A great circle is a circle around the full circumference of the Earth. The Mercator projection suggests that the "direct" route

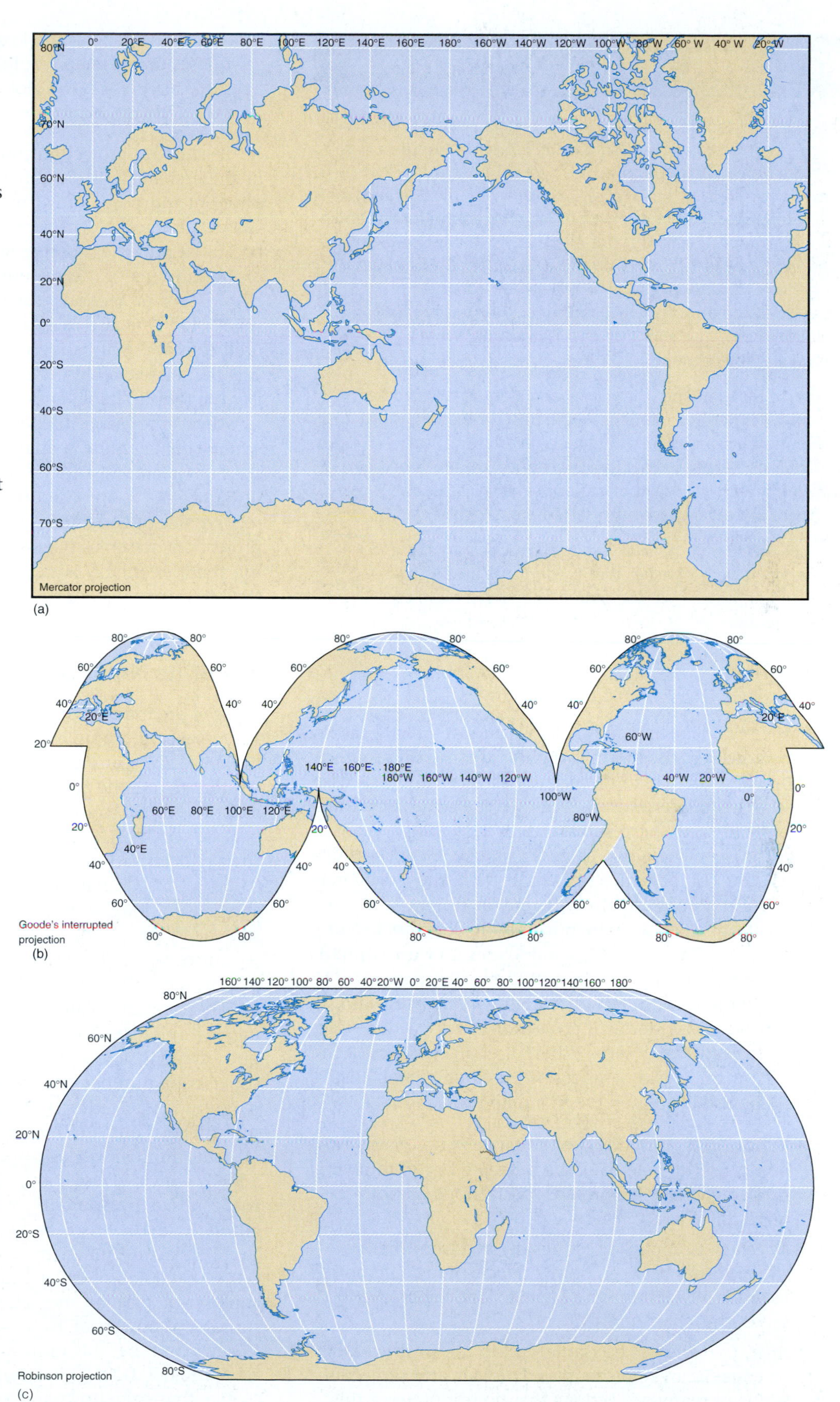

FIGURE 1.11 Typical map projections. (a) Mercator projection. Note that the map is cut off, so the higher latitudes near 90°N and 90°S are not shown. The reason is that the shape and area distortions introduced by this projection increase rapidly with latitude near the poles. (b) Goode's interrupted projection. This projection preserves relative areas and shows each of the oceans without interruption. However, it distorts the shapes of the continents. (c) Robinson projection. This projection preserves none of the four desirable characteristics perfectly, but it is a good approximation for many purposes.

CRITICAL THINKING QUESTIONS

1.1 We are concerned that the Earth's climate may change because of the increase in atmospheric carbon dioxide concentration that has resulted from human activities over the past century or more. It has been suggested that, since trees absorb carbon dioxide and release oxygen, we can solve the problem by replanting forests and planting many more trees in our urban areas. If we were to plant trees wherever they would grow on the entire planet, would this be enough to reverse the trend of increasing carbon dioxide concentration in the atmosphere? Explain the reasons for your answer. If you are not able to answer this question, what information would you need to do so?

1.2 If the universe is expanding, the galaxy and stars must have been much closer together at some time in the past. Therefore, if life had developed on many different planets, even if they were around different stars, we would have historical evidence of communication with life forms on other planets. Discuss the reasons (there are several) why this statement is incorrect.

1.3 A number of developing nations, especially those in Africa and South America, have suggested that world maps used in all textbooks and atlases should use a projection other than the Mercator projection. What do you think is their reason for this suggestion?

between San Francisco and Tokyo is almost directly east to west. However, the normal flight path for this route is almost a great-circle route that passes very close to Alaska and the Aleutian Islands. You can see the great-circle route and why it is the shortest distance by stretching a piece of string between these two cities on a globe. Over short distances the compass-direction route and the great-circle route are not substantially different in distance, and the simplicity of using a single unchanging compass heading makes navigation easy. Ships and planes now use computers to navigate over great-circle routes that require them to fly or sail on a continuously changing compass heading.

Although the Mercator projection is very widely used, other projections preserve different characteristics of the real spherical world. The following four characteristics would be useful to preserve in a map projection:

- *Relative distances.* Measured distance between any two points on the map could be calculated by multiplying by a single scale factor. Most projections do not preserve this characteristic.
- *Direction.* Compass directions derived from the straight lines between points on the map could be used for navigation. The Mercator projection preserves this characteristic.
- *Area.* Two areas of the same size on the Earth would be equal in area on the map. The Mercator projection is one of many projections that do not preserve this characteristic.
- *Shape.* The general shape of the oceans and land masses on the map would be similar to the shapes of these features on the globe. Most projections do not preserve this characteristic.

Because no projection (only a globe) preserves all four of these characteristics, different projections are used for different purposes. In this textbook, for example, a Goode's interrupted projection (**Fig. 1.11b**) is used often because it preserves relative areas correctly and shows each of the three major oceans uninterrupted by the edge of the map. Other projections, such as the Robinson projection (**Fig. 1.11c**), are also used in this textbook. The projection used for each map is identified in each figure. **Table 1.1** lists the relative ability of each of these projections to satisfy the four desirable map characteristics described here.

Always consider the characteristics of a projection when you examine a map. For example, compare the sizes of Greenland and South America in parts (a) and (b) of **Figure 1.11.** The Mercator projection exaggerates the relative size of Greenland, which the Goode's interrupted projection preserves. Generations have grown up thinking that Greenland is much bigger than it is, thanks to the Mercator projection.

Scientific Notation and Units

Some of the numbers mentioned in this chapter have been very large. For example, the history of the Earth presented in **Figure 1.5** spans nearly 5 billion (5,000,000,000) years. On that figure, the abbreviation MYA was used to represent millions of years ago. In ocean sciences, such large numbers are common, as are very small numbers. For example, the concentration of lead in seawater is about 0.000,000,000,5 gram per kilogram of water. In order to avoid using long strings of zeros, scientists use a type of shorthand for numbers like these. For example, the age of the universe is $5 \cdot 10^9$ years, and the concentration of lead in seawater is $5 \cdot 10^{-10}$ grams per kilogram.

These numbers might look odd, but they are really quite simple. The numbers are written with powers of 10. The • symbol can be stated as "to the power of," and the superscript number represents the number of orders of magnitude (factors of 10) by which the first digit must be multiplied. Thus, $5 \cdot 10^9$ is 5 to the power of 9 factors of 10 (the number is multiplied by 1 and nine zeros are added, which gives 5,000,000,000). The number is said to be raised by 9 orders of magnitude (powers of 10). Similarly, $5 \cdot 10^{-10}$ is 5 to the power of negative 10 (the number is multiplied by 0.000,000,000,1, which gives 0.000,000,000,5. Notice that the negative sign in the superscript part of the scientific notation denotes the number of powers of 10 by which the number is *reduced*. When a number has more than one nonzero digit, such as 5,230,000,000, the scientific notation is $5.23 \cdot 10^9$. Notice that this shorthand system makes it easy to compare two very large or two very small numbers. In our

TABLE 1.1 Characteristics of Selected Map Projections

Projection	*Preserves Relative Distances*	*Preserves Directions*	*Preserves Relative Areas*	*Preserves Shape*	*Areas Most Distorted*	*Common Uses*
Mercator	Poor	Excellent	Poor	Fair	Mid to high latitudes	Navigation charts, world maps, maps of limited areas
Miller's cylindrical	Good	Good	Excellent	Poor	High latitudes	World maps
Robinson	Good	Fair	Good	Fair	High latitudes, 4 "corners"	World maps
Mollweide	Good	Fair	Good	Poor	High latitudes, 4 "corners"	World maps
Goode's interrupted	Good	Poor	Very good	Poor in most areas	Relative positions of continents	World maps, global oceans
Conic	Good	Excellent	Very good	Excellent	Distortion equally distributed	Maps of the U.S., individual continents
Polar azimuthal	Fair	Excellent	Good/poor	Good	Outer edges	Maps of polar regions

example, it is easy to see that $5 \cdot 10^9$ and $5.23 \cdot 10^9$ are not very different without having to count all the zeros. This text makes extensive use of scientific notation. Appendix 1 provides a simple conversion table, if you need it.

Like other scientists, oceanographers use a variety of scientific units to identify such parameters as length, speed, time, and so on. To avoid problems with unit conversions, such as miles to kilometers, and to make the comparison of data easier, scientists have developed an International System of Units (abbreviated SI) that is steadily progressing toward being used universally, although some other units are still widely used. There are only seven base units:

- *Length:* meter, symbol m
- *Mass:* kilogram, symbol kg
- *Time:* second, symbol s
- *Electric current:* ampere, symbol A
- *Thermodynamic temperature:* kelvin, symbol K
- *Amount of a substance:* mole, symbol mol
- *Luminous intensity:* candela, symbol cd

All other units are "derived" units. For example, volume is measured in cubic meters, or m^3, and speed is measured in meters per second (m/s or $m \cdot s^{-1}$). SI units are used in much of this text. However, as is still common in science, they are not used exclusively, because some of them are, as yet, unfamiliar even to many members of the scientific community. Appendix 1 includes more information about SI units and a table of all scientific unit abbreviations used in this text.

CHAPTER SUMMARY

Introduction. The oceans influence our daily lives in many different ways. Understanding how they affect us requires an interdisciplinary approach that includes knowledge of the geology, physics, chemistry, and biology of the oceans and of how these aspects interact.

Most of the world's population lives near the oceans or a river that connects to the oceans. The oceans provide many resources, especially food and transportation corridors, but they are also susceptible to pollution, habitat damage, and other impacts of human activities. The oceans have been a source of fascination for humans throughout human history and have inspired generations of artists, poets, and writers. Recreational uses of the oceans and awareness of the unique nature of marine ecosystems and species have grown explosively in the past several decades.

The Oceans and the Earth's Environment. The Earth's climate is controlled by complex interactions between land, atmosphere, and ocean. Human activities have caused many changes in the global environment, including the oceans. The changes that might occur in the Earth's climate due to the release of greenhouse gases into the atmosphere are of particular concern. Exchanges between the oceans and the atmosphere affect the fate of these greenhouse gases.

Our Earth in the Universe. The universe is unimaginably large compared to the Earth and to distances, areas, or volumes within human experience. The universe formed from a single point about 15 billion years ago in a Big Bang explosion that left an expanding cloud of hydrogen and helium. Stars and galaxies condensed from this cloud. Young stars burn hydrogen by nuclear fusion to form helium. Small stars burn only hot enough to produce carbon and lighter elements by fusion reactions when their hydrogen is depleted. Eventually they become red giants and then eject their outer layers to become white dwarfs. Massive stars burn much hotter. When fusion reactions have converted the core of a massive star almost entirely to iron, the star explodes in a supernova and collapses to become a neutron star or black hole. Supernova explosions are so intense that they form radioactive isotopes and the heavier elements.

The solar system was formed about 5 billion years ago from a nebula that contained heavy elements created in a supernova. The solar wind, consisting of subatomic particles that stream outward from the sun, swept most of the light elements out from the inner part of the forming system. Thus, the inner planets were left to form primarily from the heavier elements and become rocky, while the outer planets became gas giants of mostly light elements, primarily in the form of ammonia and methane.

Formation of the Earth, Moon, and Oceans. During their formation, the inner planets, including the Earth, were hot, and they were rearranged by density stratification into layers that included an iron/nickel core and a rocky exterior. Atmospheres developed later from gases released in volcanic activity, and water was condensed from these atmospheres when the planets cooled. The Earth was the only planet where a balance of factors resulted in surface temperatures that allowed liquid water to form our oceans. During the millions of years it took to form the oceans, repeated evaporation and rainfall dissolved salts from the Earth's crust, and these accumulated to make the oceans salty. The moon was created early in the Earth's history when a planet-sized object collided with the Earth.

Oceans and the Origins of Life. Many organic compounds probably were formed in the early oceans, but it is not known how these simple compounds became much more complex to create the first life form. However, it is thought that all early life was found in the oceans. The first known life was bacteria-like, existed about 3.6 million years ago, and must have been chemosynthetic, as there was no oxygen in the early atmosphere. Eventually microorganisms developed that used photosynthesis, which splits water into hydrogen (which is utilized by the organism in its metabolism) and oxygen (which is released). About 1 to 2 billion years ago, photosynthetic organisms added oxygen to the atmosphere until it reached its current concentration of about 20%. Chemosynthetic organisms either died out or were restricted to limited oxygen-free environments, and species developed that depended on respiration using oxygen from their environment. Today, all species of animals and most other species respire and need oxygen in their environment. The first primitive higher animals, invertebrates such as sponges, appeared about 700 million years ago. Fishes developed about 500 million years ago. The first land plants and animals developed later, about 430 and 220 million years ago, respectively.

How to Study Ocean Data. Maps and charts are used extensively to display ocean science data. Each graph, map, or chart may represent data in a different way using different distortions of the real world. Graphs may have axes that do not start at zero or are nonlinear. The choice of axis can substantially change how the same set of data is perceived at first glance.

Contour plots are used extensively to represent the distribution of properties on a two-dimensional surface, such as the Earth's surface. The relative distance between two contours on a contour plot is a measure of the gradient in the property, but the absolute distance can be affected by the choice of contour interval. Most contour plots in this text are color-shaded between the contours. Red always represents the highest value of the parameter plotted, grading through the spectrum to blue as the lowest value.

Maps and charts are used to represent horizontal distributions. Profiles are vertical cross sections through the Earth or the oceans and generally display contours to show the vertical distribution of properties. Most profile plots in this text are greatly vertically exaggerated. This exaggeration is necessary because the depth of the oceans, or the thickness of the atmosphere or the Earth's crust, are extremely small compared to the width of the oceans or land masses.

The Earth is spherical, and a system of latitude and longitude has been developed to identify specific locations on this sphere. Latitude is referenced to the circle around the Earth at the equator, which is designated as 0° latitude. Other latitudes are expressed by the angle between a line from the Earth's center to the location in question and a line from the Earth's center to the equator. There are 90° of latitude; the North Pole is at 90°N and the South Pole at 90°S. Longitude is referenced to the prime meridian, a line designated as 0° that is drawn through the North and South Poles and that passes through Greenwich, England. Locations on the continuation of that same circle connecting the poles on the side of the Earth away from Greenwich are designated as both longitude 180°W and 180°E. Other longitudes lie between 0° and either 180°W or 180°E, and they are determined by the angle and direction between a line drawn between the Earth's center and the prime meridian and a line between the Earth's center and the specific location. One degree of latitude is the same distance regardless of the latitude or longitude. The distance represented by 1° of longitude varies from a maximum at the equator to zero at the poles.

Maps and charts must represent the spherical surface of the Earth or oceans in only two dimensions. No two-dimensional projection can correctly maintain relative distances, compass directions, relative areas, and the proper shape of features on a sphere. Therefore, all maps and charts distort one or more of these characteristics. The Mercator projection, which is the most widely used in atlases, depicts the distance between degrees of longitude as the same, regardless of latitude. Thus, this projection preserves the relative directions between locations, but it distorts all three of the other relationships. The distortion is not important in regional maps of low latitudes, but it becomes greater at high latitudes and for global maps. Ocean scientists and this text use a variety of other projections for specific purposes. For example, Goode's interrupted projection is often used because it generally preserves relative areas and distances and it can be drawn to represent all the major oceans without having to split them at the edge of the map.

To represent very large or very small numbers, scientists use a scientific notation based on powers of 10. A standard system of scientific units (SI) is now in place and is becoming more widely used, but it is not yet universal. Both scientific notation and SI units are used extensively in this text.

STUDY QUESTIONS

1. What causes a greenhouse effect?
2. What causes stars to burn and emit light?
3. Why don't stars burn forever?
4. Could a planet with a rocky crust and molten iron/nickel core have been formed around one of the first stars created after the Big Bang?
5. Why do the atmospheres of the outer planets consist mainly of methane and ammonia?
6. Why weren't the first living organisms capable of photosynthesis?
7. What are the essential things to look at the first time you study a graph?
8. What information can we get from studying a contour plot?
9. What is vertical exaggeration in a profile?
10. Are lines of longitude parallel to each other?
11. What are the four characteristics we would like a map projection to preserve?
12. On a Mercator projection map of the world, which countries or areas appear larger than they really are compared to other areas on the map?

KEY TERMS

You should recognize and understand the meaning of all terms that are in boldface type in the text. All those terms are defined in the Glossary. The following are some less familiar key scientific terms that are used in this chapter and that are essential to know and be able to use in classroom discussions or exam answers.

bathymetric (p. 13)
chemosynthetic (p. 11)
chronometer (p. 16)
contour (p. 13)
crust (p. 9)
fossil (p. 11)
galaxy (p. 7)
greenhouse effect (p. 5)
invertebrates (p. 12)
isobar (p. 13)
isopycnal (p. 13)
isotherm (p. 13)
latitude (p. 15)
longitude (p. 15)
nebula (p. 7)
photosynthetic (p. 11)
respiration (p. 12)
solar wind (p. 9)
steady state (p. 12)
stratification (p. 9)
supernova (p. 8)
topographic (p. 13)

CRITICAL CONCEPTS REMINDER

CC1 **Density and Layering in Fluids** (p. 9). The Earth and all other planets are arranged in layers of different materials sorted by their density. To read **CC1** go to page 2CC.

CC5 **Transfer and Storage of Heat by Water** (p. 10). The heat properties of water are a critical element in maintaining a climate on Earth that is suitable for life as we know it. To read **CC5** go to page 15CC.

CC9 **The Global Greenhouse Effect** (pp. 5, 10). Perhaps the greatest environmental challenge faced by humans is the prospect that major climate change may be an inevitable result of our burning of fossil fuels. The burning of fossil fuels releases carbon dioxide and other gases into the atmosphere, where they accumulate and act like the glass of a greenhouse retaining more of the sun's heat. To read **CC9** go to page 22CC.

CC10 **Modeling** (pp. 7, 12). Complex environmental systems, including the oceans and atmosphere, can best be studied by using conceptual and mathematical models. To read **CC10** go to page 26CC.

CC11 **Chaos** (p. 7). The nonlinear nature of many environmental interactions makes complex environmental systems behave in sometimes unpredictable ways. It also makes it possible for these changes to occur in rapid, unpredictable jumps from one set of conditions to a completely different set of conditions. To read **CC11** go to page 28CC.

CC14 **Photosynthesis, Light, and Nutrients** (p. 11). Chemosynthesis and photosynthesis are the processes by which simple chemical compounds are made into the organic compounds of living organisms. The oxygen in Earth's atmosphere is present entirely as a result of photosynthesis. To read **CC14** go to page 46CC.

CHAPTER 2

CHAPTER OUTLINE

CRITICAL CONCEPTS USED IN THIS CHAPTER

The dawn of ocean sciences. HMS *Challenger* is made fast to St. Paul's Rocks on the Mid-Atlantic Ridge—a hazardous task—during the first part of its 1872–1876 expedition.

History and Challenges of Ocean Studies

Until 1930, humans had never descended into the oceans deeper than about 150 m, and thus they had never visited the seafloor below that depth, an area representing about 60% of the Earth's surface. In addition, humans had never traveled below the thin surface layer of the oceans, whose depth averages almost 3.9 km and extends to as deep as 11 km. The first man to descend beyond 150 m was William Beebe. In 1934, in his sealed metal sphere, or "bathysphere," hanging on a cable several hundred meters below the ocean surface, Beebe wrote:

> *I have seen and felt the heat of molten, blazing stone gushing out of the heart of our Earth; I have climbed three and a half miles up in the Himalayas and floated in a plane still higher in the air, but nowhere have I felt so completely isolated as in this bathysphere, in the blackness of ocean's depths. I realize the unchanging age of my surroundings; we seemed like unborn embryos with unnumbered geological epochs to come before we should emerge to play our parts in the unimportant shifts and changes of a few moments in human history. Man's recent period of strutting upon the surface of the earth would have to be multiplied half a million times to equal the duration of existence of this old ocean.*
>
> —William Beebe, ***Half Mile Down,*** published in 1934 by Harcourt, Brace, New York, (pp. 173–74)

By capturing the immensity of the oceans and their history in this passage, Beebe places in perspective the brief moment in time that

humans have had to study the oceans. Still to this day, humans have visited only a tiny fraction of the seafloor and an even tinier fraction of the vast volume of deep-ocean waters.

The oceans have played a key role throughout human history. They have provided sustenance for humans living in villages and cities, both ancient and modern. They have facilitated long-distance trade from the time of the ancient Phoenicians to the mercantile system of the eighteenth century to the modern-day global economy; served for transport of armies, including those of Marc Antony, William the Conqueror, and Cortez; and provided the arena for countless naval battles. They currently offer a source of recreation for hundreds of millions of people. Perhaps more important to humans than any of these functions is the oceans' key role in mediating global climate and local weather. This role is just beginning to be understood.

Given the enormous importance of the oceans to human existence, one might expect our knowledge and understanding of them to be extensive and thorough. However, the oceans, which cover approximately 70% of the Earth's surface to an average depth of 4 km, are largely a vast domain that remains unvisited and unexplored. Amazingly, until the 1930s no human had ever descended below 150 m.

In this chapter we will review some of what is known about the oceans. We will also examine why the oceans are a difficult environment in which to conduct research. Compared to other sciences, oceanography is young, and new, sometimes startling discoveries are made almost every month.

2.1 WHAT IS OCEANOGRAPHY?

Until recent decades, oceanography has been primarily an observational discipline with greater similarity to geography than to other sciences. For many thousands of years, study of the oceans was limited almost entirely to exploring the surrounding lands and to mapping the oceans and such shallow underwater obstructions to navigation as **reefs.** Such studies were important to facilitate trade and travel between known landmasses and islands. Considerable information was also gathered about the best places to find fish and **shellfish,** especially in the coastal oceans. However, for many centuries this biological information was documented only in the oral histories of local fishers.

Aside from limited studies by scholars of a few ancient civilizations, almost all systematic studies of the oceans have been carried out within the past 200 years. During these two centuries, mapping of the oceans has continued with ever-increasing sophistication (**Table 2.1**). Geological oceanographers have mapped seafloor **topography** and sediments. Physical oceanographers have mapped distributions of salt content, temperature, and **currents**. Chemical oceanographers have mapped distributions of chemicals in seawater and in seafloor sediment. Biological oceanographers have mapped distributions of plant and animal species. Mapping the oceans has been an enormous task that is far from complete, for reasons we examine later in this chapter.

During the past several decades, oceanographers have moved beyond simply mapping the oceans to studying the processes that occur in them. As revealed in this text, the geological, physical, chemical, and biological processes in the oceans and overlying atmosphere are complicated and intimately interdependent. Consequently, oceanography (now often called "ocean sciences") must be interdisciplinary.

To perform their work properly, oceanographers must understand all basic ocean processes, although they often specialize in one aspect. In oceanography, much information must be integrated to describe complicated processes. Therefore, the scientist must take a step beyond many of the traditional sciences that rely on reductive methodology, in which investigators reduce the complexity of a problem by studying only part of it (e.g., studying the effect of varying temperature on a single **species** rather than the combined effects of all environmental variables and competition from other species). Oceanographers must integrate and compare many observations of the ever-changing marine **environment,** often by using computer models (**CC10**), as well as performing reductive methodology experiments in controlled laboratory or field conditions.

2.2 EXPLORATION AND MAPPING

Through most of human history, the oceans have been studied primarily to facilitate seaborne travel and transport. Hence, early studies were directed primarily toward exploring and mapping the oceans.

Prehistory

We do not know when humans first began to use the oceans as a food source and for transportation—the two ocean uses that remain the most valuable to us today. However, humankind has used these resources for tens of thousands of years. The bones of marine fishes discarded by Stone Age people have been found in settlements in Europe that date back to about 40,000 BCE. Artifacts, rock carvings, and paintings tell us that cave-dwelling people were using harpoons, fishhooks, nets, and primitive boats by the late Stone Age, or about 6000 BCE. The first boats were probably built much earlier, perhaps for fishing on lakes or in the shallow coastal ocean, then later for transportation and colonization. The first boats or rafts may even have been made and used by *Homo erectus*, our immediate predecessor in the evolutionary tree. Recent evidence shows that *Homo erectus* lived on the island of Flores, Indonesia, at least 750,000 years ago. To reach Flores, *Homo erectus* must have crossed a wide, deep water strait that acted as a barrier to the migration of most other species.

Early human boats were made of wood, reeds, animal skins, or tree bark. These materials usually do not survive the centuries for archaeologists to study. Indirect evidence suggests that such boats carried the Polynesians across the expanses of the Pacific Ocean to begin their colonization of that ocean's many small islands around 4000 BCE. Much earlier use of boats for transportation is likely but has not been proven.

Ocean Exploration in Early Civilizations

Recorded history tells us that early development of ocean transportation and trade centered in the Mediterranean Sea, although the Polynesian and Micronesian cultures in the Pacific Ocean may have pursued these activities during the same period.

The Mediterranean. The Minoan civilization, which prospered on the island of Crete in the Aegean Sea from about 3000 BCE, is considered the first recorded civilization to have used boats extensively for transport, trade, defense, and conquest. The Minoans' influence extended throughout the many islands of the Aegean, and legend records that a Minoan navy fought and controlled pirates in the region. Although seafaring capability continued to develop in the Mediterranean in both the ancient Greek and Egyptian civilizations, the Phoenicians, who inhabited areas that are now parts of Israel, Lebanon, and Syria, were the greatest of all the early Mediterranean seafarers. From about 1100 to 850 BCE, the Phoenicians were a great sea power, voyaging throughout the Mediterranean to Spain, Italy, North Africa, and even the British Isles, where they traded for tin. The Phoenicians also claimed to have made a 3-year voyage around the entire continent of Africa, but that claim has not been confirmed.

Much of the information needed to navigate from one port to another was a closely held secret of the navigator's art. However, maps that showed the shapes and sizes of **coastlines** and seas were made several thousand years ago. In about 450 BCE, the Greek historian Herodotus drew a map that is surprisingly recognizable as a generally accurate map of the Mediterranean and Red Sea region (**Fig. 2.1**). The most famous of ancient

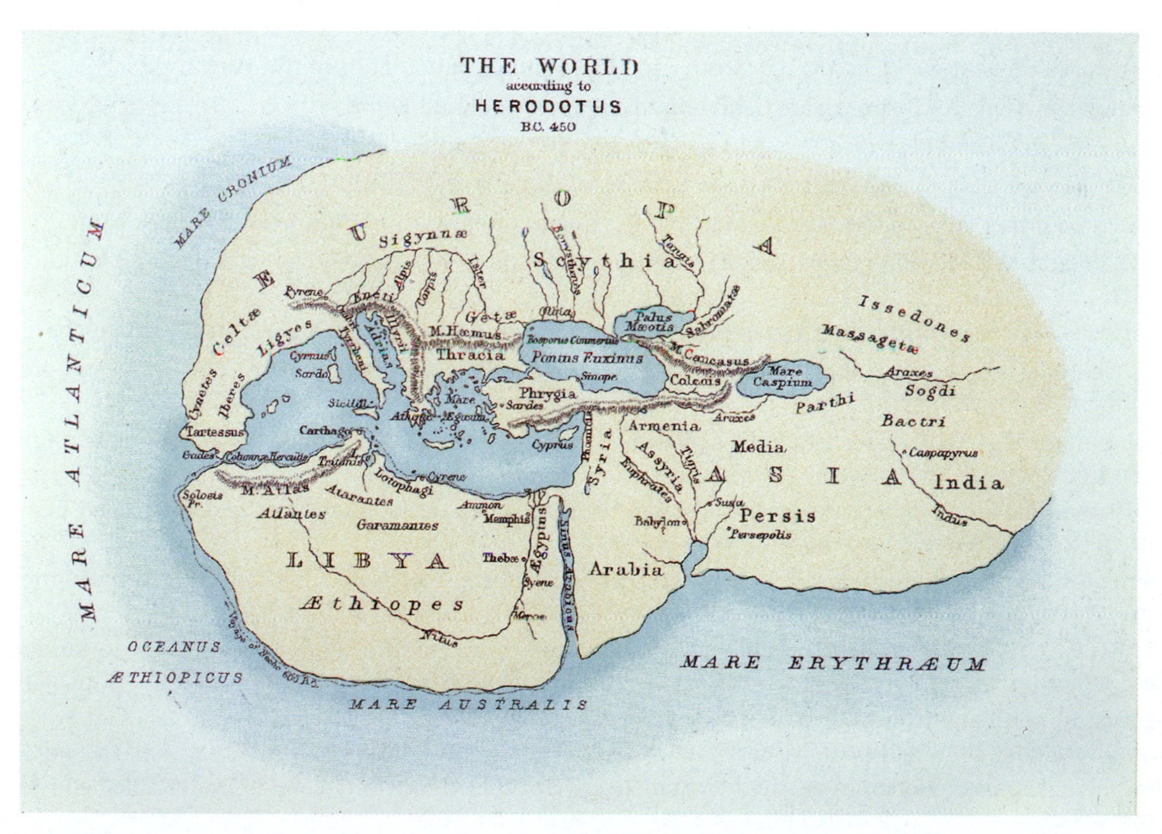

FIGURE 2.1 Map of the known world drawn by Herodotus in approximately 450 BCE. The oceans surrounding the known landmasses were thought to extend to the edge of the world. Compare this map carefully to a current Mercator projection, and you will see that the map drawn by Herodotus somewhat distorts longitude but quite accurately reproduces latitudes.

TABLE 2.1 Some Milestones in the History of Ocean Study

ca. 4000 BCE The Egyptians developed shipbuilding and ocean-piloting capabilities.

2000–500 BCE The Polynesians voyaged across the Pacific Ocean and settled all the major islands.

ca. 1100–850 BCE The Phoenicians explored the entire Mediterranean Sea and sailed into the Atlantic to Cornwall, England.

450 BCE The Greek Herodotus compiled a map of the known world centered on the Mediterranean region.

325 BCE The Greek Pytheas explored the coasts of England, Norway, and perhaps Iceland. He developed a means of determining latitude by measuring the angle between the North Star and the horizon, and he proposed a connection between the phases of the moon and the tides.

325 BCE Aristotle published *Meteorologica,* which described the geography of the Greek world, and *Historia Animalium,* the first known treatise on marine biology.

276–192 BCE The Greek Eratosthenes, a scholar at Alexandria, determined the circumference of the Earth with remarkable accuracy, using trigonometry and measurements of the angle of the sun's rays at two locations: Alexandria and Syene (now called Aswan), Egypt.

54 BCE – 30 CE The Roman Seneca devised the hydrologic cycle to show that, despite the inflow of river water, the level of the ocean remains stable because of evaporation.

ca. 150 CE The Greek Ptolemy compiled a map of the Roman World.

673–735 CE The English monk Bede published *De Temporum Ratione,* which discussed lunar control of the tides and recognized monthly tidal variations and the effect of wind drag on tidal height.

982 CE Norseman Eric the Red completed the first transatlantic voyage, discovering Baffin Island, Canada.

995 CE Leif Eriksson, son of Eric the Red, established the settlement of Vinland in what is now eastern Canada.

1452–1519 Leonardo da Vinci observed, recorded, and interpreted characteristics of currents and waves, and he noted that fossils in Italian mountains indicated that sea level had been higher in the ancient past.

1492 Christopher Columbus rediscovered North America, sailing to the islands of the West Indies.

1500 Portuguese navigator Pedro Álvars Cabral discovered and explored Brazil.

1513 Juan Ponce de León described the swift and powerful Florida current.

1513–1518 Vasco Núñez de Balboa crossed the Isthmus of Panama and sailed in the Pacific Ocean.

1515 Italian historian Peter Martyr (Pietro Martire d'Anghiera) proposed an origin for the Gulf Stream.

1519–1522 Ferdinand Magellan embarked on a circumnavigation of the globe; Juan Sebastián de Elcano completed the voyage.

1569 Gerardus Mercator constructed a map projection of the world that was adapted to navigational charts.

1674 British physicist-chemist Robert Boyle investigated the relations among temperature, salinity, and pressure with depth and reported his findings in *Observations and Experiments on the Saltiness of the Sea.*

1725 Italian naturalist-geographer Luigi Marsigli compiled *Histoire Physique de la Mer,* the first book pertaining entirely to the science of the sea.

1740 Swiss mathematician Leonhard Euler calculated the magnitude of the forces that generate ocean tides and related them to the attractive force of the moon.

1770 Benjamin Franklin published the first chart of the Gulf Stream, which was used by ships to speed their passage across the North Atlantic Ocean.

1768–1771, 1772–1775, 1776–1780 Captain James Cook commanded three major ocean voyages, gathering extensive data on the geography, geology, biota, currents, tides, and water temperatures of all the principal oceans.

1802 American mathematician-astronomer Nathaniel Bowditch published the *New American Practical Navigator,* a navigational resource that continues to be revised and published to this day.

1807 President Thomas Jefferson mandated coastal charting of the entire United States and established the U.S. Coast and Geodetic Survey (now the Office of Coast Survey).

Note: CE = of the common era; BCE = before the common era. These designations are equivalent to the more traditional abbreviations AD (anno Domini) and BC (before Christ), respectively.

TABLE 2.1 Some Milestones in the History of Ocean Study (continued)

1817–1818 Sir John Ross ventured into the Arctic Ocean to explore Baffin Island, where he successfully sounded the bottom and recovered sea stars and mud worms from a depth of 1.8 km.

1820 London physician Alexander Marcet noted that the proportion of the chemical ingredients in seawater is unvarying in all oceans.

1831–1836 The epic journey of Charles Darwin aboard HMS *Beagle* led to a theory of atoll formation and later the theory of evolution by natural selection.

1839–1843 Sir James Ross led an expedition to Antarctica, recovering samples of deep-sea benthos down to a maximum depth of 7 km.

1841, 1854 Sir Edward Forbes published *The History of British Star-Fishes* (1841) and then his *Distribution of Marine Life* (1854), in which he argued that sea life cannot exist below about 600 m (the so-called azoic zone).

1855 Matthew Fontaine Maury compiled and standardized the wind and current data recorded in U.S. ship logs and summarized his findings in *The Physical Geography of the Sea.*

1868–1870 Sir Charles Wyville Thomson, aboard HMS *Lightning* and HMS *Porcupine,* made the first series of deep-sea temperature measurements and collected marine organisms from great depths, disproving Forbes's azoic zone theory.

1871 The U.S. Fish Commission was established with a modern laboratory at Woods Hole, Massachusetts.

1872–1876 Under the leadership of Charles Wyville Thomson, HMS *Challenger* conducted worldwide scientific expeditions, collecting data and specimens that were later analyzed in more than 50 large volumes of the *Challenger Reports*.

1873 Charles Wyville Thomson published a general oceanography book called *The Depths of the Sea.*

1877–1880 American naturalist Alexander Agassiz founded the first U.S. marine station, the Anderson School of Natural History, on Penikese Island, Buzzards Bay, Massachusetts.

1884–1901 USS *Albatross* was designed and constructed specifically to conduct scientific research at sea and undertook numerous oceanographic cruises.

1888 The Marine Biological Laboratory was established at Woods Hole, Massachusetts.

1893 The Norwegian Fridtjof Nansen, aboard the *Fram,* which had a reinforced hull for use in sea ice, studied the circulation pattern of the Arctic Ocean and confirmed that there was no northern continent.

1902 Danish scientists with government backing established the International Council for the Exploration of the Sea (ICES) to investigate oceanographic conditions that affect North Atlantic fisheries. Council representatives were from Great Britain, Germany, Sweden, Finland, Norway, Denmark, Holland, and Russia.

1903 The Friday Harbor Oceanographic Laboratory was established at the University of Washington in Seattle.

1903 The laboratory that became the Scripps Institution of Biological Research, and later the Scripps Institution of Oceanography, was founded in San Diego, California.

1912 German meteorologist Alfred Wegener proposed his theory of continental drift.

1925–1927 A German expedition aboard the research vessel *Meteor* studied the physical oceanography of the Atlantic Ocean, using an echo sounder extensively for the first time.

1930 The Woods Hole Oceanographic Institution was established on the southwestern shore of Cape Cod, Massachusetts.

1932 The International Whaling Commission was organized to collect data on whale species and to enforce voluntary regulations on whaling.

1942 Harald Sverdrup, Richard Fleming, and Martin Johnson published the scientific classic *The Oceans*, which is still an authoritative source.

1949 The Lamont Geological Observatory (later renamed Lamont-Doherty Earth Observatory) at Columbia University in New York was established.

1957–1958 The International Geophysical Year (IGY) was organized as an international effort to coordinate geophysical investigation of the Earth, including the oceans.

1958 The nuclear submarine USS *Nautilus* reached the North Pole under the ice.

TABLE 2.1 Some Milestones in the History of Ocean Study (continued)

1959 Bruce Heezen and Marie Tharp published the first comprehensive map of the ocean floor topography.

1959–1965 The International Indian Ocean Expedition was established under United Nations auspices to intensively investigate Indian Ocean oceanography.

1960 The bathyscaphe *Trieste* reached the bottom of the deepest (10,915 m) ocean trench (Mariana).

1964–1965 Hot, high-salinity brines and unusual black ooze sediments were discovered at the bottom of the Red Sea.

1966 The U.S. Congress adopted the Sea Grant College and Programs Act to provide nonmilitary funding for marine science education and research.

1968, 1975 The U.S. National Science Foundation organized the Deep Sea Drilling Program (DSDP) to core through the sediments and rocks of the oceans. Reorganized in 1975 as the International Program of Ocean Drilling, this program continues today.

1970 The U.S. government created the National Oceanic and Atmospheric Administration (NOAA) to oversee and coordinate government activities related to oceanography and meteorology.

1970s The United Nations initiated the International Decade of Ocean Exploration (IDOE) to improve scientific knowledge of the oceans.

1972 The Geochemical Ocean Section Study (GEOSECS) was organized to study seawater chemistry and investigate ocean circulation and mixing and the biogeochemical recycling of chemical substances.

1977 The research submersible *Alvin* makes the first visit to a hydrothermal vent and discovers an entirely new ecosystem based on chemosynthesis.

1978 *Seasat-A*, the first oceanographic satellite, was launched, demonstrating the utility of remote sensing in the study of the oceans.

1980s The Coordinated Ocean Research and Exploration Section program (CORES) was organized to continue the scientific work of the IDOE into the 1980s.

1980s The satellite-based Global Positioning System (GPS) was developed. It was made available for public use in 1983.

1990 The *JOIDES Resolution* drilling vessel retrieved a sediment sample estimated to be 170 million years old.

1990–2002 The World Ocean Circulation Experiment (WOCE) was conducted, in which 40 nations studied ocean circulation and its interaction with the atmosphere.

1992 The *TOPEX/Poseidon* satellite, which maps ocean surface currents, waves, and tides every 10 days, was launched.

1994 The Law of the Sea Treaty entered into force.

1995 The remotely controlled unmanned Japanese submersible *Keiko* set a new depth record of 10,978 m in the Challenger Deep of the Mariana Trench.

2000 The first autonomous float–based global ocean observation system (the *Argo* float) was put into operation, deployed as part of the Global Climate Observing System/Global Ocean Observing System (GCOS/GOOS).

2003 The one-thousandth *Argo* float (of a planned 3,000) was deployed.

mapmakers was the Greek geographer Ptolemy, who lived around 150 CE. Ptolemy's maps, although little known until the end of the Dark Ages, about 1400 CE, were the basis of most maps until the 1500s.

Ptolemy's maps are remarkably detailed and accurate in their reproduction of north-to-south positions (**latitude**). However, they are distorted by substantial errors in east-to-west positions (**longitude**). For example, Ptolemy shows the east-to-west length of the Mediterranean to be about 50% too long in relation to its north-to-south dimension. These errors are a consequence of the mapmaker's inability to measure time accurately. Without precise and accurate time measurements, it was not possible to establish longitude correctly, as is discussed in Chapter 1. The east–west distortion of maps was not corrected until about the 1760s, when the first

FIGURE 2.2 A typical Polynesian outrigger canoe. This design has been used for centuries throughout much of Polynesia and Micronesia. This canoe was photographed in Papua New Guinea.

practical **chronometers** were developed that could keep accurate time on a ship.

In the 2000 years following the Phoenician era, ocean exploration slowed, and much of what may have been learned was lost in the turmoil of the Dark Ages. We know that philosophers did make several significant observations, especially during the early part of this period. In the sixth century BCE, Pythagoras declared the Earth to be spherical. In the fourth century BCE, Aristotle concluded that total rainfall over the Earth's surface must be equal to total evaporation, because the oceans did not fill up or dry out. In the same century, another Greek, Pytheas, sailed out of the Mediterranean to Britain, Norway, Germany, and Iceland. He developed a simple method of determining latitude by measuring the angle between the horizon and the North Star—a method still used today. Pytheas also proposed the concept that **tides** were caused by the moon.

In the third century BCE, Eratosthenes, a Greek studying at the library in Alexandria, Egypt, calculated the circumference of the Earth along a circle through the North and South Poles. His value of 40,000 km was very close to the 40,032 km that has been determined by extremely precise and sophisticated modern methods. Approximately 200 years later, Poseidonius incorrectly recalculated the circumference of the Earth to be about 29,000 km. Ptolemy accepted Poseidonius's incorrect value for his maps, and the error was not corrected for centuries. In fact, this error led Christopher Columbus to believe he had reached Asia when he arrived in the Americas in 1492.

Micronesia and Polynesia. About 4500 years ago, at about the same time that ocean voyaging and trade were beginning to expand in the Mediterranean, the large islands of the far western Pacific and Micronesia were colonized. This expansion, which originated on mainland Asia and is thought to have proceeded through Taiwan, may have been made possible by the development of the outrigger canoe. After an apparent pause of about 1000 years, Polynesians colonized the islands of the western and central tropical Pacific as far east as Samoa, Tahiti, and the Marquesas, which they reached in about 200 BCE. Unlike Mediterranean sailors, who could follow the coastline, the Polynesians had to navigate across broad expanses of open ocean to colonize the Pacific islands. The Polynesians crossed the ocean in double-hulled sailing canoes made of wood and reeds. A larger hull provided living space for up to 80 sailors plus plants and animals; a smaller hull or outrigger stabilized the vessel. The double-hulled design is still used in much of Polynesia (**Fig. 2.2**). Later still (about 300–1000 CE), Polynesians colonized the Pacific islands north to Hawaii, south to New Zealand, and east to Easter Island.

The Polynesians were arguably the greatest navigators in history. They successfully navigated across open oceans using only their memorized knowledge of the stars, winds, wave patterns, clouds, and seabirds. This knowledge was passed down through generations of the families of navigators, who were justifiably venerated in Polynesian culture. Both the Polynesians and Micronesians created some crude maps from wood sticks or rattan, but most Polynesian navigators did not use or need such maps. Some of the ancient Polynesian navigational knowledge persists today in the oral histories and experiences of just a few remaining descendants of the early navigators. Efforts have been made recently to sustain and preserve this knowledge through a series of ocean voyages across the Pacific in reconstructed replicas of ancient Polynesian boats, such as the *Hokule'a* (**Fig. 2.3**).

The recent journeys have proved that the ancient Polynesian navigation techniques were, and still are, remarkably accurate and reliable. There is great interest in discovering and documenting all the observational clues the navigators used. Much of the Polynesian navigational art is mysticized and undocumented, but the navigators must have known principles of wave shapes and directions, cloud formations, and other ocean phenomena that even now are not fully understood by modern scientists. For example, Polynesian navigators can deduce the location of an island hundreds of miles away by observing wave patterns created in the wake of the island. Modern science has been able to document such large-scale wave patterns around islands only since satellite observations of the sea surface became possible.

The Dark Ages and the New Era of Discovery

During the Dark Ages, when ignorance and anarchy engulfed Europe, ocean exploration by Europeans was limited. At that time, however, the Arabs were developing extensive seaborne trade with East Africa, India, and Southeast Asia. They were the first to use the **monsoons** to their advantage: voyaging from East Africa to Asia during the northern summer, when monsoon winds blow from the southwest, and returning in winter, when the winds reverse and blow from the northeast. The Vikings also made great ocean voyages during the Dark Ages, reaching Iceland in the ninth century and Greenland and Newfoundland late in the tenth century.

The middle and late 1400s brought the dawn of a new era as Europeans set out on many ambitious voyages of discovery. The Canary Islands off Northwest Africa were explored in 1416; and the Azores, in the middle of the Atlantic Ocean, were discovered in about 1430. The southern tip of Africa was rounded by Vasco da Gama in 1488; Columbus rediscovered the Americas in 1492; and the world was first circumnavigated by Ferdinand Magellan's expedition of 1519–1522, although Magellan himself was killed in the Philippines during the voyage.

Systematic mapping of the oceans began during the middle of the second millennium. After the mid-1400s, exploration of the oceans was incessant and maps improved rapidly. However, most exploration was undertaken only to discover and colonize lands that could be reached by crossing the oceans. Systematic study of the oceans is generally not considered to have begun until 1768, when Captain James Cook began the first of three voyages to explore and map the Pacific Ocean (**Fig. 2.4**). Captain Cook was the first navigator to carry an accurate chronometer to sea. He therefore was able to measure longitude precisely for the first time and improve existing maps dramatically. On his voyages, Cook made many measurements of water depth, or **soundings**, which he accomplished with a lead weight attached to a rope. He measured depths to 200 fathoms (366 m). He also made many accurate measurements of water temperature, currents, and wind speed and direction, and he documented the occurrence and general characteristics of **coral reefs**. On his final voyage, Captain Cook died at the hands of Polynesian natives at Kealakekua Bay, Hawaii.

FIGURE 2.3 The *Hokule'a.* This vessel is a reproduction of the double-hulled voyaging canoes that the Polynesians used to explore and colonize the Pacific.

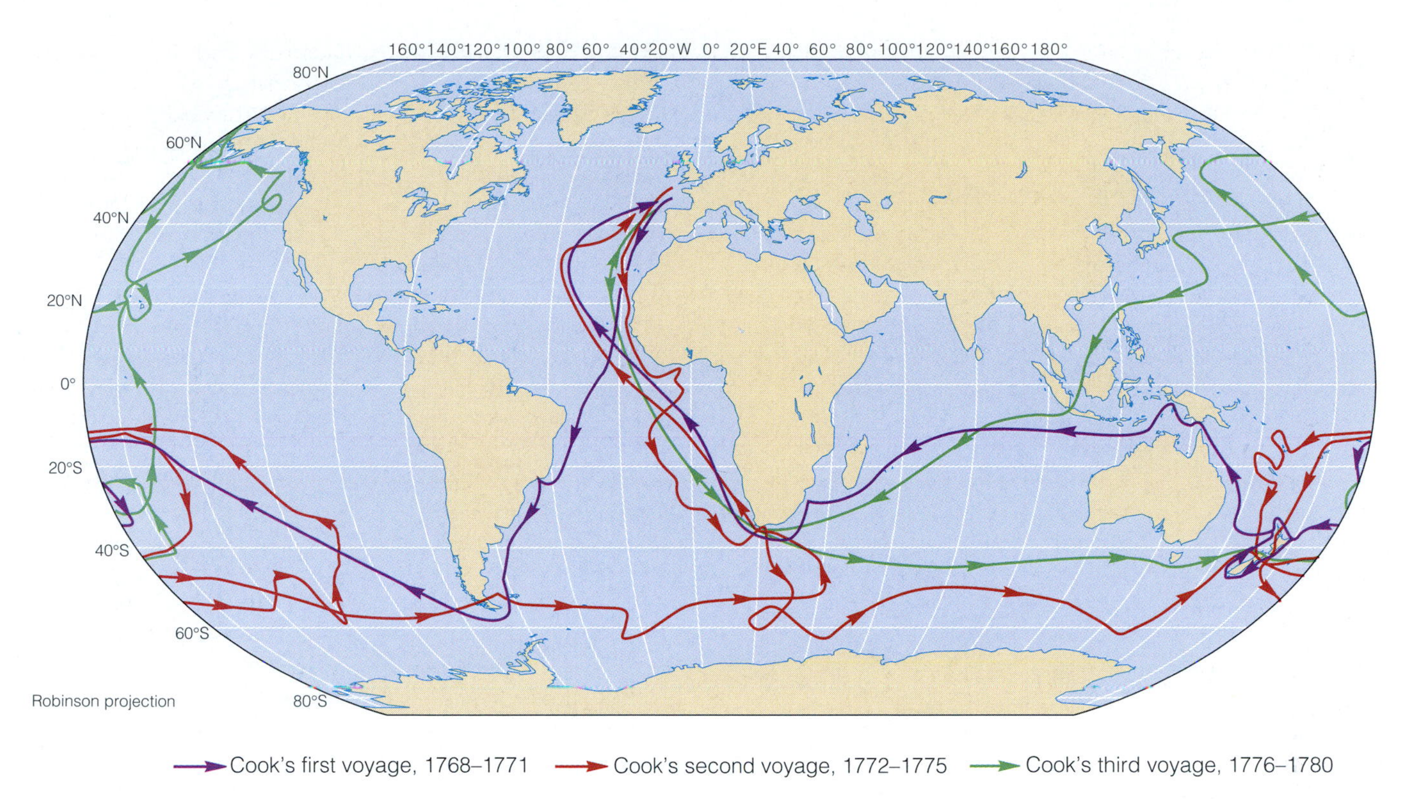

FIGURE 2.4 Routes of the three voyages of Captain Cook: 1768–1771, 1772–1775, and 1776–1780. Cook visited all the major oceans, traveling south into the waters near Antarctica and as far north as the Bering Sea. Places named after Captain Cook include Cook Inlet, Alaska; Cook Strait between the North and South Islands of New Zealand; and a group of Polynesian islands in the tropical Pacific near Tahiti.

At the same time that Cook was exploring the Pacific, Benjamin Franklin, then deputy postmaster for the American colonies, learned that mail ships that sailed the northerly route from Europe to the colonies took 2 weeks longer than ships that sailed the longer southerly route. Franklin's interest in that phenomenon led him to have many discussions with ships' navigators, from which he concluded that ships coming from the north were sailing against a great ocean current. From his observations, he and his cousin Timothy Folger were able in 1770 to develop a remarkably accurate map of the Gulf Stream current (**Fig. 2.5**).

The Birth of Oceanography

By 1800, several seafaring nations had established government offices with primary responsibility for producing charts that could be used by mariners to navigate safely and avoid reefs and **shoals**. Matthew Fontaine Maury, a U.S. Navy officer in charge of the Depot of Naval Charts, made a particularly significant contribution to the intensive ocean mapping efforts of the nineteenth century. Maury gathered data on wind and current patterns from numerous ships' logbooks, and he published his detailed findings in 1855 in a volume entitled *The Physical Geography of the Sea*. Maury also initiated cooperative efforts among seafaring nations to standardize the means by which meteorological and ocean current observations were made. Because of his many contributions, Maury has often been called the "Father of Oceanography."

FIGURE 2.5 Map of the Gulf Stream originally drawn by Benjamin Franklin and Timothy Folger in 1770.

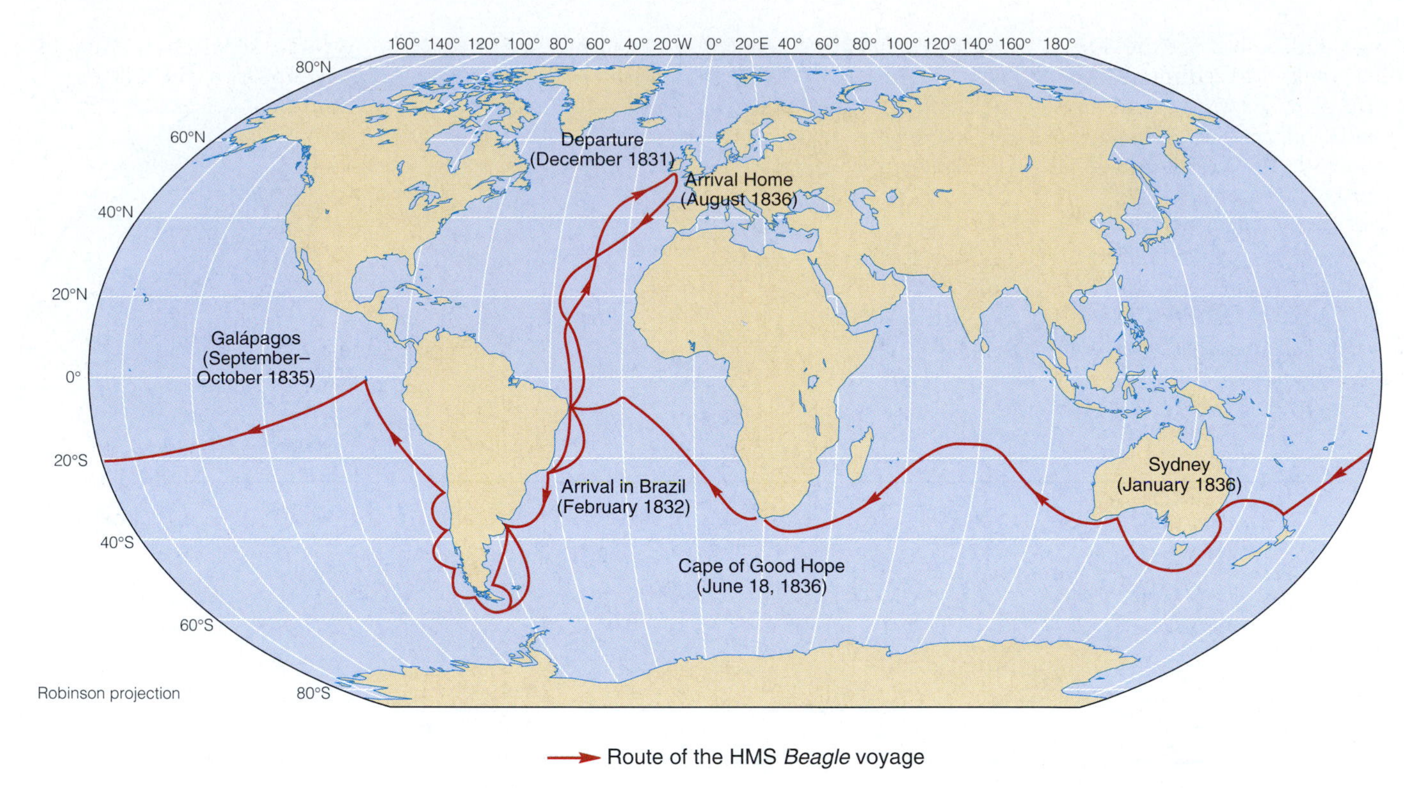

FIGURE 2.6 The voyage of HMS *Beagle* and Charles Darwin, 1831–1836. More than half of the voyage was spent in the vicinity of South America.

The Beagle. In 1831, a 5-year epic voyage of discovery was begun that forever changed the way humans view their world (**Fig. 2.6**). The major objective of the voyage of HMS *Beagle* was to complete a **hydrographic** survey of the Patagonia and Tierra del Fuego coastal regions to improve maps used by ships sailing between the Pacific and Atlantic Oceans around the tip of South America. The *Beagle* also visited the Galápagos Islands off the **coast** of Peru, crossed the Pacific to New Zealand and Australia, and returned to England across the Indian Ocean and around the southern tip of Africa.

The *Beagle* expedition has become famous primarily because of the young naturalist who traveled on the ship, observing the plant and animal life of the many places where the *Beagle* touched land. This naturalist was, of course, Charles Darwin. Darwin's observations on the *Beagle* expedition were the basis for his book *Origin of Species*, in which he proposed the revolutionary theory of natural selection. Although the observations that led to the theory of evolution were the most famous of his findings on the *Beagle* expedition, Darwin also made several other major discoveries. For example, he proposed a theory to explain the formation of coral **atolls** that is still accepted (Chap. 4). In addition, he made a startling observation on one of his land excursions during the voyage: Darwin climbed high into the Andes Mountains, which run along the west coast of South America. At the top of these mountains, he found **fossils** in the rocks that were undeniably the remains of marine creatures. Darwin concluded correctly that the rocks had originated beneath the sea. Thus, he deduced that continents were not permanent and unchanging, as was widely accepted at the time, but that they must move, at least vertically. That observation remained almost unnoticed among Darwin's works until the twentieth century, when the theories of **continental drift** and **plate tectonics** were developed (Chaps. 4, 5).

The Challenger. During the 1860s, British vessels performing investigations preparatory to laying a transatlantic telegraph cable brought up living creatures in mud samples from the bottom of the deep sea. At the time, the prevailing scientific opinion was that the deep ocean was devoid of life, because of the high pressures and low temperatures. Thus, the discovery of life on the deep-ocean floor was perhaps as dramatic a finding as the discovery of life on Mars would be if it were to occur today. The discovery of life in the deep sea led to the true birth of oceanography as a modern science, an event that can be traced precisely to the years 1872 through 1876. It was during these years that HMS *Challenger* sailed the world's oceans as the first vessel outfitted specifically so that its crew could study the physics, chemistry, geology, and biology of the oceans (**Fig. 2.7**). The *Challenger* was a sail-powered navy corvette with an auxiliary steam engine. For its scientific expedition, sponsored by the Royal Society of London, the corvette's guns were removed and replaced with laboratories and scientific gear. Included was equipment for measuring the ocean depths, collecting rocks and sediment from the ocean floor, and collecting seawater and organisms from depths between the ocean surface and the seafloor. The scientific additions crowded the vessel and left only spartan living quarters for the crew and the six scientists.

The *Challenger* sailed 127,500 km across the oceans during its 5-year expedition to study the North and South Atlantic Ocean, the North and South Pacific Ocean, and the southern part of the Indian Ocean. The expedition made hundreds of depth soundings using a lead weight on a hemp line that was hauled in by hand over a steam capstan. With this crude equipment, a single deep sounding required an entire day. Despite the extreme difficulty and tedium of obtaining deep soundings, the *Challenger* expedition was able to measure a depth of 8,185 m in the Mariana Trench east of the Philippines.

The *Challenger* also conducted

- Hundreds of observations of ocean water temperature, both near the surface and at depth
- More than 100 dredge samples of rocks and sediment from the seafloor
- One hundred and fifty open-water net trawls for fishes and other organisms
- Numerous samplings of seawater
- Many readings of ocean current velocity and meteorological conditions
- Countless visual observations of fishes, **marine mammals,** and birds

(a)

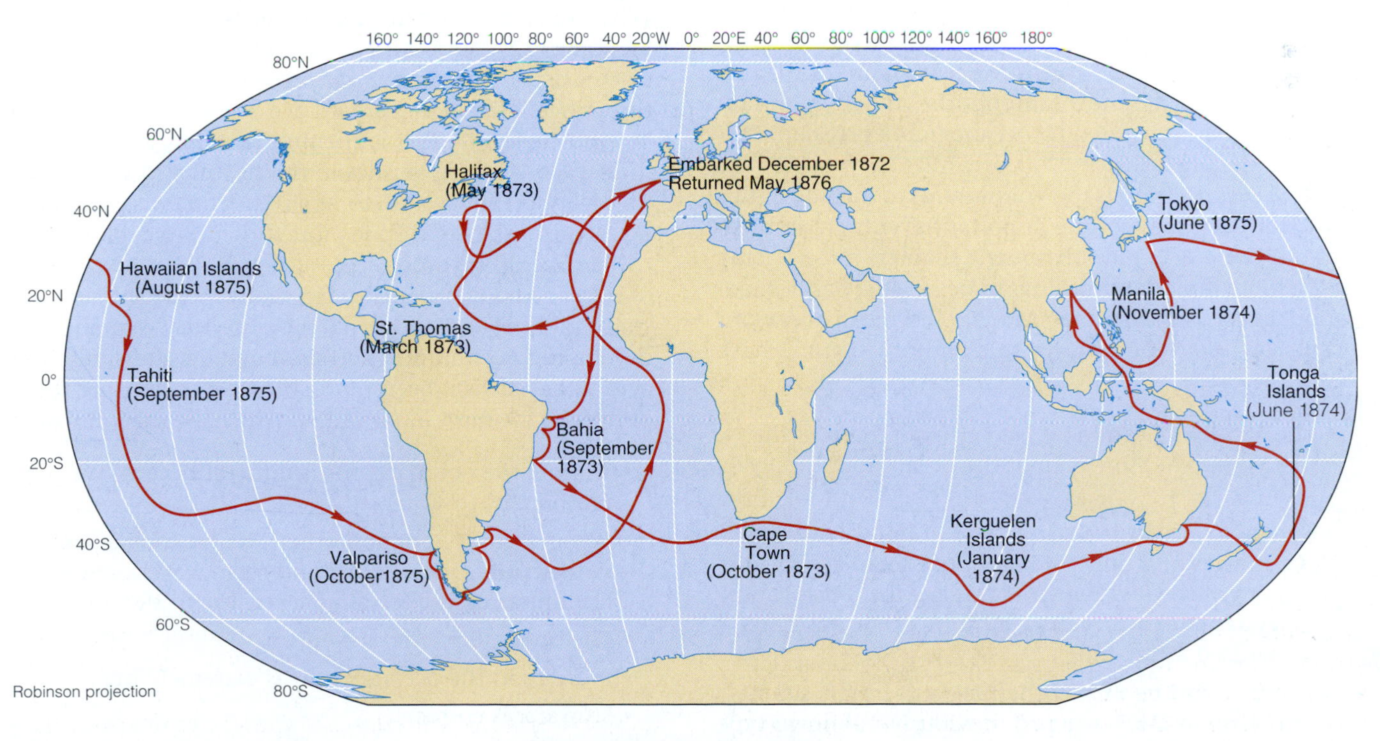

(b)

FIGURE 2.7 The *Challenger* expedition. (a) A painting of HMS *Challenger.* (b) Route of the voyage of the *Challenger,* 1872–1876, which sailed through all the major oceans, but not into the high-latitude regions.

The expedition brought back a wealth of samples and new scientific information about the oceans, including the identification and classification of almost 5000 previously unknown species of marine organisms. The quantity of data and samples obtained was so great that a special government commission was established to analyze the information. Indeed, the 50 volumes of research reports generated by the expedition were not completed until decades after the ship returned to England. The volumes contained so much information that they provided the foundation on which almost all major disciplines of oceanography were later built.

The Modern Era

In the more than 100 years since the *Challenger* expedition, oceanographers have traveled the seas in research ships ever more frequently, and with observation and sampling equipment of ever-increasing sophistication. Ocean research is still performed from research vessels in much the same way that it was in 1872. However, during the twentieth century, oceanographic research expanded to include exploration and study using **scuba** and manned **submersibles**, as well as unmanned observation using robotic vehicles and instrument packages either free-floating or attached to cables moored to the seafloor. Many of these instruments now report their data by **acoustic** links to the surface, and by radio signals sent through satellites to land-based facilities. In addition, oceanographers now make many observations from aircraft and satellites using remote sensing techniques. The introduction in 1978 of satellites specifically designed to look at ocean processes was a particularly important milestone because it allowed almost simultaneous observations to be made across an entire ocean basin for the first time. Satellites are now among the most important observing platforms used by oceanographers (Chap. 3).

The detailed history of oceanography since the *Challenger* expedition is too voluminous to include in this text. However, some of the important events are summarized in **Table 2.1**, and subsequent chapters review many of the more important findings of ocean study and exploration in the modern era. Among the most important events or discoveries during this author's lifetime have been

- The 1959 publication by Heezen and Tharp of the first comprehensive map of the ocean floor topography (Chap. 3).
- The discovery in 1964 and 1965 of hot, high-**salinity brines** and unusual black ooze sediments at the bottom of the Red Sea. This was the first observation of **hydrothermal vents,** although that was not recognized at the time.
- The first visit to a hydrothermal vent by the submarine *Alvin* in 1977, where a previously unknown type of **ecosystem,** based on **chemosynthesis** and sustaining hundreds of previously unknown species was discovered.
- The steady accumulation of evidence that has led to the acceptance of Wegener's 1912 theory of continental drift (Chap. 5). Much of the crucial evidence supporting this theory was gathered by the Deep Sea Drilling Program, which for the first time obtained samples from deep within ocean sediments.
- The initiation of satellite observations of the world ocean that allowed, for the first time, a comprehensive snapshot view of the oceans. This view did not have to rely on many individual observations from ships and enabled the observation of large-scale dynamic processes as never before.
- The development of autonomous robotic measuring devices, including *Argo* floats, which allow the gathering of data of unprecedented detail about processes that occur below the surface layer and that cannot be observed by satellite sensors.

Each of these can arguably be claimed to be as important a breakthrough or advancement for ocean sciences as, for example, Darwin's observations on the voyage of the *Beagle* were to biology. We live in an exciting age of discovery for ocean sciences and many more surprises, advancements, and revelations are to come.

2.3 DIFFICULTIES OF STUDYING THE OCEAN ENVIRONMENT

Why did it take so long to discover the most fundamental secrets of the oceans? Why did we not know of the existence of the immense mountain chains passing through all the oceans until we were already looking beyond the Earth and launching satellites and humans into space? The answer lies in the hostility of the oceans to oceanographers and to their instruments. In many ways, the ocean depths are more difficult to explore than the surface of the moon.

Oceanographers encounter many problems as they study the oceans. A principal focus of oceanographers in developing techniques and instruments to study the oceans has always been, and still is, to overcome these problems, the most important of which are the following:

- Visiting the ocean depths is difficult because we cannot breathe in water.
- Water absorbs light and other **electromagnetic radiation,** such as radar and radio waves, severely limiting their use for remote sensing in the oceans.
- The oceans are extremely deep.
- Pressure in the ocean depths is extremely high.
- Seawater is corrosive.
- The sea surface is dynamic.

"Seeing" through Ocean Water

Compared to the atmosphere, water is a much more efficient absorber of electromagnetic radiation, including radio and radar waves and ultraviolet, infrared, and visible light. In all but the shallowest areas, the seafloor cannot be seen by the naked eye or with any type of optical telescope. Even in the clearest ocean water, we see at best a distorted image of the seafloor, and only where the maximum depth is a few tens of meters at most. Because we cannot see the seafloor, mapping the ocean floor was more difficult than mapping the surface of the moon. Only in the 1920s did oceanographers discover that sound waves could be used as their "eyes" to see the seafloor. Oceanographers also discovered that the magnetic and **gravity** fields of the seafloor could be sensed through the depths of ocean water. Even so, our ability to study the deep ocean is still limited by its lack of transparency to electromagnetic radiation. For example, radar and other instruments carried on satellites can produce extraordinarily detailed maps of the planet's land surface in a matter of days, but they cannot map the seafloor directly because most electromagnetic radiation cannot penetrate the depths of the oceans. However, satellite sensors can map the seafloor topography indirectly by making very precise measurements of sea surface height (Chap. 3). Satellite instruments can also be used to produce excellent maps of ocean surface features, including wave patterns, sea surface temperatures, and the abundance of plant life in the near-surface waters.

Inaccessibility

The average depth of the oceans is 3800 m, and the greatest depth is 11,040 m. These depths are farther below sea level than the average and greatest elevations of the land are above sea level. The average land elevation is 840 m, and the maximum elevation, at Mount Everest, is 8848 m. Most of the ocean floor is as remote from sea level as the highest mountain peaks are. Put another way, most commercial airplanes fly roughly as high above the land as the deepest parts of the ocean are below the sea surface.

To take a sample of the deep-ocean waters or sediment, oceanographers must lower instruments or samplers on a wire and then haul them back up to the ship. Because the oceans are so deep, the process of lowering an instrument or sampler, probing or sampling the water column or seafloor, and retrieving the instrument is extremely time-consuming. Sometimes many hours are spent getting a single sample of mud or bottom water at one place on the ocean floor. In contrast, a scientist studying the land can collect many samples of rock, soil, plants, and animals much more efficiently.

In addition to performing the time-consuming process of lowering and retrieving instruments, oceanographic research vessels, most of which travel at only about 20 km•h^{-1}, consume large amounts of time and fuel going to and returning from sampling locations far from land. Until the relatively recent development of the satellite-based Global Positioning System (GPS) discussed in Chapter 3, navigation far from land was difficult. Location or relocation of a specific sampling site was much more difficult than on land. Because research vessels operating in the open ocean can cost tens of thousands of dollars a day to operate, the large amount of time needed to sample the deep oceans means that few samples can be collected during any oceanographic cruise and that each sample is very expensive to obtain. Therefore, samples of the seafloor, oceanic waters, and organisms living in the oceans have been obtained only at intervals of tens of kilometers throughout most areas of the oceans, especially the deep oceans.

Pressure

The pressure of the atmosphere at sea level is about 1.03 kg•cm^{-2}, or 1 atmosphere (atm). On a journey to space, a space capsule is subject to a 1-atm pressure change because the atmospheric pressure in outer space is effectively zero. Therefore, manned spacecraft must have hulls that can withstand a 1-atm pressure difference. Because most electronic equipment can operate without any problem at zero atmospheric pressure, unmanned satellites need no protection against pressure differences. In contrast, on a journey into the oceans, the pressure increases by 1.03 kg•cm^{-2} (or an additional 1 atm) for each 10 m of depth. Hence, the pressure at 100 m is 11 times as high as the pressure at sea level (1 atm of air pressure plus 10 atm of water pressure).

CRITICAL THINKING QUESTIONS

2.1 During the early part of the Dark Ages, observations about the oceans were made by people we now call "philosophers." Today those who study the oceans are known as "scientists," not philosophers. Do you agree with this distinction? Why? What are some differences between philosophy and science?

2.2 In 1492, Columbus used Poseidonius's highly inaccurate calculation for the circumference of the Earth to identify the New World as Asia. Recall that Eratosthenes had calculated the circumference of the Earth with great accuracy well before Poseidonius. For what reasons do you think Columbus used the later, incorrect circumference?

2.3 Darwin's observations about evolution are well known. His ideas of continent movement (mountains once having been under the sea, for example) are just now being noticed. Why do you suppose this is?

In the deepest part of the oceans, at 11,000 m, the pressure is a truly astounding 1101 times as high as atmospheric pressure, or over 1100 $kg \cdot cm^{-2}$, which is more than a tonne of pressure per square centimeter. Therefore, manned submersibles designed to dive to the greatest depth of the oceans would need hulls capable of withstanding a greater than 1000-atm pressure difference. Most submersibles are not designed to dive that deep, but even shallow dives to 1000 m require hulls that can withstand a greater than 100-atm pressure difference. Submarines and submersibles must have hulls of thick metal, and viewing ports of thick, durable glass or plastic. Deep-diving manned submersibles must be massive, even when made of strong, light materials, such as titanium. In addition, submersible hulls must withstand the metal-fatiguing stresses of repetitive pressurization and depressurization. These requirements make deep-diving submersibles almost prohibitively expensive.

Conductivity, Corrosion, and Fouling

Seawater poses a problem for unmanned instrument packages because most of these rely on electrical components. Such components will not work if immersed in seawater, because seawater conducts electricity and causes short-circuiting. Oceanographic instruments must be placed inside watertight containers called "housings" that must be able to withstand oceanic pressures, because the interiors of the housings normally remain at atmospheric pressure.

Seawater is extremely corrosive, as divers and other water sports enthusiasts quickly discover when they forget to wash their equipment with freshwater. Therefore, all wires, cables, sampling devices, and instrument housings must be protected. Iron and most steels corrode quickly in seawater, so special marine-grade steel or other materials must be used to minimize corrosion. These materials were not available to early oceanographers, who used more expensive and heavier materials, such as brass and bronze. Steel is still the best material available for wires to lower and raise most instrument packages or samples. Even so, the most corrosion-resistant steel wires must usually be further protected by a coating of grease or plastic. Most measurements of trace metals and organic compounds dissolved in seawater were useless until recently because of contaminants from corroding wire and metal sampler parts, and from the grease.

In addition to corrosion problems, a variety of marine organisms **foul** instruments that are left in the ocean to record data for days, weeks, or months, as is necessary for some studies. Some marine organisms, such as **barnacles**, quickly adhere to and colonize the surface of virtually any solid material. Instruments that rely on freely moving parts or on a clean surface-to-seawater contact can quickly be rendered inoperable by such biological fouling.

Wave Motion

Perhaps the most obvious difficulty faced by oceanographers is that the ocean surface is dynamic, and research vessels therefore cannot provide a stable platform on which to work. The perils of working on a rolling and pitching vessel are many. First and foremost, oceanographers must battle seasickness. In addition, they suffer mental and physical fatigue and disorientation caused by working long hours at odd times of day on an unstable platform (many shipboard research activities are continuous 24 hours a day). All oceanographers treasure the rare days of calm seas. Besides the personal hardships, dangers and difficulties are associated with the deployment and retrieval of often extremely heavy instrument packages over the side of a research vessel. The sight of heavy equipment swinging wildly on a wire from a crane over the deck of a ship when seas are rough is indeed frightening. Hanging over the side of a ship in a storm to clamp instruments that must be attached at certain intervals to a heaving wire is an experience few people would relish.

Less obvious than seasickness and the perils of equipment deployment and retrieval, but just as difficult, are the problems associated with using scientific instruments and performing scientific experiments in shipboard laboratories. Most scientific instruments are delicate and made to be used in a normal vibration-free and motion-free laboratory environment. The lurching, pounding, and vibrating to which such equipment is subjected at sea quickly expose any weaknesses. Equipment often must be specially designed or modified to operate reliably at sea. In addition, all equipment must be clamped or tied down in bad **weather**.

Logistics

A profusion of other, lesser problems is associated with studying the oceans. For example, on a research vessel hundreds of miles, and therefore days and tens of thousands of dollars, away from port, broken equipment cannot be taken to a repair shop, a technician cannot be called in, and spare parts cannot be picked up at a store. Oceanographers and research vessel crews have become skilled and ingenious at using available materials to fix equipment at sea. Nevertheless, even the greatest ingenuity sometimes fails, and research efforts must be postponed until the next cruise to the appropriate location, which may be several years later. Such postponements can also be caused by bad weather that slows or prevents work at sea, although most ocean research cruises are planned to allow some leeway for bad-weather days.

In Chapter 3, just a few of the many and varied techniques, instruments, and samplers used by oceanographers of yesteryear and today are briefly reviewed. As you study Chapter 3, keep in mind the difficulties of working in and on the oceans.

CHAPTER SUMMARY

What Is Oceanography? Oceanography is an interdisciplinary science divided into subdisciplines of physical, chemical, geological, and biological oceanography.

Exploration and Mapping. The oceans have been used for fishing for more than 40,000 years. Systematic ocean exploration began between about 6000 and 2000 years ago and was concentrated in the Mediterranean, Polynesia, and Micronesia. Subsequently, between about 2000 and 600 years ago, exploration apparently slowed for centuries, except in the Mediterranean.

After about 1400 CE, European and Mediterranean nations mounted many ocean expeditions. Scientific study of the oceans burgeoned. The voyages of Captain Cook (1768–1780), Charles Darwin and the *Beagle* (1831–1836), and HMS *Challenger* (1872–1876) produced some of the most important early systematic oceanographic studies.

Difficulties of Studying the Ocean Environment. Because water effectively absorbs all electromagnetic radiation, oceanographers can "see" through the ocean waters only by using sound waves, which travel through water with relatively little absorption.

The oceans average almost 4 km in depth. Observation and sampling instruments must be lowered through the water column—a time-consuming and expensive process. Until recently, precise navigation was extremely difficult away from sight of land, and resampling a specific location was nearly impossible.

Submersibles and instruments lowered through the ocean water column must be able to withstand very great pressure changes and the corrosiveness of seawater. They must also avoid fouling by organisms. Electrical equipment must be isolated from contact with seawater because seawater is an electrical conductor.

Oceanographers and their instruments must operate reliably on research ships that vibrate from engine noise, and that roll and pitch with wave action. Faulty equipment must be repaired at sea with available parts and personnel.

STUDY QUESTIONS

1. Why did oceanography develop as an interdisciplinary science?
2. Why was the art of navigation developed to a much greater degree by the early Polynesian and Micronesian civilizations than by the contemporary European civilizations?
3. Why were early mapmakers able to measure distances accurately in a north–south direction but not in an east–west direction?
4. Why was the *Challenger* expedition so important to the development of oceanography?
5. Why do we still know less, in some ways, about the floor of the oceans than we do about the surface of the moon?

KEY TERMS

You should recognize and understand the meaning of all terms that are in boldface type in the text. All those terms are defined in the Glossary. The following are some less familiar key scientific terms that are used in this chapter and that are essential to know and be able to use in classroom discussions or exam answers.

acoustic (p. 34)
chronometer (p. 29)
electromagnetic radiation (p. 34)
foul (p. 36)
hydrographic (p. 32)
latitude (p. 28)
longitude (p. 28)
sounding (p. 30)
submersible (p. 34)
topography (p. 24)

CRITICAL CONCEPTS REMINDER

CC10 **Modeling** (p. 24). Complex environmental systems including the oceans and atmosphere can best be studied by using conceptual and mathematical models. Many oceanographic and climate models are extremely complex and require the use of the fastest supercomputers. To read **CC10** go to page 26CC.

CHAPTER 3

CHAPTER OUTLINE

CRITICAL CONCEPTS USED IN THIS CHAPTER

Recent developments in scuba diving are revolutionizing marine sciences. The strange-looking backpack on the diver shown in the image on the far right is a scuba system that removes carbon dioxide so that the gases can be rebreathed many times and uses a computer to control the mixture of air, oxygen, and helium that the diver breathes. This unit allows divers to make very deep dives and return to the surface safely without the many hours of decompression that are usually needed. The spectacular nudibranch shown in the photo to the right is one of many never-before-seen species found on a number of very deep dives made with the "rebreather" in Papua New Guinea. It was collected from a depth of about 130 m. Scuba is normally limited to a maximum depth of about 50 m.

Studying the Oceans

I was a graduate student at sea on one of my first oceanographic research cruises. We worked a 4-hour-on, 8-hour-off watch system day and night, hanging instrument packages or sampling bottles on a wire and lowering them into the ocean only to haul them back aboard later—sometimes hours after we started, if we were taking deep samples. We used our 8 hours off to eat, sleep, and analyze the samples brought aboard. I worked the "dog" watch—4:00–8:00 AM and 4:00–8:00 PM—something I have always elected to do because I can be on deck for both dawn and sunset. These are my favorite times of day at sea because they are different every day but always awesome, regardless of the weather.

One morning, about 5:00 AM with just the slightest hint of dawn's early light on the horizon, I was out on a platform hanging from the side of the vessel, retrieving water-sampling bottles sequentially from a wire as the bottles were hauled back to the surface. The weather was reasonably calm—a welcome state of affairs, since working from these sampling platforms in bad weather can be scary, with the platform alternating between awash and as much as several meters above the water as the research vessel rolls or pitches back and forth. I was just about to unfasten a bottle from the wire, when I caught sight of an unusual motion a few meters away from the vessel's side, just within the arc of light from the floodlight we were using. Suddenly the surface of the water heaved, and a large black form emerged. As I watched, a column of water and air spurted out of the object with a noise that I had never heard before. Then, in just a few seconds, the object was gone beneath the waves again. This was my first, and still closest ever, encounter with a whale—a moment I shall never forget. The experience is a clear reminder that, although studying the oceans involves many hours of tedious work and observation, there are many rewards.

Throughout history, humans have demonstrated an irresistible drive to explore their surroundings. Civilizations have flourished and declined on all major continents. Most of these civilizations explored distant lands by crossing the seas, but only relatively recently have ocean exploration and research been conducted systematically. In this chapter, the tools and techniques used by oceanographers are briefly described. Some of these are now obsolete, but only by understanding how ingenious oceanographers need to be to overcome the difficulties of studying the oceans can we appreciate why we still know so little about them. This understanding is also necessary for a full appreciation of the immense value of recent technologies, such as the Global Positioning System (GPS), advanced satellite sensors, and computers that can rapidly analyze and graphically portray massive amounts of data.

3.1 THE UNSEEN DOMAIN

By the middle of the nineteenth century, very few land areas remained unexplored by Western civilization. A wealth of knowledge had been obtained about the sizes, locations, and shapes of the landmasses, and their mountain ranges, plains, rivers, and lakes. However, there still was almost no knowledge of what lay below the surface of the oceans. The oceans were known to cover two-thirds of the Earth's surface and to be very deep, with the exception of limited areas around continents, islands, and **reefs.** It was also known that the near-surface waters of the seas abounded with fishes and other creatures, and that much of the seafloor was covered by various types of **sediment.** However, as late as the mid-nineteenth century, the deepest ocean waters were believed to lie stagnant and unmoving in the ocean basins, and the deep waters and deep-sea floor were thought to be devoid of life.

This state of ignorance began to change in the latter half of the nineteenth century, when the first telegraph cables were laid on the seabed and the *Challenger* undertook its expedition. Despite subsequent intensive studies, the oceans remained largely a mystery for decades. For example, not until the 1950s did oceanographers begin to realize that an immense chain of undersea mountains runs through all the oceans. The first comprehensive and generally accurate **bathymetric** map was produced in 1959, and corrections and refinements continue today.

3.2 BATHYMETRY

People have long needed to map seafloor **topography** to navigate safely past obstructions such as submerged rocks and reefs. In addition, they have long been curious about the depth of water over the seafloor. Measurement of ocean water depth is called bathymetry.

Soundings

For centuries, the only way to explore any seafloor deeper than the few tens of meters that could be reached by pearl divers or in crude diving bells was by lowering a line into the water with a lead weight attached. The length of line payed out (on a ship, letting out line is referred to as "paying out") before the weight hit the bottom indicated the water depth—a measurement called a **sounding.** This method led to the unit of depth called the "fathom," which was used almost exclusively for nautical charts until it was supplanted in recent decades by the meter. The fathom was originally 5 1/2 feet, or the length of line between the outstretched arms of the man hauling the sounding line back aboard the ship. Originally, all depths were measured as a count of the number of such lengths of line that were hauled back aboard after the sounding weight hit the bottom. Later the fathom was changed to exactly 6 feet, and sounding lines, particularly those used for deeper soundings, had knots or ribbons tied at measured intervals to improve the sounding accuracy.

A modification of sounding with line and lead weights was used to collect samples of bottom sediment. The bottom of the lead weight was hollowed out and fitted with a lump of tallow. When the tallow hit the bottom, a small amount of bottom sediment adhered to it, unless the weight hit a rocky bottom. Early nautical charts included a description of the type of seafloor based on such samples. Seafloor composition was categorized as sand, silt, or mud, with the mud color sometimes noted. Many modern nautical charts, particularly those of shallow coastal waters, still include that information.

Sounding technology saw a major technological advance during the 1885 voyage of the USS *Tuscarora.* The *Tuscarora* was studying possible routes for a telephone cable between America and Japan. On the voyage, the sounding line was replaced by a single strand of piano wire with a weight attached at its end. The wire was deployed from a drum and hauled back by a winch. Because hauling with a winch was much faster than hand-hauling, the *Tuscarora* could make several deep soundings each day. By 1910, fewer than about 6000 soundings had been made in depths greater than 1000 fathoms (1800 m). Hence, fewer than 6000 depth measurements had been made in an area that represents about 40% of the Earth's surface, an area nearly 30 times as large as the combined surface area of the 48 contiguous states of the United States. Knowledge of the topography of the deep-ocean floors was only rudimentary, even as late as the beginning of World War I. Consider how good our maps of the mountains, plains, and river valleys of the United States would be if the country were covered in cloud and had been studied solely by lowering a wire through the clouds to the ground at only 200 locations.

Sounding Errors and Problems

Taking soundings using a line or wire is a tedious process that poses additional problems. Many of these must still be overcome when wires are being used to lower instrument packages. For example, determining when the weight on the wire has reached the bottom is difficult in deep water. Several kilometers of line or wire is sufficiently heavy to continue to pull more wire from the drum, even when the weight at the end has hit the bottom. Watching for a reduction in how fast the wire pays out or for a slackening of the tension in the wire, each of which occurs when weight is reduced by bottom impact, can sometimes help to overcome this problem. However, these techniques are very difficult, especially in bad **weather,** even on modern research vessels with electronic wire-tension measurement systems. As the ship rolls, the head of the crane or the A-frame over which the wire is payed out moves up and down in relation to the sea surface. When a considerable length of wire has been payed out, its weight and drag in the water prevent it from moving up and down with the ship's roll. As a result, it stretches and contracts, and the wire tension fluctuates. Therefore, wires must be several times stronger than would be necessary to carry only their own weight and the weight of any instruments attached to them.

When wires are used to lower instruments, **currents** and the action of the wind on the research vessel can also cause problems. Because the ship is slowly blown along the sea surface by the wind, it tends to move sideways from the weight or instrument package on the wire that is far below. The drag of the wire and its weight tend to prevent the wire from following the ship's sideways motion. Therefore, the wire does not drop vertically to the bottom (**Fig. 3.1a**). If subsurface currents flow in directions different from that of the research vessel's wind drift, or from that of currents at other depths, the wire's vertical path through the water can have a complex *S* shape or other curve (**Fig. 3.1b**). Because a wire never falls through a deep water column vertically, more wire than the actual depth of water beneath the research vessel must be let out if the end of the wire is to reach the seafloor. Therefore, all line and wire soundings in deep water were incorrect. The actual depth was always less than the measured depth.

A more practical problem associated with lowering instruments or wires over the side of a research vessel is that the wind or currents can blow the vessel over the top of the wire or bend the wire under the vessel. The wire can end up stretched tight across the ship's hull as it passes underneath. Continuing to lower or raise the wire in this situation could damage the ship's hull or break the wire. Therefore, most oceanographic research vessels are specifically designed to be capable of turning slowly around the wire without moving forward. This ability is usually provided by a bow thruster propeller located in a tunnel near the bow of the ship. The propeller is set at right angles to the ship's normal direction of travel and can be used to push the bow of the ship to one side or the other.

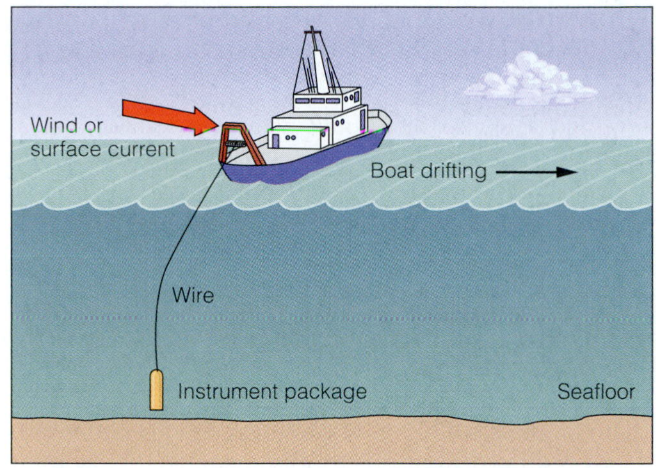

(a)

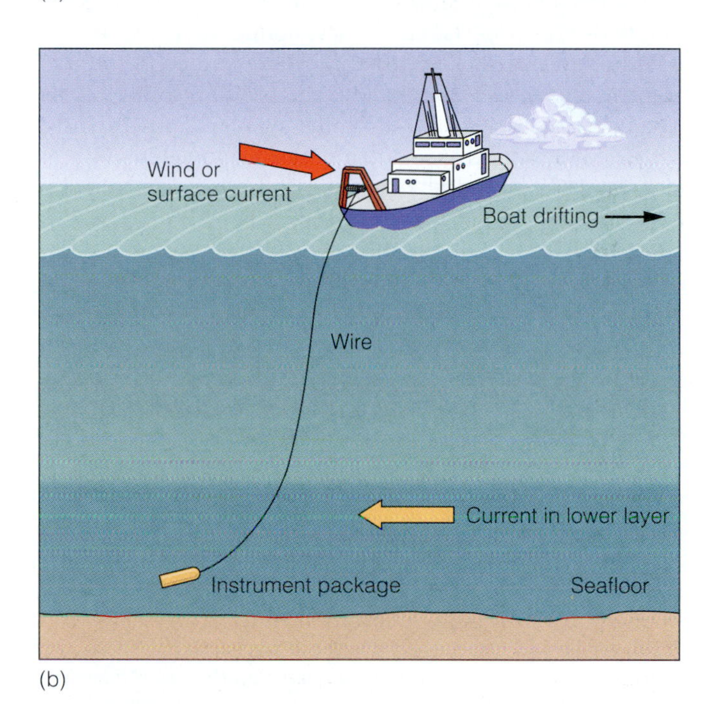

(b)

FIGURE 3.1 Winds and currents can deform the vertical path of an instrument package or sounding weight on the end of a wire. (a) Currents in the surface layer or winds that blow the vessel across the ocean surface can "tow" the instrument package or weight through the water. Therefore, a length of wire that is greater than the depth must be payed out for the instruments to reach the seafloor. (b) Currents in lower layers can further complicate the path of the wire, particularly in very deep water where several kilometers of wire must be payed out for the instrument package to reach the seafloor.

Echo Sounders

In response to the sinking of the *Titanic* in 1912, Reginald Fessender, a former assistant to Thomas Alva Edison, invented a device that could detect an iceberg almost 5 km away by sending a sound signal through the water and detecting the return echo. That sound navigation and ranging equipment, which became known as **sonar,** was quickly developed into a device for hunting submarines. After the invention of sonar, it was a simple matter to orient the sound source to point vertically downward (**Fig. 3.2**) and to detect the echo from the seafloor. The speed of sound in seawater is known, as are the relatively small changes in this speed with **salinity** and temperature. Hence, if the distribution of salinity and temperature with depth is known from other measurements, the depth of the water below a ship can be determined by measurement of the time taken for the sound to travel to the seafloor and back. The first truly successful echo sounder depth recordings were made in the North Sea in 1920 by the German scientist Alexander Behm. Subsequently, knowledge of bottom topography developed rapidly as echo sounding equipment was improved and installed in more vessels.

The great advantage of echo sounders was that they could obtain essentially continuous records of the water depth below a moving ship. By the mid-twentieth century, every research vessel was equipped with an extremely precise echo sounder called a "precision depth recorder" (PDR). Standard operating procedures on most research vessels required that the PDR be operated continuously, and that depths and precise ship positions be recorded while the vessel was under way. PDR depth measurements, although much more accurate than soundings, also have limitations. Because the PDR records only the depth of water directly under the ship's track, depths between two ship tracks still must be inferred by interpolation. Unless depth recordings from many ship tracks cover a given area of ocean floor, major features such as undersea hills and mountains may be overlooked. In addition, the PDR measures only the depth of the closest echo from under the vessel (**Fig. 3.2**). Even modern PDRs with very narrow beam widths receive echoes from a relatively large area of ocean floor. Hence, a nearby hill can cause the depth to be recorded as shallower than it really is, and narrow valleys and depressions can be missed completely (**Fig. 3.2**).

The depth information generated by research vessels after World War II had to be plotted by hand, which proved to be a daunting task. Of course, depths are now recorded electronically and processed by computers, but computers did not become practical for use at sea until the 1970s. The first truly comprehensive map of the ocean floor was completed by Bruce Heezen and Marie Tharp in 1959. Compiling it was an enormous and tedious undertaking that involved matching depth and position (navigation) data from thousands of hours of PDR recordings made by many vessels without the aid of computers. Navigation errors were corrected by comparison of the depths of each ship, recorded where ship track lines crossed. Depth data were then entered painstakingly on a blank map and carefully **contoured,** and a three-dimensional representation was drawn exactingly by hand. The map generated by this massive project was truly revolutionary. To understand this, look at a typical atlas map that shows all the oceans as a featureless uniform blue expanse, then look at the map on the inside front cover of this book. This revelation of previously unseen seafloor topography that rivals the greatest mountain chains and other features of the continents was, for oceanographers and others, like being introduced to an entirely new planet. Heezen and Tharp, and the many other people who spent years gathering data or otherwise helping to create the map, made a contribution to human knowledge that today remains startling, profound, and beautiful.

Since 1959, PDR surveys of the oceans have continued on almost all research vessels during their entire time at sea. Nevertheless, most of the ocean floor is still very poorly mapped by this technique. In fact, vast areas of the deep oceans are mapped by PDRs at a level of detail equivalent to mapping the United States only by measuring elevations along

FIGURE 3.2 Echo sounders measure the depth of water beneath a vessel by measuring how much time a sound pulse takes to travel from the vessel to the seafloor and back. Because sound travels in seawater at about 1500 m•s^{-1}, a sound pulse takes 2 s to return to the research vessel when the water depth is 1500 m. The sound pulses spread out over a narrow angle as they travel downward from the vessel. Thus, particularly where the depth is great, they are reflected off a large area of seafloor. Because the first part of the echo to return is used to measure the depth, measured depths are often inaccurate.

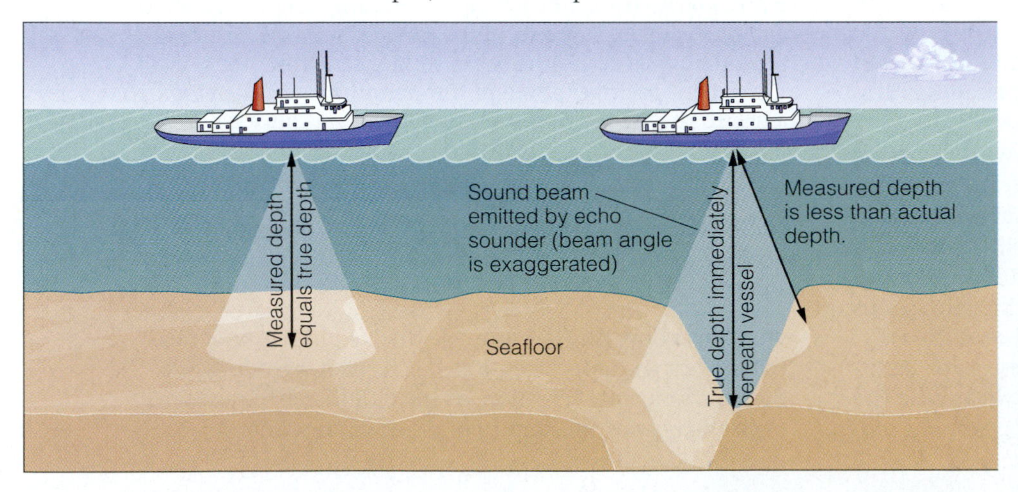

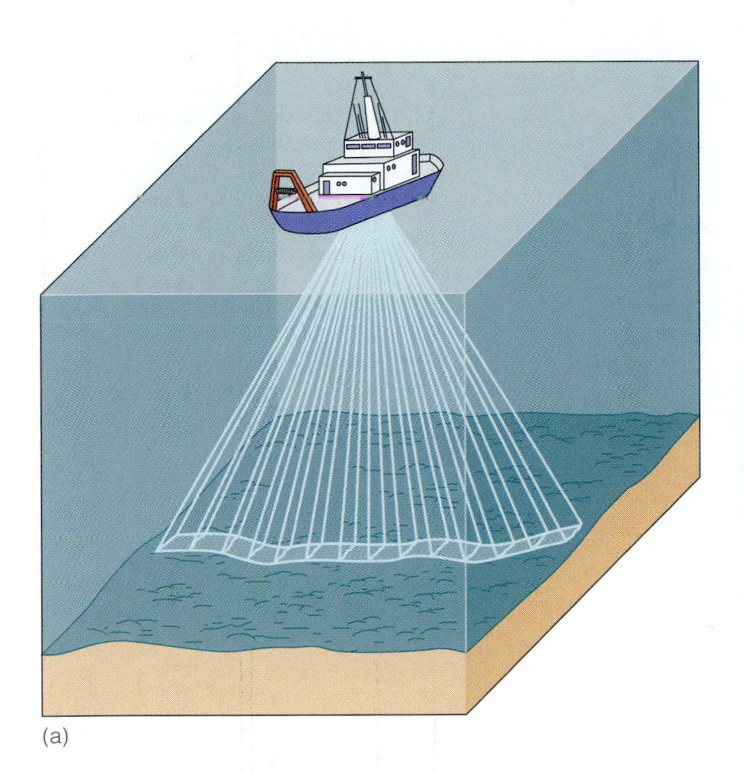

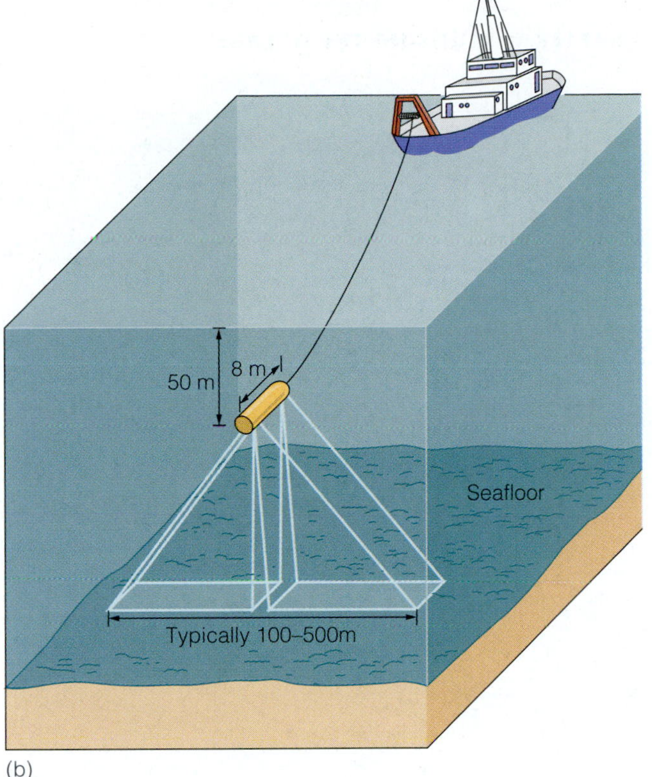

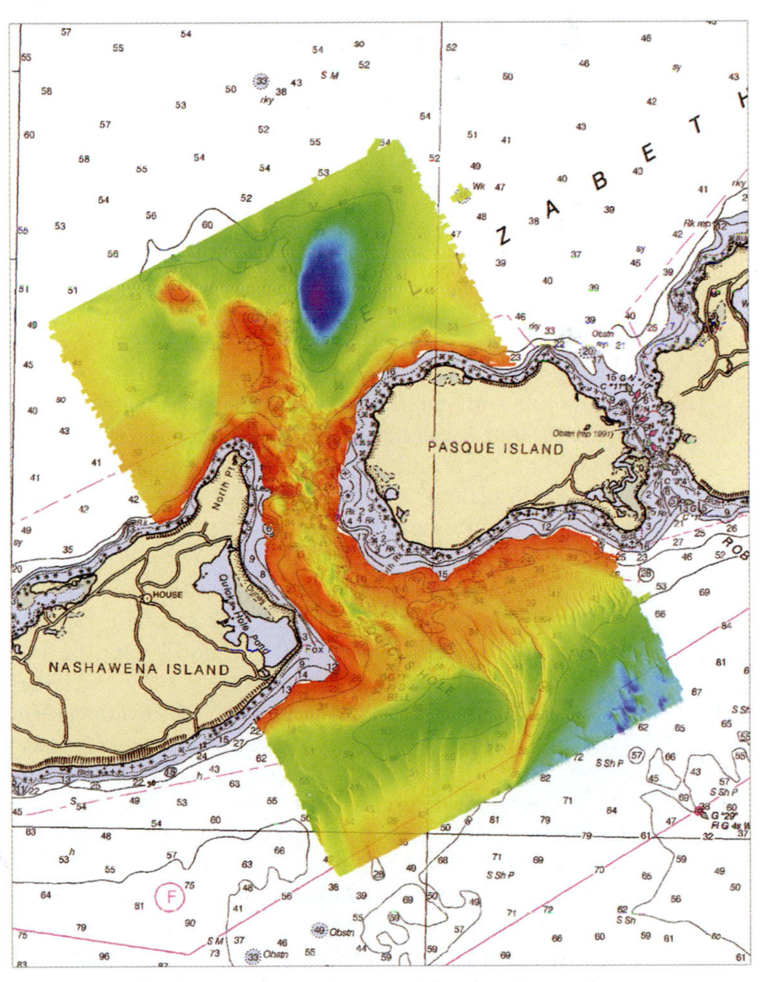

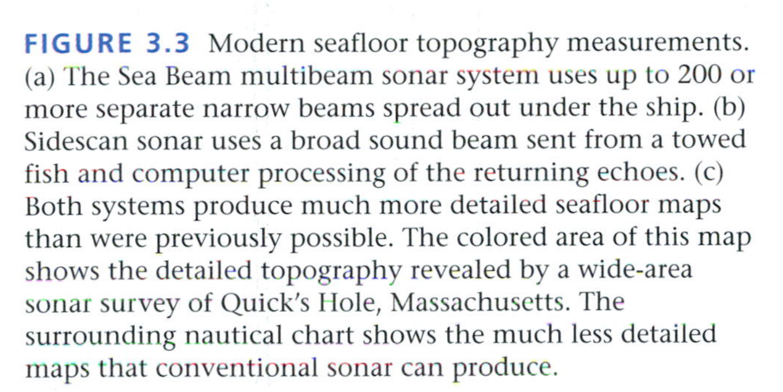

FIGURE 3.3 Modern seafloor topography measurements. (a) The Sea Beam multibeam sonar system uses up to 200 or more separate narrow beams spread out under the ship. (b) Sidescan sonar uses a broad sound beam sent from a towed fish and computer processing of the returning echoes. (c) Both systems produce much more detailed seafloor maps than were previously possible. The colored area of this map shows the detailed topography revealed by a wide-area sonar survey of Quick's Hole, Massachusetts. The surrounding nautical chart shows the much less detailed maps that conventional sonar can produce.

the interstate highways. Look at a road atlas and think about what topography we would have missed if we had mapped the United States from topographic data taken only along this highway system.

Wide-Area Echo Sounders

In the 1960s, echo sounding was revolutionized by the simultaneous development in the United States and England of somewhat different approaches to determining the seafloor topography within a wide swath under, and to either side of, a vessel track. In the American system, called "multibeam sonar" or "swath," up to 200 or more narrow sound beams are broadcast in a fan pattern beneath the ship (**Fig. 3.3a**), and the depth (corrected for the angle) is recorded for each beam. In the British system, called "sidescan sonar," two wider sound beams are broadcast at an angle, one to each side and downward from a streamlined instrument enclosure called a "fish" towed underwater behind the vessel (**Fig. 3.3b**). In this

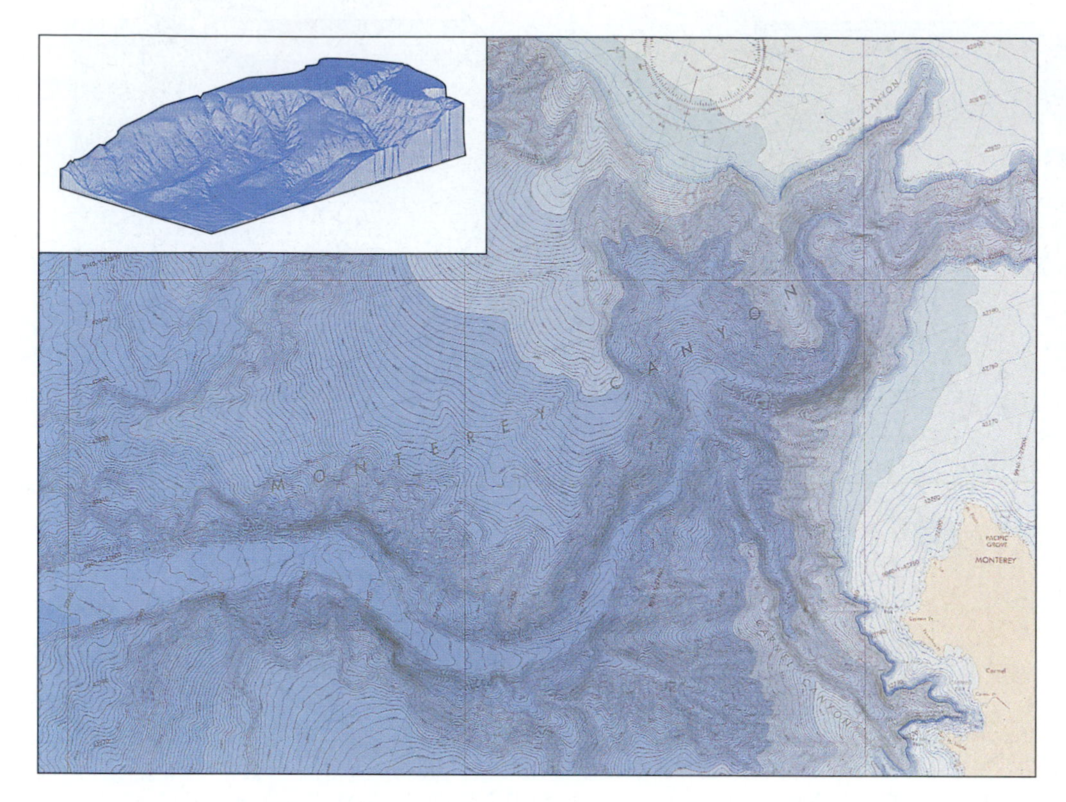

FIGURE 3.4 Bathymetric map of the Monterey Canyon region in central California, produced with data from multibeam sonar surveys. Compare the level of detail of this map with that of older navigation maps that you may have seen for other areas. The inset shows a 3-dimensional rendering of the submarine canyon that crosses the lower half of the larger map.

system, echoes from different broadcast angles within each beam return to the fish at different times, depending on the angle and therefore the distance from the fish. Within the prolonged returning sequence of echoes from each outgoing pulse, the intensity of the echo received varies according to distance and the bottom topography. Strong portions of the returning echo sequence indicate that the bottom is sloped up toward the fish. Weak echoes indicate a slope away from the fish. These represent the front and back sides of the hill, respectively.

The two wide-area echo sounding systems have somewhat different uses because each is better suited to mapping certain types of bottom terrain in certain depths of water. Each system requires powerful and sophisticated computer technology to process the signals received. Both methods provide dramatically improved maps and reveal previously unknown canyons, valleys, hills, and other features of the seafloor (**Fig. 3.3c**). The detailed charts made by these systems are so good that the U.S. Department of Defense sought unsuccessfully to classify its seafloor maps as secret. It apparently believed that the maps might give an enemy information about places where submarines could hide on the seafloor. Spectacularly detailed charts of coastal waters developed with the new echo sounding systems are now being produced as each area is surveyed. The extraordinarily precise and detailed map in **Figure 3.4** was very important in 1989 when the Loma Prieta earthquake occurred just north of Monterey, California. Using this map with postearthquake surveys and **submersible** observations, oceanographers were able for the first time to study submarine effects of an earthquake, such as the many small mud slides that occurred.

Ocean Topography from Satellites

Although new echo sounding systems are now yielding details of seafloor topography, the few suitably equipped research vessels will take many years or decades to map a significant proportion of the oceans. In contrast, satellites have been able to survey the world's oceans with unprecedented comprehensiveness in just a few days.

The *Seasat* satellite launched in 1978 made the first satellite-based maps of the ocean surface, from which the topography of the seafloor can be deduced. The sea surface topography was measured by a radar altimeter carried on board the satellite. *Seasat*'s successor, *Geosat,* carried a more accurate radar altimeter that measured the height of the sea surface with a precision of 3 to 6 cm. The height of the sea surface can be used to map the seafloor topography because it is affected by the depth of the ocean below it. Rock is denser than water, so rocks have a slightly higher gravitational attraction than water. Therefore, the sea surface is slightly higher over an undersea mountain than it is over a deeper area (the mountain "pulls" water toward it from the sides to create "mounds"). *Geosat* and several satellites that followed it made it possible to map the entire world's ocean floor in a matter of days, leading to the discovery of hundreds of previously unknown **seamounts, fracture zones,** and other topographic features in areas where depth recordings had not been made by ships (**Fig. 3.5**).

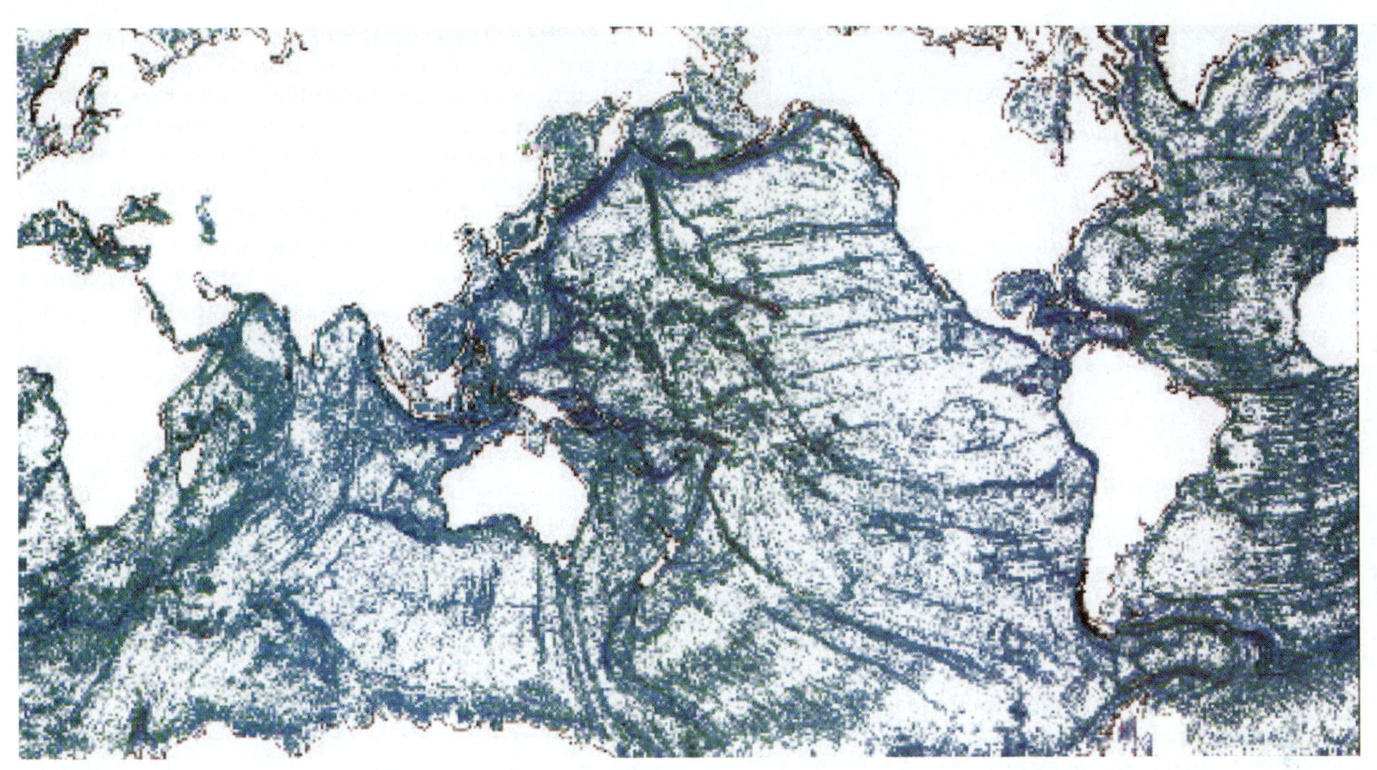

FIGURE 3.5 Ocean surface height measured by the *Geosat* satellite. The difference in sea surface level is only a few centimeters or meters between an area above an undersea mountain and an area above a nearby valley or plain. However, *Geosat* measurements are so precise that the resulting chart of sea surface height reveals all the larger features of the ocean bottom topography.

Geosat data, data from other satellites, and data from shipboard PDR surveys have all been combined to provide the most detailed view possible of the ocean floor. The map on the inside front cover is an example. However, the resolution of these maps is still such that features as large as 10 to 15 km across may be missing. Limitations inherent to both satellite and sonar bathymetry, some of which we have described, make it extremely unlikely that substantially better resolution can ever be obtained except where detailed wide-area sonar surveys are made.

In contrast to ocean floor mapping, the mapping of the planets and their moons was completed quickly and easily by scientists using modern computers aboard the *Mariner, Voyager,* and *Mars Global Surveyor* spacecraft. In addition, the resolution and comprehensiveness of our maps of Mars, Venus, and the Earth's moon are much better than our seafloor maps. This comparison illustrates the extraordinary difficulty of studying the oceans.

3.3 SEAFLOOR SEDIMENTS

The seafloor is covered by sediment ranging in thickness from zero on a small fraction of the ocean floor to several kilometers. As Chapter 8 discusses, many secrets of the Earth's history are to be found in these sediments. Because the sediments slowly accumulate layer upon layer, history is preserved in sequence; sediment becomes older at progressively greater depths below the seafloor. The upper few to tens of centimeters of sediment are especially important because many living organisms inhabit these sediments, and because processes that affect the fate of chemicals in and on sediment particles are also concentrated in this zone. Therefore, oceanographers are interested in obtaining two basic types of sediment samples:

- Samples that contain an undisturbed sequence of the layers of sediment from the sediment surface down as far as necessary to cover a long period of history
- Large samples of the top few tens of centimeters of sediment, within which organisms live

Aside from samples taken in very shallow water, the earliest sediment samples retrieved from the oceans were those that adhered to the lump of tallow at the end of a sounding line. Such samples were useful only for a gross characterization of sediment color and the size and type of the sediment grains. When he explored Baffin Bay in search of a Northwest Passage in 1820, Sir John Ross had his blacksmith construct a "deep-sea clam." That device,

CRITICAL THINKING QUESTIONS

3.1 The process of lowering and retrieving instruments to the ocean floor often takes much longer than is planned at a particular site. What might be some of the reasons for such delays?

3.2 You are planning the deployment, lowering, and retrieval of an instrument package to sample the seafloor in the deepest part of the Atlantic Ocean. How do you decide where to sample, and what do you need to do to ensure that the samples are taken at this precise location?

3.3 You are shown an instrument package that weighs several hundred pounds and is sitting on the deck of a research ship. You are told to deploy the package from the ship, lowering it to the seafloor, at a depth of 2 km. What do you need in order to do this, and how can you arrange to do this safely if the ship is in an area of strong wave action?

the forerunner of grab samplers used today, collected several kilograms of greenish mud containing living worms and other animals from depths of almost 2000 m.

Grab Samplers and Box Corers

In its basic design, a grab sampler consists of a sealed metal container, usually with two halves that open at a top hinge like a clamshell (**Fig. 3.6a**). The sampler is lowered with the clamshell jaws open, and it sinks into the sediments when it hits the bottom. A mechanism causes its two halves to close as it is pulled up, thus grabbing a sample of sediment.

Grab samplers are relatively light and simple to operate, but they can disturb and partially mix the sediment they retrieve. To minimize disturbance, box corers often are used. A box corer consists of a supporting framework that is lowered to sit on the seafloor, a heavily weighted box with an open bottom that sinks about 20 to 30 cm into the sediment, and a blade that slices under the box from the side to hold the sediment in the box when the frame is retrieved (**Fig. 3.6d**). Because they have a wide opening, box corers can collect large amounts of sediment, but they often weigh several hundred kilograms and are difficult to deploy from research vessels.

Gravity and Piston Corers

To take deeper sediment samples, other corers are used. All of these corers consist of a tube that is open at the bottom end like an apple corer. The tube is forced vertically into the sediment. When the tube is pulled out of the sediment, the core is usually held inside by a core catcher (**Fig. 3.6b,c**). There are two basic types of corers: the **gravity** corer and the piston corer. Gravity corers are allowed to fall freely on the end of a cable; they strike the bottom with great force and are driven into the sediment by weights mounted at the top of the core tube (**Fig. 3.6b**). Piston corers, and sometimes gravity corers, are attached to a release mechanism at the end of a cable (**Fig. 3.6c**). In a piston corer, the action of the piston helps the core slide into the core barrel so that longer cores can be obtained, and it helps minimize vertical distortion and disturbance of the core's sediment layers.

In very sandy sediment or other special situations, gravity and piston corers cannot penetrate. Special corers must be used that force the core barrel into the sediment, either by vibrating it mechanically or by forcing air or water down the outside of the barrel to blow the sediment away.

The smallest corers weigh a few tens of kilograms and take short cores (up to about half a meter long); the largest weigh several tonnes and can take cores more than 50 m long. Deploying and retrieving one of the largest corers, which may be over 50 m in total length, is an exacting task on any research vessel.

Unfortunately, and to the frustration of researchers who may have waited hours for the sampler to be lowered and retrieved if the water is deep, grab samplers and corers often fail to penetrate or close properly. There are several reasons for such failures. For example, the sampler might impact the seafloor at too great an angle if currents distort the wire from the vertical as discussed earlier, or the sampled material might jam in the closure mechanism. Even a tiny opening in the closure mechanism can cause the sample to be lost as the sampler is hauled back through the water column.

Drilling Ships

To study older layers, scientists must explore deeper in the ocean sediment than the 10 to 15 m reached by corers. In addition, the bedrock beneath the ocean sediment holds valuable clues to the history of the Earth and its oceans. In 1968, the United States began using a unique drilling ship, the *Glomar Challenger*, to sample the deeper sediment layers and rocks. The ship was capable of drilling holes in the ocean floor in water as deep as 6 km and could collect drill-core samples from depths up to about 2000 m below the seafloor. The Deep Sea Drilling Program (DSDP) used the vessel to obtain more than a thousand cores from throughout the world's oceans. These core samples helped to confirm the theories of **seafloor spreading** and **continental drift** (Chaps. 4, 5).

In 1983, the DSDP was succeeded by the Ocean Drilling Program (ODP), an international cooperative program funded jointly by the United States, Canada, West Germany, France, Japan, the United Kingdom, Australia, and, at one time, the Soviet Union. The *Glomar Challenger*

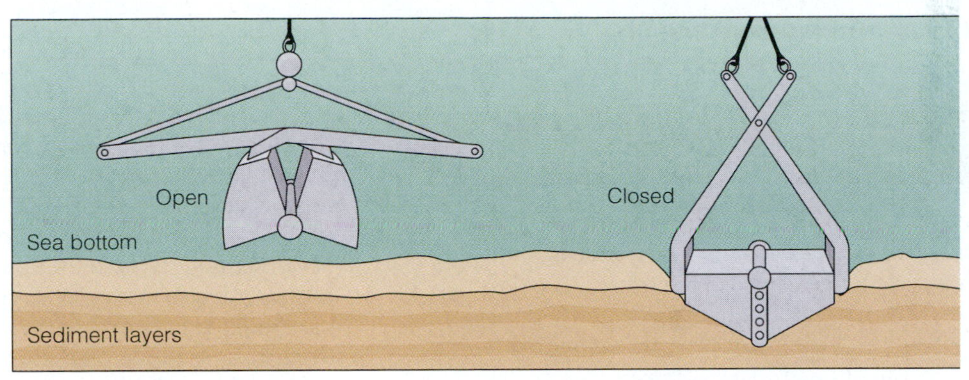

(a) Grab sampler

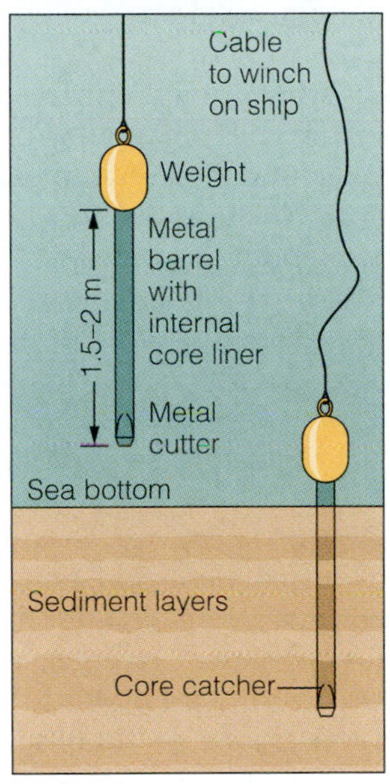

(b) Gravity corer

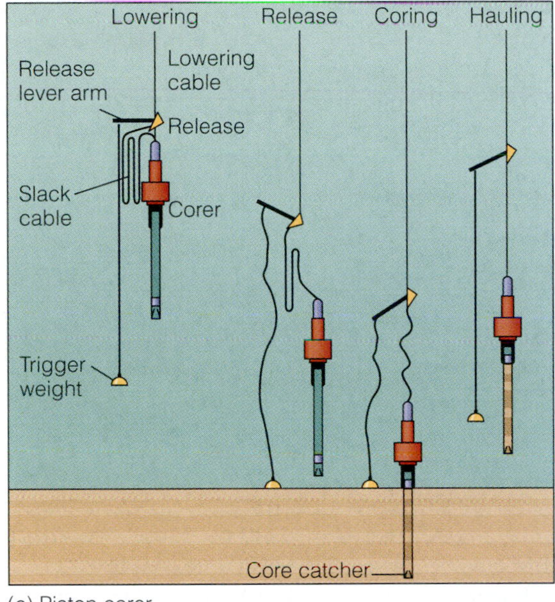

(c) Piston corer

FIGURE 3.6 Methods of sampling seafloor sediments. (a) Grab samplers are used to obtain samples of surface sediments. Gravity (b) and piston (c) corers are used to obtain long vertical samples of soft sediments. Most corers have a core catcher that consists of a ring of flexible metal leaves around the inside of the core tip. The leaves are easily pushed aside as the corer enters the sediments, but they are bent backward to close off the tube as the corer is pulled out. Typically, gravity corers are used to obtain cores up to about 2 m long, and cores from piston corers may be longer than 50 m. Corers are usually suspended from a release arm, which is activated by a trigger weight that hangs below. When the trigger weight hits the bottom, the weighted corer falls freely into the sediments. For piston corers, a loop of cable between the trigger assembly and a piston inside the core barrel becomes taut as the tip of the corer reaches the sediment surface. The corer body is forced into the sediment by its weight, but the cable holds the piston, which causes a small reduction of the water pressure on the sediment. The reduction in pressure helps the sediment core to slide into the barrel, increases core length, and minimizes vertical distortion of the sediment layers. (d) Box corers are used to collect large-volume undisturbed samples of the upper few centimeters of sediment. The box corer is forced into the sediments by its weight and is then closed off by a spade arm that cuts through the sediment and under the box.

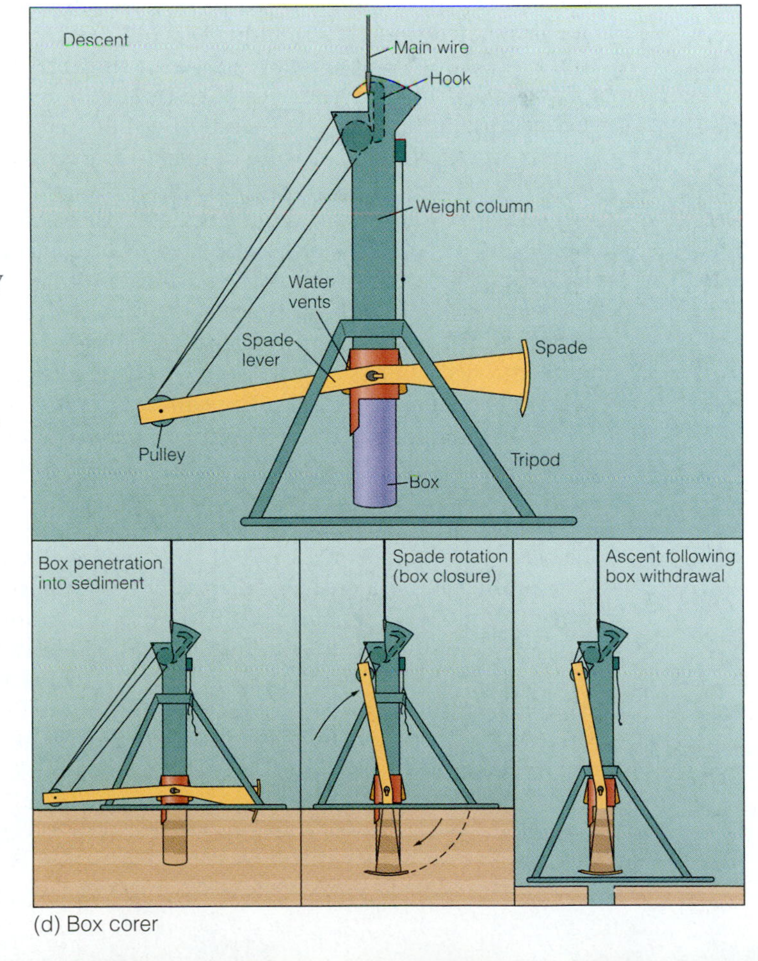

(d) Box corer

CRITICAL THINKING QUESTIONS

3.4 Before reading this chapter, did you know that the pressure changes associated with space travel are much less than those associated with travel into the depths of the oceans? Besides pressure changes, what other challenges can you think of that travel in space and in the deep oceans have in common?

3.5. Many oceanographers who first saw the Heezen and Tharp world ocean floor map in the 1970s and 1980s felt that they were being introduced to an entirely new planet. Why was this so? When did you first see a map of the ocean floor such as this? Why do you think many people are still completely unaware that there are long mountain chains on the ocean floor?

3.6 Suppose you are looking for the lost city of Atlantis, which, according to legend, sank beneath the waves. You have been told the approximate geographic location of Atlantis. What information do you need to select a sampling device that is capable of locating the city?

was replaced by the more sophisticated *JOIDES Resolution*. In 1990, in a water depth of 5700 m, 2500 km south of Japan, the *JOIDES Resolution* drilled through 200 m of recently formed volcanic rock and 460 m of sediment lying below it. A sample of sediment was retrieved that was estimated to be 170 million years old and is believed to be the oldest remaining ocean floor sediment.

In 2003, the ODP was succeeded by the International Ocean Drilling Program, led jointly by the United States and Japan. This program will eventually operate multiple vessels, including a new Japanese vessel called *Chikyu*, which is expected to start drilling in 2008. The *Chikyu* will be capable of drilling holes as deep as 6 km beneath the seafloor, far surpassing the 2-km limit of the *JOIDES Resolution*. The *Chikyu* will also be able to drill in shallower water, and it will have the ability to prevent blowouts (uncontrolled releases of gas or oil) if it penetrates formations that contain oil and gas. This will allow *Chikyu* to drill in many locations where the *JOIDES Resolution* could not.

Seismic, Magnetic, and Gravity Studies

Because obtaining corer or drill-core samples is time-consuming and expensive, oceanographers can sample directly only a small number of locations. Fortunately, certain remote sensing techniques can provide information about the sediment in areas where actual seafloor samples are not available. Seismic profiling and the measurement of magnetic and gravitational fields are the principal techniques used to obtain such information.

Like sonar measurements, seismic profiling uses a sound wave or shock wave. The sound wave passes through the ocean water into the sediment and is partially reflected at each depth where the type of sediment changes or a volcanic rock layer begins (**Fig. 3.7**). Because ocean sediments are built up layer upon layer, many such echoes, each of which corresponds to the top of a layer, are reflected from within most ocean sediments. Returning echoes are monitored with a string of **hydrophones,** devices that record sound waves, towed behind the research vessel. The sound waves received by each of the hydrophones have traveled different distances within the sediment layers and so are received at different times (**Fig. 3.7a,b**). Seismic profiles reveal structural features of the sediment that are hidden below the seafloor, including **faults,** tilting of the layers, and buried mountain tops. Modern seismic profiling systems use multiple sound sources and multiple strings of hydrophones towed behind the research vessel. Using powerful computers to analyze the resulting data, these systems can produce detailed three-dimensional images of the sediments or rock beneath the seafloor (**Fig. 3.7c,d**).

Because sound travels at different speeds in different types of sediment and rock, seismic profiles can aid in determining the nature of each layer of sediment. Distinct sediment layers in many parts of the ocean can be traced for hundreds of kilometers in all directions. Hence the layers found in seismic maps can often be correlated to the layers of sediment and rock retrieved by drilling or coring.

Sediment and rock on or below the seafloor also can be studied by precise measurement of changes in gravitational-field or magnetic-field strength. Tiny changes in magnetic-field strength are detected by instruments towed behind a research vessel as it passes over seafloor sediment and rocks that have variable magnetization (Chap. 5). Gravitational-field strength also changes slightly at different locations because the Earth is not perfectly round, and because denser sediment and rocks exert a slightly greater gravitational pull than less dense or lighter sediment at the same depth. Therefore, extremely small changes in gravitational-field strength detected by instruments carried on research ships provide information about the sediment and rock below, especially the presence of mountains of volcanic rock overlain by less dense sediment.

Dredges

Although most ocean floor is covered by thick sediment, rocky outcrops also are present, and some areas of the seafloor are partially covered with **manganese nodules** and **phosphorite nodules** (Chap. 8). Because corers and drills are not able to sample such surface rocks, dredges are commonly used. A dredge consists of a net of

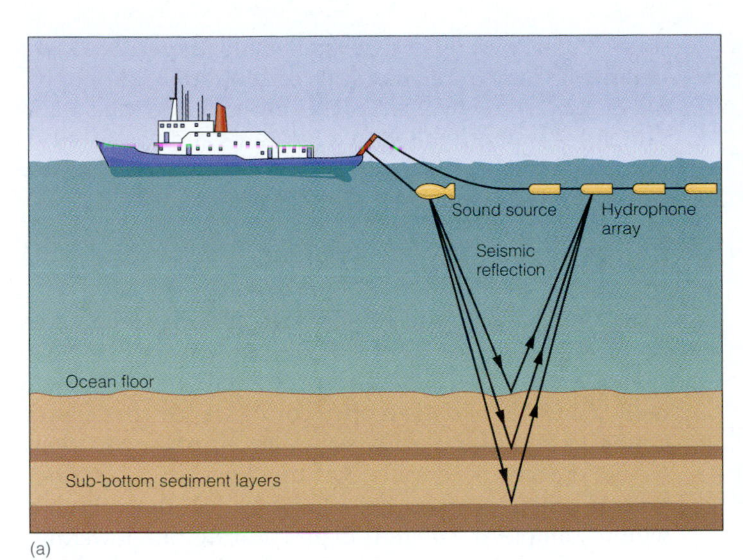

(a)

FIGURE 3.7 Seismic profiling. Two techniques are used: (a) seismic reflection, in which sound pulses penetrate the seafloor and are reflected off sedimentary layers; and (b) seismic refraction, in which sound pulses are refracted (bent) at interfaces between layers of sediment and then travel back out to be received by a second ship. At one time, the sound source was provided by dynamite sticks thrown overboard to explode just under the water surface. However, the sound source typically now used is an air gun that periodically fires a burst of compressed air into the water at a depth of about 15 to 35 m. The sound waves are generated by the collapse of the air bubbles produced by each burst. (c) A typical seismic profile shows rugged seafloor topography overlaid by sediment layers. (d) A 3-dimensional seismic image of an ancient reef sructure buried beneath the sediments off Nova Scotia, Canada. The light colored 2-dimensional sections across the center of the image and along the right side show just how much more information the 3-dimensional image can convey.

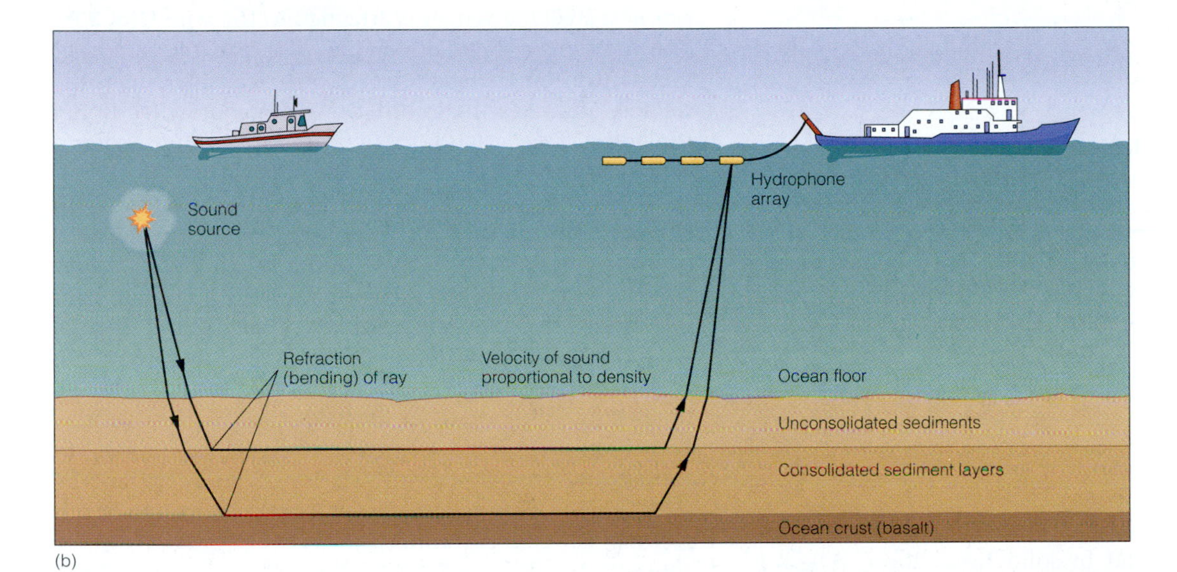

(b)

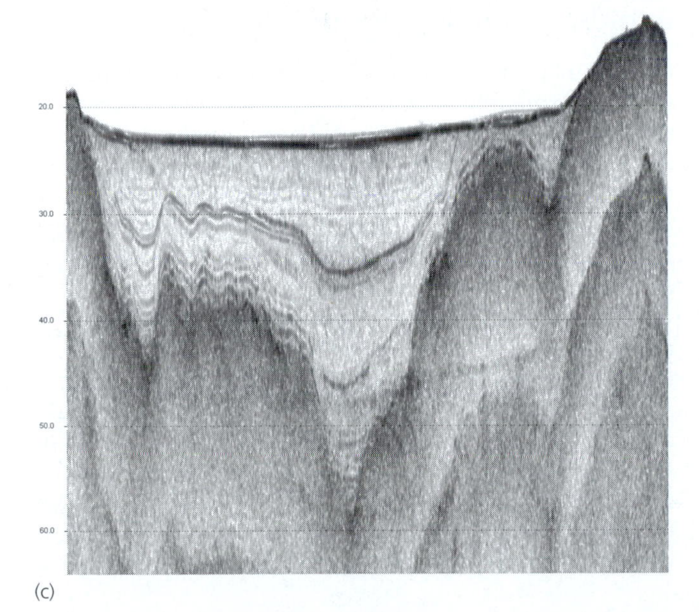

(c)

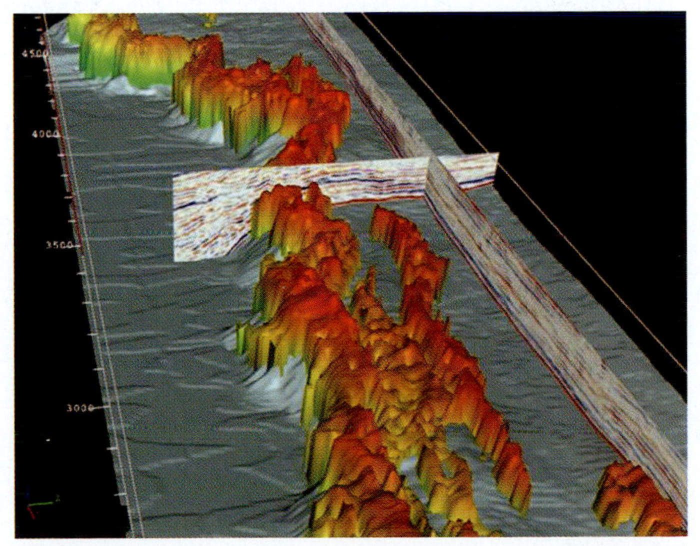

(d)

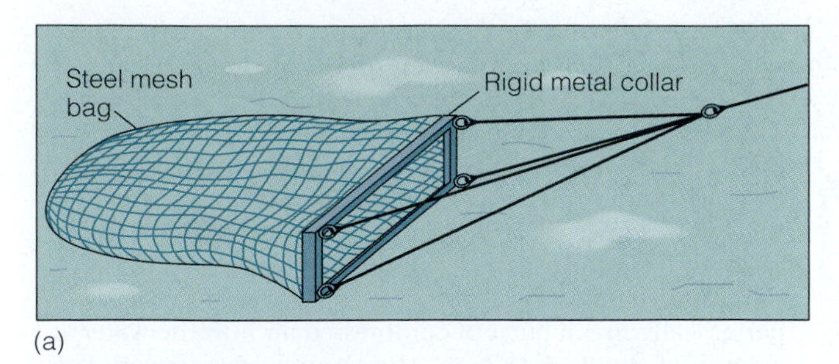

(a)

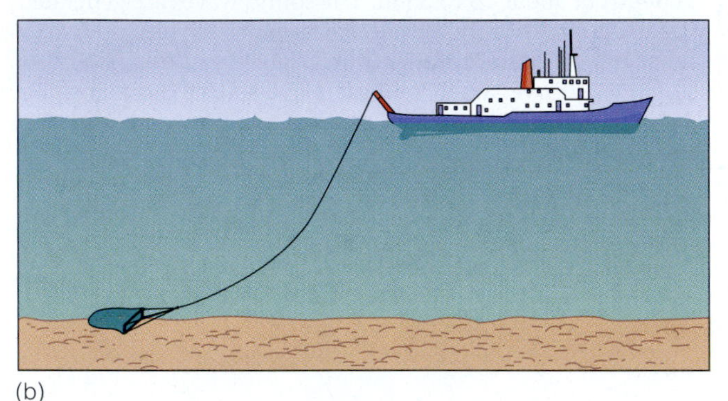
(b)

FIGURE 3.8 (a) Rock dredges are used to obtain samples of rock from seafloor where there is little or no sediment cover and to collect nodules that lie on the sediment surface. (b) The dredge is lowered on a cable and towed across the seafloor. Sediment passes into and through the net, while nodules and loose rocks are retained. The metal collar breaks off rocks from a rocky seafloor.

metal chain or nylon mesh with one end held open by a strong, rigid, metal frame (**Fig. 3.8**).

Dredges are often massive because the rocks in many areas must be broken off of solid **lava** flows. When a dredge snags on such rocks, considerable force is needed to break the rock and release the dredge. Therefore the cable used for lowering and towing dredges must be extremely strong. Steel cables 1 cm or more in diameter are sometimes required. In very deep water, 10,000 m or more of cable weighing several tonnes is required. Because the drum on which the heavy dredge cable is stored and the winch needed to drive the drum are huge, only a few of the largest research vessels can deploy the largest dredges. Submersibles or ROVs, discussed later in the chapter, now perform some of this type of sampling.

3.4 CHEMICAL AND PHYSICAL OCEANOGRAPHY

Oceanographers are interested in understanding temporal and **spatial** variations (that is, variations over time and space) in the concentrations of the many chemicals dissolved in seawater or associated with **suspended sediment.** Because most chemical concentrations cannot be measured directly by instruments lowered into the ocean, samples of water from selected locations and depths must normally be collected and brought back to a research vessel or onshore laboratory.

Sampling Bottles

In all but shallow water, where samples can be pumped up through a hose, water samples are collected in specially designed bottles. Usually the sampling bottles are designed to descend in an open configuration that allows continuous flushing with seawater, as otherwise they would quickly be crushed by the increasing pressure. Several sampling bottles can be attached at intervals on a wire to sample at different depths during one lowering. When the bottles are at the requisite depths, a brass or stainless steel messenger is attached to the wire (**Fig. 3.9**). The messenger slides down the wire and hits a trigger mechanism on the shallowest bottle, causing the bottle to close and releasing another messenger attached under the bottle. Thus, each bottle releases a new messenger to slide down the wire and close the next-deeper bottle.

For many years, most water sampling was done with Nansen bottles (**Fig. 3.9**). These have now been replaced by newer designs developed primarily to collect larger samples or to avoid sample contamination. Because concentrations of some important trace metals and organic compounds in seawater are extremely low (Chap. 6), oceanographers must often collect large volumes (sometimes tens or hundreds of liters) of seawater per sample, and then use sophisticated chemical techniques to extract and concentrate the chemicals before even the most advanced and sensitive analytical instruments can measure the concentration.

Avoiding Sample Contamination

Contamination of the sample must be avoided during its journey from the ocean depths to the laboratory. Contamination comes from many sources, including the metals, plastics, and other materials of the sampling bottle, the metal of the **hydrographic** wire, and the grease that covers the wire to protect it from corrosion. Dust, oil, and vapors in the ship and laboratory atmosphere are other potential **contaminants.**

An especially difficult contamination problem is caused by the thin **surface microlayer** (about 0.1 mm thick or less) that covers all the oceans. The microlayer always contains higher concentrations of many chemicals than the seawater below, and it can be further contaminated by discharges, such as oily cooling water from the research vessel, and by paint and corrosion chips from the vessel's hull. Until recently, all sampling bottles remained open as they were lowered through the sea

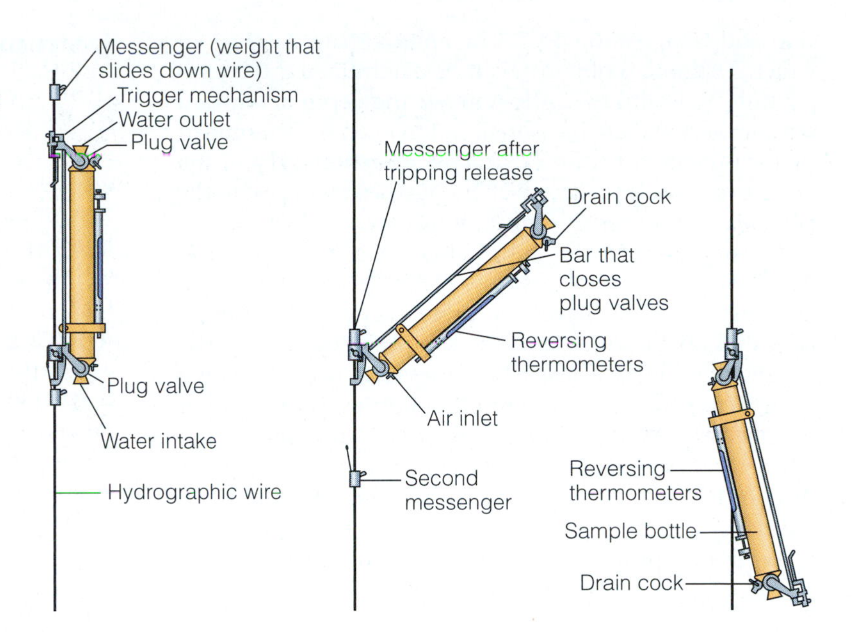

FIGURE 3.9 Nansen water-sampling bottle. When the messenger hits the trigger mechanism, the top of the bottle is released from the wire, the bottle falls into an inverted position, and a mechanical linkage closes valves at each end of the bottle. The trigger mechanism also releases a second messenger, which slides down the wire to the next sampling bottle.

surface at the side of the research vessel, and the surface microlayer deposited a film on the inside of the sampling bottle that significantly contaminated the sample. Several ingenious sampler designs, including the GoFlo bottle (**Fig. 3.10**), prevent surface microlayer contamination. The GoFlo bottle is carefully cleaned and sealed in the closed position on the research vessel before being lowered through the surface. A pressure-sensitive mechanism opens the bottle automatically once it is a few meters below the surface. The bottle can then be closed at the required sampling depth.

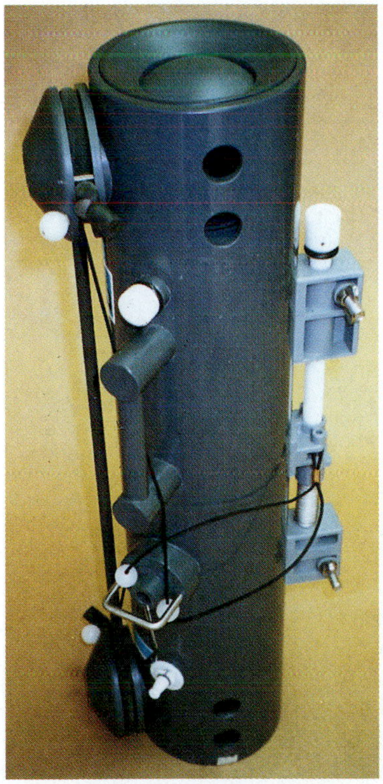

FIGURE 3.10 A GoFlo water-sampling bottle. This ingeniously designed bottle is lowered through the sea surface in a closed configuration to avoid contamination. At a depth of a few feet, a pressure-sensitive trigger opens both ends of the bottle. The bottle is closed again by means of a messenger or other triggering device when the bottle is at the required sampling depth.

For chemical parameters not affected by surface microlayer contamination, one of many other sampler designs can be used. Samplers that collect 10 liters of seawater are used routinely in many applications, and very large samplers that collect hundreds of liters are used for special analyses.

Determining the Depth of Sampling

Seawater in the oceans forms a series of horizontal layers, each of which is often only a few meters thick (Chap. 10). Water moves great distances within these layers, but it mixes only slowly with the water in the next layer above or below. Chemical oceanographers usually want to sample within each of the layers, and sometimes at closer depth intervals across the interfaces between layers. Unless the precise depths of the layers are known, selecting the exact spacing of water-sampling bottles along the wire is impossible. In addition, even if samplers are placed correctly along the wire, the depth at which each sampler is closed can be affected by curvature of the wire caused by currents or vessel drift (**Fig. 3.1**).

Density increases with depth in the oceans, so depth may be estimated by measuring density. Temperature and salinity are the two primary parameters that determine the water density, so density may be calculated and depth estimated by measurement of these two parameters. Salinity, temperature, and density relationships are discussed in

CC6 and Chapters 6 and 7. The approximate depth from which a sample is obtained can be determined if the temperature and salinity of the water at the depth at which a sampler is closed are measured. Salinity can be determined by measurement of the **electrical conductivity** of the sample of seawater collected after it has been returned to the laboratory. However, the temperature of the sample changes as it is retrieved so the temperature must be recorded at depth when the sampler is closed. This temperature measurement can be achieved by a "reversing thermometer" mounted on the outside of the sampler (Fig. 3.11a). The thermometer on a Nansen bottle reverses with the bottle itself (Fig. 3.9), but other sampling bottles, which do not themselves reverse, have an externally mounted thermometer rack that rotates mechanically through 180° as the bottle is closed (Fig. 3.12). To accurately determine the depth of sampling, two reversing thermometers can be used, one protected and one unprotected from pressure changes with depth (Fig. 3.11b).

Instrument Probes and Rosette Samplers

Using water sampling bottles and reversing thermometers to determine salinity, temperature, and depth is tedious and only provides data at those depths at which samples are taken. This method was replaced in the 1970s by instrument packages containing electronic sensors attached to a wire that make measurements continuously as they are lowered and raised. These packages are called **CTDs,** for conductivity (electrical), temperature, and depth (or sometimes "STDs," for salinity, temperature, and depth), because electrical conductivity is measured to determine salinity (Chap. 7). CTDs have many advantages over sampling bottles, especially their ability to read salinity and temperature continuously as a function of depth. Such sensors, or probes, have enabled oceanographers to observe small-scale variations in the layered structure of ocean water that cannot be discerned from widely spaced bottle samples. The wire used to lower CTDs can both supply electrical power to the sensor pack-

FIGURE 3.11 (a) Reversing thermometers like these are attached to a water-sampling bottle in a rack. (b) When the bottle is closed, the rack and thermometers are turned upside down (reversed). The protected thermometer records the true temperature, while the unprotected thermometer records a temperature that is slightly too high because the mercury reservoir is squeezed slightly by the increased pressure at depth. The temperature difference between the thermometers can be used to determine the depth at which the thermometers were reversed. Reversing thermometers are still used to calibrate CTDs, which are instrument packages that measure conductivity, temperature, and depth.

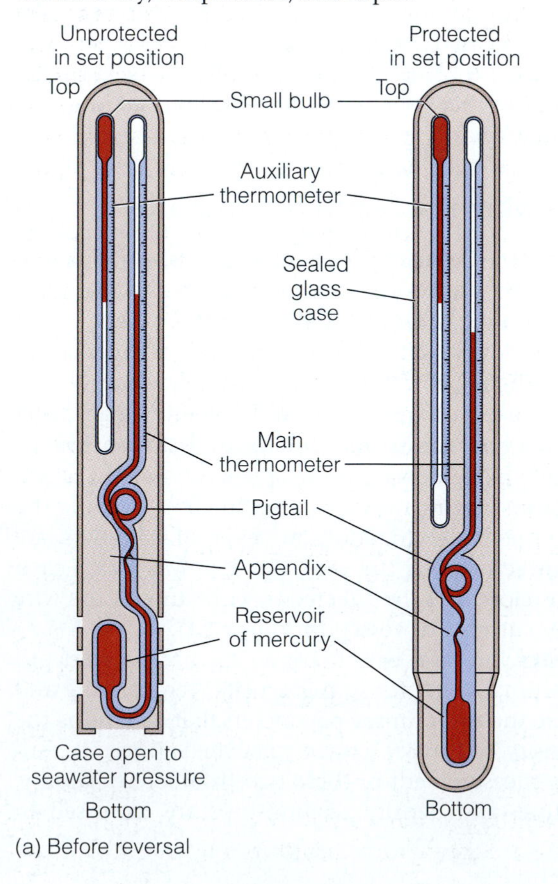

(a) Before reversal

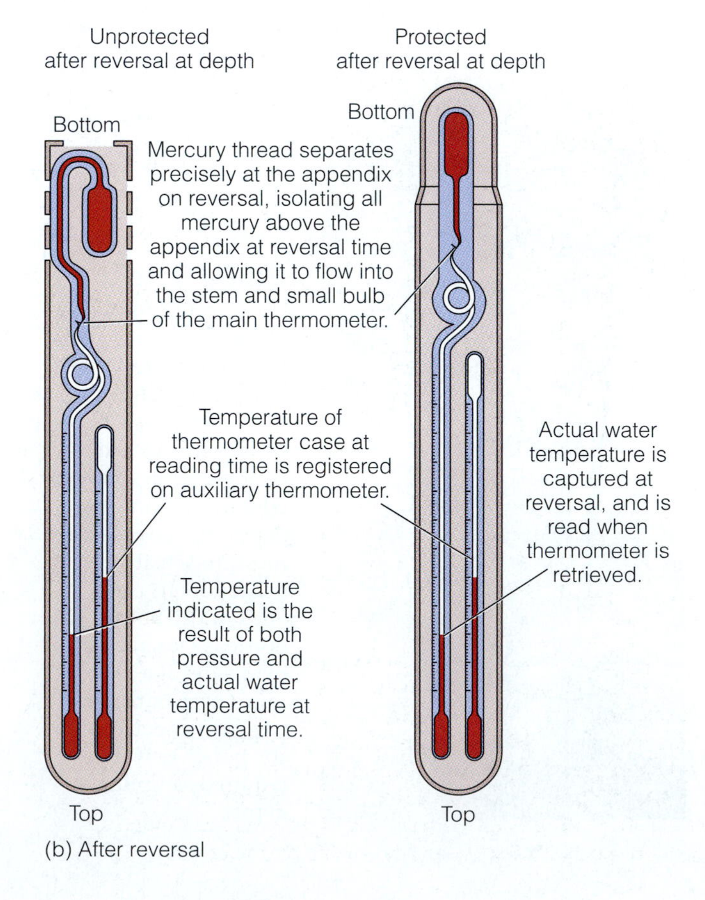

(b) After reversal

age and return the sensor signal to a processing unit in the ship's laboratory. Therefore, these CTDs enable scientists on a research vessel to see instantly the variations of salinity and temperature with depth. The special CTD wires, which have electrical conductors running throughout their length (often tens of thousands of meters), are much more expensive than the simple steel wire used previously, but the advantages justify the extra cost and complexity.

Although oceanographers no longer need to use water-sampling bottles to determine salinity, temperature, and depth, samples of water still must be collected and returned to the laboratory for analysis of most dissolved chemicals. The dissolved constituents and the importance of variations in their concentrations are discussed in Chapters 6, 14, and 19. If the sampling bottles are mounted around a CTD, samples can be taken at precise locations within the various water layers. Each bottle can be closed when the CTD readings in the shipboard laboratory show the bottle to be at the appropriate depth. For this purpose, a rosette sampler (**Fig. 3.12**) is used that consists of a rack for mounting 12 sample bottles around the CTD sensor package and an electronically operated trip mechanism. Signals sent down the support wire to the trip mechanism close each sampling bottle individually.

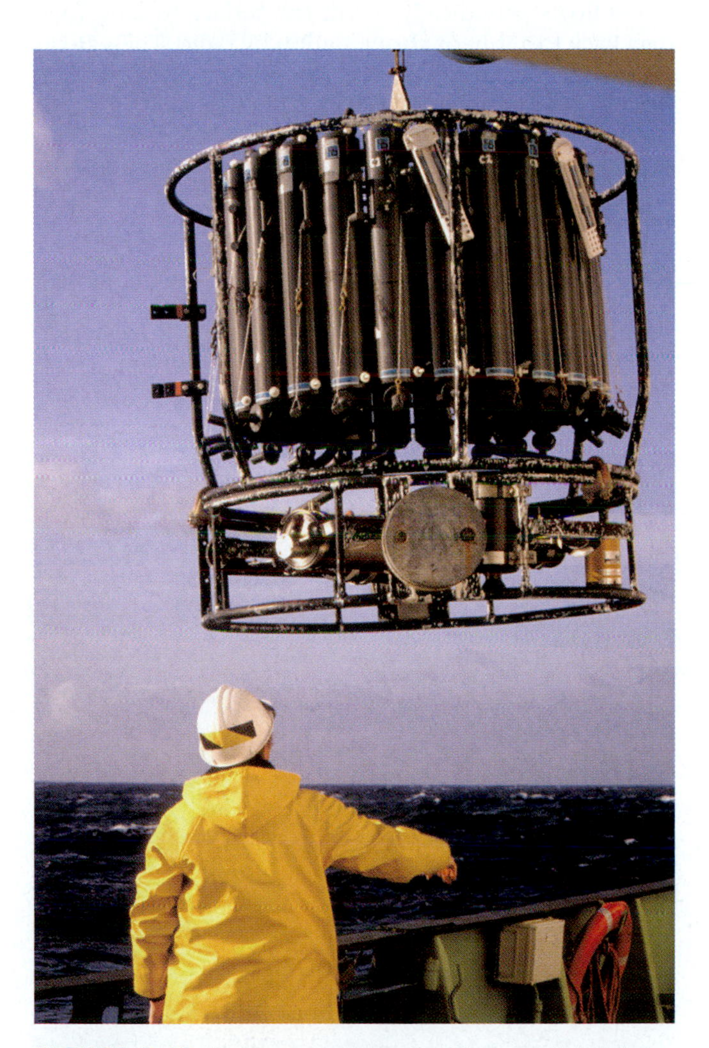

FIGURE 3.12 Rosette water sampler. The sample bottles, some fitted with thermometer racks, are arranged in a rosette pattern around an electronic device that enables researchers on the ship above to close individual sample bottles at selected depths through an electrical signal sent down the hydrographic cable. The round frame below the water-sampling bottle rack usually contains a CTD and possibly other sensors, such as turbidity-measuring devices.

There is great interest in developing sensors to add to the CTD that would continuously measure dissolved concentrations of important chemicals. Dissolved oxygen, **pH,** and **turbidity** are among the relatively few parameters for which such sensors exist. To date, none of the other sensors are as reliable and sensitive as the salinity, temperature, and depth sensors are. If a range of chemical sensors can be developed, they will probably revolutionize our understanding of the interactions among biological processes, water movement and mixing, and ocean chemistry.

Measuring Currents

Oceanographers are interested in studying the movement of water in currents that are present throughout the ocean depths. The speed and direction of currents can be computed from salinity and temperature distributions in the oceans (Chap. 10), and water movements can also be studied by the use of chemical or **radioactive tracers** dissolved in the water. These indirect methods, especially when combined with mathematical modeling techniques (**CC10**), are extremely valuable for studying the movements of water averaged over large distances (tens of kilometers or more) and long periods of time (months or longer).

Because tracers are much less useful in studying the small-scale details of current distributions and their variability with time, a variety of systems are used to measure currents directly. There are three basic types:

- *Passive devices* flow with a current wherever it goes and periodically or continuously report their position.
- *Current meters* are anchored and periodically or continuously measure the speed and direction of water flowing past them.
- *Remote sensing systems* can measure currents at various depths beneath a moving research vessel or at various depths from a fixed mooring.

Drifters, Drogues, and Floats. The simplest method of measuring currents in the **coastal zone** is to use drifters. Drifters are designed to float on the surface of the water or are weighted and designed to sink very slowly to the seafloor, where they are easily picked up and moved by even the gentlest bottom current (**Fig. 3.13a**). Drift cards (**Fig. 3.13b**) are thrown overboard in large numbers at a

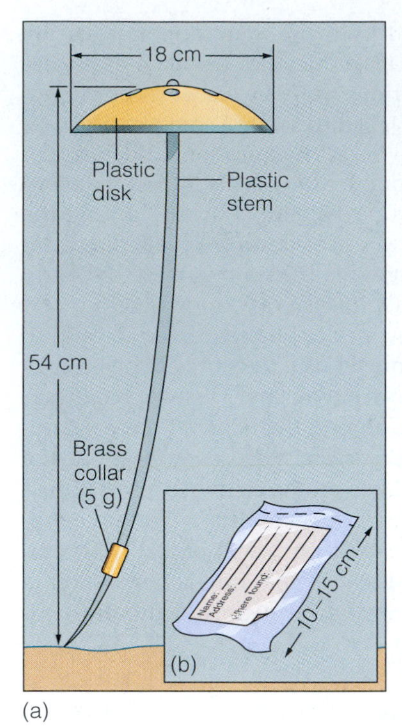

FIGURE 3.13 Surface and seabed drifters. (a) The seabed drifter is weighted such that its bottom tip drags along the seafloor as the plastic disk section "catches" the current and moves with it. (b) The surface drift card is usually a simple postage-paid return postcard sealed in plastic that offers a small reward to any finder who sends it back to the researcher and reports where (usually on a beach) and when it was found. The card floats flat on the water surface and moves with surface currents.

fixed location. Like a message in a bottle, they drift on the surface with the ocean currents until they wash up on a **beach**. Each card is numbered and bears a message asking whoever finds it to return it to the oceanographer with details of where and when it was found. The finder is often paid a small reward for return of the card. Drift cards are a very inexpensive means of gaining information about mean current directions, especially in coastal regions, where currents are often very complex and variable. Drift cards are particularly valuable in studies of the probable fate of wastes discharged or oil spilled at specific locations. Surface currents can also be studied with floats, whose movements can be followed visually, by radar bounced off a reflector, or by radio signals received from a transmitter mounted on the float.

Current speed and direction both vary with depth in the oceans. Currents flowing below the surface can be studied with two types of passive systems. In shallow coastal waters, parachute drogues are often useful. A drogue consists of a parachute attached to a weight and to a measured length of wire. The other end of the wire is attached to a float, usually with a radar reflector mounted on it. The parachute opens and is pulled down to a known depth by the weight (**Fig. 3.14**). Once at the assigned depth, the parachute drogue "sails" in the current, dragging the surface float behind it. Such devices cannot be used in deep water layers, because of the excessive drag of the long wire.

Currents in deep waters can be measured with neutrally **buoyant** floats, often called "Swallow floats" after their inventor, John Swallow. They are usually self-contained instrument packages, mounted in one or more hollow tubes and weighted so that their density is precisely the same as the density of the seawater at the depth at which the float is to operate (**CC1**). Swallow floats are deployed from a research vessel and sink through the water column until they reach a predetermined depth. The float contains a pinger, an electronic system that produces short sound pulses, or "pings." The sound emitted by the pinger is followed from listening stations aboard research vessels or onshore. The position of a float is determined by triangulation from two or more listening stations. Such floats have been followed for months as they move in the complex currents and **eddies** of the ocean depths (Chap. 10). Modern versions of the Swallow float are called "autonomous floats," and they are capable of moving back and forth vertically through the water column, recording conductivity and temperature data. These floats periodically revisit the surface to send their data back to a ship or **shore** station by radio. These floats do not have a means of recording their exact position during their time below the surface. Global Positioning

FIGURE 3.14 Parachute drogue. The parachute is deployed at a selected depth (usually less than 100 m), where it sails along with the current, dragging the surface float after it. The research ship follows the movements of the float by visual observations, with radar, or through a global positioning system unit and radio transmitter mounted on the float.

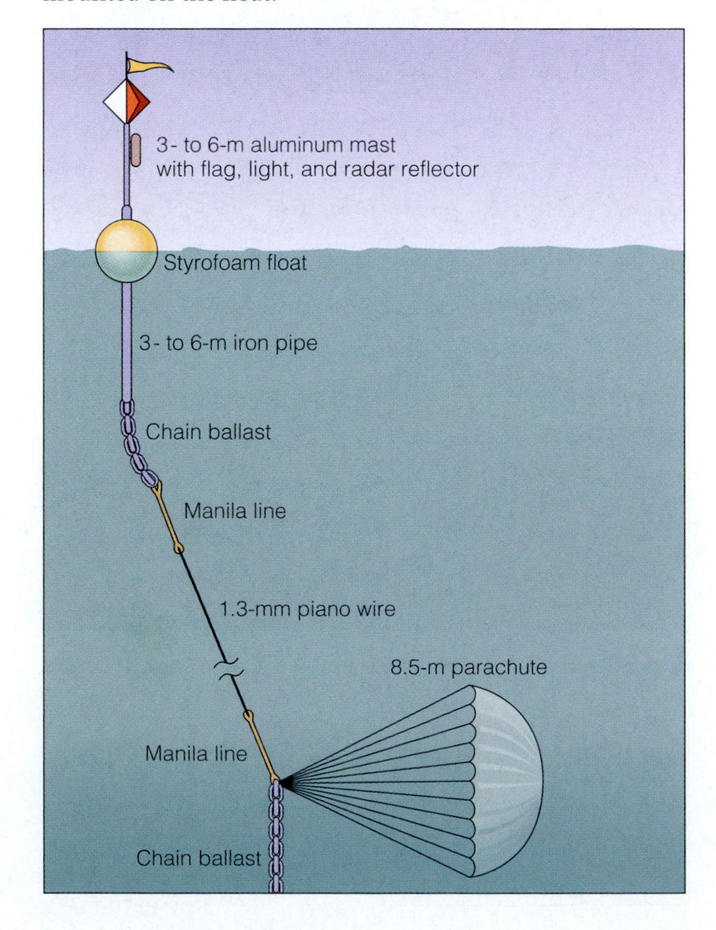

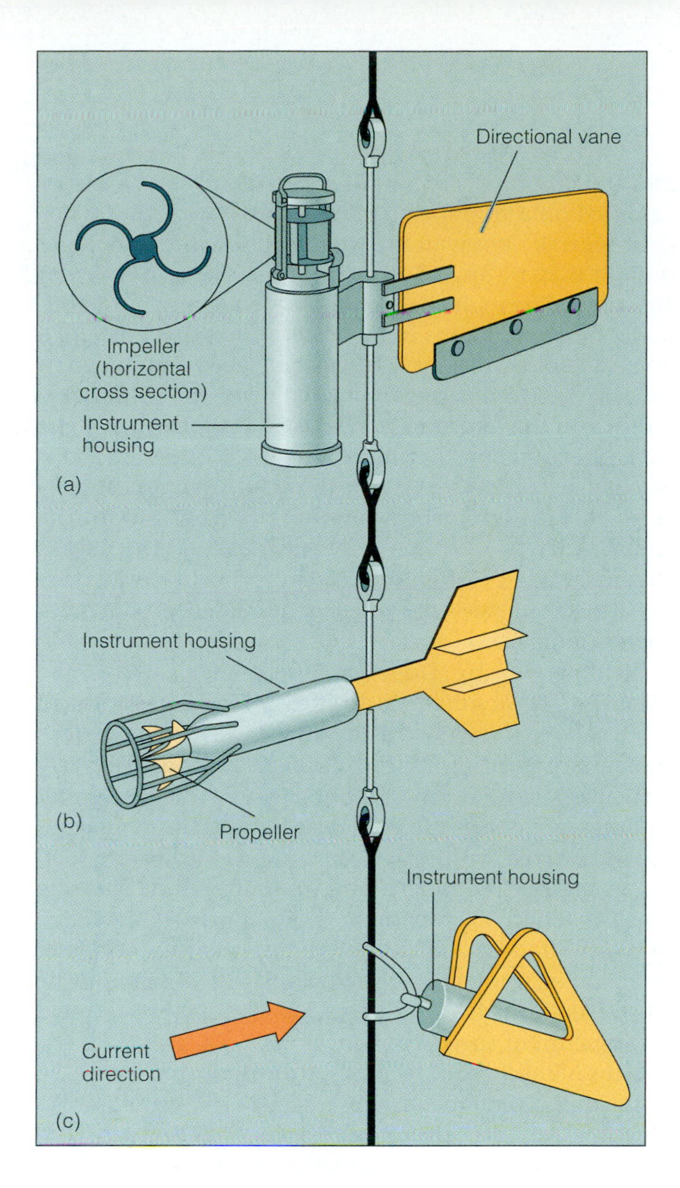

System (GPS) sensors are small enough that they can be mounted in the floats, but GPS and other electromagnetic wave–based positioning systems cannot work below the ocean surface because ocean water effectively absorbs this radiation. However, current speed and direction, averaged over the depths visited by the floats and the time interval between surface visits, can be calculated from the precise locations of the floats measured each time the float surfaces to report its data.

Mechanical Current Meters. Many current meters remain in a fixed location, measuring the rate and direction of the water flowing past them. Three types of such meters are common: (1) those having an impeller whose axis is oriented vertically (**Fig. 3.15a**); (2) those having a propellor oriented to face the current (**Fig. 3.15b**); and (3) those that rely on the current to tilt the meter body at an angle from its normal vertical position (**Fig. 3.15c**). In each case the meter is oriented to align itself with the current direction by one or more vanes or fins, just like a weather vane. Current meters are usually suspended at intervals below the surface on wire moorings (**Fig. 3.15d**). The entire mooring, including the float at the top of the wire, is often deployed well below the water surface. The string of meters is left to record currents for days, weeks, or even months, and then recovered by the release of weights from the bottom of the wire with an **acoustic** signal sent from the recovery ship to an acoustic release. This method of deployment is often necessary to avoid having the meters cut free or stolen by curious ship crews or by misguided fishers concerned about their nets catching on the mooring.

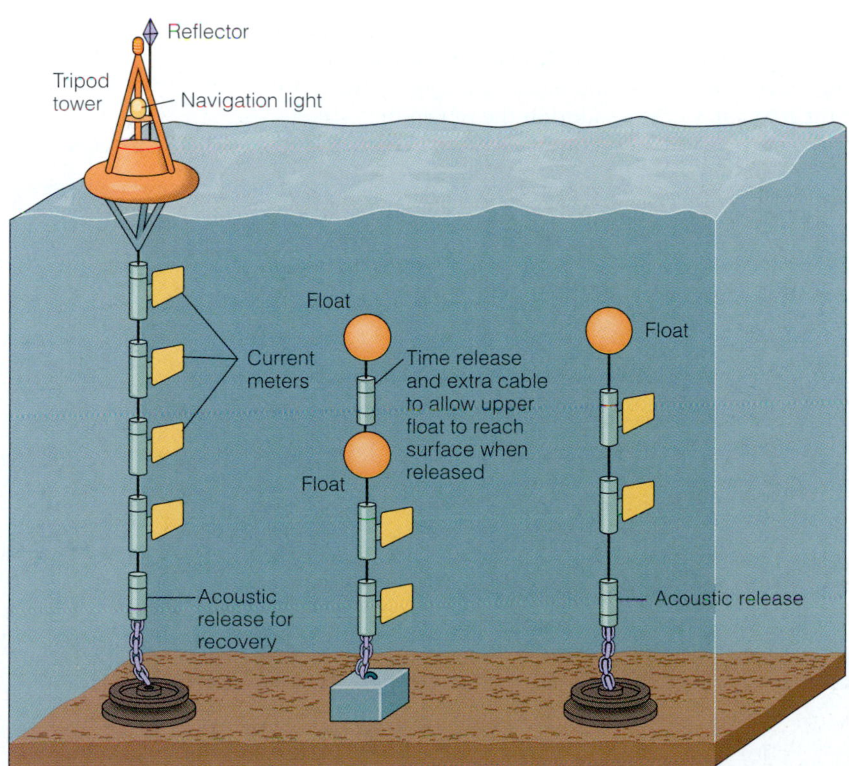

FIGURE 3.15 Typical designs for current meters are (a) a meter that measures current speed by the rate of rotation of a rotor that behaves much like a waterwheel (this type of rotor is known as a "Savonius rotor" or an "impeller"), (b) a meter that measures current speed by the rate of rotation of a propeller, and (c) a meter that measures current speed by the angle to which the meter is pushed by the current. Vanes or fins are used to orient the meter in the current. Current speed and direction, measured with a magnetic compass inside the instrument housing, are usually recorded on tape in the meter, and the records are read when the meter is retrieved. (d) Current meters are typically deployed in vertical strings, such as the three different configurations shown here, all moored to the seafloor. They are often left in place for weeks or months and then retrieved, usually by means of a release mechanism that is activated by a timer or an acoustic signal from the recovery ship. Often the current-meter moorings are entirely below the surface to discourage theft and to reduce navigation hazards.

Current meters deployed on moorings normally contain internal devices that continuously record current speed and direction. Current direction is recorded by continuous readings of the position of a magnetic compass mounted inside the meter body. Depending on the meter design (**Fig. 3.15a–c**), current speed is measured by the speed of rotation of the impeller or rotor or by the angle of the meter's tilt from the vertical.

Most current meters are sophisticated and expensive electronic instruments. However, currents have been measured with much simpler systems in some shallow coastal waters. The most ingenious and least expensive system consists of a sealed glass bottle partially filled with sealing wax and containing a magnetic needle attached to a piece of cork. The neck of the bottle is tied with string to a LifeSaver candy, and a weight, such as a rock, is attached to another string, which is also tied to the candy. The bottle is heated on board the ship to melt the wax, and the "current meter" is dropped into the water, where the weight pulls it to the bottom. Once on the seafloor, the bottle on its string is held at an angle by the current, the wax solidifies in the cold water, and soon the candy dissolves, releasing the buoyant bottle from the weight to float back to the surface. The wax surface, which was horizontal when the wax solidified, lies at an angle to the bottom of the bottle. That angle is a measure of the current speed. The bottle also records the current direction because the magnetic needle, which was facing north as it floated in the wax, is locked in place as the wax sets.

Acoustic Current Meters. Remote sensing current meters have been developed that simultaneously measure current speed and direction at multiple depths. Such meters can be mounted looking upward (**Fig. 3.16a,b**), or mounted on a ship looking downward (**Fig. 3.16c**). They send several narrow-beam sound pulses into the water column that are angled away from the meter in different directions. The returning echoes, which come from particles at different depths in the water column, are recorded from each of the beams. The current speed and direction are calculated from the Doppler shift of the sound **frequency** in the returning echoes. The Doppler shift is the same phenomenon that makes the pitch of a train whistle change as it passes. When the train is approaching—or, in this context, when current is flowing toward the acoustic meter—the pitch is increased. When the train is moving away—or when the current is flowing away from the acoustic meter—the pitch is decreased producing a deeper (more bass) tone.

Remote sensing Doppler current meters have the advantage of simultaneously measuring currents at all depths above or below the meter and within the meter's maximum operating range of 100 m or more. They are particularly useful in locations where current-meter moorings are not possible, such as in busy shipping lanes and **estuaries** with very fast currents. However, such meters are not yet widely used, in part because they are expensive and also because they must collect large amounts of data, including sound intensity, Doppler shift, and instrument orientation and tilt, which must be processed through complex computer analyses to yield data on current speed and direction.

Another technique, acoustic **tomography,** is capable of making simultaneous observations of water movements or currents within large areas of the ocean. The system is the acoustic equivalent of the computerized

FIGURE 3.16 Doppler acoustic current meters can be (a) mounted on the seabed, (b) moored in mid water, or (c) mounted in a ship's hull facing downward. The sound beams sent out by the meter are reflected off particles in the water, and the frequency of the sound in the returning echo is changed according to the direction and speed of the particles that move with the current.

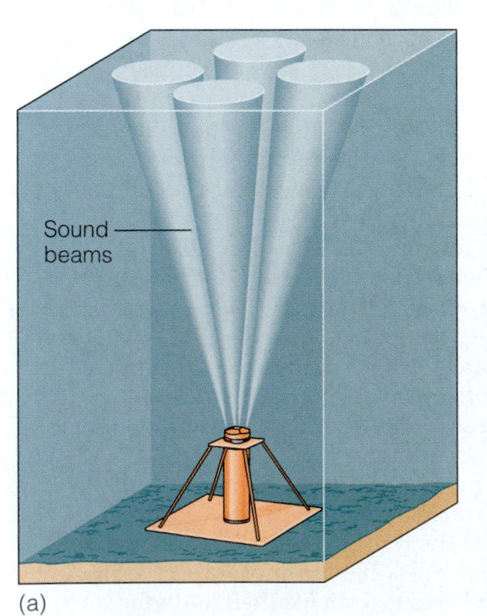

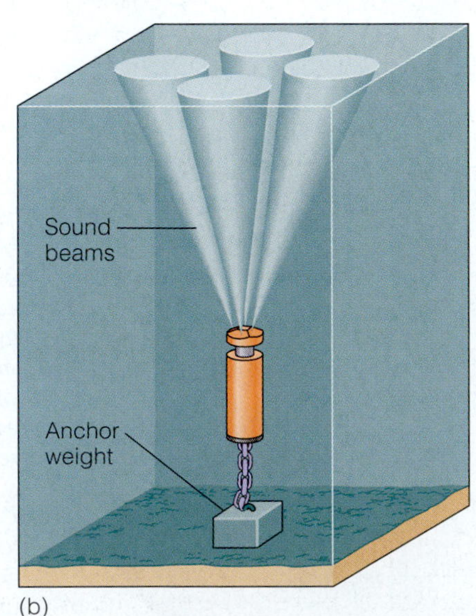

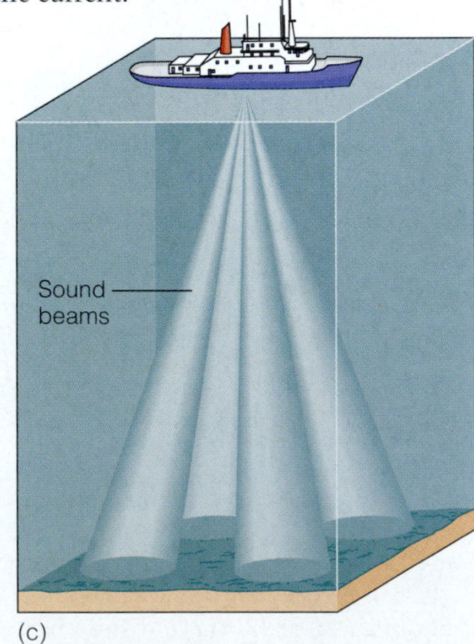

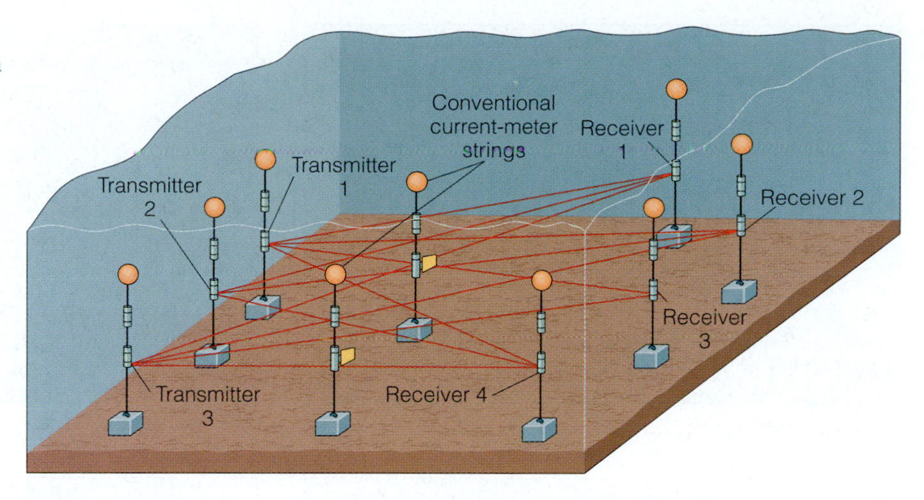

FIGURE 3.17 A typical acoustic tomography array. Sound pulses sent from each transmitter are received by each of the receivers.

axial tomography (CAT) scan used to produce three-dimensional images of internal parts of the human body. The acoustic tomography system consists of several sound sources and receivers moored at different locations within a study area that can be hundreds of kilometers across (**Fig. 3.17**). Sound is emitted by each source and received by each receiver, so numerous pathways of sound traveling between sources and receivers are possible. The sound velocity along each of the pathways is affected by such water characteristics as depth, temperature, and salinity. When currents flow within the study area, they change the distribution of temperature and salinity and therefore the average speed of sound along the source–receiver pathways. Powerful computer data analysis techniques are used to convert the small variations in travel times between each source and receiver into a detailed picture of the water property distributions and movements within the array boundaries.

3.5 LIVING ORGANISMS IN THE SEA

Almost all terrestrial **species** spend their entire lives within a narrow zone that extends from a few tens of meters above the tops of the trees to a few meters below the Earth's surface. Therefore, terrestrial organisms are relatively easy to study, as they are readily accessible and visible to the scientist. In contrast, ocean life is present throughout the thousands of meters depth of ocean waters (**pelagic** species) and for several meters, or more, into the sediment (**benthos,** or **benthic** species).

Special Challenges of Biological Oceanography

Although pelagic marine life is concentrated in the upper few hundred meters of the sea, scientists cannot readily enter the oceans to observe it. Therefore, most marine biological studies depend on methods of capturing the undersea creatures and bringing them back to a research vessel or onshore laboratory. Collecting biological samples from a given ocean location and depth is a daunting task that poses five major problems, which add to the general problems of working in the oceans discussed previously:

- Pelagic life is widely dispersed in three dimensions. Therefore, large volumes of water must be sampled or searched before representatives of all, or even most, species in any part of the ocean are captured.
- The species that make up ocean life range in size from microscopically small **bacteria, archaea,** and **viruses** to giant whales. Therefore, no single method of sampling can capture all important species in a particular study location.
- Many species of ocean life actively swim and avoid any sampling device lowered into their **environment.**
- Many species are extremely delicate and are literally torn apart by contact with sampling devices that are dragged through the water.
- Many species that live in the cold, dark, high-pressure, deep-ocean environment cannot survive the changes in these parameters as they are brought to the surface.

Nets, Water Samples, and Traps

The sampling devices most often used to collect pelagic species are nets towed through the water. Nets of different overall sizes and mesh sizes must be used for different types of organisms (**Fig. 3.18**). Small nets with very fine mesh are used to sample **phytoplankton** and **zooplankton,** and progressively larger nets and meshes are used to capture fishes and **invertebrates.** Net samples may be collected by simply lowering a net from a stationary ship and hauling it back on board. However, biologists usually want to collect samples from a specific depth, so most sampling with nets is performed by towing a weighted net slowly behind the research vessel.

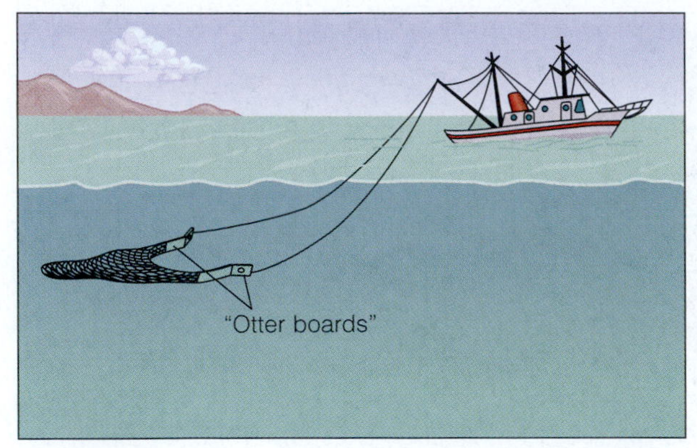

(a) Otter trawl

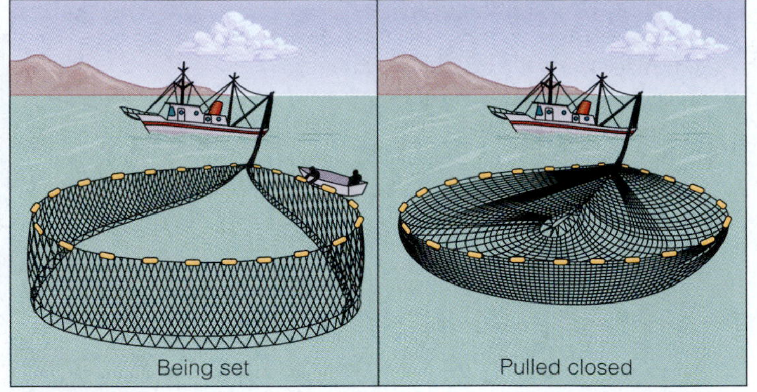

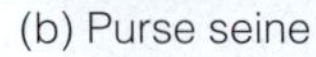

(b) Purse seine

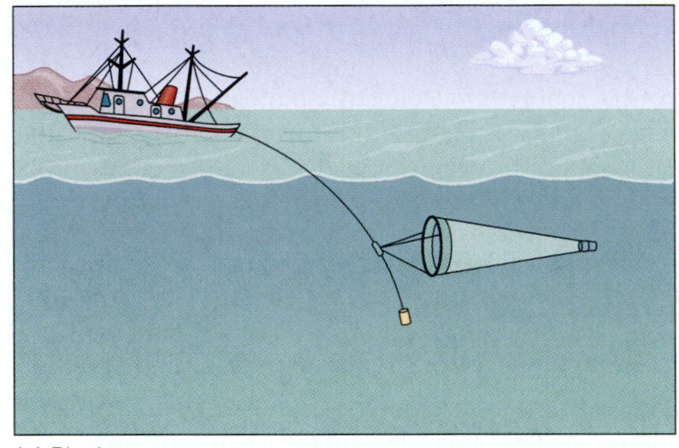

(c) Plankton net

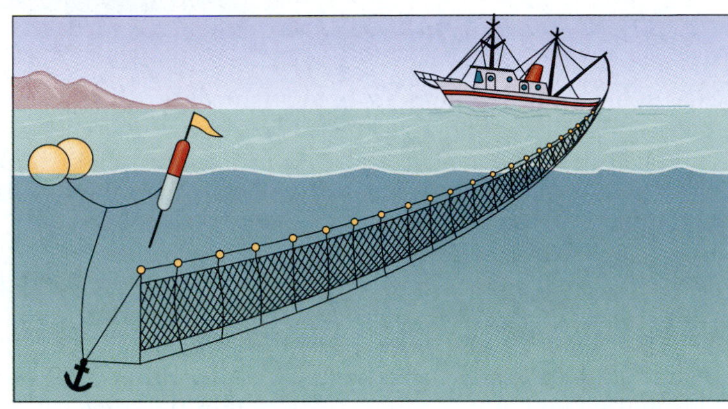

(d) Drift net

FIGURE 3.18 Many different types of nets are used to sample pelagic and benthic organisms. (a) The otter trawl uses two "otter boards" that "sail" in the current created by the movement of the vessel as the net is towed. The sailing motion of the otter boards forces the end of the net to stay open. (b) The purse seine is towed quickly around a school of fish and then drawn up under the fish, preventing their escape. (c) Plankton nets are extremely fine mesh nets with a jar or bucket attached at the narrow end to collect plankton as the net is towed through the water. Some plankton nets can be opened and closed on commands from the surface vessel in order to sample only selected depths and locations. Some have current meters attached to measure the volume of water sampled as it flows through the net. (d) Drift nets may be deployed for miles across the ocean, where they catch anything that is too large to swim through the net. Even large animals, such as dolphins and sharks, may become entangled and die.

Nets can be let out on a known length of cable over the stern of the ship, towed for a time, and then retrieved. However, without instrumentation the biologist does not know the exact depth of the net tow, how much of the catch came from shallower depths as the net was let out or retrieved, or how much water passed through the net. The amount of water passing through the net during a tow must be known if the concentration or population density of the species caught is to be determined. Many sampling nets now have various types of recording-depth gauges to measure the depth at which the net is towed. Modern nets can also be designed to be kept closed until reaching the desired sampling depth, and then opened, towed, and closed again before retrieval. In addition, many nets carry flow meters similar to impeller-driven current meters (**Fig. 3.15a**) to measure the distance the net is towed while it is sampling. To calculate the volume of water sampled by the net, the measured tow distance is multiplied by the cross-sectional area of the net opening. Some nets are equipped with video cameras to record species that evade the net.

The smallest pelagic organisms in the oceans (bacteria, archaea, viruses, and phytoplankton) can be sampled by collecting water in sampling bottles such as those described earlier. **Plankton** and microorganisms are often sampled by surrogate parameters. For example, investigators often estimate the total **biomass** of **photosynthetic** plankton by measuring the total **chlorophyll** (**CC14**) concentration. In addition, they can estimate the concentrations of some other groups of microorganisms by measuring the concentrations of other pigments similar to chlorophyll or, most recently, by measuring the concentrations of specific types of **genetic** material.

Sharks, large fishes, and squid that avoid nets are sampled with fishing lines and lures. Fishes and other large

FIGURE 3.19 Lobster pots are designed such that the lobsters can crawl into the pot to eat the bait left inside, but then cannot escape.

invertebrates that live on the seafloor are sampled with trawl nets towed across soft sediment bottoms (**Fig. 3.18a**) or by dredges similar to those used to collect rocks from the seafloor (**Fig. 3.8**). Smaller benthic organisms are collected by means of sediment collection devices such as grab samplers and box corers. On the ship, the finer sediment is washed through one or more sieves to isolate such organisms.

Fishes and invertebrates that are particularly adept at eluding nets can be caught in a variety of traps. The animals are attracted to the traps with bait or lures. In the dark deeper waters or at night, they may also be attracted by light. Many types of traps have been used. Almost all allow the organism to enter the trap easily in pursuit of the bait or lure, but make it difficult or impossible for the organism to escape. The lobster pot is the best known of such traps (**Fig. 3.19**).

Fragile Organisms

Organisms that are fragile and easily damaged by nets are particularly difficult to collect. Such organisms are numerous in the oceans. Many float freely in near-surface open-ocean waters (Chap. 14), and many live in **communities** at **hydrothermal vents** (Chap. 17). Fragile organisms can be collected only by being captured carefully in closed jars or bottles.

Fragile organisms from the upper layers of the open ocean can be collected by divers. The divers drift or swim until they encounter an interesting organism, and then they guide the organism into an open jar that they seal immediately. Fragile or elusive small fishes and other organisms can also be sampled by a simple syringelike device called a "slurp gun." The diver places the open mouth of the gun near the organism and simply sucks it into the slurp gun body by withdrawing a plunger. Slurp guns are particularly useful for collecting small organisms from within cracks and holes in **coral reefs,** and they are the most environmentally sound means of collecting small tropical fishes for aquariums.

In waters deeper than **scuba** divers can reach, fragile organisms must be collected from research submersibles. The collection methods are the same as those used by divers, but the jars, slurp guns, or other collecting devices must be manipulated by remote control.

Migrations and Behavior

Although the occurrence or concentration of a species within a given area is an important parameter, it tells us little about the organism's life cycle—including, for example, its migration, feeding, and reproduction. Observations by divers, submersibles, and remote cameras can provide some of this information, but only for species that do not migrate large distances or through great depths during the activities under observation. However, many species, including whales and other **marine mammals,** turtles, and many fish species, do sometimes swim large distances and dive to substantial depths to feed or reproduce. Some of these behaviors can be studied remotely. For example, **schools** of fish can sometimes be followed by sonar, and whales can be tracked by their songs, which can be heard at great distances. However, these are at best inadequate techniques. Recent advances in the miniaturization of electronics have led to the creation of a number of small sensing packages that can be attached to, for example, the skin of a whale to gather data on time, depth, and temperature. The instrument package stays attached for some time and then is released, floats back up to the surface, and is either recovered or sends its data to shore by radio. These instrument packages are becoming steadily more sophisticated and now may include other sensors or video cameras. One particularly interesting package contains a microphone and recording device. Attached to whales, this device has been used to record the changes in diving behavior caused when a whale swims into the sound field generated by a surface sonar unit.

3.6 SCUBA, MANNED AND UNMANNED SUBMERSIBLES

Many tasks can be performed only, or are performed best, when oceanographers are able to see what they are sampling or measuring. For example, some areas of the seafloor are characterized by jumbled rock formations and highly variable assemblages of organisms. In such areas, geological oceanographers can best identify and collect rocks and sediment of interest if they can visually inspect and select samples. Similarly, biologists can ensure that they are collecting the important species if they are able to see and select the organisms to be sampled. For many species, particularly those that do not survive transfer to an aquarium or laboratory, biologists can study feeding, defense, movement, and reproduction only if they can observe the species in its natural **habitat.**

Scuba and Habitats

In shallow waters, sampling is often performed by scuba divers. However, because of the dangers of the bends and nitrogen narcosis, scuba divers cannot descend safely below about 90 m and they can remain underwater for only a short time.

Scuba equipment supplies a diver with air at the pressure that corresponds to the diver's depth. Thus, the pressure of the air breathed increases with depth. Divers are susceptible to two dangerous syndromes. The first of these, nitrogen narcosis, occurs when high-pressure air causes high concentrations of nitrogen to build up in the bloodstream. This causes symptoms very similar to those of alcohol intoxication—with possible consequences similar to drinking and driving. The second ailment, the bends, is a different life-threatening medical problem. It is caused by breathing high-pressure air at depth for too long and then returning to the surface too quickly. As the depth and immersion time of a dive increase, nitrogen continues to dissolve into the diver's bloodstream and then transfers into other body tissues. If the diver returns to the surface too rapidly, the excess nitrogen in the blood and tissues cannot escape quickly enough and forms damaging gas bubbles in the body.

One way of extending the length of time that scuba divers can remain at depth safely is to use an underwater habitat (**Fig. 3.20**). Scientists can live for days or weeks in such habitats, which are pressurized and anchored on the seafloor. The scientists can safely make multiple scuba excursions to research sites at approximately the same depth as the habitat, as long as they return to the habitat and not to the surface. Although underwater habitats have many valuable uses, particularly for behavioral and other biological studies, their utility is limited. They are expensive to maintain and operate, they cannot readily be moved to new research sites, and they cannot significantly increase the maximum depth at which scientists can work using scuba. In addition, to avoid the bends, scientists who live in habitats for a week or more must spend several days in a decompression chamber at the end of their stay, even if the habitat is only a few meters deep.

Manned Submersibles

Most of the ocean floor and almost all of the ocean volume are too deep for scuba divers to reach. Marine scientists have used a variety of research submersibles to visit and work at greater depths. Manned oceanographic research submersibles are small submarines usually designed to carry no more than two or three scientists. They enable the scientists to observe and photograph

FIGURE 3.20 Underwater habitats, such as *Aquarius,* shown here in the Florida Keys, enable scientists to live and work up to several tens of meters underwater for several days without incurring decompression sickness. However, at the end of their stay they must undergo very long decompression periods in a decompression chamber.

organisms and the seafloor. Equipment is often attached to the outside of the submersible to collect organisms, rocks, sediment, and water samples that can be selected visually by the submersible's occupants.

The first recorded successful use of a submarine was in 1620. The vessel, built by a Dutchman, Cornelius Drebbel, had a waterproof outer skin of leather and was propelled by 12 oarsmen. It was reported to be able to stay as deep as 4 or 5 m underwater for several hours. Between 1620 and the 1930s, many different submarines were developed, but they were used primarily as warships. Today, the vast majority of submarines are still warships that are not suited to, or used for, oceanographic research.

During their early development, research submersibles had very limited capabilities because they were built primarily to transport explorers who sought to dive to ever-greater depths. Such explorers usually performed scientific observations as only a secondary interest. Early submersibles, called "bathyscaphes" or "bathyspheres," provided little more than windows and lights for their occupants to view the oceans and seafloor, and most were not equipped to collect samples. In addition, most early submersibles could only sink to the seafloor and then drift with the currents or move short distances. Therefore, positioning a submersible precisely at a previously selected site on the seafloor was not possible.

The exploration phase of research submersibles climaxed in 1960 when the bathyscaphe *Trieste* visited the deepest part of the ocean, the Mariana Trench, 10,850 m below the ocean surface. Since 1960, many new submersibles have been designed, but almost none of these are capable of reaching the depths achieved by the *Trieste.* Most are designed for much shallower dives. Recent advancements in submersible design have centered on improving the submersible's ability to find precise locations on the seafloor, travel across the seafloor during a single dive, and collect samples at selected locations. Modern submersibles are strange-looking vessels with one or more protruding robot arms and a variety of baskets, other sample-collecting devices, video cameras, and measuring instruments that hang from the hull within reach of a robot arm (**Fig. 3.21**). They also carry powerful lights because sunlight does not penetrate more than a few hundred meters of seawater. Submersibles are now used extensively for a variety of undersea observations, particularly observations of underwater volcanic features and of hydrothermal vents and their unique biological communities (Chaps. 8, 17).

Unfortunately, submersibles have disadvantages that limit their usefulness. Most submersibles must have a large surface vessel to carry or tow them to the research site, with a large crew to launch, retrieve, maintain, and repair them. Because submersibles are very expensive, the cost of each dive is extremely high in relation to the cost of other oceanographic research efforts. Submersibles have very small interior crew spaces, so dives of several hours are an uncomfortable ordeal for the pilot and passengers. The discomfort, the need to carry air-recycling systems for the crew compartment, and the limited battery power available to operate the motors, life-support system, and scientific equipment are all factors that limit the maximum duration of each dive. Limited dive duration restricts the area that can be studied on each dive, particularly in deep waters where the submersible may spend several hours descending and ascending.

FIGURE 3.21 A typical submersible, such as the *Johnson Sea Link* shown in this photograph, consists of a glass or plastic sphere inside a hull. The scientists descend within the sphere, which is provided with air and heat. Motors, articulated arms, and many different designs of samplers and cameras are mounted outside the sphere. These are used to provide propulsion and collect samples.

CRITICAL THINKING QUESTIONS

3.7 A research paper published in 1970 reported that the concentrations of several trace metals (zinc, copper, and iron) in North Atlantic Ocean waters appear to vary within a narrow range, but also appear to vary randomly with depth and location. Should you believe these data? What would you look for in the research paper to help you decide?

3.8 A grab sampler is lowered to the seafloor at a location that has never before been sampled, and it comes back closed but empty. What are the possible reasons for this result?

3.9 Undoubtedly, there are many species presently unknown to science that live in the deep oceans. It has been speculated that most of these are free-swimming in the water column and do not live sedentary lives in or on the sediment. Do you think this is a reasonable hypothesis? Why or why not?

Remotely Operated Vehicles

Many of the limitations of submersibles can be overcome by use of a remotely operated vehicle (ROV). The basic ROV consists of a television camera mounted on a frame or sled. An electric motor, which drives a propeller, and

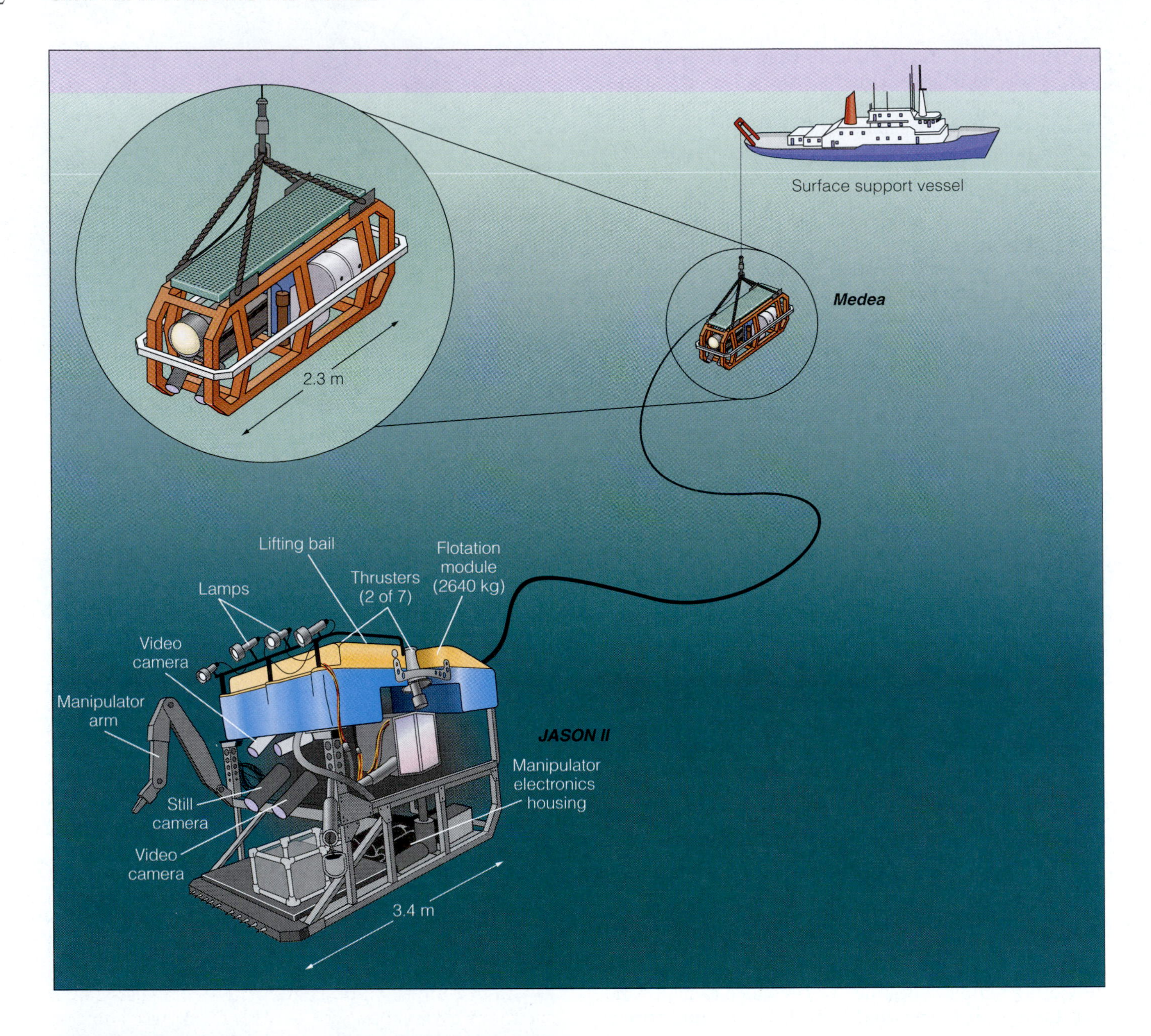

FIGURE 3.22 The *Jason II* ROV and how it is deployed as part of the *Jason II/Medea* system. The many motors, instruments, sampling arms, lights, and cameras that are mounted on *Jason II* and on *Medea* are operated remotely from a surface vessel. *Jason II* is connected to *Medea* by a 35-m-long neutrally buoyant cable and is thus isolated from any ship motions that are transmitted down the cable to *Medea*.

electronically controllable steering devices are also mounted on the sled. The sled is attached to a surface ship by a cable through which power is supplied from the ship, video images are transmitted to the shipboard laboratory, and steering signals are sent to the sled. ROVs are much less expensive than submersibles and can spend many more hours underwater. They can be deployed from much smaller research vessels than the ones that carry submersibles; they do not require the difficult, expensive, and time-consuming procedures and equipment needed to protect the lives of submersible crews and passengers; and they can allow several scientists at the same time, rather than the one or two in a submersible, to view the seafloor and to direct sampling activities.

The sampling and measurement operations performed with submersibles involve remote manipulation of robot arms and other equipment located outside the submersible. Because nearly identical equipment can be mounted on ROV sleds, they can perform the same tasks as submersibles, but they can be operated remotely by a pilot sitting comfortably in front of a video monitor in a ship's laboratory. In fact, if steerable video cameras are mounted on the ROV, the shipboard ROV operator has a better view of the environment than a submersible operator has, because the submersible operator's vision is often restricted to several small, fixed window ports. In addition, an ROV enables several scientists of different disciplines to take part in each "dive" and to participate in deciding when and where samples should be collected. Although they are less glamorous than research submersibles and are not yet developed fully for all their potential uses, ROVs may eventually replace almost all submersibles for ocean science.

Perhaps the most famous, although scientifically not very useful, achievement of modern research submersibles was the successful exploration of the wreck of the *Titanic* and the recovery of some of its artifacts. What is not widely known is that the manned submersible dives to the *Titanic* were made possible by the work of an advanced ROV, the *Argo*. It performed the lengthy and difficult search for the wreck and the video exploration of the wreck and surrounding debris field. This preliminary work made it possible to send submersibles some months later to explore the wreck further and collect artifacts. Another ROV, *Jason Jr.*, was used in the subsequent exploration of the *Titanic* by the submersible *Alvin*. *Jason Jr.* was attached to the *Alvin* rather than to a surface ship.

The *Jason II/Medea* system, developed since the exploration of the *Titanic*, is among the most sophisticated ROV systems yet built (**Fig. 3.22**). *Medea* is attached to a surface research vessel by a long cable. It carries lights, a video camera, and other instruments to survey the seafloor as it is positioned just above it. *Medea* can operate to depths of about 6500 m. *Jason II* is an ROV attached to *Medea* by a cable 35 m long. Once *Medea* has located an interesting area of seafloor, *Jason II* can be sent out to make a much more detailed survey of the area surrounding *Medea*'s location. *Jason II* can be maneuvered very precisely because it is not connected directly to the cable attached to the ship above. Consequently, *Jason II* is isolated from the sometimes substantial ship motions that are partially transferred down the cable to *Medea*, which enables *Jason II* to more easily locate and collect samples on the seafloor. The *Jason II/Medea* system can remain at work continuously on the seafloor for periods of a week or more. Such systems can locate or relocate themselves on the seafloor with a precision of a few meters by analysis of the travel times of sound pulses transmitted from the surface vessel, and from transponders mounted on the vehicles and anchored to the seafloor in a triangle surrounding the study area.

FIGURE 3.23 An Argo float system about to be deployed from a research vessel to begin its long unattended journey through the ocean depths.

Even ROVs may eventually become obsolete as the development of autonomous underwater vehicles (AUVs) proceeds. Today AUVs are limited in their abilities but are developing rapidly. An AUV is a vehicle that moves through the ocean with no human occupants and with no direct connection to surface vessels or submersibles. It performs exploration and sampling tasks according to preprogrammed instructions or through limited communication of data and instructions to and from remote operators. The simplest example of an AUV may be *Argo* (not the same as the ROV *Argo* mentioned earlier) (**Fig. 3.23**) and other similar floats. These are autonomous instrument packages programmed to alternately descend and ascend through the water column collecting data on the properties of **water masses.** They transmit their data to shore by radio each time they reach the surface. The floats are part of the Global Climate Observing System/Global Ocean Observing System (GCOS/GOOS), which by 2005 had deployed more than two thousand of these floats throughout the world oceans. Future AUVs will require the development and application of sophisticated new subsea robotics and other technologies for underwater observation, sampling, and communications.

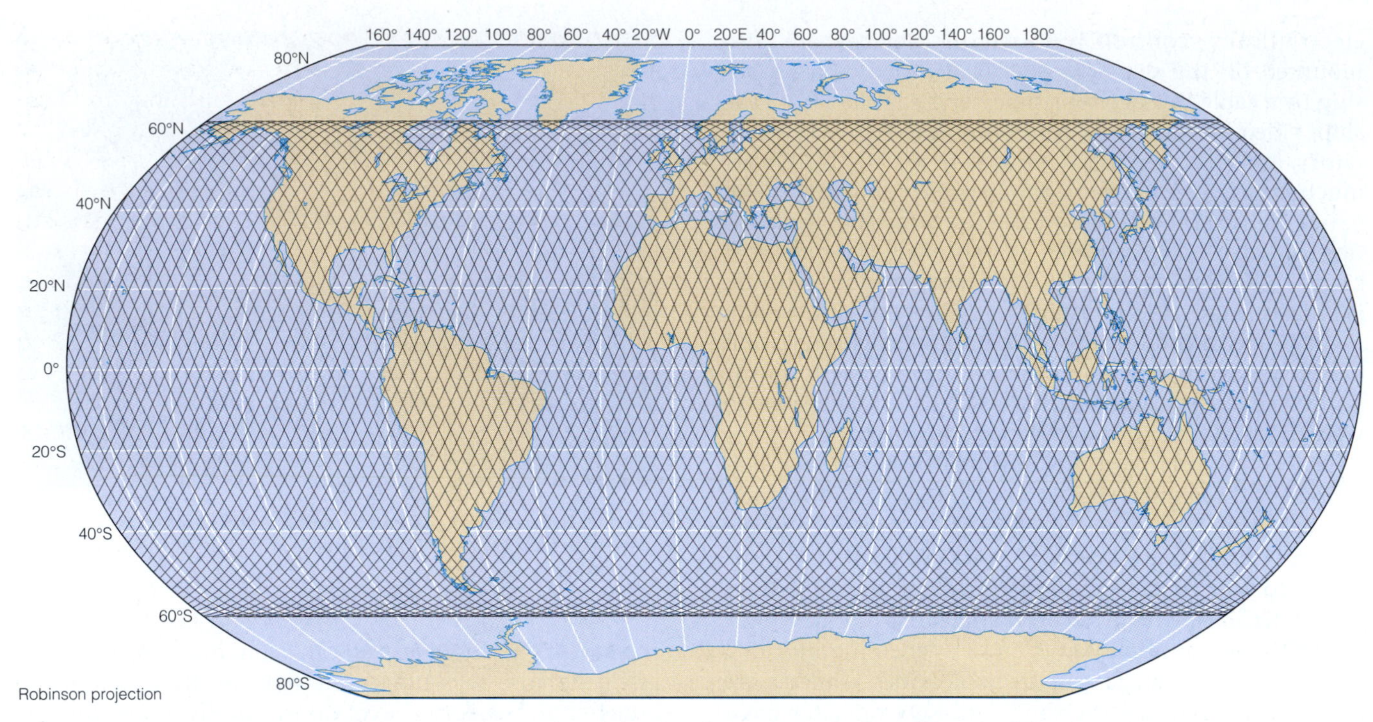

FIGURE 3.24 The *TOPEX/Poseidon* satellite travels in an orbit that is typical of satellites used for ocean observations: 1366 km high and inclined at 66° to the equator. This orbit allows coverage of 95% of the ice-free oceans every 10 days. The satellite's position is known to within 10 cm from the Earth's center. The network of curved lines on the map schematically illustrates the path that the satellite follows in each 10-day period.

3.7 SATELLITES

Satellite observations have truly revolutionized marine sciences. From a satellite, the entire surface of the world's oceans can be surveyed in just a few days (**Fig. 3.24**). In addition, large areas of the oceans can be surveyed comprehensively every few days or even hours. Such **synoptic** observations from satellites have enabled oceanographers to observe day-to-day, month-to-month, and year-to-year changes in ocean phenomena, including surface currents, surface water temperatures, waves, phytoplankton and suspended-sediment concentrations, and ice cover. Before satellites, such temporal changes could be observed only at fixed locations or within small regions because surface vessels cannot survey large regions quickly or often. Ships may take several weeks to complete a survey of several hundred square kilometers, whereas ocean conditions can change very rapidly, for example during storms. Hence, observations obtained at the beginning and end of a ship survey often describe very different states of the ocean. The degree to which satellite observations have provided a new view of the oceans is demonstrated by the number of figures in this text that present satellite data to describe important ocean processes.

Satellite observations of the oceans are made with a variety of devices, including radar, **lasers,** and color- and infrared-sensing scanners. Although satellites are excellent platforms from which oceanographers can make rapid and comprehensive observations of the oceans, their value is limited because the sensors they carry rely primarily on **electromagnetic radiation,** which does not penetrate ocean water effectively. Therefore, most satellite observations record features and processes of only the upper few meters of the oceans, and they provide little information about processes below the upper layer.

Satellites have made a special contribution to oceanography as aids to navigation. Until satellite navigation systems were developed, research ships far from land could not determine their locations with great accuracy. When clouds prevented star and sun sightings for several days, a vessel's position could be determined only with an uncertainty that often exceeded several kilometers. Research ships and other vessels now carry satellite navigation equipment that can locate the vessel with an accuracy of 200 m or less. Several navigation satellites, which are in precisely measured orbits around the Earth, send out radio signals that are received by the ship as the satellite passes overhead. A computer automatically analyzes the satellite radio signal and calculates the ship's position. The most recent and accurate satellite navigation system is the Global Positioning System (GPS), the military version of which can determine a position anywhere on the Earth's surface to within a few meters. Precise navigation enables oceanographers to locate and map small features of the seafloor and to revisit precise locations on subsequent research cruises.

Routine satellite observations of the oceans have many practical uses other than navigation and scientific research. For example, they are used by fishers to locate the best fishing grounds on the basis of surface water temperatures and by ships' captains to save fuel by avoiding storms or currents. They are also used to forecast weather, especially **hurricanes** and other storms, and to determine iceberg locations. The growing list of parameters that can be determined by satellites include currents and circulation patterns, ocean surface temperature and salinity, surface wind patterns, **wave heights** and sea state, bathymetry, chlorophyll and plankton concentrations, distribution of river plumes, concentration of suspended sediment, beach **erosion,** and **shoreline** changes. Satellite observations of oceanic and atmospheric temperatures and other parameters will be particularly important in monitoring the enhanced **greenhouse effect** and its potential impact on world **climate,** sea level, and biological processes, and in improving our ability to understand and predict such phenomena.

3.8 COMPUTERS AND MODELING

The development of computers has revolutionized ocean sciences just as it has other aspects of human society. Ocean sciences are particularly dependent on computers for two important functions: analysis and display of the geographic and depth variations in important parameters, and data analysis using mathematical models.

Almost all studies conducted by ocean scientists involve parameters or processes that vary both geographically and with depth. Data of importance to ocean processes must therefore be organized in a three-dimensional framework and displayed visually for analysis and interpretation. Many of the illustrations in this text display data on maps or on vertical map sections. As we discussed earlier in this chapter, the relatively simple seafloor map that Heezen and Tharp generated by hand before the advent of computers was a massive undertaking. Computerized mapping now makes the development of such maps much simpler and quicker, even given the massive amounts of data that can be generated by satellites and automated sensors. Most computerized mapping is now done using Geographic Information Systems (GIS). GIS is designed to store data for any parameter that is collected at a specific location and time—for example, salinity, temperature, phytoplankton concentration, light intensity. The GIS system stores each data point referenced to its geographic coordinates (three-dimensional to include depth, where appropriate) and collection time. Simple-to-use software tools allow data for multiple parameters to be overlaid on each other (spatially and temporally) for comparison and analysis, and for these results to be displayed in a variety of ways.

Ocean sciences are heavily dependent on mathematical modeling, described in **CC10**, because the processes studied occur on very variable time and space scales, and many processes, especially biological parameters and interactions, are characterized by very limited data. Global climate models, used to investigate ocean-atmosphere interactions, greenhouse-induced global climate change, global weather patterns, and many other questions, rely particularly heavily on computers. As explained in Chapter 10, the scale of water motion in the ocean is one-tenth the scale of atmospheric motions that contribute to weather and climate. Furthermore, oceans are more complex vertically than the atmosphere. As a result, if the ocean part of the global model is to operate effectively, it must be split into model cells that are smaller and separated vertically, as well as horizontally. Thus, the ocean component of these models can require as much as four orders of magnitude more computer time than the atmospheric component requires. For these models, this is significant because it can take months of time on the fastest supercomputers to run the models through just a few years of simulated climate.

CHAPTER SUMMARY

Mapping the Ocean Floor. Until 1920, the only way to determine the depth of water under a ship was to lower a weight on the end of a line or wire until it touched bottom, and then measure the length of wire let out. This method was tedious and inaccurate, especially in deep waters. It was also prone to errors because it was difficult to determine when the weight reached the seafloor and because the path of the line or wire through the water column was distorted by currents.

In 1920, echo sounders were developed that measured the time taken for a sound ping to travel from ship to seafloor and back. Precision echo sounders were operated continuously on all research ships, and the data were used in the 1950s to produce the first detailed maps of seafloor topography.

Newer echo sounders send sound pulses spread over a wide area to either side of a research vessel, and the multiple echoes are processed by computer. The data are used to create detailed three-dimensional seafloor maps. Satellites can measure sea surface height so accurately that the data can be used to map the seafloor based on slight variations of sea surface height caused by gravity differences over the varying seafloor topography.

Seafloor Sediments. Samples of seafloor sediment are obtained for several purposes by a variety of techniques.

Samples of the upper few tens of centimeters of sediment are used to study recent sediment processes or the biology of the sediment. They are usually obtained with grab samplers or box corers.

Long vertical cores are obtained to study the history of sediment that is accumulated layer by layer. They are obtained by corers or by drilling. Drilling ships are very expensive to operate but can retrieve very long cores. Samples of rocks from the seafloor are obtained by dredges.

Indirect methods of studying sediments include gravity and magnetic-field strength measurements and seismic techniques in which sound pulses are echoed off layers of buried sediment.

Chemical and Physical Oceanography. Water samples must be collected to determine most parameters of seawater chemistry and to study microscopic organisms. Samples are collected in bottles lowered to the desired depth, where they are closed and then retrieved. Sample contamination from the sampler materials and wire and from surface waters is often a problem because of the very low concentrations of some constituents. Some parameters, notably pressure (depth), conductivity (salinity), and temperature can be determined ***in situ*** by electronic sensors.

Currents are measured by tracking drifters, drogues, or floats as they move with the current; by the use of fixed current meters that measure the speed of rotation of a rotor or the tilt of the meter in the current; and by acoustic remote sensing of the movements of suspended sediment.

Living Organisms in the Sea. Sampling marine organisms is difficult because pelagic species are widely dispersed, species vary greatly in size, some species avoid samplers, and some delicate species are damaged by samplers. Pelagic species are collected by nets, water-sampling devices, traps, and fishing lines. Benthic species are collected by grab samplers, dredges, and trawl nets. Some species can be collected or observed only by the use of scuba or submersibles.

Scuba, Manned and Unmanned Submersibles. Marine organisms in shallow waters, less than a few tens of meters deep, can be observed and sampled best by scuba divers. Scuba divers can spend only a few hours a day underwater unless they live in expensive underwater habitats. Marine organisms in deeper waters can be observed and sampled from manned submersibles, but these vessels are very expensive and can remain submerged for only a few hours. Unmanned remotely operated vehicles (ROVs) using television cameras and robot arms are more economical and can remain submerged longer.

Satellites. Satellite-mounted sensors can be used to measure parameters and characteristics of near-surface waters, including temperature, phytoplankton concentrations, concentrations of suspended sediment, and surface currents and waves. Satellite observations can cover large areas almost simultaneously and repeat these observations frequently, which is not possible with research ships or fixed monitoring systems. Seafloor topography can be observed by very precise satellite radar measurements of sea surface elevation. Satellite navigation systems have dramatically improved oceanographers' abilities to map and return to specific features of the oceans.

Computers and Modeling. Computers have contributed substantially to the development of ocean sciences, primarily by organizing and displaying geographic data and by performing mathematical modeling of complex systems.

STUDY QUESTIONS

1. What are the principal difficulties encountered in determining ocean depth and mapping the seafloor?
2. Why do oceanographers still use many different types of sampling equipment to obtain samples from the seafloor?
3. If you were designing a sampler to obtain samples of water from deep in the oceans, what factors would you need to consider?
4. Why do we believe there are many species of marine animals living in the deep sea that have never been seen or sampled?
5. Do you think manned submersibles are necessary for ocean study? Explain your answer. Describe the types of studies for which they are most useful.
6. Will satellites ever completely replace research vessels? Why or why not?
7. Why is accurate navigation important to ocean studies? Discuss the reasons.
8. What methods are used to obtain information about the sediments on and below the seafloor?
9. What methods are used to obtain information about the chemistry of and organisms living in the water column of the deep oceans?
10. Why is it easier to study ocean sediments than to study ocean water chemistry or organisms that live in the deep-ocean water column?

KEY TERMS

You should recognize and understand the meaning of all terms that are in boldface type in the text. All those terms are defined in the Glossary. The following are some less familiar key scientific terms that are used in this chapter and that are essential to know and be able to use in classroom discussions or exam answers.

acoustic (p. 55)
bathymetry (p. 40)
benthic (p. 57)
benthos (p. 57)
CTD (p. 52)
current (p. 41)
electromagnetic radiation (p. 64)
frequency (p. 56)
hydrographic (p. 50)
hydrophone (p. 48)
hydrothermal vent (p. 59)
pelagic (p. 57)
phytoplankton (p. 57)
plankton (p. 58)
sediment (p. 40)
sonar (p. 42)
sounding (p. 40)
submersible (pp. 44, 60–61)
suspended sediment (p. 50)
synoptic (p. 64)
tomography (p. 56)
topography (p. 40)
tracer (p. 53)
zooplankton (p. 57)

CRITICAL CONCEPTS REMINDER

CC1 **Density and Layering in Fluids** (p. 54). Water in the oceans is arranged in layers according to water density. To read **CC1** go to page 2CC.

CC6 **Salinity, Temperature, Pressure, and Water Density** (p. 52). Sea water density is controlled by temperature, salinity, and to a lesser extent, pressure. Density is higher at lower temperatures, higher salinities, and higher pressures. Movements of water below the ocean surface layer are driven primarily by density differences. To read **CC6** go to page 16CC.

CC10 **Modeling** (pp. 53, 65). Complex environmental systems, including the oceans and atmosphere, can best be studied by using conceptual and mathematical models. To read **CC10** go to page 26CC.

CC14 **Photosynthesis, Light, and Nutrients** (p. 58). Chemosynthesis and photosynthesis are the processes by which simple chemical compounds are made into the organic compounds of living organisms. The oxygen in the Earth's atmosphere is present entirely as a result of photosynthesis. To read **CC14** go to page 46CC.

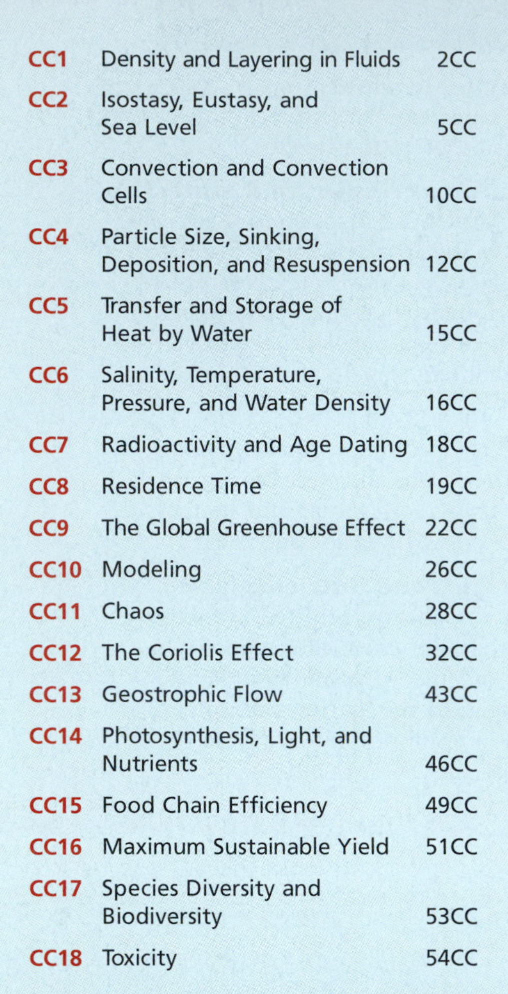

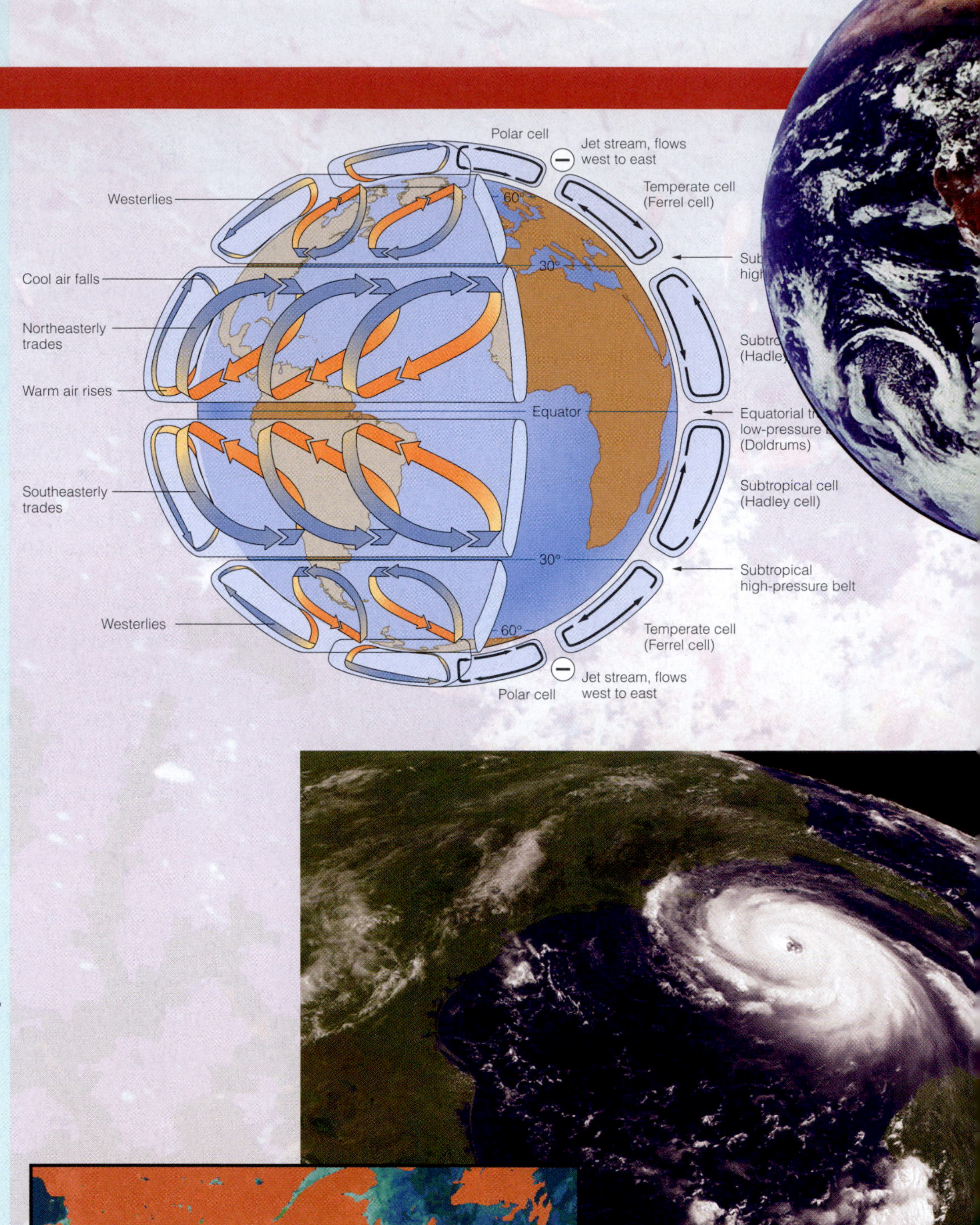

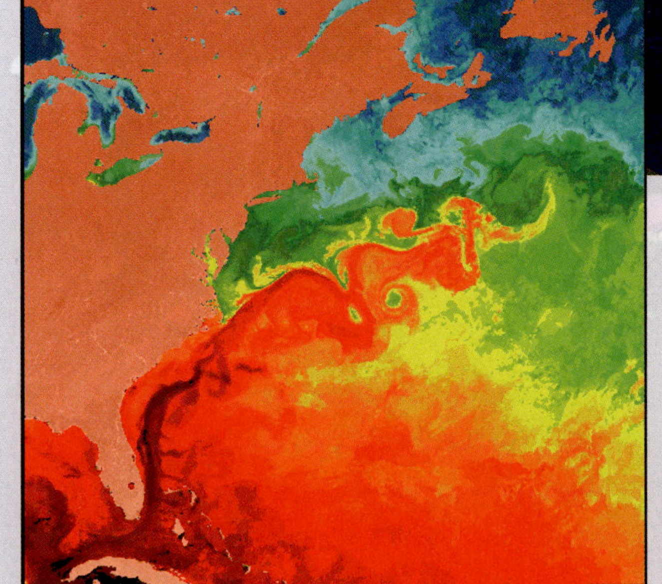

A number of key concepts are fundamental to many different ocean processes. These fundamental concepts are collected in this "Critical Concepts" section. You will find their application is critical in many different chapters. For example: the composite satellite image at top center shows cloud formations that transport heat **(CC5)**; the figure at top left shows how this heat transport creates convection cells **(CC3)**; at bottom left, a satellite image of sea surface temperatures shows transport of heat by ocean currents, here the Gulf Stream, **(CC6)**; the bottom center image shows Hurricane Katrina whose energy, like all hurricanes, was derived from ocean water **(CC5)** through convection processes **(CC3)**; the bottom right figure shows how the Coriolis effect causes moving air and water to appear to be deflected by Earth's rotation **(CC12)**; the top right figure shows the nature of convection cells **(CC3)**. All the complex air and water motions seen in figures on this page are guided by convection **(CC3)** and the Coriolis effect **(CC12)**, and most reflect movement of air or water in Geostrophic Flow **(CC13)**. The great complexity of motions seen in the satellite images results from Chaos **(CC11)**.

Critical Concepts

The Critical Concepts are the heart of the science that you must learn in order to understand the basic processes of the oceans. They are also concepts you will need to know before you can make sense of many issues that you will encounter in your future work or that will affect your life.

Some of these concepts are quite difficult to fully understand, especially without the use of mathematics at a level that is not appropriate to an introductory text like this one. Therefore, the Critical Concepts that follow are structured to allow easy learning of the essential nature of each concept. This essential nature is encapsulated in a section entitled "Essential to Know." In this section, the essential aspects of each Critical Concept are listed as brief statements of all the key facts and characteristics of the concept. The information in these "Essential to Know" lists can be learned quickly and then used for easy reference while you study the areas of ocean sciences in which they are applied. This information is all you need to know to fully understand and enjoy the broad introduction to ocean sciences that this textbook provides. However, a fuller understanding of these Critical Concepts might substantially enhance your introductory learning experience and will be necessary if you choose to study any aspect of ocean sciences at a more advanced level. Therefore, for each Critical Concept, the "Essential to Know" section is followed by a section entitled "Understanding the Concept," which provides a more detailed explanation.

In the "Understanding the Concept" sections, the essential nature of the concept is explored in more depth, and some aspects of the scientific basis for, and general applications of, the concept in ocean sciences are described. You will need to read this material carefully and to make constant reference to any figures noted. You will likely need to reread this material, perhaps several times, to fully understand it. However, you should remember that, as long as you learn the "Essential to Know" material, you will be able to understand the rest of this textbook and to apply these concepts appropriately in class discussions, written projects, and exams.

CRITICAL CONCEPT 1

Density and Layering in Fluids

ESSENTIAL TO KNOW

- Fluids are separated by gravity to form layers, each layer having lower density than the layer below it.
- Less dense fluids rise through more dense fluids.
- More dense fluids sink through less dense fluids.
- A fluid is neutrally buoyant when the fluid above has lower density and the fluid below has higher density.
- A neutrally buoyant fluid will spread horizontally to form a layer.
- A fluid in which each higher layer is of lower density than the layer below it is stably stratified. There are no density-driven vertical motions within such a stably stratified fluid.
- The depth range in a fluid within which there is a marked vertical gradient of density is called a "pycnocline." A pycnocline inhibits vertical mixing between fluid above and below the pycnocline layer.

UNDERSTANDING THE CONCEPT

All substances exist in one of three physical states: solid, liquid, or gas (Chap. 7). Fluids include both liquids and gases. Seawater and the atmosphere are both fluids that flow in response to forces such as **gravity.** Some parts of the Earth's interior are liquid. Other parts, including the **asthenosphere** and **mantle** (Chap. 4), although they are solid, can flow extremely slowly at the high temperatures and pressures present within the Earth. These solids behave like fluids over the million-year timescales in which geological processes occur.

When two fluids are brought in contact, gravity acts to distribute them vertically according to their **density.** The density of a fluid is the mass of the fluid divided by its volume. Pure liquid water has an absolute density (specific gravity) of 1000 $kg \cdot m^{-3}$ at 4°C and 1 atmosphere (atm) pressure. Generally, density is stated as a relative density (r) that is the ratio of the fluid's specific gravity to the specific gravity of pure water at 4°C and 1 atm pressure. Thus, pure water at 4°C and 1 atm has a relative density of 1.

The effects of gravity and other forces on fluids of different densities account for many of the processes studied in this text, including the circulation of ocean water and atmosphere, and the formation of ocean basins and continents.

Many fluids do not mix with one another. Examples include oil and water, and water and air. When two such fluids are placed in a container, the fluid with the higher density migrates to the bottom, leaving the less dense fluid above it. If the less dense fluid is a gas such as air, it fills the upper part of the container not occupied by the denser fluid. If the less dense fluid is a liquid such as oil, it lies in a separate layer on top of the denser fluid (**Fig. CC1.1a**). Even if we vigorously shake such mixtures, they quickly separate again into two layers after we stop shaking.

Gravity is the principal force that separates fluids of different densities. Consider a container full of water with an oil layer on top into which we introduce a small drop of water (**Fig. CC1.1b**). Because water is denser than oil, the drop of water has a greater mass than the same volume of oil. The gravitational force between any object on the Earth's surface and the Earth increases in proportion to the object's mass. Therefore, the gravitational attraction force between the Earth and the water drop is greater than that between the Earth and the oil, and the water droplet is pulled down by the Earth's gravity more than the oil is. Because the oil can flow, the water droplet sinks and the oil flows around it. When the water droplet reaches the top of the water layer, it is surrounded by a fluid of the same density, and the gravitational attraction on the water in the drop is now equal to the gravitational attraction on all the other water at its level. Therefore, the droplet sinks no farther. However, it does mix with the surrounding water by **diffusion.**

Gravitational sorting of fluids into layers according to density is called **stratification.** This process works equally well with more than two fluids. For example, if we mix mercury, water, oil, and air, the mixture will separate with water on top of the mercury, oil on top of the water, and air above the oil (**Fig. CC1.1c**). This process does not work quite the same way with solids, because they cannot flow.

Density stratification can occur not only between fluids that do not mix, but also between fluids that do mix. For example, if we add cream carefully to a cup of cold coffee, it remains as a floating layer on top of the coffee until we stir the coffee and thoroughly mix the two layers. After two liquids are thoroughly mixed, the mixture has uniform density, and the two layers do not re-form.

Density stratification can also occur in a single fluid, such as pure water or air. Just as cream will form a layer on coffee, a less dense layer of any fluid will remain on top of a more dense layer of the same fluid until they are mixed (**Fig. CC1.2**). In the natural **environment,** we often find fluids with multiple layers, each of slightly higher density from the shallowest to the deepest layer (**Fig. CC1.2**). When the density of the fluid increases progressively with depth, no layer has a tendency to sink or rise and the fluid is said to exhibit stable stratification. Ocean waters are arranged in layers of different density in this way (Chap. 10) because the density of water varies with temperature and **salinity** (**CC6**). The atmosphere is also vertically stratified (Chap. 9), and the components of the Earth's interior are arranged similarly (Chap. 4).

If we add a fluid to a column of other fluids, each of which has a different density but none of which mix, the added fluid will sink or rise until it reaches a depth at which the fluid immediately above it is of lower density and the fluid immediately below it is of higher density. At this equilibrium level, the introduced fluid is neutrally **buoyant,** so it will neither sink nor rise, but will spread out to form its own layer (**Fig. CC1.2**). For example, if we introduce oil underwater to an air-water layered system, the oil will rise to the surface of the water and spread out. This is exactly what happens when sunken ships release their oil.

When a fluid of slightly different density is added to a density-stratified column of the same fluid or of a fluid with which it mixes, some mixing will occur as the added fluid sinks or rises to its equilibrium level. We can see this effect if we pour cream from a height into a cup of coffee. Both the cream and the coffee are primarily water, but different chemicals are dissolved or suspended in each, so they differ in density. If we pour the cream in carefully, it will be partially mixed as its momentum carries it down into the coffee and as it subsequently rises because of its lower density. The cream, mixed with some coffee, will rise to form a layer on top of the coffee and will then mix only very slowly (by molecular diffusion) unless we stir

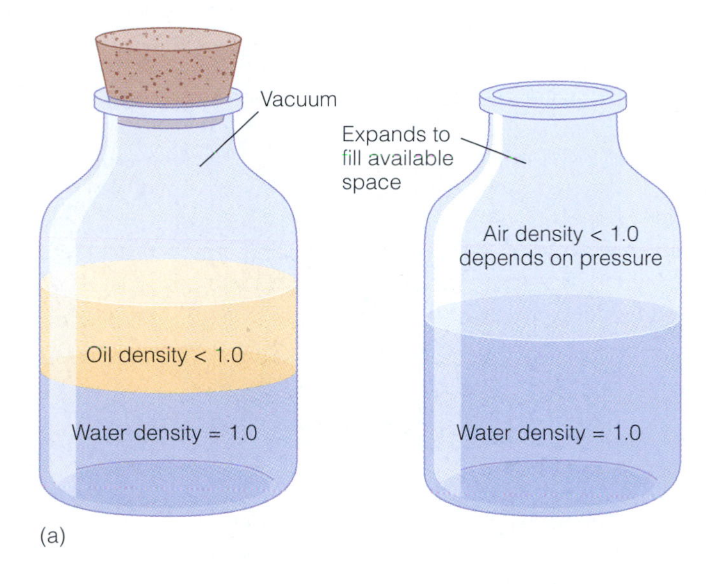

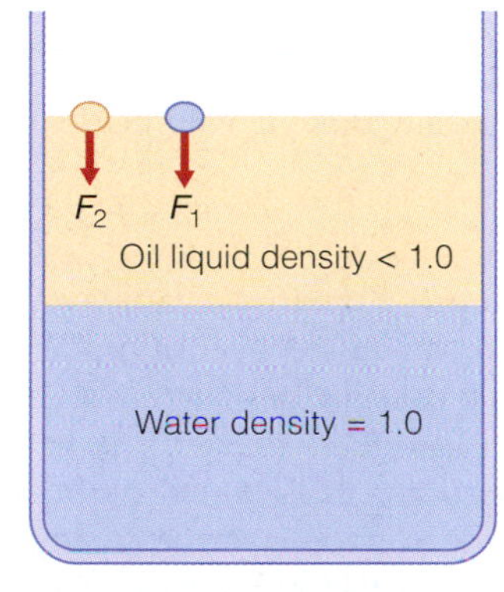

(b)

F_1 = Earth's gravitational attraction force on water droplet

F_2 = Earth's gravitational attraction force on oil droplet

$$F_1 = K \cdot \frac{m_w \cdot m_{Earth}}{r^2}$$

$$F_2 = K \cdot \frac{m_o \cdot m_{Earth}}{r^2}$$

Where: r = the distance from Earth's center to the center of the droplet
m_{earth} = mass of Earth
K = gravitational constant
m_w = mass of water droplet
m_o = mass of equal volume oil droplet

Since $m_w > m_o$ then $F_1 > F_2$

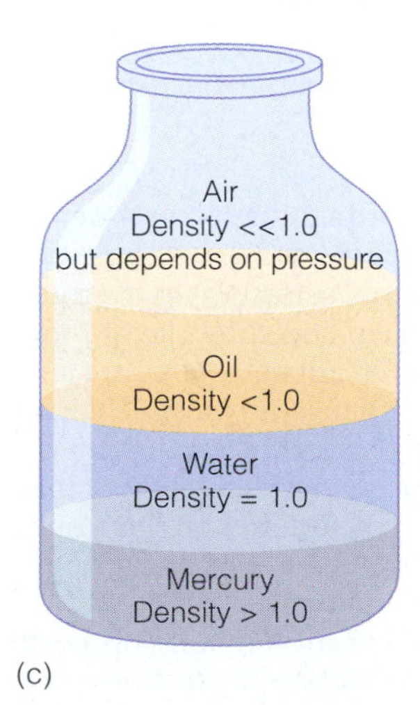

FIGURE CC1.1 Fluids (liquids and gases) that do not mix with each other are separated into layers in such a way that the density of each layer is greater than that of the layer above. (a) Oil has a lower density than water and forms a separate layer that floats on the water. If this layer is very thin, it is called a "slick." Air, which has a much lower density than either water or oil, forms a layer (the atmosphere) above the water. (b) Fluids are separated according to density because a more dense fluid has a greater mass and is subject to a greater gravitational force than the same volume of a less dense fluid. (c) Air, oil, water, and mercury separate into four layers, with density increasing toward the Earth's center (that is, with depth).

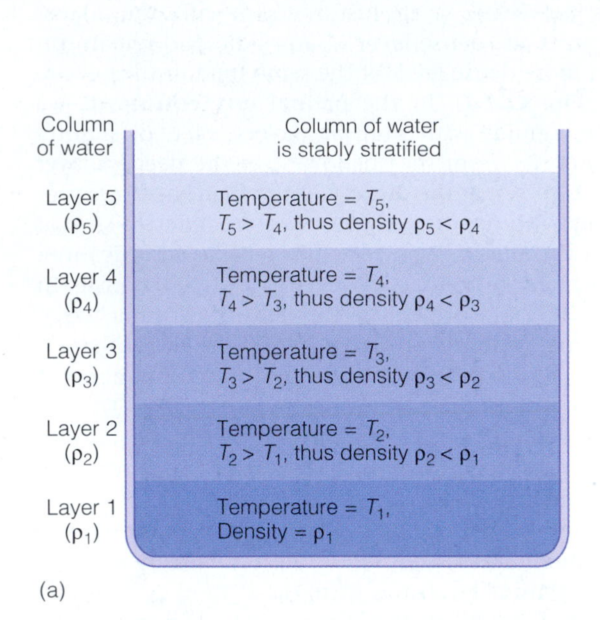

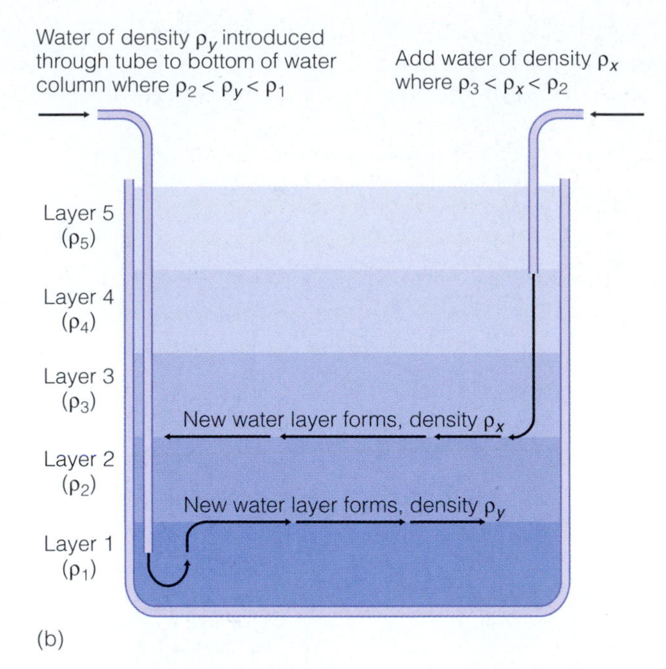

FIGURE CC1.2 Water may form distinct layers, each of which has a different density because it has a different temperature or salinity. (a) A water column is stably stratified if each successive layer has a higher density than the one above it and, thus, no water sinks or rises. (b) If water is introduced to a stratified water column, it will sink or rise until it reaches an equilibrium level at which the density of the water above it is lower and the density of the water below it is higher. It will then spread out to form a new layer at this depth. Some mixing of the introduced water with other water layers will occur as it sinks through layers of lower-density water or rises through layers of higher-density water. However, when the vertically moving volumes of water are very large (as they are in the oceans), such mixing is limited.

the mixture (and create **turbulent** diffusion). When the volumes of fluid in each layer are very large, and when the vertical motions are slow, as they are in the oceans, mixing is very limited. The oceans are vertically stratified (Chap. 10), and density-driven vertical motions of **water masses** move large volumes of water vertically to find their equilibrium level with little mixing between the ascending or descending water masses and the layers of water through which they rise or sink.

Our cream and coffee experiment can illustrate another important concept. After we carefully place a layer of cream on the coffee, there is a distinct cream layer of lower density and a distinct lower coffee layer of higher density. Between these layers is a thin region where coffee and cream are mixed in varying amounts. The vertical distribution of density in this experiment is shown in **Figure CC1.3**. Density is uniform in the surface layer of cream, becomes progressively higher (this is referred to as a "density gradient") in the transition layer of mixed cream, and is uniform in the lower layer of coffee. The transition region in which there is a vertical density gradient is called a **pycnocline.** More energy would be needed to

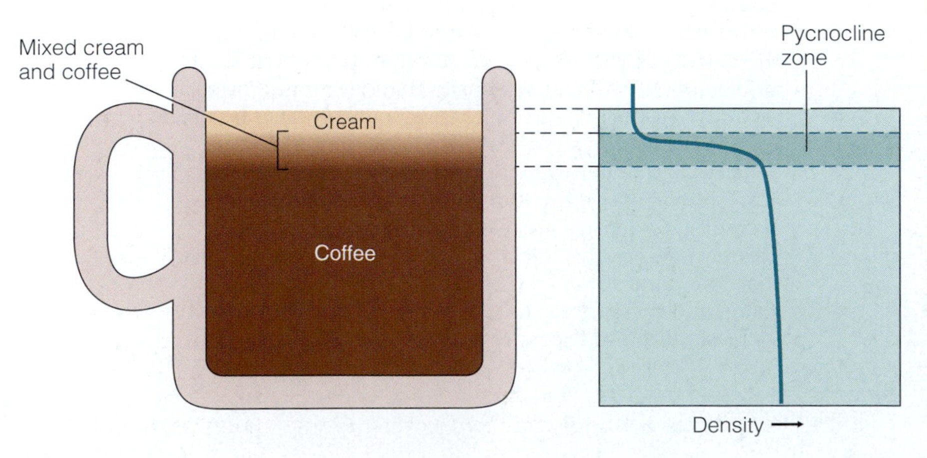

FIGURE CC1.3 If cream (which has a lower density than coffee) is introduced at the surface of a cup of cold coffee, the cream forms a surface layer overlying the coffee. Between the cream and coffee layers, there is a zone within which the cream and coffee are mixed in proportions that change progressively with depth from pure cream to pure coffee. In this zone between the coffee and cream layers, there is a rapid change in their respective proportions, which results in a sharp increase in density with depth. This zone, in which density changes rapidly with depth, is called a "pycnocline zone." Vertical mixing between the two layers takes place only slowly across a pycnocline because the lower-density upper layer (cream) has no tendency to sink and the higher-density layer (coffee) has no tendency to rise.

move a molecule of water vertically through this pycnocline than would be needed to move a molecule the same vertical distance in either the cream or the coffee. Thus, pycnoclines act as barriers to vertical mixing.

The strength of the density gradient in a pycnocline determines how strong a barrier to vertical mixing it is. Try the experiment again, but use milk instead of cream. Milk has a slightly higher density than cream (because it contains less fat). It is almost impossible to pour the milk into the coffee without the two mixing, even though milk does have a lower density than water. In this case the density gradient, and therefore the pycnocline between milk and coffee, is very weak.

Pycnoclines form in many parts of the oceans, particularly where warm surface water overlies colder deep water, and they are important because they inhibit vertical mixing. The density of fluids is altered as heat is gained from the sun and as heating or cooling occur by **conduction** and radiation at surfaces. Density is also altered by changes in pressure and by changes in fluid composition, such as the addition of salts to water (Chap. 7) and of water vapor to air (Chap. 9). The factors that affect the density of air are discussed in **CC5** and Chapter 9, and the factors that affect the density of water are examined in **CC6** and Chapter 7. **CC3** explains more about the vertical motions created by such density changes.

CRITICAL CONCEPT 2

Isostasy, Eustasy, and Sea Level

ESSENTIAL TO KNOW

- The level at which a solid floats in a liquid is determined by the relative densities of the solid and the liquid.
- Solids of equal density float in a liquid with the same percentage of their volume submerged.
- Two solids of different density will float in a liquid in such a way that the solid with the greater density has a greater fraction of its volume submerged.
- Lithospheric plates float on the asthenosphere. Because oceanic crust is denser and thinner than continental crust, the surface of the oceanic crust is always lower than the surface of the continental crust.
- Changes occur in the density and thickness of lithospheric plate sections as a result of various processes. The affected plate section rises or falls in response to such changes in a process called "isostatic leveling."
- Isostatic leveling is slow. Equilibrium may not be achieved for millions of years after an event, such as the start or end of an ice age, that changes the distribution of glaciers.
- Sea level changes on a section of coast as the section of the plate rises or falls isostatically. Isostatic sea-level changes occur at different rates on different coasts.
- Eustatic sea-level changes are caused by changes in the volume of ocean water or in the volume of the ocean basins that occur as a result of a variety of plate tectonic and climatic change processes.
- Eustatic changes of sea level take place simultaneously and uniformly throughout the world. Eustatic equilibrium is easily and quickly attained.
- Changes of sea level on different sections of the world's coasts occur at different rates, and even in different directions, because of the interaction of eustatic and isostatic changes.

UNDERSTANDING THE CONCEPT

Solid objects will sink through a fluid in which they are placed if the **density** of the solid is higher than that of the fluid. For example, a rock thrown into a pond sinks to the bottom. Solid objects whose density is less than that of the fluid will float. For example, wood or Styrofoam will float on water, and helium balloons will float on air.

When a solid object floats on a liquid, the depth to which the object is immersed is determined by its density and the density of the liquid. A simple principle describes this relationship. Called Archimedes' principle, it states that the floating solid will be immersed in the liquid deep enough that the water it displaces has exactly the same total mass as the solid.

We can understand Archimedes' principle by considering a rectangular block of wood floating on water (**Fig. CC2.1**). If the wood has a density of 0.5, or half the density of water, the wood floats with exactly half of its volume underwater and half exposed. If the wood has a density of only 0.1, or one-tenth the density of water, the wood floats with only one-tenth of its volume underwater

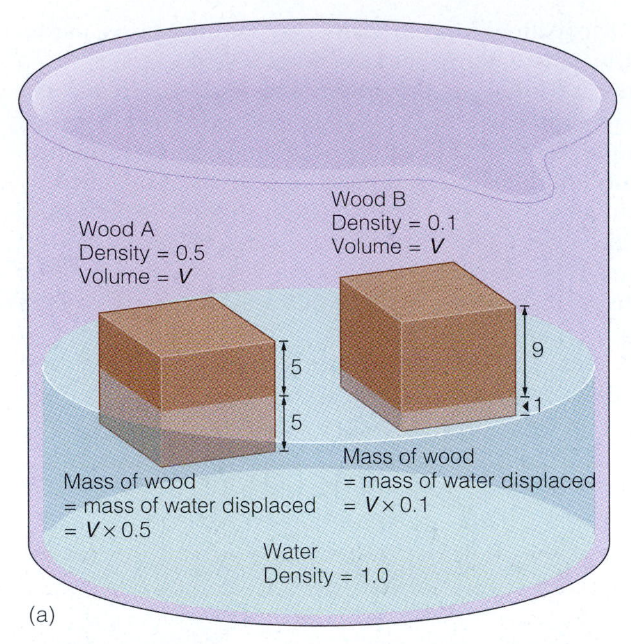

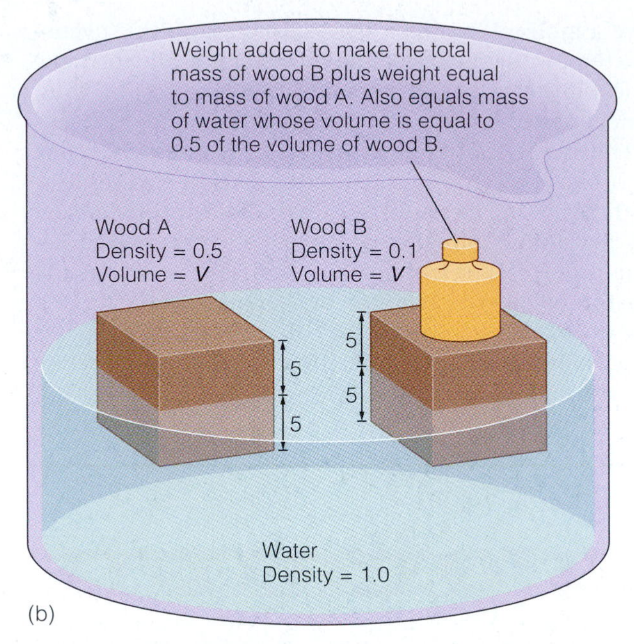

FIGURE CC2.1 Archimedes' principle determines how much of a floating solid's volume is below the surface. The solid floats at a depth such that the weight of the volume of water or other fluid that it displaces (that is, the volume of the part of the solid that is below the water or fluid surface level) equals the total weight of the solid. (a) Lower-density materials will float higher than higher-density materials. (b) Adding weight to a floating, low-density solid increases its mass and causes it to float lower and thus displace more water. The volume of water displaced will have a weight that is equal to the total weight of the floating solid plus the added weight that it supports.

and nine-tenths exposed (**Fig. CC2.1a**). If we place a weight on top of the less dense piece of wood, the wood is forced to sink into the water. It sinks until the mass of the total volume of water displaced is equal to the total mass of the wood plus the weight (**Fig. CC2.1b**).

If two pieces of wood have the same density but different thicknesses, the thicker piece will float with its upper surface higher above the water surface than the thinner piece, and it will extend deeper into the water (**Fig. CC2.2**). The two pieces have the same percentage of their individual volumes submerged. If we have two pieces of wood of identical dimensions but different densities, the wood with higher density will float lower in the water (**Fig. CC2.1a**). This effect can be observed in a river, lake, or ocean. For example, most logs or driftwood pieces float with much of their volume immersed below the waterline. In contrast, Styrofoam and balsa wood seem almost to sit on the water, because their low density means that only a small percentage of their volume is immersed.

Lithospheric plates float on the fluid surface of the **asthenosphere** in the same way that wood floats on water. Therefore, a **buoyancy** equilibrium is established in which the plates float freely on the material beneath them at a level that is determined by their density and thickness. This level is called **isostasy,** or "isostatic equilibrium."

Continental **crust** has a lower density than oceanic crust (Chap. 4). Therefore, lithospheric plates covered with continental crust float with a smaller percentage of their volume "submerged" than plates covered with oceanic crust. If the plates with continental crust and oceanic crust were the same thickness (**Fig. CC2.3**), the upper surface of the continental crust plate (the land surface) would be higher than the surface of the oceanic crust plate (the seafloor). In addition, the **lithosphere**-asthenosphere boundary would be shallower (farther from the Earth's center) under the continents than under the seafloor. However, lithospheric plates with oceanic crust are thinner than those with con-

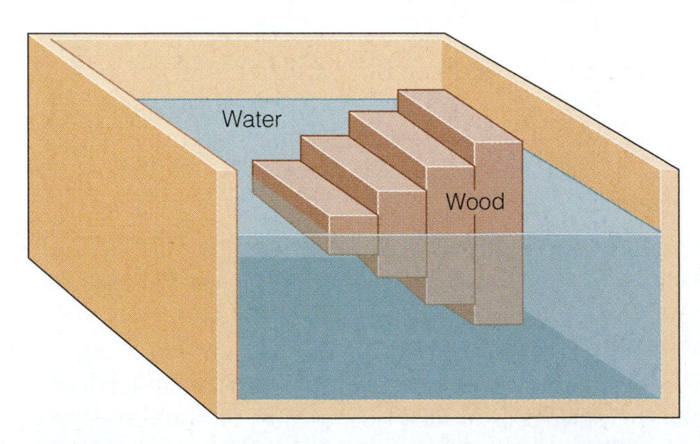

FIGURE CC2.2 Solids of the same density but of different thicknesses always float with the same proportion of their volume below the surface. Thicker blocks float with their upper surface higher above the surface of the fluid on which they float, and their lower surface deeper below.

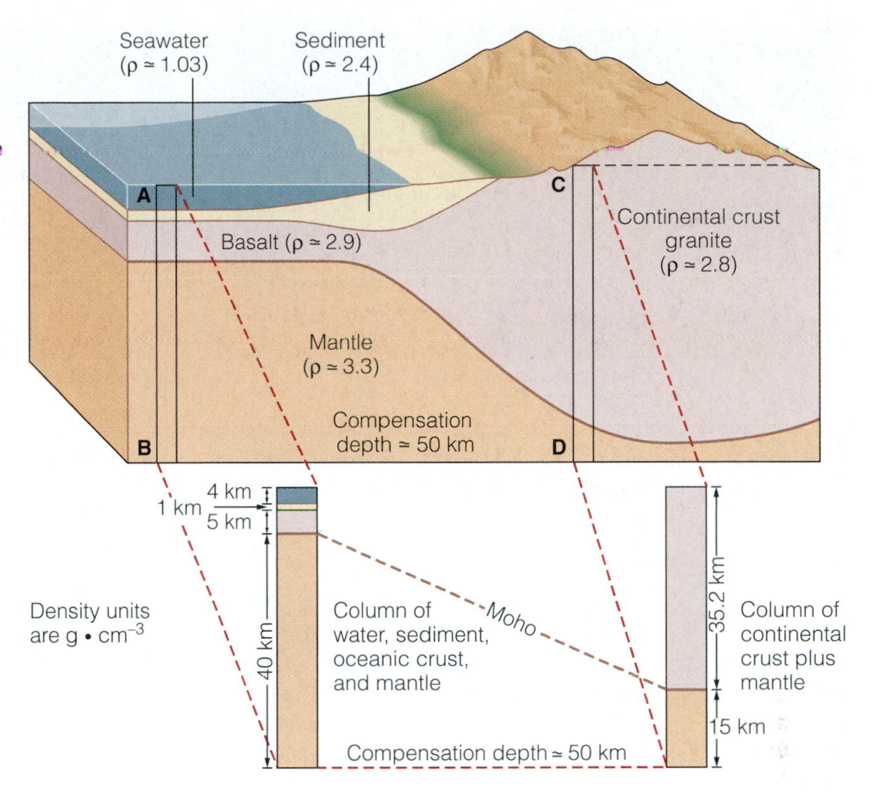

FIGURE CC2.3 Isostasy is the condition in which blocks of lithosphere float on the asthenosphere at the equilibrium level determined by Archimedes' principle. Above the compensation level (the depth at which the asthenosphere behaves as a fluid such that it can be displaced by floating lithospheric plates), the total weight of a column of continental crust plus mantle at isostasy will equal the total weight of a column of water, sediment, oceanic crust, and mantle. They do not quite do so in this figure, because the numbers are rounded.

tinental crust. Because the two types of crust are of different thicknesses, the lithosphere-asthenosphere boundary is shallower below ocean crust than it is below continental crust (**Fig. CC2.3**), and the continent surface is more elevated above the oceanic crust surface than it would be if ocean and continental crust were the same thickness.

If the density and thickness of the lithospheric plates were invariable, the continents and seafloor would always remain at a fixed equilibrium height above the asthenosphere. However, both the density and the thickness of lithospheric plates are altered by a variety of processes. First, the crust can be heated, which reduces its density; or cooled, which increases its density. Second, the thickness of the crust can be increased by the formation of mountains or volcanoes or by the **deposition** of large amounts of **sediment,** or it can be thinned by the **erosion** of mountains or stretching of the lithospheric plate. Third, the thickness of the lithospheric plate can be increased by the cooling and solidification of **mantle** material. This material is added to the bottom of the plate. Conversely, the plate can be thinned by the heating and melting of mantle material on the underside of the plate. Finally, the volume and mass of the continental crust can be increased by the development of **glaciers,** or decreased if glaciers melt, or altered by several other means, such as the growth of **coral reefs**.

If the density of a section of lithospheric plate is increased, it sinks until it reaches its new equilibrium level in a process known as **isostatic leveling** (**Fig. CC2.4**). Similarly, crust whose density has decreased will rise isostatically. Changes in thickness of the crust will also cause isostatic leveling (**Fig. CC2.4**).

Isostatic leveling is very slow because the asthenosphere is very **viscous** and flows extremely slowly to accommodate the rising or sinking lithospheric plate. The processes that cause changes in isostatic level are often localized to certain sections of a lithospheric plate and its overlying crust. Therefore, one section of a plate may be rising while another is sinking. Lithospheric plates can bend to accommodate this process. When the section of a lithospheric plate that supports the edge of a continent changes its isostatic level, sea level rises or falls along this **coast** but does not necessarily change on other coasts.

Sea level can be altered not only by isostatic leveling of the continents, but also by processes that increase or decrease either the volume of ocean water or the volume of the ocean basins. Changes in water volume or in ocean basin volume cause changes in sea level that occur simultaneously and uniformly along all the world's coasts. Simultaneous worldwide changes in sea level are called **eustatic** changes (**Fig. CC2.5b**).

Eustatic sea-level changes can be caused by several processes. For example, the volume of water in the oceans can be increased by the addition of water from melting glaciers or by elevation of the average temperature of ocean water that causes the water to expand. Correspondingly, the volume of ocean water can be decreased by increased **glaciation** or by a decline in average ocean water temperatures.

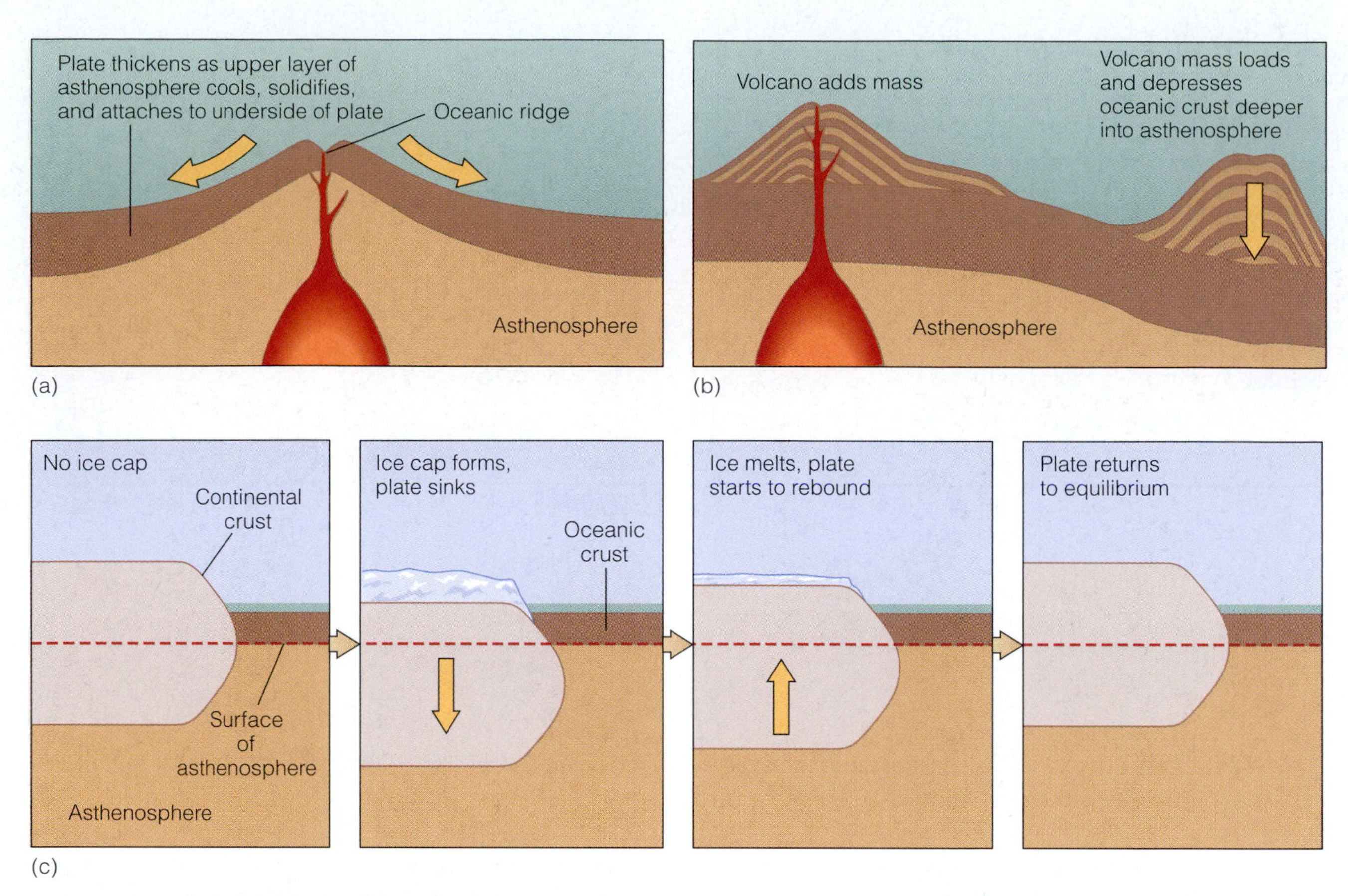

FIGURE CC2.4 Isostatic leveling takes place in response to crustal density changes, plate thickening, and plate loading processes. (a) As oceanic crust moves away from an oceanic ridge, it cools, its density increases, and it sinks lower in the asthenosphere. The plate also thickens as mantle material solidifies and is accreted to the cooling underside of the plate. (b) Volcanoes formed at hot spots add mass to the crust. The increased mass causes the plate to bend and sink lower under the volcano, especially after the old volcano moves away from a hot spot and cools, thus increasing its density. (c) A section of continental crust on which glaciers are formed will sink to a new isostatic equilibrium because of the additional weight of the glacier (like the wood in Figure CC2.1b). If the ice melts, the continent will rise in a process called "isostatic rebound" until it reaches its new equilibrium level. Isostatic changes take place very slowly compared to the climate changes that can cause large variations in the area of the continents that is covered by glaciers.

The volume of the ocean basins (actually the volume available for seawater to fill below a fixed level, independent of vertical movements of the continents) can be decreased if a larger percentage of the ocean basin floor is occupied by **oceanic ridges.** During the parts of **spreading cycles** when continents are being broken up, such as at present, there are many continents moving apart and many oceanic ridges between them. Hence, the percentage of the ocean floor covered by oceanic ridge is larger than it is during periods of reassembly of the continents. As a result, the ocean basin volume tends to be diminished, and sea level is correspondingly high.

Changes in the rate of **seafloor spreading** within a spreading cycle also affect sea level. When seafloor spreading is fast, the young, warm oceanic crust extends far from each oceanic ridge. Therefore, the volume of the ocean basin is reduced, sea level is higher, and more continental crust is covered by oceans. When seafloor spreading is slow, a smaller amount of young, warm oceanic crust is produced, the volume of the ocean basins is increased, and sea level is lower.

During periods when the Earth's continents are assembled in supercontinents such as Pangaea, the number of oceanic ridges and the rate of seafloor spreading are reduced. Therefore, young, warm oceanic crust covers a smaller percentage of the ocean floor, the ocean basin volume is increased, and sea level is lower.

The processes, just described, that affect the volume of the ocean basins are complicated and interact with each other as the various **plate tectonic** processes simultaneously change the thickness, relative abundance, and isostatic level of the continental and oceanic crusts.

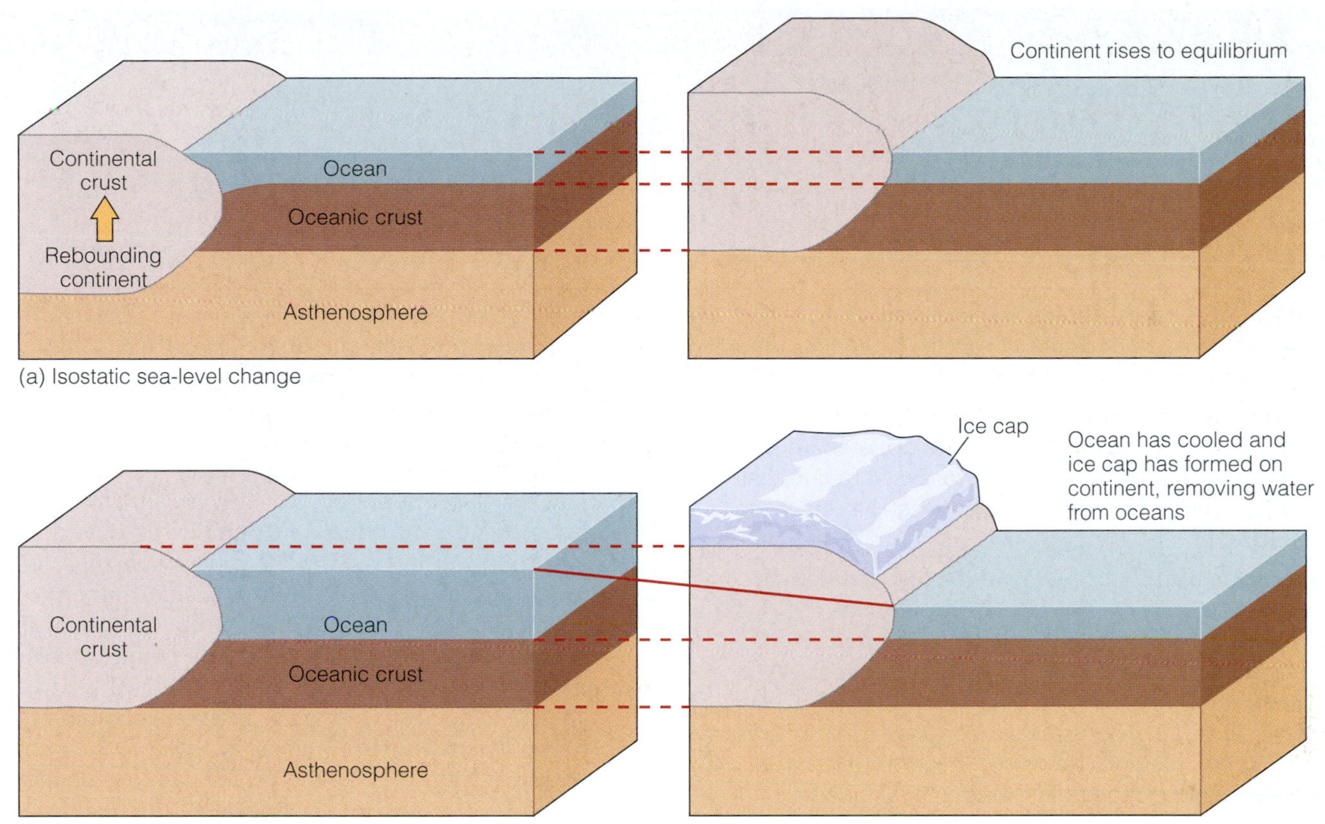

FIGURE CC2.5 (a) Isostatic sea-level changes take place when the land rises or falls while the seafloor and ocean depth remain the same. In this case, a continent that was previously depressed by the weight of an ice sheet rises slowly, exposing more of the continent to the atmosphere. (b) Eustatic sea-level changes take place when the volume of water in the oceans changes. In this case, the ocean has cooled and the water has been transferred to an ice cap on the land, thus lowering the sea level. Notice that the continent has not sunk lower in the asthenosphere in this diagram. In the situation depicted, the continent would eventually sink isostatically, but isostatic changes are much slower than eustatic changes.

At the same time, **climate** changes, which themselves may be linked to tectonic processes, also change the volume of water in the oceans.

Processes that alter sea level occur on a variety of different timescales, from centuries to millennia. Eustatic equilibrium is reached quickly in response to such changes, whereas isostatic equilibrium is attained very slowly. Sea level may be rising, static, or falling on any individual section of the world's coasts as various eustatic and isostatic changes work in concert or in opposition in different regions.

At present, there is concern that human enhancement of the **greenhouse effect** (Chaps. 2, 9, **CC9**) will lead to global warming. One fear is that if warming occurs, it will increase the volume of ocean waters as the average temperature increases and the water expands, and as possible partial melting of the polar ice sheets contributes additional water. The result would be a eustatic rise in sea level, which would have severe consequences because many coastal regions and cities would be inundated by the sea. Therefore, measuring the current rate of eustatic sea-level change is considered to be of critical importance.

Measurement of the rate of eustatic change of sea level is complicated by the interaction of isostatic and eustatic changes that occur simultaneously on any given section of coast. For example, the sea-level change measured on the northeast coast of the United States will be the net result of several processes: rising sea level due to any greenhouse warming effect, rising sea level due to progressive cooling and isostatic sinking of the continental crust, and falling sea level due to isostatic rise caused by the geologically recent melting of the **ice age** glaciers that once lay on this crust. As a further complication, human activities such as **groundwater** and oil withdrawal cause the subsidence of certain sections of coast. Because of these complicated interactions, we will be able to measure any greenhouse-induced sea-level rise only by studying sea level on many different coasts.

CRITICAL CONCEPT 3

Convection and Convection Cells

ESSENTIAL TO KNOW

- Vertical motions in a fluid can be caused by temperature or compositional changes that alter the density of parts of the fluid.
- When density is continuously decreased within one layer in a fluid (often its lower boundary with another fluid or solid) and continuously increased in a higher layer (often the fluid surface), a convection cell is established.
- In a convection cell, a plume of the lowered-density fluid rises within the fluid until it reaches an equilibrium level where the surrounding fluid is of equal density. If the rising fluid has a lower density than all fluid layers above it, it rises to the surface. At its equilibrium level (or the surface), the fluid spreads out. If it is cooled or otherwise altered so that its density increases, the fluid eventually becomes sufficiently dense that it sinks back to its original level, where it replaces rising fluid and reenters the cycle.
- Some convection cells have toroidal circulation, in which a rising column of fluid spreads out laterally in all directions at the top of the convection cell and returns by sinking in a ring-shaped band surrounding the rising plume. This pattern may be reversed with a column of sinking fluid surrounded by a ring of rising fluid.
- In other convection cells, both the rising fluid and the sinking fluid form an elongated curtain. The simplest example of such a convection cell is cylindrical in shape.
- The rising plume is called an "upwelling plume," and the sinking plume is called a "downwelling plume."
- At the top or bottom of a convection cell, the fluid flows horizontally away from a location at a divergence. The fluid flows toward a location at a convergence.
- At the top of a convection cell, divergences are regions of upwelling and convergences are regions of downwelling. At the bottom of the convection cell, divergences are regions of downwelling and convergences are regions of upwelling.
- Several convection cells may be formed side by side within a fluid. Convergences and divergences always alternate across the top or bottom of the cells.

UNDERSTANDING THE CONCEPT

CC1 explains how **density** differences cause **stratification** in fluids, in which successive layers have higher density with depth. If the density of the layers in a vertically stratified fluid does not change, the system is stable and vertical motions do not occur.

Three stratified fluids are important to oceanographers: the Earth's interior layers, the oceans, and the atmosphere. Heat is introduced into the Earth's layers by energy released during **radioactive** decay in the core and **mantle** (Chap. 4). In the oceans, a small amount of heat is introduced by conduction from the mantle through the seafloor, and a much larger amount by solar heating of surface water. Heat is also exchanged (gained in some areas and lost in others) between oceans and atmosphere by **conduction,** radiation, evaporation, and precipitation (Chaps. 7, 9). The atmosphere exchanges heat with the oceans and land, gains heat from solar radiation, and loses heat by radiation to space (Chap. 9).

The heat transfer processes between sun, atmosphere, ocean, and land all vary with time and with location on the Earth. Variable heat transfers produce density changes in the mantle, ocean, and atmosphere and cause the stratification to become unstable. Changes in composition can also alter density. For example, changes in the **salinity** of ocean water (Chap. 7) or in the water vapor pressure of air (Chap. 9) alter the fluid density. When stratification becomes unstable, vertical motions called **convection** occur.

We can understand convection best by using a simple analogy from the kitchen: the motions of water in a saucepan heated on the stove (**Fig. CC3.1**). Initially, when we place the saucepan on the burner, all the water within the pan is at room temperature. No vertical motion occurs, because the water is uniform in density. When we begin to add heat to the base of the saucepan, water at the bottom of the pan is heated and water above remains at room temperature. As bottom water is heated, its density is decreased and it rises through the overlying water to the surface. If you watch a saucepan of water carefully as you start to heat it, you can see the vertical water movements as swirls of motion.

When the water is heated above room temperature, it begins to lose heat through the surface and sides of the saucepan (**Fig. CC3.1**). Heated water rises to the surface, spreads out, and cools. As more warmed water rises to the surface, it forces the cooler surface water toward the sides

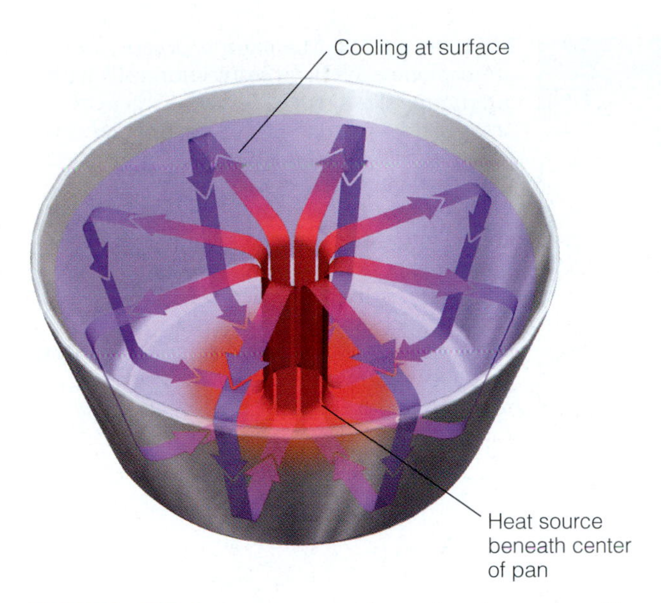

FIGURE CC3.1 In a saucepan heated at the center of its bottom, water will establish a circulation in which heated water rises to the surface, spreads toward the pan's sides and cools, and then sinks back to the bottom. This toroid-shaped convection cell is established because the water density is decreased when it is heated and increased when it is cooled. The idealized toroidal convection cell depicted would, in practice, normally be distorted because saucepans are usually heated across their entire base, not just at the center, and by turbulence in the flow patterns caused by variations in the rate of heating and cooling.

of the saucepan. Because heat is lost through the water surface, the surface water that is displaced to the edges of the pan is cooled slightly and has slightly higher density than the newly warmed water rising to the surface at the center of the pan. Accordingly, warmer water continuously flows toward the edge of the pan, cooling as it flows. As we continue to heat the saucepan, a continuous recycling motion is established. Water heated at the bottom of the water column rises at the center of the pan and is replaced by cooler water flowing from the sides of the pan. The heated water rises to the surface, then spreads out, cools, and sinks down the sides of the pan to be heated again (**Fig. CC3.1**). The closed cycle of circulation of a fluid driven by warming at a depth that causes the fluid to rise, cooling at a level of lower density, and sinking back to its original depth to rejoin the cycle is called a **convection cell.**

Convection is not as uniform as depicted in **Figure CC3.1**. The heating of the saucepan bottom and the heat loss through the water surface and pan sides are variable. Therefore, the location of the rising plume of the convection cell fluctuates, and more than one rising plume often is present. In addition, the water flow is **turbulent,** not smooth. Turbulence can be seen in the internal swirling, churning motions in fast-running streams.

If we set our stove on low heat, the water in the saucepan does not boil and convection cell motion continues indefinitely. Heat is added continuously and is lost at the same rate that it is added. In nature, most convection cells operate in this way.

Convection cells transport heat vertically within a fluid. Convection also establishes areas across the surface and bottom of the cell where warmer fluid (in the center of the saucepan) and cooler fluid (at the edges of the saucepan) are concentrated. These features of convection cells are important in many ocean processes, including **plate tectonics** (Chap. 4), ocean water circulation (Chap. 10), and atmospheric circulation (Chap. 9). The upper and lower boundaries of convection cells are often the interfaces between two fluids or between a fluid and a solid (e.g., the ocean surface and the seafloor). However, convection cells may develop between any two density surfaces within a vertically stratified fluid.

The convection cell just described is toroidal (doughnut-shaped), with a concentrated rising plume of heated water at its center and a ring of sinking water surrounding this plume (**Fig. CC3.1**). In another configuration of convection cells, both the rising plume and the sinking plume are extended laterally. In its simplest form, this type of convection cell is cylindrical in shape (**Fig. CC3.2**). Because fluid cannot both rise and sink at the same location, convection cells must be arranged so that the rising and sinking plumes alternate (**Fig. CC3.2**). The areas where fluid rises are **upwelling** zones and areas where it sinks are **downwelling** zones. Upwelling and downwelling areas must alternate across the top or bottom of the fluid. We cannot go from an upwelling zone to another upwelling zone without passing through a

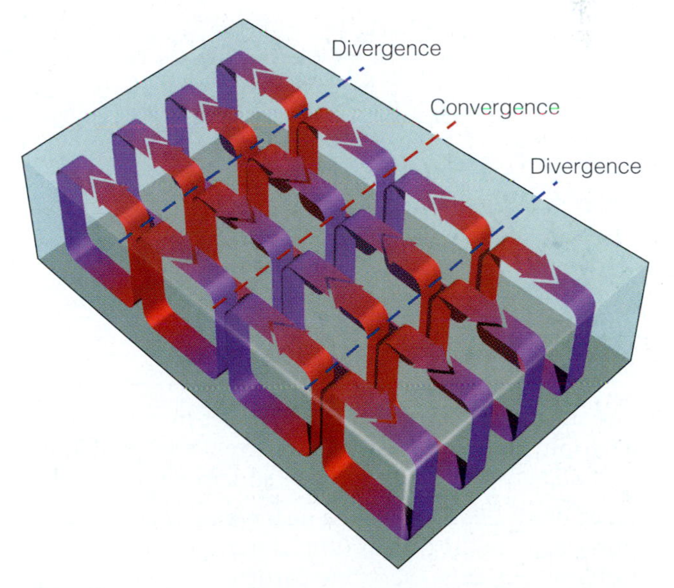

FIGURE CC3.2 Cylindrical convection cells. Note that convergences and divergences alternate between adjacent cells.

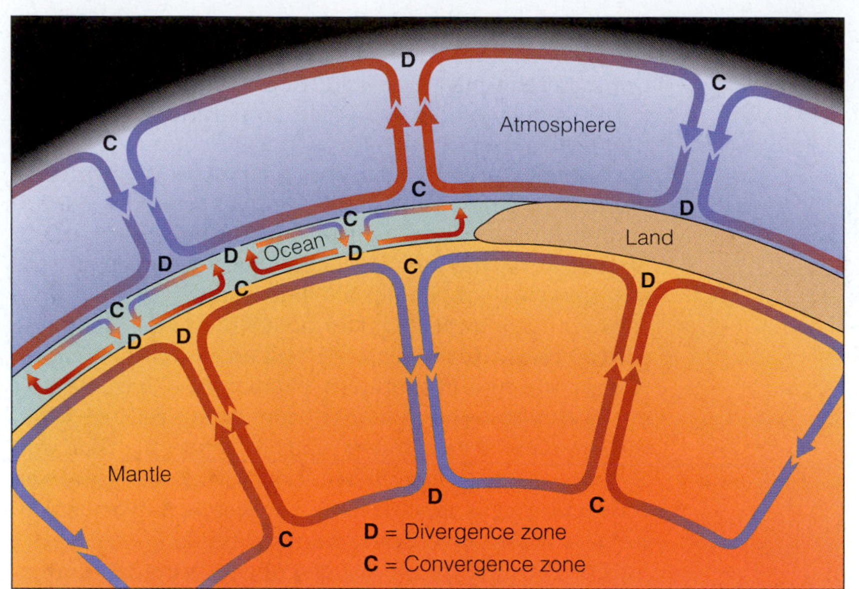

FIGURE CC3.3 The mantle, oceans, and atmosphere each have convection cells with alternating divergences and convergences across their upper and lower boundaries. In the atmosphere, we are generally concerned with the lower (sea-level) boundary of the convection processes where divergences are downwelling zones and convergences are upwelling zones. In the mantle and oceans, we are generally concerned with the upper (sea surface and upper surface of the asthenosphere) boundaries of the convection processes where divergences are upwelling zones and convergences are downwelling zones.

downwelling zone. The locations of areas of upwelling and downwelling in the atmosphere, ocean, and Earth's mantle are important to many aspects of the geology, chemistry, **climate**, and biology of the Earth.

At the top of a convection cell, fluid flows horizontally away from upwelling areas as it is displaced by upwelled fluid. Such areas are called **divergences** (**Fig. CC3.3**). Similarly, fluid flows into downwelling areas to replace downwelled fluid. Such areas are called **convergences** (**Fig. CC3.3**).

This text considers primarily processes that occur at the top of convection cells (e.g., in the upper mantle and ocean surface waters), where upwelling zones are divergences and downwelling zones are convergences (**Fig. CC3.3**). When we consider atmospheric circulation, we are concerned primarily with the bottom of the convection cell at the Earth's surface, where upwelling zones are convergences and downwelling zones are divergences (**Fig. CC3.3**).

CRITICAL CONCEPT 4

Particle Size, Sinking, Deposition, and Resuspension

ESSENTIAL TO KNOW

- The sinking rate of suspended particles in the ocean is determined primarily by particle size. Large particles sink quickly, and small particles sink more slowly.
- Large particles are deposited near where they are introduced to the oceans (unless transported by turbidity currents). Very fine particles may be transported long distances by currents before they eventually settle on the seafloor.
- Particles can be resuspended after they reach the seafloor if current speed is sufficiently high.
- Because fine-grained particles are cohesive, high current speed is necessary to resuspend them from some fine-grained sediments.
- Particles may be alternately deposited and resuspended many times where current speeds are variable.
- The particle size of grains within a sediment is determined by the range of current speeds and the size range of particles transported into the area. Areas with high maximum current speed generally have coarse-grained sediments. Fine-grained sediments are present only where minimum current speed is low and maximum current speed is not extremely high.

UNDERSTANDING THE CONCEPT

Solid particles are transported through the water by **currents** in the same way that dust particles are carried through the air by winds. Just as dust particles settle when the air is calm, waterborne particles, called **suspended sediment,** sink to the seafloor when currents are slow or absent.

Particles sink through the water in response to **gravity.** However, particles do not all sink at the same rate, because of **friction** between them and the water molecules they must push aside. The frictional resistance to a sinking particle increases with **viscosity.** The viscosity of water is low in comparison with, for example, that of molasses or motor oil.

Large, dense particles are not slowed significantly by friction as they fall through the water column. However, viscosity becomes more important as particle size decreases. To understand why smaller particles are more affected by viscosity, we must consider three factors. First, the gravitational attraction on a particle falling through the water column is directly proportional to its mass (for particles of the same **density,** it would be directly proportional to the volume). Second, because viscous friction occurs at the particle surface, particles with larger surface area are subject to greater friction. Third, the ratio of the surface area of a particle to its volume generally increases as particle size is reduced. The ratio of viscous friction to gravitational force therefore increases as particle size is reduced. Consequently, for small particles such as those in the suspended sediment, the settling rate of the particle decreases progressively (the particle sinks more slowly) as the particle size decreases.

There are some exceptions to this rule. First, the density of the particle is important. Less dense particles sink more slowly than denser particles. This fact explains how some marine organisms avoid sinking (Chap. 16). Second, the shape of the particle is important. A parachute or feather falls more slowly through the atmosphere than a rock or rice grain because the surface area of a parachute or feather is large in relation to its volume and weight. Thus, air resistance or viscous friction is enhanced. Similarly, some marine organisms have body designs that maximize their surface area to prevent them from sinking (Chap. 16).

Most suspended particles (mineral grains) in the oceans, other than organic **detritus,** are of similar density. Hence, they sink at rates that are determined primarily by their diameter. **Table CC4.1** reports typical sinking or settling rates for particles of varying sizes. Sand-sized particles sink rapidly, and smaller particles sink much more slowly. Organic detritus particles are generally of lower density than mineral grains and sink more slowly.

Particle-sinking rates are modified by the presence of currents. As current speed increases, **turbulence** increases and particle sinking is retarded. This effect can be seen in a glass of orange juice. If the glass sits undisturbed for a while, the particles of orange pulp settle to the bottom of the glass. If the orange juice in the glass is gently stirred to keep the liquid in motion, the pulp will not settle. The effect of currents on particle-sinking velocities depends on both current speed and particle size. For a given particle size, sinking rate is reduced as current speed increases until the turbulence is sufficient to prevent the particle from sinking at all. The current speed at which particles no longer sink varies with particle size. This relationship is shown in **Figure CC4.1a**. At high current speeds, only large particles settle to the seafloor, and smaller particles remain in suspension. At lower current speeds, smaller particles settle to the seafloor.

Once a particle has settled to the seafloor, it can be **resuspended** if the current speed at the seafloor is sufficiently high. **Figure CC4.1b** shows the current speeds needed to resuspend **sediment** particles of different sizes from sediments that consist primarily of grains of that size range. For all particle sizes, the speed needed to resuspend a sediment particle is higher than the speed needed to prevent it from sinking (**Fig. CC4.1a,b**). High current speeds are necessary to resuspend large particles, and somewhat lower current speeds are necessary to prevent the same particle from sinking, once resuspended.

With decreasing particle size, both the resuspension speed and the speed needed to prevent sinking decrease at approximately the same rate until silt or clay size is reached. The current speed needed to resuspend these smaller particles increases with decreasing particle size, so

TABLE CC4.1 Settling Velocity of Particles of Average Density and Approximately Spherical Shape When No Current Is Present

	Sand	*Silt*	*Clay*
Particle diameter	0.1 mm	0.01 mm	0.001 mm
Settling velocity	2.5 $cm{\cdot}s^{-1}$	0.025 $cm{\cdot}s^{-1}$	0.00025 $cm{\cdot}s^{-1}$
Time to settle 4 km (average ocean depth)	1.8 days	6 months (185 days)	50 years

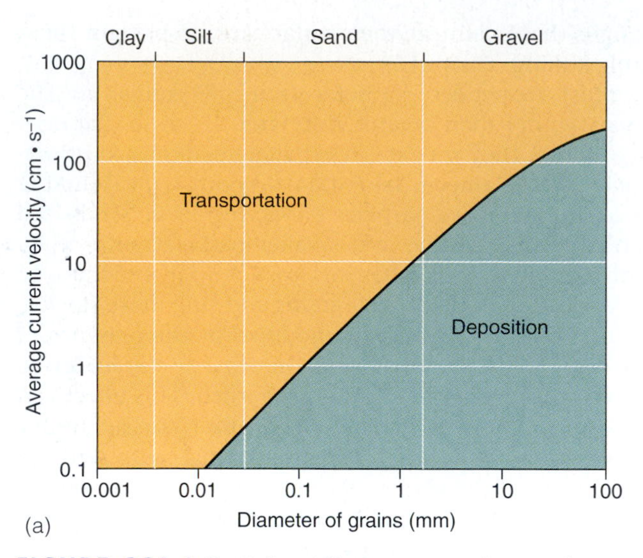

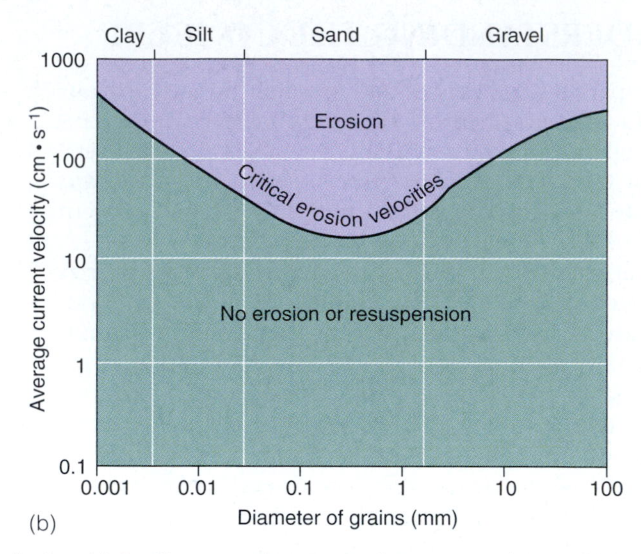

FIGURE CC4.1 Particle settling, resuspension, and current velocity. (a) Settling rate changes with current velocity. (b) Particle size and current velocity affect particle resuspension.

the speed needed to resuspend the finest sediment particles is higher than that required to resuspend sand-sized sediments. The reason is that fine sediment particles are irregular in shape and have a very large surface area in relation to their volume (size). The grains are so close to each other that they are held by **electrostatic** attraction between the individual particle surfaces. In addition, they tend to lock together because of their irregular shapes. Organic matter also may help the particles stick to each other. Because they cling together so tightly, fine-grained sediments are said to be **cohesive.**

Figure CC4.2 shows how suspended sediments of various **grain sizes** behave at different current speeds. The settling and resuspension threshold lines on this chart are the same as those in **Figure CC4.1**. Three areas are shown in **Figure CC4.2**: one where conditions are such that a particle would sink to the seafloor; another where conditions are such that particles would remain suspended; and a third where conditions are such that particles would not only remain suspended, but would also be resuspended if they were deposited on the seafloor.

Referring to **Figure CC4.2**, consider an ocean area with no local sources of particles, where current speed ranges from A to B. No sediment of grain size larger than d_2 is brought into this area by currents. Particles of grain size d_1 or smaller are not deposited in this area. At the highest current speed, B, no particles are resuspended. Therefore, only particles within the limited size range d_1 to d_2 are present in these sediments.

If current speed in **Figure CC4.2** ranged from A to C, particles up to size d_3 could be brought into the area and deposited. However, particles smaller than d_4 that were brought into the area would be alternately resuspended when current speed was high and deposited when current speed was low. These particles eventually would be removed from the area by mixing. The sediment then would consist of coarse-grained particles with diameters between d_3 and d_4. Particles smaller than d_1 would be transported through this area and would not settle.

If current speed in **Figure CC4.2** ranged from A to D, particles between sizes d_1 and d_5 could enter the area and be deposited, but finer-grained particles would not be deposited. Particles between sizes d_6 and d_7 could be resuspended and removed, but fine-grained sediments between sizes d_6 and d_1, once deposited, could not be resuspended. Particles of size range d_4 to d_7 also would accumulate. However, in most locations, fine-grained particles far outnumber larger particles in the suspended sediment. Therefore, at this site the sediments would be

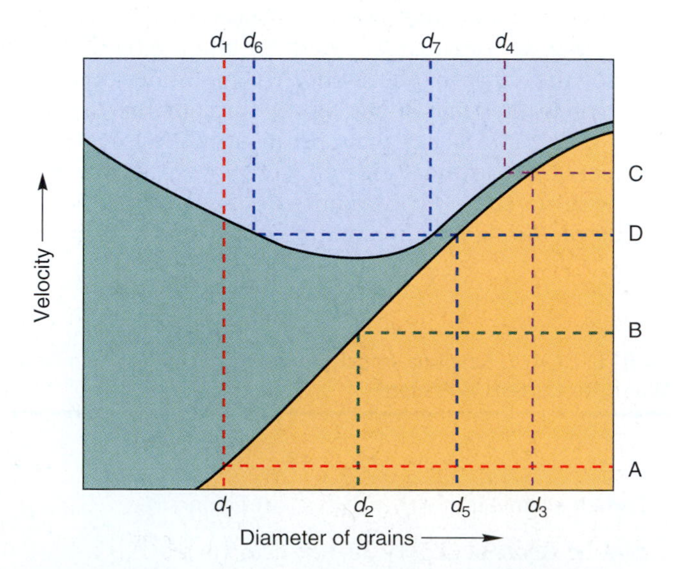

FIGURE CC4.2 Depiction of why most sediments are sorted by grain size, as explained in the text. Note how grain size sorting is determined by the range of wave and/or current speeds at a particular point on the seafloor.

fine-grained muds consisting primarily of grains between d_1 and d_6 in size.

From **Figure CC4.2**, we can conclude that large-grained particles cannot be carried far from areas where they are introduced to the oceans unless maximum current speeds are very high (but they could be transported by **turbidity currents;** Chap. 8). In contrast, fine-grained particles tend to be carried long distances before settling in an area where the minimum current speed is low, and they cannot accumulate in any area where the minimum current speed is high. The diagram further demonstrates that the sediment particles that accumulate in a given location are restricted to a range of sizes determined by the current regime.

CRITICAL CONCEPT 5

Transfer and Storage of Heat by Water

ESSENTIAL TO KNOW

- Water has a high heat capacity. More heat is needed to raise the temperature of water by 1°C than is required to do the same for almost any other substance.
- Large amounts of heat can be absorbed and stored in the waters of the mixed layer (approximately the upper 100 m) of the oceans without causing major water temperature changes.
- Heat can be transported with water masses and released at another location by radiation, conduction, and evaporation. This is a mechanism by which heat is transported from the tropics to higher latitudes.
- Water has a high latent heat of fusion. More heat is needed to melt ice than is needed to melt almost any other substance.
- Ice in the polar regions acts as a thermostat because of water's high latent heat of fusion. Large amounts of heat can be lost or gained seasonally by the conversion of ice to water, or vice versa, without changing the temperature of the ocean surface water.
- Water has a high latent heat of vaporization. More heat is needed to vaporize or evaporate water than is needed to vaporize any other substance.
- Water evaporated from the ocean surface contains large quantities of heat as latent heat of vaporization. This heat energy is transported with the air mass and then released to the atmosphere when the water vapor condenses. This is another mechanism by which heat is transported from the tropics to higher latitudes.

UNDERSTANDING THE CONCEPT

Water has a very high **heat capacity** (Chap. 7, **Fig. 7.2**). This means that it takes much more heat energy to increase the temperature of water by 1°C than it does to cause the same temperature change in other substances, such as rocks. Consequently, in regions where solar radiation is intense, the surface layer of the oceans can absorb large quantities of heat while the water temperature undergoes little change.

Most of the solar heat absorbed by the oceans is absorbed initially in the upper few meters of water because solar energy is absorbed rapidly and does not penetrate far. The heat energy is distributed rapidly by **turbulence** throughout the upper **mixed layer** of the oceans stirred primarily by winds (Chap. 10). Thus, solar heat is distributed and stored in large volumes of water. In contrast, solar heat reaching the land is transferred downward only slowly by **conduction** (a very inefficient heat transfer mechanism) and is mostly radiated back into the atmosphere and space at longer **wavelengths**.

Because large quantities of heat can be stored in the surface layers of the oceans, heat can be transported with ocean **currents**. Subsequently, the heat energy can be released to the atmosphere. Thus, the high heat capacity of water facilitates the transfer of heat from tropical regions to colder regions near the poles (Chap. 9). The mild **climate** of parts of Europe in comparison with the climate of eastern North America at similar **latitudes** is partially the result of this mechanism. Heat is transported from the tropics into European seas by the warm Gulf Stream water and then released to the atmosphere (Chap. 10).

Water also has a high **latent heat of fusion** and a high **latent heat of vaporization.** This means that a large amount of heat energy is needed to convert ice to water, or water to water vapor (Chap. 7, **Fig 7.2**). Heat energy added to ice to form water, or to water to form water vapor, is released when water freezes or water vapor condenses. Thus, heat can be stored and transferred from one location to another if the conversions between ice and water, or between water and water vapor, occur at different times or places.

For example, heat from the ocean surface can be used to vaporize water. The water vapor can be transported through the atmosphere for thousands of kilometers until it is recondensed and its heat energy released to the atmosphere.

The high latent heat of fusion is important in polar regions. If water had a lower latent heat of fusion, the extent of the Arctic Ocean ice and the Antarctica ice sheet would vary more between summer and winter than it does at present. The presence of ice moderates the climate in these regions. As heat is lost in winter and gained in summer, most of the loss or gain goes to the freezing or melting of ice, and the ocean water temperature remains at the freezing point. Thus, the heat stored in the water by melting of ice in summer is returned to the atmosphere in winter as the water refreezes. In this way, the polar ice acts virtually as a thermostat that prevents air temperature from being much higher in summer and much lower in winter.

The high latent heat of vaporization is important globally because it allows large amounts of heat energy to be transferred from the oceans to the atmosphere. Once in the atmosphere, the heat is redistributed by atmospheric circulation. Water need not boil for water molecules to be transferred from the liquid to the vapor phase (Chap. 7). At any temperature, water molecules are transferred from ocean to atmosphere by evaporation at the sea surface, but the rate of evaporation generally increases with increasing water temperature. When a water molecule is evaporated from the oceans, it carries with it its latent heat of vaporization, which is more than 1000 times the amount of heat needed to raise the temperature of the same amount of water vapor by 1°C. Once evaporated, the water carries its latent heat through the atmosphere until the air cools, causing water vapor to condense to liquid water (clouds and rain).

When water molecules condense, they release their latent heat of vaporization to the surrounding air molecules. Because the heat capacity of all atmospheric gases is less than that of water, one water molecule can release enough heat to increase the temperature of a larger number of molecules of atmospheric gases by one or more degrees. Thus, evaporation and condensation of small amounts of water can transfer large amounts of heat energy from the oceans to the atmosphere. This mechanism is responsible for transporting heat from the equator toward the poles, thus moderating climate differences that otherwise would be extreme with increasing latitude (Chap. 9). In addition, this mechanism is responsible for the more moderate climate of coastal areas than of inland areas of the same latitude. It is also the driving energy of air movements in the atmosphere, including storms and **hurricanes** (Chap. 9).

CRITICAL CONCEPT 6

Salinity, Temperature, Pressure, and Water Density

ESSENTIAL TO KNOW

- The density of water increases as the salinity increases.
- The density of seawater (salinity greater than 24.7) increases as temperature decreases at all temperatures above the freezing point.
- The density of seawater is increased by increasing pressure. Density changes about 2% because of the pressure difference between the surface and the deep seafloor. The effect of pressure on density usually can be ignored because most applications require density comparisons between water masses at the same depth.
- Water of salinity less than 24.7 has an anomalous density maximum. Pure water has its maximum density at about 4°C, but the maximum density of water occurs at lower temperatures as salinity increases. Between 4°C and the freezing point, the density of pure water decreases as temperature decreases.
- The relative importance of changes in temperature and salinity in determining seawater density varies with water temperature. Temperature variations are more important in warm ocean waters, whereas salinity variations are more important in cold ocean waters.

UNDERSTANDING THE CONCEPT

Below the ocean surface layers, movements of **water masses** are caused predominantly by differences in **density** between water masses (**CC1**, **CC3**). For this reason, water density, which depends on temperature, the concentration of dissolved salts, and pressure, is among the most important properties of seawater.

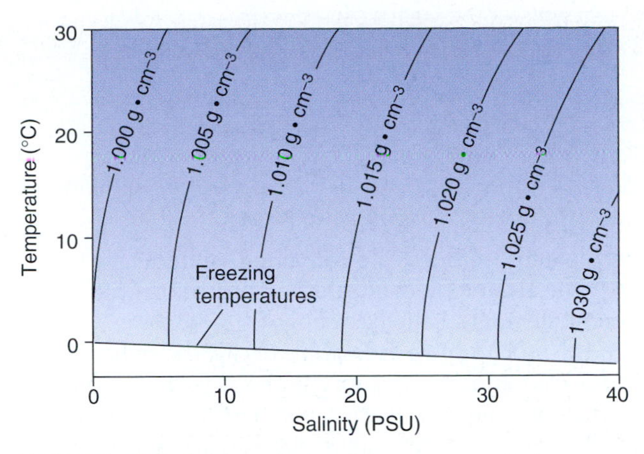

FIGURE CC6.1 The relationships among salinity, temperature, and density are complex. Generally, density increases as temperature is reduced or as salinity is increased. This figure is reproduced with additional information in Chapter 7 (Fig. 7.5a).

Increasing pressure compresses liquids so that more mass is squeezed into a smaller volume. Thus, density increases with increasing pressure. However, the molecules of water can be forced together only slightly, even by large pressure increases. Water is therefore almost incompressible, and the effect of pressure on water density is small. The density of seawater is only about 2% greater at the deepest depths of the oceans than it is at the surface. Consequently, the effects of pressure on density generally are ignored because oceanographers usually are interested in density differences between water masses at the same depth.

The density of water increases as **salinity** increases. The reason is that most **ions** have higher density than water molecules have, and dissolved substances reduce the clustering of water molecules.

Decreasing temperature generally causes liquids to contract. Therefore, decreasing temperature generally increases density. However, the behavior of pure water is an exception to this rule. Between its boiling point and 4°C, water behaves normally: density increases as temperature decreases. However, between about 4°C and its freezing point, pure water decreases slightly in density as the temperature decreases. In other words, water density has a maximum at about 4°C (Chap. 7).

As salinity increases, the temperature at which the density maximum occurs decreases. At salinity 24.7 and higher, the water density maximum is at the freezing point. The salinity of open-ocean waters is generally above 24.7. Therefore, in contrast to pure water and water with salinity less than 24.7, open-ocean seawater increases in density as temperature decreases at all temperatures above its freezing point.

The relationships among seawater density and pressure, temperature, and salinity are complex and are discussed in more detail in Chapter 7 and **Figure CC6.1**. However, they are illustrated by the following summary:

- Pure water at 4°C and 1 atmosphere (atm) pressure has, by definition, a relative density of exactly 1.
- Density varies almost linearly with salinity. Density increases by about 0.00080 (±0.00004) for each unit of salinity increase, where salinity is measured in practical salinity units, abbreviated PSU. The density of "average" seawater is about 1.028 at salinity 35, 0°C, and 1 atm pressure, which is about 3% more than that of pure water at 4°C.
- Density varies almost linearly with pressure. Density increases by about 0.00045 for every 100 m of depth in the oceans.
- The rate of change of density with temperature is a function of both temperature and salinity. Density increases as temperature decreases as follows:

Salinity	*Temperature (°C)*	*Density Increase per 1°C Reduction*
0	1	–0.00007[a]
0	10	0.00009
0	20	0.00021
0	30	0.00030
20	1	0.00001
20	10	0.00014
20	20	0.00024
20	30	0.00033
40	1	0.00017
40	10	0.00018
40	20	0.00027
40	30	0.00034

[a]The density of pure water decreases with decreasing temperature below 4°C.

- The relative importance of temperature and salinity in determining density varies with water temperature. In cold waters, variations in salinity are more important than variations in temperature. In warm waters, variations in temperature are more important.

The density of seawater is altered continuously by solar heating, which changes temperature, and by evaporation, precipitation, and ice formation and melting, which alter salinity. These processes take place at the ocean surface. Vertical motions of water masses in the oceans are caused primarily by sinking of high-density surface water formed by cooling or evaporation (increased salinity).

The vertical motions created by changes in water density are discussed in **CC3** and Chapter 10.

CRITICAL CONCEPT 7

Radioactivity and Age Dating

ESSENTIAL TO KNOW

- Atoms of an element usually have several isotopes that differ in atomic weight but are essentially identical in chemical properties.
- Some individual isotopes of many elements are radioactive. Others are stable.
- Radioisotopes called "parent isotopes" decay by losing part of the atom to become a different isotope, called a "daughter isotope." Often the daughter is an isotope of a different element.
- Radioactive decay releases heat and radioactive particles, including alpha particles, beta particles, and gamma rays.
- Each radioisotope decays at a certain fixed rate, expressed by the isotope half-life.
- One half-life is the time it takes for exactly half of the atoms of a particular radioisotope in a sample to decay.
- The age of a rock, archaeological artifact, or skeletal or undecomposed remains of an organism can often be determined by measurement of the concentration of both the parent and daughter isotopes in the sample.
- For such age dating to be accurate, no daughter isotope must have been present when the sample was formed. In addition, no parent or daughter isotope atoms must have been gained or lost by the sample after it was formed. These conditions are often not met.
- Different pairs of parent and daughter isotopes must be used for dating samples of different ages. Parent isotopes with long half-lives are useful only for dating older samples.
- Radioisotope dating is often used to calibrate other less expensive dating techniques, such as fossil dating, that reveal only the relative dates of samples from within a group of samples under study.

UNDERSTANDING THE CONCEPT

Atoms of any element may occur in several different forms called **isotopes.** Isotopes differ from each other in atomic weight, but they are almost identical in their chemical properties. Some isotopes are stable, and others are radioactive. **Radioisotopes** (but not stable isotopes) called "parent isotopes" are converted, or decay, by losing part of the atom to become a different isotope (often of a different element), called a "daughter isotope."

In the process of **radioactive** decay, the radioisotope releases heat and one or more types of radioactive particles, including alpha particles, beta particles, and gamma rays. Radioactive decay takes place at a fixed constant rate, different for each radioisotope. This rate is expressed by the radioisotope **half-life.** One-half of the atoms of a specific radioisotope in a sample will decay during one half-life for that isotope, regardless of how many atoms were present initially. Thus, after one half-life, one-half of the atoms originally present have decayed. After two half-lives, one-half of these remaining atoms have also decayed, and one-quarter of the original number of atoms of the isotope remain. After three half-lives, one-eighth of the original atoms of the isotope remain; after four half-lives, one-sixteenth of them remain; and so on.

The age of various materials, including rocks, archaeological artifacts, and the remains of living organisms, can be determined by radioisotope dating. To use this technique, the fraction of the atoms of certain radioisotopes that remain undecayed from the time the rock or other sample was formed must be determined. We can do this by measuring the concentrations of both the parent and daughter isotopes in the sample. If the sample is one half-life old, the parent and daughter will be present in equal concentrations. After two half-lives, there will be three times as many daughter atoms as parent atoms. After three half-lives, there will be seven times as many daughter as parent atoms, and so on for other ages.

Radioisotope dating depends on several critical assumptions. We must usually assume that no daughter isotope was present when the sample was formed and that neither parent nor daughter has been added or removed, except by radioactive decay, since that time. These assumptions are often incorrect and are always difficult to test because samples undergo physical and chemical changes over time that can add or remove elements. Therefore, to be certain of the measured age, often we must date a sample by two or more different methods.

A particular radioisotope is not useful for age dating if only a very small fraction of the original parent atoms have decayed or, conversely, if almost all the original parent atoms have decayed. Hence, the ideal radioisotope for dating depends on the sample's age. Several of the most frequently used radioisotopes and their daughters and half-lives are listed in **Table CC7.1**. We can see that

TABLE CC7.1 Some Radioisotopes Used for Age Dating and Their Half-Lives

Parent Isotope	*Stable Daughter Isotope*	*Half-Life[a] (years)*
Carbon-14	Nitrogen-14	5560
Uranium-235	Lead-207	700 million
Potassium-40	Argon-40	1.3 billion
Thorium-232	Lead-208	1.4 billion
Uranium-238	Lead-206	4.5 billion

[a]Dating methods are most accurate when used on samples with ages between 0.5 and 3.5 times the half-life.

carbon-14 is the most useful radioisotope for dating recent samples (if they contain carbon), whereas uranium-235 is more suitable for older materials. The other radioisotopes listed are suitable only for dating ancient samples.

The very sensitive and accurate measurements needed to perform radioisotope dating can be expensive. In addition, the dates determined must be verified because the necessary assumptions for this technique are not always true. For these reasons, radioisotope dating is often supplemented by relative dating techniques based on magnetic properties or **fossils** (Chap. 8). In some studies, radioisotope dating is performed on only a few representative samples, and the dates of these samples are used to calculate absolute dates for samples dated by relative dating methods.

CRITICAL CONCEPT 8

Residence Time

ESSENTIAL TO KNOW

- Substances, including water, move through different parts of the oceans, atmosphere, and biosphere at different rates.
- The concentration of a substance in a part of the ocean, atmosphere, or biosphere is determined by how fast, on average, a molecule of the substance moves from input to output within that particular segment of the environment. This rate is measured as a residence time.
- Residence time is calculated by dividing the total quantity of a substance within an environmental segment (or "box") by the rate of either its input or its removal from the segment.
- Residence time can be used to estimate the concentration increases that will result from increased discharges of a substance to a segment of the oceans. It is particularly useful for pollution assessments in estuaries and coastal waters. Long residence time and small estuary water volume lead to high concentrations of a contaminating substance.
- In certain situations we can use residence times to estimate the relative magnitudes of inputs to, or outputs from, a particular ocean segment without measuring them directly.

UNDERSTANDING THE CONCEPT

Marine and atmospheric sciences involve studies of the movements of elements, water, and other substances within the oceans and atmosphere, and between the oceans, atmosphere, biosphere, and solid Earth. These movements are components of complex **biogeochemical cycles** (Chap. 6). The complexity of chemical movements within these cycles makes it very difficult, if not impossible, to study a cycle as a whole. Accordingly, the

environment must be divided into segments for most studies.

A segment of the environment is chosen by a characteristic that distinguishes it from adjacent segments in some way. A segment, which for convenience is usually called a "box" or "compartment," can be defined as almost any part of the Earth, ocean, or atmosphere. For example, a box could be defined as the entire world ocean, or as a single basin such as the Mediterranean Sea, or as a simple square section of open ocean that is identified only by boundaries defined by an oceanographer. A box could also be the entire ocean **biomass** or the members of a particular fish **species** within a specified geographic area. Boxes often are defined so that the **residence times** of substances can be investigated.

The substance of interest enters a box through certain identifiable pathways (inputs) and leaves through other identifiable pathways (outputs). Substances passing through the box can move quickly from input to output, or they can remain within the box for long periods of time before exiting. The average time that an atom or molecule of a relevant substance remains in the box is an important characteristic of the system and is expressed as its residence time.

To calculate the residence time in a defined box, one must measure the total quantity of the substance in the box, and either its rate of input to or its rate of output from the box. For example, if we want to determine the residence time of sodium in the Mediterranean Sea, we estimate the total quantity of sodium by measuring the sodium **ion** concentration at different places, averaging the result, and multiplying by the volume of this sea. The rate at which sodium is entering or leaving the Mediterranean also must be measured. All of the sea's inputs or outputs must be considered, including river **runoff,** atmospheric fallout and rain, and exchange of water between the Mediterranean and the Atlantic Ocean through the Straits of Gibraltar, between the Mediterranean and the Black Sea through the Bosporus, and between the Mediterranean and the Red Sea through the Suez Canal. We could measure the total quantity of water and its inputs and outputs from the Mediterranean in the same way, but in this case we would have to consider an additional output: evaporation.

Once we know the total amount of the substance of interest in the box and the rate of its input or output, we can calculate a residence time if we assume that the system we are studying is in a **steady state** (in other words, if we assume that the amount of substance in the box remains stable over time). Because the Earth's biogeochemical cycles have had billions of years to reach equilibrium, we can usually accept that assumption when we are studying large boxes, such as an entire ocean basin. However, the smaller the box, the less likely the assumption is to be true. Although the Earth's biogeochemical cycles are in approximate steady state when averaged

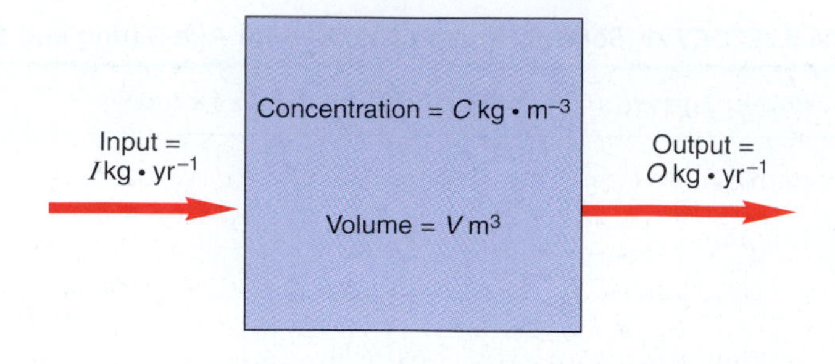

Mass of substance in box = $M = C \times V$
If M is constant
$I = O$
Residence time (yr) = $M/I = M/O$
or M (in kg) divided by I or O in kg • yr^{-1}

FIGURE CC8.1 This simple model shows how we calculate residence time by dividing the mass of the substance in the environmental segment ("box" or "compartment") by either the rate of input or the rate of removal (output) of the substance from the box.

globally, these processes can vary substantially at any one place and time because of such changes as variations in the year-to-year rate of river flow.

If the total amount of substance in the box remains stable over time, the rate of input must equal the rate of output (**Fig. CC8.1**). The residence time is calculated as the total quantity of the substance in the box, divided by either the rate of output or the rate of input. Measured in years or another time unit, residence time is equivalent to the time it would take the outputs to remove all the substance from the box if all inputs were stopped, and vice versa.

The concept of residence time has many uses. First, it provides a method of estimating the rates of some processes that are difficult to measure directly. For example, if we can measure the concentration of a particular element in Mediterranean seawater and the rate of input of this element from rivers and adjacent seas, we can calculate the rate at which this element is being removed to the **sediment.** We did not include **sedimentation** (or evaporation and salt precipitation) in our preceding examination of sodium cycling in the Mediterranean, because for sodium (as opposed to other elements), the rate of removal to the sediment is extremely slow at present, so it is negligible in relation to other inputs and outputs.

A second use of residence time is in determining the probable fate of **contaminants** released to the oceans. If we release contaminants whose residence time is long, they will remain in the oceans for a long time and may cause **pollution.** If contaminants have short residence times, human additions to the oceans will be quickly removed (usually to the ocean sediment). Residence times of the elements in the world oceans are discussed in Chapter 6.

One important characteristic of biogeochemical cycles is that mechanisms for removal of a substance from a box are generally related to the concentration of the substance in the box. If the concentration increases, the rate of output will also increase. For example, if the concentration of a substance in the Mediterranean is increased, the rate of loss of the substance from the Mediterranean will increase as seawater flows out of this sea to an adjacent sea. Similarly, higher concentrations lead to faster uptake by **biota** and faster removal to sediment, although there is often not a simple linear relationship in these instances.

An increase in the rate of input to a box generally causes the concentration in the box to increase until the output rate increases to match the elevated input rate. If the expected increase in input is known, sometimes the elevation in concentration that this would cause can be calculated and predicted. This method is extremely important in determining the potential of rivers, bays, or segments of coastal ocean to be affected by contaminant inputs. A hypothetical example illustrates how this assessment can be made. **Figure CC8.2** shows the equations we could use for an **estuary** into which is discharged river water that carries a natural concentration (C) of a substance. An industrial complex is proposed that will discharge additional amounts of this substance to the estuary. For simplicity, we will assume that the substance of interest is not removed to sediment within the estuary and that it can leave the estuary only dissolved in seawater. From the simple calculations in **Figure CC8.2**, we see that the incremental concentration (δC) that can be expected to occur in the estuary after the additional input (δI) begins is easily estimated if the residence time, volume of the estuary, and magnitude of δI are known. The important observation to be made about this result is that the concentration increase caused by an increased input of a contaminant substance is greater if the residence time (T) in the discharge water body is long in relation to its volume (V).

This finding is important because some water bodies, particularly estuaries, may have long residence times and small volumes. Such locations are not good choices for waste discharge. San Francisco Bay is a particularly good example. It has two segments: South San Francisco Bay, which has a water residence time of weeks to months; and North Bay, which has a similar volume but a water residence time of only a day or two. Inputs of treated municipal wastewater to South San Francisco Bay have caused serious water quality problems, whereas much larger inputs of treated municipal wastewater and industrial wastewater to North Bay have caused comparatively minor problems. The difference is the longer residence time in South San Francisco Bay.

Residence time calculations can be more complicated when there are multiple, separated inputs or outputs in the area of interest. For example, the calculation in **Figure CC8.2** is more complicated if a fraction of the contaminant is removed to sediments within the estuary. Nevertheless, residence time observations and calculations can be useful tools in such situations. If we can measure only inputs and concentrations of the contaminant in the estuary in **Figure CC8.2**, the calculated residence time will not tell us how much of the substance is flushed out to the ocean or how much is removed to the sediment in the estuary. We need not perform the relevant calculations here, but if we measure the inputs of the substance, its concentrations, and the residence time of water (not the residence time of the substance) in the estuary, we can calculate how much of the substance is removed with the water to the ocean and how much is retained in the estuary sediments and/or biota.

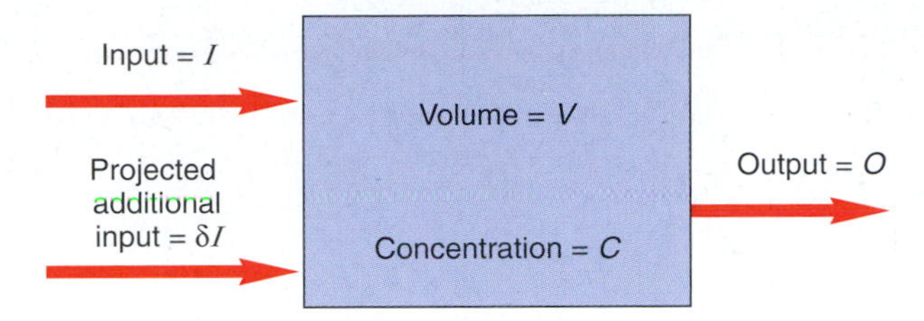

Residence time in the "box" $T = \dfrac{V \times C}{I}$

or $C = \dfrac{T \times I}{V}$ (equation 1)

Residence time remains the same when input is increased.

Residence time $T = \dfrac{V \times C}{I} = \dfrac{V\,(C + \delta C)}{(I + \delta I)}$

Where δC is the increase in concentration that occurs in the box when the input is increased by δI.

Thus, $\dfrac{T}{V} \times (I + \delta I) = C + \delta C$

or $\dfrac{T \times I}{V} + \dfrac{T \times \delta I}{V} = C + \delta C$

substituting from equation 1

$C + \dfrac{T \times \delta I}{V} = C + \delta C$

or $\dfrac{T \times \delta I}{V} = \delta C$

or the incremental increase in concentration in the box is equal to the residence time multiplied by the incremental increase in input rate, divided by the volume of the box.

FIGURE CC8.2 A simple calculation, which is based on the residence time of a substance in an estuary or similar water mass, can be used to predict the effects of a new source of contamination that increases the input rate of the substance to a defined segment of the environment, or "box."

CRITICAL CONCEPT 9

The Global Greenhouse Effect

ESSENTIAL TO KNOW

- The average temperature of the Earth's atmosphere is determined by the balance between solar energy that penetrates the Earth's atmosphere and is absorbed by gases, liquids, and solids, and energy that is lost to space by the Earth's radiation.
- The sun's radiated energy is concentrated in the visible portion of the electromagnetic spectrum, whereas the Earth's radiated energy is concentrated at longer wavelengths in the infrared portion of the spectrum.
- The Earth's atmosphere absorbs energy in the infrared and ultraviolet portions of the spectrum more effectively than it does in the visible portion of the spectrum.
- Because the Earth's atmosphere absorbs solar radiation less effectively than it absorbs energy reradiated to space by the Earth, the Earth's climate is warmer than it would otherwise be. This is the greenhouse effect.
- The efficiency of absorption of infrared energy by the atmosphere is increased by higher concentrations of certain gases, including carbon dioxide, methane, chlorofluorocarbons, nitrogen oxides, and ozone, which are called "greenhouse gases."
- Concentrations of each of these greenhouse gases in the atmosphere have been increased and continue to be increased as a result of human activities. Carbon dioxide concentrations have risen more than 35% since 1800.
- Radiation of heat from the Earth to space has probably been reduced and will continue to be reduced as a result of the absorption of radiated infrared energy by the increased gas concentrations. In contrast, the amount of solar radiation passing through the Earth's atmosphere has not been reduced by such absorption. If this were the only change occurring, the average temperature of the Earth's atmosphere would increase. This is the global climate change hypothesized to occur as a result of enhancement of the greenhouse effect by greenhouse gas releases.
- The increase in concentrations of greenhouse gases in the atmosphere and the consequent reduction in the Earth's radiative energy loss to space have many complex effects on the atmosphere, oceans, and biological systems. Many of these changes also have consequences that, on their own, would lead to warming or cooling of the Earth's atmosphere. These positive and negative feedbacks would respectively enhance or offset predicted global warming.
- Most analyses and models of this system predict that the net effect of greenhouse gas release will be a global warming of 2°C to 6°C in the next century, but the magnitude of such warming is not certain. Indeed, the net climate change may be global or regional cooling. The only certainty is that the increases in greenhouse gases will cause global climate changes of currently unknown magnitude.
- The consequences of global warming would be damaging to human civilization. Possible changes include flooding of coastal cities due to rising sea level and drastic changes in rainfall patterns that would lead to the loss of agricultural land.

UNDERSTANDING THE CONCEPT

We experience temporal and **spatial** fluctuations of temperature in the Earth's atmosphere as **weather** and seasonal change. However, the average temperature of the Earth's atmosphere remains almost constant from year to year. Changes in average temperature of several degrees have occurred in the past (**Fig. CC9.1**). These historical changes, although generally small, profoundly affected the Earth's **climate.** For example, they caused **ice ages** to begin and end.

There is concern that changes in the chemistry of the Earth's atmosphere due to human activity may cause the Earth's average temperature to rise as much as 2°C or 3°C within the next decade or so. This predicted change is usually called the **greenhouse effect** or "global warming." The fear is that the Earth's average temperature could change faster than it has ever changed during human history, and that the extent of the predicted change could be great enough to have major effects on the Earth's climate. Rapid climate changes would have numerous damaging effects on nature and on humans. Consequently, there is considerable interest in determining whether climate change will occur, how soon it might occur, and whether the problem can be reduced by altering human activities.

The only significant source of heat energy to the atmosphere is solar radiation. Several other sources, including heat **conducted** from the Earth's interior, tidal **friction,** and heat released by burning **fossil fuels,** are all negligibly small in comparison with solar input. Part of the solar radiation reaching the Earth is reflected back to space (primarily by clouds or snow). The rest is absorbed by the land, oceans, or atmosphere. Heat is

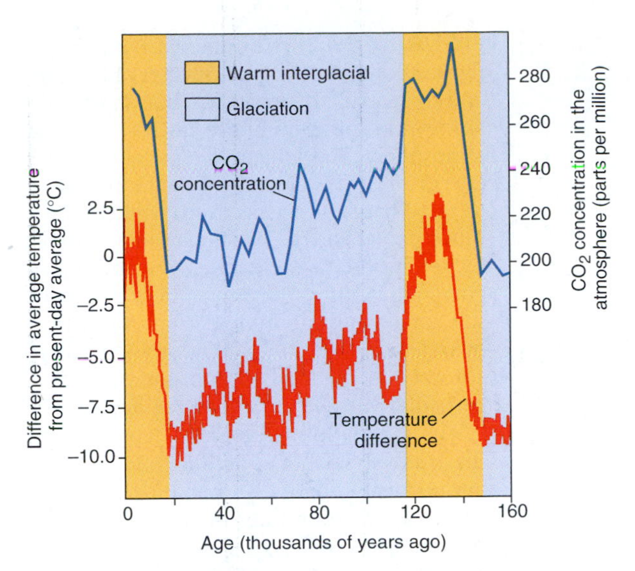

FIGURE CC9.1 Diagram showing the close relationship between variations in average air temperature and atmospheric carbon dioxide concentration during the past 160,000 years. These data were obtained from an ice core drilled into the Antarctic ice cap. Notice that the Earth's average temperature at present is approximately at its highest level since more than 120,000 years ago.

transferred between oceans and atmosphere and between land and atmosphere, and it is radiated back to space by land, ocean, and atmosphere (**Fig. 9.5**). This system of energy transfer is discussed in Chapter 9. One important feature is that the amount of solar energy that reaches the Earth must equal the amount of energy reflected or radiated back to space. If the amount of incoming energy did not equal the amount of outgoing energy, the Earth's atmosphere would either gain or lose heat continuously, and the atmospheric temperature would increase or decrease continuously.

To understand how the chemistry of the atmosphere can affect the balance between incoming and outgoing radiation, we must understand the nature of **electromagnetic radiation** (**Fig. 7.9**). All objects radiate electromagnetic energy, but the **wavelength** range emitted by an object depends on its temperature. The hotter the object, the shorter the wavelengths of electromagnetic energy that it radiates. The sun, which is very hot, radiates much of its energy in the visible and ultraviolet wavelengths, whereas the Earth and oceans, which are much cooler, radiate energy only in the much longer wavelengths of the infrared region of the spectrum.

We can understand this relationship of radiation wavelength to temperature from our everyday experience. If you turn the burner of an electric stove on at full power and hold your hand close to it, you quickly feel the heat that it radiates. You see no immediate change in the metal of the burner, because the burner is radiating all of its energy in long wavelengths in the nonvisible (infrared) part of the electromagnetic spectrum. As the burner heats further, it begins to glow red. At this higher temperature, it is radiating energy at wavelengths in the visible red and near-infrared part of the spectrum. If you could heat the burner even further, it would radiate at progressively shorter wavelengths until it appeared white, like the sun or the filament of a lightbulb.

Electromagnetic radiation from the sun must pass through the atmosphere before reaching the Earth's surface. Similarly, longer-wavelength radiation emitted by the Earth, ocean, and clouds must pass through the atmosphere as it is radiated to space. You can feel the radiated heat by placing your hand near a rock or paved surface immediately after dark in summer. As the longer-wavelength radiation from the Earth passes through the atmosphere, it is partially absorbed by atmospheric gases, but these gases do not absorb all wavelengths of electromagnetic radiation equally. Each gas absorbs some wavelengths more effectively than others.

The **absorption spectrum** of the atmosphere and the gases that contribute the strongest absorption in parts of that spectrum are shown in **Figure CC9.2**. Compare this absorption spectrum with the **emission spectra** of the sun and the Earth (**Fig. CC9.2**). We see that the sun's radiated energy is strongly absorbed by the atmosphere at ultraviolet wavelengths, but there is little absorption in the visible region of the spectrum where the sun's radiant energy is concentrated. In contrast, there are strong absorption bands in the infrared portion of the spectrum in which most of the Earth's radiation is emitted.

The atmosphere does not absorb all radiation that passes through it, even at wavelengths where the atmosphere is shown to have an absorption band (**Fig. CC9.2**). Only a fraction of the energy at such a wavelength is absorbed as it passes through the atmosphere. The rest is transmitted either to the Earth's surface or to space. The fraction absorbed is a function of the number of absorbing molecules of the gas that lie in the radiation's path as it passes through the atmosphere. If the atmospheric concentration of a gas or gases that absorb at a particular wavelength is increased, the fraction of the energy absorbed at this wavelength will increase. Consequently, the fraction transmitted to the Earth's surface or to space is decreased.

Carbon dioxide does not absorb strongly in most of the visible wavelengths that correspond to the sun's energy spectrum (**Fig. CC9.2**). However, it does absorb strongly throughout almost the entire wavelength range of the Earth's radiation. If the carbon dioxide concentration in the atmosphere is increased, there will be little effect on the quantity of solar energy passing through the atmosphere to reach the Earth's surface. However, the quantity of the Earth's radiant energy that is transmitted through the atmosphere to space will significantly decrease. Because less energy will now escape to space but the same amount will be received from the sun, excess energy will be retained in the atmosphere and global climate will warm. This process is known as the "greenhouse

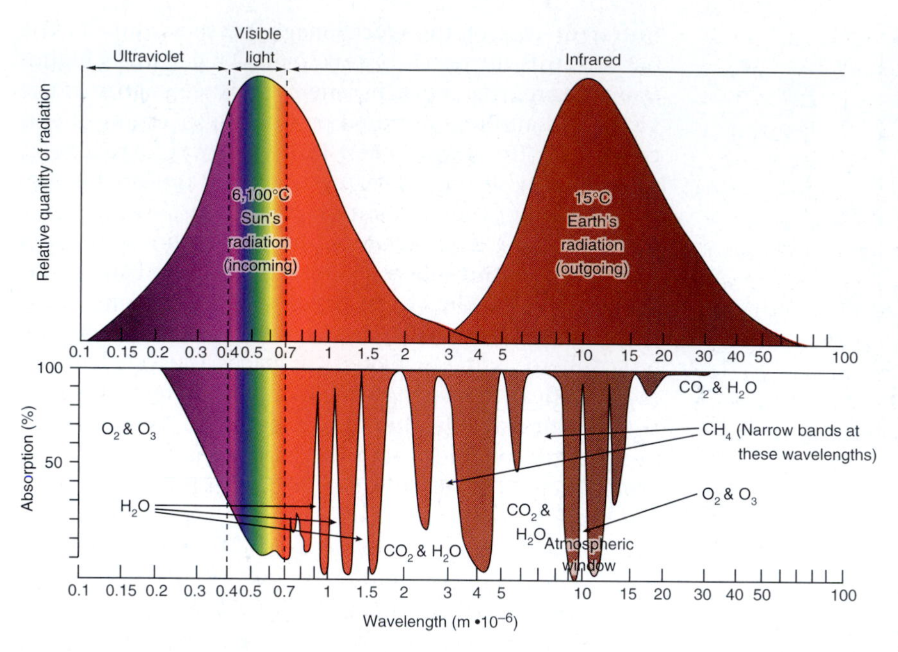

FIGURE CC9.2 The upper part of this figure shows the emission spectra of the sun and the Earth. The lower part shows the absorption spectrum of the Earth's atmosphere. The sun emits most of its energy in the visible, ultraviolet, and near-infrared parts of the spectrum, whereas the much cooler Earth emits radiation at much longer, infrared wavelengths. The amount of energy that enters the atmosphere from the sun equals the amount lost from the Earth by radiation. The sun's radiation is partially absorbed by the Earth's atmosphere, particularly in the ultraviolet region. The Earth's radiation is also partially absorbed by the atmosphere, but at the much longer infrared wavelengths. Carbon dioxide and water vapor strongly absorb radiation at these longer wavelengths, but they do not absorb strongly at most wavelengths of the incoming solar radiation. Other greenhouse gases also absorb radiation primarily in the far-infrared wavelengths.

effect." The name *greenhouse effect* refers to the similarity of this phenomenon to the effects of the glass of a greenhouse window. Glass transmits visible light but absorbs or reflects infrared energy radiated by objects within the greenhouse.

The greenhouse effect is actually misnamed. It should be called the "enhanced greenhouse effect" because carbon dioxide and other gases that occur naturally in the atmosphere already cause a greenhouse effect. The temperature at the Earth's surface would be much lower than it is now if there were no natural greenhouse effect.

We focus on carbon dioxide as a greenhouse gas because its concentration in the atmosphere increased about 35% between 1800 and 2005, and is now increasing at a faster rate each year (**Fig. 1.1**). This increase is the result of burning fossil fuels and destroying forests (vegetation uses atmospheric carbon dioxide and releases oxygen). Several other greenhouse gases may also be important, including methane, **chlorofluorocarbons** (freon and similar compounds, often called "CFCs"), nitrogen oxides, and **ozone.** Each of these gases preferentially absorbs the Earth's long-wavelength radiation more than the sun's shorter-wavelength radiation. In addition, each of these gases is released to the atmosphere in substantial quantities as a result of various human activities, including fossil fuel burning, industry, and farming. Although the quantities of carbon dioxide released to the atmosphere are much larger than the quantities of these other gases released, the other gases absorb long-wavelength radiation more effectively. Therefore, global climate change studies must also take into account these other gases.

We know that the carbon dioxide concentration in the atmosphere has increased in the past two centuries. The concentrations of other greenhouse gases have undoubtedly also increased, although we do not know by how much. Because more of the Earth's radiated heat energy is trapped and prevented from escaping to space while the amount of solar radiation reaching the Earth's surface has not been substantially changed, the Earth's atmosphere should be heating up. This is why the expected greenhouse effect has been called the "global warming problem." However, the atmospheric heat balance is more complicated than the simple input and output of radiated energy. Many other processes can affect the balance between heat absorption and heat loss by the atmosphere, and most of these processes in turn can be affected by changes in atmospheric temperature or chemistry.

These secondary effects can either add to the warming that would be expected from the greenhouse effect or act in the opposite direction and cause a tendency for the Earth's atmosphere to cool. Hence, they provide what are known as positive or negative **feedbacks** (often called "feedback loops") to the primary greenhouse effect. Positive feedbacks would cause faster atmospheric warming than would occur if just the simple enhanced greenhouse effect were operating. Negative feedbacks would cause slower atmospheric warming. If negative feedbacks were great enough, the net effect of the release of greenhouse gases could actually be a colder climate.

Most models of the complicated atmospheric heat balance predict that negative feedbacks will not dominate, and the net result of the enhanced greenhouse effect will be global warming. This conclusion is supported by the very good correlation in the Earth's past between high atmospheric carbon dioxide concentrations and high average climatic temperatures (**Fig. CC9.1**). However, there is considerable scientific uncertainty about the many feedback mechanisms, their magnitude, and how quickly they will respond to the extremely rapid (in geological time) increase of carbon dioxide concentrations in the atmosphere. Although most scientists believe that the release of greenhouse gases will cause global warming (estimated by models to be an increase of 2°C to 6°C during the next century), this outcome is not certain. What is certain is that release of greenhouse gases has altered the atmospheric heat balance and that the alterations will alter the Earth's climate. For this reason, scientists now refer to the issue not as the "greenhouse effect" or "global warming," but instead as "global climate change."

To see the complexity of the feedback mechanisms involving ocean–atmosphere interactions, we can briefly examine two such mechanisms, neither of which is currently well understood. First, consider a positive feedback. The oceans contribute the greenhouse gas methane to the atmosphere (Chap. 6). The methane is a by-product of the biological activity in the oceans. The rate of production of methane and its release to the atmosphere could increase if biological productivity in the oceans increased. About one-half of all the carbon dioxide released by human fossil fuel burning has now become dissolved in the oceans. Higher dissolved carbon dioxide concentrations may result in higher rates of **photosynthetic** plant growth and production in the oceans, which may lead to an increase in methane release from the oceans. Because methane is a greenhouse gas, it would tend to add to the global warming effect of the atmospheric carbon dioxide releases.

Second, consider a negative feedback. Clouds reflect back to space a fraction of the sunlight that hits their upper surfaces. The extent of the Earth's cloud cover is determined largely by the rate of evaporation of water from the ocean surface. The primary effect of greenhouse gases is to increase atmospheric, and thus ocean surface, temperatures. Higher water temperatures would lead to increased evaporation. Consequently, increasing global greenhouse gases may cause the extent of cloud cover to increase, and more solar radiation may be reflected to space to offset the additional heat trapped in the atmosphere by the greenhouse gases. This negative feedback is a particularly good example of the complexity of the climate change question, because additional clouds will not necessarily result in a greater reflection of solar heat to space. Certain types of clouds at different levels in the atmosphere are better reflectors than others. In addition, clouds not only reflect but also absorb incoming solar energy, and they both absorb and reflect the radiation outgoing from the Earth. Each of these factors may vary continuously as the rate and location of evaporation at the ocean surface and the ocean and atmospheric circulation change continuously in response to each other.

Of the many negative and positive feedbacks, some may respond almost immediately to carbon dioxide concentration changes in the atmosphere, and others may take centuries or millennia to respond to such changes. A number of the more important feedbacks involve interactions between the atmosphere and the oceans, and they are subject to intensive research because they must be understood if we are to model and predict global climate changes.

None of the mathematical models that were used up to 1991 were able to include any of the relevant interactions between the atmosphere and oceans. The reasons were that too little was known about these interactions to include them in a meaningful way, and even the supercomputers that were available would have been overwhelmed by the additional computations needed to include them. As computational power and knowledge of ocean–atmosphere interactions improve, and as these feedbacks are included in the global climate models, the models may dramatically change predictions of the climate changes that we can expect as a result of the enhanced greenhouse effect. Nevertheless, it will be many years, or even decades, before these complex processes will be understood well enough that we can reliably predict the future of the Earth's climate.

Fortunately, evidence is strong that the presence of many feedbacks within a complex system tends to produce stable situations that are difficult to change. The feedbacks generally tend to return the system toward a stable configuration. However, most of the feedbacks are nonlinear responses (in other words, the response does not vary in exact proportion to the change in the stimulant). Hence, the ocean–atmosphere is a **chaotic** system (**CC11**).

One important characteristic of chaotic systems is that they tend to remain relatively stable when disturbed, until the disturbance passes a critical point at which the system suddenly shifts to a completely different, but also relatively stable, state. Possibly the beginnings and endings of ice ages are driven by such chaotic dynamics. Could it be that enhancement of the greenhouse effect will not significantly affect the Earth's climate until a critical point is reached at which the climate will change abruptly to something as different as an ice age? Only decades of research or time can provide an answer.

The consequences of global climate change caused by the release of greenhouse gases are unknown and difficult to predict. However, they are likely to be severe. For example, one effect might be thermal expansion of ocean water as its temperature increases. The resultant sea-level rise would inundate many coastal cities in low-lying areas, such as Florida. Partial melting of the polar ice sheets might also contribute to sea-level rise. In addition, rainfall patterns might be changed. The bands of rainfall that now sustain the breadbaskets of the United States and Europe (see Chapter 9) might move northward, which would cause these regions to become deserts like those of North Africa. Because **hurricanes** are fueled by the heat energy of ocean water, another effect of a warmer atmosphere and ocean might be an increase in the frequency and intensity of hurricanes. Hurricanes might also be able to sustain damaging winds while traveling farther away from the equator than they do at present.

CRITICAL CONCEPT 10

Modeling

ESSENTIAL TO KNOW

- The Earth can be considered a system in which things that can range from the planet to atoms interact with each other according to physical laws.
- To understand environmental processes, we must create a conceptual or mathematical model of the system (or parts of the system) that describes the components of the system and their interactions. The parts are often called "cells," "boxes," or "compartments."
- There are many types of interaction. Examples are heat transfer between ocean and atmosphere, and the effect of changes in food supply on the population of an animal species.
- Conceptualized models must be parameterized and tested. Parameterization includes applying data to each model cell and developing equations to describe the flow of the modeled parameter between each adjacent pair of cells. Testing consists of running the model to see how it fits measured data. Parameters are adjusted between tests until the model fits the data set.
- Models must be validated by being applied to a data set that was not used in testing the model. This is often difficult to do for models for which the environmental data sets are sparse.
- Sensitivity analysis is performed by changing model inputs or assumptions and observing the changes in output. This type of analysis can help identify the parameters or processes that most affect the system's behavior and, thus, are the most important for field studies or monitoring.
- To facilitate model creation, interactions are often assumed to be linear. Interactions are linear if one component of a system varies in direct proportion to a change in the other. However, most, if not all, natural systems include many nonlinear interactions.
- The nonlinear nature of natural systems limits the degree to which models can be used to make accurate predictions of future changes in the modeled system. Models of systems that are nonlinear in nature can be used only to predict a range of likely future conditions, and these predictions become more uncertain with increasing time into the future.

UNDERSTANDING THE CONCEPT

The Earth, oceans, atmosphere, and the **biota** that inhabit these **environments** are all intimately linked in a system governed by physical laws. Each atom interacts with many others according to these laws, and it is the sum of these interactions and the atoms themselves that constitute the system. No matter how much information we obtain or how many computers we construct and program, describing this system accurately in all its details is impossible. Therefore, we must create a highly simplified system that replicates the real system in as much detail and as accurately as possible. We can create this model system either conceptually in our minds or mathematically by listing the components of the system and the mathematical relationships between each connected pair of these components.

Models that represent the entire Earth can be constructed at only a very gross level of detail. Hence, we generally construct models of parts of the system (subsystems) that we believe to be somewhat distinct from the larger system. We then represent the rest of the system by inputs and outputs to the subsystem. The descriptions of **plate tectonics,** atmospheric and ocean circulation, and marine biology, as well as other discussions in this textbook, are conceptual models of the real world.

Many different types of interactions must be described to construct models of the environment. Each model must take into account different interactions depending on which subsystem is being modeled. For example, models of atmospheric circulation must include a representation of the processes of heat transfer between ocean and atmosphere, land surface and atmosphere, and sea ice and atmosphere, and a representation of the **convective** processes due to differential heating of air masses at different locations. Similarly, models of the changes of fish populations over time must represent the effects of varying food supply and varying abundance of predatory **species** on the growth, survival, and reproduction rates of the species involved. Models, including the two just mentioned, can be much simplified or extremely complicated, especially if they represent many different processes within a subsystem.

One example of a highly simplified model, and of how it can be useful, is the model describing the possible effects of changing the rate of input of a **contaminant** to an **estuary** or other body of water that is illustrated in **Figures CC8.1 and CC8.2**. At the other extreme, in coupled ocean-atmosphere global **climate** models, which are used to investigate, among other things, the likely consequences of the **greenhouse effect**, the ocean is typically segmented into "cells" that are formed by division of the ocean depths into 20 to 40 horizontal layers, or slices, and subdivision of those slices into segments of

2° to 4° of **latitude** and **longitude.** The atmosphere is typically represented by up to 18 horizontal levels subdivided at 3° to 4° intervals of latitude and longitude. The result is models that can have several billions or tens of billions of adjoining cells. For each interface between two cells, the model must have equations, based either on observed data or on physical laws, that reflect the rate of transport of water, heat, water vapor, and other parameters between the two cells as the properties of the water and air in each cell change.

These massive global climate models can be run on only the fastest computers, now operating at 100 to 200 teraflops (a teraflop is a measure of computer speed and is equivalent to a trillion floating point operations per second). Even with these fast computers, running a model through as few as 100 years of climate can take a month or more of computer time, and the output from such a model run can consist of tens of terabytes of data. A terabyte is 1024 gigabytes, or 1,099,511,627,776 (more than a million million) bytes.

Development and application of models involve a number of fundamental steps. The first step is to conceptualize the system under study as a series of separate but interacting cells (often called "compartments" or "boxes"). Most often these cells are **spatial** areas, but they can be, for example, individual species within an **ecosystem.** The diagrams of the phosphorus and nitrogen cycles in **Figures 14.5 and 14.6** are models in which the cells are both spatial (positioned within the atmosphere, **mixed layer**, or deep layer) and representative of the different chemical forms (e.g., dissolved ammonia, nitrate, nitrite, dissolved organic nitrogen, nitrogen in animal tissues, nitrogen in **bacteria** and **detritus**). **Figures 14.9, 14.10, and 14.11** are models in which the cells are individual species in a **food chain** or **food web.**

In developing a conceptual model, it is important to know exactly what question is being asked, so that the model can be as simple as possible. For example, if we were investigating the fate of **anthropogenic** nitrogen compounds released to the atmosphere, the model in **Figure 14.6** would be adequate if we were interested only in how much of the nitrogen was transported to the deep layer and how fast. However, a much more complicated (and computer-intensive) model would be needed if we were interested in how the nitrogen released from North America was distributed among the oceans.

The second step in the modeling process is parameterization. In this process, values of the modeled parameter (e.g., nitrogen concentration, **biomass,** or heat content) must be assigned to each cell, and equations must be written to express the flows of the parameter between each pair of adjacent cells. These flow equations can be complex if they are based on physical principles. For example, heat transfer between two ocean water cells is a combination of heat **conduction, diffusion** and mixing across the boundary, and water transport between the cells. Usually a single equation is written for such flows on the basis of observed data. Often the observed data are very limited or available only for a small number of the cells, particularly in complex models such as global climate models. The model is then tested by being run with the parameters identified to see if it reproduces the data. For example, the global climate model would be examined for how closely it reproduces changes in temperature data at locations for which real-world data are available. Next, the parameters of the model are adjusted to improve the match to the real-world data. The model may be tested and parameters adjusted in this way many times.

The third step in model development is validation to determine whether the model actually does represent the way the natural system functions. In this step, the model must be tested with a data set that is completely different from the data used for the parameterization step. This is important because the process of parameterization forces the model to fit the input data. Thus, the model parameters may represent only the input set of conditions rather than the underlying processes involved.

Validating some models is relatively easy. For example, global climate models can be parameterized with decades of climate data and then validated by the use of data from a completely different time period. This is possible because relatively detailed historical atmospheric and ocean climate data are available for a number of decades in the past. Validating other models, especially complex ecological models, is much more difficult because the amount and historical coverage of the existing data are often very limited and parameterization requires using most or all of the available data. Models that cannot be properly validated may still be useful for many research studies, especially sensitivity analysis, which is described in the next paragraph, but they are inherently unreliable if used to extrapolate beyond the existing data set, and they cannot be relied upon for prediction.

The most familiar use of models is to predict future conditions, particularly **weather** and climate. However, the most important use of models is to determine the parameters or processes that most affect the system's behavior and, thus, are the most important for field studies or monitoring. To do this, we subject the models to sensitivity analysis. Sensitivity analysis is a simple, albeit sometimes tedious, process of repeatedly running the model, each time making small changes in a different input parameter or combination of input parameters (the initial value in a cell, or the coefficients of the equations relating flow between cells). This analysis can reveal which input parameters produce substantial changes in the model outputs and which produce only much smaller changes. The former are sensitive parameters that must be accurately determined (or formulated for a flow equation) because small errors will greatly affect the accuracy of the model's representation of the real world. They are also probably the most important parameters in controlling change in the real-world system that is being modeled. As a result, the most sensitive parameters of a model are often the most important parameters to focus on in field research and monitoring programs.

Models that have been adequately tested and validated can be used as predictive tools. The best-known examples of such uses of models are the weather predictions seen nightly on television, the predictions of the future paths of **hurricanes,** long-term (months) climate predictions that are of particular importance to farmers, and predictions of future climate change associated with the past and future releases of greenhouse gases. These predictions are all based on weather and climate models that have been parameterized, tested, and validated with extensive historical data. However, as we all know, the predictions are always somewhat general in nature and are often inaccurate.

For example, as hurricanes travel across the Caribbean, several days away from the United States, the model predictions of the storm's path are shown as a wedge with the future position more and more uncertain the further out in time the prediction is. The predicted range of possible landfalls of the storm is narrowed as the storm approaches, but the predicted impact point often shifts and is no longer even within the range of the impact areas predicted several days earlier. In addition, predictions of strengthening or weakening of the storm as it approaches over the ocean are often incorrect, and sometimes the storm intensity suddenly and unexpectedly increases or decreases. Similar observations could be made about other model predictions.

Why are models not better predictive tools? There are several reasons. First, the models, although very sophisticated, are not the real world, and they do not account for variability within the cells of the model. Cells often represent a large segment of the oceans or atmosphere, within which there is heterogeneity and variability that is not taken into account in the model. Second, the equations describing flow between cells are "fitted" to the testing and validation data and do not precisely describe all the factors that can affect these flows. Third, and this is the most important reason, natural systems include many nonlinear processes, and thus, these systems are **chaotic,** as discussed in more detail in **CC11**.

Even the most sophisticated models cannot make detailed and accurate long-term predictions for chaotic systems. In these systems, models will at best be able to accurately predict a range of likely future conditions. This is indeed exactly what hurricane path predictions do. These predictions are generally quite accurate for approximately a 24-h period ahead of the prediction, but they become progressively less accurate as the predictions extend out in time. Similarly, weather predictions are often quite accurate up to about 24 h ahead, but they become much less accurate farther out in time. In fact, there is a theoretical limit to future predictions in chaotic systems. Because this limit is only several days for weather predictions, these predictions will never be accurate for more than a few days into the future.

CRITICAL CONCEPT 11

Chaos

ESSENTIAL TO KNOW

- Some interactions in natural systems are linear (or almost linear over the range of values present in the environment), but many are nonlinear because the components do not vary in direct proportion to each other. For example, doubling the food supply may cause the population of an animal species to double if food is scarce, but it may result in little change in population if food is already abundant.
- Linear systems are predictable. If we increase one parameter (e.g., the available food supply), the system will adjust to equilibrium (the population will rise to a new stable value that is related in direct proportion to the new food supply).
- Nonlinear systems behave unpredictably. Changing one parameter may lead to a new equilibrium (which may be drastically different from the original system) or may cause the system to oscillate in a chaotic way.
- Complex nonlinear systems with many interactions are chaotic and unpredictable, but they tend to oscillate irregularly within an identifiable range of conditions. For example, weather is inherently unpredictable, but climate (range and average conditions) does not generally change from year to year.
- Complex nonlinear systems may respond to unusual changes in one or more parameters by "jumping" to oscillate around a different set of average conditions.
- Natural systems include many interactions, at least some of which are nonlinear. The Earth, atmosphere, and ocean processes are therefore chaotic.
- Because natural systems are chaotic, their exact state cannot ever be forecast (predicted) accurately. These systems are inherently variable, and they may undergo sudden changes to new average conditions.

UNDERSTANDING THE CONCEPT

In most cases, we conceptualize the world around us as behaving according to simple rules. For example, it seems reasonable that if we doubled the temperature difference between ocean water and the overlying air, we would find that the rate of heat transfer between the two also doubled; or if we doubled the rate of food production, we would find that the animal **species** eating this food doubled in population. These are linear relationships in which one parameter changes in direct proportion to another. Relationships need not be one to one (as just described) to be linear. The only requirement is that a certain amount of change in one parameter will always result in the change of another parameter by the same amount, even if one parameter increases and the other decreases (**Fig. CC11.1**).

Linear relationships are simple to conceptualize and to express mathematically. They are also intuitively attractive. For example, the hypothesized linear effect of changing food production rate on a consumer population seems perfectly reasonable at first. However, it does not represent the true behavior of the systems. To understand why, think about what would happen if we were to drastically increase or decrease the food supply for an animal species that has no predators. If we drastically increased the food supply, the population would increase, but at some level the species would become so crowded that some other factor (e.g., breathing space) would prevent it from increasing further. If we drastically decreased the food supply, the population would decline. Eventually it would reach such a low level that the species would not be able to reproduce successfully because, for example, the survivors would have to spread so far apart to obtain sufficient food that there would be too few available mates. Because the hypothesized relationships do not maintain proportionality if the value of one parameter is changed too much, they are nonlinear.

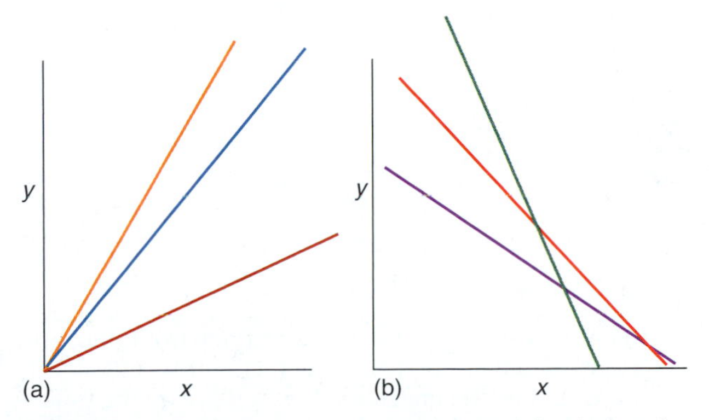

FIGURE CC11.1 Linear relationships can be depicted by the straight line on a graph that plots one variable (x) against another (y). The value of one variable may (a) increase with an increase in the other variable, or (b) decrease with an increase in the other variable.

Nonlinear relationships are depicted in **Fig. CC11.2**. Many, if not all, of the relationships that control the real world are nonlinear, and this fact critically affects our ability to model the real world. Fortunately, many relationships, such as the one just discussed, may be linear or nearly so over a wide range of values of the two parameters (**Fig. CC11.2**). For instance, the food–population relationship discussed earlier may be nearly linear within the range of food production rates that are observed in the real world. In these circumstances, we can use linear relationships in our models and the models will be similar enough to the real world to be useful.

Not until the 1970s did scientists begin to examine the behavior of mathematical models that incorporate nonlinear relationships. What they found was startling. Models that use only linear relationships, no matter how complicated, always reach a new equilibrium if one of the parameters is changed (unless parameters are constantly changing). However, even the simplest of nonlinear models involving only two parameters may never reach an equilibrium. The system may oscillate in a seemingly random, or **chaotic,** way, never exactly repeating itself.

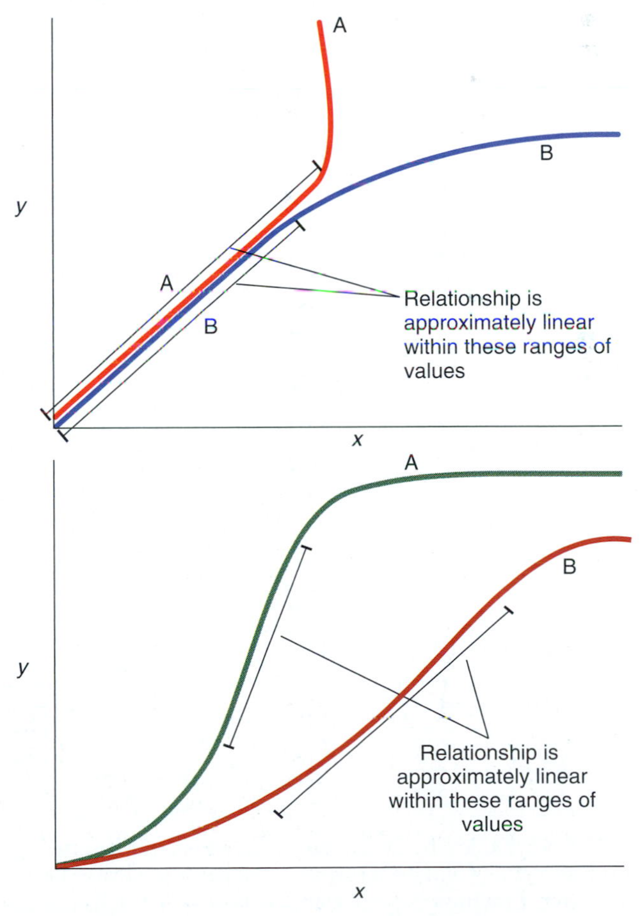

FIGURE CC11.2 Nonlinear relationships can be identified by the curvature of a graph that plots one variable (x) against another (y). Although the curvature can be complex, there is often a range of the two values within which the relationship is approximately linear.

The mathematics of nonlinear systems can be complicated. Fortunately, we can understand chaotic behavior by considering a simple system called the "Lorenzian waterwheel," named for Edward Lorenz, who first investigated the chaotic behavior of this system (**Fig. CC11.3**). In this waterwheel, the buckets are pivot-mounted so that they always stay upright (like the chairs in a Ferris wheel). Each bucket has a small hole through which water can escape (but not drip into a bucket below). This system can be modeled by three simple but nonlinear equations, and it has also been constructed and tested as an actual physical model. If we pour water into the top bucket, the waterwheel's behavior depends on how fast the water is poured, how fast each bucket empties, and how quickly the wheel is turning (which determines how long each bucket is in the right part of the rotation to be filled).

What do you think happens if we pour water into the Lorenzian waterwheel at different rates but leave everything else unchanged? If we pour the water in at a very slow rate, we might expect it to escape from the top bucket fast enough that the bucket does not fill and the waterwheel will not turn. This is indeed what happens. If we pour the water faster, the top bucket will partially fill and set the wheel in motion, bringing the next bucket under the pouring point to fill in turn and continue the rotation (**Fig. CC11.3b**). As each bucket moves through the rotation, it will progressively empty. We might expect that, if we maintain a stable rate of filling, the rotation will be smooth and the rate of rotation will increase as the fill rate increases. This is what happens when the fill rate is relatively slow. However, at faster fill rates something interesting happens.

Consider the wheel rotating relatively slowly. One or more successive buckets fills almost to the top as the wheel rotates. Because these buckets are heavy, they accelerate the rotation as they move toward the bottom point of the wheel, causing buckets behind them to fill for a shorter time. The full buckets can pass through the low point in the wheel before they have emptied, whereas the following buckets are filled less and empty completely. Once past the low point in the rotation, a once-full bucket (now partially full) may be heavier than the once partially filled buckets (now empty) behind it, so it now tends to slow the rotation of the wheel (**Fig. CC11.3c**).

We can see that the rotation rate of the wheel will vary constantly and the direction of rotation may even

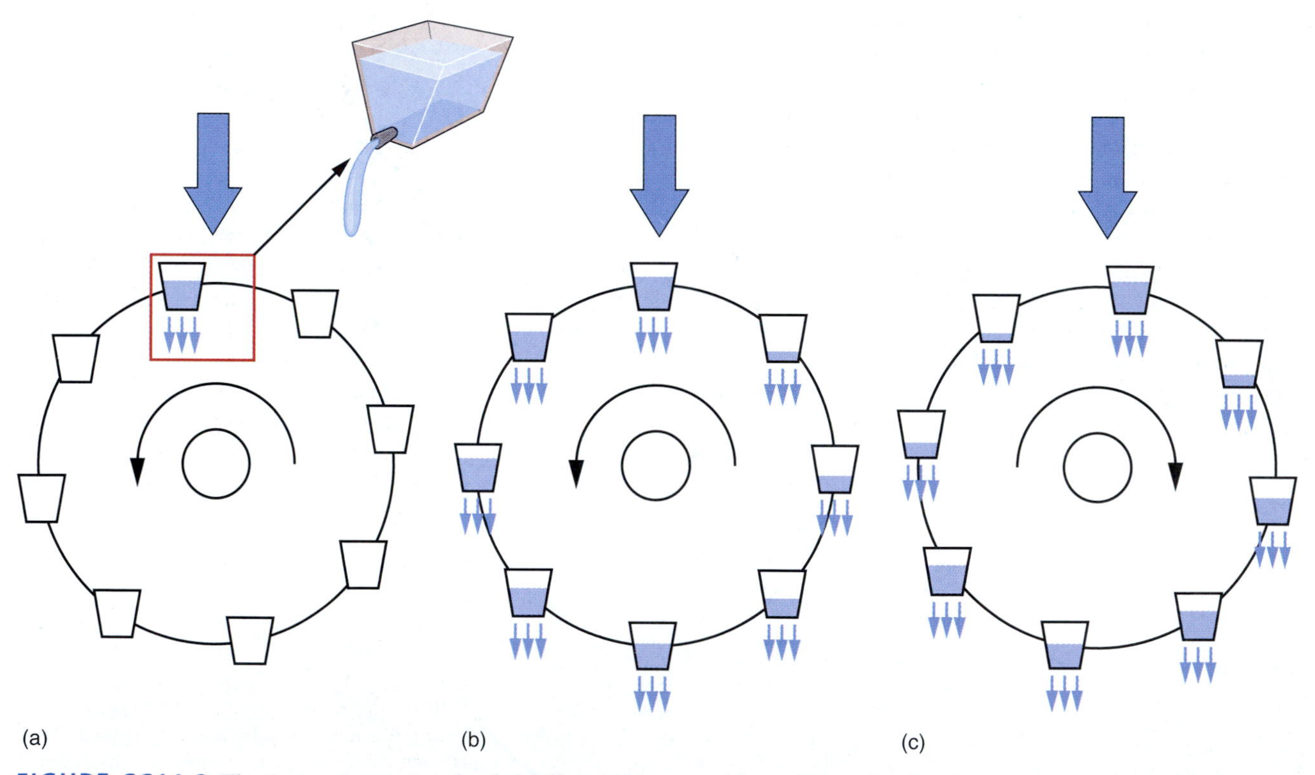

FIGURE CC11.3 The Lorenzian waterwheel. (a) If water is poured into the top bucket faster than it can escape from the bucket, the wheel will be set in motion. (b) If the water flow to the buckets is maintained, the wheel may settle into an equilibrium in which it rotates at constant speed. This can happen when, for example, the buckets are just empty by the time they reach the top and are refilled. (c) The acceleration and speed of the wheel depend on the amount of water in each bucket, which depends on the length of time that each bucket spends under the fill point, which depends on the speed of rotation of the wheel. These are nonlinear relationships, and the wheel may never settle into a constant rotation rate, even if the water input rate remains constant. Furthermore, the rotation rate may vary constantly, and the direction of rotation may even be reversed.

reverse. However, what is most surprising about the motion is that, even though we do not vary the fill rate, the wheel's rotation does not, as we might intuitively expect, settle into a regular back-and-forth oscillation. Instead, the rotation varies continuously back and forth in a pattern that is never repeated, a "chaotic" situation. This happens because of the nonlinearity in relationships, such as the relationship between the rate of rotation and the amount of filling of each bucket.

Because natural systems involve many nonlinear relationships, they can act like the Lorenzian waterwheel: sometimes they can appear stable or change smoothly to a new equilibrium if one of their important parameters changes, or they can oscillate chaotically. For example, a **convection cell** is similar to the waterwheel, although inverted. If heat is supplied very slowly to the lower part of a fluid and lost slowly from the top, the heat can be transferred by **conduction** within the fluid and there is no movement. At somewhat higher heating or cooling rates, smoothly flowing convection cells may form (**CC3**), and at even higher rates the convection cell motion becomes **turbulent** and chaotic.

Even seemingly simple biological systems can exhibit this type of behavior. For example, as discussed previously, populations respond in a nonlinear way to changes in food availability. Population increases as food is increased, but as population increases, other factors eventually limit and reduce the population. Depending on the nature of these relationships, an increase in food supply can lead to a new, higher population of the consumer species, chaotic variations of this population, or even collapse and extinction if, for example, a small increase in population drastically reduces reproduction rates because of overcrowding.

Natural systems generally depend on many interactions, some of which are nonlinear. As a result, most natural systems are chaotic, and this fact has many implications for science. Since the 1970s, a new scientific subdiscipline has developed to study chaos. The term *chaos* has now generally been replaced by *complexity*. *Complexity* is often the preferred term because certain characteristics of complex nonlinear systems can be deduced that enable us to understand their behavior and how their component parameters might vary under certain circumstances, even though the precise future status of the system cannot be predicted. We do not need to fully understand the behavior of nonlinear systems unless we become research scientists, but several characteristics of these systems are important to all of us.

Nonlinear system models are extremely sensitive to very (infinitely) small differences in initial conditions. This characteristic was discovered by Edward Lorenz when he was using a computer model of **weather** patterns, and these observations are what started the systematic study of chaos. Lorenz found that if he started his mathematical model several times from the same set of input parameters (e.g., temperature and **humidity** distribution), the predicted weather exhibited variations that, although complicated, were reproduced exactly each time the model was run.

Lorenz then decided to start his model at a time interval later than the starting point of his previous model runs. He started this new set of calculations with the values of input parameters that his previous model runs had predicted would occur at this time interval. Unexpectedly, he found that his model quickly began to predict weather patterns that differed from those that the model had predicted for this same time period when the model had been started at the earlier time (**Fig. CC11.4**). The reason for this difference, Lorenz found, was that the nonlinear equations in his model generated completely different weather patterns when the later-start model run used input data that were rounded from the values produced in the earlier-start model runs. When he ran his model from the beginning, the computer maintained the values of each parameter at each time interval to six decimal places (e.g., 0.267902). When he started his model in the middle, he rounded the input data figures he used to three decimal places (e.g., 0.268).

This sensitive dependence on initial conditions has often been called the "butterfly effect." This term originated from the example that a butterfly stirring the air one day in Beijing could alter the storm systems in New

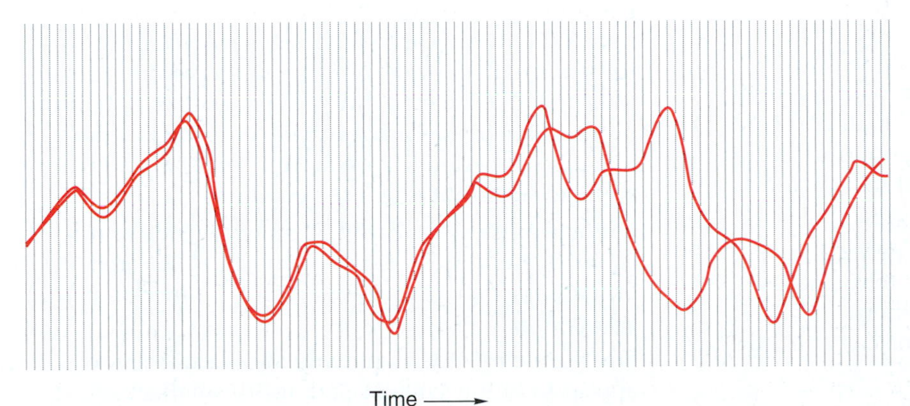

FIGURE CC11.4 These two plot lines depict the results obtained from the same nonlinear mathematical model that simulates weather patterns. The only difference between the two model runs is that the starting-point data were rounded to a smaller number of decimal places for one model run than for the other. The patterns grow farther and farther apart until they lose any resemblance to each other. Many natural systems show this sensitive dependence on initial conditions. As a result, it is impossible to measure and model such systems well enough to predict their precise future behaviors.

York the next month. As a consequence of the butterfly effect, models of nonlinear systems will never be able to predict future behavior exactly. In addition, the farther in the future that a model predicts, the less likely it is to be accurate. For example, weather forecasts will never be much better than they are today. We will never be able to predict successfully the exact time or place a thunderstorm will occur on the next day, nor predict with certainty whether it will rain a few days in the future. Similarly, we will never be able to forecast or predict the exact future status of any complex nonlinear system, including **plate tectonic** movements, **ecosystem** dynamics, or fish and wildlife population variations.

Fortunately, although many complex nonlinear systems never reach an equilibrium, they do tend to oscillate chaotically within a range of conditions (values of the component parameters) that is predictable. For example, weather is chaotic and unpredictable, but **climate** (the average and range of temperature, rainfall, etc., at a specific location and time of year) can be predicted with some confidence because it changes little from year to year. Hence, models of complex systems (**CC10**), if they adequately resemble the real world, can be used with reasonable success to predict the average conditions and range of variations that will occur in the future.

Unfortunately, although complex nonlinear systems do tend to oscillate chaotically within a range of conditions that is predictable, this dynamic "equilibrium" can be disturbed by small changes in the component characteristics. In some instances, if a critical value of one or more components is changed, the system can "jump" from one set of average conditions to a completely different set of average conditions around which it oscillates chaotically. Thus, a system may appear to suddenly change drastically, even though none of the components of the system changed substantially before the "jump" occurred. For example, a very small increase in a parameter, such as the average temperature of the oceans and atmosphere, could have little effect on climate until a critical point was reached. At that point, the Earth's climate could suddenly become much warmer or colder.

We now know that this type of sudden climate change has occurred in the past (Chaps. 9, 10, **Fig. 9.21**). Similarly, sudden changes in ecosystems, such as drastic declines or **blooms** of some species, may be a natural consequence of the nonlinearity of nature. Sudden changes in other complex systems, such as ocean circulation and the motions of tectonic plates, may have occurred in the past for this same reason. It is important to realize that sudden changes in natural systems have occurred in the past and will occur again in the future. Furthermore, no matter how well we are able to model such systems, it will be very difficult to reproduce such changes faithfully, and virtually impossible to develop an accurate predictive capability that will alert us to such changes before they occur.

CRITICAL CONCEPT 12
The Coriolis Effect

ESSENTIAL TO KNOW

- When set in motion, freely moving objects, including air and water masses, move in straight paths while the Earth continues to rotate independently.
- Because freely moving objects are not carried with the Earth as it rotates, they are subject to an apparent deflection called the "Coriolis effect." To an observer rotating with the Earth, freely moving objects that travel in a straight line appear to travel in a curved path on the Earth.
- The Coriolis effect causes an apparent deflection of freely moving objects to the right in the Northern Hemisphere and to the left in the Southern Hemisphere. The deflection is said to be *cum sole*, or "with the sun."
- The Coriolis deflection is greatest at the poles and decreases at lower latitudes. There is no Coriolis effect for objects that move directly east to west or west to east at the equator.
- Regardless of their speed, freely moving objects at the same latitude appear to complete a circle and return to their original location in the same period of time. This period, called the "inertial period," is 12 h at the poles and increases progressively at lower latitudes. The inertial period is 24 h at 30°N or 30°S, and it approaches infinity near the equator.
- For objects moving within 5° on either side of the equator, the Coriolis effect often can be ignored because the deflection is small and the inertial period is very long.
- Freely moving objects moving at the same speed appear to follow circular paths with smaller radii at higher latitudes.

- Freely moving objects at the same latitude appear to follow paths with larger radii if they are moving at higher speeds.
- The magnitude of the Coriolis deflection (the rate of increase of distance from a straight-line path) increases with increasing speed.

UNDERSTANDING THE CONCEPT

From common experience we know that freely moving objects, such as bullets, move in a straight line unless acted on by an external force. However, all is not what it seems. If we very carefully examine the bullet's flight, we find that its path is deflected very slightly to one side (in addition to the downward deflection due to **gravity**). Curiously, this deflection is always to the right in the Northern Hemisphere and the left in the Southern Hemisphere. This apparent deflection, called the **Coriolis effect**, occurs because we live on a rotating Earth, and it occurs for all objects that move freely without connection to the solid Earth surface, including bullets, thrown footballs, and moving air and **water masses**.

Some simplified examples help to explain how the Coriolis deflection works.

Merry-Go-Round without Gravity

Consider a merry-go-round rotating counterclockwise when viewed from above (**Fig. CC12.1a**). This is equivalent to the Earth seen from above the North Pole (**Fig. CC12.1c**). Imagine that you are an observer sitting on the merry-go-round and that someone at the center of the merry-go-round throws a ball outward. As the ball travels through the air in a straight line, the merry-go-round continues to turn, carrying you around with it (**Fig. CC12.1a**). You do not see the ball travel in a straight line. Instead, the ball appears to travel in a curve deflected to the right (**Fig. CC12.1b**). If the merry-go-round spins clockwise (equivalent to viewing the Southern Hemisphere from above the South Pole; **Fig. CC12.1d**), the apparent deflection of the ball is to the left (**Fig. CC12.1e**).

If you sit on the merry-go-round away from its center of rotation and throw a ball from here at a target, the ball again appears to be deflected to the right. However, it misses the target, even though it flies in a straight line. The reason is that the ball was moving with the merry-go-round when it was thrown. While in the observer's hand, the ball is constrained to move in a circle with the merry-go-round, but when released, it is no longer constrained. In fact, if the ball were simply released on the merry-go-round instead of thrown, it would roll off in the direction it was moving at the instant it was released. The same thing occurs when we twirl an object on the end of a piece of string and then let go. The object flies off in a straight line whose direction is determined by where it was in its circle of rotation when it was released (**Fig. CC12.2a**).

If a ball is thrown from a rotating merry-go-round, it is given straight-line motion by the throw. However, because of the rotation of the merry-go-round, it is already in motion in a direction tangential to the ball's circle of rotation at the point it was released. These two straight-line motions (components) are combined, so the ball's actual direction is a straight line between the directions of the two components (**Fig. CC12.2b**). The ball's direction and speed are easily calculated from the two components, which are called "vectors" (**Fig. CC12.2c**).

Centripetal Force

Spinning objects, like the object on the end of a twirled string, fly off if released. This tendency is often called "centrifugal" force. However, centrifugal force does not exist. Circling objects tend to fly off when released because the force restraining them in their circular path is no longer present. A force that acts toward the center of rotation, called **centripetal force,** must be applied to maintain any orbiting object in its circular path. This force may be exerted in several ways. For example, **friction** between the observer and merry-go-round keeps the observer from flying off, whereas tension in the string prevents a twirled object from leaving its circular path. Gravity is the centripetal force that prevents us from flying off the Earth.

Centripetal force must increase as the rate of rotation increases. For example, the amount of friction needed to keep an observer on the merry-go-round becomes greater as the merry-go-round speeds up. The rate of rotation is measured by the angular velocity (the rate of rotation measured in degrees of angle per unit time; **Fig. CC12.3a**). For a constant angular velocity, **orbital velocity** increases at points more distant from the center of rotation (**Fig. CC12.3a**). At a fixed distance from the center of rotation, angular velocity increases if orbital velocity increases, and decreases if orbital velocity decreases (**Fig. CC12.3b**).

Merry-Go-Round with Gravity

Now imagine a merry-go-round with gravity that acts toward a point located below its center of rotation (**Fig. CC12.4a**). This merry-go-round is equivalent to a small circular area of the Earth's surface centered at the North Pole with gravity acting toward the Earth's center (**Fig. CC12.4b**).

If a ball is placed at any point on the merry-go-round other than the center, a small component of the gravitational force pulls it toward the center of rotation (**Fig. CC12.4c**). If the merry-go-round and ball are rotating and the ball is released to roll freely, the ball will move outward, away from the center of rotation, just as the object on the end of a twirled string flies outward if released. However, except at the center, the component of gravity that acts toward the center of rotation (**Fig. CC12.4c**) pulls the ball toward this center. If the ball is at

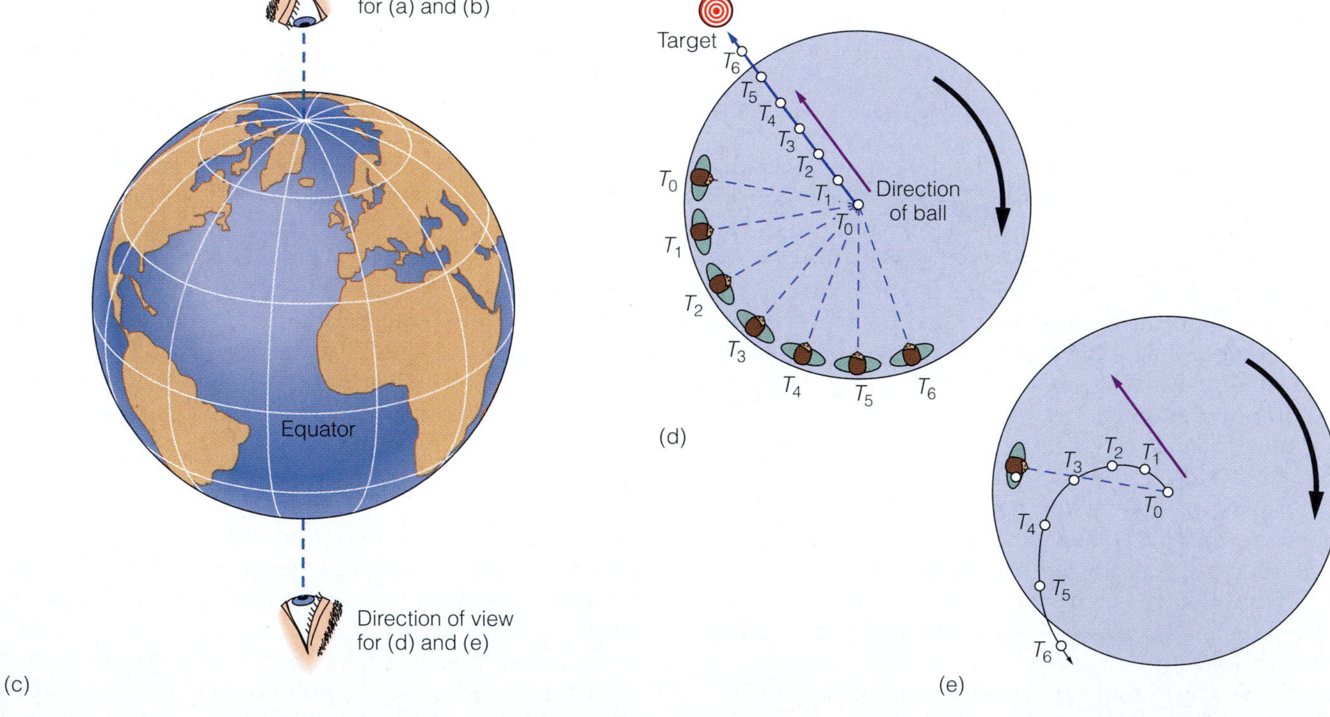

FIGURE CC12.1 (a) A ball thrown outward toward a target from the center of rotation of a merry-go-round has zero orbital velocity due to the merry-go-round rotation, and will travel outward in a straight line to hit the target. However, an observer sitting away from the axis of rotation moves around with the merry-go-round such that the observer's angle of vision to the ball and target change. (b) If the observer is not aware of the merry-go-round rotation (just as we are not aware of the Earth's rotation), the observer will see the ball appear to move in a circular path, curving to the right. To observe this phenomenon, rotate part (a) to place the observer at the observing location in part (b) for each of the time intervals T_1, T_2, and so on. (c) Looking down on the merry-go-round depicted in parts (a) and (b) is equivalent to looking down on the Earth from above the North Pole. If we were to look down from above the South Pole, the merry-go-round would appear to rotate clockwise instead of counterclockwise. (d) If the merry-go-round rotates clockwise (the equivalent of the Southern Hemisphere), the ball still moves in a straight line. However, (e) the apparent deflection is in the opposite direction, and the observer sees the ball curve to the left.

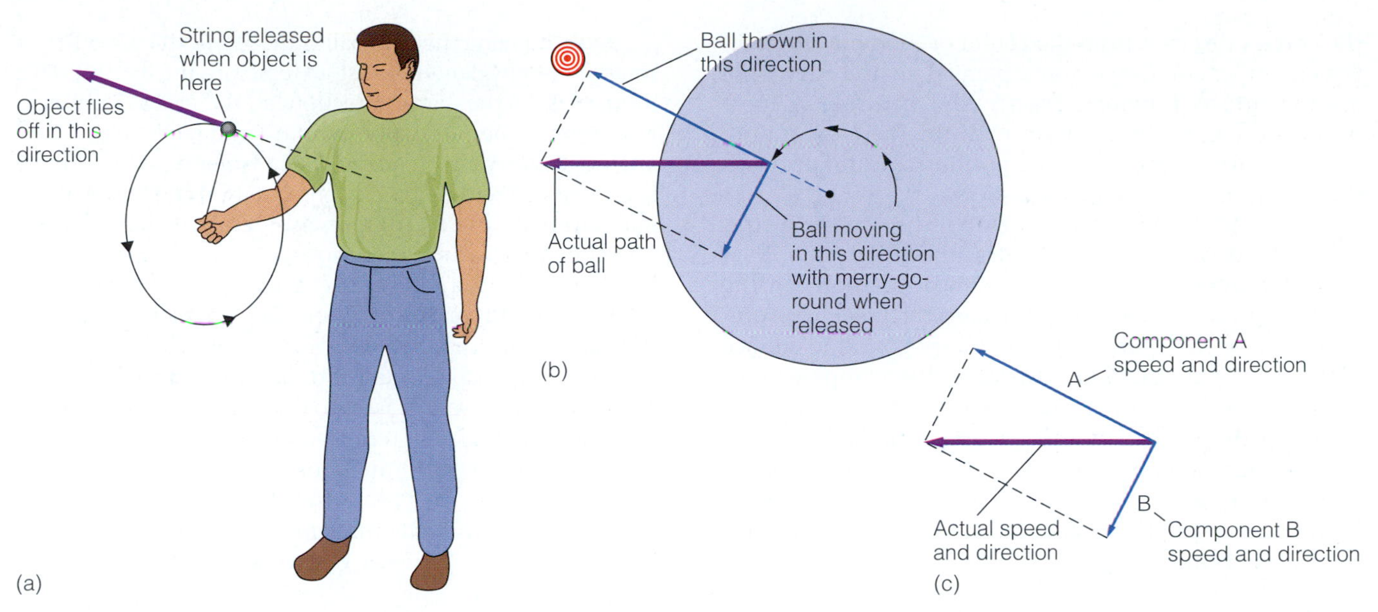

FIGURE CC12.2 (a) A centripetal force is needed to keep an orbiting object in its orbit. When an object is twirled on the end of a string, we supply this force with our muscles through tension in the string. If we release this tension by letting go of the string, the object will fly outward. (b) When a ball is thrown outward from a rotating merry-go-round, it flies off in a straight line (just as the object twirled on the end of a string does), but it also has an additional velocity (speed and direction) imparted to it by the throw. The actual velocity with which the ball moves is a combination of the two velocities. (c) We can easily determine the actual velocity of the ball by geometrically combining the two component velocities. The lines A and B are drawn in the respective directions of the two velocities. The length of each line is proportional to the speed imparted in that direction. The actual direction of the ball's motion can be determined by drawing lines (the dotted lines) parallel and equal to each arrow from the head of the other to form a parallelogram (in this case it is a rectangle). The diagonal of this parallelogram shows the actual direction of motion, and its length shows the speed.

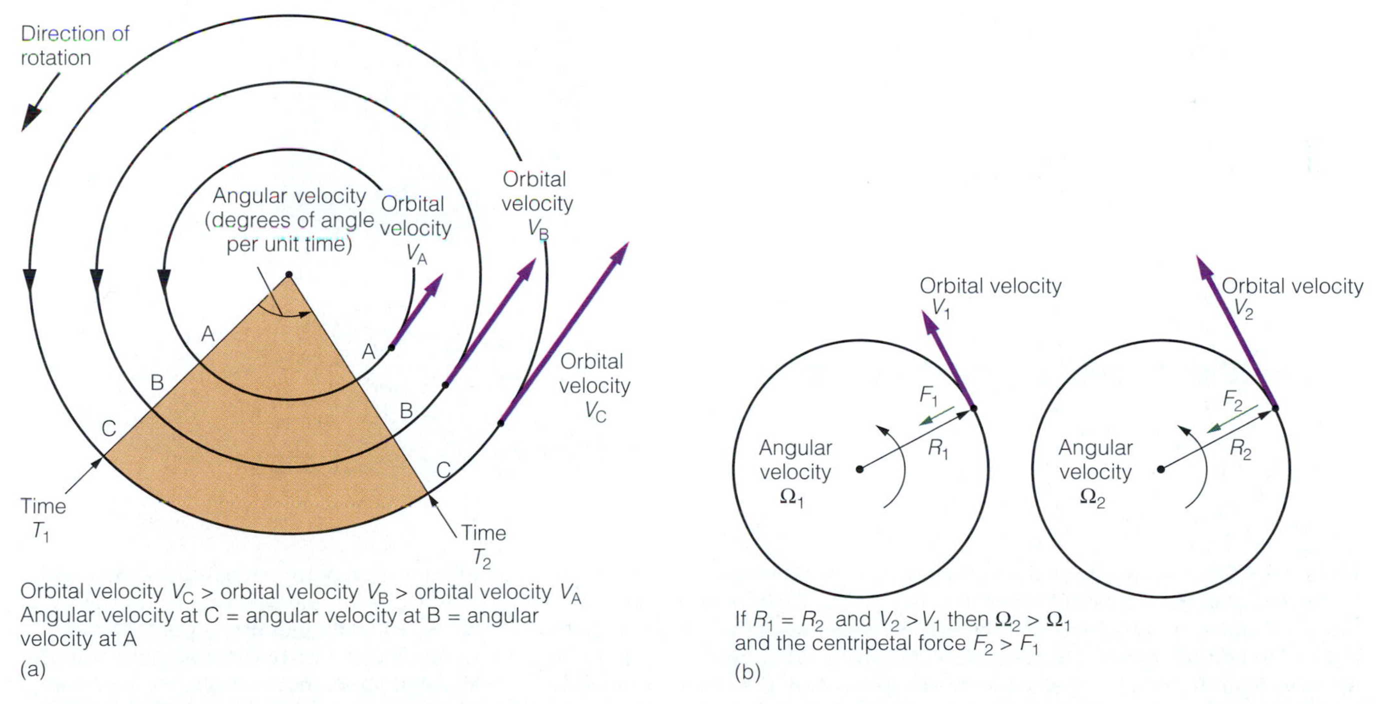

FIGURE CC12.3 The velocity of an orbiting object can be expressed as either angular velocity or orbital velocity. (a) The angular velocity is the number of degrees of angle moved through in unit time. The orbital velocity is the distance the object moves along its orbital path per unit time. On a rotating solid object (like a merry-go-round or the Earth), the orbital velocity of a point increases with distance from the center of rotation, and the angular velocity is the same for all points regardless of location. (b) Angular velocity, orbital velocity and centripetal force are related. For two objects at the same distance from the center of rotation, if the angular velocity (rate of rotation) is increased, the orbital velocity is also increased and a larger centripetal force is needed to maintain the object in orbit. For example, the tension in a string becomes greater as we twirl it and its attached object faster.

the correct distance from the center of rotation, the component of gravitational force parallel to the merry-go-round surface provides exactly the centripetal force needed to balance its rotation with the merry-go-round. Thus, the ball remains in this position and rotates with the merry-go-round (**Fig. CC12.4d**).

If we set the ball in motion with respect to the merry-go-round, this balance will be disturbed. For example, if the ball is moved toward the center of rotation (on the Earth, equivalent to moving it toward the pole), the component of gravity parallel to the merry-go-round surface is reduced (**Fig. CC12.4c**). Because this component of gravity is now less than the centripetal force needed to maintain the ball in its smaller orbit, the ball tends to move back away from the center of rotation. If the ball is moved away from the center of rotation, a small additional gravitational attraction tends to move the ball back toward the center.

Next, imagine that the ball is moved in the direction of rotation (equivalent to west to east on the Earth). This motion is in the same direction as the ball's path as it rotates with the merry-go-round, so the ball's orbital and angular velocity are increased, and a larger centripetal force is needed to keep the ball in this faster orbit. However, the gravitational force component remains unchanged. Therefore, the ball tends to move outward. Similarly, if the ball is moved in a direction exactly opposite that of the merry-go-round (east to west), the angular velocity is decreased and the ball tends to move inward. Thus, motions of the ball directed with or against the direction of rotation (east to west or west to east) are turned away from or toward the center of rotation, respectively.

No matter which direction the ball is started in motion, it will move either toward or away from the center of rotation and this movement will be counteracted by an imbalance between gravity and centripetal force.

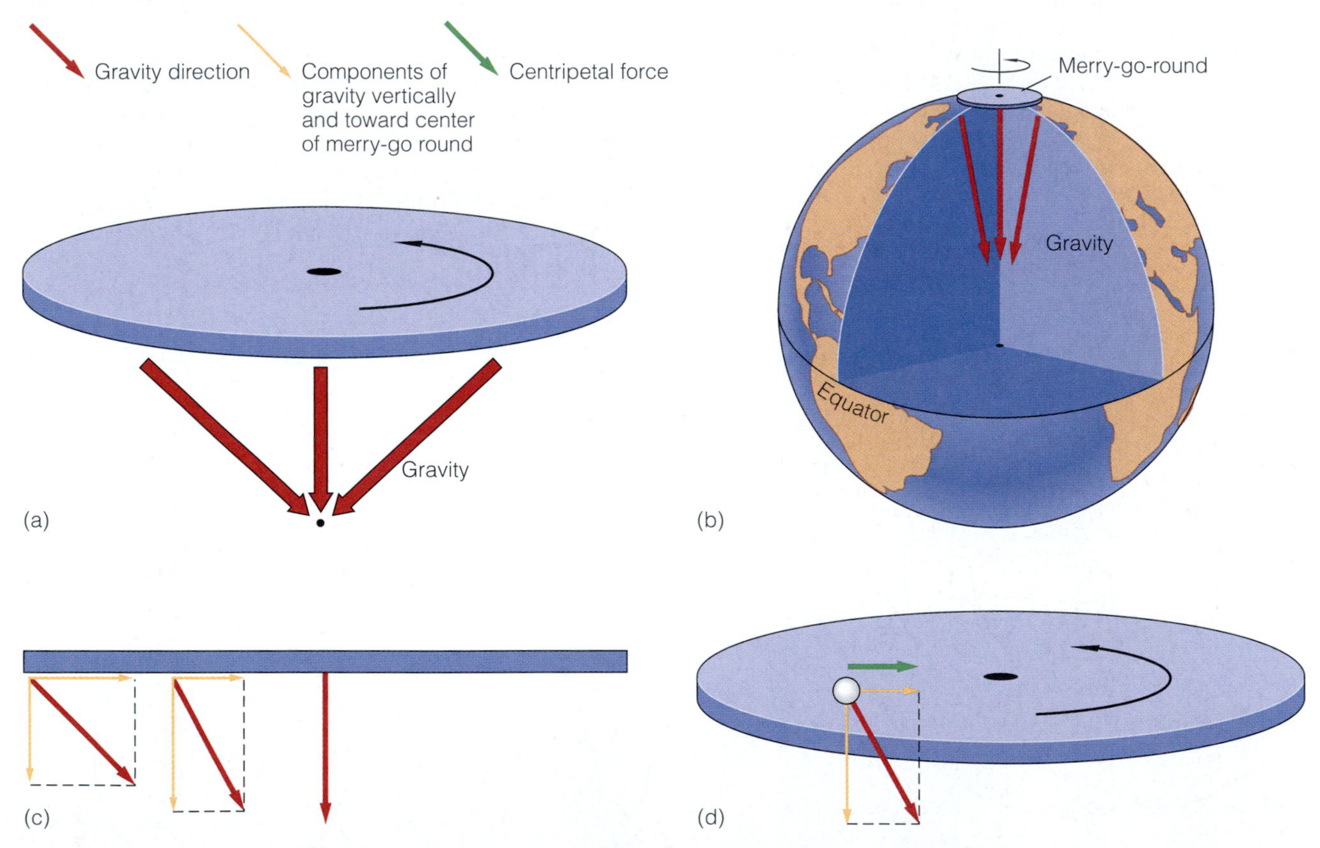

FIGURE CC12.4 (a) A hypothetical merry-go-round with a gravitational force acting toward a point beneath its center. This hypothetical merry-go-round is like a section of the Earth near the pole (b). However, in this example the distance to the center of gravity is greatly reduced to help us understand the effects of changes in the directions of gravity and centripetal force with latitude on the Earth's surface. (c) The gravitational force is the same strength (or very nearly so) at all points on the merry-go-round, but it is directed vertically down only at the center of rotation. At all other locations, it can be resolved into two components (the yellow lines), one parallel to the merry-go-round surface and another vertically downward. The component parallel to the surface increases in strength with increasing distance from the center of rotation. (d) If we place a ball on the surface of the merry-go-round, hold it in place so that it rotates with the merry-go-round, and then release it, it will tend to fly outward like an object twirled on a string. However, there is a component of gravity that acts toward the merry-go-round center. This component provides a centripetal force that, if exactly balanced with the rotation rate of the ball on the merry-go-round (its orbital velocity), will prevent the ball from flying outward.

The counteracting force will grow as the ball moves farther inward or outward. This force will slow the ball's inward or outward motion and eventually reverse its direction. The ball will oscillate toward and away from the center of rotation (**Fig. CC12.5a**) in a motion similar to that of a pendulum (**Fig. CC12.5b**).

As the ball oscillates back and forth, it is still moving in its circular orbit with the merry-go-round. We can

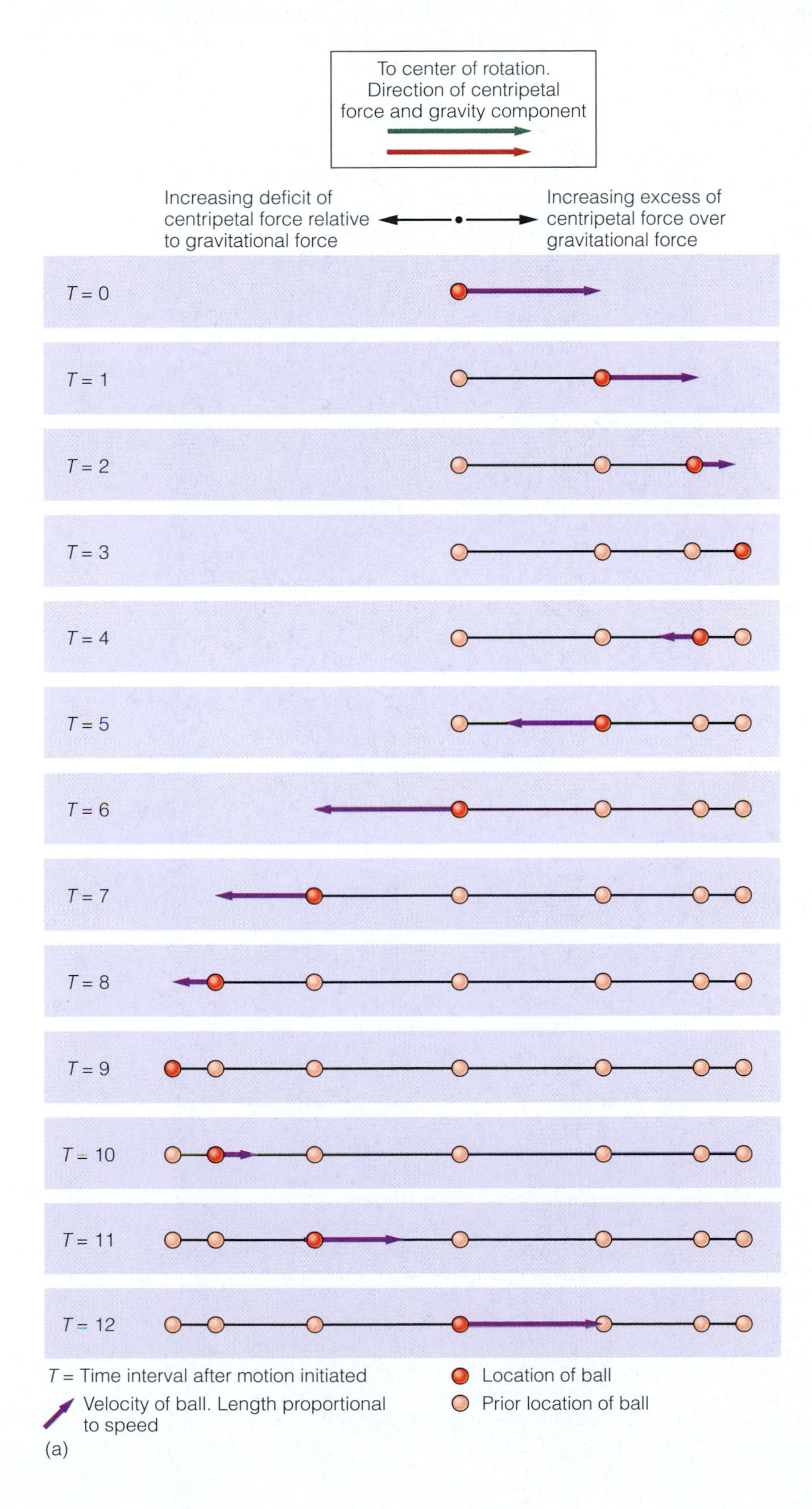

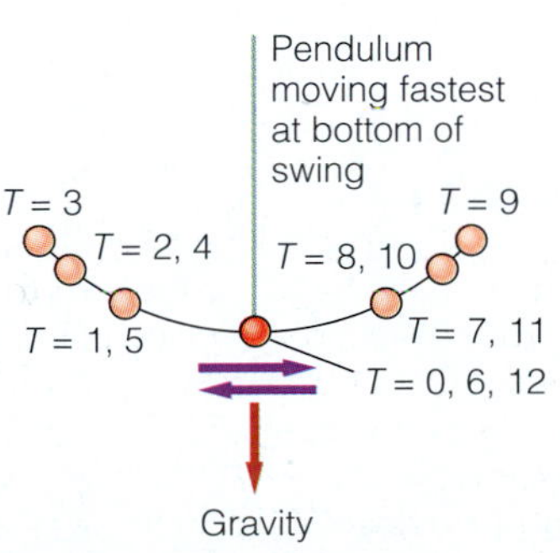

FIGURE CC12.5 (a) If a ball is set in motion toward the center of a hypothetical merry-go-round with a gravitational force that acts toward a point beneath its center (see Figure CC12.4), the ball will oscillate back and forth. As it moves toward the center, the component of gravity in this direction decreases while the required centripetal force increases (because the ball retains its orbital velocity but is now in a smaller-radius orbit). The imbalance between centripetal force and gravitational attraction slows and then reverses the ball's direction of motion. The ball is accelerated until it passes through its original location relative to the center of rotation, but it has enough momentum to continue outward. As the ball moves farther away from the center, the component of gravity increases while the required centripetal force decreases. As a result, there is an excess of gravitational force over centripetal force, and the ball is slowed and its motion eventually reversed back toward the center of rotation. (b) These oscillating motions are very similar to the movements of a pendulum.

show mathematically that the ball would always go through one complete oscillation from its extreme innermost point outward and back to this innermost point in exactly the time it takes for the merry-go-round to complete one revolution. You need not be concerned with this calculation, but the result is important because we can now see what the path of the ball set in motion would be (**Fig. CC12.6**). To an observer anywhere on the rotating merry-go-round, the ball would appear to travel in clockwise circles. It would complete a circle twice for each time the merry-go-round completed one revolution (the equivalent of two circles every 24 h on the Earth).

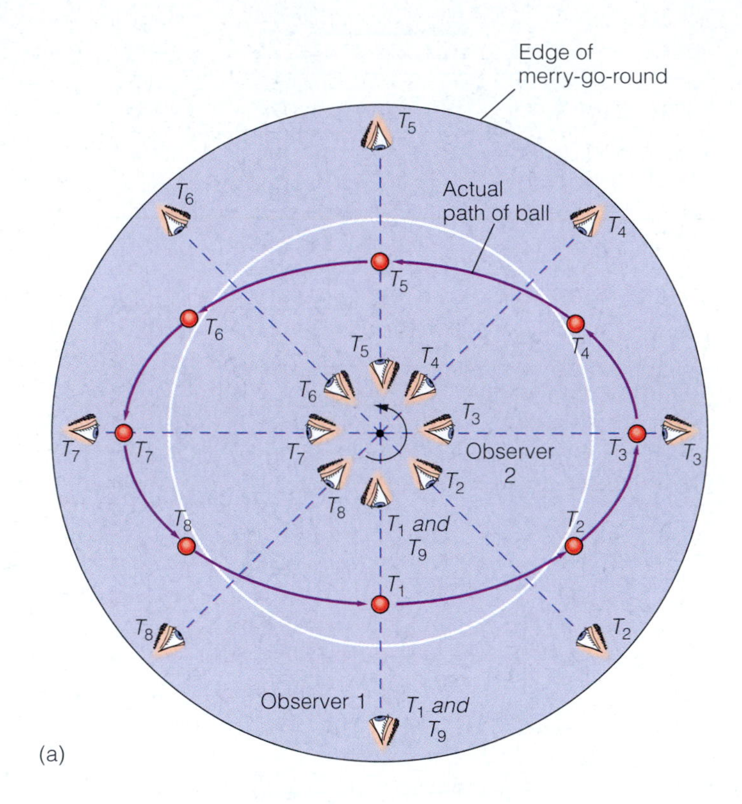

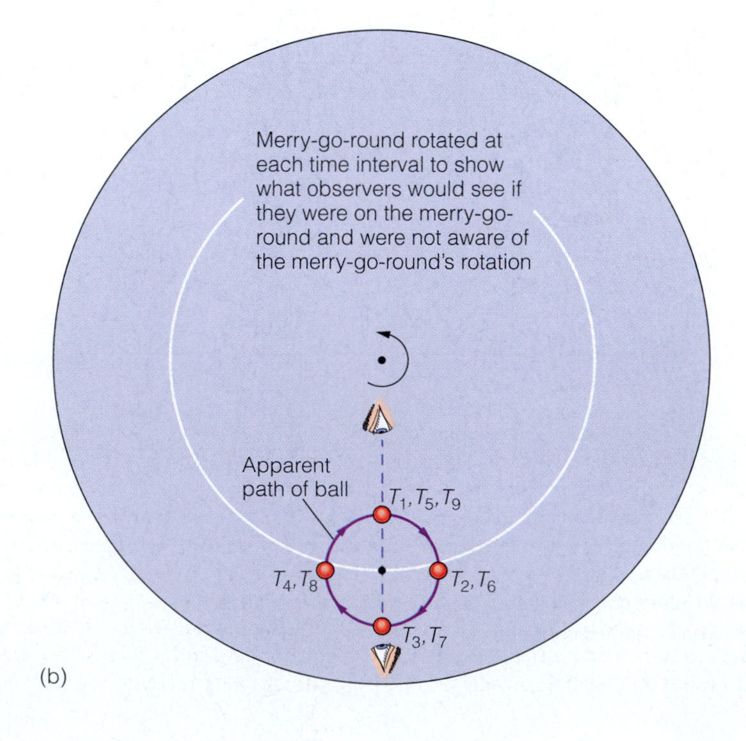

FIGURE CC12.6 (a) If a ball is set in motion toward the center of a hypothetical merry-go-round with a gravitational force that acts toward a point beneath its center (see Figure CC11.4), this motion is added to the orbital motion that the ball already has due to the rotation of the merry-go-round. The added motion and the interaction of centripetal force and gravity cause the ball to oscillate back and forth toward and away from the center as shown in Figure CC11.5. When the rotational motion is added, the actual path of the ball on the merry-go-round is an ellipse, and it returns to its original position after exactly one revolution. (b) To an observer at any location on the rotating merry-go-round, the ball appears to move in a circle. Follow the ball's direction and distance from each observer's eye in part (a) of this figure and see how they plot in a circle in part (b).

The Earth, Centripetal Force, and Gravity

The Earth is more complicated than a merry-go-round because it is a sphere, although not quite an exact sphere. Gravity acts toward the Earth's center, whereas centripetal force acts toward the center of rotation (perpendicular to the Earth's axis of rotation). Therefore, gravity and centripetal force act in the same direction only at the equator, and the difference between their directions increases from zero at the equator to 90° at the poles (**Fig. CC12.7a**).

Centripetal force can be resolved into two components: one parallel to the Earth's surface oriented north–south, and the other oriented toward the Earth's center (**Fig. CC12.7a**). The centripetal force needed to keep an object on the Earth's surface, and variations in this force needed to balance even very large changes in an object's speed relative to the Earth's surface, are extremely small in comparison with gravity. Therefore, variations in the component of centripetal force that acts toward the Earth's center (the same direction as gravity) are easily compensated by gravity and the **pressure gradient** (**Fig. CC12.8**). In other words, moving objects whose angular velocity is increased are immeasurably reduced in weight, whereas those whose angular velocity is reduced are immeasurably heavier.

If the Earth were perfectly spherical, the component of centripetal force parallel to the Earth's surface would not be compensated by gravity, because gravity would act perpendicular to the Earth's surface at all locations. Consequently, freely moving objects would "slide" across the surface toward the equator (**Fig. CC12.7a**). This explains why the Earth is not a perfect sphere, but is instead an oblate spheroid (the shape you would get if you squeezed the Earth at its poles to deform it like a squeezed basketball). The Earth's diameter as measured from the North to the South Pole is 12,714 km, slightly less than its 12,756-km diameter measured at the equator. Because the Earth is not a perfect sphere, gravity acts at a very small angle to the surface (**Fig. CC12.7b**), and there is a small component of gravity parallel to the Earth's surface. This gravity component compensates for the component of centripetal force parallel to the Earth's surface (**Fig. CC12.7b**), so freely moving objects do not slide toward the equator.

If an object is set in motion relative to the Earth, the object's angular velocity will change just as it did for the ball on the merry-go-round. On the Earth, the vertical component of the altered centripetal force is compensated by gravity and the Earth's pressure gradient. However, if the angular velocity of an object is altered, the balance between the small component of gravity acting parallel to the Earth's surface and the component of centripetal force acting parallel to the Earth's surface is upset. The

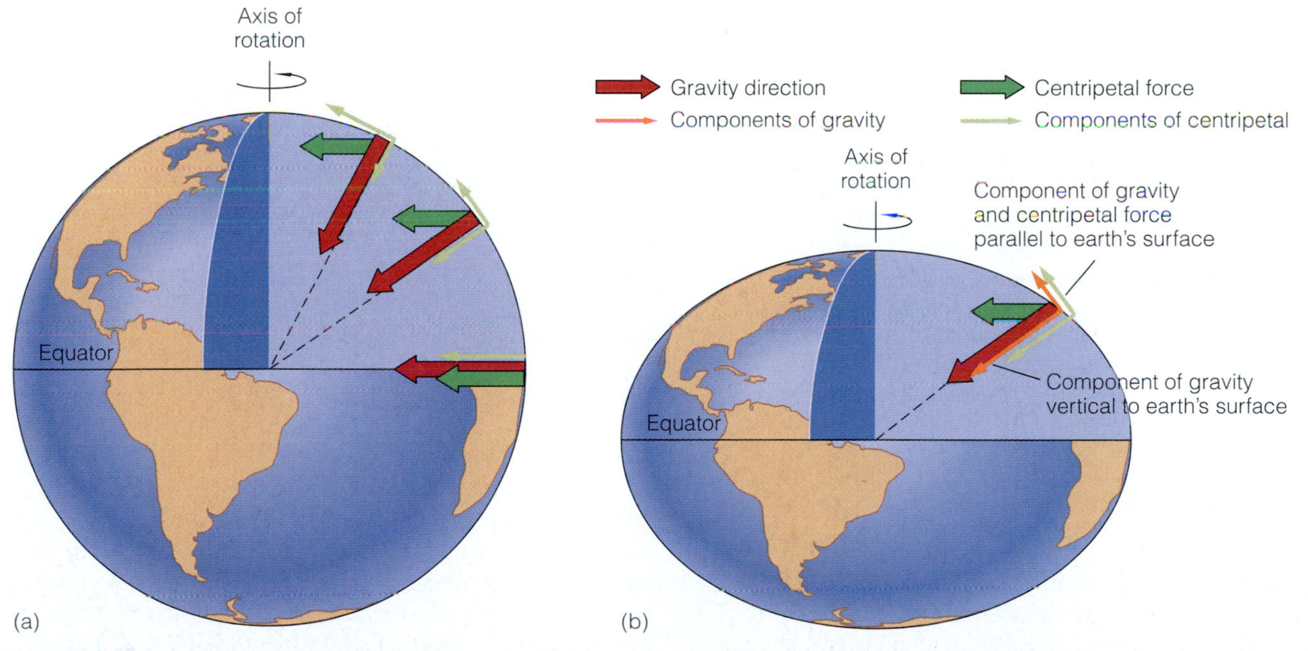

FIGURE CC12.7 (a) If the Earth were exactly spherical, there would be no component of gravity parallel to the Earth's surface at any latitude. However, to keep an object in orbit at the Earth's surface there would have to be a component of a centripetal force parallel to the Earth's surface, everywhere except at the equator. Because there would be no gravitational force component to provide this component of centripetal force, an object anywhere on the Earth's surface except at the equator would move into a wider orbit (in which the required centripetal force was lower). Thus, the object would slide toward the equator. (b) In response to the force imbalance that would be created if the Earth were spherical, the Earth and all other planets that were at one time totally molten or gaseous were not formed as perfect spheres. Instead, the Earth is an oblate spheroid, a squashed spheroid shape such as shown (but much exaggerated) in the figure. Because the Earth is an oblate spheroid, gravity does not act exactly perpendicular to the Earth's surface, except at the equator and poles. Consequently, at all points other than these, there is a small component of gravity parallel to the Earth's surface that balances the required centripetal force for an object at that point.

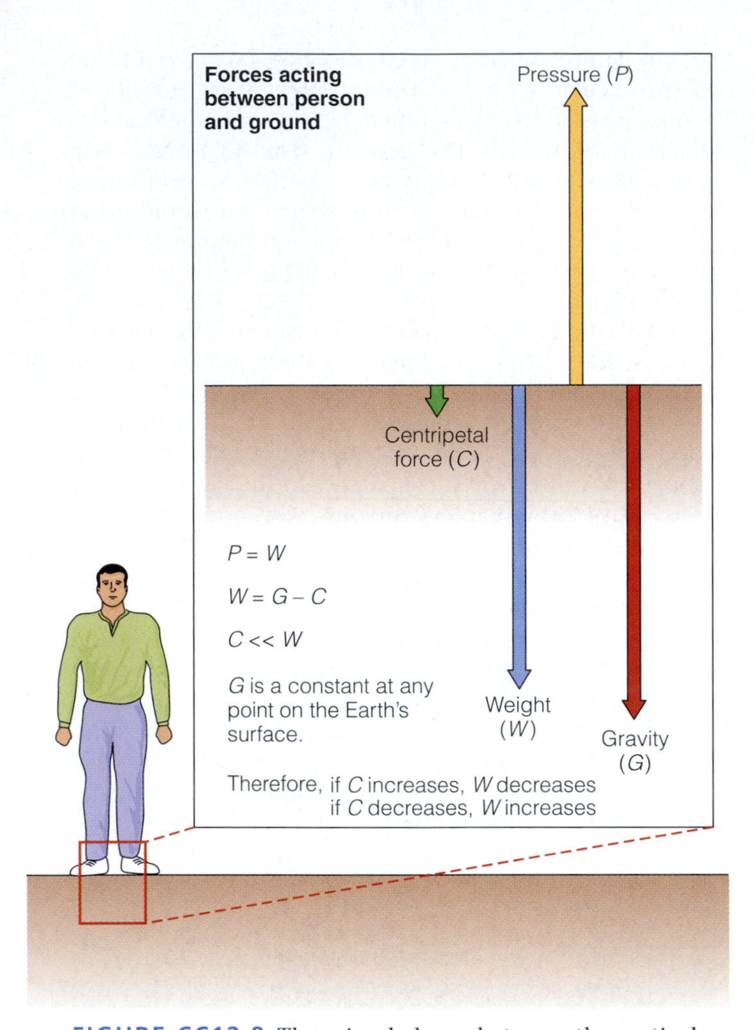

FIGURE CC12.8 There is a balance between the vertical components of gravity (which is invariable at any given location on the Earth), centripetal force, and the Earth's pressure gradient for objects at the Earth's surface. The pressure gradient force is equal to the object's weight. The pressure gradient force and the object's weight must be very slightly smaller than the gravitational force because the gravitational force must provide the required centripetal force to maintain the object in its orbit with the rotating Earth. Thus, the object's weight equals the gravitational force minus the very much smaller centripetal force. The magnitude of the centripetal force has been greatly exaggerated in this figure.

component of gravity parallel to the surface cannot compensate for changes in the corresponding component of centripetal force (unless the Earth changes its shape). Consequently, altering the angular velocity of a freely moving object on the Earth's surface causes it to move across the surface, toward or away from the pole.

Motion on the Earth: North and South

Now let's look at motions on the Earth's surface. First, consider a projectile fired toward the north in the Northern Hemisphere. This projectile has a component of motion from west to east imparted to it by the Earth's rotation. The projectile retains this west-to-east velocity, but the velocity of the Earth rotating underneath it decreases as the projectile moves north (**Fig. CC12.9a**). The projectile continues eastward while the Earth below it moves eastward progressively more slowly. Thus, the projectile "leads" the Earth's rotation at an increasing rate and follows a path that appears to an observer to be a curve deflected to the right (**Fig. CC12.9b**). In addition, because gravity prevents the projectile from maintaining its original distance from the Earth's axis of rotation, the projectile's angular velocity increases as it moves north. This requires increased centripetal force and creates a force imbalance that "pushes" the projectile back to the south away from the Earth's axis of rotation.

Similarly, a projectile fired southward in the Northern Hemisphere passes over an Earth that moves eastward progressively faster under the projectile, and it "lags" the Earth's rotation (**Fig. CC12.9c**). It also appears to be deflected to the right and is "pushed" back north because it has a lower angular velocity. In the Southern Hemisphere, the situation is similar, but the deflection is to the left (**Fig. CC12.9b,c**).

Motion on the Earth: East and West

When an object is set in motion in a west-to-east or east-to-west direction, it moves either with or against the Earth's rotation (**Fig. CC12.10a**). Thus, the object's angular velocity is increased or decreased. When it is set in motion west to east, angular velocity is increased and therefore the object tends to move outward away from the pole. A very small reduction in the object's weight allows gravity to compensate for the increased centripetal force and prevents the object from flying upward away from the Earth's surface. However, there is no compensating force to provide the needed increase in the component of centripetal force across the Earth's surface, and the object moves toward the equator, or to the right in the Northern Hemisphere and left in the Southern Hemisphere (**Fig. CC12.10b**).

When an object is set in motion east to west, the angular velocity is decreased and the centripetal force needed to maintain its orbit is slightly decreased. The object's weight increases slightly and prevents it from moving vertically downward. However, the reduction in the component of centripetal force parallel to the Earth's surface remains unbalanced and the object is moved across the Earth's surface toward the pole. Again, the deflection is to the right in the Northern Hemisphere and to the left in the Southern Hemisphere (**Fig. CC12.10b**).

If the object is moving either west to east or east to west exactly at the equator, there is no component of centripetal force parallel to the Earth's surface, a small adjustment in weight can totally compensate the changes in centripetal force, and there is no deflection (**Fig. CC12.10b**).

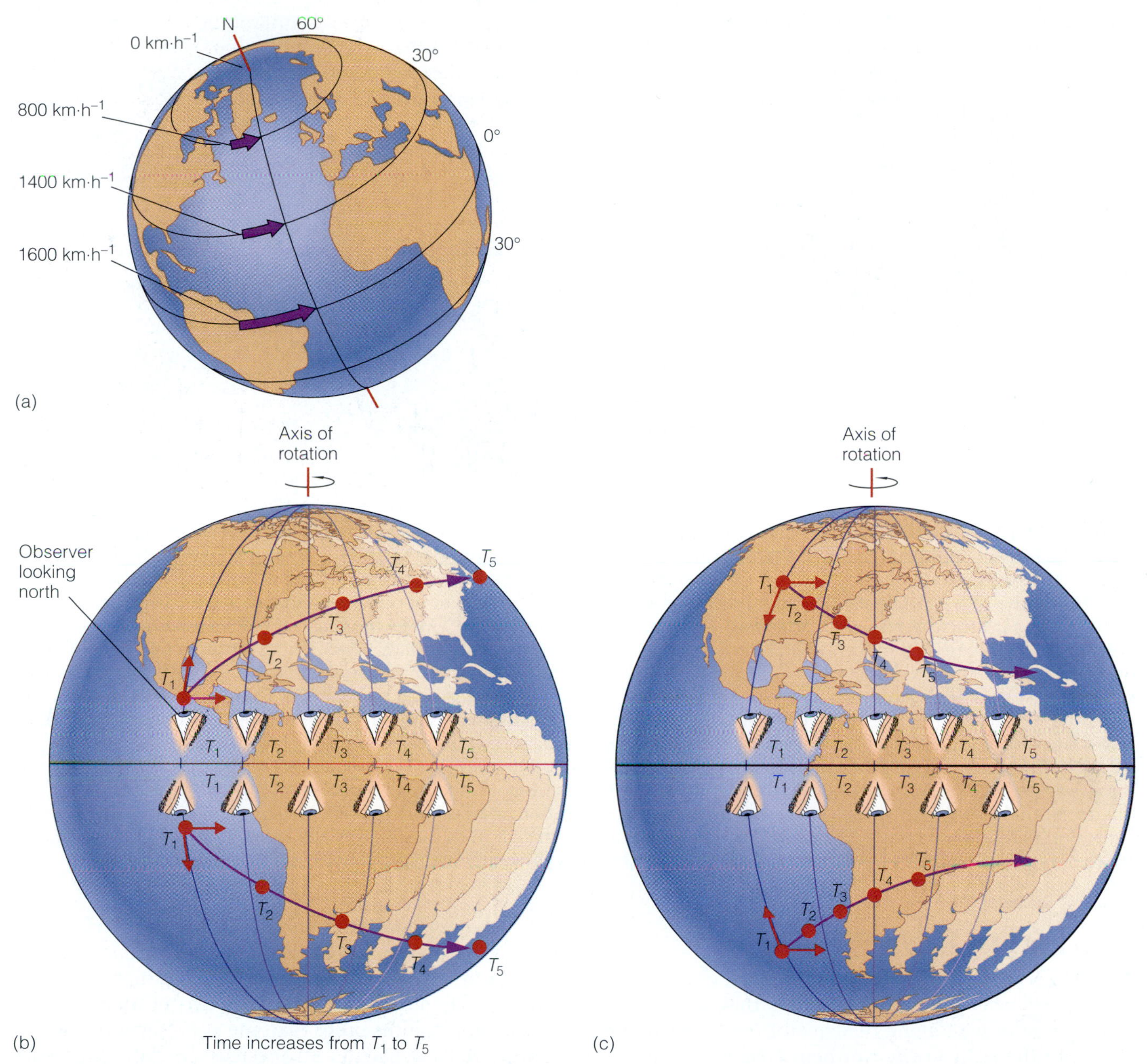

FIGURE CC12.9 (a) The orbital velocity due to the Earth's rotation of points on the Earth's surface decreases as latitude increases. (b) If an object is started in motion directly toward the pole in either hemisphere, it moves progressively into latitudes in which the Earth's surface is moving more slowly in its rotational orbit. However, if the object is moving freely (that is, if it is not attached by strong frictional forces to the solid Earth), it retains the orbital velocity that it had when it was started in motion. Thus, in addition to its motion to the north (or south in the Southern Hemisphere), the object continues to move to the east at its original orbital velocity and progressively moves ahead of the latitude from which it was set in motion. To an observer moving with the Earth's surface, the result is an apparent deflection in the path of the object, to the right in the Northern Hemisphere and to the left in the Southern Hemisphere. (c) If an object is started in motion directly toward the equator in either hemisphere, it moves progressively into latitudes in which the Earth's surface is moving more quickly in its rotational orbit. Thus, in addition to its motion to the south (or north in the Southern Hemisphere), it continues to move to the east at its original orbital velocity and progressively falls behind the latitude from which it was set in motion. The apparent deflection of the path of the object is, once again, to the right in the Northern Hemisphere and to the left in the Southern Hemisphere.

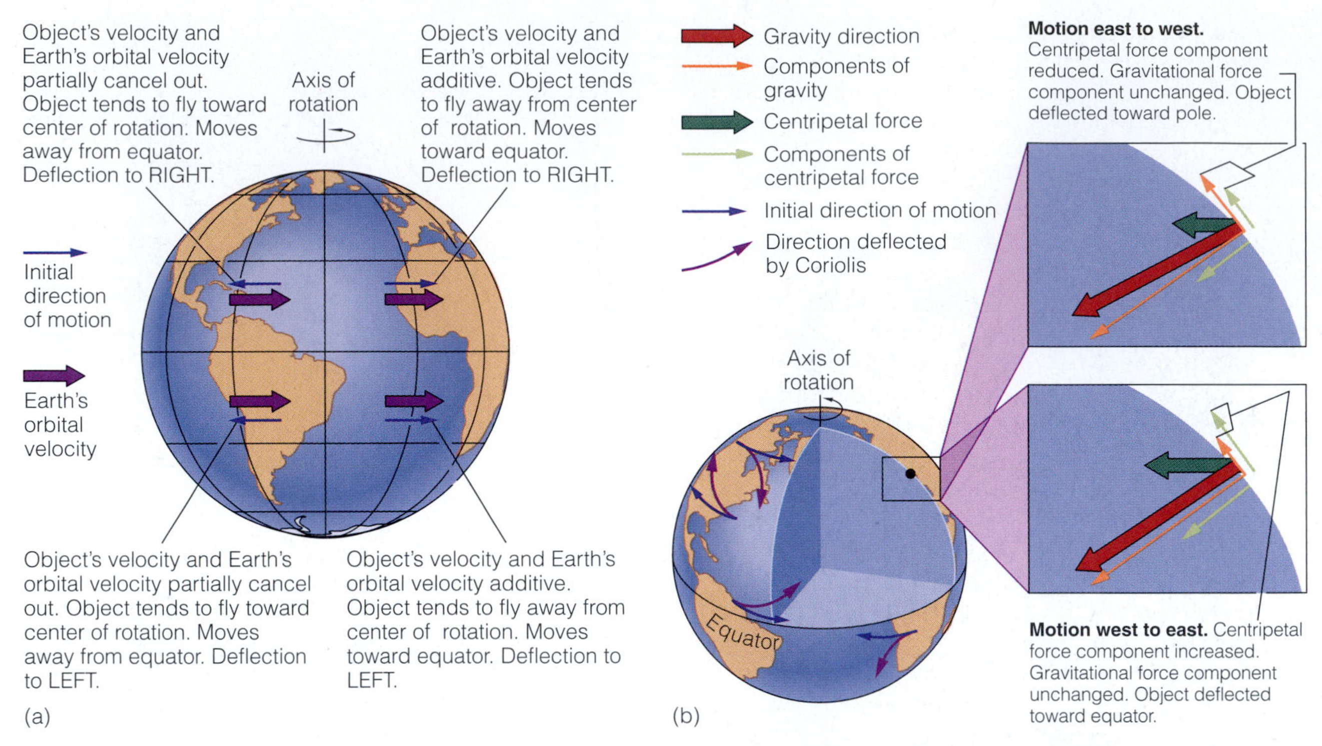

Object's velocity and Earth's orbital velocity partially cancel out. Object tends to fly toward center of rotation. Moves away from equator. Deflection to RIGHT.
Axis of rotation
Object's velocity and Earth's orbital velocity additive. Object tends to fly away from center of rotation. Moves toward equator. Deflection to RIGHT.
Initial direction of motion
Earth's orbital velocity
Object's velocity and Earth's orbital velocity partially cancel out. Object tends to fly toward center of rotation. Moves away from equator. Deflection to LEFT.
Object's velocity and Earth's orbital velocity additive. Object tends to fly away from center of rotation. Moves toward equator. Deflection to LEFT.
(a)
Gravity direction
Components of gravity
Centripetal force
Components of centripetal force
Initial direction of motion
Direction deflected by Coriolis
Axis of rotation
Equator
Motion east to west. Centripetal force component reduced. Gravitational force component unchanged. Object deflected toward pole.
Motion west to east. Centripetal force component increased. Gravitational force component unchanged. Object deflected toward equator.
(b)

FIGURE CC12.10 (a) When objects are started in motion across the Earth's surface in an east-to-west direction, their velocity is in the opposite direction to that imparted by the orbital velocity of the Earth's surface due to the Earth's rotation. As a result, the effective orbital velocity of the object is reduced, and the centripetal force needed to maintain the body in its orbit with the Earth's spin is decreased. The slight decrease in the vertical component of centripetal force is readily compensated by a negligibly small increase in the object's weight (change in pressure gradient). (b) In contrast, no such compensation is possible for the component of centripetal force that acts parallel to the Earth's surface. The slight excess of gravitational force over the component of centripetal force parallel to the Earth's surface needed to maintain the body in its orbit with the Earth's spin deflects the object's path toward the axis of rotation or toward the pole in each hemisphere. Thus, the deflection is to the right in the Northern Hemisphere and to the left in the Southern Hemisphere. Similarly, (a) when objects are started in motion across the Earth's surface in a west-to-east direction, their velocity is in the same direction as that imparted by the orbital velocity of the Earth's surface due to its rotation. (b) The effective orbital velocity of the object is increased, but the centripetal force provided by the component of gravity parallel to the Earth's surface is too small to maintain this orbital velocity. Therefore, the object's path is deflected away from the axis of rotation or away from the pole in each hemisphere. This deflection is also to the right in the Northern Hemisphere and to the left in the Southern Hemisphere. For ease of illustration, the magnitude of centripetal force has been greatly exaggerated relative to the force of gravity in this diagram.

Coriolis Effect Characteristics

All objects moving freely relative to the Earth are deflected by the Coriolis effect unless they are moving directly east or west along the equator. The deflection is always to the right in the Northern Hemisphere and to the left in the Southern Hemisphere. However, it is easier to say that the direction of the deflection is ***cum sole,*** which means "with the sun." To an observer in the Northern Hemisphere looking toward the equator and thus toward the arc of the sun across the sky, the sun moves across the sky from left to right. In the Southern Hemisphere, the sun moves from right to left.

Freely moving objects appear to move in circular paths on the Earth's surface. These circular paths are distorted if objects move large distances across the Earth's surface. This is because the Earth is equivalent to a flat merry-go-round only at the North or South Pole, where the apparent rotation occurs across a surface that is perpendicular to the axis of rotation. At all other points on the Earth, the motion occurs on a surface that is inclined at an angle other than 90° to the axis of rotation. As a result, at all points other than the poles, a component of the Coriolis deflection is directed vertically downward and "blocked" by adjustments of the object's weight. Consequently, the rate of deflection is reduced.

The proportion of centripetal force directed vertically downward increases from zero at the poles to 100% at the equator (**Fig. CC12.7a**). Therefore, the Coriolis deflection is at a maximum at the poles, is reduced progressively with decreasing **latitude**, and reaches zero at the equator. In addition, the time necessary for a freely moving object to complete a circle increases with decreasing latitude. This time period, called the "inertial period," is 12 h at the poles, increases to 24 h at 30° latitude, and is infinity at the equator.

The diameter of the circle in which freely moving objects appear to move is determined by both the object's speed and its latitude. Because freely moving objects at the same latitude will complete a circle in the same amount of time, a faster moving object must move in larger circles. A bullet travels in a circle of huge diameter and along only a very tiny part of its circular path in the extremely short time it remains airborne. Consequently, it appears to travel in a "straight" line. Air masses in the atmosphere and water masses in the oceans move much more slowly and travel in paths of much smaller radius.

For objects moving at the same speed, the inertial period and the diameter of the apparent circular path decrease with increasing latitude. Therefore, the Coriolis effect "increases" with increasing latitude. Because the diameter of the circle in which even a slowly moving object appears to travel is very large near the equator, the Coriolis deflection is often considered negligible at latitudes within 5° of the equator.

Objects moving at different speeds at the same latitude are deflected through the same angle in equal times, but the faster object is deflected farther from its original path (**Fig. CC12.11**). Thus, the Coriolis deflection "increases" with increasing speed.

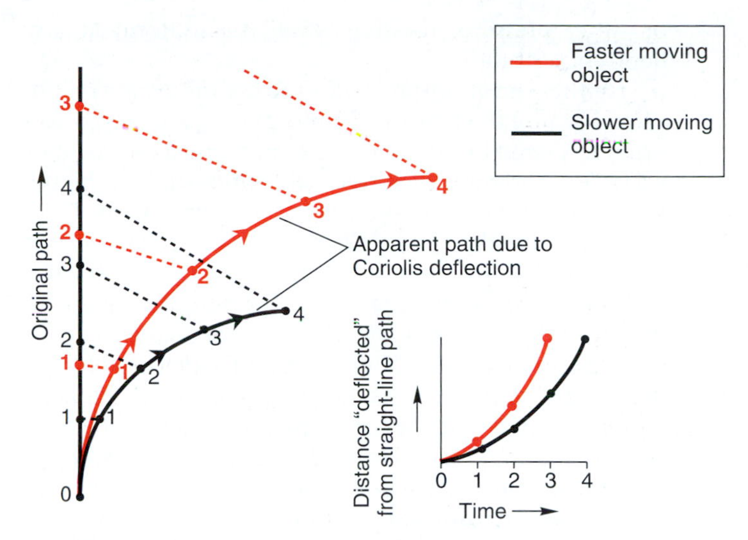

FIGURE CC12.11 The distance between the location of an object deflected by the Coriolis effect and the location it would be expected to occupy if it were not deflected (that is, if it followed a straight-line path) increases more quickly with time if the object's speed is greater. Thus, the magnitude of the Coriolis deflection is said to "increase" with increasing speed of the object.

CRITICAL CONCEPT 13

Geostrophic Flow

ESSENTIAL TO KNOW

- Horizontal pressure gradients exert a force that accelerates fluid molecules in the direction of pressure decrease on the gradient. The acceleration increases as the strength of the the pressure gradient increases.
- The Coriolis effect deflects fluids that flow on a pressure gradient until they flow across the gradient. The flow then continues along a line of constant pressure (isobar) as a geostrophic flow.
- Geostrophic flow conditions occur when the pressure gradient force is balanced by the Coriolis deflection.
- Geostrophic wind and current speeds are determined by the steepness of the pressure gradient. Wind or current speed increases as the steepness of the gradient increases.
- Geostrophic wind or current speed and direction can be determined from isobaric maps. The direction of flow is parallel to the isobars, and the speed is higher where the pressure gradient is steeper (isobars are closer together).
- Geostrophic winds and currents flow counterclockwise around low-pressure zones and clockwise around high-pressure zones in the Northern Hemisphere. In the Southern Hemisphere, they flow clockwise around low-pressure zones and counterclockwise around high-pressure zones.

UNDERSTANDING THE CONCEPT

The atmosphere and ocean waters are **stratified** fluids (**CC1**), in each of which **density** decreases with increasing distance from the Earth's center. The only exceptions are in limited areas where stratification is unstable. It is in these areas of unstable stratification that density-driven vertical motions of the fluid occur.

If the oceans or atmosphere were at equilibrium, density would be uniform at any one depth or altitude. The vertical density gradient would be the same everywhere, so the total weight of the overlying water and/or air column at a specific depth or altitude would be the same everywhere. Because this total weight determines atmospheric or water pressure, pressure would be uniform at any depth or altitude.

Neither the atmosphere nor the oceans are at equilibrium, because the density of atmospheric gases and ocean water is altered locally by changes in such factors as air or water temperature, dissolved salt concentration in the water, and water vapor pressure in the air (Chaps. 7, 9). Consequently, the vertical distribution of density in both oceans and atmosphere varies from place to place, and there are horizontal variations of pressure at any given height in the atmosphere or depth in the oceans.

Surfaces that consist of points of equal altitude or depth are referred to as horizontal or level surfaces. However, these surfaces are actually spherical because the Earth is a sphere. Horizontal **pressure gradients** develop in the oceans as a result of density differences between **water masses** and also because winds tend to move ocean surface waters, causing the water to pile up in some locations (Chap. 10).

Where there is a horizontal pressure gradient, the fluid is subject to a force that tends to accelerate molecules from high-pressure areas toward low-pressure areas. The acceleration is greater when the pressure gradient is steeper. We perform one of the simplest demonstrations of acceleration along a pressure gradient every time we open a soda bottle or can. Once they are free to do so, the gas molecules in the high-pressure zone within the bottle are accelerated toward the lower pressure outside.

In our soda bottle experiment, the density gradient is extremely steep because the pressure difference between the air outside the bottle and the gas in the bottle is large and the distance between the high-pressure zone inside the bottle and the low-pressure zone of the surrounding air is very short. When we open the bottle, gas molecules in the high-pressure zone are accelerated very rapidly and must move only a short distance to reach the low-pressure zone. As a result, the pressure equalizes almost instantaneously. In contrast, the pressure differential between horizontally separated high- and low-pressure zones in the oceans and atmosphere is very small, and these zones are separated by much greater distances. Consequently, the accelerations produced by atmospheric and oceanic horizontal pressure gradients are small. In addition, air or water does not flow directly from the high-pressure zone to the low-pressure zone, because the air and water molecules are subject to the **Coriolis effect** (**CC12**) once they have been set in motion.

To understand how motions induced by the pressure gradient and Coriolis effect interact, consider the following facts. Freely moving objects, including air or water molecules, actually travel in straight paths unless acted on by another force, such as the pressure gradient. The Coriolis deflection is only a perceived deflection seen by an observer on the rotating Earth. However, from our rotating frame of reference, the freely moving object is deflected ***cum sole.*** In **Figure CC12.11**, we can see that the magnitude of the deflection and the rate of increase both increase with time. These are the characteristics of an acceleration in the direction 90° *cum sole* to the direction of motion, and this "acceleration" increases with the object's speed (**Fig. CC12.11**).

We can now examine what happens when a fluid begins to flow in response to a horizontal pressure gradient. The fluid is accelerated down the pressure gradient from the high-pressure region toward the low-pressure region. As it begins to move, it is deflected by the Coriolis effect. The faster it moves, the greater is the Coriolis deflection (**CC12**). The direction of motion thus turns away from the direct path down the pressure gradient, to the right in the Northern Hemisphere and to the left in the Southern Hemisphere. As the moving fluid is deflected, the pressure gradient acceleration continues to act toward the low-pressure region while the Coriolis deflection continues to act at 90° *cum sole* to the changed direction of flow. The speed increases as the molecules are accelerated, thus increasing the Coriolis deflection.

The molecules continue to accelerate while the direction of flow is progressively turned *cum sole* until the flow is directed across the pressure gradient along a **contour** of equal pressure (**Fig. CC13.1**). When the flow is in this direction, the pressure gradient acceleration and Coriolis deflection act in opposite directions and balance each other. The speed of the fluid at this balance point is determined by the steepness of the pressure gradient because stronger pressure gradients cause greater accelerations that must be balanced by greater Coriolis deflections, which increase with increasing speed (**CC12**).

This type of flow, in which the pressure gradient and Coriolis deflection are balanced, is called **geostrophic** flow, and the moving air or water masses are geostrophic winds or geostrophic **currents**. The most important features of geostrophic winds or currents are that the flow is directed along contours of equal pressure within a pressure gradient, and that wind or current speed is determined by the steepness of the pressure gradient.

Geostrophic flows are almost never aligned exactly along the contours of equal pressure on the pressure gradient. The reason is that pressure gradients are continuously changing in response to changes in the factors that create them, such as wind stress on the oceans and **convection** in the atmosphere. In addition, **friction** between moving air masses and the ground or ocean surface, or between ocean water and the seafloor, reduces the Coriolis deflection. Thus, geostrophic winds at the surface and currents near the seafloor do not flow exactly along contours of equal pressure. Instead, they are aligned generally along these contours but slightly offset toward the center of low-pressure zones and away from the center of high-pressure zones.

Horizontal pressure gradients are usually shown as maps of lines of equal pressure called **isobars** (Appendix 1). Most **weather** maps in newspapers and on television

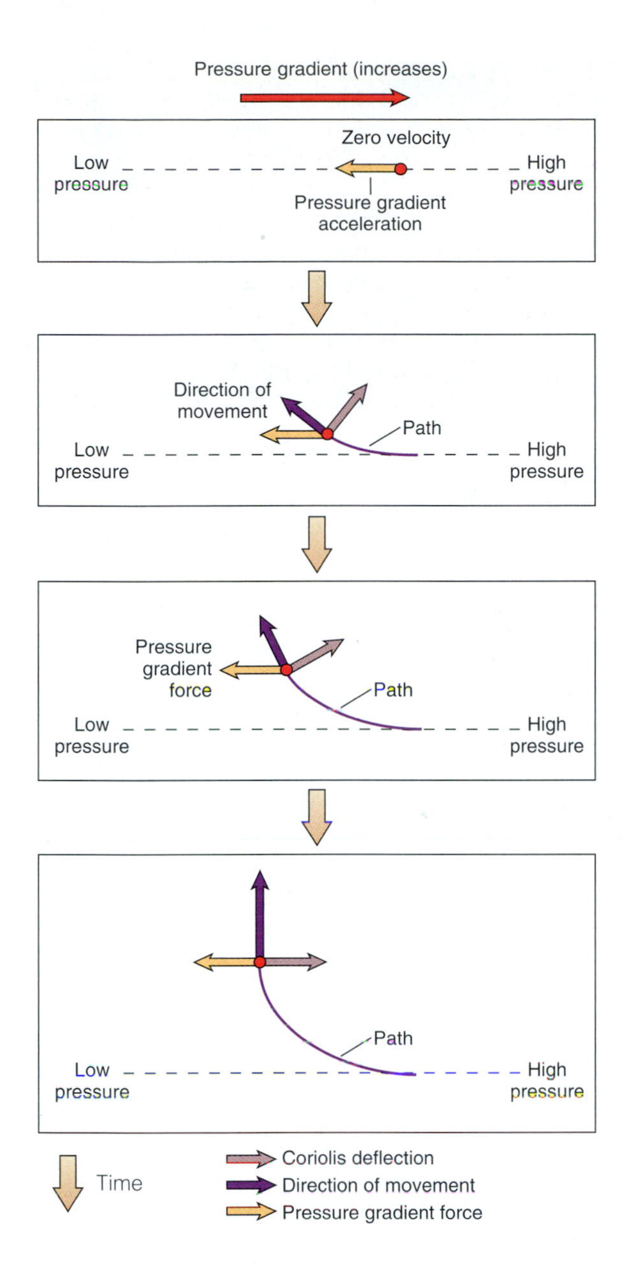

FIGURE CC13.1 When a current is initiated on a horizontal pressure gradient, the initial direction of motion is directly down the gradient. However, as it moves, the water mass is deflected by the Coriolis effect. It is accelerated and deflected until it is flowing directly across the pressure gradient and the pressure gradient force is balanced by the Coriolis deflection. This is a "geostrophic" current. If the pressure gradient is steeper, the acceleration is greater and the geostrophic current is faster, but it is still balanced by the greater Coriolis deflection associated with the higher speed.

are of this type. The isobaric maps in **Figure CC13.2** show high- and low-pressure zones and the pressure gradients between these zones. The pressure difference between adjacent isobars is the same for all adjacent isobars at all locations on each map. The gradient is steeper where the isobars are closer together. Therefore, the spacing of the isobars reveals the steepness of the pressure gradient (change in pressure per unit distance).

Most wind and ocean current systems are geostrophic, and their direction of flow is almost parallel to the pressure contours. As a result, pressure contour maps can be used to estimate both the speed and direction of winds or currents. The wind or current direction is parallel to the isobars. Because the Coriolis deflection is to the right in the Northern Hemisphere, winds and currents in this hemisphere flow counterclockwise around low-pressure zones (**Fig. CC13.2a**) and clockwise around high-pressure zones (**Fig. CC13.2b**). In the Southern Hemisphere the deflection is to the left, and winds and currents flow clockwise around low-pressure zones (**Fig. CC13.2c**) and counterclockwise around high-pressure zones (**Fig. CC13.2d**). The wind or current speed is determined by the steepness of the pressure gradient. Accordingly, wind or current speeds are higher where isobars are closer together and slower where they are more widely separated (**Fig. CC13.2**).

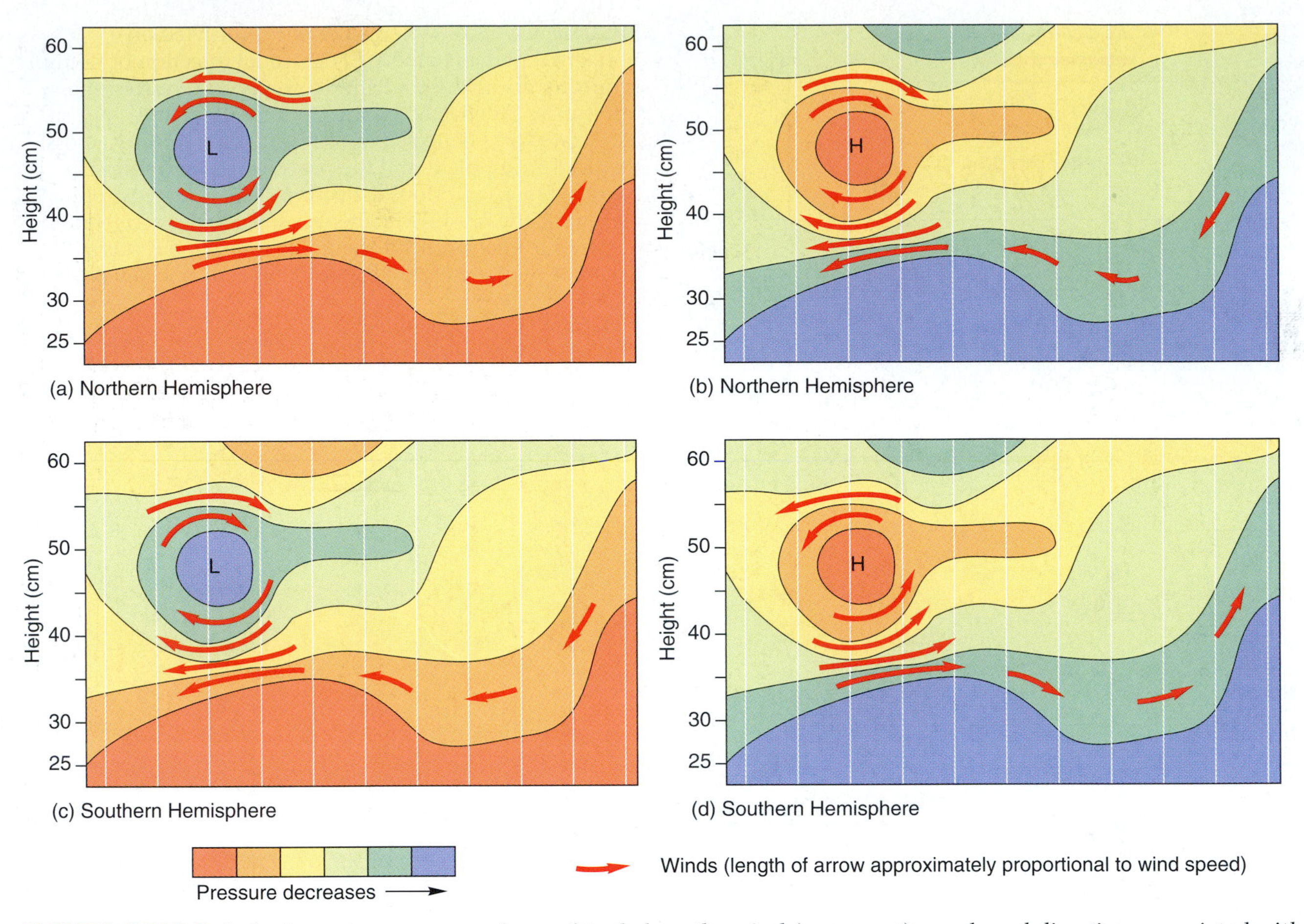

FIGURE CC13.2 Isobaric contour maps can be used to deduce the wind (or current) speeds and directions associated with the isobars. The contour map examples in this figure show circulation at (a) a low-pressure zone in the Northern Hemisphere, (b) a high-pressure zone in the Northern Hemisphere, (c) a low-pressure zone in the Southern Hemisphere, and (d) a high-pressure zone in the Southern Hemisphere. Note that the directions of rotation are opposite for the two hemispheres. Winds blow counterclockwise around a low-pressure zone and clockwise around a high-pressure zone in the Northern Hemisphere, and in the reverse directions in the Southern Hemisphere.

CRITICAL CONCEPT 14

Photosynthesis, Light, and Nutrients

ESSENTIAL TO KNOW

- Most organic matter in the Earth's ecosystems is created by photosynthesis.
- Plants perform photosynthesis through a complex series of reactions that occur in pigmented cells called "chloroplasts."
- A nitrogen-containing pigment called "chlorophyll *a*" plays the central role in transferring light energy to the photosynthesis process. Chloroplasts contain accessory pigments that capture light energy and transfer it to chlorophyll *a*.
- Accessory pigments collect light energy at wavelengths at which chlorophyll *a* absorbs light poorly. Each species has a different suite of accessory pigments optimized to collect light energy available in its specific habitat.
- Plant photosynthesis uses light energy, carbon dioxide, and water to produce oxygen (which is released) and relatively simple organic compounds.

- Most organic compounds produced by photosynthesis are used in respiration of the plant cells. Only a small proportion enter biochemical cycles and are converted to more complex organic molecules.
- Only a very small percentage of the light energy used in photosynthesis is used to produce biomass.
- Nitrogen and phosphorus, among other elements, are present in key compounds involved in photosynthesis. Hence, depletion of available nitrogen and phosphorus stops production of these compounds and limits primary productivity.
- Bacterial photosynthesis produces an insignificant amount of organic matter in comparison with plant photosynthesis in almost all of the ocean environment. Bacterial photosynthesis uses light energy, carbon dioxide, and a hydrogen donor, such as H_2S or H_2, to create relatively simple organic compounds. No oxygen is released.
- Chemosynthetic bacteria synthesize organic compounds by using chemical energy from the oxidation of compounds such as H_2S and H_2 instead of light energy.

UNDERSTANDING THE CONCEPT

For life to exist, carbon, hydrogen, oxygen, nitrogen, phosphorus, and many other essential elements must be synthesized into organic compounds. Most organic matter on the Earth was initially converted from inorganic matter to organic compounds by **photosynthesis.** Exceptions include a few simple organic compounds created by electrical discharges in the atmosphere and the **chemosynthetic communities** described in Chapters 14 and 17.

Photosynthesis is a complex process whereby plants and certain **bacteria** use light energy to convert carbon dioxide to organic compounds. Plant photosynthesis is much more important than bacterial photosynthesis.

The first step in photosynthesis is the capture of light energy by plants or bacteria. Light is captured in complex cell structures called "chloroplasts" located near the plant surface. The chloroplasts contain **chlorophyll** *a* and, in many **species,** the related pigments chlorophyll *b* and chlorophyll *c*. There are also a number of "accessory pigments," including carotenoids, xanthophylls, and phycobilins. Each plant species has its own unique combination of pigments. Accessory pigments capture light energy and transfer it to chlorophyll *a*, which then performs the next step in the synthesis. Chlorophylls *b* and *c* can be characterized as accessory pigments because they, too, collect light energy and pass it on to chlorophyll *a*. The transfer of energy from accessory pigments to chlorophyll *a* is an extremely efficient process. In some cases, transfer efficiency approaches 100%.

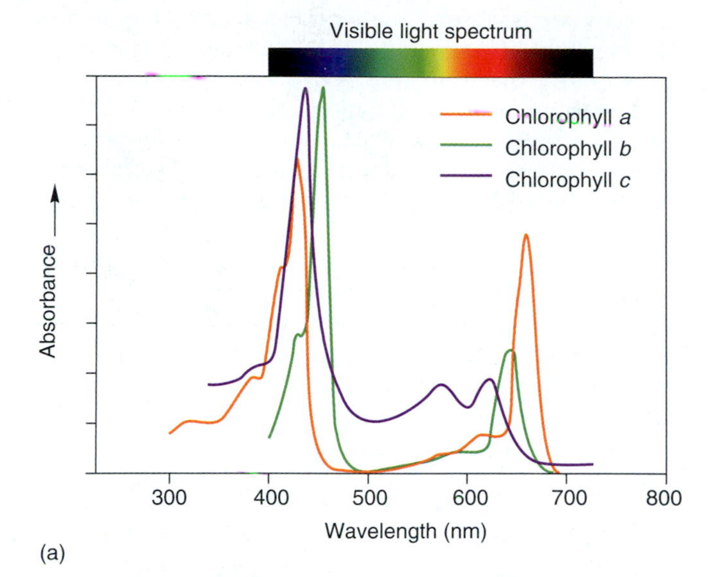

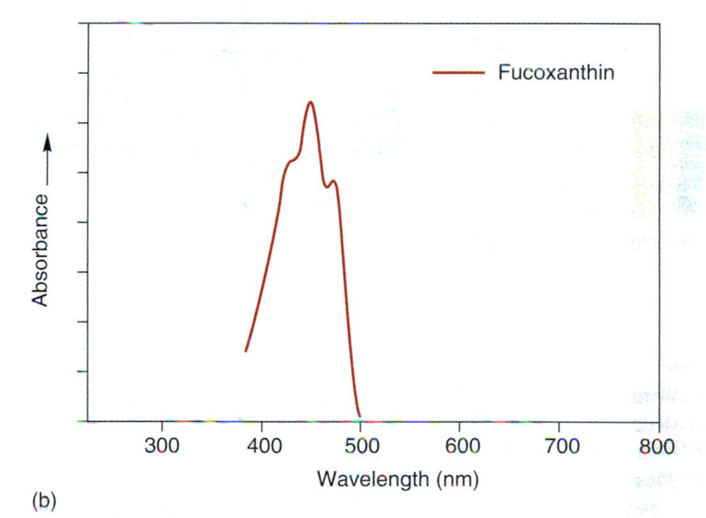

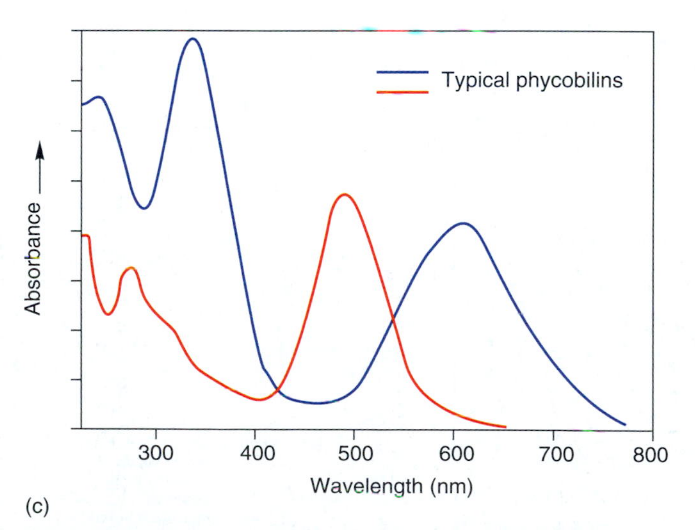

FIGURE CC14.1 Absorption spectra of photosynthetic pigments. (a) Chlorophylls *a*, *b*, and *c*. (b) Fucoxanthin (a typical xanthophyll). (c) Typical phycobilins. Note that the other pigments absorb energy in parts of the spectrum that are not absorbed efficiently by chlorophyll *a*.

Why are accessory pigments needed? Chlorophyll *a* absorbs light effectively only toward the red and violet ends of the spectrum (**Fig. CC14.1a**). To capture light energy from the central blue–green–yellow part of the spectrum, other pigments are needed. Carotenoids, which include the group of pigments called xanthophylls, absorb energy primarily in the blue and blue-green parts of the spectrum (**Fig. CC14.1b**), whereas phycobilins absorb energy primarily in the green, yellow, and ultraviolet parts (**Fig. CC14.1c**). Chlorophylls *b* and *c* also absorb in different parts of the spectrum than chlorophyll *a* (**Fig. CC14.1a**). Thus, the chloroplast contains an array of accessory pigments to capture as much of the available light energy as possible.

Red and blue-violet light are absorbed more effectively by seawater than are green and blue light (**Fig. 7.10**). Hence, accessory pigments are especially important in marine plants that must photosynthesize below the immediate surface layer. **Absorption spectra** of seawater from the open ocean and from coastal waters are different (**Fig. 7.10**), and the **wavelength** distribution of light changes with depth. Species of marine plants that live at different locations and depths have different accessory pigment compositions to optimize their capture of available light energy.

Chlorophyll *a* (**Fig. CC14.2**) is a complex molecule that contains not only carbon (C), hydrogen (H), and oxygen (O) atoms, but also a magnesium atom (Mg) and, most importantly, nitrogen atoms (N). When dissolved nitrogen is depleted, nitrogen-containing compounds, including chlorophyll *a*, cannot be synthesized and **primary production** is limited or stopped (Chap. 14). Thus, nitrogen as nitrate (NO_3^-), nitrite (NO_2^-), or ammonia (ammonium **ion** [NH_4^+] and NH_3) is essential to photosynthesis.

After chlorophyll *a* has captured light energy, it can transfer an excited electron (an electron to which excess energy has been added) to other molecules, which transfer the energy through yet other molecules in a complex series of steps that are not fully understood. During these transfers, the raw materials of photosynthesis—water and carbon dioxide—are brought together and converted into organic compounds.

Some photosynthesis reactions can take place only in the presence of light. The complex process that occurs can be simplified and summarized as follows. Light energy is used to split a water molecule into a hydrogen atom with an excited electron and a hydroxyl ion, each of which immediately reacts with other molecules in the chloroplast. The hydroxyl ion is then converted to oxygen and water, which are released, and the hydrogen atom with its excited electron is combined with NADP (nicotinamide adenine dinucleotide phosphate) to form NADP-H. At the same time, additional light energy is transferred to ADP (adenosine diphosphate) to form ATP (adenosine triphosphate) (**Fig. CC14.3**). ATP is a versatile energy-storing molecule that can donate the considerable energy associated with one of its phosphate bonds by releasing or transferring this phosphate group, which changes ATP back into ADP. ATP is an energy provider in both photosynthesis and **respiration.** Both ATP and ADP contain phosphorus, which is therefore also an essential **nutrient** for photosynthesis.

Once light energy is stored in ATP and NADP-H, the remaining steps in photosynthesis can continue in the dark. However, if light is removed for an extended time, the store of these molecules in the chloroplast is used up.

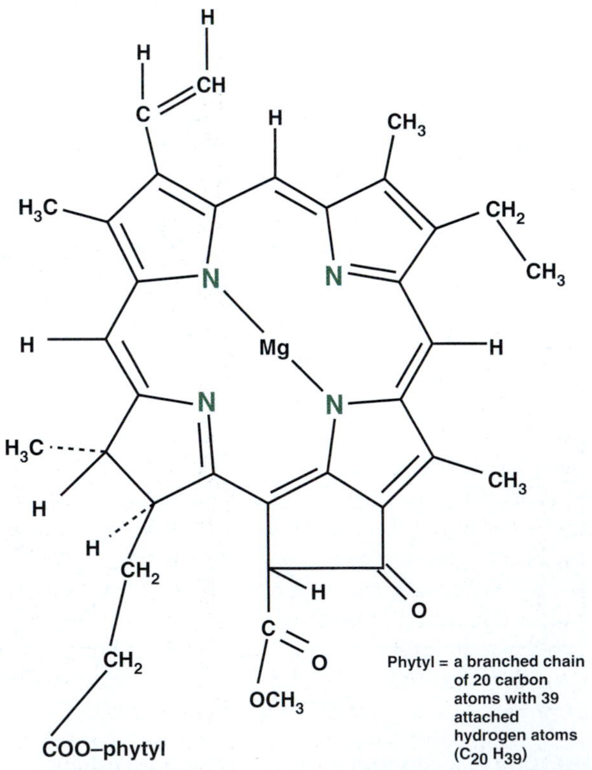

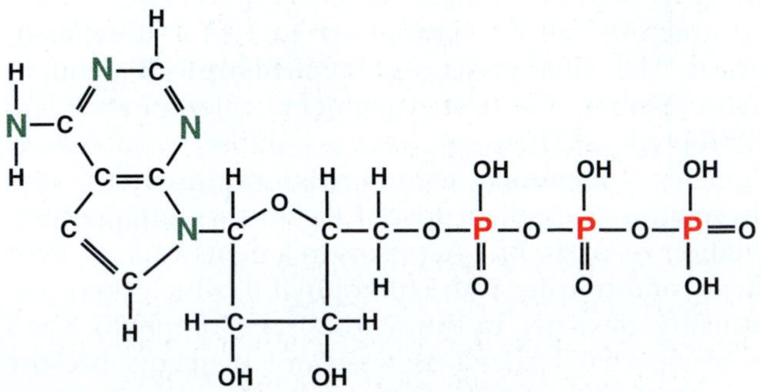

FIGURE CC14.2 The molecular structure of chlorophyll *a*, which is needed for photosynthesis. The molecule is complex and contains four nitrogen atoms.

FIGURE CC14.3 The molecular structure of ATP, a chemical essential to life, is complex. The molecule contains both phosphorus (P) and nitrogen (N) atoms.

The photosynthesis steps that can occur in the dark use hydrogen from NADP-H and energy from ATP to reduce carbon dioxide. Oxygen is removed from carbon dioxide and released, whereas the carbon is combined with other carbon atoms to form organic compounds.

There are several different pathways for these reactions, each of which produces a different organic compound or compounds. One of the most common pathways leads to the production of glucose ($C_6H_{12}O_6$). Glucose and other organic molecules that are formed by photosynthesis are subsequently used by plant cells as basic building blocks for the vast array of other organic compounds created by living organisms in their **biochemical** cycles. Large amounts of glucose and the other compounds created by photosynthesis are also converted back into carbon dioxide in the respiration processes of the plant. The overall efficiency of the transfer of light energy through the process of photosynthesis to create new **biomass** is extremely low, a few percent at most.

The process of primary production can be depicted by the following, much simplified equation:

$$6\ CO_2 + 6\ H_2O \xrightarrow{\text{visible light}} C_6H_{12}O_6 + 6O_2$$

Bacteria photosynthesize in a somewhat different way. First, bacteria contain a unique form of chlorophyll called "bacteriochlorophyll." In addition, they do not use water to provide hydrogen for photosynthesis, and they do not release oxygen. Bacterial photosynthesis can be depicted by the following simplified equation:

$$6\ CO_2 + 6\ H_2[A] \xrightarrow[\text{far red light}]{\text{visible light}} C_6H_{12}O_6 + 6[A]$$

$H_2[A]$ is a hydrogen donor and can be H_2S, H_2, or various organic compounds. Bacterial photosynthesis does not contribute significantly to ocean primary production, because **environments** where both light and a suitable hydrogen donor are available are severely limited.

In bacterial photosynthesis, the needed energy is provided by light. However, it is noteworthy that many of the molecules that replace water as the hydrogen donor in bacterial photosynthesis can also combine with oxygen to provide the energy needed to synthesize organic compounds in chemosynthesis (Chap. 14).

CRITICAL CONCEPT 15

Food Chain Efficiency

ESSENTIAL TO KNOW

- All organisms require food for respiration, growth, and reproduction.
- All organisms lose some of the food as waste products.
- The rate of biomass production by an organism that is available as food for the next trophic level equals the total amount of food ingested, minus the proportion used for respiration and reproduction and the proportion lost as waste products.
- The percentage of ingested food converted to new biomass is the food chain (or trophic) efficiency. It is usually approximately 10% (ranging from 1% to 40%), regardless of species or food source.
- Approximately 90% of the available food biomass is used in respiration and reproduction or lost as waste products at each trophic level in a food chain.
- Long marine food chains that lead to consumers at higher trophic levels, such as tuna, utilize primary production hundreds of times less efficiently than short marine food chains, such as those that support anchovies or sardines, or most terrestrial food chains used to produce food for human consumption.

UNDERSTANDING THE CONCEPT

All organisms require food, which they can obtain by performing **photosynthesis** or **chemosynthesis** or by ingesting organic matter. In all **species**, food is distributed and used in four ways: (1) to provide energy through respiration for the life processes of the organism, (2) to provide the basic materials for production of the myriad compounds contained in any organism's body, (3) to provide the basic materials needed to produce offspring (including eggs and sperm), and (4) as waste products that are **excreted.** Hence, not all food is incorporated into new body tissues or **biomass.** The amount converted to new biomass equals the total food intake minus the amounts used for **respiration** and reproduction and lost as waste. Food used for reproductive processes does create new biomass, but the vast majority of eggs, **larvae,** and juveniles of most species are consumed by predators. Consequently, most new organic matter produced during reproduction does not contribute to the species' adult biomass.

The percentage of food that is ingested by a particular species and converted to new biomass is often referred to as **food chain efficiency** or "trophic efficiency." The total biomass of a species often can be estimated if the total amount of available food and the food chain efficiency are both known. For example, studies of Loch Ness have determined that it would be virtually impossible for a viable population of "monsters" to live there, because the total amount of food available in the loch would not be enough to sustain even one very large animal, unless it converted all available food into new biomass at an impossibly high food chain efficiency.

The efficiency of the conversion of food to biomass is an important parameter in managing food supplies for human populations. For example, much of the corn fed to beef cattle is used by the cattle in respiration, reproduction, and waste generation. Hence, although beef has a higher protein content than corn and therefore is a desirable food, the same amount of corn could feed more people if eaten directly than if first fed to cattle, which are then eaten. This concept is particularly important in many parts of the world where famine persists, but where cattle are raised on corn and other grains that could be consumed directly by people.

Terrestrial **food chains** used for human food are generally very short. Humans consume primarily plant material, which is at the **primary production** or first **trophic level,** or plant-fed animals, which are at the second trophic level. Ocean food chains that lead to human foods are, in many cases, much longer. For example, tuna feed generally at the fifth trophic level (**Fig. 14.10**). The overall efficiency of the food chain that leads from primary production (**phytoplankton**) to tuna is very low because each step in the food chain is inefficient.

Food chain efficiency is difficult to measure exactly because it varies with many parameters, such as the amount of food available, the quality of the food, the nature and timing of the reproductive cycle, and the level of physical activity needed to avoid predators. As an example of this variability, consider that two human children of the same height may ingest approximately the same amount of food, but one may be much thinner than the other. We might conclude that the thinner child has a "faster metabolism." What we mean is that the thinner child uses more food for respiration (because of either greater physical activity or a **genetic** disposition to utilize food less efficiently in respiration) or that the thinner child sends a greater proportion of food to waste (because of genetic differences in the ability to digest and assimilate food). Of course, if the food ingested by these two children were not equal in quantity, the thinner child could also be thinner simply because of a limited food supply and eating less. Another illustration of the variability of food chain efficiency is the difference in efficiency between children, whose biomass is increasing, and adults, whose biomass stays relatively constant, even though they may ingest approximately the same amount of food as children.

Food chain efficiency is variable. It has been determined that, if averaged over a population and over time, this efficiency ranges from about 1% to 40% and, in most instances, is approximately 10% between any two trophic levels. Hence, on average, all species above the primary producer trophic level convert only about 10% of their food into new biomass, which then is available to be consumed at the next trophic level. The following relationships summarize marine food chains:

1 kg of biomass at trophic level 1
(phytoplankton)

produces

0.1 kg of biomass at trophic level 2
(**zooplankton**)

produces

0.01 kg of biomass at trophic level 3
(e.g., small fishes, **baleen** whales)

produces

0.001 kg of biomass at trophic level 4
(e.g., larger fishes such as mackerel, squid)

produces

0.0001 kg of biomass at trophic level 5
(e.g., predatory fishes such as tuna)

produces

0.00001 kg of biomass at trophic level 6
(e.g., killer whales)

In this example, certain types of organisms are designated as feeding at a particular trophic level. However, some of them, such as killer whales, may feed at more than one trophic level.

Many seafood species captured and eaten by people feed at a much higher trophic level than terrestrial animals used as human food. Consequently, human use of the primary production of the oceans is grossly inefficient in comparison with human use of terrestrial primary production.

CRITICAL CONCEPT 16

Maximum Sustainable Yield

ESSENTIAL TO KNOW

- Fishing initially reduces the size of a fish stock. However, because the reduction results in greater food availability for the fishes remaining, the rate of production normally increases.
- The additional biomass produced represents an excess over that needed to maintain the population. This excess can be harvested safely.
- As fishing increases, stock size is reduced to a critical level known as the "maximum sustainable yield," at which the production and reproduction rates of the population are just sufficient to balance the removal rate due to predation and fishing. If fishing yield is increased and continued beyond the maximum sustainable yield, the population can no longer sustain itself and collapses.
- Maximum sustainable yield is difficult to establish because fish stocks vary as a result of year-to-year climate-induced changes and changes caused by other factors, such as diseases.
- Maximum sustainable yield also depends on the age structure of the population and on the degree of age selectivity in fishing methods used. Harvesting older fishes tends to increase the sustainable yield because the remaining younger fishes are faster-growing, but it also tends to reduce the breeding population because the younger fishes are sexually immature.
- Maximum sustainable yield is usually established by using one year's data for stock size and reproductive success to project the survival of adults and young into the next year. Unexpected events, such as disease outbreaks, can render such estimates inaccurate and inadequately protective.
- Most fisheries are managed at a yield 20% to 40% below the estimated maximum sustainable yield to allow a safety margin. This safety margin may not always be adequate, but any safety factor means that fishes that could be harvested will not be.

UNDERSTANDING THE CONCEPT

Although fisheries and **shellfisheries** provide only a small fraction (about 1%) of the world's total human food supply, they provide a much larger percentage of its required protein. In many areas, the oceans are the only significant source of protein.

The world's total fish catch is currently approximately 80 million tonnes per year, about one-third of the estimated global annual fish production. We do not know at what level increased fishing would cause severe damage to the ocean **ecosystem,** but it is believed that damage would be likely if the global fishery catch reached 45% to 50% of total fish production—only about a 35% increase from the current level of fishing. The oceans do not appear to have the potential to solve the world's food supply problem, although some optimistic biologists believe that the global fishery catch could be raised by a factor of as much as 10. These optimistic estimates are based on the assumption that existing fisheries can be exploited to the maximum possible extent, that many new stocks of fishes will be discovered, and that many currently underutilized exotic **species,** including **invertebrates** such as **sea cucumbers** and **sea urchins,** will be fully utilized. Unfortunately, fisheries throughout the globe are already showing signs that overexploitation has depleted the stocks of many fish species, and many fisheries have been closed or severely restricted in attempts to reverse this damage.

If fisheries are to continue sustaining human populations in areas that historically have been dependent on them, and if seafood is to continue supplying the same or a larger proportion of human food, the world's fisheries must be managed carefully. The goal of such management must be to catch the maximum amount of seafood that can be taken from the oceans without damaging individual species or marine ecosystems. To fully meet this objective, we would need to harvest species selectively at low **trophic levels** instead of currently consumer-desired species, such as tuna. Furthermore, we would have to learn to farm the sea as we do the land, eliminating undesirable plants and animals from ocean farms.

Neither approach is likely to be acceptable, at least in the foreseeable future. Consequently, the principal approach of fishery management, which will probably continue for many years, is to manage fishing of each individual species that is targeted by fishers. The specific objective is to maximize the total amount of the species caught (the yield) while ensuring that the **standing stock** does not decline to levels that cannot sustain the yield in the future. Thus, the goal is to manage each species to ensure the **maximum sustainable yield.**

To determine the maximum sustainable yield, we must have a good understanding of the species' life cycle. Consider how fish stocks respond to fishing. The stock (or **biomass**) of a fish species is limited primarily by its food supply and its predators. If human or other predators harvest more of the species than is normally taken by its natural predators, more food will be available for others of the species. If excess food is available, the species will reproduce and grow to use this food until the population is again at a size where food availability limits further growth. Exceptions occur when the excess food is consumed by competitor species, but in most cases, if we harvest a species and reduce its stock somewhat, the total production of the species will increase. If we harvest an amount equal to this increased production each year, the population will remain stable, but at a lower number than before we began to harvest. If we harvest more than the amount by which the production is increased, the stock will progressively decline.

As a stock declines with increased fishing, a critical level, generally 35% to 70% of the original stock, is reached at which the remaining stock becomes so small that it is barely able to grow and reproduce fast enough to replace fishes removed from the population by predation and fishing. If fishing continues to increase beyond this critical point, the stock will begin to decline precipitously (**Fig. CC16.1a**), and unless the catch is reduced, the stock will collapse to a very low abundance and will not recover. The critical level at which the species is just able to replace the stock lost to fishing is the maximum sustainable yield.

One of the most important characteristics of a fishery is the "fishing effort," which is the number of boat or person days of fishing expended. As a typical fishery develops, fishing effort increases as more boats and fishers target the resource species. At first, yield increases rapidly, but as the maximum sustainable yield is approached, fishing effort increases faster than yield because stock size is reduced (**Fig. CC16.1a**). Each boat must fish longer to catch the same amount of fish. Consequently, either the cost of each fish caught rises, or the catch and income of each fisher declines. Therefore, for economic reasons the optimum harvest level of a species may be well below its maximum sustainable yield. Fishery management often controls fishing effort to control yield (**Fig. CC16.1b**).

If the maximum sustainable yield is exceeded in an uncontrolled fishery, fishing effort often rises dramatically as stocks decline. Fishers targeting the species try to protect their livelihoods by increasing their efforts in an attempt to maintain their individual historical catch levels. This dynamic occurs both with technologically advanced fishers targeting a regional resource, such as North Atlantic cod, and with subsistence fishers in island communities whose increasing populations cause an increased fishing effort for local **reef** fishes.

Changes in the age structure of populations that are subject to fishing also affect maximum sustainable yield because the age structure affects the population's ability to reproduce. Like children, young fishes increase their biomass with time faster than adults do, even if they consume similar amounts of food. If larger adult fishes of a

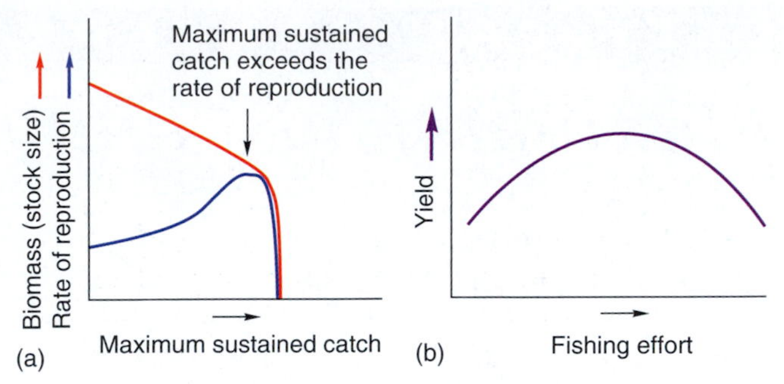

FIGURE CC16.1 Conceptual model showing (a) the changes in fish stock size with increasing fishing, and (b) fishing yield as a function of fishing effort. Note that moderate fishing normally can produce high sustained yields because the rate of reproduction initially increases as the stock size is reduced and more food becomes available to younger fishes. However, if the fishing effort and yield are increased further, the stock size is reduced to such a level that there are no longer enough individuals to sustain a high rate of reproduction, and the stock collapses.

particular stock are preferentially targeted by fishers, the excess food and decreased competition from larger fishes will enable a greater number of young fishes to grow successfully. Because young fishes gain weight or biomass faster than adult fishes do, a greater number of young fishes in the population will cause an increase in the rate of biomass production, even if the available food supply remains the same. If large fishes are preferentially removed, the stock gains more biomass per year and the maximum sustainable yield is increased. Hence, fishery management commonly attempts to control the size of harvested fishes by establishing minimum mesh sizes for nets, requiring the release of captured small individuals, or other methods.

Although it is generally advantageous to reduce the average age of a fish population to increase the maximum sustainable yield, the stock can be adversely affected if too many large fish are removed. Only the larger fish of most species are of reproductive age. Therefore, reducing the numbers of large fishes also reduces the breeding population.

Clearly, establishing maximum sustainable yield is a difficult task that requires knowledge of stock size, age structure, and reproductive process for each species. It is complicated further by the natural year-to-year variability of fish populations and the requirement that maximum sustainable yield be established before, or early in, a given year. The natural variability of fish populations from year to year is very large, at least for some species, because of **climate** variations, disease, and other factors. For example, a relatively small change in water temperature may alter the timing of a **phytoplankton bloom.** In turn, this change may cause an entire year class of **larvae** to die if the larvae are dependent on the timely availability of this food. In addition, because diseases are present in fish populations, the equivalent of epidemics can occur and decimate the stock. Therefore, fish stocks can vary dramatically from year to year and on even shorter timescales, and the maximum sustainable yield will vary accordingly.

If the size and condition of the stock were continuously monitored and known, the maximum sustainable yield could be continuously adjusted to accommodate changes. However, data cannot be gathered and analyzed quickly enough to do this. Consequently, maximum sustainable yields generally are estimated from the previous year's data. Estimates of the stock of adults and number of juveniles that enter the population are made each year. In addition, estimates are made of the survival rates of both adults and juveniles through the coming winter. These data are used to project what the stock will be in the following year, and the projection is used to estimate maximum sustainable yield and set fishing limits for the following year. If an unexpected event occurs that adversely affects the population after the maximum sustainable yield has been estimated and fishing limits established, the permitted yield may be high enough to damage the stocks. If the unexpected decrease in the stock is recognized early enough in the fishing season, emergency measures can be taken to reduce fishing efforts. However, the stock size is often not well known until after the fishing season ends.

To account for uncertainties and variability in maximum sustainable yield, most fisheries set the permissible catch 20% to 40% below the estimated maximum sustainable yield. This practice leads to conflicts because some people feel that this safety margin is not enough to ensure protection of the stock and others feel that part of the resource is being wasted because the maximum sustainable yield is not fully used.

CRITICAL CONCEPT 17

Species Diversity and Biodiversity

ESSENTIAL TO KNOW

- *Species diversity* is a well-defined term that expresses a combination of species richness (number of species) and evenness (degree to which the community has balanced populations with no dominant species).
- High species diversity is generally equated with healthy ecosystems, but there are exceptions.
- *Biodiversity* is a poorly defined term that refers to a combination of genetic diversity (genetic variation within populations of a species), species diversity (species richness and evenness within a community), ecosystem diversity (variations in communities of species within an ecosystem), and physiological diversity (variations in feeding, reproduction, and predator avoidance strategies within a community or ecosystem).

UNDERSTANDING THE CONCEPT

The term **biodiversity** entered into common usage when the global community of nations recognized the need to preserve species from extinction and unique **habitats** or **ecosystems** from destruction. The term has no precise definition but is based on the much more precisely defined technical term **species diversity,** which has been in use for many years.

Species diversity is a measure of species richness and evenness. *Richness* describes the number of individual species, whereas *evenness* expresses the degree to which a **community** has balanced populations in which there is not a small number of numerically dominant **species.** Several statistical indices have been developed to express both species richness and evenness as a single species diversity number.

Species richness is important because the larger the number of species present in a community, the more robust the community is considered to be. If there are many species, a community disturbance (such as a change in temperature or **salinity**) may lead to exclusion or even extinction of one or a few species, but most species are likely to survive. Evenness is important because if the community is dominated by only a few species and these species are excluded or rendered extinct by a disturbance, competition by remaining species to become the new dominant species may cause severe ecosystem instability. Alternatively, an uneven community may reflect effects of a disturbance, such as **pollution,** that favors dominance by tolerant species.

Although communities generally are thought to be desirable and stable if they contain many species and have strong evenness, those characteristics may not always be ideal. For example, **coral reefs** that are undisturbed by storm wave damage for many decades become dominated by just a few species of hard **corals,** and most other hard coral species are excluded by competition with these dominant species. Thus, the beautiful untouched hard coral communities present in a few sheltered locations have low species richness and species are unevenly distributed. These **reefs** have low species diversity (at least of corals).

In contrast, a coral reef that has been damaged several years earlier by a **hurricane** has more species of corals (and possibly other species) and greater evenness, or higher species diversity. Few of us would consider this damaged **environment** an ideal situation, but the periodic disturbance may be necessary to restore and preserve the diversity of coral reefs since high species diversity is generally considered a positive attribute of any ecosystem.

During the hundreds of millions of years that life has existed on the Earth, virtually all the species that have lived have become extinct and been replaced by others. Because this natural process still continues today, none of the species now on the Earth are likely to be living a few million years in the future. Human disturbances, including pollution, have caused many species to become endangered or extinct. The rate of species extinction due to human disturbances is estimated to be many times faster than natural extinction, and possibly is much faster than the rate at which new species can evolve to replace those that are lost.

Only recently have scientists begun to realize that probably the best way to control the extinction rates of species is to control losses of their habitat and disturbances of their community structure. In response to the need to consider habitat and community structure in protecting individual species, a new concept was developed called "biodiversity." Biodiversity is still often interpreted as species richness, but it should include many attributes of natural ecosystems. For example, biodiversity can be separated into four components:

- ***Genetic** diversity*, or variation in the genes within a species or population. High genetic diversity is thought to maximize the potential for new species development. It is also thought to maximize the potential for species survival when the species is subjected to an environmental disturbance because some members of the species may be more resistant.
- *Species diversity*, or species richness and evenness. High species diversity is thought to maximize the stability of the ecosystem and its resistance to environmental disturbances.
- *Ecosystem diversity*, or variation in the communities of species within an ecosystem. High ecosystem diversity is thought to reflect the availability of a wide variety of **ecological niches.** If the range of niches is large, environmental disturbance is unlikely to alter the ecosystem in such a way that more than a few species lack a niche within which to survive.
- *Physiological diversity,* or variation in physiological adaptations to feeding, reproduction, and predator avoidance within a community or ecosystem. High physiological diversity is thought to reflect a greater ability of the community within an ecosystem to adjust to environmental disturbances and maintain its stability.

A global agreement was established in 1993 to preserve biodiversity. Because *biodiversity* is a poorly defined term that encompasses many different characteristics of organisms in their living environment, this agreement will continue to be very difficult to translate into management actions. Protection of biodiversity may mean very different things to different nations or individuals.

CRITICAL CONCEPT 18

Toxicity

ESSENTIAL TO KNOW

- A toxic chemical is a substance that can cause death or adverse sublethal effects in organisms exposed to it at a concentration above a critical threshold.
- Toxic chemicals have many possible sublethal adverse effects on organisms, such as inhibition of the ability to photosynthesize or feed. The most important sublethal effects appear to be those that interfere with reproductive success.
- All toxic chemicals have a concentration threshold below which they have no sublethal or lethal toxic effects. Hence, for every chemical in the ocean environment, there is a concentration below which it is environmentally safe.
- Many substances that are toxic at high concentrations are also essential to life, and the growth of marine organisms may be inhibited if they are not present above a certain concentration. For these substances, there is an optimum range between the minimum

concentration that an organism requires to supply its needs and the concentration above which the substance is toxic.

- The range of optimum concentration can be large or small. The concentration at which toxicity occurs can be many times higher than or very close to the natural range.
- Anthropogenic inputs of toxic substances can be assimilated safely in the oceans if the amount introduced does not cause concentrations to exceed the threshold at which sublethal toxicity occurs. The quantity that can be safely assimilated is different for each substance and determined in part by its sublethal toxicity threshold and background concentrations.
- Sublethal or lethal toxicity threshold concentrations are difficult to determine because they vary among species, among substances, and with other factors, such as physical stresses and synergistic and antagonistic effects of other chemical constituents.
- Marine organisms bioaccumulate most toxic substances. Bioaccumulation occurs when the concentration in the organism is higher than the environmental concentration but the concentrations are in equilibrium.
- A few toxic substances are biomagnified in marine organisms. Biomagnification occurs when the organism retains all the toxic substance to which it is exposed in its food or environment and does not lose any of the substance, even if its environmental concentration decreases.
- Carcinogenicity, mutagenicity, and teratogenicity can be considered to be lethal or sublethal effects, but there is probably no concentration threshold below which there is no effect. Each exposed organism has a small probability of suffering an effect at any specific concentration. More individuals within a population suffer the effect as concentration increases.
- Carcinogenic, mutagenic, and teratogenic substances occur naturally. Anthropogenic inputs will increase the incidence of the effects of such substances. However, at least for some of these substances, anthropogenic inputs would need to be large before the increased incidence would be significant or measurable in comparison with the natural incidence of effects.

UNDERSTANDING THE CONCEPT

A toxic chemical is a substance that can cause death or adverse sublethal effects in organisms, including human beings, that are exposed to the substance at or above certain concentrations. If an organism is exposed to a concentration that is lethally toxic, the organism dies. If an organism is exposed to a concentration that is sublethally toxic, the organism is not killed, but it is disadvantaged in some way.

Examples of the many possible sublethal effects are partial inhibition of **photosynthesis** in plants and partial inhibition of feeding, hunting, or prey avoidance capabilities in animals. However, the most important sublethal effects are generally believed to be those that reduce reproductive success. Reproductive success can be reduced by partial inhibition of egg or sperm production, reduction in the probability of fertilization when egg and sperm meet, reduction in the survival ability of **larvae** or juveniles, and in other ways. Because of the wide range of possible sublethal effects caused by toxic chemicals, identifying and measuring such effects is often impossible either in the **environment** or in laboratory experiments.

Effects of any toxic chemical vary with its concentration (**Fig. CC18.1**) and with individual **species.** Above a certain concentration, the toxic chemical produces death. At somewhat lower concentrations, the same substance may produce sublethal adverse effects. Below an even lower concentration, the substance causes no sublethal effects and does not harm the organism. This concentration is the sublethal toxicity threshold (**Fig. CC18.1a**). Any concentration below the sublethal toxicity threshold could be considered an acceptable concentration. However, this is not true for naturally occurring toxic substances, because some are also essential **nutrients.** These substances have an optimum concentration range below the sublethal toxicity threshold, within which they do not harm the organism (**Fig. CC18.1b**). However, below this optimum range, some essential function is limited by a lack of the substance, and the organism is disadvantaged (**Fig. CC18.1b**).

All toxic substances have an acceptable or optimum concentration range within which they do not harm organisms or the **ecosystem.** Therefore, **anthropogenic** additions of toxic chemicals to the oceans are not always harmful. Such inputs can be safe if the **assimilative capacity** or acceptable concentration range in the environment is not exceeded. The assimilative capacity is determined by the sublethal toxicity threshold of the most sensitive species.

The relationships shown in **Figure CC18.1** generally apply to all toxic chemicals other than **carcinogens, mutagens,** and **teratogens.** However, concentrations at which the inhibition of growth ends (for essential nutrients), sublethal toxicity begins, and lethal toxicity occurs can all vary by orders of magnitude for different species, and even for different life stages of a species, exposed to a specific toxic substance.

Most importantly, the optimum concentration range for essential nutrients may vary dramatically. For example, zinc is a toxic but essential chemical that has a wide optimum concentration range. In certain parts of the oceans, zinc may be the nutrient most likely to be depleted and hence most likely to restrict the growth of certain organisms. Because it has a large optimum concentration range, zinc becomes sublethally toxic only at concentrations that are many orders of magnitude above those normally present in the oceans. In contrast, copper has a very narrow optimum concentration range. Like zinc, copper may be growth-limiting to certain species in some parts of the oceans. However, at concentrations only a few times higher than the limiting concentration and within the range often present in **contaminated** coastal waters, copper can reduce the rate of photosynthesis by certain species of **phytoplankton.** Hence, copper is sublethally toxic at concentrations very close to those that occur naturally in the oceans.

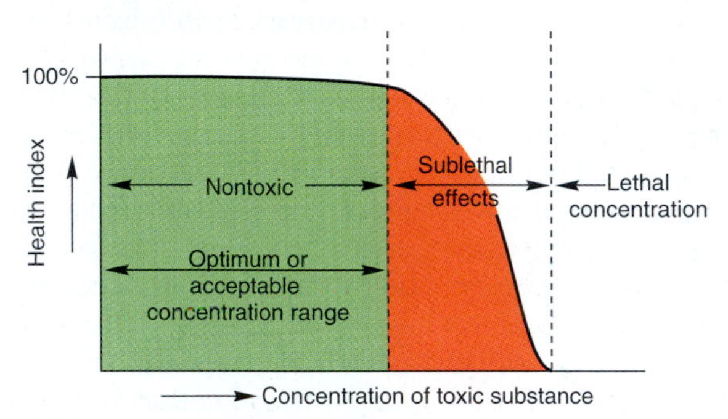

(a) Nonessential element or compound

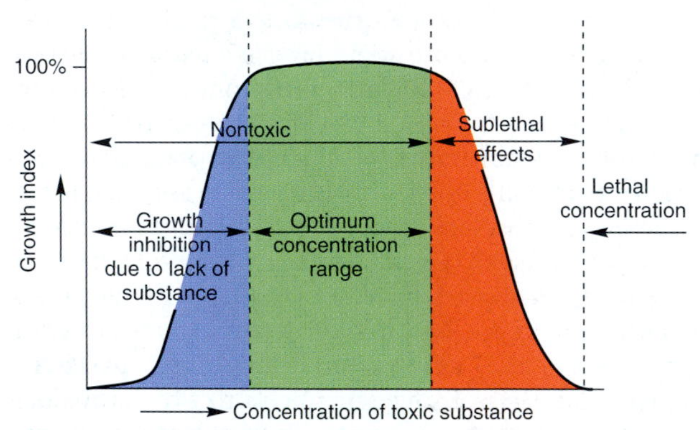

(b) Essential element or compound

FIGURE CC18.1 Conceptual model of toxicity, showing the general behavior of two different types of toxic substances: (a) elements or compounds that are not essential to life, and (b) elements or compounds that are essential to living organisms for their growth and reproduction, but that become toxic to the organisms if their concentration is too high. The health index plotted on the y-axis for nonessential substances in (a) can be many different factors, such as reproductive success or incidence of disease. For essential substances (b), the index of growth similarly can be such factors as primary production rate or biomass. The concentrations at which transitions occur among growth inhibition, the optimum or nontoxic range, sublethal concentrations, and lethal concentrations vary among species, among life stages (e.g., juveniles and adults), and among toxic substances. They can even vary among individuals of a species or for a single individual, depending on the level of environmental stress that the individual is experiencing as a result of other factors.

Acceptable or optimum concentration ranges and the range of concentrations present naturally must both be considered in assessments of whether anthropogenic discharges might create **pollution** problems. For example, much smaller quantities of copper than of zinc can be discharged safely to the oceans.

Concentrations that represent the transitions between growth inhibition, nontoxicity, sublethal toxicity, and lethal toxicity, are very difficult to determine for several reasons. First, members of a species differ somewhat in susceptibility to a given toxic chemical. Second, different species and life stages of a single species may have dramatically different susceptibilities to a particular toxic chemical. Third, complicated antagonistic and synergistic effects occur among toxic substances and other chemicals in the environment. Synergistic effects increase toxicity. For example, high concentrations of one trace metal can increase toxicity of another trace metal. Antagonistic effects are the opposite; high concentrations of one trace metal may reduce the toxicity of another metal. Fourth, the susceptibility of a species to a toxic chemical may be altered by stress due to other environmental factors, including temperature, **salinity,** and the extent of competition from other species. Fifth, the toxicity of many substances is determined by their chemical form. For example, **ionic** copper is substantially more toxic than copper **complexed** with organic matter.

Because of the variability among chemicals, among species, among species' life stages, among chemical states, and with synergistic, antagonistic, and stress factors, it is not possible to determine a single concentration at which a toxic chemical is lethal or sublethal in the environment. For this reason, toxicologists often measure lethal toxic concentrations of chemicals on selected species in the laboratory by using **bioassays**. They assume that sublethal effects will occur at concentrations below the lethal concentration. They also assume that some species are more sensitive to the toxic chemical than the tested species is. A "safe" concentration in the environment is then estimated. Generally, a concentration two orders of magnitude below the minimum lethal concentration is assumed arbitrarily to be safely below the level at which any sublethal adverse effects occur. However, in some cases, such as copper, the optimum concentration range (**Fig. CC18.1b**) is less than two orders of magnitude below the lethal concentration threshold. Therefore, this approach may set the "safe" concentration below naturally occurring levels and sometimes even below the level needed to sustain growth.

Bioassays are conceptually simple. The test species is exposed usually for 96 h (a duration chosen because one set of tests can be set up and run in a 5-day workweek) to several different concentrations of the toxic substances in seawater. The test species is also exposed to control seawater with no toxic substance added. After exposure, the number of organisms that have died or exhibited a sublethal effect (e.g., failure of an **embryo** to develop or of a larval stage to **metamorphose** to the next life stage) are counted. Generally, no death or sublethal effect occurs in the control seawater (otherwise the experiment is not acceptable, because a factor other than the tested contaminant has caused a confounding effect). If the test substance is toxic, some or all individuals will have died or shown the sublethal effect in the test with the highest contaminant concentration. A progressively smaller proportion of the test population usually is affected at lower contaminant concentrations. From these tests, the contaminant concentrations at which one-half of the test organisms are killed or suffer the observed sublethal effect are calculated. These values are called the LC_{50} and EC_{50}, respectively.

Bioassay tests must be interpreted carefully because they do not use the most sensitive organism. Without testing all species, there is no way to determine which is the most sensitive. In fact, because different life stages of some species have different levels of sensitivity (juvenile stages are generally more sensitive than adults), all life stages would have to be tested. In addition, most marine species are difficult to keep alive in the laboratory because this alien environment stresses the organisms. Species used for bioassays must be stress-tolerant, and hence they are also likely to be tolerant of toxicant stress.

Many factors, such as salinity, temperature, light intensity, water chemistry, **sediment** characteristics, food supply, and competition or cooperation with other individuals or species, are different in bioassays than in the environment. The additional stresses of the test environment may make test organisms more susceptible to toxicant stress.

Bioassays are often conducted with samples of wastes in the form discharged (e.g., sewage or industrial **effluent**). Although this approach takes into account synergism or antagonism between components of the waste, there remains the question of whether additional such effects occur when the waste is mixed in seawater and subject to different conditions in the environment.

Most toxic chemicals are more highly concentrated in tissues of marine species than in the surrounding seawater. This fact has led to a popular belief that toxic chemicals are taken up continuously by marine species and that concentrations of the toxins in their tissues increase progressively throughout their lifetime, regardless of how the concentration changes in their environment. This may be true for some species and some toxic chemicals, but it is not true for most species and most toxic substances. Most toxic chemicals are **bioaccumulated** by most marine species, but are not **biomagnified.**

Bioaccumulation occurs when the concentration of a compound is regulated by equilibria between the organism and its food and/or surrounding water. The toxic substance is taken in by the organism from its food and/or directly from the water, but the organism is capable of **excreting** some of the toxic substance either directly back to the water or through its urine or feces. Hence, if the food or seawater concentration of the toxic substance increases, the concentration in the organism will increase, but if the food or seawater concentration decreases, the concentration in the organism will also decrease. However, it may take some time for equilibrium to be reached.

Certain toxic substances are more dangerous than others in marine ecosystems because they biomagnify. Biomagnification occurs when organisms at each **trophic level** in a **food chain** retain all or almost all of the toxic substance ingested in food. Consequently, the concentration of the toxin increases at each trophic level.

Certain chemicals are carcinogenic (cancer-causing), mutagenic (causing **genetic** changes in the offspring by altering the parental **DNA**), or teratogenic (causing abnormal development of the **embryo**). Although each can be considered a sublethal effect, many scientists do not believe that these effects follow the toxicity-concentration relationships shown in **Figure CC18.1**, but rather that they may occur at any concentration of these substances.

If this hypothesis is correct, a single molecule of such a chemical could cause cancer in an exposed individual. However, if many individuals were each exposed to a single molecule, only a few (if any) would be expected to develop cancer. If the same group of individuals were each exposed to two molecules of the chemical, twice as many would be expected to develop cancer. This type of effect is said to be "probabilistic." The probability that any one individual will contract cancer from a given concentration of the chemical is reduced proportionally as the concentration is reduced, but it never reaches zero.

There are many naturally occurring carcinogens, teratogens, and mutagens, but only a very small percentage of marine populations suffers adverse effects as a result of these naturally occurring compounds. Because such a small fraction is affected, great numbers of individuals would have to be monitored to estimate accurately the percentage of a population that suffers such effects. Furthermore, determining the extent to which a higher incidence of such effects might be caused by human contamination of the oceans, or parts thereof, would be even more difficult.

If the probabilistic-effect hypothesis is correct, any quantity of anthropogenic input of a carcinogen, mutagen, or teratogen will increase the incidence of such effects in marine species. The increase may be extremely small in comparison with the incidence of such effects due to the naturally occurring concentrations, in which case many scientists believe adverse effects on the marine ecosystem would be negligible. Hence, some believe that it would be environmentally safe and acceptable to dispose of limited quantities of anthropogenic carcinogens, mutagens, and teratogens in the ocean.

CHAPTER 4

CHAPTER OUTLINE

CRITICAL CONCEPTS USED IN THIS CHAPTER

An eruption of Pu'u O'o, the active vent of the Kilauea Volcano in Hawaii. This hot-spot volcano erupts relatively smoothly, but it was ejecting cinders and ash into the atmosphere that were "raining" on the airplane windshield like a hailstorm when this photograph was taken. The shiny surface below the vent mouth is a crust of solidified lava that conceals a rapidly flowing molten river of lava. The river of lava extended several kilometers down the slopes of Kilauea, where it destroyed several homes shortly after this photograph was taken.

Plate Tectonics: Evolution of the Ocean Floor

On January 3, 1983, the Kilauea Volcano on the Big Island of Hawaii erupted. I was scheduled to visit Hawaii in May of that year. Kilauea had erupted a number of times during the previous decade, but all those eruptions had ended after just a day or two, except for one that lasted about 2 weeks and another for almost a month. As a result, I was very disappointed because I believed I would not have the opportunity to see this eruption. To my surprise, the eruption continued through May, and I was fortunate to be able to fly around and over Kilauea during a very active period. At that time, lava was spewing from the vent to feed several fast-running rivers of glowing-red, molten lava. It is impossible to describe the raw power and majesty that I felt privileged to experience. Even from several hundred feet above the volcano, the airplane engine noise was drowned out by the roar of the venting lava and the explosive sounds created as ash and gases mixed with the lava were ejected into the air. This was a sound I will never forget. As I watched and listened in awe, I could feel the radiated heat through the plane's windows. The plane was buffeted about in the thermal air currents rising from the vent and pelted by cinders and ash ejected to an altitude well above where we were flying.

This experience, matched only by witnessing the 1989 Loma Prieta earthquake and its destruction of freeways and buildings in the San Francisco Bay area, has given me a deep appreciation for the almost unimaginable forces that accompany plate tectonics. Even these experiences cannot but hint at the scale of plate tectonic processes.

For example, the Kilauea eruption that started in 1983 has continued without interruption for more than two decades. In fact, it is still continuing in 2006 as I write these words. Is this length of continuous eruption unique? Not at all! For example, Kilauea was in a nearly continuous eruption for over 100 years between 1823 and 1924. Indeed, it is believed that Kilauea has been an active volcano for most of the 50,000 to 100,000 years since it first emerged above the ocean surface and for the 200,000 to 500,000 years before that as it was steadily built up from the ocean floor—an enormous mountain-building project that truly dwarfs the human scale.

The ocean floor has irregular and complex **topographic** features—including mountain ranges, plains, depressions, and plateaus—that resemble topographic features on land (**Inside front cover map**). During the period of human history, those features have remained essentially unchanged aside from local modifications due to **erosion**, **sedimentation**, volcanic eruptions, and **coral reef** formation. However, the human **species** has existed for only a very brief period of the planet's history (**Fig. 1.5**). In the billions of years before humans appeared, the face of the Earth was reshaped a number of times.

4.1 STRUCTURE OF THE EARTH

The continents and ocean basins are not permanent. Instead, the location, shape, and size of these features are continuously changing, albeit imperceptibly slowly on the timescale of human experience. The global-scale processes that reshape the face of the planet are, as yet, far from fully understood. However, we do know that these processes, called **plate tectonics**, originate deep within the Earth.

Layered Structure of the Earth

The Earth consists of a spherical central core surrounded by several concentric layers of different materials (**Fig. 4.1**). The layers are arranged by **density**, with the highest-density material at the Earth's center and the lowest-density material forming the outer layer, the Earth's **crust**. This arrangement came about early in the Earth's history, when the planet was much hotter than it is today and almost entirely fluid. The densest elements sank toward the Earth's center, and lighter elements rose to the surface (**CC1**). The Earth's center is still much hotter than its surface. Heat is generated continuously within the Earth, primarily by the decay of **radioisotopes** (**CC7**).

The core, which is about 7000 km in diameter, is composed primarily of iron and nickel, and is very dense. It consists of a solid inner core and a liquid outer core. The **mantle**, which surrounds the core, is composed of material that is about half as dense as the core. The upper mantle, known as the **asthenosphere**, is thought to consist of material that is very close to its melting point and "plastic," so it is capable of flowing very slowly without fracturing. The best examples of such materials in everyday life are glass and **glacial** ice. Although glass appears to be a solid, examination of glass windowpanes that have been in place for centuries reveals that the glass is thicker at the bottom than at the top. The glass is flowing slowly downward in response to **gravity**. Similarly, ice within glaciers also flows slowly. Recent evidence indicates that not only the asthenosphere, but also much of the deeper mantle that generally appears solid, may be capable of flowing very slowly. The ability to flow is a critically important factor in the tectonic processes shaping the Earth's surface.

The asthenosphere is surrounded by the **lithosphere**, the outermost layer of the Earth, which varies from just a few kilometers in thickness at the **oceanic ridges** to 100 to 150 km in the older parts of the ocean basins and up to 250 to 300 km under the continents. The lithosphere consists of the mostly rigid outer shell of the mantle plus the solid crust that lies on the mantle. The lithosphere is less dense than the asthenosphere and essentially floats on top of the plastic asthenosphere. The oceans and atmosphere lie on top of the lithosphere. Pieces of lithosphere are rigid, but they move across the Earth's surface and in relation to each other as they float on the asthenosphere. They can be many thousands of kilometers wide, but they are generally less than 200 km thick. Hence, being platelike, they are called **lithospheric plates** (or "tectonic plates" or just "plates"). Processes that occur where the plates collide, where they move apart, or where plates "slide" past each other are the principal processes that create the mountains and other surface features of the continents and the ocean floor.

Lithosphere, Hydrosphere, and Atmosphere

Relative to the Earth, the lithosphere is thin, rather like the skin on an apple (**Fig. 4.1**). At the top of this thin layer are the mountains, ocean basins, and other features of the Earth's surface. From the deepest point in the ocean to the top of the highest mountain is a vertical distance of approximately 20 km. This 20-km range is very small compared to the Earth's radius, which is more than 6000 km. Consequently, the planet is an almost smooth sphere (from space it looks smoother than the skin of an orange) on which the mountain ranges and ocean depths are barely perceptible.

There are two types of crust—oceanic and continental—both of which have a substantially lower density than the upper mantle material on which they lie. Oceanic

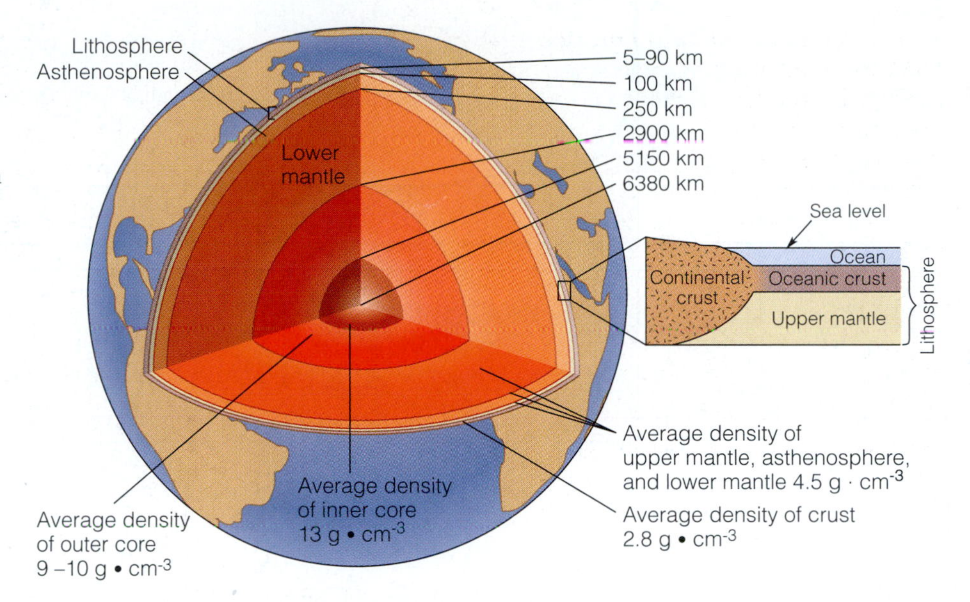

FIGURE 4.1 A cross section of the Earth showing its layers. Note that the thickness of the lithosphere has been greatly exaggerated in this diagram. If it were drawn to the correct scale, the lithosphere would appear as just a thin line at the Earth's surface.

crust has a higher density than continental crust (**Fig. 4.2**). According to the principle of **isostasy** (**CC2**), lithospheric plates float on the asthenosphere at levels determined by their density. Consequently, if the continental and oceanic crusts were of equal thickness, the ocean floor would be lower than the surface of the continents. However, oceanic crust is much thinner (typically 6–7 km thick) than continental crust (typically 35–40 km), which causes an additional height difference between the surface of the continents and the ocean floor (**Fig. 4.3**).

The density difference between continental and oceanic crust is due to differences in their composition. Both are composed primarily of rocks that are formed from cooling **magma** and consist mainly of silicon and aluminum oxides. However, continental crust is primarily granite, and oceanic crust is primarily gabbro or **basalt**, both of which have higher concentrations of heavier elements, such as iron, and thus a higher density than granite.

Surrounding each continent is a **continental shelf** covered by shallow ocean waters. The continental shelf is an extension of the continent itself, so this portion of the continental crust surface is submerged below sea level. Both the width and the depth of the shelf vary, but it is generally in waters less than 100 to 200 m deep and ranges in width from a few kilometers to several hundred kilometers. The continental shelf slopes gently offshore to the **shelf break**, where it joins the steeper **continental slope** (**Figs. 4.3, 4.4**). The continental slope generally extends to depths of 2 to 3 km. Seaward of the base of the slope, the ocean floor either descends sharply into a deep-ocean **trench** (**Fig. 4.4a**) or slopes gently seaward on a **continental rise** that eventually joins the deep-ocean floor (**Figs. 4.3, 4.4b**). Much of the deep-ocean floor is featureless flat **abyssal plain**, but other areas are characterized by low, rolling **abyssal hills**.

A layer of **sediment** lies on top of the oceanic crustal rocks, constituting part of the crust. The thickness of the sediment is highly variable, for reasons discussed in this chapter and Chapter 8.

The **hydrosphere** consists of all water in the lithosphere that is not combined in rocks and minerals—primarily the oceans and the much smaller volume of freshwater (**Table 6.1**). The oceans cover all the oceanic

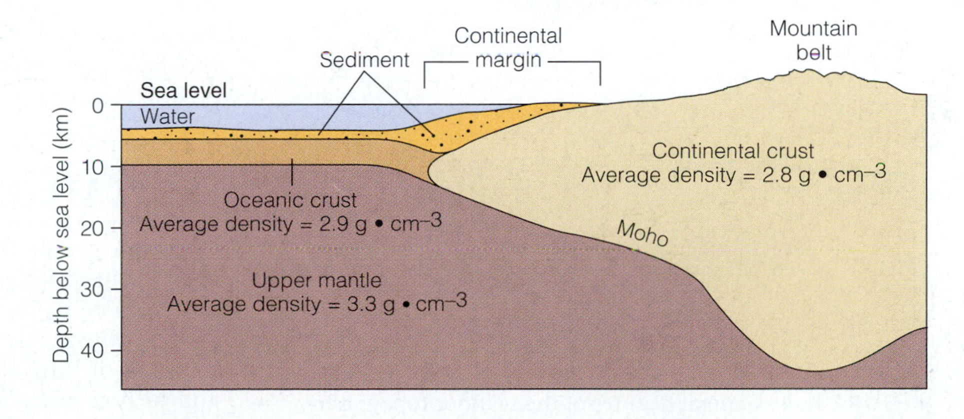

FIGURE 4.2 Structure of the lithosphere. Continental crust is typically 35 to 40 km thick, whereas oceanic crust is typically only 6 to 7 km thick. Continental crust has a lower density than oceanic crust, but both continental and oceanic crust have a considerably lower density than upper mantle material.

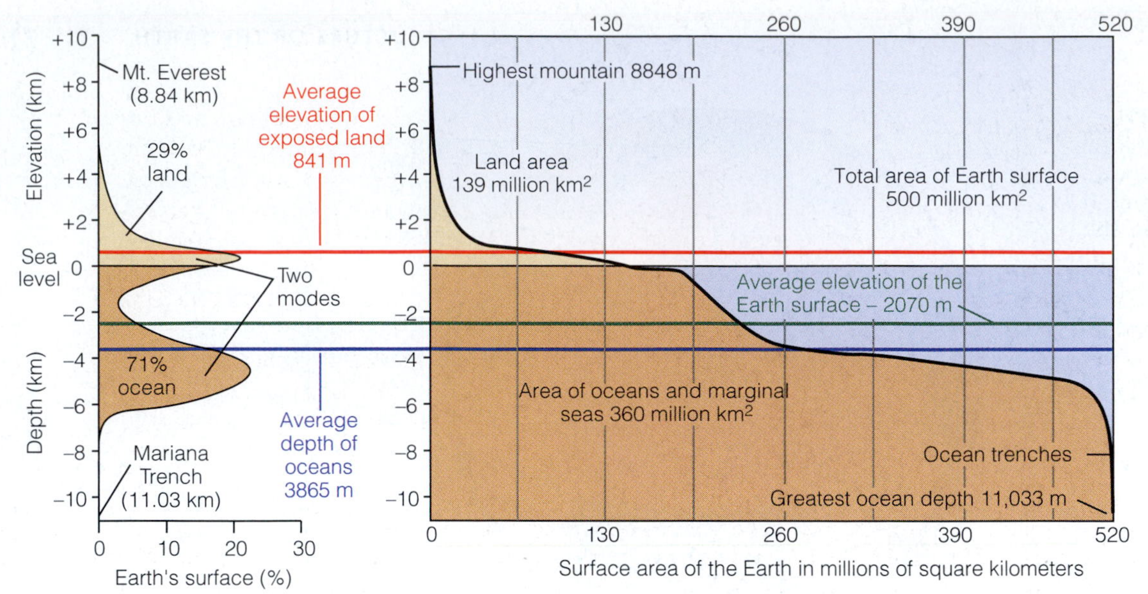

FIGURE 4.3 Relative heights of parts of the Earth's surface. The average elevation of the continents is 841 m above sea level, and the average depth of the oceans is 3865 m. Almost 30% of the Earth's surface is above sea level. Much of the surface of the continental crust is below sea level.

crust and large areas of continental crust around the edges of the continents—all of which total more than two-thirds of the Earth's surface area.

The atmosphere is the envelope of gases surrounding the lithosphere and hydrosphere and is composed primarily of nitrogen (78%) and oxygen (21%). Although these gases have much lower densities than liquids or solids, they are dense enough to be held by the Earth's gravity. Less dense gases, including hydrogen and helium, are so light that they tend to escape from the Earth's gravity into space. Although the light gases were present in large quantities when the Earth first formed, only trace concentrations are present in the atmosphere today. The atmosphere is discussed in Chapter 9.

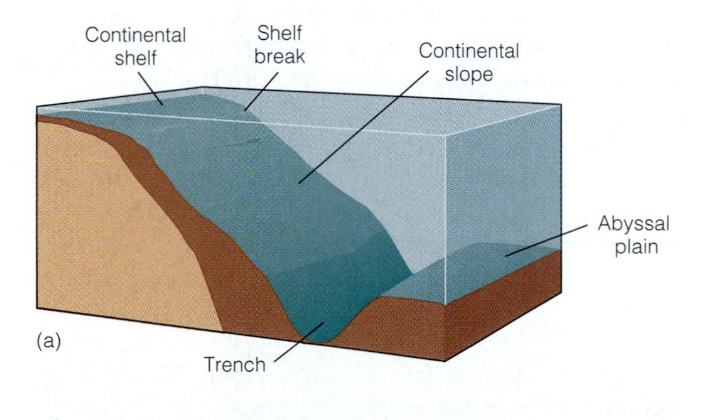

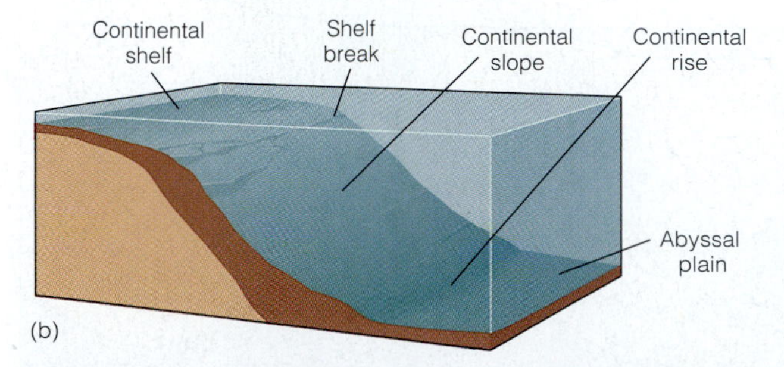

FIGURE 4.4 General features of the seafloor topography between the continents and the deep ocean floor.

Studying the Earth's Interior

The processes that occur beneath the lithospheric plates are very difficult to study because they occur deep within the Earth beneath kilometers of crustal rocks and upper mantle. Scientific drilling is currently limited to depths of only a few kilometers, and thus, samples of materials below the crust have not been obtained directly. Studies of the processes occurring below the crust must rely on indirect observations, such as examination of volcanic rocks, studies of earthquake-energy transmission through the Earth, and studies of meteorites. Meteorites are examined because they are believed to represent the types of material that make up the Earth's core and mantle.

A technique called "seismic **tomography**" now enables scientists to use earthquake waves to study the Earth's internal structure in more detail than ever before. It has yielded several intriguing findings about the Earth's interior. For example, features that resemble mountains and valleys have been found on the core–mantle transition zone. The mountainlike features extend downward into the molten core and are as tall as those found on the Earth's surface. Some of these features may be sedimentlike accumulations of impurities that float upward out of the liquid nickel/iron core. Computer tomography is described in Chapter 5.

4.2 PLATE TECTONICS

Over the geological timescales of the Earth's history, oceans and continents have been formed and have disappeared to be replaced by others through the movements of the lithospheric plates. We now have a basic understanding of how and why the lithospheric plates move and of how plate tectonics has continually shaped and changed the face of the planet.

Driving Forces of Plate Tectonics

The movements of the lithospheric plates are thought to be driven by heat energy transferred through the mantle by **convection** (**CC3**). The mantle has areas where its constituent material is **upwelled** and **downwelled**. As discussed later in this chapter, the convection processes in the mantle are complex, probably **chaotic** (**CC11**), possibly multilayered, and as yet very poorly understood. However, we do know that upwelling material has been heated by the Earth's core and **radioactive** decay within the mantle itself. As a result, it is slightly less dense than the material through which it rises. The upper mantle loses some of its heat by **conduction** outward through the Earth's cooler crust. As the upper material cools and contracts, its density increases. When its density exceeds the density of the mantle material below, the cooled material will tend to sink (downwell).

The precise nature and locations of the **convection cells** in the Earth are not yet known. At least two types of circulation have been hypothesized (**Fig. 4.5**). A shallow circulation is believed to occur within the upper 650 km of the mantle, and a much deeper circulation is believed to extend down almost 3000 km to the core–mantle

FIGURE 4.5 A generalized depiction of the convection processes thought to occur in the Earth's mantle. The view here is from below the Southern Hemisphere. Not quite half of the Earth has been removed by slicing across the Earth at an angle to the equator. This slice removes much of the Southern Hemisphere, including Antarctica, Australia, much of the southern Atlantic and Indian Oceans, and most of the Pacific Ocean as far north as Hawaii and Indonesia. Look at a globe to see this better. Although it is known that the convection processes beneath some hot spots extend to the core–mantle boundary and that some subducted blocks of lithosphere may sink to this boundary, the details of mantle convection circulation are not yet known. Convection cells must exist in the liquid outer core, but almost nothing is known about this circulation. The solid inner core rotates a little faster than the rest of the Earth, making an additional revolution about every 400 years. Scientists believe the faster rotation of the inner core may help explain the periodic reversal of the Earth's magnetic field and the plate tectonic processes that cause the continents to periodically gather together and then spread apart. However, much research is needed before these processes can be understood.

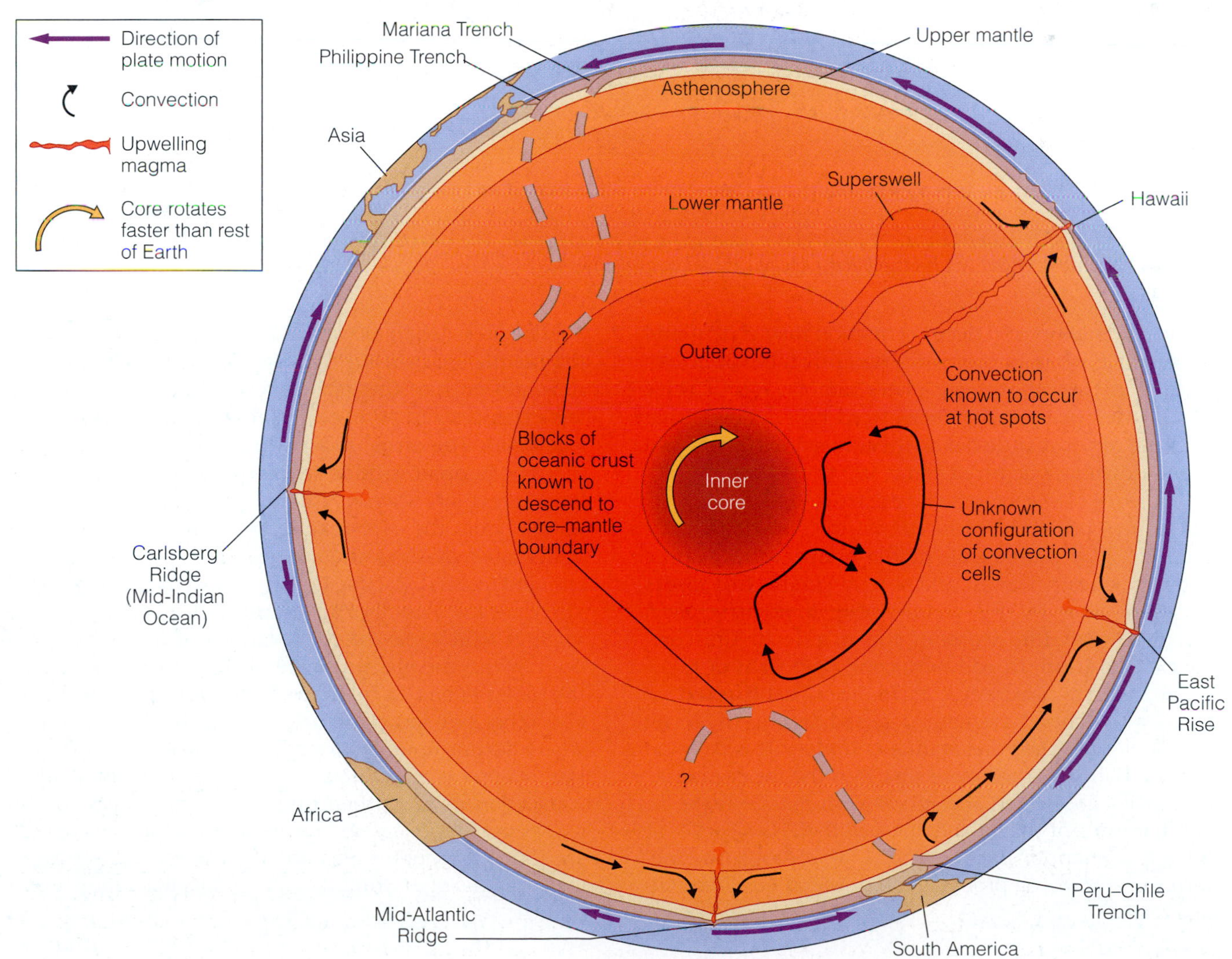

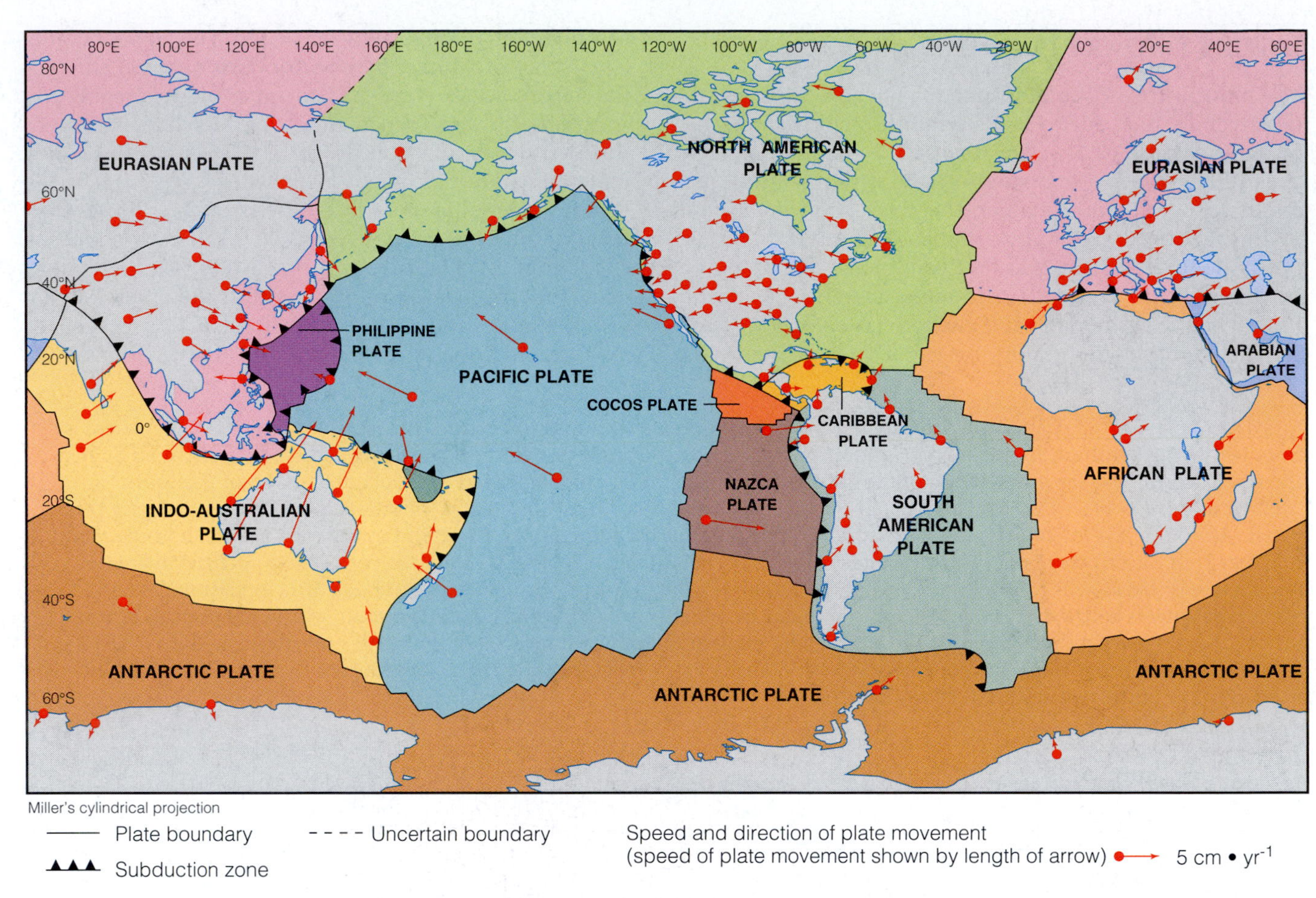

FIGURE 4.6 The major lithospheric plates and their current directions of motion relative to each other. Continental North America lies on the North American Plate, which extends to the center of the Atlantic Ocean. At the Pacific coast boundary of this plate, the North American and Pacific Plates are sliding past each other. The directions of motion show that the Atlantic Ocean is becoming wider, while the Pacific Ocean is becoming smaller.

boundary. Shallow and deep convection may coexist and interact in as yet unknown ways. Recently it has been proposed that the lithospheric plates ride on top of three **mesoplates** (*meso* is the Greek word for "middle") that lie in the upper mantle below the lithospheric plates and extend down to a depth of about 100 to 300 km.

Evidence that supports the existence of deep convection comes primarily from computer tomography. For example, the plumes of higher-temperature mantle material beneath Hawaii and several similar **hot spots** (discussed later in this chapter; **Fig. 4.10**) have been traced through the depth of the mantle to the core-mantle transition. In addition, cool, rigid slabs of lithosphere, which apparently have been **subducted** at deep-sea trenches, have been detected deep in the lower mantle close to the core–mantle boundary. It is believed that the slabs of lithosphere sink into the lower mantle, where they are heated and mixed with mantle material, and that some of this mixed material then rises back into the upper mantle. A single cycle of cooling, sinking, warming, and rising probably takes several hundred million years.

The areas where downwelling is thought to occur in the mantle are beneath the deep-ocean trenches, which surround most of the Pacific Ocean, but they are found less extensively in other oceans (**Inside front cover map, Fig. 4.10**). The areas under which mantle upwelling is thought to occur include two **superplumes**, which are large areas of warmer mantle material that "swell" upward from the core–mantle boundary. These areas lie under much of Africa and a large region of the southwestern and central Pacific Ocean, respectively (**Fig. 4.5**). There are also a number of volcanic hot spots in the crust, some of which may be the locations of upwelling from deep within the mantle. Some of these may originate in the superplumes. As we shall see in this chapter, there are also other hot spots and areas along the oceanic ridges where magma upwells through the lithosphere. However, this magma is now thought to originate from shallow convection in the asthenosphere, or upper mantle. The convection processes in the mantle and asthenosphere are still very poorly understood, and new discoveries or explanations of existing data are published almost every month.

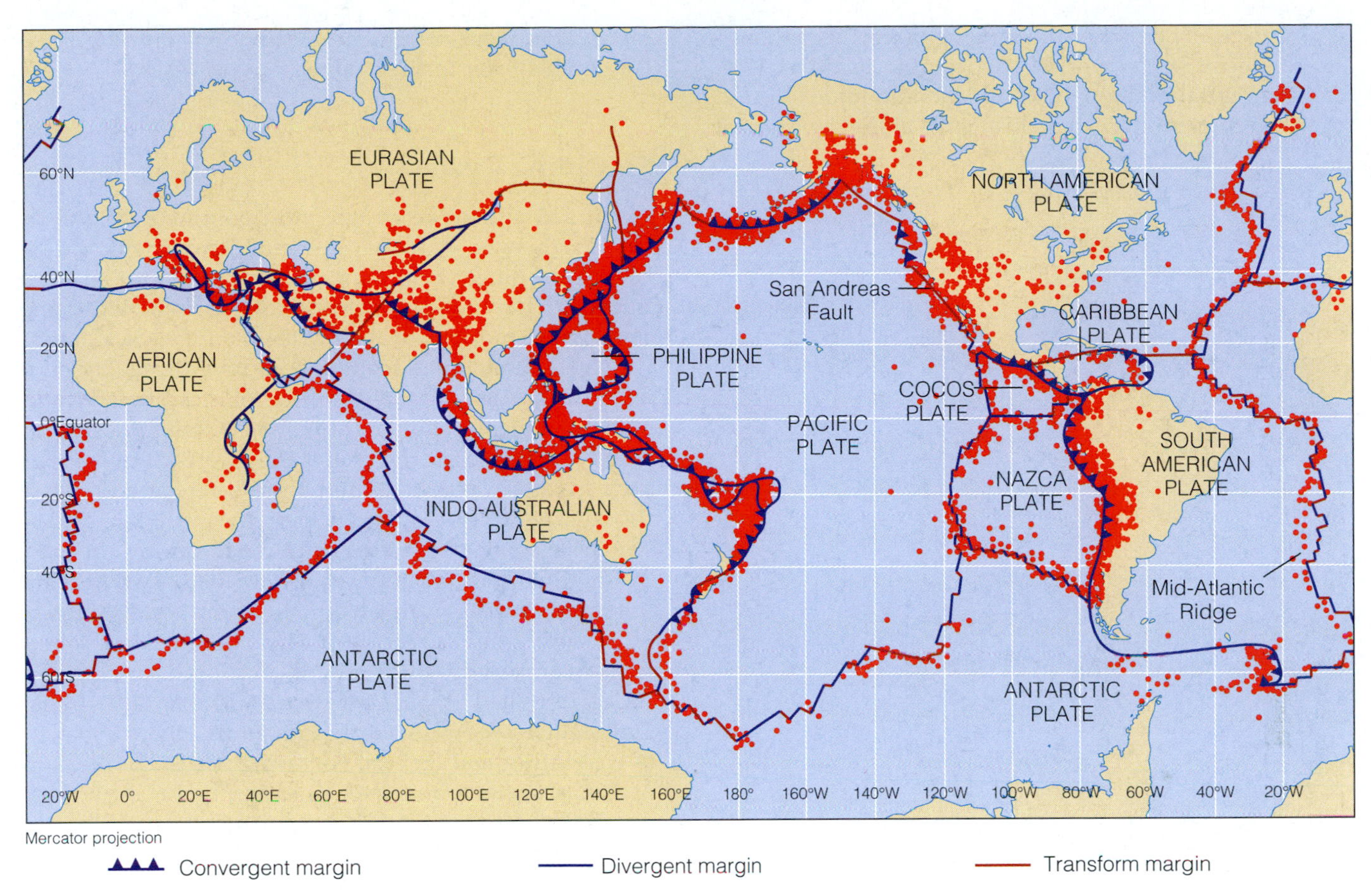

FIGURE 4.7 Locations of the world's earthquakes. Compare this figure with Figure 4.6. Earthquakes are concentrated along plate boundaries and appear to be more frequent at the plate boundaries where plates are converging.

The forces that actually move the lithospheric plates across the Earth's surface act in response to the mantle convection, but they, too, are far from fully understood. The plates are believed to be dragged across the surface as slabs of cold, dense lithosphere as their edges sink at **subduction zones**. The mechanism is somewhat like that of an anchor that, when dropped into the water, will drag down the line attached to it, even if, by itself, this line would float. The descending slabs of lithosphere create an additional effect similar to the vortex created as a ship sinks that drags floating materials down with it. Finally, the plates may move in response to gravity. As we shall see later, oceanic crust near the oceanic ridges (also called "mid-ocean ridges"), where new crust is formed, floats higher than other oceanic crust. The oceanic crust cools, becomes more dense, and sinks as it moves away from the oceanic ridges toward the subduction zones. Thus, the oceanic crust lies on a slope, albeit a very shallow one, between the oceanic ridges and subduction zones and thus has a tendency to "slide" downhill (in response to gravity), with newer, higher crust "pushing" older, lower crust down the slope.

Present-Day Lithospheric Plates

Seven major lithospheric plates are generally recognized—the Pacific, Eurasian, African, North American, South American, Indo-Australian, and Antarctic Plates—as well as several smaller plates (**Fig. 4.6**). However, some plate boundaries are not yet fully defined, and additional small plates undoubtedly exist that currently are considered to be parts of larger plates.

All plates are in motion, and over geological time, both the direction and speed of movement can vary. **Figure 4.6** shows the directions in which the plates are moving relative to each other at present. The movement of the plates is extremely slow, just a few centimeters per year, or about the rate at which human fingernails grow. Despite their slowness, plate movements can completely alter the face of the planet in a few tens of millions of years. As they move, plates collide, separate, or slide past each other, and sometimes they fracture to form smaller plates. At plate edges, the motion is not smooth and continuous; it occurs largely by a series of short, sharp movements that we feel as earthquakes. Most earthquakes occur at plate boundaries (**Fig. 4.7**).

Interactions between the moving plates are responsible for most of the world's topographic features. Most major mountain ranges, both on land and undersea, and all the deep-ocean trenches are aligned along the edges of plates. How plate movements create topographic features is examined later in the chapter.

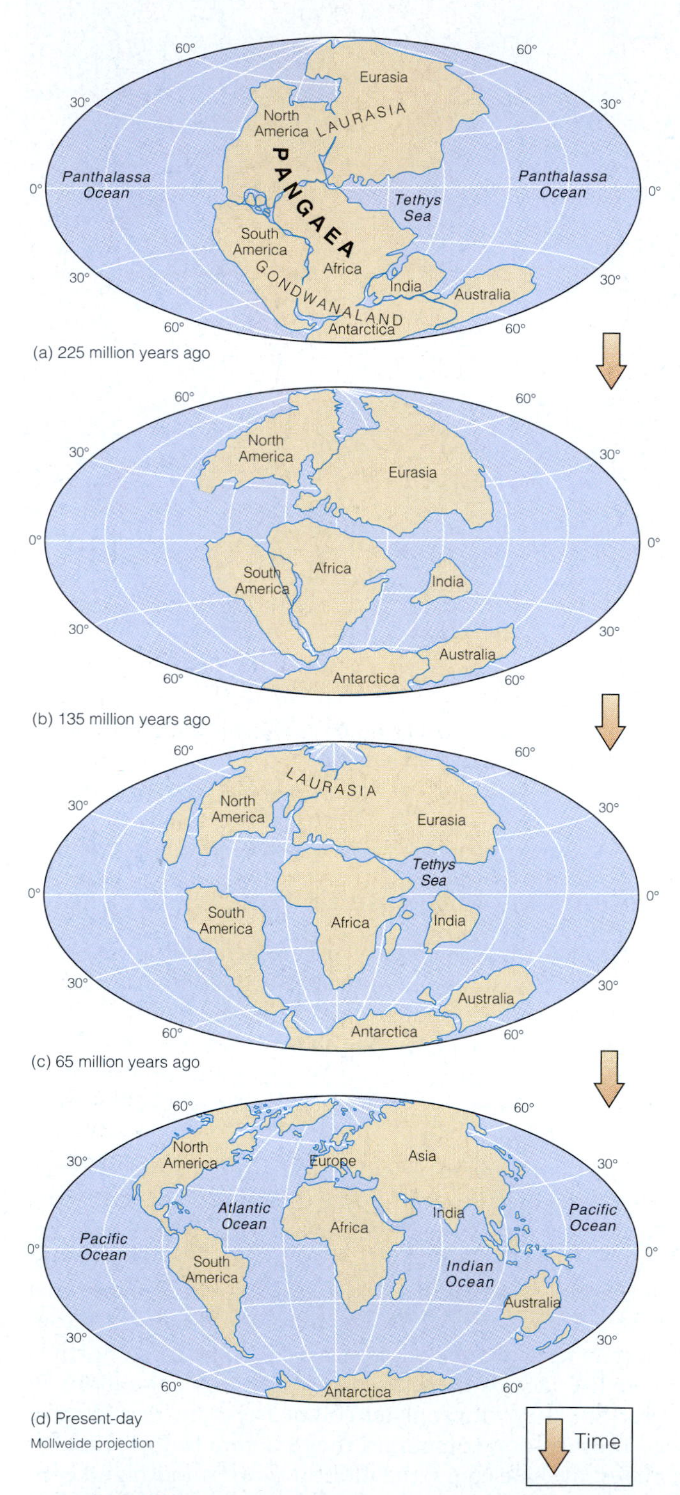

FIGURE 4.8 A reconstruction of the movements of the continents over the past 225 million years, which is the length of the present spreading cycle.

Spreading Cycles

About 225 million years ago, most continental crust on the Earth was part of one supercontinent called "Pangaea." Since then, Pangaea has broken up and the fragments have spread apart to their present locations. The relative motions of the fragments of Pangaea during the past 225 million years have been investigated by studies of the magnetism and chemistry of rocks, **fossils**, and ocean sediments, which can reveal subtle clues as to when and where on the planet they were formed. From such studies we can determine from where and how fast the continents have drifted (Chap. 5).

The history of **continental drift** during the past 225 million years is quite well known (**Fig. 4.8**). Initially the landmasses now known as Eurasia and North America broke away from Pangaea as a single block. North America later broke away from Eurasia, and South America and Africa separated from Antarctica, Australia, and India. Only much later were North and South America joined, as were Africa and Eurasia. Following the initial breakup of Pangaea, India and later Australia broke away from Antarctica. Since its break from Antarctica, India has moved northward to collide with Asia. This northward movement has been much faster than the rate of movement of other plates.

Although the history of the continents before the breakup of Pangaea is much less understood, Pangaea is known to have been assembled from a number of smaller continents that came together several hundred million years ago. The largest piece of Pangaea, called "Gondwanaland," appears to have remained intact for more than a billion years. The other pieces of continental crust probably broke apart, spread out, and then re-joined to form a supercontinent several times before the formation of Pangaea. Therefore, we refer to the past 225 million years as the "current **spreading cycle**."

As many as 10 spreading and assembling cycles may have occurred during the Earth's history. Each continent is itself a geological jigsaw puzzle, consisting of pieces assembled and broken apart during previous spreading cycles. Today, for example, many areas in the interior of continents show geological evidence that they were subduction zones during earlier spreading cycles (**Fig. 4.9**). Subduction zones form at plate boundaries where oceanic crust is downwelled into the asthenosphere, so these ancient subduction zones could not have been formed where they now lie in the middle of continents. There are several types of plate boundaries, each with its own characteristic topographic features. The various types are identified in the next section, and each type is then further described in subsequent sections of this chapter.

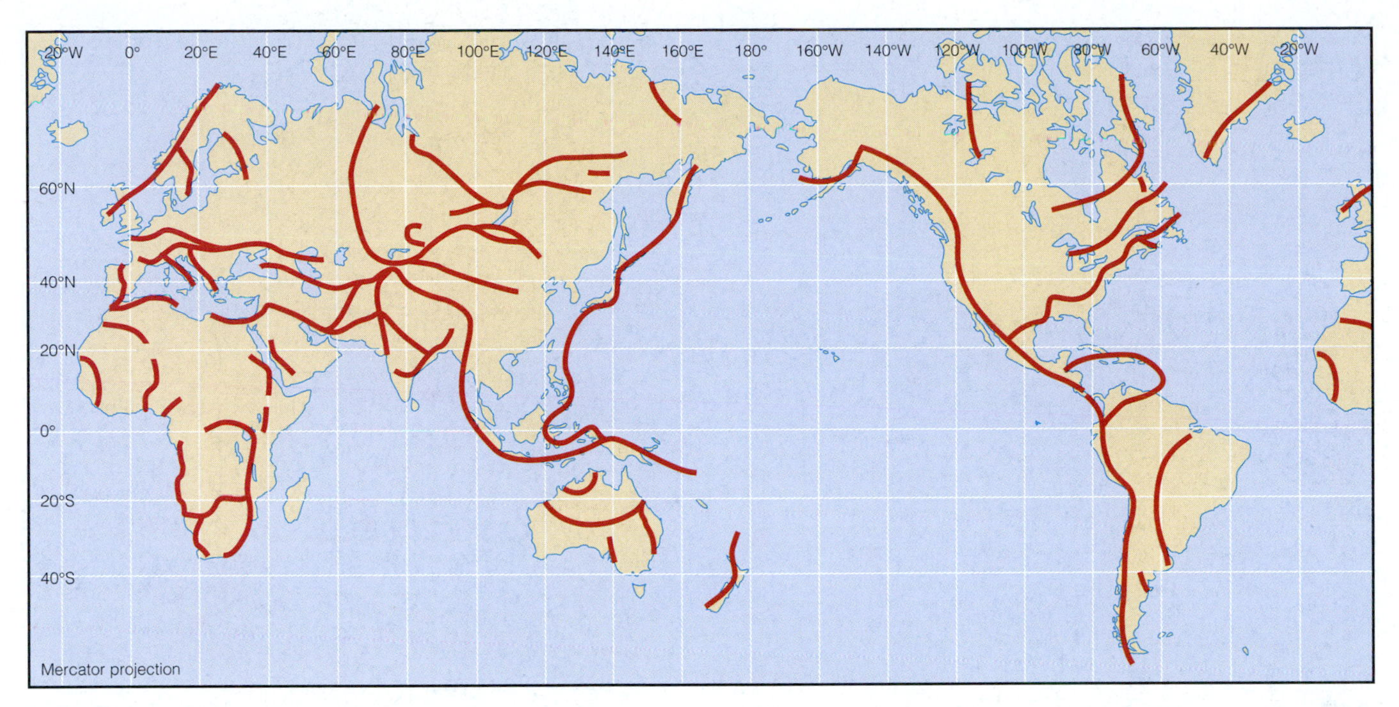

FIGURE 4.9 Sutures and active subduction zones. Sutures are locations where plates once converged and, thus, are the remnants of subduction zones. The oldest sutures that have been identified are in the interior of continents and are more than 570 million years old.

4.3 PLATE BOUNDARIES

One of three actions occurs at each plate boundary: two lithospheric plates collide with each other (**convergent plate boundary**), pull away from each other (**divergent plate boundary**), or slide past each other (**transform plate boundary**). Each action deforms the Earth's crust in a different way, creating characteristic topographic features. The behavior of the Earth's crust at a plate boundary depends on the type of crust at the edge of each of the adjacent plates and the directions of movement of the plates in relation to each other.

There are three types of convergent plate boundaries. In one type, the edge of one **converging** plate is oceanic crust and the edge of the other is continental crust (e.g., the Pacific **coast** of South America). In the second type, both plate edges have oceanic crust (e.g., the Aleutian and Indonesian island arcs); and in the third type, both plate edges have continental crust (e.g., the Himalaya Mountains that divide India from the rest of Asia). At the first two types, in which at least one of the plates has oceanic crust on its converging edge, oceanic crust is downwelled (subducted) into the mantle (**CC3**, **Fig. 4.5**), and these boundaries are called subduction zones.

Divergent plate boundaries are locations where lithospheric plates are moving apart. There are two types. One type, called an **oceanic ridge**, is a plate boundary where both plates have oceanic crust at their edges (e.g., the Mid-Atlantic Ridge and the East Pacific Rise). The second type, called a **rift zone**, is a location where a continent is splitting apart (e.g., the East African Rift Zone). If the **divergence** at a rift zone continues long enough, a new ocean will form between the two now separated sections of the continent. The gap that would otherwise be created as two plates move apart at a divergent plate boundary is filled by magma upwelling from below.

There are few long sections of transform plate boundary on the present-day Earth. The principal such boundary is the Pacific coast of North America, primarily California, where the Pacific Plate slides past the North American Plate. Much shorter sections of transform plate boundary occur at intervals along all plate boundaries.

At certain locations along plate boundaries, three plates meet (**Fig. 4.6**) in areas called "triple junctions." There are of two types of triple junction: stable and unstable. Stable triple junctions can persist for long periods of time, although their locations may migrate. Unstable triple junctions are those in which the relative motions of the plates cannot be sustained over time because of their geometry. The reason for this is a little complicated, but for a triple junction to be stable, the geometry must be such that each plate can continue to move at the same rate and in the same direction as a whole plate (the plate cannot "bend"). However, the rate at which the plate is subducted or added to by oceanic ridge spreading can be different along its boundaries with each of the other two plates it contacts. Triple junctions where three divergent plate boundaries intersect, as they do near Easter Island in the Pacific Ocean, are always stable. Most other triple junctions are unstable. Unstable triple junctions exhibit

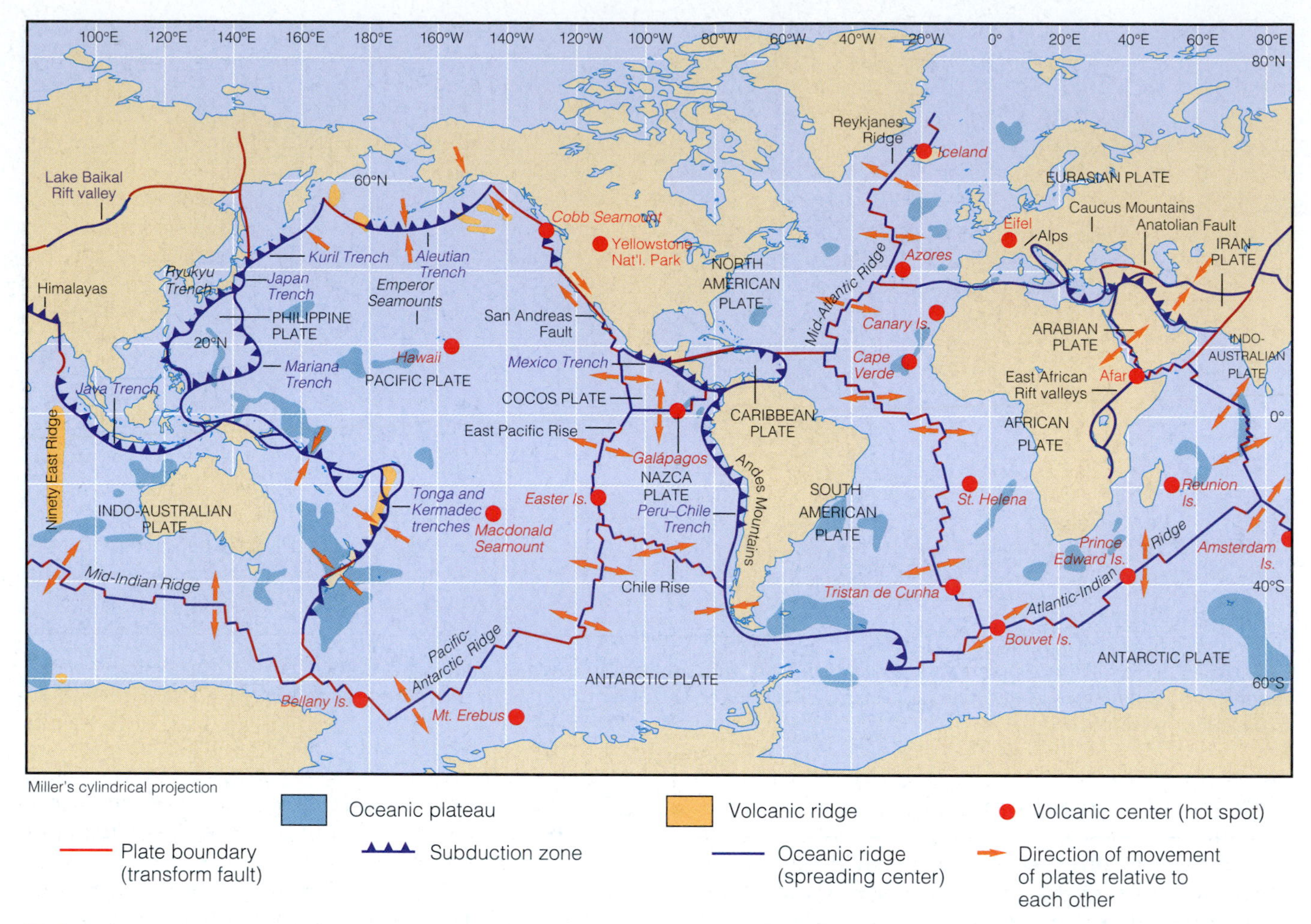

FIGURE 4.10 The locations of hot spots, subduction zones (trenches), and oceanic ridges. Hot spots occur both at plate boundaries and within plates. Subduction zones occur mostly in the Pacific Ocean, and oceanic ridges are interconnected throughout all of the world's oceans.

enhanced and complex tectonic activity, and their locations move as the interacting plates move relative to each other. An example of an unstable triple junction is the junction where three convergent plate boundaries (the Ryukyu, Japan, and Mariana trenches) intersect south of Japan (**Fig. 4.10**).

4.4 CONVERGENT PLATE BOUNDARIES

At two of the three types of convergent plate boundaries, lithosphere is downwelled (subducted) at a subduction zone. Lithosphere (sediments, oceanic crust, and solid upper mantle layer) that enters a subduction zone was formed at an oceanic ridge, mostly millions of years earlier (up to about 170 million years ago). Lithosphere slowly cools during the millions of years it travels horizontally at the top of the mantle away from where it was formed. As it cools, its density increases so that, by the time it enters a subduction zone, its density exceeds that of the mantle material beneath it and it can sink into (and through) the asthenosphere.

Why does the lithosphere sink through the asthenosphere only at subduction zones where two plates meet? It would seem that it should sink sooner, since, before it reaches the subduction zone, its density is already higher than that of the underlying asthenosphere. There are two reasons it does not sink sooner. First, the difference in density between lithosphere supporting old, cool oceanic crust and the warmer asthenosphere below is very small. Second, the resistance to flow is high because the lithosphere material is stretched out as a flat plate across the top of the asthenosphere (like a canoe paddle that meets strong water resistance when held flat for the thrust stroke but slices through the water easily when turned sideways). In a subduction zone, the edge of the subducting plate is bent downward (equivalent to turning the paddle blade sideways), thus allowing it to sink much more easily.

On today's Earth, most subduction zones surround the Pacific Ocean (**Fig. 4.10**). At each of these subduction zones, it is the Pacific Plate that is being subducted. The Pacific Ocean floor is being destroyed by subduction faster than new seafloor is created at its ridges, so the Pacific Ocean is becoming smaller. In contrast, the Atlantic and Indian Oceans are becoming larger. The oldest remaining seafloor sediment yet found (170 million years old) was retrieved from a hole drilled south of Japan by the International Ocean Drilling Program. All oceanic crust that existed prior to 170 million years ago and the associated sediment are believed to have entered subduction zones, where they were either destroyed or added to the edges of continents.

Subduction Zones Next to Continents

The subduction zones that are located along the **coastlines** of continents mark convergent plate boundaries where a plate that has oceanic crust at its edge converges with a plate that has continental crust at its edge (**Fig. 4.11**). In this convergence, the lithosphere of the plate supporting oceanic crust is thrust beneath the plate that supports the less dense continental crust. As the oceanic crust sinks beneath the continental crust, a subduction zone is formed with a characteristic deep trench parallel to the coast.

The edge of a continent that is carried into the plate boundary at a subduction zone is squeezed and thickened, forming a chain of coastal mountains along the edge of the continent. Because ocean sediments and sedimentary rocks have lower densities, like the continental rocks from which much of the sediment originated (Chap. 8), they are relatively **buoyant** as they are dragged downward into the subduction zone by their underlying oceanic crust. As a result, these materials tend to be "scraped off" the oceanic plate rather than subducted with it. The sedimentary materials are further compressed, folded, and lifted as more such material accumulates from continuing subduction of additional oceanic crust. Thus, some of the marine sediments from the subducting oceanic crust are collected in the subduction zone and contribute to the formation of the coastal mountains. This process explains why Darwin found marine fossils high in the Andes Mountains (Chap. 2).

Volcanoes. Another process that occurs at subduction zones is the formation of a line of volcanoes located a few tens or hundreds of kilometers from the plate boundary inland on continental crust on the plate that is not subducted. The volcanoes are formed when oceanic crust and some associated sediment are subducted beneath the edge of the continental plate. As this material is subducted deeper into the Earth, it is heated by the **friction** of its movement and by the hotter mantle material below. Oceanic crustal rocks have a lower melting point than mantle material has. In addition, subducted sediments and oceanic crust contain substantial amounts of water. Hence, much of the oceanic crustal rock and some of the surrounding mantle material melt, and the melted material forms pockets of magma mixed with heated water and sediment. The magma rises to form volcanoes near the edge of the continental plate (**Fig. 4.11**). As magma erupts, pressure decreases, and water and other constituents that are released by the heating of subducted crust and sediment become gaseous and expand rapidly. The result is that these eruptions are often explosive. The

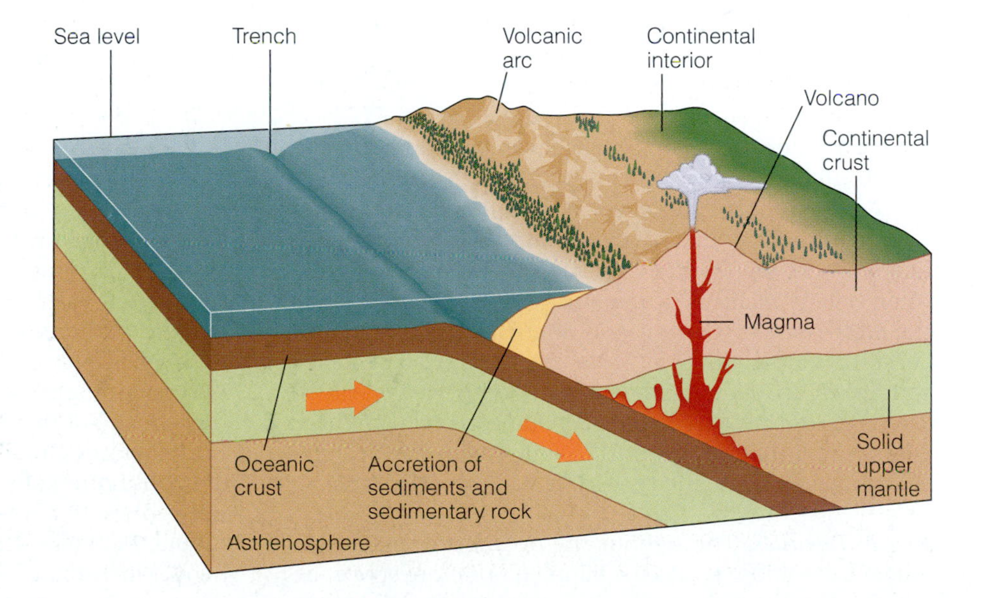

FIGURE 4.11 Subduction zone at a continental margin. A plate with oceanic crust is subducted beneath a plate with continental crust. This process forms a trench, a chain of volcanoes near the edge of the continent, and a chain of coastal hills or mountains, which are formed by compaction of the continental crust edge and accretion of compacted sediments and sedimentary rock from the subducting plate.

(a) (b)

FIGURE 4.12 (a) Mount St. Helens, in Washington State, is a subduction zone volcano that erupted violently in 1980. (b) The top of the mountain was blown away during the eruption, leaving behind a large crater.

1980 explosion of Mount St. Helens in Washington State is an example (**Fig. 4.12**).

Most active subduction zones in which the two plates have continental and oceanic crust at their respective edges are in the Pacific Ocean, and therefore they are often called "Pacific-type margins." Such margins include the west coast of Central and South America, the west coast of North America from northern California to southern Canada, and the coast of Southcentral Alaska. These areas have the well-developed ocean trench, coastal mountain range, and inland chain of volcanoes that characterize Pacific-type plate margins (**Fig. 4.13**).

Exotic Terranes. Throughout the oceans are small areas where the seafloor is raised a kilometer or more above the surrounding oceanic crust. Called **oceanic plateaus**, these areas constitute about 3% of the ocean floor (**Fig. 4.10**). Some oceanic plateaus are extinct seafloor volcanoes, others are old volcanic ridges, and still others are fragments of continental crust called "microcontinents." Small extinct undersea volcanoes can be broken up and subducted with oceanic crust. However, the larger oceanic plateaus are too thick to be subducted. Therefore, when sections of oceanic plateau enter a subduction zone, their rocks become welded onto the edge of the continent to form **exotic terranes** (**Fig. 4.14**). Exotic terranes may also be formed when islands, including those originally formed at **magmatic arc** subduction zones (see the next section), are carried into a subduction zone where they impact the edge of a continent. Much of the Pacific coast of North America consists of exotic terranes.

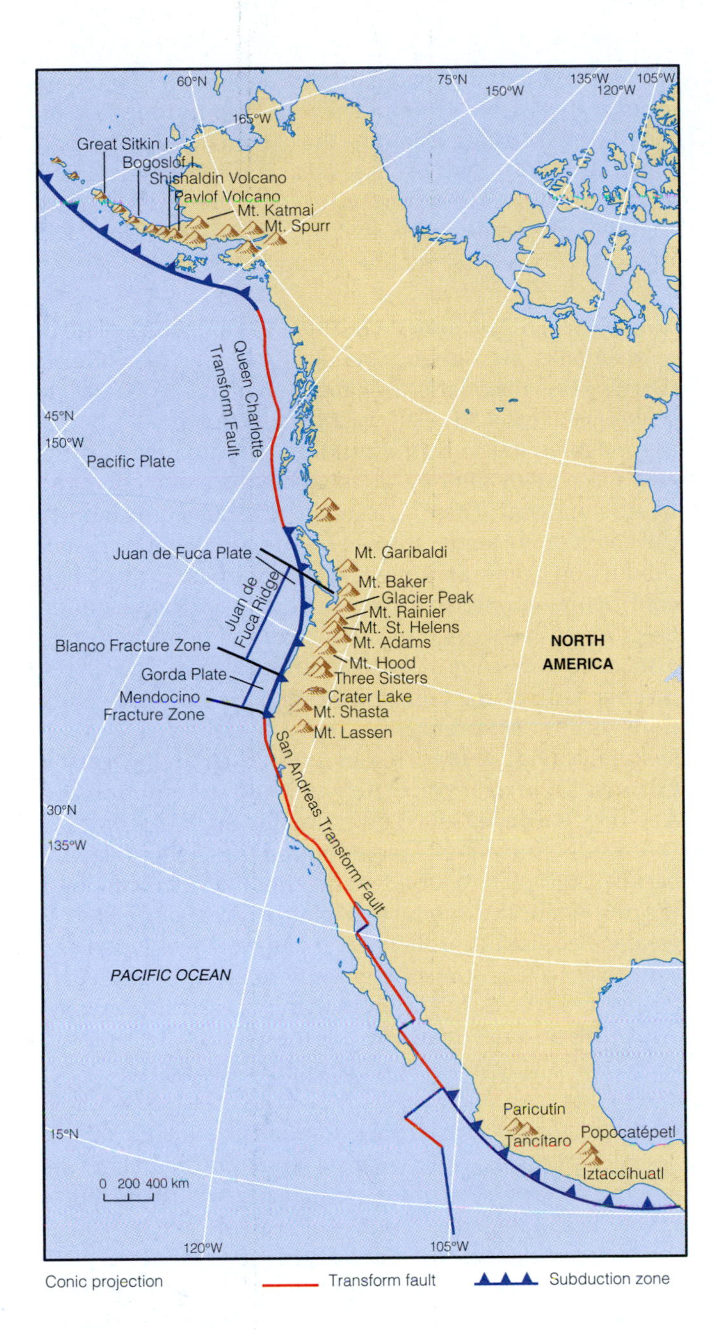

FIGURE 4.13 Stretches of the west coasts of Mexico, California, Oregon, Washington, Canada and Alaska are characterized by trenches and subduction zone volcanoes. The active volcanoes associated with the subduction of the Juan de Fuca and Gorda Plates under the North American Plate do not extend farther south than Mount Lassen in northern California. To the south, the San Andreas Fault, which can be traced from Cape Mendocino through southern California and into northern Mexico, is a transform fault and not a subduction zone. Continuing southward, active volcanoes are again found in central Mexico, where the transform faults of California end and the subduction zone between the Cocos Plate and the North American Plate begins.

CRITICAL THINKING QUESTIONS

4.1 The structure of the Earth and the processes of plate tectonics depend on the way that fluids and solids of different density interact. List examples of everyday situations in which the difference in density between substances or within a substance determine their behavior. One such example is oil floating on top of water.

4.2 If continental crust and oceanic crust were able to flow like the material in the asthenosphere, how might the Earth be different? How would the processes of plate tectonics be different?

4.3 Most ice ages in the Earth's history have lasted about 50 million years. The most recent ice age began 2 to 3 million years ago. About 10,000 years ago, the glaciers that extended into much lower latitudes than they do today began to melt and the Earth's climate warmed. Does this evidence indicate that the most recent ice age is over? Explain the reasons for your answer.

4.4 Is it possible that plate tectonic motions occur at present on any planet or moon in our solar system other than the Earth? Why or why not?

Subduction Zones at Magmatic Arcs

At convergent plate boundaries where both plates have oceanic crust at their edges, a chain of volcanoes erupts magma to form an island chain parallel to the subduction zone. Because the most prominent examples of this type of plate boundary on the present-day Earth are curved (arced), these boundaries are called magmatic arcs (or "island arcs"). Processes that occur at such plate boundaries are similar to processes that occur at subduction zones at a continent's edge. However, because no continental crust is present, the oceanic crust of one plate (which has higher density) is subducted beneath the oceanic crust of the other plate (which has lower density). Because oceanic crust cools with age, and density increases with decreasing temperatures, older oceanic crust has a higher density and is subducted beneath younger crust. For example, at the Aleutian Islands plate boundary, the older Pacific Plate is being subducted below the Bering Sea portion of the younger North American Plate.

The subduction of oceanic crust at magmatic arcs forms a trench system that parallels the plate boundary (**Fig. 4.15**). Sedimentary materials from the subducting plate accumulate on the edge of the nonsubducting plate, just as they do at subduction zones at a continent's edge. They may form a chain of low **sedimentary arc** islands joined by an underwater ridge called an "outer arc ridge." Behind the ridge is an outer arc basin, an area of the nonsubducting plate where the crust is affected little by the subduction processes.

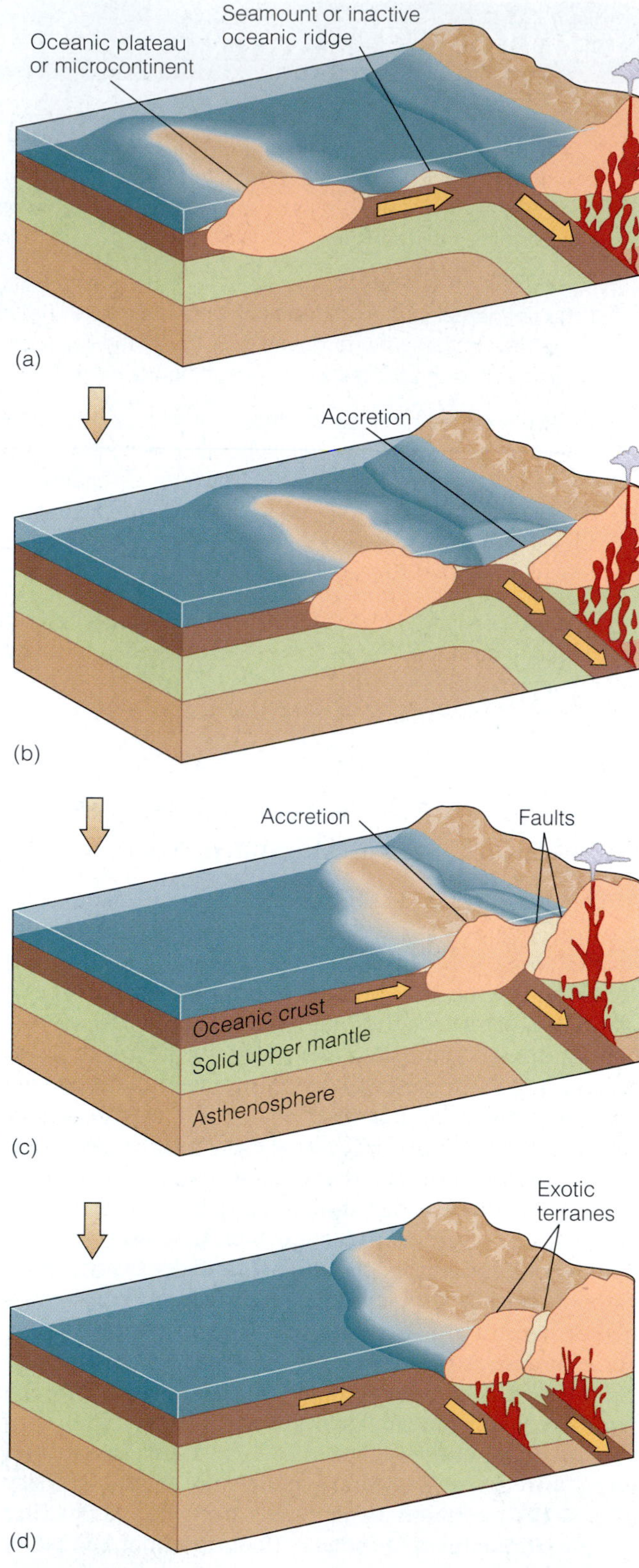

FIGURE 4.14 Formation of exotic terranes. Oceanic plateaus, inactive oceanic ridges, and volcanoes are scraped off the oceanic crust as it enters a subduction zone. The scraped-off material forms new continental crust that is welded onto the edge of the continent.

Most present-day magmatic arc subduction zones are around the Pacific Ocean. They include Indonesia (**Fig. 4.16**), the Mariana Islands, the Aleutian Islands, and Japan (**Fig. 4.6**). Each of these areas has the characteristic trench and magmatic arc, some have the low outer arc islands, and some have shallow, sediment-filled back-island basins, as described in the next section.

Magmatic Arc Volcanoes. On the nonsubducting plate of a magmatic arc subduction zone, a line of volcanic islands (the magmatic arc) forms parallel to the plate boundary (**Fig. 4.15**). The islands are constructed by rising magma, which is produced by the sinking, heating, and subsequent melting of subducted oceanic crust and mantle material. These volcanic islands are equivalent to the chain of volcanoes formed on the continental crust of subduction zones at the edges of continents. Like their continental counterparts, magmatic arc volcanoes often erupt explosively because subducted sediments and their associated water are also heated and erupted with the magma. One of the best known of these explosive eruptions is the 1883 eruption of Krakatau (Krakatoa) in Indonesia (**Fig. 4.16**). This eruption altered the Earth's **climate** for several years afterward, and the eruption and resulting **tsunami** killed an estimated 36,000 people.

At all convergent plate boundaries, the distance between the plate boundary and the line of volcanoes is shorter where the subducting plate's angle of descent is greater. Because lithosphere cools and its density increases as it ages, older lithosphere tends to sink more steeply (faster) into the asthenosphere than younger lithosphere. Thus, the distance between the trench and the volcano chain is less where the subducting lithosphere is older.

Back-Arc Basins. Sometimes a subducted plate's rate of destruction can exceed the rate at which the two plates are moving toward each other, particularly where old lithosphere is sinking steeply into the asthenosphere. Under these circumstances, the edge of the nonsubducted plate is stretched, which causes a thinning of the lithosphere at its edge. The thinning may create a **back-arc basin** (sometimes called a "back-island basin") behind a magmatic arc (**Fig. 4.15**). In extreme cases, the lithosphere is stretched and thinned so much that magma rises from below to create new oceanic crust in the basin.

Back-arc basins are generally shallow seas with a floor consisting of large quantities of sediment eroded by wind and water from the newly formed mountains of the magmatic arc and from nearby continents. The Mariana Trench subduction system, where the Pacific Plate is being subducted beneath the Philippine Plate, provides an example of a back-arc basin. The back-arc basin lies between the Mariana Trench–Island subduction zone and the Philippine Trench (**Fig. 4.10**). About 200 km to the west of the Mariana subduction zone is a now-inactive oceanic ridge that is the former location of back-arc spreading (**Inside front cover**).

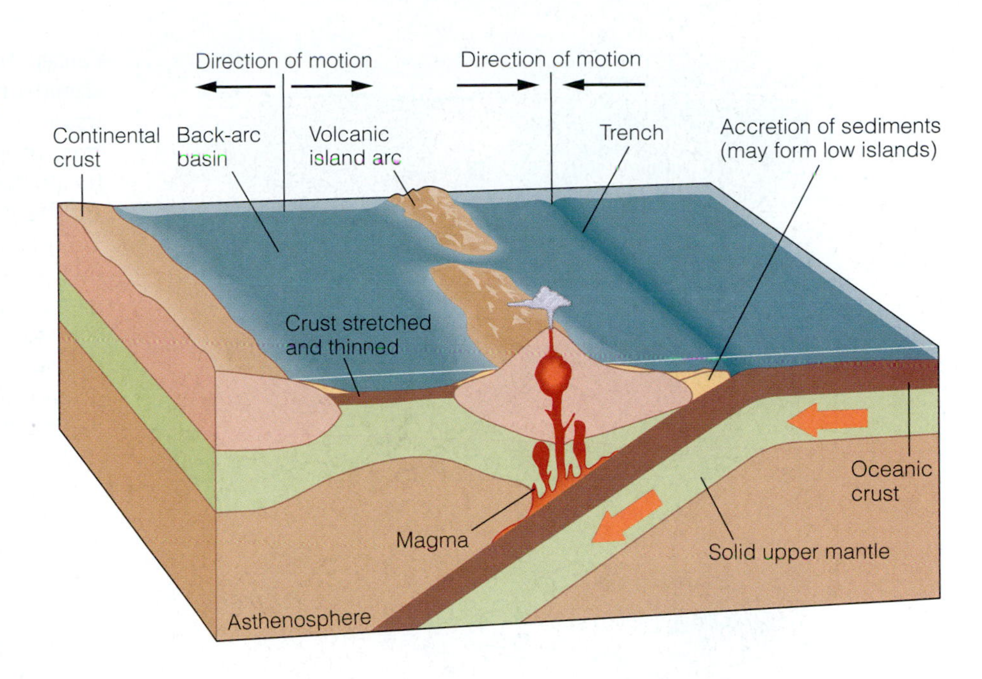

FIGURE 4.15 At magmatic arc subduction zones, as the plate with the higher-density oceanic crust is subducted, a chain of volcanic islands (magmatic or island arc) is formed along the edge of the nonsubducting plate. Sometimes low sedimentary islands also form between the island arc and the trench. At some magmatic arcs, such as shown here, the oceanic plate subducts fast enough that the adjacent plate with a continental crust edge cannot move toward the plate boundary as fast as the subduction occurs. In this situation, a trench and island arc form outside a back-arc basin, which is created between the continent edge and the subduction zone. The directions of plate motion shown are relative to each other. In some cases, the trench may migrate seaward, while the continent migrates more slowly in the same direction so that a back-arc basin is still formed.

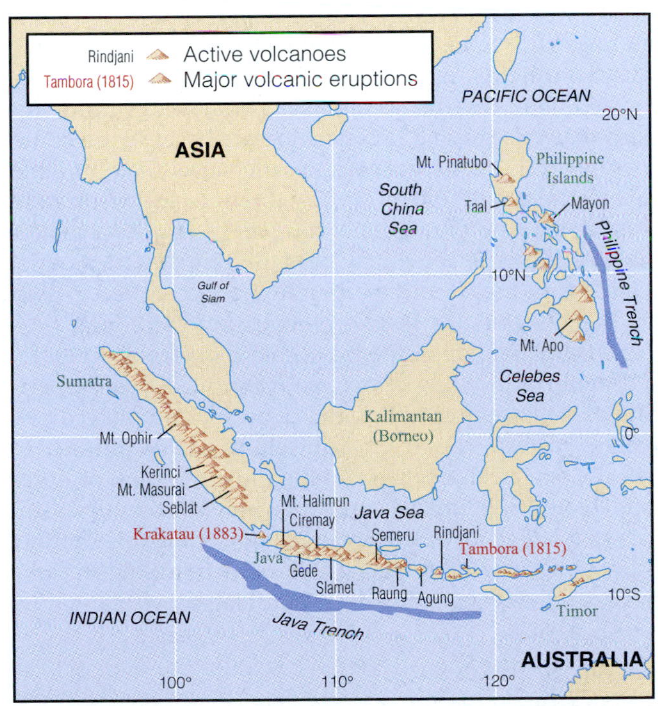

FIGURE 4.16 The outer arc ridge, volcanoes, and back-arc basins of Indonesia and the Philippines. The numerous active volcanoes on the major islands of Indonesia, including Sumatra and Java, are evidence of the very active nature of this oceanic convergence. The eruptions of Krakatau in 1883 and Tambora in 1815 were two of the largest eruptions of the past several centuries.

Collisions of Continents

Continental collision plate boundaries form at the convergence of two plates that both have continental crust at their edges. When the lithosphere of one plate meets the edge of the other plate, neither is sufficiently dense to be dragged into the asthenosphere and subducted. Therefore, the continental crusts of the two plates are thrust up against each other. As the collision continues, more continental crust is thrust toward the plate boundary, and the two continents become compressed. One continent is generally thrust beneath the other, lifting it up. The forces created and the energy released by such a collision are truly immense. The collision transforms the newly joined continent by raising a high mountain chain along the plate boundary. The effect of such collisions is not unlike the effect of a head-on collision of two cars, in which each car's hood is compressed and crumpled upward, and one car may ride partially under the other.

Collision plate boundaries are relatively rare on the Earth today, but the geological record indicates that they were more common in the past, such as when Pangaea was formed from preexisting continents. The most prominent continental collision plate boundaries today are the ones between India and Asia and between Africa and Eurasia (**Fig. 4.10**). Older examples, formed before the present spreading cycle, include the Ural Mountains of Russia and the Appalachian Mountains of North America.

The ongoing India–Asia collision is a particularly vigorous collision. It began about 40 million years ago when the edges of the continental shelves of the two conti-

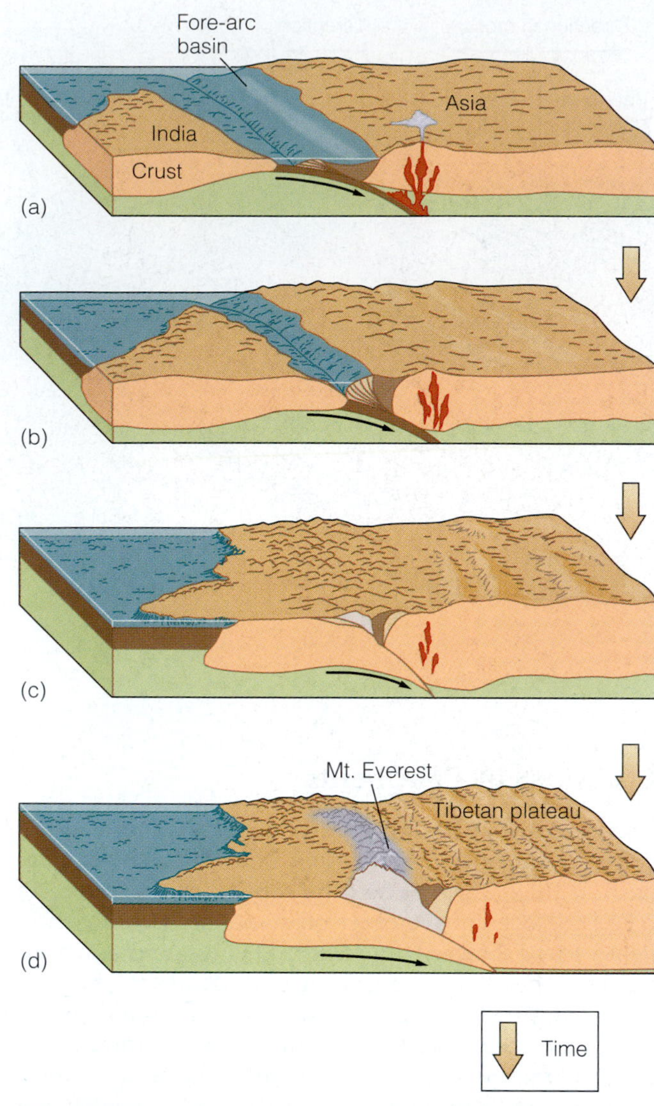

FIGURE 4.17 The continental collision boundary between India and Asia. About 50 million years ago, the northern margin of the Indian continental crust began a collision with the southern margin of the Asian continental crust. India is being thrust under the Asian continent as the collision continues. The Himalayas and the high Tibetan Plateau were both created by the collision. The tops of the Himalaya Mountains are formed from sedimentary rocks scraped off the oceanic crust of the ocean floor destroyed as the continents first came together.

nents first collided (**Fig. 4.17**). The movement of India toward Asia before the collision was very fast in relation to the speed at which other plates are known to move. The high speed explains the extreme violence of the India–Asia collision, during which India has been thrust under the Asian continent, creating the Himalaya Mountains and compressing and lifting the vast high steppes of Asia. The Himalaya Mountains continue to be uplifted, and the many powerful earthquakes felt throughout the interior of China and Afghanistan are caused by earth movements that release the compressional stress continuously built up by the India–Asia collision. This collision may also be compressing and fracturing the Indo-Australian Plate across its middle, as its northern section is slowed by the collision and its southern section continues northward.

A similar continental collision began about 200 million years ago between Africa and Eurasia. The convergence is slower than the India–Asia convergence, so the collision is less violent. Nevertheless, the Africa–Eurasia collision is responsible for building the Alps and for the numerous earthquakes that occur on the Balkan Peninsula (parts of Turkey, Greece, Albania, Bosnia and Herzegovina, Bulgaria, Croatia, Macedonia, Romania, Slovenia, Serbia, and Montenegro) and the adjacent region of Asia (parts of Turkey, Armenia, Georgia, Azerbaijan, and Iran).

4.5 DIVERGENT PLATE BOUNDARIES

The processes by which divergent plate boundaries are formed are not fully understood. For many years, the plates were believed to be driven apart by the upwelling of magma from below. It is still believed that rift zones in the continents may be at least partially caused by upwelling of magma accumulated under the continents, and that this magma may accumulate because the continental crust acts as an insulator and causes heat to build up in the mantle below it. However, it is now thought that the most prominent rift zone on the Earth today, the East African Rift Zone, may be caused by upwelling associated with the warm superplume that is believed to exist in the lower mantle beneath the African continent.

At oceanic ridges, the plate boundary is continuously formed by upwelling of magma from the asthenosphere, but this upwelling is not the cause of the plate divergence. Instead, the magma upwelling occurs in order to fill the gap created when sinking (downwelling) of lithosphere at a subduction zone elsewhere on one (or both) of the plates moves the entire plate toward this subduction zone, "pulling" the plate away from the oceanic ridge and creating a divergent boundary.

Oceanic Ridges

When two plates are pulled apart (diverge), the gap between them is filled by upwelled magma. Hot, low-density magma rises through the splitting crust and spreads on the seafloor from a series of volcanoes aligned along the gap. This is the primary process responsible for building the connected undersea mountain chains of the oceans (**Inside front cover map**) and is often called **seafloor spreading**. The mountain chains are known as oceanic ridges and oceanic rises. Most are in the approximate center of their ocean basins and follow the

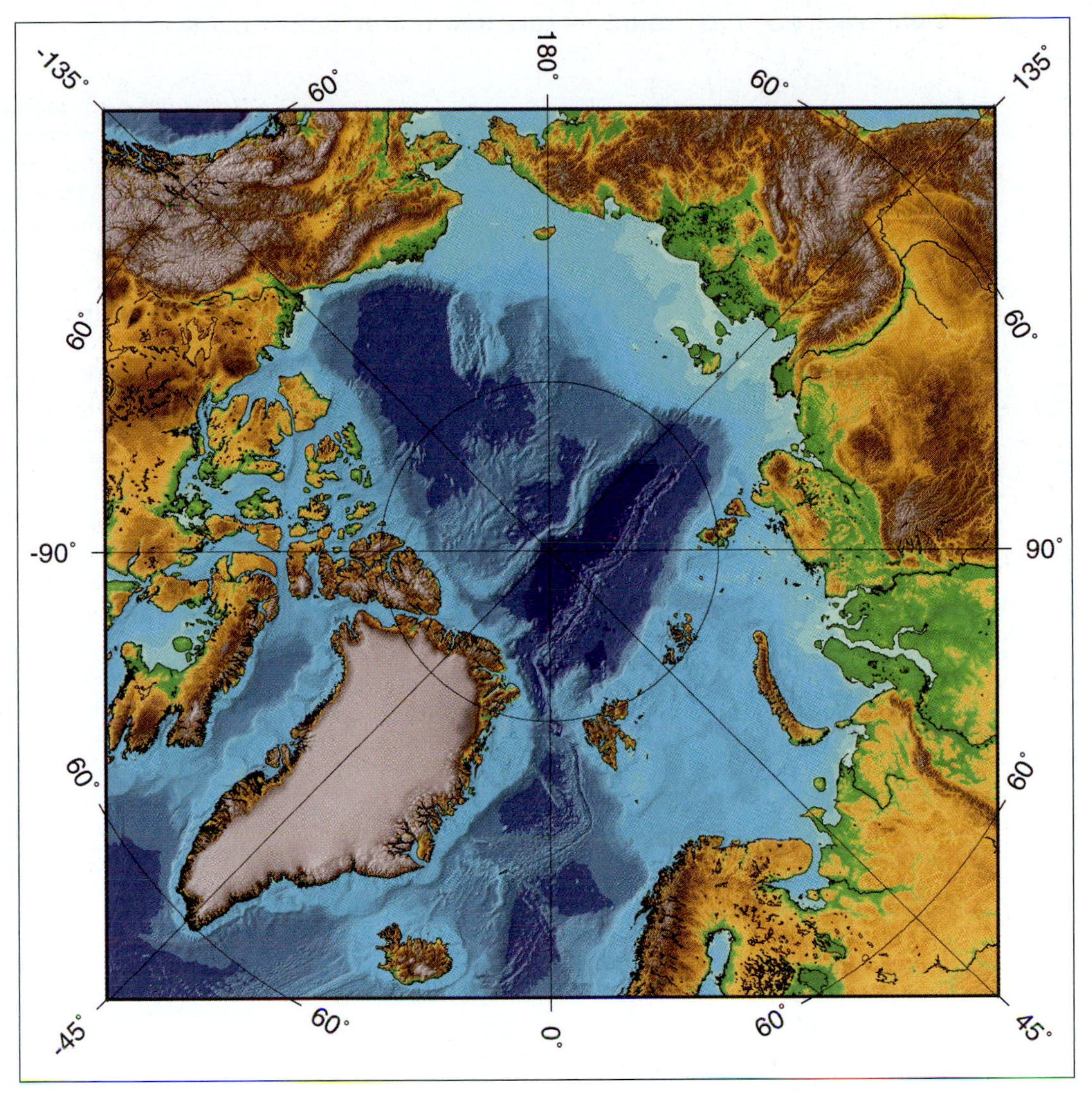

FIGURE 4.18 Topography of the Arctic Ocean floor. The Mid-Atlantic Ridge extends into the center of the Arctic Ocean. Shallow sills restrict water circulation between the Arctic Ocean and other oceans. The sill between the Arctic Ocean and the Bering Sea is especially shallow, and the circulation between the Arctic and Pacific Oceans is more restricted than the circulation between the Arctic and Atlantic Oceans.

general shape of the coastlines on either side of the ocean (**Fig. 4.10**). An example is the Mid-Atlantic Ridge, which bisects the Atlantic Ocean from near the North Pole to the southern tip of South America (**Fig. 4.18, Inside front cover map**).

As the plate edges separate at an oceanic ridge, new oceanic crust is formed by a sequence of processes that causes oceanic crust to be layered. Molten magma upwells from below and forms a chamber of molten magma below the seafloor. Some magma cools and solidifies on the sides of this chamber to form a rock called "gabbro." This becomes the lowest of four layers that are found consistently in almost all oceanic crust. The next layer up is formed by magma that rises through vertical cracks and solidifies in wall-like sheets called "dikes." The third, and initially upper, layer is composed of pillow basalt, which is formed when erupting **lava** is cooled rapidly by seawater. The final layer of oceanic crust is composed of sediments deposited from the ocean water column.

New crust formed at an oceanic ridge or rise is hot and continues to be heated from below as long as it remains above the divergence zone. Because it is hot, it is less dense than older, cooler crust. Beneath the new crust, the solid mantle portion of the lithosphere also is thin. Therefore, the new crust floats with its base higher on the asthenosphere (**CC2**), and consequently, the seafloor is shallower. As the new crust is moved progressively farther from the divergent plate boundary, it cools by conduction of heat to the overlying water and becomes denser. In addition, the lithosphere slowly thickens as mantle material in the asthenosphere just below the base of the lithosphere cools, solidifies, and is added to the lithospheric plate. Therefore, the aging crust on the plate sinks steadily to float lower on the asthenosphere. Thus, oceanic ridge mountains slowly move away from the divergence zone and sink, while new mountains form at the divergence zone to take their place.

Oceanic Ridge Types. There are three basic types of oceanic ridges distinguished by spreading rate, the rate at which the two plates are being pulled apart. Two of these types constitute a majority of the length of the world's oceanic ridge system. The third, where the spreading rate is very slow, is less well studied.

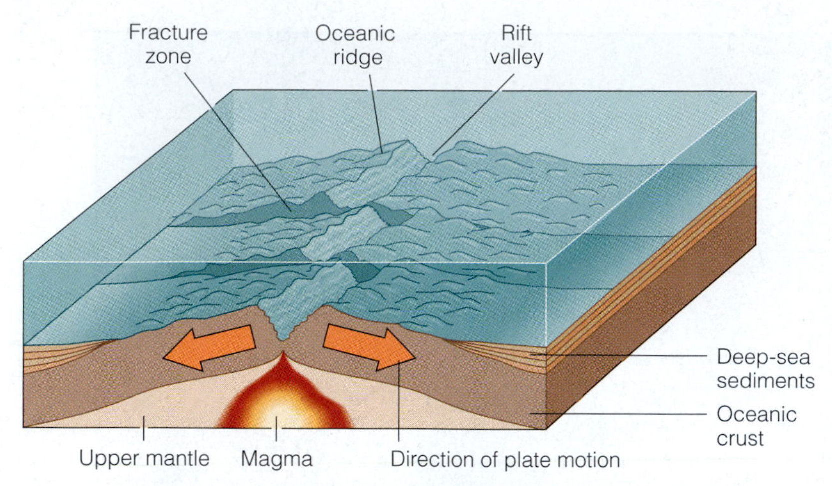

FIGURE 4.19 Oceanic ridge divergent plate boundaries are characterized by a ridge with a central rift valley and by fracture zones. On slowly spreading ridges, such as the Mid-Atlantic Ridge, the rift valley may be 30 km wide and up to 3 km deep, and the ridge is rugged with steep slopes. On rapidly spreading ridges, such as the East Pacific Rise, the rift valley is typically 2 to 10 km wide and only 100 m or so deep, and the ridge is broad with low-angle slopes and relatively smooth topography.

The two faster-spreading types of oceanic ridges have certain common characteristics. Each has a string of volcanoes aligned along the axis of the ridge. The volcanoes create a continuously spreading layer of new oceanic crust on the ocean floor. The most common type of oceanic ridge, exemplified by most of the Mid-Atlantic Ridge, has a well-defined, steep-sided **central rift valley** extending down its center (**Fig. 4.19, Inside front cover map**). The valley is the gap formed by the two plates pulling apart.

CRITICAL THINKING QUESTIONS

4.5 Are there planets or moons on which there were probably active plate tectonic motions in the past but not at present? If so, which planets or moons are the most likely to have had such motions? Why?

4.6 Using the surface topography maps of the planets and moons that are now available, how would you investigate whether, in fact, there had been plate tectonic activity on these planets?

4.7 Volcanoes occur at intervals along the length of divergent plate boundaries, but not along the length of transform faults such as the San Andreas Fault. Why?

4.8 Evidence exists that there are extinct volcanoes in California arranged roughly parallel to the San Andreas Fault. What are the possible explanations for their origin? How would you determine which of the possible explanations was correct?

Numerous earthquakes occur beneath the valley, and volcanic vents erupt magma to create the new oceanic crust. Central rift valleys are generally 25 to 50 km wide and 1 to 2 km lower than the surrounding peaks. The oceanic ridge mountains stand 1 to 3 km above the surrounding ocean floor and are extremely rugged.

The second type of oceanic ridge, exemplified by the East Pacific Rise, is much broader and less rugged than the Mid-Atlantic type, and the central rift valley is absent or poorly defined (**Inside front cover map**). The lack of a central rift valley is thought to be due to a faster spreading rate. The less rugged topography of this type of ridge is reflected in the seafloor of the surrounding plate, which is flatter than seafloor surrounding more slowly spreading ridges.

The third type of oceanic ridge occurs where the spreading rate is extremely slow, and it is poorly studied. This type of oceanic ridge has very few and widely spaced volcanoes. In addition, much of the new seafloor generated at these ridges is missing the characteristic basalt layer found elsewhere. This layer is thought to be missing because magma at these slowly spreading ridges has time to cool and solidify before it rises to the seafloor. The entire process appears to be different at these very slowly spreading ridges than at other oceanic ridges. At very slowly spreading ridges, the seafloor is apparently just cracked apart between volcanoes, and warm but solid rock then rises to become new seafloor. The Gakkel Ridge beneath the Arctic Ice Cap, northeast of Greenland, is an example. By some estimates, up to 40% of the oceanic ridge system worldwide may be of this extremely slowly spreading type.

Oceanic Ridge Volcanoes. Where the peaks of oceanic ridge volcanoes approach the ocean surface, as they do near Iceland, eruptions can be violent. In such cases, numerous small steam explosions are created at the volcano vent when hot erupting magma and seawater come in contact (you may have seen video footage of such eruptions). However, most eruptions at oceanic ridge volcanoes are quiet and smooth-flowing, despite the magma–seawater contact. There are several reasons for the lack of explosiveness. First, unlike the magma of subduction zone volcanoes, the magma rising into oceanic ridge volcanoes does not include sediments and water. Second, water movements carry heat away from the erupting magma, minimizing or preventing the production of steam. Third, the boiling point of water is several hundred degrees Celsius at the high pressures present at the depths of most oceanic ridge volcanoes.

Because they are not explosive and they occur deep within the ocean, most oceanic ridge volcanic eruptions occur without being noticed. Nonetheless, they can be at least as large and as frequent as eruptions on land. For example, a sidescan **sonar** survey of part of the East Pacific Rise in the early 1990s revealed a huge lava field formed by an eruption believed to have occurred within the previous 20 years. The lava field covers an area of

220 km² and has an estimated average thickness of 70 m. The volume of lava is estimated to be about 15 km³—enough to cover 1000 km of four-lane highway to a depth of 600 m, or to repave the entire U.S. interstate highway system 10 times. The lava field is the largest flow known to be generated during human history (much larger lava fields from prehistoric eruptions are known). Only a very small fraction of the 60,000 km of oceanic ridge has been studied by sidescan sonar surveys or by **submersibles**. Hence, the huge young lava field on the East Pacific Rise may be dwarfed by undiscovered flows from other undersea eruptions.

Another example of unseen volcanic activity under the oceans is provided by studies of the Juan de Fuca Ridge, which is about 500 km off the Oregon coast (**Fig. 4.13**). In 1990, a 20-km-long string of at least 10 volcanoes was discovered on the ridge. Some are more than 200 m high and about 1 km wide. All of these volcanoes apparently were formed between 1981 and 1990, because they were not found in 1981 surveys of the area.

Rift Zones

Rift zones occur where a continent is being pulled or is breaking apart. If the rifting continues long enough, an ocean basin is created between two separate, smaller continents. A rift zone may be created when a continental crustal block remains in one place for a prolonged period. In such instances, heat beneath the continent is partially trapped by the insulating effect of the continental rocks. The mantle temperature may increase until the rock at the base of the lithosphere melts, at which time hot magma could rise and split the continental block. Rift zones may also be formed by variations in the convection processes operating within the mantle that cause the mantle upwelling to become located under the continent (**CC3**). As yet, we have no detailed understanding of the relative roles of mantle convection and the insulating effect of continental crust, or of how these processes interact to split continents and create new oceans. In contrast, the processes that occur once a rift zone has been initiated are much better understood, and they are discussed next.

New Oceans. A new ocean may form in a location where the temperature of the mantle below the continent becomes elevated (**Fig. 4.20**). The process may occur as follows: As the mantle temperature increases, density

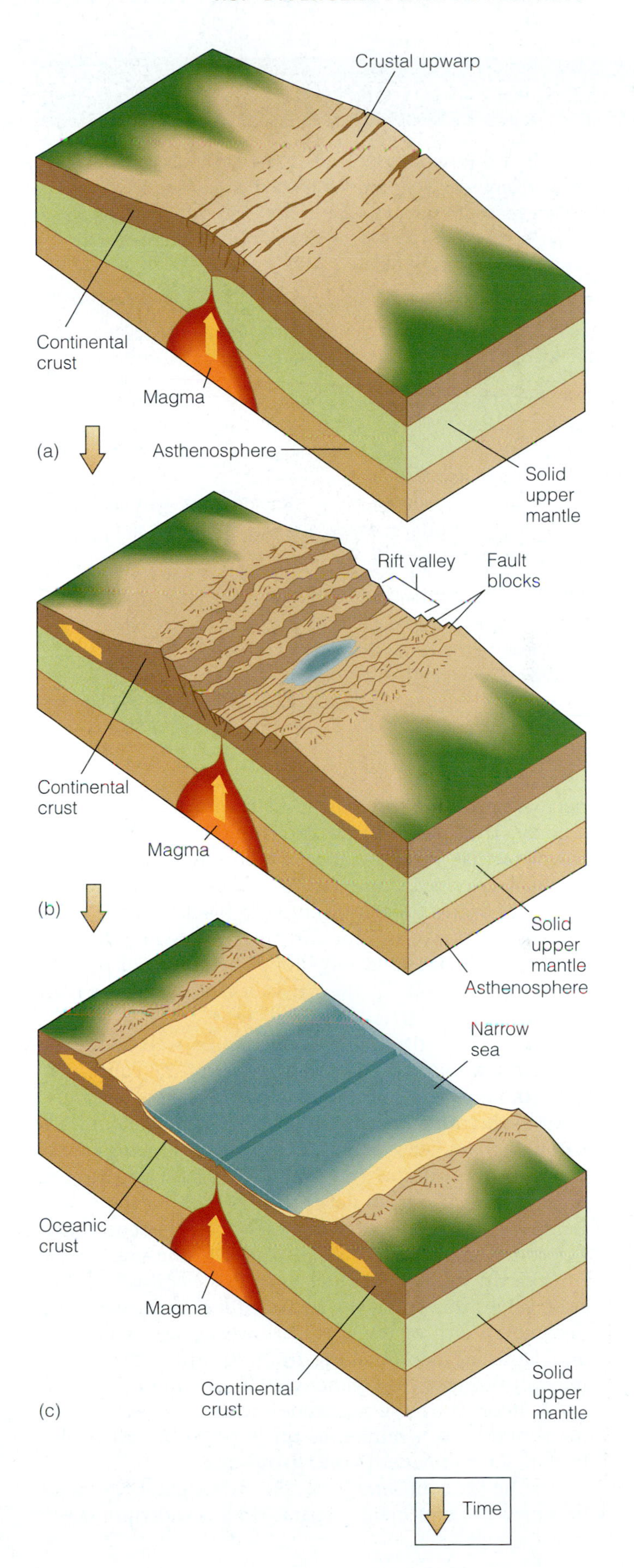

FIGURE 4.20 The history of new ocean formation at a rift zone. (a) Rising magma beneath a continent pushes the crust upward and creates cracks and fractures. (b) The crust is stretched and thinned. Collapses occur along a series of fault blocks, leading to the formation of a rift valley. Lakes may be formed in the rift valley. (c) The rift widens, allowing ocean water to enter and creating a narrow sea, such as the Red Sea. The rift continues to widen, and an oceanic ridge is formed down its center, eventually creating a new ocean.

CRITICAL THINKING QUESTIONS

4.9 If the Atlantic Ocean stopped expanding and started to contract, what would happen to New York City? Where would it be located a few million years after this change of direction occurred? Would it still be on land? Where would the nearest mountains be? Where would the coastline be?

4.10 There have apparently been several spreading cycles in the Earth's history. What might cause the continents to repeatedly collect together, then break apart, only to collect together again?

4.11 One of the hypothesized effects of the enhanced greenhouse effect is that sea level will rise. Will sea level rise, and if so, why? What will be the causes of sea-level change if climate changes? How many of these causal factors can you list? Which of these factors would be the most important? Would any of them tend to lower sea level? What effects would a sea-level rise of 1 m or 10 m have on humans?

4.12 In the 1995 movie *Waterworld,* melting of all the ice in glaciers and the polar ice caps caused the oceans to expand and cover all the land surface on the Earth except for a few small islands. Is this possible? Why or why not?

decreases, causing the crust to be thrust upward to form a dome. The upward thrust stretches the crust and causes fractures that extend outward from the center of the dome. As such a fracture widens, the lithosphere beneath the bulge begins to melt. Eventually, blocks of crust break off and slip down into the rift valley formed by the fractures. This occurs unevenly along the length of the rift, with some sections of the rift valley being deeper than others.

Initially the rift valley floor may be well above sea level. In wet climates, the deeper areas of the valley may fill with rainwater, forming long narrow lakes. Volcanoes form in the rift valley and on its sides as magma upwells through cracks left by the fracturing and slipping blocks of crust. The sides of the rift move steadily apart, and magmatic rocks accumulate at the bottom of the valley. Because magmatic rocks, when they cool, are denser than the continental rocks being displaced to the sides of the rift, the rift valley floor sinks until it is eventually below sea level. At this point, the rift may fill with seawater and become an arm of the ocean. However, new volcanoes and landslide debris falling from the rift valley flanks may fill and temporarily increase the elevation of the rift valley floor. Thus, the connection with the ocean may be made and broken numerous times before the rift valley becomes a "permanent" **marginal sea**.

As the rift continues to widen, the original continent becomes two continents separated by a widening ocean with an oceanic divergence at its center. The rift valley provides a route through which the excess heat built up below can be released. If the quantity of heat is limited, the rifting process may stop before an ocean is formed. However, if excess heat is supplied at a high rate, the spreading may continue and a new ocean may be formed.

Of the several rift zones on the Earth today, the East African Rift is among the newest (**Inside front cover map**). The African continent appears to have remained at its present location for more than 100 million years. It is elevated several hundred meters higher than most of the other continents and is believed to be directly above a mantle superplume upwelling. The East African Rift displays a rift valley, rift valley lakes, and rift volcanoes typical of newly formed **continental divergent plate boundaries**. The Red Sea is probably a more advanced rift zone, where a connection with the ocean, perhaps a permanent one, has already been made.

4.6 TRANSFORM FAULTS AND FRACTURE ZONES

Transform faults are formed where two plates slide by each other. Numerous earthquakes occur along such **faults** as the edges of the two plates periodically lock and then break loose and slide past each other. The best-known transform fault is the San Andreas Fault in California, where a long section of the Pacific Plate is sliding northward past the North American Plate (**Fig. 4.13**). Most transform faults are much shorter than the San Andreas Fault.

Motions of plates at transform faults do not create the dramatic trenches, volcanoes, and mountain ranges of other types of plate boundaries. However, the action of one plate against another creates complicated stresses near the edge of each plate, produces faults along which earthquakes occur, and forms low hills or mountains. Mountain building may be enhanced at bends on the plate boundary. One such bend on the San Andreas Fault is at the Santa Cruz Mountains south of San Francisco and was the site of the October 1989 Loma Prieta earthquake.

Numerous short transform faults are formed at divergent and convergent plate boundaries. As two plates move apart or together, they also rotate in relation to each other because the plates are moving on a spherical Earth (**Fig. 4.21**). The rate of spreading and creation of new oceanic crust will vary along the plate boundary when such rotation occurs. To accommodate the varying rates of spreading or subduction between two rigid plates, new plate material must be added or destroyed at different rates in adjacent segments of the plate boundary. This is accomplished by the creation of transform faults between short sections of the plate boundary (**Fig. 4.22**). In the transform fault region, the edges of the two plates slide past each other, one side moving slightly faster in

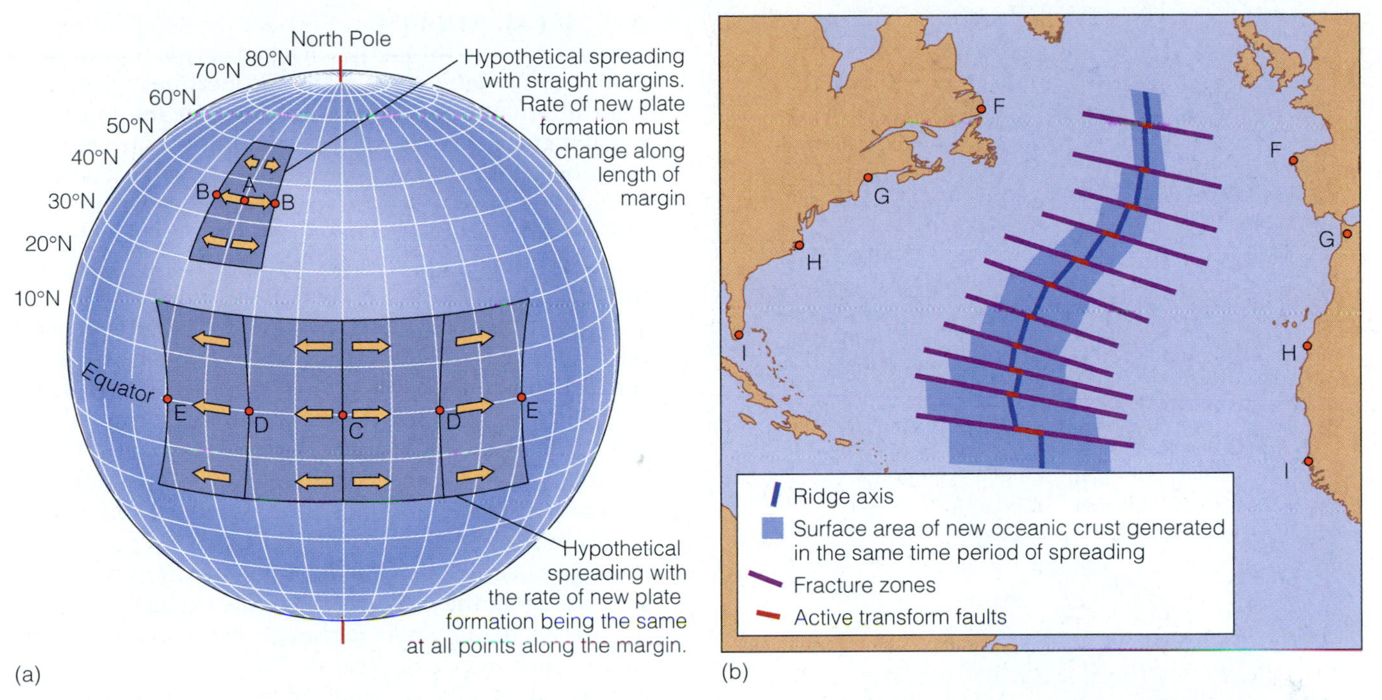

FIGURE 4.21 (a) Because lithospheric plates have rigid shapes and move on a spherical surface, they must rotate with respect to each other, and the rate of formation or destruction of crust at any plate boundary must vary along the plate boundary's length as shown by the spreading from A to B in this figure. The principle is similar to the movement of your eyelids as you open and close your eyes. The eye is exposed or covered more rapidly at the center than at the two corners. The rate of spreading cannot be equal along the plate boundary as the plates would need to bend and deform as shown in the sequence C–D–E. (b) Each of the points labeled with the same letter on either side of the Atlantic Ocean were joined at one time. Since the pairs are now separated by different distances, crust must have been formed at different rates along the oceanic ridge. Because the lithosphere is rigid and cannot stretch along its edge, variation in the rate of formation or destruction of crust is accommodated by the formation of transform faults between segments of plate boundary. You can demonstrate why this is necessary with a simple test using this book. With the book open on a flat surface, place one hand on the lower right corner of the right-hand page and press down so that it cannot move. With your other hand, lift all the pages (without the hard cover) at the top right corner slightly and grasp them between your finger and thumb. Hold these pages tightly together so that they cannot slide across each other, and try to lift and rotate these pages toward the lower left corner. Now loosen your grip on the top corner of the pages and try the same movement allowing the pages to slide across each other. The resistance you felt and the distortion of the page edges is no longer there. Although the motions in this test are different from the motions of lithospheric plates on a sphere, the principle is the same. The book pages, when allowed to, slide across each other—the equivalent of transform faults.

one direction than the other is moving in the opposite direction.

To see how transform faults accommodate the changing rates of plate creation or destruction explained in **Figure 4.21**, we can examine the process at oceanic ridges (**Fig. 4.22a**). The transform fault is the segment of the plate boundary where the line of the oceanic ridge is offset. Along this segment, the sections of crust newly formed on each of the two plates slide past each other. However, once the spreading motion has transported the new oceanic crust past the central rift valley of the adjacent segment, it is no longer adjacent to the other plate. Instead, it is adjacent to another piece of its own plate. Because the two sides of this join are now parts of the same plate and moving in the same direction, the two sides do not slide past each other. These areas are called **fracture zones**. The topographic roughness formed at the transform fault remains after the fracture zone has moved away from the plate boundary. The **Inside front cover map** shows numerous transform faults and associated fracture zones that cut perpendicularly through the oceanic ridges. At each transform fault, the line of the ridge **crest** is offset.

Some transform faults connect divergent and convergent plate boundaries, and others occur in subduction zones. The seafloor at subduction zones is covered by deep sediments that flow into any depressions and blanket the topography. Therefore, transform faults in subduction zones are less well defined by topography than their oceanic ridge counterparts are.

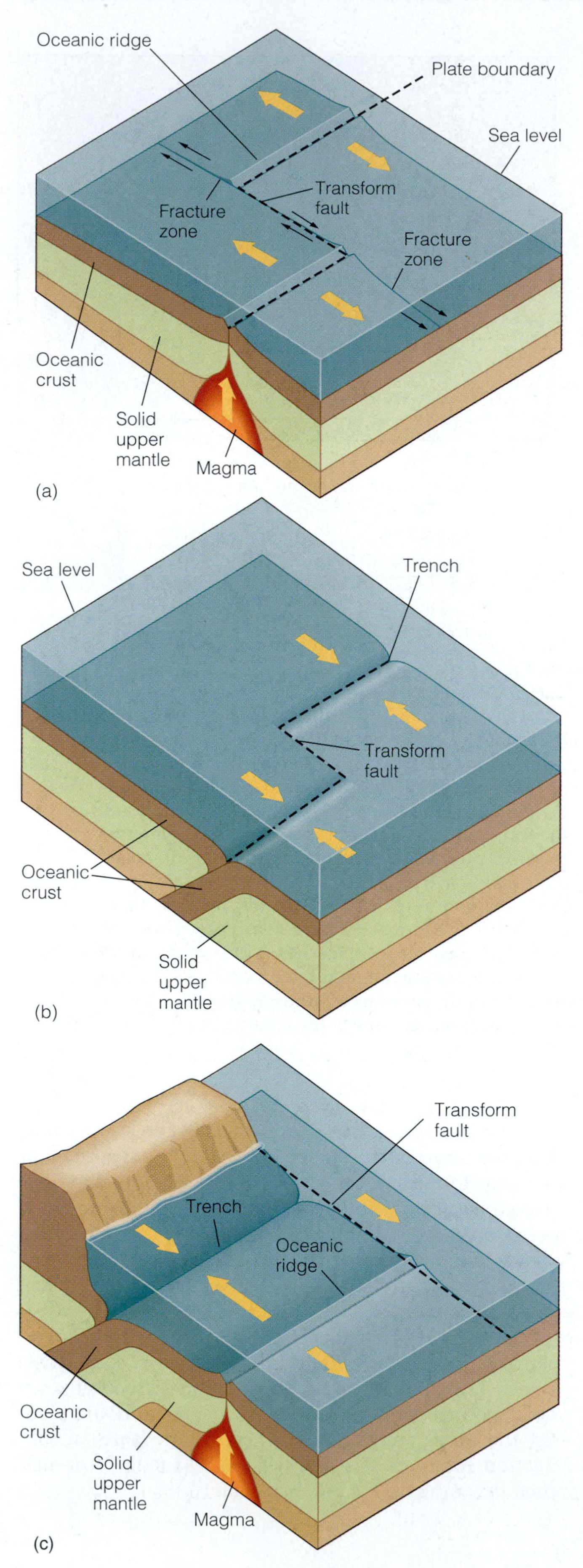

4.7 HOT SPOTS

Scattered throughout the world are locations where heat from the mantle flows outward through the crust at a much higher rate than in the surrounding crust. Some of these hot areas are within plates, and some are at plate boundaries. Most of them are beneath oceanic crust, but others are under continental crust (**Fig. 4.10**). Most hot locations occupy only a very limited area at the Earth's surface and are, therefore, called "hot spots."

The Earth's hot spots are not yet well understood. Many, but not all, are now known to be located where convection plumes upwell from the deep mantle (**Fig. 4.5**, **CC3**). Some of these may be plumes that extend all the way to the boundary between the core and mantle. Others may originate from one of the two known superplumes in the lower mantle. Yet others, such as that at Yellowstone Park, seem to be linked to shallow convection processes in the upper mantle only. Finally, some locations where the hot area under the lithosphere is more extensive may not themselves be hot spots (although hot spots may occur within such regions). Instead, they may be locations where continents have remained for prolonged periods of time. At such locations, the continents may act as a blanket to trap the heat flowing upward through the upper mantle, and the heat may melt and thin the crust. The East African Rift may be an example. However, one of the known superplumes is located beneath the region in which the East African Rift occurs, and this rift may be the result of several interacting processes.

Most oceanic hot spots are characterized by active volcanoes that rise through the ocean floor to form islands. Examples of islands located at hot spots are Iceland and the island of Hawaii. Both have active volcanoes that generally erupt relatively quietly, without explosions, and bring copious amounts of magma to the surface (**Fig. 4.23**). The magma solidifies and steadily builds the island.

The quantity of lava produced by hot-spot volcanoes is so large that such volcanoes are the tallest topographic features on the Earth. For example, the island of Hawaii rises about 5500 m from the seafloor to the ocean surface and then rises another 4205 m above the ocean surface to its highest mountain peak. Its total elevation, about 9700 m, is gained over a horizontal distance of less than 200 km and far exceeds the 8848-m elevation of Mount Everest. The enormous weight of Hawaii and the adjacent

FIGURE 4.22 Adjacent plates move past each other at a transform fault. (a) Most transform faults connect two segments of an oceanic ridge. The plates slide past each other in the region between the two ridge crests. This is the transform fault. Beyond this region, the two ridge segments are locked together and form a fracture zone. A transform fault can also connect (b) two adjacent segments of a trench or two trenches, or (c) an oceanic ridge and a trench.

FIGURE 4.23 The 1983 eruption of Pu'u O'o, the active vent of Kilauea Volcano in Hawaii. Kilauea is a hot-spot volcano that ejects small amounts of ash and large amounts of lava in a relatively smooth, nonexplosive manner. This eruption of Kilauea was still continuing in 2006.

islands has depressed the Pacific Plate through **isostatic leveling** (CC2). As a result, the seafloor around the islands has a broad moatlike depression some 500 m deeper than the surrounding seafloor. The downward deformation of the Pacific Plate extends about 300 km from Hawaii. Surrounding the "moat" region is another broad area where the seafloor is deformed upward as the crust is compressed outward by the island rising through its center.

Until recently, it was thought that hot spots remain fixed in place with respect to the Earth's rotational axis for tens or hundreds of millions of years as the lithospheric plates move over them. There is now evidence that hot spots do migrate independent of each other and of the lithospheric plates.

As a lithospheric plate moves over (or in relation to) an oceanic hot spot, the most recently formed volcanic island moves away from the hot spot. With the migration of the island from the hot spot, another segment of plate is brought over the hot spot, new volcanic vents open in it, and eventually another island may be formed. Continued plate movement takes this island away from the hot spot, and yet another island is formed, and so on. The results of that process can be seen in the ocean floor topography as a trail of islands and **seamounts** (undersea cone-shaped mountains). In the Pacific Ocean, the chains of islands and seamounts that align northwest of Macdonald Seamount, Easter Island, and Hawaii are all remnants of their respective hot spots (Fig. 4.10).

The island and seamount trails provide a history of the rate and direction of movement of the plate relative to the hot spot. Of the Hawaiian Islands, the island of Hawaii is the youngest and has several active volcanoes, including Kilauea (Fig. 4.23) and Mauna Loa. About 20 km to the southeast of Hawaii, a seamount called Loihi has been built and continues to grow by volcanic action (Fig. 4.24). Its current peak is about 970 m below sea level, but if volcanic activity continues at its present rate, Loihi may become the next Hawaiian island 10,000 to 40,000 years from now.

Radioisotope dating (CC7) of the volcanic rocks in the Hawaiian Island–Emperor Seamount chain shows each island or seamount to be progressively older with distance northwest from Hawaii (Fig. 4.24a). For example, Oahu was formed about 2 to 3 million years ago and has only inactive, although not yet necessarily extinct, volcanoes. As hot-spot volcanic islands migrate away from the hot spot, their volcanoes become inactive, the islands cool and sink isostatically, and they are subjected to erosion (Fig. 4.24b).

Some hot spots lie on divergent plate boundaries where two plates are moving apart. A volcanic island formed at such a location is steadily broken apart as the plates diverge. Iceland is a good example. Each side of the island, with its cooling volcanoes, migrates away from the hot spot on its respective plate. Evidence of this process can be seen also in the ocean floor topography. For example, the Icelandic Ridge stretching between Greenland and Europe consists of sediment-buried remnants of volcanoes that occupied the Icelandic hot spot when the Atlantic Ocean was narrower. Similarly, the Walvis Ridge and Rio Grande Rise in the South Atlantic Ocean are remnants of the Tristan da Cunha hot-spot volcanoes (Fig. 4.10).

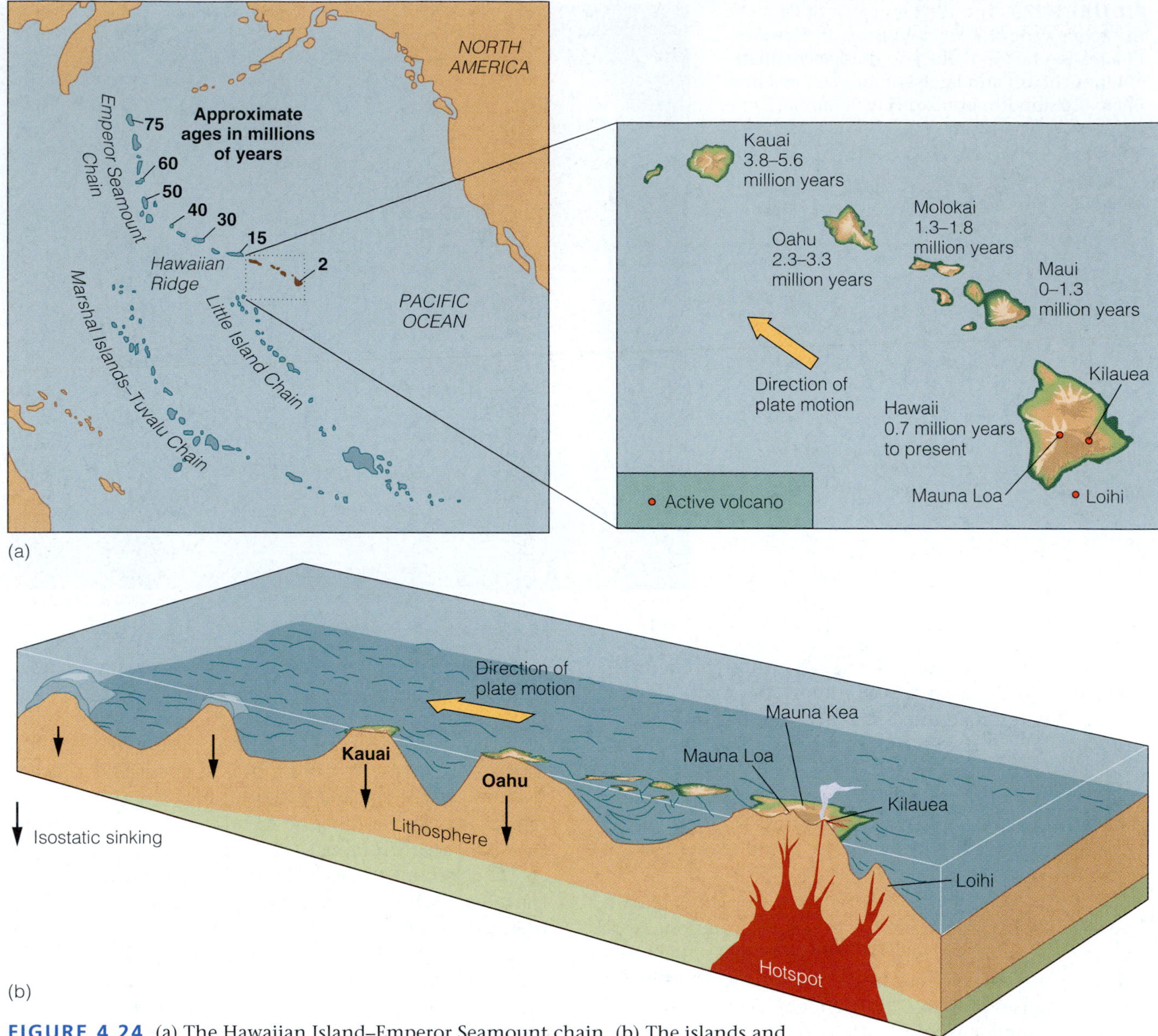

FIGURE 4.24 (a) The Hawaiian Island–Emperor Seamount chain. (b) The islands and seamounts are formed at the hot spot. As the plate moves relative to the hot spot, new islands are formed, and the older islands move off the hot spot, cool, sink isostatically, and are eroded.

A Lesson about Science

The rate of Pacific Plate movement appears to have varied relatively little during the past 50 million years. However, a distinct change in the direction of the Hawaiian Island–Emperor Seamount chain occurs at islands formed about 50 million years ago (**Fig. 4.24**). Seemingly identical changes of direction are also seen in the Easter Island and Macdonald Seamount chains.

Studies of the change in direction of the Hawaiian Island–Emperor Seamount chain provide a lesson for this author and most of the ocean and geological science community. This story provides you, the readers of this text, with a perfect example of why critical thinking skills are so important. It also illustrates that the material in this book represents the most recent scientific consensus regarding what is known about our ocean world but that consensus is always evolving.

As recently as 1998 (when the first edition of this text appeared), the consensus view of the scientific community was that hot spots remained fixed in place for tens or hundreds of millions of years as the lithospheric plates moved over them. At that time the scientific consensus was also that the distinct change in the direction of the Hawaiian Island–Emperor Seamount chain indicated an abrupt (in geological time) change in the Pacific Plate's direction of motion. Experts in the field agreed that the available data fit this conclusion. For example, the ages of the islands in the chain had been measured and did increase with distance from the hot spot. Other hot-spot trails were also found on the Pacific Plate, and they mirrored both the direction and the change in direction of the Hawaiian Island chain.

Fortunately, some scientists continued to think critically, even when the available data seemed to fit these explanations very well. These scientists were uncomfortable that nobody could explain how or why the direction of the Pacific Plate movement changed so abruptly. Thus, when the opportunity presented itself, they examined cores drilled into three seamounts of the Emperor chain that are located within part of the chain that was formed before the "change in direction of motion." They analyzed these cores for many different parameters, but the key data were obtained by examination of the **paleomagnetic** signatures of the magmatic rocks that form these seamounts. These data allowed the researchers to determine the rocks' paleolatitudes (the **latitudes** at which these rocks had formed by cooling and solidification of liquid magma). Paleomagnetism, and how it can be used to determine paleolatitude, are described in Chapter 5.

What they found was that these islands had not been formed at the same latitude that Hawaii now occupies. Instead, each older seamount had been formed progressively farther north. How could this be if hot spots are fixed in place? Of course, it cannot. Most scientists had accepted the hypothesis that hot spots remain fixed in place relative to the Earth's axis of rotation. That hypothesis now appears to be wrong. This hot spot must have moved south during a period about 50 to 80 million years ago, when the Emperor seamounts were formed. The lesson here is that, when a new hypothesis is proposed, there are almost always a few pieces of data missing, or data that do not exactly fit the "facts" of the hypothesis. These minor inconsistencies are the clues to how a hypothesis might be wrong.

Where does this new evidence leave our understanding of hot spots and how they fit into the jigsaw puzzle of plate tectonics? First, it raises many more questions. For example, why are there almost identical changes in direction in other Pacific hot-spot trails? Did all the Pacific hot spots move in the same direction at the same time? Scientists must answer these questions, and many more, if we are to better understand plate tectonic processes. Indeed, a recently developed hypothesis suggests that three mesoplates exist in the upper mantle, and that these mesoplates move independently of the lithospheric plate motions. Perhaps the hot spots move with these mesoplates, which could explain why the hot spots move and change direction together. Perhaps the existence of mesoplates will be verified and further studies of them will provide a fuller understanding of plate tectonics, or perhaps an entirely different explanation will be discovered for the next edition of this text.

4.8 PLATE INTERIORS

Oceanic crust is formed and destroyed by the processes that occur at divergent and convergent plate boundaries and hot spots. Its topography is modified by various processes over the tens or hundreds of millions of years between its creation and destruction. The edges of the continents, which are formed at divergent plate boundaries and altered at convergent plate boundaries, also are modified by a variety of processes between the time they are formed and the time they enter convergent plate boundaries.

Oceanic Crust

As it moves away from an oceanic ridge, lithosphere cools, becomes denser, and sinks steadily deeper into the asthenosphere. Therefore, the seafloor becomes lower with increasing distance from the oceanic ridge. As the cooling crust sinks, it is progressively buried by a continuous slow "rain" of solid particles through the water column that accumulate as sediments on the seafloor (**Figs. 4.19, 4.25**). Because sediments tend to accumulate faster in topographic lows, the original topography of the rugged oceanic crust that is formed at oceanic ridges is progressively buried and smoothed as the crust moves away from the oceanic ridge. The effect is very similar to that of snowfalls. If left undisturbed, a few centimeters of snow obscures features, such as street curbs and potholes, and softens larger features by mounding around them. As more snow accumulates, larger features are buried, and even cars in a parking lot may be difficult to find. Similarly, the lower topography of the oceanic crust is completely obscured after it has traveled a few hundred kilometers away from the ridge (remember that oceanic crust takes millions of years to move such distances).

The higher topography of the oceanic crust survives the sedimentation as rounded hills or mountains rising above the surrounding flatter areas. The largest topographic features of the oceanic ridges are commonly cone-shaped volcanoes. Their conical form is preserved and even enhanced by sediment accumulation. Therefore, much of the deep-sea floor is characterized by cone-shaped abyssal hills and mountains (called "seamounts"), even in regions far from the oceanic ridges (**Inside front cover map**).

Some seamounts have flat tops and are called "tablemounts" or **guyots**. Volcanic mountain cones that have sufficient elevation when first formed at the oceanic ridge or at hot spots emerge above sea level as islands. The island tops are eroded by wind, water, and waves much faster than the volcano can cool and sink isostatically. Thus, before such volcanoes sink isostatically below the surface, their tops are totally eroded away. The eroded flat tops are preserved once the volcanoes are completely submerged because erosion is extremely slow under the ocean surface away from winds and surface waves. The Hawaiian Island–Emperor Seamount chain exemplifies the various stages of this process (**Fig. 4.24**).

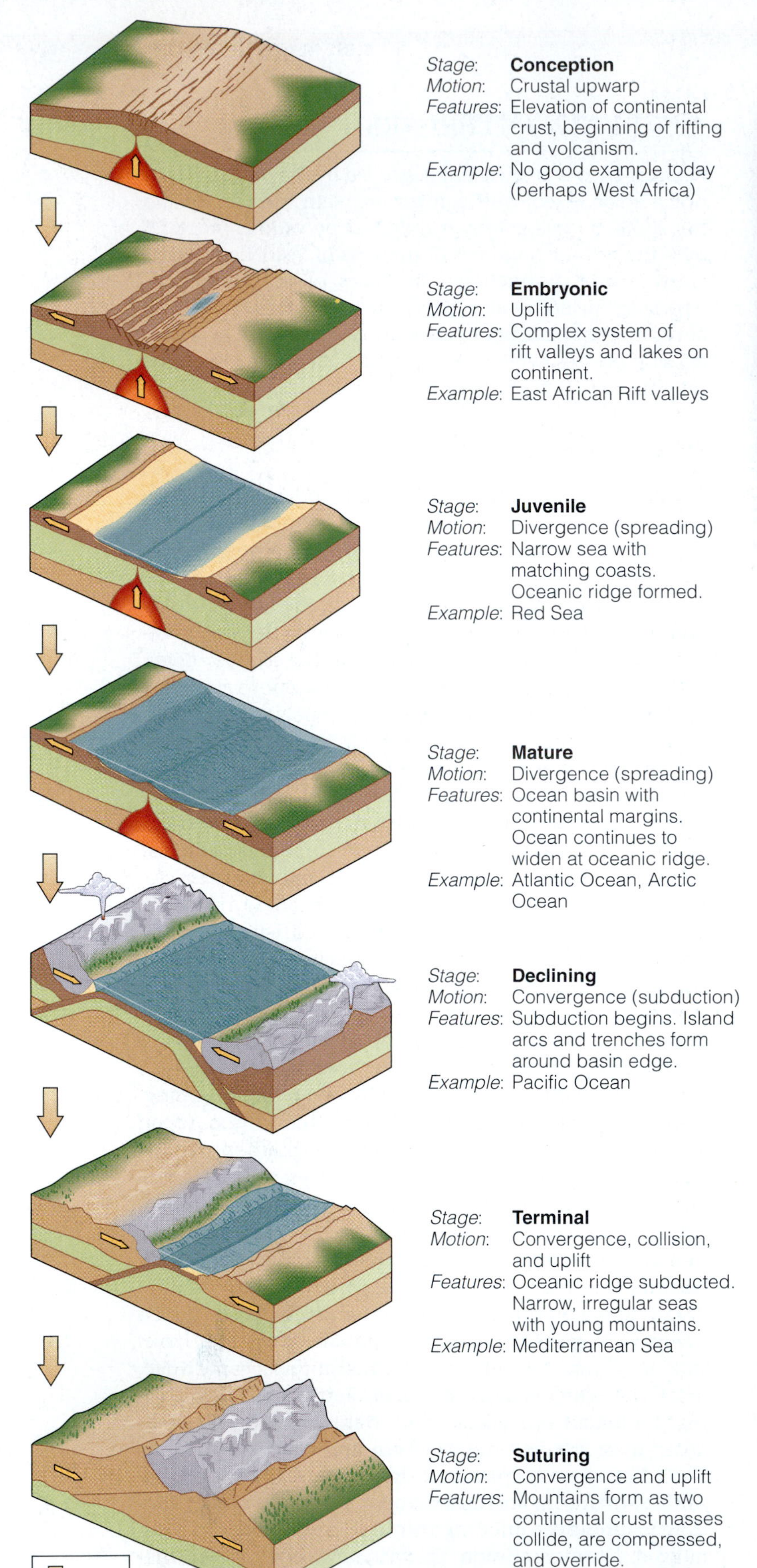
Stage: Conception
Motion: Crustal upwarp
Features: Elevation of continental crust, beginning of rifting and volcanism.
Example: No good example today (perhaps West Africa)
Stage: Embryonic
Motion: Uplift
Features: Complex system of rift valleys and lakes on continent.
Example: East African Rift valleys
Stage: Juvenile
Motion: Divergence (spreading)
Features: Narrow sea with matching coasts. Oceanic ridge formed.
Example: Red Sea
Stage: Mature
Motion: Divergence (spreading)
Features: Ocean basin with continental margins. Ocean continues to widen at oceanic ridge.
Example: Atlantic Ocean, Arctic Ocean
Stage: Declining
Motion: Convergence (subduction)
Features: Subduction begins. Island arcs and trenches form around basin edge.
Example: Pacific Ocean
Stage: Terminal
Motion: Convergence, collision, and uplift
Features: Oceanic ridge subducted. Narrow, irregular seas with young mountains.
Example: Mediterranean Sea
Stage: Suturing
Motion: Convergence and uplift
Features: Mountains form as two continental crust masses collide, are compressed, and override.
Example: India–Eurasia collision. Himalaya Mountains
Time

FIGURE 4.25 The history of an ocean from its creation at an oceanic ridge to its disappearance at a continental collision.

FIGURE 4.26 A coral atoll with its shallow interior lagoon (Kayangel Atoll, Palau).

Some seamounts form the submerged base of nearly circular coral reefs called **atolls** (**Fig. 4.26**). **Reef-building corals** grow only in shallow water (less than several hundred meters), and most species inhabit the warm tropical oceans (Chap. 17). In such regions, a coral reef may become established around islands that are formed when the tops of oceanic ridge or hot-spot volcanoes extend to or above sea level (**Fig. 4.27**). The coral reef continues to build upward from the flanks of the volcano as the volcano sinks isostatically. If the upward growth rate of the reef is fast enough to match the sinking rate of the volcano, the top of the live coral remains in sufficiently shallow water to continue growing.

The coral first forms a **fringing reef** adjacent to the island (**Fig. 4.27a**). As the volcano continues to sink, this reef continues to grow upward and becomes a **barrier reef** separated from the island by a **lagoon** (**Figs. 4.27b, 4.28**). Eventually the volcano completely sinks and leaves only an atoll (**Figs. 4.26, 4.27c**). On the historic *Beagle* voyage of 1831 to 1836, Charles Darwin visited Keeling Atoll and several reef-fringed islands of the South Pacific. Even though Darwin had no knowledge of plate tectonics, he proposed an explanation for the formation of atolls that is very similar to our understanding of that process today.

Continental Edges

We have seen how the processes at convergent, divergent, and transform fault plate boundaries form and shape the edges of continents. However, fewer than half of the coasts (or margins) of today's continents lie at plate edges. Most are located in the middle of the lithospheric plates. Most of these coastlines were formed ini-

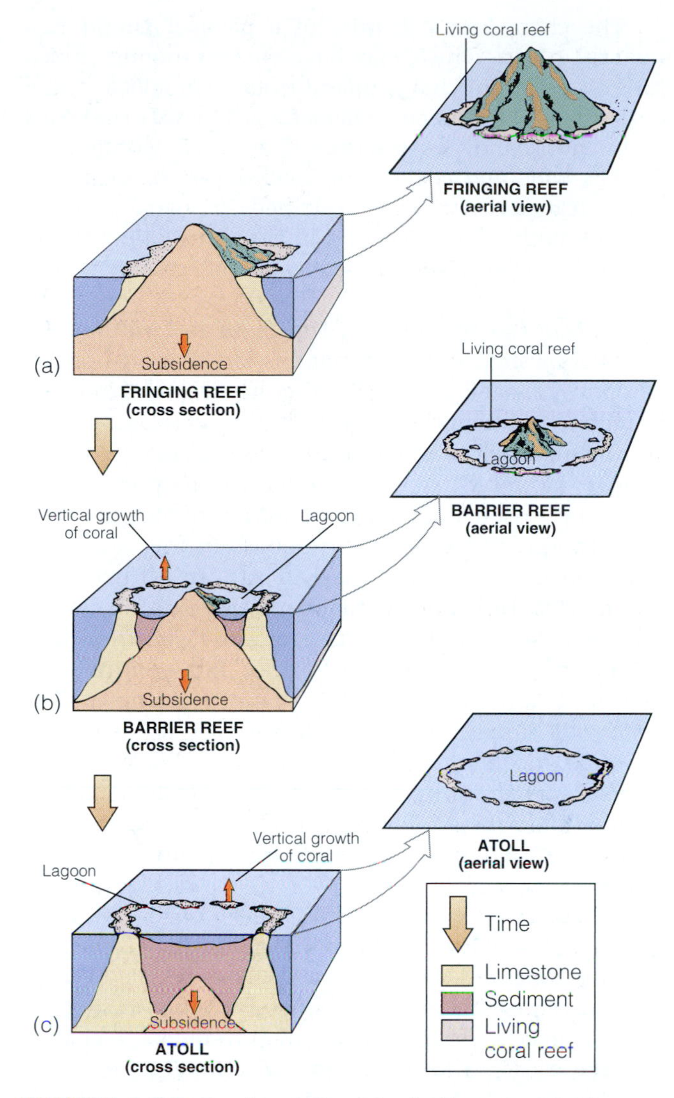

FIGURE 4.27 Coral reefs form in shallow water in the tropics and subtropics. Reefs are well developed around volcanic islands created at hot spots because there is little runoff of freshwater and sediment, both of which inhibit healthy growth of coral reefs. Coral reefs evolve as volcanic islands are formed and then sink isostatically as they move off the hot spot and cool. (a) Fringing reefs are formed around the perimeter of rising or static volcanic islands. (b) When the island sinks, a barrier reef is formed as the fringing reef grows upward. (c) Eventually the island sinks completely beneath the surface, leaving an atoll where the barrier reef continues to grow upward.

FIGURE 4.28 This volcanic island in the Society Islands of French Polynesia is surrounded by a well-developed barrier reef.

tially at the rift zones created as Pangaea broke apart. Because there are few earthquakes or volcanoes at such **continental margins**, they are called **passive margins**. The Atlantic coasts of North America and western Europe are examples.

Passive Margins. Development of a passive margin begins as the edge of a new continent is formed at a mature continental rift valley (**Fig. 4.29**). The new edge is isostatically elevated because of heating in the rift zone. Consequently, rivers drain away from the edge toward the interior of the continent. Shallow seas form in the rift as blocks of the continental crust edge slide down into the rift zone. Because few rivers carry sediment into the seas, the **turbidity** of the water in these seas is low. Consequently, the light needed for **photosynthesis** (**CC14**) penetrates deep into the clear waters, and **primary productivity** (growth of marine plants; see Chap. 14) is high. High productivity leads to large quantities of organic matter that may accumulate in the sediment of the new marginal seas. Because the seas are shallow and the rift valley is narrow, they may be periodically isolated from exchange with the large ocean basins. Under these conditions, thick salt deposits may be formed as seawater evaporates and its dissolved salts precipitate (Chap. 8).

As the passive margin moves away from the divergent plate boundary, which by that time has developed an oceanic ridge, both continental crust and oceanic crust cool and subside isostatically. Eventually, the edge of the continent subsides sufficiently that the direction of the slope of the landmass is reversed. Then the rivers that flowed away from the margin during its early history instead flow into the ocean at the margin. They bring large quantities of sediment from the land, which increases turbidity and siltation in the coastal ocean, thus reducing light penetration and primary productivity. Both the marginal seas and their deposits of organically rich sediment eventually are buried deeply as they subside farther below sea level. When the organic sediments are sufficiently deep to be heated and subjected to high pressure, the organic matter may be converted to the **hydrocarbon** compounds of oil and gas. If the overlying rocks are permeable, the oil and gas migrate toward the surface. Where the overlying rocks are not permeable, they trap the oil and gas to form reservoirs. Reservoirs formed at passive margins provide most of the world's oil and gas.

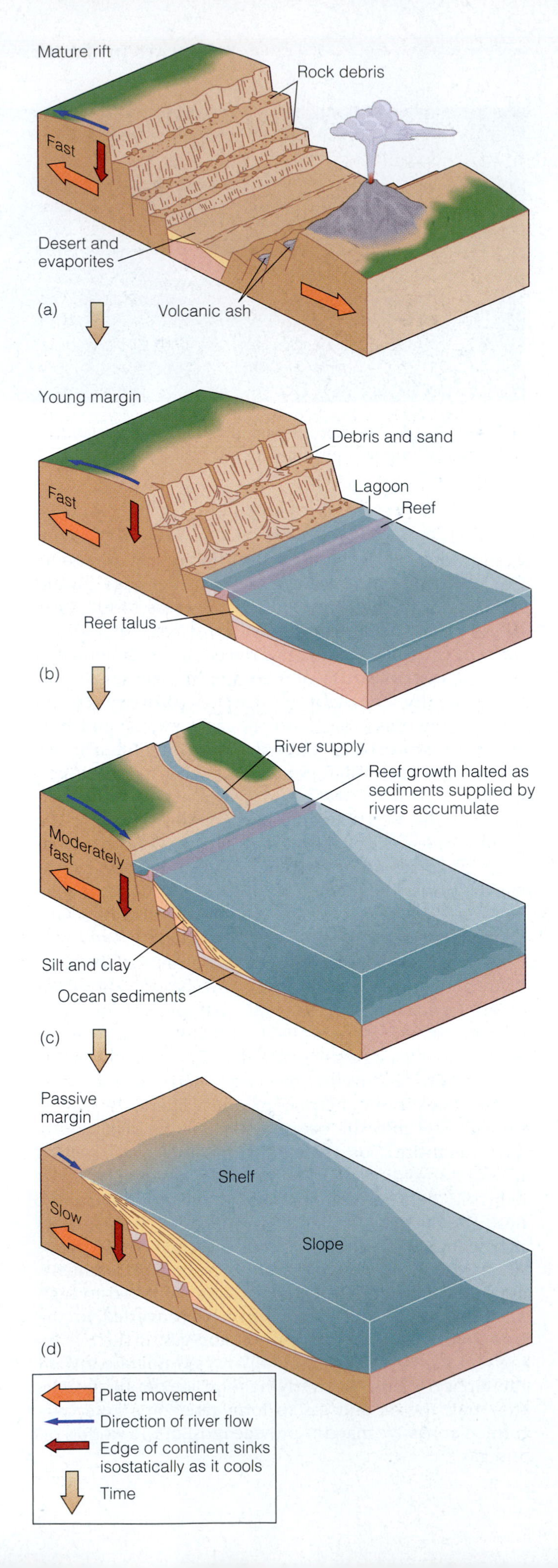

The characteristic feature of a passive margin is a **coastal plain**, which may have ancient, highly eroded hills or mountain chains inland from it. In addition, the coastal region is generally characterized by **salt marshes** and many shallow **estuaries**. Offshore, the continental shelf is wide and covered by thick layers of sediment. These features have been modified in many areas by changes both in sea level and in the distribution of glaciers that, in turn, were caused by climate changes.

The Fate of Passive Margins. The depth and width of the continental shelf at passive margins can be influenced by isostatic changes caused by the heating of relatively immobile continental blocks (**CC2**). For example, the continental shelf of much of the west (Atlantic) coast of Africa is narrower than those of many other passive margins. The reason is that the entire continent has been lifted by isostatic leveling in response to the heat accumulation that is causing the East African Rift to form. Africa's Atlantic coast eventually may become a new subduction zone, if the Atlantic Ocean crust near Africa cools sufficiently and if the East African Rift continues to expand. Another possibility is that the rifting in East Africa will simply stop. Many factors will influence the outcome, including how much heat is available to drive the East African rifting process and to continue spreading at the Mid-Atlantic Ridge, how cold and dense the Atlantic Ocean crust is near West Africa, and the driving forces on other plates.

Whatever happens in Africa, the probable ultimate fate of some passive margins is that they will become subduction zones. This will be the outcome if the oceanic crust at the passive margin cools sufficiently to subduct into the mantle, causing the plate movements to change and the ocean to begin to close. Eventually, the ocean could be destroyed as two continents meet at a collision margin.

FIGURE 4.29 A passive margin develops after the initial stages of creation of a new ocean. (a) A rift forms at a continental margin. Layers of volcanic ash, desert sands, and evaporites form as lakes develop and dry up in the rift valley. (b) As the new ocean forms, reefs and high-productivity lagoons form along the ocean edge because rivers flow inland and turbidity is low in such young oceans. (c) As the margin moves away from the new oceanic ridge, it cools and subsides. Rivers start to flow toward the new ocean and deposit lithogenous sediments in the previously productive lagoons and reefs. Productivity in the coastal waters, especially coral reef growth, is suppressed by high turbidity caused by the river inputs of suspended sediment. (d) Sediment layers build up on the passive margin to form a broad sediment-covered continental shelf. The buried lagoon and reef sediments often form oil reservoirs.

4.9 SEA-LEVEL CHANGE AND CLIMATE

Although tectonic processes and sedimentation are the principal agents that shape the seafloor topography, changes in climate also affect certain features, particularly the continental shelves and coasts. Substantial climate changes have been observed during recorded human history. For example, much of North Africa and the Middle East, which is now a region of desert and near-desert, was a fertile region with ample rainfall during Greek and Roman times. However, such short-term changes are small in comparison to changes that occur during the hundreds of millions of years of a tectonic spreading cycle.

Climate Cycles

Over the past 1000 years, the Earth's average temperature has varied by about 0.5°C. During the past 10,000 years, it has varied within a range of about 2°C to 3°C (**Fig. 4.30**). However, in the immediately preceding 2- to 3-million-year period, the Earth's temperatures were as much as 10°C below what they are today. That period is often called an **ice age**. During this ice age, the Earth's climate alternated between periods of glacial maxima (when temperatures were about 10°C below those of today) and interglacial periods (when temperatures were close to those of today). During glacial maxima, the polar ice sheets extended to much lower latitudes than they do now. Several earlier major ice ages lasted about 50 million years. Longer periods with warmer climates have occurred between ice ages. The most recent ice age may not yet be over, and the relatively warm climate of the past 10,000 years may simply represent an interglacial period.

Climate varies on timescales ranging from a few years to tens or hundreds of millions of years. Variations over the past 18,000 years are shown in **Figure 4.30**. These shorter timescale variations are superimposed on longer-term variations. Understanding the causes and consequences of the historical variations is important. Such knowledge may provide the key to understanding the effects and consequences of the extremely rapid (on geological timescales) global warming that has been predicted to occur and that may already have begun as a result of human enhancement of the atmospheric **greenhouse effect**.

Eustatic Sea-Level Change

When the Earth's climate cools for a long time, ocean water cools and contracts and polar ice sheets expand as water is transferred from oceans to land in the form of glaciers and snow. Both of these processes cause the sea level to drop. Conversely, warm climate periods tend to heat and expand ocean water, and melting continental ice returns water to the oceans, raising the sea level. Such changes of sea level, called **eustatic** changes, occur synchronously throughout the world. In contrast, sea-level changes caused by isostatic movements of an individual continent affect only that continent (**CC2**).

During the initial breakup of Pangaea, the climate was relatively warm, and it remained so until about 10 to 15 million years ago. During the warmest part of this period, about 75 million years ago, sea level was considerably higher than it is today, and as much as 40% of the Earth's present land area was below sea level. For example, a shallow sea extended from the Gulf of Mexico far north into Canada and covered what is now the land between the Rocky Mountains and the Appalachian Mountains. In contrast, at the peak of the most recent glacial period, about 20,000 years ago, sea level was at least 100 m below its present level. Sea level has risen and fallen by various amounts many times as Pangaea has broken apart in the present spreading cycle, and coastlines have migrated back and forth accordingly.

Sea-Level Change and Continental Margin Topography

Erosion by rivers carves out valleys, and the rivers transport the eroded sediment downstream, where it is deposited in lower-lying areas or in the coastal oceans. Erosion by waves and winds at coastlines (Chap. 13) also tends to reduce topography and deposit the eroded sediment in the shallow waters of the continental shelves. In contrast, at water depths of more than a few meters, erosional forces in the oceans are generally reduced and sediment accumulation is dramatically reduced, except in proximity to rivers that transport massive sediment loads to the ocean (Chap. 8).

As a result of sea-level oscillations, the area between the edge of the continental shelf and an elevation several tens of meters above the present sea level has been subjected to alternating cycles of wind, river, and wave erosion at some times and sediment **deposition** at others. These processes have substantially modified the topography. The effect is most apparent at passive margins. On most such margins, there is evidence of sea incursion and erosion throughout the area between the continental shelf edge and areas far inland from the current coastline. This evidence includes buried deposits that contain freshwater and shallow-water marine organisms and flat or low-relief topography.

The continental shelf, which has been progressively covered by the rising sea during the past 15,000 years, is cut across by numerous **shelf valleys**. Most of the valleys were carved out by rivers during the last ice age,

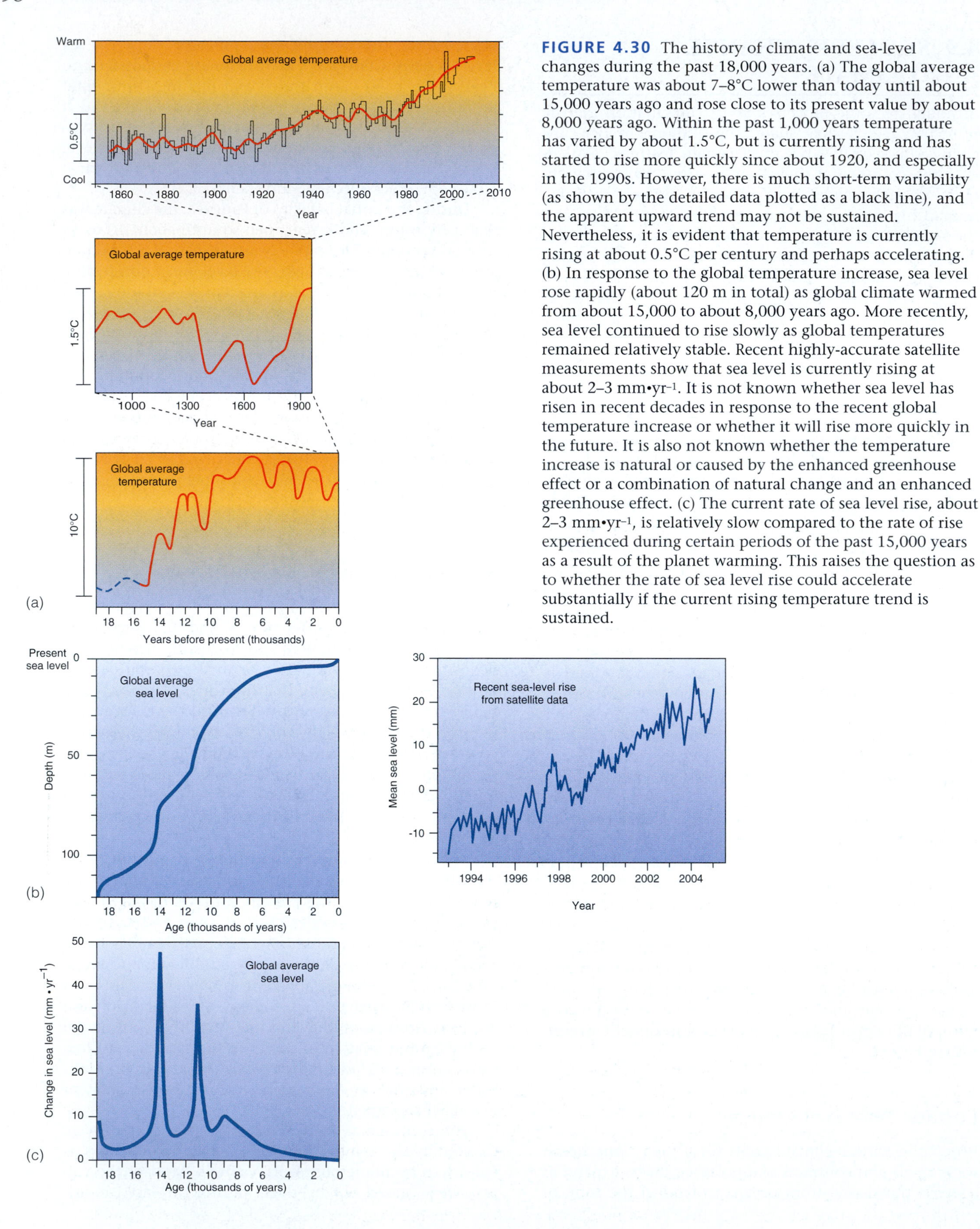

FIGURE 4.30 The history of climate and sea-level changes during the past 18,000 years. (a) The global average temperature was about 7–8°C lower than today until about 15,000 years ago and rose close to its present value by about 8,000 years ago. Within the past 1,000 years temperature has varied by about 1.5°C, but is currently rising and has started to rise more quickly since about 1920, and especially in the 1990s. However, there is much short-term variability (as shown by the detailed data plotted as a black line), and the apparent upward trend may not be sustained. Nevertheless, it is evident that temperature is currently rising at about 0.5°C per century and perhaps accelerating. (b) In response to the global temperature increase, sea level rose rapidly (about 120 m in total) as global climate warmed from about 15,000 to about 8,000 years ago. More recently, sea level continued to rise slowly as global temperatures remained relatively stable. Recent highly-accurate satellite measurements show that sea level is currently rising at about 2–3 mm•yr^{-1}. It is not known whether sea level has risen in recent decades in response to the recent global temperature increase or whether it will rise more quickly in the future. It is also not known whether the temperature increase is natural or caused by the enhanced greenhouse effect or a combination of natural change and an enhanced greenhouse effect. (c) The current rate of sea level rise, about 2–3 mm•yr^{-1}, is relatively slow compared to the rate of rise experienced during certain periods of the past 15,000 years as a result of the planet warming. This raises the question as to whether the rate of sea level rise could accelerate substantially if the current rising temperature trend is sustained.

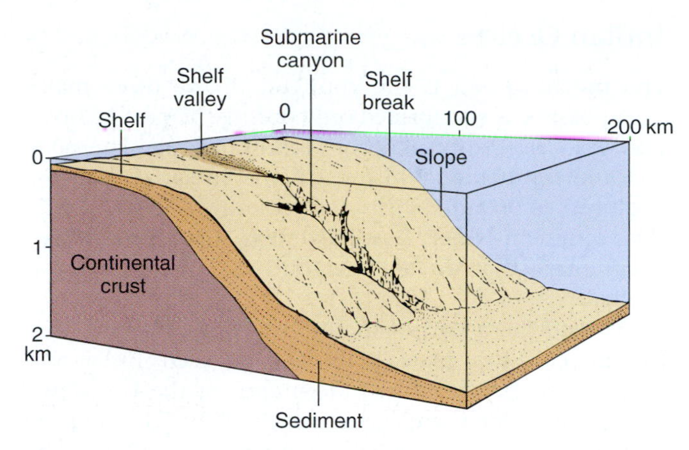

FIGURE 4.31 The continental shelf is often cut across by shelf valleys, some of which connect to steep-sided submarine canyons that cut across the continental slope.

when sea level was lower. Many **submarine canyons** are extensions of shelf valleys (**Fig. 4.31**).

Isostatic Sea-Level Change

In addition to eustatic processes, isostatic processes induced by changing climate can affect coastal topography. During an ice age, massive ice sheets accumulate over many parts of the continental crust. The weight of the ice forces the continental crust to sink lower into the asthenosphere (**CC2**). This process also depresses (lowers the level of) the continental shelves. For example, the continental shelf of Antarctica currently is depressed about 400 m lower than most other shelves by the weight of its ice sheets. At the end of an ice age, when the ice melts, a depressed section of a plate slowly rises until it again reaches isostatic equilibrium. However, isostatic leveling is much slower than eustatic sea-level change. Many areas of continental crust, including the coasts of Scandinavia and the northeastern United States and Canada, are still rising in response to the melting of glaciers that occurred several thousand years ago.

Glaciers shape topography in other ways as well. They often cut narrow, steep-sided valleys. Many valleys left after glaciers melted have been submerged by rising sea level to become deep, narrow arms of the sea known as **fjords** (Chap. 13).

Sea Level and the Greenhouse Effect

Although oscillations of sea level are normal occurrences in geological time, human civilization has emerged during a long period of relatively stable sea level (**Fig. 4.30**). If the predicted greenhouse warming of the planet by as much as 2°C during the first half of this century does indeed occur, the higher average global temperature is expected to cause the ocean water to warm and expand, and the current polar ice sheets to melt. As a result, sea level could rise by several meters or more. A sea level rise of only a few tens of centimeters would inundate vast areas, including major coastal cities. Therefore, the history of coastal modification during periods of sea-level rise is of more than academic interest.

4.10 PRESENT-DAY OCEANS

Having learned about processes that create, shape, and destroy ocean floor topography, we can look with greater understanding at the present-day oceans, which are all connected. The Atlantic Ocean connects with both the Indian and Pacific Oceans near Antarctica. The Arctic Ocean connects with the Pacific Ocean only where shallow water covers the continental shelf in the Bering Strait between Alaska and Siberia. In contrast, the Arctic Ocean is connected with the Atlantic Ocean by the deeper and much wider passage through the Nansen Fracture Zone between Greenland and Spitsbergen, Norway (**Fig. 4.18**). The somewhat deeper connection between the Arctic and Atlantic Oceans allows somewhat restricted water transfer between the two basins (Chap. 10).

Pacific Ocean

The Pacific Ocean is the world's largest and, on average, deepest ocean. It is almost completely surrounded by narrow continental shelves and deep trenches of subduction zones. It has many islands, including volcanic islands and atolls formed at hot spots, islands in magmatic and sedimentary arcs, and islands that appear to be small pieces of continent. Aside from hot spots and limited areas of new spreading, notably the East Pacific Rise, the Pacific Ocean crust is generally old, cold, and dense. Hence, it floats low in the asthenosphere.

The oldest oceanic crust found in the Pacific Ocean is not substantially older than the oldest oceanic crust in other oceans. This is because all of the Pacific Ocean's older crust has been subducted. Passive margins of continents that once surrounded the Pacific were converted to subduction zones long ago. Oceanic crust that once was present near the passive margins also has been subducted. The sediment that once covered the subducted crust was either subducted or compacted into rock and added to island arcs or continents as exotic terrane.

When the Pacific Ocean margins changed from passive margins to subduction zones as the ocean stopped expanding hundreds of millions of years ago, a ring of coastal mountains formed on the surrounding continents. The mountains prevented most rivers on these continents from draining into the Pacific. In addition, trenches in which river-borne sediments accumulated were formed around much of the Pacific. In other areas, magmatic arcs formed, creating marginal seas where sediments discharged by rivers were trapped. All oceanic crust now found in the Pacific Ocean, other than crust formed in marginal basins that are not part of the Pacific Plate, was created after these mountain chains, trenches, and magmatic arcs were formed. Hence, since its formation, the oceanic crust of the Pacific Plate has remained remote from any major rivers and the large amounts of sediment that they can contribute (Chap. 8), and the **sedimentation rate** on this crust has been slow for hundreds of millions of years. Because it is not significantly older than the crust in other oceans and because the sedimentation rate is low, the Pacific Ocean floor has a relatively thin cover of sediment. As a result, the abyssal plains are not as flat as parts of the Atlantic Ocean's abyssal plains.

Atlantic Ocean

The Atlantic is a long, narrow ocean that is often considered to include the Arctic Ocean because the Mid-Atlantic Ridge stretches essentially from the North Pole to Antarctica. The Atlantic Ocean is expanding in an east–west direction. An oceanic ridge, the Mid-Atlantic Ridge, runs down its center. Most of the Atlantic coasts of Europe, Africa, and North and South America are passive margins. The Atlantic's oldest oceanic crust is at two subduction zones: the Puerto Rico–Cayman Trench at the entrance to the Caribbean Sea, and the South Sandwich Trench located northeast of the Antarctic Peninsula.

The Atlantic Ocean has fewer island arcs and hot spots than the Pacific has. Greenland, the world's largest island, is part of the North American continent and is joined to Canada by the submerged continental shelf of Baffin Bay. Continental shelves are broader in the Atlantic Ocean than in the Pacific Ocean, and they constitute a larger percentage of the total area of the Atlantic than of the Pacific. The Atlantic also has several shallow marginal seas, including the Baltic, Mediterranean, and Caribbean Seas and the Gulf of Mexico. Thick salt deposits and oil and gas reservoirs typical of developing passive margins are present in many locations around the edges of the Atlantic Ocean. Such deposits formed early in the ocean's history, during periods when the new ocean and its marginal seas were isolated from the rest of the world ocean.

The Atlantic Ocean floor has fewer seamounts and a smoother abyssal plain than the Pacific Ocean floor, in part because of the relative dearth of hot-spot volcanoes in the Atlantic. In addition, topography is buried by very large quantities of sediment deposits that are found on the Atlantic Ocean crust, except on the newly formed oceanic ridge mountains. The large quantities of sediments are derived from the great quantities of sediment that rivers have transported into the Atlantic as the passive margins on both sides of the ocean have eroded. The Atlantic continues to receive freshwater **runoff** and some sediment from vast drainage areas of Europe, Africa, and the Americas, particularly in equatorial regions. Two large rivers, the Amazon and the Congo, empty into the equatorial Atlantic. They contribute about one-fourth of the total worldwide river flow to the oceans.

Indian Ocean

The Indian Ocean is the youngest of the three major ocean basins; it was formed only during the past 125 million years by the breakup of Gondwanaland. The plate tectonic features and history of the Indian Ocean are complex. For example, it is not yet known how and why the prominent Ninety East Ridge (**Fig. 4.10, Inside front cover map**) that divides the Indo-Australian Plate was formed.

The northern part of the Indian Ocean is dominated by the collision plate boundary between India and Eurasia. The newly formed mountains of the Himalayas are readily eroded, and large quantities of sediment are transported into the Arabian Sea and the Bay of Bengal by many rivers. Those rivers include three that are among the world's largest: the Indus, which empties into the Arabian Sea; and the Ganges and Brahmaputra, both of which empty into the Bay of Bengal. The enormous quantities of sediment flowing from India and the Himalayas have accumulated to form massive **abyssal fans** and extensive, relatively shallow abyssal plains in large areas of the northern Indian Ocean.

To the east of the India–Eurasia collision, along the northeastern edge of the Indian Ocean, is the very active Indonesian subduction zone. Its trenches and island arcs extend between mainland Asia and Australia. To the west of the India–Eurasia collision, the northern Red Sea is a rift zone that becomes an oceanic ridge system at its southern end. This oceanic ridge extends toward the south and into the central Indian Ocean, where it divides. One of the ridges that originates at that point extends southwest around Africa, and the other extends southeast around Australia. The Indian Ocean is opening in a complex manner as the African, Antarctica, and Indo-Australian plates move apart.

Passive margins are present along most of West Africa, most of Australia, and the coast of India. There are few islands in the Indian Ocean other than the many islands that form the Indonesian arc, where subduction is occurring along an **oceanic convergent plate boundary**. Madagascar, which is now part of Africa, appears to be a fragment of Pangaea that broke away from India when India began to move rapidly northward toward Asia, long after the initial breakup of Pangaea itself.

Marginal Seas

Several arms of the major oceans are partially isolated from the major ocean basins by surrounding landmasses. Such marginal seas are of four types:

One type consists of shallow submerged areas of continental crust. Examples include the Baltic Sea, the North Sea, Baffin Bay, and Hudson Bay.

A second type is formed in back-arc basins behind subduction zones, and marginal seas of this type often contain deep areas where oceanic crust is present. The marginal sea is separated from the open oceans by islands of the magmatic arc and a submerged ridge that connects the islands. Many such seas contain thick sediment deposits derived primarily from erosion of the newly formed islands of the arc. Examples include the Java Sea behind the Indonesian Arc, the South China Sea behind the Philippines, and the Aleutian Basin or Bering Sea behind the Aleutian Island chain.

The third type of marginal sea is a narrow remnant of an old closing ocean. The Mediterranean Sea, which lies between the converging African and Eurasian Plates, is a good example. Such seas may have thick sediments.

Finally, some marginal seas are the long and narrow arms formed as a continental divergent plate boundary develops into a new ocean. The primary example is the Red Sea.

CHAPTER SUMMARY

The Earth and Plate Tectonics. The Earth consists of a solid inner core, liquid outer core, plastic mantle, and solid overlying lithosphere. The mantle, especially the upper mantle or asthenosphere, is close to its melting point and can flow like a fluid, but very slowly. The thin lithosphere consists of continental or oceanic crust overlying a layer of solidified mantle material and is separated into plates that float on the asthenosphere. About 225 million years ago, all the continents were joined. Since then they have been separated by plate tectonic movements.

Plate Boundary Processes. Lithospheric plates may pull away from (diverge), collide with (converge), or slide past each other. Oceanic crust is created at divergent plate boundaries and destroyed at convergent plate boundaries. Divergent plate boundaries are the oceanic ridges and areas where continents are being pulled apart. Oceanic ridges are undersea mountain chains with many active volcanoes, and they are offset at transform faults.

Convergent plate boundaries are downwelling zones where old oceanic crust is subducted. Subduction zones at the edges of continents are characterized by an offshore trench and coastal mountains formed by compression of the continental crust plate and accumulation of sediment scraped off the subducting oceanic crust plate. Subducted and heated crust melts and magma rises to form volcanoes on the continental crust plate. Subduction zones at which oceanic crust is at the edge of both plates are characterized by a trench and a magmatic arc (and sometimes a separate sedimentary arc) of islands on the nonsubducting plate. A back-arc basin is present if the subduction rate is high and the nonsubducting plate is stretched. A collision where two continents meet at a convergent plate boundary is characterized by mountain chains created by compression of the continental crusts of the two colliding plates.

Hot Spots. Hot spots cause persistent volcanic activity. Some are situated over zones where upwelling convection extends throughout the mantle. Lithospheric plates move independently of any movements of most hot spots. As the lithospheric plate and/or hot spot move with respect to each other, hot-spot trails of islands and seamounts are formed.

Plate Interiors. As new oceanic crust moves away from a divergent plate boundary, it cools, sinks isostatically, and is buried by sediment. Edges of continents that are not at plate boundaries are known as passive margins and are characterized by a flat coastal plain, shallow estuaries and swamps, and a wide, heavily sediment-covered continental shelf.

Sea-Level Change and Climate. The Earth's climate is naturally variable. When the average surface temperature changes, eustatic changes of sea level occur globally. When the Earth warms, sea level rises as ocean water expands thermally, and vice versa. At the Earth's warmest temperatures, the oceans covered as much as 40% of the present land surface area. At its lowest temperatures, sea level was at least 100 m lower than it is today, and most of the continental shelves were exposed.

Isostatic leveling causes sea level to change in relation to the local coast. If continental crust is weighted by ice during a glacial period, or if its temperature falls (density increases), it sinks. If crust loses weight (as it does when ice melts during warm periods) or warms, it rises. However, isostatic leveling is very slow.

Present-Day Oceans. The Pacific is the largest and oldest ocean. It is ringed by subduction zones and has many volcanic islands and atolls formed at hot spots and magmatic arcs. Because few rivers drain directly into it and sediments are trapped in subduction zones and marginal seas, its seafloor has a relatively thin sediment cover.

The Atlantic Ocean is widening as lithospheric plates move apart at the Mid-Atlantic Ridge. It has few islands and broad continental shelves. Compared with the Pacific Ocean, it has more rivers and thicker average sediment cover.

The Indian Ocean is the youngest ocean. It has a complex oceanic ridge system, few islands, and thick sediment cover, especially in the north, where major rivers empty from the new, easily erodable Himalaya Mountains created at the India–Eurasia continental collision.

There are four types of marginal seas: shallow seas where continental crust is submerged, long narrow seas where continents are breaking apart, seas between continents that are moving toward a future collision, and back-arc basins behind subduction zones.

STUDY QUESTIONS

1. Describe the Earth's mantle. How do we know what the mantle is made of and how it behaves?
2. What are the differences between continental crust and oceanic crust, and why are these differences important?
3. What is a hot-spot trail? How do hot-spot trails show that lithospheric plates move across the Earth's surface?
4. What three types of motion occur at plate boundaries?
5. List the types of convergent plate boundaries. Describe the characteristics and locations of volcanoes associated with convergent plate boundaries. Why are there few or no volcanoes at convergent plate boundaries where two continents collide?
6. What processes occur at oceanic ridges to form their mountainous topography and the fracture zones that cut across them?
7. Describe the changes in seafloor depth and sediment cover with increasing distance from an oceanic ridge. What causes these changes?
8. Why are passive margins described as "passive"? What are their characteristics?
9. Distinguish between isostatic and eustatic processes that cause sea level to change. How do these processes complicate efforts to measure changes in global sea level by measuring sea-level heights at various coastlines?
10. If there were no ice cap on the Antarctic continent, which coast of the United States—the California coast, the Pacific Northwest coast, or the Mid-Atlantic coast—would the Antarctic coast and continental shelf resemble? Why?
11. Describe and explain the principal differences in geography and seafloor topography between the Atlantic Ocean and the Pacific Ocean.

KEY TERMS

You should recognize and understand the meaning of all terms that are in boldface type in the text. All those terms are defined in the Glossary. The following are some less familiar key scientific terms that are used in this chapter and that are essential to know and be able to use in classroom discussions or exam answers.

abyssal hill (p. 71)
abyssal plain (p. 71)
asthenosphere (p. 70)
atoll (p. 94)
back-arc basin (p. 82)
barrier reef (p. 94)
central rift valley (p. 86)
coastal plain (p. 96)
continental collision plate boundary (p. 83)
continental divergent plate boundary (p. 88)
continental drift (p. 76)
continental shelf (p. 71)
continental slope (p. 71)
convection (p. 73)
convergence (p. 77)
convergent plate boundary (p. 77)
crust (p. 70)
divergence (p. 77)
divergent plate boundary (p. 77)
downwelled (p. 73)
eustasy (p. 96)
exotic terrane (p. 80)
fracture zone (p. 89)
fringing reef (p. 94)
guyot (p. 93)
hot spot (p. 74)
hydrosphere (p. 71)
ice age (p. 97)
isostasy (p. 71)
isostatic leveling (p. 91)
lithosphere (p. 70)
lithospheric plate (p. 70)
magma (p. 71)
magmatic arc (p. 80)
mantle (p. 70)
marginal sea (p. 88)
mesoplate (p. 74)
oceanic plateau (p. 80)
oceanic ridge (pp. 70, 77)
passive margin (p. 95)
plate tectonics (p. 70)
rift zone (p. 77)
seafloor spreading (p. 84)
seamount (p. 91)
sediment (p. 71)
sedimentary arc (p. 81)
sedimentation rate (p. 100)
shelf break (p. 71)
spreading cycle (p. 76)
subducted (p. 74)
subduction zone (p. 75)
superplume (p. 74)
topographic (p. 70)
transform fault (p. 88)
transform plate boundary (p. 77)
trench (p. 71)
upwelled (p. 73)

CRITICAL CONCEPTS REMINDER

CC1 **Density and Layering in Fluids** (p. 70). The Earth and all other planets are arranged in layers of different materials sorted by their density. To read **CC1** go to page 2CC.

CC2 **Isostasy, Eustasy, and Sea Level** (pp. 71, 85, 91, 96, 97, 99). Earth's crust floats on the plastic asthenosphere. Sections of crust rise and fall isostatically as temperature changes alter their density and as their mass loading changes due to melting or to the formation of ice stemming from climate changes. This causes sea level to change on the coast of that particular section of crust. Sea level can also change eustatically when the volume of water in the oceans increases or decreases due to changes in water temperature or changes in the amount of water in glaciers and ice caps on the continents. Eustatic sea level change takes place synchronously worldwide and much more quickly than isostatic sea level changes. To read **CC2** go to page 5CC.

CC3 **Convection and Convection Cells** (pp. 73, 77, 87, 90). Fluids that are heated from below, such as Earth's mantle, or ocean water, or the atmosphere, rise because their density is reduced. They continue to rise to higher levels until they are cooled sufficiently, at which time they become dense enough to sink back down. This convection process establishes convection cells in which the heated material rises in areas of upwelling, spreads out, cools, and then sinks at areas of downwelling. To read **CC3** go to page 10CC.

CC7 **Radioactivity and Age Dating** (pp. 70, 91). Some elements have naturally occurring radioactive (parent) isotopes that decay at precisely known rates to become a different (daughter) isotope, which is often an isotope of another element. This decay process releases heat within the Earth's interior. Measurement of the concentration ratio of the parent and daughter isotopes in a rock or other material can be used to calculate its age, but only if none of the parent or daughter isotopes have been gained or lost from the sample over time. To read **CC7** go to page 18CC.

CC11 **Chaos** (p. 73). The nonlinear nature of many environmental interactions makes complex environmental systems behave in sometimes unpredictable ways. It also makes it possible for these changes to occur in rapid and unpredictable jumps from one set of conditions to a completely different set of conditions. To read **CC11** go to page 28CC.

CC14 **Photosynthesis, Light, and Nutrients** (p. 95). Chemosynthesis and photosynthesis are the processes by which simple chemical compounds are made into the organic compounds of living organisms. The oxygen in Earth's atmosphere is present entirely as a result of photosynthesis. To read **CC14** go to page 46CC.

CHAPTER 5

CHAPTER OUTLINE

CRITICAL CONCEPTS USED IN THIS CHAPTER

Plate tectonic processes affect us all. This freeway collapse occurred in 1989, in Oakland, California, after an earthquake of relatively modest magnitude occurred about 60 miles away. It is normal for one or more earthquakes of approximately this magnitude or greater to occur, on average, every few years somewhere along the Pacific coast transform fault plate boundary.

Plate Tectonics: History and Evidence

In 1939, a renowned and respected geologist, Andrew Lawson, was shown a movie and working model that demonstrated how thermal convection in the Earth's crust could build mountains. He is reported to have responded,"*I may be gullible. I may be gullible! But I am not gullible enough to swallow this poppycock*" **(as quoted in *Challenger at Sea: A Ship That Revolutionized Earth Science*, 1992, by Kenneth Hsü. Published by Princeton University Press, Princeton, NJ, p. 57). Like many other geologists of the era, Lawson simply refused to concede that the mountains and continents were not fixed and immovable, despite a mounting volume of evidence to support the theory of continental drift (which has now evolved into the theory of plate tectonics). The theory of continental drift was proposed in 1912 by Alfred Wegener, who was not a geologist and, therefore, not to be believed by those who really knew geology. In fact, suggestions that mountains were not fixed and that continents moved on a fluid inner layer of the Earth had been made much earlier by, for example, Benjamin Franklin, who in 1782 stated,**

> *Such changes in the superficial parts of the globe seemed to me unlikely to happen if the earth were solid to the center. I therefore imagined that the internal parts might be a fluid more dense, and of greater specific gravity, than any of the solids we are acquainted with; which therefore might swim in or upon the fluid. Thus, the surface of the globe would be a shell, capable of being broken and disordered by the violent movements of the fluid on which it rested.*

(Benjamin Franklin, Letter of September 22, 1782, to Abbe Soulavie, quoted in *Benjamin Franklin* [1938, 1987 edition], by Carl Van Doren, published by Bramhall House, New York, p. 660)

There was also ample evidence to support the theory that mountains were not fixed in place. For example, on his 1831–1836 expedition on the *Beagle*, Charles Darwin observed corals high in the Andes Mountains and concluded that these mountains had once been below the sea surface. For all their eminence in other fields, though, Franklin and Darwin were not geologists, and their observations and those of many others were therefore scorned or simply ignored.

In this chapter, we will learn more about how the theory of plate tectonics was developed through the scientific method, which requires that the preponderance of evidence support a theory before its acceptance. The primary lesson we learn is that scientific theories must be extensively validated to be accepted, but even upon acceptance they are, and should be, subject to challenge, review, and revision. This aspect of the scientific method is now well entrenched. One might also believe that dogmatic denials of evidence because the evidence does not fit a favored theory or because the source of the evidence was not an acknowledged expert in the field are a thing of the past, as science is now much more advanced. But is this true? Human nature changes slowly, and I wonder whether this is indeed the case. Which outrageous theories out there today will become the "continental drift" equivalent of the future?

The concepts of **plate tectonics** described in Chapter 4 are now almost universally accepted, even though many uncertainties remain about the details of the processes involved. The hypothesis that the continents are not fixed was not accepted until about the 1970s. For several decades before that, this hypothesis had been dismissed as ridiculous by many geologists, even though much of the evidence that we now believe shows that the continents have moved was already available.

This chapter briefly reviews the history of plate tectonic theory and the many pieces of evidence that have resulted in its acceptance. The chapter provides insight into the complexity of Earth sciences. It also illustrates the way in which many disconnected studies in biology, physics, chemistry, and geology can become the crucial pieces in solving a larger puzzle. Scientific studies that have no apparent application to a particular problem may suddenly provide the key to a better understanding of the world around us, and of how we can best manage and sustain human interactions with this **environment**.

5.1 METHODOLOGY OF EARTH SCIENCES

An idealized view of science is that knowledge is gained in a stepwise process. Each incremental step involves establishing a hypothesis, reviewing available data, designing and performing tests, and conclusively verifying or disproving the hypothesis on the basis of the test results. This simple process is the accepted scientific method, and it works well in many areas of science. However, environmental systems are extremely complex, because many different interlinked processes control the world we see. For example, the geology at any point on the Earth's surface may be determined by a host of processes that operate on timescales ranging from minutes and hours to tens of millions of years. Such processes include volcanism, earthquakes, marine **sedimentation**, **erosion**, **weathering**, **glaciation**, sea-level changes, and **climate** changes. It is rarely possible to isolate and study the effects of only one such process. Therefore, usually we must study the net result of all the interlinked processes. Identifying and understanding the relative influence of each process is like trying to solve a giant multidimensional jigsaw puzzle from which many pieces are missing.

Because of the extreme complexity of environmental systems, studies in Earth sciences, including ocean sciences, generally must be framed as a series of alternative hypotheses, and many different pieces of evidence must be gathered to address these hypotheses. Almost without exception, no single piece of evidence is ever conclusive. In many instances, data will support, or can be explained by, two apparently contradictory hypotheses. For example, Darwin's observations of marine **fossils** high in the Andes support a hypothesis that these mountains were not always elevated and were raised up from the sea. They also support a hypothesis that the fossils were placed on top of the mountains during a time when sea level was higher and the sea covered the mountain range.

Given that conclusive data are rarely available to prove or disprove alternative hypotheses in Earth and environmental sciences, acceptance or rejection of a particular hypothesis and its elevation to the status of well-established scientific theory (an idea that has a large body of observational evidence to support it and that has come to be accepted by most scientists in the field of study) is based on the weight of available evidence. This standard is similar to the "preponderance of the evidence" standard used in the American judicial system. Individual scientists may choose to give more weight to certain pieces of evidence than others, in the same way that judges and jurors do. Often, accepted hypotheses are reluctantly rejected only when an overwhelming preponderance of new evidence becomes available. Accordingly, Earth sciences tend to move forward by periodic drastic revisions of accepted theory when the overwhelming weight of contrary evidence finally becomes too great to ignore. The development and acceptance of the hypothesis that the continents move is a classic example of this process. After the initial hypothesis was proposed, many years passed and numerous individual pieces of indirect evidence accumulated before the scientific community accepted the idea that continents move.

5.2 HISTORY OF PLATE TECTONIC THEORY

Before the theory of continental drift was formally proposed in 1912, the continents and oceans were universally believed to be fixed and immovable. Geologists knew that ocean fossils could be found at the tops of mountains and that many mountains were formed from uplifted seabed, as shown by tilted layers of sedimentary rock (**Fig. 5.1**). They also knew that sharply defined changes in fossil types within a landmass indicated that the landmass consisted of several pieces that at one time had been separated but now were welded together. However, such evidence was believed to strongly support the prevailing hypothesis that the Earth was only a few thousand years old and that all of its features had been created by catastrophic changes wrought during its early history. Indeed, some people felt that such evidence also supported a hypothesis, based on the teachings of the Bible, that a single catastrophic flood had created many of the features.

During the nineteenth century, on the basis of geological studies and Darwin's theory of evolution, the Earth was determined to be probably many millions of years old. However, these findings simply led to minor modifications of the prevailing theory. For example, landforms were acknowledged to be modified by volcanoes and erosion, but the continents were still believed to be fixed and immovable.

Wegener's Observations and Theory

In the early 1900s, Alfred Wegener, a meteorologist and polar explorer, was intrigued by certain puzzling features of the continents. His initial interest was based on an observation that the **coastlines** of South America and Africa matched in shape and could be closely fitted together. This was not a new observation. About 400 years earlier, the similarity had attracted the attention of Leonardo da Vinci and a Dutch cartographer, Abraham Ortelius. In a 1596 publication called *Thesaurus Geographicus*, Ortelius suggested that the continents had once been joined but had been separated by earthquakes and floods. Furthermore, in the 1800s several scientists hypothesized that the Atlantic Ocean had been created when a cataclysmic event of unknown origin separated the continents early in the Earth's history.

Wegener's observations went beyond earlier speculation that the continents fit together. For example, he observed that mountain ranges in South America and Africa lined up with each other when the continents were reconstructed. Fossils in the rocks and geological evidence of ancient **glaciers** also lined up and matched at the appropriate locations on the edges of the fitted continents.

Wegener expanded his interest to the other continents and published a detailed account of his observations and theory in 1915, 3 years after first having proposed the outlines of the theory. He hypothesized that about 200 to 250

FIGURE 5.1 Sedimentary rock layers tilted from their original horizontal orientation. This formation is on the coast of northern California and consists of sedimentary rock added to the continent as the continental edge was uplifted by tectonic processes.

million years ago, all the continents were joined as parts of a single supercontinent, which he named "Pangaea." He further hypothesized that Pangaea had been surrounded by a single world ocean, called "Panthalassa." In retrospect, Wegener's 200-to-250-million-year estimate for the onset of the breakup of Pangaea was surprisingly accurate, considering that it was based on the very poor methods available at the time for estimating the age of the Earth and its oceans, rocks, and fossils. Dating (age estimation) techniques are discussed in CC7.

Although Wegener's hypothesis explained many otherwise unexplainable geological features, his theory was ignored and even scorned by the geological science community for decades. Some of the scorn was probably due to the fact that Wegener was a meteorologist and not a geologist. Many scientists were, and still are, jealous of their own expertise and "turf." Accordingly, geologists may have been unwilling to listen to the ideas of someone with no formal qualifications in their discipline. Furthermore, in the early twentieth century, no geologist was able to hypothesize a mechanism that would provide the driving force to split Pangaea and set its continents in motion, as Wegener's hypothesis required.

The Evidence Accumulates

Wegener's theory, which became known as the "**continental drift theory**," made little progress in acceptance until the 1950s, when a variety of new evidence began to accumulate. First, it became clear that the maximum age of the ocean basins was less than 200 million years, much less than the age of the continents, where rocks about 4 billion years old had been found. Second, studies of **paleomagnetic** properties of rocks and **sediments** indicated that the continents must have moved, unless there had been more than one north magnetic pole on the Earth at certain times in the past. Many people in the scientific community began to suspect that parts of Wegener's hypothesis might indeed be correct, and serious study of his theory began.

Some of the most important new evidence resulted from rapidly intensifying studies of the oceans. During the 1950s, the great mountain chains of the **oceanic ridges** were discovered and found to consist of newly formed volcanic rock. These extensive ridges were also found to be the sites of many shallow earthquakes, which indicated motion of the **crust** and, more important, a tension or pulling apart of the seafloor at the ridges.

A Proposed Mechanism

In 1960, Harry Hess, a Princeton geologist and oceanographer, first proposed the theory later known as **seafloor spreading**. He suggested that the Earth's outer shell moves about in pieces as it floats on the material below. He hypothesized that such motions had caused Pangaea to break apart, the continents to drift toward their present locations, and new oceans to form between them. Hess also theorized that ocean floor is created at oceanic ridges and destroyed at deep-ocean **trenches**. At that time, he had little evidence to support these hypotheses, and he referred to them as "geo-poetry."

The Magnetic Anomaly Clue

During the 1950s, magnetometers were towed routinely behind research ships to measure small changes in the Earth's magnetic field. The magnetometer readings revealed a striped pattern of alternating bands of slightly higher and slightly lower magnetism parallel to the oceanic ridge (Fig. 5.2). Such deviations from the mean value are known as anomalies. Although the anomalies were known to be related to changes in the magnetic field of the seafloor, the striped patterns were a mystery for a number of years. Ultimately, however, they became possibly the most important single piece of evidence in convincing the majority of geologists of the truth of Wegener's hypothesis.

In 1963, British geophysicists Drummond Matthews and Fred Vine hypothesized that the **magnetic anomaly** stripes were a historical record of the creation of new seafloor. From studies of terrestrial rocks in the early 1900s, the Earth's magnetic field was known to reverse periodically but irregularly. Matthews and Vine suggested that each adjacent magnetic anomaly stripe corresponded to a period of normal or reversed magnetic field. Later this idea was applied to estimate the age of the seafloor and demonstrate that the oldest part of the Atlantic Ocean floor was only about 170 million years old. Matthews and Vine's findings are discussed later in this chapter.

Acceptance of the Theory

The magnetic anomaly data was perhaps the key piece of evidence that, when added to the known fossil and geological data, finally became overwhelming. In the mid 1960s, almost 50 years after it was first proposed, Wegener's theory was accepted by the scientific community as a whole, although skepticism persisted. During the past several decades, new evidence has accumulated rapidly to support the theory of moving **lithospheric plates** and the idea that ocean floors are continuously born and destroyed. Within the past few years, satellite and other very precise distance measurement techniques have enabled scientists to observe directly the slow creep of lithospheric plates across the Earth's surface.

The following sections review many of the pieces of evidence for plate tectonic theory. Each piece of evidence is important, but only when taken together do they provide irrefutable proof that continents move and overwhelming evidence to support the hypothesis that they

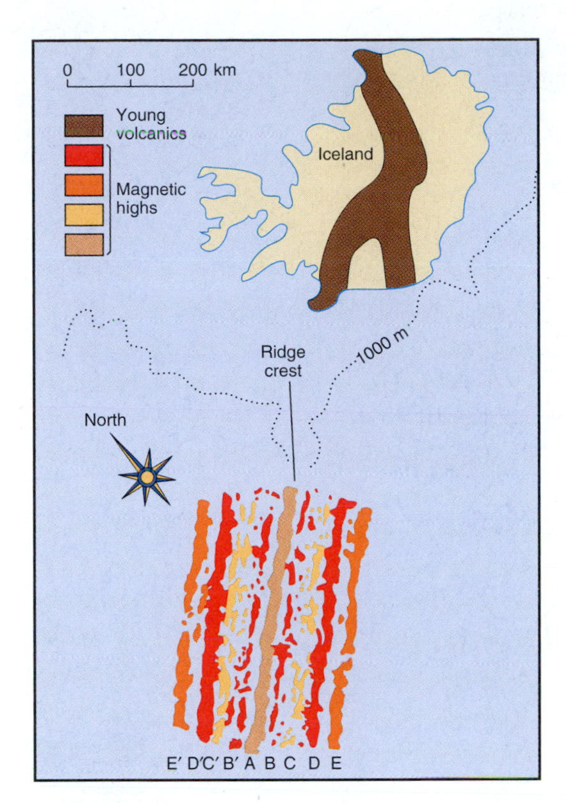

FIGURE 5.2 The striped magnetic anomaly pattern on the ocean floor on either side of an oceanic ridge. The stripe marked A is the crest area. Stripes B and B' and C and C' are separated parts of stripes formed at the ridge crest during progressively older periods between which the Earth's magnetic field reversed. This section of seafloor is south of Iceland. Iceland is part of the oceanic ridge and has a line of active volcanoes through its center.

are assembled and reassembled as they drift across the Earth's surface and that oceans are created and destroyed.

5.3 FIT OF THE CONTINENTS

The first piece of evidence that the continents might drift was the observation that the coastlines of the continents, particularly South America and Africa, closely match each other in shape.

Gaps and Overlaps in the Fit

Alfred Wegener's reconstruction of Pangaea from the present continents was very similar to the current view of that reconstruction, which is shown in **Figure 5.3**. However, because Wegener used the coastlines of the continents to fit the pieces, his reconstruction was not perfect and left substantial gaps in many places. The imperfections were cited by skeptics as evidence that Wegener's theory was wrong. However, most of the imperfections were eventually found to be the result of an incomplete understanding of the true edges of the continents. The edge of the continental crust is not the coastline, since the **continental shelf**, which is covered by shallow ocean waters, is also continental crust and therefore part of the continent (Chap. 4).

In 1965, Sir Edward Bullard, a British geophysicist, resolved many questions about gaps in the Pangaea reconstruction. He reconstructed the continents using depth **contours** that marked the edge of the continental shelf. Today this type of reconstruction is usually based on the 500-m depth contour. This contour corresponds to the edge of the deepest continental shelf, which is around Antarctica, and therefore lies on the **continental slope** around most continents. Reconstructions that include the continental shelves fill most of the gaps in

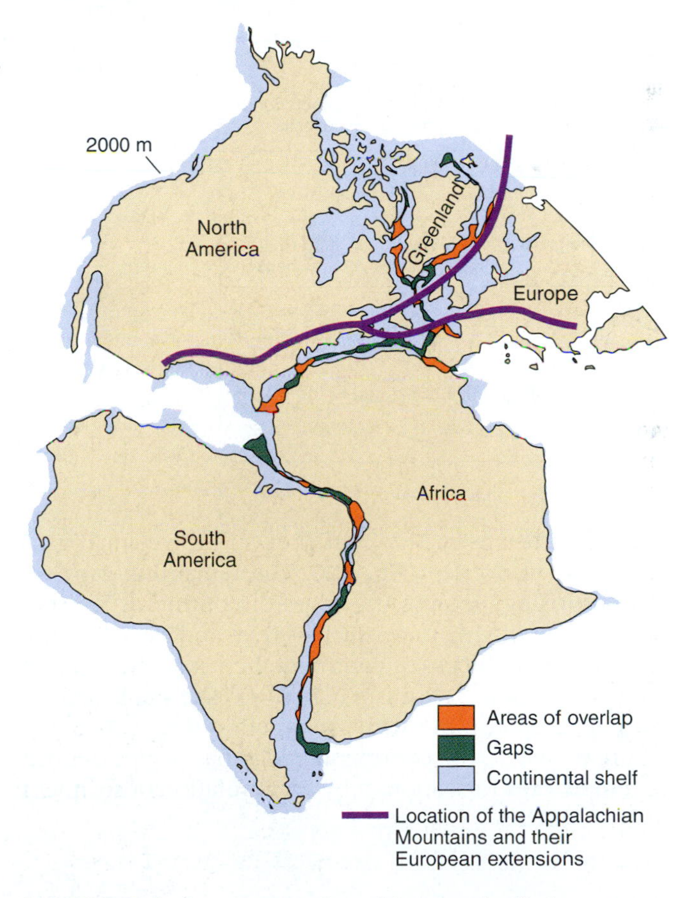

FIGURE 5.3 A reconstruction of Pangaea by fitting continents at the shoreline and along the continental slope at a depth of 2000 m. Ancient mountain chains such as the Appalachians and their extensions in Europe are aligned in a way that suggests they were separated as Pangaea broke apart. By looking just at the coastlines, you can see how Wegener's original construction would have had substantially greater areas where the fit was not exact.

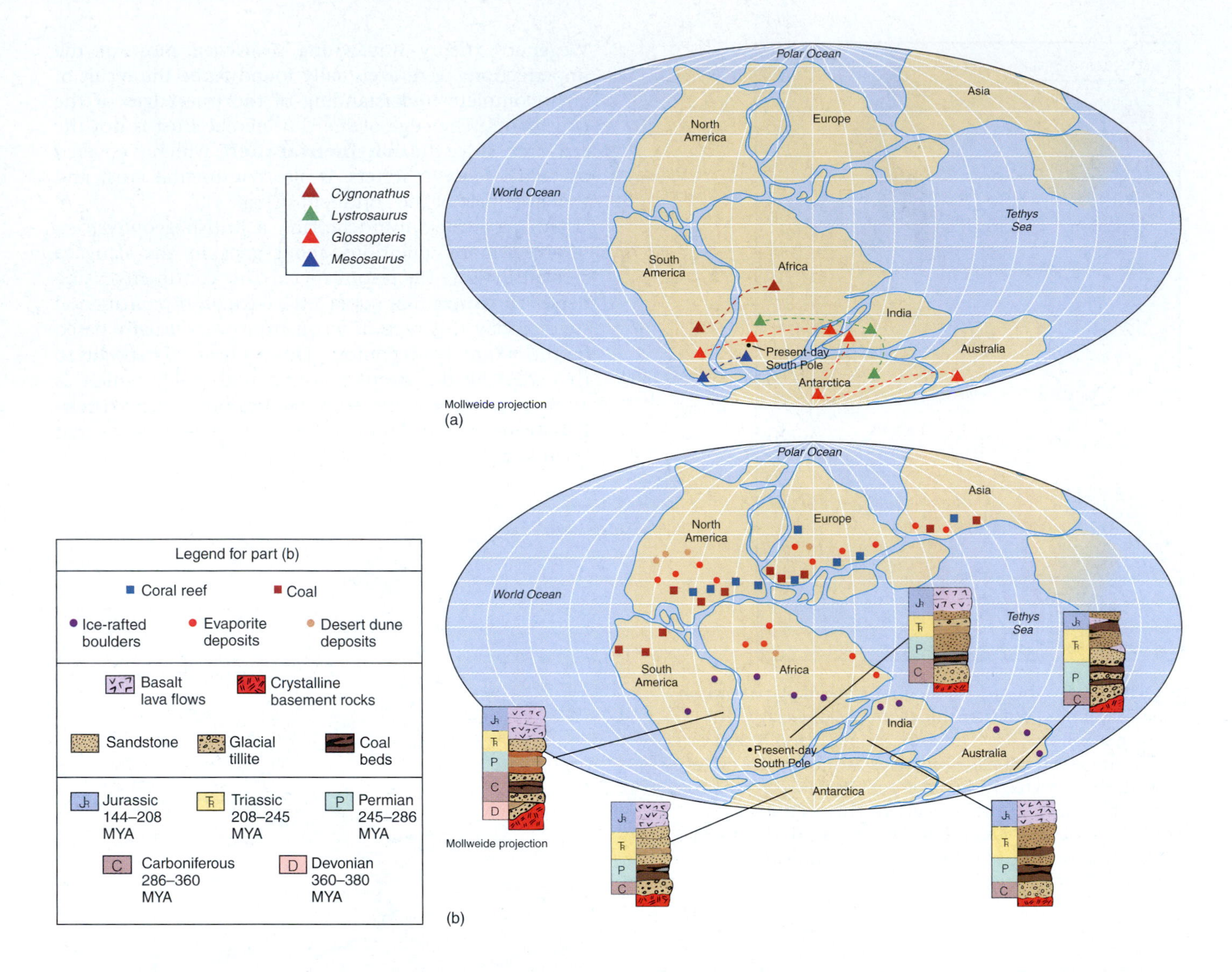

Wegener's reconstruction, but they still leave small gaps and areas of overlap (**Fig. 5.3**). The remaining gaps are not surprising, because the edges of continents are subjected to modifying forces such as river and wave erosion, the collapse of blocks of continental crust at the edges of some plates, and volcanism (Chap. 4). The small areas of overlap, seen also in this reconstruction (**Fig. 5.3**), are at locations where the continental edge has been extended by **coral reef** formation or by accumulation of sediment at river **deltas**.

5.4 GEOLOGICAL SIMILARITIES BETWEEN PREVIOUSLY JOINED EDGES OF CONTINENTS

If the continents were originally joined to form Pangaea, evidence of geological formations that were split apart should be found on each of the two new continents formed from the split. Certain pieces of such evidence were readily available to Wegener.

Mountain Chains

Some mountain chains or their eroded remnants end at the edge of one continent and reappear on the edge of another continent, exactly where they should if the continents were joined in Pangaea. For example, the Appalachian Mountains of the United States run through New England, the Canadian Atlantic Provinces, and Newfoundland and terminate abruptly at the Atlantic Ocean. Mountain chains of the correct age originate on the **coasts** of Ireland and Brittany (Northwest France) exactly where they should according to Wegener's reconstruction (**Fig. 5.3**). These mountains were formed at a **subduction zone** during a previous **spreading cycle**. They were incorporated into the interior of Pangaea when the supercontinent formed, and they subsequently split apart during its breakup.

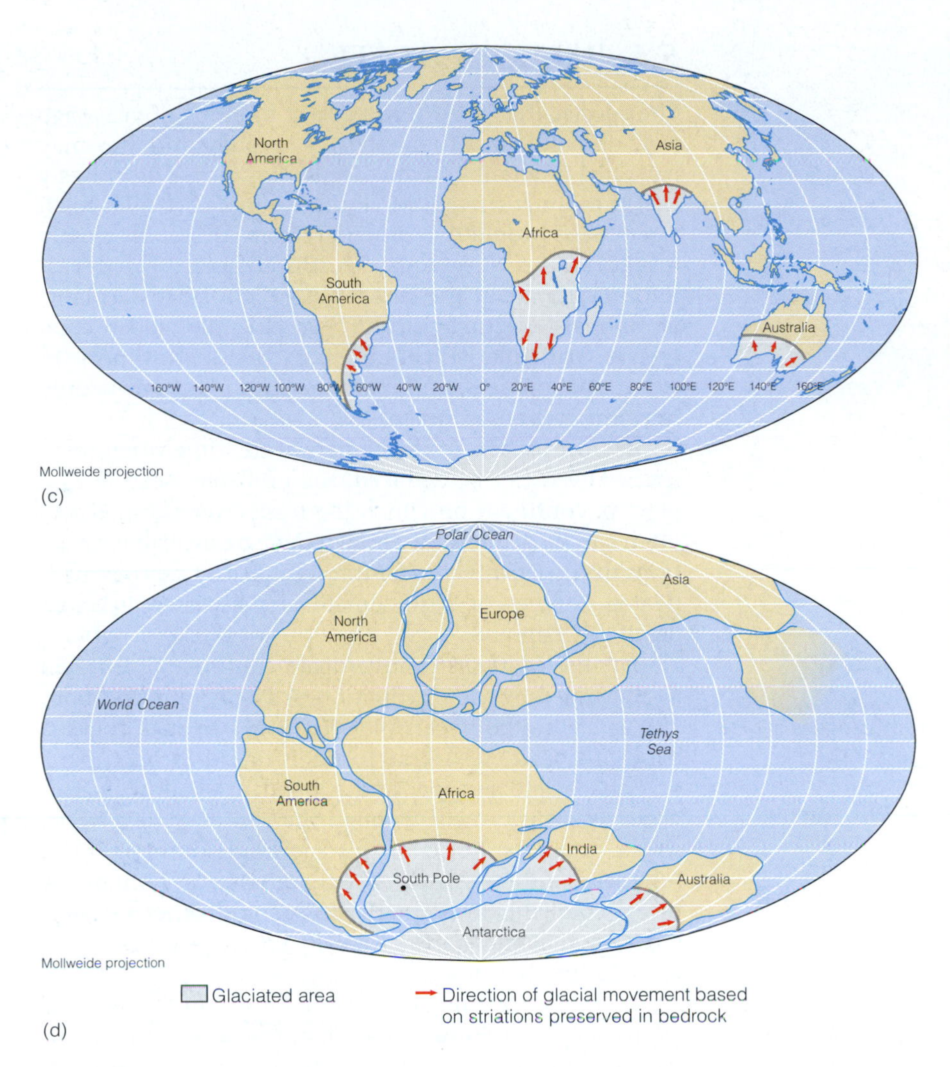

FIGURE 5.4 Fossil, geological, and glacial evidence for the existence of Pangaea and the present spreading cycle. (a) The fossils of some species that lived more than 200 million years ago are found on widely separated continents, indicating that these landmasses were joined at that time. In contrast, more recent fossils show that the species have evolved separately on each continent. (b) The distributions of 200-million-year-old and older rocks suggest that the continents must have been joined prior to that time. (c) The directions of flow of glaciers approximately 200 million years ago, as deduced from scour lines in the rock. (d) A reconstruction of Pangaea, showing that the directions of glacier movement would have been away from the South Pole approximately 200 million years ago.

Rocks and Fossils

Wegener found similarities in fossil and rock types between continent edges previously joined in Pangaea. More recent and extensive studies confirm extremely strong similarities in the fossil and mineral geological record of continents now located on opposite sides of an ocean (**Fig. 5.4a,b**).

The similarity in fossil assemblages, particularly fossils of terrestrial **species**, at locations that once were joined in Pangaea is not surprising. Because these areas were parts of a common landmass, **flora** and **fauna** were of a uniform type within each climatic region until the continents parted. As the pieces of Pangaea broke off, plant and animal life on the two sides of each separation became isolated from each other. Flora and fauna evolved separately and differently on each of the new continents in response to their individual changing climates. For this reason, fossil assemblages in rocks that formed since Pangaea first started to split apart provide a history of the breakup. Fossil information, combined with other data, can be used to estimate the approximate time at which each piece of Pangaea broke off and began to drift away (**Fig. 5.5**).

The distributions of various characteristics of 200- to 300-million-year-old rocks, including coal deposits, desert sandstones, windblown sand, and salt deposits, parallel the distributions of fossil assemblages. These geological features also connect in a consistent way in the reconstruction of Pangaea (**Fig. 5.4b**).

Glaciers

Large numbers of massive glaciers were present about 250 to 300 million years ago on the Gondwanaland section of Pangaea (**Fig. 4.8**), including the areas now known as Antarctica, Australia, southern Africa, southern South America, and India (**Fig. 5.4c**). The locations of these glaciated areas on Pangaea can be reconstructed (**Fig. 5.4d**). The scour lines left on the rocks by glaciers reveal

FIGURE 5.5 The continents broke away from Pangaea at different times. The earliest new ocean was the southern North Atlantic, followed by the Southern Ocean and the South Atlantic. The Red Sea is very young and still forming a new ocean, and the Gulf of California is even younger. The numbers on this figure represent the time (in millions of years ago, MYA) when Pangaea or its fragments broke apart at various locations shown by the arrows.

not just the presence of the glaciers, but also their flow direction. The flow directions for glaciers on each of the continental pieces that once formed the glaciated Gondwanaland section consistently align with each other in the reconstruction. The position and direction of the glaciers suggest that Gondwanaland was primarily in the Southern Hemisphere 250 to 300 million years ago. The area where Antarctica and South Africa joined in Gondwanaland was then at the approximate location of the South Pole.

Reef Corals

About 300 million years ago, Gondwanaland and Laurasia (**Fig. 4.8**) were separated by the shallow equatorial Tethys Sea, which later closed as Laurasia and Gondwanaland came together to form Pangaea. Tropical **reef** corals apparently grew in abundance in the Tethys Sea, and tropical terrestrial flora and fauna covered the southern half of Laurasia. Fossils of Tethys Sea **corals** and tropical Laurasian flora are now found in 300-million-year-old bands of rocks across North America, North Africa, and Eurasia. The fossil assemblages align well in a reconstruction of Pangaea (**Fig. 5.4a**).

5.5 PALEOCLIMATE

The species that lived hundreds of millions of years ago were different from their relatives of today, and they may have had different climatic needs. However, it is unlikely that the climatic needs of all species in a particular fossil assemblage would be different from those of closely related species alive today. Fossils therefore can reveal information about the climate in which the fossil plants or animals originally lived. For example, most coral species, particularly reef-building corals, grow only in warm, shallow seas, and many tropical plants grow only in warm, wet environments.

Because aspects of the climate at the time when fossil species lived can be deduced, the probable location of a piece of continent on which the fossils were formed also often can be deduced. If enough data are available, variations in the Earth's climate through history can be investigated. Studies of past climate are discussed in more detail in Chapter 8.

The use of fossils to investigate continental drift can be illustrated by the distribution of fossils in rocks about 280 to 350 million years old (**Fig. 5.4a**). In such rocks, a band of tropical terrestrial fossils is found through parts of Canada and Europe, and a band of shallow, warm-water fossils is present through the United States, North Africa, and Asia. Rocks of this age from southern South America, southern Africa, Australia, India, and Antarctica reveal cold-climate terrestrial fossils and other physical evidence that these areas were covered by extensive glaciers during that time.

Two explanations are possible for the distribution of fossils shown in **Figure 5.4a**. First, the continents might have remained in the same position during the past 300 million years. If that were true, the polar ice sheet would have spread from the South Pole as far north as the equator (in Africa) 300 million years ago. In addition, tropical temperatures would have been present only in a narrow band of **latitude** centered about 30°N, and temperatures near the North Pole would have been much higher than those anywhere in the Southern Hemisphere. This bizarre set of climatic conditions could not occur under any circumstances that we can currently believe might have existed. The alternative explanation is that 300 million years ago, the continents were assembled as shown in **Figure 5.4**, with all the glaciated regions joined. The supercontinent must have been mostly in the Southern Hemisphere, with the pieces of North America and Eurasia that are now at about 30°N located on or close to the equator.

Additional information about paleoclimates can be obtained from the composition of rocks. For example, deposits of coal, desert sands, and salt are each indicative of a particular climate. The distribution of these deposits confirms the climate and location of Gondwanaland that was deduced from fossils (**Fig. 5.4a,b**).

5.6 ISOSTASY AND THE MOHOROVIČIĆ DISCONTINUITY

Early in the twentieth century, a Croatian geophysicist, Andrija Mohorovičić, discovered an abrupt change in the velocity of seismic waves at a depth of about 30 to 40 km below the mean elevation of the continent surface and about 6 km below the mean depth of the ocean floor. The abrupt velocity change indicates that the rocks below this boundary have a composition very different from that of the rocks in continental or oceanic crust. The boundary is called the "Mohorovičić discontinuity" or simply the "Moho." The Moho marks the boundary between the crust and the **mantle**. From the location of the Moho, we can deduce that the continental crust averages 30 to 40 km in thickness, and the oceanic crust about 6 km in thickness.

The mean elevation of the top of the oceanic crust is distinctly different from that of the top of the continental crust. Whereas 75% of the continental crust surface is between sea level and 1 km in elevation, 75% of the oceanic crust surface is between 3 and 6 km below sea level (**Fig. 4.3**). Continental crust has a **density** of 2.7 to 2.8 $g \cdot cm^{-3}$, and oceanic crust has a density of approximately 3.0 $g \cdot cm^{-3}$. If the continental and oceanic crust float on the underlying mantle in response to **gravity**, this arrangement, with the surface of the thicker and lighter continental crust elevated above the surface of the thinner and denser oceanic crust, can be explained by **isostasy** (**CC2**).

We now know that the Earth's crust apparently lies on top of slabs of solid upper mantle, which are tens of kilometers thick, and together the crust and solid upper mantle form the lithospheric plates. It is these plates that are "floating" on the **asthenosphere**. We also know that the continents and oceans are continuously adjusting their heights in a process called **isostatic leveling**. They do so in response to a variety of factors, including changes in density associated with the heating or cooling of sections of the crust and with the thickening or stretching of the crust where plates collide or tear apart. Sections of the plates are also slowly but continuously adjusting their isostatic levels in response to changes in mass loading on the crust, by addition of material through volcanic eruptions, addition or loss of ice loads as glaciers are formed or melt, and removal of material by erosion.

The vertical movement of isostatic leveling is very slow, compared to many of the processes that cause the density or mass loading to change, so individual pieces of oceanic or continental crust may not be at isostatic equilibrium. However, if the lithospheric plates are floating on asthenosphere material that flows slowly like a fluid, the average elevations of the oceanic and continental crusts should be close to the equilibrium elevations corresponding to their average densities and thicknesses. In fact, calculations based on these averages do show that the continental crust and oceanic crust are in the approximate relative vertical positions that they would be expected to occupy at isostatic equilibrium (**CC2**). These calculations provided important evidence in the development of plate tectonic theory by demonstrating that the lithospheric plates could be floating on an underlying fluidlike material. The knowledge that the plates were floating became the basis for understanding the process by which continents could be made to drift across the face of a seemingly solid Earth.

CRITICAL THINKING QUESTIONS

5.1 What is the scientific method? Give two examples of things that you have studied in school or college (other than the theory of plate tectonics discussed in this chapter) that were accepted as fact only after scientific studies had provided several pieces of information proving that previous ideas about our world were wrong. In each case, what were the most important pieces of information? Describe what you know about the nature of the important scientific studies that provided the necessary new information in each case.

5.2 This text emphasizes the interrelated nature of physical, chemical, geological, and biological processes. The amount of information developed by Earth scientists is increasing at an ever faster rate as methods of observing the Earth improve rapidly. As a result, most university curricula now require students to concentrate on one of the subdisciplines of science. In particular, students of the physical sciences often learn little about biological sciences, while students of biology often learn little about the physical sciences.

(a) Discuss whether the trend toward specialization in a subdiscipline is a desirable direction, and explain why or why not.

(b) In college, should some individuals be trained to be specialists, while others remain less well trained in specific disciplines but are better trained to integrate information across disciplines?

(c) How would you do this?

(d) Should colleges also seek to ensure that all students have an understanding of the basic processes of all Earth science subdisciplines and of how they interrelate? Why or why not?

(e) Comment on whether your college curriculum addresses the needs identified in your answer to part (d) effectively or not. How could the curriculum be improved in this respect?

5.3 Most earthquakes occur at plate boundaries but some, often very intense, earthquakes occur in the middle of plates. For example, in 1811 and 1812, two earthquakes occurred with epicenters located near New Madrid, Missouri, that were strong enough to ring church bells in Montreal, more than 1500 km away. Why do you think earthquakes occur in the middle of lithospheric plates, and why are they felt so far away?

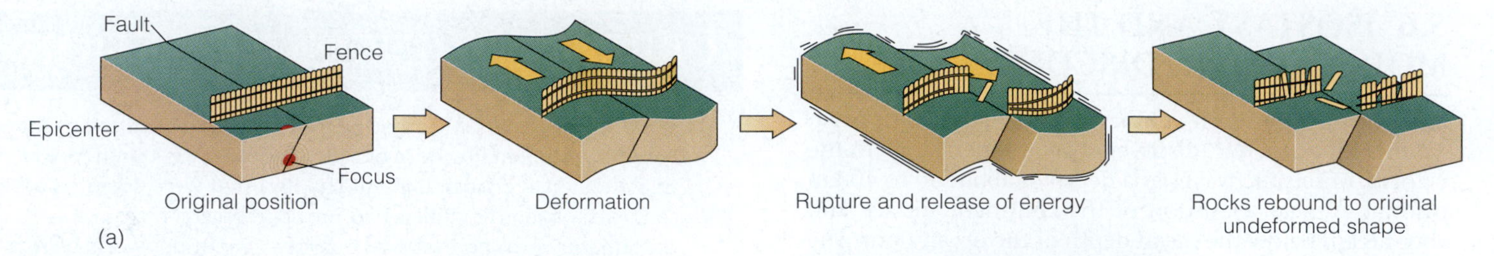

FIGURE 5.6 (a) Rocks store energy as they are deformed by movement on either side of a fault. Eventually they rupture, release the stored energy in the form of earthquake waves that radiate outward in all directions, and then rebound to their former shape, although they are now displaced from the rocks on the other side of the fault, all within a few seconds. The earthquake focus may be many kilometers below the Earth's surface. (b) This fence was displaced 2.5 m during the 1906 San Francisco earthquake.

5.7 EARTHQUAKES

Since antiquity, certain areas have been known to be likely to have earthquakes, and earthquakes have been known to be associated with abrupt earth movements that can raise, lower, or move one section of land in relation to another. Earthquakes occur when accumulated stress in solid rock leads to its fracture, resulting in a rapid vibrating motion of the rock on either side of the fracture (**Fig. 5.6**). Earthquakes were of considerable interest and importance to continental drift studies because they were recognized as possibly being caused by movements of the continents.

Earthquake Locations

The location of an earthquake within the Earth is known as the **focus** (plural *foci*). Earthquake foci occur as deep as 650 km below the Earth's surface. The location on the surface of the Earth directly above an earthquake focus is known as the "epicenter." When the epicenter locations of earthquakes were first plotted at about the end of the eighteenth century, several major bands of concentrated earthquake activity were evident. For example, earthquakes were concentrated around the American and Asian rims of the Pacific Ocean, through the Mediterranean and Middle East, and in the Himalaya and Tibetan Plateau region of Asia (**Fig. 5.7**). However, the bands of undersea earthquakes within the ocean basins were not known until sensitive **seismographs** became available in the middle of the twentieth century.

During the 1940s and 1950s, earthquake studies began to reveal well-defined narrow bands of earthquake occurrences running in the ocean basins (**Fig. 4.7**). At the same time, oceanographers were mapping the locations of the oceanic ridge mountain chains for the first time. They quickly saw that many earthquakes were located along the **crests** of the oceanic ridge mountain chains. Indeed, earthquake observations were used to identify the probable locations of oceanic ridges in poorly surveyed areas of the oceans. The close correspondence between earthquake locations and the oceanic ridges became an important piece of evidence that led to the development of the seafloor spreading theory. This theory provided a crucial link, because it identified a possible mechanism whereby the continents could be moved apart as in Wegener's theory.

Other studies revealed that the oceanic ridges are covered by recent volcanic **lava** flows, and that they are areas where the heat flow through the seafloor from the Earth's interior is greater than in the rest of the ocean. That information was the basis for Harry Hess's 1960 proposal that oceanic crust is formed by **magma** that is **upwelled** at the oceanic ridges and destroyed at the trenches.

Although some earthquakes are known to occur in the center of lithospheric plates, the vast majority of all earthquakes occur at plate boundaries (compare **Figs. 4.6 and 4.7**). That fact supports the hypothesis that pieces of crust move across the Earth's surface as separate rigid plates that are modified primarily at their edges, where they meet and interact with other plates.

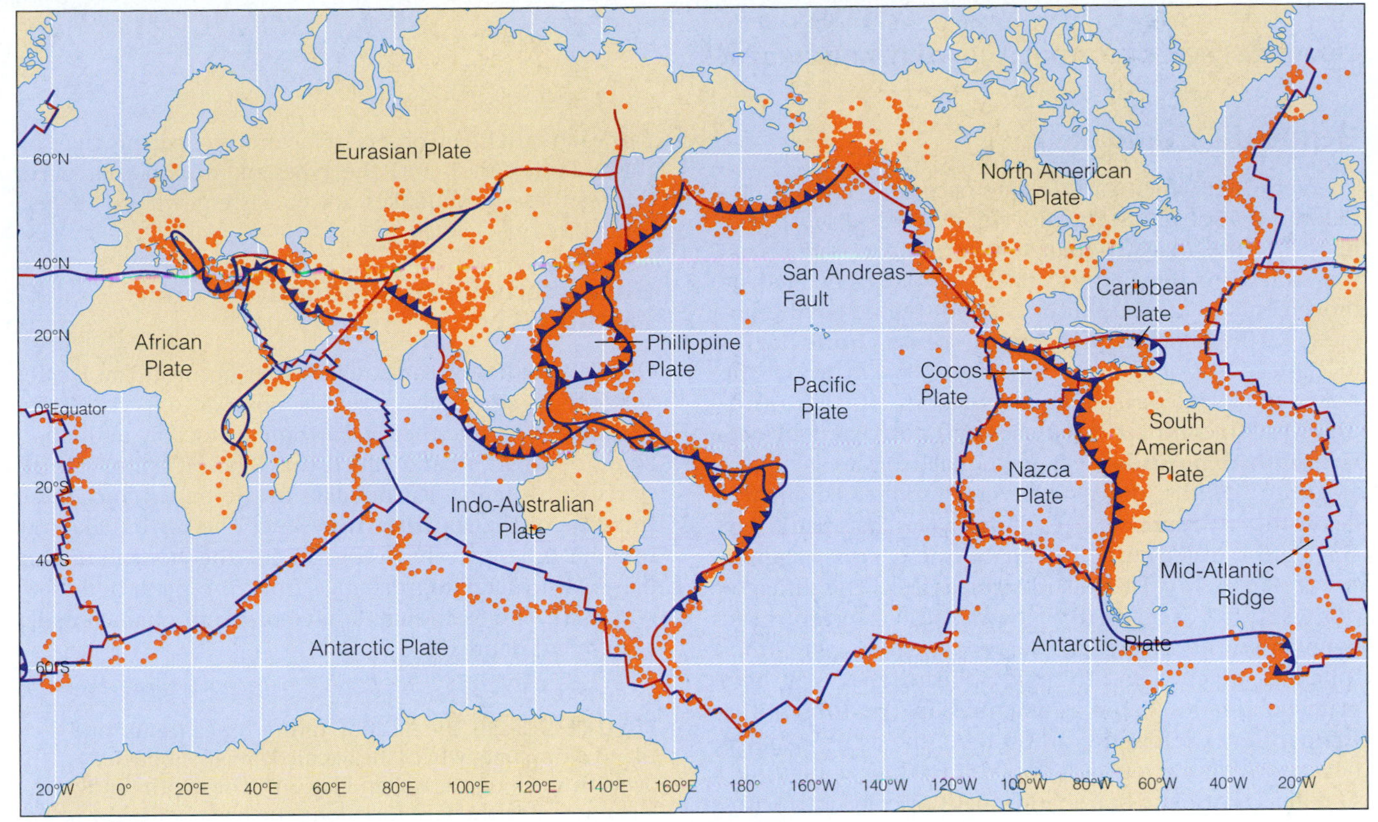

Mercator projection

(a)

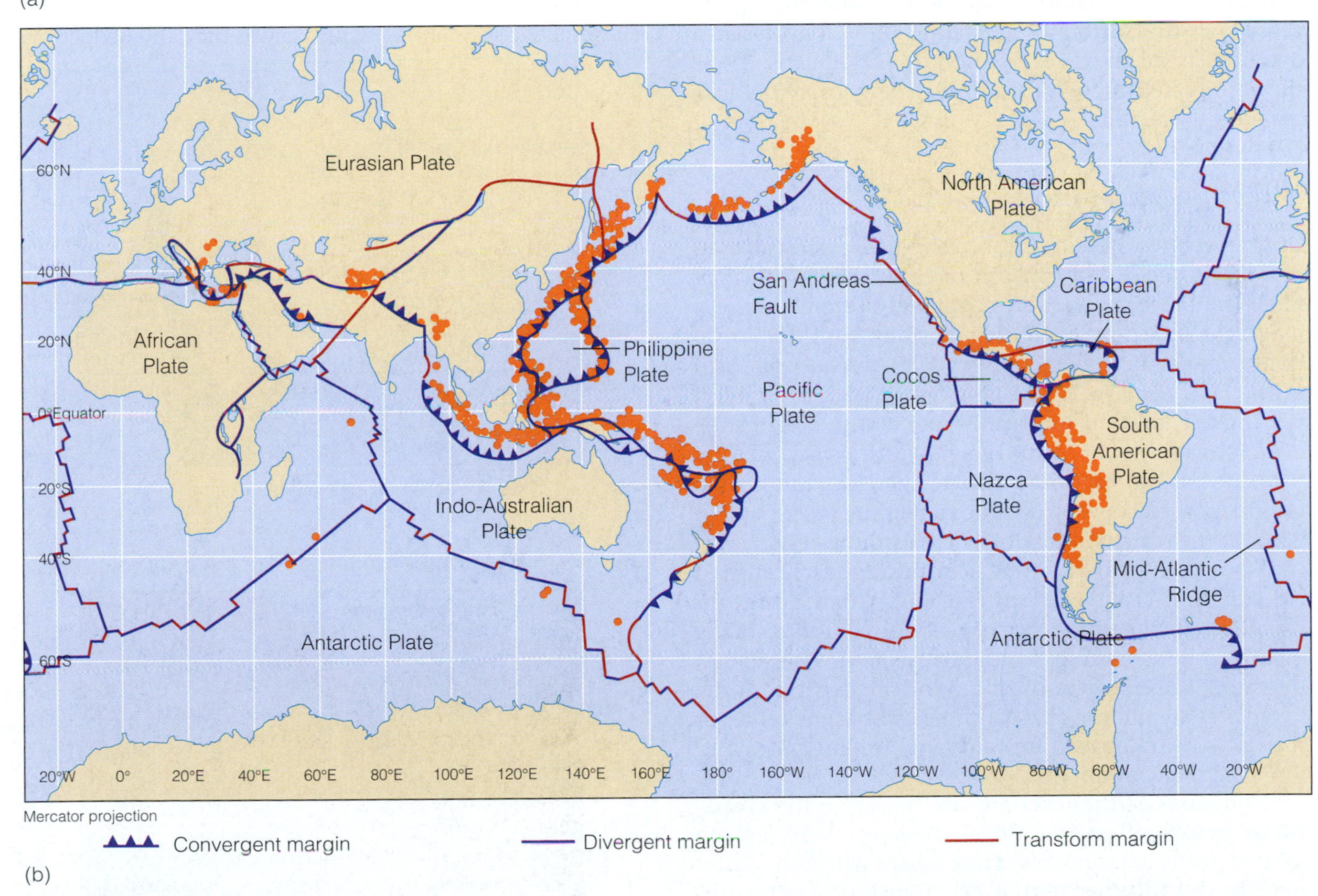

(b)

FIGURE 5.7 (a) Earthquakes occur primarily at subduction zones, oceanic ridges, and transform faults. (b) Earthquakes that have foci deeper than 100 km below the Earth's surface are restricted almost exclusively to subduction zones such as the trenches that ring the Pacific, or to continental collision plate boundaries, such as the Himalayas and the eastern Mediterranean.

Depth of Earthquake Foci

The depths at which earthquakes occur provide substantial information about the movements of the lithospheric plates. Although earthquakes have dramatic effects on the surface of the Earth's crust, they are generated by movements that are usually centered well below the surface. Earthquakes at **divergent plate boundaries** (oceanic ridges) have foci less than 100 km below the crust surface. In contrast, at **convergent plate boundaries** (subduction zone trenches and **continental collision plate boundaries**), earthquake foci are generally at greater depths, down to several hundred kilometers below the crust surface. Deep earthquakes, below 100 km, are almost exclusively limited to these **convergence** zones (**Fig. 5.7b**). This distribution reflects the different processes that occur at **divergent** and convergent plate boundaries. At divergent plate boundaries, the lithosphere is stretched and thinned. All earthquakes at these plate boundaries occur as fractures in the thin brittle **lithosphere**. Below the lithosphere, the asthenosphere flows slowly but freely and does not fracture as solids do. At convergent plate boundaries, slabs of cold brittle lithosphere are **subducted** deep into the mantle. As the slabs descend, they continue to fracture under the stresses caused by the surrounding mantle until they reach a depth of as much as several hundred kilometers. There they are heated enough that their rocks become fluidlike and do not fracture.

Directions of Motion

Until recently, seismological techniques were able to record the depth and location of earthquakes with only moderate precision. Newer techniques use multiple seismographs linked by computers to study both the arrival times and detailed characteristics of the waves from each earthquake. Such measurements can be used to establish very precise locations of earthquakes and to describe the directions and types of movements of the rocks on the two sides of the rupturing **fault**. Therefore, substantial information can now be obtained about the processes that occur at the plate boundaries where earthquakes occur.

Many earthquakes on the oceanic ridges are centered either in or near the **central rift valley** in the center of the ridge. Such earthquakes show both vertical displacement and horizontal separation between the two sides of the fault. These directions of motion are expected if the two plates are pulling apart and the crust at the center of the rift is being pushed upward by upwelling magma to form the ridge mountains.

Numerous earthquakes are also found on the **transform faults** of oceanic ridges. However, very few are located in the adjacent **fracture zones** where the plates have locked together (**Fig. 4.22**). Transform fault earthquakes show evidence of a horizontal component of displacement between the two sides of the fault as they slide past each other. The observed locations of earthquakes on the oceanic ridges and the types of motion that occur during these earthquakes are consistent with plate tectonic theory.

Characteristics of Subduction Zone Earthquakes

Detailed studies of earthquakes that occur at the Earth's subduction zones also provide observations consistent with plate tectonic theory. Earthquake foci at subduction zones occur in a band called the "Benioff zone," which starts with shallow earthquakes in the trench and under the edge of the **sedimentary arc**. The earthquake foci become progressively deeper under the nonsubducting plate with increasing distance from the plate boundary (**Fig. 5.8**). The earthquakes have different characteristics at different depths.

FIGURE 5.8 (a) The location of the Tonga trench and island arc system, where the Pacific Plate is subducted beneath the Indo-Australian Plate. (b) The depths and locations of earthquake foci along the plate boundary in relation to the trench and island arc. The earthquake foci trace the depth and direction of the movement of lithospheric slabs of the Pacific Plate as they are subducted.

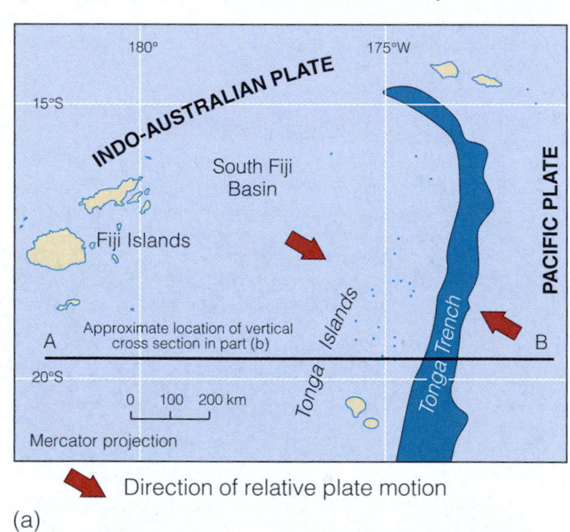

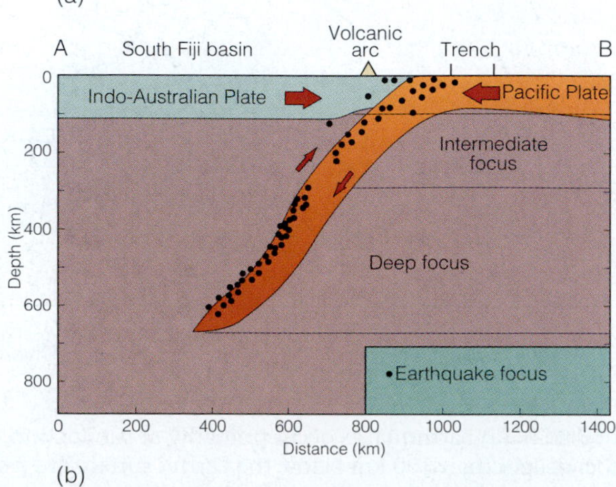

In the trench, earthquakes are produced by the initial bending of the plate: the plate fractures as its upper section is stretched and its lower section is compressed by the bending. The characteristics of earthquakes at intermediate depths (under the island arcs) indicate that one plate slides past the other. Farther under the nonsubducting plate, deeper earthquakes indicate extension or compression of the descending plate. Extension (stretching) of the descending plate occurs if part of it is descending rapidly into the mantle and pulling the rest of the plate behind it. In some subduction zones, this extension can lead to a fracture and complete separation of the descending plate. The section of subducting plate that has broken off then descends into the mantle far more rapidly than the remaining plate behind it. Compression of the descending plate occurs when the descent is slow and the lowest part is impeded by the mantle, thus "backing up" the rest of the plate descending behind it. The deepest earthquakes, at approximately 670 km, are generated primarily by compression. Hence, the mantle may be more rigid at this depth.

Computer Tomography

Computer **tomography** has recently been applied to earthquake waves. The technique uses earthquake waves passing through the interior of the planet to draw pictures of the Earth's interior in the same way that CAT (computerized axial tomography) scanners use X-rays to draw three-dimensional pictures of the internal organs of the human body. Earthquakes create several types of seismic waves, which behave differently as they pass through the Earth (**Fig. 5.9a**). For example, compressional (pressure) waves pass through both liquids and solids, whereas shear waves do not pass through liquids. In addition, the velocity of earthquake waves varies with depth below the Earth's surface and between different types of rock. Hence, the times of arrival of the various waves caused by an earthquake can be monitored at several locations, and the data can be processed by computer to draw a three-dimensional picture of the density of rocks and the shape and size of the Earth's liquid core (**Fig. 5.9b,c**).

Computer tomography has dramatically improved our ability to study the Earth's interior and the processes of plate tectonics. For example, the seismic tomography map in **Figure 5.10** shows warm upper mantle under many known spreading centers and cold upper mantle under most subduction zones. Unfortunately, computer tomography data can be very difficult to interpret.

Computer tomography studies have provided information that confirms the theory of plate tectonics in its essentials, while suggesting substantial modifications in our interpretation of the details of plate tectonic processes deep in the Earth. For example, until the 1990s, most scientists believed that the **convection** processes (**CC3**) that drive plate motions were restricted to the asthenosphere, and they thought that oceanic ridges were the primary areas of upwelling in the mantle convection system. However, computer tomography has traced the origins of some **hot spots**, such as the Hawaiian Islands, to an area below the asthenosphere that is close to, or at, the core–mantle boundary. Furthermore, evidence has been found that slabs of lithosphere have descended deep into the lower mantle from subduction zones.

These observations led to a revised plate tectonic hypothesis that suggests that the processes driving plate tectonics are not restricted to the crust and shallow

CRITICAL THINKING QUESTIONS

5.4 If you took a single sample of the rocks of a lava flow and found that its paleomagnetic field was oriented exactly horizontal and parallel to the Earth's surface, what would this tell you about the rock? Why?

5.5 You have two colleagues who live in other parts of the continent and who have each studied one rock from their own local area. One reports that his rock has a paleomagnetic field exactly like your rock (exactly horizontal and parallel to the Earth's surface), but the other reports that her rock's paleomagnetic field is oriented about 10° to the left of the horizontal direction to the pole and has a dip angle of 25°. What additional information would you want to have about both your rock and the two rocks studied by your colleagues in order to interpret and explain the different paleomagnetic fields in these rocks?

5.6 Do you believe everything that you learn from science textbooks? Explain why or why not. Describe at least one alternative explanation for at least two pieces of scientific evidence presented in this chapter that is not discussed in the chapter or in Chapter 4.

5.7 At one time, everyone believed the Earth was flat. At one time, everyone believed the continents did not move. Discuss one example of something that we now take for granted about the Earth and our environment that may prove to be just as wrong in the future.

5.8 If oceanic crust were less dense than continental crust, but both were still less dense than mantle material, how might the world be different? Explain your answer by describing what the Earth's surface would probably look like.

5.9 Look at the Ninety East Ridge, which divides the Australian Plate (**Fig. 4.10**, **Inside front cover map**). Examine the distribution of earthquakes (**Fig. 5.7**) and volcanoes (**Fig. 5.11**). Hypothesize alternative explanations for how and why this ridge was formed.

mantle. Instead, the driving processes include convective activity that extends from the core–mantle boundary to the base of the lithosphere. Some features of this undoubtedly complex and **chaotic** convection are reasonably well understood. For example, upwelling from deep within the mantle is believed to occur at some hot spots and **downwelling** to occur at subduction zones. The oceanic ridges are believed to be zones where shallow upwelling fills gaps between plates, which are being pulled apart by the deep convection motions. However, very few details of convection processes that connect the oceanic ridges, subduction zones, and hot spots, and that control movement of the lithospheric plates, are currently known.

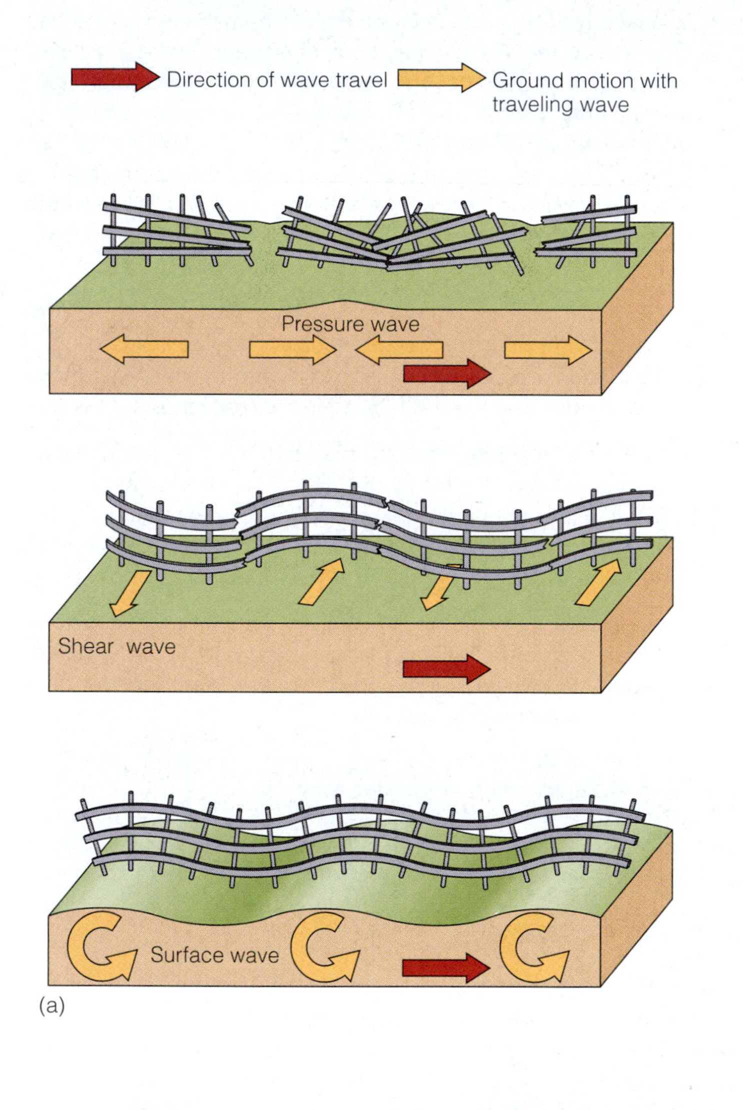

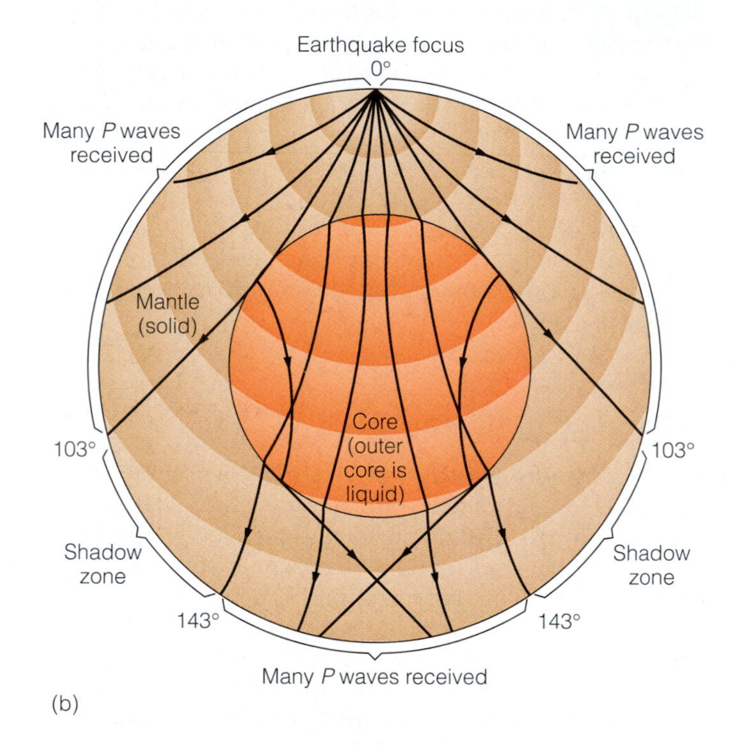

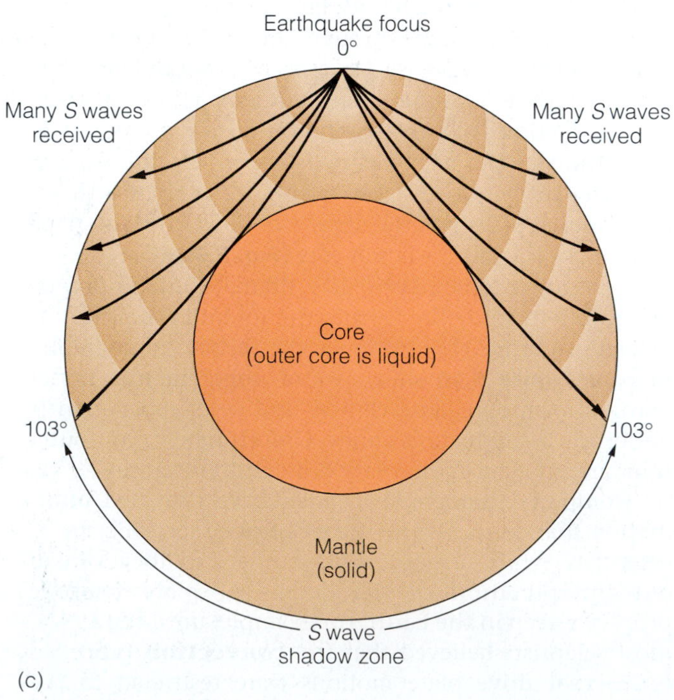

FIGURE 5.9 (a) Three types of earthquake waves are illustrated. Pressure waves alternately compress and then expand the ground in the direction of wave travel, shear waves cause the ground to sway back and forth across the direction of wave travel, and surface waves cause oscillating up and down movements of the ground similar to ocean waves. (b) Pressure (*P*) waves can travel through both solids and liquids. Therefore, if strong enough, they can be detected on the side of the Earth opposite the earthquake epicenter. (c) Shear (*S*) waves cannot travel through liquids, so they do not pass through the Earth's core and are not detected on the other side of the Earth. Surface waves travel only across the Earth's surface.

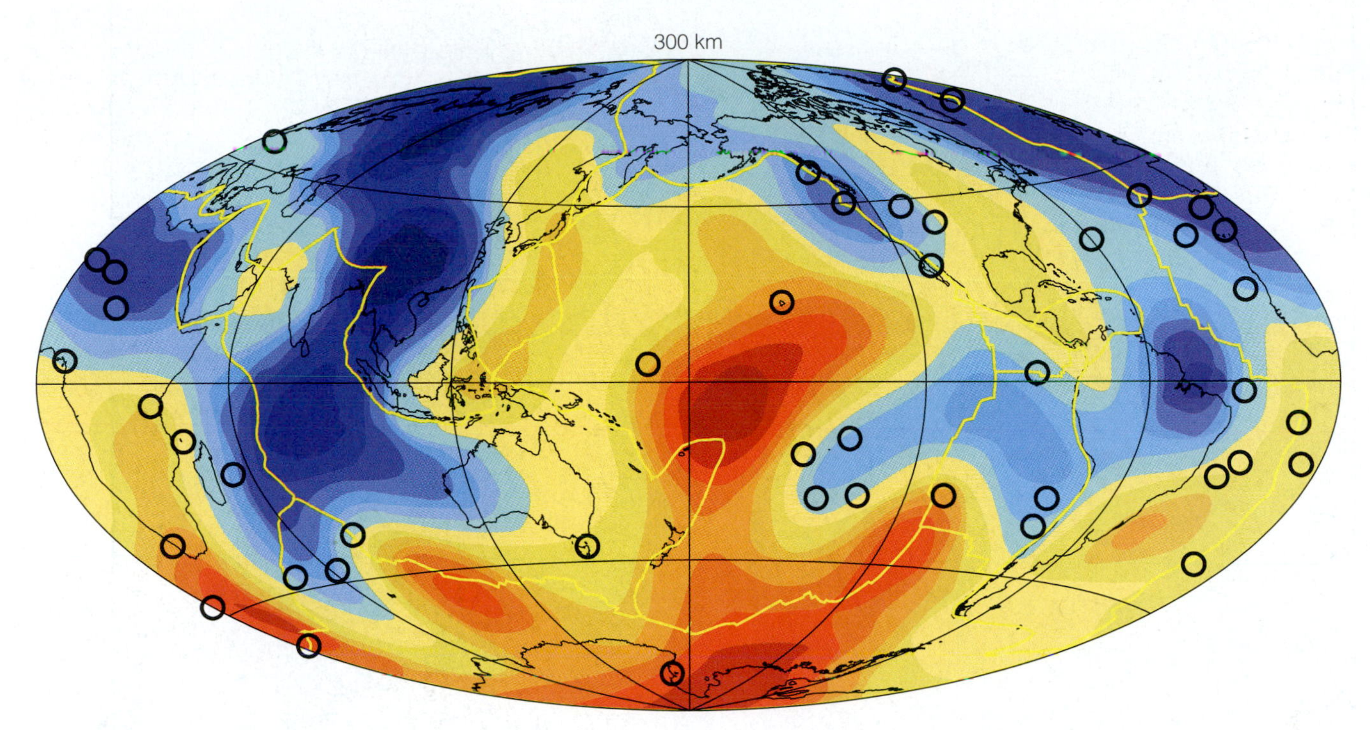

FIGURE 5.10 Heterogeneity of the mantle at a depth of 300 km, as revealed by computer tomography. Earthquake waves travel more slowly through warmer, less rigid materials (shown in red) and faster in cooler, more rigid materials (shown in blue). Yellow lines represent plate boundaries, and black circles represent the surface locations of volcanic hot spots. Note that most hot spots at 300 km are displaced from their surface locations or do not appear to extend to this 300 km depth, so upwelling must not be exactly vertical through the mantle and some hot spots may be evidence of only shallow convection processes.

5.8 VOLCANOES

Volcanoes are present in areas where molten rock or magma upwells through the crust to the Earth's surface. Both the location of volcanoes and the type of magma they produce provide information about the processes of plate tectonics.

Oceanic Ridge and Hot-Spot Volcanoes

The seafloor spreading theory requires that new oceanic crust be created at divergent plate boundaries as the crust of the two lithospheric plates is pulled apart. Hence, the material that constitutes the new ocean crust must be supplied by upwelling from below. During the 1950s and 1960s, the oceanic ridges were found to be composed of volcanic rocks of recent origin. This finding, which indicated that the seafloor at oceanic ridges was indeed formed by upwelling of magma in volcanoes, became an important piece of evidence supporting the concept of seafloor spreading. We now know that volcanic eruptions occur frequently along oceanic ridges and that most eruptions occur in the central rift valley. Those facts indicate that oceanic ridges are locations where magma upwells through the crust.

Subduction Zone Volcanoes

Most of the world's volcanoes, other than those on divergent plate boundaries or those at hot spots, are within 100 or 200 km of deep-ocean trenches. Where the trenches are adjacent to the edge of a continent, the volcanoes are on only the continent side of the trench and lie in a chain that parallels the trench (**Fig. 4.11**). Where the trench is not adjacent to the edge of a continent, the volcanoes are on a chain of volcanic islands or **seamounts**, which also parallel the trench (**Figs. 4.15, 4.16**). The location of such volcanoes and their characteristics provide evidence for, and can be explained by, plate tectonic processes. All are on the nonsubducting plate and are formed when the oceanic crust of the subducted plate descends into the mantle at an angle under the nonsubducting plate (**Fig. 4.11**).

Lava Types

Different types of magma are formed and erupt as lava from different types of volcanoes. The three principal types of lava are **basaltic**, andesitic, and rhyolitic. Basaltic lava has a chemical and mineral composition similar to that of ocean crust and has a very low water content. Rhyolitic lava has a chemical composition similar to

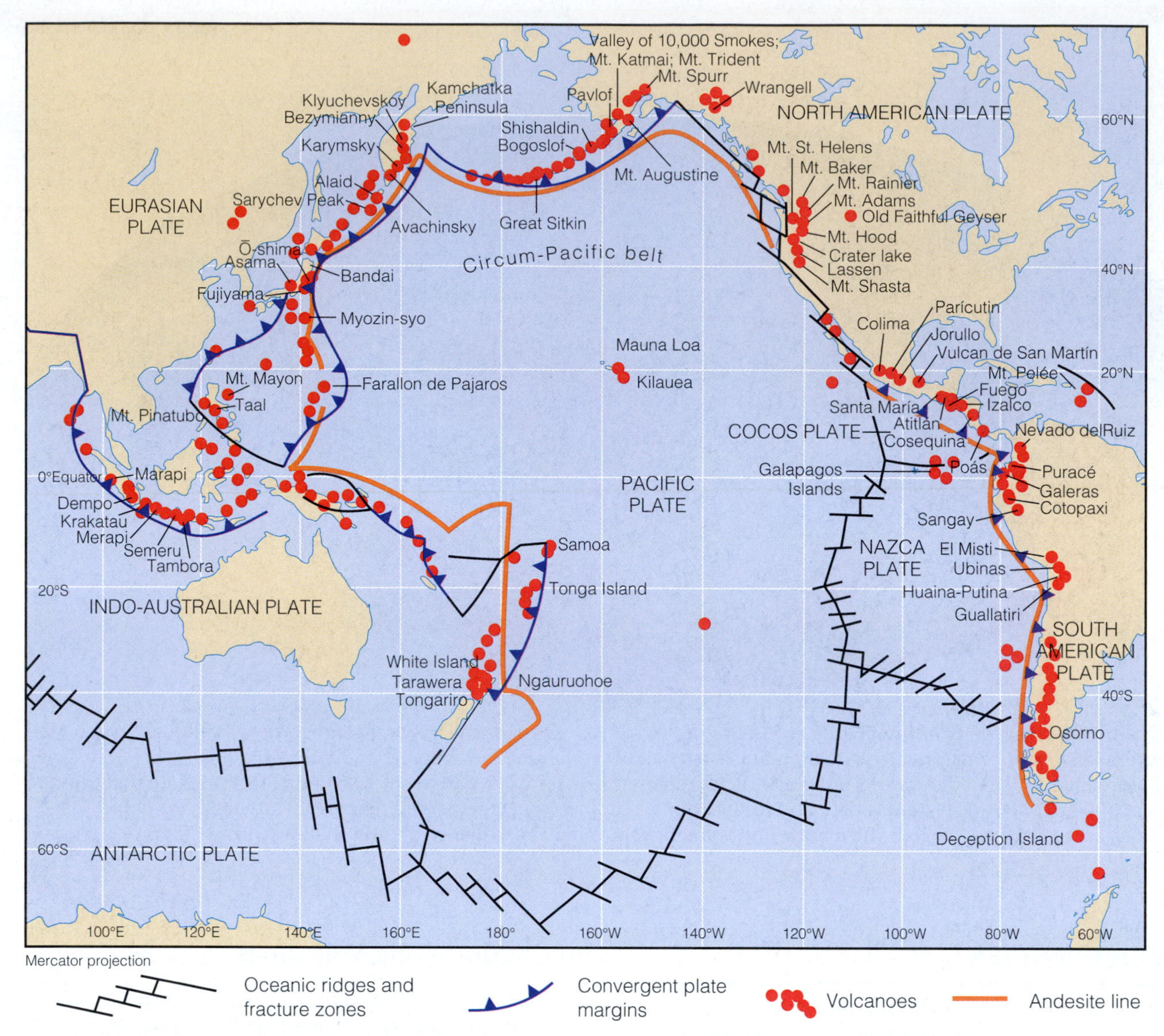

FIGURE 5.11 Location of volcanoes and the andesite line in the Pacific Basin. None of the volcanoes that are inside the andesite line erupt andesitic lava. Note that many more volcanoes are located along the oceanic ridges but they are, as yet, not mapped or named.

that of continental crust, and andesitic lava is intermediate in composition between basaltic and rhyolitic lava.

The distribution of volcanoes that erupt the different types of lava suggests that basaltic lava is produced by melting of mantle material, andesitic lava by melting of subducted ocean crust, and rhyolitic lava by melting of continental rocks. Hot-spot and oceanic ridge volcanoes erupt only basaltic lava and are found in regions of upwelling from the mantle. Volcanoes at **oceanic convergent plate boundaries** can erupt both basaltic and andesitic lava, which would represent a mixture of melted ocean crust from the subducted plate and molten mantle material that rises with it through the crust. Volcanoes on the continents can erupt all three types of material, with rhyolitic lava being formed, for example, where a hot spot or developing **continental divergent plate boundary** heats and melts the continental crust.

All volcanoes on the Pacific Plate, including oceanic ridge and hot-spot volcanoes, erupt only basaltic lava. The Pacific Ocean crust is composed of those basaltic rocks. Pacific-region volcanoes that erupt andesitic lava are found only outside the andesite line (**Fig. 5.11**), which closely corresponds to the edge of the Pacific Plate. The andesitic volcanoes are all on the adjacent plates. At these locations, the subducting Pacific Plate descends at an angle beneath the adjacent plate to the approximate depth of the asthenosphere, where the subducted oceanic crust is heated sufficiently to partially melt.

Chemical and mineralogical studies and laboratory experiments have provided additional evidence for the origin of the three types of lava. Basaltic lava is believed to be formed by melting of mantle material in the absence of water (within the magma itself). Rhyolitic lava is formed by melting of continental crust in the presence of water. Andesitic lava is probably formed by melting of oceanic crust rocks at high pressures in the asthenosphere in the presence of water. Lava from subduction zone volcanoes may also include a contribution from mantle material melted from the mantle surrounding the subducted plate. These findings are consistent with plate tectonic theory and the known distribution of volcanoes.

5.9 PALEOMAGNETISM

All magma contains at least small quantities of iron oxides. When magma cools to its solidifying point, small crystals of a mineral called "magnetite" are formed. The crystals are weakly magnetic because they contain iron oxide. Iron oxide molecules are dipolar. In molten magma, these molecules rotate and move about freely. When the magma solidifies, the heat energy is still sufficient to enable the molecules some freedom to rotate. As the rock cools further, however, the declining heat energy approaches the magnitude of the weak energy of the Earth's magnetic field. Gradually, as heat energy is lost, the molecules respond to the Earth's magnetic field by becoming parallel with each other and with the magnetic field (just as a compass needle becomes oriented to the magnetic field). When the temperature falls below about 350°C to 500°C, the molecules are "locked" in place and can no longer rotate in response to the Earth's magnetic field. If the rocks are then moved, the iron oxide molecules generally remain oriented as they were when the rock solidified. The result is that the rock retains a magnetic record of the direction in which it was oriented when it solidified. That record is called "paleomagnetism." Basalt (rock formed from basaltic magma) is rich in iron, so it is particularly effective in preserving the magnetic field at the time of its formation, but other rocks and some sediment also provide readable paleomagnetic records.

Reading the Paleomagnetic Record

The paleomagnetic record would be of little use to us if the locations of the Earth's north and south magnetic poles at the time when the rocks were formed were not known. Fortunately, the magnetic poles appear to have been close to the Earth's north and south poles of rotation for billions of years, even though they wander slightly. However, the Earth's magnetic field completely reverses periodically for reasons we do not yet fully understand. During a reversal, the north magnetic pole is located close to the geographic South Pole and the south magnetic pole close to the geographic North Pole.

Some areas of rock may be tilted or rotated by movements associated with mountain building or earthquakes. However, many rocks of a given age within a continent will not be subjected to such small-scale motions for hundreds of millions of years or more. Therefore, if we take samples of rocks of a certain age from many different nearby locations on a continent, the paleomagnetic record will identify their original orientations to the magnetic pole. The location of the pole itself has not changed significantly, and if the paleomagnetic direction of the rocks does not now point approximately to the Earth's pole, the reason must be that the continent has rotated (**Fig. 5.12a**). The paleomagnetic record thus reveals how continents have rotated with time. If rocks of the same age from many different locations on a continent are examined, the approximate distance of the continent from the magnetic pole when the rocks were formed also can be calculated.

During the 1950s, paleomagnetic data were gathered from several continents and analyzed. At that time, scientists did not know that the magnetic pole had remained close to the geographic pole for billions of years. For convenience in plotting and interpreting their paleomagnetic data, they initially assumed that the continents had always been where they are now. They assumed that the paleomagnetic data showed the changing location of the magnetic pole in relation to the fixed continental locations. The data plots were called "polar wandering curves" (**Fig. 5.13**). Recently formed rocks from all continents indicate that the magnetic pole was located where it is now relative to the continent when the rocks were formed. However, data from rocks of increasing age showed that the magnetic pole apparently had wandered across the face of the planet. Further, when polar wandering curves were plotted for two continents, the indicated locations of the magnetic pole in relation to the continents did not match in any but the most recent rocks.

Such findings have only two possible explanations: either the continents have remained fixed and there were two (or more) north magnetic poles in the past, or there was only one north magnetic pole and the continents have moved. The first explanation is believed to be impossible according to the known laws of physics. Therefore, in the 1950s many scientists became convinced that continents did indeed drift. Acceptance of the continental drift theory was not universal, as many geologists refused to believe the paleomagnetic data. In support of their skepticism, they cited the many rock samples whose paleomagnetic data were not consistent with the systematic pattern of polar wandering for the continent on which they were located. We now know that the inconsistent data were due to methodological errors and to localized changes in the orientation of some rocks that occurred after they were formed.

Dip Angle and Paleolatitude

We normally hold a compass horizontally so that its needle can rotate to align north and south. However, if we turn the compass on its side and orient it toward the pole, we find that the Earth's magnetic field is not parallel to the ground (except at the magnetic equator). In the Northern Hemisphere, the compass needle points at an angle toward the ground (the **dip angle**) in the direction of the north magnetic pole (**Fig. 5.12b**). The dip angle increases as we move toward the magnetic pole. At the magnetic pole, the

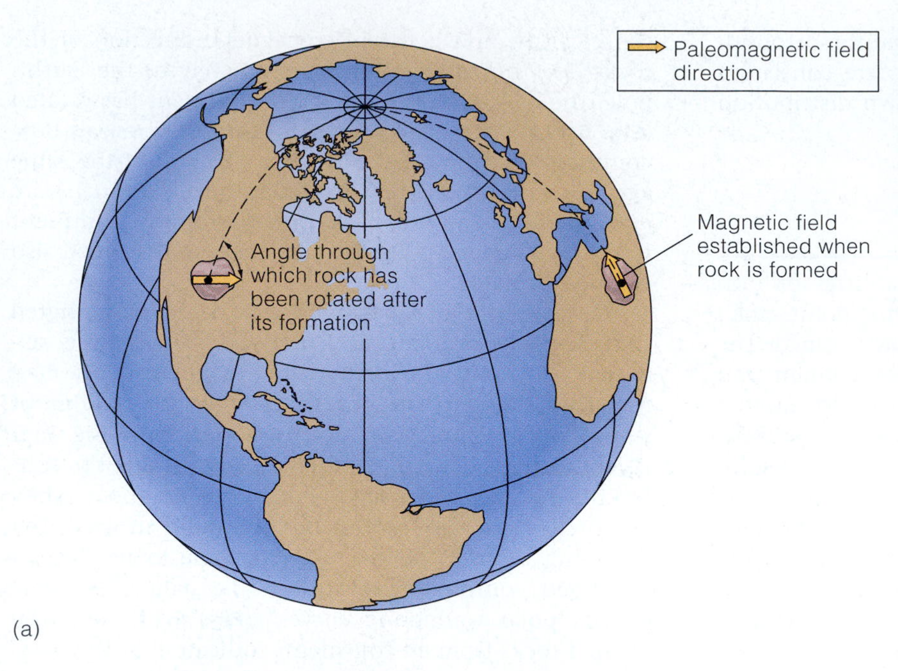

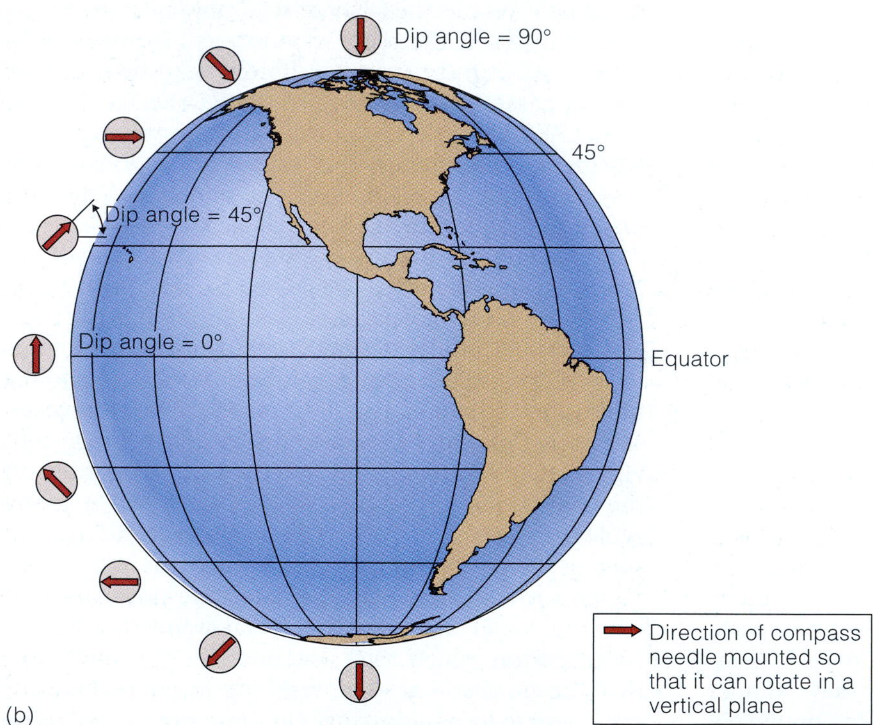

FIGURE 5.12 The principles of paleomagnetism. (a) If the compass direction of a rock's magnetization is not directed toward the pole, either the pole has moved or the rock has been rotated since it was formed. (b) The dip angle is the angle between the rock's direction of magnetization and the Earth's surface (if the rock has not been tilted since being formed). Because the dip angle changes with latitude, the original dip angle of a rock's magnetization reveals the latitude at which it was formed.

needle points at a 90° angle directly into the ground. At the magnetic equator, the dip angle is zero and the needle is parallel to the ground. Because the magnetic and geographic poles are always reasonably close together, the dip angle of the compass needle is a measure of latitude. Larger angles correspond to higher latitudes.

Magnetic materials in magma orient both horizontally and vertically to align with the Earth's magnetic field when the magma solidifies. Therefore, rocks carry paleomagnetic information about both their horizontal orientation with respect to the pole at the time they were formed, and the latitude at which they were formed, the paleolatitude. Reading the paleolatitude information can be complicated because rocks are subject to tilting during mountain building and other processes after they are formed. However, other information can be used to measure tilting of the rock after it solidified, such as the angle of tilt of overlying or underlying sedimentary rock layers. Hence, if many samples are collected from different locations within a continent, the latitudes at which each continent lay when rocks of different ages were formed can be determined.

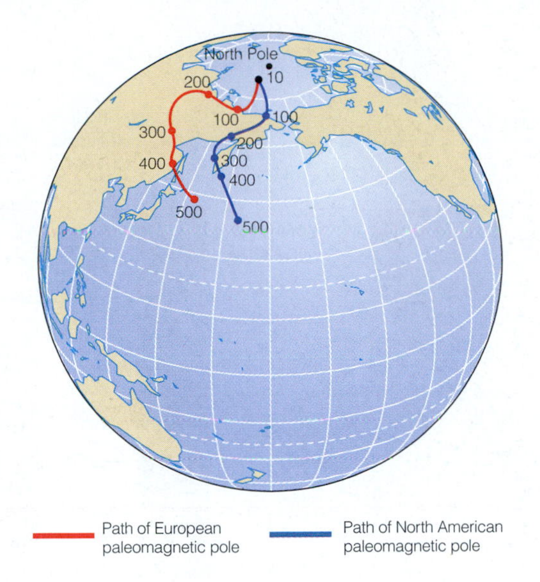

FIGURE 5.13 The apparent paths of polar wandering for North America and Europe.

Although paleomagnetic information can reveal the latitude and orientation of a continent at different times in the past, it does not place the continent at a unique location on the Earth, because no **longitude** information is obtained. Data for many locations and ages must be compared for the longitudinal history of plate motions to be determined.

Paleolatitude data for the Indian continent illustrate how such data are used to investigate both the direction and the speed of movement of the continents (**Fig. 5.14**). The data reveal that India broke away from the remains of Pangaea about 100 million years ago, then traveled rapidly northward and collided with Eurasia. The collision began about 40 to 50 million years ago when the bases of the continental shelves of India and Eurasia first met.

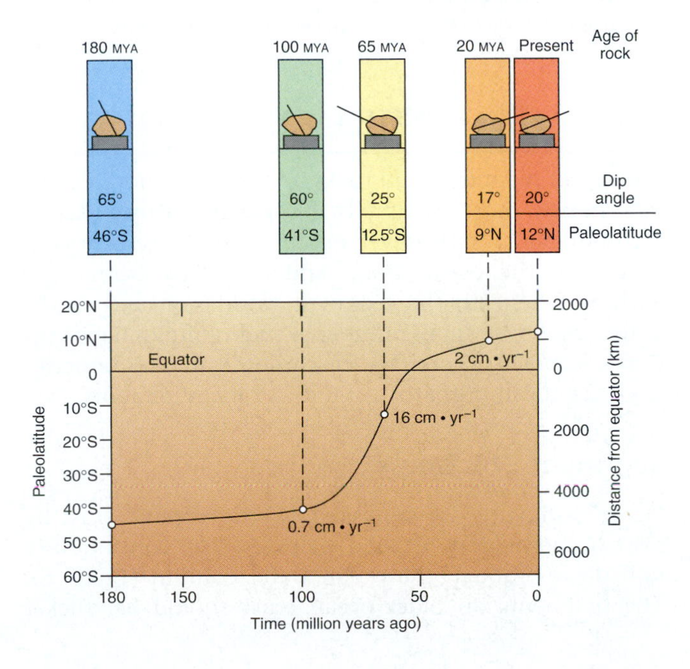

FIGURE 5.14 An example of how dip angles measured for rocks of different ages from a single continent can be used to deduce the continent's changes of latitude as it moved on its tectonic plate. The data show the movements of India. Notice that India's northward motion accelerated about 100 million years ago, but slowed about 50 million years ago. That reduction in the rate of movement to the north was probably caused by the beginning of the collision between the continental crusts of India and Asia.

5.10 OCEAN FLOOR MAGNETIC ANOMALIES

The strength of the Earth's magnetic field measured at the ocean surface varies slightly from one place to another. The variations are called "magnetic anomalies."

Confusion and Enlightenment

When they began to collect data on magnetic anomalies, oceanographers did not have a specific purpose in mind, but they thought the measurements might provide some information about the composition of seafloor rocks. When they plotted their measurements on maps, what they found was entirely unexpected and, for a number of years, unexplained. On the seafloor was a series of zebra-like stripes of positive and negative magnetic anomalies (**Fig. 5.2**). The positive and negative anomalies were areas of the ocean floor where the Earth's magnetic-field strength was alternately slightly higher and slightly lower than average, respectively.

This curious set of data was eventually explained by Drummond Matthews and Fred Vine. They knew from measurements on land that the Earth's magnetic field had reversed at irregular intervals of between tens and hundreds of thousands of years (**Fig. 5.15**). At each reversal, the north and south magnetic poles exchanged location. Matthews and Vine theorized that lava on the ocean floor was magnetized when it solidified, and that the stripes represented rocks formed during successive periods of normal and reversed orientation of the Earth's magnetic field. When the Earth's magnetic field is normal, the new ocean floor rock is magnetized in the same direction as the present-day field. Therefore, the magnetism of this rock is added to the Earth's magnetic field and causes a positive (normal) magnetic anomaly. Conversely, when the Earth's magnetic field is reversed, the new rock is magnetized in the reverse direction, and its magnetic-field direction is opposite to that of the Earth's present-day field. The opposite field of this rock slightly reduces the Earth's magnetic field and causes a negative magnetic anomaly.

Matthews and Vine observed that symmetrical patterns of alternating positive and negative anomaly stripes are present in both directions from the center of the

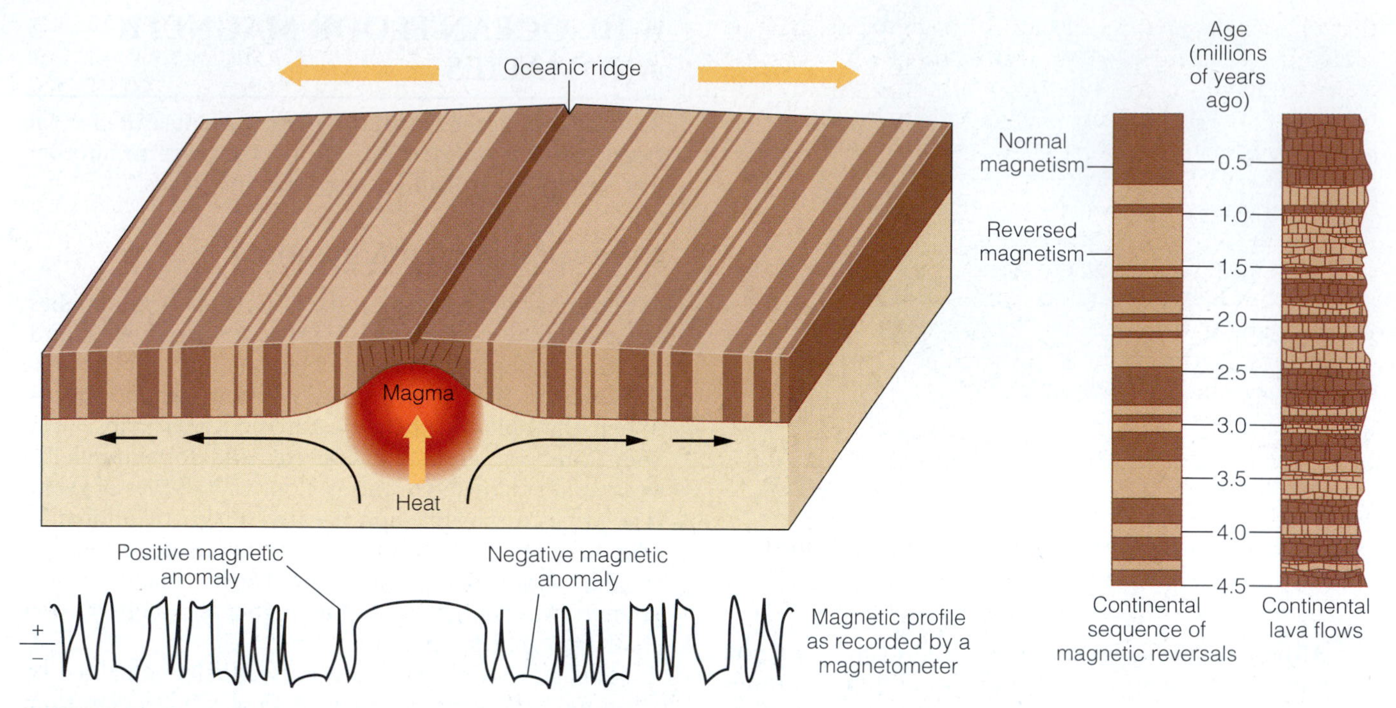

FIGURE 5.15 The sequence of magnetic-field reversals preserved in the magnetic anomalies of the oceanic crust on either side of the oceanic ridges is identical to the sequence of magnetic reversals known from continental lava flows. As new crust is formed at the oceanic ridge center, it is magnetized in the direction of the Earth's magnetic field. When the Earth's magnetic field reverses, new crust with the reversed magnetic field is formed at the ridge center and splits and separates the crust formed immediately before it. The result is a stripe of opposite magnetic anomaly direction on each side of the ridge.

oceanic ridge. Further, the stripes parallel the ridge crest. They explained these observations by suggesting that new ocean crust is formed continuously at the oceanic ridge. As the new ocean floor is formed, it is magnetized in the direction of the Earth's magnetic field at that time. The new crust cools and solidifies as it ages, and its original magnetic field is locked in place so that it cannot reorient when the Earth's magnetic field reverses.

As seafloor spreading continues during times of normal orientation of the Earth's magnetic field, new crust is formed continuously at the oceanic ridge and a widening band of normally magnetized seafloor is created. When the Earth's magnetic field reverses, all new crust is magnetized in the reverse direction and a new stripe is created at the ridge. The new stripe is formed in the middle of the previous stripe, and the two halves of the previous stripe move laterally in opposite directions away from the center of the ridge as seafloor spreading continues (**Fig. 5.15**). When the Earth's magnetic field reverts to normal orientation, the stripe with reversed magnetism is split in two and carried away from the ridge.

A Lesson about "Useless" Data

Matthews and Vine's elegant explanation for the magnetic anomaly stripes of the ocean floor was probably the evidence that finally removed almost all doubt within the scientific community that the continents drift. Notably, this crucial piece of information came from a data set gathered at substantial cost, even though it had little or no immediate purpose or application. Such is the essential nature of oceanography and other Earth and environmental sciences. All data are valuable because a seemingly nonuseful data set often becomes the key to solving a major problem or mystery of the science.

5.11 AGE OF THE OCEAN FLOOR

The age of sediments and rocks on the ocean floor can be determined most accurately by **radioisotope** dating techniques (**CC7**). However, obtaining the necessary samples from the ocean floor, and especially from deep within or beneath the sediment, is difficult and costly. Only a relatively small number of such samples have ever been recovered. Therefore, the earliest estimates of ocean floor age relied primarily on indirect measurements.

Sediment Thickness

Ocean sediments are deposited on the oceanic crust by slow accumulation of solid particles that continuously rain down through the ocean water column (Chap. 8). The sediments on older ocean crust should be thicker than those on younger crust because they have been accumulating for a longer time. However, many other factors affect sediment thickness in different parts of the oceans (Chap. 8). Oceanographers quickly learned that the oceanic ridges have little or no sediment cover and that the thickness of sediment generally increases with distance from the ridges. Initially this observation was made from dredge, core, and grab samples, which indicated only that sediment covered a greater proportion of any given area of the seafloor at a distance from the oceanic ridges than on or near the ridges themselves. More detailed information was obtained later by seismic studies and by scientific seafloor drilling projects (Chap. 2).

Magnetic Anomalies

The magnetic anomaly stripes on the ocean floor are like tree rings. By determining the time at which each magnetic reversal occurred, we can determine the age of the ocean floor. Initially, determining the times when the Earth's magnetic field reversed was a tedious and expensive task. It involved matching of magnetic analyses with radioisotope dating of some rock samples, and relative dating of many other rock samples by studying their fossil assemblages (**CC7**). However, the chronological record of magnetic reversals is now reasonably well known (**Fig. 5.16**). Accordingly, the age of any part of the ocean floor can be estimated easily and inexpensively from available or readily obtainable ocean magnetic anomaly data.

The early maps of magnetic anomalies were not very useful, because the history of magnetic reversals was not well known. However, the maps did confirm that ocean floor becomes older with distance from the oceanic ridges. The data also enabled scientists to estimate that no ocean floor more than about 200 million years old (close to the 170 million years we now accept as correct) could still exist and suggested that the oldest ocean floor is at subduction zones. These findings supported the concept that new ocean floor is created at oceanic ridges and destroyed at trenches.

Deep-Sea Drilling

In 1968, the scientific drilling ship *Glomar Challenger* began operations. During the next several years, the vessel obtained the data that many scientists believe finally

FIGURE 5.16 The age of the oceanic crust of the world's oceans, as established from magnetic anomalies. The youngest crust is adjacent to the divergent plate boundary of the oceanic ridges. Age increases away from the ridge to the oldest crust, which is entering subduction zones. The youngest crust is also the warmest and has the lowest density. Crust cools, increases in density, and sinks isostatically with increasing age.

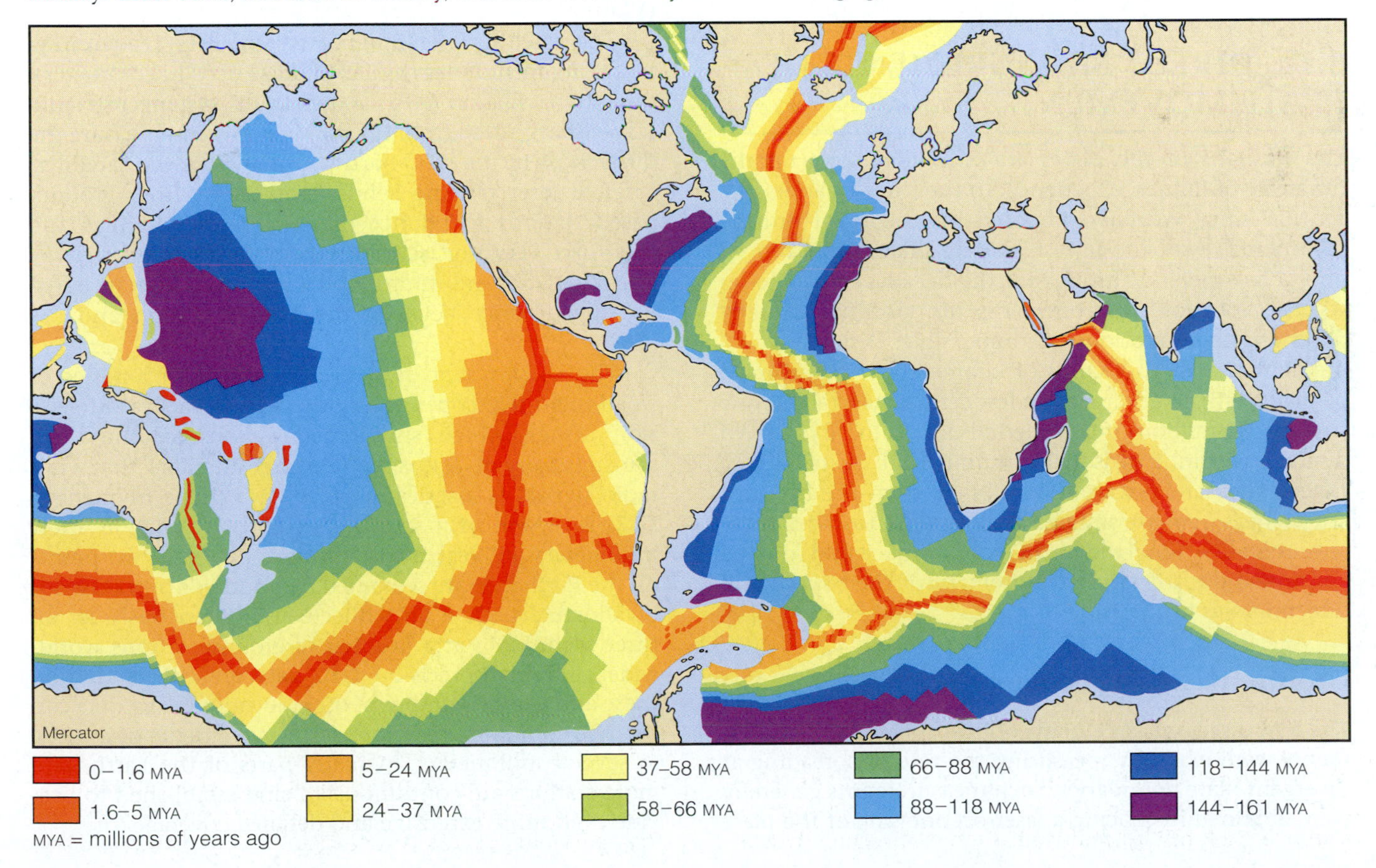

established seafloor spreading and continental drift beyond any reasonable doubt. The *Glomar Challenger* was capable of drilling into the ocean floor sediment in waters as deep as 6000 m. It drilled many holes into the ocean floor and collected samples of the sediment through which it drilled. Some holes were drilled down to the oceanic crust bedrock, and when the drill could penetrate it, samples of the crust itself were also obtained.

One of the *Glomar Challenger*'s first tasks was to drill a series of holes in the North Atlantic Ocean floor in a line from the center of the oceanic ridge toward the North American continent. Samples of bedrock, or more often sediment lying immediately above the bedrock, were dated by radioisotope techniques and known ages of fossils (**CC7**). These studies proved that sediments are thin near the oceanic ridge and become thicker with distance from the ridge. They also demonstrated that the surface sediment at each sampling site has been recently deposited, and that the buried sediment immediately above the crust becomes progressively older with distance from the ridge. More recent worldwide drilling studies by the *Glomar Challenger* and its successor, the *JOIDES Resolution*, have confirmed that the oldest ocean crust known to remain today is about 170 million years old. The data gathered by the two drilling vessels again confirm the theory that new ocean crust is created at oceanic ridges and destroyed at subduction zones.

5.12 DIRECT MEASUREMENTS OF PLATE MOVEMENT

The motions of the plates are extremely slow on the timescale of human experience. In the three-quarters of a century since Wegener proposed his theory of drifting continents, even the fastest-moving plates have moved only a few meters. Throughout the decades of debate on whether or not the continents are drifting, actually observing these motions was impossible. Distance measurements across thousands of kilometers of ocean must be made with an accuracy of a few centimeters (about one part in 100 million) if continental drift is to be observed directly within a reasonable number of years. Before **lasers** and satellites were developed, even the best distance measurement systems were unable to meet these requirements. Movements of continents could not be confirmed directly, and measurement of their speed and direction was not possible.

Ground-Based Measurements

The first direct measurements of lithospheric plate movements were made by measuring the rate of spreading at divergent plate boundaries. Such measurements generally were obtained by placing a laser on one side of the plate boundary and a reflector on the other side. The time taken for the laser light to travel across to the mirror and back was measured precisely. The travel time is a measure of the distance between laser and mirror. Experiments were performed at several locations, including the rift valleys that pass through the middle of Iceland and of the Red Sea. The studies revealed movements of approximately the expected speed at divergent plate boundaries. However, most of the measurements were barely accurate enough to demonstrate that motion was occurring, and many scientists thought the measured motions might reflect only local earth movements rather than motions of drifting plates.

Satellite Measurements

Commercial satellite navigation systems (GPS; Chap. 2) are not accurate enough to measure continental drift directly. An accuracy of a few meters is more than good enough for terrestrial navigation, but not for the most exacting scientific needs. However, the most precise GPS measurements, when used in conjunction with two other recently developed sophisticated measurement systems, are capable of measuring the speed and direction of movement of the lithospheric plates. The two other systems are satellite laser ranging using Laser Geodynamics Satellites (LAGEOS) and very long baseline interferometry (VLBI).

The satellite laser ranging system uses **frequency**-matching techniques similar to those used in commercial satellite navigation (GPS) systems. GPS systems use radio waves transmitted from several satellites and measure the differences in the signal arrival times from the satellites. Satellite laser ranging sends laser beams from ground-based stations to bounce off mirrors on satellites and then uses very precise methods of determining the frequency match and timing of the returning light to measure the distance between the ground station and the satellite.

The VLBI system (**Fig. 5.17**) consists of a series of small radio telescopes, many the size of backyard television dishes, which simultaneously observe the radio transmissions of a quasar. (Quasars are stars that emit radio signals as a series of pulses at a single frequency.) VLBI is capable of measuring a change as small as a few millimeters per year in the distance between two points on the Earth that are separated by hundreds or thousands of kilometers. Data are collected simultaneously by all three techniques at a number of ground stations worldwide, and high-precision GPS data are collected from these and a large number of other locations. The data can then be combined to detect earth movements of as little as several millimeters on many parts of the Earth, and more stations are continuously being established to provide even more extensive and detailed coverage.

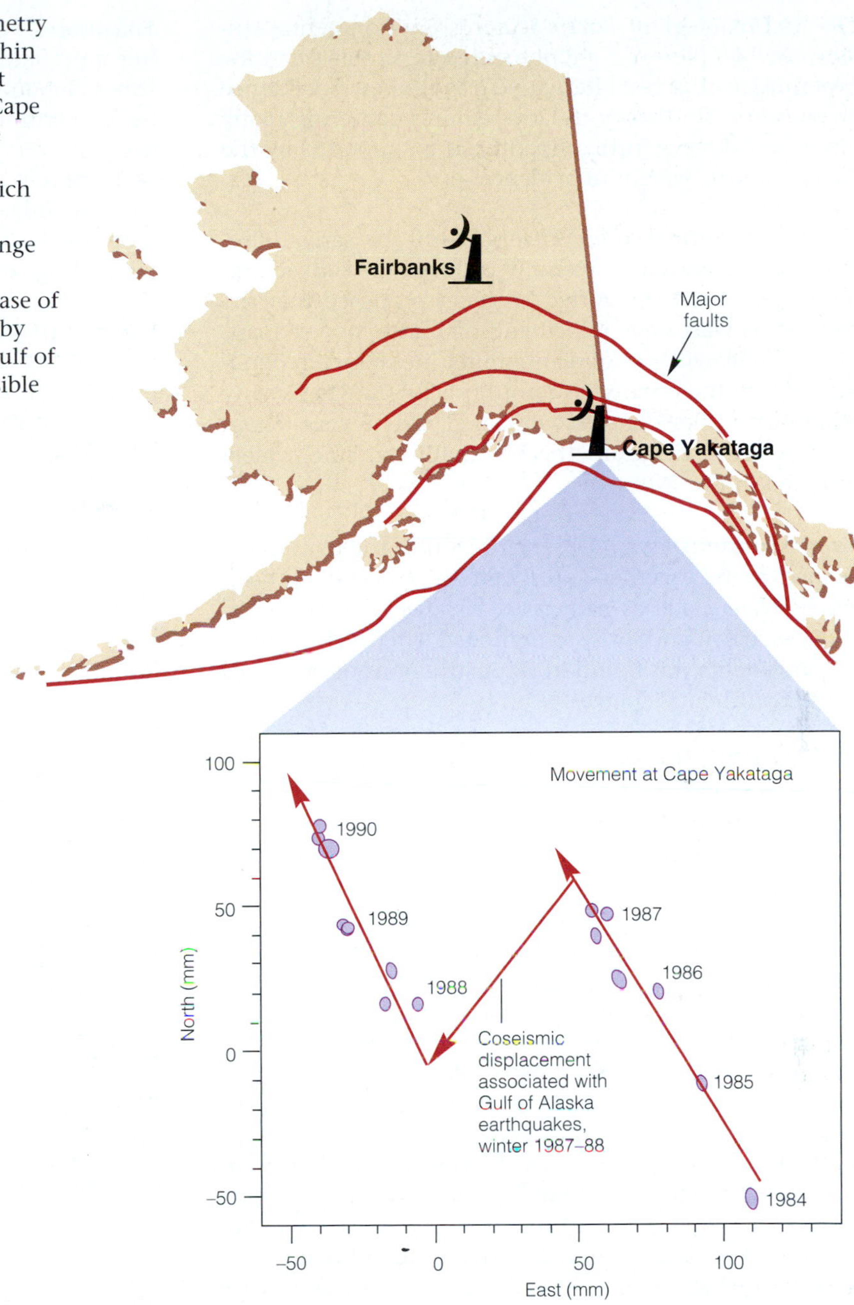

FIGURE 5.17 Very long baseline interferometry (VLBI) observations show the movements within the North American Plate in Alaska. The chart plots the relative locations of Fairbanks and Cape Yakataga during the years 1984 to 1990. Cape Yakataga appears to creep slowly closer to Fairbanks as the North American Plate on which they lie is compressed by the northward movement of the Pacific Plate. The sharp change in relative positions that occurred during the winter of 1987 to 1988 was related to the release of some of the compressional stress in the plate by the major earthquakes that occurred in the Gulf of Alaska that winter. The ellipses show the possible range of error of each location measurement.

Emerging Details of Plate Movements

Since the 1980s, direct measurements have revealed relative motions of lithospheric plates that are consistent with those suggested by plate tectonic theory. Measured speeds and directions of plate motions are similar to estimates based on geological records.

Extremely precise measurements are beginning to reveal details of plate motions (**Fig. 5.17**). Plates or sections of a plate apparently move in a combination of slow creeping motions and short sharp movements usually associated with earthquakes. The sharp earth movements may not cause any disturbance that can be observed on the ground and can occur in a given area even though an earthquake may be centered tens or hundreds of kilometers away.

Increasingly sophisticated measurement systems are now beginning to reveal a wealth of information about plate motions. Observations are identifying areas of compression or tension that build up and are released as the plates move. Eventually such knowledge may help in the development of a reliable method to predict the year or month and location of probable earthquakes.

CHAPTER SUMMARY

The Methodology of Earth Science. Environmental systems are complex and are not suited to simple scientific experiments that test alternative hypotheses. Acceptance or revision of a theory usually requires gathering many pieces of evidence until a hypothesis is supported by the overwhelming weight of evidence.

History of Plate Tectonic Theory. Until the early 1900s, the continents were universally believed to be fixed and immovable. In 1912, Alfred Wegener proposed that the continents were once joined and had later moved apart. He provided evidence of many similarities between opposite sides of the oceans. Not until the 1950s, when many other pieces of evidence accumulated, did his theory gain support. Plate tectonic theory finally became widely accepted in the 1960s.

Fit of the Continents. Observations that the continents seemed to fit together as though they had once been joined were made as early as the sixteenth century. In 1965, such evidence was strengthened when a closer fit of the continents was found in reconstructions of the edges of the continental shelves.

Geological Similarities between Previously Joined Edges of Continents. At the places where continents are hypothesized to have been joined, mountain chains run off one edge and resume on the other, now separated, edge. Rocks, fossils, and minerals on the two continents are found to be identical at appropriate locations.

Paleoclimate. Fossils of tropical coral reef species are found in 300-million-year-old rocks across North America, North Africa, and Eurasia, far from the tropics. This and similar distributions of fossils provide further evidence that the continents have moved.

Isostasy and the Mohorovičić Discontinuity. A seismic discontinuity is present about 30 to 40 km below the continents and about 6 km below the seafloor. Calculations based on the density and thickness of continental and oceanic crust show that these depths are consistent with isostasy, the theory that lithosphere floats on underlying asthenosphere.

Earthquakes. Earthquakes are concentrated in the central rift valleys or axes of oceanic ridges, in transform faults, and in subduction zones. Only shallow earthquakes occur at oceanic ridges, consistent with thin lithosphere and upwelling. Both shallow and deep earthquakes occur at subduction zones. Earthquakes at various depths in subduction zones show characteristics of plate bending, plate extension, and plate compression that are consistent with processes occurring when slabs of cold lithosphere are subducted into the mantle. Computer tomography has revealed upwelling plumes in the mantle and cold slabs of crustlike material descending through the mantle at subduction zones.

Volcanoes. Volcanoes are located along oceanic ridges, at hot spots, and on the nonsubducting plate at subduction zones. Subduction zone volcanoes erupt lava characteristic of continental crust or of mixed continental and oceanic crust. They often erupt explosively, indicating a high content of water from subducted marine sediments. Oceanic ridge and hot-spot volcanoes usually do not erupt explosively and they expel lava of a composition resembling oceanic crust.

Paleomagnetism. When rocks form, they are magnetized in the direction of the Earth's magnetic field, which dips progressively toward the Earth's center with increasing latitude. If the rocks are later moved or rotated, their original latitude and orientation toward the pole can be deduced from their magnetic field. Such paleomagnetic information for rocks of various ages and for continents indicate that the continents must have moved progressively.

Ocean Floor Magnetic Anomalies. The magnetism of seafloor rocks and sediment causes small anomalies in the Earth's magnetic field. Magnetic anomalies have a striped pattern that is symmetrical on either side of the oceanic ridges. The pattern was caused by periodic reversals in the Earth's magnetic field. Stripes represent rocks formed at the divergent plate boundary when the Earth's magnetic field was alternately normal and reversed. The pattern is symmetrical on either side of the ridge because each new stripe was split by new oceanic crust that was created after a field reversal.

Age of the Ocean Floor. The relative age of the seafloor can be determined from magnetic anomaly data and from sediment thickness. Absolute ages of seafloor sediment have been measured for cores drilled deep into the sediment. The age of sediment immediately above the crustal rocks increases with distance from the oceanic ridges. The oldest sediment ever found in the oceans (about 170 million years old) lies on the oldest oceanic crust next to the subduction zones in the Northwest Pacific.

Direct Measurements of Plate Movement. Direct measurements of the speed and direction of tectonic plate movements were first made using lasers to measure the distance across spreading centers. Now they are made routinely using a combination of global positioning satellites, satellite laser ranging systems, and multiple radio telescopes that simultaneously monitor the same star.

STUDY QUESTIONS

1. How are ocean sciences different from the traditional sciences of chemistry and physics?
2. Why did geologists not believe for centuries that the continents move?
3. Why is the seafloor lower than the surface of the continents?

4. Fossils of species that lived only in sediments of the abyssal parts of the oceans occur in rocks that lie in the middle of the continents. How did they get there?
5. What principal characteristics of the two coastal regions of previously joined continents can show that they once were joined?
6. Which are the most violent earthquakes—those at subduction zones or those at oceanic ridges? Explain why.
7. Most hot spots create volcanoes that rise several kilometers from the seafloor and often create islands, whereas volcanoes at oceanic ridges do not generally build mountains as high above the seafloor. Explain why.
8. Would the dip angle of newly magnetized rocks be changed after the Earth's magnetic field reversed? If so, explain how.
9. Which oceanic crust is younger—that just south of India or that immediately adjacent to the Mid-Atlantic Ridge? Explain why.
10. Direct measurements of the movement of individual locations near the edges of tectonic plates show that they both creep slowly and move sharply when earthquakes occur. Explain how the slow creeping motion may occur. Is that motion the same at all points within a given tectonic plate? Explain why or why not.

KEY TERMS

You should recognize and understand the meaning of all terms that are in boldface type in the text. All those terms are defined in the Glossary. The following are some less familiar key scientific terms that are used in this chapter and that are essential to know and be able to use in classroom discussions or exam answers.

asthenosphere (p. 113)
basaltic (p. 119)
central rift valley (p. 116)
continental drift (p. 108)
continental shelf (p. 109)
contour (p. 109)
convection (p. 117)
convergent plate boundary (p. 116)
dip angle (p. 121)
divergent plate boundary (p. 116)
downwelling (p. 118)
focus (p. 114)
fracture zone (p. 116)
frequency (p. 126)
hot spot (p. 117)
isostasy (p. 113)
isostatic leveling (p. 113)
lithosphere (p. 116)
lithospheric plate (p. 108)
magma (p. 114)
magnetic anomaly (p. 108)
oceanic ridge (p. 108)
paleomagnetism (p. 108)
radioisotope (p. 124)
seafloor spreading (p. 108)
seamount (p. 119)
seismograph (p. 114)
subducted (p. 116)
subduction zone (p. 110)
tomography (p. 117)
transform fault (p. 116)
trench (p. 108)
upwelling (p. 114)
weathering (p. 106)

CRITICAL CONCEPTS REMINDER

CC2 **Isostasy, Eustasy, and Sea Level** (p. 113). Earth's crust floats on the plastic asthenosphere. Sections of crust rise and fall isostatically as temperature changes alter their density and as their mass loading changes due to melting or to the formation of ice stemming from climate changes. This causes sea level to change on the coast of that particular section of crust. Sea level can also change eustatically when the volume of water in the oceans increases or decreases due to changes in water temperature or changes in the amount of water in glaciers and ice caps on the continents. Eustatic sea level change takes place synchronously worldwide and much more quickly than isostatic sea level changes. To read **CC2** go to page 5CC.

CC3 **Convection and Convection Cells** (p. 117). Fluids that are heated from below, such as the Earth's mantle, rise because their density is reduced. They continue to rise to higher levels until they are cooled sufficiently, at which time they become dense enough to sink back down. This convection process establishes convection cells in which the heated material rises in areas of upwelling, spreads out, cools, and then sinks at areas of downwelling. Areas of downwelling are the subduction zones, while areas of upwelling include the oceanic ridges and hot spots. To read **CC3** go to page 10CC.

CC7 **Radioactivity and Age Dating** (pp. 108, 124, 125, 126). Some elements have naturally occurring radioactive (parent) isotopes that decay at precisely known rates to become different (daughter) isotopes, which are often isotopes of another element. Measurement of the concentration ratio of the parent and daughter isotopes in a rock or other material can be used to calculate its age, but only if none of the parent or daughter isotopes have been gained or lost from the sample over time. Radioisotope age dating is used in conjunction with other dating methods including variations in fossil assemblages and magnetic field properties. To read **CC7** go to page 18CC.

CHAPTER 6

CHAPTER OUTLINE

CRITICAL CONCEPTS USED IN THIS CHAPTER

All known life forms on Earth depend on water and its unique properties. Although it is well known that plants and animals must take up or ingest water to survive, water is also important in many other ways. For example, the seawater in this photograph at Deacon's Reef, Papua New Guinea, is transparent to visible light. As a result, light energy can penetrate into the ocean, where it is used for photosynthesis. Many of the chemical elements and compounds invisibly dissolved in the seawater are essential to life. Water is important in many other subtle ways as well. For example, the organisms living in this reef ecosystem, and the terrestrial plants and animals that live on nearby land, thrive in a climate in which the temperature changes little between day and night. If water did not have a high heat capacity, these organisms would experience daily temperature variations and extremes comparable to those that occur in the middle of the largest deserts.

Water and Seawater

Day after day, day after day,
We stuck, nor breath nor motion,
As idle as a painted ship
Upon a painted ocean.

Water, water, every where,
And all the boards did shrink,
Water, water, every where,
Nor any drop to drink.

—Samuel Taylor Coleridge,
"The Rime of the Ancient Mariner,"
1798

The section of this famous poem quoted here speaks of the anguish of the ancient mariner becalmed and suffering from raging thirst after all the drinkable water brought aboard his vessel at the last landfall had been consumed. One of the great ironies of history is that many a sailor has died of thirst while surrounded by nothing but water as far as the eye can see. Less than 1% of the Earth's water is found in the liquid form with a low enough concentration of dissolved salts that it may be drinkable by humans and other land mammals. Even some of this water is contaminated with pathogens or traces of toxic chemicals that make it unfit to drink. Clean, salt-free drinking water is just as fundamental to our survival as food is. For anyone living in a drought-stricken land or sailing the ocean, drinkable water is perhaps the most precious commodity that exists.

Although most of us are aware that water is essential to life as we know it, the range of unique properties of water that make life possible is less well known. For example, the properties of water

- Help to create and control **climate** and **weather**
- Influence the formation and modification of the land and seafloor
- Enable essential chemicals to be transported to and within living organisms
- Control many features of our physical **environment,** such as rain, snow, and the waves on oceans and lakes
- Underlie the functioning of many aspects of modern society, ranging from cooling systems for automobile engines and power plants, to ice cubes that keep drinks cold in summer

The dissolving power of water, the composition of seawater, and the processes that add or remove dissolved substances in ocean water are important to the studies of **sediments** (Chap. 8), life in the oceans (Chaps. 14, 16, 17), and **pollution** (Chap. 19). In this chapter, we will investigate the extraordinary properties of water that permit it to support life on our planet, and we will discuss how water's dissolving ability affects the chemical composition of ocean water. In the following chapter we will examine the physical properties of water.

6.1 ORIGINS AND DISTRIBUTION OF THE EARTH'S WATER

Early in its history, the Earth was hot and mostly molten. Heavy elements such as iron and nickel migrated toward the Earth's center, while lighter elements, such as silicon, aluminum, and oxygen, moved upward toward the surface. The lightest elements, which included hydrogen and oxygen, and compounds of light elements, including carbon dioxide, methane, and water vapor, migrated upward to form an atmosphere. The lightest gaseous elements, hydrogen and helium, were largely lost to space. After the Earth cooled, the **crust** solidified and water vapor in the atmosphere condensed into liquid water.

In the billions of years since the Earth's water first condensed, the temperature of the atmosphere apparently has changed relatively little. Hence, liquid and solid water probably were present throughout that period in oceans and ice sheets. Volcanic activity releases water that was trapped deep within the Earth's interior as the Earth solidified. However, the rate at which water is released by volcanic activity is believed to be small in comparison with the volume of the oceans and ice sheets. Therefore, the amount of water at the Earth's surface probably has remained relatively unchanged for many millions of years.

Uniqueness of the Earth

The Earth is the only planet in the solar system with liquid water on its surface. The outer planets—Saturn, Jupiter, Uranus, and Neptune—and their moons are cold, and any water present is ice. The surface of Mars is currently too cold to have liquid water, although it may have been warmer in the past. Certain features of the Martian surface suggest that liquid water was present at one time in its history, and we now have direct evidence from instruments on robotic vehicles on the Martian surface that this was almost certainly the case. The surfaces of Mercury and Venus are hot and could have water only as a vapor. In addition, the mass and therefore the **gravity** of Mercury, Venus, and Mars are sufficiently small that much of the water formed during their early history has probably escaped into space.

Water is the only known substance that is present in all three physical forms—liquid, solid (ice), and gas (atmospheric water vapor)—within the range of temperatures and pressures found on the Earth's surface and in its atmosphere. The presence of water in all three forms and the conversion of water from one form to another are important to the maintenance of climatic conditions within the range we consider acceptable for human life and for the stability of the Earth's **ecosystems.**

Distribution of the Earth's Water

More than 97% of the world's water is in the oceans (**Table 6.1**). However, the tiny fractions of water in the atmosphere and in freshwater systems (lakes, rivers, streams, and **groundwater**) are disproportionately important to humanity. Freshwater may be the most precious and scarce natural resource supporting human civilization. Transport of water vapor from the oceans through the atmosphere to land as rain and snow determines the availability of the freshwater resource. The transfer of water between oceans and atmosphere determines critical aspects of our climate. In Chapter 9 we learn more about the importance of the hydrologic cycle whereby water is exchanged among the oceans, atmosphere, freshwater, groundwater, and polar ice sheets.

TABLE 6.1 The World's Water

Location	*Percentage of total* (1.4×10^9 *km*^{3})	
Oceans		**97.2**
Freshwater		
Ice caps and glaciers	2.38	
Surface waters (lakes, rivers, streams)	0.022	
Air and soil moisture	0.001	
Groundwater	0.40	
Total freshwater		**2.8**

6.2 THE WATER MOLECULE

To understand the unusual properties of water, we need to understand the structure of atoms and molecules, particularly the water molecule.

Atoms and Electrons

Atoms consist of negatively charged electrons orbiting in shells that surround a nucleus. The nucleus contains neutrons that have no charge and protons that are positively charged. The electrical charge of a proton is equal and opposite to that of an electron, and the number of protons in each atom is equal to the number of electrons. Hence, atoms are electrically neutral.

Each electron shell is capable of holding only a certain number of electrons. For example, the innermost shell holds a maximum of two electrons, and the second shell holds a maximum of eight. Elements that have their outermost shell filled are noble gases (helium, neon, argon, krypton, xenon), which are inert (that is, they do not react chemically with most atoms).

The atoms of most elements have an outermost electron shell that is not completely filled with the maximum number of electrons. However, it is energetically favorable for each shell to be filled. Therefore, atoms of some elements, such as chlorine or oxygen, that have a nearly full outer electron shell, have a tendency to gain electrons in order to fill the shell. Atoms of other elements, such as sodium, that have outer electron shells less than half full, have a tendency to lose one or more electrons, producing an empty outer shell.

Chemical Bonds between Atoms

To fill their outer electron shells, two or more atoms of different elements can combine to create a molecule of a chemical compound. This can happen in two ways. First, one atom can donate one or more electrons to an atom of another element that has an incomplete outer shell. For example, a sodium atom can lose an electron, and a chlorine atom can gain this electron. Sodium thus becomes a positively charged sodium **ion,** and chlorine becomes a negatively charged chloride ion. Because the two ions have opposite electrical charges, they are **electrostatically** attracted to each other and are bonded together by this attraction (**Fig. 6.1**). This type of bond is called an **ionic bond.**

The second way in which atoms of different elements can share electrons is called a **covalent bond.** One or more outer-shell electrons spend part of their time in the outer shell of each of the two atoms that are bonded. For example, each of two hydrogen atoms can share its single electron with an oxygen atom. To describe such electron sharing simply, each hydrogen electron orbits both its own hydrogen atom nucleus and within the outer shell of the oxygen atom electron cloud. At the same time, two

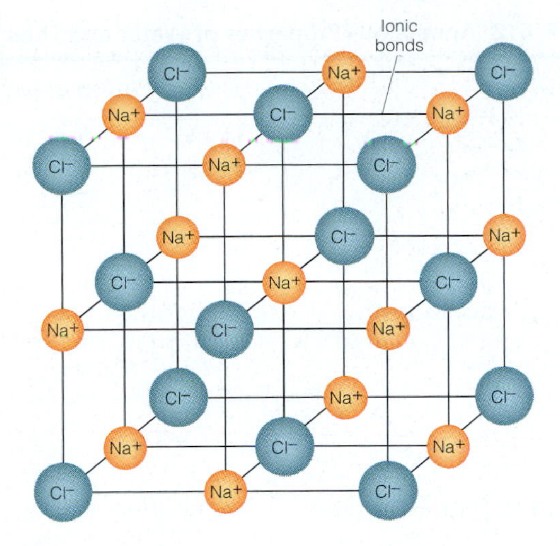

FIGURE 6.1 The sodium and chlorine atoms of common salt are held together in a salt crystal by ionic bonds. The ionic bond is an electrostatic attraction between the positively charged sodium ion (Na^+) and the negatively charged chlorine ion (Cl^-).

electrons from the outer shell of the oxygen atom orbit within their own shell and around the hydrogen atoms (**Fig. 6.2a**). In this way, the oxygen has the extra electrons it needs to fill its outer electron shell for part of the time, and each of the two hydrogen atoms has two electrons to fill its outer electron shell for part of the time. Because the outer shell of each atom in the molecule is full at least part of the time, the bonded atoms form a stable molecule. Most covalent bonds are stronger than most ionic bonds. Consequently, atoms in a covalently bonded molecule usually are more difficult to break apart than those that are bonded by ionic bonds.

Van der Waals Force and the Hydrogen Bond

Molecules are electrically neutral. However, there is a weak attractive force between all molecules called the **van der Waals force.** It is caused by the attraction of the protons of one atom to the electrons of another. An additional attractive force or bond, called the **hydrogen bond,** is present between the molecules of water. In other chemical compounds, the hydrogen bond is either absent or weaker than it is in water. The strength of the hydrogen bond is what gives water most of its anomalous properties (**Table 6.2**).

Like other molecules, the water molecule is electrically neutral. However, the arrangement of atoms and electrons in the water molecule is such that the side of the molecule away from the hydrogen atoms has a small net negative charge, whereas the two areas where the hydrogen atoms are located have a small net positive charge (**Fig. 6.2b**). Molecules that behave as though they have a positive and a negative side are called "polar mol-

TABLE 6.2 Anomalous Properties of Water and Their Importance

Property	*Special Characteristics of Water*	*Importance*
Heat capacity	Higher than that of any solid or liquid other than ammonia.	Water moderates climate in coastal regions. Currents transport large amounts of heat. Ocean water temperatures are relatively invariable in comparison with terrestrial temperatures.
Latent heat of fusion	Higher than that of any substance other than ammonia.	When ice forms, most of the energy lost is released to the atmosphere, and ice absorbs large amounts of heat in melting. Ice therefore acts as a thermostat to keep high-latitude water and atmosphere near freezing point all year.
Latent heat of vaporization	Higher than that of any other substance.	Heat is transported from the low-latitude ocean by evaporation and atmospheric circulation and released to the atmosphere through precipitation at higher latitudes.
Thermal expansion	Pure water has a density maximum at 4°C. The temperature at which the maximum occurs decreases as salinity increases; and there is no maximum in sea water.	Freshwater and low-salinity seawater stay unfrozen under ice in winter in lakes and estuaries.
Solvent property	Can dissolve more substances than any other liquid.	Water dissolves minerals from rocks and transports them to the oceans. Water is the medium in which the chemical reactions that support life occur.
Surface tension	Higher than that of any other known liquids except mercury.	Surface tension controls the formation of droplets in the atmosphere and bubbles in the water. Some organisms use surface tension to anchor themselves to or walk on the surface.
Physical states	The only substance present as a gas, liquid, and solid within the temperature range at the Earth's surface.	Water vapor evaporated from the ocean helps transport heat from warm low latitudes to cold high latitudes. Liquid water also contributes to that transport through ocean currents. In polar regions, the presence of both ice and water moderates climate.

ecules." The reasons for the polarity are complicated, but they can be understood through a simplified example. The six electrons in the outer shell of the oxygen atom are arranged in pairs. The two hydrogen atoms are covalently bonded to the electrons in one of these pairs. This leaves two pairs of unshared orbiting electrons on the side of the oxygen atom opposite the hydrogen atoms, causing this side of the oxygen to have a negative charge bias. On the hydrogen side of the molecule, hydrogen electrons are shared with, and actually spend more time on, the oxygen atom, leaving the positively charged nucleus of each hydrogen atom "exposed." Hence, this side of the molecule has a positive charge.

Because the water molecule is polar, the negatively charged side of one molecule is attracted to the positively charged side of an adjacent molecule. This attractive force is the hydrogen bond (**Fig. 6.2c**). It is relatively strong, but not as strong as ionic or covalent bonds. The relative strengths of bonds between atoms and molecules are listed in **Table 6.3**. The relative strength of a bond is an indication of how much energy is needed to break that bond. Substantially more energy is needed to break hydrogen bonds than is needed to counter van der Waals forces between molecules. The relatively high strength of the hydrogen bond is responsible for the anomalous properties of water.

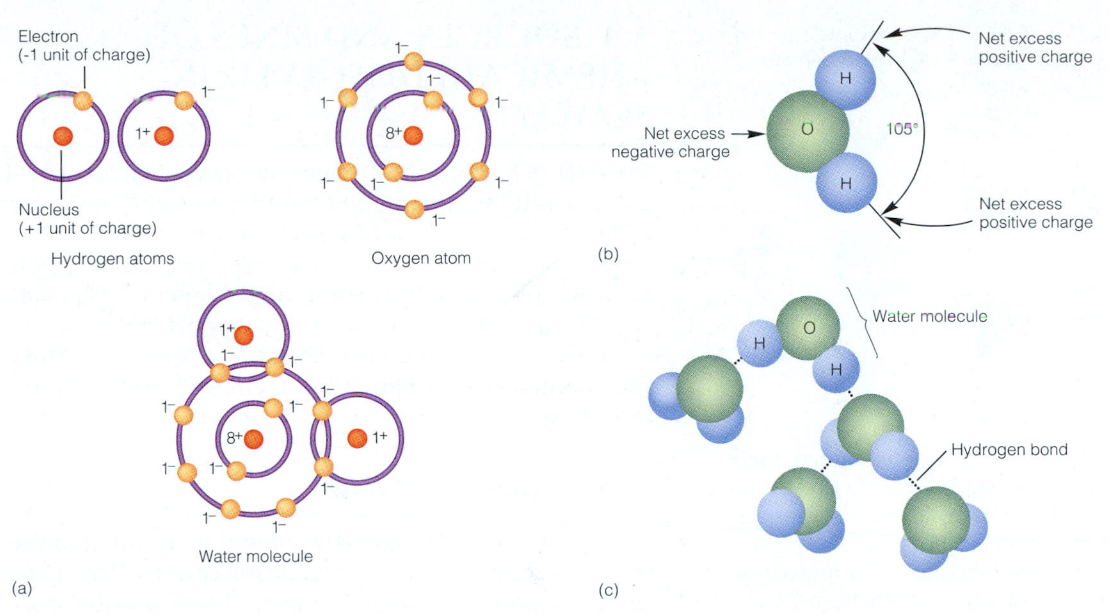

FIGURE 6.2 The water molecule consists of two hydrogen atoms and one oxygen atom. (a) The oxygen atom has six electrons in its outer shell. It lacks two electrons that would be needed to complete this shell. Each hydrogen atom has one electron and needs one more to complete its single shell. In a water molecule, the electrons from the hydrogen atoms and the six electrons from the outer shell of the oxygen atom are shared and orbit around both the oxygen and the hydrogen atoms. This is a covalent bond in which electron clouds from the bonded atoms partially overlap. (b) Both hydrogen atoms in a water molecule are on one side of the oxygen atom and are separated by an angle of 105°. The hydrogen electrons orbit partially around the oxygen atom, leaving the positive nucleus of each hydrogen atom partially "exposed" and giving this side of the water molecule a slight net positive charge. Similarly, the electrons orbiting the opposite side of the oxygen atom give that side of the molecule a slight net negative charge. Hence, water is a polar molecule. (c) Hydrogen bonds are formed between water molecules when the positive side of one molecule is attracted to the negative side of another.

TABLE 6.3 **Relative Strengths of Bonds between Atoms and Molecules**

Type of Bond	*Approximate Relative Strength*
Van der Waals force (between molecules)	1
Hydrogen bond (between molecules)	10
Ionic bond (between atoms)	100
Covalent bond (between atoms)	Usually >1000

6.3 THE DISSOLVING POWER OF WATER

Water can dissolve more substances and greater quantities of these substances than any other liquid. Its unique dissolving power is related to the polar nature of the water molecule. Many inorganic chemical compounds—for example, sodium chloride (NaCl)—have ionic bonds. In such compounds, the two ions (Na^+ and Cl^- in the case of NaCl) are held together by the attraction of the two opposite electrical charges. If such a compound is placed in water, the electrostatic attraction is greatly reduced and the ionic bonds are broken by a process called **hydration.** Hydration occurs when a positive ion (e.g., Na^+) is surrounded by water molecules oriented with their negative sides toward the positive ion (**Fig. 6.3**). Conversely, a negative ion (e.g., Cl^-) is surrounded by water molecules oriented with their positive sides toward the negative ion. Thus, each ion becomes free to move independently of the other ion, and the compound dissolves. A variety of ionically bonded inorganic com-

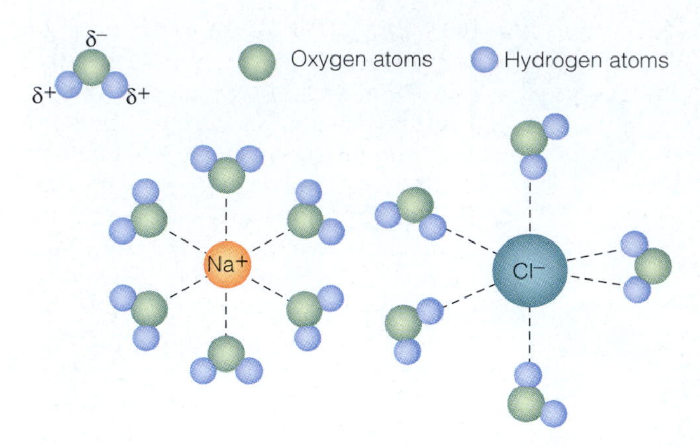

FIGURE 6.3 Compounds that are ionically bonded, such as common salt, are readily dissolved in water because both the positively charged ion (the sodium cation, Na^+) and the negatively charged ion (the chloride anion, Cl^-) are hydrated. Each ion is surrounded by polar water molecules oriented so that their positive sides (hydrogen atoms) face the anion and their negative sides (oxygen atom opposite the hydrogen atoms) face the cation. Hydration reduces the attraction between the cation and the anion and promotes their dispersal (solution) in the water.

pounds, or "salts," are dissolved in seawater. Many organic compounds also can be ionized and dissolved. Covalent compounds are generally less soluble in water than ionic compounds, although many covalent compounds, such as silica (SiO_2, quartz, sand), dissolve in small quantities.

Water's exceptional dissolving power is the reason why most elements are present in seawater, even though many are in very low concentrations. Most of the dissolved substances occur as either **cations** (positively charged ions) or **anions** (negatively charged ions), each of which has a surrounding sphere of properly oriented water molecules. Ions dissolved in seawater are the source of elements needed for the growth of marine plants, on which most ocean life depends (Chap. 14).

The dissolving power of water is also important to terrestrial life. Land plants obtain many of their needed elements from solution in water through their root systems. Animals, including humans, obtain many of the elements and other chemicals needed for their **biochemical** systems by dissolving these substances from their food through the digestion process. In fact, almost all processes that support the growth and function of living organisms on our planet depend on the dissolving power of water. Life as we know it would not exist if water did not have the ability to dissolve and separate the ions, enabling them to be moved in solution into and within living tissue.

6.4 SOURCES AND SINKS OF CHEMICALS DISSOLVED IN SEAWATER

Seawater is a solution of many different chemical compounds: cations, anions, and both organic and inorganic compounds that are not ionized. Concentrations of the compounds in ocean water are determined by their behavior in global **biogeochemical cycles** (**Fig. 6.4**) and by their abundance in the materials of the Earth's crust. Compounds are both added to ocean water from sources such as river **runoff** and removed from seawater to sinks such as the seafloor sediments.

Biogeochemical Cycles

Continental rocks are **weathered** and transported to the oceans both as particles and as dissolved ions. The particles are deposited as ocean sediment. Many dissolved elements are used in biological processes and subsequently are incorporated in seafloor sediment in the form of the **hard parts** and **detritus** of marine organisms. Elements dissolved in seawater also can be precipitated directly or on particulate matter or they may be **adsorbed** by (adhere to) mineral grains and detritus and thus be removed to the sediment.

Over millions of years, ocean sediments are compacted by the weight of overlying sediments. Water is squeezed out and minerals precipitate between the grains, cementing the grains together to form sedimentary rock. Vast volumes of sedimentary rock and sediment eventually enter a **subduction zone.** Some of this

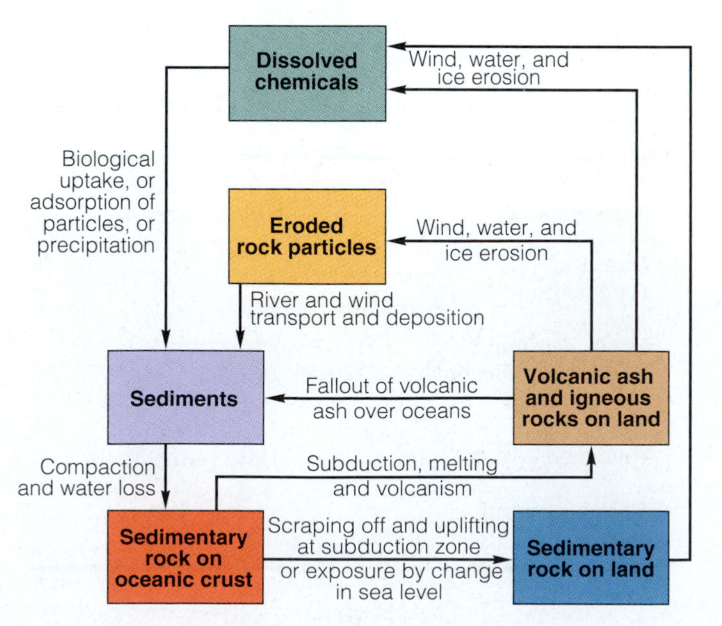

FIGURE 6.4 A simplified biogeochemical cycle.

material is scraped off the descending **lithospheric plate** edge and raised to form **sedimentary arc** islands, **continental margin** rocks, and **exotic terranes** (Chap. 4), where it is again weathered and **eroded,** thus continuing the biogeochemical cycle. Much of the sedimentary rock and sediment is **subducted** into the **mantle,** where some of it is melted and ejected through volcanoes as ash or **lava.** This material either reenters the oceans and returns to the ocean sediment directly, or collects on the land, where it is weathered and eroded to start a new cycle.

Biogeochemical cycles are actually much more complicated than **Figure 6.4** indicates. For example, many elements used by marine organisms are rapidly recycled to seawater solution by the decomposition of detritus. Similarly, some elements from oceanic crustal rocks enter seawater in fluids discharged by **hydrothermal vents** at **oceanic ridges** (Chap. 17).

Steady-State Concentrations

The concentration of each element in ocean water is determined by the rate at which it enters the ocean water and the rate at which it is removed. If the rate of input exceeds the rate of removal, the concentration will rise, and vice versa. Although global biogeochemical cycles include processes that take hundreds of millions of years, these cycles have been ongoing for billions of years and are thought to be at an approximate **steady state.** Steady state is achieved when the total quantity of an element in each of the compartments (square boxes) in **Figure 6.4** remains approximately the same over time (its input rate equals its removal rate).

To see how a steady state is achieved, consider what would happen if an element's rate of input to the oceans were increased. The total quantity of an element within the global biogeochemical cycle does not change. Therefore, as the input to the oceans increases and the total quantity of the element in the oceans increases, the total quantity of the element in the other compartments must decrease. Consequently, less of the element is in the compartments that provide inputs to the oceans, and its input must decrease. If an element's input to the oceans (or any other compartment) increases, so do the rates of the various removal processes.

Here's a simple analogy: If we pour orange juice (the input) into a glass full of water and stir continuously, the glass will overflow. At first, the overflowing liquid (the output) is almost pure water, with only traces of orange juice mixed in it. The input of orange juice has increased, but neither the concentration in the glass nor that in the output reaches a new steady level instantly. As we continue to pour orange juice into the glass, the concentration of juice in the glass increases progressively to full concentration. The concentration of juice in the output also increases progressively in response to the change in concentration in the glass. When the concentrations of orange juice in the glass and in the output match the concentration of orange juice in the input, the system has reached a new steady state. Subsequently, unless we change the concentration of orange juice in the input, or alter the output in some way (such as by allowing the orange juice pulp to settle and accumulate at the bottom of the glass), the input equals the output and the concentration of orange juice within the glass does not vary. Biogeochemical cycles are more complicated than our simple analogy because they have many inputs and outputs, and multiple "glasses" that empty into each other. However, they reach steady state in a similar way, and this steady state can be disturbed only by changes in one or more inputs or outputs.

Because the global biogeochemical cycles are approximately at steady state (aside from changes caused by **contaminants;** see Chap. 19), the rate of input of most elements to the oceans is approximately equal to the rate of removal.

Residence Time

The concentrations of elements in seawater are determined largely by the effectiveness of the processes that remove them from solution. Concentrations are high if the element is not removed rapidly or effectively from solution in ocean waters and if it is abundant in the

CRITICAL THINKING QUESTIONS

6.1 Continental rocks and soils are constantly leached as freshwater dissolves ions and gases in rocks and transports them to the oceans. However, it is thought that the concentration and composition of dissolved salts in the oceans has remained essentially unchanged for hundreds of millions of years. Explain how this stability is possible and explain why the hundreds of millions of years of leaching have not depleted the concentrations of such soluble elements as sodium and potassium in terrestrial rocks.

6.2 In this chapter the concept of steady-state processes is illustrated by an analogy of pouring orange juice into a glass filled with water.

(a) List some processes that occur in our homes, industry, or economy that are steady-state processes in the sense that the outputs change in response to changes in the inputs (or vice versa).

(b) How quickly does this response lead to a new steady state in the examples you have given, and how would changes in inputs or outputs in your example produce temporal changes equivalent to the changes in "concentration" of the "substance" in the "box" discussed in CC8?

TABLE 6.4 Crustal Abundance, Oceanic Residence Time, and Seawater Concentration of Several Elements

Element	*Crustal Abundance (%)*	*Oceanic Residence Time (years)*	*Concentration in Seawater ($mg \cdot kg^{-1}$)*
Na (sodium)	2.4	60,000,000	10,780
Cl (chlorine as Cl^{-1})	0.013	80,000,000	19,400
Mg (magnesium)	2.3	10,000,000	1,280
K (potassium)	2.1	6,000,000	399
S (sulfur as sulfate, $SO4_4^{2-}$)	0.026	9,000,000	898
Ca (calcium)	4.1	1,000,000	412
Mn (manganese)	0.5	7,000	0.00002
Pb (lead)	0.001	400	0.0000027
Fe (iron)	2.4	100	0.0003
Al (aluminum)	6.0	100	0.00003

Earth's crust. The effectiveness of removal is expressed by the **residence time** (CC8), which is a measure of the mean length of time an atom of the element spends in the oceans before being removed to the sediment.

The relationships of crustal abundance, residence time, and concentration in seawater for several elements are shown in Table 6.4. Note that the effectiveness of the removal processes, as expressed by residence time, is the principal determinant of concentration.

Constancy of Ocean Water Composition

Scientists believe that the proportion of each element in continental rocks, sediments, subducted sediments, and the atmosphere has remained approximately constant for more than 500 million years. Hence, the concentrations of dissolved constituents in seawater have similarly remained almost constant. Even for those **nutrient** elements, such as nitrogen and phosphorus, that vary substantially in concentration both seasonally and geographically (Chap. 14), seawater composition is constant when considered as an annual and regional average. Marine life prospers only when important nutrient elements are above critical concentrations and when toxic elements (including some of the nutrient elements) are below concentrations that are harmful. Thus, the evolution of life in the oceans has depended on the constancy of seawater composition (Chap. 14, CC14, CC18).

6.5 SALINITY

Many of the properties of water are modified by the presence of dissolved salts. The total quantity of dissolved salts in seawater is expressed as **salinity.** Until the early 1980s, salinity was expressed in grams of dissolved salts per kilogram of water or in parts per thousand, for which the symbol is ‰ (note that this is different from the percent symbol, %). The symbol ‰ is read as "per mil." Open-ocean seawater contains about 35 g of dissolved salts per kilogram of seawater and thus has a salinity of 35‰.

The original approach to measuring salinity was to evaporate water and weigh the salt residue. This tedious procedure was inaccurate because some dissolved ions, such as bromide and iodide, decomposed in the process and the elements were lost as gases. Various other methods have been used to determine salinity, including measuring the chloride concentration (which is closely related to total dissolved solids because seawater follows the principle of constant proportions, as discussed in the next section).

The most precise and widely used method of salinity determination is the measurement of **electrical conductivity.** Salinity is measured by comparison of the conductivity of two solutions, one of which has a precisely known salinity. Until the early 1980s, the comparison was made with a standard seawater whose salinity was determined precisely by a reference laboratory in Copenhagen, Denmark, and later in England. This method worked well but became very difficult as oceanography grew and the reference laboratory had to supply standard water samples to hundreds of laboratories worldwide. For this reason, and to improve the precision of salinity measurements, salinity has been redefined as a ratio of the electrical conductivity of the seawater to the electrical conductivity of a standard concentration of potassium chloride solution.

Because salinity is now defined as a ratio of electrical conductivities, it is no longer measured in parts per thousand but is expressed in "practical salinity units" (PSU). The average seawater salinity is now expressed as 35 without the ‰ symbol. Seawater with a salinity of 35 PSU does have a concentration of almost exactly 35 g of dissolved salts per kilogram. All salinity values in this text are stated without the ‰ symbol, as is consistent with currently accepted practice. However, the symbol is still used in some publications.

Very small salinity changes can significantly alter seawater **density.** Consequently, salinity often must be measured to ±0.001, and to achieve that precision, conductivity must be measured to ±1 part in 40,000. Such precision is readily achievable in the laboratory, and compact rapid-reading conductivity sensors achieve nearly that precision when mounted in the **CTD** probes described in Chapter 3.

6.6 DISSOLVED CHEMICALS IN SEAWATER

The chemicals dissolved in seawater include most elements, a variety of naturally occurring and human-made **radionuclides** (**radioisotopes**), and numerous organic compounds. Elements in solution are generally ionized. Many ions are compound ions, such as nitrate (NO_3^{2-}) and phosphate (PO_4^{3-}), in which atoms of one element are combined with atoms of other elements. Hence, although we talk about dissolved nitrogen or phosphorus, we generally use the term "dissolved constituents," not "dissolved elements." Dissolved constituents are designated major, minor, or trace according to their concentration. Trace constituents include organic compounds, radionuclides, and trace elements. The term "trace elements" is widely used, although it is not accurate, because dissolved trace elements are present in seawater as simple or compound ions. Dissolved gases are generally considered separately.

Concentrations of dissolved constituents of seawater are usually measured in parts per million ($mg \cdot kg^{-1}$) or parts per billion ($\mu g \cdot kg^{-1}$). One part per million is roughly equivalent to one teaspoonful mixed into 5000 liters of water, or enough water to fill more than 14,000 cans of soda. One part per billion is roughly equivalent to mixing one teaspoonful into 5,000,000 liters of water, enough water to fill about five Olympic-size swimming pools. Some dissolved trace metal and organic constituents of seawater occur at concentrations in the parts per trillion range. One part per trillion is equivalent to one cent in 10 million dollars or one second in 31,700 years.

Major Constituents

The major dissolved constituents have concentrations greater than 100 parts per million by weight. The six major constituents (**Table 6.5**) are chlorine, sodium, magnesium, sulfur (as sulfate), calcium, and potassium. They occur as the ions identified in **Table 6.5.** Together, these six ions constitute 99.28% of all dissolved salts in the oceans (**Fig. 6.5**), and sodium and chloride (the constituents of table salt) alone constitute more than 85%.

With the exception of calcium, which is used by marine organisms to build calcium carbonate hard parts of their bodies (Chap. 8), the major constituents are not utilized significantly in biological processes and do not interact readily with inorganic particles. Therefore, the primary process by which they are removed from the oceans is by the precipitation of salt deposits. This process occurs in shallow, partially enclosed **marginal seas** or embayments where the rate of removal of water by evaporation far exceeds its replacement by precipitation, river flow, and mixing with the open ocean (Chap. 8).

Inputs of major constituents from rivers and other sources are very small in comparison with the quantities in the oceans. Because concentrations of these constituents are not affected significantly by inputs or removal processes, their residence time is very long and the relative concentrations of these major constituents do not vary significantly. This principle of constant proportions is a cornerstone of chemical oceanography. The ratios of major constituent concentrations in seawater vary significantly only in enclosed seas, where evaporation leads to salt precipitation (Chap. 8), and in **estuaries,** where the ocean water is mixed with substantial quantities of river water. River water has a much more variable composition of the constituents.

Minor Constituents

The minor constituents of seawater include bromine, carbon, strontium, boron, silicon, and fluorine. They have concentrations between 1 part per million and 100 parts per million (**Table 6.5**, **Fig. 6.5**). Nitrogen and oxygen are not considered to be among the minor constituents,

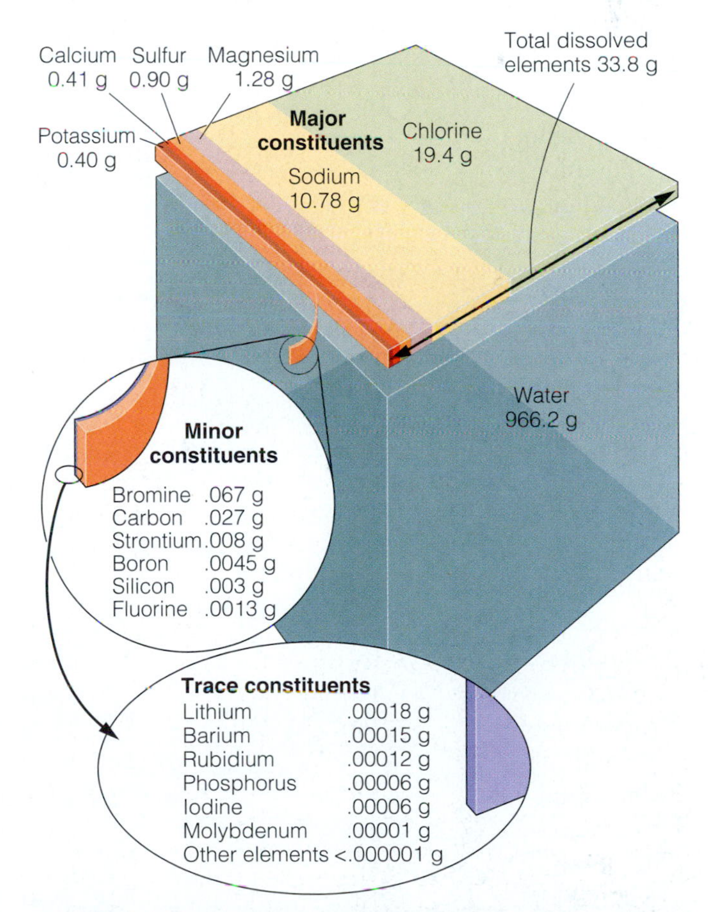

FIGURE 6.5 The dissolved elements in 1 kg of seawater. The total mass of dissolved elements is less than the total mass of dissolved salts (salinity) because the mass of dissolved salts includes the H^+ and OH^- ions associated with the dissolved species of some elements in seawater (see Table 6.5).

TABLE 6.5 Concentrations and Speciation of the Elements in Seawater (Salinity 35)

Element	*Chemical Symbol*	*Concentration* ($mg{\cdot}kg^{-1}$)	*Some Probable Dissolved Species*
Chlorine	Cl	1.94×10^4	Cl^-
Sodium	Na	1.078×10^4	Na^+
Magnesium	Mg	1.28×10^3	Mg^{2+}
Sulfur	S	8.98×10^2	SO_4^{2-}, $NaSO_4^-$
Calcium	Ca	4.12×10^2	Ca^{2+}
Potassium	K	3.99×10^2	K^+
Bromine	Br	67	Br^-
Carbon	C	27	HCO_3^-, CO_3^{2-}, CO_2
Nitrogen	N	8.3	N_2 gas, NO_3^-, NH_4
Strontium	Sr	7.8	Sr^{2+}
Boron	B	4.5	$B(OH)_3$, $B(OH)_4^-$, H_2BO_3
Oxygen	O	2.8	O_2 gas
Silicon	Si	2.8	$Si(OH)_4$
Fluorine	F	1.3	F^-, MgF^+
Argon	Ar	0.62	Ar gas
Lithium	Li	0.18	Li^+
Rubidium	Rb	0.12	Rb^+
Phosphorus	P	6.2×10^{-2}	HPO_4^{2-}, PO_4^{3-}, H_2PO_4
Iodine	I	5.8×10^{-2}	IO^{3-}, I_2
Barium	Ba	1.5×10^{-2}	Ba^{2+}
Molybdenum	Mo	1×10^{-2}	MoO_4^{2-}
Uranium	U	3.2×10^{-3}	$UO_2(CO_3)_2^{4-}$
Vanadium	V	2.0×10^{-3}	$H_2VO_4^-$, HVO_4^{2-}
Arsenic	As	1.2×10^{-3}	$HAsO_4^{2-}$, $H_2AsO_4^-$
Nickel	Ni	4.8×10^{-4}	Ni^{2+}
Zinc	Zn	3.5×10^{-4}	$ZnOH^+$, Zn^{2+}, $ZnCO_3$
Krypton	Kr	3.1×10^{-4}	Kr gas
Cesium	Cs	3×10^{-4}	Cs^+
Chromium	Cr	2.1×10^{-4}	$Cr(OH)_3$, CrO_4^{2-}
Antimony	Sb	2.0×10^{-4}	$Sb(OH)_6^-$
Neon	Ne	1.6×10^{-4}	Ne gas
Copper	Cu	1.5×10^{-4}	$CuCO_3$, $CuOH^+$
Selenium	Se	1×10^{-4}	SeO_3^{2-}
Cadmium	Cd	7×10^{-5}	$CdCl_2$
Xenon	Xe	6.6×10^{-5}	Xe gas
Aluminium	Al	3×10^{-5}	$Al(OH)_4^-$
Iron	Fe	3×10^{-5}	$Fe(OH)_2^+$, $Fe(OH)_4$
Manganese	Mn	2×10^{-5}	Mn^{2+}, $MnCl^+$
Yttrium	Y	1.7×10^{-5}	$Y(OH)_3$
Zirconium	Zr	1.5×10^{-5}	$Zr(OH)_4$
Thallium	Tl	1.3×10^{-5}	Tl^+
Tungsten	W	1×10^{-5}	WO_4^{2-}
Niobium	Nb	1×10^{-5}	Not known
Rhenium	Re	7.8×10^{-6}	ReO_4^-

TABLE 6.5 (Continued)

Element	*Chemical Symbol*	*Concentration* ($mg \cdot kg^{-1}$)	*Some Probable Dissolved Species*
Helium	He	7.6×10^{-6}	He gas
Titanium	Ti	6.5×10^{-6}	$Ti(OH)_4$
Lanthanum	La	5.6×10^{-6}	$La(OH)_3$
Germanium	Ge	5.5×10^{-6}	$Ge(OH)_4$
Hafnium	Hf	3.4×10^{-6}	Not known
Neodymium	Nd	3.3×10^{-6}	$Nd(OH)_3$
Lead	Pb	2.7×10^{-6}	$PbCO_3$, $Pb(CO_3)_2^{2-}$
Tantalum	Ta	$<2.5 \times 10^{-6}$	Not known
Silver	Ag	2×10^{-6}	$AgCl_2^-$
Cobalt	Co	1.2×10^{-6}	Co^{2+}
Gallium	Ga	1.2×10^{-6}	$Ga(OH)_4$
Erbium	Er	1.2×10^{-6}	$Er(OH)_3$
Ytterbium	Yb	1.2×10^{-6}	$Yb(OH)_3$
Dysprosium	Dy	1.1×10^{-6}	$Dy(OH)_3$
Gadolinium	Gd	9×10^{-7}	$Gd(OH)_3$
Praseodymium	Pr	7×10^{-7}	$Pr(OH)_3$
Scandium	Sc	7×10^{-7}	$Sc(OH)_3$
Cerium	Ce	7×10^{-7}	$Ce(OH)_3$
Samarium	Sm	5.7×10^{-7}	$Sm(OH)_3$
Tin	Sn	5×10^{-7}	$SnO(OH)_3^-$
Holmium	Ho	3.6×10^{-7}	$Ho(OH)_3$
Lutetium	Lu	2.3×10^{-7}	$Lu(OH)_3$
Beryllium	Be	2.1×10^{-7}	$BeOH^+$
Thulium	Tm	2×10^{-7}	$Tm(OH)_3$
Europium	Eu	1.7×10^{-7}	$Eu(OH)_3$
Terbium	Tb	1.7×10^{-7}	$Tb(OH)_3$
Mercury	Hg	1.4×10^{-7}	$HgCl_4^{2-}$, $HgCl_2$
Indium	In	1×10^{-7}	$In(OH)^{2+}$
Rhodium	Rh	8×10^{-8}	Not known
Palladium	Pd	6×10^{-8}	Not known
Platinum	Pt	5×10^{-8}	Not known
Tellurium	Te	5×10^{-8}	$Te(OH)_3$
Bismuth	Bi	3×10^{-8}	BiO^+, $Bi(OH)^{2+}$
Thorium	Th	2×10^{-8}	$Th(OH)_4$
Gold	Au	2×10^{-8}	$AuCl_2^-$
Ruthenium	Ru	5×10^{-9}	Not known
Osmium	Os	2×10^{-9}	Not known
Radium	Ra	1.3×10^{-10}	Ra^{2+}
Iridium	Ir	1.3×10^{-10}	Not known
Protactinium	Pa	5×10^{-11}	Not known
Radon	Rn	6×10^{-16}	Rn gas

Note: Even for the more abundant constituents, concentration may vary slightly. For the rarer elements, the listed concentrations are uncertain and may be revised as analytical methods improve.

(a)

(b)

FIGURE 6.6 There are many different inputs of trace elements to the oceans. Two of the most important are inputs from volcanoes and rivers. (a) Inputs from volcanoes can occur through atmospheric dust clouds formed by explosive eruptions such as the Mt. St. Helens eruption shown in Figure 4.12a or directly from the many active undersea and other volcanoes whose lava flows directly into the sea. This photograph shows lava from the Kilauea volcano on Hawaii as it flows directly into the ocean on the south shore of the island. (b) Trace elements are also introduced to the oceans by river discharges, both in solution and associated with suspended particles. This photograph shows a turbid plume of water discharged into the Arctic Ocean by the MacKenzie river, which flows through the Northwest Territories of Canada.

because they are present in seawater primarily as dissolved molecular oxygen (O_2) and nitrogen (N_2) gases. Dissolved gases are considered separately later in the chapter. Together, the six major and six minor constituents constitute more than 99.6% of all the dissolved solids. Several minor constituents, notably carbon and silicon, are utilized extensively in biological processes. Therefore, the concentrations of these constituents in ocean waters are variable, changing in response to uptake by marine organisms and release during decay of organic matter. Such variations are discussed in Chapter 14.

Trace Elements

Other than the 12 major and minor constituents and the dissolved gases, all the elements listed in **Table 6.5** have concentrations in seawater of less than 1 part per million. Together, these trace elements constitute only about 0.4% of the total dissolved solids in seawater. Most trace elements are used extensively in biological processes or attach easily to particles that remove them from seawater. Many are introduced in significant quantities by hydrothermal vents, undersea volcanoes, decomposition of organic matter, atmospheric sources such as volcanic gases, river outflows, and release from seafloor sediment (**Fig. 6.6**). Concentrations of the various trace constituents vary substantially in different parts of the ocean in response to variations in the input and removal processes.

Many trace elements are essential minerals for marine life, others are toxic, and many are essential for marine life at low concentrations but toxic at higher concentrations (**CC18**). Measurements of the concentrations of trace metals and observations of their **spatial** and temporal variations in the oceans are important to marine biological and pollution studies.

Many trace elements that are essential (e.g., iron, zinc, and copper) or toxic (e.g., lead and mercury) have seawater concentrations of about one part per billion (10^9) down to less than one part in 10^{13} (**Table 6.5**). Marine chemists have great difficulty measuring such extremely small concentrations. For many elements, the amount that dissolves from surfaces of samplers and sample bottles or is contributed by dust in laboratory air can be greater than the quantity in the water sample itself.

Radionuclides

A variety of naturally occurring and human-made radionuclides are present in seawater at extremely low concentrations (**CC7**). **Radioactivity** emanating from individual **isotopes** can be measured at exceedingly low levels, and sample contamination is generally a less critical problem than it is for trace metals. Therefore, determining even extremely small concentrations of radionuclides in seawater is relatively easy. In addition, many radionuclides are introduced to the oceans at known locations, and thus radionuclides can be used to trace the movements of ocean water (Chap. 10).

Radionuclides behave in biogeochemical cycles in ways that are virtually identical to those of the stable (nonradioactive) isotopes of their elements. Therefore, radionuclide distributions in seawater, sediments, and marine organisms, and movements of radionuclides among them, are used to infer movements of stable isotopes through the marine biogeosphere.

Organic Compounds

Thousands of different dissolved organic compounds are present in seawater. They include naturally occurring compounds such as proteins, carbohydrates, **lipids, amino acids,** vitamins, and petroleum hydrocarbons, as well as synthetic contaminants, including **DDT** and **PCBs** (polychlorinated biphenyls). The number of organic compounds is so large that probably less than 1% of them have been identified. Most organic compounds are very difficult to study because they are present in the parts per trillion concentration range (one teaspoonful in 5000 Olympic-size swimming pools). However, we believe that marine plants and perhaps some animals cannot grow successfully without dissolved compounds, such as vitamins (Chaps. 14, 15). We also know that certain organic compounds are highly toxic, even at concentrations in the parts per trillion range.

Most organic compounds in seawater are naturally occurring compounds. Many are produced by marine organisms and released to seawater, either through **excretions** (similar to urine) or by being dissolved after the death and decomposition of the organism. Organic compounds are also transported from land to the ocean in rivers and through the atmosphere. In some areas, especially where runoff from **mangrove** swamps or **salt marshes** carries large quantities of organic matter, the concentrations of colored dissolved organic compounds (sometimes called "gelbstoff") are high enough to make the water appear brownish yellow.

Some dissolved organic compounds are removed from solution by attachment to particles that sink to the seafloor. However, most such compounds are taken up by marine organisms or decomposed to their inorganic constituents in the water column, primarily by **bacteria** (Chap. 14).

Dissolved Gases

Gases are free to move between the atmosphere and the oceans at the ocean surface. The net direction of the exchange is determined by the **saturation solubility** and concentration of the gas in seawater. The saturation solubility is the maximum amount of the gas that can be dissolved in water at a specific temperature, salinity, and pressure. If seawater is undersaturated, a net transfer of gas molecules into the water occurs. If seawater is oversaturated, the net transfer is from the water into the atmosphere.

The atmosphere is composed primarily of nitrogen (78%) and oxygen (21%) and contains several other minor gases. Carbon dioxide constitutes about 0.037% of all atmospheric gases. The distribution of gases dissolved in ocean waters is very different (**Table 6.6**). The oceans have proportionally more oxygen and less nitrogen than

TABLE 6.6 Distribution of Gases in the Atmosphere and Dissolved in Seawater

	Percentage of Gas Phase by Volume		
Gas	*Atmosphere*	*Surface Oceans*	*Total Oceans*
Nitrogen (N_2)	78	48	11
Oxygen (O_2)	21	36	6
Carbon dioxide[a]	0.037	15	83

[a]CO_2 in the atmosphere, CO_2 plus HCO_3^- plus CO_3^{2-} in the oceans.

the atmosphere because of differences in the saturation solubility of these gases. In addition, the ratios of carbon dioxide concentration to oxygen and nitrogen concentrations are much higher in seawater than in the atmosphere because carbon dioxide reacts with water in a complicated way to produce highly soluble carbonate (CO_3^{2-}) and bicarbonate (HCO_3^-) anions.

Oxygen and Carbon Dioxide. Gases can be exchanged between the ocean and the atmosphere only at the ocean surface. In the water column beneath the ocean surface, the proportions of dissolved gases are changed primarily by biochemical processes. In the shallow **photic zone** where light penetrates, carbon dioxide is consumed and oxygen released during **photosynthesis.** However, in the much larger volume of deep-ocean waters where no photosynthesis occurs, the dominant process affecting dissolved gas concentrations is the consumption of oxygen through **respiration** and decomposition. The excess carbon dioxide produced by respiration and decomposition can escape to the atmosphere only at the ocean surface, and there is no mechanism for resupplying dissolved oxygen to the water below the photic zone. Therefore, deep-ocean water is depleted of oxygen and stores large quantities of carbon dioxide (Table 6.6).

The total quantity of carbon dioxide dissolved in the oceans is about 70 times as large as the total in the atmosphere. The processes that control oxygen and carbon dioxide concentrations at different depths and locations within the oceans are discussed in more detail in Chapter 14. The role of the oceans in absorbing and storing carbon dioxide is critical to the fate of carbon dioxide that has been, or will be, released by industrialized civilization, and to the severity of global climate change due to the enhanced **greenhouse effect** (CC9).

The saturation solubility of gases varies with pressure, temperature, and salinity. It increases with increasing pressure and decreasing temperature. Therefore, seawater at depth in the oceans can dissolve much higher concentrations of gases than surface seawater can. There is continuous movement of water from ocean surface layers to the deep layers and eventually back to the surface (Chap. 10). As water moves through the ocean depths, concentrations of dissolved gases are changed by biochemical processes and by mixing with other **water masses** that have different concentrations of the gases. Geological processes, including the release of gases from undersea volcanoes and from decaying organic matter in sediments, can also change the concentrations of some gases, but those processes are generally of minor significance.

When water that is saturated with carbon dioxide sinks below the surface layer, gases can no longer be exchanged with the atmosphere. However, carbon dioxide is released into deep-ocean waters by respiration and by decay of organic matter. The added gas remains **dissolved** because the saturation solubility is increased by the higher pressures.

The increase in **solubility** caused by increased pressure is what keeps carbon dioxide dissolved in carbonated sodas. The carbon dioxide is dissolved in the soda under increased pressure at the bottling plant and then sealed in its container at the higher pressure. When the container is opened, the internal pressure is released and the carbon dioxide bubbles out because the concentration exceeds the saturation solubility at the lower pressure.

The saturation solubility of gases generally increases as temperature decreases and is generally lower at ocean water salinities than in pure water (Fig. 6.7). Oxygen concentrations are higher in cold surface waters near the polar regions than in tropical waters. In tropical waters, the oxygen concentration is low enough that, under certain circumstances, it can be inadequate for the respiration needs of some marine species. For the same reason, tropical waters are more vulnerable to marine pollution by oxygen-consuming waste materials, such as sewage (Chap. 19).

Other Gases. In addition to the major atmospheric gases, several other gases are present in seawater (Table 6.7). With the exception of sulfur dioxide, these gases are produced primarily by marine organisms, so surface waters are oversaturated and net movement of the gases is into the atmosphere (Table 6.7). The quantities of such gases supplied to the atmosphere by the oceans are relatively small in comparison with those from other sources. However, the ocean concentrations of methane, for example, must be taken into account in global climate change studies because atmospheric methane contributes significantly to the greenhouse effect (CC9). Atmospheric sulfur dioxide comes primarily from **fossil fuel** burning, industrial processes, and volcanoes. Sulfur dioxide can be converted to sulfuric acid, the principal component of **acid rain.** The oceans act as a sink for this air contaminant by absorbing sulfur dioxide and the sulfate ions present in the runoff from acid rain.

pH and Buffering

The acidic or **alkaline** property of water is expressed as **pH,** which is a measure of the concentration of hydrogen ions (H^+). The pH increases as the hydrogen ion concentration decreases, and it is measured on a logarithmic scale of zero to 14: pH 0 is the most acidic, pH 14 is the most alkaline, and pH 7 is neutral. Pure water (free of dissolved carbon dioxide) is neutral, pH 7. Seawater normally is about pH 8, mildly alkaline. The near-neutral pH of natural waters is very important to aquatic biology. For example, persistent acid rain, which can be about pH 5, can severely damage or destroy aquatic life in lakes by reducing the natural pH of the lake water. In addition, the numerous marine species with calcium carbonate hard body parts cannot construct these parts if the water is even mildly acidic.

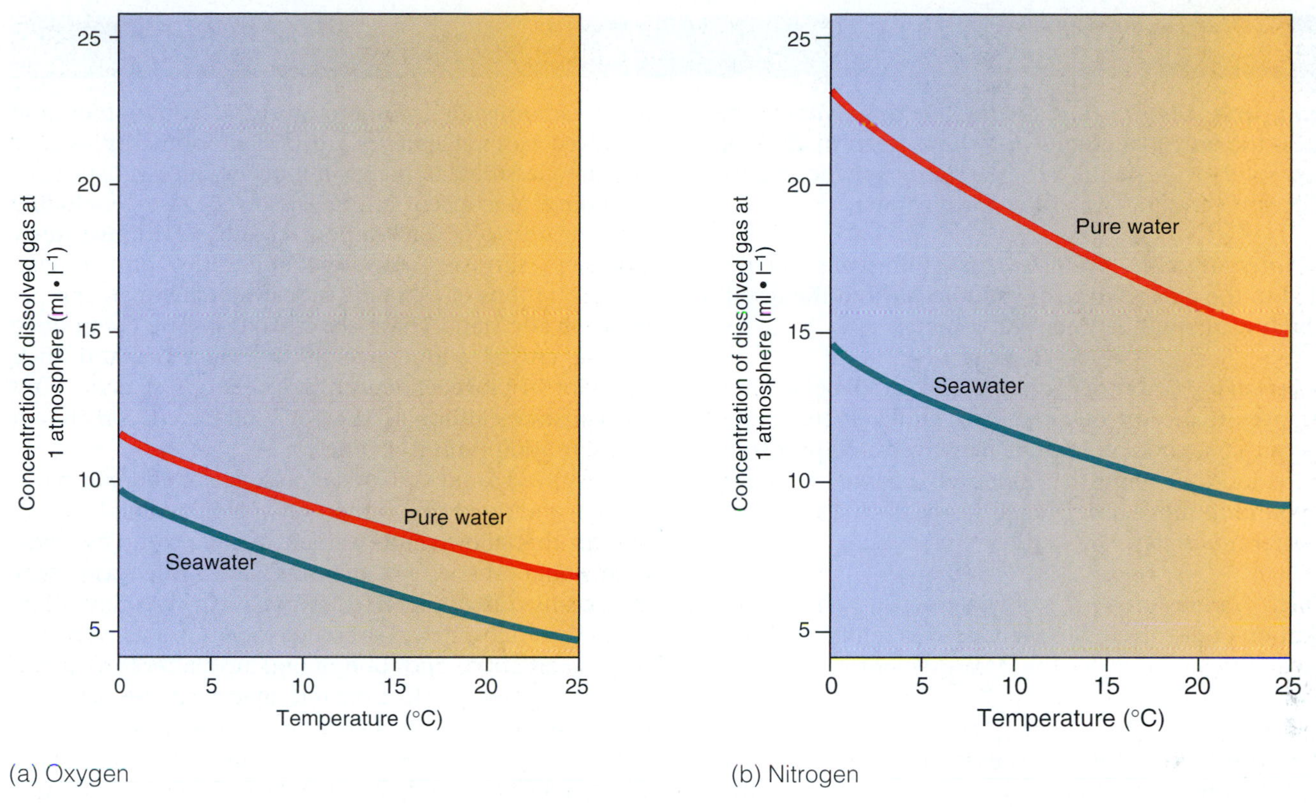

FIGURE 6.7 Solubility of oxygen (a) and nitrogen (b) in pure water and seawater. Note that the solubility of each gas is reduced as temperature increases and is lower at seawater salinity than in freshwater.

Seawater pH is buffered (maintained within a narrow range of pH 7.5 to 8.1) through the reactions of dissolved carbon dioxide. Dissolved carbon dioxide combines with water to form carbonic acid (H_2CO_3). The carbonic acid partially dissociates (separates into ions) to form a hydrogen ion and a bicarbonate ion (HCO_3^-), or two hydrogen ions and a carbonate ion (CO_3^{2-}). Carbon dioxide, carbonic acid, bicarbonate, carbonate, and hydrogen ions coexist in equilibrium in seawater:

$$CO_2 \text{ (gas)} + H_2O \rightleftharpoons H_2CO_3 \rightleftharpoons H^+ + HCO_3^- \rightleftharpoons 2H^+ + CO_3^{2-}$$

The addition of acid to seawater increases the number of hydrogen ions, which reduces the pH of the seawater and forces the equilibrium to shift so that less carbonate and bicarbonate are present. This shift reduces the number of hydrogen ions present, which offsets the reduction of pH. The opposite shift occurs if alkali is added to seawater. Such buffering capacity partially protects the ocean waters from pH changes that otherwise might result from acid rain or from acidic or alkaline industrial **effluents.**

TABLE 6.7 Estimated Flux of Gases between Oceans and Atomosphere

Gas	*Total Transfer* ($g \cdot yr^{-1}$)	*Direction of Net Transfer*
Sulfur dioxide (SO_2)	1.5×10^{14}	Atmosphere to ocean
Nitrous oxide (N_2O)	1.2×10^{14}	Ocean to atmosphere
Carbon monoxide (CO)	4.3×10^{13}	Ocean to atmosphere
Methane (CH_4)	3.2×10^{12}	Ocean to atmosphere
Methyl iodide (CH_3I)	2.7×10^{11}	Ocean to atmosphere
Dimethyl sulfide $(CH_3)_2S$	4.0×10^{13}	Ocean to atmosphere

CHAPTER SUMMARY

Origins and Distribution of the Earth's Water. The Earth is the only planet known to have water in all three states: solid (ice), liquid, and gas (atmospheric vapor). The oceans contain 97% of the Earth's water.

The Water Molecule. Water has many unusual properties related to the polar nature of its molecule and the resulting hydrogen bond between water molecules.

The Dissolving Power of Water. Water can dissolve more substances than any other known liquid. Water molecules can surround and hydrate negatively and positively charged ions, allowing the oppositely charged ions of a compound to move independently within the surrounding water molecules.

Salinity. Various properties of water depend on the concentration of dissolved salts, expressed as salinity. Salinity is usually determined by the measurement of electrical conductivity.

Composition of Seawater. Seawater is a complex solution containing almost all naturally occurring elements, many radionuclides, numerous organic compounds, and dissolved gases. Elements are present primarily as ions and are classified as major, minor, or trace constituents according to concentration. Dissolved gases, including oxygen, nitrogen, and carbon dioxide, can move freely between surface seawater and the atmosphere. Concentrations of elements in seawater are in an approximate steady state. These concentrations are determined by the crustal abundance of the elements and their behavior in biogeochemical cycles. Ratios of major ions are constant, although the total dissolved constituent concentration (salinity) varies.

Dissolved carbon dioxide is present as carbonate, bicarbonate, and molecular carbon dioxide in a complex equilibrium that allows more carbon dioxide to be dissolved than would otherwise be the case. This is one reason why the oceans contain about 70 times as much carbon dioxide as the atmosphere. Carbon dioxide is produced by respiration and decomposition of organic matter throughout the ocean depths. Most organic matter in the oceans is produced through photosynthesis, which uses carbon dioxide and releases oxygen. Photosynthesis requires light energy and occurs only in upper layers of ocean waters.

KEY TERMS

You should recognize and understand the meaning of all terms that are in boldface type in the text. All those terms are defined in the Glossary. The following are some less familiar key scientific terms that are used in this chapter and that are essential to know and be able to use in classroom discussions or exam answers.

adsorbed (p. 136)
alkaline (p. 144)
anion (p. 136)
biogeochemical cycle (p. 136)
cation (p. 136)
contaminant (p. 137)
covalent bond (p. 133)
density (p. 138)
detritus (p. 136)
electrical conductivity (p. 138)
eroded (p. 137)
excretion (p. 143)
groundwater (p. 132)
hard parts (p. 136)
hydration (p. 135)
hydrogen bond (p. 133)
hydrothermal vent (p. 137)
ion (p. 133)
ionic bond (p. 133)
isotope (p. 143)
nutrient (p. 138)
pH (p. 144)
photic zone (p. 144)
photosynthesis (p. 144)
radionuclide (radioisotope) (p. 139)
residence time (p. 138)
respiration (p. 144)
runoff (p. 136)
salinity (p. 138)
salt marsh (p. 143)
saturation solubility (p. 143)
sediment (p. 132)
solubility (p. 144)
spatial (p. 142)
steady state (p. 137)
van der Waals force (p. 133)
water mass (p. 144)
weathered (p. 136)

STUDY QUESTIONS

1. What are the principal sources and sinks of elements dissolved in seawater? Explain why the totals of these sources and sinks must be approximately balanced.
2. Why are some elements present in high concentrations in seawater, whereas others are present in much lower concentrations?
3. Why is the solubility of carbon dioxide in seawater so high?

CRITICAL CONCEPTS REMINDER

CC7 **Radioactivity and Age Dating** (p. 143). Seawater contains naturally occurring and man-made radioisotopes of many elements at very low concentrations. These are useful for tracing movements of the elements through biogeochemical cycles. To read **CC7** go to page 18CC.

CC8 **Residence Time** (p. 137, 138). The residence time of an element in seawater is the average length of time atoms of the element spend in the oceans and it depends, to a large extent, on the rate of processes that remove them from solution. Residence time is a major factor in determining the concentration of elements in seawater. To read **CC8** go to page 19CC.

CC9 **The Global Greenhouse Effect** (p. 144). The oceans play a major part in studies of the greenhouse effect because the oceans store large amounts of carbon dioxide and are a source of other greenhouse gases including methane. To read **CC9** go to page 22CC.

CC14 **Photosynthesis, Light, and Nutrients** (p. 138). Chemosynthesis and photosynthesis, the processes by which simple chemical compounds are made into the organic compounds of living organisms, depend on the availability of a number of dissolved nutrient elements. To read **CC14** go to page 46CC.

CC18 **Toxicity** (pp. 138, 142). Many dissolved constituents of seawater become toxic to marine life at concentrations above their natural levels in seawater. A number of these constituents are nutrients that can limit primary production if their concentrations are too low. Thus, life depends on the relatively invariable concentrations of dissolved chemicals in the oceans. To read **CC18** go to page 54CC.

CHAPTER 7

CHAPTER OUTLINE

CRITICAL CONCEPTS USED IN THIS CHAPTER

Explorers from another star system would probably call Earth "The Water Planet" as more than 70 percent of our blue globe is covered by ocean and seas. A further 3 percent of the Earth is permanently covered by ice sheets and glaciers, and about the same amount of area is seasonally covered by snow. Our atmosphere also contains large amounts of water vapor as shown by the swirling clouds that can be seen from space. The existence of liquid water is essential to life as we know it and the existence of water in all three forms water vapor, liquid water, and ice is critical to the maintenance of our planet's climate.

Physical Properties of Water and Seawater

All of us grow up being aware that water is important to our lives. We know that if there were no water to drink, we could not survive. However, most of us have no idea about just how many ways water affects our lives. Most of us are totally unaware that water is a substance with many properties that are unique among all chemical substances and that it is these properties that make life as we know it possible on the Earth. Perhaps one of the least-known special properties of water is its extraordinary ability to conduct sound with very little absorption. Did you know that some species of whales sing underwater, probably as a means of communicating with each other? Whales, and possibly many other ocean animals, are able to communicate over distances of hundreds, perhaps even thousands, of kilometers through sound and have probably been doing so for millions of years. Compare that with humans, who could not communicate over long distances by sound or by any other means until Samuel Morse perfected the first operational telegraph and the age of electronic communication began in 1835, less than 200 years ago.

Chapter 6 described the polar nature of the water molecule and explained how this property gives water the ability to dissolve more chemical substances than any other solvent. The polar nature of the water molecule also gives water a number of unique physical properties (Table 6.2).

A basic understanding of the physical properties of water and of how these properties are modified by changes in temperature, pressure, and the concentration of dissolved salts is fundamental to ocean sciences. Water's unique physical properties (Table 6.2) are critically important to processes discussed in many other chapters in this text, especially ocean–atmosphere interactions (Chap. 9) and the circulation of water in the oceans (Chap. 10). The behavior of sound and of **electromagnetic radiation**, including visible light, as they pass through seawater is important to ocean life (Chap. 14) and contributes to the difficulties faced by oceanographers in studying the oceans (Chaps. 2, 3). Light and sound transmission also influence the survival and reproduction strategies of marine **species** (Chap. 17). In this chapter we will examine the most important physical properties of water and some of the ways that they influence the Earth and our **environment.**

7.1 HEAT PROPERTIES OF WATER

Of the properties of water, perhaps none are more important than the unique responses of water, ice, and water vapor to the application and removal of heat.

Heat Energy and Phase Changes

All substances can exist in three different phases: solid, liquid, or gas. In a solid, attractive forces between molecules (**van der Waals forces** and **hydrogen bonds**, if present) are strong enough to ensure that the molecules stay firmly fixed in place relative to each other, even though the molecules (or atoms in a pure element) vibrate. Heating the solid increases the strength of these vibrations. When the heat energy of each molecule is sufficient to overcome most of the attractive forces, the solid melts and becomes liquid. In a liquid, the molecules have enough energy to vibrate, rotate, and translate (temporarily move about in relation to each other). However, they do not have enough energy to escape completely from the attractive forces of their neighbors. If more heat is added to a liquid, each molecule eventually has enough energy to break free of the attractive forces. The compound then becomes a gas, in which each molecule is free to move about by itself. If heat is removed, the process is reversed; gas becomes liquid, and then solid, as heat is progressively lost.

Molecules of different chemical compounds have van der Waals forces of different strengths. As a result, each compound requires a certain characteristic quantity of heat energy to convert from solid to liquid or from liquid to gas. The stronger the attractive force between the molecules, the more heat energy each molecule must have to break free of this force. Therefore, the temperature at which a solid melts (the melting, or freezing, point) and the temperature at which a liquid vaporizes (the boiling, or condensation, point) increase as the van der Waals attractive force increases.

Freezing and Boiling Points

To convert solid water (ice) to liquid water and liquid water to gaseous water (water vapor), the heat energy supplied must overcome both the van der Waals attractive force and the much stronger attractive force of the hydrogen bond. Therefore, the amount of heat energy that each molecule of water must have to become free to rotate and move in relation to adjacent molecules (that is, to change from ice to water) is much greater than it would be if the hydrogen bond were not present. Similarly, the amount of heat energy that each water molecule must have to free itself completely and enter the gaseous state is much greater than it would be without the hydrogen bond.

The extra energy needed to break the hydrogen bond causes water's freezing point and boiling point temperatures to be anomalously high. If there were no hydrogen bond, water would freeze at about –90°C and boil at about –70°C, and all water on the Earth would be gaseous. There would be no oceans or life as we know it. Comparison of the boiling and freezing points of water with those of other hydrogen compounds that are formed with elements that, like oxygen, have two electrons short of a full outer electron shell illustrates the anomalous nature of water (Fig. 7.1). Molecules in which hydrogen is combined with elements whose atoms are similar in size to oxygen but have a different number of outer-shell electrons (hydrogen fluoride, HF; ammonia, NH_3) are also polar. They, too, have hydrogen bonds and anomalous melting and boiling points (Fig. 7.1).

As discussed later in this chapter, adding salt to water raises the boiling point and lowers the freezing point.

Heat Capacity and Latent Heat

In addition to the freezing and boiling points, the numerical values of three other related heat properties of water are anomalously high because of the hydrogen bond: the **heat capacity**, the **latent heat of fusion**, and the **latent heat of vaporization**. These properties express the quantity of heat per specified quantity of a substance needed to raise the temperature, to convert solid to liquid, and to convert liquid to gas, respectively. Heat is a form of energy, and as such, it is quantified by a unit of energy. The SI unit of energy is the **joule** (abbreviated J). However, this SI unit is not yet universally used. Many students will be more familiar with the **calorie** (abbrevi-

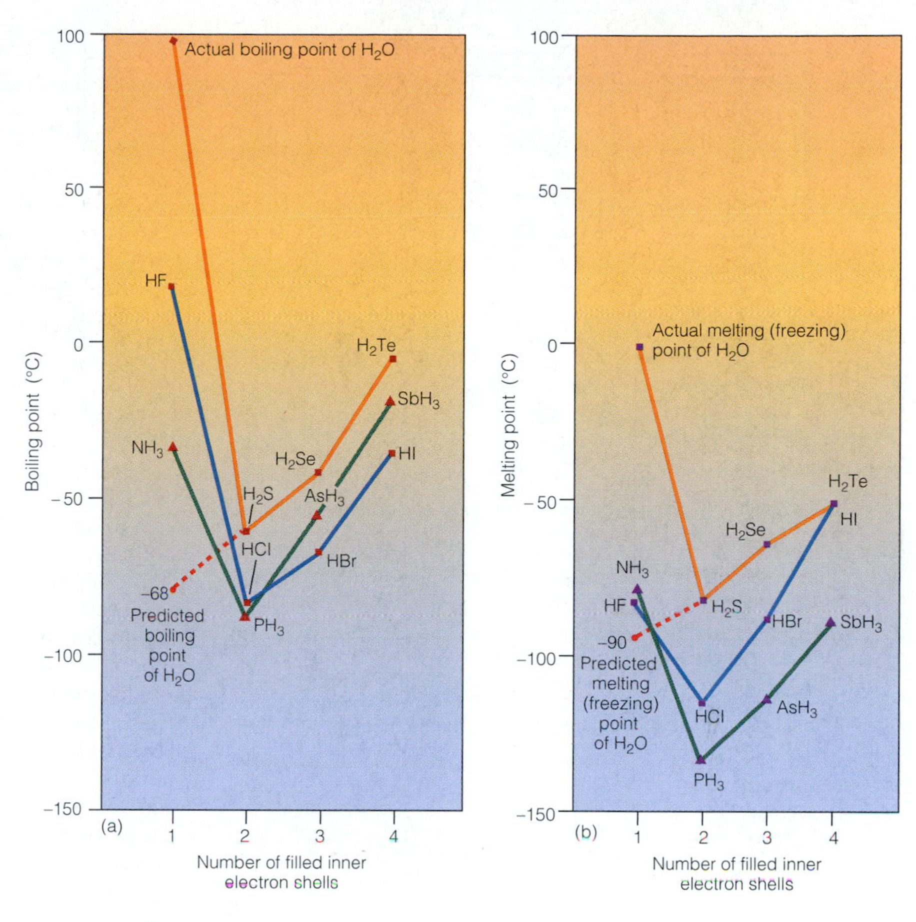

FIGURE 7.1 Freezing and boiling points of hydrogen compounds. The lines on the chart connect compounds of hydrogen with elements that have the same number of electrons in their outer shell but a different number of filled inner shells. If there were no hydrogen bond, both the boiling point and the melting point of compounds would be expected to increase progressively along these lines as the number of filled shells increased. However, water (H_2O), hydrogen fluoride (HF), and ammonia (NH_3) all have strong hydrogen bonds between their polar molecules, whereas there is no appreciable hydrogen bond in the compounds of hydrogen that are formed with sulfur (S), selenium (Se), tellurium (Te), chlorine (Cl), bromine (Br), iodine (I), phosphorus (P), arsenic (As), and antimony (Sb). In this latter group of compounds, the hydrogen bond is virtually absent because the atom combined with the hydrogen is much larger, and, therefore, the polarity of the molecules is much less. Since the hydrogen bond must be broken to convert solid to liquid or liquid to gas, the boiling points (a) and melting (freezing) points (b) of water, hydrogen fluoride, and ammonia are much higher than would be expected if they were not held by hydrogen bonds.

ated cal) as the unit of energy. In this text, we will use the calorie, and we will also identify the equivalent value in joules, since the joule will become universally accepted and used within the next few years.

To understand the concepts of heat capacity and latent heat, we can consider the sequence of events that occurs when we add heat to ice (**Fig. 7.2**). Within the solid ice, the individual molecules vibrate, but the vibrations are suppressed by the hydrogen bonds. Adding heat energy to the ice increases the intensity of vibration of each water molecule and increases the temperature of the ice. The temperature of 1 g of ice is increased by 1°C when approximately 0.5 calorie (2.1 J) is added. As heat continues to be added, the temperature continues to rise until the ice reaches its melting point. Heat that raises (or lowers) the temperature of a substance to which it is added (or removed) is called **sensible heat**. The quantity of heat needed to increase the temperature of a specified quantity of a substance is called the "heat capacity." For ice, the heat capacity is approximately 0.5 $cal \cdot g^{-1} \cdot °C^{-1}$ (2.1 $J \cdot g^{-1} \cdot °C^{-1}$). That is, one-half of a calorie (2.1 J) is needed to raise the temperature of 1 gram of ice by 1°C.

After the ice has reached the melting point temperature, the temperature of the solid–liquid (ice–water) mixture does not increase again until an additional 80 calories (334 J) of heat energy has been added for each gram of ice (**Fig. 7.2b,c**). That is, as heat is added, the ice is progressively converted to water. Heat added to a substance that does not raise its temperature but instead changes its state (solid to liquid or liquid to gas) is called "latent" heat. The heat added to melt a specified quantity of ice is the latent heat of fusion (melting). The term *latent* is used because the heat is stored in the molecules of the liquid water and is released when the water is refrozen (fused). Hence, the latent heat of fusion of ice is 80 $cal \cdot g^{-1}$ (334 $J \cdot g^{-1}$).

If we continue to add heat after all the ice is converted to water, the heat is again taken up as sensible heat and the temperature of the water rises. The heat capacity of water is about twice that of ice, or 1 $cal \cdot g^{-1} \cdot °C^{-1}$. That is, 1 calorie (4.2 J) of heat energy must be added to raise the temperature of 1 gram of water by 1°C.

After 100 $cal \cdot g^{-1}$ (418 $J \cdot g^{-1}$) of heat energy has been added to the water (starting as melted ice water at 0°C), the water temperature reaches the boiling point. With the continued addition of heat, the temperature does not rise again until an additional 540 $cal \cdot g^{-1}$ (2,260 $J \cdot g^{-1}$) of heat energy has been added and all the water has been converted to a gas. Hence, the latent heat of vaporization of water is 540 $cal \cdot g^{-1}$ (2,260 $J \cdot g^{-1}$). Finally, if we add more heat to the gaseous water (water vapor), its temperature rises by about 1°C for each additional 0.5 $cal \cdot g^{-1}$ (2.1 $J \cdot g^{-1}$), so the heat capacity of water vapor is about 0.5 $cal \cdot g^{-1} \cdot °C^{-1}$ (2.1 $J \cdot g^{-1} \cdot °C^{-1}$).

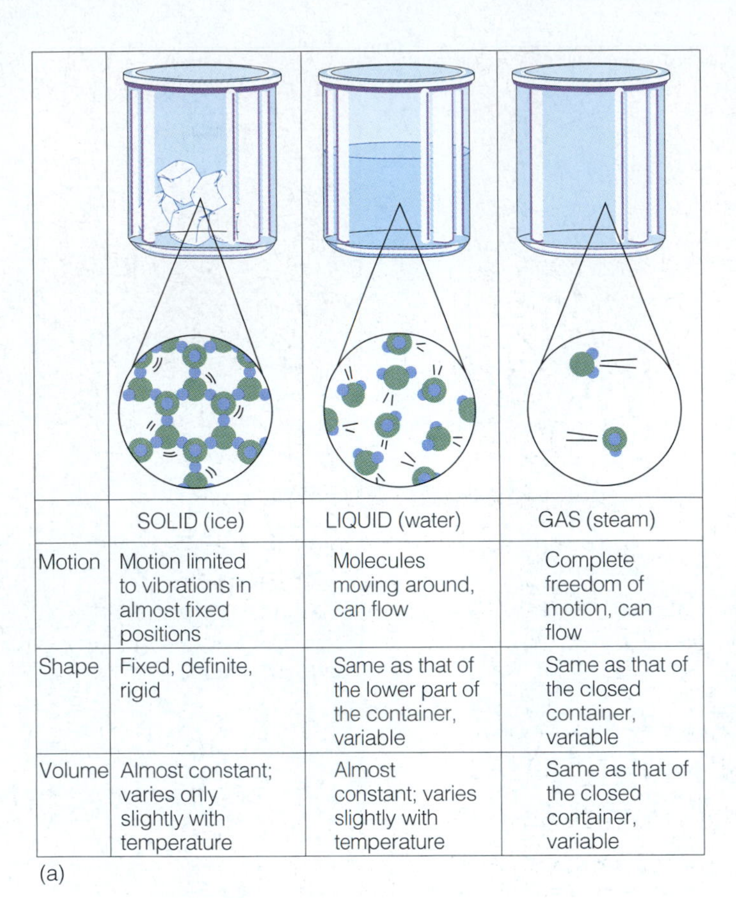

	SOLID (ice)	LIQUID (water)	GAS (steam)
Motion	Motion limited to vibrations in almost fixed positions	Molecules moving around, can flow	Complete freedom of motion, can flow
Shape	Fixed, definite, rigid	Same as that of the lower part of the container, variable	Same as that of the closed container, variable
Volume	Almost constant; varies only slightly with temperature	Almost constant; varies slightly with temperature	Same as that of the closed container, variable

(a)

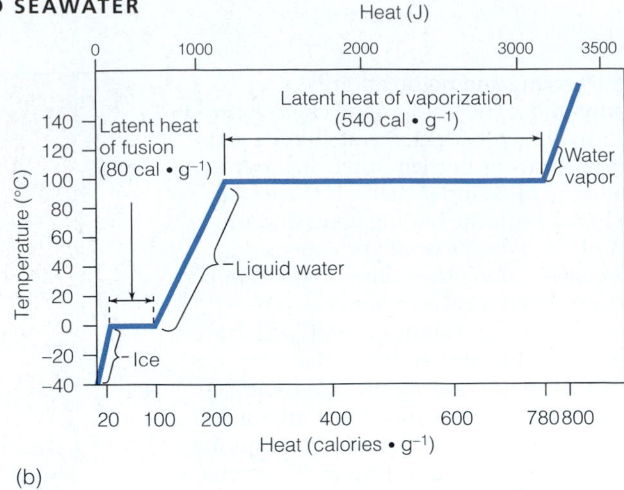

(b)

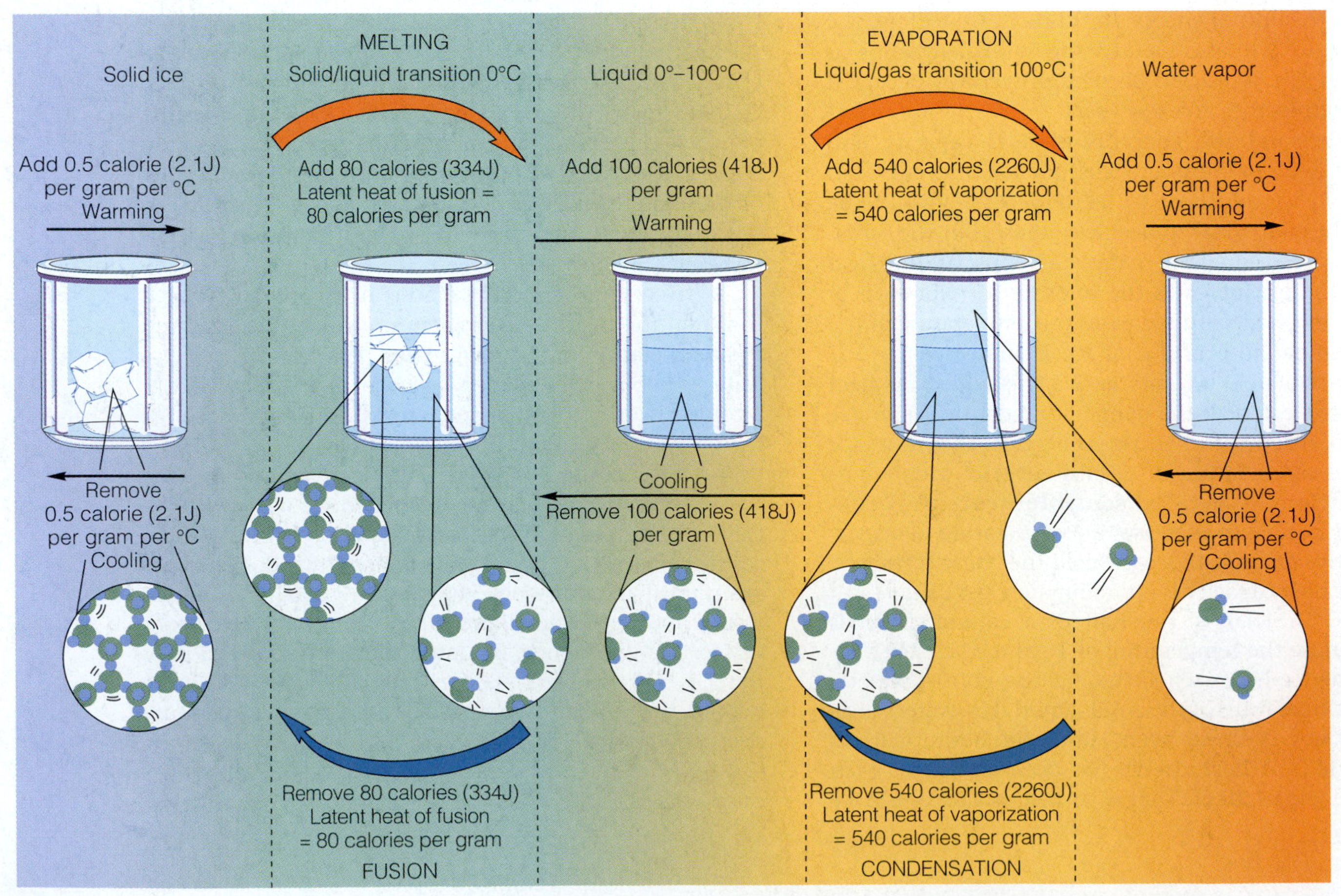

(c)

FIGURE 7.2 (a) Characteristics of solid, liquid, and gaseous water. (b, c) Water's latent heat of fusion (melting) is the amount of heat that must be added to convert ice to water at the freezing (melting) point temperature. Water's latent heat of vaporization is the amount of heat that must be added to water to convert water to water vapor at boiling point temperature. Both of these are high because of the need to provide energy to overcome the hydrogen bond attraction. As latent heat is added to change the phase, there is no change in temperature. In the reverse processes, the same amounts of latent heat that were added are released when water vapor condenses and when water freezes. The heat capacity of water is the amount of heat required to raise the temperature of 1 g of liquid water by 1°C and is measured in $cal{\cdot}g^{-1}{\cdot}°C^{-1}$ ($J{\cdot}g^{-1}{\cdot}°C^{-1}$). The same amount of heat is released as water cools.

The latent heats and heat capacity of water are very high primarily because the hydrogen bond is stronger than van der Waals forces. More heat energy is needed to overcome the attraction between molecules in solid, liquid, and gaseous forms of water than in substances whose molecules are bound only by van der Waals forces. Consequently, the heat capacity of liquid water is the highest of all liquids other than liquid ammonia and is higher than that of all solids; the latent heat of fusion of water is the highest of all substances other than ammonia; and the latent heat of vaporization of water is the highest of all known substances.

Implications of the High Heat Capacity and Latent Heats of Water

The anomalously high heat capacity and latent heats of water have many implications. In our everyday experience, we rely on the high latent heat of fusion to keep our iced drinks cold. When we add ice cubes to a cold drink, the drink remains cold for a long time, until all the ice has melted. The drink, a mixture of ice and water, remains cold as it gains heat from its surroundings, until it has absorbed 80 calories (334 J) for each gram of ice, converting the ice to water. Subsequently, when all the ice has melted, the drink warms quickly because each calorie (each 4.2 J) gained can raise the temperature of 1 g of water by 1°C.

The high heat capacity of water is illustrated by the behavior of hot drinks. A hot drink cools much more slowly than, for example, an empty cup or glass taken out of a hot dishwasher. The reason is that more heat must be lost per unit weight of water than of any other substance (other than liquid ammonia) to lower its temperature by a given number of degrees.

The high heat capacity of water allows large amounts of heat energy from the sun to be stored in ocean waters without causing much of a temperature change. The heat is released to the atmosphere when atmospheric temperatures fall. Because water has this heat-buffering capability, coastal locations have milder **climates** than inland locations (Chap. 9). The heat-buffering capacity of the oceans and the transfer of latent heat between ice, oceans, and atmosphere are important to global climatic control and to predictions of the effects of releases of gases that contribute to the **greenhouse effect** (Chap. 9).

In the polar ocean and coastal regions, water's high latent heat of fusion acts to control the air and water temperatures in much the same way that ice cubes cool an iced drink. During winter, latent heat is released to the atmosphere as water freezes to form more sea ice. Conversely, in summer, heat added to the polar regions is used to melt the ice. Because the heat lost or gained is latent heat, not sensible heat, the ocean surface water and sea ice remain at or close to the freezing point throughout the year. Because air temperatures are partially controlled by ocean temperatures, the annual climatic temperature ranges in polar ocean regions are small (Chap. 9, **CC5**).

Evaporation

Liquids, such as water, can be converted to a gas at temperatures below their boiling point by a process known as "evaporation." For example, puddles of cold water can evaporate from the ground after a rainstorm. Although the average energy level of water molecules remains the same unless there is a change of temperature, the energy level of individual molecules varies around this average. Some molecules that temporarily possess higher amounts of energy can overcome their hydrogen bonds and escape completely from the liquid phase (evaporate).

Water evaporating from the oceans carries its latent heat of vaporization into the atmosphere and releases it when the water subsequently condenses as rain. Because the latent heat of vaporization is high, water can transport large quantities of heat energy from ocean to atmosphere, where that energy can be redistributed geographically. For example, areas of the Earth where evaporation is slow and precipitation is high are warmed, and areas where evaporation is high are cooled (Chap. 9, **CC5**).

Water's high latent heat of vaporization facilitates cooking and can be observed in a kitchen. If we heat a saucepan of water on a stove, the water can be brought to a boil quickly, but then, even if we do not change the heat setting, it takes much longer for the pan to boil dry. It takes only 100 cal•g^{-1} (418 J•g^{-1}) to heat water from its freezing point to its boiling point, but more than five times as much, 540 cal•g^{-1} (2260 J•g^{-1}), to vaporize or convert the water to steam. Water's high latent heat of vaporization is also apparent when we bake food for substantial lengths of time at temperatures well above 100°C and find that the food is still moist.

Because evaporating molecules gain the energy to escape by colliding with molecules that remain behind in the liquid, the average energy level (temperature) of the water molecules that remain behind is reduced. This is why we feel colder when we first climb out of a swimming pool or shower. While we are wet, water molecules evaporate, using heat energy gained from our skin and the air and leaving behind colder skin. After toweling off, we have less water on our skin and the amount of evaporation is reduced. Another example of this phenomenon is the wind chill factor used in **weather** forecasts. Increasing the speed of wind blowing across a wet or damp surface, such as the skin, increases the evaporation rate and thus increases the rate of surface cooling.

At temperatures below the boiling point, the average water molecule has less energy than it does at the boiling point. Therefore, more heat must be supplied to evaporate each molecule of water than is needed to vaporize it at the boiling point, and the latent heat of vaporization increases with decreasing temperature. For example, the latent heat of vaporization at 20°C is 585 cal•g^{-1} (2449 J•g^{-1}), approximately 45 cal•g^{-1} (189 J•g^{-1}) higher than at 100°C.

7.2 EFFECTS OF PRESSURE, TEMPERATURE, AND DISSOLVED SALTS ON SEAWATER DENSITY

Many of the movements of **water masses** in the oceans are driven by differences in **density**. Solid objects that have higher density than water sink, and those that have lower density rise and float. Liquid water can also rise or sink if its density is different from that of the surrounding water (**CC1, CC3**). Water density is controlled by changes in pressure, temperature (**CC6**), and concentration of dissolved constituents (**salinity**).

Pressure

Because water molecules can be forced together only slightly as pressure increases, water is virtually incompressible, and its volume decreases and density increases only very slightly with pressure. Therefore, pressure changes are not as important to controlling water density in the oceans as changes in temperature and salinity are. However, the pressure at the greatest depths in the oceans is more than 1000 times as great as atmospheric pressure. This pressure change causes seawater density to be approximately 2% higher than it is in shallow water at the same temperature and salinity. This small difference must be taken into account in some research studies. Like all other gases, water vapor is compressible, and its density varies substantially with pressure.

Temperature

Increasing temperature adds energy that enables the molecules of solids and liquids to vibrate, rotate, and/or translate more vigorously. Thus, the average distance between the molecules generally increases. As the same

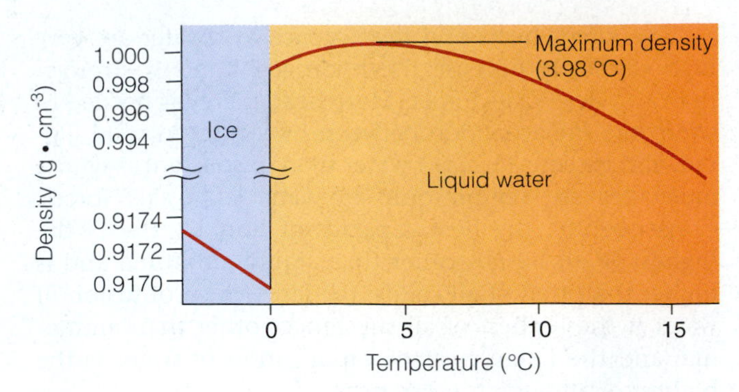

FIGURE 7.3 Density of pure water plotted against temperature. Water with no dissolved salts has a maximum density at 3.98°C. At lower temperatures, water density decreases until the freezing point is reached. There is a discontinuity in the density scale. The density of ice is much less than that of liquid water because of the open structure of the ice crystal lattice. When solid ice and liquid water occur together at 0°C, the ice floats on the water because its density is less than that of the water.

number of molecules occupies a larger volume (the material expands), increasing the temperature causes the density to decrease. Ice is no exception to this rule. However, pure water (but not seawater, as discussed later) behaves anomalously, because liquid water has a density maximum at 4°C. Between 0°C and 4°C, water density actually increases with increasing temperature (**Fig. 7.3**). In the rest of liquid water's temperature range—4°C to 100°C—pure water behaves normally and density decreases with increasing temperature.

The reason for the anomalous effect of temperature on water density is the hydrogen bond. Water molecules form clusters in which the molecules are arranged in a latticelike structure. The atoms in the cluster are held in place by hydrogen bonds (**Fig. 7.4**). The structure is similar to that of ice, and the molecules of water in a cluster occupy a larger volume than molecules that are not clustered. The ordered clusters remain together for only a few ten-millionths of a second, but they are continuously forming, breaking, and re-forming. Both the number of clusters present at any time and the number of molecules in each cluster increase as the temperature decreases (more unbroken hydrogen bonds are present). Because clustered molecules occupy a greater volume than unclustered molecules, an increase in the number of clusters and in the number of molecules per cluster decreases the density of the water. Above 4°C, there are too few clusters to counteract completely the normal temperature effect on density. Below 4°C (actually 3.98°C), however, clustering increases fast enough that the decrease in density caused by the clustering is faster than the increase caused by the normal temperature effect.

CRITICAL THINKING QUESTIONS

7.1 If water's heat capacity and latent heat of vaporization were much lower, would we still have rain, and would puddles of rainwater behave differently? Explain your answer. Describe some of the things that would be different in your everyday life.

7.2 If we look through the water of a swimming pool, we can see the bottom. If the water surface is calm, the tiles and lane markings appear clear and sharp. However, if the surface is disturbed, even if only by tiny ripples, these markings appear indistinct and seem to shimmer. Why?

7.3 Describe all the ways that you can think of in which the Earth would be different if the water molecule were not polar.

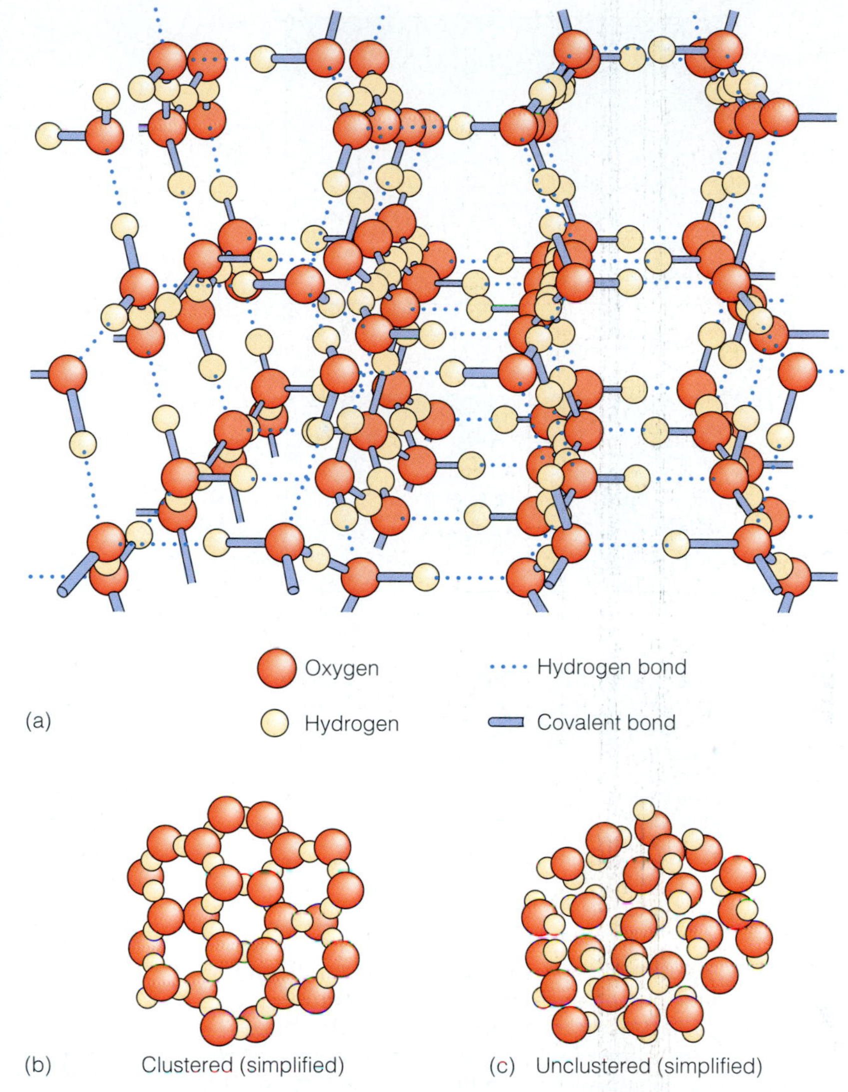

FIGURE 7.4 The hydrogen bond plays a major role in the properties of ice and water. (a) Molecules in the ice crystal lattice are bonded to each other by hydrogen bonds and arranged in a hexagonal lattice. The resulting structure is very open, which explains the low density of ice. (b) Water molecules constantly form clusters that are temporarily held together by hydrogen bonds. The clusters consist of chains or networks of water molecules arranged in hexagons. The representation here shows this hexagonal form but, for simplicity, does not show the hydrogen bonds. (c) When water molecules are unclustered, the molecules are closer to each other than they are in clusters. Clusters are more likely to form and may persist longer at low temperatures, which is why water's density decreases below 3.98°C.

Dissolved Salts and Density

Salts dissolved in water increase the water density for several reasons. First, the **ions** or molecules of most substances dissolved in seawater have a higher density than water molecules. Dissolved substances also reduce the clustering of the water molecules, further increasing the density, particularly at temperatures near the freezing point.

Combined Effects of Salinity and Temperature

The density of ocean waters is determined primarily by salinity and temperature. **Figure 7.5a** shows the relationships among the salinity, temperature, and density of seawater. Raising the temperature of freshwater from 4°C to 30°C (the range between the temperature of maximum density and the highest temperature generally found in surface waters) decreases its density by about 0.0043 (from 1.0000 to 0.9957), or about 0.4%. At a constant temperature, changing the salinity from 0 to 40 (approximately the range of salinity in surface waters) changes the density by about 0.035, or about 3.5%. These observations suggest that salinity is more important than temperature as a determinant of density. This is often true in rivers and **estuaries** where the water has a wide range of salinity, but the range of salinity in open-ocean waters is much smaller. In fact, 99% of all ocean water has salinity between 33 and 37, and 75% has salinity between 34 and 35 (**Fig. 7.6**). Similarly, 75% of ocean water has a temperature between 0°C and 5°C, and the rest has a much wider temperature range, between about –3°C and 30°C.

With the exception of water discharged by **hydrothermal vents**, the highest temperatures in the oceans

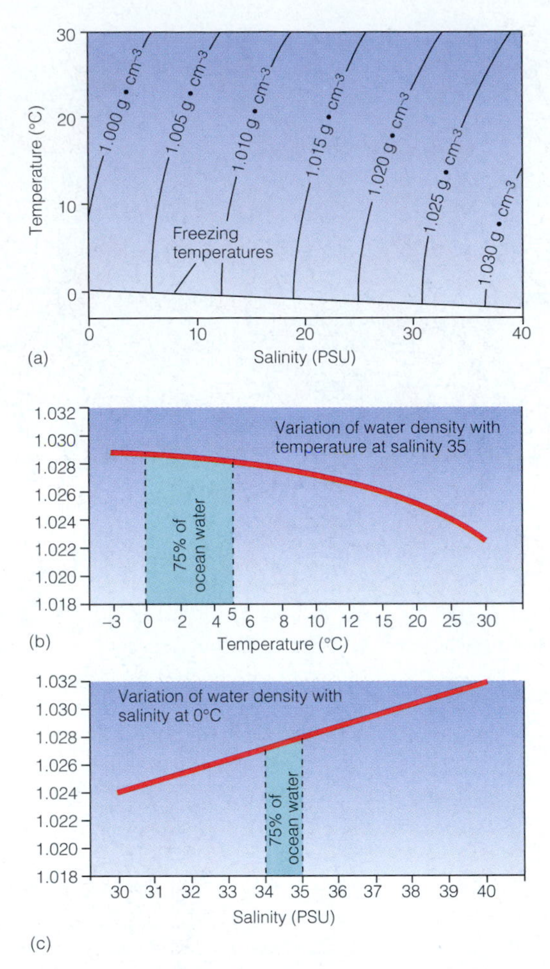

FIGURE 7.5 The effects of temperature and salinity on the density of water. (a) The relationships among salinity, temperature, and density are complex. The effects of temperature and salinity individually are more obvious if we examine (b) the relationship of temperature (–3°C to 30°C) to density at a fixed salinity, and (c) the relationship of salinity (30 to 40 PSU) to density at fixed temperature. The shaded sections of the plots in (b) and (c) represent the narrow ranges of salinity and temperature present in most ocean water.

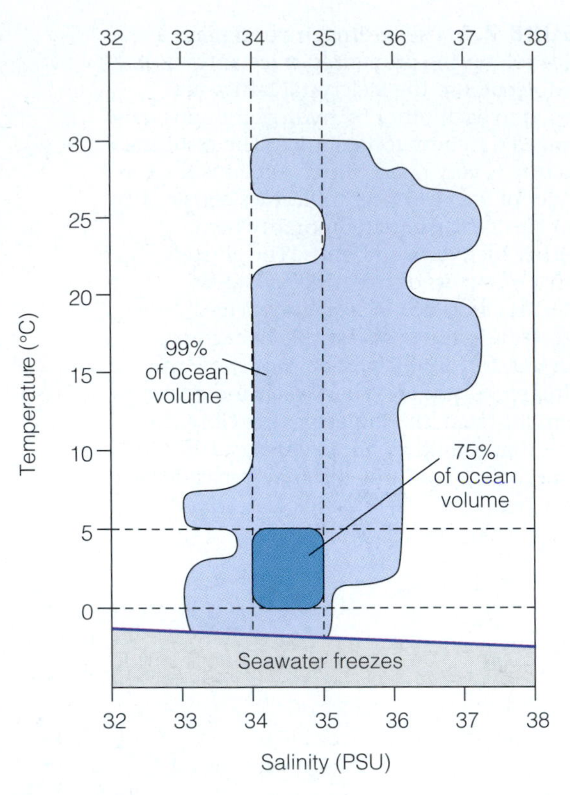

FIGURE 7.6 Temperature and salinity characteristics of ocean waters. Ninety-nine percent of all ocean water has salinity and temperature within the range shown by the larger shaded area. Seventy-five percent of all ocean water has the much narrower range of salinity and temperature within the inner darker-shaded area.

are in surface waters in tropical regions. **Figure 7.5b,c** relates salinity, temperature, and seawater density, and shows that, in most of the ocean, temperature and salinity are of approximately equal importance in determining the density of ocean waters. However, their relative importance varies with location and depth. For example, temperature changes are more important to density variations in the tropical water column, where salinity variation is relatively small but temperature variation with depth is relatively large. In contrast, salinity is more important in some high-**latitude** regions, where salinity variations are relatively large as a result of high volumes of freshwater **runoff** and the formation and melting of ice, but temperatures are generally uniform and near the freezing point.

7.3 ICE FORMATION

When surface water is cooled to the freezing point, the behavior of ice and water in lakes and oceans is determined by two factors: the relative density of ice and water, and the variations of water density with salinity at temperatures near the freezing point.

Dissolved Salts and Freezing

Increasing the salinity lowers the freezing point of water (**Fig. 7.7**). This is why salt is used to de-ice roads. Increasing the concentration of dissolved salts also inhibits the clustering of water molecules and therefore lowers the temperature at which the density maximum occurs (**Fig. 7.7**). As salinity increases, the temperature of the water density maximum decreases more rapidly than the freezing point temperature. Therefore, as salinity increases, the difference between the temperature of maximum density (4°C in pure water) and the freezing point is narrowed. At a salinity of 24.7, the density maximum and the freezing point are the same temperature (–1.36°C). At any salinity above 24.7, the freezing point is reached before a density maximum occurs. Since open-ocean water generally has a salinity above 24.7, seawater does not have an anomalous density maximum.

Because freshwater has a maximum density at 4°C and seawater has no maximum density, lakes and rivers behave differently from the oceans when freezing. As winter begins, the surface water of a freshwater lake cools and its density increases. This cold, dense surface water sinks and displaces bottom water upward to be cooled in its turn. This **convection** process is called "overturning" (**CC3**). Convection continues until all the water in the lake is 4°C and is at the maximum density. As the surface layer of water cools below 4°C, its density no longer increases. Instead, it actually decreases. Therefore, this water does not sink. The water in this surface layer continues to cool until it reaches 0°C and then the water freezes. The resulting ice has a lower density than water and floats. With further cooling, the lake water continues to lose heat, but because of the density maximum there is no convection, so the heat is lost by **conduction** through the overlying ice layer. As heat continues to be lost, the water immediately under the ice cools to 0°C and freezes. Therefore, the ice layer grows progressively thicker from underneath. Ice is not a very good conductor of heat. In addition, a large amount of heat must be lost to convert water to ice because of the high latent heat of fusion. Consequently, the ice layer on a lake surface will grow to a thickness of at most only a meter or two, depending on the length and severity of the winter. If the lake is deep enough, a pool of liquid water at 4°C will remain below the ice of the lake throughout the winter. Many freshwater organisms that cannot tolerate being frozen survive the winter in such pools of water.

In contrast to freshwater, as seawater (salinity greater than 24.7) cools, its density increases continuously until it reaches its freezing point of about –2°C. Cooling water continues to sink away from the surface until ice forms. This is why most of the deep water in the oceans is cold, generally below 4°C (Chap. 10). As seawater freezes, most of the dissolved salts are excluded from the ice and remain in the unfrozen water. Therefore, salinity increases in the remaining water, and as a result, the freezing point decreases (**Fig. 7.7**). If freezing occurs quickly, or if winds break up the surface ice and splash water onto the ice surface, isolated pockets or pools of liquid water can be present within the ice. The water in the pockets has high salinity, and its temperature may be several degrees below the freezing point of seawater with normal salinity. Some microscopic plants and animals survive the Arctic and Antarctic winters in such isolated pockets of liquid inside the sea ice.

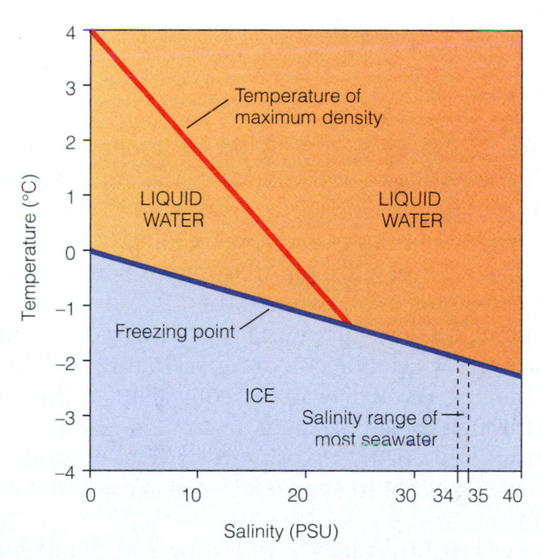

FIGURE 7.7 Increasing the concentration of dissolved salts in water reduces both the freezing point and the temperature at which the anomalous density maximum occurs. However, the temperature of maximum density is reduced more rapidly than the freezing point as salt concentration increases. Consequently, seawater (99% of which has salinity greater than 33) has no anomalous density maximum.

Density of Ice

Most substances have higher density in the solid phase than in the liquid phase. Water is an exception to this rule. Ice at 0°C has a density of 0.917 $g \cdot cm^{-3}$, about 8% less than the density of water of the same temperature. Hence, water expands when it freezes. The reason is that the ice crystal has an open lattice structure, and fewer water molecules are packed into the volume that they would occupy as water (**Fig. 7.4**). In winter, expansion of water as it freezes bursts water pipes, breaks up rocks, and causes potholes in street surfaces when small cracks are filled with water and then freeze.

Ice floats because it is less dense than water. If ice did not float, it would sink to the bottoms of ponds and lakes as they froze, and surface water would continue to freeze and sink until no liquid water was left. Hence, aquatic organisms would have no liquid water refuge below the ice in which to survive the winter. In addition, sea ice formed in the polar regions would sink to the floor of the oceans and accumulate. Because there would be no way to bring the ice to the surface to remelt it, the oceans would have a warm-water surface layer over a solid ice layer extending to the seafloor.

7.4 SURFACE TENSION AND VISCOSITY

As we have seen, hydrogen bonding plays an important role in several thermal properties of water. Hydrogen bonding also contributes to the high **surface tension** of water and its relatively high **viscosity**. Surface tension and viscosity are major factors in determining the

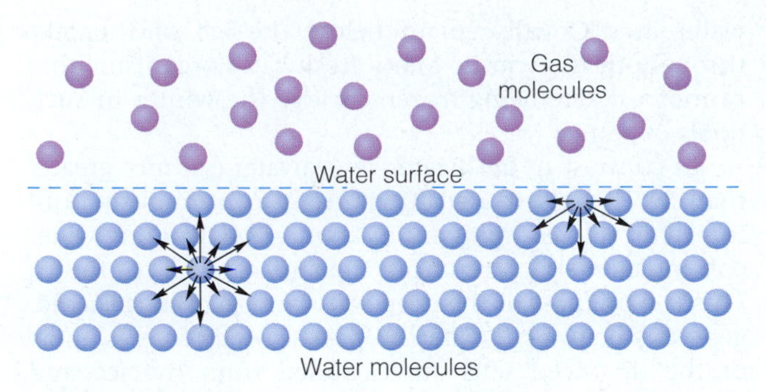

FIGURE 7.8 All molecules in a liquid are attracted to each other by van der Waals forces. Liquids have surface tension because the molecules at the surface are attracted more strongly to other molecules of the liquid than to the gas molecules above, which are more dispersed and thus, on average, farther away. Molecules at the surface have a net inward attraction that makes the surface behave somewhat like the rubber of an inflated balloon. The surface tension of water is particularly high because the water molecules are attracted to each other by both van der Waals forces and the much stronger hydrogen bond.

behavior of waves and controlling the processes by which water molecules are transported across biological membranes.

Surface Tension

Molecules within a liquid are subject to attractive forces that pull them toward all adjacent molecules. Moreover, molecules at the surface of a liquid are attracted much more strongly to their neighboring liquid molecules than to the more remote gas molecules above the surface (**Fig. 7.8**). Therefore, all molecules on the surface are pulled toward the interior of the liquid. The attractive force tends to minimize the number of molecules at the surface and creates a surface tension that pulls the liquid surface into a configuration with the minimum possible surface area. Because of the strong hydrogen bond attraction between water molecules, the surface tension of water is higher than that of any known liquid other than mercury.

Surface tension pulls water droplets into a spherical shape because a sphere has the lowest possible ratio of surface area to volume. We can see the spherical shape in beads of water on a newly waxed automobile, and in drips from a leaky faucet.

If we pour water into one side of a saucepan (or ocean), **gravity** distributes the water so that all parts of the water surface are at the same level. Similarly, if we disturb a surface—say, by sloshing the saucepan—the water will quickly return to its original state, in which its surface is flat and horizontal. Surface tension "pulls" the surface into the minimum area possible, which in the saucepan is a flat horizontal plane. For this reason, the high surface tension of water critically affects the generation and dissipation of small waves on the ocean surface (Chap. 11).

If an object is to break through a water surface, it must be heavy enough to push aside the water molecules and break the surface tension. Any object with a density greater than that of water can be made to float on the water surface if its weight per unit of water surface area is too low to break the surface tension. For example, a steel sewing needle or razor blade can be positioned to float on the water surface. Similarly, insects such as the water strider can walk on water because of its high surface tension.

The high surface tension of water has significant effects on the formation and behavior of gas bubbles and spray, which are created in the oceans primarily by breaking waves. Once they have formed in water, small gas bubbles cannot easily break through the ocean surface to escape to the atmosphere. When bubbles do break through the surface, a large number of small water droplets are ejected into the air. Because of their high surface tension, the droplets do not readily combine into larger drops that would fall back to the water surface faster. The high surface tension of water helps to retain gas bubbles in the water and water droplets in the air, enhancing the efficiency of gas transfer between oceans and atmosphere. Because small droplets have large surface areas in relation to their volume, evaporation from the oceans is also enhanced.

Because of the high surface tension of water, many small aquatic organisms, such as mosquito **larvae**, can anchor themselves from below in a very thin surface water layer called the **surface microlayer**. The surface microlayer, discussed in Chapter 16, is just a few molecules thick.

Viscosity

Viscosity is a measure of a liquid's internal resistance to flow or the resistance of the liquid to the movement of an object through it. A low-viscosity liquid flows easily, whereas a high-viscosity, or "viscous," liquid flows slowly. Water has a lower viscosity than many liquids, such as honey or motor oil. However, the viscosity of water is high enough to provide substantial resistance to minute organisms that swim or sink in water. To these tiny organisms, moving through water is equivalent to our swimming through tomato ketchup. In fact, many microscopic organisms rely on the viscosity of water to prevent them from sinking out of the upper layers of ocean water where they live.

Water viscosity varies with salinity and temperature. Viscosity increases as temperature decreases (slightly less than 1% per °C within the normal range of ocean water temperatures) because more structured water molecule clusters are present at lower temperatures. The clusters do not move out of the way as easily as individual molecules do. Hence, microscopic organisms in warm tropical waters have more difficulty in avoiding sinking than do organisms in cold polar waters. Many tropical microorganisms have compensated by evolving elongated spines or frilled appendages that increase "drag" as they move through the water. Related polar species generally do not have these ornate features.

Viscosity also increases as salinity increases, because the component ions of the dissolved salts are very effective in surrounding themselves with water molecules to form clusters that resemble water molecule clusters. However, the change in viscosity caused by dissolved salts is small within the range of salinity in freshwater and the oceans. Viscosity increases by less than 1% from freshwater to salinity-35 seawater at the same temperature.

7.5 TRANSMISSION OF LIGHT AND OTHER ELECTROMAGNETIC RADIATION

Light is a form of electromagnetic radiation emitted by the sun and stars. However, visible light occupies only a small segment of the very wide electromagnetic radiation spectrum (**Fig. 7.9**). Ultraviolet light, X-rays, and gamma rays have progressively shorter **wavelengths** than visible light, whereas infrared light, microwaves, and radio waves have progressively longer wavelengths.

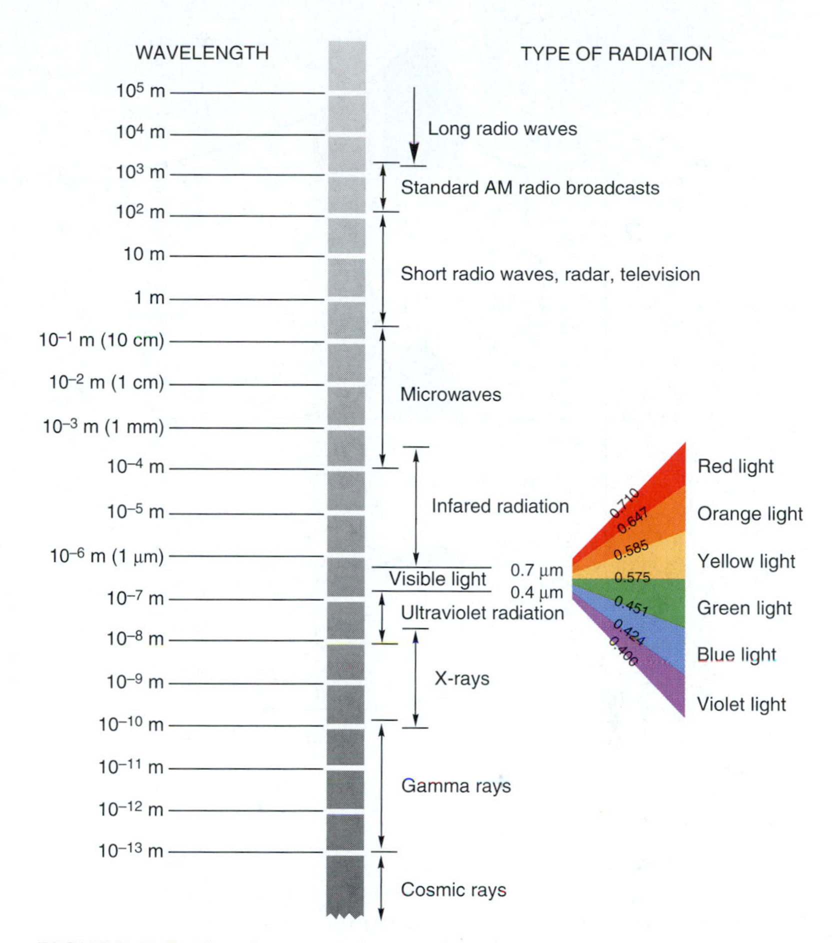

FIGURE 7.9 The electromagnetic radiation spectrum.

Absorption

Water absorbs electromagnetic radiation, but the depth of water penetrated varies with the intensity of the radiation and with wavelength. At most wavelengths, absorption is so effective that, even at very high intensity, the radiation can penetrate only a few centimeters or meters of water. Absorption is less effective within the narrow range of wavelengths of visible light and for very long-wavelength radio waves. However, even at the wavelengths of lowest absorption, the most intense radiation (e.g., **lasers**) cannot penetrate more than a few tens or hundreds of meters of water. For this reason, oceanographers cannot use electromagnetic radiation, such as radar, radio waves, or laser light, to "see" through the oceans in the same way that radar can see through clouds to the Earth's surface or even the surface of Venus.

Because radio waves and other forms of electromagnetic radiation do not penetrate far through water, they cannot be used for underwater communication. For example, long-range communication with submarines is a considerable problem. Very low-**frequency** radio waves can penetrate a few tens of meters into the ocean and are used to communicate with submarines near the surface. Because intense visible light can penetrate up to several hundred meters below the ocean surface, satellite-mounted lasers also are used to communicate with submarines at shallow depths. The inability of electromagnetic radiation, such as radar, to penetrate ocean water is the principal reason why submarines are among the best places to hide strategic missiles. Detection of submarines relies primarily on sound waves, which are transmitted easily through ocean water.

The greater transmissibility of visible light through water (compared to other electromagnetic wavelengths) is critically important to life in the oceans (Chaps. 14, 15). Like life on the land, most life in the oceans depends on plants that need light to grow by **photosynthesis** (**CC14**).

Some wavelengths of visible light are better absorbed by water than others (**Fig. 7.10a**). Wavelengths at the red end of the visible spectrum are the most rapidly absorbed, and violet wavelengths also are absorbed relatively quickly; blue and green light penetrate farthest. Infrared radiation (wavelengths longer than red light) and ultraviolet radiation (wavelengths shorter than violet light) are effectively absorbed.

As light penetrates the ocean, it is absorbed not only by water molecules, but also by suspended particles. Accordingly, light penetrates deeper in clear ocean waters than in high-**turbidity** coastal waters with high **suspended sediment** loads or high **plankton** concentrations (**Fig. 7.10b–d**).

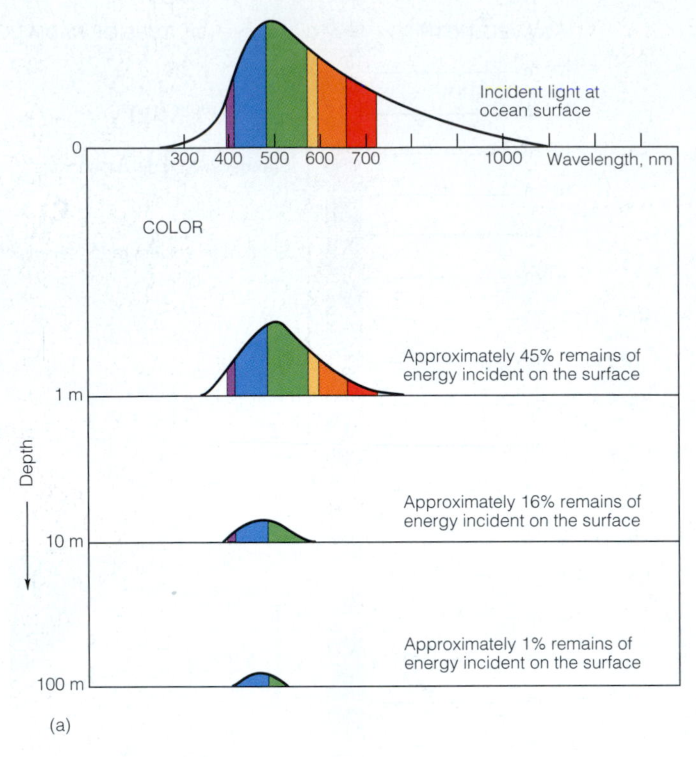

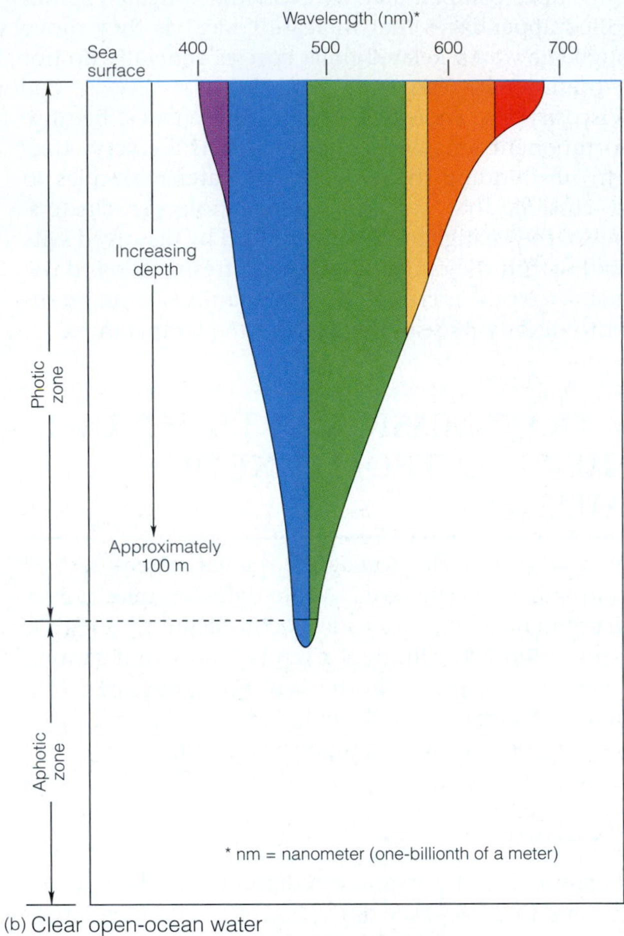

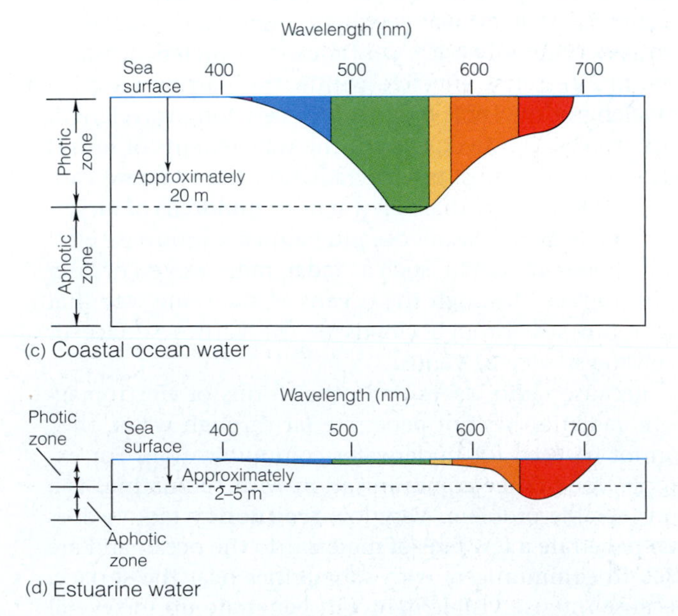

FIGURE 7.10 Light penetration and absorption in ocean waters. (a) Absorption of electromagnetic energy by ocean water. Note that energy in the infrared (about 700–1000 nm) and ultraviolet (about 300–400 nm) wavelengths is absorbed very rapidly compared to energy in the visible light wavelengths (400–680 nm). Also, energy at longer wavelengths of visible light (red/orange) is absorbed more quickly than energy at shorter (blue/green) wavelengths. The remaining parts of the figure show the depth at which light of visible wavelengths is reduced to less than 1% of the intensity at the surface in (b) clear open-ocean water, (c) coastal-ocean water, and (d) estuarine water. The aphotic zone begins at the depth where the ambient light intensity is approximately 1% of that at the surface. Note that the photic zone depth is much reduced in coastal and estuarine waters. Below the photic zone, light is insufficient for phytoplankton to produce enough organic matter by photosynthesis to survive. Note also that the wavelength range of light reaching the lower part of the photic zone moves from the blue/green part of the spectrum toward the red end of the spectrum in coastal and estuarine waters. This is the reason why open-ocean waters usually appear blue while coastal and estuarine waters do not.

Scattering and Reflection

We are all familiar with the idea that light is reflected by smooth "shiny" surfaces such as the silvered surface of a mirror. Light is also reflected when it encounters a rough surface. However, individual light rays meet different parts of the rough surface at different angles, and as a result, they are reflected in different directions. When this occurs, the light is said to be **scattered**.

As light penetrates the ocean water column, scattering occurs when some of the photons of light bounce off water molecules, molecules of dissolved substances, or suspended particles. The light scatters in all directions, but some is directed upward and is said to be **backscattered**. In very clear ocean waters with few suspended particles or plankton, most of the red and yellow light is absorbed rather than scattered (**Fig. 7.10b–d**). Light of blue wavelengths has a greater probability of being backscattered because blue light travels through a greater length of the water column. Therefore, most of the backscattered light in clear ocean waters is of blue wavelengths, and it is this backscattered light that gives the ocean its deep blue color.

In waters with more suspended particles or plankton, there is a higher probability of backscattering of all visible wavelengths. Hence, in more turbid waters, the backscattered light comprises a much wider range of wavelengths, and the water appears not blue, but green or brownish. The specific color is determined by the nature of the suspended particles. All particles absorb light more effectively at some wavelengths, and their color is determined by which wavelengths of the light they do not absorb. Light of these nonabsorbed wavelengths is scattered and, therefore, constitutes the wavelengths we see when looking at the particles. Most particles in the oceans are either green, such as many **phytoplankton** cells, or brown, such as sand and other mineral particles. Ocean waters with large phytoplankton populations (Chap. 14) tend to appear greenish, whereas coastal waters with large loads of suspended mineral grains appear muddy brown. Therefore, color variations of the ocean surface indicate the quantity of particulate material and phytoplankton in the near-surface water. Ocean surface color measured by sophisticated sensors aboard satellites (**Fig. 7.11**) is used to investigate distributions of the living and nonliving particles in the oceans (Chaps. 8, 14).

The underwater world seen in movies or video is a blaze of colored creatures. However, **scuba** divers see such colors only in very shallow water. Just a few meters below the surface, the red and yellow wavelengths of natural sunlight are absorbed, so bright red fishes appear black and most objects appear blue or bluish green. The true colors are revealed only when lights with a full spectrum of visible wavelengths are shone on the marine life. For this reason, underwater photographers or videographers must carry powerful lights.

Scuba divers can observe scattering and absorption of light by particles by noting changes in the visibility and lighting of the water. In turbid water with more particles, daylight does not penetrate very far. In addition, at shallow levels to which some natural light penetrates, horizontal visibility is reduced. Just as daylight is blocked from penetrating vertically into the water by scattering and absorption, light traveling from an object horizontal to the diver is absorbed and scattered. If the absorption and scattering are substantial, the light from distant objects does not reach the diver and the objects cannot be seen. When light is scattered equally in all directions, a scuba diver may not be able to visually distinguish between up and down.

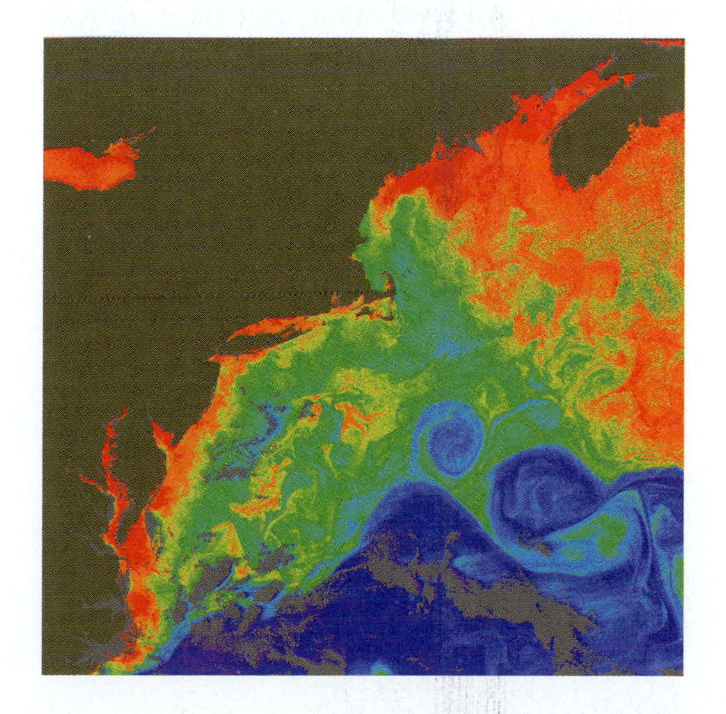

FIGURE 7.11 This image of the Northwest Atlantic Ocean shows data obtained from a satellite-mounted instrument called the Coastal Zone Color Scanner, which measures the intensity of light reflected or backscattered by the Earth's surface. The data are from a narrow wavelength band, in which the intensity of the light received by the scanner from the ocean surface increases with increasing concentrations of chlorophyll-containing phytoplankton in the photic zone. Intensities of light in this wavelength band are depicted by the false colors in this image. Red colors show the areas of highest chlorophyll and phytoplankton concentration; and progressively lower concentrations are depicted by the orange, yellow, green, and blue shades. Note that the highest concentrations of chlorophyll are found in the coastal region and that the image reveals the complex patterns associated with different water masses.

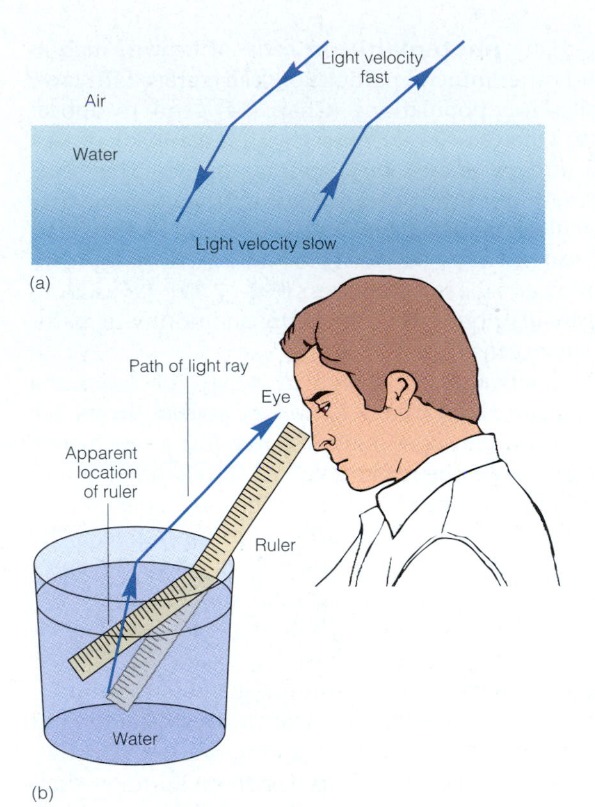

FIGURE 7.12 (a) Light rays are bent, or refracted, when they pass between air and water. (b) Inserting a ruler or stick through a water surface at an angle illustrates the distortion in the way refraction makes objects appear. The ruler looks as if it is bent at the point it enters the water.

Refraction

The speed of light in seawater is less than that in air. As light waves pass from air into water or from water into air, they are **refracted** (their direction is changed because of the change in speed) (**Fig. 7.12a**). An easy way to see refraction is to sight down the length of a ruler and dip the end of the ruler into a tumbler of water (**Fig. 7.12b**). The tip of the ruler appears to be bent upward as it passes into the water. Refraction causes fishes in an aquarium to appear bigger and closer than they really are. Similarly, fishes and other objects seen through a scuba mask appear larger and closer than they are.

Refraction between air and water increases as salinity increases. Therefore, measurements of refraction are often used to determine approximate salinity.

7.6 TRANSMISSION OF SOUND

Electromagnetic radiation travels both through a vacuum and through substances such as air and water. Sound is fundamentally different because it cannot be transmitted through a vacuum. The sound effects in space movies are pure fantasy, because sound cannot travel across the vacuum of space. Sound is transmitted as a vibration in which adjacent molecules are compressed in sequence. One molecule is pushed into the next, increasing the "pressure" between the two molecules. The second molecule pushes on a third, increasing the pressure between the second and third molecules and relieving the pressure between the first two molecules, and so on. Hence, sound waves are pressure waves transmitted through gases, liquids, and solids. The pressure waves can range from very high frequencies, which the human ear cannot hear (ultrasound), to low frequencies below the audible range that we sometimes feel as vibrations from our surroundings.

Sound waves are absorbed, reflected, and scattered as they pass through water, but much less than electromagnetic radiation is. Therefore, sound travels much greater distances in water and sound waves are the principal tools of communication and remote sensing for both oceanographers and marine animals.

Sound Velocity

The speed of sound is about four times greater in water than in air. In seawater, the speed of sound increases with increases in salinity, pressure, and temperature (**Fig. 7.13a,b**). However, the changes due to salinity variations are relatively small within the range of salinities in ocean waters. Thus, sound velocity generally decreases with depth in the upper layers of the oceans, where the temperature change is large, but then increases again with depth in the deep layers, where temperature variations are small and pressure changes are more important (**Fig. 7.13c**). Sound velocity in the oceans can be determined if the salinity, temperature, and depth (pressure) are known.

Sonar

Sound velocity is important because sound waves are used to measure distances in water. Sound pulses sent through water bounce off objects and the seafloor. If the sound velocity at all points within the sound path is known and the time taken for the sound to go out and the echo to return is measured, the distance to the object or seafloor can be calculated. This is the principle of **sonar**.

Sonar systems send and receive sound pulses to measure ocean depths or to measure the distance and direction of objects that reflect sound, such as submarines. **Schools** of fish, concentrations of tiny animals called **zooplankton** on which fishes feed, and concentrations of suspended mineral grains also can be detected with sonar.

Different frequencies are employed for different applications. For example, low frequencies are used for deep-ocean **sounding**. Higher frequencies allow smaller particles (or animals) to be detected, but higher-frequency sound is more effectively absorbed by water.

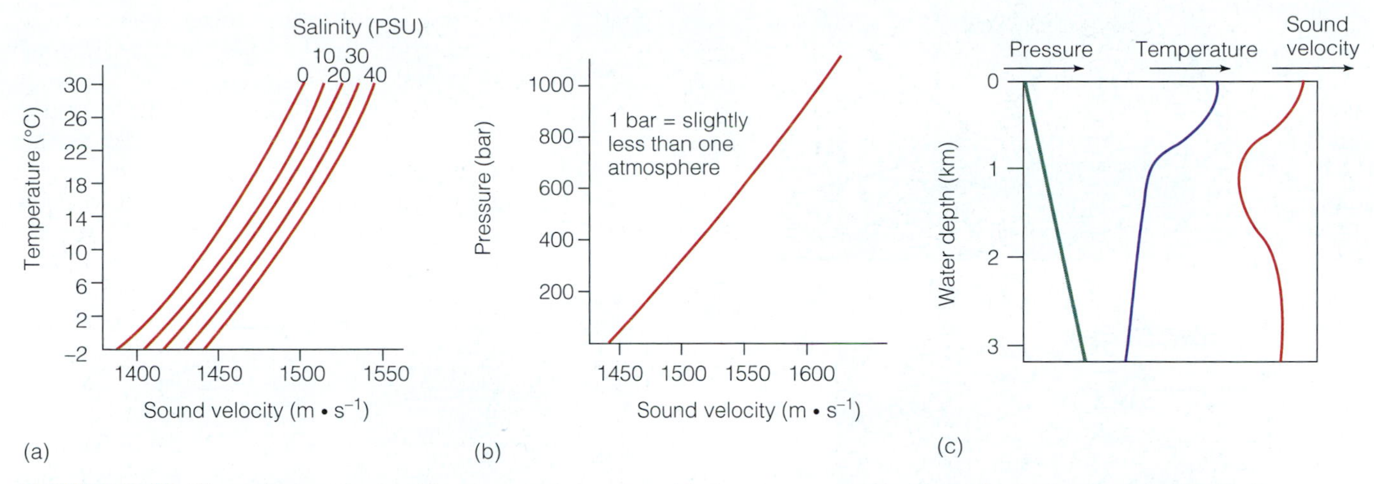

FIGURE 7.13 The velocity of sound in ocean water varies with temperature, pressure, and salinity. Sound velocity increases with (a) increasing temperature or salinity and (b) increasing pressure. Salinity is comparatively less important in determining sound velocity than temperature or pressure. (c) Variations of pressure and temperature with depth in the oceans produce changes in the sound velocity that are sometimes complex. Generally sound velocity decreases with depth until it reaches a minimum at about 1000 m (1 km) and then increases as depth increases.

Therefore, higher frequencies are used in applications involving small objects in shallow water, such as finding fishes and tracing plumes of particles from river discharges or ocean dumping. Dolphins use their natural sonar to locate objects such as fishes, and they can tune the sound frequencies they use: low frequencies for long distance, and higher frequencies for more detailed echo "vision" at shorter distances.

Sound Refraction and the Sound Channel

Just as light is refracted as it passes between air and water (**Fig. 7.12a**), sound waves are refracted as they pass between air and water (**Fig. 7.14a**), or through water in which the sound velocity varies. At any specific ocean depth, the parameters that determine sound velocity are fairly constant over great distances (Chap. 10). Hence, sound waves that travel horizontally are little affected by refraction. Sound waves that travel vertically are also not affected significantly by refraction, because they pass perpendicularly to the horizontal layers of varying sound velocity. In contrast, sound waves that pass through the ocean at any angle other than horizontal or vertical are refracted in complicated curving paths. The paths are determined by the initial direction of the sound waves and their varying velocity in the waters through which they pass (**Fig. 7.14b**).

Two important consequences of sound refraction in the oceans are of interest. First, sonar sound pulses sent out by a vessel at any downward angle not close to the vertical are refracted away from a zone called the "shadow zone" (**Fig. 7.14b**). The zone starts at a depth where sound velocity is at a maximum, being slower at both shallower and deeper levels. Such a maximum is present within submarine operating depths in many parts of the oceans. Hence, submarines can hide from sonar detection if they can remain in the shadow zone. This is one reason why the navies of the world continuously monitor changes in the temperature and salinity of the upper few hundred meters of the ocean throughout their operating areas. Small changes in these characteristics can change the location and effectiveness of the shadow zone. Because sound travels in complicated curved paths between the sources and echoing objects, knowledge of salinity and temperature distributions is essential in determining the sound velocity distribution and estimating the location and depth of an object detected by sonar.

The second important consequence of sound refraction in the oceans is that sound emitted at or about the depth of a sound velocity minimum is focused in what is called the "sound channel" (**Fig. 7.14c**). With the exception of sound waves traveling out vertically (or nearly so) from such a source, all sound waves are refracted back and forth as they pass alternately up and down across the depth of the velocity minimum. A sound velocity minimum, and therefore a sound channel, is present at a depth of several hundred meters throughout most of the world's oceans. Sound normally spreads out spherically (in all directions), and the loss in intensity (attenuation) due to the spreading is proportional to the square of the distance from the source (**Fig. 7.14d**). In contrast, sound is focused in the sound channel and spreads cylindrically, so that the attenuation is proportional only to the distance from the source. Hence, sound can travel very large distances within the sound channel with relatively little attenuation. For example, sounds of small underwater

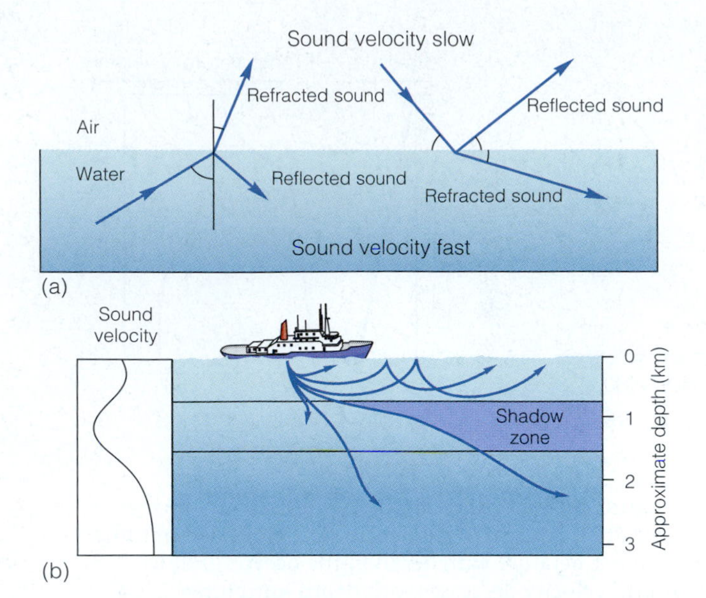

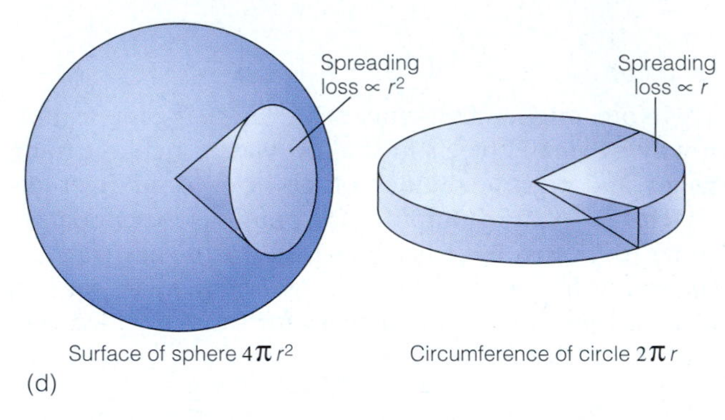

FIGURE 7.14 (a) Sound of certain wavelengths may be reflected when it encounters an interface between two water layers in which the sound velocity is different. Sound of other wavelengths can pass through such an interface and be refracted. (b) Sound can be reflected and refracted in complex ways. A sonar pulse sent from the surface can be refracted in such a way that there is a shadow zone within the depth range around the sound velocity minimum. Because sound emitted by a surface vessel cannot enter the zone, submarines hide in this shadow zone. (c) A sound generated at the depth of minimum sound velocity can be refracted back and forth within the low-velocity sound layer as it travels out from the source. This layer is called the "sound channel." Sound can be transmitted very long distances horizontally within the sound channel with little loss of intensity. (d) Outside the sound channel, sound spreads spherically, and its intensity is reduced in proportion to the square of the distance from the source. Within the sound channel, sound spreads cylindrically and loses intensity only in proportion to the distance from the source.

explosions in the sound channel near Australia have been detected as far away as Bermuda in the North Atlantic Ocean.

Acoustic Thermometry

The vast distance over which sound can travel in the sound channel is the basis of an experiment, begun in 1991, that may provide a means to determine whether enhancement of the greenhouse effect is causing global climate change. Determining whether the Earth's average surface and/or atmospheric temperature has recently changed or is now changing as a result of greenhouse effect enhancement is very difficult. Changes of only one-tenth of a degree per year in the global average temperature could drastically change the Earth's climate in just a few decades. However, because of the high temporal and **spatial** variability of temperatures in the ocean, land surface, and atmosphere, the best measurements currently possible from conventional thermometers or satellites are not accurate enough to identify changes in the global average temperature of less than about one degree. In contrast, changes of only a few thousandths of a degree per year in the average temperature of ocean water can be identified by measurement of the travel times of sounds over distances of thousands of kilometers in the sound channel. This method of monitoring is known as "**acoustic** thermometry."

The principle of acoustic thermometry is simple and was proven in 1991 with a preliminary experiment, the Acoustic Thermometry of Ocean Climate (ATOC) program. A large underwater transducer (a sound producer like a loudspeaker) is lowered into the sound channel. In the preliminary experiment, the transducer was placed in the ocean near Heard Island in the southern Indian Ocean (**Fig. 7.15**). The transducer transmits low-frequency sound pulses into the sound channel. At several locations tens of thousands of kilometers away, sensitive **hydrophones** detect the incoming sound and precisely record its arrival time. The travel time is measured to within a fraction of a second and used to compute the average speed of sound between transducer and receiver.

The speed of sound in seawater increases with increasing temperature. Therefore, if the travel time between a transducer and a listening point in the system decreases from year to year, it will indicate that the average temperature within the sound channel between those two points has increased (unless the average salinity or average depth of the sound channel has changed). An experimental acoustic array with the sound source located just north of the Hawaiian Islands and receiving stations surrounding the Pacific Ocean basin at distances of 3000 to 5000 km from this source has demonstrated that average ocean temperature measurements can be made with a precision of about 0.006°C. Results from this North Pacific Acoustic Laboratory (NPAL) program have revealed surprisingly large seasonal variations of average temperatures, but the program has not yet continued long enough to reveal any long-term trends. Establishment of long-term acoustic thermometry monitoring arrays has been substantially delayed by controversies concerning the possible effects of the sounds produced on marine species, especially whales, dolphins, and other **marine mammals**. Intensive studies of such possible effects have led to the development of substantial safeguards and improvements that are expected to prevent or minimize any adverse impacts on ocean life.

An acoustic thermometry monitoring program, if fully implemented, should provide an extremely sensitive measurement of climate change. Currently, arrays are planned or are in continuing test operations in both the Pacific and the Indian Oceans.

Ocean Noise

Scuba divers know that the undersea world is not silent. Besides the sounds of breathing regulators and boat propellers, an observant scuba diver hears a variety of other noises. They often sound like a well-known snap-crackle-pop cereal. The natural sources of sound in the oceans are many and varied, but generally they fit into three categories: noises from breaking waves and bursting air bubbles, noises from vessels and other human mechanical equipment, and biological noises. Each of the sources produces sounds across a wide range of frequencies.

Biological sounds are produced by many species and through various means. For example, whales and dolphins use sound to locate objects, such as prey, and to communicate with each other. Dolphins may also use intense bursts of sound to confuse or stun their prey. Certain **crustaceans** (shrimp, crabs, and lobsters) make clicking noises as they close their claws, and certain fishes make sounds by inflating and deflating the **swim bladder** (a gas sac used to control **buoyancy**).

Natural sounds and the sounds of human activities, both of which are always present in the ocean, constitute a background noise above which the sonar sound probes used by oceanographers or navies must be strong enough to be heard. Submarines can be detected easily if they send out sonar signals to locate vessels that are searching for them, so instead they use sensitive directional microphones to detect the noise of ships' propellers.

Because sound can travel very long distances in water, sensitive microphones can pick up sounds made hundreds or even thousands of kilometers away. Using sensitive microphone arrays, navies obtain data that are analyzed by supercomputers to separate the sounds of submarine or ship engines and other ship noises from background noise. This technology has become highly sophisticated. In some cases, the listening arrays can identify a specific submarine or surface vessel by its own specific sound "fingerprint" and can determine the vessel's location precisely and track its movements across oceans. This capability has been extended so that the system can now identify and track individual whales.

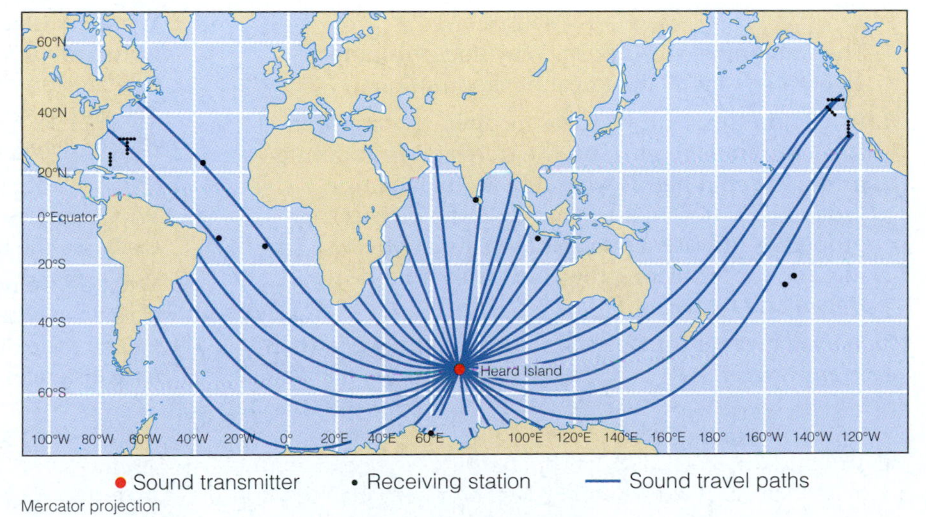

FIGURE 7.15 The Heard Island experiment. Sound pulses were emitted from a source in the sound channel at Heard Island, and the time of travel to the various receiving sites (shown by the black dots) was measured very precisely. Because the travel time of sound would change if the average temperature of the ocean water in the sound channel changed, arrays such as this can be used to monitor changes in the average temperature of ocean waters at the depth of the sound channel—a measurement that cannot be made from satellites.

CHAPTER SUMMARY

Water has many anomalous physical properties related to the polar nature of its molecule and the resulting hydrogen bond between water molecules.

Heat Properties of Water. Water has anomalously high freezing and boiling points. Hence, water can be present in solid, liquid, and vapor phases in the range of temperatures found at the Earth's surface. Water also has anomalously high heat capacity, latent heat of fusion, and latent heat of vaporization. These anomalous properties enable water to store and transport large amounts of heat through the oceans and atmosphere. Without these properties, the Earth's climate would be much more extreme.

Effects of Pressure, Temperature, and Dissolved Salts on Seawater Density. Pure water has an anomalous density maximum at 4°C, but the maximum disappears with increasing salt concentration and is not present in seawater of normal salinity. Seawater density increases with increasing salinity and with decreasing temperature. Water expands as it freezes, so ice is less dense than water and floats. As seawater freezes, the salts are excluded and left in solution.

Surface Tension and Viscosity. Water has an anomalously high surface tension. This property affects the behavior of the sea surface, including the creation and dissipation of small waves and the formation and behavior of gas bubbles and spray droplets in breaking waves. Water has a high viscosity relative to many other liquids. This high viscosity allows microscopic organisms that are more dense than water to remain in the surface layers of the oceans, because water provides a relatively high resistance to their motion including sinking.

Transmission of Light and Other Electromagnetic Radiation. Water effectively absorbs electromagnetic radiation. Absorption is less effective at the wavelengths of visible light. Visible light can penetrate up to several hundred meters into clear ocean waters and is the energy source for photosynthesis. Blue and green wavelengths of light are absorbed less than red, yellow, and violet wavelengths.

Suspended particles absorb, scatter, and reflect light. The depth to which light penetrates to support photosynthesis is reduced as the concentration of suspended particles increases.

Transmission of Sound. Sound is absorbed much less effectively by water than electromagnetic radiation is. Sound travels long distances through water with little loss of intensity, and it is used by oceanographers and marine animals for communication and remote sensing in the oceans. The principal use of sound by oceanographers is to measure ocean depth and seafloor **topography** with sonar.

Sound velocity varies slightly with salinity, temperature, and depth (pressure). Very accurate measurements of ocean depth require detailed knowledge of salinity and temperature variations in the water column. Sound pulses can be used to make very precise measurements of year-to-year changes in average ocean water temperature based on the time sound takes to travel from a source to receiving microphones thousands of kilometers away.

STUDY QUESTIONS

1. The water molecule is polar. Why does this property give water a high latent heat of vaporization?
2. In Florida citrus groves, growers often try to encourage ice crystals to grow on the trees and fruit during unusually cold nights when the air temperature drops below freezing. Explain how the physical properties of water and ice make this an effective method for protecting the citrus fruit from being frozen and damaged.
3. Why is it easier for a person to float in the ocean than in freshwater?
4. If ice were dense enough that it always sank, how would lakes behave in winter? How might the oceans be different?
5. What is surface tension, and why is this property of water important in the ocean? Why is the high surface tension of water important in our everyday life? The surface tension of water is reduced by the addition of detergents. How and why is that effect useful to us?
6. Why is it dark deep in the oceans? Why can we not easily see the bottoms of streams and rivers?
7. What do we need to know about the properties of the ocean water beneath a ship if we want to use sonar to measure the depth accurately? Why does sound travel long distances in ocean water?

KEY TERMS

You should recognize and understand the meaning of all terms that are in boldface type in the text. All those terms are defined in the Glossary. The following are some less familiar key scientific terms that are used in this chapter and that are essential to know and be able to use in classroom discussions or exam answers.

acoustic (p. 164)
backscattered (p. 161)
calorie (p. 150)
conduction (p. 157)
convection (p. 157)
density (p. 154)
electromagnetic radiation (p. 150)
estuary (p. 155)
frequency (p. 159)
heat capacity (p. 150)
hydrogen bond (p. 150)
hydrophone (p. 164)
ion (p. 155)
joule (p. 150)
latent heat of fusion (p. 150)
latent heat of vaporization (p. 150)
plankton (p. 159)
refracted (p. 162)
salinity (p. 154)
scattered (p. 161)
sensible heat (p. 151)
sonar (p. 162)
sounding (p. 162)
surface tension (p. 157)
suspended sediment (p. 159)
turbidity (p. 159)
van der Waals forces (p. 150)
viscosity (p. 157)
water mass (p. 154)
wavelength (p. 159)

CRITICAL CONCEPTS REMINDER

CC1 Density and Layering in Fluids (p. 154). Water in the oceans is arranged in layers according to the water density. Many movements of water masses in the oceans are driven by differences in water density. To read **CC1** go to page 2CC.

CC3 Convection and Convection Cells (pp. 154, 157). Fluids, including ocean water, that are cooled from above or heated from below, sink or rise because their density is increased or reduced respectively. This establishes convection processes that are a primary cause of vertical movements and the mixing of ocean waters. To read **CC3** go to page 10CC.

CC5 Transfer and Storage of Heat by Water (p. 153). Water's high heat capacity allows large amounts of heat to be stored in the oceans and released to the atmosphere without much change of ocean water temperature. Water's high latent heat of vaporization allows large amounts of heat to be transferred to the atmosphere in water vapor and then transported elsewhere. Water's high latent heat of fusion allows ice to act as a heat buffer reducing climate extremes in high latitude regions. To read **CC5** go to page 15CC.

CC6 Salinity, Temperature, Pressure, and Water Density (p. 154). Sea water density is controlled by temperature, salinity, and to a lesser extent pressure. Density is higher at lower temperatures, higher salinities, and higher pressures. Movements of water below the ocean surface layer are driven primarily by density differences. To read **CC6** go to page 16CC.

CC14 Photosynthesis, Light, and Nutrients (p. 159). Photosynthesis, the major process in the production of living matter, depends on the availability of light. Thus, the transmission of light through ocean water limits the depth to which photosynthesis may occur. To read **CC14** go to page 46CC.

CHAPTER 8

CHAPTER OUTLINE

CRITICAL CONCEPTS USED IN THIS CHAPTER

This beautiful cove and beach on the south shore of Hawaii are known locally as the "Green Sand Beach." The shoreline consists of volcanic ash that is easily eroded. The eroded cliffs supply the source of the sand on this beach and those nearby, as well as on the seafloor on the narrow continental shelf. The minerals are not weathered and have a distinctly greenish hue that, unfortunately, does not show well in photographs but that can be seen more easily in the inset close-up image of the sand grains.

Ocean Sediments

One of my first encounters with ocean sediments took place in the Atlantic Ocean just a few kilometers from the Straits of Gibraltar, the Pillars of Hercules of ancient legend. The research vessel crew had just deployed and retrieved a 15-m-long gravity corer that weighed more than a tonne, in itself an exciting and somewhat dangerous event to watch. The sediment core retrieved was over 12 m long, and in its plastic tube core liner, it had to be cut into sections before it was moved into the onboard laboratory. Once inside, each segment was carefully cut down the length of the tube, separating the core in two so that the sediments were exposed along the length of the core. At first glance, it was clear that the sediments were not uniform, because there were a number of distinct layers at intervals along the core length.

As I looked at the core, I thought about the history that it could tell. In that area, each centimeter of core represented approximately 100 years of history. Just below the sediment surface at the top of the core was a layer 1 or 2 cm thick that contained much larger particles than the sediment either above or below this layer. I was told the large particles were "clinker," the solid material left behind when coal was burned and thrown overboard by the many coal-fired ships that plowed these waters throughout much of the 1900s. Further down the core were distinct layers of sand grains interrupting the much finer mud that made up most of the core. Each of these sand grain layers represents a slump of materials from shallower areas near the shore called "turbidity currents."

Turbidity currents can be triggered by earthquakes and can cause tsunamis. Could the exceptionally large turbidity current layer that was found about 30 to 35 cm below the sediment surface in this core have occurred during the time of the Greek philosophers, or during the Minoan civilization? Could it have been associated with a large

tsunami? Perhaps this event was responsible for the legend of Atlantis. Lower still, there was an even larger turbidite layer. The turbidity current that created that layer must have occurred before the dawn of civilization. And what of the other segment of the core 12 m below the surface sediments? That sediment may have been deposited more than 100,000 years ago. What stories could these sediments tell of changes in the Earth's climates and of the evolution of species in the oceans, if we could learn to read their record?

Why should we be interested in marine **sediments**? Approximately 70% of the Earth's surface is covered by the oceans, and almost all of the seafloor beneath these oceans is covered by sediments. These sediments provide a home for many living species and contain a wealth of information about the history of the Earth's **climate,** oceans, and continents. Sediments accumulate layer by layer and include the remains of dead organisms. By studying changes in the type and chemical composition of the organisms, and the chemistry of sediment in each layer beneath the sediment surface, we can learn a great deal about the Earth's history and the evolution of life.

This chapter reviews the origin and types of particles present in ocean sediments and the mechanisms by which they enter the oceans and are transported to the seafloor. It also describes how the type of sediment that accumulates at any location can change with climate and **plate tectonics** and examines how the sediments can be used to investigate the Earth's history.

8.1 SEDIMENTS AND BIOGEOCHEMICAL CYCLES

Sediments are important components of **biogeochemical cycles** (Chap. 6), and they act as both sinks and sources of elements dissolved in seawater. Human civilization has now altered the **steady state** (Chap. 6) of the biogeochemical cycles of many elements by mining rocks and burning **fossil fuels;** by deforestation, which increases **erosion;** and by releasing **contaminants** that cause **acid rain** and promote **leaching.** Many human activities have increased the rate at which some elements enter the ocean and atmosphere, which will probably cause concentrations of these elements in seawater to rise, even if the rate of removal to the sediment increases to match the increased input rate (**CC8**).

For many elements, the rate of input to the oceans has varied over millions of years. Periods of intense volcanic activity or changes in land erosion rates related to sea-level and **glaciation** changes, for example, affect the rate of input to the oceans. Without a full understanding of these processes, we cannot understand or predict how releases of elements by human activities will affect the ocean **environment.** Biogeochemical cycles are difficult to study because of their many complex interactions and because some of the key processes occur on timescales of tens of millions of years. Fortunately, seafloor sediments preserve a history of the biogeochemical cycles and their variations over the past 170 million years (the age range of all surviving seafloor; Chaps. 4 and 5). Sedimentary rocks on land preserve even older history. Studying the jigsaw puzzle of information contained in ocean sediment is essential to understanding the consequences of human activities, including **pollution** (Chap. 19).

8.2 CLASSIFICATION OF SEDIMENTS

Sediments are separated into categories to facilitate comparisons and communication among scientists. There are a number of classification schemes. The most widely used schemes are based either on the range of sizes of the particles (grains) that make up the sediment or on the origin of the predominant particles in the sediment. Each type of scheme has its advantages, as we shall see in this chapter.

Classification by Grain Size

The size of the grains in marine sediments varies from boulders tens of centimeters or more in diameter to grains that are so small (less than 0.001 mm in diameter) that they cannot be distinguished except under the most powerful electron microscopes. Various classification schemes have been used, but generally sediments are described as clay, silt, sand, or gravel (**Fig. 8.1**). Sediments with the finest **grain size** ranges—silt and clay—are often classified together as mud.

Sediments with a wide range of grain sizes can be found in different areas of the oceans or at different depths within the layers of sediment at any given location. However, most marine sediments that have accumulated at the same time and in the same area of the oceans consist of only grains that are within a narrow range of grain sizes. Sediments that consist mostly of grains that are within a narrow range of sizes are said to be well **sorted.** For example, most sediment on the deep-ocean floor consists of well-sorted mud, and many areas near **beaches** consist of well-sorted sand. Sediments that consist of a wide range of grain sizes are said to be poorly sorted. For example, some undersea landslide deposits are poorly sorted, consisting of a mixture of mud, sand, and gravel. Poorly sorted sediments are found in only limited

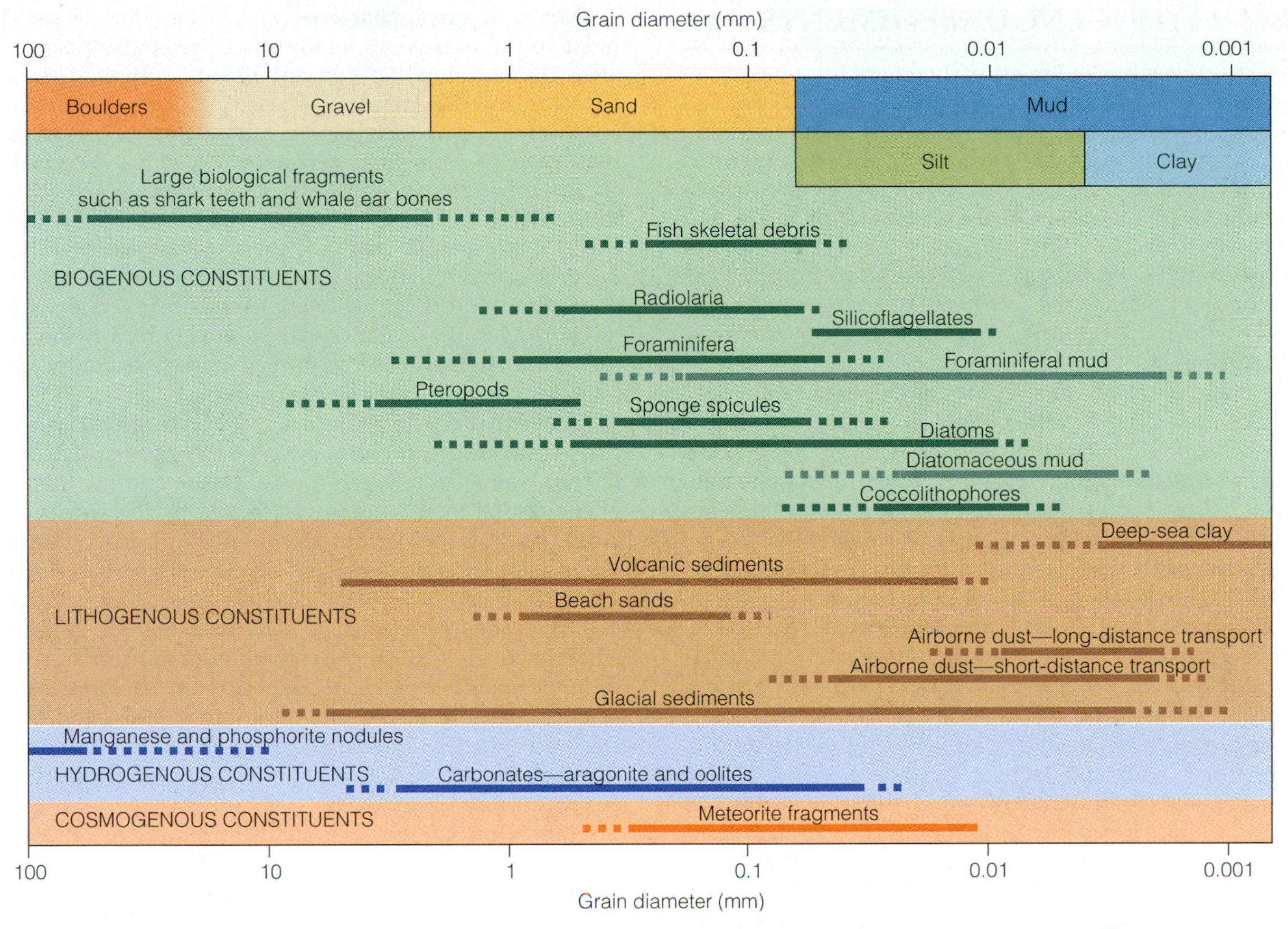

FIGURE 8.1 Various types of particles, each of which has its own characteristic size range, are commonly found in marine sediments. The particles in foraminiferal and diatomaceous muds are, on average, smaller than the sizes of the foraminifera and diatoms from which they are derived because the hard parts of these organisms are partially dissolved and broken mechanically during transport to, and burial in, the sediments.

areas in the oceans. Transport mechanisms that cause sediment sorting are discussed in **CC4**, and sorting of beach sand is discussed in Chapter 13.

Classification by Origin

Sediments can be classified as **lithogenous, biogenous, hydrogenous,** or **cosmogenous** to indicate the origin of the dominant type of particle. Lithogenous particles consist of fragments of rocks and minerals **weathered** and eroded from rocks on land, and fragments of ash and rock from terrestrial and undersea volcanic eruptions. Biogenous particles are the remains of organisms. Because most organic matter is decomposed rapidly, biogenous particles are predominantly the mineralized **hard parts** of marine organisms. They include, for example, the hard parts of certain **phytoplankton** and **zooplankton** species, **mollusk** shells, fish bones, and whale and shark teeth. Hydrogenous particles are formed by precipitation of dissolved inorganic chemicals from seawater to form solids. Cosmogenous particles are meteorite fragments that have passed through the atmosphere. Each category of particles can be subdivided into several different types. For example, biogenous particles can be classified according to the types of organisms from which they are derived (**Fig. 8.1**).

The distribution of sediment particles in the oceans is determined by the size range and origin of each type of particle, the susceptibility of each type of particle to decomposition or dissolution in seawater, and the mechanisms transporting the particles. The distribution of sediments is discussed later in this chapter. However, to understand the distribution of sediments, we need first to understand the origins and sources of the particles that make up sediments and the processes that transport and transform the particles in the ocean.

8.3 LITHOGENOUS SEDIMENTS

Most lithogenous sediment particles are the products of weathering and erosion of terrestrial rocks by water and wind. For this reason, sediments with a high proportion of lithogenous particles are often called **terrigenous** sediments. Rock fragments are continuously flaked off of solid rock by the action of running water or waves, as a result of freezing and thawing of ice, or by the actions of plant roots or animals. Once they are broken into fragments, rock particles are further weathered to smaller particles and transported by streams, rivers, waves, winds, and **glaciers.**

Rocks of the landmasses are composed of a large variety of minerals. The minerals are slowly altered and partially dissolved by reactions with oxygen, carbon dioxide, and water. During such chemical weathering, many minerals partly dissolve, leaving behind resistant minerals including quartz, feldspars, and clay minerals, which are all **siliceous** minerals. Clay minerals are layered structures of silicon, aluminum, and oxygen atoms. Some clay minerals also contain iron or other elements.

Five natural transport mechanisms bring lithogenous sediment to the oceans: freshwater **runoff,** glaciers, waves, winds, and landslides. Humans have added another mechanism: ships. Until recently, all vessels simply threw trash and garbage overboard. Cans, bottles, plastics, and clinker (cinders from coal-burning ships) can now be found in sediments throughout the oceans, particularly under major shipping lanes (Chap. 19). The various transport mechanisms introduce different size ranges of sediment particles to the oceans, and each mechanism has a different distribution of input locations.

Transport by Rivers

Rivers vary substantially in terms of their volume of flow, velocity, and **turbulence.** In the fast-flowing upper parts of rivers where they cut through steeply sloped mountain valleys, strong turbulence can be created, and as a result, substantial quantities of rock can be eroded. Such turbulent rivers carry a wide size range of particles. As rivers reach flatter land near the coast, turbulence decreases, and the largest particles are deposited in this region (CC4). Consequently, under normal flow conditions, rivers transport mostly fine-grained particles to the oceans. In contrast, when heavy rains flood rivers, both the volume of the river discharge and the river velocity increase dramatically, and great quantities of deposited sediment, including larger particles, are **resuspended** from the riverbed. River inputs are continuous, but the rate of input, particularly of the larger particles, peaks during major flood events. The peaks can be dramatic. Rivers may transport more sediment to the ocean in a few days after an unusually massive storm, such as a **hurricane,** than they do during several years or longer of normal conditions.

Not all rivers discharge significant amounts of sediment to the oceans. On many coasts, rivers flow across areas that are now flat **coastal plains,** although they follow valleys that were cut by glaciers or rivers when sea level was lower. As sea level rose, many of the valleys were inundated to form long **estuaries** in which **current** speeds are relatively low. Most particles transported by the rivers collect in the estuaries and do not reach the ocean. For example, many of the rivers of the Atlantic coast of North America pass through such estuaries and do not transport much sediment to the ocean. Unless sea level changes, the estuaries will eventually fill with sediment, and **deltas** may be formed by accumulation of sediment discharged at the river mouths.

Rivers that discharge onto active **continental margins** at **subduction zones** are generally short and drain limited land areas between the **coastline** and the ridges of the coastal mountain ranges created by the **subduction** process. The Pacific coasts of North and South America are examples (Fig. 8.2). Because they drain limited land areas, such rivers generally carry only a small load of eroded rock particles. The load may be greater where there are gaps in the coastal mountains, but even so, only small quantities of **suspended sediment** may be transported to the oceans. For example, the considerable suspended sediment load from the rivers draining the Sierra Nevada in California is deposited in the northern part of San Francisco Bay.

Approximately 90% of all lithogenous sediments reach the ocean through rivers, and 80% of this input is derived from Asia. The largest amounts are from four rivers that flow into the Indian Ocean (Fig. 8.2). The Ganges, Brahmaputra, and Irrawaddy discharge into the Bay of Bengal, and the Indus discharges into the Arabian Sea. Most of the other rivers that transport large amounts of suspended sediment into the ocean empty into **marginal seas.** Examples are the Chang (Yangtze) River (to the East China Sea), the Huang (Yellow) River (to the Yellow Sea), and Mekong (to the Andaman Sea)—all in Asia—and the Mississippi (to the Gulf of Mexico) in the United States. With the exception of the Amazon's input to the central Atlantic, river discharges of lithogenous sediment to the Pacific and Atlantic Oceans are very limited.

Erosion by Glaciers

Ice is squeezed into cracks in the rocks over which glaciers flow, causing pieces of rock to break off of the floor and sides of the glacial valley. This process of erosion is extremely effective, as can be seen from the deep **fjords** that the abundant glaciers cut during the last **ice age** (Fig. 13.4). The rock eroded by a glacier is bulldozed, dragged, and carried down the glacial valley with the glacier, and deposited at its lower end. At present, glaciers reach the sea only at certain places in high **latitudes,** including Antarctica, Alaska, and Greenland. These glaciers release most of the eroded rock into the water close

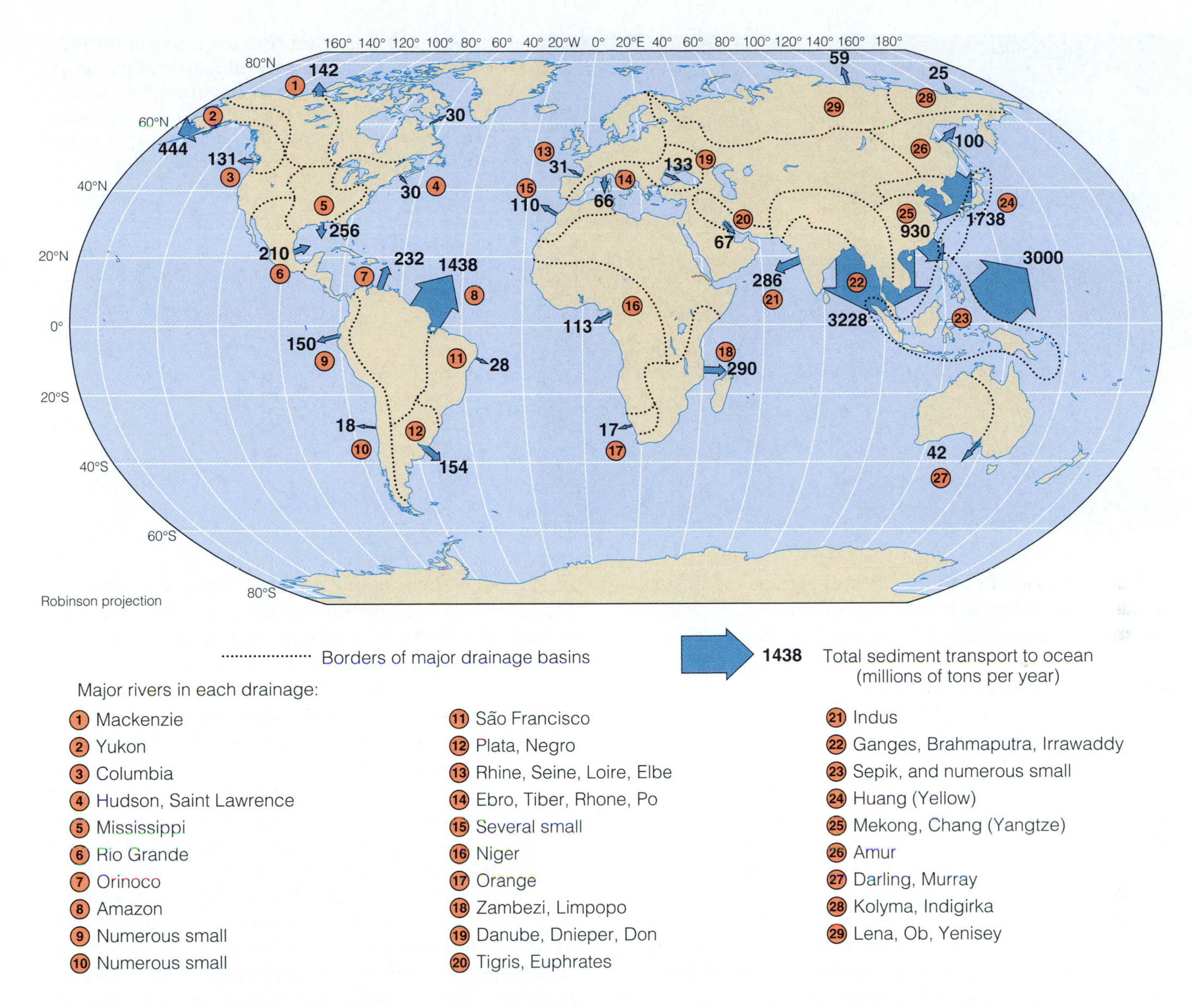

FIGURE 8.2 The major river inputs of terrigenous sediments to the oceans. These inputs are dominated by runoff from the Asian continent to the marginal seas of Southeast Asia. The Amazon is the only very large source of input to the Atlantic Ocean.

to where they enter the sea. Some icebergs that break off the ends of glaciers may be transported by ocean currents hundreds of kilometers or more before they melt. Rock particles are transported with the ice and can be released far from the glacier, but the amount of sediment transported in this way is small.

Glaciers can transport rock fragments that range from the finest grains to house-sized boulders. The larger boulders, pebbles, and sand grains are almost all deposited in ocean sediments close to the end of the glacier, but finer-grained material can be transported by currents to more remote **deposition** sites.

Some glaciers release large amounts of "glacial flour," particles so finely ground that they remain suspended for weeks or months, even in lakes or slowly flowing streams where turbulence is very low. Where glaciers discharge into lakes or fjords with long **residence times** (CC8), glacial flour can become so concentrated that the water is distinctly milky (Fig. 8.3).

Erosion by Waves

Waves continuously erode many coastlines (Chap. 13). The eroded rock particles are similar to those introduced to the oceans by rivers. However, particles eroded by waves generally have a larger proportion of unweathered mineral grains, unless the eroding coastline is composed of sedimentary rock (Chap. 13).

FIGURE 8.3 Bear Glacier enters a lake in British Columbia, Canada. Note the milkiness of the water due to the large quantities of fine particles (glacial flour) released from the melting glacial ice. When glaciers enter the oceans, much of the fine-grained material coagulates because of the high salinity and is deposited quickly.

Wave erosion of coastlines creates particles of all sizes, from large boulders that fall from undercut cliff faces to the finest clay particles. Waves sort the particles, transporting small ones offshore while leaving larger ones on or close to the **shore** (Chap. 13).

Transport by Winds

Dust particles can be transported very long distances through the atmosphere by winds before they are deposited on the ground or ocean surface. Dust particles in the atmosphere fall to the ground and are resuspended by air currents in much the same way that particles are deposited and resuspended by ocean currents (**CC4**). Normal winds can transport only very fine particles, but, as we can observe on a windy day at a dry sand beach, high winds can also transport larger grains.

When dust particles fall on the ocean surface, they become suspended sediment in the water. Consequently, there is a continuous flux of dust particles from the land to the oceans. The flux of airborne dust to the oceans is particularly strong in some locations, such as the northern subtropical Atlantic Ocean, where the prevailing winds blow from the Sahara out across the ocean. Dust clouds that stretch thousands of kilometers over the ocean can be seen in this area in satellite images (**Fig. 8.4**). Filters used by scientists to collect airborne particles in Florida and the Bahamas are often rust-red because they have collected large quantities of the red, clay-sized particles from the Sahara Desert that are transported across the Atlantic Ocean by winds. Windblown dust also is transported from the Gobi Desert in Asia over the North Pacific Ocean. These lithogenous clay-sized particles form a significant fraction of the deep-ocean sediment between 20°N and 30°N in the Pacific Ocean, where the accumulation rate of other particles is very low.

Although the amount of dust in the atmosphere is usually small, the continuous flux of small quantities through the atmosphere adds up to a substantial input to the oceans over geological time. Consider how much dust collects on a piece of furniture in only a few days. Multiply that amount by billions to see how much dust can reach the ocean surface over a period of millions or tens of millions of years.

Fine particles are deposited on all parts of the ocean surface at a relatively uniform rate, although rates of deposition are greater in areas downwind from deserts. High storm winds carry larger particles, up to fine sand size, over the oceans. Such particles are deposited relatively close to their coast of origin because most storm winds do not persist for long periods or blow over great distances.

Although winds generally transport only smaller particles, larger lithogenous particles do enter the atmosphere as a result of volcanic eruptions, especially the explosive eruptions of **convergent plate boundary** volcanoes (Chap. 4). Explosive eruptions can instantaneously fragment and blast upward large quantities of rock, which are then carried by winds as ash and cinders so that they rain out over a very large area. The principal fallout occurs immediately around and downwind of the volcano, where the larger particles drop. Eruptions produce particles of all sizes, including tiny particles that can be carried thousands or tens of thousands of kilometers in the atmosphere before being deposited. The most violent eruptions throw large quantities of ash into the upper atmosphere, where swift winds, such as the **jet stream,** can transport them one or more times around the planet before they finally fall. Volcanic ash from the largest eruptions can remain in the upper atmosphere for years and affect the Earth's climate by reducing the amount of solar energy that passes through the atmosphere to the Earth's surface (**CC9**).

Such ash can also reduce the **ozone** concentration in the Earth's **ozone layer** (Chap. 9).

Recent explosive eruptions that have ejected large quantities of ash into the atmosphere include the Mount Pinatubo eruption in the Philippines in 1991 and the Mount St. Helens eruption in Washington in 1980. The Pinatubo eruption completely buried the U.S. Clark Air Base with ash, even though the base was tens of kilometers from the volcano. The 1980 eruption of Mount St. Helens is estimated to have ejected about 1 cubic kilometer (km^3) of ash into the atmosphere, and Mount Pinatubo about four or five times that amount. These eruptions were dwarfed by two Indonesian eruptions: Krakatau in 1883 and Tambora in 1815, which ejected an estimated 16 and 80 km^3 of ash, respectively. The eruption that created the Long Valley **caldera** in California 700,000 years ago may have ejected 500 km^3 of ash. Much of the ash ejected by these eruptions would have fallen on the ocean's surface or been washed into the ocean in runoff and then transported to the sediments. Distinct ash layers are found in some ocean sediments corresponding to these and other eruptions.

Transport by Landslides

Landslides occur when loose soil or rock moves down a slope under the force of **gravity** in a process known as "mass wasting." Water, ice, and winds may be involved in loosening the soil or rock, but the rock or soil is transported by gravity, not by the water, ice, or wind. Some landslides occur on the slopes of a **shoreline** and may carry rock and soil particles of a wide range of sizes directly into the ocean. Most such material is initially deposited, but it then becomes subject to erosion and transport by waves. Mass-wasting processes also occur on steep slopes on the ocean floor. **Slumps** or **turbidity currents** (discussed in more detail later in this chapter) can occur as a result of accumulation of sediment on a **continental shelf** or at the head of a **submarine canyon,** or accumulation of volcanic rock on the side of a submerged or partly submerged volcano. These events can transport substantial amounts of lithogenous sediments into the deep-ocean basins.

8.4 BIOGENOUS SEDIMENTS

Almost all life in the oceans is sustained by the conversion of dissolved carbon dioxide to living organic matter by **photosynthetic** organisms. Photosynthesis takes place only in the upper part of the water column (at most a few hundred meters deep) where sufficient light is present (Chap. 14, CC14). With only minor exceptions, the photosynthetic organisms of the oceans are small (less than about 2 mm), as are most of the animals that feed on them (Chap. 14). Very few of these microscopic organisms live out a full natural life span, because most are consumed by larger animals. Those that are consumed are only partially digested by the larger animals, and their hard parts are not normally decomposed or digested as they pass through the **food web.** Undigested food particles are packaged together in animal guts to form fecal

FIGURE 8.4 This August 2004 satellite image shows a huge yellowish-brown, windblown dust plume from the Sahara Desert. The plume extends northward from the African continent. After traveling approximately 1000 km over the Atlantic Ocean, passing over the Canary Islands en route, the plume turns westward toward North America as seen near the top of this image.

material that is **excreted** in the form of **fecal pellets.** Fecal pellets, although small, are much larger than the microscopic organisms of which they are made up. Therefore, they tend to sink toward the seafloor relatively rapidly, as do bodies of larger animals that have not been ingested by another animal (CC4). Most life in the deep-ocean waters and on the seafloor is sustained by the rain of this organic-rich **detritus** (Chap. 15).

Recently it was discovered that a small **planktonic tunicate** called a "giant larvacean" may contribute a substantial proportion of the total detritus rain through a previously unknown mechanism. The giant larvacean secretes an intricate netlike structure to capture its food. The larvacean lives inside the structure, which, consequently, has been called a "house." When the web becomes clogged with particles, which happens about once a day, the larvacean simply releases the house and secretes a new one. The abandoned houses are large enough that they sink fairly rapidly, and this source may then contribute a substantial proportion of the detritus that reaches the seafloor, at least in some areas.

As detritus falls through the water column, it is either utilized by animals or decomposed by **bacteria** and **fungi,** and decomposition continues on the seafloor. In all but a few locations where the rain of detritus is extraordinarily heavy or where there is insufficient oxygen to support bacterial decomposition, the organic matter in detritus is essentially completely decomposed. Consequently, little organic matter is incorporated in the accumulating bottom sediment in most areas. By contrast, many marine species have hard parts that are not decomposed as they fall through the water column, and these materials constitute the overwhelming majority of inputs of biogenous particles to ocean sediment.

The hard parts, shells, or skeletons of marine organisms are either **calcareous** or siliceous. Calcareous organisms have hard parts of calcium carbonate ($CaCO_3$); siliceous organisms have hard parts of silica in the form of opal, which has the same basic composition as glass (SiO_2). Ocean surface waters are populated by very large numbers of calcareous and siliceous organisms, most of which are smaller than a few millimeters. Hence, a continuous rain of both calcareous and siliceous particles falls from the surface layers. This material does not rain uniformly on the sediment surface, because productivity and types of organisms vary between ocean areas, and because calcium carbonate and silica dissolve in seawater at slow but variable rates.

The two major factors that determine the accumulation rate of biogenous sediment are the rate of production of biological particles in the overlying water column and the rate of decomposition or dissolution of these particles as they fall to the seafloor. The percentage of biogenous material in the sediment is determined by these two factors plus the rate of accumulation of other particles.

Regional Variations of Biogenous Particle Production

Chapters 14 and 15 examine the factors that determine **primary productivity** rates and the types of organisms that inhabit various regions of the oceans. The distributions of productivity, **biomass,** and dominant organism type (e.g., calcareous or siliceous) in the **ecosystem** are major factors in determining the nature of sediments that accumulate in any given area.

In high latitudes and areas of **coastal upwelling** (Chap. 15), siliceous **diatoms** are the dominant photosynthetic organisms (Fig. 8.5). Diatoms are among the largest of the phytoplankton (up to about 2 mm) and yield relatively large siliceous particles. At lower latitudes and in the open ocean, many of the dominant photosynthetic organisms have no hard parts. However, one group of very tiny photosynthetic organisms called **coccolithophores** (Fig. 8.6) may grow in abundance. They are covered by a number of small calcareous plates that contribute extremely fine particles to the sediment. The white chalk cliffs of Dover in England consist primarily of coccolithophore plates.

Zooplankton with calcareous shells include **foraminifera** and **pteropods** (Fig. 8.7), many species of which are present throughout the oceans. Animals with siliceous hard parts are less common, but the microscopic **radiolaria** (Fig. 8.8) have intricate silica shells and are abundant in tropical waters that have high primary productivity.

FIGURE 8.5 Several species of diatoms (Chrysophyta) photographed at 25 times magnification. Diatoms are algae that are a preferred food source for many small marine animals and animal larvae. They are photosynthetic, and each one is covered by a hard silica frustule.

FIGURE 8.6 Coccolithophore (*Coronosphaera mediterranea*) photographed at aproximately 5500 times magnification. Note the covering of intricate calcareous plates.

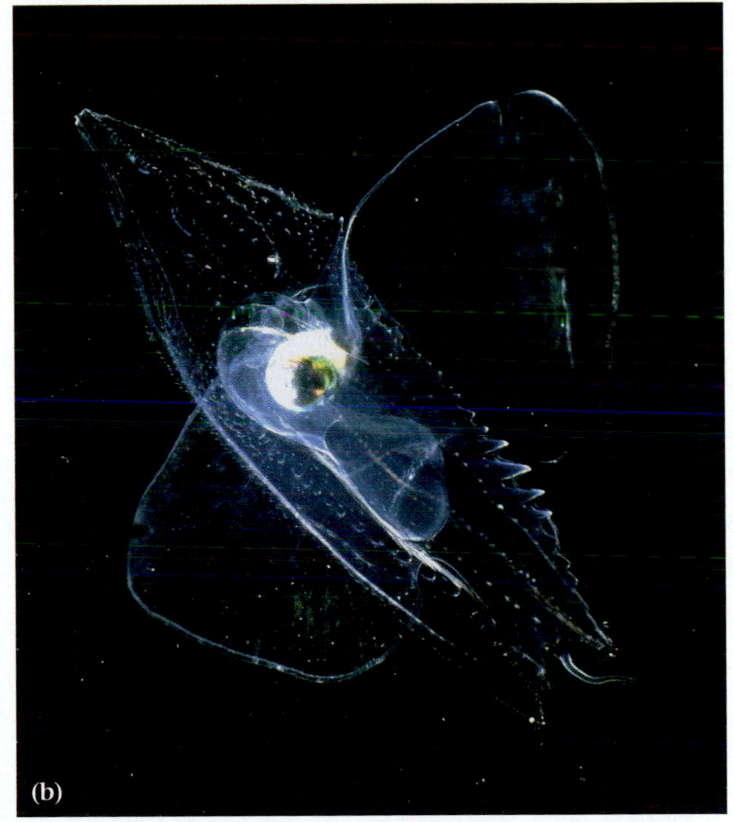

FIGURE 8.7 (a) Most foraminifera are pelagic (live in the water column). However, the species shown in this photograph (the round platelike objects that are about 1 cm in diameter, *Marginopora vertebralis,* Papua New Guinea) is benthic and lives on the coral reef surface. Several of the individuals shown here have died and lost their green pigmentation, revealing the calcareous hard parts that can survive to become sediment particles. (b) A pteropod, or sea butterfly—*Cymbula* sp. (Venus slipper). The animal "flaps" its two "wings" (modified foot) to propel itself through the water. It lives free-swimming in the oceans and eats plankton.

FIGURE 8.8 Radiolaria (photographed here at 7 times magnification) are small organisms that have intricate silica shells, which make a major contribution to the sediments. They occur in abundance in some tropical waters where primary productivity is high in the surface water layer.

Dissolution of Biogenous Particles

Seawater is undersaturated with silica. Therefore, siliceous particles dissolve as they fall through the water column to the seafloor. The rate of dissolution is very slow at all depths, but it decreases with depth throughout the upper 2 km of the water column (**Fig. 8.9**). Thin siliceous hard parts are almost always totally dissolved and reach the sediment only in areas where siliceous organisms grow in great abundance. For example, many radiolaria have thin, intricate shells that are dissolved relatively easily and reach the sediment in substantial quantities only in highly productive tropical waters where they abound.

The production of fecal pellets can enhance the accumulation of siliceous and calcareous particles. When packaged in relatively large fecal pellets, such particles fall more quickly to the seafloor and are partially protected from dissolution until the organic matter in the pellet disintegrates.

Many diatoms have a relatively thick siliceous **frustule** (hard part), much of which reaches the seafloor, where it continues to dissolve until buried by other particles. Diatom frustules are present in the sediment in locations where their production rate and the rate of sediment accumulation (the **sedimentation rate**) are high.

Calcareous particles also dissolve in seawater, but their behavior is different and more complex than that of siliceous particles. In the upper water layers, where water temperatures are high and pressures low (Chap. 10), seawater is generally **supersaturated** with calcium carbonate. Consequently, calcareous shells dissolve only very slowly or not at all in near-surface waters (**Fig. 8.9**). The **solubility** of calcium carbonate increases with pressure and with decreasing temperature. Therefore, solubility tends to increase with depth.

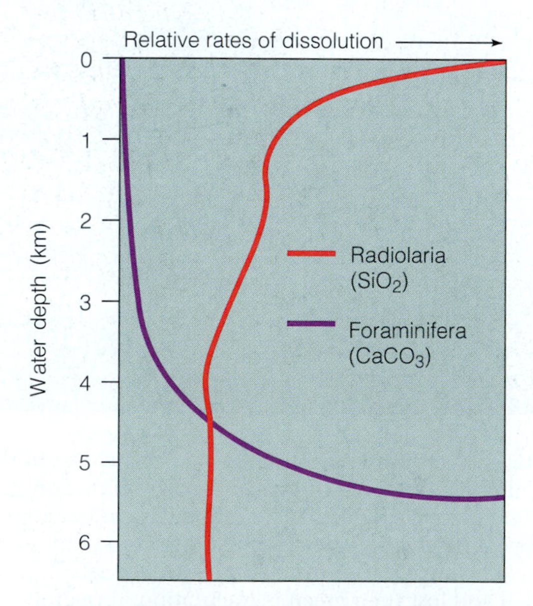

FIGURE 8.9 Changes in the dissolution rate of siliceous shells (radiolaria) and calcareous shells (foraminifera) with depth. Siliceous shells are more soluble in surface layers; calcareous shells are more soluble at greater depth.

Not all calcareous hard parts are made of the same mineral. Calcium carbonate shell material is of two distinct forms: **calcite** and **aragonite**. Some types of animals, including most foraminifera, have calcite shells, whereas others, such as pteropods, have aragonite shells. Aragonite dissolves much more readily in seawater than calcite. Therefore, pteropod shells are dissolved completely at shallower depths than foraminiferal shells. Where pteropods are more abundant than foraminifera, shallow-water sediments can consist of predominantly pteropod particles. Sediments at intermediate depths are predominantly foraminiferal, and sediments at greater depths have almost no calcareous component (**Fig. 8.10**).

Carbonate Compensation Depth

Below a certain depth, seawater is undersaturated with respect to calcium carbonate, and calcareous debris starts to dissolve. Below this level, the degree of undersaturation and hence the dissolution rate of calcium carbonate increase with depth. Eventually, at the **carbonate compensation depth (CCD),** the dissolution rate is fast enough to dissolve all of the calcium carbonate before it can be incorporated in the sediment.

The CCD depends not only on pressure and temperature, but also on other factors, especially the concentration of dissolved carbon dioxide. An increase in carbon dioxide concentration lowers the **pH,** making the water more acidic, and thus increases calcium carbonate solubility.

Carbon dioxide solubility in seawater increases as temperature decreases. The deepest water layers of the oceans were formed by the sinking of cold water in certain high-latitude regions (Chap. 10). Thus, the deep-water layers contain relatively high concentrations of dissolved carbon dioxide. In addition, deep waters below the warm surface water layer accumulate dissolved carbon dioxide as animals **respire** and bacteria and other **decomposers** convert organic matter to energy and carbon dioxide (Chap. 14). Therefore, older deep water (water that has been away from the surface longer) has a higher carbon dioxide concentration than newer deep water. The deep water of the Pacific Ocean is older than the deep water of the Atlantic Ocean (Chap. 10). Consequently, calcium carbonate is more soluble and the CCD is shallower in the Pacific Ocean because of its higher dissolved carbon dioxide concentrations. The CCD is at approximately 4000 m in the Atlantic, at about 2500 m in the South Pacific, and at less than 1000 m in the North Pacific, which has the oldest deep waters with the highest carbon dioxide concentrations.

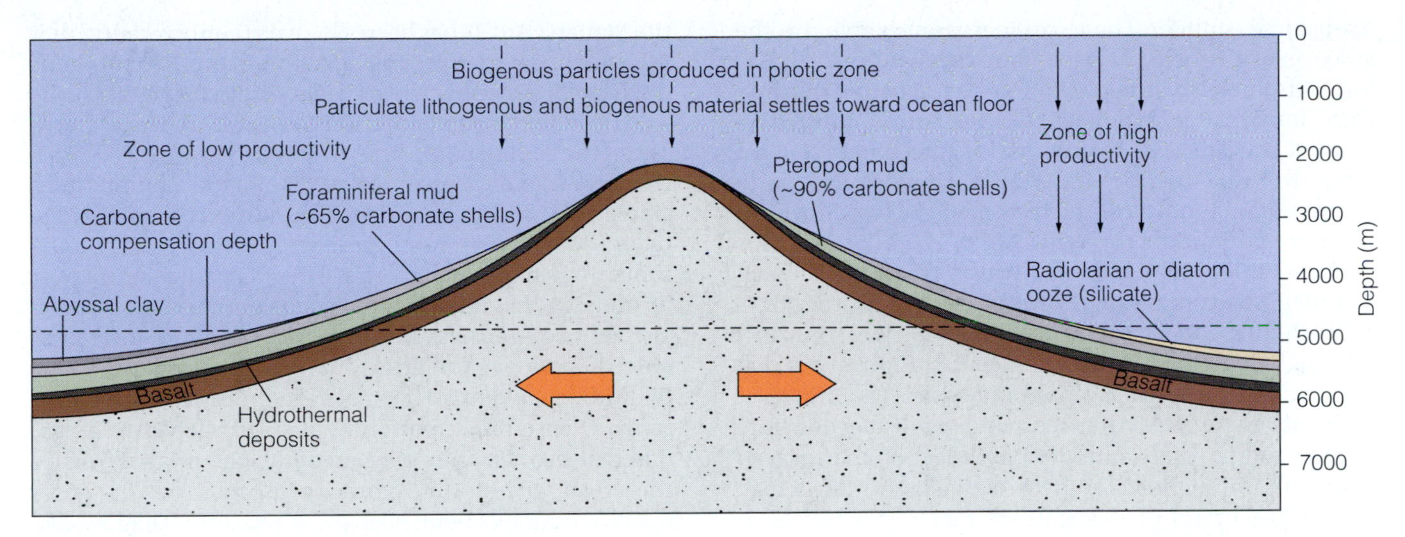

FIGURE 8.10 Idealized depiction of layered sediment accumulation on an actively spreading oceanic ridge. There is a thin, patchy cover of predominantly hydrothermal sediments at the center of the oceanic ridge. On the flanks of the ridge, which are shallower than the carbonate compensation depth (CCD), calcareous sediments dominated by pteropods and foraminifera are deposited. However, pteropods dominate the surface sediments only at depths much shallower than the CCD because their shells are made of a more soluble form of calcium carbonate than those of foraminifera. As the new oceanic crust moves away from the spreading center and sinks isostatically, it is covered by layers of hydrothermal sediments, pteropod and foraminiferal oozes, and then siliceous sediments. These siliceous sediments can be either biogenous, if diatomaceous or radiolarian production in the overlying water column is high, or fine-grained lithogenous clays. In the Pacific Ocean, the lithogenous sediments are primarily deep-sea clays because subduction zones prevent inputs of turbidites. In the Atlantic Ocean, they are generally a mixture of deep-sea clays and turbidites.

Carbonate Compensation Depth and the Greenhouse Effect

The carbonate compensation depth (CCD) does not remain fixed but changes with climate, particularly with changes in the carbon dioxide concentrations of ocean water. An understanding of these changes is necessary to predict the fate of the excess carbon dioxide released to the atmosphere by the burning of fossil fuels (Chap. 9, CC9).

Atmospheric carbon dioxide, carbon dioxide dissolved in seawater (primarily as carbonate and bicarbonate **ions**), and carbon locked up in calcareous sediments or sedimentary rocks are all involved in the same biogeochemical cycle. Some of the excess carbon dioxide released to the atmosphere will be absorbed by ocean water and calcareous sediments, but how much? Will the higher carbon dioxide concentrations lower seawater pH and move the CCD to shallower depth, causing more calcium carbonate to dissolve (Chap. 6)? Will this extra carbonate cause more carbon dioxide to be released to the atmosphere when the deep waters are brought to the surface? How fast will such changes take place? At present, these and many related questions cannot be answered with confidence. However, because atmospheric carbon dioxide and the CCD have varied substantially throughout the Earth's history, we may be able to find answers in the record preserved in deep-sea sediments.

8.5 HYDROGENOUS SEDIMENTS

The input of hydrogenous sediments to the ocean floor is small in comparison with the inputs of biogenous and lithogenous material. Nevertheless, hydrogenous material can be an important component of the sediment, particularly in some areas where the accumulation rate of biogenous and lithogenous material is low.

Seawater is generally undersaturated with most dissolved substances. Consequently, hydrogenous sediments are formed under special conditions where the chemistry of ocean water is altered. Various types of hydrogenous sediment form under different sets of conditions. The types include **hydrothermal minerals, manganese nodules, phosphorite nodules** and crusts, carbonates, and **evaporites.**

Hydrothermal Minerals

Because the Earth's **crust** is thin at the **oceanic ridges,** more heat flows through the seafloor from the **mantle** at the ridges than elsewhere. The excess heat drives a series of **hydrothermal vents** that discharge hot water (sometimes several hundred degrees Celsius) into the cold surrounding seawater. The hot water is devoid of dissolved oxygen when discharged, and it contains high concentrations of metal sulfides, such as iron sulfide and manganese sulfide. These sulfides are soluble in the absence of dissolved oxygen, but they are oxidized to form insoluble hydrous oxides as the vent plumes mix with much larger volumes of the surrounding oxygenated seawater. The hydrous oxides precipitate to form a cloud of fine, metal-rich particles. These particles sink to the seafloor and accumulate as metal-rich sediment in the area surrounding the vent.

In addition to iron and manganese, hydrothermal vent deposits contain high concentrations of other metals, including copper, cobalt, lead, nickel, silver, and zinc. Although some particles accumulate to form metal-rich sediment near the vents, many fine particles formed at the vents are probably transported large distances before settling and provide an input of hydrogenous particles to sediments throughout the deep-ocean basins. This may also be the origin of the material that forms the manganese nodules described later in this chapter.

The mechanism of hydrothermal vent circulation is not fully understood. The excess heat flow at the ridge is believed to heat water that enters cracks in the newly formed volcanic rocks. The heated water rises (**CC1**) through cracks in the seafloor and is replaced by cold seawater that seeps through cracks in the rocks and sediment on the sides of the ridge (Chap. 17). As the seawater migrates through the sediment and rocks, it is heated, its oxygen is depleted by reaction with sulfides, and its pH drops so that it leaches salts and metal sulfides from the sediment and surrounding rock before it is discharged at the vent. The vents are, in essence, parts of convection cells (**CC3**, **Fig. 17.16**).

Hydrothermal vents were not discovered until the late 1970s. They are surrounded by commercially attractive deposits of metal-rich sediment, and they also support unique **communities** of organisms, many of which are found nowhere else on the Earth (Chaps. 14, 17). Hydrothermal vents are limited to small areas on the ridge, often separated by substantial distances, but they are probably present throughout the oceanic ridge system. Whether the metal-rich sediment is abundant enough to be commercially exploitable is not known. In addition, whether the unique biological communities could be safeguarded if the minerals were commercially exploited is uncertain.

Test mining of hydrothermal minerals has been done along the axis of the Red Sea where a new oceanic ridge is forming. Because the deep waters of the Red Sea do not mix with open-ocean waters, the hydrothermal minerals have collected in huge quantities on the Red Sea floor. Hot water of very high **salinity** discharged by the hydrothermal vents also has accumulated in the deep basin because of its high **density** (**CC6**).

Undersea Volcano Emissions

Undersea volcanic activity is not limited to the oceanic ridges. Volcanoes also erupt under the sea at some **hot spots** and at some convergent plate boundaries, such as the Mariana Arc. It has been discovered only recently that sulfur-rich water, comparable to the hydrothermal fluids discharged at oceanic ridge vents, is discharged into the water column at some locations on these undersea volcanoes. The sediments in the areas surrounding these discharges sustain populations of biological communities similar to those found surrounding oceanic ridge hydrothermal vents, and these sediments are likely to be similarly rich in some metals. There is particular interest in studying these newly discovered communities and discharges because these undersea volcanoes are located at much shallower depths than most oceanic ridge hydrothermal vents. This makes them easier to reach for study. Even more importantly, these shallower vents are different because they discharge into the upper layers of the ocean where photosynthesis also occurs. This raises many questions about how these relatively shallow discharges and communities may affect and/or interact with the chemistry and biology of the surrounding waters.

Manganese Nodules

Manganese nodules are dark brown, rounded lumps of rock, many of which are larger than a large potato. Enormous numbers of nodules litter parts of the deep-ocean floor (**Fig. 8.11a**). They are potentially valuable as a mineral resource because they usually consist of about 30% manganese dioxide, about 20% iron oxide, and up to 1% or 2% other metals that are much more valuable, including copper, nickel, and cobalt.

Manganese nodules are probably formed by various mechanisms and from various sources of manganese, iron, and other elements. These chemicals may be from the dissolved ions in river water, from seawater leaching of volcanic materials in the oceans, or from the hydrothermal vents on the oceanic ridges. At present, the major source is thought to be the oceanic ridge hydrothermal vents. Particles of **colloidal** size (submicroscopic) are hypothesized to form at the hydrothermal vents, be transported throughout the oceans, and accumulate by **adsorption** on the surface of the nodule, perhaps aided by bacterial action.

Most nodules initially form around a large sediment particle (e.g., a shark's tooth) and are then built up in layers (**Fig. 8.11b**) at the very slow rate of about 1 to 10 mm per million years. Nodules are most common where the sedimentation rate is extremely slow. In areas where sediment accumulates rapidly, nodules are buried by new sediment before they have time to grow. Where nodules do form, occasional disturbance of the nodule by marine organisms is hypothesized to be necessary to prevent the nodule from being buried.

The areas of greatest accumulation of manganese nodules are the deepest parts of the Pacific Ocean (**Fig. 8.11c**), where the sedimentation rate is very low. The greatest density of manganese nodules is in the region of the Pacific Ocean south of a line between Hawaii and southern California and north of 10°N latitude.

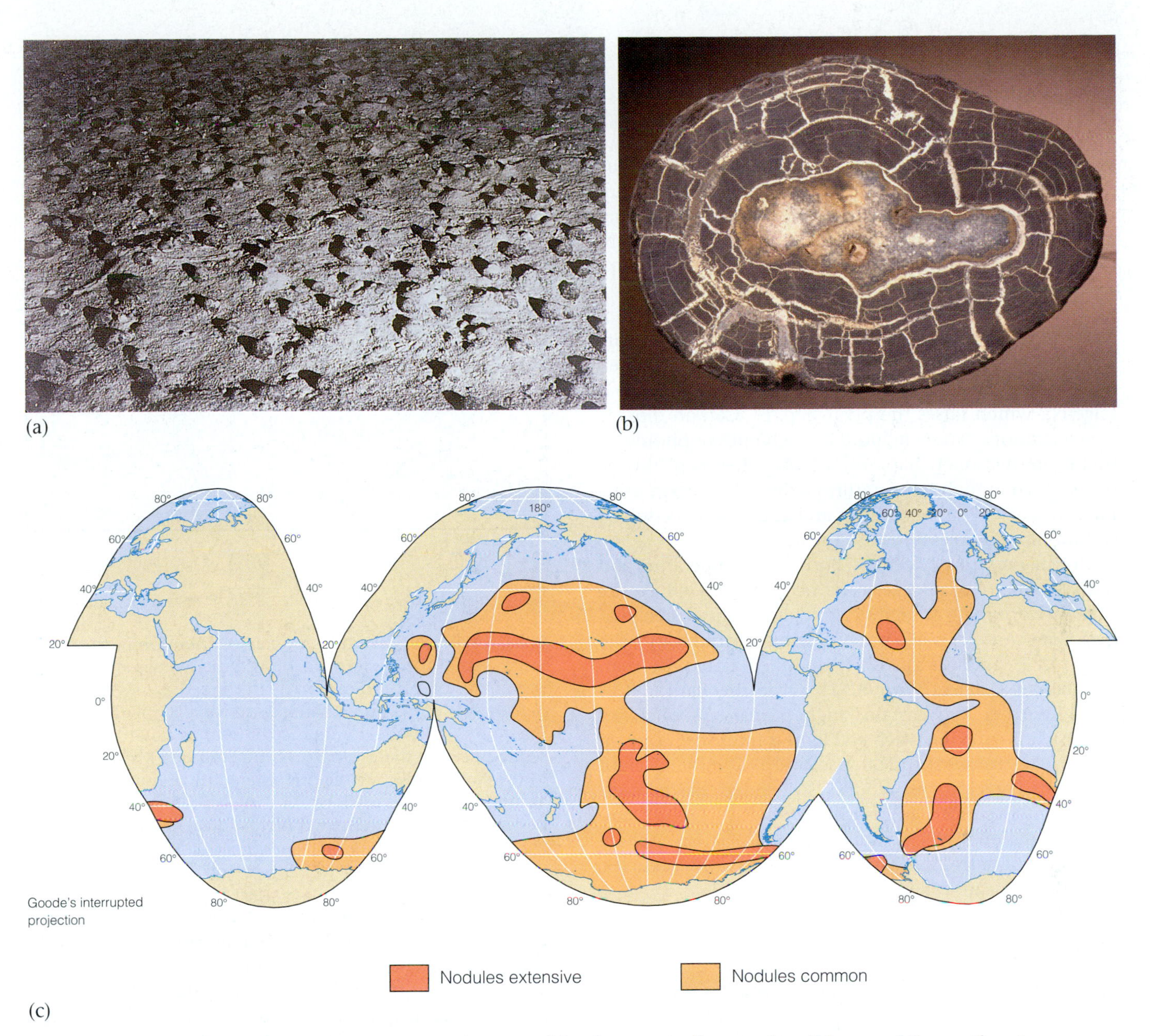

FIGURE 8.11 (a) Manganese nodules cover certain areas of the deep-ocean floor, such as this area of the northeastern Atlantic Ocean. (b) A sectioned manganese nodule shows that the nodule is made up of concentric layers. (c) The areas of the greatest density of manganese nodules are on the abyssal floor of the North Pacific Ocean between about 10°N and 20°N and in the central South Pacific Ocean.

Phosphorite Nodules and Crusts

Phosphorite is a mineral composed of up to 30% phosphorus. Phosphorite nodules form in limited areas of the continental shelf and **continental slope** and on some **seamounts.** Phosphorite nodule formation apparently requires low dissolved oxygen concentrations in bottom waters and an abundant supply of phosphorus. These conditions are present in **upwelling** regions where productivity is high. In such regions, the decomposition of falling detritus depletes oxygen in the bottom waters and supplies relatively large quantities of phosphorus, which is released as phosphate when the organic matter decays.

Phosphorite nodules grow slowly (1 to 10 mm per 1000 years) by a mechanism not yet fully understood. In contrast to manganese nodules, they do not form concentric layers but instead grow only on the underside, accumulating phosphate released by the decomposition of organic matter. Because very large deposits of phosphates are available on land, phosphorite nodules are unlikely to become commercially valuable.

Carbonates

Many **limestone** rocks lack **fossils.** Some consist of biogenous sediments that have lost their fossils through diagenetic changes (discussed later in this chapter), but others consist of hydrogenous sediments formed by direct precipitation of calcium carbonate. Calcium carbonate precipitates from seawater under conditions that apparently were widespread in the oceans at different times in the past, but now are present only in very limited regions, such as the Bahamas.

Calcium carbonate precipitation is more likely when water temperatures are high, because calcium carbonate solubility decreases with increasing temperature, as well as when the concentration of dissolved carbon dioxide is lowered, which raises the pH (Chap. 6). Carbon dioxide concentrations can be reduced by high rates of photosynthetic production (Chap. 14, **CC14**). They can also be reduced when high temperatures reduce the carbon dioxide solubility of surface waters and allow some dissolved carbon dioxide to escape into the atmosphere.

In the present-day oceans, both high temperature and high productivity appear to be necessary to reduce the calcium carbonate solubility enough to cause precipitation. For example, warm water flowing north through the Straits of Florida (Chap. 10) is upwelled across the shallow Bahama Banks (**Fig. 8.12**). Over the banks, it is heated further and sustains high primary productivity because it is **nutrient**-rich and light is intense (Chap. 14, **CC14**). This causes the pH to rise, and as a result, calcium carbonate precipitates around suspended particles to form white, rounded grains called "ooliths." Ooliths collect in shallow-water sediment in many areas. Where they are especially abundant in relation to other particles, the sediments are called "oolite sands."

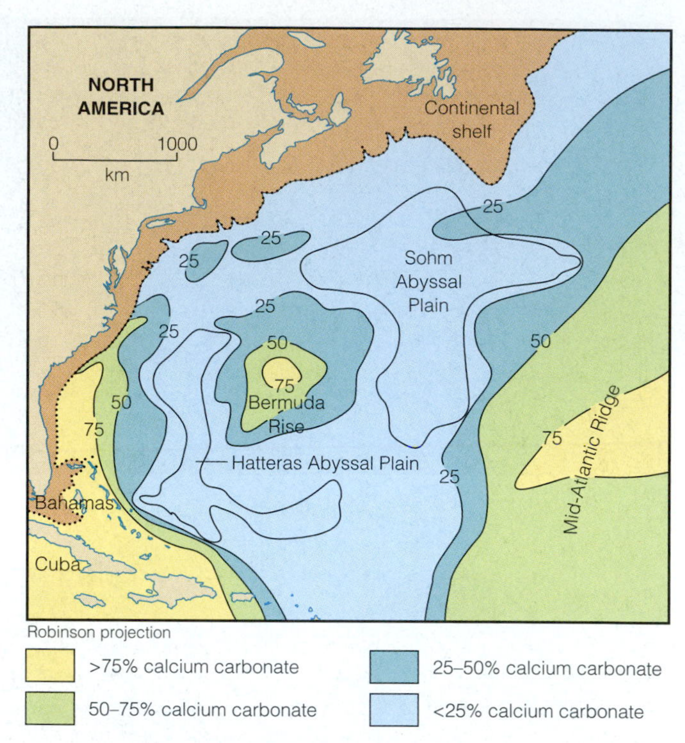

FIGURE 8.12 Calcium carbonate distribution in the sediments of the western North Atlantic. Values are the percentage of calcium carbonate in the sediment. The high percentages of calcium carbonate on the Bahama Banks are due primarily to the dominance of hydrogenous carbonate sediments. The shallow parts of the Bermuda Rise and the Mid-Atlantic Ridge are regions where lithogenous sediment accumulation rates are low, allowing biogenous carbonate particles to dominate the sediments in these regions.

Evaporites

In marginal seas in arid climates, the evaporation rate of ocean water exceeds the rate of replacement by rainfall and rivers (Chap. 9). Under such conditions, salinity progressively increases until the seawater becomes saturated with salts. As evaporation continues, salts, called "evaporites," are precipitated in succession: calcium and magnesium carbonates first, then calcium sulfate, and finally sodium chloride. Only a few marginal seas are evaporating in this way today, and even in these areas salt precipitation occurs only on tidal flats or in shallow embayments around the perimeter of the sea. Areas where evaporites form today include the Dead Sea, Red Sea, and Persian Gulf. Such seas must have been more abundant in the past because salt layers are present at many locations deep within ocean sediments, as well as in previously submerged parts of the continents.

Seismic profiles reveal several extensive layers of buried evaporite sediments, some more than 100 m thick, beneath the floor of the Mediterranean Sea. For such layers to have formed, the entire Mediterranean must have evaporated almost to dryness. This process apparently occurred several times when global sea level was lower and the connection between the Mediterranean and the Atlantic Ocean was broken at the **sill** that now lies underwater across the Straits of Gibraltar.

8.6 COSMOGENOUS SEDIMENTS

Most meteorites that enter the Earth's atmosphere are destroyed before reaching the surface. As they burn up, the meteorites form particles that rain out onto land and ocean and are eventually incorporated into sediments. These cosmogenous particles contribute only minor amounts of material to ocean sediments in comparison with biogenous, lithogenous, and hydrogenous sources. Nevertheless, the total input of meteorite dust to ocean sediments is estimated to be tens of thousands of tonnes each year.

Two types of meteorites hit the Earth: stony and iron-rich. The dust from the meteorites is found in sediment as cosmic spherules, which are spherical particles that show the effects of having been melted in the atmosphere. Stony meteorites form silicate spherules that are difficult to distinguish from lithogenous sediment grains. Most iron-rich spherules are about 0.2 to 0.3 mm in diameter and are magnetic, which makes them easier to find.

Recent studies have focused on finding the impact craters and debris from large meteorites that were not destroyed in the atmosphere but instead struck the oceans. A particularly large meteorite may have landed in the Caribbean Sea or Gulf of Mexico 65 million years ago. That impact is discussed later in this chapter.

8.7 SEDIMENT TRANSPORT, DEPOSITION, AND ACCUMULATION

Sediment particles are transported by ocean currents and waves in the same way that dust and sand are blown around by winds. High winds and fast water currents both cause particles to be suspended and carried until the wind or current speed diminishes. Particles are then deposited on the ground or ocean floor, but they may be picked up again if the wind or current speed increases. **CC4** explores the relationship between particle size and sinking rate, between current speed and sinking rate, and between current speed and resuspension of deposited particles.

Large particles sink rapidly, and high current speed is necessary to prevent them from being deposited. Once deposited, large particles are not resuspended unless current speed is considerable. Many large particles are transported to the ocean by rivers during periods of peak river flow that follow flooding rains. When such particles reach the ocean, where currents are weaker, they are either deposited or resuspended and transported by waves. In areas where wave energy is limited, the large particles are not resuspended, but they remain in sediment at the river mouth.

Waves can generate orbital water motions that have higher speeds (**orbital velocity**) than ocean currents have. As a result, they can resuspend large particles in waters that are shallow enough for the wave energy to reach the seafloor, and sand-sized particles can be resuspended and transported in the nearshore zone. Waves can transport sand long distances along the coast in shallow water (Chap. 13). However, the speed of water motion in waves is reduced with depth below the surface (Chap. 10). Sand-sized particles that are transported offshore are deposited when they reach depths at which the wave orbital velocity is no longer high enough to resuspend them. Smaller sand-sized particles that can be resuspended at lower water speeds are deposited in deeper waters than larger particles. Thus, waves tend to sort sand-sized particles in nearshore sediments: larger particles in shallow water and progressively smaller particles with increasing water depth.

Clay-sized particles are generally **cohesive,** but they tend to remain in suspension once resuspended, unless current speed is reduced substantially (**CC4**). The peak water velocity in waves is often sufficient to resuspend most cohesive sediments, and fine particles accumulate in nearshore areas only if these areas are well protected from waves and have slow currents. Such areas include both **wetlands** and fjords. Fine particles are transported by currents until they reach a low current area, where they are deposited permanently. The finest particles may be transported many thousands of kilometers and for many years before being deposited in the deep oceans, but they are often combined into clumps by **electrostatic** attraction or packaged into fecal pellets. The larger conglomerated particles are deposited more rapidly.

CRITICAL THINKING QUESTIONS

8.1 Fifty centimeters deep in a 10-m-long sediment core taken from the center of an abyssal plain in the Pacific Ocean, a thin layer of sediment is found that has a much larger grain size than the sediment either below or above it.
(a) Without knowing anything more about the material, give the possible explanations for this layer.
(b) How could you determine which one of these alternative explanations was most likely?

8.2 If you were dating different levels within a sediment core and found a layer that was older than both the layer above it and the layer below it, how would you explain this?

8.3 Offshore oil drilling now takes place in very deep waters on the continental slope and may eventually take place at abyssal depths. In many cases, plans call for oil and gas produced from these wells to be collected through short pipelines from several wells in a small area and then shipped ashore on tankers.
(a) Why might it not be a good idea in some deep locations to build pipelines on the seafloor to transport the oil ashore (ignoring economic reasons)?
(b) On what parts of the continental slope, if any, do you think pipelines should not be built?

8.4 Some sedimentary rocks found in the interior of continents are composed of very fine-grained silicate mineral particles with no calcium carbonate and low metal concentrations. Other continental sedimentary rocks are composed of coarse sand grains with little calcium carbonate, and yet others are composed of fine-grained material that is principally calcium carbonate. Describe the characteristics of the locations at which each of these three rock types were originally accumulated as sediments.

Aside from an occasional large particle (e.g., shark's tooth or whalebone), deep-ocean sediments remote from coastal sediment inputs are fine-grained, because almost all large particles are deposited in nearshore sediments and most particles introduced directly to the deep ocean are small. For example, most marine organisms are microscopic and therefore produce small particles. Meteorite dust particles, dust particles carried by winds, and most hydrogenous particles are also small.

Turbidity Currents

In some deep-ocean sediments, particularly those on **abyssal plains** adjacent to continental slopes, layers of coarse-grained sediments (sand and gravel) can be found. Layers of such sediments are separated by layers of the fine-grained sediments that normally accumulate on the deep-ocean floor. The coarse-grained sediments are transported downslope to the abyssal plain by turbidity currents.

Turbidity currents are similar to avalanches. Snow accumulates on a mountainside until it becomes unstable, breaks loose, and tumbles down the slope as an avalanche. Similarly, sediments accumulate on the continental shelf edge and slope until they become unstable, break loose, and flow down the continental slope as turbidity currents. Turbidity currents can be triggered by disturbances such as earthquake vibrations or sudden large discharges of sediment by rivers in the same way that avalanches can be triggered by noises or storm winds. Pockets or layers of methane or **methane hydrates** (Chap. 18) formed by the decay of organic matter in the sediments may rupture and be released when a turbidity current occurs. This may have the effect of providing a "lubricated" layer on which the turbidity current travels, which may enhance both the size and velocity of the turbidity current.

Once a turbidity current has been triggered, the disturbed sediments flow down the continental slope. As it gathers momentum, the turbidity current entrains more sediment and water, just as an avalanche gathers more snow.

Turbidity currents can flow down the continental slope at speeds of 70 $km \cdot h^{-1}$ or more, which are sufficient to suspend and retain large particles in suspension. The velocity is reduced when the turbidity current reaches the abyssal plain, and the entrained sediment particles are deposited from the suspended sediment cloud. The larger particles are deposited first, followed by progressively smaller particles, as speed and turbulence diminish. The result is the formation of a **graded bed** of sediments on the abyssal seafloor within which the largest grains are at the bottom and grain size progressively decreases upward (**Fig. 8.13**). Such graded beds, called "**turbidite** layers," can be meters thick. In many locations, a number of them are separated by layers of finer sediment. The fine-grained sediment layers were deposited slowly in the years, centuries, or millennia between successive turbidity currents.

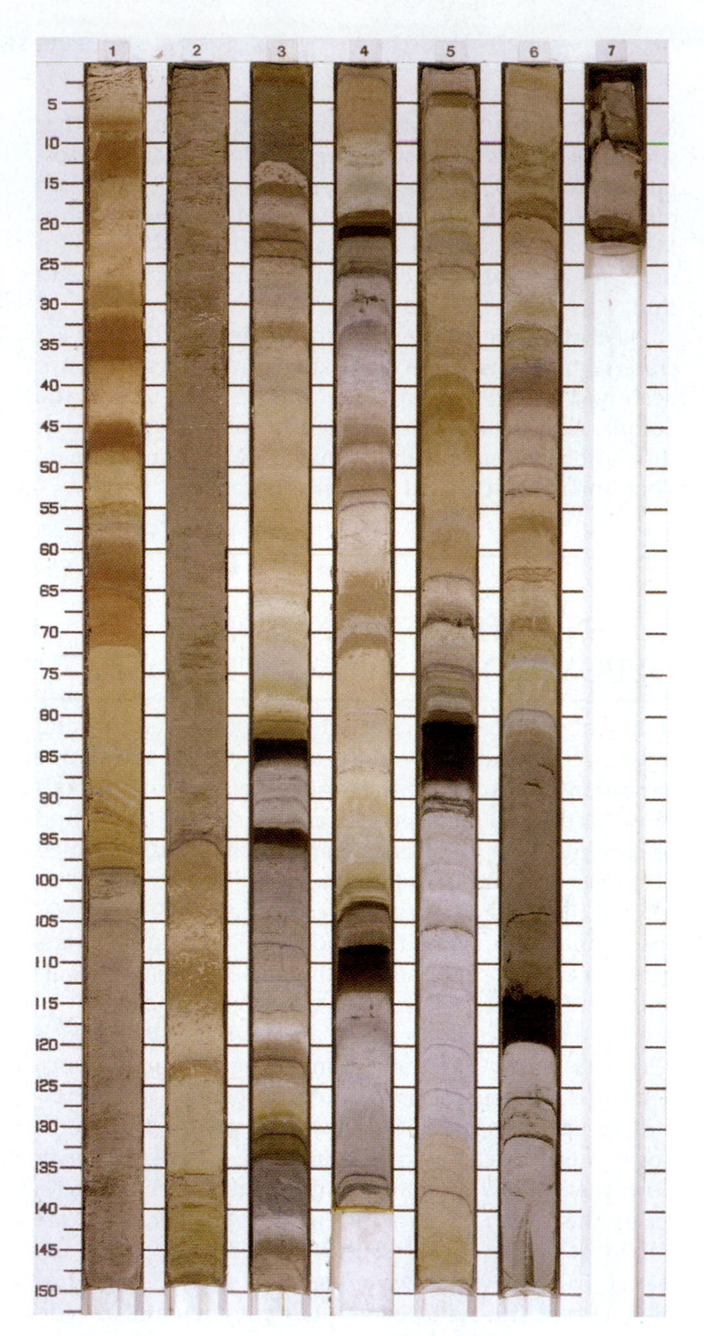

FIGURE 8.13 There are a number of turbidite layers in this Deep Sea Drilling Program core taken at the base of the continental slope in the North Pacific Ocean off Vancouver Island, British Columbia. Note the sharp change in grain size at the bottom of each turbidite layer and the gradual reduction in grain size above this boundary. The convex shape of the boundaries between layers is due to distortion of the core during the drilling process.

Turbidity currents can travel long distances on the abyssal plain before finally depositing all their terrigenous sediment load. Turbidity current deposits are generally thickest at the lower end of submarine canyons, where they form **abyssal fans** that decrease in thickness and median grain size in a seaward direction.

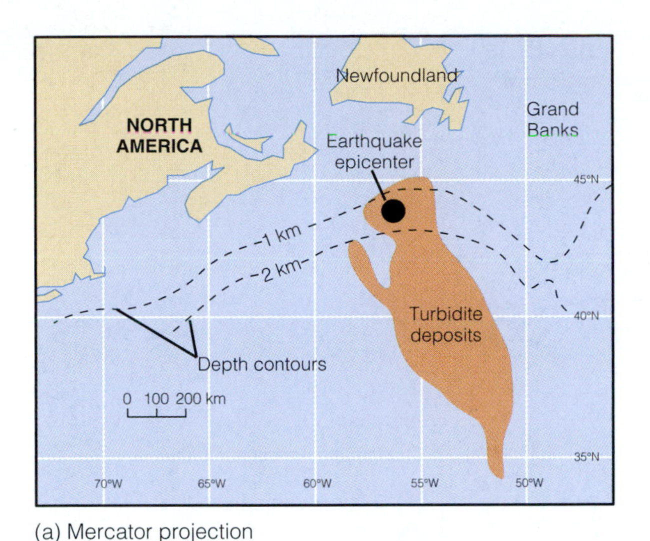

(a) Mercator projection

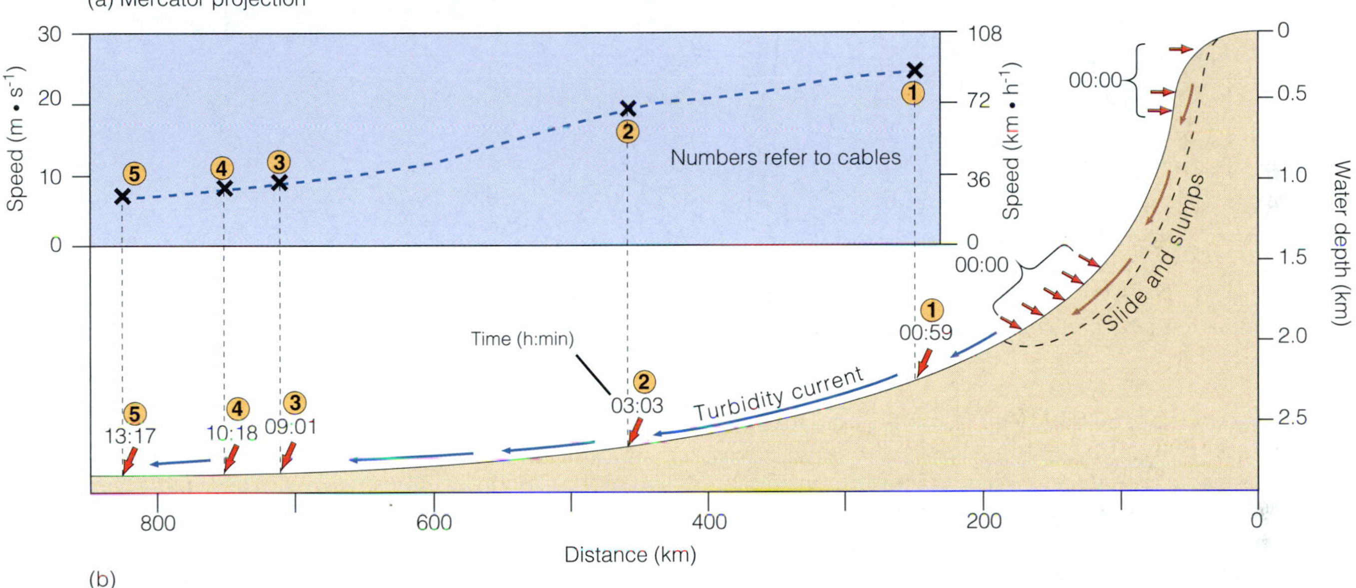

(b)

FIGURE 8.14 Submarine cables in the North Atlantic were severed by an earthquake-triggered turbidity current in 1929. (a) The earthquake epicenter was located on the continental slope. Turbidity currents caused by the earthquake spread turbidite sediments across large areas of the deep-sea floor extending more than 1000 km from the source. (b) The turbidity currents broke several submarine telephone cables at precisely known times. From these data, it was found that the turbidity currents had reached speeds in excess of 70 km•h^{-1}.

Where a deep-ocean **trench** is present at the base of the continental slope, turbidity currents are intercepted by the trench and do not reach beyond it to the abyssal plain. This is one reason why the **topography** of the Atlantic Ocean abyssal floor is flatter than that of the Pacific Ocean. The Atlantic Ocean floor topography is buried by turbidites that have been able to reach the oceanic ridge since the Atlantic Ocean first began to form. Because most of the Pacific Ocean floor was created after subduction zones formed around this ocean, turbidite deposits in the Pacific are rare.

Turbidity currents must be common events in the oceans, particularly where continental shelves are narrow and terrigenous sediment inputs are high. They are difficult to observe because they last only a few hours at most and occur unpredictably, although some may cause tsunamis. Nevertheless, their destructive power must be respected. In fact, this destructive power is what led to the first quantitative observation of the speed and geographic extent of turbidity currents. In 1929, an earthquake occurred in the Atlantic Ocean off Nova Scotia. The turbidity current triggered by the earthquake plunged down the continental slope, and snapped and buried several undersea telephone cables (**Fig. 8.14**). Because the precise times at which successive cables broke were known, the peak turbidity current speed was later estimated to be at least 70 km•h^{-1}. This turbidity current, like many others, moved primarily down a submarine canyon and traveled more than 600 km before slowing and depositing a turbidite layer across the abyssal plain. Because turbidity currents often flow down submarine canyons, it has been suggested that their scouring action maintains or even creates the canyons.

By studying the layers of sediments, scientists have determined that extremely large turbidity currents have occurred in the past and, therefore, may occur again. For example, about 20,000 years ago, a turbidity current occurred in the western Mediterranean that was estimated to have deposited 500 km^3 of sediments on the deep-sea floor, enough material to cover all of Texas with nearly 2 m of mud and sand. Another slide off the coast

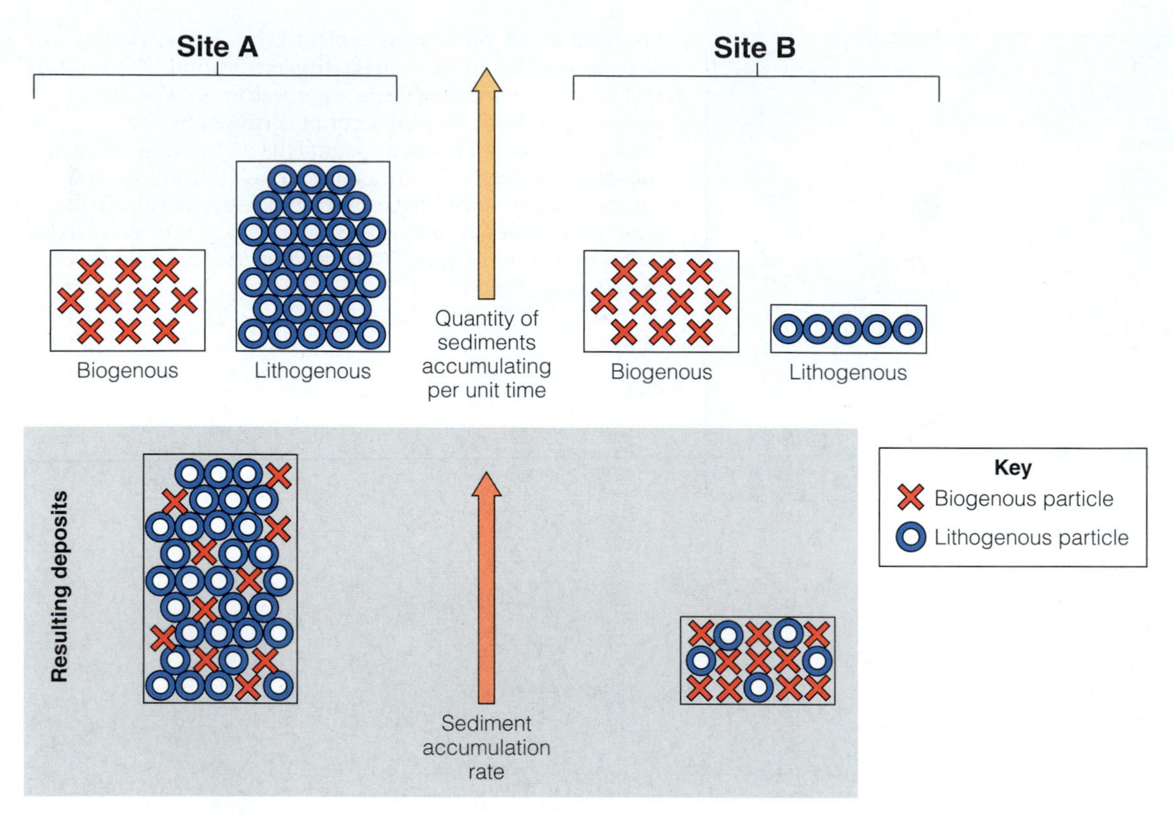

FIGURE 8.15 The relationships between the accumulation rates of different types of particles, and the accumulation rates and characteristics of the sediments.

of Norway occurred 30,000 to 50,000 years ago, involved more than twice the volume of sediment as the Mediterranean example, and left a scar in the continental slope that is larger than the state of Maryland.

Turbidity currents carry shallow-water organisms to great depths, where some, particularly microorganisms, may survive and adapt. Their occurrence also may affect the food web. After the 1989 Loma Prieta earthquake in California, dense **schools** of fish were observed feasting on the sediment-dwelling organisms exposed on the floor and sides of Monterey Canyon, where the sediment slumped and presumably caused a turbidity current.

Recently found evidence suggests that massive slides may occur when parts of oceanic ridge mountains collapse, carrying volcanic rocks far from the ridge. Such avalanches may cause intense turbidity currents.

Accumulation Rates

All sediments are mixtures of particles from many different sources. The accumulation rate and type of sediment are determined by the relative quantities of particles from each source that are deposited at each location. For example, **Figure 8.15** schematically illustrates the accumulation of sediments at two hypothetical locations. The input rate of biogenous particles to the sediment is the same at each location, but the input rate of lithogenous sediment is much higher at the first location than at the second. The sediments that accumulate at Site A are predominantly lithogenous, whereas the sediments at Site B are predominantly biogenous. However, the accumulation rate is much faster at A than at B. Because sediments are characterized by their predominant material, the Site A sediment would be called "lithogenous sediment," and the Site B sediment would be called "biogenous sediment," or **ooze.** An ooze is a sediment that contains more than 30% biogenous particles by volume. Where one type of organism is responsible for most of the biogenous particles, the ooze can be named after this type: pteropod ooze, radiolarian ooze, diatom ooze, foraminiferal ooze, and so on.

Lithogenous particles are the dominant input to ocean sediments. Most lithogenous material is discharged to the oceans from land as relatively large particles and deposited near river mouths and glaciers and in estuaries and wetlands. Sediment accumulation rates in these nearshore regions range from about 100 cm per 1000 years up to extreme rates such as the 7 m per year found in the delta of the Fraser River in British Columbia, Canada. Somewhat smaller but still large quantities of lithogenous sediment are transported offshore and are deposited on continental shelves. Such sediment can also reach the deep-ocean floor in areas where the continental shelves are narrow, or as turbidites. Many continental shelves are areas of high bio-

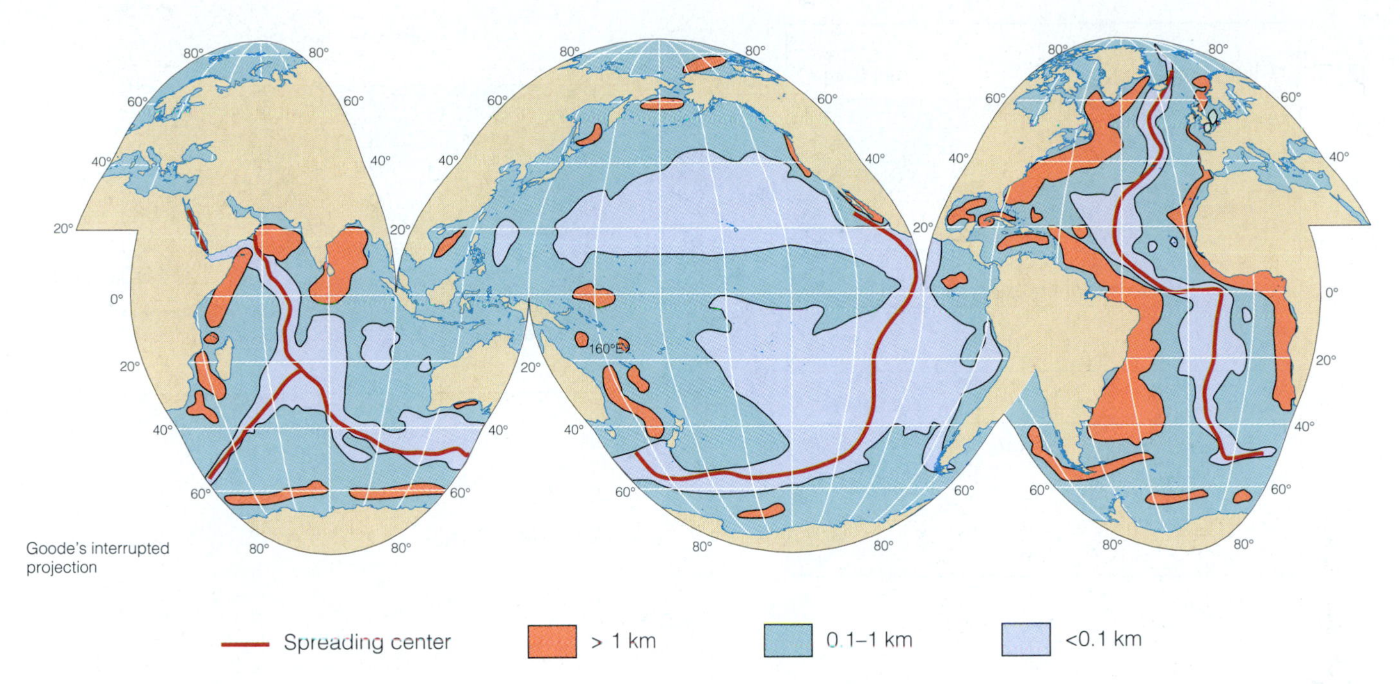

FIGURE 8.16 Thickness of accumulated sediments on the world's oceanic crust.

logical productivity (Chap. 15). Accordingly, sedimentation rates on the continental shelf and slope and within marginal seas, such as the Mediterranean, generally are about 10 to 100 cm per 1000 years.

In the deep oceans remote from land, lithogenous inputs are much reduced, and biogenous material, especially calcium carbonate, is dissolved before it can settle and be buried. Therefore, sedimentation rates in the deep-ocean basins are very low, approximately 0.1 cm per 1000 years. Under highly productive areas or on shallow seamounts or **oceanic plateaus** remote from land, the increased rate of **sedimentation** of biogenous material raises the sedimentation rate by about an order of magnitude, to approximately 1 cm per 1000 years.

Although sediments accumulate very slowly on a human timescale, they accumulate to substantial thickness in some places (**Fig. 8.16**). At a sedimentation rate of 0.5 cm per 1000 years, sediments approaching 1 km thick (850 m) can accumulate in 170 million years, which is the approximate age of the oldest oceanic crust.

8.8 CONTINENTAL MARGIN SEDIMENTS

Because most biogenous, hydrogenous, and cosmogenous sediments consist of silt- or clay-sized particles, they do not readily accumulate on continental shelves where currents are swift. Consequently, continental shelves are covered with lithogenous sediment of larger grain sizes, except in those areas where currents are slow or production of biogenous particles is very high.

On many continental shelves, lithogenous sediments accumulate continuously as new sediment is supplied by rivers and coastal erosion. The particles are sorted by grain size, and grain size generally decreases with distance from shore. However, on some continental shelves the supply of lithogenous material is very limited. For example, along the Mid- and North Atlantic coasts of the United States, most river-borne sediment and sediment eroded from the shore is trapped in estuaries and coastal **lagoons.** Consequently, little lithogenous sediment is transported offshore beyond the **longshore drift** system (Chap. 13). If the shelf is wide and currents are generally swift on the outer part of the shelf, little or no new sediment is supplied to the seafloor. Such areas of the continental shelf are not bare rock, but are covered by generally coarse-grained sediments, known as **relict sediments,** that were deposited under a different set of ocean conditions (**Fig. 8.17**).

Relict sediments, including those on the outer continental shelf of the U.S. east coast, contain terrestrial fossils and shells of organisms, such as oysters, that live only in water less than a few meters deep. These materials are too big to have been moved by currents but now are under 100 m or more of water. The reason is that the relict sediments were deposited several thousand years ago when the sea level was lower and the coastline was near what is now the edge of the continental shelf.

The distribution and age of relict sediments are determined by the sea level. Much of the detailed history of

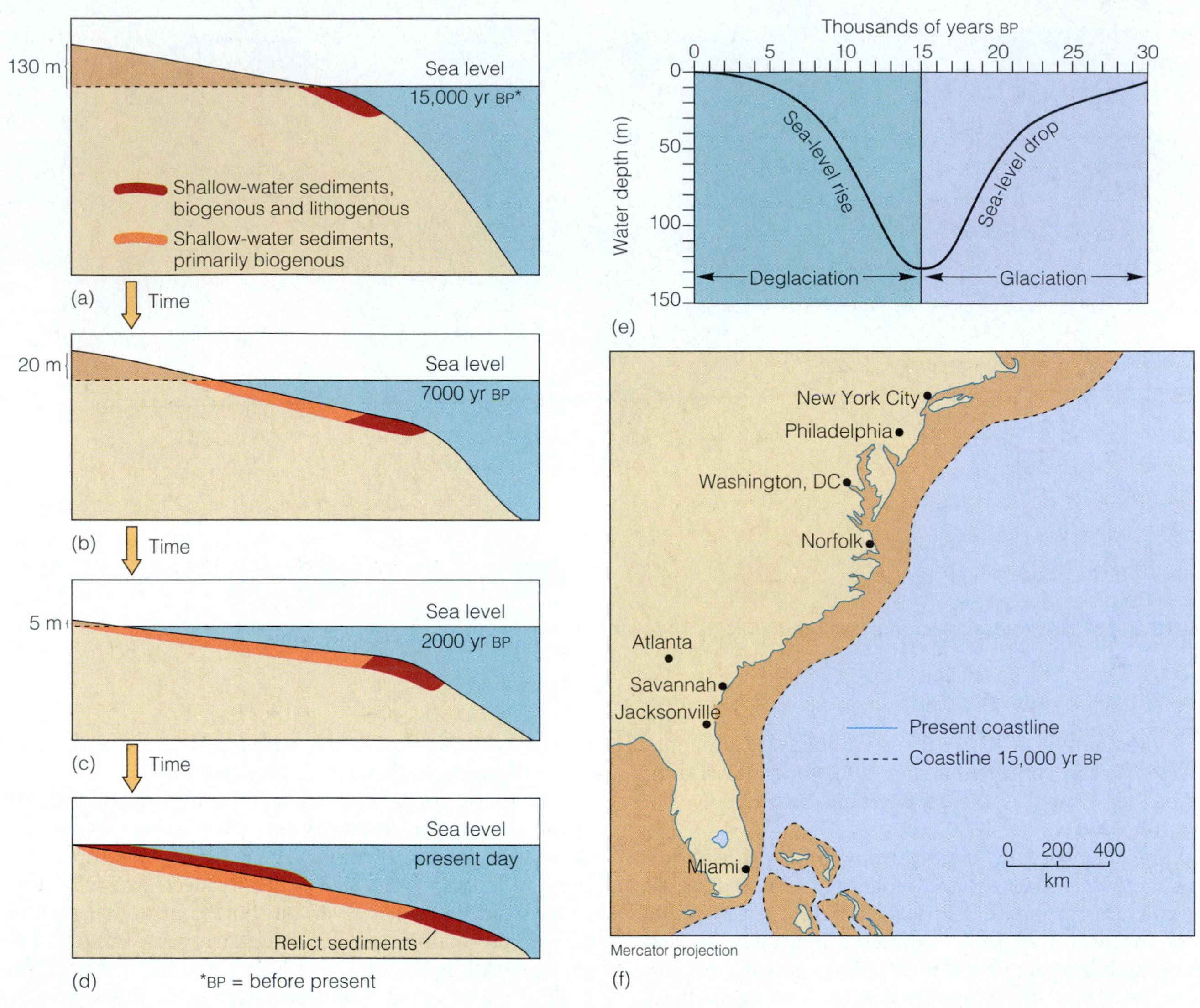

FIGURE 8.17 Relict sediments. (a) Sediments characteristic of shallow water were deposited on the outer continental shelf when sea level was lower. (b) As sea level rose, the relict sediments were not buried by more recent sediments, probably because of the rapid rise of sea level and the trapping of lithogenous sediments in newly submerged river valleys. (c) Shallow-water sediments formed close to the new shoreline as it migrated inland. (d) As sea level rose more slowly, the sediments close to the now slowly retreating shoreline were buried by terrigenous and shallow water biogenous sediments. (e) The history of sea-level change during the past 30,000 years shows the rapid rise between 15,000 and 5000 years ago and the much reduced rate of rise in the past 5000 years. (f) Locations of the Atlantic coastline of North America 15,000 years ago and today. Relict sediments are found primarily near the coastline location of 15,000 years ago.

recent sea-level change has been learned through studies of the location of relict sediments. Sea level has been rising from its most recent low point about 15,000 years ago to the present day, although not at a uniform rate.

8.9 DISTRIBUTION OF SURFACE SEDIMENTS

Surface sediments are the materials currently accumulating on top of older sediment. The older sediment buried below the surface layer may have a different character, if it was deposited under different conditions.

The distribution of **pelagic** surface sediments and their current rate of accumulation are shown in **Figure 8.18**. Continental margin sediments are not shown, because they are lithogenous except in areas dominated by **coral reefs** and in some areas of calcium carbonate precipitation. Each type of sediment and the factors that govern its occurrence are described briefly in the sections that follow.

Radiolarian Oozes

Under a region of high productivity that extends in a band across the deep oceans at the equator (Chap. 15), surface sediments are fine muds that consist primarily of radiolarian shells. Although both calcareous and siliceous organisms grow in abundance in this upwelling region, calcareous material dissolves and does not accumulate in sediments below the CCD. Radiolaria are prolific in tropical waters, and the rate of input of radiolarian shells to the sediment is much higher there than in other deep-sea mud areas. Because siliceous material dissolves slowly, many radiolarian shells are buried in the surface sediment before they fully dissolve.

The rate of input of fine-grained lithogenous particles that are resistant to dissolution is similar in this area to the rate of input of this type of particle in the adjacent areas where deep-sea mud accumulates. However, the rate of accumulation of radiolarian shells, and therefore the overall sedimentation rate, is much higher in these deep-sea mud areas. Therefore, the deep-sea lithogenous mud particles are "diluted" to become a minor component in radiolarian oozes.

The Atlantic Ocean has no band of radiolarian ooze because the accumulation rate of lithogenous sediment is much higher there than in the tropical Pacific and Indian oceans, the seafloor is relatively shallow, and the CCD is relatively deep. The radiolarian content of tropical Atlantic sediments is diluted and masked by the greater contributions of calcareous and lithogenous sediment. Hence, the overall sedimentation rate in the tropical Atlantic Ocean exceeds the sedimentation rate in the tropical Pacific and Indian oceans.

Diatom Oozes

Diatoms dominate the siliceous phytoplankton in upwelling areas except in the tropical upwelling zone where radiolaria dominate. Upwelling and abundant diatom growth occur in a broad band around Antarctica and in the coastal oceans along the west coasts of continents (eastern ocean margins) in subtropical latitudes (Chaps. 10, 15). Diatom oozes dominate in these locations if the sedimentation rates of lithogenous and calcareous particles are relatively slow.

Aside from large coastal inputs from Antarctica's glaciers, sedimentation in the Southern Ocean is limited because of the paucity of landmasses and river inflows. Hence, the accumulation rate of lithogenous particles is slow in the band of high diatom productivity that surrounds Antarctica, and the sediments are dominated by diatom frustules.

Terrigenous sediment inputs are also limited on the west coasts of North and South America because of the limited drainage areas of coastal rivers and the trapping of terrigenous sediment in the trenches that parallel these coasts. Because the seafloor is deep in the Pacific coast upwelling regions and the CCD is relatively shallow in the Pacific Ocean, calcareous particles are dissolved and do not accumulate in sediments. Diatom frustules therefore dominate the sediments in the north and south subtropical Pacific offshore basins. In contrast, on the eastern ocean boundaries in the Atlantic Ocean (southern Europe and West Africa) and Indian Ocean (Australia), in areas where the seafloor is relatively shallow, calcareous particles accumulate faster than diatomaceous particles, and the sediments are calcareous.

Diatoms are very abundant in Northern Hemisphere cold-water regions (Bering Sea and North Atlantic), just as they are near Antarctica. However, lithogenous sediment inputs are greater in the Northern Hemisphere than in the Southern Ocean, and much of the seafloor is above the CCD. Therefore, diatomaceous particles are diluted by calcareous and lithogenous particles, and diatomaceous oozes accumulate only in limited deep areas of the Bering Sea and northwestern Pacific Ocean.

Calcareous Sediments

In areas where the seafloor is shallower than the CCD, calcareous sediment particles accumulate faster than deep-sea clays. Therefore, where the seafloor is shallower than the CCD and terrigenous inputs are limited, calcareous particles are a major component of the sediment. Calcareous sediments are present on oceanic plateaus and seamounts and on the flanks of oceanic ridges.

Deep-Sea Clays

Few large particles are transported into the deep oceans remote from land unless they are carried by turbidity currents (mostly in the Atlantic Ocean). Most deep-ocean areas far from land sustain relatively poor productivity of marine life (Chap. 15). Because of the great depth, the relatively small quantities of biogenous material that fall toward the seafloor are mostly dissolved before reaching, or being buried in, the sediment. This is especially true for calcareous organisms in areas where the seafloor is below the carbonate compensation depth (CCD). Therefore, the particles deposited on the deep-sea floor far from land are mainly fine lithogenous quartz grains and clay minerals.

Deep-sea clays (sometimes called "red clays") are reddish or brownish sediments that consist predominantly of very fine-grained lithogenous material. Everywhere but at high latitudes and in a high productivity band extending across the equatorial Pacific and Indian Oceans, they cover deep-ocean floor that is remote from land. The reddish or brownish color is due to oxidation of iron to form red iron oxide, which we know as rust, during the slow descent of the particles to the seafloor. Oxidation continues on the seafloor until the particles are buried several millimeters deep. The sedimentation rate is very low, so burial is very slow and the interval during which oxidation occurs is very long.

Siliceous Red Clay Sediments

In the deep basins of the North and South Pacific, South Atlantic, and southern Indian Oceans, transitional areas

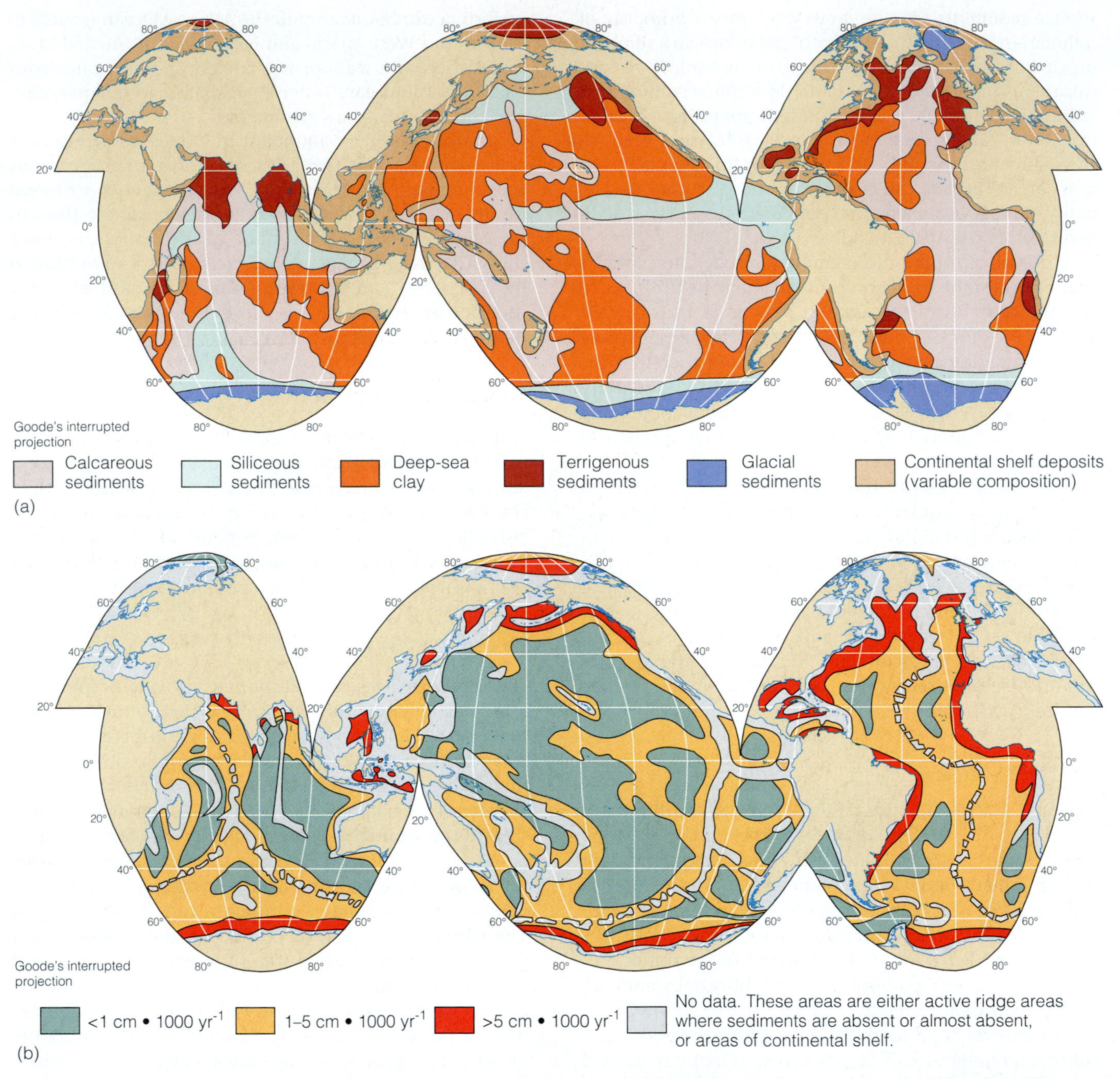

FIGURE 8.18 Present-day sedimentation. (a) The distribution of present-day surface sediments of the world oceans. (b) Approximate sedimentation rates in the present-day oceans. Note the higher rates of sedimentation near the continents. Surface sediments in these areas are predominantly terrigenous. Sediments are predominantly red clays in areas where the sedimentation rate is lowest. Note the generally slower rate of sediment accumulation on the abyssal floor of the Pacific Ocean compared to that of the Atlantic Ocean.

are present between the central deep basin, whose surface sediments are red clays, and higher latitudes, where surface sediments are diatom oozes. Sediments in these transitional areas are mixtures of deep-sea clay and diatomaceous sediments in varying proportions. The transitional sediments are often classified with deep-sea clays and often called "deep-sea muds."

Ice-Rafted Sediments

In the Arctic Ocean, northern Bering Sea, and a band immediately surrounding Antarctica, the sediment consists largely of lithogenous material carried to the oceans by glaciers. It includes sand, pebbles, and even boulders at distances of up to several hundred kilometers offshore from the glaciers. Some lithogenous material is released directly to the ocean when the edge of the ice shelf or glacier melts. Some is transported by icebergs, also called "ice rafts," that break off the glaciers. The material released and deposited as the ice melts is known as glacial, or "ice-rafted," debris. Because glaciers can carry huge quantities of eroded rock, the sedimentation rate can be very high. Biogenous particles therefore are diluted by and mixed with much larger volumes of ice-rafted lithogenous material.

Terrigenous Sediments

Sediments dominated by terrigenous particles are present near the mouths of major rivers and may extend hundreds of kilometers offshore. Such sediments are deposited in the northern Arabian Sea, the Bay of Bengal, the Gulf of Mexico, and the Atlantic Ocean near the mouth of the Amazon River in Brazil. Terrigenous sediments transported by turbidity currents are present at the foot of the continental slope in the North Atlantic. On the North Pacific coast of North America, rivers carry large sediment loads from the glaciers in the mountains of British Columbia and Alaska. Many of the glaciers terminate and melt before reaching the ocean. Ice rafts are few, so coarse sand, pebbles, and boulders are not transported beyond the immediate vicinity of the glacier termination. However, large quantities of glacially ground sand, silt, and mud are transported to the ocean by rivers, and offshore by waves and currents.

Hydrothermal Sediments

The central basin of the Red Sea is the only area known to have hydrothermal mineral deposits as the dominant sediment type. The other locations where such sediments accumulate are small and scattered throughout the oceans, mostly on the oceanic ridges. The extent of such deposits is not known, because only a very tiny fraction of the oceanic ridge system has been surveyed by methods that reveal hydrothermal vents and their associated sediments. Vents have been found on oceanic ridges in all oceans and may be present at intervals of a few kilometers to hundreds of kilometers along the entire oceanic ridge system.

8.10 THE SEDIMENT HISTORICAL RECORD

Sediments accumulate continuously by deposition of new layers of particles on top of previously deposited layers. Therefore, individual layers that make up the sediment below the sediment surface were deposited progressively earlier as depth below the sediment surface increases. The type of particles deposited at a specific location and time depends on many factors, such as the proximity to land, whether rivers or glaciers flowed into the adjacent coastal ocean, the type and amount of dust in the atmosphere, and the composition and production rates of organisms in the water column. These factors do not remain constant over geological time, because they are affected by plate tectonic movements, climate changes, and changes in volcanic activity, all of which may be related to each other in complex ways. Such changes, singly or in combination, can move coastlines, change seafloor depth, raise mountains, alter rivers, create and melt glaciers, change atmospheric dust composition by creating or reducing deserts or by altering the amounts of volcanic material injected into the atmosphere, and change the temperature, salinity, and nutrient distributions in the oceans. Each of the changes affects the type of sediment that accumulates. Hence, buried sediments may be very different from those currently accumulating.

Buried sediments provide information about the conditions at the time they were buried. Ocean sediments therefore can provide a history of plate tectonics and climate changes covering about 170 million years, the approximate age of the oldest remaining oceanic crust. To read this history, we must be able to determine the date at which a particular layer of sediment was deposited. Then we must examine the composition of the sediment particles to reconstruct the sedimentation conditions at the time the layer was formed.

Stratigraphy is the study of the Earth's history through investigation of the sediment layers beneath the ocean floor (and on land where sediments have been compressed and converted into sedimentary rock). Stratigraphic studies of ocean sediments usually involve the examination of sediment cores (Chap. 3), which are sectioned into slices 1 or 2 cm thick (**Fig. 8.19**). Each slice represents the accumulation during a period of thousands or tens of thousands of years, depending on the sedimentation rate.

The stratigraphic record can be complicated to read. Gaps may appear in the record if previously deposited sediment was eroded because of a temporary increase in current speed. Furthermore, adjacent layers may have had substantially different sedimentation rates. Biological activity also may have altered the historical record. Bottom-dwelling organisms rework, or mix, the upper few centimeters or tens of centimeters of sediment as they consume the organic matter, feed on other sediment dwellers, or build burrows. Their churning of the sediments is called **bioturbation.** Despite these limitations, the stratigraphic record provides vital information about the Earth's history, including previous climate changes. Understanding climate changes will enable us to understand and predict future changes, such as may result from enhancement of the **greenhouse effect.**

FIGURE 8.19 Sediment cores, such as this deep-sea drilling core, are first cut down the middle and then segmented into small slices, each representing the accumulation of sediments during a period of hundreds or thousands of years in the past.

Sediment Age Dating

A fundamental requirement of stratigraphic studies is that the age (date of deposition) of the sediment layers be determined. Methods for determining sediment ages include **radioactive** dating (**CC7**), magnetic dating, and biological dating.

Radioactive dating of sediments is difficult because often the assumptions on which the technique is based are not met. Therefore, researchers commonly use several different methods to be certain of a measured age, including dating of sediments based on their magnetic properties. Chapter 5 describes the periodic reversals of the Earth's magnetic field and how **magnetic anomalies** are measured to determine the age of the oceanic crust. In much the same way, the magnetic properties of sediment vary with depth below the sediment surface, changing at each depth in the sediment column that corresponds to a magnetic reversal.

Many of the marine species whose remains make up the biogenous fraction of older sediments have become extinct, have been replaced by new species, or have evolved into substantially different forms. Consequently, the ages of sediments at different locations often can be matched by comparison of the species compositions of organisms found in their biogenous fractions. If the sediment at a specific depth at one sampling site contains the fossil remains of all the same species that the sediment at a different depth at another sampling site contains, the sediments are likely to be the same age. Thus, we can date sediments in relation to each other and determine relative sedimentation rates.

Diagenesis

Interpreting the stratigraphic record requires an understanding of **diagenesis,** the physical and chemical changes that occur within sediment after it is buried. In some cases, during the millions of years they lie buried, minerals are altered from one form to another as weathering processes that began on land or in suspended sediment continue. However, the more important diagenetic changes occur in the **pore water** (interstitial water) that is trapped between the sediment grains.

Chemicals that dissolve in pore waters during diagenesis include silica and calcium carbonate, which continue to dissolve from buried siliceous and calcareous material. Also during diagenesis, other substances are released by continuing oxidation of sediment organic matter, and these too may be dissolved in the pore water. Chemicals dissolved from the sediment particles during diagenesis migrate upward by **diffusion** toward (and in some cases into) the overlying water. As sediments accumulate, their weight compacts the underlying sediment. In the process, pore waters are slowly squeezed out and migrate upward through the sediment.

As organic matter decomposes in the sediment, dissolved oxygen in the pore water is consumed and oxidation continues with oxygen from nitrates and then sulfates (Chap. 14). The sulfates are reduced to sulfide, and the presence of sulfide dramatically changes the solubility of many elements. Iron, manganese, and several other metals form sulfides that are much more soluble than their **hydrated** oxides, which form in oxygenated water. In sediment that has enough organic matter to produce sulfides, iron and manganese are dissolved in pore waters that migrate upward. The metal sulfides migrate with the pore water until it approaches the sediment surface, where it encounters dissolved oxygen diffusing down into the sediment or introduced by bioturbation. The iron and manganese then are oxidized and redeposited on sediment grains. As sediment accumulates, these elements may be dissolved and moved upward continuously, accumulating in a layer near the sediment surface where sulfide and oxygen mix.

In contrast to iron and manganese, the sulfides of many elements (e.g., copper, zinc, and silver) are less soluble than their hydrated oxides. These metals are not dissolved in pore waters that contain sulfide, so they do not migrate up toward the oxygenated surface sediments and overlying seawater. On the contrary, these elements can diffuse with oxygenated seawater downward into the sediment, where they can be converted to sulfides and buried with the accumulating sediment. The complex migrations of elements within pore waters and the associated changes in the nature of the sediment particles must be understood if we are to deduce the nature of the sediment particles at the time of their deposition and thus read the stratigraphic record.

Diagenetic processes are also important to life. For example, the nutrients nitrate, phosphate, and silicate are transported to the sediments in detritus that falls to the sediment surface. The nutrients are released to the pore water as organic matter is oxidized and they then diffuse back into the water column, where they can be reused by living organisms.

Tectonic History in the Sediments

Much of the history of tectonic processes can be revealed by studies of the changes in sediment characteristics with depth below the sediment surface. For example, consider the sediment layers that have accumulated at locations between the oceanic ridges and continents (**Fig. 8.10**). Remember that the seafloor is progressively older with increased distance from the oceanic ridge (Chaps. 4, 5). However, the older seafloor was once at the oceanic ridge and sank **isostatically** (**CC2**) as it cooled, so it becomes progressively deeper. Remember also that all sediments are mixtures (**Fig. 8.15**).

As another example of how sediment deposits can reveal the history of plate tectonic movements, the age of a buried calcareous sediment layer can reveal the times during which the seafloor was at depths shallower than the CCD. In reality, reading the sediment history is complicated because important factors, such as the CCD level, have undoubtedly varied over the millennia. In addition, the locations at which certain types of organisms are abundant may have changed. Currently, siliceous diatoms abound near the poles, and siliceous radiolaria dominate near the equator. However, the distributions of these organisms might have been very different at times in history when the climate was different from the present-day climate.

Climate History in the Sediments

The past 170 million years of the Earth's climate history are preserved in ocean sediments, primarily in biogenous particles. Each marine species that contributes calcareous and siliceous material to ocean sediments has an optimal set of conditions for growth. Although nutrient concentrations, light intensity, and other factors are important, temperature is crucial for most species. Hence, fossils in sediments can be used to determine the geographic variations in ocean water temperatures.

Because different species of foraminifera thrive at different temperatures, the species composition of foraminiferal fossils in buried oozes can be used to study past climate. However, because some species have evolved or become extinct, the temperature tolerances of fossil species often are assumed to be the same as those of modern species. This assumption may not be correct.

Past climate information also can be obtained from measurements of the ratio of oxygen **isotope** concentrations in calcareous sediments. The basis for this method is somewhat complicated. Oxygen has three isotopes: oxygen-16, oxygen-17, and oxygen-18. Almost all ocean water molecules consist of either an oxygen-16 atom combined with two hydrogen atoms, or an oxygen-18 atom combined with two hydrogen atoms (the concentration of oxygen-17 is very small). Oxygen-16 is lighter than oxygen-18. Consequently, water containing oxygen-16 is lighter than water containing oxygen-18. The lighter water molecules evaporate more easily than the heavier molecules. Hence, when seawater evaporates, the water vapor is enriched in oxygen-16. When the Earth's climate cools, more water is precipitated and stored in the expanding glaciers and polar ice sheets. Because this water was evaporated from the oceans, oxygen-16 is transferred preferentially to the ice, the ratio of oxygen-16 to oxygen-18 in ocean water decreases, and sea level falls. Organisms that live in seawater incorporate the oxygen-16 and oxygen-18 of seawater into their calcium carbonate body parts in a ratio that is determined partly by the ratio of these two isotopes in the seawater. Therefore, the ratio of oxygen-16 to oxygen-18 in calcareous sediment can be used to deduce climatic characteristics and sea level at the time of deposition. The ratio is high when the climate is warm and sea level is high, and low when sea level is lowered by evaporation and more water is stored on land as ice and snow.

In the same way that water molecules containing oxygen-16 and oxygen-18 evaporate at slightly different rates, molecules containing these oxygen isotopes react at slightly different rates in the chemical processes that produce the calcium carbonate of calcareous organisms. The ratio of oxygen-16 to oxygen-18 in the calcareous parts differs between species, but for a single species it depends on the ratio in the seawater and the seawater temperature. At any given time, the oxygen isotope ratio of ocean surface water is relatively uniform throughout the oceans. Hence, calcareous remains of the same species deposited in different parts of the ocean at the same time have different ratios of oxygen-16 to oxygen-18 because the water temperatures were different where the particles were deposited. A low ratio of oxygen-16 to oxygen-18 indicates colder water.

Studies of sediments aimed at revealing the Earth's climate history have become more important because of growing concern about human enhancement of the greenhouse effect. An understanding of past climates can help us understand future climate change due to both natural causes and human activities. Stratigraphic studies are becoming ever more sophisticated and are now beginning to reveal details of ancient ocean current systems and ocean chemistry, including carbon dioxide concentrations. Such information is continually improving our understanding of the complicated **feedback** mechanisms between the atmosphere and oceans (Chap. 9, CC9) that will determine the future effects of greenhouse gases on our planetary environment.

Support for Extinction Theories in the Sediments

About 65 million years ago, the last of the dinosaurs became extinct. Fossil and sediment records show that, at the same time, more than half of all species of marine animals became extinct. Although other extinctions have been found at different times in the sedimentary record, the dinosaur extinction was among the most dramatic. Several theories have been proposed to explain that mass extinction, but two appear to be better supported than others. One is that the extinction was caused by an intense period of volcanic activity, and the other is that it was caused by the impact of a giant meteorite striking the Earth. In either case, large amounts of dust and gas would have been released to the atmosphere, altering global climate and causing the extinctions. For example, the 1883 eruption of Krakatau, a relatively small eruption in comparison with some that may have occurred in the past, is known to have affected the world climate for a decade.

At present, neither of the extinction theories can be proved or refuted. However, recent evidence from places like the Chicxulub, Mexico, area support the meteorite theory and provide a good illustration of how past events can be deduced from sediments.

FIGURE 8.20 The location of the Chicxulub crater, which is thought to be the impact site of a massive meteorite that may have been responsible for climate changes that, in turn, led to the extinction of the dinosaurs. The crater is at least 180 km in diameter, but it may be twice that size.

Evidence of Impact at Chicxulub

The area that is now the Yucatán Peninsula of Mexico was underwater 65 million years ago at a depth of about 500 m. Magnetic surveys in this area reveal a circular feature (**Fig. 8.20**), 180 km in diameter, that may be the buried remains of the impact crater of a huge meteorite. Studies of sedimentary deposits of the appropriate age from many areas surrounding this circular feature, now called Chicxulub, show an unusual series of layers at the level that corresponds to deposition 65 million years ago.

Sediments below the strange layers are fine-grained biogenous oozes of the type normally deposited at a depth of approximately 500 m. The unusual layers above these normal sediments have characteristics that could be explained by the sequence of events that followed a massive meteorite impact. The deepest and oldest strange layer consists primarily of rounded, coarse grains several millimeters in diameter. These grains were not rounded by water weathering. They are glassy tektites, which form when rocks are melted by meteorite impacts, ejected into the atmosphere where they resolidify, and fall back to the Earth. The next higher layer consists of coarse-grained sediments that contain fossilized wood and pinecone fragments. These materials are not normally found in marine sediments. It is hypothesized that this layer was created when the meteorite impact generated a huge **tsunami** (Chap. 11) that smashed onto land, tore rocks and trees loose, and carried them back to sea as it receded. The tsunami may have been so large that it sloshed back and forth across the Gulf of Mexico for several days.

Above the two older extraordinary layers is a series of layers of progressively finer-grained sand mixed with other mineral particles that are not normally found in marine sediments. These progressively finer-grained layers may consist of beach sands and eroded soil particles suspended and transported to the deep water by waves generated by the meteorite impact. Above these finer-grained layers, the sediments finally grade into normal biogenous oozes similar to the ones that underlie the unusual layers. As wave energy decreased, progressively smaller particles would have been deposited in the same way that turbidites are. Much of the finer-grained material overlying the graded sequence may have come from the fallout of dust ejected into the atmosphere during the impact. The impact hypothesis is further supported by the fact that the fine-grained layer contains a high concentration of iridium, an element that is very rare on the Earth but is present in much higher concentrations in meteorites.

Many variations of the preceding interpretation of events that may have deposited the anomalous sediment layers 65 million years ago in the Gulf of Mexico are already emerging. One suggestion is that the tsunami may not have been caused directly by the impact of the meteorite on the ocean. Instead, the impact may have caused a magnitude-11 earthquake (about a million times more powerful than the 1989 Loma Prieta earthquake near San Francisco), which could have sent giant turbidity currents down the continental slope and thus generated massive tsunamis. The history of any such mudslides should be preserved in the deep-sea sediments and surely will be sought by oceanographers. The odd sediment layers in the Gulf of Mexico will undoubtedly be studied extensively, and their meaning debated for many years.

Evidence of Other Impacts

Researchers believe that there may have been a number of other large impacts similar to the Chicxulub event at different times in the past, and they are searching for evidence of the impact craters. A number of locations have been identified as possible impact sites, including an area off the northwestern coast of Australia, where there may have been an impact about 250 million years ago, and an area at the mouth of the Chesapeake Bay, where an impact may have left an 85-km-wide crater about 35 million years ago. Decades of careful research will undoubtedly be needed to demonstrate that meteorite impacts did or did not occur in these areas and others. If confirmed, additional research will be needed to investigate and identify the effects that those impacts may have had on the biosphere. Eventually we may be able to establish with some certainty whether the extinction of the dinosaurs and other extinctions in the Earth's past were caused by meteorite impacts.

CHAPTER SUMMARY

Classification of Sediments. Sediments can be classified by their predominant grain size or by the origin of the majority of their particles. Grain size ranges usually are classified as gravel, sand, silt, and clay, in order of decreasing size.

Lithogenous Sediments. Lithogenous sediment particles are primarily chemically resistant minerals produced by weathering of continental rocks. They are transported to the oceans by rivers, glaciers, waves, and winds. Lithogenous sediments are deposited in thick layers beyond the mouths of some rivers, but many rivers trap sediments in their estuaries. Fine-grained dust, especially from deserts, can be transported by winds for thousands of kilometers before settling on the sea surface and then sinking to the seafloor. Some volcanic eruptions eject large quantities of ash that reach ocean sediment in a similar way.

Biogenous Sediments. Biogenous sediments are the remains of marine organisms. In most cases, the organic matter is decomposed before the hard inorganic parts are deposited and buried.

Some marine species have calcareous or siliceous hard parts, both of which dissolve in seawater. Most siliceous diatom frustules dissolve very slowly and become included in sediments. The intricate siliceous shells of radiolaria generally dissolve before being buried in sediments in all but tropical waters where radiolaria are abundant. Because the solubility of calcium carbonate in seawater increases with pressure, the rate of dissolution of calcareous hard parts increases with depth. Calcareous shell material has two forms: calcite and aragonite. The aragonite of pteropods is dissolved much more easily than the calcite of foraminifera. Pteropod oozes are present only in relatively shallow water, whereas foraminiferal oozes predominate at intermediate depths. Below the carbonate compensation depth, little or no calcareous material survives dissolution to be incorporated in the sediment. The CCD varies between oceans because of differences in dissolved carbon dioxide concentrations, and it also varies historically with climate changes.

Hydrogenous Sediments. Hydrogenous sediments, which are less abundant than terrigenous (lithogenous) or biogenous sediments, are deposited by the precipitation of minerals from seawater. They include manganese nodules that lie on the sediments in areas of the deep ocean, phosphorite nodules and crusts on continental shelves beneath oxygen-deficient water, calcium carbonate deposits in a few shallow areas where water temperature is high and dissolved carbon dioxide concentration is low, and evaporites (salt deposits) in coastal waters with limited water exchange with the open ocean, low rainfall, and high evaporation rate. Hydrogenous sediments also include metal-rich sediments accumulated by the precipitation of minerals from water discharged by hydrothermal vents.

Cosmogenous Sediments. Cosmogenous sediment particles are fragments of meteorites or tektites created by meteorite impacts. They constitute only a tiny fraction of the sediments. Many are found as cosmic spherules.

Sediment Transport and Deposition. Large particles are deposited quickly, unless current speeds are high, and they are difficult to resuspend. Large particles collect close to river mouths, glaciers, and wave-eroded shores. Because orbital velocity in waves is higher than ocean current speeds, large particles can be transported in the nearshore zone but cannot be transported far from shore. Smaller particles may be carried long distances before being deposited.

Turbidity currents that resemble avalanches carry large quantities of sediment down continental slopes and onto the abyssal plains. Graded beds of turbidites are present in deep-ocean sediments, and they are common on the abyssal floor near the continents, except where there are trenches.

Sediment Accumulation Rates. Sediments are mixtures of particles of different origins. Sediment characteristics are determined by the relative accumulation rate of each type of particle. Sediment accumulation rates range from about 0.1 cm per 1000 years in the deep oceans to more than 1 m per year near the mouths of some rivers.

Continental Margin Sediments. Because fine-grained sediments are transported off the continental shelf by currents, shelf sediments are generally sand- and silt-sized lithogenous particles. Relict sediments, deposited when sea level was lower and containing remains of terrestrial and shallow-water organisms, are present in some areas on the continental shelf.

Distribution of Surface Sediments. Deep-sea clays are present far from the continents on the abyssal plains. Radiolarian oozes occupy a band of high primary productivity that follows the equator, but not in the Atlantic, where the radiolarian shells are diluted with lithogenous sediment. Diatom oozes are present in high-latitude areas and in other upwelling areas where primary productivity is high and lithogenous inputs are low. Sediments dominated by terrigenous particles are deposited near the mouths of major rivers. Calcareous sediments dominate where the seafloor is shallower than the CCD and lithogenous inputs are low. Hydrothermal sediments are deposited in some areas on oceanic ridges.

The Sediment Historical Record. Sediments accumulate continuously and provide a record of the Earth's history. The stratigraphic record can be complicated because sediment layers may be disturbed or chemically and physically altered by diagenesis. Stratigraphic studies of ocean sediments have revealed much of the Earth's climate and tectonic history during the past 170 million years. For example, such studies have provided information about the impact of a large meteor that may have contributed to the extinction of the dinosaurs and other species 65 million years ago.

KEY TERMS

You should recognize and understand the meaning of all terms that are in boldface type in the text. All those terms are defined in the Glossary. The following are some less familiar key scientific terms that are used in this chapter and that are essential to know and be able to use in classroom discussions or exam answers.

abyssal fan (plain) (p. 184)
adsorption (p. 180)
biogenous (p. 171)
biogeochemical cycle (p. 170)
bioturbation (p. 191)
calcareous (p. 176)
carbonate compensation depth (CCD) (p. 178)
coastal plain (p. 172)
coccolithophores (p. 176)
cohesive (p. 183)
colloidal (p. 180)
continental shelf (margin, slope) (pp. 175, 181)
cosmogenous (p. 171)
decomposer (p. 178)
delta (p. 172)
deposition (p. 173)
detritus (p. 176)
diagenesis (p. 192)
diatom (p. 176)
diffusion (p. 192)
electrostatic (p. 183)
erosion (p. 170)
evaporite (p. 179)
excreted (p. 176)
fecal pellets (p. 176)
fjord (p. 172)
foraminifera (p. 176)
fossil (p. 182)
frustule (p. 178)
glacier (glaciation) (pp. 170, 172)
graded bed (p. 184)
grain size (p. 170)
hard parts (p. 171)
hydrogenous (p. 171)
hydrothermal vent (p. 179)
lagoon (p. 187)
leaching (p. 170)
limestore (p. 182)
lithogenous (p. 171)
manganese nodules (p. 179)
marginal sea (p. 172)
mollusk (p. 171)
ooze (p. 186)
pelagic (p. 188)
phytoplankton (p. 171)
pore water (p. 192)
pteropod (p. 176)
radiolaria (p. 176)
relict sediment (p. 187)
residence time (p. 173)
resuspended (p. 172)
runoff (p. 172)
sedimentation rate (p. 178)
shore (shoreline) (pp. 174, 175)
siliceous (p. 172)
slump (p. 175)
sorted (p. 170)
steady state (p. 170)
submarine canyon (p. 175)
suspended sediment (p. 172)
terrigenous (p. 172)
turbidite (p. 184)
turbidity current (p. 175)
turbulence (p. 172)
weathered (p. 171)
wetland (p. 183)
zooplankton (p. 171)

STUDY QUESTIONS

1. Sediments are deposits of particles of many different sizes from many different sources. Why can we classify sediments by their grain size?
2. List the four principal types of sedimentary particles classified by their origin. Can we classify sediments from all locations according to these four groups? Explain why or why not.
3. A sample of the surface sediments on the continental shelf a kilometer offshore contains large amounts

of organic matter compared to other continental shelf sediments. What can you conclude or hypothesize about the characteristics of this region?

4. What conditions are necessary for deposits of metal-rich sediments to be formed at hydrothermal vents?
5. Why would you not expect to find turbidites near the East Pacific Rise?
6. Why do we find relict sediments that have only a small amount of overlying sediment on the continental shelf off the northeastern United States, but not off the coast of Louisiana?
7. Describe the characteristics of surface sediments on the deep-sea floor. Why do they have these characteristics and how do they vary with location?
8. If you drilled into the sediments of the deep-sea floor just beyond the continental rise of the northeastern United States, what layers of sediment would you be likely to encounter with depth?
9. A 30-cm-long sediment core has a high concentration of organic matter throughout its length, but oxygen is not depleted in the pore waters. What might explain this?

CRITICAL CONCEPTS REMINDER

CC1 **Density and Layering in Fluids** (p. 180). Fluids, including the oceans, are arranged in layers sorted by their density. Heat sources under the seafloor can heat and reduce the density of any water present, and this heated water rises through the seafloor and up into the water column. To read **CC1** go to page 2CC.

CC2 **Isostasy, Eustasy, and Sea Level** (p. 193). Earth's crust floats on the plastic asthenosphere. Sections of crust rise and fall isostatically as temperature changes alter their density. Oceanic crust cools progressively after it is formed and sinks because its density rises. Thus, the seafloor becomes deeper with increasing distance from an oceanic ridge. To read **CC2** go to page 5CC.

CC3 **Convection and Convection Cells** (p. 180). Fluids that are heated from below, including water within sediments and cracks in the seafloor, rise because their density is decreased. This establishes convection processes. Heated water rising through the seafloor is replaced by colder water that seeps or percolates downward into the seafloor to replace it. To read **CC3** go to page 10CC.

CC4 **Particle Size, Sinking, Deposition, and Resuspension** (pp. 171, 172, 174, 176, 183). Suspended particles (either in ocean water or in the atmosphere) sink at rates primarily determined by particle size: large particles sink faster than small particles. Once deposited, particles can be resuspended if current (or wind) speeds are high enough. Generally large particles are more difficult to resuspend, although very fine particles may be cohesive and also difficult to resuspend. Sinking and resuspension rates are primary factors in determining the grain size characteristics of beach sands and sediments at any given location. To read **CC4** go to page 12CC.

CC6 **Salinity, Temperature, Pressure, and Water Density** (p. 180). Sea water density is controlled by temperature, salinity, and to a lesser extent pressure. Density is higher at lower temperatures, higher salinities, and higher pressures. Heated water discharged by hydrothermal vents has a high enough salinity in some areas of very limited mixing that it is more dense than the overlying seawater and collects in a layer next to the seafloor. To read **CC6** go to page 16CC.

CC7 **Radioactivity and Age Dating** (p. 192). Some elements have naturally occurring radioactive (parent) isotopes that decay at precisely known rates to become a different (daughter) isotope. Measuring the concentration ratio of the parent and daughter isotope can be used to calculate the age of the various materials since they were first formed, but only if none of the parent or daughter isotope are gained or lost from the sample during this time period. This condition is usually not met in sediments and radioisotope age dating is difficult so other dating methods including variations in fossil assemblages, and magnetic field properties are used extensively. To read **CC7** go to page 18CC.

CC8 **Residence Time** (pp. 170, 173). The residence time of seawater in a given segment of the oceans is the average length of time the water spends in the segment. In restricted arms of the sea or where residence time is long, very fine grained particles from glaciers and rivers can become concentrated in the water. Residence time is also a major factor in determining the change in concentration of an element in seawater if its inputs to the oceans change. To read **CC8** go to page 19CC.

CC9 **The Global Greenhouse Effect** (pp. 174, 179, 194). The oceans play a major part in studies of the greenhouse effect as the oceans store large amounts of carbon dioxide both in solution and as carbonates in sediments, formed at shallow enough depths that the carbonates are not dissolved. To read **CC9** go to page 22CC.

CC14 **Photosynthesis, Light, and Nutrients** (pp. 175, 182). Chemosynthesis and photosynthesis are the two processes by which simple chemical compounds are made into the organic compounds of living organisms. Many sediments have high concentrations of particles that originate from photosynthetic organisms or species that consume these organisms. To read **CC14** go to page 46CC.

CHAPTER 9

CHAPTER OUTLINE

CRITICAL CONCEPTS USED IN THIS CHAPTER

The oceans and atmosphere are intimately linked in many ways. Top: A waterspout, or tornado, is shown as it touches down in a strong thunderstorm near Grand Cayman Island in the Caribbean Sea. Bottom: A major hurricane approaches the state of Florida in this satellite image. Right: The global wind patterns in September 1978 measured by satellite sensors. The white arrows show the wind direction and the colors indicate the wind speed, with red showing the strongest winds and blue the weakest. These atmospheric phenomena, ocean thunderstorms, hurricanes, and the global winds, are all driven by heat energy released by ocean surface waters.

Ocean-Atmosphere Interactions

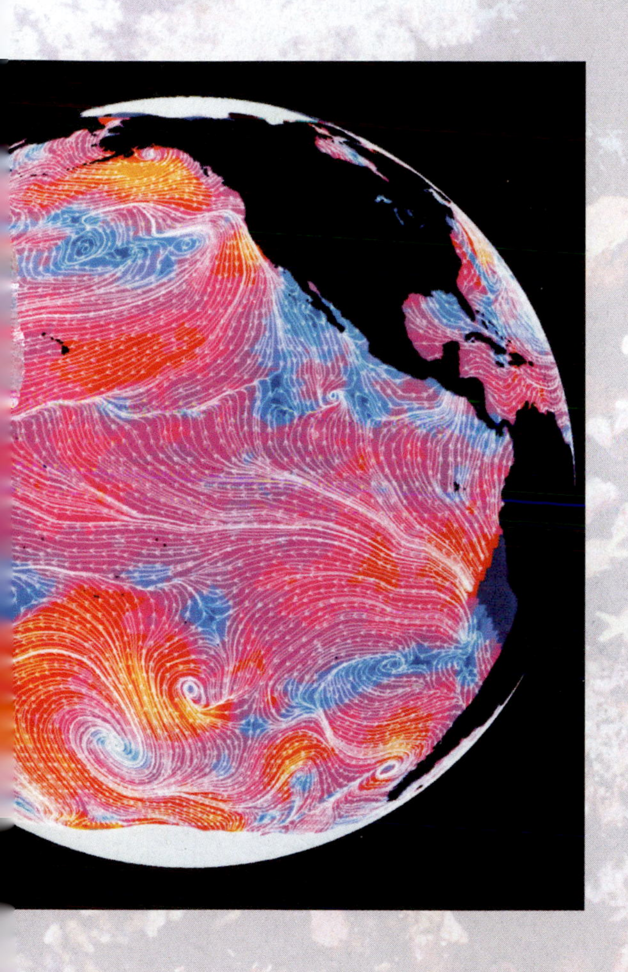

In the nineteenth century, scientists seeking to explain the Earth's ice ages determined that carbon dioxide and other atmospheric gases cause a "greenhouse effect" and hypothesized that lower-atmospheric concentrations of these gases may have been responsible for the ice ages. In 1895, a Swedish chemist, Svante Arrhenius, predicted that carbon dioxide emissions from human industry were likely to lead to a global warming. In 1939, Guy Stewart Callendar, a British steam engineer, concluded that both carbon dioxide concentrations and temperature had risen since the Industrial Revolution. Most scientists found Arrhenius's predictions and Callendar's observations to be implausible, until the 1950s, when the possibility of global warming was rediscovered. In the early 1960s, measurements of atmospheric carbon dioxide by Charles Keeling revealed that its concentration was increasing quickly, and researchers began intensive studies of the greenhouse effect and possible climate changes due to industrial emissions. This issue has now become perhaps the most important environmental issue facing humanity.

Why are we discussing climate change, an atmospheric pollution problem, in a text on ocean sciences? Consider the words of marine scientist Rachel Carson (who is famous for her book ***Silent Spring,*** which alerted the world to the dangers of DDT and other toxic chemicals in our environment):

> *For the globe as a whole, the ocean is the great regulator, the great stabilizer of temperature. It has been described as "a savings bank for solar energy, receiving deposits in seasons of excessive*

insolation and paying them back in seasons of want." Without the ocean, our world would be visited by unthinkably harsh extremes of temperature.

—RACHEL CARSON,
***The Sea Around Us*, published in 1950 by Oxford University Press, New York, (p. 170)**

This quote refers to the importance of the intimate dance performed between our atmosphere and oceans. The capacity of the oceans to retain and release heat energy is but one facet of that dance, which mediates our climate, weather, and environment.

The Earth's **climate** (average temperature, rainfall, etc., for prior years) and **weather** (temperature, rainfall, etc., occurring at a specific time) are determined by the distribution of heat and water vapor in the atmosphere. The oceans play an important role in controlling this distribution.

Ocean–atmosphere interactions are also important because ocean **currents** are generated by winds or by **density** differences between **water masses**. Seawater density is determined primarily by changes in temperature and **salinity** that occur at the ocean surface. Surface water temperature is controlled by solar heating and radiative cooling, and salinity is altered by evaporation and precipitation.

The oceans contain many times more heat energy than the atmosphere does. This is because water is denser than air, and the total mass of ocean water is more than 200 times the total mass of air in the atmosphere. In addition, water has a much higher **heat capacity** per unit mass than air (or the rocks and soil of the land). Just the top few meters of ocean water have the same heat capacity as the entire atmosphere.

The oceans contain more than 97% of the world's water. The ocean surface covers more than 70% of the Earth's surface. The transfers of heat and water vapor between the oceans and the atmosphere that occur at this ocean surface are the main driving forces that determine the world's climate.

Studies of the complex interactions between the oceans and atmosphere have been important since oceanographic science began. However, such studies have received added attention because of potential climate changes as a result of the enhanced **greenhouse effect** (**CC9**). We cannot understand such climate changes without understanding ocean–atmosphere interactions.

9.1 THE ATMOSPHERE

Gases are highly compressible. Even small changes in pressure produce measurable changes in volume. As pressure increases, the gas molecules are forced closer together and the volume is reduced. In the atmosphere, pressure decreases rapidly with altitude (**Fig. 9.1**). Consequently, the density of the atmosphere decreases progressively with increasing height above the Earth's surface. As a result, the atmosphere is **stratified** throughout its more than 100-km height (**CC1**).

Temperature and vapor pressure (water vapor concentration) also affect the density of air, but much less than the changes in pressure with altitude. The atmosphere is steeply stratified near the Earth's surface because of the rapid pressure change with altitude. As a result, density-driven vertical motions of air masses in the lower atmosphere are limited to only a few kilometers above the Earth's surface.

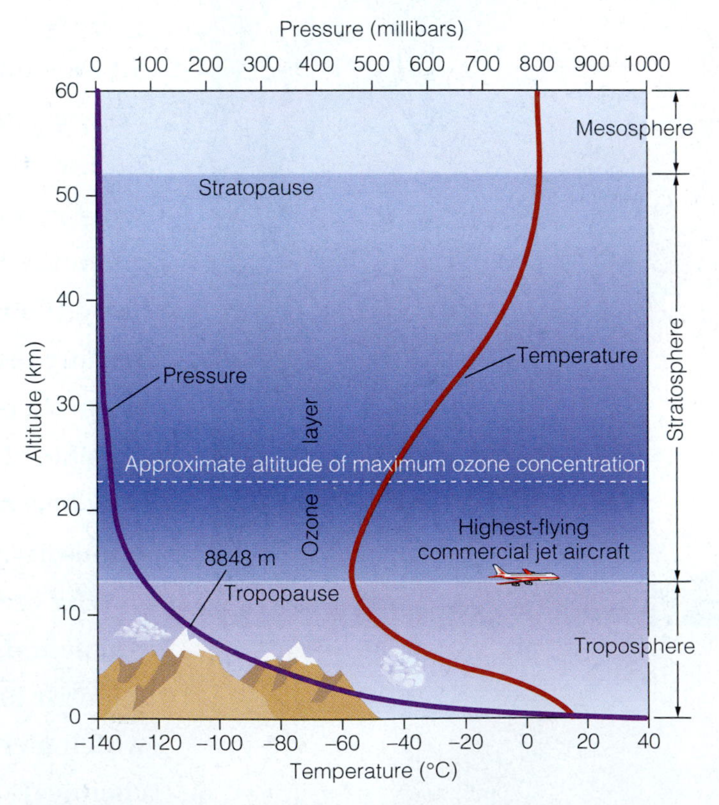

FIGURE 9.1 Structure of the atmosphere. The atmospheric circulation that determines the Earth's climate and weather takes place within the troposphere. Movements of gases between the troposphere and the stratosphere are limited and slow. Ozone is present throughout the stratosphere, but its concentration peaks at an altitude of about 22 km.

Atmospheric Structure

The atmosphere is separated vertically into three distinct zones: **troposphere**, stratosphere, and mesosphere. The troposphere lies between the Earth's surface and an altitude of about 16 to 18 km near the equator and less than 10 km over the poles (jet airplanes fly at an altitude of about 10 to 12 km). Vertical movements of air masses are caused by density changes and occur mainly in the troposphere. Although temperature decreases with altitude within the troposphere, density also decreases because the reduction in atmospheric pressure with altitude more than offsets the effect of lowered temperature (**Fig. 9.1**). This chapter pertains primarily to the troposphere because the atmospheric motions created by heat transfer from oceans to atmosphere are restricted to this zone.

Between the top of the troposphere and an altitude of approximately 50 km, is the stratosphere. The air in the stratosphere is "thin" (that is, molecules are much farther apart than they are at ground level, because pressure is low), vertical air movements are primarily very slow **diffusion**, and temperature increases with altitude (**Fig. 9.1**). Within the stratosphere is a region called the **ozone layer**, where oxygen gas (O_2) is partially converted to **ozone** (O_3) by reactions driven by the sun's radiated energy.

Ozone Depletion

The concentration of ozone in the Earth's ozone layer decreased progressively during the 1980s and 1990s, particularly in the region around the South Pole, where the depletion is so great that it has become known as the "ozone hole" (**Fig. 9.2**). The decrease varies seasonally and is greatest in each hemisphere during that hemisphere's spring.

The ozone layer absorbs much of the sun's ultraviolet light, so depletion of the ozone layer could have severe adverse consequences for people and the **environment**. Ultraviolet light causes sunburn, eye cataracts, and skin cancers in humans; inhibits the growth of **phytoplankton** and possibly land plants; and may have harmful effects on other **species**.

The effects of ozone depletion on the environment are largely unknown. Studies have shown that the growth rate of phytoplankton in the Southern Ocean around Antarctica is significantly reduced (by 6 to 10%) in areas under the ozone hole. All life in the Antarctic, including whales, penguins, and seals, is ultimately dependent on phytoplankton for food (Chaps. 14, 15). Consequently, besides the direct effects of the increased ultraviolet radiation, ozone depletion may adversely affect species by reducing their food supply.

There is still some uncertainty about the causes of

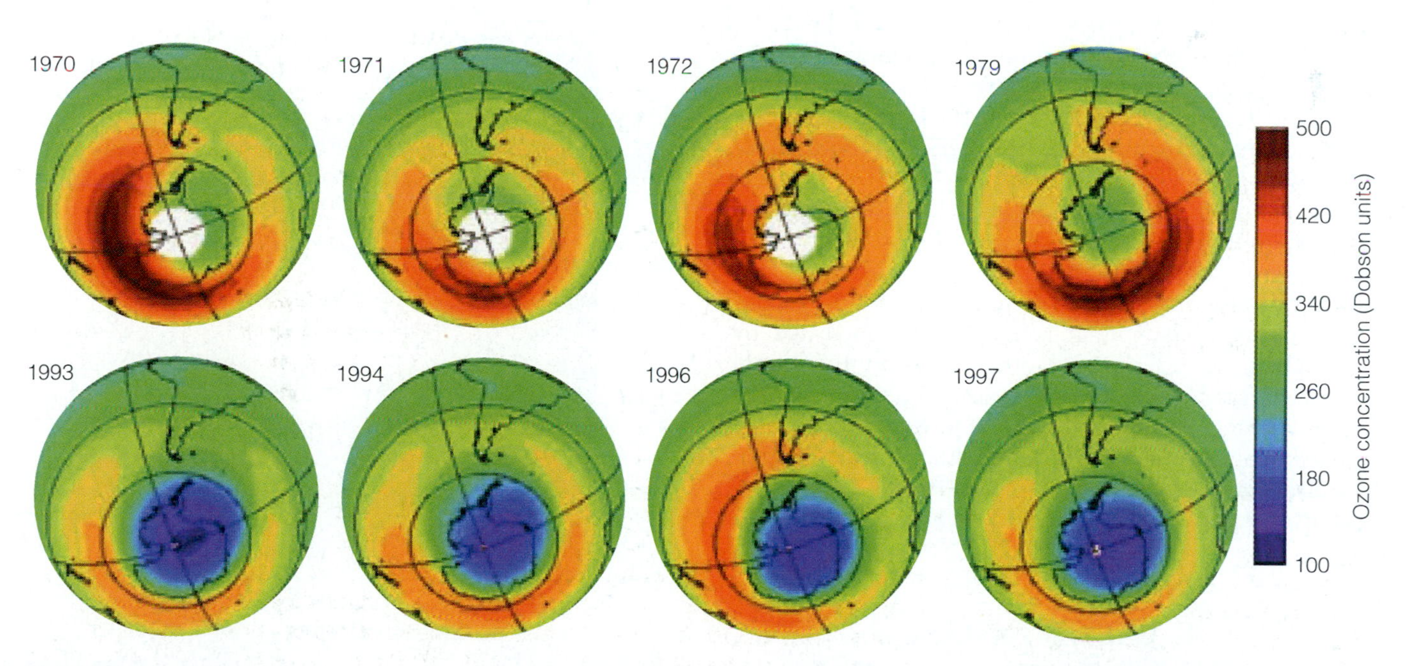

FIGURE 9.2 Average total amount of ozone above the Earth's surface over Antarctica during the month of October, as measured by a series of satellite instruments. The small circular areas that have no color directly over the pole in the 1970, 1971, and 1972 data are areas where no data were obtained. Nevertheless, it is clear that there is a broad region of very low ozone levels over Antarctica during the 1990s that was not present in the 1970s. This is the Antarctic ozone hole.

ozone depletion, but the generally accepted view is that it is caused primarily by gases called **chlorofluorocarbons (CFCs)** released to the atmosphere. CFCs are synthetic chemicals used in many industrial applications. Until just a few years ago, almost all refrigerators and air conditioners used substantial quantities of a mixture of CFCs called "freon."

CFCs are highly resistant to decomposition and may remain in the atmosphere for decades. The long **residence time** (CC8) allows the CFCs to diffuse slowly upward until they reach the ozone layer. There they are decomposed by ultraviolet light, and their chlorine is released. The chlorine atoms take part in a complicated series of chemical reactions with ozone molecules whereby the ozone molecule is destroyed but the chlorine atom is not. Because it is not destroyed, the chlorine atom can destroy another ozone molecule and then another. In fact, one chlorine atom, on average, can destroy thousands of ozone molecules.

Ozone depletion in polar regions is greatest in spring because stratospheric clouds of supercold ice crystals are formed at that time. These clouds remove nitrogen compounds from the ozone layer. Thus, in spring, chlorine that would normally react with the nitrogen compounds reacts with the ozone instead.

CFC use as an aerosol gas in spray cans was banned and eliminated virtually worldwide in the early 1980s. International treaties were established that called for total elimination of CFC manufacture and replacement of these compounds by other chemicals by the year 2000. Although these measures will probably eliminate ozone depletion eventually, most scientists believe that the depletion will continue for more than a decade, until the CFCs already released have diffused slowly upward through the atmosphere and been decomposed. Indeed, ozone depletion continues to be observed in both hemispheres.

Depletion of ozone in the ozone layer in the stratosphere should not be confused with the problem of elevated ozone concentrations in smog in the troposphere (lower atmosphere). Ozone is released into the troposphere by various human activities and is created by photochemical reactions of other gases in the troposphere. Ozone is one of the principal harmful components of smog. However, this ozone does not contribute to the ozone layer because it does not last long enough in the atmosphere to be transported up through the troposphere and stratosphere to the ozone layer.

9.2 WATER VAPOR IN THE ATMOSPHERE

In the troposphere, air masses move continuously, both vertically and horizontally, and these movements control the Earth's weather and climate and the winds that create ocean waves and ocean surface currents. The movements of air masses are caused primarily by changes in air mass density that occur as water vapor is added or removed.

Water Vapor and Air Density

At temperatures below the boiling point of water, water vapor in the atmosphere behaves in much the same way as sugar in water. Sugar dissolves in water, but the maximum quantity of sugar that can be dissolved is limited. More sugar can be dissolved at higher temperatures. Similarly, water vapor can be "dissolved" in air in limited quantities, and air is able to hold more water vapor at higher temperatures. The maximum amount of water that can remain in the vapor phase in air is expressed by the water vapor **saturation pressure**. The variation of water vapor saturation pressure with temperature is shown in Figure 9.3. The amount of water vapor that can be held in air at average ocean surface temperatures (approximately 20°C) is many times greater than the amount that can be held in air near the top of the troposphere (approximately –40°C to –60°C).

When water is evaporated from the sea surface, water molecules displace (move aside) oxygen and nitrogen molecules. The displaced molecules are mixed and distributed in the surrounding air mass by random motions of the gas molecules. This is important because the atmospheric pressure and, therefore, density of the air mass would be increased if this were not the case. In contrast,

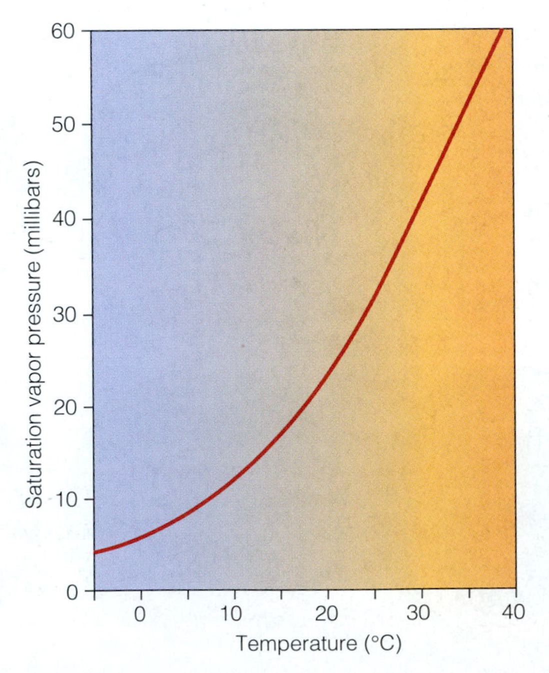

FIGURE 9.3 Variation of the water vapor saturation pressure with temperature. Air can hold less water vapor at lower temperatures. This is the reason that water condenses into rain as warm, moist air rises and cools adiabatically.

water molecules (H_2O, molecular mass 18) are lighter than either the nitrogen (N_2, molecular mass 28) or the oxygen (O_2, molecular mass 32) molecules that they displace. Hence, light molecules displace heavier molecules, and the density of air is, therefore, reduced when water vapor is added (provided there is no change in pressure). As a result, moist air is less dense than dry air at the same temperature and pressure. Therefore, an air mass to which water vapor has been added tends to rise until it reaches its equilibrium density level in the density-stratified atmosphere (**CC1**). Conversely, if water vapor is removed, air increases in density and tends to sink.

Water vapor is continuously added to the atmosphere by evaporation of water from the ocean surface and removed by condensation and precipitation. The rate of evaporation varies with location, as explained later in this chapter. Where evaporation of ocean surface water is particularly rapid, the density of the air mass at the ocean surface is reduced and the air tends to rise. In many parts of the oceans where the surface water is warmer than the overlying air, the density of the air is also reduced by warming at the ocean surface.

Water Vapor, Convection, and Condensation

If a gas expands without any external source of heat, its temperature decreases in a process called **adiabatic expansion**. As a result, when air rises in the atmosphere, the pressure decreases and the air is cooled by adiabatic expansion. Because water vapor saturation pressure is highly temperature-dependent, cooling the air also results in a decrease in the water vapor saturation pressure. When the air has risen sufficiently that the saturation pressure is reduced below the actual water vapor pressure, the air can no longer retain all of its water vapor. The excess water vapor is converted to liquid water. However, water molecules in air are generally far enough apart that they do not easily condense to form water.

Air must often become **supersaturated** before water molecules combine to form enough clusters to provide nuclei for raindrops. This process is highly variable and can be affected by many factors, including the presence of dust particles, which also may act as raindrop nuclei. In some cases, water vapor does not condense until the temperature is below the freezing point of water. Then snow or hail can form. In many cases, the water will condense to form large numbers of extremely small water droplets—droplets that are too small to fall out of the sky (**CC4**) or to combine with each other to form larger droplets. These tiny water droplets form the clouds of the atmosphere.

Water vapor contributes to vertical movements of air masses, not only because of its low molecular weight, but also because of water's **latent heat of vaporization** (Chap. 7). When warm, moist air rises through the atmosphere and cools, and water vapor condenses, the water's latent heat of vaporization is released. The heat warms the air mass and thus reduces its density, causing the air mass to rise. Release of latent heat is the major driving force for the spectacular rising plumes of moist air that form thunderhead clouds, which can reach the upper parts of the troposphere. Latent heat released to the atmosphere is also the main energy source for **hurricanes** and other storms.

Atmospheric **convection cells** are created by evaporation or warming at the sea surface that decreases the density of the air mass and causes it to rise. The rising air mass cools by adiabatic expansion, and eventually water vapor is lost by condensation. As the water vapor condenses, temperature is increased by the release of latent heat and the air tends to rise farther. Adiabatic expansion and radiative heat loss then cool the air mass and increase its density so that it sinks (**CC3**). The location, size, and intensity of convection cells determine the location and intensity of rainfall and snowfall, as well as the distribution of heat energy in the atmosphere. These factors determine climate and weather.

9.3 WATER AND HEAT BUDGETS

Averaged over the entire planet and over time periods of months or years, the amounts of heat and water that enter the atmosphere from all sources must equal the amounts removed from the atmosphere by all routes. If total inputs of heat or water did not equal total outputs, the average atmospheric temperature and/or the amount of water vapor in the atmosphere would progressively increase or decrease from year to year. In fact, the Earth's climate normally remains essentially unchanged for centuries.

In the Earth's past, very small imbalances between heat gained and heat lost have slightly changed the total amount of heat in the atmosphere and, thus, the Earth's climate. Today, atmospheric emissions of gases, such as carbon dioxide, may be upsetting or may already have upset the balance between heat gained and heat lost that has remained stable throughout the past several centuries (**CC9**).

Because the inputs and outputs of heat and water to and from the atmosphere must balance, we can construct global heat and water budgets.

Water Budget

The global water budget is relatively simple (**Fig. 9.4**). Water is evaporated from the oceans and from lakes, rivers, soils, and vegetation on the land. The amount of water evaporated from land and ocean annually is enough to cover the entire globe with water almost 1 m deep. The evaporated water eventually condenses and falls as rain or snow over both land and ocean. Approximately 93% of water evaporated to the atmosphere is evaporated from

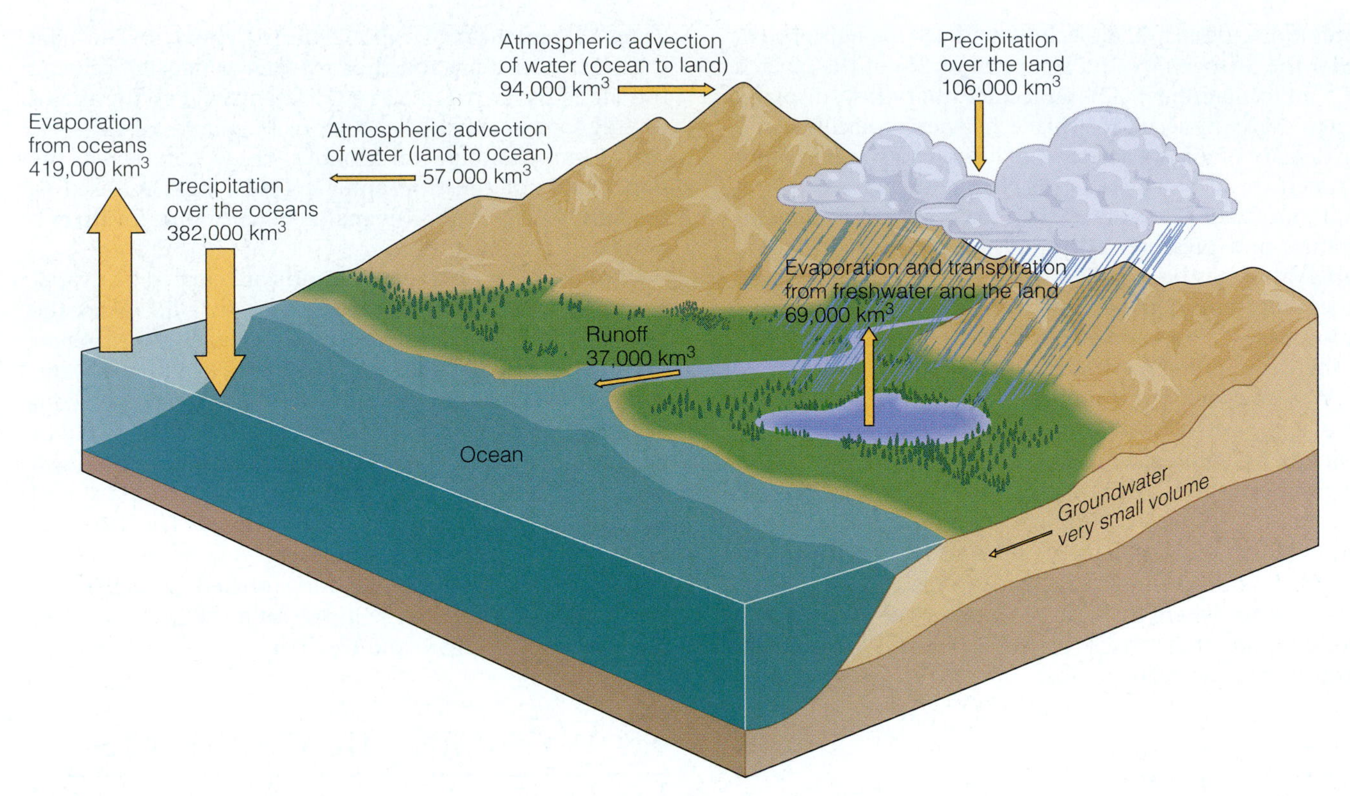

FIGURE 9.4 The Earth's water budget. The values represent the volumes of water transferred annually between the various reservoirs by the indicated routes.

the oceans. However, only about 71% of the total global precipitation (rain and snow) falls on the oceans. Thus, more water evaporates from the ocean than reenters it as rain, and more rain falls on the land than is evaporated from it. This pattern is fortunate because the excess precipitation on land is the source of the Earth's freshwater supply. The excess water flows back to the oceans through rivers, streams, and **groundwater**.

Neither the rate of water evaporation nor the amount of rainfall is the same everywhere on the oceans or on land. Clearly, deserts and tropical rain forests receive different amounts of rainfall and sustain different amounts of evaporation. The distribution of evaporation and rainfall is discussed later in this chapter.

Heat Budget

The Earth's heat budget (**Fig. 9.5**) is more complicated than its water budget. The source of heat energy for the atmosphere is the sun. Solar radiation energy is partially absorbed by water vapor, dust, and clouds in the atmosphere and converted to heat energy. Part of the sun's energy is also reflected back, or **backscattered**, to space by clouds. Solar energy that reaches the Earth's surface is partially reflected, and partially absorbed and converted to heat energy by the land and ocean waters. The Earth, ocean, atmosphere, and clouds also radiate heat energy. In fact, all bodies radiate heat. The peak **wavelength** radiated increases as the temperature of the body decreases. Because the Earth is much cooler than the sun, the wavelengths radiated by the Earth, ocean, and clouds (infrared) are much longer than the wavelengths emitted by the sun (visible and ultraviolet). The difference in wavelength and the differential absorption by atmospheric gases at different wavelengths are the basis for the greenhouse effect (**CC9**).

About 20 to 25% of the solar radiation reaching the Earth is absorbed by atmospheric gases and clouds and converted to heat energy. Almost 50% of the solar radiation is absorbed by the ocean water and the land, then transferred to the atmosphere by the **conduction** of **sensible heat**, as latent heat of vaporization, and by radiation. The net transfer averaged over the global oceans is from ocean to atmosphere. Heat transfer by conduction is relatively small, but the transfer of latent heat from ocean to atmosphere equals about one-quarter of the total solar energy that reaches the Earth. Thus, the oceans effectively capture a major portion of the sun's radiated energy and transfer much of it to the atmosphere as latent heat of vaporization. This mechanism is responsible for moderating climate differences between the Earth's tropical and polar regions.

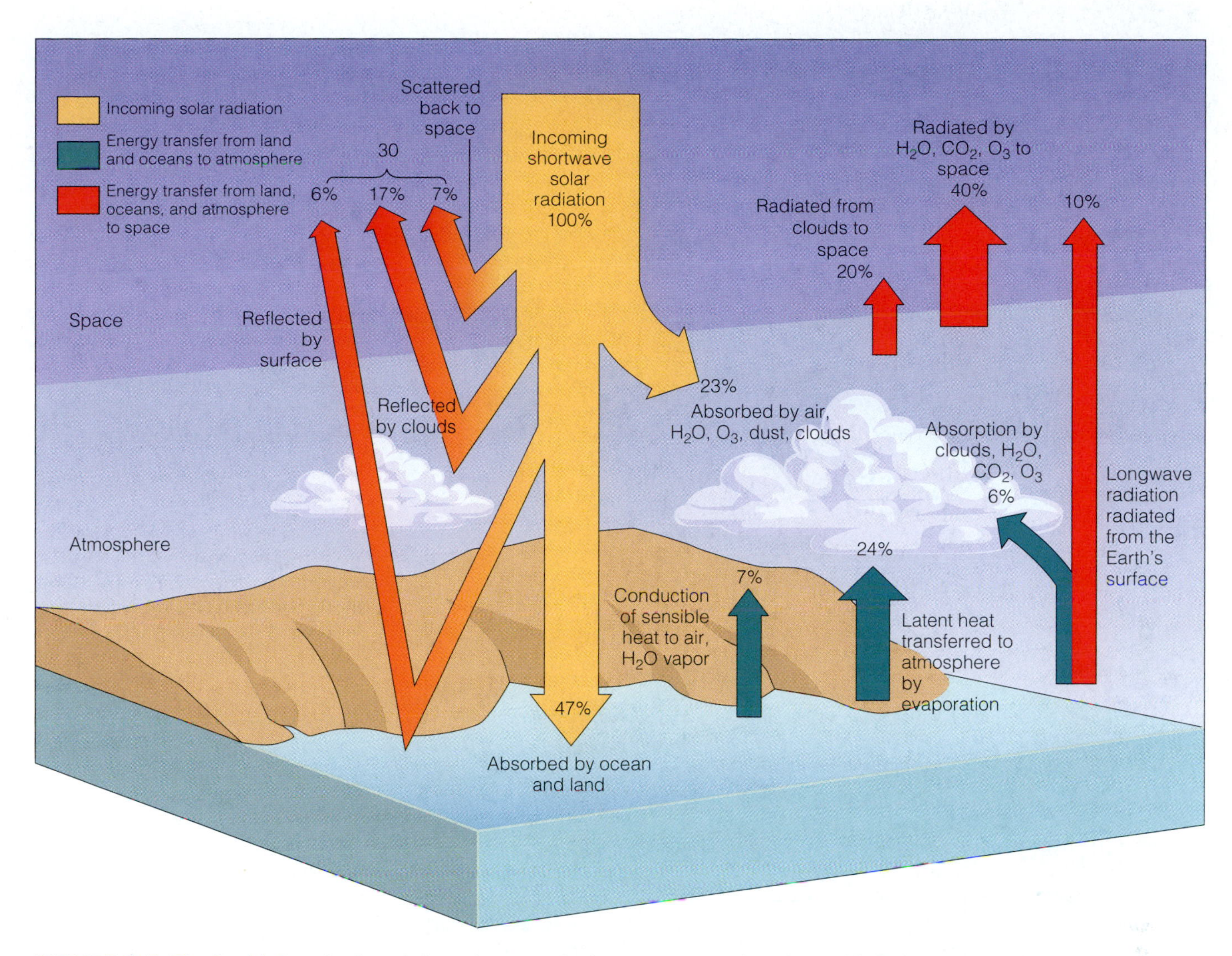

FIGURE 9.5 The Earth's heat budget. Solar radiation, which is concentrated in the visible light wavelengths, is reflected, scattered, and absorbed by the gases and clouds of the atmosphere and by the ocean and land surface. All of the absorbed energy is radiated back to space, primarily as infrared radiation. The diagram shows the percentages of the incoming solar radiation that are absorbed, reflected, and reradiated to space by the various components of the Earth–ocean–atmosphere system. Each of these may change in an unpredictable way as the greenhouse effect is enhanced by anthropogenic inputs to the atmosphere (**CC9, CC10, CC11**).

Latitudinal Imbalance in the Earth's Radiation

At the equator, the sun passes directly overhead at noon on the spring and autumnal **equinoxes** (Fig. 9.6a). The sun's radiated energy therefore passes vertically through the atmosphere, and the **angle of incidence** at the Earth's surface is 90°. At the poles on the equinoxes, the sun remains on the horizon all day, and its radiation must pass through the atmosphere parallel to the Earth's surface.

Because the poles are only about 6000 km farther from the sun than the equator is, and because the mean distance from the sun to the Earth is 148,000,000 km, the intensity of the sun's radiated energy is only about 0.01% lower at the poles than at the equator. However, the solar energy is spread over a progressively larger area with increasing **latitude** (Fig. 9.6b). Consequently, at 30°N or 30°S, the solar radiation received by an area of the Earth's surface is only 86% of that received by an equal area at the equator. The proportion at 60°N or 60°S is only 50%, and at the poles it is zero.

Nearer the poles, solar radiation must also travel a longer path through the atmosphere because of the lower angle to the Earth's surface. More of the solar radiation is absorbed as it passes through the atmosphere because of

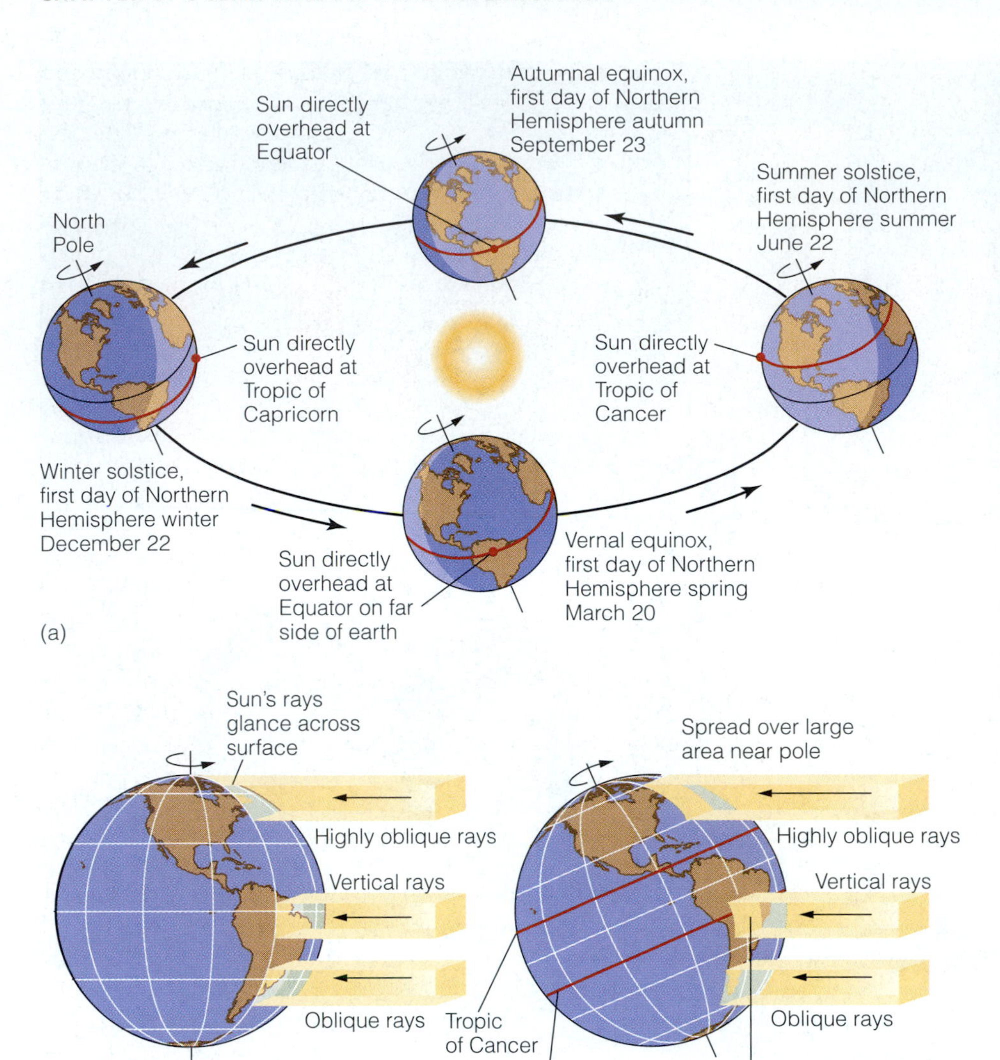

FIGURE 9.6 (a) The Earth, spinning on its tilted axis, moves around the sun once a year. In northern summer (on the right), the Earth is tilted such that the sun is directly overhead at a latitude north of the equator at a point that moves daily around the Earth. Similarly, in northern winter, this point is south of the equator. At the spring and autumnal equinoxes, the sun is overhead at the equator throughout the day as the Earth spins. (b) The angle of incidence of solar radiation received at the Earth's surface also changes with the seasons, but generally the area over which a given amount of the sun's energy is spread is less near the equator and increases toward the poles. The globe on the left shows the Earth at an equinox. The globe on the right shows the Earth at the winter (northern) solstice.

the increased path length, but this effect causes a relatively small difference in solar intensity at the Earth's surface.

The intensity of the Earth's long-wavelength radiation increases with temperature. However, the temperature difference between equatorial and polar regions is too small to produce more than a slight latitudinal variation in the Earth's longwave radiation intensity. The amount of long-wavelength radiation emitted by the Earth differs little between the equator and the poles (**Fig. 9.7**).

The total solar radiation absorbed by the Earth must equal the total long-wavelength radiation lost by the Earth when averaged over the whole Earth, but this balance does not occur at all latitudes. There is a large difference between the rates at which solar heat is absorbed at the poles and at the equator. There is a much smaller difference between the rates at which long-wavelength radiation is lost at the poles and at the equator (**Fig. 9.7**). Consequently, there must be mechanisms for transferring heat energy from low latitudes to high latitudes. If heat were not transferred, the polar regions would be much colder and the equatorial regions much warmer. Planets and moons that have no (or a limited) mechanism of heat transfer between latitudes have dramatically greater differences in surface temperature between their polar and equatorial regions than does the Earth. Hence, latitudinal heat transfer is one of the critical mechanisms that maintains the habitability of the Earth's entire surface.

Heat is transported latitudinally in two ways. First, latent heat of vaporization enters the atmosphere as water is evaporated in warm tropical regions and some of the heat is transported to higher latitudes by atmospheric convection cells. At higher latitudes, cooling causes water vapor to condense in the form of rain or snow, thus releasing latent heat to the atmosphere. Second, ocean water warmed by the sun in low latitudes is transported by ocean currents to higher latitudes, where it transfers heat to the atmosphere by conduction and evaporation. Ocean transport is more important near the equator, whereas atmospheric transport is more important near the poles.

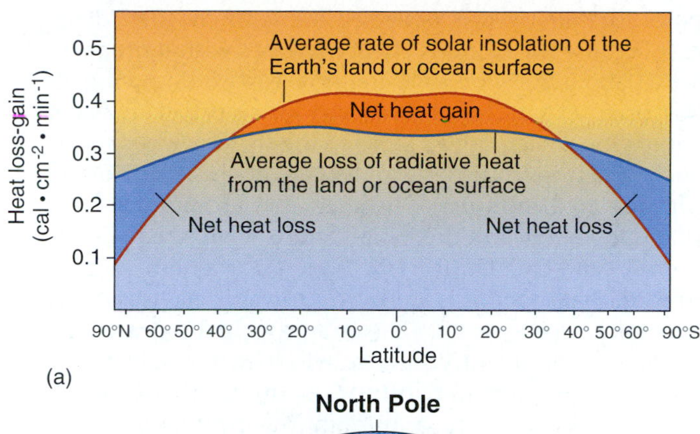

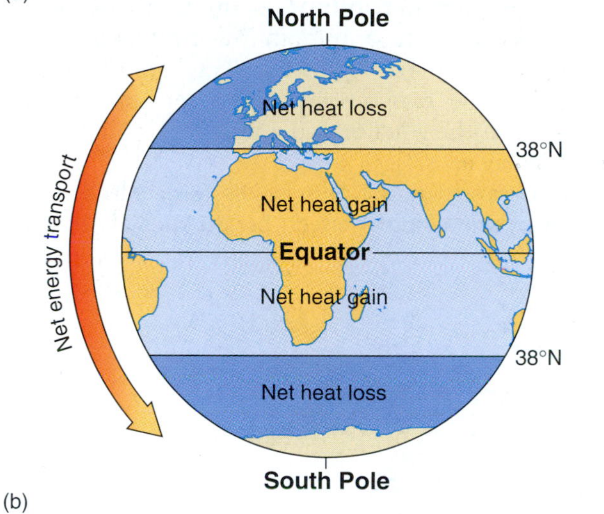

FIGURE 9.7 Variation in radiative heat loss and gain at the Earth's surface with latitude. There is a net gain between the equator and about 35° to 40°, and a net loss at higher latitudes.

9.4 CLIMATIC WINDS

Winds are extremely variable from one day to the next at any location. The day-to-day variations are part of what we call "weather." Weather events may include calm, thunderstorms, hurricanes, fog, and the other phenomena that a weather forecaster tries to predict hours or a few days before they happen. If we average the day-to-day variation over relatively large areas and over periods of several days or weeks, we find patterns of average winds, temperatures, rainfall, and so on that are relatively repeatable from one year to the next in each area. These long-term average patterns are called "climate." This chapter is concerned primarily with climatic winds, temperatures, and rainfall, although some aspects of various weather events are also examined.

Climatic Winds on a Nonrotating Earth

On a nonrotating Earth with no continents or **Coriolis effect** (**CC12**), wind patterns would be simple (**Fig. 9.8**). More heat is transferred to the atmosphere from land and oceans than is absorbed by the atmosphere directly from the sun (**Fig. 9.5**). Solar heating of the oceans is greatest at the equator. Thus, ocean water is warmer and the rate of heat transfer by evaporation of water from ocean to atmosphere is greater than at higher latitudes. The equatorial air mass would rise through the troposphere, creating a low-pressure zone and horizontal **pressure gradient** in the atmosphere at sea level. The rising air would be replaced by air flowing into the region from higher latitudes. Once it reached its equilibrium density level (**CC1**), the rising air would spread toward the poles. As the air mass moved north or south in the upper troposphere, it would progressively cool and release its water vapor as rain.

Once the air mass reached the pole, it would have sufficiently cooled and lost water vapor by condensation to become dense enough to sink and create a high-pressure zone in the atmosphere at sea level. The air would then move toward the equator along the horizontal atmospheric pressure gradient between the polar high-pressure zone and the equatorial low-pressure zone. As the air moved back toward the equator, it would be warmed and would gain water vapor from evaporation of ocean surface water. This is a simple convection cell motion (**CC3**). There would be a **convergence** at sea level at the equator, and **divergences** at sea level at the poles. Surface winds would blow from the poles toward the equator in each hemisphere. This simple two-convection-cell pattern (**Fig. 9.8**) illustrates the basic principles of atmospheric circulation, but it does not exist on the Earth, because we have so far not accounted for the fact that the planet rotates and moving air masses are deflected by the Coriolis effect.

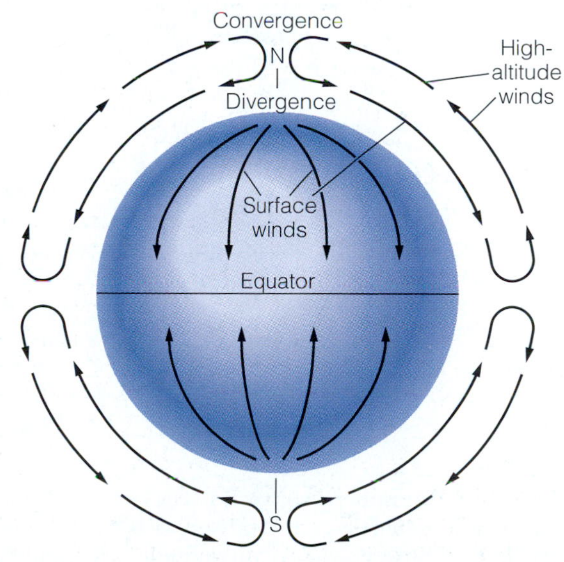

FIGURE 9.8 Atmospheric convection cells and winds on a hypothetical water-covered, nonrotating Earth.

Climatic Winds and the Coriolis Effect

Deflection of air movements by the Coriolis effect results in an atmospheric **convection** system that consists of three convection cells arranged latitudinally in each hemisphere (**Fig. 9.9**). To some extent, this real-world system has the same net result as the simple two-convection-cell system: air rises at the equator and sinks at the poles, and heat is transported from the equator to the poles.

To understand how the six-cell circulation pattern develops, consider the movements of an air mass initially located at sea level on the equator. The air mass is heated by solar radiation and has a high water vapor content because of the relatively high evaporation rate. It rises through the troposphere to its density equilibrium level and spreads out toward the poles, just as it would on a nonrotating Earth. As the air mass moves away from the equator, where there is no deflection by the Coriolis effect, it is not deflected significantly until it reaches 5° to 10° N or S (**Fig. 9.9**). As it continues to move toward the poles, the air mass is deflected increasingly toward the east because the Coriolis effect increases with latitude.

At about 30°N or 30°S, the now eastward-moving air has risen, expanded, cooled, and lost most of its water vapor. The cooler, dry air sinks to the surface, where some air moves toward the pole and some moves away from it, forming an atmospheric divergence at ground level. Air that moves toward the pole in this region enters the "Ferrel cell" circulation (**Fig. 9.9**). Air that moves back toward the equator is deflected toward the west. The deflection decreases as the air mass moves nearer the equator. Consequently, surface winds in the subtropical regions blow predominantly from the northeast in the Northern Hemisphere and from the southeast in the Southern Hemisphere (**Fig. 9.9**). These are the **trade winds**, and the convection cell air movements that occur between the equator and about 30°N and 30°S are called "Hadley cells."

Surface winds in the Ferrel cells (**Fig. 9.9**) are also deflected by the Coriolis effect, but to the east because

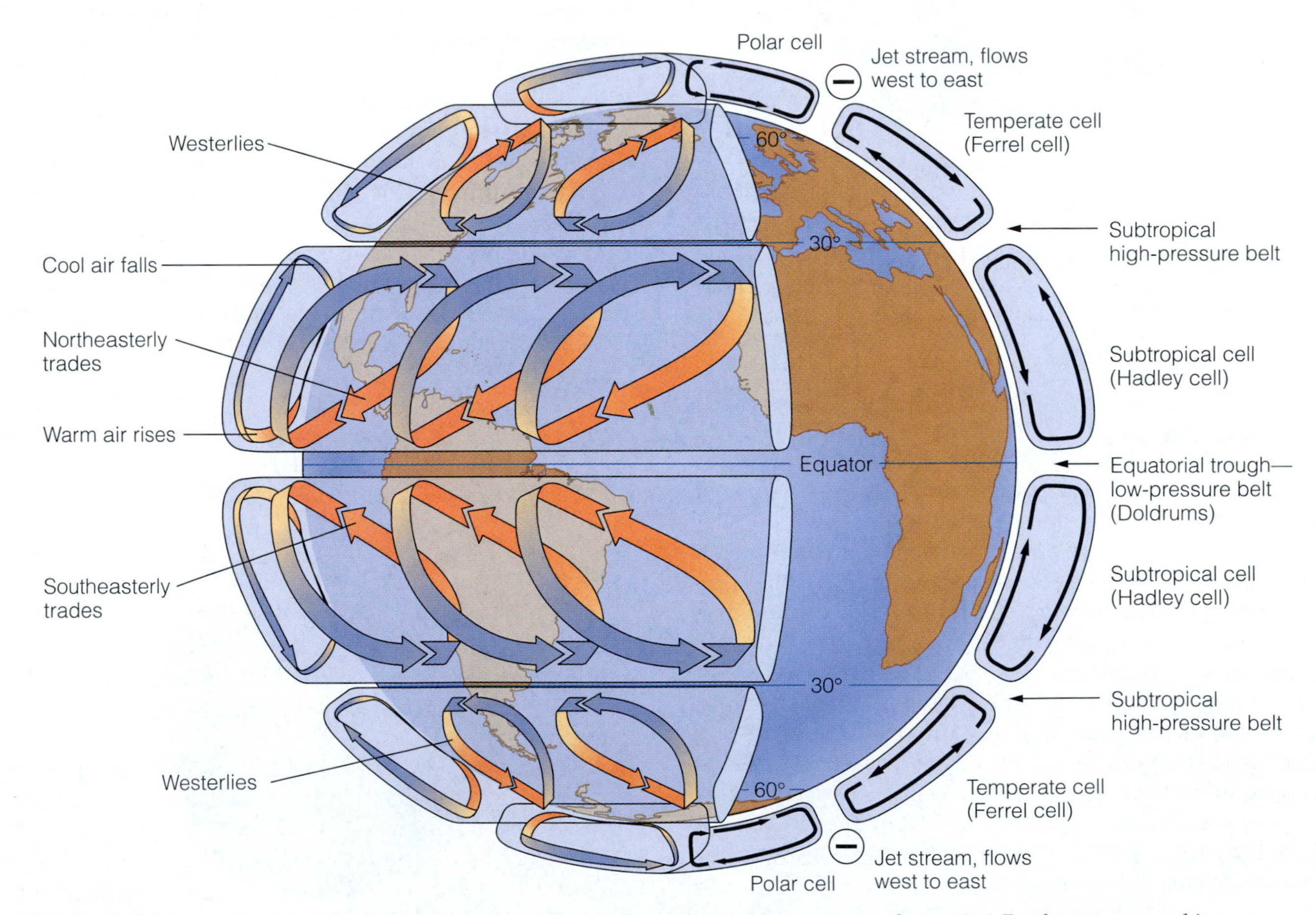

FIGURE 9.9 Atmospheric convection cells and winds on a hypothetical water-covered, rotating Earth are arranged in latitudinal bands. There are three cells in each hemisphere. Upwelling occurs at the equator and the polar fronts, and downwelling at the poles and in mid latitudes. Note how the Coriolis effect deflects air masses: air masses moving away from the equator are deflected to the east, and air masses moving toward the equator are deflected to the west. Atmospheric convection cells are arranged in this general pattern on the Earth, but they are modified substantially by the influence of the land masses.

they are moving away from the equator. Therefore, this is a region of persistent surface winds from the southwest in the Northern Hemisphere and from the northwest in the Southern Hemisphere, commonly called the **westerlies**.

Although the predominant winds are from the east in subpolar regions, air movements in the polar cells are complex because of the stronger Coriolis deflection and the presence of the **jet streams**. Jet streams are swiftly moving west-to-east air currents high in the troposphere. They are located over the boundaries between the Ferrel and polar cells, called the "polar **fronts**" (or "Antarctic front" in the Southern Hemisphere). The jet streams undergo complex **meanders** associated with movements of the polar fronts. Because the location of the jet stream in the Northern Hemisphere is an indicator of the movements of storms formed at the polar front, many weather forecasts in the United States routinely report the trajectory of the jet stream.

In any one region, the predominant wind direction and its persistence, the average extent of cloud cover, and the average rainfall are determined largely by the region's location with respect to the atmospheric convection cells (**Fig. 9.9**). In the equatorial region known as the Doldrums, where trade winds converge and atmospheric upwelling occurs, sea-level winds are light and variable. In addition, because rising moist air masses are cooled, cloud cover is persistent and rainfall high. This region is called the "**intertropical convergence** zone" because it is where two wind systems converge. Air rising from the sea surface through the troposphere causes atmospheric pressure at sea level to be low in the intertropical convergence zone. Air also rises through the troposphere at the polar front. However, unlike the intertropical convergence zone that migrates seasonally but otherwise remains generally invariable, the polar front oscillates in wavelike motions, called "Rossby waves" (Chap. 11), which have timescales of days or weeks. These waves can spawn massive storms called **extratropical cyclones**.

In the atmospheric **downwelling** (or subsidence) regions between the Hadley cells and Ferrel cells and at the poles, atmospheric pressure is high, winds are light, skies are usually cloudless, and rainfall is low. The subtropical high-pressure zones between the Hadley and Ferrel cells are called the "horse latitudes." Sailing ships were often becalmed for long periods in these regions, and horses carried on the vessels were killed to conserve water. Because of the high-pressure zone at the South Pole, much of Antarctica is a desert with very low annual precipitation. In fact, there are dry desert valleys in Antarctica with no snow cover, and the interior of Antarctica receives only about 50 mm of rain per year (desert climates are defined as having less than 254 mm of rain per year). The massive amounts of ice and snow in the thick ice sheet that covers other parts of Antarctica took millennia to accumulate, and the ice sheet exists only because Antarctic temperatures are too low for snow to melt during the summer.

Under the centers of the atmospheric cells (between convergences and divergences), winds are generally persistent, particularly in the trade wind zones. Rainfall and cloud cover are variable, and the zones are affected by periodic strong storms, especially in the higher-latitude areas of the **westerly** zones that are affected by storms formed on the polar front. Westerly surface winds under the Ferrel cells can be particularly strong, especially in the Southern Hemisphere, where few continents disturb the circulation. Sailors call the area of strong westerly winds at about 40°S the "roaring forties."

The atmospheric convection cell system described here is a simplified view of the climatic wind patterns on the Earth's surface. Complications of this simplified pattern occur because the interactions between ocean and atmosphere differ from the interactions between land and atmosphere. Complications also occur because of the seasonal movement of the Earth around the sun and certain longer-term oscillations of the ocean–atmosphere interaction, such as **El Niño**, which is described later in this chapter. Finally, local wind patterns are highly variable because weather events operate at much smaller geographic scales than do the global climatic wind patterns discussed in this section.

Seasonal Variations

Because the Earth's axis is tilted in relation to the plane of the Earth's orbit around the sun, the latitude at which the sun is directly overhead at noon changes progressively during the year. At the Northern Hemisphere summer **solstice** (June 21 or 22), the sun is directly overhead at 23.5°N, the Tropic of Cancer. At the autumnal equinox (September 22 or 23) and the spring equinox (March 20 or 21), the sun is directly overhead at the equator. At the Northern Hemisphere winter solstice (December 21 or 22), the sun is directly overhead at 23.5°S, the Tropic of Capricorn (**Fig. 9.6a**).

As the Earth moves around the sun, the latitude of greatest solar intensity migrates north and south, which results in a corresponding north or south displacement of the atmospheric convection cells. The displacement of the atmospheric convection cells is the cause of seasonal climate variations. The seasonal movement of convection cells is shown by a seasonal shift in wind bands (**Fig. 9.10**) and in surface-level atmospheric high- and low-pressure zones (**Fig. 9.11**).

Because the latitude at which lower-atmosphere air temperature is highest changes seasonally, the convection cells also migrate seasonally. Tropical lower-atmosphere air temperature is controlled primarily by radiation and evaporative heat inputs from the oceans and land. Because water has a high heat capacity, a great amount of heat energy is necessary to warm the ocean surface waters (**CC5**). Therefore, water temperature, evaporation rate, and air temperature do not increase in step with changes in solar intensity as the latitude of greatest solar intensity

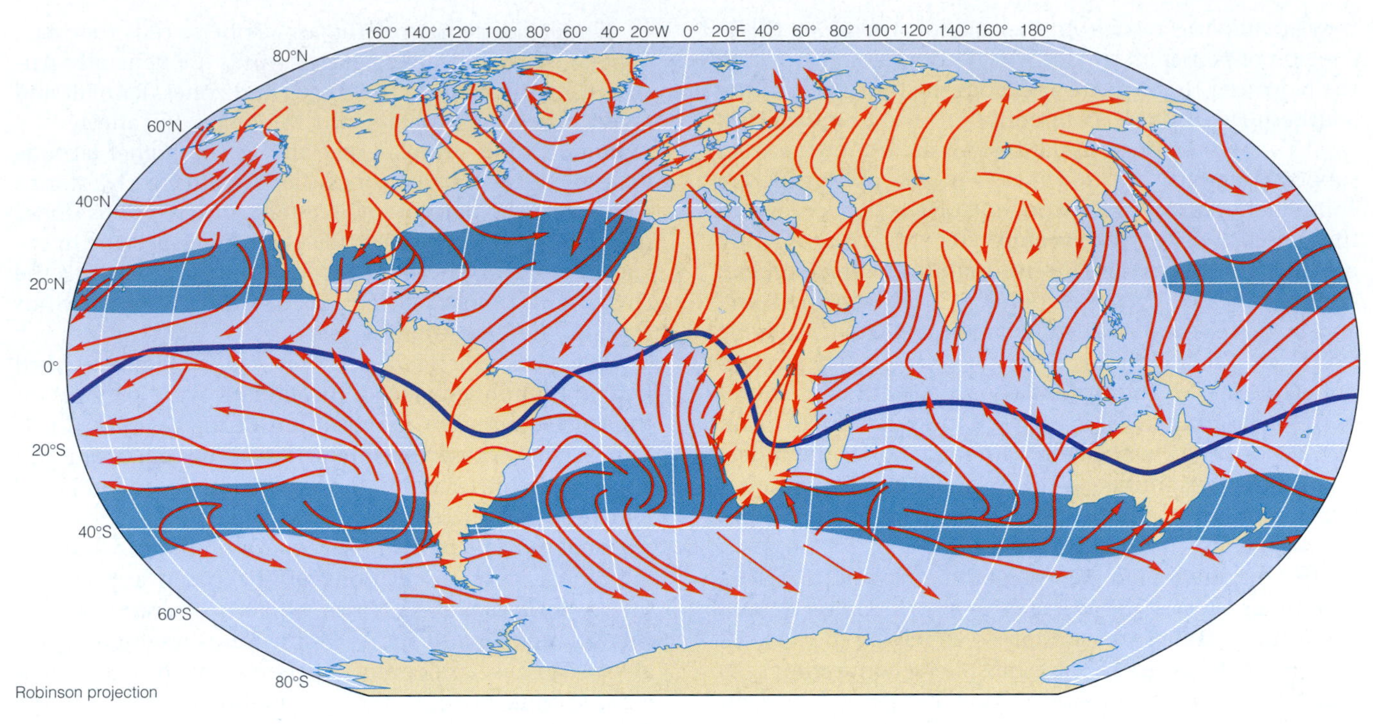

(a) Northern winter

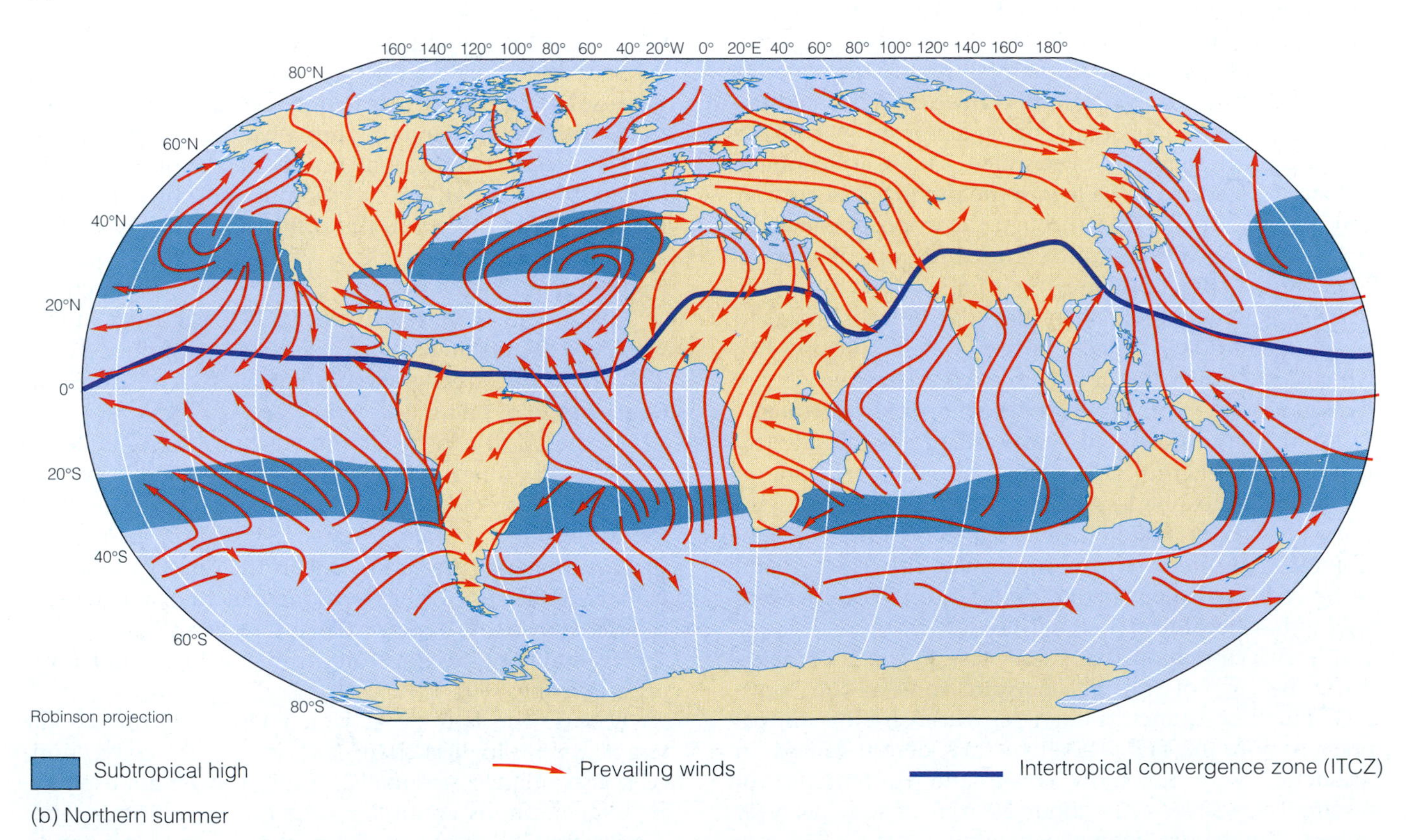

(b) Northern summer

FIGURE 9.10 Climatic winds over the ocean surface in (a) northern winter and (b) northern summer. Note the seasonal migration of the subtropical high-pressure zones, the intertropical convergence zone, and the associated trade winds. The migration is more pronounced over the major continental landmasses than over the oceans, except in the Indian Ocean, where the seasonal migration is enhanced by the strong monsoonal interaction between Asia and the Indian Ocean.

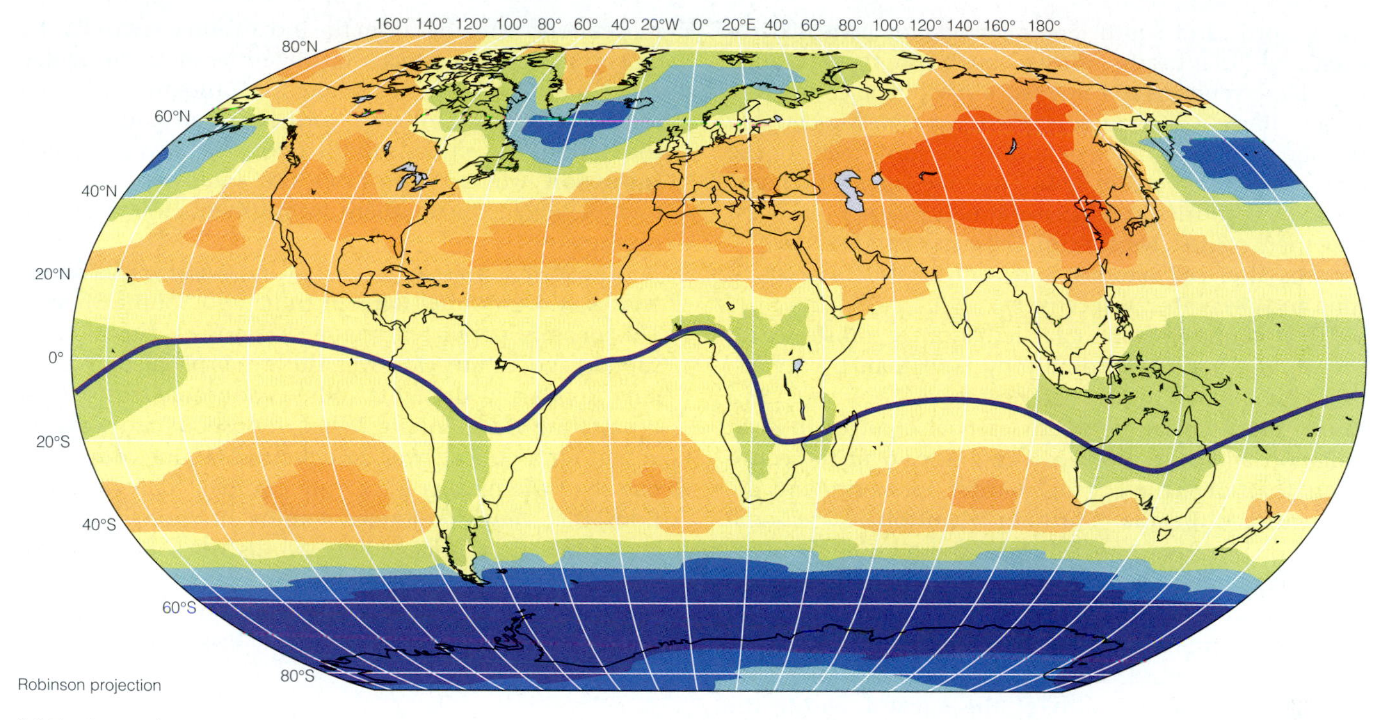

(a) Northern winter

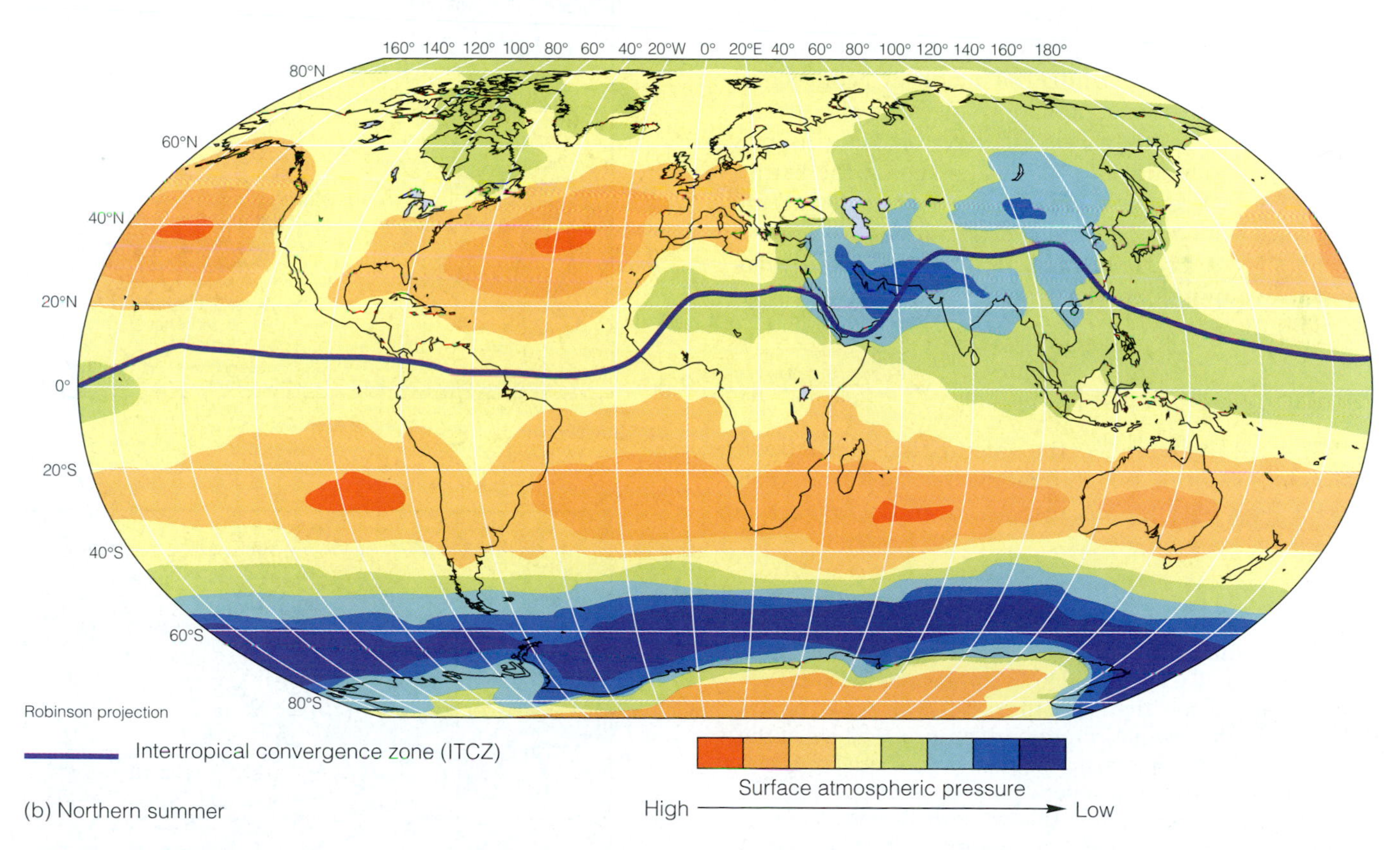

(b) Northern summer

FIGURE 9.11 Mean atmospheric pressures at sea level in (a) northern winter and (b) northern summer. The greater complexity of the geographical distribution and seasonal change of atmospheric pressure in the Northern Hemisphere is due to the concentration of the landmasses in this hemisphere. The extreme seasonal changes over the center of the Asian continent show that oceanic air masses have little influence in this region.

shifts north and south. Several weeks of increased (or decreased) solar heating must elapse before these parameters change enough to cause the atmospheric convection cells to move. Consequently, there is a time lag between the sun's seasonal movement and the movement of the atmospheric convection cells. The time lag is why many northern countries have their warmest weather in August and their coldest weather in February, approximately 2 months after the summer and winter solstices occur in June and December, respectively.

Heat captured from solar radiation is transferred to the atmosphere more rapidly by land than by oceans, primarily because of the high heat capacity of water. As a result, the time lag between seasonal changes in solar radiation and air temperature is shorter over the continents than over the oceans. Because the continents are concentrated in the Northern Hemisphere, seasonal migration of convection cells has a shorter time lag north of the equator than south of the equator. Furthermore, the center of the latitudinal wind bands, the intertropical convergence, migrates farther north of the equator in the Northern Hemisphere summer (especially over the continents) than it does south of the equator during the Southern Hemisphere summer (**Fig. 9.10**).

Seasonal changes in climate are greater in areas that are under an atmospheric convergence or divergence zone in one season and under the middle of a convection cell in another. Seasonal climate changes are also greater in higher latitudes than near the equator. The reason is that the total amount of solar heat reaching the Earth's surface per day depends on both the sun's angle and the amount of time during which the sun is above the horizon on a given day. Both the sun's intensity and the day length vary more between summer and winter at high latitudes than at low latitudes.

Monsoons

In the northern summer, the intertropical convergence moves farther north over the landmasses of Asia and Africa than it does over the oceans. Warm, moist tropical air moves north onto Asia from the Indian and western Pacific Oceans in the trade wind zone under the southern Hadley cell (**Fig. 9.12a**). Thus, in summer the southern Hadley cell trade winds extend into the Northern Hemisphere, where the Coriolis deflection is to the right. These winds, called **monsoon** winds, blow from the southwest, whereas trade winds in the Southern Hemisphere blow from the southeast.

The warm, moist air carried by monsoon winds causes storms and torrential rains in India and Southeast Asia. This densely populated region depends on the rains to support agriculture. However, the area is plagued by alternate floods and droughts. When monsoons are strong, devastating floods occur. When monsoons are weak or absent for a year or two, drought and famine occur. The nature of the multiyear, or "interannual," climate variations that cause prolonged famine or drought is just beginning to be understood. We discuss some of these variations later in this chapter.

In the northern winter, the intertropical convergence moves south over the Indian and Pacific Oceans and

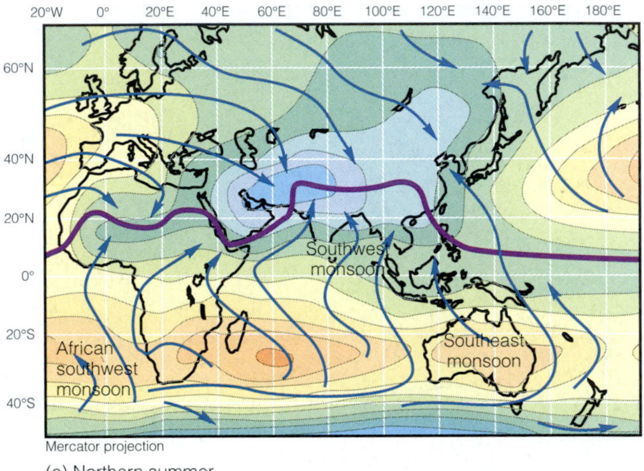

(a) Northern summer

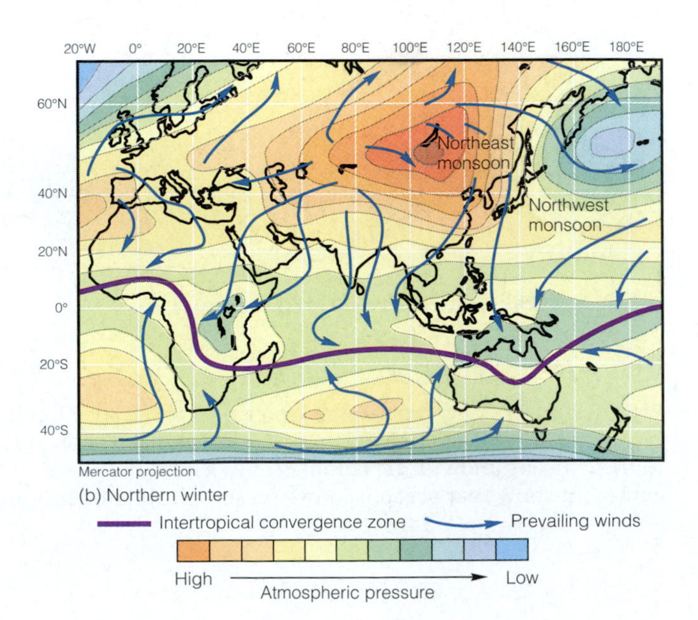

(b) Northern winter

FIGURE 9.12 Indian Ocean monsoon winds. (a) In summer, an intense low-pressure zone forms over Asia as the land is heated by the sun and loses this heat to the atmosphere more quickly than does the ocean. The heated air rises to form the low-pressure zone at a latitude where, if the Earth were covered by ocean, the Hadley and Ferrel cells would intersect and atmospheric pressure would be high. The low-pressure zone causes the atmospheric convection cells to shift to the north and bring warm, wet, southwest monsoon winds to parts of Asia. A similar, but less pronounced effect occurs over North Africa. (b) In winter, the Asian landmass loses heat more quickly than does the ocean, and an intense high-pressure zone is formed by the sinking cold, dry air mass. As this air mass spreads to the south, it shifts the convection cells south, which creates the northeast monsoons and brings dry weather to much of Asia.

Africa. In addition, the divergence between the Hadley and Ferrel cells moves far to the south over Asia in winter for the same reason that the intertropical convergence zone moves far north in summer. Thus, in winter this downwelling zone lies partly south of the Himalaya Mountains. The cool, dry downwelling air flows south over India, producing dry winters (**Fig. 9.12b**). Similar, but weaker and more complicated, seasonal monsoon wind reversals occur throughout East Asia, tropical Africa, and northern Australia (**Fig. 9.12**).

Sailors have taken advantage of the monsoons for centuries. Before Europeans explored the region, coastal traders sailed on winter trade winds from India and Southeast Asia to East Africa and Madagascar and returned on the summer monsoon winds.

9.5 CLIMATE AND OCEAN SURFACE WATER PROPERTIES

Ocean surface water salinity and temperature are controlled by solar radiation, transfer of heat and water between atmosphere and oceans, ocean currents, vertical mixing, and locally, river **runoff**.

Ocean Surface Temperatures

Figure 9.7 shows the generalized latitudinal distribution of solar radiation that reaches the Earth's surface. Ocean surface water temperatures generally decrease from the equator toward the poles (**Figs. 9.13, 9.14**), and thus, they reflect the distribution of solar radiation.

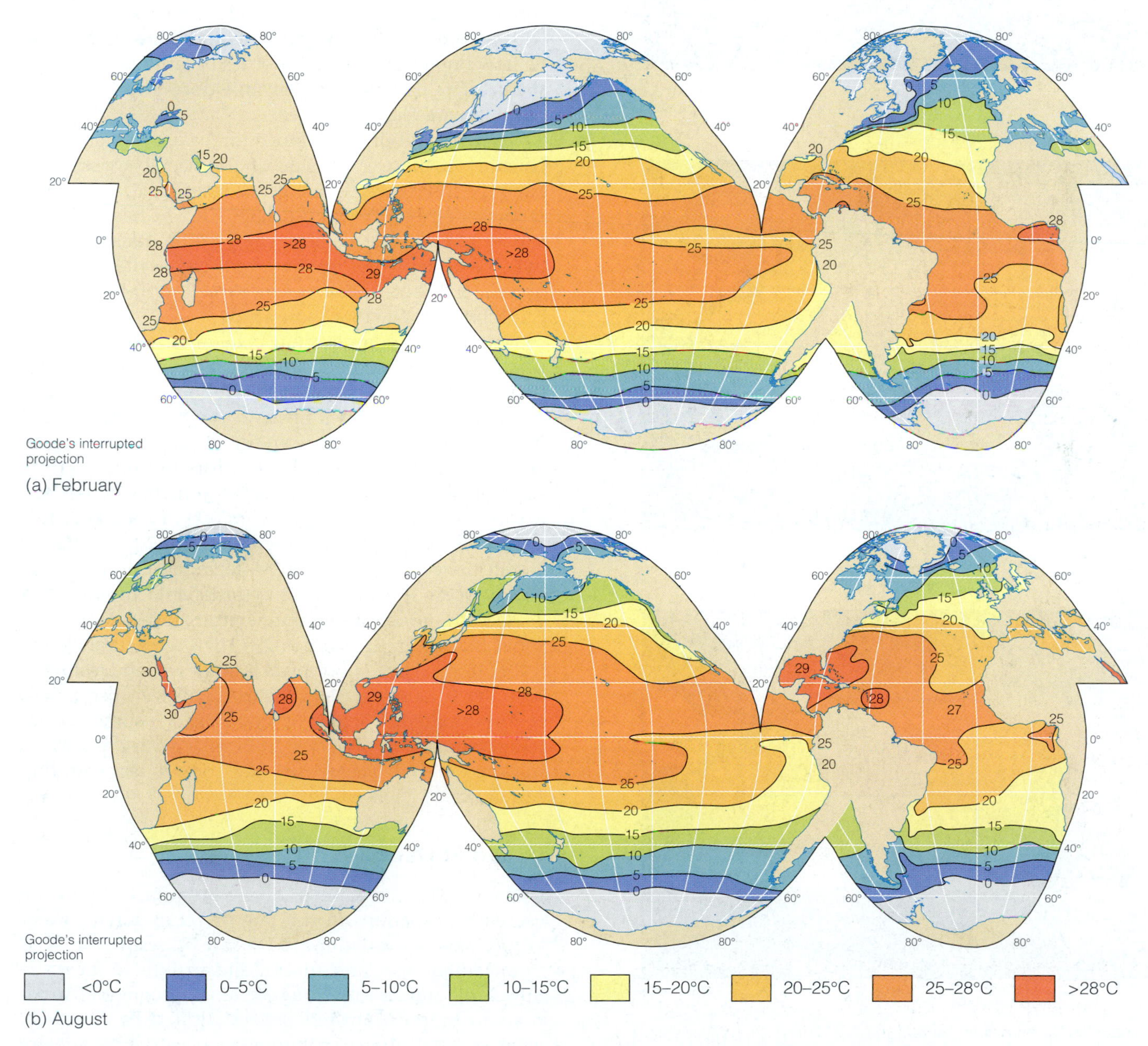

FIGURE 9.13 Average surface water temperatures in the world oceans in (a) February and (b) August.

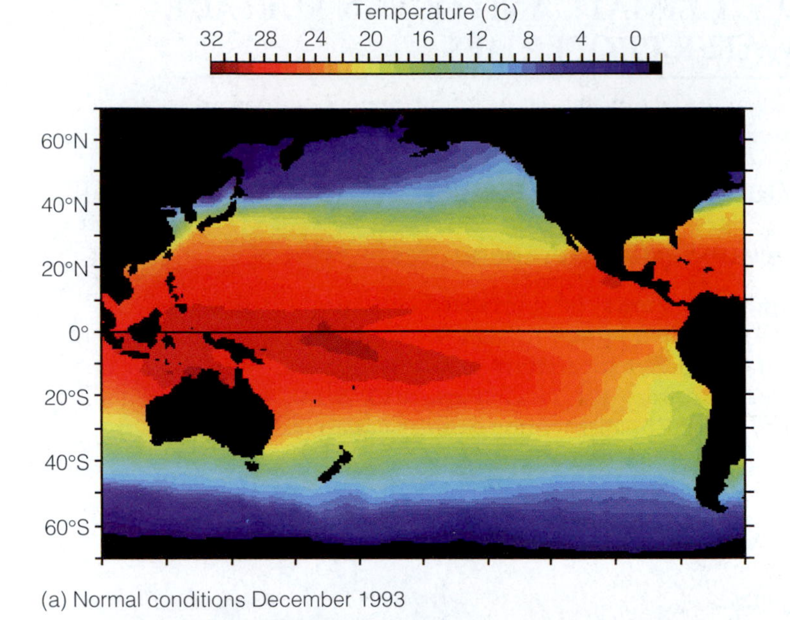

(a) Normal conditions December 1993

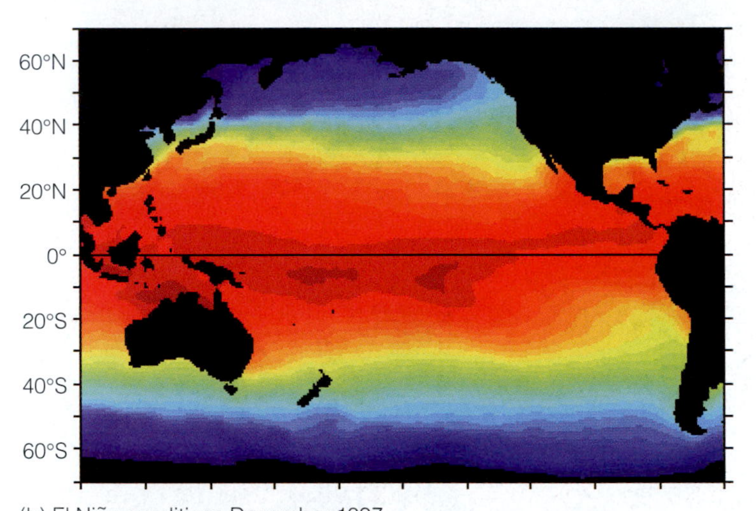

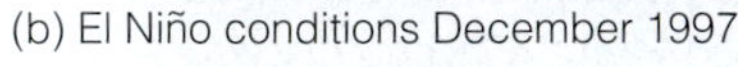

(b) El Niño conditions December 1997

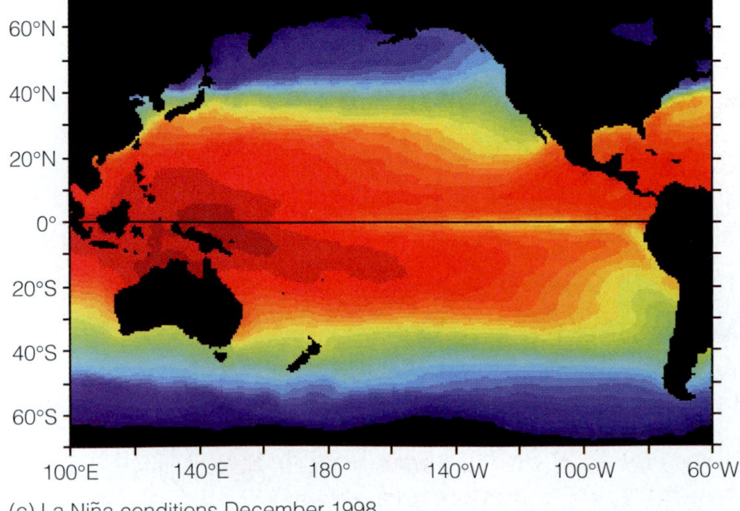

(c) La Niña conditions December 1998

FIGURE 9.14 Sea surface temperature in the Pacific Ocean, as measured by satellite-mounted sensors. (a) During non-El Niño conditions, there is a strong band of upwelling along the equator and in the Peruvian coastal zone, which can be seen in this December 1993 image by the yellow color that indicates lower surface water temperature in these areas. (b) During a strong El Niño (December 1997), there is no upwelling along the equator, and the Peruvian coastal upwelling is greatly reduced. (c) During La Niña conditions (December 1998), the upwelling area along the equator is substantially larger, and ocean water temperature around Indonesia is higher than normal because warm surface water is driven to the west.

On an Earth without ocean currents or winds, ocean surface water temperatures would decrease uniformly with latitude, and the **isotherms** in **Figure 9.13** would be horizontal lines parallel to lines of latitude. Although such a basic pattern is evident in **Figure 9.13**, complications also are readily apparent.

First, a band of low-temperature water lies along the equator on the eastern side of the Pacific Ocean. The band is a surface water divergence where cold, deep water is continuously **upwelled** (Chap. 10). The relatively low surface temperature at the equator in the Atlantic Ocean may be a result of similar, but less well-developed, upwelling. In the Indian Ocean, upwelling at the equator is inhibited by the monsoon winds.

The second complication evident in **Figure 9.13** is that isotherms in mid latitudes are inclined toward higher latitudes on the western sides of the oceans. Such a pattern is especially evident in the North Pacific and North Atlantic Oceans. It is a result of warm surface ocean currents that flow poleward from the subtropics on the western sides of the ocean basins and then across the ocean to the east. The currents are part of the **subtropical gyres** discussed in Chapter 10. In the North Atlantic Ocean the current is the Gulf Stream, and in the North Pacific Ocean it is the Kuroshio (or Japan) Current.

Ocean surface water temperature responds to seasonal changes in the sun's angle. High temperatures typical of tropical and subtropical latitudes extend to higher latitudes in summer (**Fig. 9.13b**). However, because of the heat-buffering action of ice (Chap. 7), surface water temperatures do not vary significantly with the seasons in the Arctic Ocean or the ocean surrounding Antarctica.

Ocean Surface Salinity

Differences in the salinity of ocean surface waters are caused by variations in the rate of ocean surface water evaporation, by variations in the rate of freshwater input from rainfall and land runoff, and in limited areas, by upwelling and downwelling. Because geographic variations in evaporation and precipitation rates are somewhat complex, the distribution of ocean surface water

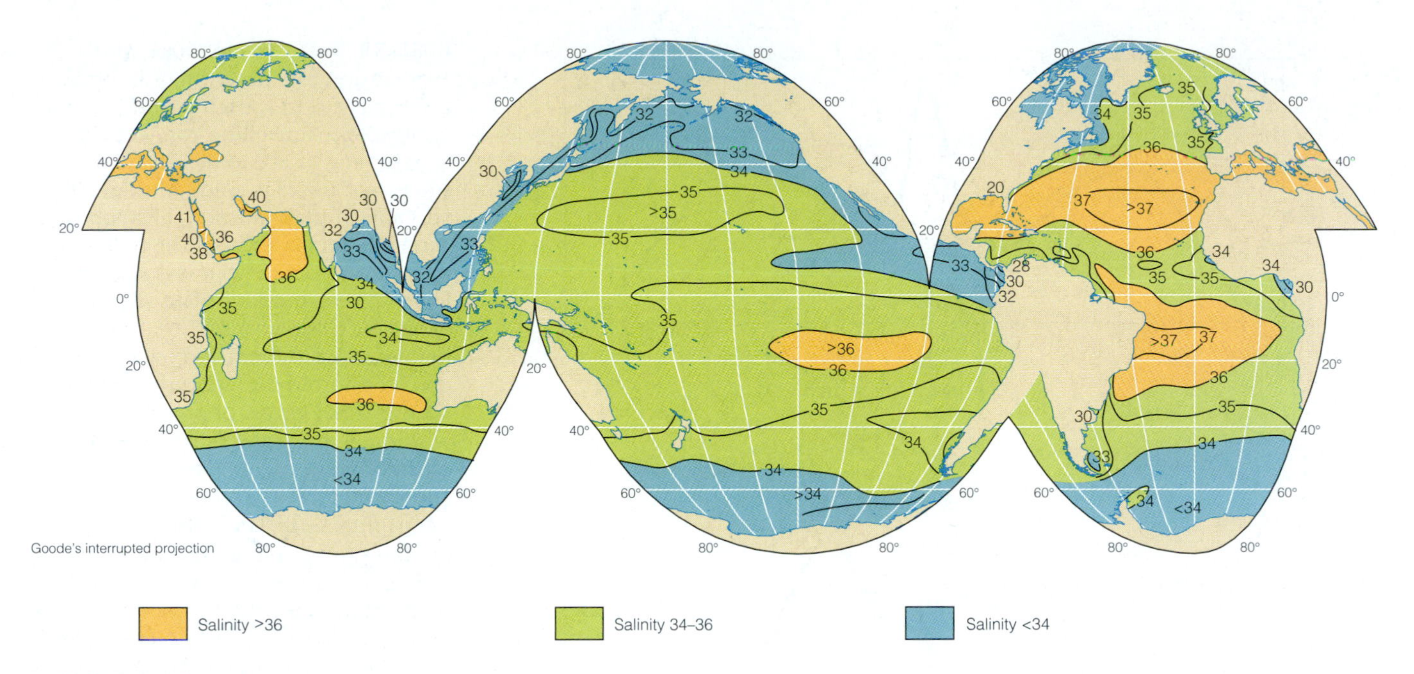

FIGURE 9.15 Surface water salinity in the world oceans.

salinity (**Fig. 9.15**) is somewhat more complicated than that of surface temperatures. Salinity is generally highest in surface waters in subtropical regions that are remote from land. It is generally lower at the equator, in polar and subpolar regions, and near continents. Surface salinity is higher in the subtropical Atlantic Ocean than in the subtropical regions of the other oceans.

The distribution of salinity is related to variations in evaporation and precipitation with latitude (**Fig. 9.16**). Precipitation is high near the equator and at mid latitudes around 40° to 60° N and S, which are the atmospheric convergence regions between the Northern and Southern Hemisphere Hadley cells and between the polar and Ferrel cells (**Figs. 9.9, 9.10**). In these regions warm, moist air converges and rises through the atmosphere. As the air rises, its water condenses to form clouds and rainfall. Precipitation is lower in subtropical latitudes (the horse latitudes) and near the poles because these regions are under the centers of the atmospheric convection cells or at divergences.

Evaporation varies with latitude in a different way than precipitation does (**Fig. 9.16**). The evaporation rate is higher at warmer ocean surface water temperatures. Hence, evaporation generally decreases progressively from equator to pole. However, the evaporation rate is lower in equatorial latitudes than in subequatorial regions for two main reasons. First, persistent cloud cover reduces the solar intensity in the equatorial region. Second, the equatorial region is persistently calm, whereas the trade wind regions have higher winds and hence greater evaporation. The higher rate of evaporation is due to both increased airflow over the water and increased surface area of water caused by waves, particularly breaking waves that create water droplets and bubbles with relatively large surface areas.

Figure 9.16b shows that precipitation exceeds evaporation across the equatorial oceans and at latitudes above about 40°N and 40°S. Surface salinity therefore tends to be lower in these regions. In contrast, evaporation exceeds precipitation in subtropical regions. Consequently, surface salinity tends to be higher in these regions (**Fig. 9.16b**).

The pattern of higher salinity in subtropical regions and lower salinity in equatorial regions and at high latitudes is observed in surface waters of the central parts of each ocean (**Fig. 9.15**). The distribution is more complex in some regions near continents because of large quantities of freshwater runoff.

The distribution of precipitation with latitude is substantially altered by the presence of landmasses (**Fig. 9.17**). The higher surface salinity in the central North and South Atlantic oceans (**Fig. 9.15**) is the result of interactions of the mountain chains of North and South America with the prevailing winds, and the outflow of high-salinity water from the Mediterranean (Chap. 10). In the westerly wind zones, moist air masses that move from the Pacific Ocean onto the American continents lose their moisture as rainfall on the west side of the Andes in South America and on the west side of the Rocky Mountains, the Sierra Nevada, and the coastal mountain ranges in North America. Consequently, the precipitation runs off into the Pacific Ocean. In contrast, the Northern Hemisphere trade winds in the Atlantic Ocean blow across the relatively narrow and low-altitude neck of Central America and, therefore, are able to carry their moisture all the way to the Pacific. The tongue of low-salinity water that reaches west into the Pacific Ocean from Central America (**Fig. 9.15**) is evidence of this transport.

The Southern Hemisphere trade winds in the Atlantic Ocean carry their moisture predominantly into the Amazon and Orinoco river basins. Both of these major

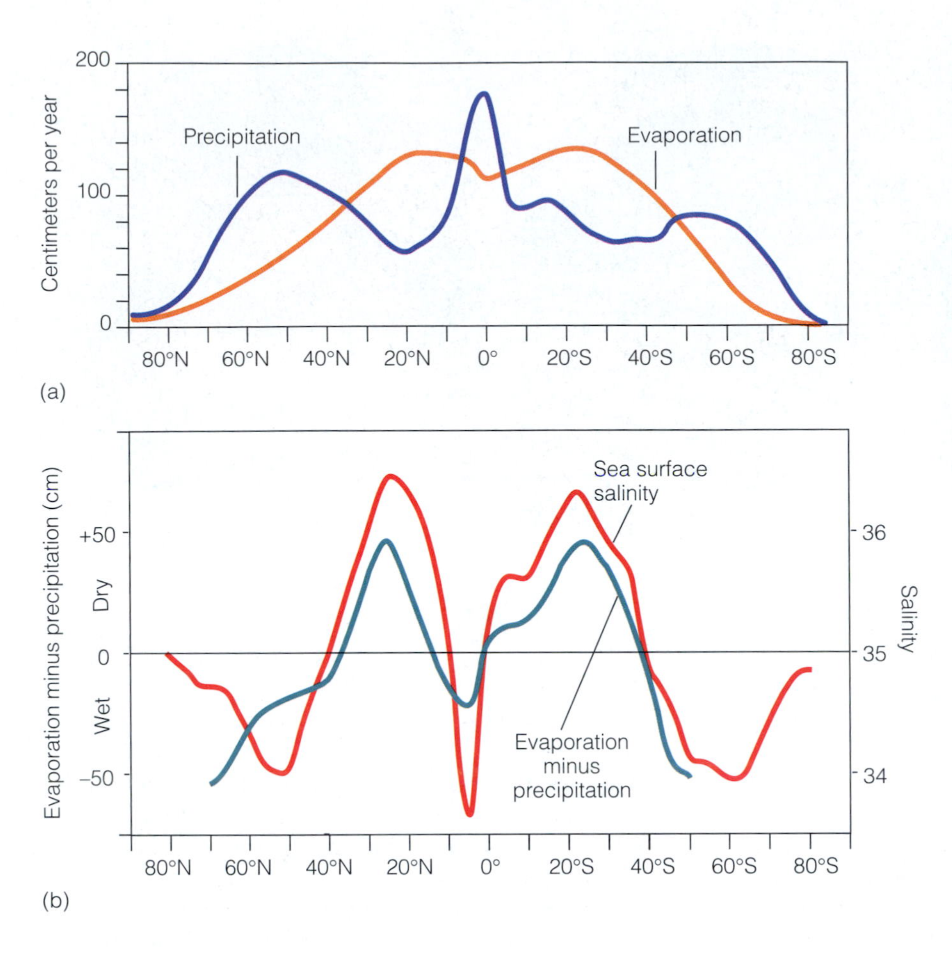

FIGURE 9.16 Precipitation and evaporation as a function of latitude. (a) Precipitation is high at the atmospheric upwelling zones at the equator and at 50° to 60° N and S, somewhat lower in mid latitudes, and very low at high latitudes. Evaporation generally increases from the poles toward the equator, but it is slightly lower at the equator because of the extensive cloud cover. Thus, there is a net excess of evaporation over precipitation in mid latitudes, whereas there is a net excess of precipitation within a few degrees north and south of the equator and at latitudes above approximately 40° in both hemispheres. (b) The salinity of surface ocean waters varies according to the difference between evaporation and precipitation. It is high where evaporation exceeds precipitation, and low where precipitation exceeds evaporation.

rivers discharge into the Atlantic Ocean equatorial region, where salinity is generally low.

Extremely high and extremely low salinities are most common in **marginal seas**, which have limited water exchange with the open oceans. Hence, if evaporation exceeds precipitation, salinity increases in such seas, and the high-salinity water is not mixed effectively with the lower-salinity water of the open ocean. Surface water salinity is high in the Mediterranean and the Arabian Gulf, and it is especially high in the Red Sea. All of these regional seas are in arid regions (**Fig. 9.17**) with low river runoff and high evaporation rates. Similarly, if precipitation and river runoff exceed evaporation, the salinity of a marginal sea is low. The best examples are the Baltic Sea and the marginal seas of Southeast Asia. In the Baltic, evaporation is low, and rainfall and runoff are moderately high (**Fig. 9.17**). In the marginal seas of Southeast Asia, rivers fed by monsoon rains discharge huge volumes of freshwater, and high rainfall rates overcome a relatively high rate of evaporation.

9.6 INTERANNUAL CLIMATE VARIATIONS

In addition to seasonal variations, there are year-to-year and multiyear oscillations of climate that are related to ocean–atmosphere interactions. Although we know relatively little about such longer-term climatic variations, oscillations have been identified in the tropical Pacific Ocean, the North Pacific Ocean, the Indian Ocean, the North Atlantic Ocean, and the Arctic Ocean. These may be linked with each other in ways we do not yet understand. The oscillation in the tropical Pacific Ocean, usually called "El Niño," is associated with climate changes affecting much of the globe and has been extensively studied.

El Niño and the Southern Oscillation

El Niño (Spanish for "the male child") has been so named because it was first observed at about Christmastime off the **coast** of Peru. The coastal and equatorial waters off Peru are normally regions where cold water from below the surface layers of the oceans is upwelled. Chapter 10 explains how winds cause this upwelling. The upwelled water contains high concentrations of **nutrients** that are essential to phytoplankton growth (Chap. 14). Upwelling areas therefore have high **primary productivity** rates, and massive **schools** of anchovy feed on the abundant phytoplankton that grow in the Peruvian upwelling region (Chaps. 14, 15). Until they were **overfished** (Chap. 18), the anchovy populations were so large that they yielded about 20% of the total world fish catch.

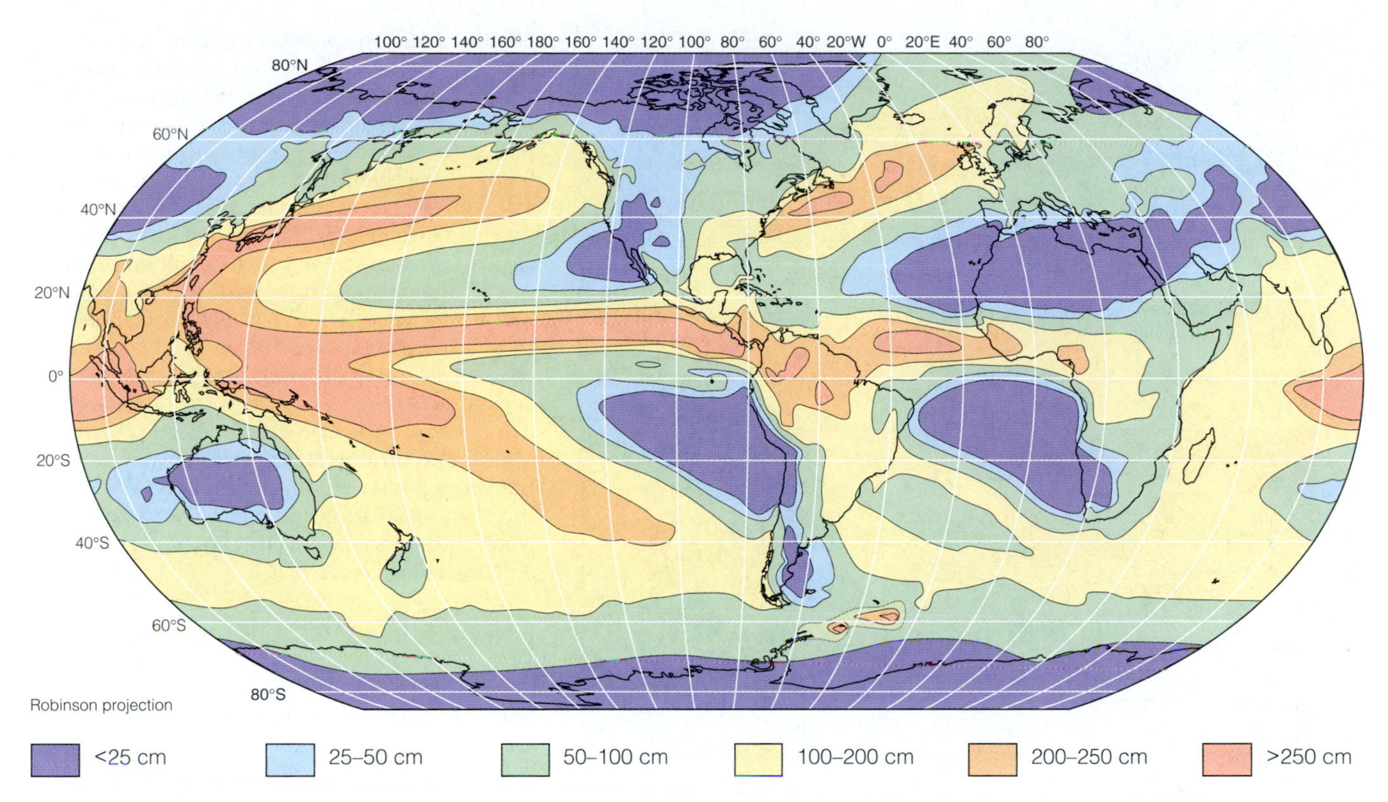

FIGURE 9.17 Distribution of mean annual atmospheric precipitation. Precipitation is high in the tropical atmospheric upwelling zone at the equator, generally low at mid latitudes in the atmospheric downwelling zones between the Hadley and Ferrel cells, higher in the atmospheric upwelling zones between the Ferrel and polar cells, and low again in the polar atmospheric downwelling zones. This pattern is more evident in the Southern Hemisphere because it is less modified by the presence of landmasses than it is in the Northern Hemisphere.

Symptoms of El Niño. The first sign of El Niño on the Peruvian coast is a temperature change in ocean surface water. Much warmer water (5°C or more warmer) displaces the normally cold surface water (**Fig. 9.14**). The less dense warmer water causes stable stratification of the water column and inhibits upwelling (Chap. 10, **CC1**), thus reducing phytoplankton populations that were nourished by the cold upwelled water. As the phytoplankton decline, the anchovy population that feeds on them plummets and the fishery collapses. El Niño recurs generally every 3 to 5 years and usually lasts several months, until the situation reverts to normal. El Niño events have been documented in historical records as far back as 1726.

El Niño is not an isolated event that occurs only off the coast of Peru. In fact, it is a complex series of cyclical changes in both ocean and atmosphere that affects much of the Earth's surface. The series of changes has been called the "El Niño/Southern Oscillation" (**ENSO**). The name reflects the fact that the changes in ocean and atmosphere involve a very large region of the Pacific, especially the region immediately south of the equator. The sequence of ocean–atmosphere interactions during an ENSO, now relatively well known, is described in the next section.

Between El Niños, a broad band of upwelling extends across the equatorial Pacific Ocean, and strong **coastal upwelling** occurs off the coast of Peru. The upwelling regions and their causes are discussed in Chapter 10. The areas of upwelling are shown by colder surface waters (**Fig. 9.14a**). Trade winds move warm surface water to the west, where it accumulates near Indonesia (**Fig. 9.14a**). An atmospheric low-pressure area is well developed in this region because of the strong evaporation from the ocean surface and the high water temperature.

The low-pressure zone normally brings plentiful rains to Indonesia and the surrounding region. In contrast, a well-developed high-pressure zone off Peru maintains low rainfall over that country, which normally has an arid climate.

Where warm water that has been pushed to the west by trade winds accumulates near Indonesia, the sea level becomes elevated by as much as a few tens of centimeters in extreme cases. For reasons not yet understood, this situation begins to change in the months of November through April. The Indonesian atmospheric low-pressure

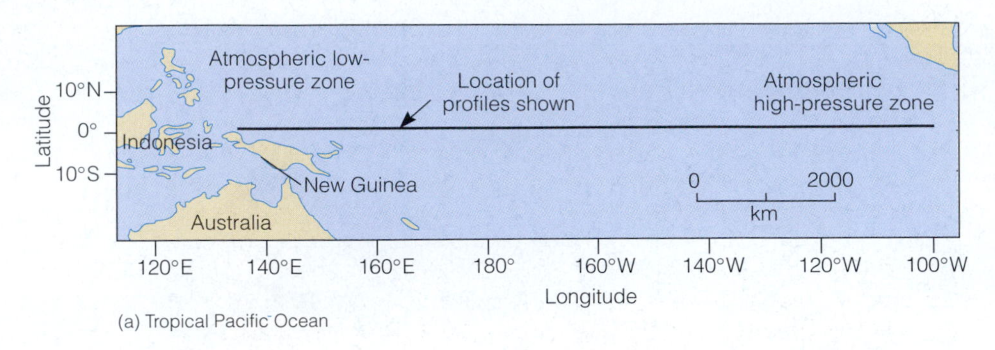

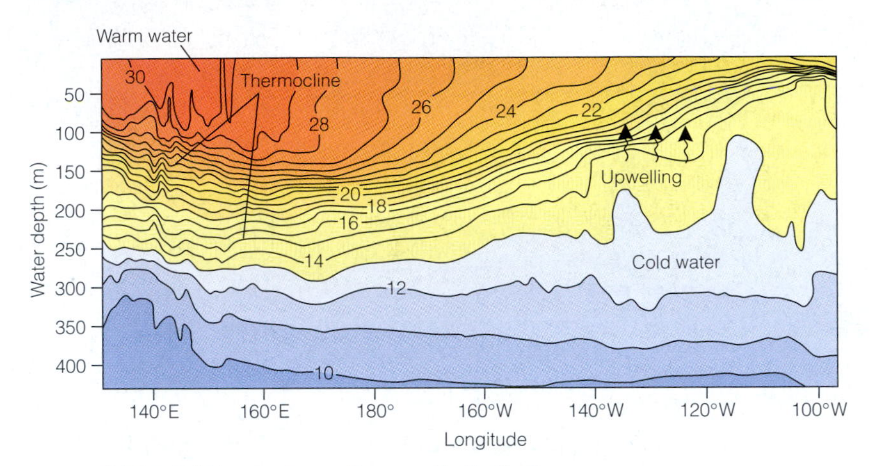

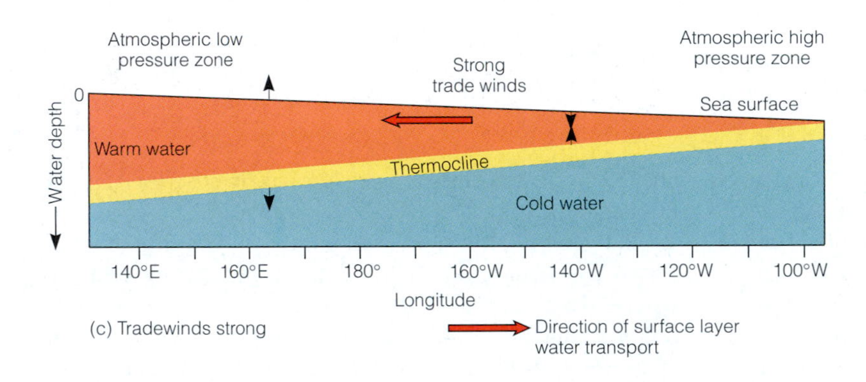

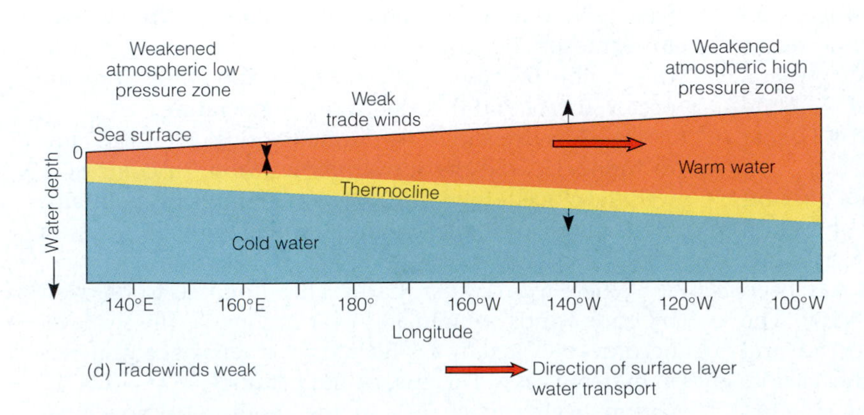

FIGURE 9.18 One possible model for the development of El Niño. (a) The area of the Pacific Ocean where El Niño develops. (b) When trade winds are strong, warm surface water is pushed westward and piles up on the western boundary of the Pacific Ocean near Indonesia. The surface layer is thinned near Peru, and upwelling occurs. (c,d) The vertically exaggerated cross sections are a simplified depiction of the interaction of trade winds with sea-level and thermocline depth in the two extreme modes of ENSO. The cross section in part (c) shows the effect of strong trade winds on the sea surface and surface water layer. This is the normal, or in extreme cases, the La Niña situation. When the trade winds diminish, surface layer water is transported eastward toward Peru, resulting in the El Niño situation shown in part (d). In an El Niño, the surface layer is thickened near Peru, and upwelling ceases.

zone and the Peruvian high-pressure zone both weaken and, in some years, actually reverse. During such periods, trade winds diminish and may reverse direction for several days. This change is sometimes preceded by two strong hurricanes, one north and one south of the equator. Trade winds do weaken in most years, and they return to normal by April, but in some years these events are followed by an El Niño.

El Niño/Southern Oscillation Sequence. ENSO follows a relatively well-known sequence that occurs after the Indonesian low-pressure and Peruvian high-pressure systems weaken. The sequence begins as the strength of the southeast trade winds lessens in response to the weakening Peruvian high-pressure zone. The southeast trade winds are the cause of the Peruvian upwelling (Chap. 10). Therefore, the upwelling is reduced or may be stopped if the trade winds abate entirely or reverse direction.

At this stage of ENSO, the sea surface near Indonesia is elevated (a few tens of centimeters) in relation to the region near Peru (**Fig. 9.18c**). The difference is normally partially compensated for (the sea surface slope is balanced) by higher atmospheric pressure near Peru (Chap. 10, **CC13**). However, the atmospheric pressure difference depends on the relative strength of the Indonesian low-pressure and Peruvian high-pressure zones. Because both are weakened at this stage of ENSO, the horizontal pressure change within the tropical

Pacific atmosphere is too small to balance the sea surface slope.

In response, warm surface water from the western tropical Pacific flows toward the east and Peru (**Fig. 9.18d**). The flow has characteristics of an extremely long-wavelength wave that travels directly from west to east along the equator, where there is no Coriolis effect (**CC12**). Therefore, the wave is free to flow across the entire ocean without deflection. This is one of the unique characteristics of ENSO. The warm surface water flows eastward across the Pacific Ocean over the cold upwelled water that is normally found in this region (**Fig. 9.14b**). The warmer, less dense surface water arrives near Peru and establishes a steep **pycnocline** in December, about 9 months after the event has started. Because steep pycnoclines are effective barriers to vertical mixing and water movements, the Peruvian coastal and tropical Pacific upwelling is further inhibited.

El Niño usually persists for about 3 months, but in extreme cases it may last 15 months or more. It ends when the reestablished trade winds again begin to drive the warm surface water of the tropical region to the west, thus restarting the upwelling process near Peru.

The oscillation set in motion by El Niño sometimes appears to "overshoot" as the system recovers. The trade winds become stronger, the water near Indonesia becomes warmer, and upwelling is stronger and water temperature lower near Peru (**Fig. 9.14c**). In addition, the low-pressure zone near Indonesia and the high-pressure zone near Peru become especially well developed. This situation is known as La Niña ("the female child"), to contrast it with El Niño ("the male child").

Effects of El Niño. The climatic effects of El Niño are felt far beyond the tropical Pacific. It appears that the extended area of warmer-than-usual tropical waters in the Pacific enhances the transfer of heat energy toward the poles. This change is manifested in many ways as the atmospheric convection cell system responds to the stimulus. In the tropical Pacific, El Niño's effects are sometimes devastating. Reduction of the warm-water pool and weakening of the atmospheric low near Indonesia brings droughts to this normally high-rainfall region.

In contrast, Peru has heavy rains and coastal flooding as sea level rises. The sea level rises because the surface water layer is warmer and less dense than normal. Consequently, the surface layer is "thicker" than the normally higher-density and cold surface water layer. Offshore from Peru, the marine **food web** collapses from lack of nutrients normally supplied by upwelled waters. In extreme cases, the result can be massive die-offs of marine organisms, which decay, strip oxygen from the water, and produce foul-smelling sulfides (Chap. 14). The principal fishery, anchovy, is not the only biological resource affected. In severe El Niños, lack of food decimates the huge colonies of seabirds that live on islands near Peru, and causes penguins and **marine mammals** to undertake unusual migrations and probably suffer population losses.

Changes in atmospheric circulation caused by El Niño also alter ocean water temperatures in areas far from the tropical Pacific. In the particularly strong 1982–1983 El Niño, subtropical fishes were caught as far north as the Gulf of Alaska. This El Niño caused droughts in Australia, India, Indonesia, Central America, west-central South America, Africa, and Central Europe. Excessive rains in many cases plagued parts of Southeast Asia, California, the east coast of the United States, and parts of South America, Britain, France, and the Arabian Peninsula (**Fig. 9.19**). La Niña, in contrast, brings flooding rains to India, Thailand, and Indonesia, and drought to Peru and northern Chile.

CRITICAL THINKING QUESTIONS

9.1 What causes air masses to move vertically in the atmosphere? Why is the vertical circulation of the lower atmosphere restricted to a layer of only about 12 km, although the entire atmosphere is more than 45 km deep?

9.2 Would the height of the restricted vertical circulation in the atmosphere be changed if the Earth's average surface atmospheric temperature increased? Why or why not?

9.3 On a nonrotating Earth, we believe there would be one atmospheric convection cell in each hemisphere, with upwelling at the equator and downwelling at the poles. On the Earth today, we have three cells in each hemisphere because of the Coriolis deflection. Since the Coriolis deflection will be reduced as the Earth's rotation is progressively slowed down, at a lower speed of rotation would you expect a system with two convection cells in each hemisphere to be established eventually? Why or why not?

9.4 It is known that in Biblical times, parts of the areas that are now desert in Eurasia (Egypt, Israel, Syria, and so on) were fertile regions with plentiful rainfall. What does this tell us about the location of the atmospheric convection cells?

9.5 If the Earth's direction of rotation on its axis were somehow reversed, would the location of the world's deserts change? Why or why not? If so, which parts of each of the continents would become deserts, and why?

9.6 If the trade winds were strengthened by global warming, how might the frequency and strength of El Niños be affected? Explain the reasons for your answer.

9.7 If there were no mountain chains in the western part of the United States, how would the climate in Nevada and the plains states be different? Explain the reasons for your answer.

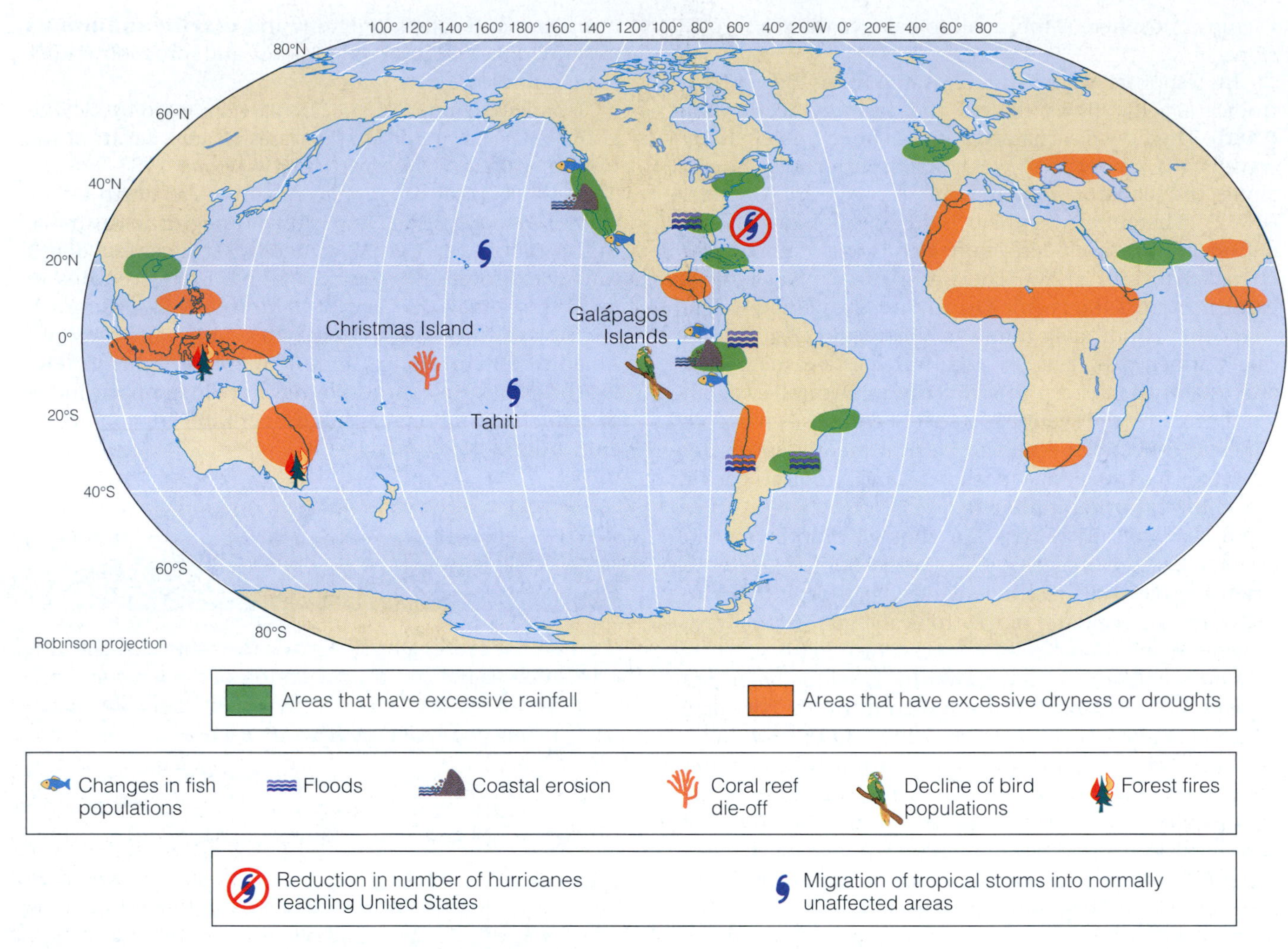

FIGURE 9.19 Strong El Niño events have worldwide effects that, in different areas, include droughts, excessive rainfall and floods, coastal erosion, tropical storms in areas where they do not normally occur, and various changes in terrestrial and marine ecosystems, some of which can be disastrous.

Modeling El Niño

El Niños have occurred regularly for centuries. La Niñas also may have occurred regularly, but they have less dramatic effects and were not recorded until scientific studies began. However, there is considerable variability from one decade to the next. In some decades, the cycle was relatively inactive; in others, it was pronounced. Between 1950 and 1980, El Niño occurred quite regularly and was followed regularly by La Niña throughout most of the period (**Fig. 9.20**). The 1980s and 1990s experienced a very active ENSO cycle, with five El Niño episodes (1982–1983, 1986–1987, 1991–1993, 1994–1995, and 1997–1998) and three La Niña episodes (1984–1985, 1988–1989, 1995–1996). Two of the El Niños (1982–1983 and 1997–1998) were the strongest of the twentieth century, and two consecutive El Niños occurred during 1991–1995 without an intervening La Niña. Because El Niño can have worldwide impacts on **ecosystems** and fisheries, and can cause widespread droughts and floods, major scientific efforts have been made to develop mathematical models (**CC10**) that can predict the occurrence of the phenomenon ahead of time. A number of models have been developed, each of which has been calibrated with the historical data. As the models have developed, they have progressively become more sophisticated and have required increasingly detailed atmospheric and oceanic data. Gathering these data is one of the most intensive and comprehensive observational programs ever undertaken in the Earth sciences.

Several models of ENSO have been developed. Their development has led to a much better understanding of the dynamics of the ENSO events. By 1997, several models were sufficiently developed to make predictions, and all of them forecast that a modest El Niño would occur in the winter of 1997–1998. The model predictions were partly correct. However, the El Niño was not a modest one. It turned out to be the most intense event on record, costing an estimated 23,000 lives and $33 billion in damage worldwide. What went wrong?

Fortunately, this El Niño was extensively observed by satellite sensors, moored instrument arrays, autonomous

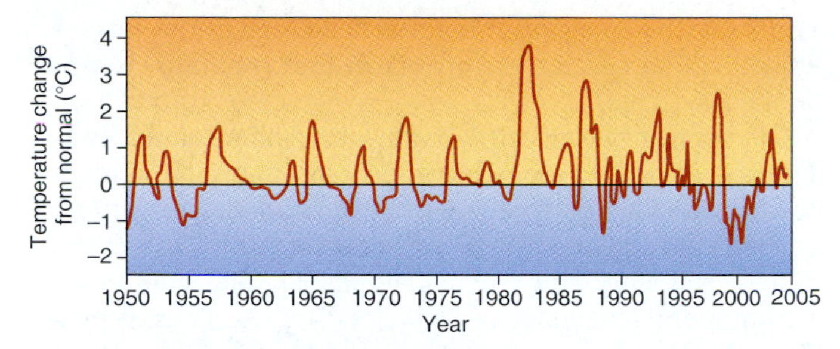

FIGURE 9.20 Changes of mean sea surface temperature off Peru from 1950 through 2004 indicate the occurrences of El Niño (above 0) and La Niña (below 0). No strong La Niña events occurred between 1975 and 1987, or between 1988 and 1996.

floats, research vessels, and aircraft. When the data were analyzed, the cause of the model failures appeared to be a burst of westerly winds in the western Pacific that occurred just as the eastward warm-water flow of the El Niño phenomenon was starting. These winds, unanticipated in the models, appear to have been timed perfectly to increase the flow of warm water across the ocean and, thus, increase the intensity of the El Niño.

Further research suggested that similar bursts of westerly winds also may have occurred just as other unusually intense El Niños had started. It has been hypothesized that these westerly winds are part of another ocean–atmosphere oscillation, called the Madden-Julian Oscillation (MJO). About every 30 days, this oscillation causes bursts of winds that are associated with an eastward-moving patch of tropical clouds. These clouds extend high into the troposphere and are easily observed from satellites. If this hypothesis is correct, the most intense El Niños may turn out to be formed only when the timing of this burst and the ENSO sequence are exactly right. The MJO cannot yet be forecast well in advance, and it may never be if the system is **chaotic** (**CC10, CC11**), which is very likely. Unfortunately, this would mean that the intensity of an El Niño could not be accurately forecast until it had already started.

Other Oscillations

ENSO is the best-known ocean–atmosphere oscillation, and it is believed to be the most influential in causing the interannual variations of the Earth's climate. A number of other oscillations are known to have an influence on climate, but primarily just in specific regions. The MJO, discussed in the previous section, is one such oscillation, but there are several others that have longer periods and greater effects on regional climates. These include relatively well-known oscillations in the North Atlantic and North Pacific oceans: the North Atlantic Oscillation (NAO) and the Pacific Decadal Oscillation (PDO), respectively. More recently, oscillations have been observed in the Arctic Ocean and in the Indian Ocean. The details of what is known about these oscillations and their effects on regional climates are beyond the scope of this text. However, the following paragraphs provide a very brief overview of each.

The North Atlantic Oscillation is a periodic change in the relative strengths of the atmospheric low-pressure region centered near Iceland and the subtropical high-pressure zone near the Azores. This pressure difference drives the wintertime winds and storms that cross from west to east across the North Atlantic. When the gradient between the two zones weakens, more powerful storms create harsher winters in Europe. The shift can affect marine and terrestrial **ecology**, food production, energy consumption, and other economic factors in the regions surrounding the North Atlantic. The periodicity of NAO is highly variable, but it ranges from several years to a decade or more. The Arctic Oscillation is closely coupled with the NAO. The Arctic high-pressure region tends to be weak when the Icelandic low is strong. The strength of the Arctic high affects climate around the Arctic Ocean, as well as the direction and intensity of ice drift.

The Pacific Decadal Oscillation is characterized by two modes. In one mode, the sea surface temperatures in the northwestern Pacific, extending from Japan to the Gulf of Alaska, are relatively warm and the sea surface temperatures in the region from Canada to California and Hawaii are relatively cool. In the other mode, these relative temperatures are reversed. In the first mode, the jet stream flows high across the Pacific and dips south over the Pacific coast of North America. Pacific storms follow the jet stream onto the continent over Washington, Oregon, or British Columbia; and California is cool and dry. Moist, relatively warm air is transported across the northern half of the United States, and as a result, relatively mild, but often wet, winters occur in this region and in states influenced by warm, moist air masses from the Gulf of Mexico. In the second mode, the jet stream and Pacific storm tracks parallel the Pacific coast into Canada and Alaska. Parts of Alaska and much of California experience a mild winter. However, the jet stream location allows cold, dry Arctic air to flow over the central United States, which pushes the Gulf of Mexico air mass south, and a generally cold winter occurs in the eastern United States. We discuss some of the effects of the PDO in more detail in the next section.

In the Indian Ocean, there is no strong oscillation comparable to ENSO in the Pacific. However, there is evidence of weaker oscillation that is similar. Although the Indian Ocean oscillation is much weaker than ENSO, it exhibits a similar "normal" situation, characterized by warmer water on the western side of the ocean and cooler water on the eastern side, and movement of warm water from west to east as the system changes to its other mode. Modeling has shown that the intense drought of 1998–2002 over a swath of mid latitudes spanning the United States, the Mediterranean, southern Europe, and Southwest and Central Asia was probably caused by a combination of the strong El Niño of 1997–1998, and the weaker similar oscillation in the Indian Ocean.

Climate Chaos?

The decade following the 1976–1977 winter was extremely unusual in the United States and elsewhere. For example, in the following eight years, five severe freezes damaged Florida orange groves. Such freezes had occurred on average only once every 10 years during the previous 75 years. In addition, waves 6 m or higher tore into the southern California **coastline** 10 times in the 4 years from 1980 to 1984. Such waves had occurred there only 8 times in the preceding 80 years.

Were all these unusual events the result of a sudden shift in climate or just due to normal year-to-year variations of the Earth's climate? Because any one such change could be just year-to-year variation, researchers examined the records of 40 different environmental variables that reflect climatic conditions around the Pacific region. The variables include wind speeds in the subtropical North Pacific Ocean, the concentration of plant **chlorophyll** in the central North Pacific Ocean, the salmon catch in Alaska, the sea surface temperature in the northeastern Pacific, and the number of Canada goose nests on the Columbia River. The 40 variables were combined to derive a single statistical index of climate for each year from 1968 to 1984.

The index showed a distinct and abrupt steplike change in value between 1976 and 1977, which is truly remarkable because such a clear signal of changed conditions is seldom, if ever, found even in any single environmental variable. Usually the many complex natural variations would obscure such a pattern. Was this change real? Did it indicate a chaotic "phase shift" between two relatively stable sets of environmental conditions? This study was perhaps the first evidence that chaotic shifts do indeed occur. It later became known that this shift was associated with the Pacific Decadal Oscillation (PDO).

In 2000, this study was repeated, but this time using as many as 100 variables: 31 time series of atmospheric or oceanic physical variables, such as indices of the sea surface temperature; and 69 biological time series, such as catch rates of salmon. The analysis was performed for the period 1965–1995, but because data were not available for all indices for each year, the analysis was done in two blocks—1965–1985 and 1985–1995—using a slightly different suite of indices for each. The results were as startling as those of the early studies. Distinct phase shifts were identified between 1976 and 1977, and between 1988 and 1989 (**Fig. 9.21**). These dates correspond with reversals of the PDO.

Further evidence that these phase shifts may involve elements of chaos comes from the observation that some of the individual biological indices that changed in 1997 did not simply reverse in 1989. Indeed, some made step increases in both years, and some made step decreases in both years. These observations indicate that some species did not return to their former numbers in the ecosystem and others were advantaged by both shifts. Thus, these shifts may be an agent of long-term change in the success of individual species within ocean and, perhaps, terrestrial ecosystems.

If changes in the Earth's ocean–atmosphere climate system and associated ecological effects are chaotic, what does this mean? Chaotic systems often remain stable, varying within a well-defined range of behaviors, until the entire system suddenly shifts to a new stable state and varies within a different range of behaviors (**CC11**). Thus, it may mean that ecosystems are constantly changing in nonpredictable ways. It may also mean that, even if no dramatic climate changes have yet occurred in response to human releases of greenhouse gases, there is no guarantee that any future changes will be small or slow. Drastic and sudden climate changes could be triggered if the greenhouse gas concentration in the atmosphere continues to increase, or such changes could already be inevitable. Decades of research on the Earth's ocean–atmosphere climate system will be needed before we can comfortably forecast future climate in a world in which we have caused an unprecedented rapid increase in the concentration of greenhouse gases in the atmosphere. Furthermore, such forecasts may never be more reliable than the local weather forecast.

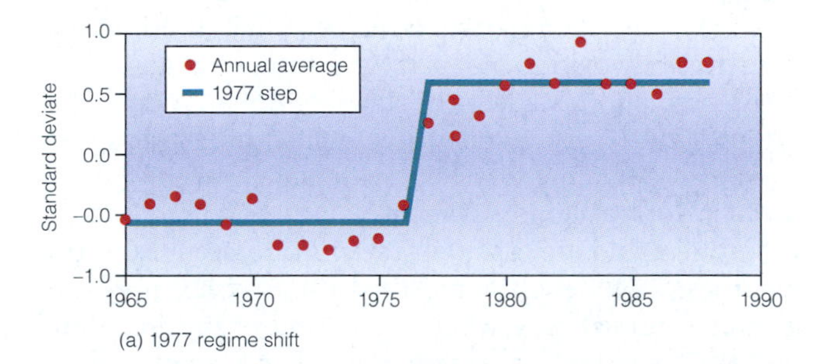

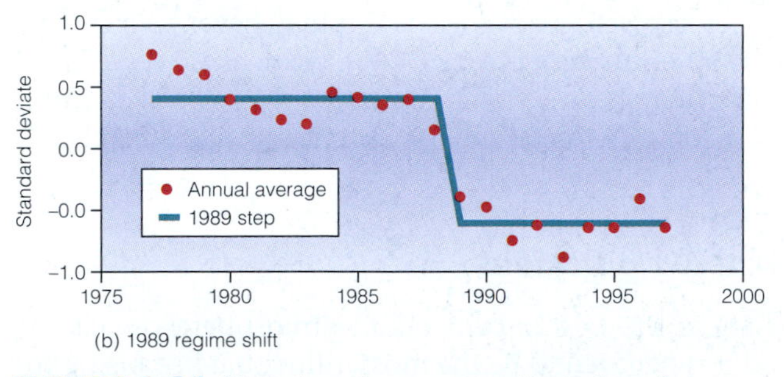

FIGURE 9.21 Variations in a composite index of environmental variables related to climate in the North Pacific Ocean. Time series were examined for 100 variables—31 climatic and 69 biological—although data were not available for all variables throughout the analysis period. (a) A regime shift occurred in the winter of 1976–1977, as shown by the step change in this function, which includes time series for 91 variables. (b) A regime shift also occurred in 1989, as shown by the step decrease in an index that includes time series for 94 variables. However, this second shift did not return the system to the same state as before 1977.

9.7 LAND–OCEAN–ATMOSPHERE INTERACTIONS

Landmasses interact with atmospheric circulation in two ways. First, they can physically block or steer air masses moving across the Earth's surface. Air masses must rise over mountain chains or flow around them. Second, the thermal properties of landmasses and of the ocean are very different because rocks and soil have lower heat capacities than water has. Consequently, land warms more quickly than the oceans under the same solar radiation intensity. Also, land cools and gives off heat to the atmosphere much more quickly than the oceans do.

Interactions of landmasses with the atmosphere and oceans exert substantial control over the climate characteristics of most of the Earth's land and much of its oceans. The effects of the interactions on climate may be extremely localized and limited to one small area, or they may range across an entire continent or ocean. On the local scale, effects include land and sea breezes and the "island effect," which are discussed later in this chapter. On a larger scale, effects include the monsoons and the climate characteristics of the interior of continents. In the following sections of this chapter we discuss how land–ocean–atmosphere interactions can occur on different geographic scales. We review first the global interactions that determine climate, then larger-scale weather events that extend or travel across large regions of a continent or ocean, and finally smaller-scale weather features that occur locally.

9.8 GLOBAL CLIMATE ZONES

Climate in a given region is defined by several factors, including the average annual air temperature, seasonal range of temperatures, range of temperatures between day and night, extent and persistence of cloud cover, and annual rainfall and its seasonal distribution. Each factor varies with latitude and location in relation to land, mountain chains, and ocean. The factors also vary from year to year. Hence, most climate zones are not separate and distinct, but merge into one another. The demarcation lines drawn between the zones in **Figure 9.22** are therefore only approximations.

Ocean Climate Zones

Climate zones over the oceans are generally latitudinal bands that parallel the average ocean surface temperatures shown in **Figure 9.13**.

Polar ocean zones are covered with ice, and ocean surface temperatures remain at or close to freezing all year (**Fig. 9.23a**) because of the heat-buffering effect of ice (Chap. 7, **CC5**). These zones have low rainfall and generally light winds (except where land and ocean interact). In subpolar zones, sea ice forms in winter but melts each year. Despite low rainfall in the polar and subpolar zones, ocean surface salinity, particularly in the Arctic Ocean, is low. Salinity is lowered by **ice exclusion** during the continual freeze-and-thaw cycle that creates sea ice. As water freezes, dissolved salts are excluded from the ice, which produces cold, high-salinity water that sinks because of its high density (**CC1**). When sea ice melts in the spring and summer, the nonsalty water released from the melting ice mixes with the upper layer of ocean water and lowers its salinity. In the Arctic Ocean, freshwater runoff from the continents also contributes to low surface water salinity.

Temperate ocean zones are located in the zones of strong westerly winds (**Fig. 9.10**). They have high rainfall (**Fig. 9.17**) and are subject to strong storms called "extratropical cyclones." These storms form especially in winter at the atmospheric polar fronts, which are locations where the polar and Ferrel cells meet (**Fig. 9.9**). The storms travel eastward and toward lower latitudes. Extratropical cyclones are described in the next section.

Subtropical ocean zones are located at the divergence between the Hadley and Ferrel atmospheric circulation cells (**Fig. 9.9**). In these zones, winds are generally light, skies are usually clear, and rainfall is low. Many of the world's most desirable **beach** vacation areas are on coasts in these zones.

Trade wind zones are generally dry with limited cloud cover, but they are subject to persistent winds. Ocean surface water evaporation is very high, and thus, salinity tends to be high (**Figs. 9.15, 9.16**). In the equatorial region, ocean surface waters are warm (**Fig. 9.13**), evaporation is high (**Fig. 9.16a**), winds are light (**Fig. 9.9**), clouds are persistent, and rainfall is high and continuous throughout the year (**Fig. 9.17**).

Land Climate Zones

Land (terrestrial) climate zones are not arranged in the same type of orderly latitudinal bands that characterize ocean climate zones (**Fig. 9.22**). The reasons for this difference include the difference in thermal properties between land and water and the effect of mountains on climatic air mass movements.

CRITICAL THINKING QUESTIONS

9.8 No hurricanes are fully formed on the North African side of the Atlantic Ocean, but many hurricanes are formed in the equivalent region off the Pacific coast of Mexico. Why?

9.9 Almost no hurricanes form in the South Atlantic Ocean. Why?

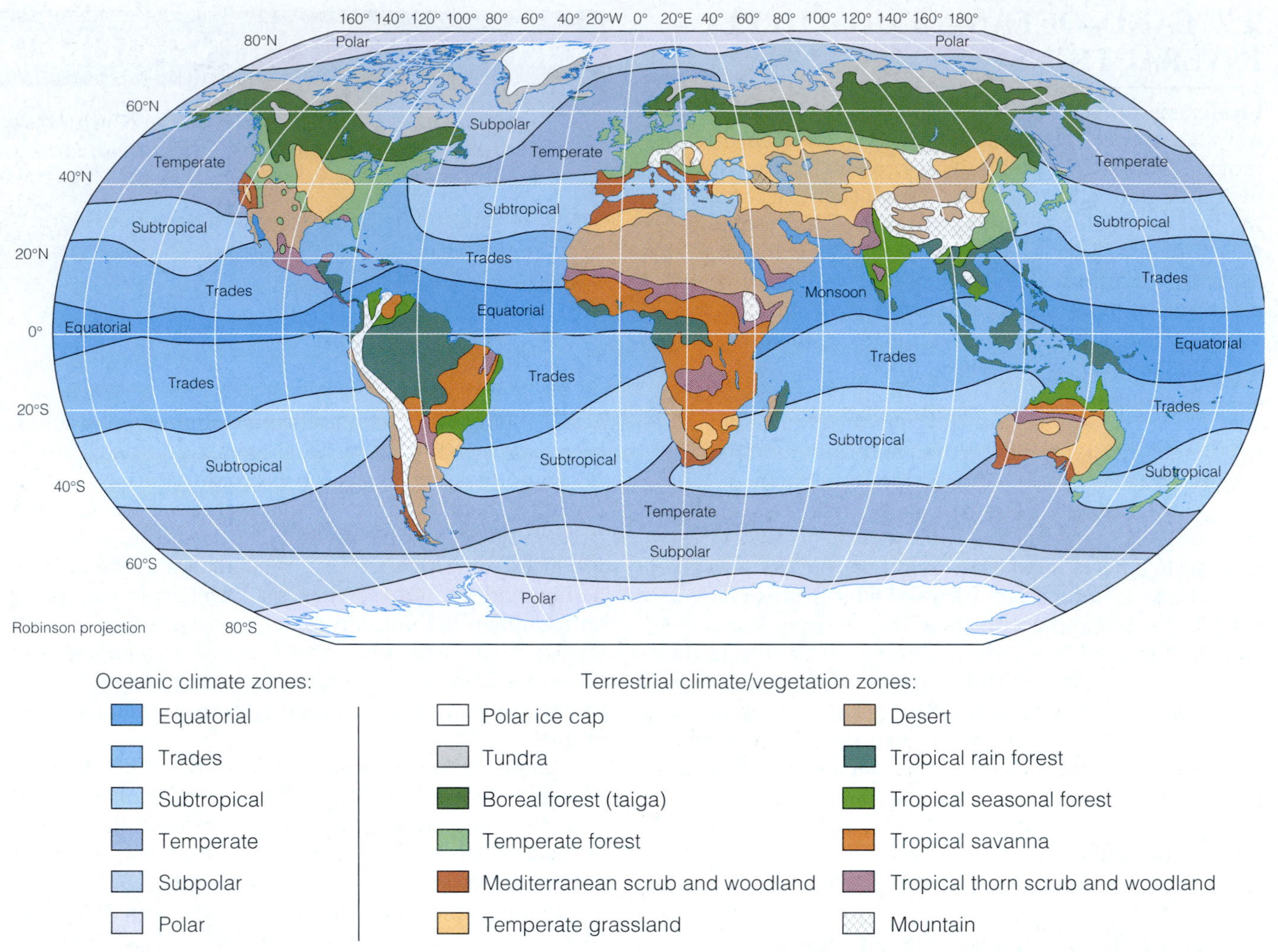

FIGURE 9.22 The climate zones of the oceans are arranged in bands that generally follow a latitudinal pattern and that are related primarily to ocean surface water temperature (see Figure 9.13). Terrestrial climate and vegetation zones follow a more complex pattern that reflects the presence of mountain chains, the prevailing direction of air mass movements, and the proximity to the oceans—all of which affect the distribution of rainfall and seasonal temperature ranges.

The average ocean surface temperature at any latitude changes little between seasons (**Fig. 9.23a**) because of the high heat capacity of water (Chap. 7, **CC5**). In contrast, soil and rocks have a much lower heat capacity. Consequently, changes in solar intensity with season cause substantial seasonal changes in average temperature at the land surface (**Fig. 9.23b**). Seasonal changes are small in equatorial regions and increase with increasing latitude. The reason is that the annual range of daily total solar energy received at the Earth's surface increases with distance between the equator and the poles.

Figure 9.23b shows the temperatures for continental locations far from the influence of the oceans. Daily and seasonal temperature changes are less in coastal locations than in regions far from the oceans at the same latitude (**Fig. 9.24**). The ocean provides a source of relatively warm air in winter and cool air in summer. Although the coastal air mass moderates climates in all coastal locations, the distance inland to which this effect reaches on a specific coast is determined by the prevailing wind direction and by the location and height of mountain ranges.

If climatic winds blow onshore, coastal-ocean air can moderate temperatures and enhance rainfall many hundreds of miles beyond the coast. For example, warm coastal air that flows eastward across Europe is unimpeded by mountains. Hence, the moderate marine climate extends farther inland in western Europe than in the western parts of other continents, such as North and South America, where coastal mountains impede the westward flow of warm, moist air from the Pacific Ocean (**Fig. 9.22**). In the latter regions, the mountain effect removes the principal source of the air mass's heat: its water vapor.

If climatic winds blow offshore, the moderating influence of the ocean is reduced. For example, Boston, Massachusetts, has a wider range of temperatures than Portland, Oregon (**Fig. 9.24**). Although both are coastal cities and are at about the same latitude, Boston is in

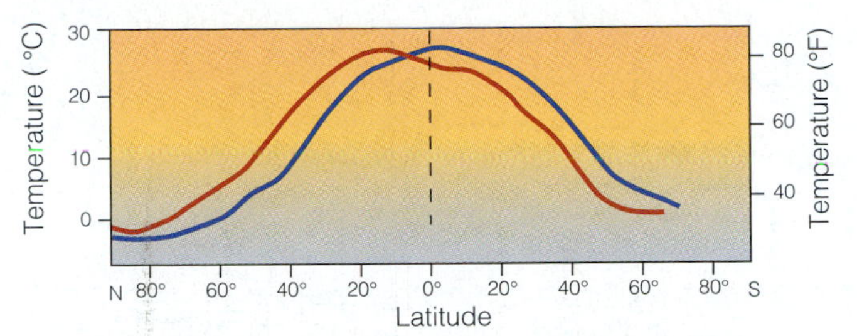

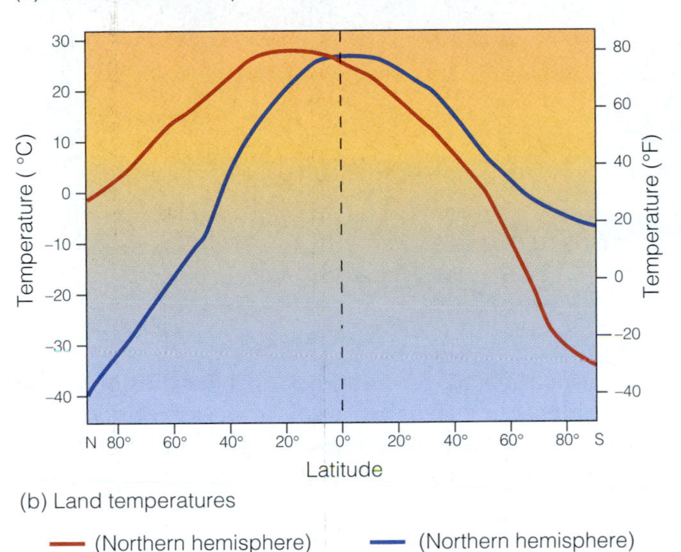

FIGURE 9.23 Seasonal and latitudinal variation in temperatures at the Earth's surface. (a) Average ocean surface water temperature. (b) Average temperature over land (or ice-covered ocean). Seasonal variation is much greater over land than over ocean, especially at high latitudes.

a region of offshore climatic winds, and Portland is in a region of onshore climatic winds.

Coastal land climates are also affected by ocean currents. For example, Scandinavia and Alaska are at about the same latitude, but winters are much more severe in Alaska than in Scandinavia. In winter, most of Scandinavia is in the westerly wind zone and the temperate ocean climate zone, whereas most of Alaska is in the polar easterly wind zone and the subpolar ocean climate zone (**Figs. 9.10, 9.22**). In the North Atlantic Ocean, the Gulf Stream current carries warm water from Florida north and west across the ocean to the seas around Scandinavia. The warm water provides heat energy through evaporation to moderate the Scandinavian atmosphere in winter. In fact, enough heat energy is provided to cause the westerly wind zone to extend into the region throughout the year (**Fig. 9.10**). A similar warm current, the Kuroshio Current, flows north and west from south of Japan toward Alaska (Chap. 10). However, this warm current is deflected by the Aleutian Island arc and does not reach the Bering Sea on the west coast of Alaska.

The terrestrial climate zones shown in **Figure 9.22** are defined primarily by seasonal temperature ranges and rainfall rates. In any given location, these parameters are controlled by the interaction of several factors, including ocean water temperatures, climatic winds, distance from the oceans, and the locations of mountain chains. A complete description of the zones is beyond the scope of this text, but the knowledge gained in this chapter can be used to explore the distribution of terrestrial climate zones depicted in **Figure 9.22**.

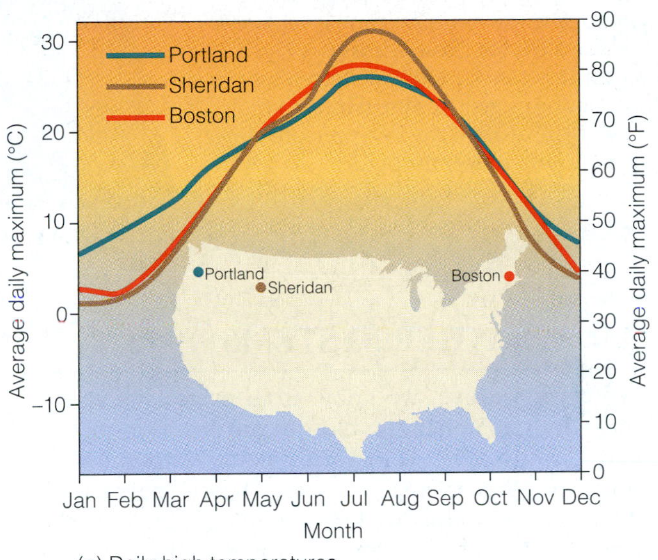

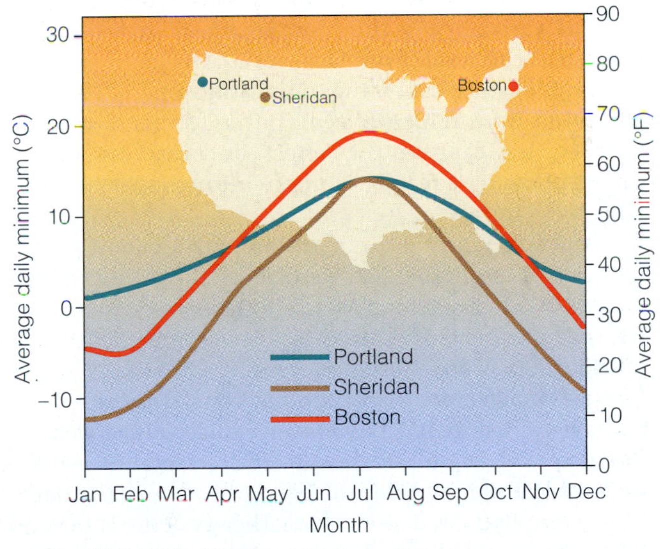

FIGURE 9.24 Annual variation in monthly average air temperature for Portland, Oregon; Sheridan, Wyoming; and Boston, Massachusetts. (a) Daily high temperatures. (b) Daily low temperatures. These cities are at approximately the same latitude; however, their climates are very different. Portland has a maritime climate with a small annual temperature range; Sheridan has a continental climate with a large annual temperature range; and Boston has a more continental climate than would be expected from its coastal location, because the climatic winds there blow from the continent toward the ocean.

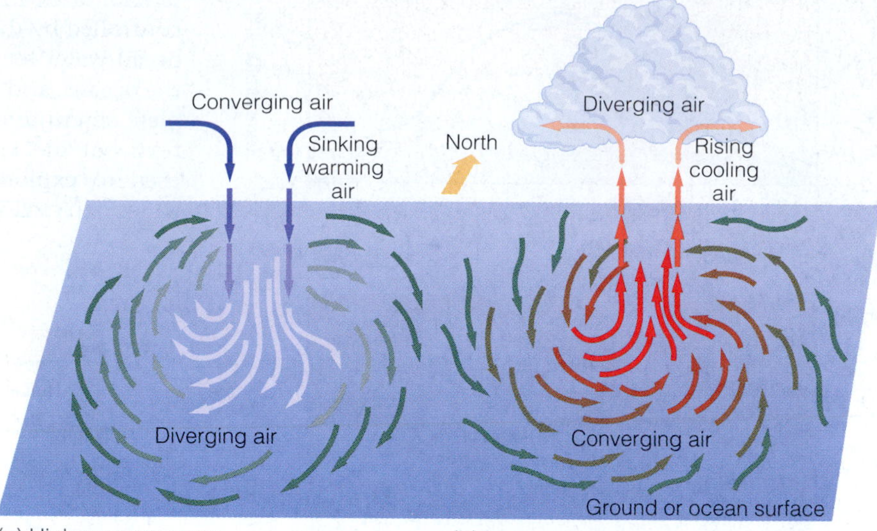

FIGURE 9.25 Geostrophic winds and zones of high and low pressure. (a) Cold, dense air sinks, creating a high-pressure zone in the lower atmosphere. The sinking air warms and flows out from the high-pressure zone. As the air flows outward, it is deflected by the Coriolis effect and forms geostrophic winds that flow clockwise around the high-pressure zone (counterclockwise in the Southern Hemisphere). (b) Warm, moist air rises, creating a low-pressure zone in the lower atmosphere. The rising air cools, and water vapor condenses, causing precipitation. The warm rising air at the center of the low-pressure zone is replaced by air drawn toward the low from across the Earth's surface. As the air flows inward, it is deflected by the Coriolis effect and forms geostrophic winds that flow counterclockwise around the low-pressure zone (clockwise in the Southern Hemisphere).

9.9 WEATHER SYSTEMS

Wherever we live, we know from everyday experience that winds do not always blow in the directions of the climatic winds depicted in **Figure 9.10**. Furthermore, contrary to the average climatic zone characteristics discussed previously, we know that some days are cloudy and rainy and others are sunny and dry. These short-term variations, called "weather," are caused by temporal and **spatial** variability in the motions of air in the atmosphere that can result from local variations in evaporation rate or in heating or cooling of the atmosphere. They can also result from **friction** as air flows over land or ocean surfaces. We can see this effect by observing how a smooth-flowing stream becomes churned as it interacts with rocks in rapids.

Atmospheric air movements are chaotic, and precise motions of any individual particle of air cannot be predicted (**CC11**). Therefore, we cannot ever hope to forecast weather precisely. Any forecast will be progressively less reliable the further ahead in time it is made. Weather forecasts made more than a few days ahead are unreliable and always will be. At best, future longer-range weather forecasts will be able to predict only trends in average weather that might occur during a period of many days.

A complete discussion of weather systems is beyond the scope of this book. However, the behavior of high- and low-pressure zones, hurricanes, and extratropical cyclones is described briefly in the sections that follow.

High- and Low-Pressure Zones

Atmospheric pressure at the Earth's surface varies at any given location from day to day and over distances of tens or hundreds of kilometers across the surface at any given time. Once created, areas of slightly higher or lower pressure, which may cover hundreds or thousands of square kilometers, can persist for days or weeks. Atmospheric high- or low-pressure zones ("highs" and "lows" in weather reporter jargon) generate winds because air tends to flow on the pressure gradient outward from a high-pressure zone and inward toward a low-pressure zone. The moving air does not flow directly away from a high-pressure zone or toward a low-pressure zone, because the moving air mass is deflected by the Coriolis effect (**CC12**).

The Coriolis effect deflects moving air masses ***cum sole*** (which means "with the sun")—that is, to the right in the Northern Hemisphere and to the left in the Southern Hemisphere. Thus, in the Northern Hemisphere, air that flows toward a low-pressure zone is deflected to the right. The deflection continues until the Coriolis effect is balanced by the pressure gradient. This balance results in a situation known as **geostrophic** flow, as discussed in **CC13**. A pattern is established in which winds flow counterclockwise around the low-pressure zone (**Fig. 9.25b**). Similarly, clockwise winds are established around a high-pressure zone (**Fig. 9.25a**). In the Southern Hemisphere, winds flow in the opposite direction: clockwise around a low-pressure zone and counterclockwise around a high-pressure zone. Friction between the moving air mass and the ground or ocean surface reduces the Coriolis deflection. Consequently, winds at the surface tend to spiral toward the center of low-pressure zones and away from the center of high-pressure zones.

Because winds near the Earth's surface are usually in approximate geostrophic balance, both wind speed and direction can be estimated from a simple **contour** map of atmospheric pressure, called an **isobaric chart** (**CC13**). Therefore, atmospheric pressure is among the most important measurements made at weather stations worldwide, and isobaric charts are a major weather forecasting tool. **Wave height** forecasts also can be made from such charts because the relationships among wave height and wind speed, duration, and distance over which the wind blows (called **fetch**) are known (Chap. 11).

High- and low-pressure zones continuously form and re-form in the atmosphere and vary substantially in size and persistence. Low-pressure zones can be formed over the ocean where slightly higher ocean surface temperatures and/or higher evaporation rates occur. The result is an air mass in the low-pressure zone that has a temperature and water vapor content that is elevated compared to surrounding air masses. As increasing temperature and water content lead to a decrease in density, this air mass rises as a plume (**Fig. 9.25b**).

Hurricanes

Hurricanes are among the most destructive forces of nature. These intense storms form mostly during local summer or fall in both the Northern and Southern Hemispheres. Hurricanes develop over the warm ocean waters of tropical regions where a low-pressure zone develops that draws air inward across the ocean surface. If conditions are favorable, the low-pressure zone deepens (that is, the pressure drops further), and a hurricane may form.

Winds in the developing hurricane are deflected by the Coriolis effect (**Fig. 9.25b**) until they flow geostrophically around the low-pressure zone. As the air flows in toward and around the low-pressure zone across the ocean surface, it is warmed, gains moisture, and tends to rise. The winds thus blow toward the center, or "eye," of the hurricane in a rising helical pattern (**Fig. 9.26**). Because winds do not reach the eye to alleviate the low pressure, the low-pressure zone continues to intensify. The horizontal pressure gradient increases, and the geostrophic wind speeds increase (**CC13**).

A well-formed hurricane has an eye, usually about 20 to 25 km in diameter, within which winds are light and there are few clouds (**Fig. 9.27**). The eye forms when the low-pressure zone becomes deep enough to draw warm, relatively dry air downward in the eye as winds are deflected from the eye in their rising helical pattern (**Fig. 9.26**). The eye is surrounded by a wall of clouds (the eye wall) where winds are moderate, but intense thunderstorms are created as the rotating air mass rises. Outside the eye wall is a region of intense winds that blow across the ocean surface (**Fig. 9.26**). The winds are counterclockwise in the Northern Hemisphere and clockwise in the Southern Hemisphere.

Hurricanes drift to the west in the prevailing trade wind direction at about 10 to 30 $km \cdot h^{-1}$, and sometimes faster. Their westward paths are deflected by the Coriolis effect, and they generally turn away from the equator. Over warm ocean water, hurricanes maintain or gain intensity as high temperatures and high winds encourage evaporation and feed more latent heat energy into the storm. When a hurricane moves over cold water or land, it quickly loses energy and releases its moisture as intense rainfall (10 to 20 billion tonnes of rainfall per day from an average hurricane).

Hurricane winds and flooding from the torrential rains can cause massive damage on coasts and far inland. However, the most destructive and life-threatening effect of a hurricane is often the **storm surge**, or wave, that it pushes ahead of it. Storm surges (Chap. 11) cause flooding of low-lying coastal areas. Hurricane Camille in 1969 took more than 250 lives. Most of the victims were swept away and drowned in the storm surge. Camille also caused almost $1.5 billion in damage to the Gulf of Mexico coast. Hurricane Hugo, which hit South Carolina in 1988, and Hurricane Andrew, which hit South Florida in 1992, caused much greater damage than Camille but took few lives. Hurricane Andrew is estimated to have caused more than $30 billion in damage. Andrew may have been the most expensive natural disaster in U.S. history until Hurricane Katrina ravaged Louisiana and Mississippi in 2005, causing more than $125 billion in damage and more than 1,000 deaths.

In recent years, satellites and other techniques have made early hurricane warnings possible. Many lives have been saved because the warnings enable people to evacuate the low-lying coastal areas that storm surges inundate. Hurricanes (called "cyclones" in the Indian Ocean) that plague Bangladesh still regularly take hundreds of lives, despite better warnings, because the local population has no evacuation route or plan. Forecasting the tracks of hurricanes to predict their landfalls and warn populations remains difficult because hurricanes can move erratically as they are steered by surface and upper-atmosphere winds.

Hurricanes form in all tropical oceans, but they are rare in the South Atlantic and eastern South Pacific (**Fig. 9.28**). Before March of 2004, only two tropical cyclones had ever been recorded in the South Atlantic Ocean, and neither had reached hurricane strength. However, in that month, Hurricane Catarina formed east of Brazil. The storm caused at least two deaths and substantial property damage when it came ashore on the southern coast of Brazil.

Hurricanes form only over surface waters with temperatures above about 26°C. Hurricanes generally do not form closer to the equator than 5°N or 5°S, because the Coriolis effect is too weak in this equatorial band. In the Indian and western Pacific Oceans, hurricanes are called "cyclones" north of the equator and "typhoons" south of the equator. They are also called "baguios" around the Philippines and "willy willys" near Australia (**Fig. 9.28**). They are known as hurricanes only in the Atlantic and eastern Pacific Oceans, but they are the same phenomenon regardless of the name.

Historical records suggest that the frequency and intensity of hurricanes may vary on a cycle several decades long. Records also suggest that we were in a relatively quiet part of the cycle during the 1970s and 1980s. Researchers therefore expect hurricanes to be more fre-

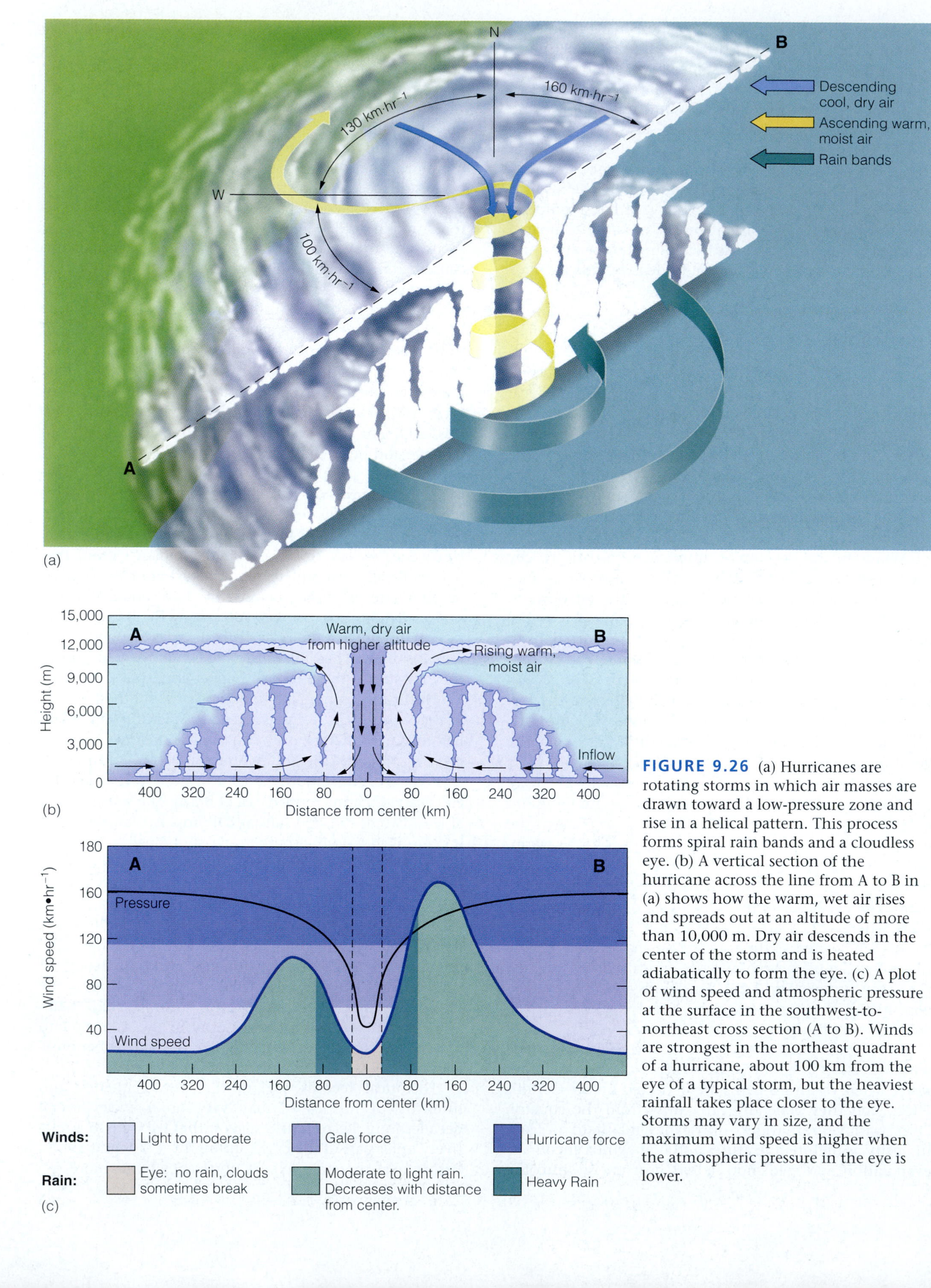

FIGURE 9.26 (a) Hurricanes are rotating storms in which air masses are drawn toward a low-pressure zone and rise in a helical pattern. This process forms spiral rain bands and a cloudless eye. (b) A vertical section of the hurricane across the line from A to B in (a) shows how the warm, wet air rises and spreads out at an altitude of more than 10,000 m. Dry air descends in the center of the storm and is heated adiabatically to form the eye. (c) A plot of wind speed and atmospheric pressure at the surface in the southwest-to-northeast cross section (A to B). Winds are strongest in the northeast quadrant of a hurricane, about 100 km from the eye of a typical storm, but the heaviest rainfall takes place closer to the eye. Storms may vary in size, and the maximum wind speed is higher when the atmospheric pressure in the eye is lower.

FIGURE 9.27 This satellite image of hurricane Katrina was obtained as the hurricane moved across the Gulf of Mexico, just a few hours before it struck land, costing many lives and causing billions of dollars of damage in Louisiana and Mississippi. In the image, the outer bands of cloud are just reaching the coast near New Orleans while the trailing edge clouds extend as far as the Yucatan Peninsula in Mexico, so this monster storm was about 1000 km in diameter.

quent and intense during the next decade or two. Such a trend may already be evident in the intense and frequent hurricanes that have occurred in all oceans in a number of years since the early 1990s. For the United States and many other countries, this prediction is important. Coastal development has been intense during the past several decades, particularly in the southeastern United States and Gulf Coast communities that are in the hurricane band. If hurricane frequency and intensity increase, coastal damage and loss of life also could increase dramatically.

The prospect of global warming also worries researchers. If ocean surface waters were to warm just a few tenths of a degree, hurricanes would probably become more frequent and intense, and they would maintain energy farther from the equator. In addition, warmer water has lower density, so sea level would rise. Large coastal cities of the northeastern United States could be exposed to the full force of the devastating storms. Perhaps worse, future hurricanes might have much higher winds than the approximate maximum of 200 $km \cdot h^{-1}$ observed in the most intense recent hurricanes. Studies of **sediment** deposited from ancient storms have provided evidence of "super" hurricanes in the Earth's past that had wind speeds probably higher than any ever observed directly, and greater than what most buildings could withstand.

Extratropical Cyclones

Powerful **cyclonic** storms form in the westerly wind regions in winter. Such storms form primarily at the polar front where cold westward-flowing polar air and warmer eastward-flowing mid-latitude air converge. Along the front between these two air masses, which are moving in opposite directions, waves periodically develop. The waves develop into cyclonic storms as the warm air flows over the top of the cold, denser air and each moving air mass is deflected by the Coriolis effect (**Fig. 9.29**). The energy source for these storms is latent and sensible heat transferred from the ocean to the warm air mass at low latitudes.

Storms that form at the polar front are called "extratropical cyclones." They can be extremely large and carry huge amounts of moisture. They move eastward with the prevailing winds. Extratropical cyclones move more slowly than hurricanes and generally have somewhat lower, although still very high, wind speeds. They can be much larger than hurricanes and can cause storm surges. Consequently, they can cause severe and extensive wave

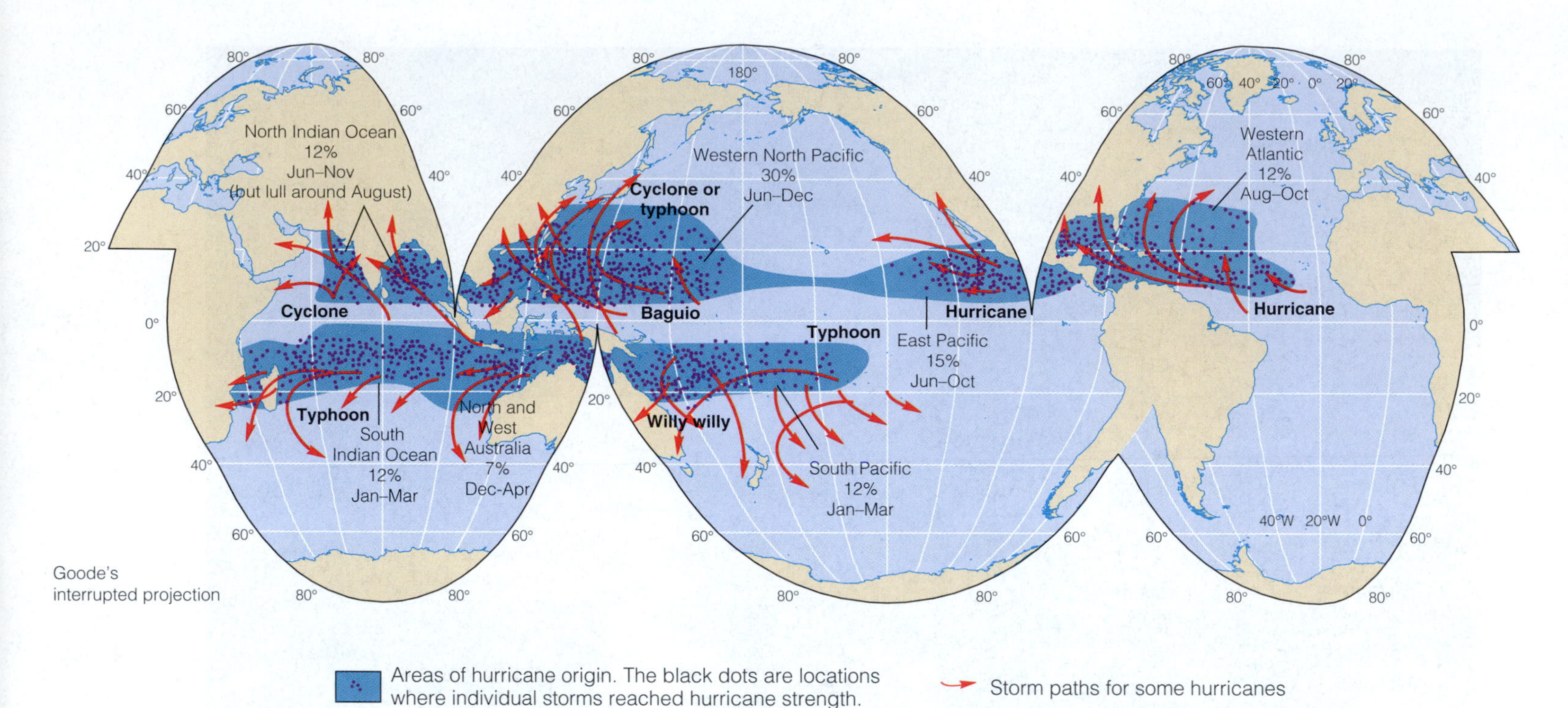

FIGURE 9.28 Hurricanes develop where the surface ocean water temperature is greater than 26°C, except in the equatorial region between 5°N and 5°S, where the Coriolis effect is too weak to generate these storms. Surface waters are too cold for strong tropical storm formation in the eastern Atlantic Ocean and below the equator in the eastern Pacific Ocean. Once formed, hurricanes travel generally east to west, curving toward higher latitudes as they are steered by the Coriolis effect and by other weather systems. Hurricanes are most frequent in the western North Pacific Ocean, where they are called "typhoons" or "baguios."

erosion on exposed **shores**. The massive Pacific storms that bring winter rains, winds, and beach erosion to California and the Pacific Northwest are extratropical cyclones.

Extratropical cyclones can also be formed by the passage of a cold front over a warm **western boundary current**. For example, the nor'easters that often strike the Mid-Atlantic coast of the United States in winter are extratropical cyclones formed over the Gulf Stream.

9.10 LOCAL WEATHER EFFECTS

Interactions among ocean, atmosphere, and land are local in geographic scale. These interactions occur on or near the coast and sometimes produce large differences in weather and climate within a coastal region.

Land and Sea Breezes

All parts of the Earth's surface are subjected to a daily cycle of heating and cooling as night follows day. Ocean surface water absorbs solar radiation during the day without a significant temperature change for three main reasons. First, water has a high heat capacity. Second, much of the solar energy penetrates beneath the surface of the water before being absorbed, particularly in low-**turbidity** waters. Third, wind stirs the upper water column, mixing and distributing the heated water. Similarly, ocean surface water cools slowly at night. Consequently, even in locations where daytime solar radiation is intense, the temperature of ocean surface water varies by at most a few degrees between day and night. The temperature of the land surface, in contrast, can vary over a much larger range **diurnally.** Anyone who has spent time in the middle of a continent, particularly in the mountains, may have observed the rapid heating and cooling of the land. A rock that is too hot to touch in the middle of a sunny day will feel cold within a few hours after sunset that evening.

Heat is transferred continuously between land and atmosphere and between ocean and atmosphere. As a result, and because of its low heat capacity, the air mass next to the Earth's surface tends to change temperature relatively quickly to match that of the land or ocean surface.

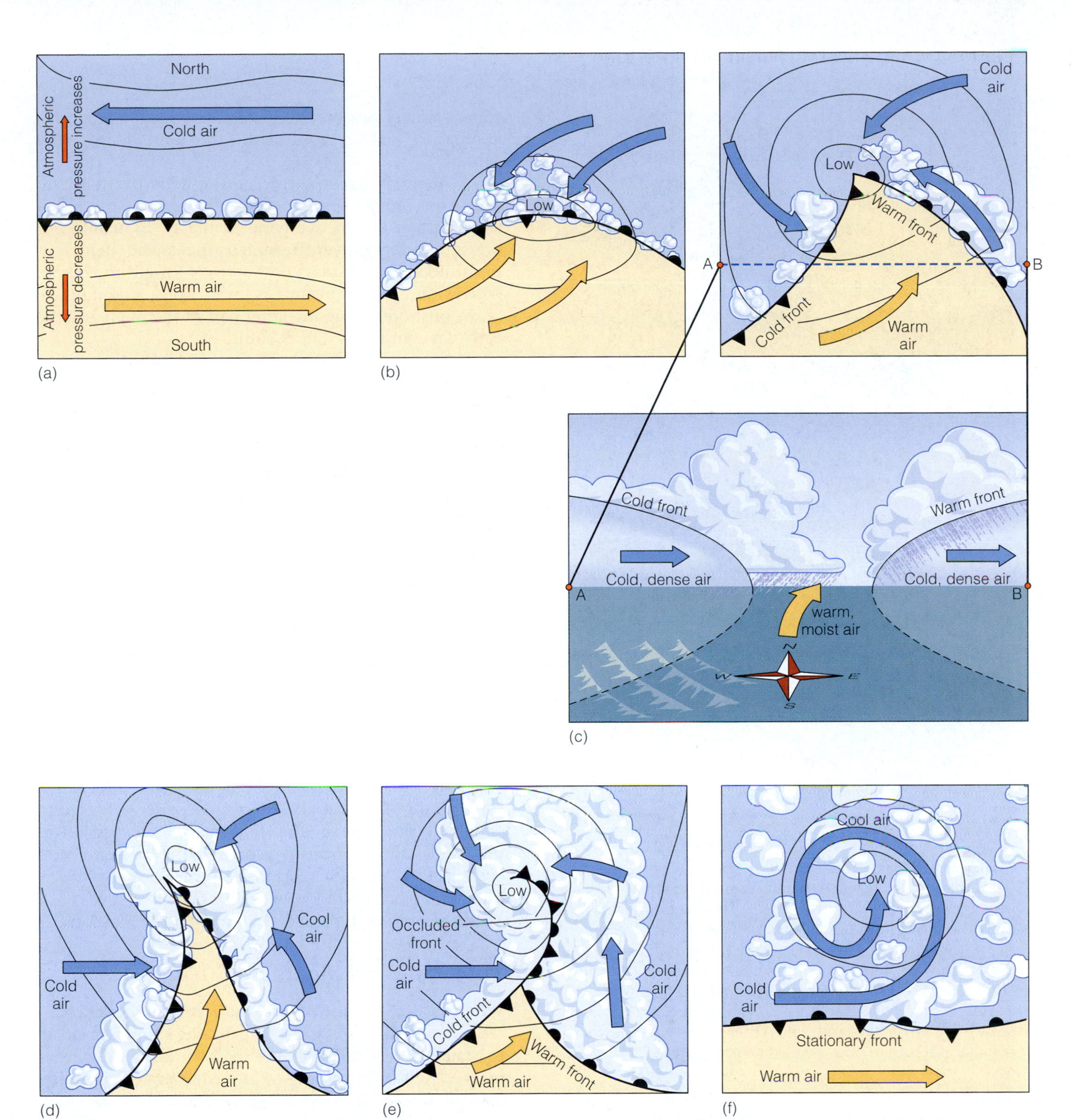

FIGURE 9.29 Formation of extratropical cyclones. All the diagrams except (c) are surface level map views. (a) The polar front separates eastward-moving warm air from westward-moving cold air. (b) Extratropical cyclones are formed when a wave develops on this front. (c) A low-pressure zone is formed as warm, moist air flows north and rises, initiating counterclockwise rotation of the air masses around the low-pressure zone (see Figure 9.25b). Once formed, the low-pressure zone may strengthen if the cold air mass flows over relatively warm ocean surface waters, is warmed, gains water vapor through evaporation, and also begins to rise. The vertical cross section shows that extensive areas of rain clouds form both where the advancing cold air mass forces warm, moist air to rise (cold front), and where warm, moist air is deflected to the right by the Coriolis effect and rises over the cold air mass (warm front). (d) Eventually the cold front catches up to the warm front as warm air continues to rise over the cold air mass. (e) This leads to the formation of an occluded front, which is the line where the warm and cold front meet. (f) The polar front re-forms and the extratropical cyclone slowly dissipates when warm air no longer flows into the central low-pressure zone, and atmospheric pressure rises. These diagrams show air masses in the Northern Hemisphere. In the Southern Hemisphere, the rotation is in the opposite direction.

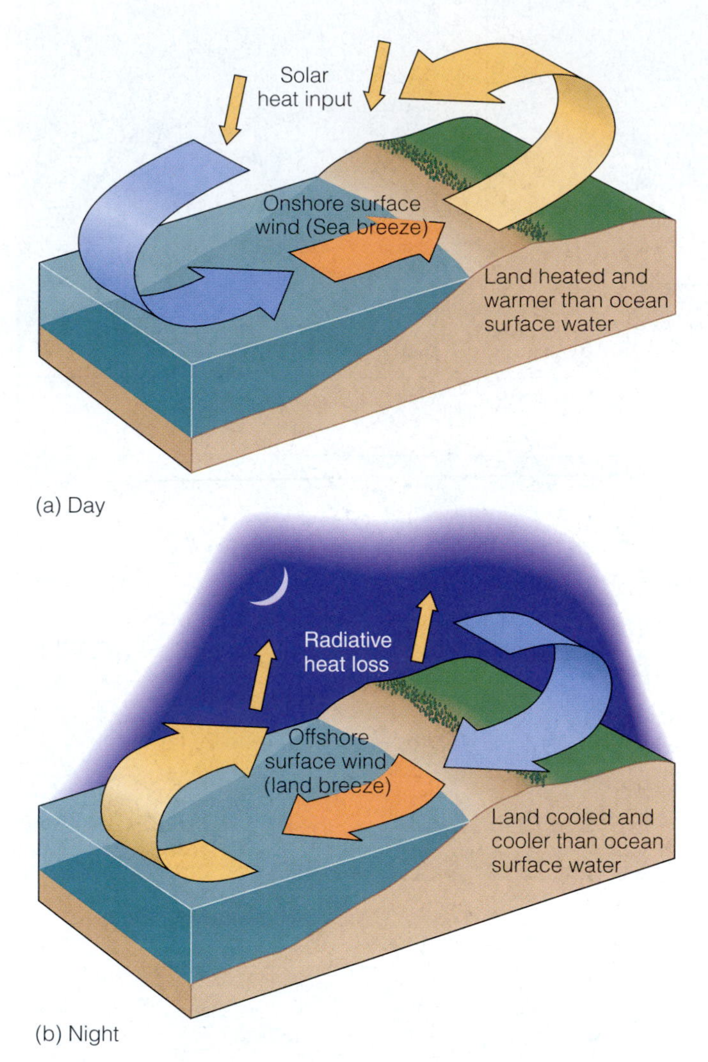

FIGURE 9.30 (a) Land heated by the sun during the day heats the ground-level air, which rises and is replaced by cooler air drawn in from the ocean. Thus, the surface wind blows from ocean to land and is called a "sea breeze." (b) At night, the land loses its heat quickly through radiation, but the ocean water remains at the same temperature. As the land surface cools to below the sea surface temperature, warmer air rises over the ocean and is replaced by cooler air drawn from over the land. Thus, the surface wind blows from land to ocean and is called a "land breeze."

During daylight hours, coastal water and its overlying air mass warm only slowly, whereas the adjacent land and its overlying air mass warm much more rapidly. Where air temperature is higher over coastal land, the less dense air mass rises, causing the atmospheric pressure to be reduced (**Fig. 9.30a**). Cooler air from over the ocean is drawn inland by the resulting atmospheric pressure gradient to replace the rising heated air. This process creates onshore winds called "sea breezes." The sea breeze blows until the solar radiation is reduced in intensity and the land has cooled to the same temperature as the coastal surface waters. Sea breezes generally weaken as nightfall approaches and are gone within an hour or two after sunset.

As night progresses, the land and its overlying air mass cool rapidly until they are colder than the ocean surface water and its overlying air mass (**Fig. 9.30b**). Air pressure increases over the land as the cooled, denser air mass sinks. This air mass now spreads toward the ocean, where atmospheric pressure is lower. Thus, a land breeze blows offshore during the latter part of the night and the early daylight hours (**Fig. 9.30b**).

The reversing local sea breezes and land breezes are present on almost all coasts, but they vary significantly in intensity with location, season, and larger-scale regional weather patterns. Probably for thousands of years, fishers in many countries have used sea breeze and land breeze patterns to their advantage, sailing offshore with the land breezes in the early morning hours and returning with the sea breezes late in the day. Many fishers still follow such a regimen, even though they have motorized vessels.

Coastal Fog

The famous fogs of San Francisco and other coastal locations are often related to sea breezes. In many coastal areas, especially where coastal upwelling occurs, the surface waters near the coast are colder than surface waters some tens of kilometers offshore. When sea breezes develop, warm moist air over the offshore water is drawn into the sea breeze system. As this air flows shoreward, it passes over cooler coastal water. If it becomes cooled sufficiently, some of the water vapor it carries is condensed to form extremely small water droplets that we call "fog" (**Fig. 9.31**). Coastal fogs of this type occur along the entire west coast of the United States and along the east coast from Maine to Canada. Similar fogs occur on the west coasts of South America and Africa and in many other locations.

The Island and Mountain Effect

When an air mass moving over the Earth's surface encounters a mountain chain across its path, the air mass is forced to rise up the mountain slope to pass the barrier. As the air mass rises, it cools because of the reduced pressure. If the air is moist before it begins to rise, and if it rises high enough, water vapor will condense to form clouds and eventually rain (**Fig. 9.32a**). Once the air mass has crossed the mountain ridge, it flows down the mountain slope. As it descends and pressure increases, the air is warmed and consequently can retain more water vapor. Because much of the water vapor originally present has been lost on the upslope of the mountains, rainfall ends and cloud water droplets are revaporized as the air descends. The **leeward** side of the mountains therefore receives very little rain.

In many coastal locations, the wind blows consistently from the ocean onto land. If a coastal or near-

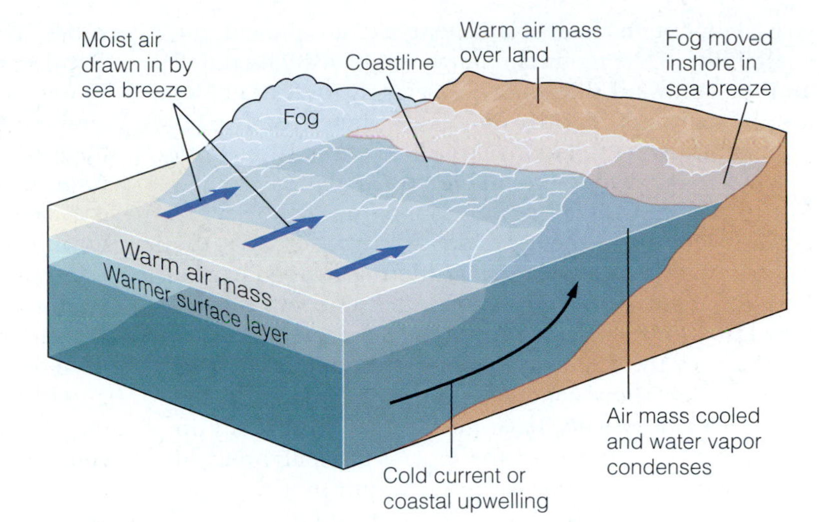

FIGURE 9.31 Coastal fog develops when moist air is drawn inland by a sea breeze or climatic winds and passes over a cold nearshore water mass. The cold nearshore water can be the result of cold-water currents or, more often, of coastal upwelling that brings cold, deep water to the surface.

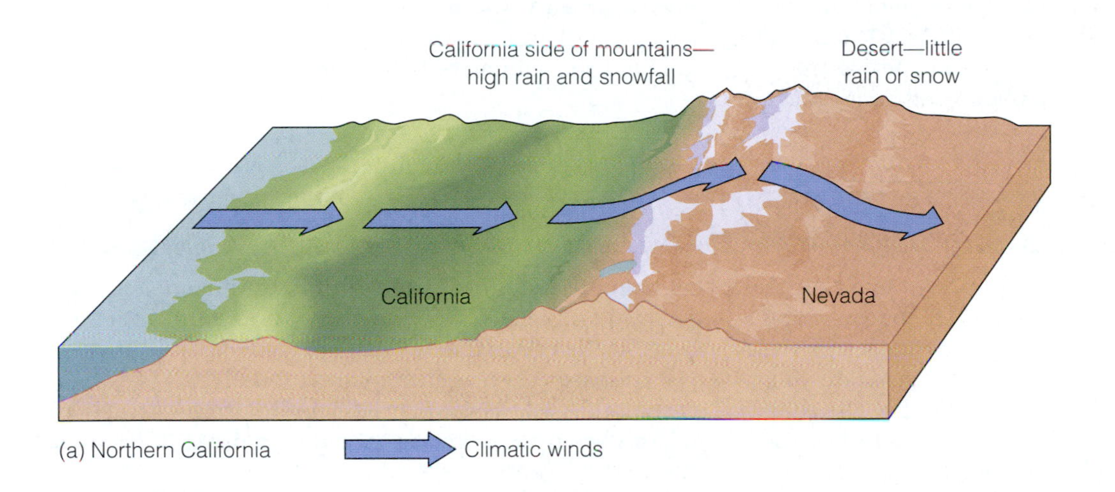

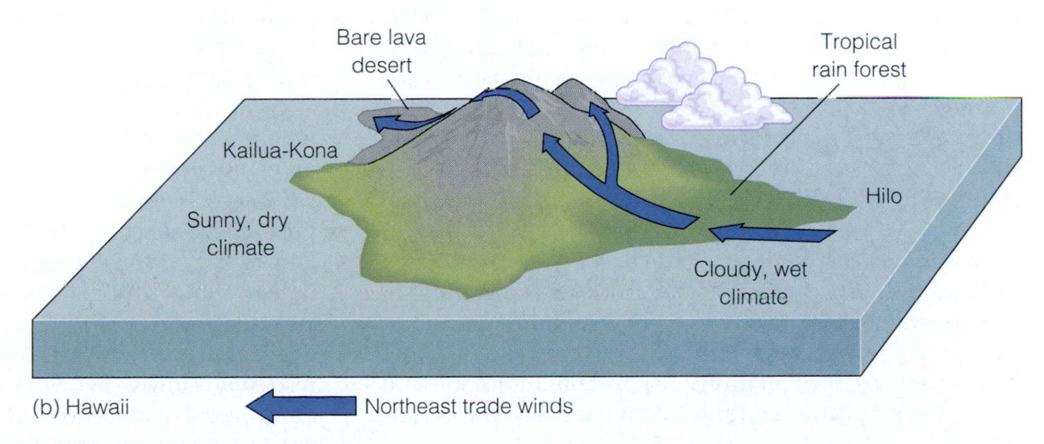

FIGURE 9.32 The mountain or island effect occurs where a moving mass of moist air, usually from over the ocean, encounters a mountain or mountainous island, rises, cools adiabatically, and loses water as rain or snow. Once over the mountains, the air mass descends, warms, and can hold more moisture. As a result, rain and snowfall stop. (a) The Sierra Nevada mountains of California intercept moist air moving inland from the ocean. Rain and snowfall are abundant on the California side of the mountains, but the Nevada side is a desert. (b) On the Hilo side of Hawaii, trade winds blowing from the northeast produce clouds, plentiful rainfall, and tropical rain forests, but the opposite side of the island, near Kailua-Kona, is sunny and has low rainfall.

coastal mountain chain of sufficient height is present, the coastal side of the mountain range will have high rainfall and the landward side will be desert. The best examples are in parts of South America, where the Andes Mountains divide the fertile coastal region of Chile from the desert or near-desert interior of Argentina, and in North America, where the west side of the Sierra Nevada chain is fertile and the east side is desert. Note that the coastal mountains of northern California are too low and have too many gaps to dry the air masses completely as they move off the Pacific toward the Sierra Nevada.

Skiers in the Sierras can use the mountain effect to decide where snow conditions are best. Resorts on the California side usually have more snow, particularly in poor snow years. However, the snow is generally wet and heavy. The resorts on the Nevada side, just over the ridge **crest**, generally get less snow. However, snow that does accumulate falls through undersaturated dry air and often has a dry powdery character that is desirable for skiing.

The mountain effect also operates on mountainous islands, where it is called the "island effect." The island effect is extremely important to the climates of many islands, particularly those in the trade wind zone, where winds blow reliably from one direction almost all year. The island of Hawaii is a particularly good example. Hawaii is composed of several volcanoes, some of which are over 4000 m high, and it is located in the northeast trade wind zone (**Fig. 9.32b**). Hilo, on the **windward** northeast coast, has rain almost every day. It is within a coastal belt of tropical rain forest with many continuously flowing streams and waterfalls. Kailua-Kona, on the leeward southwest side of the island, is hot, sunny, and arid, with only occasional showers throughout most of the year. It is surrounded by almost bare fields of solidified **lava** on which little can grow because of the low rainfall.

Other islands in the Hawaiian chain, and many other tropical Pacific island vacation spots, have windward wet climates and leeward dry climates. However, each island is somewhat different. On Oahu, for example, the mountains are relatively low and cut by passes through which clouds carry afternoon showers to Honolulu before they can dissipate. On other islands, the mountains are too low to produce the island effect. Kahoolawe, like most other low islands in the trade wind zone, is extremely arid and gets rainfall only from infrequent storms.

CHAPTER SUMMARY

Atmosphere and Water Vapor. Air masses move vertically if a change in temperature or in their concentration of water vapor causes their density to change. In the lower atmosphere, vertical movements are generally limited to the troposphere (about 12 km altitude). The addition of water vapor to air decreases its density because lighter water vapor molecules displace heavier nitrogen and oxygen molecules. Atmospheric convection is caused by heating or evaporation at the sea or land surface. Water vapor is evaporated continuously from the oceans to the atmosphere. As air rises, it expands as pressure decreases, which causes it to cool. The saturation pressure of water vapor in air decreases with decreasing temperature. Hence, as air cools it becomes supersaturated with water, which condenses to rain.

Water and Heat Budgets. Enough water is evaporated from land and oceans each year to cover the world 1 m deep. About 93% of this water comes from the oceans, but nearly 30% of the resulting precipitation falls on land. The excess of precipitation over evaporation on land enters lakes, streams, and rivers and returns to the oceans as runoff.

Almost 25% of the solar radiation reaching the Earth is absorbed and converted to heat in the atmosphere. About 50% is absorbed by oceans and land, and the rest is reflected to space. Of the heat absorbed by the oceans, about half is lost by radiation, and half is transferred to the atmosphere as latent heat of vaporization. Solar energy per unit area received by the Earth is at a maximum at the equator and decreases toward the poles. Heat radiated and reflected per unit area varies little with latitude. At the equator, more heat energy is received than is lost to space, whereas at the poles more heat is lost than is received. Heat is transferred from the tropics to polar regions by atmospheric and ocean current transport.

Climatic Winds. Heat transfer from oceans to the atmosphere causes atmospheric convection. Horizontal air movements in the convection are the winds. The atmospheric convection cell system consists of Hadley, Ferrel, and polar cells arranged between the equator and the pole in each hemisphere. Trade winds in the Hadley cell blow westward and toward the equator, and westerly winds in the Ferrel cell blow eastward and toward the pole. Between cells, winds are generally calm. Rainfall and clouds are heavy at upwelling regions and light at downwelling regions. The convection cells shift north and south seasonally as the Earth's angle to the sun changes. Because solar heat is released to the atmosphere relatively slowly by evaporation, the location of the atmospheric convection cells lags behind the location of greatest solar heating as the cells migrate seasonally.

Climate and Ocean Surface Water Properties. Ocean surface water temperatures generally decrease with latitude, but currents and upwelling distort the pattern. Surface water salinity is determined primarily by differences between evaporation and precipitation rates, except near continents where freshwater runoff is high. Salinity is highest in the subtropics and polar regions, where rainfall is low. Salinity is low at mid latitudes, where rainfall is high and evaporation less than at lower latitudes, and at the equator, where evaporation is reduced by persistent cloud cover and lack of winds. Extreme high salinity occurs in marginal seas where evaporation exceeds precipitation; and extreme low salinity, in marginal seas where precipitation exceeds evaporation.

Interannual Climate Variations. El Niño/Southern Oscillation is a complex sequence of interrelated events that occur across the equatorial region of the Pacific Ocean. Warm surface water is transported westward by the trade winds until it accumulates near Indonesia and flows back to the east along the equator, where there is no Coriolis effect. During El Niño, upwelling is stopped near Peru, with often disastrous effects on marine life. Droughts occur in locations as far away as Central Europe, and severe storms occur in California and other places. When an El Niño ends, the system may overshoot to produce La Niña, which has effects generally opposite those of El Niño, but less severe.

In addition to ENSO, interannual oscillations of ocean–atmosphere characteristics have been identified in the North Pacific, North Atlantic, Arctic, and Indian Oceans. Each of these oscillations affects regional climates, but the effects are generally smaller and less widespread than those due to ENSO. Studies of the Pacific Decadal Oscillation suggest that it, and probably other such oscillations, and their ecosystem effects are chaotic and may cause nonpredictable irreversible ecosystem change.

Global Climate Zones. Climate zones in the oceans are arranged generally in latitudinal bands. Land climate zones are more complex and depend on proximity to the ocean and the locations of mountain ranges. The ocean moderates coastal climates because surface waters do not change much in temperature either during the day or during the year. The moderating influence of the oceans can extend far into the continents in regions where the winds are generally onshore and no mountain chains block the passage of the coastal air mass inland.

Weather Systems. Winds caused by atmospheric high- and low-pressure zones blow geostrophically, almost parallel to the pressure contours because the pressure gradient and Coriolis deflection are balanced. Winds flow counterclockwise around a low and clockwise around a high in the Northern Hemisphere, and vice versa in the Southern Hemisphere. Hurricanes form around a low-pressure zone over warm water in latitudes high enough that the Coriolis effect is significant. Hurricane winds blow in toward the eye in a rising helical pattern. Winds accelerate as heat energy is added by evaporation from the ocean surface, but the hurricane loses strength when it moves over land or cool water, where its energy supply is removed. Extratropical cyclones form at the polar front as warm air flows over cold polar air and is deflected by the Coriolis effect.

Local Weather Effects. The Earth's surface heats by day from solar radiation and cools at night as the heat is lost. Primarily because of water's high latent heat, the ocean surface water temperature varies little in this daily cycle. The land temperature varies more. During the day, land next to an ocean heats, then loses some of its heat to the air. The warmed air rises and is replaced by cooler air from over the ocean, creating a sea breeze. At night, the land cools rapidly, which creates a land breeze that flows seaward to displace warmer, less dense air over the ocean. Coastal fogs form when warm, moist air from over the ocean passes over a cold coastal water mass as it enters the sea breeze system.

When moist air masses encounter mountains, they rise and cool. If they rise high enough, water vapor condenses and causes precipitation. After crossing the mountains, the air mass descends, is compressed, and warms. When it is no longer saturated with water vapor, precipitation ceases. Hence, the windward side of the mountains is wet, and the leeward side is arid.

STUDY QUESTIONS

1. Why do the CFCs released into the lower atmosphere take a very long time to reach the Earth's ozone layer?
2. Why is the amount of water vapor in air so important?
3. What are the reasons for the net heat loss to space from polar regions of the Earth and the net heat gain from the sun in the tropics?
4. Why are there three atmospheric convection cells in each hemisphere? Discuss why the cells are not centered on the equator and why they are more complicated in the Northern Hemisphere.
5. Why are persistent cloud cover and rainfall present at atmospheric convergences over the oceans?
6. How do the locations of the world's deserts relate to the locations of the atmospheric convection cells?
7. Given that the summer solstice is in June, why is August usually the hottest month in most of the United States?
8. The Hawaiian Islands lie in the northeast trade wind zone. Why do the northeast sides of the islands have wet climates and the southwest sides dry climates?
9. Antarctica is a desert, despite its ice cap, which is several kilometers thick. Why is it a desert?
10. Why are the cloud patterns that we see in weather satellite images dominated by swirls?
11. Why don't hurricanes form at the equator? Why don't they form over land?

KEY TERMS

You should recognize and understand the meaning of all terms that are in boldface type in the text. All those terms are defined in the Glossary. The following are some less familiar key scientific terms that are used in this chapter and that are essential to know and be able to use in classroom discussions or exam answers.

adiabatic expansion (p. 203)
angle of incidence (p. 205)
backscattered (p. 204)
chaotic (p. 221)
chlorofluorocarbons (CFCs) (p. 202)
climate (p. 200)
conduction (p. 204)
contour (p. 226)
convection cell (p. 203)
convergence (p. 207)
Coriolis effect (p. 207)
cum sole (p. 226)
cyclonic (p. 229)
diffusion (p. 201)
divergence (p. 207)
downwelling (p. 209)
El Niño (p. 209)
equinox (p. 205)
extratropical cyclone (p. 209)
front (p. 209)
geostrophic (p. 226)
greenhouse effect (p. 200)
groundwater (p. 204)
heat capacity (p. 200)
hurricane (p. 203)
intertropical convergence (p. 209)
isobaric chart (p. 226)
isotherm (p. 214)
jet stream (p. 209)
latent heat of vaporization (p. 203)
leeward (p. 232)
meander (p. 209)
monsoon (p. 212)
ozone layer (p. 201)
pressure gradient (p. 207)
pycnocline (p. 219)
residence time (p. 202)
runoff (p. 213)
salinity (p. 200)
saturation pressure (p. 202)
sensible heat (p. 204)
stratified (p. 200)
solstice (p. 209)
storm surge (p. 227)
supersaturated (p. 203)
trade wind (p. 208)
troposphere (p. 201)
upwelled (p. 214)
weather (p. 200)
westerly (p. 209)
western boundary current (p. 230)
windward (p. 234)

CRITICAL CONCEPTS REMINDER

CC1 **Density and Layering in Fluids** (pp. 200, 203, 207, 217, 223). Fluids, including the oceans and atmosphere, are arranged in layers sorted by their density. Air density can be reduced by increasing its temperature, and by increasing its concentration of water vapor causing the air to rise. This is the principal source of energy for Earth's weather. To read **CC1** go to page 2CC.

CC3 **Convection and Convection Cells** (pp. 203, 207). Evaporation or warming at the sea surface decreases the density of the surface air mass and causes it to rise. The rising air mass cools by adiabatic expansion, and eventually loses water vapor by condensation, which increases temperature as latent heat is released. As air continues to rise, adiabatic expansion and radiative heat loss then cool the air and increase its density so that it sinks. These processes form the atmospheric convection cells that control Earth's climate. To read **CC3** go to page 10CC.

CC4 **Particle Size, Sinking, Deposition, and Resuspension** (p. 203). Suspended particles (either in ocean water or in the atmosphere) sink at rates primarily determined by particle size: large particles sink faster than small particles. This applies to water droplets in the atmosphere. The smallest droplets sink very slowly and form the clouds, while larger droplets fall as rain. To read **CC4** go to page 12CC.

CC5 **Transfer and Storage of Heat by Water** (pp. 209, 223, 224). Water's high heat capacity allows large amounts of heat to be stored in the oceans and released to the atmosphere without much change of ocean water temperature. Water's high latent heat of vaporization allows large amounts of heat to be transferred to the atmosphere in water vapor and then transported elsewhere. Water's high latent heat of fusion allows ice to act as a heat buffer reducing climate extremes in high latitude regions. To read **CC5** go to page 15CC.

CC8 **Residence Time** (p. 202). The residence time is the average length of time that molecules of contaminants such as chlorofluorocarbons spend in the atmosphere before being decomposed or removed in precipitation or dust. Long residence time allows such contaminants to diffuse upwards into the ozone layer in the upper atmosphere. To read **CC8** go to page 19CC.

CC9 **The Global Greenhouse Effect** (pp. 200, 203, 204, 205). The oceans and atmosphere are both important in studies of the greenhouse effect, as heat, carbon dioxide and other greenhouse gases are exchanged between atmosphere and oceans at the sea surface. The oceans store large amounts of heat, and larger quantities of carbon dioxide both in solution and as carbonates. To read **CC9** go to page 22CC.

CC10 **Modeling** (pp. 205, 220, 221). The complex interactions between the oceans and atmosphere that control Earth's climate and affect the fate of Greenhouse gases can best be studied by using mathematical models, many of which are extremely complex and require massive computing resources. To read **CC10** go to page 26CC.

CC11 **Chaos** (pp. 205, 221, 222, 226). The nonlinear nature of ocean-atmosphere interactions makes at least part of this system behave in sometimes unpredictable ways and makes it possible for climate and ecological changes to occur in rapid, unpredictable jumps from one set of conditions to a completely different set of conditions. To read **CC11** go to page 28CC.

CC12 **The Coriolis Effect** (pp. 207, 219, 226). Water and air masses move freely over the Earth and ocean surface while objects on the Earth's surface, including the solid Earth itself, are constrained to move with the Earth in its rotation. This causes moving water or air masses to appear to follow curving paths across the Earth's surface. The apparent deflection, called the Coriolis Effect, is to the right in the Northern Hemisphere and to the left in the Southern Hemisphere. The effect is at a maximum at the poles, and is reduced at lower latitudes, becoming zero at the equator. To read **CC12** go to page 32CC.

CC13 **Geostrophic Flow** (pp. 218, 226, 227). Air and water masses flowing on horizontal pressure gradients are deflected by the Coriolis effect until they flow across the gradient such that the pressure gradient force and Coriolis effect are balanced, a condition called geostrophic flow. This causes ocean currents and winds to flow around high and low pressure regions in near circular paths. The familiar rotating cloud formations of weather systems seen on satellite images are formed by air masses flowing geostrophically. To read **CC13** go to page 43CC.

CHAPTER 10

CHAPTER OUTLINE

CRITICAL CONCEPTS USED IN THIS CHAPTER

These images are of the ocean surface taken from the space shuttle with a low sun angle that enhances reflections and allows features of the ocean surface to be seen. The complex swirls and patterns revealed in all these images are caused by eddies and other surface water motions, some of which come and go within a few hours. The eddy motions are too small and too short lived to be observed by oceanographers using ship-based measurements, so their existence was not known until observations were made from space. Eddies exist in all parts of the oceans. The top left image was taken off the Gulf of California. The island that can be seen in the image is Santa Catalina Island, Mexico. The center image shows part of the Gulf of Mexico off Central Florida. The upper right image was taken of part of the Southwestern Indian Ocean between Africa and Madagascar and the lower right image of the Pacific Ocean between California and Hawaii.

Ocean Circulation

Most of us are aware that the oceans are in constant motion. However, our experiences are primarily of waves and tides, and the sometimes swift, sometimes lazy currents that run in rivers and estuaries. As a result, few people have any appreciation for the monumental scale and persistence of the water motions that make up the currents of ocean circulation. The following passage conveys some small measure of understanding of the great currents that flow in the oceans.

> *Man stands with bowed head in the presence of nature's visible grandeurs, such as towering mountains, precipices, or icebergs, forests of immense trees, grand rivers, or waterfalls. He realizes the force of waves that can sweep away lighthouses or toss an ocean steamer about like a cork. In a vessel floating on the Gulf stream one sees nothing of the current and knows nothing but what experience tells him; but to be anchored in its depths far out of sight of land, and to see the mighty torrent rushing past at a speed of miles per hour, day after day after day, one begins to think that all the wonders of the earth combined can not equal this one river in the ocean.*
>
> —John E. Pillsbury, "*The Gulf Stream, 1891,*" as quoted in Appendix 10, ***Report of the Superintendent of the United States Coast and Geodetic Survey for 1890*** (pp. 461–620)

The Gulf Stream is a current that can be observed from a vessel anchored within it, as Pillsbury did, by sailors who know how much it speeds or slows their vessels. It can also be observed now in satellite images. However, currents of even greater geographic scale flow beneath the surface layer of the oceans. Unseen, they have flowed for millennia with little change. Their importance was summarized well by Rachel Carson:

There is, then, no water that is wholly of the Pacific, or wholly of the Atlantic, or of the Indian or the Antarctic. The surf that we find exhilarating at Virginia Beach or at La Jolla today may have lapped at the base of Antarctic icebergs or sparkled in the Mediterranean sun, years ago, before it moved through dark and unseen waterways to the place we find it now. It is by the deep, hidden currents that the oceans are made one.

—Rachel Carson, ***The Sea Around Us*****, published in 1950 by Oxford University Press, New York (p. 147)**

Throughout the oceans, from surface to seafloor, water is in continuous motion, ranging in scale from movements of individual molecules to oceanwide movements of water by tidal forces and winds. Oceanographers generally consider ocean water movements in three categories: waves, **tides,** and ocean circulation. Wave motions (Chap. 11) and tides (Chap. 12) are primarily short-term (lasting from seconds to hours) oscillating motions that move water in orbital paths around a single location.

Although seawater properties such as **salinity,** temperature, and chemical concentrations vary with location and depth, large volumes of water have nearly uniform properties in many areas. Such uniform volumes of water are called **water masses.** Water masses are transported across the oceans by **currents,** and vertically between depths by **convection** (**CC3**). As this circulation occurs, water masses mix with other water masses at their interface (usually a horizontal interface between layers; **CC1**). Mixing is the exchange of water between adjacent water masses, which results in a new water mass whose properties are a proportional combination of the properties of the individual water masses before mixing. Currents and mixing are the major processes that transport and distribute heat, dissolved chemicals, **suspended sediments,** and some plants and animals in the oceans.

Chapter 9 describes how ocean surface currents can affect local **climates.** This chapter examines the physical processes that control present-day ocean circulation, reviews the major features of circulation, and describes some aspects of the relationship of ocean circulation to present, as well as past, global climate. Subsequent chapters describe how ocean circulation controls and/or affects the distribution and abundance of life in the oceans. Although waves and tides have relatively little influence on large-scale ocean circulation, they contribute substantially to local currents, particularly in coastal waters and **estuaries**, and to vertical mixing, particularly in the upper several hundred meters of the water column (Chaps. 11, 12).

10.1 ENERGY SOURCES

Winds are the primary energy source for currents that flow horizontally in the ocean surface layers (less than 100 to 200 m deep). Hence, surface currents are often called "wind-driven currents" or "wind drift currents." Currents that flow deep in the oceans below the level affected by winds are generated by convection caused by variations in water **density** (Chap. 7, **CC3**). Higher-density waters sink and displace (push aside or up) less-dense deeper water. Because density is determined by temperature and salinity (**CC6**), deep-water circulation is called **thermohaline circulation**.

The primary energy source for both wind-driven circulation and thermohaline circulation is the sun. Winds are generated by density variations in the Earth's atmosphere caused by solar heating and radiative and evaporative cooling (Chap. 9). Thermohaline circulation is driven by density differences between ocean water masses caused by temperature and salinity variations. Variations in temperature and salinity are controlled, in turn, primarily by balances between solar heating and radiative cooling, and between solar evaporation and precipitation, respectively (Chap. 9).

10.2 WIND-DRIVEN CURRENTS

On a windy day at the **shore**, we can easily see that winds cause ocean water motions such as waves. Winds also create currents that can transport large volumes of water across the oceans.

Generation of Currents

When winds blow across the ocean surface, energy is transferred from wind to surface water as a result of the **friction** between the wind and the surface. The energy transferred to the ocean surface sets the surface layer of water in motion and generates both waves and currents. The process of energy transfer from winds to waves and currents is complex and depends on many factors, including wind speed, **surface tension** of the water, and roughness of the surface (that is, whether waves are already present, how high they are, and whether they are breaking). Therefore, the percentage of wind energy that is converted into **kinetic energy** of ocean currents is highly variable. With a steady wind, the speed of the surface current is generally about 1% to 3% of the wind speed, so a wind of 60 km•h^{-1} will generate a surface current of about 1 to 2 km•h^{-1}.

Surface water set in motion by the wind flows horizontally across the water below it. Because of internal friction between the surface water and the water below,

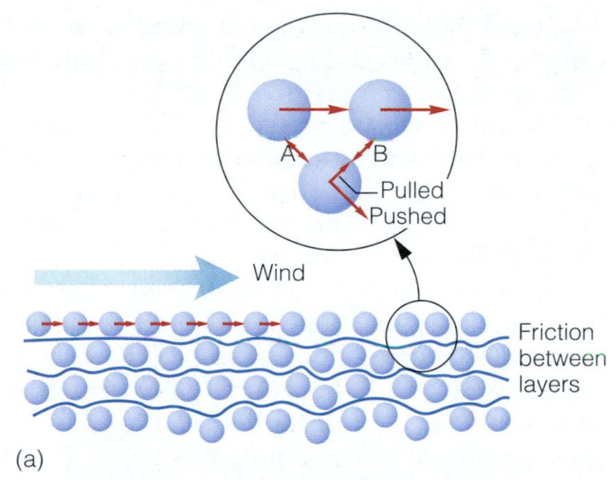

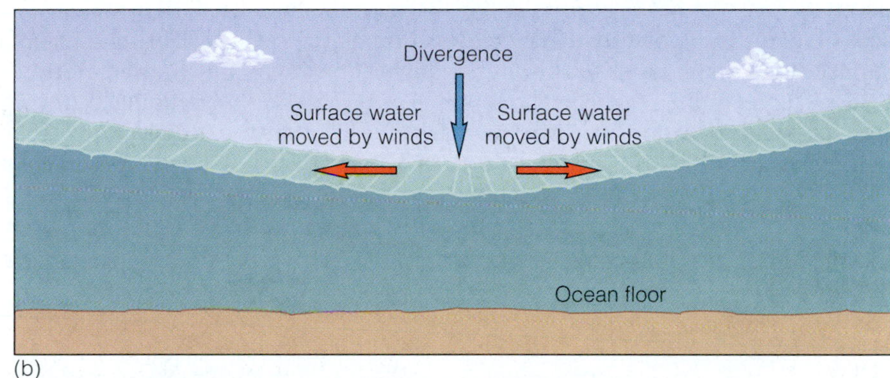

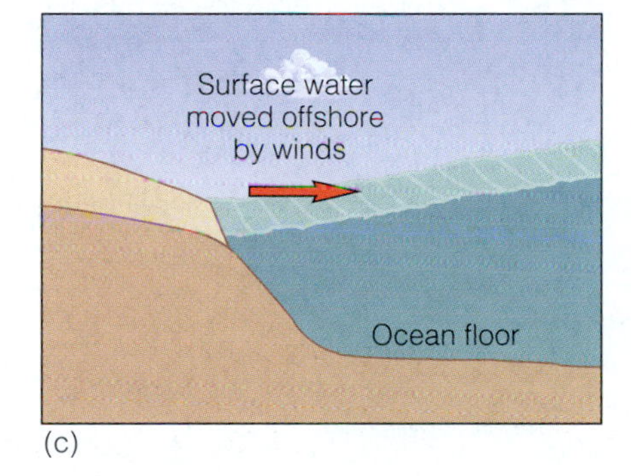

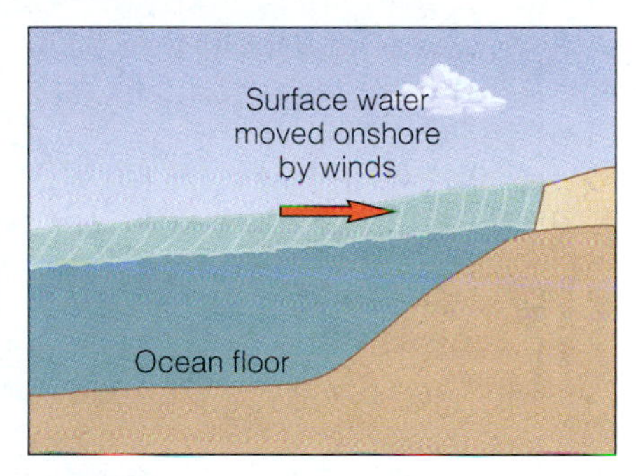

FIGURE 10.1 (a) Wind energy is converted to water movements called "currents" by friction between the moving air and the water surface. The resulting kinetic energy of the water at the surface is transferred vertically downward into the water column by friction between the water molecules. Friction occurs when the molecules of one layer are moved as shown in the enlarged diagram of three adjacent molecules. As the upper-layer molecules are moved, the distance between them and the molecules immediately below is altered. At point A, the distance between molecules is slightly decreased. As the distance decreases, the repulsive force between the molecule's electron clouds increases faster than the attractive gravitational force between the molecules (partly because the electron clouds are closer to each other than the two centers of mass are). Thus, the molecule below is "pushed" forward and downward. Conversely, the distance between molecules at the point labeled B is slightly increased, so the repulsive force between the electron clouds is reduced faster than gravitational force is reduced, and the molecule below is "pulled" forward and upward. This process transfers kinetic energy downward through successive layers of molecules. (b) Sloped sea surfaces can be created in the open ocean when winds transport the surface water layer either away from a divergence, as shown here, or toward a convergence that then becomes an elevated area of sea surface. (c) Sloped sea surfaces can also be created at a coastline when winds transport the surface water layer offshore or onshore. The sea surface slopes are greatly exaggerated in these diagrams.

wind energy is transferred downward. If we consider the water column to consist of a series of very thin horizontal layers, we can envision the moving surface layer transferring some of its kinetic energy to the next-lower layer by friction, setting that layer in motion. The second layer, in turn, transfers some of its resulting kinetic energy to the layer below it, setting that layer in motion (**Fig. 10.1a**). However, only a fraction of the kinetic energy of each moving layer is transferred to the layer below it. Consequently, the speed of a wind-driven current decreases progressively with depth. Wind-driven currents are restricted to the upper 100 to 200 m of the oceans and generally to even shallower depths.

Restoring Forces and Steering Forces

Once current motion has been started, it will continue for some time after the wind stops blowing because the water has momentum. It is like a bicycle that continues to roll forward after the rider stops pedaling. Just as the bicycle slows and eventually stops because of friction in the wheel bearings and between tires and road, ocean currents slow and would finally stop if winds stopped blowing and did not restart. The energy of a current is dissipated by friction between water layers flowing over each other. However, the frictional force between moving layers of water is small. This is one reason why the currents created by winds can flow for long periods after the winds stop and can also flow into and through regions with little wind, such as the Doldrums (**Figs. 9.9, 9.10**).

There is a second reason for the continued flow of wind-created currents after the winds stop. Transport of surface water by wind-driven currents while winds are blowing can cause the sea surface to be sloped (**Fig. 10.1b,c**). The sea surface slopes created by wind-driven transport are extremely small, no more than a few centimeters of height difference across hundreds or thousands of kilometers of ocean surface. However, when the surface is sloped, a horizontal **pressure gradient** is formed (**CC13**) that causes the water to flow from high pressure toward low pressure and tends to restore the ocean surface to a flat horizontal plane. Because the pressure gradient is aligned in the same direction as the sea surface slope, water tends to flow in the downslope direction, or "downhill."

Once set in motion, the direction and speed of any current are modified by friction and three other factors. First, any body in motion on the Earth, including water moving in currents, is subject to deflection by the **Coriolis effect** (**CC12**). Second, the presence of **coasts** can block current flow and cause the water to mound up or to be deflected. Third, current speed and direction are affected by the presence of horizontal pressure gradients.

The following sections examine how the interactions between climatic winds and the **restoring** and steering forces account for the location and characteristics of surface currents in the oceans (**Fig. 10.2**).

10.3 EKMAN MOTION

During his Arctic expedition in the 1890s, Fridtjof Nansen noticed that drifting ice moved in a direction about 20° to 40° to the right of the wind direction. The deflection occurs because water set in motion by the wind is subject to the Coriolis effect (**CC12**). The surface

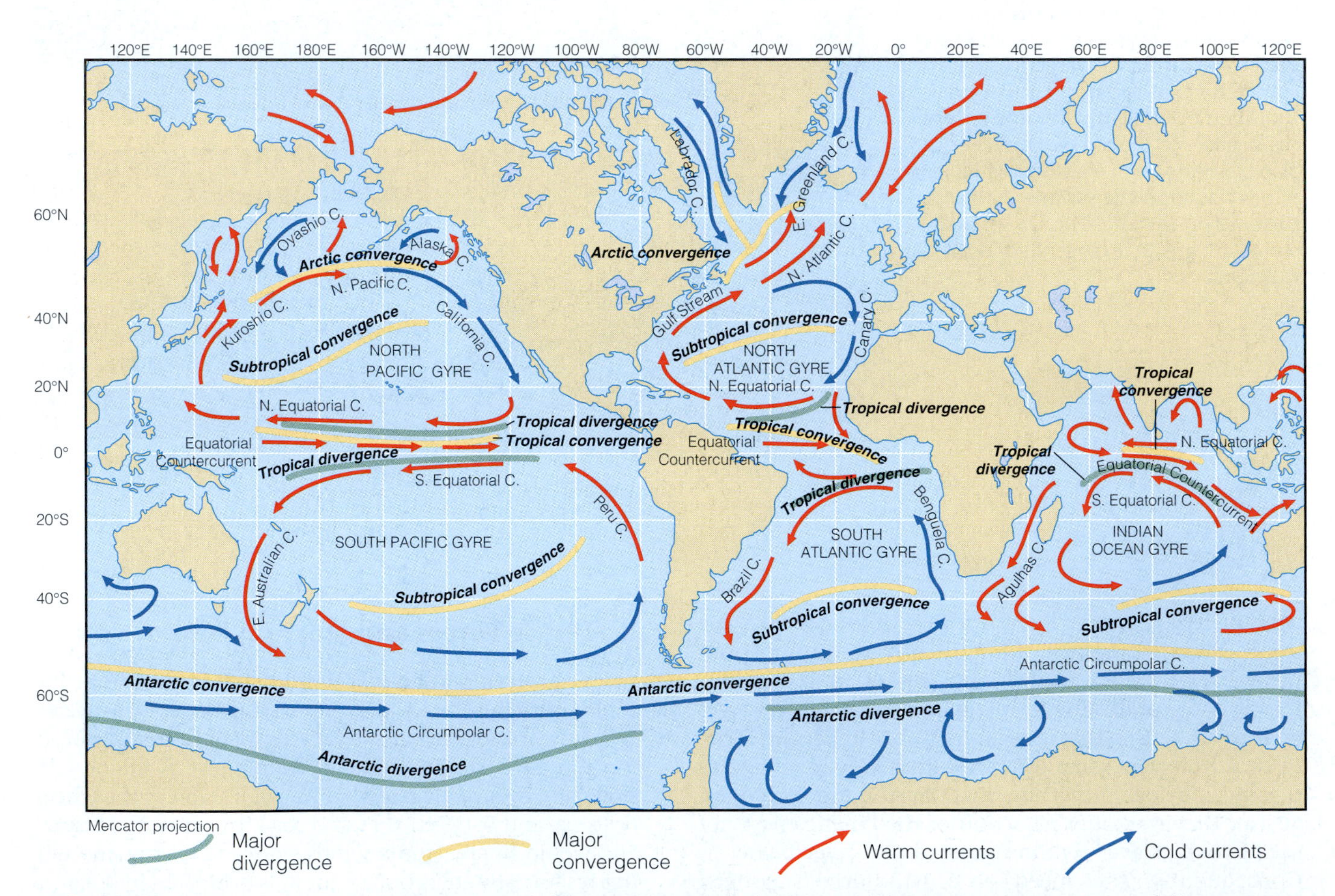

FIGURE 10.2 The wind-driven surface currents of the oceans and the locations of the major convergences and divergences are related. The surface currents form subtropical gyres both north and south of the equator in each ocean basin, and there are subtropical convergences within these gyres. Convergences are also found at higher latitudes between the subtropical gyres and the Antarctic Circumpolar Current, and between the subtropical gyres and the complicated high-latitude gyres in the Northern Hemisphere. Divergences lie on either side of the equator between the subtropical gyres and surrounding Antarctica.

current that carries drifting ice does not flow in the direction of the wind but is deflected ***cum sole*** ("with the sun")—that is, to the right in the Northern Hemisphere and to the left in the Southern Hemisphere (**CC12**). Remember the term *cum sole*. It always means "to the right in the Northern Hemisphere and to the left in the Southern Hemisphere."

The Ekman Spiral

As surface water is set in motion by the wind, energy is transferred down into the water column, successively setting in motion a series of thin layers of water (**Fig. 10.1a**). The surface layer is driven by the wind, each successively lower layer is driven by the layer above, and speed decreases with depth. In addition, each layer of water, in turn, is subject to the Coriolis effect. When it is set in motion by friction with the layer above, each layer's direction of motion is deflected by the Coriolis effect *cum sole* to the direction of the overlying layer. This deflection establishes the Ekman spiral (**Fig. 10.3**), named for the physicist who developed mathematical relationships to explain Nansen's observations that floating ice does not follow the wind direction.

Surface water set in motion by the wind is deflected at an angle *cum sole* to the wind (**Fig. 10.3**). Under ideal conditions, with constant winds and a water column of uniform density, Ekman's theory predicts the angle to be 45°. Under normal conditions, the deflection is usually less. As depth increases, current speed is reduced and the deflection increases. The Ekman spiral can extend to depths of 100 to 200 m, below which the energy transferred downward from layer to layer becomes insufficient to set the water in motion. This depth is a little greater than the depth at which the flow is in the direction opposite that of the surface water, which is called the "depth of frictional influence." At this depth, the current speed is about 4% of the surface current speed. The water column above the depth of frictional influence is known as the "wind-driven layer."

The most important feature of the Ekman spiral is that water in the wind-driven layer is transported at an angle *cum sole* to the wind direction and that the deflection increases with depth. Ekman showed that, under ideal conditions, the mean movement of water summed over all depths within the wind-driven layer is at 90° *cum sole* to the wind. Such movement is called **Ekman transport**.

For the Ekman spiral to be fully established, the water column above the depth of frictional influence must be uniform or nearly uniform in density, water depth must be greater than the depth of frictional influence, and winds must blow at a constant speed for as long as a day or two. Because such conditions occur only rarely, the Ekman spiral is seldom fully established.

Pycnocline and Seafloor Interruption of the Ekman Spiral

In most parts of the ocean, there is a range of depths within which density changes rapidly with depth (**Fig. 10.4**). Such vertical density gradients are called **pycnoclines** (**CC1**). In most parts of the open ocean, a permanent pycnocline is present with an upper boundary that is usually at a depth between 100 and 500 m. In some mid-latitude areas, another shallower seasonal pycnocline develops in summer at depths of approximately 10 to 20 m. Where a steep (large density change over a small depth increment) pycnocline occurs within the depth

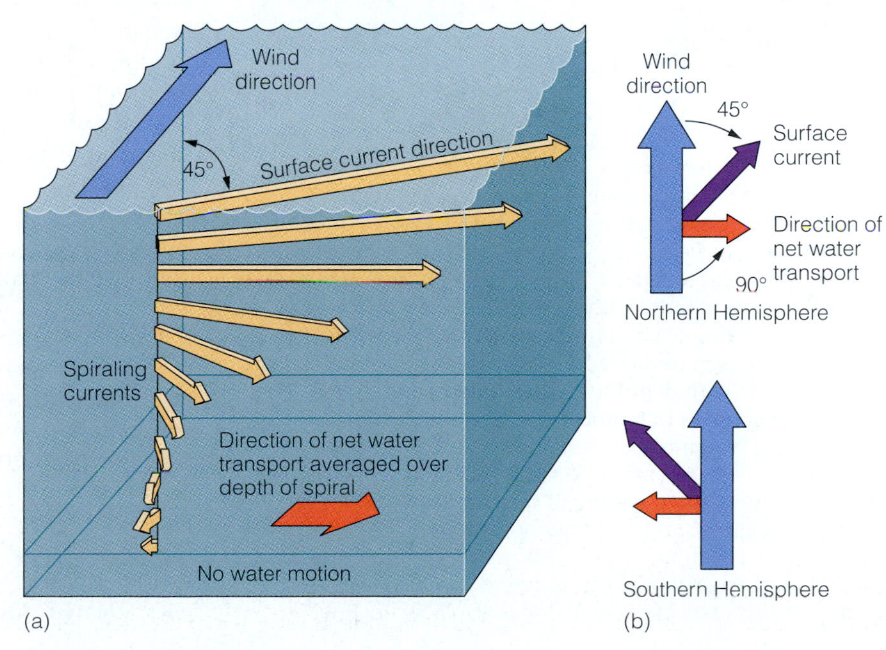

FIGURE 10.3 The Ekman spiral. (a) In the Northern Hemisphere, the Coriolis deflection causes a surface current to flow at an angle to the right of the wind direction. As the kinetic energy of the wind-driven current is transferred downward in the water column, each layer of water is deflected to the right of the direction of the layer above, producing a spiraling current called the "Ekman spiral," in which speed decreases with depth. (b) If the wind blows long enough over deep water, the Ekman spiral motion becomes fully developed. When the spiral is fully developed in the Northern Hemisphere, the surface current direction is 45° to the right of the wind direction, and the net water transport is 90° to the right of the wind direction. In the Southern Hemisphere, the deflection is to the left.

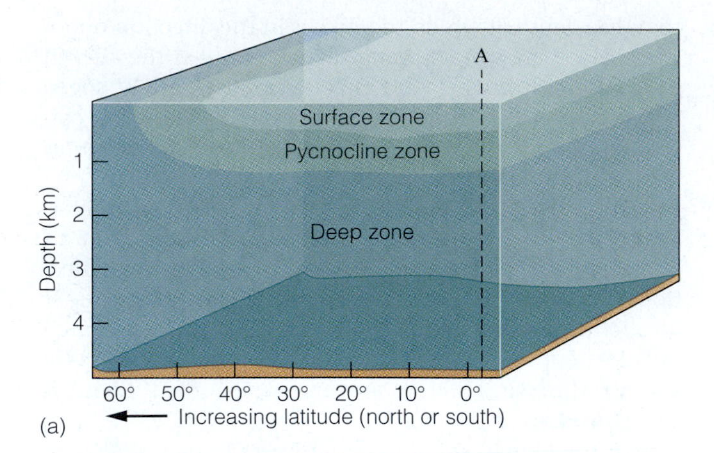

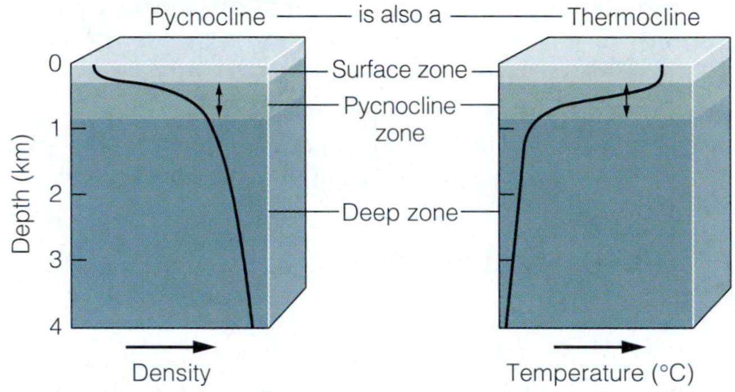

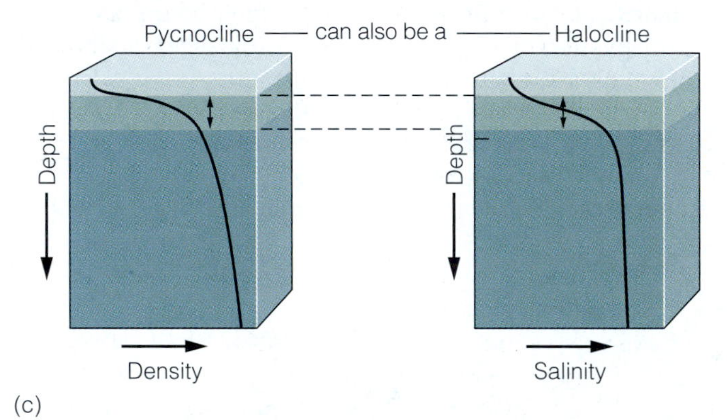

FIGURE 10.4 Density stratification in the oceans. (a) Throughout most of the open oceans (except in high latitudes where surface waters are cooled), a surface or mixed layer of relatively warm, nutrient-depleted, low-density water overlies a layer called the "pycnocline zone" in which density increases rapidly with depth. Below the pycnocline zone is a deep zone of cold, nutrient-rich water in which density increases more slowly with depth. (b) The rapid density change with depth in the pycnocline is due primarily to changes in temperature, and the pycnocline is often called a "thermocline." Shallow seasonal thermoclines can develop above the main pycnocline and in coastal waters because of solar heating of the surface water. (c) In some other regions, and particularly in shallow water near major freshwater inputs, a pycnocline may develop in which the density change with depth is due to vertical variations in salinity. In this case, the pycnocline is also a "halocline."

range of the Ekman spiral, it inhibits the downward transfer of energy and momentum. The reason is that water layers of different density slide over each other with less friction than do water layers of nearly identical density. Water masses above and below a density interface are said to be frictionally decoupled. When a shallow pycnocline is present, the wind-driven layer is restricted to depths less than the pycnocline depth and the Ekman spiral cannot be fully developed. Pycnoclines and the vertical structure of the ocean water column are discussed in more detail later in this chapter.

The Ekman spiral also cannot be fully established in shallow coastal waters, because of added friction between the near-bottom current and the seafloor. Where the Ekman spiral is not fully developed because of a shallow pycnocline or seafloor, the surface current deflection is somewhat less than 45° and Ekman transport is at an angle of less than 90° to the wind. In water only a few meters deep or less, the wind-driven water transport is not deflected from the wind direction. Instead, the current flows generally in the wind direction, but it is steered by bottom **topography** and flows along the depth **contours.**

Where Ekman transport pushes surface water toward a coast, the water surface is elevated at the **coastline** (**Fig. 10.5a**). If the Ekman transport is offshore, the sea surface is lowered at the coastline. Similarly, where Ekman transport brings surface water together from different directions, a **convergence** is formed at which the sea surface becomes elevated (**Fig. 10.5b**). In contrast, where winds transport surface water away in different directions, a **divergence** is formed at which the sea surface is depressed.

10.4 GEOSTROPHIC CURRENTS

Ekman transport of surface layer water tends to produce sloping sea surfaces by "piling up" the water in some locations and "removing" surface water from others. **Geostrophic** currents are the result of horizontal pressure gradients caused by such variations in sea surface level (**CC13**).

Ekman Transport, Sea Surface Slope, and Pressure Gradient

If winds blew continuously, Ekman transport would cause the sea surface slope to increase continuously unless other water movements caused the sea surface to return to a flat configuration. Although sloping sea surfaces do develop, the slopes are always very small—so small that we perceive the sea surface to be flat even in regions of the most persistent winds. Water movements associated with Ekman transport that tend to restore the flat sea surface configuration are driven by the horizontal pressure gradients created under sloping sea surfaces (**Fig. 10.5**).

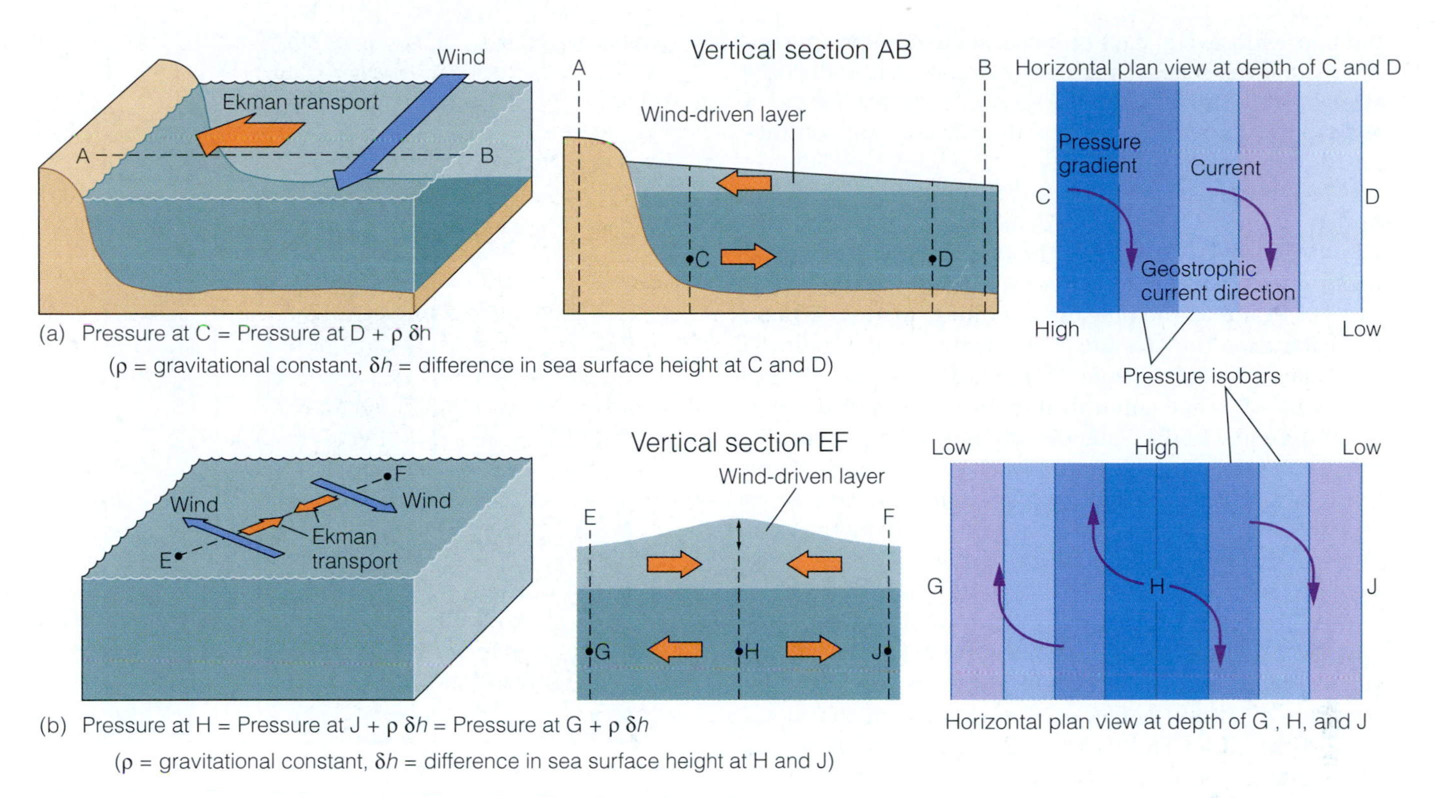

FIGURE 10.5 Ekman transport, geostrophic currents, and Ekman circulation. (a) Where winds blow surface water shoreward, a horizontal pressure gradient develops that causes water below the surface layer to flow offshore, and water to be downwelled near the coastline. This flow is deflected by the Coriolis effect such that, if the system reaches equilibrium, the lower-layer geostrophic current flows parallel to the coast (across the pressure gradient). (b) Away from the coastline, in locations where Ekman transport of surface waters tends to create a mound of surface water at a convergence, a horizontal pressure gradient develops that causes geostrophic currents below the surface layer to flow across the pressure gradient, and water downwells at the convergence. The circulation caused by the Ekman transport, downwelling, and geostrophic flow in the lower layer is called "Ekman circulation." Ekman circulation occurs in reverse at divergences or where winds move surface water offshore.

Horizontal pressure gradients develop in response to the Earth's gravitational force on the water.

The pressure gradient force tends to move water from the high-pressure region under the elevated part of the sea surface toward the lower-pressure region under the depressed part (**Fig. 10.5**). This direction is opposite that of the Ekman transport. The flow induced by the pressure gradient tends to reduce the gradient and sea surface slope.

Ekman transport occurs only in the wind-driven layer, which usually occupies only the upper few tens of meters of the water column. The mean water flow in the wind-driven layer is toward the point of highest sea surface elevation. However, the pressure gradient created when Ekman transport generates a sloping sea surface extends to depths beyond the wind-driven layer and is not restricted by shallow pycnoclines. As a result, the mean flow below the wind-driven layer is directed away from the area of maximum sea surface elevation (**Fig. 10.5**). In the area of maximum elevation, water is **downwelled** from the surface to below the wind-driven layer to join the pressure gradient flow.

Balance between Pressure Gradient and Coriolis Effect

All fluids, including air and water, are subject to the Coriolis effect when they flow horizontally (**CC12**). In the wind-driven layer, the Coriolis effect causes Ekman transport at approximately 90° *cum sole* to the wind direction. Ekman transport stores wind energy as **potential energy** associated with the elevation of the sea surface in one area relative to another. Water flows in response to the horizontal pressure gradient created by the sea surface elevation. As the water flows, it is accelerated and deflected *cum sole*. If allowed to develop long enough, the resulting currents, called "geostrophic currents" (**CC13**), are deflected to flow parallel to the contours of constant pressure (or along the side of the "hill" parallel to the sea surface height contours), and the Coriolis deflection is balanced by the pressure gradient force.

Geostrophic currents generated under ideal conditions beneath sloping sea surfaces are shown in **Figure 10.5**. These ideal conditions would be established only if winds blew at uniform speed over the entire area represented in the figure for several days or more. Such equilibrium conditions are never fully met in the oceans. Consequently, **Figure 10.5** shows only the general characteristics of geostrophic currents. In reality, the currents are continuously changing as winds vary.

Geostrophic current speed and direction are determined by the magnitude and distribution of the horizontal pressure gradient (**CC13**). The pressure gradient can be determined from the sea surface slope and the water density. For example, the pressure at point C in **Figure 10.5a** is greater than the pressure at the same depth at point D by an amount that is equal to the additional weight per unit area of water column at point C. The additional weight is equal to the water density multiplied by the difference in sea surface height between the locations. Using these simple relationships, oceanographers have been able to estimate some surface current speeds by measuring sea surface height differences very precisely with satellite radar sensors.

Sea surface slopes that are created by wind-driven currents and that sustain geostrophic currents are extremely small. For example, one of the fastest ocean currents is the Florida Current, a part of the Gulf Stream that flows around the tip of Florida from the Gulf of Mexico into the Atlantic Ocean. The Florida Current flows at about 150 $cm \cdot s^{-1}$ (5.4 $km \cdot h^{-1}$). This current flows on a sea surface with an approximate height difference of only 20 cm across about 200 km—a slope of 1 in 1 million. Most sea surface slopes are even smaller, and the geostrophic currents are correspondingly slower.

Dynamic Topography

Measuring very tiny sea surface height differences to determine current velocities would appear to be a very difficult task. However, oceanographers have learned to measure the differences indirectly. They carefully measure the density of seawater throughout the depth of the water column at different locations and calculate the **dynamic height**.

The dynamic height is the height of the water column calculated from a measured density profile of the water column between the surface and a depth below which it is assumed there are no currents. We can see how the calculation of dynamic height works if we remember that density is mass per unit volume. Consider several columns of water, each with a 1-cm^2 surface area, but at different locations (**Fig. 10.6**). If the total mass of water in each column were the same, the height of the column with a lower average density would be greater than that of the column with higher average density (**Fig. 10.6**). From several calculated dynamic heights, we can estimate the average sea surface slope between locations. By making such calculations for many different locations, we can draw a detailed map of sea surface height called a "dynamic topography map." Detailed water density distribution data can also be used to identify slopes on pycnoclines, which are then used to estimate geostrophic current speeds and directions in deep-ocean water masses.

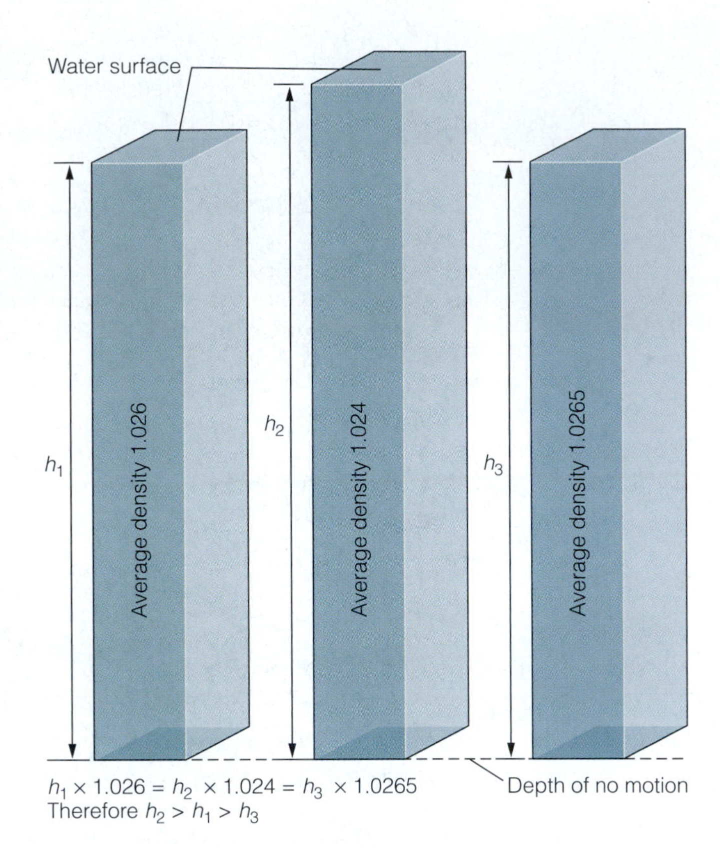

FIGURE 10.6 Calculations of dynamic height can be used to determine sea surface slopes. The height of the water column above a "depth of no motion" is calculated from measurements of the distribution of density with depth. The depth of no motion is a depth below which it can be reasonably assumed that there are no currents flowing. If the average density of the water column above the depth of no motion is known for several locations, the height of the sea surface at each location and the horizontal pressure gradients between the locations can be calculated.

This procedure for calculating dynamic height or sea surface slope is relatively simple, but it involves assumptions that are not usually absolutely correct. First, a depth of no motion in the oceans must be selected, and it must be assumed that there are no currents at or below this depth. The pressure at this depth must be the same at all locations; otherwise the horizontal pressure difference would cause water to flow horizontally. Because pressure does not vary at the depth of no motion, the mass of water above this depth must be the same at any location.

Until just the past few years, dynamic topography maps were generated from measurements made at many different individual locations at various times, sometimes months or years apart. The necessary measurements of salinity and temperature were made from research vessels, which take many weeks to complete a series of measurements at stations spaced across an ocean. As a result, until recently, dynamic topography maps represented average conditions over many weeks or months, and we knew very little about short-term (up to months) variations in geostrophic currents. Oceanographers still rely on indirect calculations of dynamic height to investigate geostrophic currents. However, the development of unattended moored arrays, autonomous floats, and satellite sensors now allows simultaneous data to be gathered at many locations. Moored vertical strings (arrays) of instruments at key locations and autonomous floats that profile the water column can collect and send back to shore-based facilities continuous data that can be used to calculate dynamic height. More than 2000 autonomous floats are now deployed, and thousands more are planned. Satellite radar sensors have now become sensitive enough to measure sea surface heights directly—and with such precision that the data can be used to draw detailed dynamic topography maps of any part of the world oceans every few days. Although the satellite data give only the dynamic height at the sea surface, these data, when combined with the data collected by moored and autonomous instruments, have drastically improved our ability to observe geostrophic currents. As a result, we are just beginning to understand their complexities and variability.

Energy Storage

Geostrophic currents flowing near the surface are generated by winds because they depend on Ekman transport to establish the sea surface slopes under which they flow. However, geostrophic currents continue to flow even if winds abate, because Ekman transport stores wind energy as potential energy by piling up the water. The stored potential energy is converted to momentum to maintain geostrophic currents during intervals when winds are light or cease.

The length of time during which geostrophic currents continue to flow after winds abate depends on the magnitude of the sea surface slope and the area over which the slope occurs. The larger the area of sloping sea surface and the greater the slope, the longer it takes for geostrophic currents to restore the sea surface after winds abate. Where local winds create sea surface slopes over small areas, geostrophic currents restore the sea surface and cease to flow within a few hours after the winds abate. This is often the case for geostrophic currents created by the interaction of storms and coastal water masses.

10.5 OPEN-OCEAN SURFACE CURRENTS

Surface currents of the open oceans (**Fig. 10.2**) are wind-driven currents initiated by Ekman transport and maintained as geostrophic currents. Although they typically extend to depths of several hundreds of meters, they may extend as deep as 2000 m in some limited cases. The most obvious features of ocean surface currents are the **gyres** in subtropical **latitudes** of each ocean. Separate **subtropical gyres** are present north and south of the equator in each ocean except in the Indian Ocean, where the Northern Hemisphere gyre would occur at a location now occupied by landmasses. Subtropical gyres in the Northern Hemisphere flow clockwise; those in the Southern Hemisphere flow counterclockwise.

At latitudes above the subtropics, ocean surface currents are more complicated. The North Atlantic and North Pacific oceans have high-latitude gyres similar to, but less well formed than, the subtropical gyres. The high-latitude gyres rotate in the opposite direction (counterclockwise) from the northern subtropical gyres. In the Southern Hemisphere, high-latitude gyres are absent because the surface current that flows clockwise from west to east completely around Antarctica is not interrupted by a landmass.

CRITICAL THINKING QUESTIONS

10.1 You might have heard it said that because of the Coriolis deflection, water going down a bathtub drain will swirl in opposite directions depending on whether the drain is located in the Northern or Southern Hemisphere. Do you think this is true? Explain why or why not.

10.2 Imagine that the viscosity of water was much higher, so that it flowed more like syrup.
(a) Would the Ekman spiral and Ekman transport be different? If so, how would they be different and why?
(b) Would there still be geostrophic currents? If so, how would they be different?

10.3 If seawater had a maximum density at 4°C, how would the deep water layers be different from those of today? Where would deep water masses be formed?

10.4 If we were to build a solid causeway across the Pacific Ocean from Seattle to Tokyo, how would the surface currents in the North Pacific Ocean be affected, and how might this affect the climate of Japan, Alaska, and California?

10.5 If the Earth were covered in water except for the North American continent, would there be geostrophic gyres? If so, where? What would the current pattern be in the Southern Hemisphere?

10.6 Discuss why you would expect the Gulf Stream to slow down and Europe to be colder if the rate of formation of North Atlantic Deep Water was reduced.

10.7 Why is there no geostrophic gyre in the Mediterranean Sea?

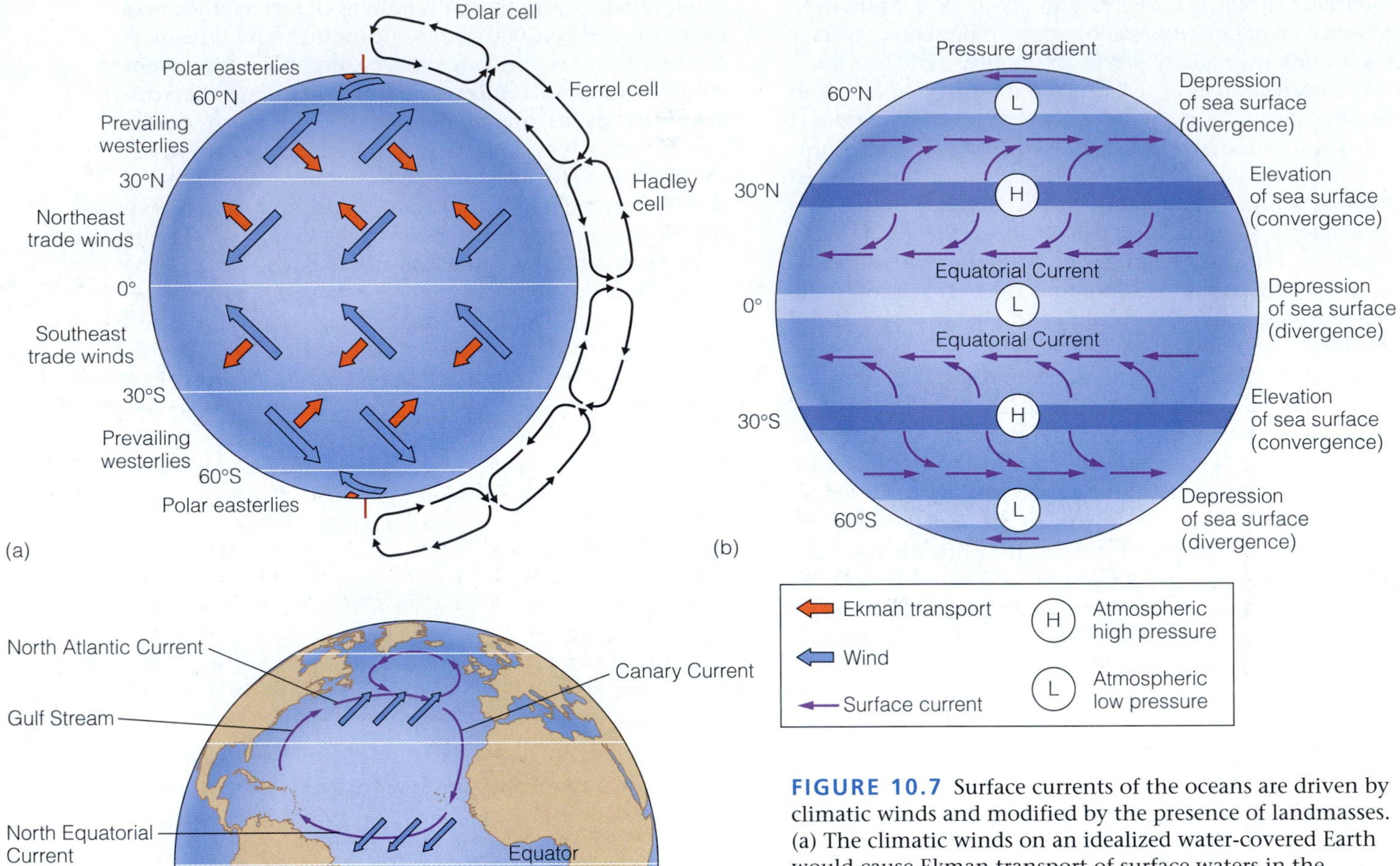

FIGURE 10.7 Surface currents of the oceans are driven by climatic winds and modified by the presence of landmasses. (a) The climatic winds on an idealized water-covered Earth would cause Ekman transport of surface waters in the directions shown. (b) Those water movements would lead to the formation of convergences and divergences, and geostrophic currents would flow on the pressure gradients created by the Ekman transport of the surface layer. These geostrophic currents would flow around the Earth from east to west, or vice versa, between the latitudes of the convergences and divergences. (c) However, landmasses block the flow of geostrophic currents and divert them to the north or south such that geostrophic gyres are formed.

Geostrophic Currents on a Water-Covered Earth

If there were no continents, geostrophic surface layer currents would be relatively simple. The climatic winds and **convection cells** that would be present on a rotating, water-covered Earth are shown in **Figure 9.9**. **Figure 10.7a** shows these idealized climatic winds, and the directions in which the wind-driven layer of ocean water would be moved by Ekman transport due to the winds. In the **trade wind** and polar easterly zones, Ekman transport is partly toward the pole and partly toward the west. In the **westerly** wind zones, Ekman transport is toward the equator and to the east. Surface water layer divergences and **upwelling** of subsurface water occur at the equator and at the atmospheric downwelling regions between the polar and Ferrel cells in each hemisphere. Surface water convergences occur at the atmospheric upwelling regions between the Hadley and Ferrel cells in each hemisphere. Thus, Ekman transport would produce a depression of the sea surface at the equator and between the polar and Ferrel cells in each hemisphere, and an elevation of the sea surface between the Hadley and Ferrel cells in each hemisphere (**Fig. 10.7b**).

The sloping sea surfaces created by climatic winds would establish horizontal pressure gradients within the water column and thus establish geostrophic currents. If an equilibrium were reached, the geostrophic currents would flow around the Earth from east to west or from west to east across the pressure gradients (**Fig. 10.7b**). Currents would flow to the west in the trade wind zones and polar regions and to the east in the westerly wind zones. Geostrophic currents on the real Earth cannot flow in this way, because they are interrupted by continents in all but the westerly wind zone around Antarctica. The Antarctic Circumpolar Current does, in fact, flow from west to east around the Earth in the Southern Hemisphere westerly wind zone (**Fig. 10.2**), just as the no-continent model suggests (**Fig. 10.7b**).

Geostrophic Gyres

The continents provide a western boundary (the Atlantic coast of North and South America, Pacific coast of Asia and Australia, and Indian Ocean coast of Africa), and an eastern boundary (the Atlantic coast of Europe and Africa, Pacific coast of North and South America, and Indian Ocean coast of Southeast Asia and Australia) for each of the three major oceans. Within each ocean, geostrophic east-to-west or west-to-east currents are established at the same latitudes and in the same locations as those in the simple ocean-covered Earth model (**Figs. 10.2, 10.7b**). Where these currents meet a continent, they are blocked and diverted to the north or south.

The westward-flowing current within the trade wind zone of the Northern and Southern Hemispheres are called the North Equatorial Current and the South Equatorial Current, respectively (**Fig. 10.2**). The North and South Equatorial currents carry large volumes of water from east to west until they meet the western boundary continent. Surface water "piles up" there, and the Equatorial currents are deflected away from the equator. The currents then flow toward higher latitudes as **western boundary currents** until they enter the westerly wind zone, where they form the west-to-east geostrophic flow. After the currents reach the eastern boundary, they are deflected primarily toward the equator. After turning toward the equator, they flow as **eastern boundary currents** until they enter the trade wind zone, re-join the Equatorial Current, and complete their circuit, or gyre (**Figs. 10.2, 10.7c**).

As western boundary currents flow toward the pole, they are subject to the Coriolis effect. Therefore, if there were no opposing horizontal pressure gradient, they would be deflected to the east. Similarly, as eastern boundary currents flowed toward the equator, they would be deflected to the west (**Fig. 10.8a**). In each case the deflection would be toward the center of the gyre. However, Ekman transport of surface water in the trade wind and westerly wind zones and the Coriolis deflection of the gyre currents, including the boundary currents, all tend to move water toward the center of the gyre. As a result, an equilibrium

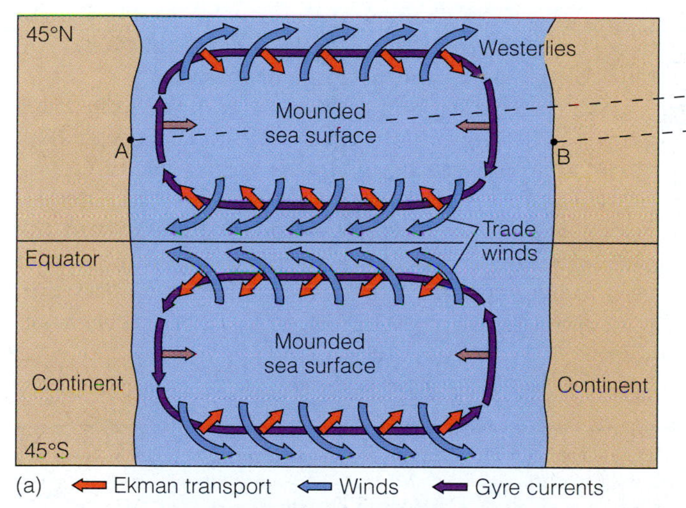

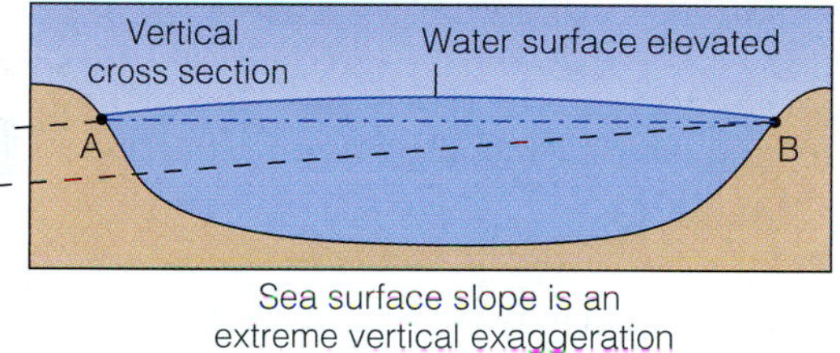

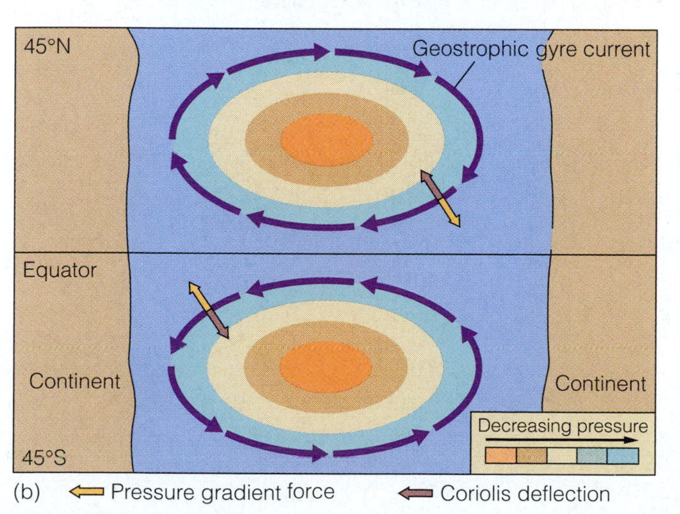

FIGURE 10.8 Geostrophic currents of the subtropical gyres. (a) Wind-driven Ekman transport due to the trade winds and westerlies, and the Coriolis deflection of the boundary currents, both act to elevate the ocean surface at the center of each subtropical gyre. (b) This elevated sea surface creates a horizontal pressure gradient that tends to move water outward. As water flows on this pressure gradient, it is deflected by the Coriolis effect and flows geostrophically around the mound. At equilibrium, the sea surface elevation does not vary, the tendency of water to be moved to the center of the mound is balanced by the pressure gradient, and the pressure gradient generates geostrophic currents. Thus, the wind energy of the trade winds and westerlies is transferred into the geostrophic currents that flow continuously and form the subtropical gyres.

situation is created in which there is a mounded sea surface at the center of the gyre (**Fig. 10.8**). Currents that make up the gyre flow around the mound in geostrophic balance between the Coriolis effect and the horizontal pressure gradient that surrounds this central location.

Hence, gyres are geostrophic currents that flow on mounded surfaces. The center of each subtropical gyre is called a "subtropical convergence" (**Fig. 10.2**). Subtropical gyres are essentially the same as the idealized gyres shown in **Figure 10.8**. However, they are modified in shape and location by the configuration of each ocean's coastlines (**Fig. 10.2**).

Geostrophic ocean gyres act like giant flywheels. They spin at an almost constant speed that represents the average wind energy input to the gyre. If winds cease or are abnormally light, the geostrophic current gyre, or flywheel, slows very gradually because **turbulence** and friction between the gyre currents and the water layers or seafloor dissipate the energy very slowly. When winds resume, Ekman transport builds the height of the mound and therefore increases the sea surface slope and speed of the gyre. This process occurs very slowly because massive volumes of water must be moved to elevate the sea surface even a millimeter or two over a mound that is hundreds or thousands of kilometers in diameter. Winds that drive the ocean gyres are variable over periods of days, but very long periods of calm or unusually high winds are rare. Because gyre current speeds change very slowly, they are relatively invariable and generally do not reflect normal wind speed variations.

Westward Intensification of Boundary Currents

Western boundary currents include the Gulf Stream and the Brazil, Kuroshio, East Australian, and Agulhas currents (**FIG. 10.2**). Eastern boundary currents include the California, Peru, Canary, and Benguela currents. Western boundary currents are narrower, faster, deeper, and warmer than eastern boundary currents (**Table 10.1**). They are generally so deep that they are constrained against the edge of the **continental shelf** and do not extend across the continental shelf to the shore.

The reasons for the westward intensification of boundary currents are quite complex. However, the major reason is related to the Earth's rotation. Western boundary currents are intensified because the strength of the Coriolis effect varies with latitude.

We can look at westward intensification in a simplified way. The Coriolis effect, which decreases from the poles to the equator, is weaker in the latitudes of the trade wind band than in the westerly wind band. Therefore, water moving eastward is deflected more quickly toward the south in the westerly wind zone than water moving westward is deflected toward the north in the trade wind zone (**Fig. 10.9**). Consequently, in the westerly wind zone, geostrophic flow tends to transport surface water toward the center of the gyre over the entire width of the ocean. In contrast, surface water transported westward in the trade wind zone tends to flow with less deflection (that is, in a circle of larger radius; **CC12**) across the ocean (**Fig. 10.9**), is constrained at the equator by the Ekman transport of the trade winds in the other hemisphere, and tends to pile up on the west side of the ocean.

As a result, the mound at the center of the subtropical gyre is offset toward the west side of the ocean (**Fig. 10.10**). The western boundary current is laterally compressed against the continent, and the sea surface slope is greater toward the western boundary than toward the eastern boundary. Greater sea surface slopes cause faster geostrophic currents. Therefore, western boundary currents are faster than eastern boundary currents. The west-

TABLE 10.1 Boundary Currents

	Western Boundary Currents	*Eastern Boundary Currents*
Northern Hemisphere examples	Gulf Stream Kuroshio Current	California Current Canary Current
Southern Hemisphere examples	Agulhas Current Brazil Current	Peru Current Benguela Current
Width	Narrow (≤100 km)	Broad (≈1000 km)
Depth	Deep (to 2 km)	Shallow (≤500 m)
Speed	Fast (>100 km•day^{-1})	Slow (<50 km·day^{-1})
Volume transport	Large (50 ×10^{6} m^{3}•s^{-1})	Small (10–15 ×10^{6} m^{3}•s^{-1})
Boundaries with coastal currents	Sharply defined	Diffuse
Upwelling	Almost none	Frequent
Nutrients	Depleted	Enhanced by upwelling
Fishery	Usually poor	Usually good
Water temperature	Warm	Cool

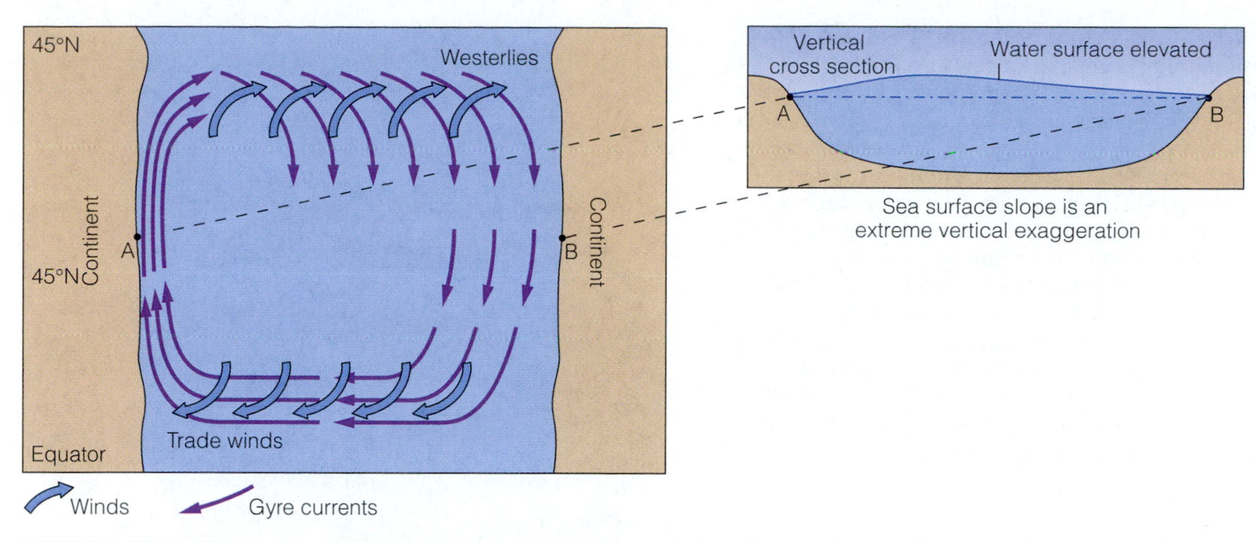

FIGURE 10.9 The westward-moving geostrophic currents of the subtropical gyres flow near the equator, where the Coriolis deflection is weak, whereas the eastward-moving currents flow where the Coriolis deflection is much greater. Hence, the westward-moving water flows across the ocean more directly, piling up water on the west side. The greater Coriolis deflection tends to turn the eastward-moving gyre currents toward the equator and "spreads" the southerly part of the gyre currents that returns water from the trade wind region to the tropics.

ward offset of the subtropical gyre mound causes the pycnocline to be depressed (deeper) on the west side of the ocean in relation to the east side (**Fig. 10.10**). Hence, western boundary currents are narrower and deeper than eastern boundary currents.

This simplified explanation serves reasonably well. However, for a more detailed understanding of westward intensification, we first need to learn about vorticity. Imagine a motionless wheel suspended horizontally above the Earth's surface at the equator (**Fig. 10.11**). To an observer on the Earth, the wheel does not appear to rotate as the Earth rotates. Now imagine that we transport this still motionless wheel to the North Pole and again suspend it horizontally above the Earth. To an observer on the Earth, the stationary wheel appears to spin, and at the same rate as the Earth spins. This effect also applies to spinning gyres of water in the oceans. The rate of apparent rotation increases with distance from the equator and is called **planetary vorticity.** Western boundary currents undergo an increase in their apparent rate of rotation as they flow poleward, whereas eastern boundary currents undergo a decrease in their apparent rate of rotation as they flow toward the equator. This is the underlying reason for the westward intensification, but it requires some explanation.

In a spinning fluid, the angular momentum must be conserved. It is not necessary to fully understand this, but you can see an application of this principle when an iceskater spins. With the arms close to the body, the skater spins quickly. If the arms are extended, the spin becomes

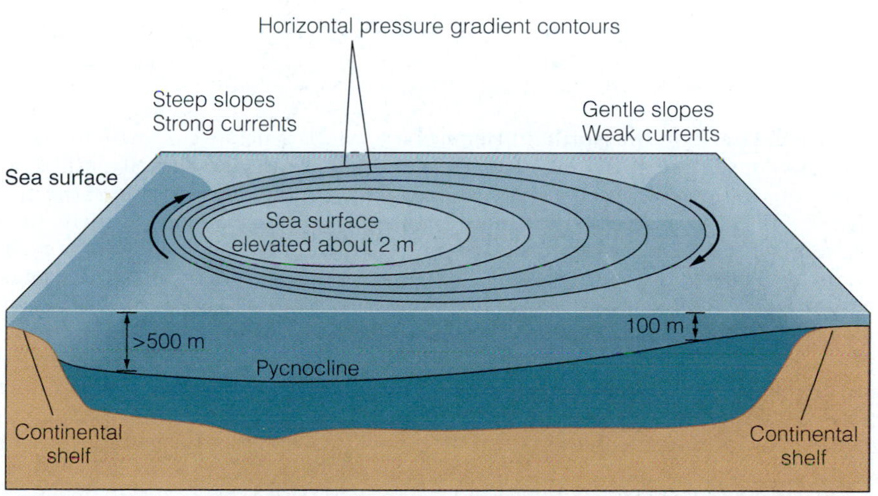

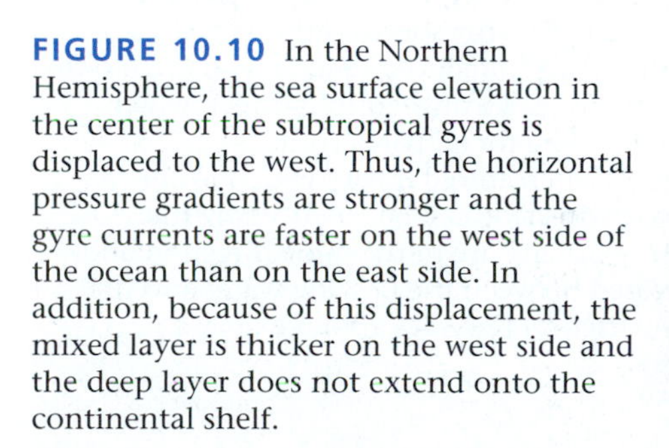

FIGURE 10.10 In the Northern Hemisphere, the sea surface elevation in the center of the subtropical gyres is displaced to the west. Thus, the horizontal pressure gradients are stronger and the gyre currents are faster on the west side of the ocean than on the east side. In addition, because of this displacement, the mixed layer is thicker on the west side and the deep layer does not extend onto the continental shelf.

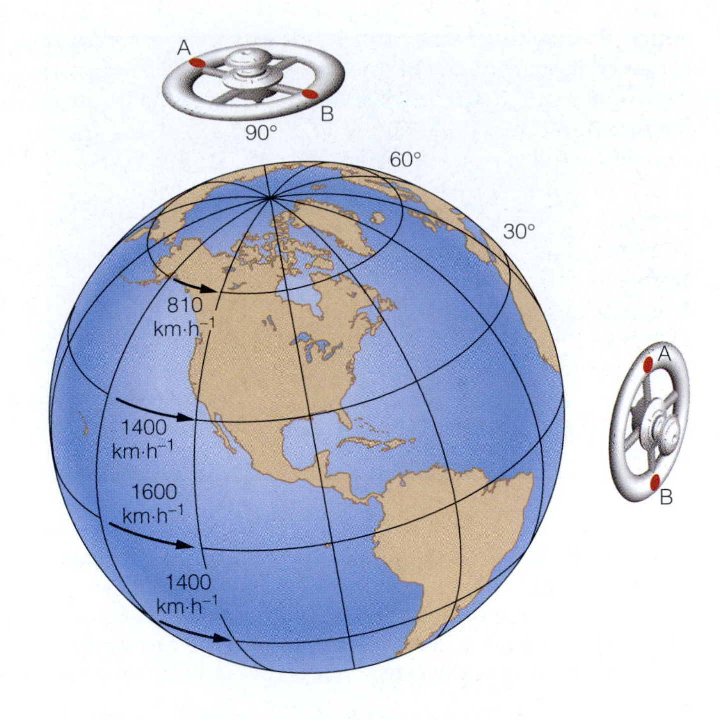

FIGURE 10.11 Vorticity (apparent rotation) increases with increasing latitude. Imagine a rotating Earth with a circular space station orbiting above a specific location on the equator but not rotating around its own axis. An observer looking up from the equator would see that the nonrotating space station stays in the same position in the sky and does not rotate. Now imagine the space station being positioned directly over the North Pole, but still not rotating about its own axis. To an observer looking up from the pole, the nonrotating space station would appear to rotate one revolution each 24 h. It is more difficult to explain what an observer would see if the space station were moved from the equator to mid latitudes. However, the observer would see it rotate, and the rotation rate would increase progressively faster with increasing latitude until, at the poles, it reached one apparent rotation per 24 h.

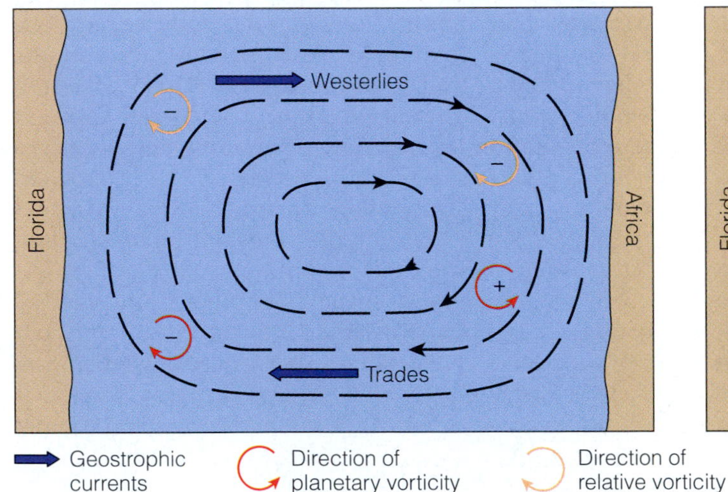

(a) Symmetrical gyre, vorticity unbalanced on western boundary

(b) Westward intensified gyre, vorticity balanced by friction on western boundary

FIGURE 10.12 In the Northern Hemisphere, winds generate clockwise rotating gyres. This rotation creates clockwise vorticity (or spin) relative to the Earth, called relative vorticity. Vorticity, called planetary vorticity, is also imparted by the variations in the magnitude of the Coriolis Effect. On the eastward side of the gyre, water flowing southward is subject to a decreasing Coriolis Effect, which induces an anticlockwise planetary vorticity (for simplicity, a reduced tendency to turn to the right or spin). On the westward side of the gyre the northward flowing water is subject to increasing Coriolis Effect, which induces clockwise planetary vorticity. However, the total vorticity, or spin, must be the same for all parts of the gyre. If it were not, then the total mass transport rate of one part of the gyre would speed up while another slowed down, which is not possible. (a) Shows that, if the gyre were symmetrical, the relative vorticity on the eastward side can be balanced by the planetary vorticity as they are opposite. On the westward side, planetary and relative vorticity are in the same direction and cannot be balanced. (b) If the gyre is offset (intensified) to the west, friction is increased between the flowing water and the seafloor of the continental slope against which the gyre current is concentrated. The friction provides a source of counterclockwise vorticity to balance the clockwise relative and planetary vorticities.

much slower, but momentum is not lost. When the arms are once more brought in, the spin becomes faster again. Of course, friction with the ice does eventually steal momentum, slowing down the spin. Angular momentum and vorticity are related, and vorticity is conserved in much the same way as angular momentum. However, the vorticity that must be conserved is a combination of the planetary vorticity and the relative vorticity. The relative vorticity is the vorticity due to the wind-driven motion of the water in the gyre. The math can be a little complicated, so the process is just summarized in **Figure 10.12**.

Figure 10.12a represents a symmetrical gyre with no westward intensification. On the east side of the gyre, the planetary vorticity is counterclockwise and the relative vorticity is clockwise. Thus, the decrease in planetary vorticity as water moves south can be compensated for by an increase in the relative vorticity (speeding up the wind-driven rotation of the gyre, which is in fact what happens as the current flows farther south into the trade wind belt). However, on the west side both the planetary vorticity and the relative vorticity are counterclockwise. The increase of planetary vorticity could be partly offset if the relative vorticity decreased, but the magnitude of the planetary vorticity at the north end of the western boundary current is much larger than the relative vorticity, so the clockwise motion would be stopped unless there was a strong source of positive (counterclockwise) relative vorticity. This can be supplied by a strong frictional stress that opposes the northward flow of the western boundary current. A strong frictional stress, in turn, can be established if the current is constrained to flow against the land boundary (**Fig. 10.12b**), which generates friction at the interface between the current and the seafloor. If the current flows fast and is narrow and deep, the area of contact with the seafloor (along the edge of the continental shelf) is maximized, and, therefore, so is the shear "frictional" stress. As a result, the westward intensification occurs in all ocean current gyres and is caused by the variation in magnitude of the Coriolis effect with latitude.

The characteristic differences between western and eastern boundary currents have major consequences for processes that sustain fisheries off the adjacent coasts (Chap. 15). The differences are also an important factor in the transport of heat poleward (Chap. 9). Warm western boundary currents are fast-flowing and deep. They have little time to cool and only a small surface area from which to lose heat as they move into higher latitudes. Therefore, heat energy carried away from the equator by western boundary currents is released to the atmosphere primarily when the water enters the westerly wind zone, where the gyre currents flow eastward. Heat transported by western boundary currents moderates the climates of regions into which they flow. For example, the Gulf Stream transports heat north and then east across the Atlantic Ocean to moderate western Europe's climate.

10.6 EQUATORIAL SURFACE CURRENTS

The Northern and Southern Hemisphere trade wind zones are separated by the Doldrums, where winds are very light and variable (**Figs. 9.9, 9.10**). In the simplified model of subtropical gyre formation discussed in the preceding section, the Northern and Southern Hemisphere trade winds produce Ekman transport of surface water to the northwest and southwest, respectively (**Fig. 10.7**). This pattern would be present if the Doldrums (the atmospheric **intertropical convergence** zone) were exactly over the equator. However, atmospheric circulation interacts in a complex manner with continental landmasses, which are concentrated in the Northern Hemisphere. One result of this interaction is that the Doldrums are displaced somewhat to the north of the equator during most of the year (**Fig. 9.10**).

Ekman Transport in the Equatorial Region

Because the atmospheric intertropical convergence zone is displaced north of the equator, part of the southeast trade wind zone extends into the Northern Hemisphere. Consequently, although the Coriolis effect is weak near the equator, trade winds are deflected to the right after they pass north of the equator. Once the Southern Hemisphere trade winds have crossed the equator into the Northern Hemisphere, they generate Ekman transport that is deflected to the right, away from the equator, in a generally northeast direction toward the Doldrums (**Fig. 10.13**). This pattern contrasts with the Ekman transport to the southwest, or away from the Doldrums, that was assumed in the simple model (**Fig. 10.7**). Ekman transport is directed to the southwest in the major portion of the southeast trade wind belt that lies south of the equator (**Fig. 10.13a**).

When the intertropical convergence zone is displaced north of the equator, the sea surface slopes up in a northward direction toward the center of the Northern Hemisphere subtropical gyres throughout the northeast trade wind zone (**Fig. 10.13b**). In addition, the sea surface slopes up in a southward direction toward the center of the Southern Hemisphere subtropical gyre throughout the portion of the southeast trade wind zone in the Southern Hemisphere. These slopes are the same as the slopes developed in the simple model of the subtropical gyres shown in **Figures 10.7 and 10.8**. However, the sea surface slopes upward toward the north in the portion of the southeast trade wind zone that lies between the equator and the Doldrums (**Fig. 10.13b**). In this region, Ekman transport is northeast, to the right of the wind, toward the Doldrums, where the sea surface level is undisturbed by local winds.

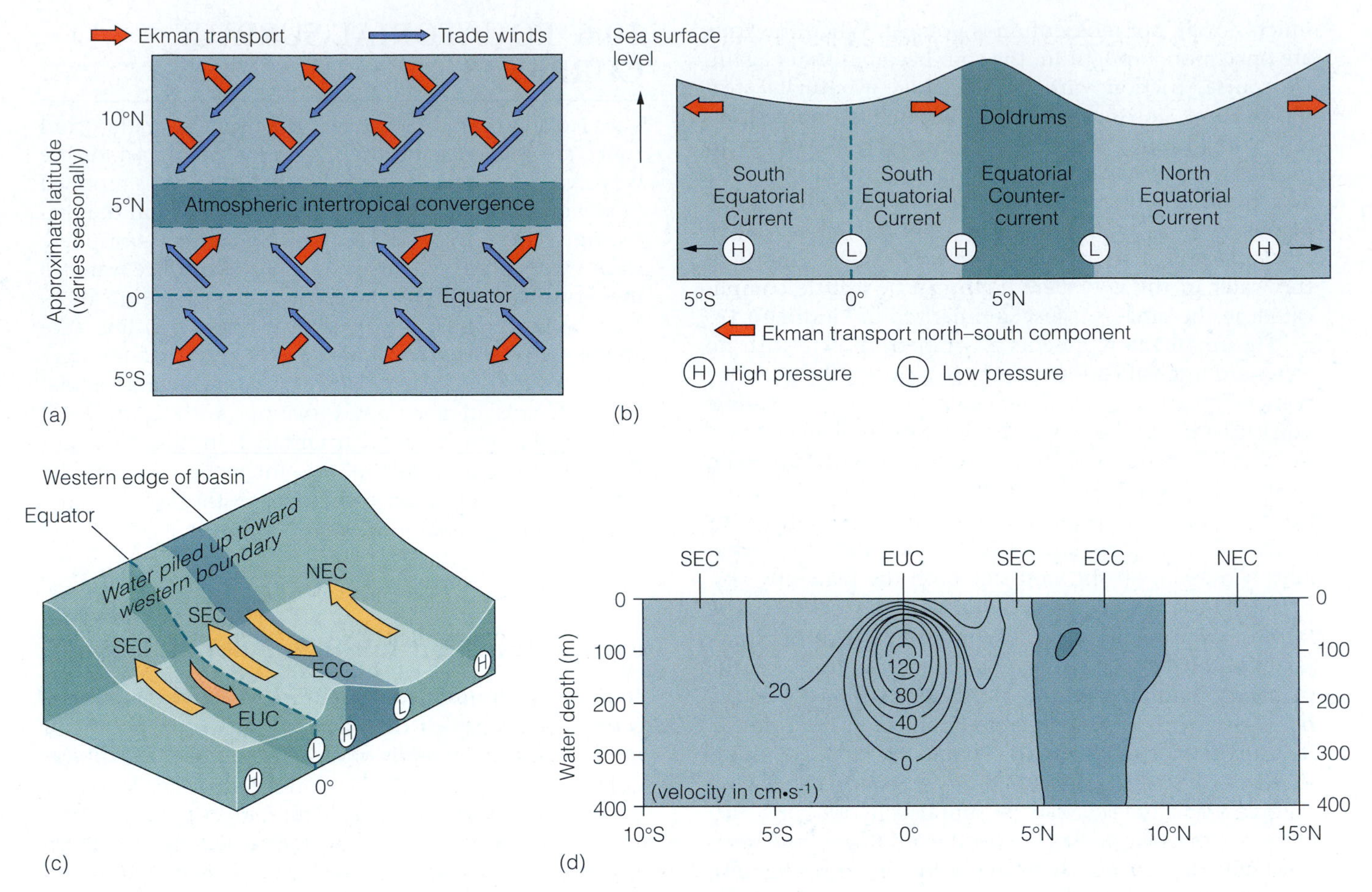

FIGURE 10.13 Currents of the Pacific Ocean equatorial region. (a) The intertropical convergence is displaced north of the equator, especially in the winter, such that the Southern Hemisphere southeast trade winds extend north of the equator. Because Ekman transport is to the right of the wind direction in the Northern Hemisphere, the pattern of Ekman transport just north of the equator is complex. (b) This pattern of Ekman transport produces complex latitudinal variation of sea surface elevation in the equatorial region. (c) The complex sea surface elevation pattern sets up a eastward-flowing Equatorial Undercurrent (EUC) between the South Equatorial Current (SEC) and the Equatorial Countercurrent (ECC). (d) The Equatorial Undercurrent is a fast current—much faster than the North and South Equatorial currents (NEC and SEC) or the Equatorial Countercurrent. The EUC flows between the surface and a depth of about 300 m, and its maximum velocity occurs at about 100 m.

Geostrophic Currents in the Equatorial Region

As a result of Ekman transport processes, a small "hill" of water is present in the Northern Hemisphere, and its **crest** is at the intersection of the Doldrums with the southeast trade wind zone (**Fig. 10.13b**). In the pressure gradient south of the crest, geostrophic flow is to the west, in the same direction as the geostrophic flow immediately south of the equator. North of the crest, in the pressure gradient within the Doldrums, geostrophic flow is to the east. This is the Equatorial Countercurrent. The arrangement of sea surface slope, pressure gradients, and geostrophic currents in the equatorial region is shown in **Figure 10.13c**. The characteristics of each of the currents are described in **Table 10.2**.

The equatorial currents of the subtropical gyres and the westward component of Ekman transport in the trade wind zones move very large volumes of surface layer water toward the west. Consequently, the sea surface slopes upward toward the western boundary. Much of the water that flows westward is either diverted north or south in the subtropical gyres or transported back toward the east in the Equatorial Countercurrents. However, water also flows back to the east in a current called the "Equatorial Undercurrent."

Equatorial Undercurrents flow at the equator, are about 400 km wide, and have a vertical thickness of about 200 m (**Fig. 10.13d**). In the Pacific, the core of the current rises from a depth of about 200 m in the west to about 40 m in the east. In the Atlantic, the core is at about 100 m depth. Equatorial Undercurrents are very

TABLE 10.2 Characteristics of Equatorial Surface Layer Currents

Current	*Sea Surface Slope*	*Pressure Gradient*	*Hemisphere*	*Coriolis Deflection*	*Direction Flow*
South Equatorial Current (SEC)[a]	Down to north	Decreases to north	Southern	To left	Westward
South Equatorial Current (SEC)[b]	Up to north	Increases to north	Northern	To right	Westward
North Equatorial Current (NEC)	Up to north	Increases to north	Northern	To right	Westward
Equatorial Countercurrent (ECC)	Down to north	Decreases to north	Northern	To right	Eastward
Equatorial Undercurrent (EUC)	Down to east	Decreases to east	Equator	None	Eastward

[a]Component of South Equatorial Current south of the equator (Figure 10.13).
[b]Component of South Equatorial Current north of the equator (Figure 10.13).

swift, with speeds comparable to that of the Gulf Stream. The Equatorial Undercurrent flows on the west-to-east pressure gradient created by the "piling up" of surface water toward the western boundary. However, the fact that it flows directly down the pressure gradient distinguishes it from other geostrophic currents, which all flow parallel to the contours of constant pressure (across the pressure gradient) to balance the pressure gradient and the Coriolis effect (**CC12, CC13**).

The Equatorial Undercurrent can flow down the pressure gradient because it flows precisely west to east at the equator, where the Coriolis deflection is zero. If the current deviates slightly to the north, the Coriolis deflection (to the right in the Northern Hemisphere), although weak, turns the current back to the south. If it deviates slightly to the south, the Coriolis deflection (to the left in the Southern Hemisphere) turns it back to the north. Thus, the Equatorial Undercurrent flows directly across the oceans for thousands of kilometers without deviating from its west-to-east direction. The Pacific Ocean Equatorial Undercurrent is particularly important for the critical role it plays in **El Niño** (Chap. 9).

10.7 HIGH-LATITUDE SURFACE CURRENTS

In the North Atlantic and North Pacific, secondary gyres occur at higher latitudes than the subtropical gyres. They are called "subpolar gyres" and rotate in the direction opposite that of the adjacent subtropical gyres. Subpolar gyres are wind-driven in the same way as subtropical gyres. However, the prevailing winds of the region, the polar easterlies, are much more variable than trade winds. Because of this wind variability and the complex shapes of the ocean basins in polar regions, the subpolar gyres are more complex and variable than the subtropical gyres.

In the Southern Hemisphere, subpolar surface currents are different from their Northern Hemisphere counterparts. The primary reason is the lack of continents in the Southern Hemisphere. Because of the wide connection between the oceans, Southern Hemisphere subpolar currents flow all the way around the Antarctic continent (**Fig. 10.2**). At high latitudes, a geostrophic current, the East Wind Drift, flows westward around Antarctica. It flows in response to the pressure gradient caused by elevation of the sea surface near the Antarctic continent due to southerly Ekman transport in the polar easterly wind belt. Farther from Antarctica, an eastward current, the Antarctic Circumpolar Current (or West Wind Drift), flows geostrophically around Antarctica under a sea surface that slopes up to the north as a result of the Ekman transport of the westerlies (**Fig. 10.2**). The Antarctic Circumpolar Current is combined with, and a part of, the eastward-moving currents of the Southern Hemisphere subtropical gyres in each ocean.

10.8 UPWELLING AND DOWNWELLING

Although wind-driven surface water motions are primarily horizontal, winds can also cause vertical water movements. Vertical movements occur because wind-driven Ekman transport drives a layer of surface water up to about 100 m deep across the underlying water layers. Winds moving surface water away from an area cause a

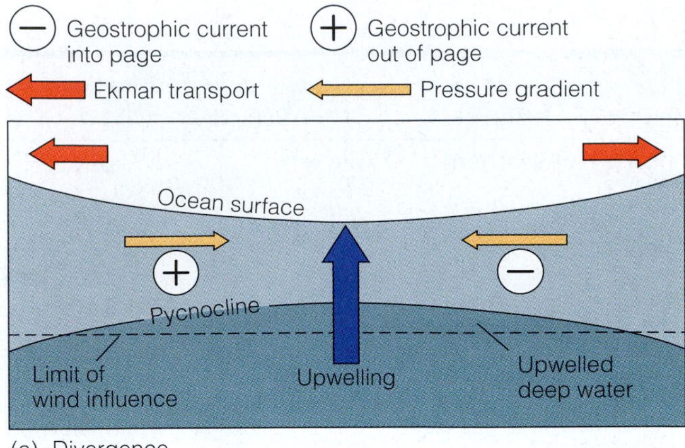

(a) Divergence

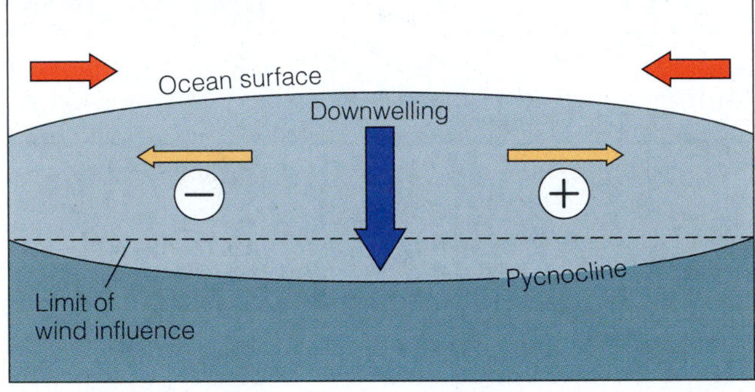

(b) Convergence

FIGURE 10.14 Divergences and convergences are areas of upwelling and downwelling in the open oceans. (a) At a divergence, the surface layer is thinned and the pycnocline depth reduced. If wind mixing extends below the pycnocline depth, cold, nutrient-rich, deep water is upwelled and mixed into the surface layer. (b) At a convergence, the surface layer is thickened, and the pycnocline is driven deeper. There is no mixing of cold, nutrient-rich subpycnocline water into surface layers, even when winds are very strong.

divergence at which the surface water is replaced by water upwelling from below (**Fig. 10.14a**). Upwelling is important because colder water upwelled from below the permanent pycnocline has high concentrations of **nutrients**, such as nitrogen and phosphorus compounds. The nutrients are needed to fertilize **phytoplankton**, the microscopic plants that grow only in near-surface waters and that are the principal source of food supporting all animal life in the oceans. Most surface waters are deficient in nutrients and cannot sustain phytoplankton growth unless nutrients are resupplied by upwelling or recycling (Chap. 14).

At a convergence, winds moving surface waters toward an area elevate the sea surface and create a high-pressure region in the subsurface water, causing downwelling (**Fig. 10.14b**). Surface water is transported downward at the convergence and then outward below the wind-driven layer under the influence of the horizontal pressure gradient. Downwelling thickens the layer of warmer surface water above the pycnocline and tends to isolate cold subpycnocline water, preventing it from mixing upward into the surface layers. Downwelling also occurs in areas where the surface water density is increased by cooling or evaporation.

Locations of Upwelling and Downwelling

Divergences are upwelling areas where **primary productivity** is high because of the supply of nutrients from below. Convergences are downwelling areas where productivity is poor. The principal open-ocean divergences include a band that parallels the equator between the North and South Equatorial Currents and a band around Antarctica between the Antarctic Circumpolar Current and the eastward coastal flow around the continent (**Fig. 10.2**). In the equatorial region, upwelling is inhibited by the flow of warm surface water from west to east in the Equatorial countercurrents. Consequently, persistent upwelling of nutrient-rich, cold subpycnocline water at the equatorial divergence is limited primarily to the east side of the Pacific Ocean off the Peruvian coast (**Fig. 9.18**). Upwelling in this region can be inhibited by the movement of warm surface water from west to east during El Niño (**Fig. 9.14b**). The principal open-ocean convergences are in the center of each of the subtropical surface current gyres (**Fig. 10.2**).

Upwelling and downwelling can occur when winds blow parallel to the coast and cause Ekman transport of the surface layer offshore or onshore. Surface water transported offshore is replaced by deeper water that moves inshore and is upwelled (**Fig. 10.15a**). Surface water transported toward shore displaces near-shore surface water downward and forces the deeper water to move offshore (**Fig. 10.15b**). These processes are particularly important in areas where there is a shallow pycnocline, with warm, nutrient-depleted water at the surface and colder, nutrient-rich water below. In such areas, offshore transport causes **coastal upwelling** that supplies large quantities of nutrients to support phytoplankton growth (Chap. 15).

Boundary Currents and Upwelling or Downwelling

The characteristics of ocean gyre boundary currents strongly influence the occurrence and persistence of coastal upwelling off a particular coast. Western boundary currents are deep and generally compressed against the continental shelf edge. Hence, a deep layer of warm surface water generally lies over the cold, nutrient-rich subpycnocline water and prevents it from extending onto the continental shelf (**Fig. 10.15b**). Because coastal winds normally move aside a surface layer only a few tens of

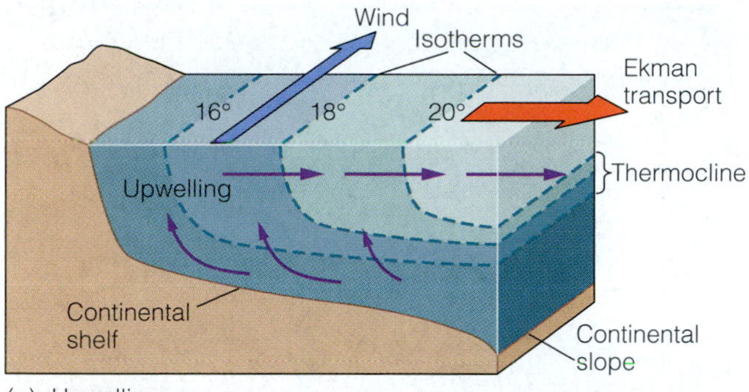

(a) Upwelling

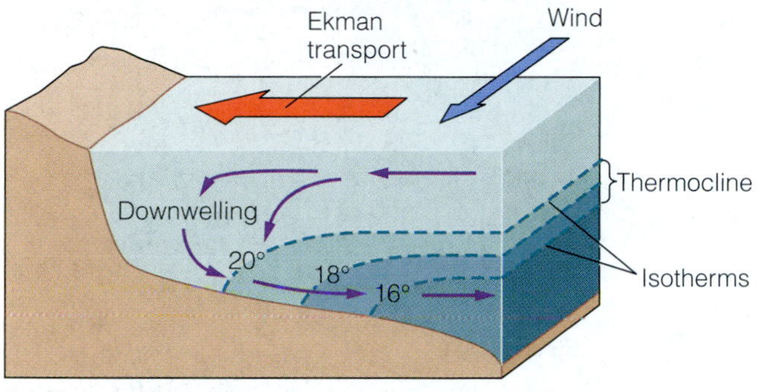

(b) Downwelling

Water mass movements →

FIGURE 10.15 Coastal upwelling and downwelling. (a) When winds produce Ekman transport of the surface water layer in an offshore direction, the pycnocline (usually a thermocline) is raised, sometimes to the surface, and cold, nutrient-rich, deep water is upwelled into the surface layer. (b) When winds produce Ekman transport of the surface water layer in an onshore direction, the pycnocline (again usually a thermocline) is depressed, shelf surface water is downwelled, and cold, nutrient-rich, deep water is forced deeper and off the shelf, so it cannot enter the surface layer, even if winds are strong.

meters deep, offshore Ekman transport in western boundary current regions leads to upwelling of warm, nutrient-poor water from within the deep surface layer (**Fig. 10.15b**). Thus, upwelling of nutrient-rich water is rare in western boundary current regions.

Eastern boundary currents are relatively shallow and wide, and they extend onto the continental shelf. Cold, nutrient-rich, deep water can migrate onto the shelf as a bottom layer below the shallow, warm surface layer (**Fig. 10.15a**). Only moderate coastal winds with offshore Ekman transport are needed to cause cold, nutrient-rich waters to upwell to the surface. Consequently, coastal upwelling is more frequent, widespread, and persistent on the eastern boundaries of the oceans (west coasts of continents) than on the western boundaries of the oceans (east coasts of continents).

10.9 COASTAL CURRENTS

Coastal currents are independent of and more variable than the adjacent oceanic gyre boundary currents. The zone in which coastal currents flow varies in width along different coasts. For example, the coastal current zone is wide off the Mid-Atlantic coast of North America because the continental shelf is wide, and the western boundary current (Gulf Stream) flows in deeper water off the edge of the shelf. In contrast, the coastal current zone off the Pacific coast of North America is narrow because the continental shelf is narrow, and the relatively shallow eastern boundary current flows over the narrow shelf almost to the coastline. Variable winds, tides, freshwater inflow from rivers, friction with the seafloor, and steering by coastline irregularities, such as capes, affect currents more in coastal waters than in the open ocean.

Coastal currents generally flow parallel to the coastline in a direction determined by the winds (and by Ekman transport). Coastal winds are more variable in speed and direction than are trade winds and other global

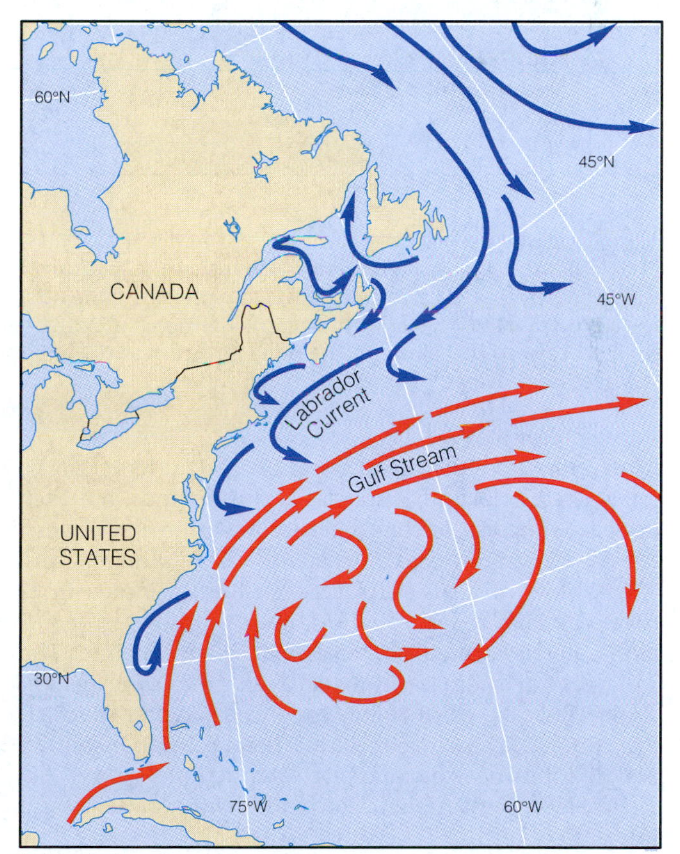

Conic projection

FIGURE 10.16 Off the Atlantic coast of North America, the coastal currents are cold-water currents that flow to the south on the continental shelf inshore from the northward-flowing Gulf Stream.

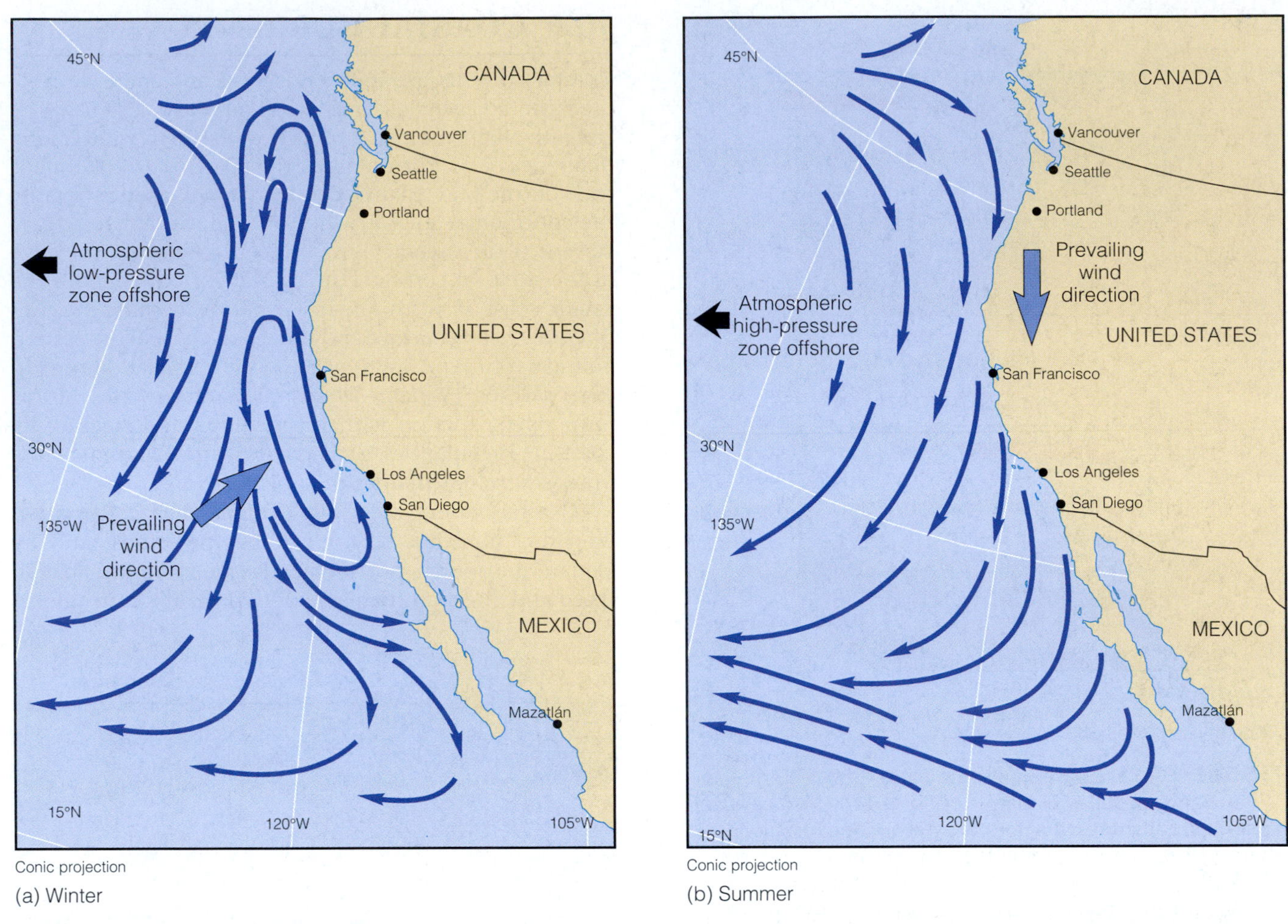

FIGURE 10.17 Coastal currents reverse seasonally off the Pacific coast of North America. (a) They flow to the north during winter, when there is an atmospheric low-pressure zone offshore and the coastal winds blow mainly from the southwest. (b) In the summer, when there is atmospheric high pressure offshore and coastal winds are mainly from the north, the coastal currents flow to the south, in the same direction as the offshore southward current that is part of the subtropical gyre in the North Pacific Ocean.

winds that drive the ocean gyres. Coastal currents can be established quickly in response to storm winds, but they can also disappear or change direction within a few hours.

The strongest coastal currents occur when strong winds blow in areas with large freshwater inputs from rivers. In such areas, a shallow, low-salinity layer of surface water is formed over a very steep pycnocline. The surface layer slides easily over layers below the pycnocline. Thus, the wind energy is concentrated in the shallow, surface-layer current, rather than being transmitted and distributed throughout a greater depth.

Because they are directed by local wind patterns and interaction with the coastline, coastal currents may flow in the opposite direction from the adjacent ocean gyre boundary currents. For example, coastal currents off the Atlantic coast of North America generally flow to the southwest from Labrador to Cape Hatteras and along the Carolina, Georgia, and Florida coasts (**Fig. 10.16**) in the direction opposite that of the adjacent Gulf Stream.

The boundary between the warm Gulf Stream and colder coastal water, called a **front**, can be clearly discerned as a distinct difference in water color. The nearshore water is greenish or brownish, whereas Gulf Stream water is a clear, deep blue. The color difference is due to the greatly reduced concentrations of suspended particles and phytoplankton in Gulf Stream water (Chap. 7).

Off the west coast of North America, the boundary between the water of the California Current (which is part of the North Pacific Ocean subtropical gyre) and the coastal water is not well defined. The main reason is that the California Current is a weak, diffuse eastern boundary current. During winter and spring, a weak and variable coastal current, the Davidson Current, flows north in the opposite direction from the adjacent boundary current. The Davidson Current is driven by the predominant winds of winter and spring that blow from the southwest. These winds produce Ekman transport of water onshore and a geostrophic flow to the north on the resulting sea surface slope (**Fig. 10.17a**). In summer and fall, prevailing winds change to blow from the north. They cause Ekman transport offshore, a generally southward-flowing coastal current, and upwelling (**Figs. 10.15, 10.17b**). The summer and fall upwelling is responsible for the very high primary productivity and abundant sea life in coastal waters off the west coast of North America.

10.10 EDDIES

Chapter 9 describes how winds are arranged in global patterns that determine climate. It also explains that winds are highly variable in any given location. The local variations are part of what we know as **weather**, and we readily associate them with the swirling patterns of clouds seen on television weather forecasts. The global ocean current systems are analogous to the climatic winds, and the oceans have their own "weather," which includes variable swirling motions and meandering fronts, just like atmospheric weather. In the oceans, the swirling motions are called **eddies**.

The principal difference between atmospheric weather and ocean current "weather" is the scale of the eddy motions. Ocean currents move much more slowly than atmospheric winds. Consequently, the Coriolis effect tends to make ocean currents flow in curving paths that have a much smaller radius than those followed by atmospheric winds (**CC12**). In other words, ocean eddies are much smaller than atmospheric eddies. Because ocean eddies are smaller, they are probably more numerous than atmospheric eddies.

Satellites can photograph some atmospheric eddies that contribute to the weather. However, such **synoptic** measurements cannot be made so easily within the body of the oceans. Therefore, we know relatively little about ocean current variability. Most of what we do know about such variability comes from recent satellite observations of surface currents.

Satellite Observations of Eddies

Satellite infrared sensors can readily detect ocean surface temperature by measuring the sea surface heat radiation. Satellite optical sensors tuned to different **wavelengths** of light can detect ocean color by measuring **backscattered** sunlight. Such satellite observations are most effective for observing ocean surface currents where temperature and/or color differ markedly between water masses. In most areas, the differences are very small. The easiest surface currents to observe by satellite are western boundary currents because the water they carry is warmer than the adjacent coastal waters. Areas near river discharges are also good for satellite observations of currents because the suspended particles in river outflow alter ocean color.

CRITICAL THINKING QUESTIONS

10.8 Global climate may change in the next few decades, with the average temperature of the ocean surface waters increasing by as much as several degrees.
(a) Hypothesize what might happen to the thermohaline circulation of the oceans as a result.
(b) Speculate on what would be the effects on the intensity, geographic distribution, and frequency of upwelling of water from below the permanent thermocline. Explain your reasons for these speculative changes.

10.9 The carbonate compensation depth (CCD) is explained in Chapter 8. From what you have learned about ocean circulation, what do you think would be likely to happen to the atmospheric concentrations of carbon dioxide if the CCD were to migrate to shallower depths? How long would you expect it to take for these changes to occur? Explain why.

10.10 If the hypothesized global warming does indeed occur, the Arctic ice pack may melt completely.
(a) How would this affect the bottom water mass of the Arctic Ocean?
(b) How might it also affect the characteristics of bottom water masses in the other oceans?
(c) What other changes in ocean circulation might be caused?
(d) Would these changes be significant to your grandchildren, or would they take thousands of years to occur? Explain the reasons for your answer.

Even in areas where satellite observations are most effective, color or temperature variations of the sea surface are so small that they must be computer-enhanced to be seen by researchers. Computer enhancement of satellite images generally assigns bright false colors to areas that are slightly different in temperature or in true color, thus making the demarcations between currents or water masses more prominent. **Figure 10.18** is a color-enhanced satellite image of the Gulf Stream region of the Atlantic Ocean.

Gulf Stream Rings

The image in **Figure 10.18** shows the warm Gulf Stream flowing northward from the tip of Florida along the edge of the continental shelf. The front between this water and the colder coastal water is very sharp. A similar front can be seen between Gulf Stream water and colder Sargasso Sea water in the interior of the North Atlantic subtropical gyre. Great complexity is apparent along these fronts. We can see numerous secondary fronts and **meanders** of the Gulf Stream. We also see isolated, almost circular areas of warm water on the coastal-water side of the Gulf Stream

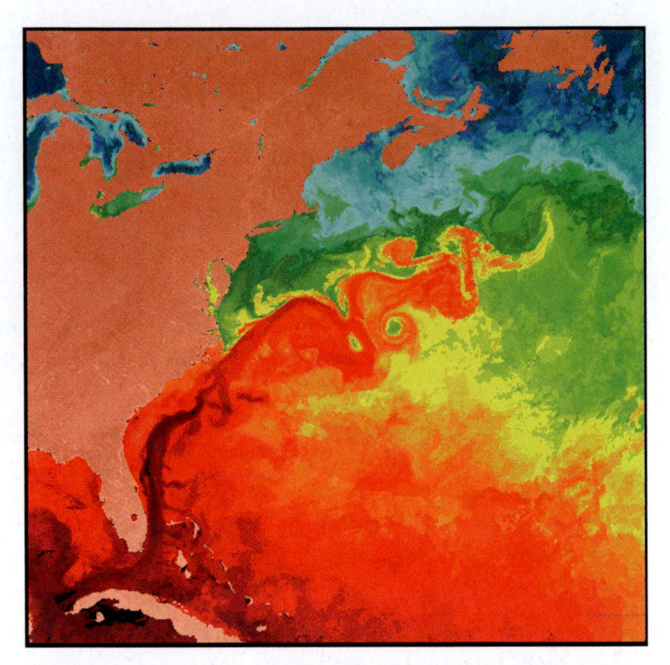

FIGURE 10.18 This image of the northwestern Atlantic Ocean shows data obtained from a satellite-mounted instrument called the Coastal Zone Color Scanner, which measures the intensity of light reflected or backscattered by the Earth's surface. The data are from the infrared wavelength band, in which the intensity of the light received by the scanner from the ocean surface increases with increasing surface water temperature. Intensities of light in this wavelength band are depicted by the false colors in this image. Shades of red denote the areas of highest temperature, and progressively lower temperatures are depicted by orange, yellow, green, and blue shades. The image shows the Gulf Stream separating the coastal water mass from the Sargasso Sea water mass. The complex meanders of the Gulf Stream and both cold-core and warm-core rings are easy to see.

and similar isolated areas of cold water on the Sargasso Sea side.

Gulf Stream meanders are continuously forming and changing shape and location. Meanders travel slowly northward along with the Gulf Stream and can become larger as they move. If a meander becomes tight enough, it can break off the Gulf Stream entirely and form a spinning ring of water. If the meander is to the coastal-water side of the Gulf Stream, a warm-core ring that spins clockwise is formed and isolated (**Fig. 10.19**). If the meander is to the Sargasso Sea side of the Gulf Stream, a cold-core ring that spins counterclockwise is formed as colder coastal water is pinched off inside the meander (**Fig. 10.19**).

The rings are from 100 to 300 km across and can extend to the seafloor. They are encircled by swiftly flowing currents that move at approximately 90 cm·s^{-1}. Both cold- and warm-core rings drift southward and slowly disappear as the core water temperature changes to match that of the surrounding water. Many rings are reattached to, and reabsorbed by, the Gulf Stream. Cold-core rings can maintain their integrity for up to several months, but warm-core rings do not generally last as long, because of the shallower water and narrow shelf.

Gulf Stream rings transport heat, dissolved substances (including nutrients), and marine organisms that are weak swimmers or that prefer the warmer or colder water. Certain locations within the complex frontal system, with its associated warm- and cold-water rings, are better than others for certain fish **species**. Hence, the location of the Gulf Stream and its rings is important to fishers. This information is also important to ships if they are to minimize fuel costs by taking advantage of, or not fighting, ocean currents. The position of the Gulf Stream front and its rings is now monitored and forecasted. Many oceangoing fishing and other vessels obtain frequent up-to-date satellite images of the Gulf Stream region while at sea, just as they receive weather forecast maps.

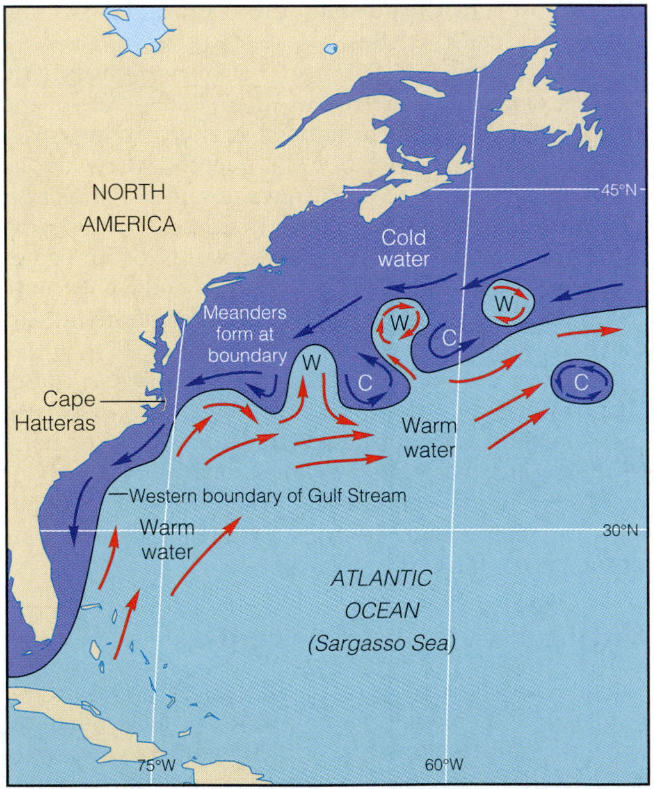

FIGURE 10.19 Gulf Stream rings. As the Gulf Stream passes north of Cape Hatteras and no longer flows along the edge of the continental shelf, meanders form. The meanders become more extreme as they move north and may eventually be pinched off to form warm- or cold-core rings. The rings generally drift south and dissipate or are reabsorbed by the Gulf Stream. Meander and ring formation is variable and complex, as shown in Figure 10.18.

Swiftly flowing meanders and rings are also present in other western boundary currents, such as the Kuroshio Current.

Mesoscale Eddies

Eddies somewhat similar to Gulf Stream rings are present throughout the oceans. They are called "mesoscale eddies." Such eddies are generally less well defined than Gulf Stream rings, and their current velocities are lower. Mesoscale eddies, which in some cases extend all the way to the deep-sea floor, are the ocean equivalent of atmospheric high- and low-pressure zones. They range in diameter from 25 to 200 km, drift a few kilometers per day, and have rotating currents of approximately 10 $cm \cdot s^{-1}$, only about 10% of the speed of Gulf Stream ring currents. In comparison, atmospheric depressions are about 1000 km across, travel about 1000 km per day, and sustain rotating winds of up to approximately 20 $m \cdot s^{-1}$.

Like Gulf Stream rings, mesoscale eddies transport and distribute heat and dissolved substances within the oceans. The distributions of some dissolved constituents are important to ocean life and to the processes that may lead to global climate change as a result of the **greenhouse effect**. Consequently, mesoscale eddies are the subject of considerable ongoing research efforts. Because mesoscale eddies are smaller and slower-moving than atmospheric eddies, and because ocean currents are difficult to measure, the tracking, understanding, and forecasting of ocean "weather," particularly that below the surface layers, will always be much more difficult than the observation and forecasting of atmospheric weather.

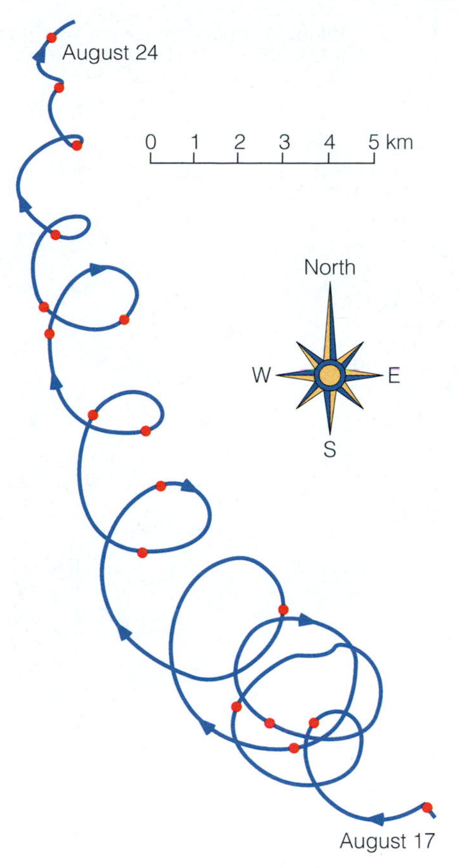

FIGURE 10.20 Inertial current in the Baltic Sea. The inertial motion is superimposed on a mean drift toward the north-northwest. The red dots indicate intervals of 12 h.

10.11 INERTIAL CURRENTS

Once established by the wind, currents will continue to flow even after the wind stops, until the momentum built up by the wind is dissipated by friction. If there is no other wind or pressure gradient force on the water after the wind stops, the water will flow in a curved path under the influence of the Coriolis effect. If the current is limited to a narrow range of latitude within which the Coriolis effect does not change significantly in magnitude (**CC12**), the water flows in circles. The circular currents are called "inertial currents." Although inertial currents flow even when there is no wind, they are ultimately wind-driven currents because winds are responsible for starting the water in motion and thus provide the momentum needed to drive the currents.

Inertial currents occur frequently throughout the oceans, but the water rarely moves in perfect circular patterns. In **Figure 10.20**, for example, a circular inertial current is superimposed on a mean drift to the north-northwest. The inertial current and mean wind-driven current were generated by winds blowing at different times, different velocities, and probably different locations. Studies of ocean currents are always complicated by the simultaneous presence of both wind-driven and inertial currents that are caused by different wind events but that are combined to form the current that is being measured at any given place and time. In fact, measured currents are even more complex because geostrophic currents, tidal currents (Chap. 12), and wave-induced **longshore currents** (Chap. 11) are also combined with inertial and other wind-driven currents.

10.12 LANGMUIR CIRCULATION

In addition to currents generated by Ekman transport, winds cause a completely different type of near-surface motion in which the upper few meters of water flow in corkscrewlike motions. These motions, called **Langmuir circulation**, are generated when the wind blows across the sea surface at speeds in excess of a few kilometers per hour.

In Langmuir circulation, water travels in helical vortices aligned in the direction of the wind (**Fig. 10.21**). The

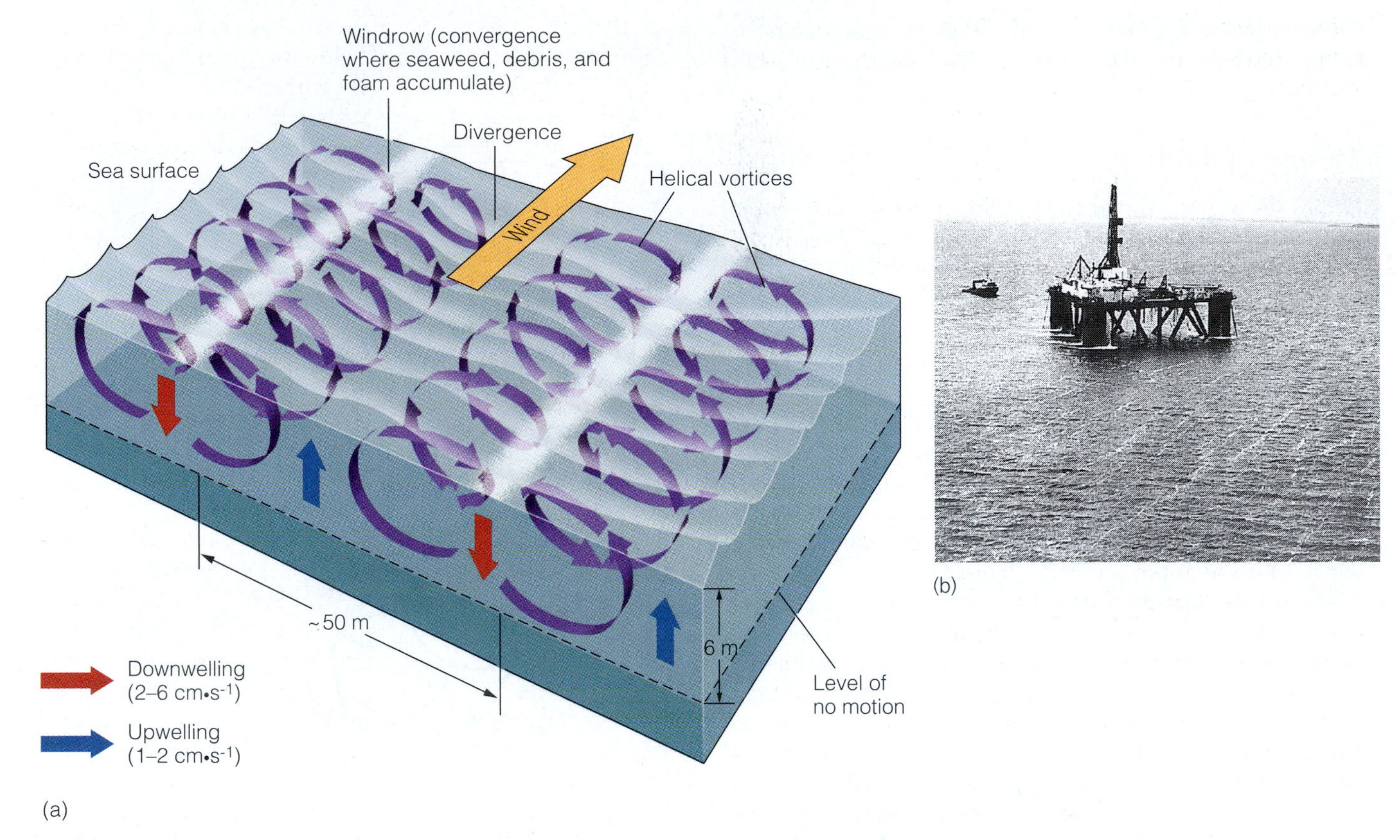

FIGURE 10.21 Langmuir circulation. (a) Moderate but persistent winds produce corkscrewlike motion of the upper few meters of water. Long linear cells of this helical motion line up alongside each other in the direction of the wind. Adjacent cells rotate in opposite directions, and alternating convergences and divergences are formed between adjacent cells. (b) Convergences between Langmuir circulation cells are visible in this photo as the windrows of foam that collect at the convergences when the wind is blowing strongly.

helical vortices are stacked side by side and rotate in alternating directions like convection cells (**CC3**). Within each cell, water moves across the surface, sinks at a line of downwelling between two cells, moves back under the surface, and rises at a line of upwelling between two cells, while still moving in the wind direction (**Fig. 10.21a**).

The width and depth of the Langmuir cells are determined by the wind speed and may be limited by shallow water or a shallow pycnocline. Typical cells are a few meters deep and a few meters to about 30 m wide. The top of the downwelling zone between two Langmuir cells is a sea surface convergence. Any oil, debris, or foam that is floating on the surface when Langmuir circulation is established is transported by the circulation to a convergence. The convergence zones are readily observable in lakes or oceans as the linear or nearly linear windrows of floating material that form when strong winds blow (**Fig. 10.21b**). The distance between two of the parallel windrows is equal to twice the width of each Langmuir cell.

The mechanism of Langmuir cell formation is not yet fully understood. However, together with wave action, this type of circulation is known to be very important in mixing the upper layers of water—a key process in the transport of dissolved gases and heat between the ocean and atmosphere (Chap. 9).

10.13 THERMOHALINE CIRCULATION

Wind-driven currents dominate water motions in the upper layer of the oceans above the pycnocline. Below the pycnocline, currents are driven by density differences between water masses. Density differences cause water masses to sink or rise to the appropriate density level (**CC1**). In areas where water masses sink or rise, the density distribution with depth, and hence the pressure distribution, is different from that in surrounding areas. Thus, horizontal pressure gradients are formed. Once a water mass has moved vertically to reach its density equilibrium level (**CC1**), it flows horizontally in response to the pressure gradient. Because seawater density is determined primarily by temperature and salinity, these water movements are called "thermohaline circulation."

Thermohaline circulation is difficult to study, and most of our knowledge of it comes from studies of density and other characteristics of the deep-ocean water masses. Much of our understanding of thermohaline circulation comes from modeling studies (**CC10**), but the models are themselves limited by the relatively small amount of data that is available to calibrate and test them.

Depth Distribution of Temperature and Salinity

Ocean waters are arranged in a series of horizontal layers of increasing density from the surface to the ocean floor (**CC1**). Throughout most of the oceans, the layers form three principal depth zones: the surface zone (usually called the **mixed layer**), the pycnocline zone, and the deep zone (**Fig. 10.4**). The surface zone, which is approximately 100 m thick, is usually of uniform or nearly uniform density. However, a seasonal pycnocline is present in the surface zone in many mid-latitude areas (**Fig. 10.22**). Water in the surface layer above the pycnocline, or above the seasonal pycnocline if it is present, is continuously mixed or stirred by winds. In the permanent pycnocline zone, water density increases rapidly with depth. This zone extends from the bottom of the surface layer, where water temperature is approximately 10°C, to a depth that varies with location between about 500 and 1000 m. In the deep layer below the permanent pycnocline, density increases slowly with depth.

The marked density increase with depth in the pycnocline zone is due to decreasing temperature, increasing salinity, or a combination of temperature and salinity changes (**Fig. 10.4**, **CC6**). Where temperature changes cause density to change with depth, the pycnocline is also a **thermocline** (**Fig. 10.4b**). Where salinity changes cause density to change with depth, the pycnocline is also a **halocline** (**Fig. 10.4c**). Haloclines are more important in nearshore waters where freshwater **runoff** produces surface waters of low salinity.

Figures 10.23 and 10.24 show the vertical distributions of temperature and salinity, respectively, in the centers of the Atlantic, Pacific, and Indian Oceans. Throughout much of the area between 45°N and 45°S, surface water salinity is actually higher than that of deep waters (**CC6**). The reason is that evaporation exceeds precipitation throughout most of the mid-latitude and subtropical-latitude open oceans (**Fig. 9.16**). If the water temperature were uniform with depth, the surface waters would be denser than those below and would sink. However, the temperature difference between surface and deep waters is more than enough to offset the density difference due to salinity.

Within the mixed layer, shallow secondary thermoclines may form during summer in areas where intense solar heating warms surface water, winds are light, and vertical mixing is limited to shallow depths (**Fig. 10.22c**).

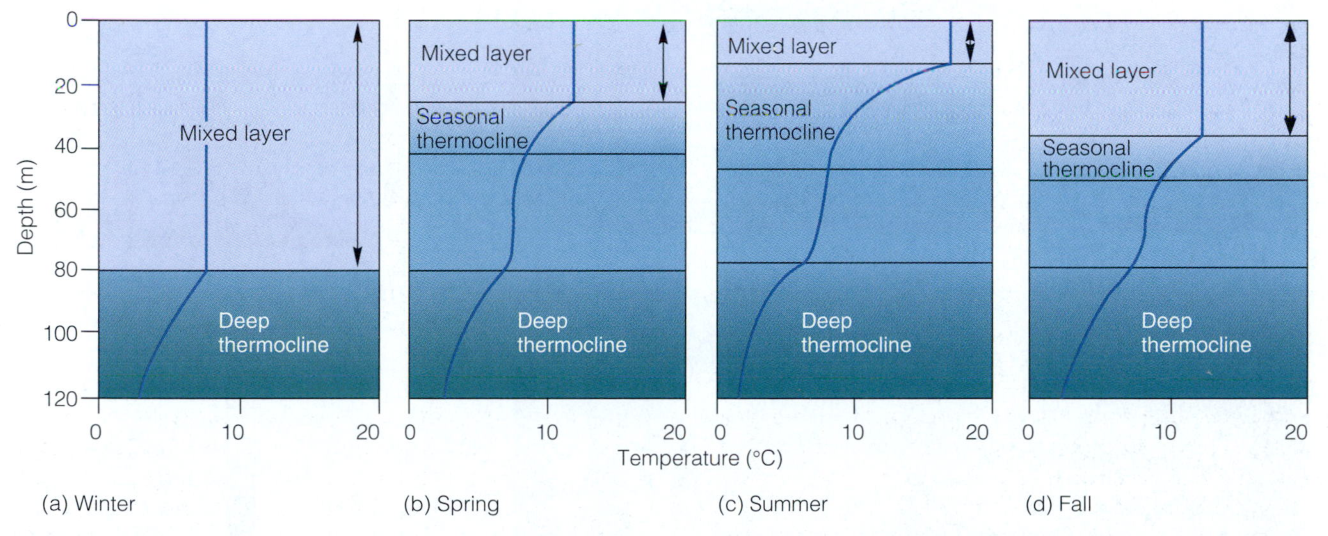

FIGURE 10.22 In many subtropical and temperate regions, a shallow summer thermocline forms and acts as a barrier to vertical mixing. (a) In winter, the water column is well mixed by the strong winter winds and by convection (due to cooling of surface water) down to the top of the permanent thermocline. (b) In spring, the surface layer is warmed and decreases in density, forming a shallow thermocline whose depth is determined by the depth of wind mixing. (c) As surface water becomes warmer and winds lighter in summer, the seasonal thermocline becomes stronger and shallower. (d) During fall, increased winds increase the depth of the mixed layer, and cooling of the surface water causes convective mixing, which eventually completely destroys the seasonal thermocline.

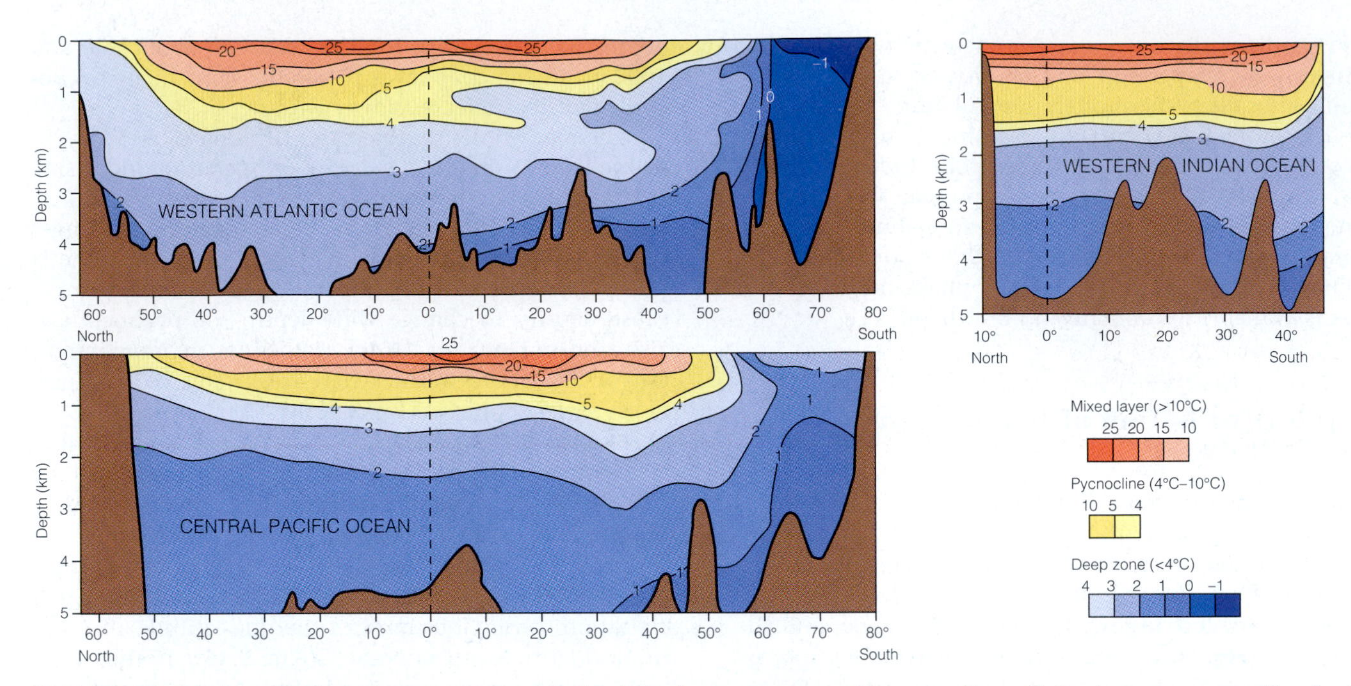

FIGURE 10.23 Distribution of temperature with depth in the open oceans. There are three distinct zones arranged by depth. In the deep zone below the pycnocline, water is uniformly cold (below 4°C). Except in high latitudes, there is a warm (>10°C) surface layer that is generally less than 1 km deep. Between the surface and deep zones, the pycnocline zone is a region in which temperature decreases rapidly with depth. Around Antarctica, in the North Atlantic Ocean, and to a lesser extent, in the North Pacific Ocean, the warm surface layer and pycnocline layer are absent because surface waters in these regions are cooled and sink to form the deep water masses.

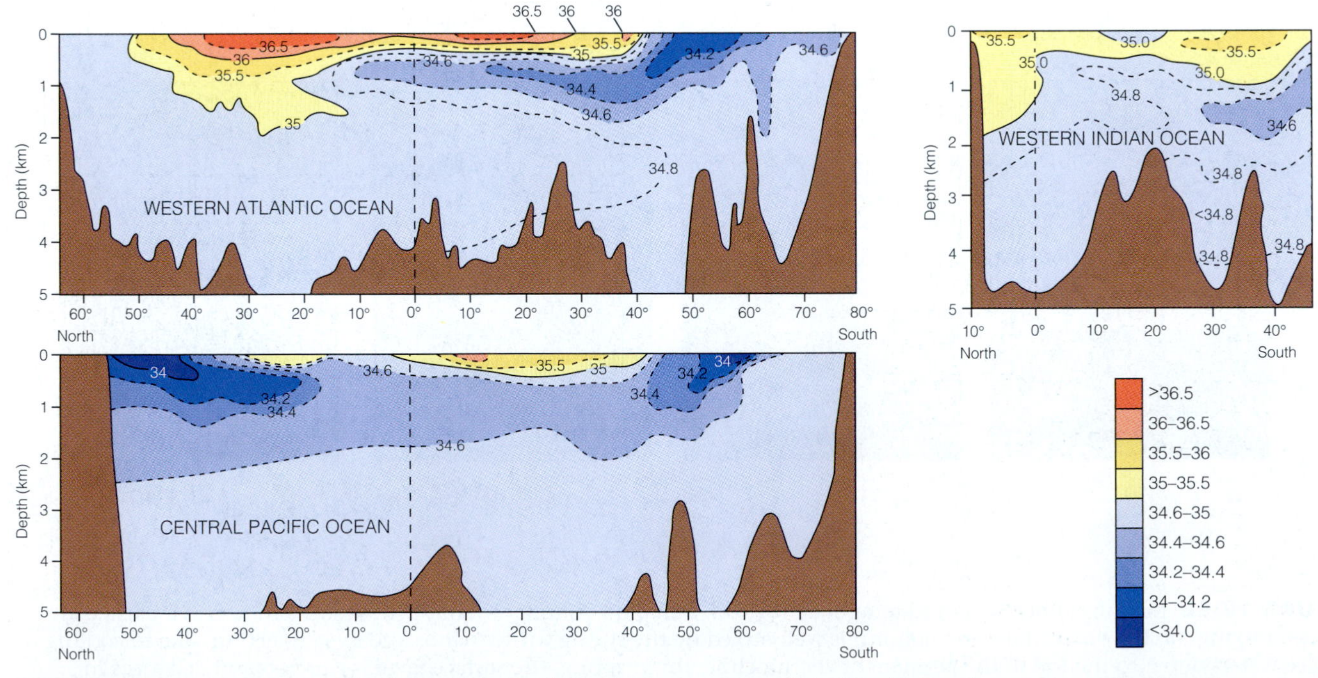

FIGURE 10.24 Distribution of salinity with depth in the deep oceans. This distribution is more complex than the distribution of temperature. The surface layer has more variable and generally slightly higher salinity than the deeper water masses have, except in high latitudes, where precipitation is high and evaporation is low. The influence of the outflow of the warm, high-salinity water mass from the Mediterranean is evident in the tongue of higher-salinity water below 1 km depth between 20°N and 40°N in the Atlantic Ocean.

Seasonal thermoclines are important to biological processes, particularly in coastal waters (Chaps. 14, 15).

The pycnocline acts as an effective barrier to vertical mixing of water masses. Where density changes rapidly with depth, large amounts of energy are needed to move parcels of water molecules either up or down across the density gradient (**CC1**). Consequently, seasonal changes in temperature and salinity caused by changes in solar intensity and rainfall rates (Chap. 9) do not generally penetrate below the mixed layer. The pycnocline also acts as a barrier to vertical mixing of dissolved gases and other chemicals. Therefore, deep water is effectively isolated from the mixed layer and atmosphere.

In high latitudes, other than in the North Pacific and Arctic Oceans, there is no pycnocline because heat lost from the oceans exceeds heat gained from solar radiation (**Fig. 9.7**). Consequently, surface waters are cooled and their density increases. If cooling is intense enough, the cooled surface water becomes more dense than the water below and sinks. Thus, in high-latitude regions without a pycnocline, deep-ocean water masses are formed as cooled surface waters sink.

The pycnocline layer and deep layer extend throughout each ocean in all but high latitudes. The thermocline layer is deeper in some areas, particularly the North Atlantic, and in some areas complicated tongues of water occur, particularly at depths within and just below the pycnocline zone. The tongues are different water masses because they have their own characteristic temperature and salinity, and they indicate the presence of deep-ocean currents. The pycnocline is shallower at the equator as a result of upwelling there.

Formation of Deep Water Masses

Water density can be increased by a decrease in temperature or by an increase in salinity (**CC6**). Salinity can be increased by evaporation or by **ice exclusion** (dissolved salts remain in solution as seawater freezes). Higher-density water sinks to the depth at which the water below has higher density and the water above has lower density (**CC1**). It then spreads laterally to form a thin layer extending out from the source area (**Fig. 10.25**).

Water within each such layer is a separate water mass within which salinity, temperature, and therefore density

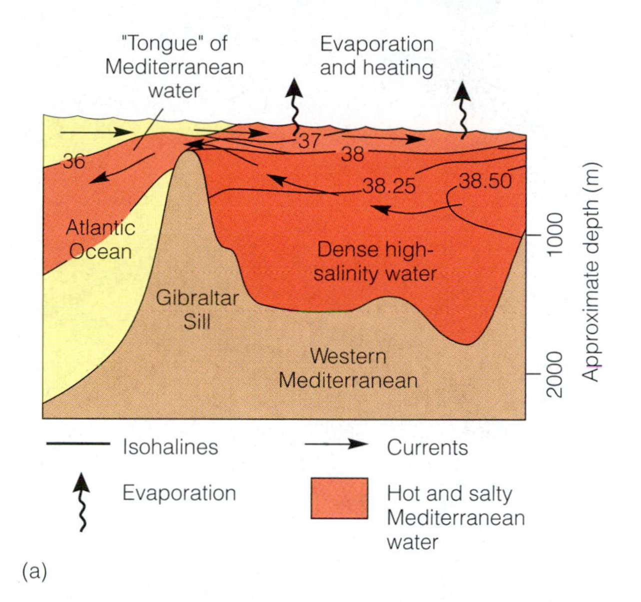

(a)

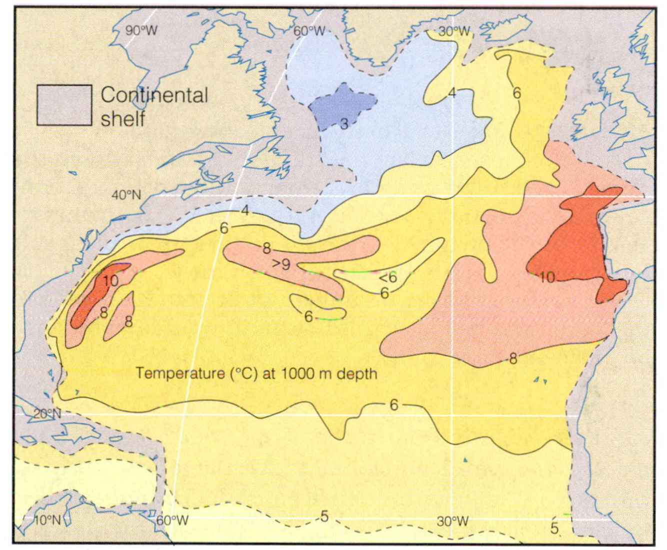

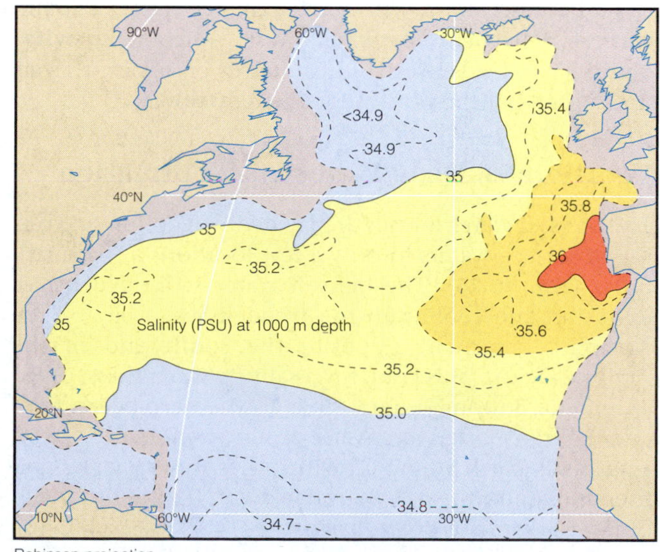

Robinson projection

(b)

FIGURE 10.25 (a) The dry, sunny climate in the Mediterranean causes warming and an excess of evaporation over precipitation. The warm, high-salinity water created there spills over the sill at the Straits of Gibraltar into the Atlantic Ocean, where it sinks to its density equilibrium level at about 1000 m and spreads out as a distinctive layer. (b) The temperature and salinity characteristics of the Mediterranean water can be detected at 1000 m throughout much of the North Atlantic Ocean.

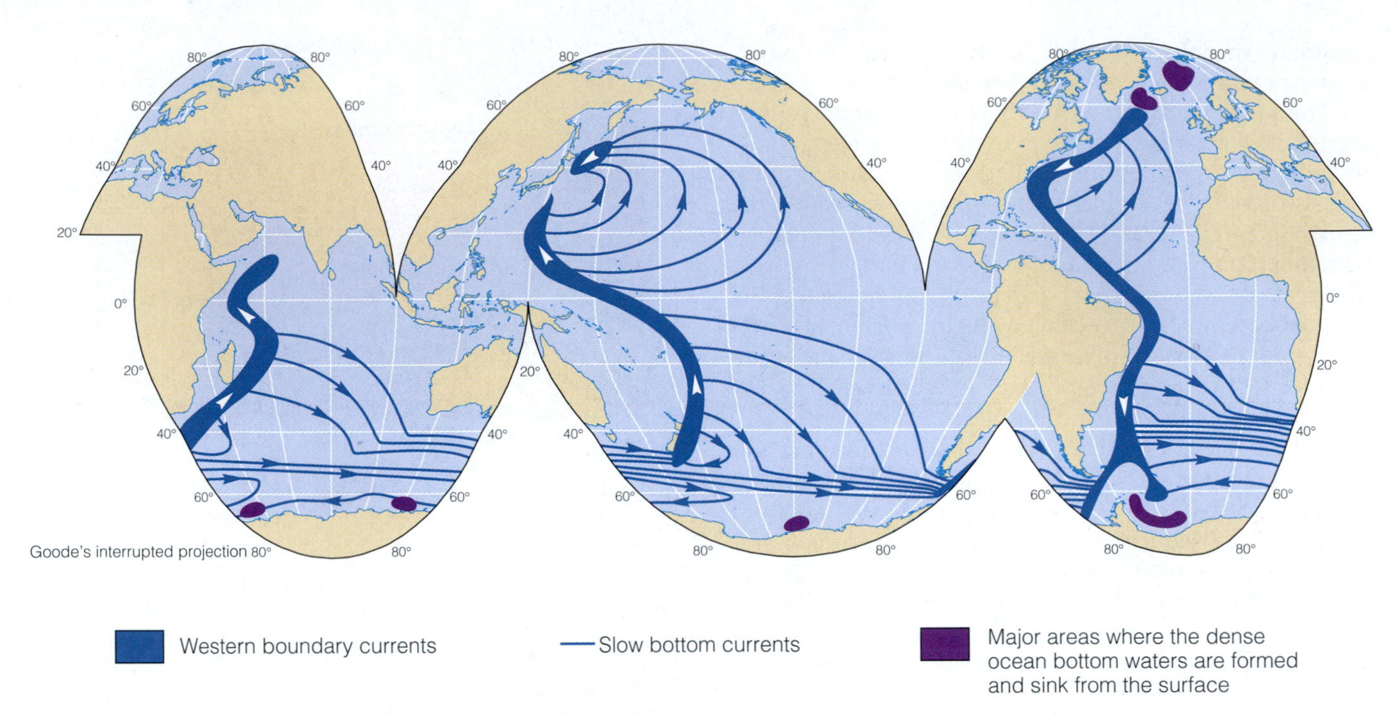

FIGURE 10.26 The densest water masses that flow along the ocean floor are created by cooling at only a very few locations. Deep water-mass circulation is still not well understood. The densest bottom water is formed in the Weddell Sea near Antarctica. It sinks, then flows around Antarctica and northward into each of the oceans. Cold, dense water is also formed near Greenland and in the Norwegian Sea. It flows south until it meets and flows over the more dense Antarctic Bottom Water (see Figure 10.27) or returns northward in the gyre circulation. Bottom currents flow in gyres in each basin and are modified by topography. Currents are intensified on the western boundaries of the oceans. The western boundary current in the Atlantic Ocean flows southward from the zone of deep water-mass formation near Greenland, and the western boundary currents in the Pacific and Indian oceans flow northward from the Southern Ocean around Antarctica. Return flows on the eastern boundary are more diffuse.

vary only slightly from those of the source water. Water masses can spread horizontally over large areas because the density differences between them inhibit vertical movement and mixing. Less energy is needed for the water mass to flow horizontally than for it to flow or mix vertically because, in a stably **stratified** water column, vertical movements of water must overcome **gravity**. Vertical mixing between water masses of adjacent layers does occur, but the process is extremely slow.

Locations of Deep Water-Mass Formation

The densest water masses are created by cooling or freezing of surface waters in only a few locations at high latitudes (**Fig. 10.26**). The densest water is formed by ice exclusion and cooling in the Weddell Sea, a bay on the Antarctic continent opposite the south end of the Atlantic Ocean. This cold, high-salinity water mass, called "Antarctic Bottom Water," sinks to the deep-ocean floor and is transported eastward around Antarctica. As it sinks, it is partially mixed with other water masses and then moves northward along the ocean bottom into each of the three major ocean basins.

No bottom water is formed at the northern end of the Indian or Pacific oceans. The Indian Ocean does not extend into high latitudes north of the equator. The Pacific Ocean is effectively separated from polar regions by the shallow **sills** between the Aleutian Islands, which mark the southern boundary of the Bering Sea, and by the shallow, narrow Bering Strait that connects the Arctic Ocean and Bering Sea. Cold, dense water formed in the Arctic Ocean cannot flow into the Pacific Ocean over the shallow areas. In the North Pacific Ocean itself, precipitation rates are high, and low surface salinity prevents the formation of deep water.

The Atlantic Ocean is partially isolated from the Arctic Ocean by shallow ridges between Scotland and Greenland, and the coldest Arctic Ocean deep water cannot enter the Atlantic Ocean readily. However, in the Norwegian Sea and particularly the Greenland Sea, intensive cooling forms North Atlantic Deep Water (NADW), which sinks and flows south in vast quantities. It is probably the most voluminous water mass, and its presence can be traced throughout much of the world's oceans. The Atlantic Ocean has water of higher average salinity than any other ocean, primarily because large amounts of high-salinity water are introduced from the Mediterranean Sea (which is in an atmospheric divergence zone where rainfall is low and evaporation exceeds precipitation; see **Fig. 10.25**). Higher salinity and temperature

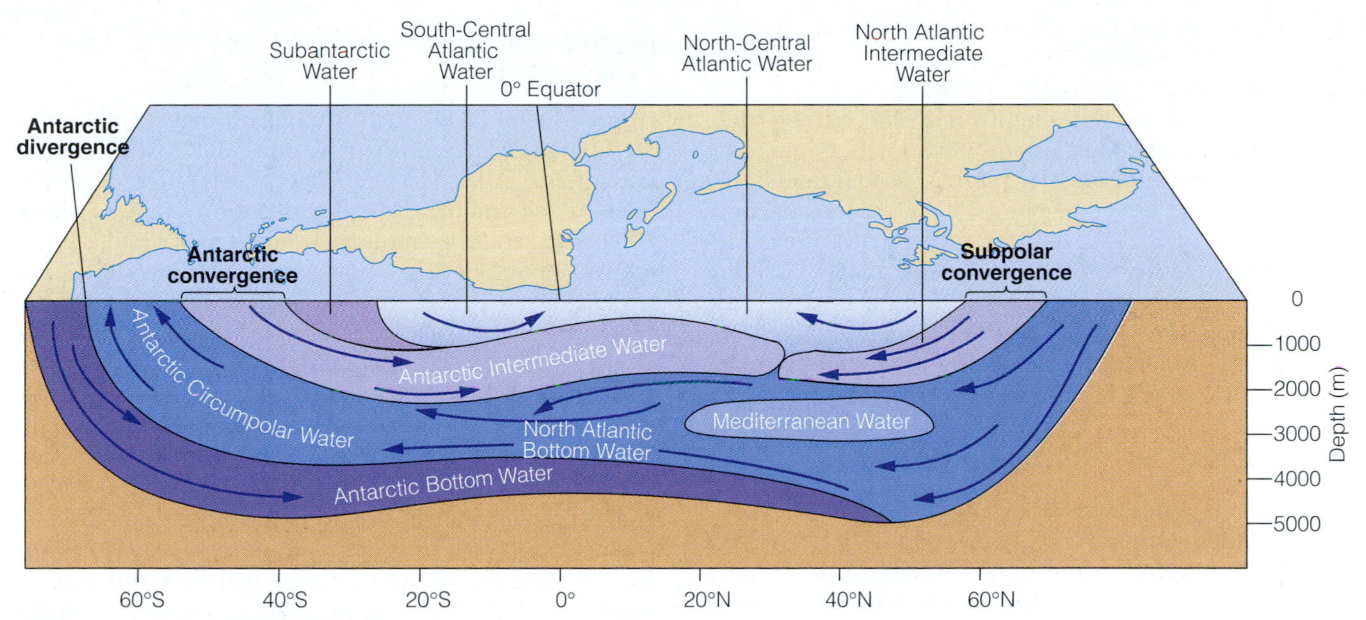

FIGURE 10.27 A vertical cross section of the Atlantic Ocean shows the various water masses that form layers at different depths. Antarctic Bottom Water is the densest water mass, and it flows northward from around Antarctica. North Atlantic Bottom Water sinks near Greenland and flows southward over the top of Antarctic Bottom Water. Intermediate-depth water masses are formed and sink at the Antarctic and subpolar convergences. The near-surface layers are more complex. Note the tongue of Mediterranean Water that spreads across the North Atlantic Ocean from the Straits of Gibraltar at about 2000 to 3000 m depth between 20°N and 55°N.

make NADW readily distinguishable from Antarctic Bottom Water. Because it is less dense than Antarctic Bottom Water, southward-moving NADW flows over the northward-moving Antarctic Bottom Water (**Fig. 10.27**).

Most of the Arctic Ocean is covered year-round by a floating ice sheet (**Fig. 10.28**). Beneath the ice, surface waters have low salinity and are separated from the deep waters by an intense halocline. The low salinity is due partly to river runoff into the Arctic Ocean and partly to ice exclusion. Salt is excluded and transported below the pycnocline during freeze-up in **brines** left by the freezing, but it is not returned in summer when seasonal ice melts. Sea ice contains very little salt, and melting of seasonal ice adds about the same volume of freshwater to the Arctic Ocean surface layer as river runoff does.

High-salinity brine is also formed by ice exclusion during winter freezing in the continental shelf regions that surround Antarctica. This brine sinks rapidly to the seafloor or through the halocline without mixing effectively with the lower-salinity surface water through which it passes. Brine may collect in pools on the continental shelf floor and then drain down the slope. Once in the deep basin below the halocline, the brine eventually mixes with the bottom water, most of which enters the Southern Ocean originally from the Atlantic Ocean, and the mixed water mass becomes Antarctic Bottom Water. Most Antarctic Bottom Water is formed in the Weddell Sea section of the Antarctic coast, located south of the

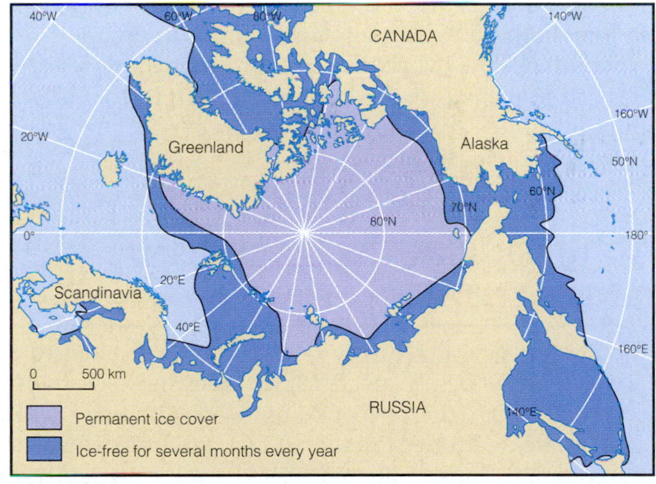

FIGURE 10.28 Most of the area of the Arctic Ocean is covered by a permanent floating ice pack. Only the areas around the edges become ice-free for a few weeks each year. The areas covered by seasonal ice in winter extend far south into the Bering Sea and the Sea of Okhotsk between Russia and Japan, and into the North Atlantic near Greenland and eastern Canada. Most northern European seas are kept ice-free year-round as a result of the warming influence of Gulf Stream water transported into the area. There is strong evidence that the area of seasonal sea ice and permanent ice pack are becoming smaller from year to year, and this may be evidence of the enhanced greenhouse effect.

southwest corner of the Atlantic Ocean near the tip of South America. However, lesser contributions are made at several other locations around Antarctica.

Deep water-mass movements are affected by the Earth's rotation and the Coriolis effect in the same way that surface currents are. Therefore, deep currents are intensified along western boundaries of the oceans and tend to flow in gyrelike motions within the northern and southern basins of each ocean (**Fig. 10.26**). Eastern boundaries generally have a weak, diffuse flow in the direction opposite that of the western boundary flow (**Fig. 10.26**). In addition, deep currents are affected by **oceanic ridges** and other bottom topography.

Water Masses at Intermediate Depths

Currents and water masses at intermediate depths are caused by the sinking of cool, higher-density water at the Antarctic Convergence, and by the sinking of warm but high-salinity and slightly higher-density water at subtropical convergences, where evaporation exceeds precipitation (**Fig. 10.27**). Evaporation exceeds precipitation in the Mediterranean Sea, and warm, high-salinity Mediterranean water discharges into the Atlantic Ocean, where it sinks to intermediate depths (**Figs. 10.25, 10.27**).

If water continuously sinks to form deep water masses, deep water must be displaced, warmed, and returned to the surface. Cold, deep water is mixed with warmer, less dense water from above by slow vertical mixing between water masses and by more turbulent mixing induced by currents, **internal waves,** and tides where they flow over rough seafloor topography. Water from intermediate depths may also return to the surface by upwelling at divergences such as the Antarctic Divergence (**Fig. 10.27**).

10.14 OCEAN CIRCULATION AND CLIMATE

Circulation of water in the oceans can be likened to a giant conveyor belt. Water is cooled near the poles; transported through the deep oceans, where it mixes with other water; returned to the surface far from where it sank; and warmed and transported back to a location where it cools and enters the cycle again.

The ocean circulation system carries heat from one part of the planet to another and transfers dissolved chemicals between surface and deep-water layers. Consequently, changes in circulation can cause and be caused by changes in climate, and they can cause changes in the distribution of dissolved chemicals that directly affect the biological **communities** of the oceans. Ocean circulation also carries some of the excess carbon dioxide from **fossil fuel** burning into the deep oceans (**CC9**). Therefore, deep-ocean circulation research is of great interest, particularly for global climate change studies.

Meridional Overturning Circulation: The Conveyor Belt

Figure 10.29 is a much simplified depiction of the Meridional Overturning Circulation (MOC), which transports North Atlantic Deep Water (NADW) through the oceans. The complete circuit takes an average of about 1000 years, and the amount of water transported is enormous—about 20 times the combined flow of all the world's rivers. This circulation is often referred to as the "conveyor belt circulation."

The MOC starts in the North Atlantic near Greenland and Iceland, where surface water is cooled by the cold air mass that flows from the Canadian Arctic. The cold, high-density water sinks to form NADW, which then flows south through the deep Atlantic Ocean, around Africa, and eventually northward into the Indian and Pacific Oceans. The water mixes progressively toward the surface. Little is known about where and how this mixing takes place, but it is thought to take place slowly throughout most of the oceans and more quickly through turbulent mixing in some areas of the Indian and North Pacific Oceans and at least some regions around Antarctica. However, eventually it mixes with surface waters and enters the mixed-layer current system. Complex exchanges and movements of surface water eventually return surface water to the Atlantic Ocean to replace the water that originally formed NADW.

Surface water in the North Atlantic Ocean is warmer than NADW. Westerly winds blow over this relatively warm water, which is transported by the Gulf Stream to the northeastern Atlantic Ocean near Europe. The westerly wind air mass releases its heat and moisture over Europe, causing Europe's climate to be extremely mild and wet in comparison with climates of other land areas at the same latitude (Chap. 9). Thus, the MOC transfers heat from the central Indian Ocean and North Pacific to the North Atlantic region near Europe. At the same time, the MOC transfers dissolved nutrients from the Atlantic Ocean to surface waters of the Indian and North Pacific Oceans.

The MOC Climate Switch

Ocean **sediment** records have shown that, during the last **ice age**, which peaked about 18,000 years ago, the MOC appears to have generally operated more slowly and weakly than it does today and to have varied in strength. The variances may have occurred abruptly at times. Some of those abrupt changes may have almost entirely turned the MOC off or back on again during the past tens of thousands of years. It is believed that such abrupt changes have, at least on some occasions, coincided with very abrupt (on the scale of decades) changes in climate. Current evidence suggests that variations in the MOC circulation, at least during the cold, ice-age period, may have been **chaotic** (**CC11**) and that ocean circulation

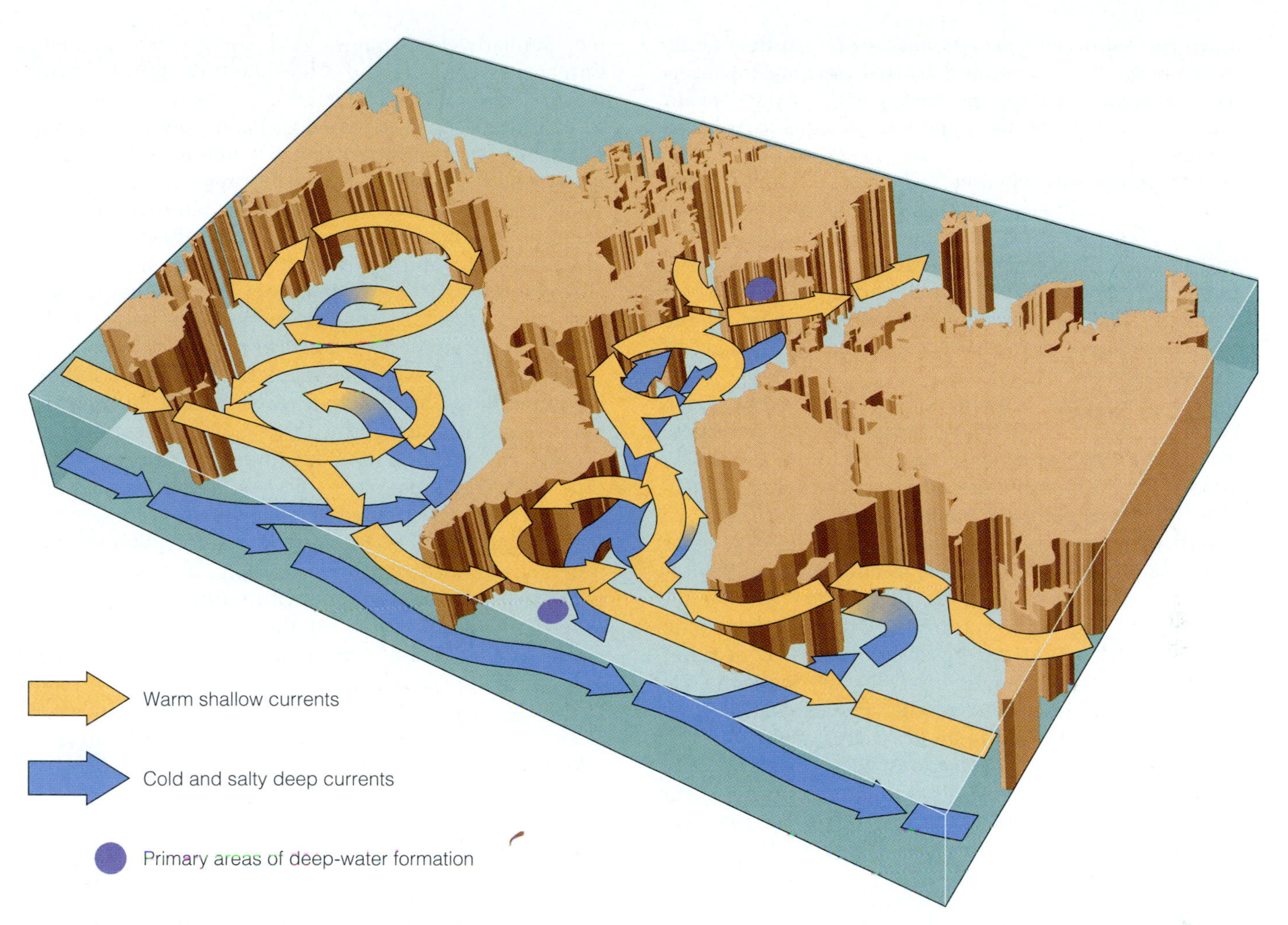

FIGURE 10.29 A simplified depiction of the Meridional Overturning Circulation (MOC), or "conveyor belt," that starts with the formation of North Atlantic Deep Water near Greenland. This water mass flows south through the Atlantic, flows over and mixes with the deep water formed near Antarctica, and is transported around Antarctica and then north into the Indian and Pacific Oceans. This deep water is gradually warmed, mixed upward, and returned to the North Atlantic in a complex and not yet well understood surface and near-surface circulation. This system is instrumental in transporting heat from low latitudes to high latitudes of the North Atlantic Ocean.

may have switched periodically between two modes. About 13,500 years ago, the ice age ended when the average temperature increased by about 6°C in as little as 100 years. The abrupt change was accompanied by a 20% increase in atmospheric carbon dioxide concentration and an increase in the intensity of the MOC.

During the 13,500 years since the last ice age ended, the Earth's climate has been warmer than at any other time during the past million or more years. However, this warm period was interrupted by a short cold period, the Younger Dryas period, that started 11,000 years ago and lasted several hundred years. During the Younger Dryas period, western Europe's climate cooled within a matter of decades and then returned just as abruptly to its former warm condition. Apparently the Younger Dryas cold climate occurred during a period when the MOC was severely slowed or stopped. The reason for the temporary slowing or stoppage of the conveyor belt is not known, but it might be related to the flow of meltwater from **glaciers**.

During the early postglacial period, about 13,000 years ago, glaciers extended far to the south, and most meltwater from the North American ice sheet probably drained down the Mississippi Valley to the Gulf of Mexico. Once the ice sheet had retreated far enough, cold meltwater probably was diverted largely to the Saint Lawrence River, which empties into the ocean where NADW is formed. The cold meltwater would have been essentially freshwater. Because of its low salinity, it would have floated on the ocean water to form a stable, low-density surface layer over a large area of the North Atlantic Ocean. The low-salinity surface layer would have acted as a virtual cap, severely restricting the formation of NADW and slowing the MOC. Within several hundred years after glacial meltwater first flooded in large volumes down the Saint Lawrence, its flow to the North Atlantic probably diminished because, by that time, most glaciers had finished melting. At that point, NADW again formed, the MOC resumed, and Europe's climate warmed rapidly.

We are not sure whether a number of abrupt climate changes that have occurred during the past 10,000 to 12,000 years coincided with significant changes in the rate of water transport within the MOC, but it appears that at least some such abrupt climate changes did coincide with changes in the MOC. We cannot yet determine whether changes in the MOC cause climate changes, or vice versa.

The annual rate of formation of Greenland Sea Deep Water, one source of NADW, decreased by as much as 90% in the 1980s and 1990s compared to the 1970s. The reasons for the decrease are unclear. However, it is likely to be related to climatic change, which, in turn, may be related to increased concentrations of greenhouse gases in the atmosphere. Whether the recent change in the formation rate of Greenland Sea Deep Water will significantly slow the MOC is not known. If it does, one possible result could be the abrupt (years to decades) onset of a cold climate period in Europe, similar to the Younger Dryas period, and to other climate changes elsewhere. However, modeling studies (**CC10**) tell us that the climate change, although serious, would not be catastrophic, or plunge the world overnight into chaos and another ice age, as depicted in the movie *The Day after Tomorrow*.

10.15 TRACING WATER MASSES

Although direct measurements have been made of deep-ocean currents, most of what we know about deep-ocean circulation was learned from studies of the horizontal and vertical distribution of water masses. Water masses are most often characterized by their temperature and salinity, but the concentrations of several dissolved constituents are also used to trace water masses (**CC6**).

Conservative and Nonconservative Properties

Conservative properties of seawater are properties that are not changed by biological, chemical, or physical processes within the body of the oceans. A conservative property can be changed only by mixing with other water masses that have different values of that property, or by processes that occur at the ocean surface, or where rivers or other sources, such as **hydrothermal vents**, enter the oceans. For example, salinity, a conservative property, can be changed at the ocean surface by evaporation or precipitation, or at river mouths by the introduction of freshwater. Small salinity changes can also occur at hydrothermal vents, where high-salinity waters are emitted. Similarly, temperature is a conservative property until water contacts the atmosphere or seafloor, with which it can exchange heat.

Dissolved constituents of seawater, such as Na^+ and Cl^-, whose concentrations are not significantly affected by biological or chemical uptake or removal from ocean water, are also conservative properties. In contrast, concentrations of constituents such as oxygen, carbon dioxide, phosphate, and many trace metals are substantially altered by biological and chemical processes that remove and release them to solution within the body of the oceans. These are nonconservative properties.

Conservative properties are particularly useful in tracing water masses because they can be used to identify the masses as they are formed, to trace their transport through the oceans, and to determine how they mix with other water masses. When two water masses mix, the value of the conservative property in the mixed water mass is determined by the proportions in which the two water masses have mixed. For example, consider the mixing of two water masses, one with salinity of 35 and the other with salinity of 37. If the salinity of the mixed water mass is 36, the mixture must consist of equal volumes of the two water masses (2 volumes of 36 = 1 volume of 37 + 1 volume of 35). If the salinity of the mixture is 35.5, the mixture must consist of three parts water of salinity 35 and 1 part water of salinity 37 (4 volumes of 35.5 = 3 volumes of 35 + 1 volume of 37).

TS Diagrams

We can identify different water masses by plotting temperature against salinity in what are called "TS diagrams" (**Fig. 10.30**). Water samples from the same water mass have the same salinity and temperature and therefore will appear as a single point on a TS diagram. Samples from water masses that have different temperature and salinity appear as separated points. When two water masses are mixed, the temperature and salinity of the mixed water will be along a straight line drawn between the two points that represent the original water masses. The location of the TS value for a mixed water sample on the line between the TS values of the two water sources indicates the relative proportions of the two original water masses in the mixture.

When three water masses mix in such a way that no unmixed sample of one of the three water masses remains, some of the mixed water samples will have TS values that are no longer on straight lines between the three original TS points (**Fig. 10.30c**). However, in such cases the temperature and salinity of the original water mass, which now exists only as a part of mixtures, can often be deduced by extrapolation of the straight-line portions of the TS curve (**Fig. 10.30c**).

A TS diagram for water samples taken at different depths at a single station in the South Atlantic Ocean is shown in **Figure 10.30d**. At this station, the principal water masses are warm, saline surface water; North

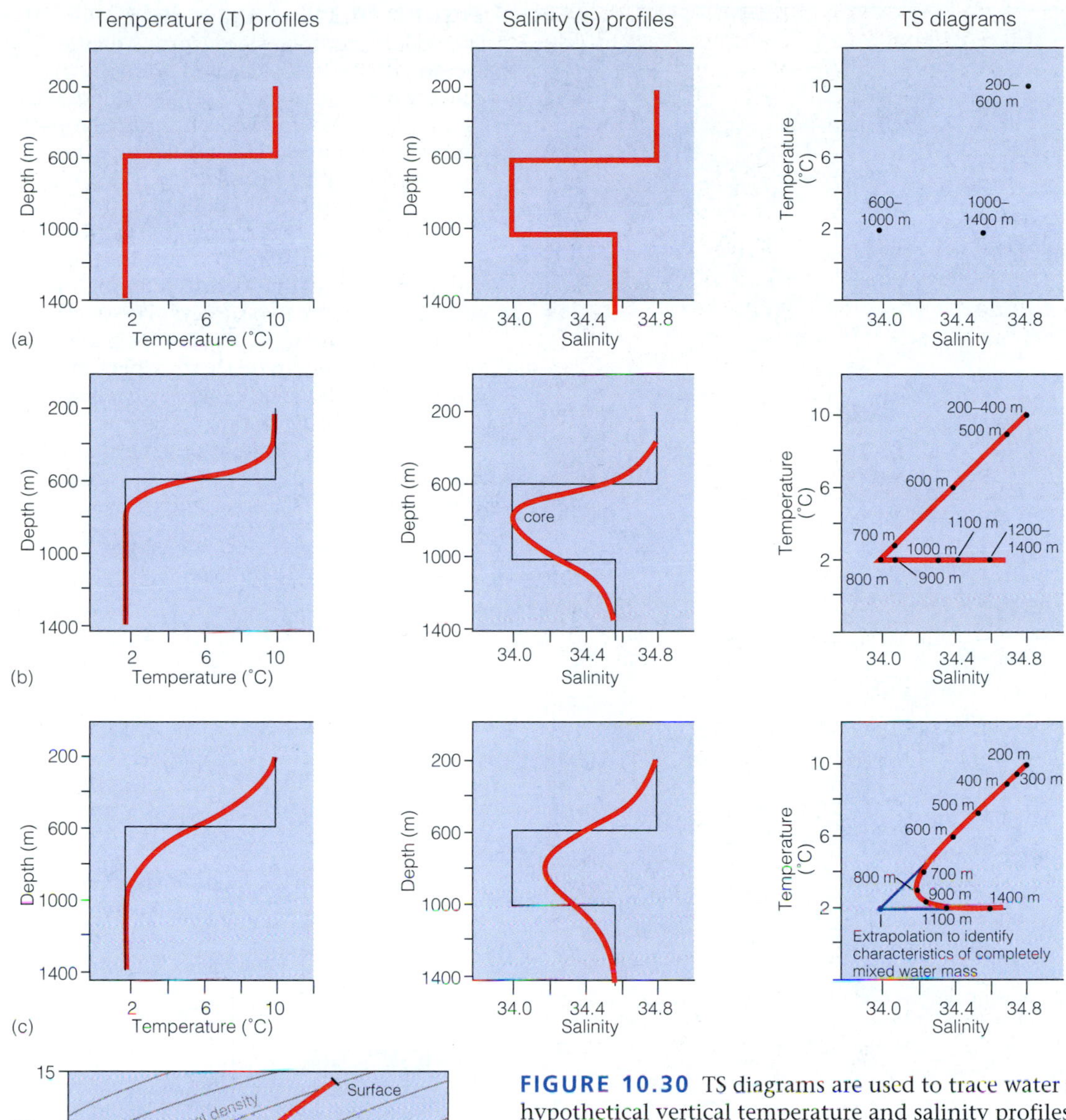

FIGURE 10.30 TS diagrams are used to trace water masses. In this figure, hypothetical vertical temperature and salinity profiles are used to construct examples of TS diagrams. (a) A simple three-layer system. All TS points plot at one of three locations that represent the three water masses. (b) If the three layers mix, but some of the unmixed mid-depth water mass is still present (at 700 m depth), the TS points plot on two straight lines that connect the three original water-mass characteristic points. (c) Even if all of the mid-depth water mass is mixed with at least some of the other two water masses, a TS plot can be used to identify the temperature and salinity characteristics of the original unmixed water mass. (d) A typical TS diagram for a station in the South Atlantic Ocean.

Atlantic Deep Water; water that is a mixture of surface water, Antarctic Intermediate Water, and North Atlantic Deep Water; and Antarctic Bottom Water. We need not be concerned with how these water masses acquired their different temperature and salinity characteristics. What is important is that, once the source properties are known, the presence of a particular water mass in the area for which the TS diagram is drawn can be determined. Note that we can detect the presence of Antarctic Intermediate Water in the diagram, even though no unmixed Antarctic Intermediate Water is present at the station. We can also calculate from the diagram the percentage of Antarctic Intermediate Water present in the mixed water at any depth. For example, at 800 m, where the proportion of this water mass is greatest, Antarctic Intermediate Water is about 55% of the mixture.

By drawing TS diagrams for many different stations across the oceans, oceanographers can trace movements of individual water masses and investigate mixing processes at different points within the oceans.

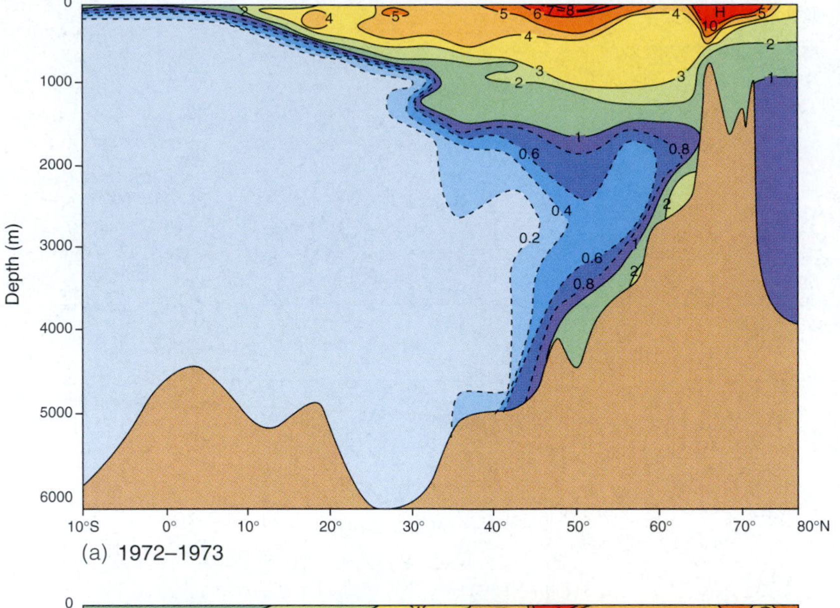

FIGURE 10.31 (a) Tritium distribution with depth in the western North Atlantic Ocean (1972–1973). Released primarily by nuclear bomb tests in the 1950s, tritium has spread throughout the ocean surface layers and is being transported steadily into the deep oceans with North Atlantic Deep Water, which is formed by the cooling of surface waters in the North Atlantic. (b) Distribution in 1981. The values are corrected for the radioactive decay that occurred during the 9 years between the two surveys. The strong source in the surface layer at 50°N is the outflow from the Arctic Ocean where the Soviet Union did most of its nuclear bomb testing. Comparing the two diagrams shows the progressive movement of tritium into the deep layer as North Atlantic Deep Water continuously forms, sinks, and moves southward.

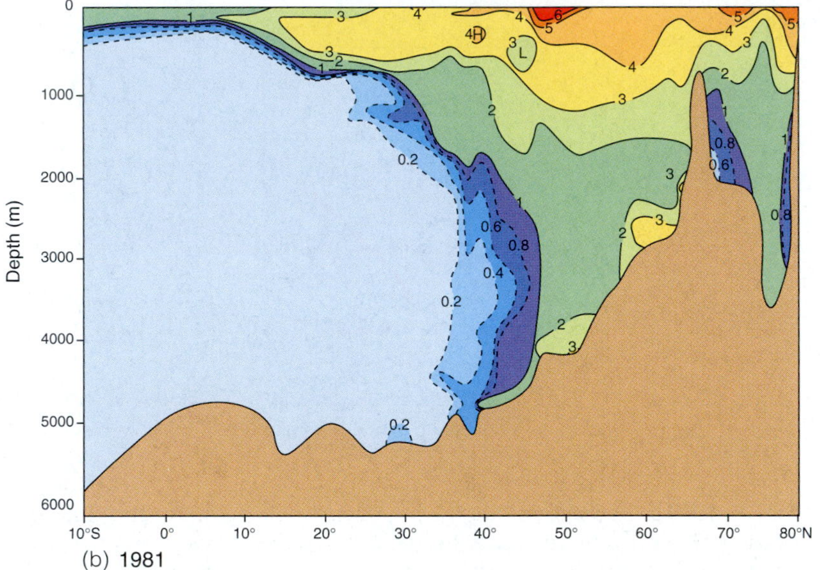

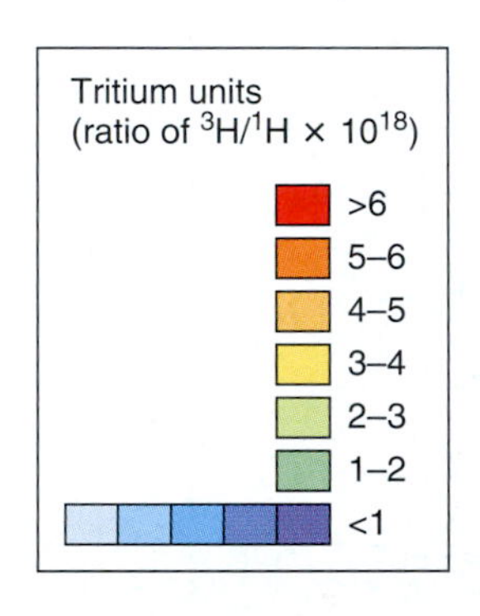

Tracers

Salinity and temperature are excellent tracers, but they tell us only which water masses are mixed and where the mixed water mass is transported to. Without other information, these tracers cannot tell us how fast the water masses move and mix. Several different dissolved components of seawater are now used in conjunction with temperature and salinity to trace ocean water masses and to provide information on their rate of movement.

For some applications, nonconservative properties such as oxygen and carbon dioxide concentrations can be used as tracers. Concentrations of these gases are altered by the decomposition of organic particles, which consumes oxygen and releases additional carbon dioxide to solution. Therefore, if we know the oxygen and carbon dioxide concentrations in water masses when they sink below the mixed layer, changes in those concentrations can provide information about the relative ages (since the time when they sank) of water masses in different parts of the oceans.

Human activities have provided a number of useful tracers during the past half century. These tracers include **radionuclides** created by nuclear weapons testing and nuclear reactors, and certain synthetic organic compounds, particularly **chlorofluorocarbons (CFCs)**. Some radionuclides that do not occur naturally are now found dissolved in ocean water in extremely small concentrations. They are derived primarily from nuclear bomb tests in the atmosphere, which peaked in the 1950s, and from more recent releases from nuclear reactors. These radionuclides are good tracers because the time and location of their introduction to the oceans is well known and they decay at known constant rates (**CC7**).

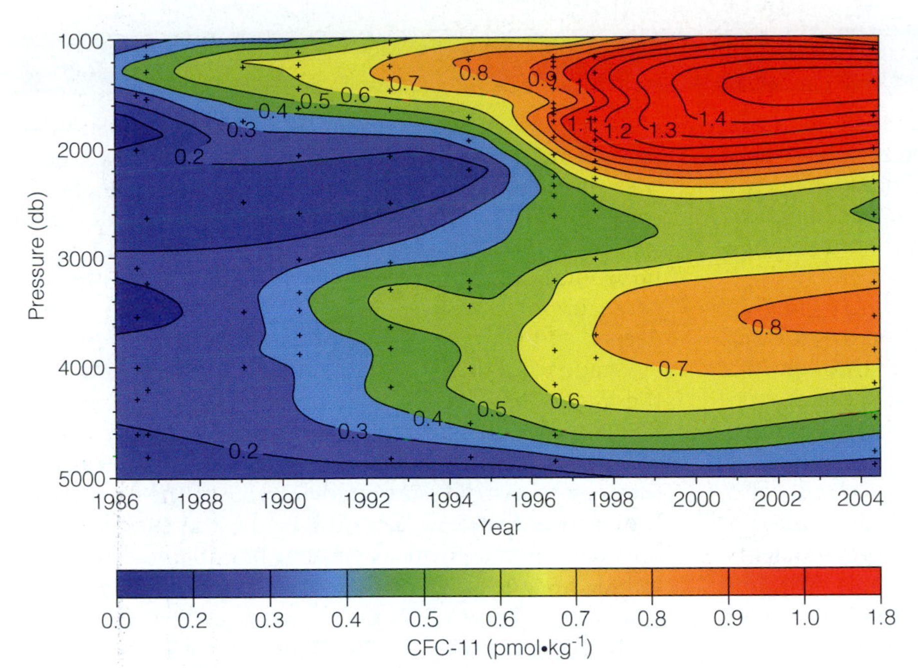

FIGURE 10.32 The concentrations of CFC-11, a specific chlorofluorocarbon compound, have been measured throughout the depth of the water column to the east of Abaco Island in the Bahamas periodically between 1986 and 2004. This figure plots the vertical distribution of the concentrations of CFC-11 in this area as a function of time. The data show the arrival at this subtropical location of CFC-11 in recently formed deep water masses originating from the high latitude, deep water mass formation areas in the polar and subpolar regions of the North Atlantic (see Figures 10.27 and 10.29). There are three distinct water layers seen arriving in this area, centered at about 1200 m, 1800 m, and 4000 m. The deepest layer is deep water formed from the outflow of the Arctic Ocean and the two shallower layers are both deep water formed near Greenland. The 1800 m layer arrived near Abaco later than the shallower layer, and appears to reflect a small reduction in the temperature of deep water formed near Greenland that had been observed near Greenland about a decade earlier.

Tritium (^3H), a **radioactive isotope** of hydrogen, is a particularly useful tracer. Tritium is produced naturally by cosmic rays in the atmosphere. Thus, it occurs naturally in ocean water, but only at extremely small concentrations. In water unaffected by human releases, approximately one of every 10^{19} atoms of hydrogen is a tritium atom. Large quantities of tritium were released during hydrogen bomb testing. Tritium reacts quickly with oxygen in the atmosphere to form tritiated water, which enters the oceans in rainfall. Because the tritium is incorporated in the water molecule itself, it is a perfect tracer for water masses. Releases of tritium during the nuclear bomb testing era of the mid-twentieth century raised the tritium concentration in surface seawater to a concentration of approximately 6 atoms of tritium in 10^{18} atoms of hydrogen. Tritium has a **half-life** of only approximately 12 years, so bomb-produced tritium will continue to be useful as a tracer for only a few decades, until all the tritium has decayed.

In the Northern Hemisphere, most tritium was produced by nuclear tests in the Soviet Arctic. Therefore, tritium concentrations are particularly high in water flowing into the Atlantic Ocean from the Arctic Ocean. This tritium is now distributed throughout the mixed layer of the oceans and is currently being transported into the deep oceans with sinking cold water masses created in polar regions. The progress of deep water-mass formation has been monitored by the use of tritium as a tracer (**Fig. 10.31**).

Chlorofluorocarbons (CFCs), which are used as refrigerants and are thought to be responsible for **ozone** depletion in the atmosphere (Chap. 9), are also excellent water-mass tracers (**Fig. 10.32**). These compounds do not occur naturally. They are dissolved at very low concentrations in seawater and are not readily decomposed by biological processes.

Unfortunately, for most water-mass tracers, ***in situ*** instrumentation is not sensitive enough, and thus, water samples must be collected for laboratory analysis. Because the cost of ship time required to collect water samples from deep within the water column is very high, sampling cannot be as intensive either **spatially** or temporally as would be desirable in most instances. Thus, many studies rely on mathematical modeling (**CC10**) using the limited data that are available.

Study of recent and ongoing processes of deep water-mass formation is extremely important to our understanding of the effects of fossil fuel burning on global climate. One estimate suggests that more than half of the carbon dioxide released by burning fossil fuels since the Industrial Revolution in the mid-1800s is now dissolved in ocean water. However, there is still considerable uncertainty about the accuracy of the estimate. Ocean water that sinks below the mixed layer carries carbon dioxide from the atmosphere into the deep oceans, where it is locked away from contact with the atmosphere for hundreds or thousands of years (or longer if it is incorporated in carbonate sediments; Chap. 8). Tracer studies will help us to determine how much carbon dioxide from fossil fuels has already been transported safely below the pycnocline and to estimate how quickly additional carbon dioxide will follow it.

CHAPTER SUMMARY

Energy Sources. Winds are the energy source for most currents in the ocean's upper few hundred meters. Deeper currents and most vertical movements of ocean water are caused by thermohaline circulation driven by differences in the density of water masses that are due to temperature and salinity differences.

Wind-Driven Currents. Winds blowing across the sea surface transfer kinetic energy to the water through friction and cause wind-driven currents with speeds generally about 1% to 3% of the wind speed. Currents continue after winds abate, until their momentum is eventually lost because of internal friction. Currents also continue because winds transport water, cause the sea surface to slope, and create horizontal pressure gradients on which water is moved. Current speed and direction are modified by friction, the Coriolis effect, horizontal pressure gradients, and land and seafloor topography.

Ekman Motion. Surface water set in motion by winds is deflected *cum sole* relative to the wind direction by the Coriolis effect. Wind energy is transferred down into the water column as each water layer transfers its energy to the layer below. If winds blow consistently for long enough over deep water, an Ekman spiral develops in which the surface current is deflected 45° *cum sole*, current speed decreases with depth, and current direction is progressively deflected *cum sole* with depth. Mean water transport (Ekman transport) in the layer within which the Ekman spiral develops is at 90° *cum sole* to the wind.

Wind-driven currents rarely reach deeper than 100 to 200 m. Where the seafloor is shallower than the Ekman spiral depth, friction reduces the surface water deflection below 45° and the mean transport below 90°. If a pycnocline is present within the Ekman spiral depth, downward transfer of wind energy is inhibited, and wind-driven currents do not occur below the pycnocline.

Geostrophic Currents. Ekman transport causes surface water to "pile up" in some areas. The resulting sloping sea surface results in a horizontal pressure gradient within the upper several hundred meters. Water flows on such gradients and is deflected by the Coriolis effect. The deflection continues until the flow is across the gradient (parallel to the pressure contours) at a speed where the Coriolis deflection is just matched by the pressure gradient force. This is geostrophic motion.

Ocean Surface Currents. Trade winds drive surface waters generally west in the Equatorial currents. Westerly winds drive surface waters generally east. Continents deflect the flow north or south. Thus, current gyres are created between trade wind and westerly wind latitudes in each hemisphere and ocean. These gyres flow clockwise in the Northern Hemisphere and counterclockwise south of the equator. Ekman transport by trades and westerlies piles up water in the gyre center and creates horizontal pressure gradients that maintain geostrophic gyre currents. Western boundary Currents are faster, deeper, and narrower than eastern boundary currents.

Trade wind-driven Equatorial Currents move water to the west. Some water is returned to the east in the Equatorial Countercurrent, which flows at the intertropical convergence and not at the equator. At the equator, the west-to-east pressure gradient causes the Equatorial Undercurrent to flow east just below the surface layer. This current is as swift as western boundary currents, is unique because it is not deflected by the Coriolis effect, and is important in El Niño.

In high latitudes of the North Atlantic and Pacific Oceans, counterclockwise subpolar gyres occur that are complex, weak, and variable. At high southern latitudes, no continents block the East Wind Drift current that flows westward around Antarctica.

Upwelling and Downwelling. Upwelling occurs where Ekman transport moves surface waters away from a divergence or coast. Downwelling occurs where Ekman transport adds to the surface layer depth at convergences or coasts. Upwelling areas are usually highly productive because upwelling can bring cold, nutrient-rich water from below into the mixed layer, where nutrients are needed by phytoplankton. Divergences and upwelling occur in part of the area between the North and South Pacific Equatorial currents, around Antarctica, and in many coastal locations. Convergences occur in the centers of ocean gyres and in many coastal regions. Coastal upwelling is more prevalent inshore of eastern boundary currents than inshore of western boundary currents.

Coastal Currents. Coastal current directions are determined by local winds and may be opposite those of the adjacent gyre currents. Coastal currents are variable and are affected by freshwater inflow, friction with the seafloor, steering by seafloor topography and coastline, and tidal currents.

Eddies. Ocean eddies are similar to, but smaller than, the atmospheric eddies seen on satellite images as swirls of clouds. Satellites can be used to observe some ocean eddies. Gulf Stream rings are eddies 100 to 300 km wide that form cold-core (counterclockwise-rotating) or warm-core (clockwise-rotating) rings when a meander of the Gulf Stream breaks off. The rings drift slowly south, may reattach to the Gulf Stream, and may last several months. Mesoscale eddies present throughout the oceans have slower currents than Gulf Stream rings, may extend to the deep-sea floor, and are 25 to 200 km wide.

Inertial Currents. Once established, wind-driven currents that continue to flow and are deflected into circular paths by the Coriolis effect are called "inertial currents."

Langmuir Circulation. Strong winds can set up a corkscrewlike Langmuir circulation aligned in the wind direction. Langmuir cells are a few meters deep and about 30 m wide, and they lie side by side. Foam or floating debris collects in the linear downwelling regions between cells.

Thermohaline Circulation. When surface water density is increased by cooling or evaporation, the water sinks and spreads horizontally at a depth where the density of the water above is higher and the density of the water below is lower. Throughout most of the oceans, the water column consists of a uniform-density mixed surface layer (about 100 m deep), a permanent pycnocline zone (extending from the bottom of the mixed layer to a depth of 500 to 1000 m) in which density increases progressively with depth, and a deep zone in which density increases slowly with depth. Most high-latitude areas where surface waters are cooled have no pycnocline. Secondary, temporary pycnoclines can develop in the mixed layer in summer. Pycnoclines act as a barrier to vertical mixing.

Deep water masses are formed by cooling and ice exclusion at high southern latitudes, in the Arctic Ocean, and at high latitudes in the North Atlantic Ocean. Antarctic Bottom Water is formed primarily in the Weddell Sea and flows north as the deepest and densest water layer. North Atlantic Deep Water is formed in the Greenland and Norwegian Seas and flows south. Water masses at intermediate depths are formed by the sinking of cooled water at the Antarctic Convergence and of warm, high-salinity water produced by evaporation at subtropical convergences and in the Mediterranean Sea. Deep and intermediate-depth water masses are subject to the Coriolis effect, and thus, currents in the deep layer are intensified along western boundaries and tend to flow in gyres. They are also affected by seafloor topography.

Ocean Circulation and Climate. Ocean circulation transfers heat from the tropics toward the poles, moderating mid- and high-latitude climates. North Atlantic Deep Water forms near Greenland and Iceland. It sinks and spreads south in the deep Atlantic and then north into the Pacific and Indian Oceans. It mixes progressively upward to the surface layer, which is warmed and eventually transferred back to the North Atlantic by the Gulf Stream. This Meridional Overturning Circulation (MOC) has varied in intensity and switched off and on in the past. Periods when it has not operated seem to be associated with colder climates, especially in Europe. The change to colder climate appears to take place abruptly when the MOC is switched off.

Tracing Water Masses. Water masses are characterized by their salinity and temperature, which are conservative properties everywhere but at the surface and in some areas of the seafloor. Nonconservative properties such as oxygen concentration, as well as concentrations of human-made radionuclides and persistent organic compounds, are also used as water-mass tracers.

STUDY QUESTIONS

1. What are the major energy sources for the surface currents and the thermohaline circulation of the oceans?
2. Why doesn't a wind-driven current flow in the same direction as the wind that causes it?
3. Wind-driven Ekman transport moves surface water across the oceans. How does this lead to geostrophic currents that continue to flow even if the winds abate? Why don't geostrophic currents flow in the same direction as the Ekman transport?
4. Subtropical gyres are present in each ocean and each hemisphere in the latitudes between the trades and westerlies. Why? Would they exist if the Earth were totally covered by oceans? Why or why not?
5. Why do the subtropical gyres rotate clockwise in the Northern Hemisphere and counterclockwise in the Southern Hemisphere?
6. Why are there two currents (the Equatorial Countercurrent and the Equatorial Undercurrent) that flow eastward near the equator? How do they behave differently from one another?
7. How are upwelling and downwelling caused in the oceans, and why are they important? Where would you expect to find upwelling and downwelling areas in the oceans? Why?
8. What factors affect coastal currents but not deep-ocean currents?
9. Why are the deepest water masses in the oceans formed at high latitudes? Why are they not formed in the Indian and North Pacific oceans?
10. Why is there no pycnocline in most high-latitude regions of the oceans?
11. Why are temperature and salinity the two most important tracers of water masses?

KEY TERMS

You should recognize and understand the meaning of all terms that are in boldface type in the text. All those terms are defined in the Glossary. The following are some less familiar key scientific terms that are used in this chapter and that are essential to know and be able to use in classroom discussions or exam answers.

brine (p. 267)
chaotic (p. 268)
chlorofluorocarbons (CFCs) (p. 272)
climate (p. 240)
coastal upwelling (p. 256)
coastline (p. 244)
convection cell (p. 248)
convergence (p. 244)
Coriolis effect (p. 242)
cum sole (p. 243)
current (p. 240)
divergence (p. 244)
downwelled (p. 245)
dynamic height (p. 246)
eastern boundary current (p. 249)
eddy (p. 259)
Ekman transport (p. 243)
El Niño (p. 255)
estuary (p. 240)
friction (p. 240)
geostrophic (p. 244)
gyre (p. 247)
halocline (p. 263)
ice exclusion (p. 265)
intertropical convergence (p. 253)
kinetic energy (p. 240)
Langmuir circulation (p. 261)
longshore current (p. 261)
meander (p. 259)
mixed layer (p. 263)
nutrients (p. 256)
phytoplankton (p. 256)
planetary vorticity (p. 251)
potential energy (p. 245)
pressure gradient (p. 242)
pycnocline (p. 243)
radioactive isotope (p. 273)
radionuclide (p. 272)
runoff (p. 263)
salinity (p. 240)
sill (p. 266)
stratified (p. 266)
subtropical gyre (p. 247)
surface tension (p. 240)
suspended sediment (p. 240)
synoptic (p. 259)
thermocline (p. 263)
thermohaline circulation (p. 240)
tracer (p. 272)
trade wind (p. 248)
turbulence (p. 250)
upwelling (p. 248)
water mass (p. 240)
western boundary current (p. 249)

CRITICAL CONCEPTS REMINDER

CC1 Density and Layering in Fluids (pp. 240, 243, 262, 263, 265). Water in the oceans is arranged in layers according to the water density. Many movements of water masses in the oceans, especially the movements of deep water masses, are driven by differences in water density. To read **CC1** go to page 2CC.

CC3 Convection and Convection Cells (pp. 240, 262). Fluids, including ocean water, that are cooled from above, sink because their density is increased. This establishes convection processes that are a primary cause of vertical movements and the mixing of ocean waters. These processes are also important in transporting and distributing heat and carbon dioxide between the atmosphere and oceans and between regions of the globe. To read **CC3** go to page 10CC.

CC6 Salinity, Temperature, Pressure, and Water Density (pp. 240, 263, 265, 270). Sea water density is controlled by temperature, salinity, and to a lesser extent pressure. Density is higher at lower temperatures, higher salinities, and higher pressures. Movements of water below the ocean surface layer are driven primarily by density differences. To read **CC6** go to page 16CC.

CC7 Radioactivity and Age Dating (p. 272). Some elements have naturally occurring radioactive (parent) isotopes that decay at precisely known rates to become a different (daughter) isotope. Radioactive isotopes, especially those that were released during the period of atmospheric testing of nuclear weapons, are useful as tracers that can reveal the movements of water masses in the ocean, especially the rates at which deep water masses are formed. To read **CC7** go to page 18CC.

CC9 The Global Greenhouse Effect (p. 268). The oceans and atmosphere are both important in studies of the greenhouse effect, as heat and carbon dioxide and other greenhouse gases are exchanged between the atmosphere and oceans at the sea surface. The oceans store large amounts of heat and carbon dioxide both in solution and in carbonates. To read **CC9** go to page 22CC.

CC10 Modeling (pp. 263, 270, 273). The complex interactions between the oceans and atmosphere can best be studied by using mathematical models. The motions of water masses within the body of the oceans, especially motions below the surface layer, are also studied extensively using mathematical models because they are extremely difficult to observe directly. To read **CC10** go to page 26CC.

CC11 Chaos (p. 268). The nonlinear nature of ocean-atmosphere interactions makes at least parts of this system behave in sometimes unpredictable ways. It also makes it possible for changes in ocean circulation to occur in rapid, unpredictable jumps between one set of conditions and a different set of conditions, and these changes can affect climate. To read **CC11** go to page 28CC.

CC12 The Coriolis Effect (pp. 242, 243, 245, 250, 255, 259, 261). Water and air masses move freely over the Earth and ocean surface while objects on the Earth's surface, including the solid Earth itself, are constrained to move with the Earth in its rotation.

This causes moving water or air masses to appear to follow curving paths across the Earth's surface. The apparent deflection, called the Coriolis effect, is to the right in the Northern Hemisphere and to the left in the Southern Hemisphere. The deflection is at a maximum at the poles, is reduced at lower latitudes, and becomes zero at the equator. To read **CC12** go to page 32CC.

CC13 **Geostrophic Flow** (pp. 242, 244, 245, 246, 255). Air and water masses flowing on horizontal pressure gradients are deflected by the Coriolis Effect until they flow across the gradient such that the pressure gradient force and Coriolis Effect are balanced, a condition called geostrophic flow. This causes ocean currents and winds to flow around high and low pressure regions (regions of elevated or depressed seasurface height) in near circular paths. The circular gyres that dominate the global circulation of ocean waters are the result of water masses flowing geostrophically. To read **CC13** go to page 43CC.

CHAPTER 11

CHAPTER OUTLINE

CRITICAL CONCEPTS USED IN THIS CHAPTER

A boat approaches the wreckage of a hotel at Ton Sai Bay on Phi Phi Island, Thailand, two days after the hotel was destroyed by a tsunami. Although this damage is significant, this same tsunami rose to much greater heights and caused far more severe damage on the Island of Sumatra, on parts of mainland Thailand, and even on the Island of Sri Lanka, which is far from the area where the tsunami was spawned (in the ocean northwest of Sumatra). All told, more than 220,000 people died in eleven different countries in southern Asia and Africa.

Waves

News Flash

[Washington, DC]—The President declared a global emergency this morning less than 24 hours after a catastrophic tsunami hit the islands of Hawaii. Almost nothing has been heard from Hawaii since a few minutes after the tsunami warning center issued a warning at 1:17 AM, and it is assumed that power has been lost throughout the islands. Sporadic radio messages have been received from the astronomical observatories on the mountain of Mauna Kea on the Big Island reporting a massive earthquake, felt at about 1:00 AM, followed by a "wall of noise" coming from lower on the mountain that lasted at least several minutes.

Several hours after the tsunami warning, the entire west coast of the United States and Canada was hit by the tsunami, which is reported to be by far the largest in human history. At some locations the waves were reported to be only a meter or two high, but at others waves as high as 50 meters were reported. In some places, including Los Angeles, many thousands are dead, missing, or injured. Loss of life would have been greater, were it not for the extensive efforts of officials to evacuate everyone to higher ground in the few short hours before the waves hit.

The President was quoted as saying that the tsunami was probably the greatest catastrophe in human history. He called for a crisis meeting of all world leaders because the scale of the disaster was beyond the capability of even the United States to respond. He stated that millions or perhaps tens of millions may have died, and many more millions were injured and in need of medical attention. The President asked for an immediate freeze on all nonessential uses of gasoline and aviation fuel because almost all West Coast refineries appear to be badly damaged.

The first satellite images of Hawaii since the disaster show that a large area of the south side of the Big Island, about 10% to 15% of its total area, has simply disappeared beneath the waves, and it is thought that the collapse of this area may be what caused the tsunami. The satellite image also shows that the tsunami must have ripped across the Hawaiian Islands up to a height of perhaps 300 m above sea level, leaving massive devastation behind it. All the airports and ports in Hawaii appear to have been destroyed or severely damaged, and rescuers will probably be able to land only by helicopter.

Reports are slowly coming in from other countries around the Pacific Rim, most reporting at least some loss of life and coastal damage. Contact has not yet been made with many of the island nations of the Pacific.

Scenario for a Hollywood disaster movie? Perhaps not! The truth is that tsunamis of this scale have been generated in the past by collapses of parts of volcanic islands, including Hawaiian islands, and will certainly happen again someday, perhaps thousands of years in the future, but perhaps much sooner.

We are very much aware of continuous wave motion anytime we sail or visit the **shore**. Even on the calmest day, the water still laps gently onto the **beach** or rocks. If we stay long enough, we also notice that in most places the water level rises and falls on the shore as the hours pass. This slow up-and-down oscillation of the sea surface is the incoming and outgoing **tide**. Because the high point of the wave (high tide) is separated by hours from the low point of the wave (low tide), we do not think of tides as waves. However, tides, which are discussed in Chapter 12, are a form of wave motion.

For all who visit or live near the shore, ocean waves are important. Waves are a source of pleasure to people who visit the shore, and, as described in Chapter 13, waves are responsible for building and maintaining the sand beaches that are essential to many of our recreational pleasures. Unfortunately, waves also drown the unwary, **erode** beaches and **coastlines**, and damage piers and other structures. Waves and tides are also important because they affect the navigation and safety of the myriad commercial and recreational vessels that ply the world oceans. This chapter describes what ocean waves are, how they are formed, and how they behave. Knowledge of the behavior of waves can enhance our enjoyment of activities such as swimming, surfing, **scuba** diving, sailing, or just standing by the water's edge watching the **surf**. It can also protect us from the dangers that waves often present.

11.1 COMPLEXITY OF OCEAN WAVES

Ocean waves are complicated motions of the water surface. The surface of the water oscillates up and down, sometimes only a few centimeters and sometimes several meters, but the oscillations are never exactly regular. Successive waves are of different **wave heights**, and the **wave period** also varies. Sometimes there may seem to be a pattern, especially on days with little or no wind. On such days the waves appear to be less complicated, and the intervals between successive waves are usually longer. However, on the open sea during a storm, waves appear to break at random times and places. In addition, waves that roll in toward shore do not all break at exactly the same distance from shore, and each breaking wave looks a little different from the preceding one.

Although they are complex, ocean waves are not difficult to understand. The reason for the complexity is that what we see at any one place and time is a combination of many different waves. The combination of two or more simple waves of different periods or of two waves traveling in different directions is called **wave interference**, and this process produces complex waveforms. We can see how waves interact and combine in a lake or even a bathtub. When we throw a stone into a lake, the waves (usually called "ripples" if they are small) radiate outward in neat, regular, circular patterns from the point of impact. When we throw two stones into a lake a few meters apart from each other, the waves radiate in regular, circular patterns from each point of impact. Where the waves from one stone meet the waves from the other stone, a complicated wave pattern develops on the water surface as the two sets of waves interact.

If we watch carefully, we may be able to see that the two sets of waves are not really altered when they meet. Each wave simply passes through the other set, maintaining its original form and direction. The complicated surface pattern arises where the two sets of waves cross each other because the vertical displacement of the surface is the sum of the displacements of each of the two sets of waves at that point in space and time. We will return to the concept of wave interference later. For now, the important point to understand is that even the most complicated seas are the sum of a number of different simple waves of different heights and periods that may come from different directions.

11.2 PROGRESSIVE WAVES

Waves that we create by throwing a rock into a lake and waves that we see on the ocean move freely on the water surface. Therefore, they are called **progressive waves** (**Fig. 11.1b**). Another type of wave, called a **standing wave**, behaves differently, as discussed later in this chapter.

If we look carefully at the ripples caused by tossing a rock into a lake, we see that a series of waves, not just a single wave, moves out from the impact point. Each of the individual waves has a rounded top where the water surface is elevated (**Fig. 11.1a**). The point of highest elevation of the wave is the wave **crest**. Between each successive pair of crests is a rounded depression of the water surface, or **trough**. Most progressive waves, whether in a lake or in the deep ocean, also have this general shape.

Waves also have several other important characteristics (**Fig. 11.1a**). The distance between two adjacent crests (or troughs) is called the **wavelength**, represented as L. Wave height (H) is the vertical distance between crest and trough, and wave **amplitude** is equal to $H/2$, the vertical distance between the crest or trough and the mean water level. Wave period (T) is the time the wave takes to move a distance equal to one wavelength. It is equivalent to the time that elapses between the arrival of two successive crests (or troughs) at a point on the surface. Wave period is measured as the number of seconds per wave. Wave **frequency** (f) is the number of wave crests (or troughs) that pass a point on the ocean surface in a given time. It is measured as number of waves (or fractions of a wave) per second. Wave period (T), usually measured in seconds, and wave frequency (f) are related by a simple equation:

$$T = 1/f$$

We can estimate frequency without a stopwatch by counting the number of waves that pass in several minutes and dividing by the elapsed time (in seconds).

Two other important characteristics of progressive waves, **wave speed** and **wave steepness**, can be determined from the wavelength, wave height, and period or frequency. Wave speed is often called "celerity" (abbreviated C) because there is essentially no net forward movement (or speed) of the water as the wave passes. As explained later in this chapter, only the wave energy and waveform, and not the water, move forward with the wave. Wave speed (C) is calculated by the following equations:

$$C = L/T$$

$$\text{or } C = f \times L$$

Wave steepness is equal to the ratio of wave height to wavelength (H/L).

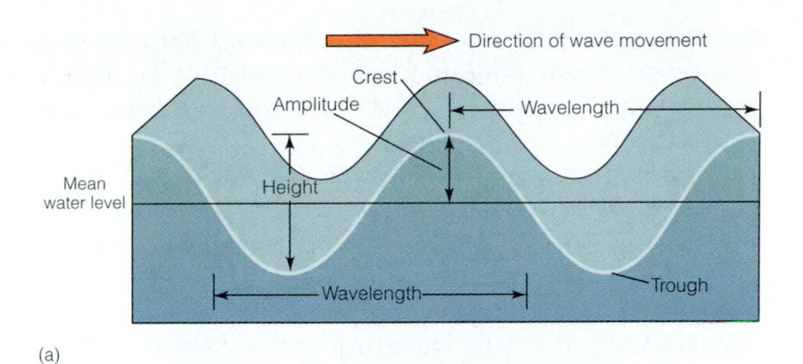

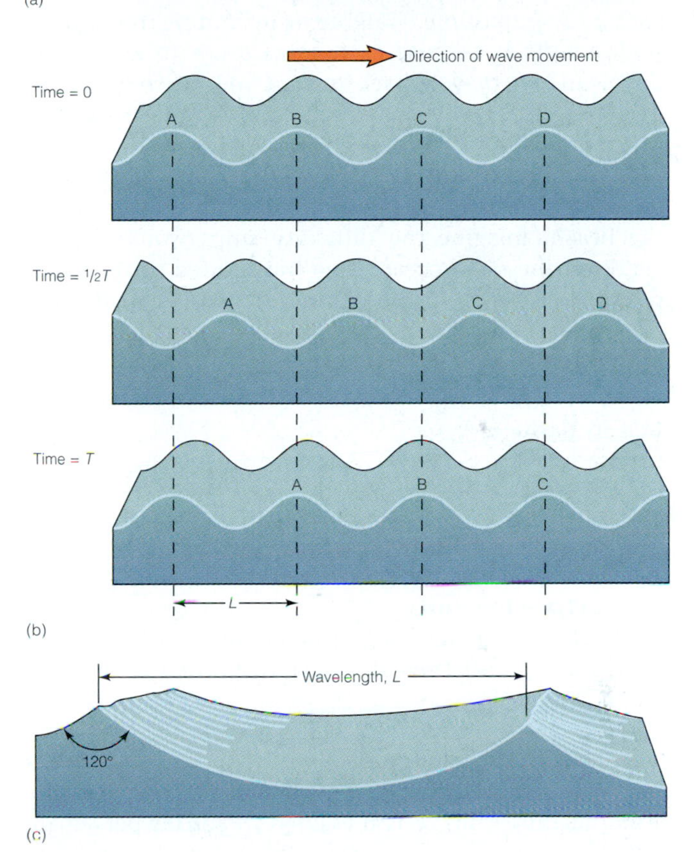

FIGURE 11.1 (a) All waves can be characterized by the presence of a crest and a trough, and by their wavelength and wave height. (b) Progressive waves travel over the ocean surface at a speed that can be characterized by their period (T) or their frequency (f), which is the inverse of the period. Most progressive waves are smooth sine waves with rounded crests and troughs. (c) As waves are built by the wind, they develop a pointed crest and become trochoidal in shape.

11.3 WHAT IS WAVE MOTION?

When a wave moves across the ocean surface without breaking, there is almost no net forward motion of the water itself. As we can easily see at the shore, objects such as logs and surfers floating outside the **surf zone** are not carried inshore with the waves. As waves pass, the floating objects ride up and down on each wave but remain at almost the same location indefinitely. Surfers must paddle forward to "catch" a wave before they can ride it inshore.

If water moved forward with the wave, the world would be an entirely different place. If you have walked or swum out through the surf on a beach, you have experienced what this would be like. When a wave breaks on the beach, the foaming water in the wave crest does move forward faster than the water under the wave is moving back away from shore. Walking or swimming through the surf is difficult because each wave tends to knock you down and carry you back toward the beach until the trough of the wave arrives and carries you seaward as water drains down the beach. However, once you have passed through the surf zone, the same waves that knocked you down in the surf no longer push you toward the beach. Imagine the difficulty ships would have if water on the open ocean were transported in the direction of the waves, in the manner that water in the surf zone is.

Wave Energy

A wave has both **kinetic energy** and **potential energy**. Kinetic energy is possessed by water molecules that are moving in the wave; and potential energy, by water molecules that have been displaced vertically against **gravity** and **surface tension**.

At the wavelengths of most ocean waves, the total energy (E) per unit area of a wave is approximately

$$E = 0.125(g\rho H^2)$$

where ρ is the absolute **density** of water (in $g{\cdot}cm^{-3}$), g is the acceleration due to gravity ($9.8\ m{\cdot}s^{-2}$), and H is the wave height (in m). E is measured in **joules** per square meter ($J{\cdot}m^{-2}$). Wave energy does increase with wavelength, but this factor becomes important only for waves of very long or very short wavelength.

Because water density changes very little in the open oceans (Chap. 7) and g is a constant, the total energy of a wave depends primarily on its height. The total energy of a wave is multiplied by a factor of 4 if the wave height is doubled.

Restoring Forces

Instead of moving forward with the wave, each water molecule within a wave moves in an orbital path. In deep water, the orbital path is circular (**Fig. 11.2**). Only the waveform and energy associated with the wave move forward. For the waveform to move forward, a **restoring force** must exist that tends to return the sea surface to its original flat configuration after the water is initially displaced.

The principal restoring forces acting on ocean waves are surface tension and gravity. Surface tension pulls the surface equally in all directions, contracting the surface to its minimum area—a flat plane (Chap. 7). A trampoline provides a good analogy for surface tension. The trampoline surface is depressed and stretched when someone lands on it, but its "surface tension" causes it to snap back to its normal flat configuration, launching the trampoliner into the air.

Gravity acts on water molecules within a wave and causes a **pressure gradient** to develop beneath the sloping surface of the waveform (**Fig. 11.3**). The water flows in response to the pressure gradients and tends to flatten the sea surface. In simpler terms, gravity causes the water to fall from the high parts of the wave to fill the depressions and restore the surface to a flat configuration.

Although interactions of restoring forces with water molecules in the wave are somewhat complicated, the

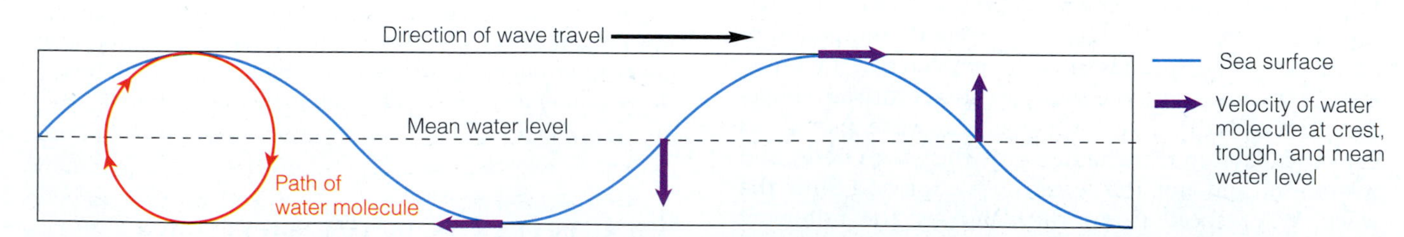

FIGURE 11.2 In a progressive wave, water at the crest (or at points directly below the crest) is moving forward but has no vertical velocity. After the crest passes, the forward velocity slows, and vertical velocity increases until, halfway between crest and trough, the velocity is entirely vertical. As the wave continues forward, the vertical velocity is slowed, and the water gains velocity in the backward direction. At the trough, the vertical velocity is zero, and the backward velocity is at a maximum. The waveform moves forward, but each particle of water moves in a circular orbit whose diameter equals the wave height (for water at the surface) and returns to its starting point after each wave. You can see this for yourself if you push down on a water bed. The wave travels across the bed, but the plastic "surface layer" must return to its original location after the wave has passed.

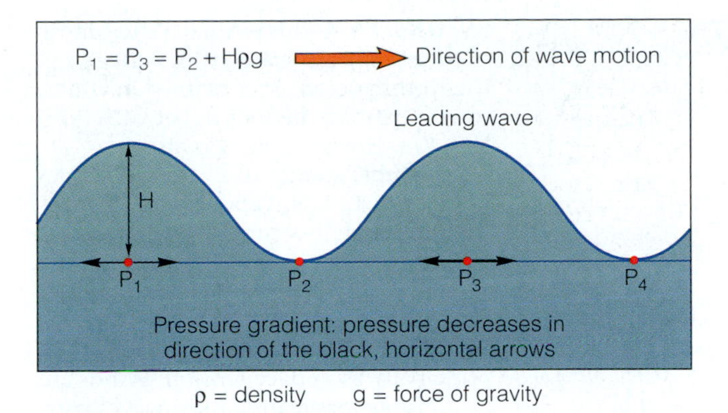

FIGURE 11.3 The horizontal pressure gradients under a wave illustrate how waves move. The pressure difference between locations directly beneath the crest and locations at the same depth beneath the trough tends to move water away from each crest toward the trough. Between two crests, the pressure gradients are symmetrical and tend to move water toward the trough. However, the water under each crest has kinetic energy because it is moving forward with the wave motion, and this energy is transferred forward from wave to wave with the wave motion. In front of the leading wave, the pressure gradient tends to move the water forward, displacing the still water surface in front, transferring some of the kinetic energy into potential energy, and beginning the wave motion. Some energy is transferred backwards from the last wave of a group of waves, in much the same way, to form a new wave at the back of the train.

principle is straightforward. Consider a water molecule located at the high point of an elevation of the sea surface. The water molecule has potential energy because it is elevated above the mean water level. The restoring forces accelerate the molecule downward, and potential energy is converted to kinetic energy.

When the molecule reaches the mean surface level, its initial potential energy has been converted to kinetic energy and its motion is vertically downward. Because it has kinetic energy, it continues its downward motion but is slowed by the restoring forces. As it slows, kinetic energy is converted to potential energy. The molecule will continue in motion until all of its kinetic energy is converted to potential energy, at which point the molecule is at the same distance below the mean surface level as its starting point was above that level.

This process explains why water moves up and down as potential energy is converted to kinetic energy and back. Why, then, does the water surface in a progressive wave move in a circular path and not simply oscillate vertically up and down, falling from crest to trough and then flowing back up to the crest to repeat the cycle? In fact, this simple vertical oscillation can occur at some locations where a vertical barrier exists at the exact location of a wave crest or trough. Where this happens, the wave is not a progressive wave but instead is a standing wave. Standing waves (discussed later in the chapter) behave differently from progressive waves.

In contrast, if no barrier is present, the leading wave of any series of waves is adjacent to an undisturbed water surface. The horizontal pressure gradient under the leading wave moves water molecules toward the undisturbed surface (**Fig. 11.3**). This movement causes the undisturbed water surface at the leading edge of the wave to be displaced. Thus, the restoring force leads to a disturbing force that transfers energy to undisturbed water ahead of the leading wave, wave energy is translated forward, and the wave is progressive.

The two restoring forces, gravity and surface tension, act on all waves. Gravity is the principal restoring force for most waves within the range of wavelengths normally present in the oceans, and such waves are often called "gravity waves." In contrast, for waves with very short periods (<0.1 s) that can have only very small wave height (for reasons discussed later), gravity is less important and surface tension is the principal restoring force. These small waves are called **capillary waves** because *capillarity* is another term for surface tension. For waves with very long periods, such as the tides, an additional restoring force, the **Coriolis effect** (**CC12**), which is not actually a "force," is important (**Fig. 11.4**).

11.4 MAKING WAVES

Once they are created, waves must have a restoring force to sustain the wave motion, but waves must be formed initially by a displacing force. Small waves can be created by small displacing forces acting over short periods of time. Large waves can be created only by large displacing forces or lesser displacing forces exerted continuously for extended periods of time.

Forces That Create Waves

Most ocean waves are caused by the interaction of winds with the water surface, but waves are also created by impacts on ocean water, by rapid displacement of ocean water, by gravitational attraction between the Earth, moon, and sun, and by the passage of vessels or marine animals through the sea surface. The ocean waves created by these various processes vary from waves with a period of less than one-tenth of a second to waves with a period of more than a day (**Fig. 11.4**).

Impacts on ocean water can be generated by earthquakes or volcanic explosions that cause the seafloor to move abruptly, or by meteorite impacts on the ocean surface. Rapid displacements of ocean water can be caused by seafloor **slumps** (landslides) and **turbidity currents**, and by collapses of coastal cliffs. Such impacts can

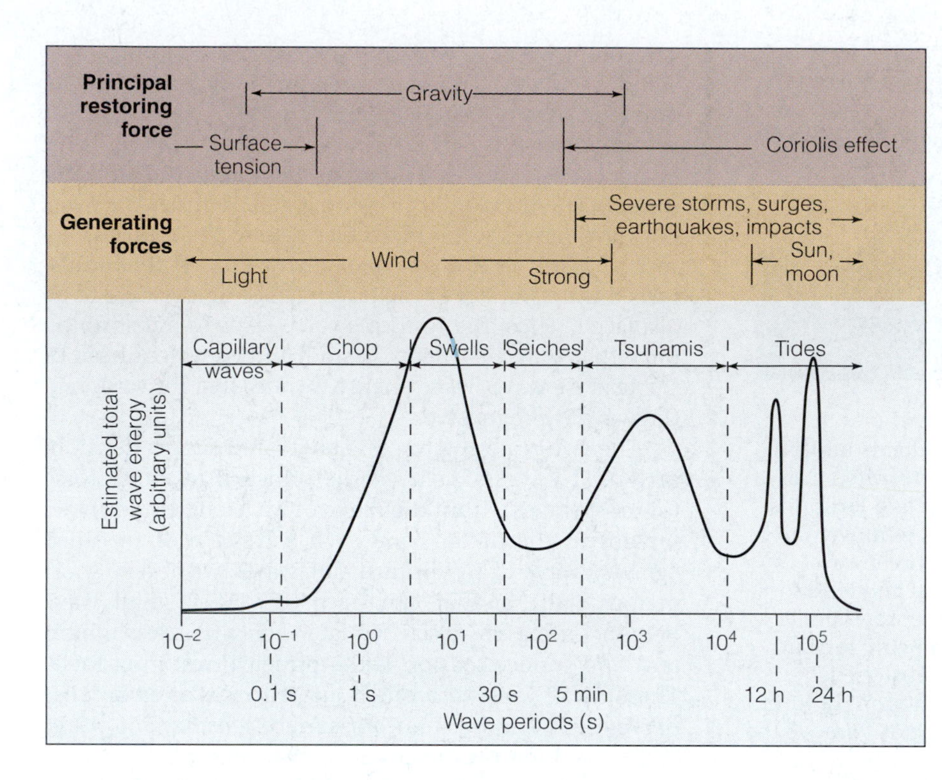

FIGURE 11.4 It is estimated that most of the energy associated with waves in the world's oceans is contained in wind waves that have periods of between 0.2 and 30 s. However, there is also considerable energy associated with tides (periods of 12 h and above) and, despite their infrequent occurrence, with tsunamis and severe storm waves (periods between about 1 minute and 1 h). The smallest waves, capillary waves, are created by winds and restored principally by surface tension. Winds are responsible for creating most waves with periods less than about 15 minutes, and gravity is responsible for restoring all except the shortest-period capillary waves within this range of periods. Longer waves are formed by earthquakes and tidal forces, and the Coriolis effect becomes the principal restoring force for the longest waves.

create waves with very long periods (typically 10 to 30 minutes) called **tsunamis** (or sometimes "seismic sea waves"). Tsunamis are occasional events and contribute only a fraction of the total wave energy of the oceans (**Fig. 11.4**), but they can be extremely destructive.

Impact and displacement waves can also be generated by comparatively minor events, such as whales and dolphins jumping out of the water and ships moving through the water leaving wakes. Waves created by these minor impacts are insignificant in number and intensity, in comparison with wind-generated waves. However, in sheltered harbors ship wakes may be the principal source of wave energy and erosion of the **shoreline**.

The gravitational attraction between the Earth, moon, and sun is the force that creates waves that we know as tides (Chap. 12). Tide waves have periods of predominantly about 12 or 24 h and contribute a significant amount of the total wave energy of the oceans (**Fig. 11.4**).

Wind Waves

As we know from experience, winds are highly variable, or gusty, from second to second. Air moving over a land or ocean surface becomes **turbulent** when more than a gentle breeze blows. Small variations of wind speed and pressure that occur as the wind blows across a smooth sea surface are believed to lead to the creation of capillary waves, which then grow larger if the wind continues to blow. Capillary waves have a maximum wavelength of 1.73 cm, rounded crests, and *V*-shaped troughs.

The mechanism by which wind generates capillary waves is not fully understood, but two factors seem to be important. First, where atmospheric pressure is slightly increased because of turbulence in the atmosphere, the sea surface tends to be slightly depressed in relation to the adjacent surface, where the pressure is lower. Thus, when winds blow across a perfectly flat sea surface, tiny areas of elevated or depressed sea surface are formed. Second, when winds blow over the ocean surface, a **shear stress** (**friction**) develops between air and water because of the velocity difference across the air–sea interface.

The formation of capillary waves increases the roughness of the surface, which in turn increases the shear stress between wind and ocean surface and allows larger waves to be built. Capillary waves also alter the sea surface, so that some areas are tilted slightly up toward the wind (the back of the wave) and others (the front of the wave) are tilted slightly away from the wind. The wind can push harder on the back of the wave than on the front, further accelerating and building the wave (**Fig. 11.5**).

Wind–wave interaction actually is more complicated than this simple description suggests. For example, a wave changes the wind flow across the water surface just as an aircraft wing alters the airflow to provide the lift necessary for a plane to fly. Once a wave is formed, the wind tends to build the wave height, primarily by uplifting the **leeward** side of the wave. In addition, differences in wave velocities cause small capillary waves to combine into the longer-period, higher waves known as

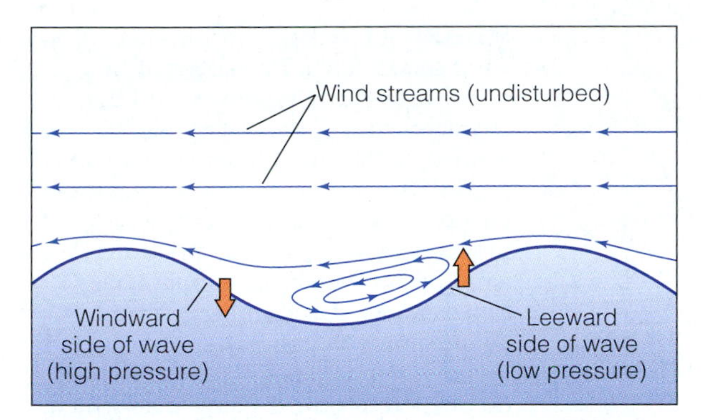

FIGURE 11.5 When the wind speed is great enough to form waves, the waves interfere with the wind flow immediately above the surface of the water. The wind is able to push harder on the elevated, windward surface of the wave, and the air pressure is slightly lower on the leeward side of the wave than on the windward side. These forces combine to increase the height of the wave.

"gravity waves." In this way, as the wind continues to blow across the sea surface, wind energy is accumulated and longer-period waves are formed that collectively represent most of the ocean wave energy (**Fig. 11.4**).

11.5 SEA DEVELOPMENT AND WAVE HEIGHT

As a wave's height is built up by winds, its shape is modified from that of capillary waves. The shape first becomes close to the smooth sine wave form shown in **Figure 11.1a,b**. Subsequently, the crest and trough are modified progressively as the wave builds, and the wave shape becomes trochoidal (pointed crests and rounded troughs; **Fig. 11.1c**). As wind energy is absorbed by waves, their height, speed, period, and wavelength are increased.

Winds never blow uniformly over the water, because they are always highly variable in both time and space. Consequently, in the area where the waves are formed by the wind, waves of many different heights, wavelengths, and even directions of travel are present at the same time. Mariners refer to this confused state as a "sea." In a "calm," there are no significant waves and the sea surface is flat. In a **swell**, waves are generally smooth, mostly of the same wavelength and from the same direction. How a sea becomes a swell is explained later in this chapter when we consider **wave dispersion**.

Waves sometimes break in deep water just as they do near the shore, but waves breaking in deep water are often called "whitecaps." Waves break because they have become too steep. When the steepness of a **deep-water wave** (H/L) reaches 1:7 (wave height equals one-seventh of wavelength), the wave becomes unstable and the crest tumbles down the forward slope of the wave, creating a breaking wave. If winds are very strong, the tops of the largest waves may be blown off by the wind.

Waves break when they have reached their maximum possible height for the wind speed. They also break when seas are developing and the waves cannot be modified quickly enough to longer wavelengths and greater heights to absorb the energy introduced by the winds. As a wave becomes oversteepened and breaks, some of its energy is dissipated by turbulence, which releases heat. Consequently, this excess energy cannot be used to increase the wave height.

Factors Affecting Maximum Wave Heights

A sea is said to be fully developed when the amount of wave energy lost to turbulence in breaking waves is equal to the difference between the total energy input from winds and the amount of wind energy needed to maintain the sea. Waves of many different heights and periods are present in a fully developed sea, but storms with stronger winds produce waves with longer maximum periods than are produced by lower wind speeds (**Fig. 11.6a,b**). The maximum height of waves created by any specific storm or series of storms depends on the wind speed and the length of time the wind blows. The maximum height also depends on the wind **fetch**, which is

CRITICAL THINKING QUESTIONS

11.1 If the surface tension of water were much smaller than it is, how would you expect this to affect the formation of waves on the ocean? How would the waves behave differently when they approached a beach?

11.2 You are sailing across the water into an almost regular swell that comes from directly ahead of you. Suddenly you notice that the same swell now appears to be coming at you from two slightly different directions at a small angle from each side of the boat's bow and the wave pattern ahead looks confused compared to the smooth swell you were sailing on a few minutes before. What are you seeing, and why? Should you be concerned?

11.3 Sailing ships exploring the oceans before there were maps were able to safely enter lagoons of atolls or those behind fringing coral reefs, even though in many such reefs there are only a few narrow entrances where the water is deep enough for safe passage.
(a) How did they do this?
(b) How might they have done this differently on days when the ocean surface was extremely calm, lacking even a gentle swell?

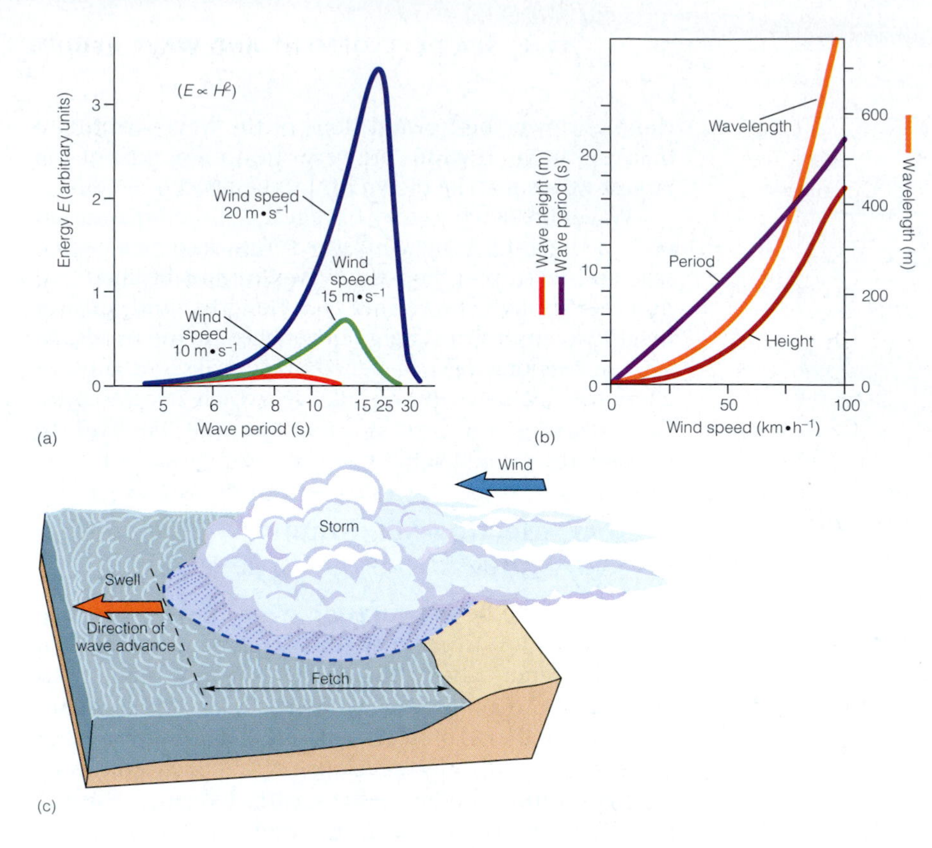

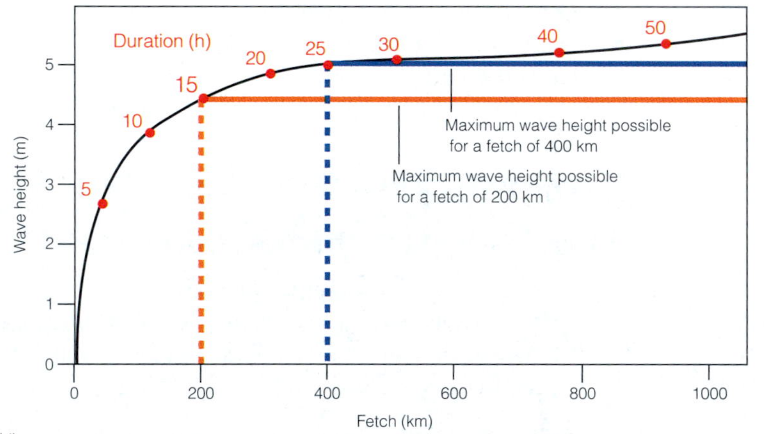

Data plotted are for a wind speed of 54 km•h^{-1}

30 ● Number of hours of continuous winds needed to reach this wave height

FIGURE 11.6 Relationships between wind speed, fetch, the length of time over which the wind blows, and the characteristics of the waves that are formed. (a) Strong (higher-speed) winds produce storms in which the total wave energy is great, and this energy is stored in waves of long periods. (b) Strong winds also produce waves that have longer wavelengths and greater wave heights, provided that the winds blow long enough and have sufficient fetch. (c) The fetch is the distance over which the winds blow across the water to create waves. Fetch can be limited by the presence of a coastline or, where the storm is completely over water, by the dimensions of the storm itself. (d) The maximum wave height that can be reached is limited by the fetch, and this maximum wave height is reached only if the winds blow for a sufficiently long time.

the uninterrupted distance over which the wind blows (**Fig. 11.6c**). In shallow water (depth less than one-half of the wavelength), water depth is a limiting factor.

Figure 11.6d shows the effect of wind duration and fetch on maximum wave height. As the wind begins to blow, waves accumulate energy from the wind. Wave height initially increases rapidly, then at a progressively slower rate. However, if the fetch is small, the maximum wave height is limited because waves relatively quickly travel out of the area where the wind is blowing. We can see the effects of a small fetch in lakes or in harbors protected from ocean waves. The fetch is restricted in such areas because the waves encounter a shoreline. No matter how strong winds are or how long they blow, only small waves can be created in harbors and all but the largest lakes.

The highest ocean waves are created where winds are strong and blow persistently over long fetches. High waves can also be generated when a series of strong storms passes in the same direction across the same fetch over a period of several days or longer. In this situation, some of the shorter-wavelength waves created by one storm do not have time to travel out of the area before the next storm arrives. Each successive storm simply builds the height of waves remaining from preceding storms.

Maximum Observed Wave Heights

Persistent winds and storm tracks are arranged in bands with an east–west orientation (Chap. 9). Therefore, calm regions alternate with stormy regions from north to south in the oceans. The calm regions limit the fetch that can occur in a north or south direction in the ocean basins, but that does not mean that waves travel only east or west. First, the **trade winds** and **westerly** winds in the east–west wind bands do not blow directly east or west. Second, storms produce rotating winds and, thus, winds and waves of all directions. Third, if waves encounter a landmass or **reef**, the direction of wave travel may be altered as discussed later in this chapter.

Not surprisingly, high waves are most common in the Pacific Ocean, because the Atlantic and Indian Oceans are much narrower and provide a more limited fetch. The longest fetches are within a band of westerly winds stretching around Antarctica. A giant wave at least 34 m high was measured by the U.S. Navy vessel *Ramapo* on February 7, 1933. Although that was the largest wave reliably reported, higher waves undoubtedly occur from time to time. It is not surprising that such higher waves have not been measured, since any sailor whose vessel is faced with a 30-m or higher wave has many concerns other than measuring wave height. Most vessels would survive the ride through even the highest wave unless the wave were extremely steep-sided and breaking. A 30-m-high steep-sided wave breaking over the deck of even the largest ocean vessel may be the last wave the vessel ever encounters. Indeed, such large waves may have caused the disappearance of many ships.

So far, the highest waves reliably reported in the Atlantic and Indian Oceans were about 15 m high. However, in September of 2004 the center of **hurricane** Ivan passed directly over six tide/wave gauges mounted at depths of 60 to 90 m on the Gulf of Mexico **continental shelf**. The highest wave measured by these gauges was 27.7 m. The area of highest winds within Ivan did not pass directly over any of the gauges, and it was calculated that the highest wave generated by this storm may have exceeded 40 m. Satellites such as the *TOPEX/Poseidon* satellite can measure ocean surface heights to an accuracy of within less than 5 cm. Unfortunately, these sensors cannot yet be used to measure the heights of individual waves, because the width of their measurement beams is large compared to the distance between ocean wind waves.

Effects of Currents on Wave Height

Steep-sided waves can be created when waves travel in the direction opposite that of a strong ocean **current**. The opposition of current and wave motion shortens the wavelength of the waves and increases their steepness, particularly on the front face of the wave. This effect is illustrated in **Figure 11.7**.

In extreme cases, waves can have almost vertical faces. Long-wavelength waves travel faster than ocean

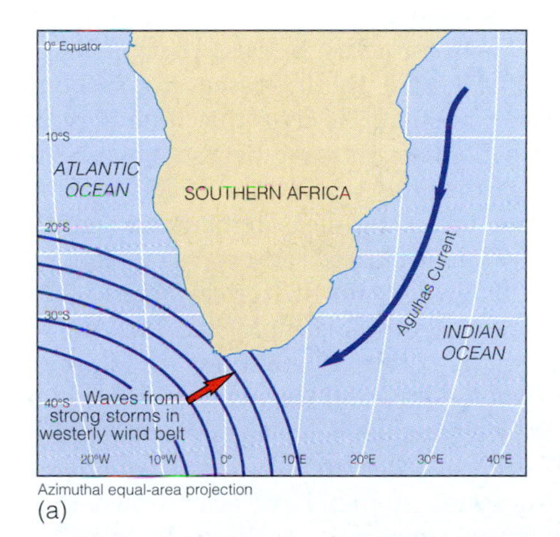

(a)

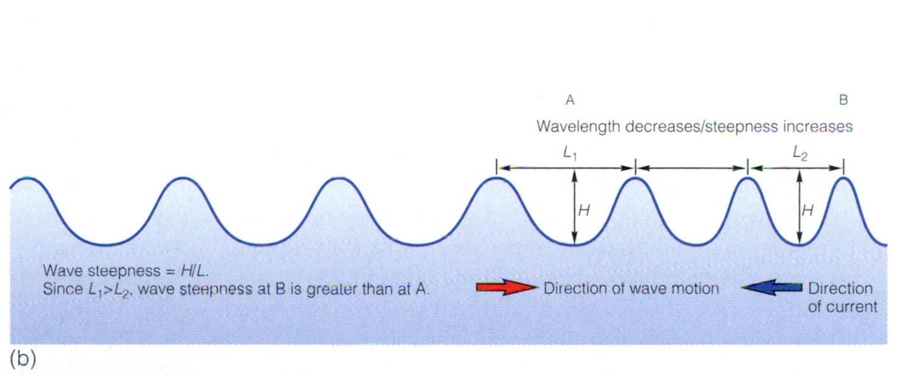

(b)

(c)

FIGURE 11.7 (a) Waves from around Antarctica travel northward and meet the strong, southward-flowing Agulhas Current. (b) As each wave encounters the current, its forward speed is decreased because the current moves surface water southward and water particles in each wave are a little farther south after each wave orbit, instead of in their original position. As a wave is slowed by this process, the following wave that has not yet encountered the current is not slowed, so its crest becomes closer to the crest of the wave preceding it. Hence, wavelength decreases and steepness increases as the figure shows. Waves are slowed by friction with the seafloor when they enter shallow water, which causes the wavelength to shorten and steepness to increase in the same manner. (c) The tanker *World Glory,* sinking after breaking its back in waves steepened by the Agulhas Current.

CRITICAL THINKING QUESTIONS

11.4 Describe how waves would be refracted if they entered a very long basin whose underwater sides described a perfect *V* shape, with the depth of the *V* slowly decreasing with distance into the basin. Draw your answer showing the shapes of several successive waves as they travel up the basin. Include wave rays in your diagram.

11.5 In some parts of the oceans, there are often strong seas with breaking waves. In other areas, the waves are usually smooth swells. In yet other areas, the sea surface is often calm, and finally, in some areas, ships may encounter large swell waves with steepened fronts. Using what you have learned in this chapter and in Chapters 9 and 10, identify on a map at least one area of the Pacific Ocean where you would expect to find each of these situations. Explain the reasons for your choices.

11.6 In strong storms at sea, ships often alter course, for safety, so that they travel directly into the direction from which the wind and waves are coming, even though this may be far from their intended course of travel. Why do you think it is safer to travel into the wind and seas than at an angle across this direction?

vessels. When a vessel encounters a large wave with a nearly vertical face, the vessel may be unable to climb up and over the wave rapidly enough to avoid what is often a fatal plunge under the wave.

Such steep waves occur often in some areas, including where the swift Agulhas Current flows south along the Indian Ocean **coast** of southern Africa (**Fig. 11.7a**, Chap. 10). Large waves built by Antarctic storms travel north into this area, which is a major shipping lane for traffic between the Indian and Atlantic Oceans and is heavily traveled by oil supertankers. Several ships have been severely damaged or sunk there by shortened and steepened waves. In 1968, for example, the tanker *World Glory* broke in two, spilled its entire cargo, consisting of 14 million gallons of oil, and sank after encountering such a wave (**Fig. 11.7c**).

Tankers are constructed with most of the hull weight concentrated in the engine room and crew quarters at the stern, and the anchor-handling and other equipment in the bow. In the middle of the ship, oil is stored in separate compartments held together relatively weakly by the hull. A tanker can break in the middle if it becomes suspended with its center section on the crest of a high, steep-sided wave or suspended between two waves with the bow section on one wave and the stern on the other. Sophisticated satellite-based radar and other sensors are providing rapidly improving wave forecasting in areas such as the Agulhas Current.

Are Wave Heights Increasing?

Higher waves possess higher energy than lower waves and can create more damage when they impact a shoreline. Therefore, it is important to know whether wave heights are increasing as a result of changing **climate** that might increase wind speeds. However, measuring wave heights is not as simple as just recording the maximum wave height at one location over a period of time. Generally, oceanographers measure a parameter called "significant wave height," which is a measure of the average height of approximately the highest one-third of all waves within a given area at a specific time. Methods of measuring wave heights have varied in the past several decades, so there are very few long time-series records of wave heights measured by a consistent method. This lack of consistent methodology makes any conclusion about long-term trends in wave height difficult.

Wave heights in the North Atlantic Ocean were measured by researchers on a lightship moored off the southwest tip of England during 1960–1985. These measurements showed not only that the maximum wave heights vary widely between years, but also that the maximum wave heights apparently increased from about 12 m to about 15 m, about 25%, over this time period. Since 1985, similar increases in wave heights have been observed in other parts of the Atlantic and Pacific Oceans by other techniques, including measurements of significant wave height by satellite-based sensors, calculation of significant wave heights from historical wind data records, and seismological data. Seismological data can be used because the microscale fluctuations in seismic records are related to the impact of waves on the coastline.

Although wave height studies have shown that wave heights are increasing in some areas of the Atlantic and Pacific Oceans, the same studies have also shown that wave heights have not increased significantly in other areas of these same oceans. At present, we do not know whether a long-term trend toward higher maximum wave heights is real. If it is real, we do not know whether it is related to climate change, whether the trend will continue, or in which areas this change is occurring. We are also unsure to what extent opposite trends or a lack of change may have occurred in other oceans.

However, we do know that any increase in maximum wave heights must be related to increased wind speeds and/or storm frequencies. Small changes in ocean surface temperature may cause changes in the ocean–atmosphere interactions that create winds (Chap. 9). Hence, if wind speeds or storm frequencies have increased, the changes may be related to enhancement of the **greenhouse effect** caused by human release of greenhouse gases

(**CC9**). Alternatively, the change in wave heights may be related to a natural long-term climatic cycle. Long-term climatic changes or cycles with periods of 30 years or more are known to occur. For example, historical records for Atlantic hurricanes suggest that they were fewer and weaker during the past 20 to 30 years than in preceding decades, but are now increasing in both number and strength again (Chap. 9).

Considerably more information will be needed to fully identify the trends in wave heights and to determine whether these trends are part of a natural climatic cycle, are related entirely to the greenhouse effect (**CC9**), or are partially natural and partially related to the greenhouse effect. These issues are important because any long-term trend of increasing wave heights will pose problems for ships and increase coastal erosion.

11.6 WAVE DISSIPATION

When the force that creates a wave is removed, the wave continues to move across the ocean surface. For example, waves created by a rock thrown into a lake continue long after the rock's impact, and large waves may crash ashore on days when there is no wind to create them. The reason is that wave energy is dissipated (lost from the wave) only very slowly.

Energy is lost from a wave because of **viscosity** (internal friction) between water molecules moving within the wave and air resistance as the wave moves across the surface, but this energy loss is very small. Frictional loss in a flowing fluid depends on the fluid's viscosity. The viscosity of water is not high (Chap. 7), so frictional loss is a very small fraction of the total wave energy in all but very short-period (<0.1 s) capillary waves. For capillary waves, viscosity is important as a means of dissipating wave energy because such waves contain only a very small amount of energy. In fact, the energy in capillary waves can be totally dissipated by viscosity in a minute or less after the wind stops. Capillary waves therefore disappear quickly when there is no wind, but larger waves do not.

Wave steepness diminishes once the wind ceases. Hence, in the open ocean, waves do not break and lose energy through turbulence unless winds blow and add excess energy to the sea. Minor exceptions occur when waves encounter winds that blow, or currents that flow, against the wave's direction of travel. Waves then slow, and wave steepness increases (**Fig. 11.7b**).

The waves created by a rock thrown into a lake spread in a circular pattern away from the point of impact. This process does not decrease the total energy of each wave. However, as the wave moves out from the impact point, it spreads over an ever-increasing area (**Fig. 11.8**), and the wave height is reduced. Storm waves also spread out from where they are generated, but some storms generate their largest waves across a broad storm **front**. The resulting waves travel out from the storm area in the same general direction, and wave energy tends to be concentrated in that direction (**Fig. 11.8b**).

Because waves lose very little energy as they travel across the ocean surface, most waves travel until they meet a shore, where their energy is converted to heat or used to erode the shore, transport sand along the coast, or create **longshore currents** (Chap. 13). Although breaking waves release a considerable amount of heat energy, this process does not cause a measurable change

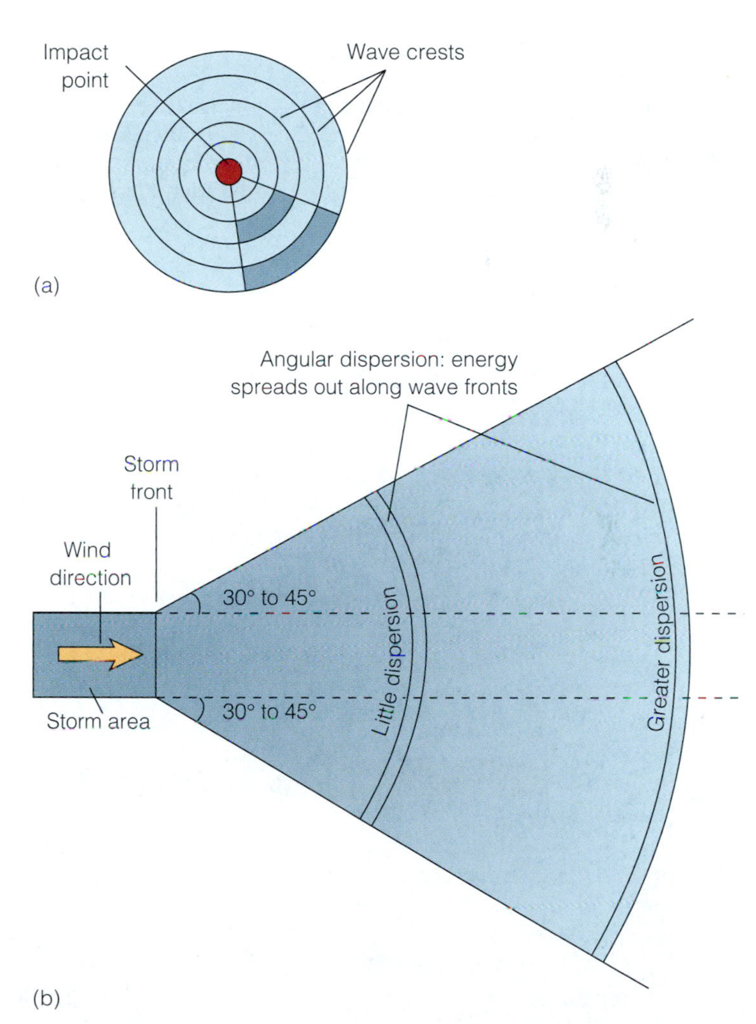

FIGURE 11.8 (a) Waves spread out uniformly in all directions from an impact point, and energy is spread along the increasing length of each wave as it moves away from this point. (b) Waves generated along a storm front travel forward in the direction of the winds in the storm, but they also spread out in a fan shape, thus dispersing (spreading) the wave energy along a progressively lengthening wave crest with distance from the storm.

in water temperature, because water has a high **heat capacity** (Chap. 7, **CC5**). In addition, heat from breaking waves is dissipated in very large volumes of water.

11.7 DEEP-WATER WAVES

Individual water molecules in waves in deep water (depth >$L/2$) move in vertical circular orbits oriented in the direction of wave travel (**Fig. 11.9**). The forward motion at the top of the orbit is slightly greater than the backward motion at the bottom of the orbit, hence there is a very small net forward movement of water in the wave (**Fig. 11.9b**). Water motion within a wave is not restricted to water depths between crest and trough. Scuba divers who have dived where long-period swell waves are present know that the surging wave motion can be felt 10 m or more below the surface, even when the surface swell waves are less than a meter high.

Water immediately below the surface moves in an orbit whose diameter is equal to the wave height, whereas water farther below the surface moves in orbits whose diameter decreases as depth increases. The decrease in orbital diameter with increasing depth depends on the wavelength. The diameter is reduced to one-half of the wave height when the depth is equal to one-ninth of the wavelength, and it is almost zero at a depth of one-half of the wavelength (**Fig. 11.9**).

FIGURE 11.9 Orbital motion of water particles beneath waves. (a) The orbits in which individual particles move within a wave decrease in diameter exponentially with increasing depth. The wave motion becomes essentially zero at a depth equal to one-half of the wavelength. (b) The particle orbits are not exactly closed. Instead each particle moves forward a very small distance (exaggerated in this figure) during each orbit. Therefore, there is a very small net mass transport of water in the direction of the wave.

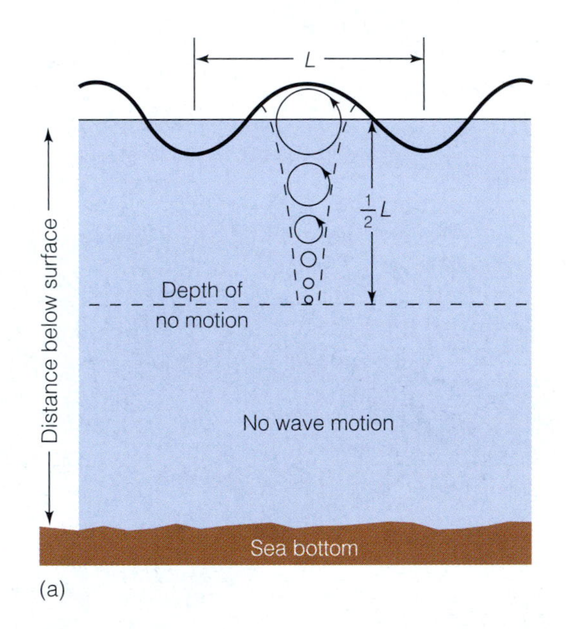

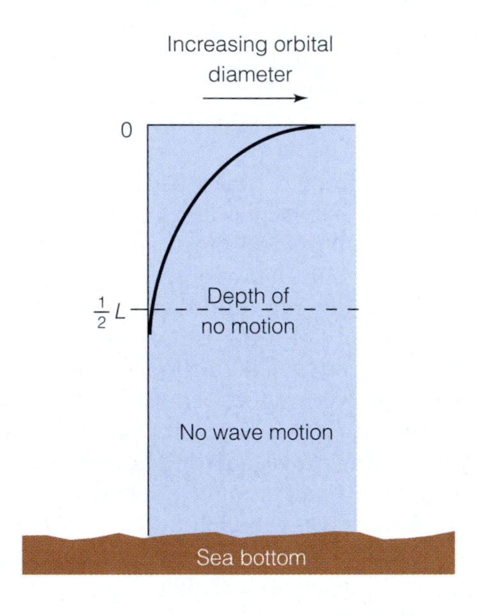

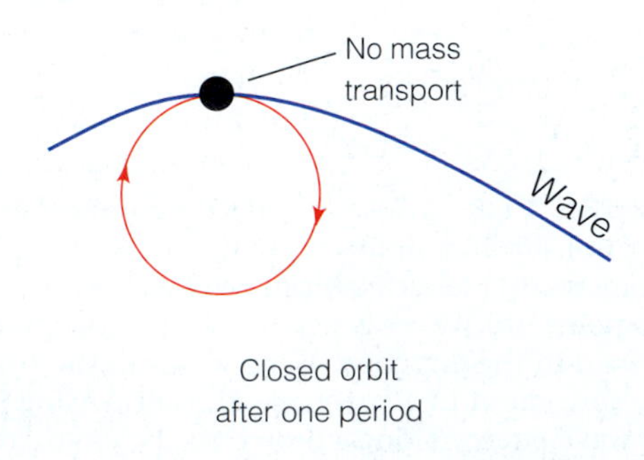

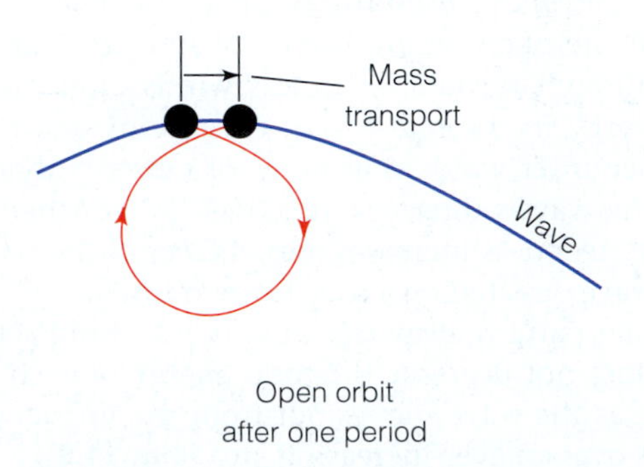

TABLE 11.1 **Wavelength, Period, Celerity, and Depth of No Motion for Deep-Water Waves**

Wavelength		*Period*	*Frequency (waves per minute)*	*Depth of No Motion (L/2)*		*Depth at Which Motion Reduced by 50% (L/9)*		*Deep Water Celerity*		
(m)	*(ft)*	*(s)*		*(m)*	*(ft)*	*(m)*	*(ft)*	*(m·s⁻¹)*	*(km·hr⁻¹)*	*(miles·hr⁻¹)*
1.6	5	1	60	0.8	2.5	0.2	0.6	1.6	5.6	3.5
6.2	20	2	30	3.1	10	0.7	2.2	3.1	11	7.0
14	45	3	20	7.1	23	1.6	5.0	4.7	17	10
25	80	4	15	13	40	2.8	8.9	6.2	22	14
39	126	5	12	20	63	4.3	14	7.8	28	17
56	181	6	10	28	90	6.2	20	9.4	34	21
77	246	7	8.6	38	123	8.5	27	11	39	24
100	322	8	7.5	50	161	11	36	12	45	28
126	407	9	6.7	63	203	14	45	14	51	31
156	502	10	6.0	78	251	17	56	16	56	35
225	724	12	5.0	113	362	25	80	19	67	42
306	985	14	4.3	153	492	34	109	22	79	49
400	1287	16	3.7	200	644	44	143	25	90	56
506	1628	18	3.3	253	814	56	181	28	101	63
624	2008	20	3.0	312	1004	69	223	31	112	70
975	3138	25	2.4	488	1569	108	349	39	140	87
1405	4521	30	2.0	703	2261	156	502	47	169	105

The speed of a wave in deep water also depends on wavelength and is approximated by the following equation:

$$C = \sqrt{gL/2\pi}$$

where g is the acceleration due to gravity. Because g (9.8 $m \cdot s^{-2}$) and π (3.142) are constants, we can simplify this equation to

$$C = 1.25\sqrt{L}$$

C is measured in meters per second ($m \cdot s^{-1}$), and L in meters (m).

Thus, the speed of deep-water waves (depth $>L/2$) increases with increasing wavelength. Because wave speed is also equal to L/T, we can calculate the wavelength of a deep-water wave if we know its period:

$$C = L/T = 1.25\sqrt{L}$$

$$L = 1.25^2 \times T^2 = 1.5625T^2$$

Table 11.1 lists wavelengths and celerities for deep-water waves of different periods. The table can be useful for scuba divers in planning a dive. If the divers count the number of waves that pass their boat per minute, they can calculate the average wave period. From **Table 11.1**, they can then determine the $L/2$ and $L/9$ depths for this wave period. If the $L/9$ depth exceeds the depth of the planned dive, the divers can expect to feel a strong wave surge throughout the dive. However, the surge will be less than that at the surface, and at depths greater than $L/2$ there will be no surge.

Wave Trains

Waves usually travel in groups called "wave trains." As the leading wave in a wave train travels, some of its energy is transferred to the molecules of undisturbed water in front of it. This energy is used to initiate orbital motion of the previously undisturbed water molecules. However, only half of the energy of the individual wave is transferred forward; the remaining energy is transferred back to the second wave in the train. By examining the pressure gradient under the wave, we can see how this happens (**Fig. 11.3**). The second wave in the train also transfers energy forward and back to the third wave. Thus, energy is lost from the front wave of the wave train. The front wave progressively loses height and eventually disappears. The energy transferred backward from the first wave is transferred progressively rearward from one wave to another, until it passes the last wave and builds a new wave at the rear of the wave train.

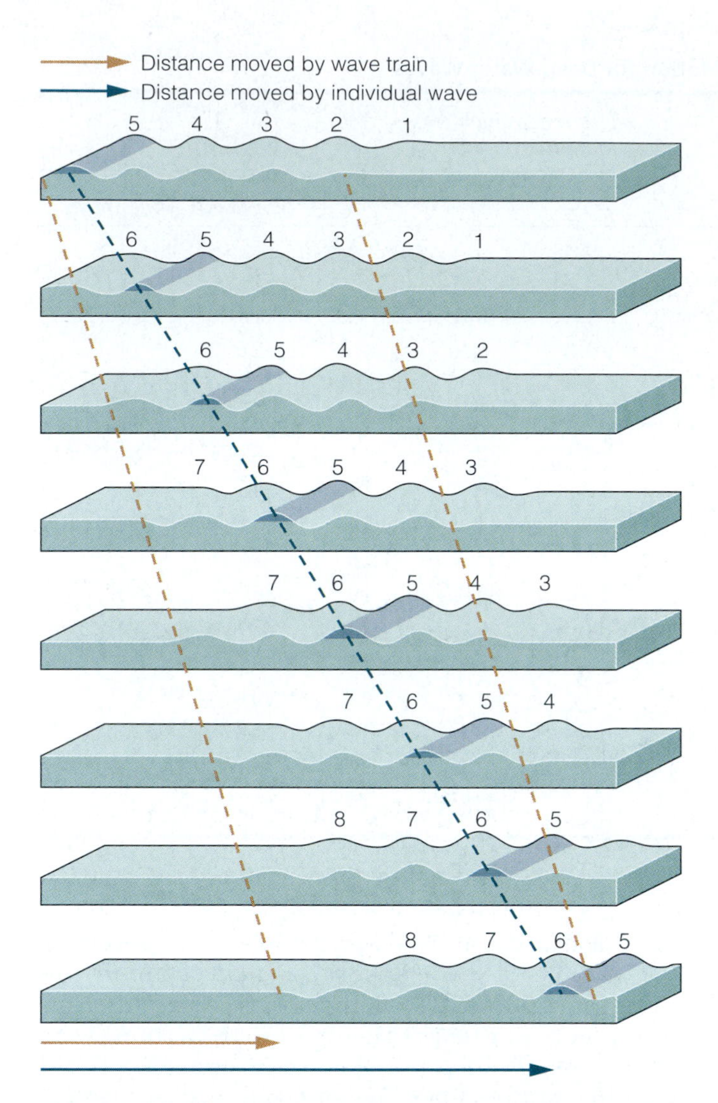

FIGURE 11.10 Waves travel in groups called wave trains. As the leading wave of the group travels forward, it transfers half of its energy forward to initiate motion in the undisturbed surface ahead and transfers the other half to the wave behind it to maintain the wave motion. The leading wave is constantly decaying (losing energy) as it transfers energy forward to displace the undisturbed water surface ahead of it. As a result, the leading wave in the wave train is constantly disappearing, while a new wave is constantly being formed at the back of the train. To visualize this process, follow wave number 5 in this diagram. Notice that the train travels at half the speed of an individual wave.

In deep water, a wave train and the energy associated with it move at one-half the speed of individual waves within the wave train. The time it takes for waves of a particular wavelength to cross the ocean must be calculated from this group speed, which is one-half of the individual wave speed specified in **Table 11.1**.

As a result of the rearward transfer of energy in the wave train, the front wave in the train continuously decays, and a new wave is continuously created at the rear of the train (**Fig. 11.10**). Individual waves form at the back of the train, progress forward as their predecessors are decayed, and eventually reach the front of the train, where they decay themselves. We can see this process by carefully watching ripples caused by a stone thrown into a lake or the waves of a ship's wake (**Fig. 11.11**).

Wave Interference

Only very rarely are waves in a specific part of the ocean regularly spaced and all of the same height. The reason for the irregularity in height and spacing is that waves of several different periods and moving in different directions pass a given location simultaneously. If we plot the sea surface height as a function of time, we get the type of complicated wave pattern shown in **Figure 11.12a**. Surprisingly, the pattern in **Figure 11.12a** is simply the sum of the five simple waves shown in **Figure 11.12b**.

When wave trains of similar wave heights but slightly different periods arrive at the same time, the sea surface will appear to alternate between periods of low and high

FIGURE 11.11 A ship's bow and stern waves travel out from the ship as wave trains. Note how the leading wave disappears as it moves away from where the ship has passed and is replaced by another wave behind it. The second wave in turn also disappears, to be replaced by another one behind it, and so on. The wave energy is slowly spread as the wave crest lengthens by dispersion, and the wave train eventually disappears.

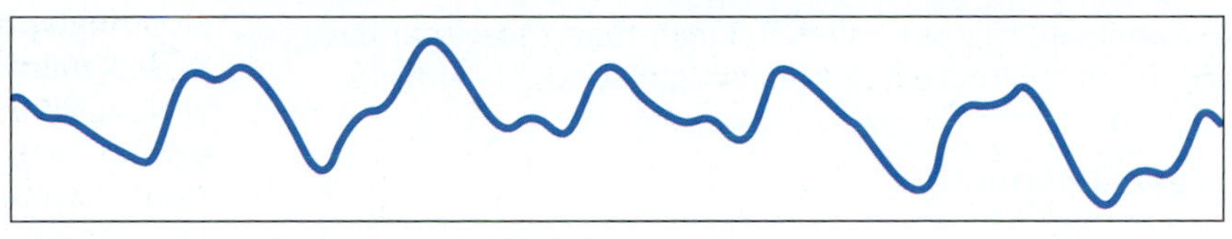

(a) Waveform generated by interference (addition) of component waves

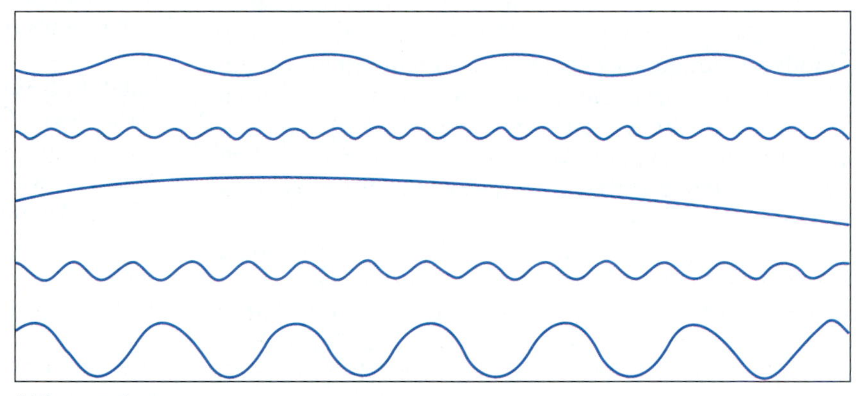

(b) Component waves

FIGURE 11.12 The waveforms observed at any point in the ocean are often complex. The complex waveform in part (a) is simply the result of interference (addition) of the five simple sine waves in part (b).

waves. During relatively calm periods (the segments where the waveform depicted in **Figure 11.13b** is almost flat), the crest of one wave arrives at almost the same time as the trough of another wave, and the elevations of the two waves cancel each other out (**Fig. 11.13**). During the intervals when the waves are high, the crests of the two waves arrive at the same time and combine to form a larger wave (**Fig. 11.13**). When waves of several different, but similar, periods arrive simultaneously, the situation is more complex. However, the process of wave addition and subtraction, called "wave interference," is what causes waves to vary in height in irregular patterns.

The irregular pattern of waves caused by wave addition is important to surfers, who await sets of big waves. Contrary to some popular beliefs, the periodicity in wave heights does not always follow a consistent pattern. Sets of waves that are especially high may appear every few minutes or only every few hours at any given location on any day. Persons who visit rocky coastlines when the sea is rough should be aware of this phenomenon. It is not safe to assume that a dry position on the rocks is beyond the reach of the waves. More than a few unfortunate individuals are washed off the rocks and drown every year because they have made that assumption. It is wise always

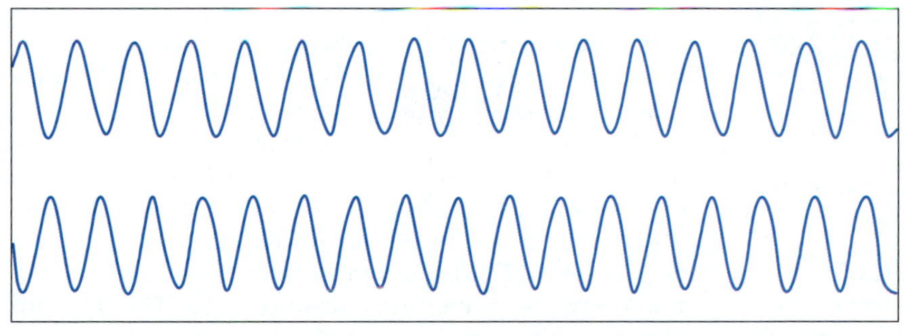

(a) Component waves

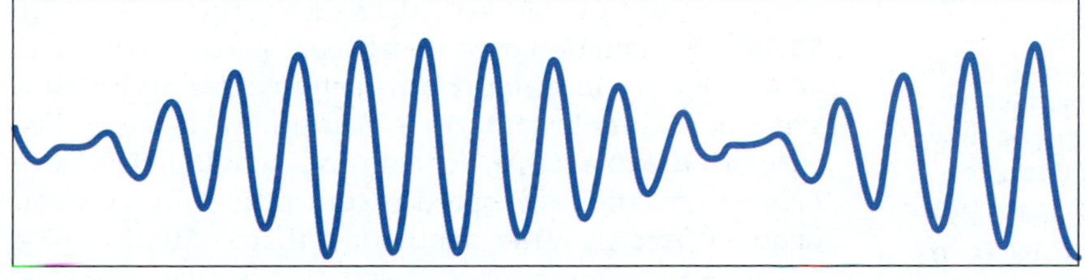

(b) Waveform generated by interference (addition) of component waves

FIGURE 11.13 When two sets of waves of very similar wavelengths arrive at the same location at the same time, their addition can produce large, long-period variations of the wave height in the resulting waveform. (a) These two simple waves have slightly different wavelengths. (b) When they are combined, they produce a very different waveform. This is why we may see periodic sets of very much higher (and lower) waves than normal arriving at the shore.

to anticipate that waves much larger than those that have arrived in the preceding hours might suddenly appear.

Wave Dispersion

Storms create waves with a wide range of periods. The speed of deep-water waves increases as period (and wavelength) increase (Table 11.1). Hence, waves with longer periods and longer wavelengths travel faster than shorter-period waves, and longer-period waves move away from the area where they were generated more rapidly than shorter-period waves do. Wave trains of different periods become separated as they move away from the storm center where they were created (Fig. 11.14). This phenomenon is called "wave dispersion" and causes waves to be sorted by period (and wavelength). Because waves are sorted by period as they are dispersed with distance from their origin, the longer-period waves from a storm arrive at locations far from the storm before the shorter-period waves do. If the storm that created the waves occurred about 100 km away, waves with a 5-s period will arrive about 3 h after waves with a 10-s period. If the storm occurred about 1000 km away, the 5-s waves will arrive about 36 h after the 10-s waves. The waves from the more distant storm will have traveled farther and been dispersed by wavelength to a greater extent (Fig. 11.14).

Normally, waves of more than one storm arrive simultaneously, so the shorter-period waves from any given storm are usually obscured by longer-period waves from another storm. The shorter-period waves can be observed if detailed wave records are analyzed mathematically.

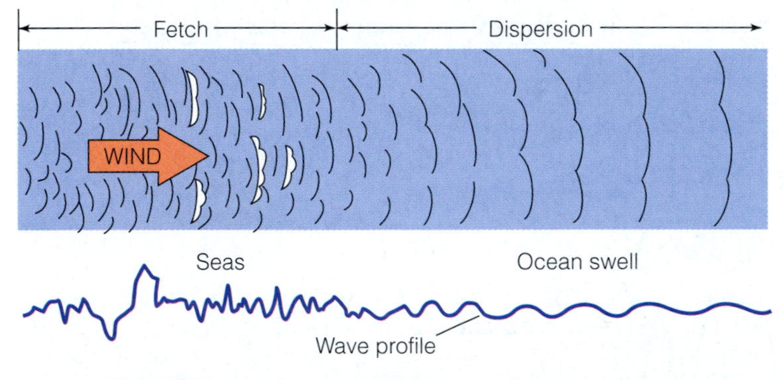

FIGURE 11.14 Within the fetch of a storm, the waves are irregular because they are the sum of many separate waves of different wavelengths, as shown in Figure 11.12. Long-wavelength waves move faster than shorter-wavelength waves, such that, as waves move away from the storm, the long-wavelength waves move ahead of the shorter-wavelength waves. This sorting of the waves by wavelengths is called "wave dispersion." Because they are separated by wavelength, the waves appear to become smoother as they travel out from the storm fetch. The greater the distance from the fetch, the greater the difference in the arrival times of waves of two different wavelengths.

Oceanographers are able to calculate the locations of storms many thousands of kilometers away by analyzing waves arriving at two or more locations. The wave dispersion at each location reveals the distance of the storm from that location, and the distances of the storm from two widely separated locations reveal the storm location. For example, storms near Antarctica can be located by wave measurements made in locations such as the United Kingdom and Brazil. Although this method is reliable, satellite observations of cloud patterns are now a much easier way to locate storms.

As waves disperse by wavelength in their direction of travel, wave steepness decreases because the waves also spread laterally and their energy is distributed over a wider area. The reduction in steepness is accompanied by a change in the wave shape, making it closer to the ideal smooth sine wave shape (Fig. 11.1a). Thus, at some distance from the area where they are generated, the waves are sorted by wavelength and form a swell—smooth undulations without sharp or breaking crests. A swell originates from a sea and is caused by the combination of wave (wavelength) dispersion and lateral dispersion of wave energy (spreading of the wave front like the spreading of a ripple as it moves away from the point of impact of a stone) as the waves travel from their origin.

11.8 WAVES IN SHALLOW WATER

On a calm-wind day at the shore, when there is a swell from a distant storm, we observe that offshore in deep water, the waves roll smoothly toward the shore without breaking. Once they reach shallow water, the same waves form breakers. Clearly something happens in shallow water that alters the behavior of waves.

Interaction with the Seafloor

When waves enter water shallower than half their wavelength, they begin to interact with the seafloor. The wave motion becomes inhibited because the solid seafloor prevents water molecules from moving in circular orbits. Molecules immediately above the seafloor can move only back and forth. As a result, the orbits of water molecules within the wave are distorted into elongated ellipses (Fig. 11.15). Compression of the orbits and friction between water and seafloor slow the forward motion of the wave.

As water depth decreases between $L/2$ and $L/20$, the forward speed of the wave decreases progressively (Fig. 11.16). The equation used to calculate the speed of waves when they are in water depths between $L/2$ and $L/20$ is somewhat complex. In these depths, waves are called **intermediate waves**. In contrast, at water depths of $L/20$ or less, the wave speed is controlled only by water depth. Waves in water depths less than $L/20$ are called

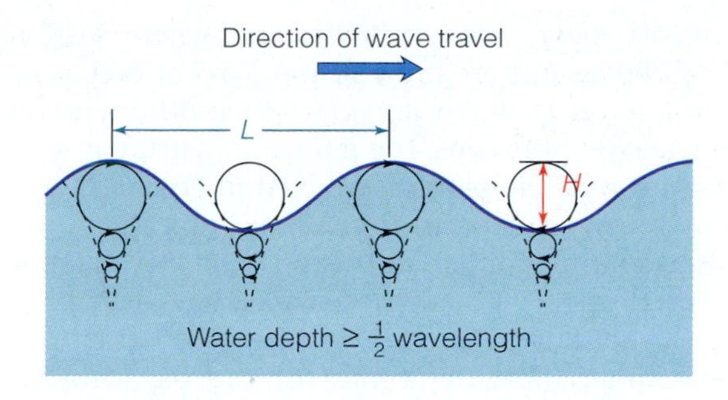

(a) Deep-water wave

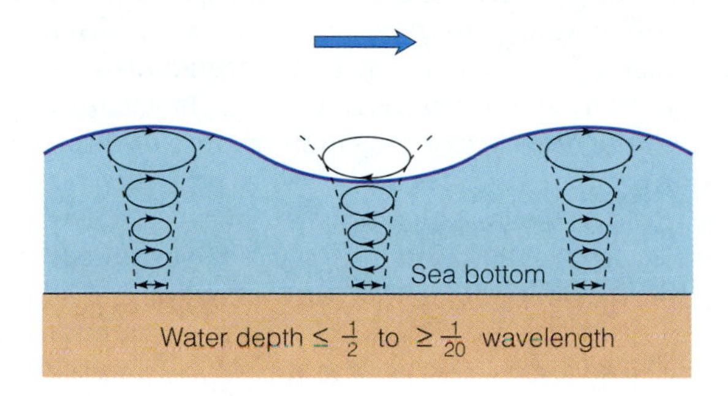

(b) Intermediate wave

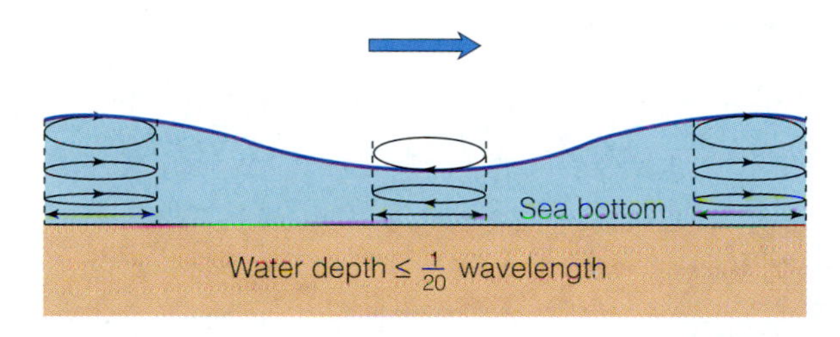

(c) Shallow-water wave

FIGURE 11.15 Effects of water depth on waves. (a) When waves are traveling across the ocean where the water depth is greater than one-half of the wavelength, the orbits of water particles in the wave are circular and are determined only by the wavelength. (b) Where the water depth is less than one-half but more than one-twentieth of the wavelength, the wave motion is slowed and distorted by friction with the seafloor, and the particle orbits are flattened into ovals. (c) Where the water depth is less than one-twentieth of the wavelength, the particle orbits are flattened so much that they become nearly horizontal, back-and-forth motions. In this case the wave speed is determined only by the depth.

shallow-water waves. The speed of shallow-water waves is given by the equation

$$C = \sqrt{gD} = 3.13\sqrt{D}$$

where C is the celerity ($m \cdot s^{-1}$), D is the depth (m), and g is the acceleration due to gravity ($m \cdot s^{-2}$).

Thus, all waves, regardless of their period, travel at the same speed when they are in water of the same depth, provided that the depth is less than $L/20$. Because wavelength and period are related, longer-period waves will become shallow-water waves in deeper water than short-period waves do (because $L/20$ is greater when L and T are larger).

As waves enter shallow water and are slowed, their period does not change. Because wave speed is the ratio of wavelength to period ($C = L/T$) and the period does not change, the wavelength must decrease as the speed decreases (**Fig. 11.16**). We can easily see why the wavelength decreases if we consider what happens to two successive waves following each other into shallow water. The first wave enters shallow water and is slowed before the second wave. The second wave comes closer to the first wave because it does not slow until it also reaches the shallow water. As they move into shallower water, both waves continue to be slowed, but the first wave continues to be slowed sooner than the second wave, which is always in deeper water. Therefore, the wavelength decreases progressively as waves move into shallower water and wave speed continues to decrease (**Fig. 11.16**). As wavelength decreases, wave steepness increases (**Fig. 11.7**).

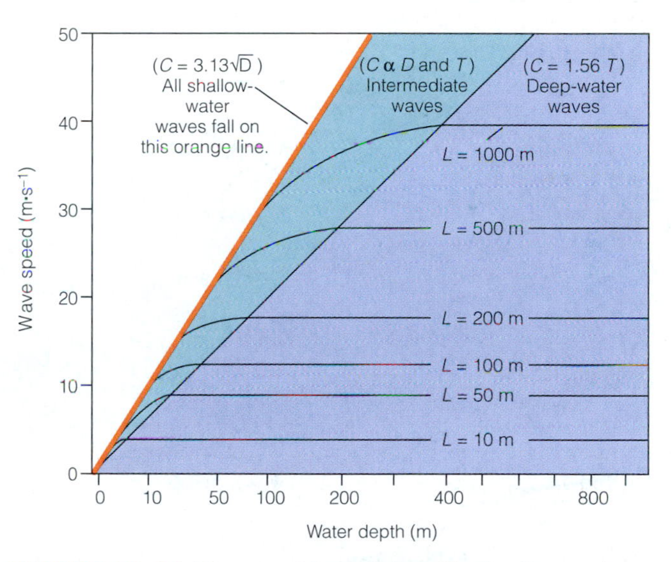

FIGURE 11.16 The speed (or celerity, C) of a deep-water wave is constant and depends on the wavelength (or wave period). For example, a deep-water wave of wavelength 1000 m has a speed of almost exactly 40 $m \cdot s^{-1}$. The speed of shallow-water waves is determined only by depth and decreases as water depth decreases. To determine the speed of a shallow-water wave in this figure, simply find the speed on the orange line that corresponds to the water depth. The speed of intermediate waves is determined by both water depth and wavelength. For example, the speed of an intermediate water wave of wavelength 1000 m decreases from almost exactly 40 $m \cdot s^{-1}$ when the water is 400 m deep, to a little more than 30 $m \cdot s^{-1}$ when the depth equals one-twentieth of the wavelength (50 m) and the wave becomes a shallow-water wave.

Wave Refraction

Waves usually approach a shoreline at an angle. Consequently, because one end of the wave crest line enters shallow water and slows while the rest of the wave is still in deeper water and traveling at its original speed, the wave is **refracted**, or bent (**Fig. 11.17a**). Lines perpendicular to the crest lines of successive waves in **Figure 11.17a** show the path that a specific point on the wave crest follows as it moves inshore. These lines are called **wave rays** or "orthogonals." As the wave continues to move inshore, more of the wave enters shallow water and slows. However, the section of the wave that first entered shallow water is still moving into progressively shallower water and slowing further, so wave refraction continues.

Note that the wave rays bend toward shallower water. As the refraction process continues to bend the wave, the wave crests tend to become aligned parallel to the shore. The refraction is seldom complete, so a wave crest rarely reaches the shore or breaks at precisely the same time along its entire length. However, even if waves approach the shore at a large angle, they are refracted and always reach the shore almost parallel to it (**Fig. 11.17a**).

If waves move into shallow water where seafloor ridges or depressions are present, the parts of each wave where the water is shallower slow, while other parts in deeper water do not. Thus, the refraction pattern is modified from the simple pattern depicted in **Figure 11.17a**. In many areas, seafloor **topography** immediately offshore is a continuation of land topography. For example, many coastlines have horseshoe-shaped bays where valleys reach the shore (**Fig. 11.17b**). The valley often continues offshore from the center of the bay, and the seafloor falls away in a valley-shaped depression between submerged ridges that extend out from the bay's headlands. Most bays of this type have a horseshoe-shaped sandy beach at the center and rocky headlands at each end (**Fig. 11.17b**). As we shall see, the headlands are rocky, and the bay's interior is sandy because of the wave refraction.

Figure 11.17b shows the locations of successive wave crests as waves move toward a bay. As a wave approaches shore, the first part of the wave to encounter shallow water will be that over the undersea ridge extending out from the headland closest to the direction from which the wave arrives. While this part of the wave slows, the rest of the wave continues at its original speed. Then

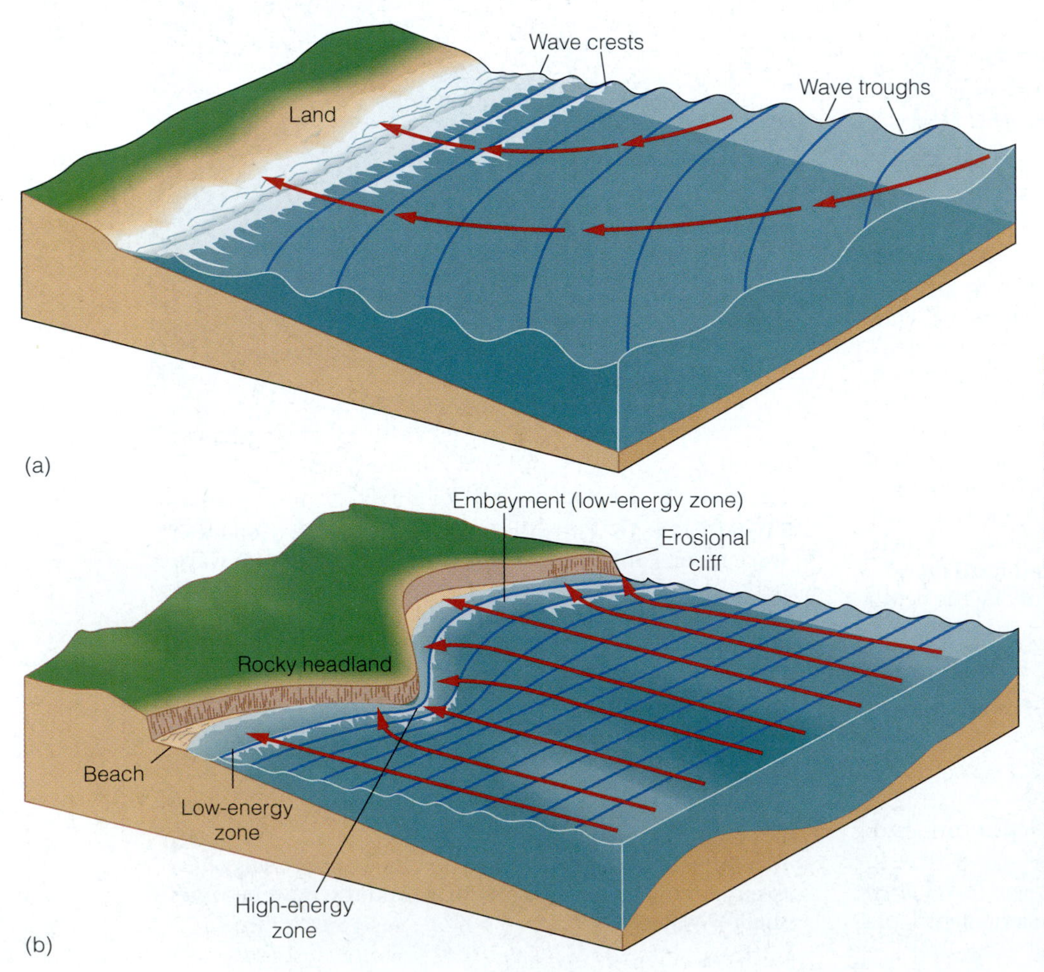

FIGURE 11.17 Wave refraction. (a) Waves generally approach the coastline at an angle. Consequently, part of the wave front reaches shallow water and is slowed before the rest of the wave. It continues to slow as the other parts of the wave move more rapidly toward shore until they, too, are progressively slowed. As a result, the wave front is bent and becomes aligned more closely parallel to the beach. The path followed by any point on the wave crest is called a "wave ray." (b) As a wave approaches a bay flanked by headlands whose topography extends onto the offshore seafloor, the first parts of the wave to be slowed are those over the underwater extensions of the headlands, and the last part to be slowed is that over the deeper water at the center of the bay. Thus, the wave is refracted, and wave energy is concentrated on the headlands and spread out inside the bay.

another part of the wave slows where it encounters the submerged ridge extending from the second headland. Thus, while parts of the wave are slowed at each headland, the rest of it continues at its full speed into the center of the bay. As the wave front enters the bay, the center section travels farther in deep water than the parts that enter the bay at each side. The wave is refracted so that it breaks at almost the same time along the entire length of the beach. Note again that the wave rays always bend toward shallower water.

Refraction redistributes the wave energy. In deep water, the wave has the same amount of energy per unit length along the entire length of its wave crest. As the wave is refracted in a bay, the total length of the wave crest is increased (**Fig. 11.17b**), and the same wave energy is distributed over this greater length. Lengthening of the wave crest also results in a lowering of the wave height. In contrast, at a headland, refraction reduces the length of the wave crest and, consequently, increases the wave energy per unit area and the wave height.

Refraction focuses wave energy on headlands while spreading wave energy along the beach within the bay. This is why we generally swim on the beach near the center of the bay and not at the headlands. The same wave that breaks gently at the middle of the beach smashes violently at the headland. This is also the reason we should exercise caution while walking on rocks of a headland. The headland is where waves are highest and crash most violently ashore. In addition, on narrow headlands where the bottom topography is appropriate, a single wave with the right period that approaches from the right direction may be refracted to hit the headland simultaneously from two directions (**Fig. 11.17b**).

The wave refraction that concentrates wave energy at headlands and spreads energy within a bay determines the character of the shore at these locations. At the headlands, where wave energy is focused, sand does not accumulate, because it is carried away by the wave action, and the shore is steadily eroded. In contrast, within the bay, gentler wave action transports sand toward the shore, where it builds and maintains the beach (Chap. 13).

In areas where the seafloor has complex topography with offshore rocks, ridges, and depressions, wave refraction can be far more complicated than the simple patterns depicted in **Figure 11.17**. Sometimes we can tell where such underwater features are by carefully watching wave refraction patterns from a beach or headland.

Breaking Waves

When waves enter shallow water and interact with the seafloor, their height is altered by the interaction. At first, the wave height is reduced slowly as the water depth decreases below $L/2$. This loss in wave height is caused by flattening of the orbital paths of water molecules in the wave (**Fig. 11.15b**). However, in water depths of about $L/10$ and less, the trend reverses and wave height increases rapidly as water depth decreases (**Fig. 11.18**). The reason is that wavelength decreases as water depth decreases, each wave is "squeezed" by its neighbors, and kinetic energy is converted to potential energy.

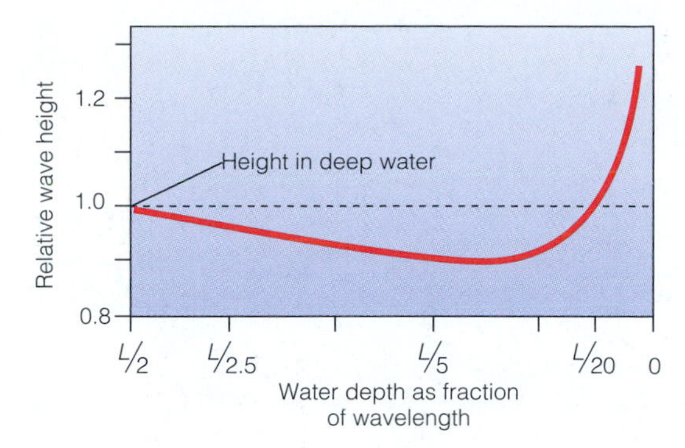

FIGURE 11.18 Wave height changes as a wave moves into progressively shallower water. The wave height declines slowly until the water depth is about one-tenth of the wavelength, then increases rapidly as depth decreases. Most waves become unstable and break before their height is much greater than it was in deep water (usually before the wave height exceeds 1.1 times its height in deep water). However, very long-wavelength waves, such as tsunamis, may be dramatically increased in height before they break.

Wavelength decreases faster than wave height as water depth decreases below $L/2$. Hence, wave steepness increases (H decreases but L decreases more rapidly, and H/L therefore increases). The speed and wavelength continue to decrease in water shallower than $L/20$ as the wave becomes a shallow-water wave (**Fig. 11.16**), but wave height increases (**Fig. 11.18**), wave steepness increases more rapidly as depth decreases (H increases, L decreases, and H/L increases rapidly), and wave shape is much modified.

Wave steepness increases until the wave becomes unstable and breaks. Wave heights and wavelengths vary among waves reaching the shore at any one time. Waves that follow each other therefore become unstable and break at different depths and thus different distances from shore. The area offshore within which waves are breaking is called the "**surf zone**."

Waves rarely approach the shore from a direction exactly perpendicular to the shoreline, and the seafloor rarely has exactly the same slope along the entire shoreline, so almost always one part of a wave breaks before another. A wave often breaks progressively along the crest of the wave as the crest moves progressively along its length into water shallow enough to cause the wave to break.

Waves break in different ways that depend on several factors, including wave period, wave height, and the slope of the ocean floor. These factors determine how quickly the wave becomes oversteepened and unstable.

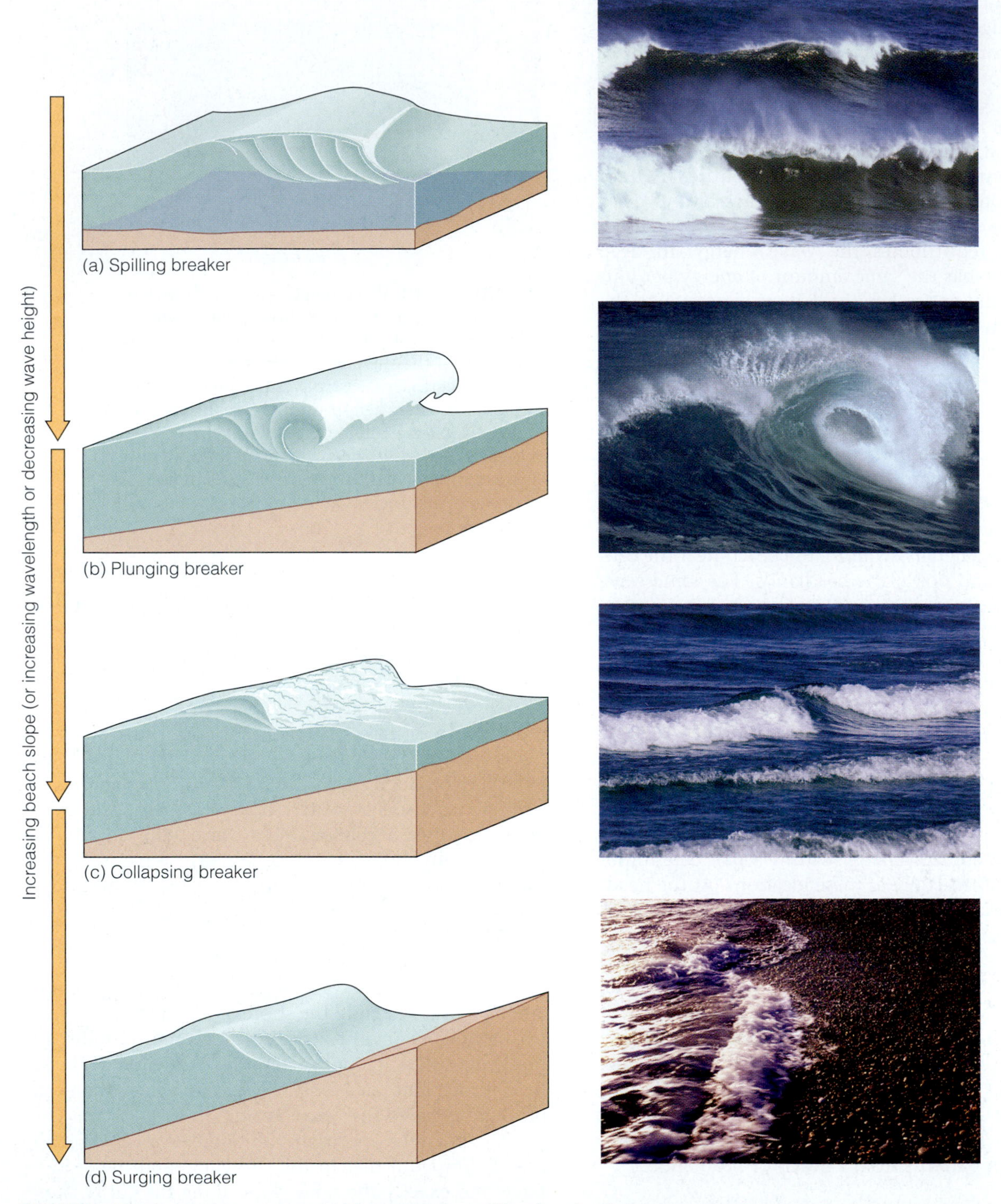

FIGURE 11.19 Breaking waves. (a) The peak of a spilling breaker becomes oversteepened, and it breaks by spilling water from the crest down the front face of the wave. (b) The peak of a plunging breaker travels fast enough to outrun the rest of the wave, and the crest curls over and smashes down on the front of the wave. (c) The base of the front side of a collapsing breaker becomes unstable before the crest, and the front face of the wave collapses. (d) A surging breaker never really breaks. The front face of the wave simply surges up and down on the beach, and the wave energy is partially reflected back out to sea.

FIGURE 11.20 This surfer is riding in the tube formed by a plunging breaker in the north end of the Banzai Pipeline, the famed North Shore surfing area of Oahu, Hawaii.

Spilling breakers are formed when the seafloor over which the wave is traveling is almost flat. When a wave reaches a water depth of about 1.2 times its wave height, it becomes unstable. Where the seafloor has very little slope, the crest begins to tumble down the forward face of the wave (**Fig. 11.19a**). The forward face fills with churning, turbulent water and air bubbles that we see as a white foam. As the wave continues inshore, wave steepness increases slowly because of the almost flat seafloor, but as water spills off the wave crest it reduces wave height and steepness. The spillage occurs at a fast enough rate to maintain the wave at its critical steepness against the tendency for the shallowing water to increase the steepness. Thus, spilling breakers break progressively as they travel inshore. Wave height is reduced progressively by the spilling action as the water depth decreases, until finally the turbulent wave crest encounters the seafloor and wave motion is ended.

Plunging breakers are the spectacular curling waves that many surfers covet (**Fig. 11.20**). They are formed when the seafloor slope is moderately steep. When the wave reaches the depth at which it becomes unstable, the bottom part of the wave is slowed more quickly than the upper part can slow or spill. Thus, the bottom of the wave lags behind as water in the wave crest outruns it while still traveling in its wave orbit. As a result, the wave crest curls over in front of the wave and plunges downward until it crashes into the trough preceding the wave (**Fig. 11.19b**).

Collapsing breakers, which are relatively rare, occur where the seafloor has a steep slope and the lower part of the wave is slowed so rapidly that the leading face of the wave collapses before the crest arrives. As the wave breaks, foam and bubbles are concentrated at the base of the forward face of the wave, and the crest collapses behind, usually with little splash (**Fig. 11.19c**).

On very steep shores, waves may appear not to break at all. The waves simply surge up and down a very steep beach with little bubble production (**Fig. 11.19d**). Surging breakers are rare, but the very small, gentle waves that lap onto some beaches in very calm conditions behave somewhat like surging breakers because even a flat beach has significant slope in relation to the tiny wave height. Surging breakers do not break up in foam and bubbles, because most of the wave energy is reflected by the steep shore face.

Waves are reflected from vertical or nearly vertical solid objects such as cliffs or **seawalls** with little loss of energy, just as light is reflected by a mirror. When waves are reflected, they pass through and interfere with the incoming waves. Wave patterns created by such reflection can be very complex and are of great concern to engineers who build harbors, marinas, and other coastal structures. Waves are also diffracted by solid objects. Diffraction occurs when part of a wave is blocked by a solid object and the edge of the remaining wave spreads out after passing the object (**Fig. 11.21**).

Often we see waves breaking at some distance from shore. Reduced in height, the same waves continue toward shore until they break again near the water's edge. This pattern is an indication that the seafloor is shallower in the offshore surf zone than it is just inshore of that area. In tropical locations, the shallow area is commonly a fringing **coral reef** (Chap. 13). In other locations, it is a **longshore bar** (Chap. 13). **Fringing reefs** and longshore bars are aligned parallel to many coastlines. They help to protect coasts from erosion by waves because they dissipate some wave energy, particularly from the highest and most energetic waves, before the waves reach the shore.

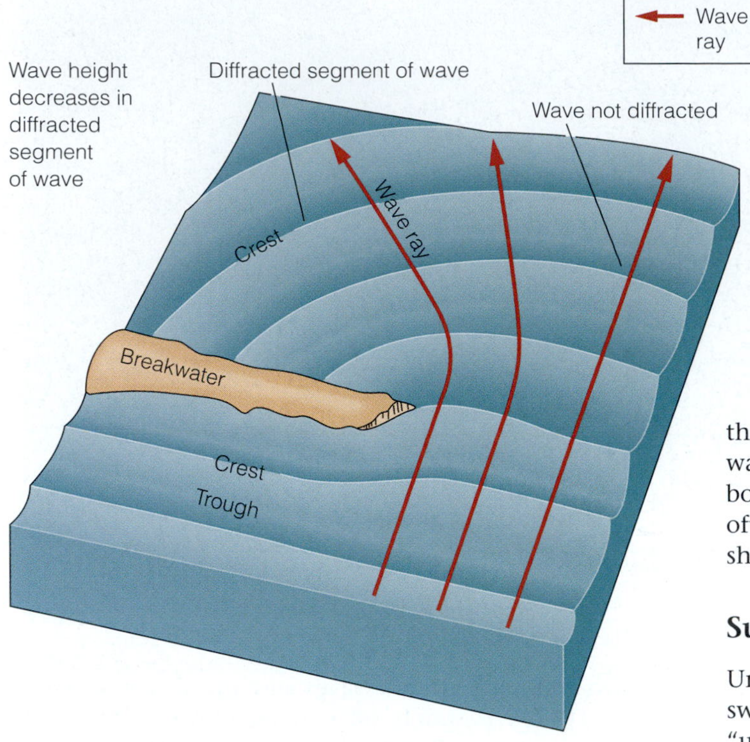

FIGURE 11.21 A wave is diffracted when the wave crest is intercepted by a solid object, such as a breakwater. The part of the wave that passes by the end of the breakwater is bent such that the wave spreads out behind the breakwater.

Surfing

To surf a wave, surfers must propel themselves forward on the board to join the wave motion just at the point where the wave becomes unstable and begins to break. They must position themselves on the forward face of the chosen wave and point the board "downhill." If the surfers "catch" the wave correctly, they will move forward with the wave. In the correct location, surfers are balanced such that their tendency to fall down the wave because of gravity is just offset by the pressure on the board from the water pushing upward in its orbital motion. Therefore, they must place the board where water is moving upward into the breaking crest. Plunging breakers (**Figs. 11.19b, 11.20**) provide unique conditions for surfing because most of the front face of the wave is traveling upward and then forward to join the curling crest.

Many surfers believe the best surfing is found where waves approach the shore from a large angle. In such places, refraction does not align the waves exactly parallel to the shore before they break. A wave breaks progressively along the crest as each section reaches shallow water. Surfers simply ride laterally along the wave crest. They remain on the section of wave crest where the forward face of the wave is moving upward in its last orbital motion before breaking. Where plunging breakers break in this way, surfers can ride under the wave crest in the tube formed where the crest plunges forward ahead of the lower part of the wave (**Fig. 11.20**).

Humans are not the only animals that enjoy riding the orbital motion of waves. Porpoises ride ships' bow waves by positioning themselves where the water in the bow wave is moving forward in its orbital motion. They often ride this way for hours, "pushed" along by the ship.

Surf Drownings and Rip Currents

Unfortunately, many people drown while playing or swimming in the surf. Drownings are often blamed on "undertow" that supposedly sucks unwary individuals underwater away from the beach. However, "undertow" does not exist. People drown because of two simple phenomena, each of which can be easily avoided or escaped.

When large waves break very close to the beach, the force of a wave can easily knock over a person who walks out into the surf. Once knocked over by a wave, an unwary individual may be washed seaward as the water flows back down the beach from the collapsed wave. The person is then at a place where the next wave crashes down. This wave washes the person first toward and then away from the beach as it breaks and the water runs back. In these circumstances, many people simply panic and drown because they cannot recover their balance and stand up.

If you find yourself in this situation, you will not drown if you follow two simple rules. First, grab a breath each time a wave has passed, before you are hit by the next one. Second, use the waves to your advantage. Let a wave carry you inshore. When your inshore motion ceases, do not try to stand up. Instead, either dig your hands and feet into the sand or start crawling toward the beach as you grab that next important breath. As each new wave arrives, let go and allow it to carry you farther inshore. If you take this approach, you will soon be sitting safely on the beach telling others how much fun you had playing in the surf. If you are a good swimmer, an alternative is to swim out beyond the breaking waves, where you can float comfortably and rest before swimming back to shore or calling for help.

The second major cause of beach drownings is **rip currents** (often incorrectly called "rip tides"). Waves create a

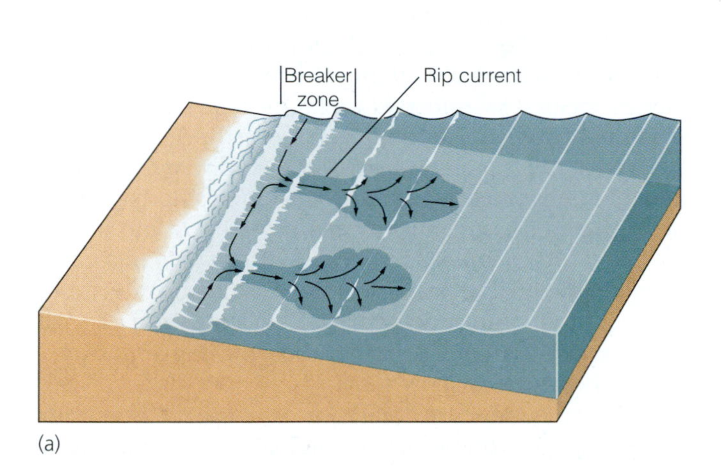

(a)

(b)

FIGURE 11.22 (a) Rip currents are formed at intervals along a beach. They are fast, narrow currents that return water offshore after it has been transported onto the beach by the forward transport of waves. (b) Rip currents are sometimes visible because of foam or discoloration of the water caused by sediments suspended by the swiftly moving current.

small but significant net movement of water in the direction of wave travel (**Fig. 11.9b**). This net forward movement is increased in the surf zone because water in the wave crest outruns water in the wave trough as the wave breaks. Accordingly, in the surf zone, water is continuously transported toward the beach. However, it cannot simply accumulate on the beach, but must flow back through the surf zone. When the waves approach the beach at an angle, as they usually do, water that is transported onto the beach is also transported along the beach in the direction of the waves. Eventually the water encounters an area where it can flow back out through the surf zone against the waves more easily than it can continue to accumulate and flow along the beach (**Fig. 11.22a**). In such areas, wave heights are generally somewhat smaller because a depression or shallow channel runs offshore from the beach.

The corridors in which water returns from the beach to deeper water are often narrow and may be spaced well apart. The return flow through these corridors consists of all the water that was transported onshore over a broader width of beach (**Fig. 11.22**). Because large amounts of water are moving through the narrow corridors across the surf zone, the flow rate or current through the return corridors can be fast. These return flows are rip currents.

Swimmers who enter a rip current will find themselves carried rapidly out to sea by the current. A rip current can be so fast that even the most accomplished swimmer cannot swim against it back to the beach. Tragedies occur when a panicked swimmer tries to fight the rip current, becomes exhausted, and drowns. However, even poor swimmers can easily avert such tragedies if they understand that rip currents are narrow, often only a few meters wide. A swimmer who meets a rip current should simply turn and swim parallel to the beach. A few strokes will bring the swimmer out of the rip current to an area where it is easy to swim safely ashore. A less desirable alternative is simply to ride with the rip current until it stops, usually only 200 or 300 m offshore, and then signal for help.

The locations of rip currents cannot always be seen easily, particularly by a swimmer in the water, but they are most likely where depressions run down the beach into the surf. Rip currents may reveal themselves in plumes of increased **turbidity** (**Fig. 11.22b**), reduced wave heights in the surf zone, or, occasionally, lines of floating debris or foam moving offshore.

11.9 TSUNAMIS

Although tsunamis are relatively rare and most tsunamis are small enough to cause no harm, a large tsunami can be devastating, as, for example, the December 2004 tsunami that caused the deaths of many thousands of people around the Indian Ocean. Far bigger tsunamis, like the hypothetical one described in the introduction to this chapter, have occurred in the past and will occur again some day.

CRITICAL THINKING QUESTIONS

11.7 In extreme storms, ships may turn and travel in the same direction as the wind and waves are traveling, even if doing so takes them in almost the opposite direction from their intended course. Why do you think this might be safer than traveling into the wind and waves?

11.8 Are there internal waves in the atmosphere? If so, where would you be most likely to find them? If there were internal waves in the atmosphere, would you expect their wavelengths to be longer or shorter than those of internal waves in the oceans? Why?

Tsunamis can be generated when an earthquake or volcanic eruption moves a section of seafloor or coast, when a meteorite or coastal land collapse impacts the sea surface, or when an undersea landslide takes place. Each of these events can produce an abrupt displacement of the ocean water that causes the water column to oscillate, generating waves that generally have much longer periods than wind waves. Tsunamis are not related to tides, but they are often incorrectly called "tidal waves." *Tsunami* is a Japanese word that is translated as "harbor wave," but tsunamis are definitely not restricted to harbors.

Not all undersea earthquakes, volcanic eruptions, or landslides cause tsunamis. Tsunamis are most likely to be created when such seismic events cause a section of seafloor to move vertically or to slump. The sudden vertical movement either pushes up the seafloor and overlying water to form a wave crest or lowers the seafloor and overlying water to form a wave trough. The event may cause only one, almost instantaneous, vertical movement of the seafloor. Nevertheless, a series of several waves is created as the water oscillates before returning to a level configuration. This series resembles the series of waves created by the impact of a stone thrown into a pond. Hurricanes and **extratropical cyclones** can cause a different phenomenon, called a **storm surge**, that elevates the sea surface and may cause flooding and damage similar to that caused by tsunamis. These storm surges are not considered to be wind-driven waves, although they behave in a similar manner. They are caused by elevation of the sea surface created by the lower atmospheric pressure at the center of the storm (Chap. 9).

Tsunamis consist of a series of waves with extremely long wavelengths (typically 100 to 200 km) and periods (typically 10 to 30 minutes). Only a very small fraction of the ocean basins is deeper than 6 km (**Fig. 4.3**), and half of the ocean floor is less than 4 km deep. Hence, the water depth is almost always less than one-twentieth of the tsunami's wavelength, which causes tsunamis to behave as shallow-water waves. The speed of a shallow-water wave is determined by the water depth. In water 4 km deep, tsunamis travel at approximately 200 $m \cdot s^{-1}$ (720 $km \cdot h^{-1}$), or nearly the speed of a jet airliner. Because tsunamis are shallow-water waves, their speed changes with depth (**Fig. 11.16**), and they are refracted as they pass over seafloor topography. Tsunamis can be spread out or focused by undersea ridges or depressions, just as wind waves are as they approach a coastline.

When the tsunami is over deep-ocean waters, its wave height rarely exceeds 1 or 2 m. Therefore, ships at sea are not affected by tsunamis. Indeed, it is almost impossible to detect a tsunami at sea, because of its very long wavelength and limited wave height. A tsunami raises and lowers a ship only a meter or two, and each rise and fall takes several minutes. Tsunamis become dangerous only when they enter shallow water.

As a tsunami enters shallower water, it slows and its wavelength is reduced, but its period is unchanged, as is true for any other wave (**Fig. 11.16**). As a wave slows, the wave height increases. Because water depth is very small in comparison to a tsunami's wavelength, wave height builds very rapidly (**Fig. 11.18**). The tsunami does not break, because its wavelength is so long that even a large increase in wave height does not produce steep, unstable waves. Nevertheless, the leading edge of the tsunami wave can produce tremendous surf as it flows turbulently across the shore and coast.

The tsunami reaches the shore as a wave that can be tens of meters high and can take 5 to 10 minutes to pass from trough to crest. As the tsunami moves inshore, sea level rises several meters above normal, and enormous quantities of water are transported onshore and into any **estuaries** or rivers. The water simply keeps pouring onshore for several minutes as the wave crest approaches. The enormous energy stored in the wave is released as the water in the wave flows turbulently onto land and past any structures it encounters. Very strong currents and the equivalent of large breaking waves are generated as the water is concentrated in flows through harbors and channels and between structures. Buildings and trees can be destroyed, and boats and debris can be carried far inshore to be left stranded when the wave recedes. If the ocean is not calm when the tsunami arrives, wind waves will add to the tsunami wave and may contribute to the destruction as they break far inland from the normal surf zone.

The impact that creates a tsunami may cause either a trough or a crest to be formed initially. At the coast, the first indication of the tsunami's arrival may be a rise in the normal level of the sea or a recession of the sea that lasts several minutes and exposes large areas of seafloor that are not normally exposed. Regardless of whether a trough or crest arrives at the coast first, the tsunami will consist of several waves following each other. After the first wave, successive waves may be larger, but they eventually decrease in height. Waves can continue to arrive for 12 h or more. Many drownings occur when curious sightseers or beachcombers walk out onto a beach exposed by a tsunami trough and are caught by the following crest. Despite its 10- to 20-minute period, the crest of a tsunami moves onshore much too fast for someone on the beach to escape after the water begins to rise.

In 1883, the Indonesian island volcano Krakatau erupted with an extremely violent explosion that almost instantly blew a large fraction of the island's mass into the air. The explosion and the subsequent collapse of the remaining sides of the volcano into the underwater **caldera** (crater) created by the explosion caused a tsunami with waves of unusually long periods (estimated to be as much as 1 to 2 h). On the island of Java, about 60 km away from the eruption, the tsunami hit with waves about 30 m high. The waves destroyed many structures and carried a ship more than 3 km inland, where it was stranded almost 10 m above sea level. Krakatau's tsunami killed an estimated 35,000 people and was observed by water-level recorders as far as 18,000 km away in Panama.

In 1946, an earthquake in the Aleutian Trench off Alaska caused a tsunami at Scotch Cap, Alaska, that destroyed a concrete lighthouse 10 m above sea level, killed the lighthouse operators, and tore down a nearby radio mast mounted 33 m above sea level. About 5 hours later, the tsunami heavily damaged the Hawaiian Islands and swept away 150 people to their deaths. In response to this type of disaster, a tsunami warning system was developed for the Pacific Ocean. As a result, when a tsunami hit Hawaii in 1957 with waves larger than those in 1946, no lives were lost.

On December 26, 2004, a great earthquake struck in the **subduction zone** just offshore from the Indian Ocean coastline of the northern part of the island of Sumatra in Indonesia. The earthquake generated a tsunami that was estimated to be 30 m high when it hit the coast of Sumatra. The tsunami spread across the Indian Ocean, smashing into Thailand to the east and India and Sri Lanka to the west. Continuing across the Indian Ocean to the west, the tsunami then impacted the western coastline of Africa. More than 220,000 people were killed or disappeared, with the majority of deaths occurring in Indonesia. Many deaths also occurred in Thailand, Sri Lanka, and India; and several deaths were also reported in Somalia, more than 5000 km from where the earthquake occurred. In addition to the deaths, hundreds of thousands were rendered homeless, and the damage to property is estimated to have been many billions of dollars.

No doubt most of you who are reading this will have seen videos of the December 2004 tsunami. Although some of the videos are horrific, they are also instructive. Many of the video records illustrate well that a tsunami does not arrive as a single short-lived wave. Clearly to be seen in these videos is the turbulent front edge of the advancing wave followed by the mass of water in the wave continuing to pour ashore in a relentless current for several minutes before finally receding. Eyewitness accounts also attest to the fact that several waves came ashore, 10 to 20 minutes apart, and that, in some places, the crest of the first wave arrived without warning, whereas in others, the trough arrived first, providing some warning as the sea receded rapidly and much farther out than normal. The videos also show that it is not possible to outrun a tsunami wave. However, there were many survivors, some of whom survived because they ran to higher ground at the first sign that a tsunami might hit.

We all should learn from this event. It is imperative to seek higher ground immediately if we are on or near the shoreline and we either feel an earthquake or see the ocean rapidly recede in an abnormal fashion. We must also leave immediately if we see a massive wave moving toward shore, but unfortunately, fleeing might not guarantee safety. Finally, if a tsunami has occurred, even a small one, we must stay well away from the shore for at least 12 h.

Although the 2004 Indian Ocean tsunami was an event of epic human scale, it is dwarfed by giant tsunamis that occurred before recorded history and will probably occur again. For example, Chapter 8 describes the monstrous tsunami that is thought to have been created when a meteorite hit the Earth at Chicxulub, Mexico. In addition, studies of the Hawaiian Islands have shown that huge tsunamis may be created when large segments of volcanic islands break loose and slump to the ocean floor (Chap. 13). A slump of 350 km^3 of the island of Hawaii's coastline about 120,000 years ago led to the deposit of marine **fossils** in a location on the island that was at that time at least 5 km inland and at least 400 m above sea level. Blocks of **coral** were apparently swept to a height of 325 m on the island of Lanai, perhaps by this same tsunami.

Tsunamis are most common in the Pacific Ocean because it has many **subduction zones** in which earthquakes are likely to cause vertical seafloor movements. However, tsunamis may occur wherever vertical movements of the land or seafloor occur. Tsunamis are most likely to be damaging on island or continental coasts with narrow continental shelves because much of the wave energy can be dissipated as a tsunami moves over the shallow waters of a wide continental shelf. A tsunami may cause no damage at all on **atolls** or other islands that have no continental shelf, because the tsunami has no opportunity to build in height before it reaches the coast. **Table 11.2** lists 11 destructive tsunamis that have occurred around the Pacific Ocean since 1990.

The tsunami warning system, headquartered in Hawaii, provides warning of possible tsunamis whenever an earthquake occurs in the Pacific that might cause such

TABLE 11.2 Some Destructive Tsunamis since 1990

Date	*Location*	*Maximum Wave Height (m)*	*Number of Fatalities*
September 2, 1992	Nicaragua	10	170
December 12, 1992	Flores Island, Indonesia	26	>1,000
July 12, 1993	Okushiri, Japan	31	239
June 2, 1994	East Java, Indonesia	14	238
November 14, 1994	Mindoro Island, Philippines	7	49
October 9, 1995	Jalisco, Mexico	11	1
January 1, 1996	Sulawesi Island, Indonesia	3.4	9
February 17, 1996	West Papua, Indonesia	7.7	161
February 21, 1996	Northern coast of Peru	5	12
July 17, 1998	Papua New Guinea	15	>2,200
December 26, 2004	Sumatra, Indonesia	30	>220,000

waves. Often tsunami warnings are not followed by a dangerous series of tsunami waves. The reason is that the wave refraction patterns are complex and different for each tsunami. Therefore, even when a tsunami is actually observed near its source, predicting whether the tsunami will significantly affect any section of coast at a more remote location is very difficult. Even though some warnings may prove false, heeding such warnings is always wise.

In Hawaii and Oregon, tsunami warnings are broadcast through sirens at points along the coast and through the emergency radio and television broadcasting system. In other parts of the United States, warnings are disseminated through the emergency broadcast system. Always leave the beach for higher ground immediately after an earthquake is felt, because a "mild" earthquake may actually be a major earthquake centered some kilometers offshore, and a tsunami warning may be issued too late.

11.10 INTERNAL WAVES

Wind waves are oscillations that occur at the ocean surface where there is a sharp density discontinuity between water and the considerably less dense overlying atmosphere. Waves can be created at any density interface with a sharp density gradient where a fluid of low density overlies a fluid of higher density. Such sharp vertical density gradients, called **pycnoclines**, occur in the water column in many areas of the oceans (Chap. 10).

If a pycnocline is displaced up or down, the displacement will set the pycnocline in motion and create **internal waves**. Relatively little is known about internal waves, but their principal causes are apparently tidal motions and shear stress between layers as the layers flow across one another in wind-driven currents or in the water-mass movements within the body of the ocean waters that are called **thermohaline circulation**. Occasionally, where the pycnocline is shallow, internal waves can be generated by propellers or bow waves of ships.

The wave height of internal waves can be great (up to 100 m) in comparison with that of surface waves. The reason is that a given amount of energy causes waves of larger amplitude where the density difference across the wave interface is smaller. The density difference across the pycnocline is much smaller than that across the sea surface. Internal waves travel approximately one-eighth as fast as surface waves with similar period, and they typically have periods of 5 to 8 minutes and wavelengths of 0.6 to 1 km. They have much less energy than surface waves, but they can be very important to sea life, submarines, and offshore oil platforms.

Internal waves behave just like surface waves as they enter shallow water and interact with the seafloor. They slow, their wavelength is reduced, and eventually their wave height increases until they break. Because of their long wavelength, internal waves generally break on the outer part of the continental shelf. In some locations, including the continental shelf offshore of New York and New Jersey, breaking internal waves may mix **nutrient**-rich, cold water from below the pycnocline into warmer, nutrient-poor surface waters. This mechanism may supply nitrogen, phosphorus, and other nutrients to **phytoplankton** and thereby support larger populations of fishes and other animals than would otherwise be present (Chaps. 14, 15). Internal waves may also move **sediments** up and down canyons.

Submarines must be wary of internal waves because such waves can substantially change the submarine's depth in an uncontrollable and unpredictable way. Internal waves are speculated to have caused the sinking and loss of the nuclear-powered submarine USS *Thresher,* with its crew of 129, in 1963. Drilling and oil production platforms also can be affected by internal waves. They are designed to withstand surface waves and are constructed on a tower resting on the seabed or on pontoons that float well below the depth of significant surface wave motion. In 1980, a series of internal waves moved a production platform to face almost 90° from its original direction, although the platform did not topple.

11.11 ROSSBY AND KELVIN WAVES

Rossby waves, often called "planetary waves," are caused by the interaction of the latitudinal gradient in the Coriolis effect (**CC12**) with the **geostrophic** flow of water around a sea surface or with the flow of air around a depression or elevation of atmospheric pressure (Chaps. 9, 10, **CC13**). These waves occur both in the ocean and in the atmosphere. Their generation is a somewhat complex process, but in the ocean they have wavelengths hundreds to thousands of kilometers long and wave heights of just a few centimeters. Because they are so long and low, they were not observed directly in the oceans until satellites were able to measure sea surface height very precisely and to do so across wide areas of ocean within short periods of time.

Rossby waves move only from east to west. However, they can appear to move from west to east if they are carried by a much faster-moving air or water-mass movement. For example, satellite images show atmospheric Rossby waves as the by now familiar large-scale **meanders** of the **jet stream** and the high- and low-pressure zones we see on **weather** maps. These features generally move from west to east with the main flow of air masses in the atmosphere, but the Rossby wave embedded in this flow travels from east to west. Rossby waves move at speeds that vary with latitude, but the speeds of oceanic Rossby waves are on the order of only a few kilometers per day. Thus, for example, a single wave can take months or even years to cross the Pacific Ocean at mid latitudes.

There are two types of Kelvin waves: coastal and equatorial. Coastal Kelvin waves have sufficiently long periods and slow speeds that they are deflected into and constrained to move along a boundary by the Coriolis effect. Coastal Kelvin waves thus flow along the coastal margin in a counterclockwise direction in the Northern Hemisphere and a clockwise direction in the Southern Hemisphere. At the equator, where the Coriolis effect changes sign, there is a special case where Kelvin waves flow directly along the equator from west to east. Equatorial Kelvin waves travel about three times as fast as Rossby waves, so they can cross the Pacific Ocean in about 70 days.

Rossby and Kelvin waves and their effects are poorly understood, but they are believed to affect phytoplankton distribution in the open oceans. They are also believed to affect weather—dramatically in some instances—by creating periodic but irregular variations in the location of ocean currents, such as the Gulf Stream, and in surface water temperatures (and thus atmospheric energy) in various ocean regions. Kelvin and Rossby waves are also associated with the development and relaxation of the El Niño/Southern Oscillation (**ENSO;** Chap 9).

11.12 STANDING WAVES

Standing waves are completely different from the various types of progressive waves described previously in this chapter. Standing waves are sometimes called "stationary waves" or "seiches." Their crests and troughs alternate at fixed locations, and they do not progress across the water surface.

A simple standing wave can be produced in a rectangular cake pan partially filled with water. If you slowly lift one end of the pan a few centimeters and then set it down quickly, the water will slosh back and forth from end to end of the pan. As you tilt the pan, the water surface remains level, but as you set it down the water must move quickly so that the surface returns toward the level position. However, the surface overruns the level position, and water continues to flow until the surface is tilted in the direction opposite the tilt that occurred when you first set the pan down. An oscillating motion of the water surface continues back and forth (**Fig. 11.23a**) as the wave energy is reflected off either end of the pan until friction slows and finally stops the motion. The wavelength of the standing wave equals twice the length of the pan, and a standing wave is formed rather than progressive waves because the wave is blocked by (cannot progress past) the ends of the pan.

The water surface at the center of a standing wave does not move up and down. This position is called a **node** (**Fig. 11.23**). At the **antinodes** (**Fig. 11.23**), which are at the ends of the standing wave, water within the waves can move only up and down; hence, there are no horizontal currents. Because water must be moved from one end of the wave to the other as it oscillates, water at the node moves back and forth horizontally. The current flows in one direction at the node as the wave moves in that direction, and then reverses when the wave moves back. Horizontal or nearly horizontal reversing currents occur at all points within the wave, except at the antinodes. The maximum speed of these currents is highest at the node and decreases with distance from the node until it is zero at the antinodes.

Standing waves are important within lakes and restricted ocean basins. They form especially easily in

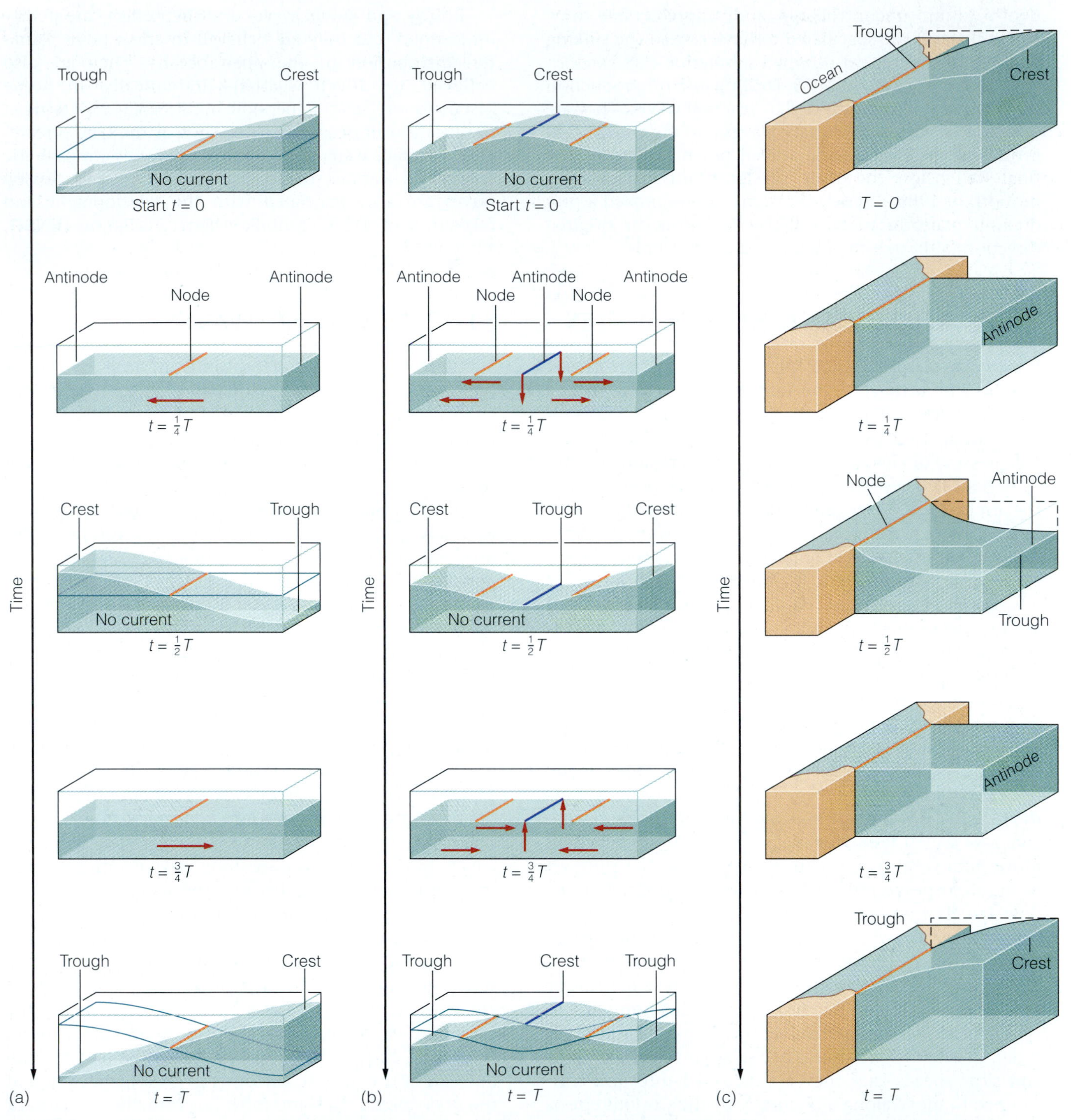

FIGURE 11.23 The motion of water in standing waves. The standing-wave motion is a seesaw oscillation of the surface. Standing waves have one or more nodes (where there is no change in the water surface height but there are oscillating horizontal currents) and one or more antinodes (where there are no currents but the surface oscillates up and down). (a) A standing wave with a single node. (b) A standing wave with two nodes. (c) A standing wave that is essentially a half wave with a node at the basin entrance and only one antinode, at the closed end of the basin.

basins with steep sides because the sides reflect wave energy and little energy is lost by friction with the seafloor. Standing waves are refracted like progressive waves. Because water moves horizontally except at the antinodes, standing waves in very large basins, such as the Great Lakes of North America, are deflected by the Coriolis effect. The influence of the Coriolis effect on a standing tide wave is examined in Chapter 12.

The standing wave just described is essentially one-half of a wave because the crest and trough of the wave are at opposite ends of the basin. Standing waves with more than one node can be established in certain basins (**Fig. 11.23b**).

When a progressive wave of exactly the right period arrives at the entrance to a basin of exactly the correct length, a standing wave is established in the basin. Each successive crest of the standing wave must arrive at the basin entrance at the same time that a crest of a progressive wave arrives from the other direction. This type of **tuned** oscillation is often important in tidal motions within enclosed bays or estuaries (Chap. 12). The oscillation is tuned in much the same way that organ pipes are tuned to specific wavelengths of air oscillations that create different musical notes. The pure note in an organ pipe is a single-node standing wave, whereas **harmonics** are standing waves that have two or more nodes within the pipe.

CHAPTER SUMMARY

Complexity of Ocean Waves. Waves usually seem to come in haphazard sequences of heights and periods. The reason is wave interference, the combination of waves of several different periods and from different directions.

Wave Motion. Water within a progressive wave moves in almost circular orbits whose diameters decrease with depth and are almost zero at a depth of half the wavelength. There is a very small net forward motion of water with the wave. Waves are created when the water surface is displaced. Once started, the wave motion is sustained by a restoring force that tends to return the sea surface to a flat, level state. Gravity is the principal restoring force for most ocean waves, and surface tension is important for very short-period capillary waves. The Coriolis effect affects very long wavelength waves, such as tides.

Tides are created by gravitational attractions of the sun and moon. Some ocean waves are caused by earthquakes or the passage of vessels. Most ocean waves are created by winds. Winds blowing over a calm water surface create tiny capillary waves that increase sea surface roughness. The wind increases the wave height and causes capillary waves to be combined into waves with longer wavelengths.

Capillary waves are dissipated quickly as energy is lost through water viscosity. Most ocean waves are gravity waves and lose little energy until they reach the shore, break, and dissipate much of their energy as heat.

Wave Breaking and Wave Height. Winds are usually variable, causing seas with waves of many different wavelengths and heights. If winds are strong and blow long enough, waves begin to break when their height is one-seventh of their wavelength. Maximum wave heights are determined by the wind strength and duration and by the fetch. Wave heights up to 34 m have been reported. Wave period and wave steepness increase when waves travel directly into a current.

Wave Trains and Wave Dispersion. Waves usually travel in groups called "trains" that move at half the speed of the individual waves. The leading wave in the train is continuously destroyed and a new one created at the rear. Wave trains having waves of different wavelengths combine to produce the confused sea state that is characteristic of storms. The speed of deep-water waves increases with wavelength, so waves of different wavelengths move away from a storm at different speeds. Thus, waves are sorted by wavelength dispersion as they move away from a storm. As wavelength dispersion occurs, wave steepness declines and a swell is formed.

Waves in Shallow Water. When waves enter water shallower than half their wavelength, the wave orbital motion encounters the seafloor. As a result, wave orbits are flattened and both wave speed and wavelength are reduced, while period remains unchanged. Waves continue to slow as they enter progressively shallower water until the water depth is less than one-twentieth of the wavelength, after which the speed is determined only by depth. Waves are refracted in shallow water because any part of the wave in shallower water is slowed more than any part in deeper water. Refraction tends to turn waves to align parallel to the shore, focuses wave energy on headlands, and spreads energy within bays. Waves are reflected off barriers such as seawalls, and they are diffracted when part of the wave passes a barrier that blocks the rest of the wave.

As waves enter shallow water, wave height slowly decreases until water depth is about one-tenth of the wavelength and then increases rapidly as depth decreases further. Waves break when they become oversteep. There are four types of breakers: spilling breakers, in which the crest tumbles down the front of the wave; plunging breakers, in which the crest outruns the bottom of the wave and curls over before crashing down; collapsing breakers, in which the front face of the wave collapses;

and surging breakers, in which the water simply surges up and down the shore.

Waves transport water forward onto the beach. Water returns offshore through narrow rip currents at intervals along the beach. Rip currents are a major cause of drownings, but even weak swimmers can survive them if they are aware that rip currents are narrow.

Tsunamis and Storm Surges. Tsunamis are trains of very long-wavelength (100 to 200 km) waves with periods of 10 to 30 minutes created by earthquakes or other such events that abruptly displace a section of seafloor. Tsunamis travel at speeds in excess of 700 $km{\cdot}hr^{-1}$ but behave as shallow-water waves because their wavelength far exceeds the ocean depths. Tsunamis are rarely more than a meter or two in height until they enter shallow coastal waters, where their height builds rapidly. Tsunamis can do tremendous damage because they pour onshore for 10 minutes or more before the wave crest passes.

Hurricanes and other major storms can cause storm surges in which the water level ahead of the storm is elevated, and this storm surge can cause storm waves to reach far inland of the normal high water line.

Rossby and Kelvin Waves. Rossby waves are very low-amplitude waves that move slowly from east to west, have wavelengths of tens or hundreds of kilometers, and are responsible for periodic but irregular variations in weather along their track. Kelvin waves flow along the coastlines of the ocean basins—counterclockwise in the Northern Hemisphere, clockwise in the Southern Hemisphere, and directly west to east at the equator.

Internal Waves. Internal waves form on pycnoclines. They have long periods and wavelengths, have large heights in comparison with surface waves, and break over the continental shelf, where they enhance the vertical mixing of water.

Standing Waves. Standing waves are formed in basins where the ends of the basin prevent progressive waves from passing. At the node of a standing wave, the water surface does not move vertically but there are horizontal reversing currents. At the antinodes, water motion is vertical and there are no horizontal currents.

STUDY QUESTIONS

1. Why do successive waves that arrive at the beach have different heights?
2. You are standing at the end of a pier with a stopwatch. How would you measure the average wavelength of the waves passing by?
3. What is a restoring force, and why is it necessary if waves are to develop?
4. List the factors that determine the maximum wave height in a sea. How do they differ in the Atlantic and Pacific Oceans?
5. Why do deep-water waves sometimes break instead of just getting bigger?
6. Why is it not a good idea to enter a narrow harbor mouth when the tide is flowing out of the entrance and large waves are entering the harbor mouth from offshore?
7. Why do waves disperse as they move out from a storm? What would you see if you stood at the beach and observed the waves for many hours after the first waves from a distant storm arrived on an otherwise calm sea?
8. What are wave trains? Why do they move more slowly than individual waves?
9. What are the principal differences between internal waves and surface waves? Why do these differences exist?
10. What are rip currents, and why do they occur?
11. Why do tsunamis travel almost as fast as jet aircraft? Why would you not even notice a passing tsunami if you were on a boat far out to sea in deep water?
12. Describe standing waves. How do they differ from progressive waves? How do they resemble progressive waves?

KEY TERMS

You should be able to recognize and understand the meaning of all terms that are in boldface type in the text. All of those terms are defined in the Glossary. The following are some less familiar key scientific terms that are used in this chapter and that are essential to know and be able to use in classroom discussions or on exams.

amplitude (p. 281)
antinode (p. 305)
Coriolis effect (p. 283)
crest (p. 281)
current (p. 287)
deep-water wave (p. 285)
erosion (p. 280)
fetch (p. 285)
frequency (p. 281)
harmonics (p. 307)
intermediate wave (p. 294)
internal wave (p. 304)
jet stream (p. 305)
joule (p. 282)
kinetic energy (p. 282)
longshore current (p. 289)
node (p. 305)
potential energy (p. 282)
pressure gradient (p. 282)
progressive wave (p. 281)
pycnocline (p. 304)
refracted (p. 296)
restoring force (p. 282)
rip current (pp. 300–301)
shallow-water wave (p. 295)

shear stress (p. 284)
standing wave (p. 281)
storm surge (p. 302)
surf (p. 280)
surf zone (pp. 281, 297)
surface tension (p. 282)
swell (p. 285)
trough (p. 281)
tsunami (p. 284)
tuned (p. 307)
turbulent (p. 284)
viscosity (p. 289)
wave dispersion (p. 285)
wave height (p. 280)
wave interference (p. 280)
wave period (p. 280)
wave ray (p. 296)
wave speed (p. 281)
wave steepness (p. 281)
wavelength (p. 281)

CRITICAL CONCEPTS REMINDER

CC5 **Transfer and Storage of Heat by Water** (p. 290). Water's high heat capacity allows large amounts of heat to be stored in the oceans with little change in temperature. Thus, when waves break and release their energy as heat, they cause very little change in the temperature of the water. To read **CC5** go to page 15CC.

CC9 **The Global Greenhouse Effect** (pp. 288–289). Major climate and climate related changes may be an inevitable result of our burning fossil fuels. The burning of fossil fuels releases carbon dioxide and other gases into the atmosphere, where they accumulate and act like the glass of a greenhouse retaining more of the sun's heat. One of these climate related changes is the possibility that average wave heights in the oceans will increase, which would add to existing safety issues for ocean vessels and increase shoreline erosion rates. To read **CC9** go to page 22CC.

CC12 **The Coriolis Effect** (pp. 283, 305). Water masses move freely over the Earth surface while the solid Earth itself is constrained to move with the Earth's rotation. This causes moving water masses, including some long period waves, to appear to follow curving paths across the Earth's surface. The apparent deflection, called the Coriolis effect, is to the right in the Northern Hemisphere and to the left in the Southern Hemisphere. It is at a maximum at the poles, reduces at lower latitudes, and becomes zero at the equator. To read **CC12** go to page 32CC.

CC13 **Geostrophic Flow** (p. 305). Water and air masses flowing on horizontal pressure gradients are deflected by the Coriolis effect until they flow across the gradient such that the pressure gradient force and Coriolis effect are balanced, a condition called geostrophic flow. The interaction of geostrophic currents and the change of the magnitude of the Coriolis effect with latitude are the cause of Rossby waves that flow east to west across the oceans and atmosphere. The meanders in the jet stream seen in many weather maps are Rossby waves. To read **CC13** go to page 43CC.

CHAPTER 12

CHAPTER OUTLINE

CRITICAL CONCEPTS USED IN THIS CHAPTER

These two sets of photographs were taken from the same location 6 h apart at the port of Anchorage, Alaska, at approximately high tide and low tide. The tidal range this day was almost 11 m. For scale, note the cargo containers on the stern of the ship. They are the type hauled on 18-wheel trucks or flatbed railcars.

Tides

We had been blessed by several days of good weather during which we collected sediment and water samples from the southern end of Cook Inlet in Alaska. We were assessing the extent of oil contamination from natural sources and oil rigs, and even though we were on a small vessel with a crew of just three, the samples represented tens of thousands of dollars of effort. Our samples had to be transported to Seattle for analysis. They would be useless unless they reached the laboratory in less than 48 h, and a strong Gulf of Alaska storm had moved into the area much faster than had been forecast. Our allotment of time to use the vessel was over, and the vessel was committed to other programs for the remaining few short weeks of summer. If we had to replace these samples, we would have to wait at least a year, and in Cook Inlet the chances of having such good weather again were very small. Our original plan to have the samples picked up by a float plane would not work in this weather. We were anchored in the shelter of Augustine Island only about 60 miles from Homer and the airport from which we needed to ship the samples, but those 60 miles were to our east, right across the mouth of Cook Inlet, which was exposed to the storm's waves. The waves were up to 5 to 7 m high and very steep-sided. It would be suicide to try to sail across the direction of travel of those waves, as we would most certainly capsize and sink. No scientific samples, no matter how important and valuable, were worth that risk. The vessel's captain, who had many years of experience at sea, told us there was no way we could get back to Homer safely until the storm abated.

As night fell, I sat at the chart table feeling defeated. At daybreak, the storm continued unabated, but by then I had developed a plan and had convinced the captain that it would get us safely to Homer. We left our shelter in the morning at the precise time that my plan required and headed north-northeast, about 60° away from the direction

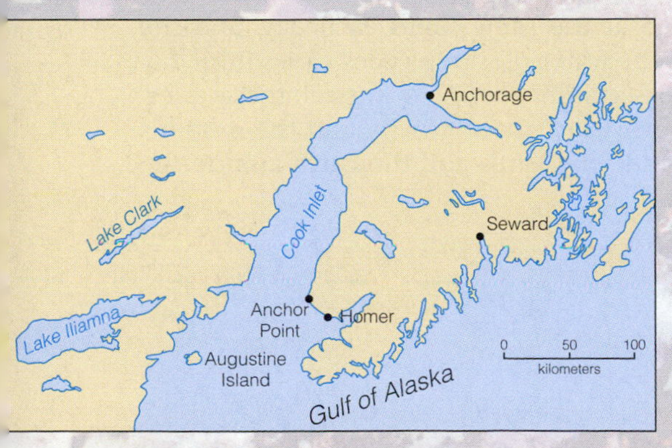

of Homer, our intended destination. About 10 h later we were safely in the Salty Dog Saloon in Homer, telling our story to the locals, our samples already on an airplane bound for Seattle. Our captain, who had thought we would be forced to turn back and would not reach Homer that day, told me, "Doug, you are a seaman." For generations, most of my family have been seamen, and I feel an intimate bond with the sea, so this was perhaps the greatest compliment anyone could ever give me.

However, my plan was not based on some innate knowledge of the oceans or experience in these waters. Instead, it was based on several scientific and practical concepts that you will learn from this text. First, I knew that it was safe for us to sail in the same direction that the steep-sided waves were traveling and that the waves were steepened when they were opposed by a strong current (Chap. 11). Next, I knew that there were strong tidal currents in Cook Inlet, and that tidal currents reverse during the tidal cycle, but at times that cannot necessarily be predicted from the times of high and low tide (details about that in this chapter). Finally, I knew that Cook Inlet is an estuary, and that the residual freshwater flow to the ocean is concentrated on the west side of the inlet, where we were, and, thus, waves would be steeper than on the other side of the estuary.

What was my plan? Sail with the waves during the time they were opposed by the tidal current, gaining distance to the north so that, when the tidal currents reversed, we could turn to sail back across the inlet. Because we would then be farther north, our course across the inlet would not be across the direction of wave travel, but instead at a much safer angle to it. We could reach relative shelter from the wind on the opposite side of the inlet north of Homer before the tidal currents reversed again. The only "data" that I needed to look at to develop this plan were the weather map and forecast, a prediction of which direction the wind was likely to shift to (Chap. 9), tide prediction tables (which all vessels carry), and a summary of the little information available on the tidal-current cycle in the inlet. As I expected, the plan worked perfectly. Our first several hours traveling in the wave direction were rough and uncomfortable, but the rest of the trip was comparatively smooth and easy.

Most of us are familiar with the water movements that slowly expose and then cover the seaward part of the **shore** during a day. This rise and fall of sea level is called the **tide**. If we observe tidal motions long enough at one location, we will see that they are periodic. At some locations, the tide rises and falls twice during a day; at others, it rises and falls only once. The times of high and low tide vary predictably, and the height of the tide also changes somewhat from day to day. Indeed, the **tidal range** is often very different for the two tides on a single day. In addition, we may observe that the sea-level change between high and low tide is large in some places, while in other places there appears to be no tide at all.

Tides are important to mariners because many harbors and channels are not deep enough for vessels to navigate at low tide. Tides are an energy source that has been harnessed for electricity generation in some parts of the world. They are also important to many marine creatures, especially **species** that live in the **intertidal zone** between the **high-tide line** and **low-tide line** and must cope with alternate periods of immersion in water and exposure to air.

Tides cause tidal **currents** that can be very swift in coastal waters and within harbors and **estuaries**. In estuaries, tidal currents reverse direction as the tide rises and falls. When the current flows in from the sea, it is a **flood** current. When it flows out, it is an **ebb** current. The reversals between flood and ebb currents do not necessarily occur at high or low tide. However, where only one tide occurs each day, there is one flood and one ebb; and where two tides occur each day, there are two floods and two ebbs.

It has been known for more than 2000 years that tides are related to the movements of the sun and moon. However, the relationship was not explained until 1686, when Sir Isaac Newton published his theory of **gravity**. Tides are a simple phenomenon in principle, yet they can be extraordinarily complex.

12.1 TIDE-GENERATING FORCES

Tides continuously raise and lower the sea surface. The vertical movement of the water surface is as much as several meters, twice a day in some areas. Clearly large amounts of energy are necessary to move the water involved in this process. The energy is supplied by the gravitational attraction between the water and the Earth, moon, and sun—but how? Tides are cyclical, but they do not come and go at the same times each day, and the height to which the water rises or recedes varies from day to day. Why does this happen? To answer these questions, we must first learn about gravity and the motions of the moon around the Earth and the Earth around the sun.

Gravity

Newton's law of gravitation states that every particle of mass in the universe attracts every other particle of mass with a force that is proportional to the product of the two masses and inversely proportional to the square of the distance between the two masses. The gravitation force (*F*) can be expressed as

$$F = G \times \frac{M_1 \times M_2}{D^2}$$

where *G* is the gravitational constant (a constant defined by Newton's law and equal to $6.673 \times 10^{-11}\ m^3 \cdot kg^{-1} \cdot s^{-2}$, M_1 and M_2 are the two masses, and *D* is the distance between them.

For approximately spherical objects, such as planets, the entire mass can be considered to be at the object's center of gravity, and the distance *D* is measured between the centers of the two masses. The gravitational force increases as either mass increases or the distance between the two objects decreases.

We are familiar with gravity because it is the force that attracts all objects on the Earth's surface, including ourselves, to the much greater mass of the Earth itself. In addition, the Earth and every object on it are subject to gravitational attractions toward the sun, the moon, other planets, and every other celestial body. However, these gravitational forces are extremely small in relation to the Earth's gravitational pull. Although the sun, stars, and some planets each have a mass much greater than the Earth's mass, they are much farther away from us than we are from the Earth's center. For example, gravitational attractions that the sun and moon exert on any object at the Earth's surface are approximately 0.06% and 0.0003% of the Earth's gravitational attraction, respectively. Gravitational attractions of other celestial bodies are much smaller.

Although they are almost negligible in comparison with the Earth's gravitational pull, gravitational forces exerted by the moon and sun on the Earth are the cause of the tides. To understand how these forces cause the tides, we must consider the characteristics of orbits in which celestial bodies move.

Orbital Motions and Centripetal Force

Any two bodies that orbit together in space, orbit around their common center of mass. This works like a seesaw. If the two riders are of equal weight, the seesaw balances when the riders are equal distances from the center. Similarly, two planetary bodies of equal mass would orbit around a point exactly midway between them. However, if one rider is heavier than the other, the heavier rider must sit closer to the center of the seesaw to make it balance. Similarly, two planetary bodies of unequal masses orbit around a point that is closer to the more massive of the two bodies. The Earth's mass is 81 times that of the moon. Therefore, the common point of rotation is about 4700 km from the Earth's center, or approximately 1640 km below the Earth's surface (**Fig. 12.1**). Similarly, because the sun has far greater mass than the Earth, the common point of rotation of the Earth and the sun is deep within the sun.

As two planetary bodies orbit each other, they must be constrained in their orbit by a **centripetal force** that prevents each from flying off in a straight line (**CC12**). The centripetal force is supplied by the gravitational attraction

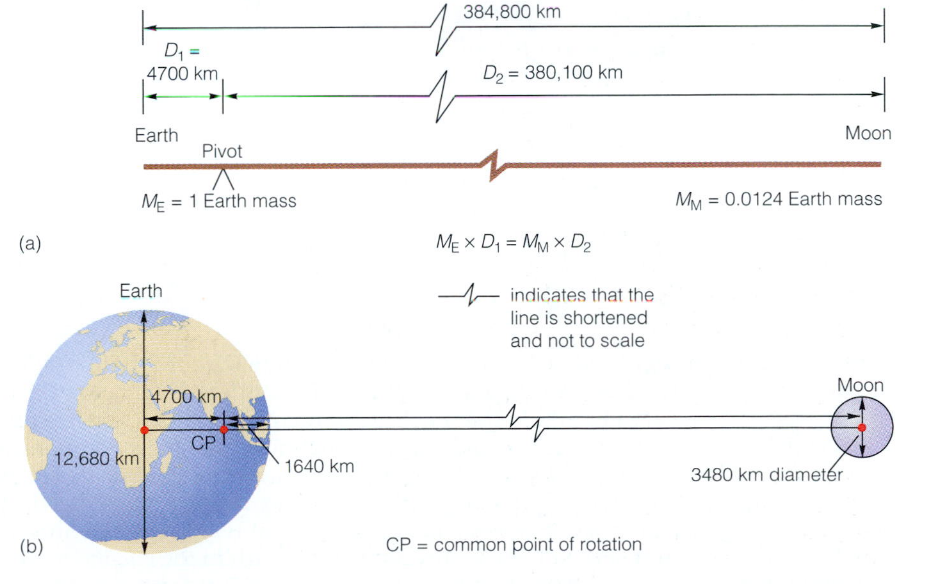

FIGURE 12.1 Any two bodies in space orbit around a common point of rotation. Just as, on a seesaw, the heavier person must sit nearer the pivot point to provide balance, the common point of rotation for two bodies orbiting in space must be closer to the more massive body. (a) Thus, the Earth's center must be much closer to the common center of rotation than the moon's center is. (b) Because the Earth is much more massive than the moon, their common point of rotation is 1640 km below the Earth's surface.

between them. The centripetal force required is the same for all parts of each of the two bodies. Because centripetal force varies with distance from the center of rotation (**CC12**), this may be difficult to understand. However, within either of the orbiting bodies, all particles of mass move at the same speed in circular paths of the same diameter. We can understand this motion if we examine the motions of the two objects as they orbit each other without considering other motions, such as the Earth's spin on its axis or orbits that involve a third body.

Figure 12.2 shows a nonspinning planet (a hypothetical "Earth") orbiting with another planetary object (a hypothetical "moon"). In this diagram, the common center of rotation, which is also the center of mass of the two bodies, is located beneath the surface of the larger planetary body (the hypothetical Earth). Careful examination of **Figure 12.2** reveals that, as the two bodies rotate around each other, each particle of mass on the planet moves in a circular orbit, and all of the orbits have the same diameter. However, as **Figure 12.2** shows, the centers of the circular orbits are all displaced from the common center of rotation, except the orbit of the planet's center of mass. Because the particles of mass in the planet move in circles of the same diameter and complete the circles in the same amount of time, all particles require the same centripetal force to maintain their orbits. This is true no matter where the common center of rotation is. It is true for the Earth–moon system, where the center of rotation is inside the Earth, and for the Earth–sun system, where the center of rotation is inside the sun.

The Balance between Centripetal Force and Gravitational Force

The total gravitational force between the moon and the Earth (or the sun and the Earth) must be equal to the total centripetal force needed to maintain these bodies in their orbits, or else the orbits would change. Although all particles of mass within the Earth are subject to the same centripetal force, the gravitational force on each particle of mass within the Earth varies with its distance from the moon or sun. For example, the distance from the moon's center to the Earth's center is approximately 384,800 km,

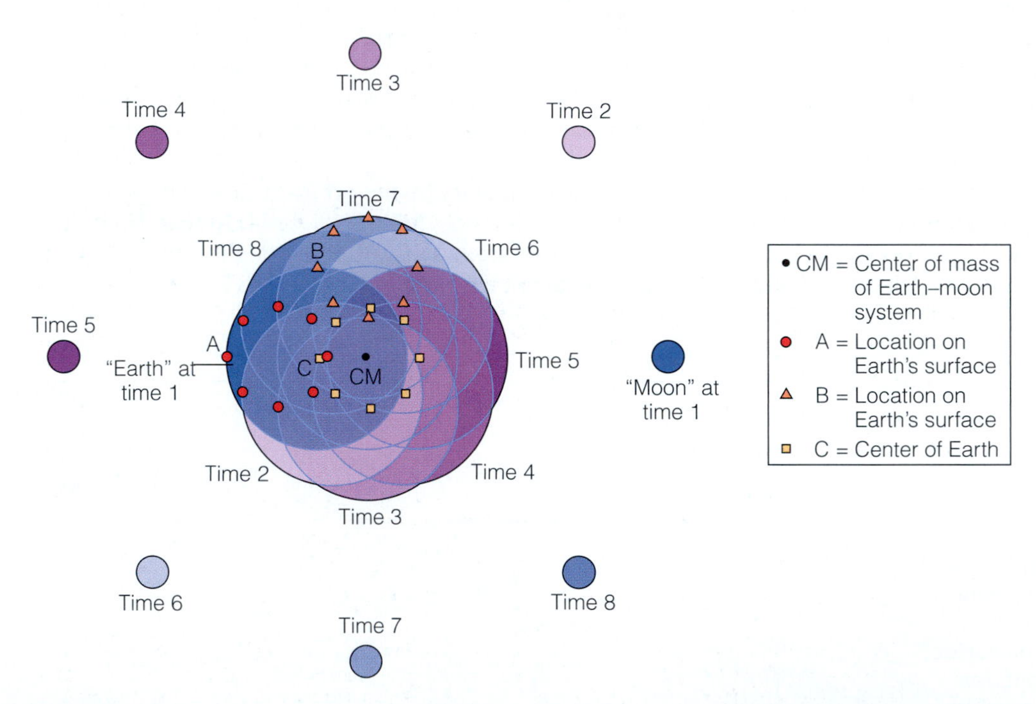

FIGURE 12.2 This figure is a representation of the location of the Earth in relation to the moon as they orbit around their common center of mass (CM), which is inside the Earth. In this representation the Earth is assumed to be nonrotating. As the two bodies rotate around their common center of rotation, points A and B, which are at the Earth's surface, and point C, which is the center of the Earth, would move through the color-coded locations shown in the figure. Notice that all points on or within the Earth move in circular paths that have the same diameter. This is also true for all points on or within the moon, or on or within any other orbiting bodies. Because the orbital diameter and rotation rate are always the same, the centripetal force needed to keep each body in orbit is equal at all points within each of the two bodies. Note that a centripetal force is also needed to maintain parts of the Earth in the circular motion of the Earth's spin, but this is a separate motion with a different distribution of its centripetal force and does not affect the centripetal force between the Earth and moon.

and the Earth's diameter is approximately 12,680 km. Therefore, a point on the Earth's surface nearest to the moon is only 378,460 km from the moon's center, and a point farthest from the moon is 391,140 km from the moon's center. Because the gravitational force is inversely proportional to the square of the distance, the change in the moon's gravitational force between the point on the Earth nearest to the moon and the Earth's center is $378{,}460^2$ divided by $384{,}800^2$, or about 1 to 1.034. Hence, the moon's gravitational attraction is about 3% greater at the Earth's surface nearest the moon than it is at the Earth's center (the average gravitational attraction between moon and Earth). Similarly, the moon's gravitational attraction at the Earth's surface farthest from the moon is about 3% less than the average gravitational attraction.

The average centripetal force for each orbiting body must equal the average gravitational force if the orbit is to remain stable. Because the centripetal force is the same at all points on the Earth and gravitational force is not, there is a net excess gravitational force on the side of the Earth nearest the moon, and a net deficit of gravitational force on the side farthest from the moon (**Fig. 12.3**). These imbalances are tiny in relation to the Earth's gravitational force and are easily compensated in the solid Earth by small changes in **pressure gradient** (the gradient of pressure within the atmosphere, ocean, or solid Earth that increases toward the Earth's center). Pressure within the Earth (and other objects) represents the force needed to resist the gravitational force tending to pull all material toward the Earth's center). For people, the compensation in the pressure gradient causes a change in weight. Each of us weighs about 0.000,000,1 kg (0.1 mg) less when the moon is over the opposite side of the Earth and when it is directly overhead than when the moon is just below the horizon—much too small a difference to notice.

Unlike solid objects, ocean water can flow in response to the imbalances between gravitational force and centripetal force. These flows are the tides. For convenience, we refer to the net excess of gravitational force over centripetal force as the "tidal pull" or "tide-generating force."

Distribution of Tide-Generating Forces

At the point on the Earth's surface nearest the moon, there is a slight tidal pull. Water is pulled upward toward the moon at this point. However, the Earth's gravity, which is comparatively very strong, acts directly opposite the tidal pull at this location. Therefore, the tidal force at this point (and at the opposite side of the Earth, which is farthest from the moon) has little effect, just as the tidal force has only an extremely small effect on our weight.

At other points on the Earth's surface, the tidal pull is exerted in a direction different from that of the Earth's gravity because the direction between that point on the Earth's surface and the moon, and the direction between that point and the Earth's center of mass, are aligned at an angle. Thus, at every point except directly under the moon and exactly on the other side of the Earth, the tidal pull has a component that acts parallel to the Earth's surface (**Fig. 12.4**). This component of tidal pull cannot be compensated by the Earth's gravity and therefore causes water to flow in the direction of the force.

The net result of the tidal forces acting on the Earth's oceans is to move water toward the points nearest to the moon and farthest from the moon (**Fig. 12.4b**). This movement creates bulges of elevated water surface at these points, and a depression of the water surface in a ring around the Earth halfway between these points. The diagrams in this book necessarily greatly exaggerate the

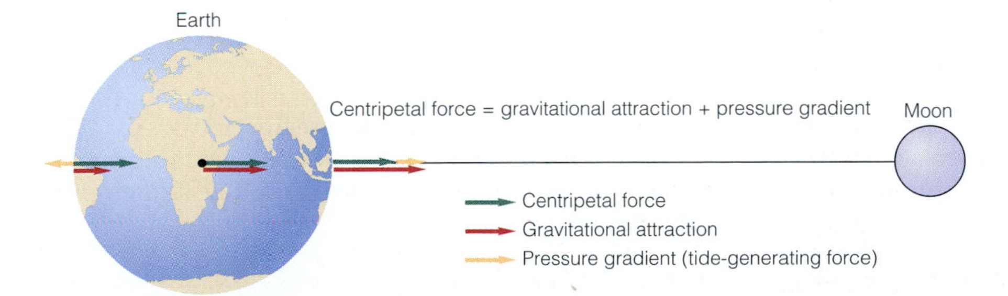

FIGURE 12.3 The total gravitational force between the Earth and moon (or sun) must equal the total centripetal force needed to maintain the two bodies in their common orbit. However, because the gravitational force varies slightly at different points on the Earth, whereas the centripetal force is the same at all points, there is a slight imbalance between them everywhere except at the exact center of mass of the Earth. On the side of the Earth nearest the moon, the gravitational force due to the moon is slightly higher than it is at the Earth's center. The slight excess of gravitational force over centripetal force at this point is easily compensated by the pressure gradient. Similarly, there is a slight excess of centripetal force over gravitational force on the Earth at the point directly away from the moon. Thus, objects at points directly under the moon or directly opposite the moon weigh very slightly less than they do at other points on the Earth.

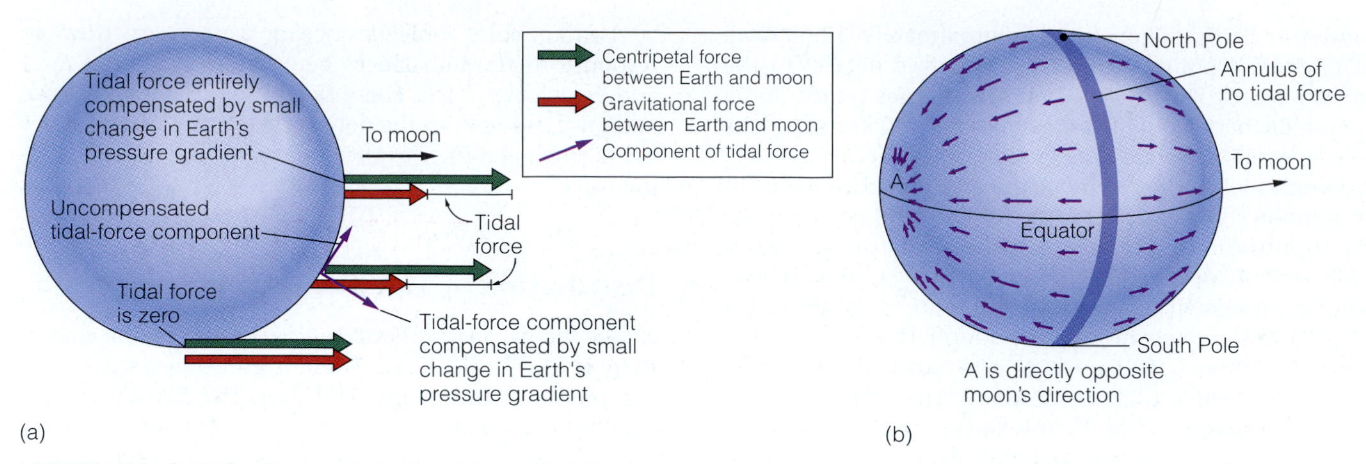

FIGURE 12.4 Horizontal tidal forces. (a) Components of gravitational attraction, centripetal force, and the Earth's pressure gradient. The vertical component of the tidal force is easily compensated by a minute change in the pressure gradient. (b)The horizontal tidal force is zero at a point directly under the moon (or sun) and a point located exactly on the opposite side of the moon. It is also zero in a ring (annulus) around the Earth that is equidistant from these two points. The horizontal tidal force increases to a maximum and then decreases between the point directly under the moon and the annulus where the force is zero. The same is true between the point directly away from the moon and the annulus, but the tidal force is in the opposite direction.

bulges. Even on an Earth totally covered by oceans, the bulges would be less than a meter high.

The tide bulges are oriented directly toward and away from the moon. Because the Earth is spinning, the point on the Earth facing the moon is continuously changing. Therefore, as the Earth spins on its axis and the bulge remains aligned with the moon, but the locations on the Earth that are under the tide bulge change.

Relative Magnitude of the Lunar and Solar Tide-Generating Forces

Although all planets and stars exert tidal forces on the Earth, only tidal forces of the moon and the sun are significant. Tidal forces are caused by the small difference in gravitational force from one side of the Earth to the other. This difference depends on the distance between the Earth's center and the other body's center, and on the other body's mass. Although we need not perform the calculation here, we can show mathematically that the magnitude of the tidal force is proportional to the mass of the attracting body (sun or moon) and inversely proportional to the cube of the distance between the centers of the two bodies (r^3). This means that the tidal force is much more dependent on the distance between bodies than it is on mass.

For the Earth–moon system, the tide-generating force (F_M) is given by

$$F_M = K \times \frac{\text{(Mass of moon)}}{\text{(Earth-to-moon distance)}^3}$$

and for the Earth–sun system, the tide-generating force (F_S) is given by

$$F_S = K \times \frac{\text{(Mass of sun)}}{\text{(Earth-to-sun distance)}^3}$$

where K is a constant that is always the same for tidal forces between the Earth and any other planetary body.

The sun's mass is about 27 million times greater than the moon's. However, the sun is about 390 times farther from the Earth. The ratio of solar to lunar tidal force can be calculated with the following equations:

$$\begin{aligned}\frac{F_S}{F_M} &= \frac{\text{(Mass of sun)}}{\text{(Mass of moon)}} \times \frac{\text{(Earth-to-moon distance)}^3}{\text{(Earth-to-sun distance)}^3}\\ &= \frac{27{,}000{,}000}{1} \times \frac{(1 \times 1 \times 1)}{(390 \times 390 \times 390)}\\ &= 0.26\end{aligned}$$

Despite the sun's much greater mass, its tide-generating force is only about 46% of the moon's tide-generating force because the sun is much farther away from the Earth.

12.2 CHARACTERISTICS OF THE TIDES

Tides are measured by a variety of gauges that continuously monitor the sea surface height. The tide is described by a plot of the water surface height as a function of time, called a "tidal curve" (**Fig. 12.5**). The tidal curves in **Figure 12.5** show that tidal motions are similar to the **progressive waves** described in Chapter 11. In some locations, the tidal curve resembles a simple progressive wave. In other locations, it resembles the complex waveforms produced when waves of different **wave periods** and **wave heights** interfere. Tidal curves are, in fact, the net result of several tide waves of different periods, as explained in the next section.

The most important characteristics of a tidal curve are the times and relative elevations of high tide and low tide, and the tidal range, which is the difference between the height of the high tide and that of the low tide (**Fig. 12.5**). Because the tide is a wave, the tidal range is an expression of the tide wave height.

Diurnal, Semidiurnal, and Mixed Tides

Tides are classified into three general types based on the number and relative heights of the tides each day at a given location. **Diurnal tides** (or daily tides) have one high tide and one low tide in each tidal day, which equals approximately 24 h 49 min (**Fig. 12.5a**). The next section explains why the tidal day is 49 min longer than the solar day. **Semidiurnal tides** (or semidaily tides) have two high and two low tides each tidal day, and thus a period of 12 h 24½ min. For pure semidiurnal tides, the two high tides (and the two low tides) each day are equal in height (**Fig. 12.5b**).

Mixed tides also have two high tides and two low tides each tidal day, but the heights of the two high tides (and/or of the two low tides) in each tidal day are different (**Fig. 12.5c**). Mixed tides have a higher high water (HHW) and a lower high water (LHW), as well as a higher low water (HLW) and a lower low water (LLW), each day. **Figure 12.5c** shows that the relationship of these four daily extreme levels varies greatly among different locations.

Mixed tides are the most common. On the North American continent, tides are mixed along the entire Pacific **coast**, on the Atlantic coast of Canada north of Nova Scotia, in the Caribbean Sea, and in parts of the Gulf of Mexico (**Fig. 12.6**). Tides are semidiurnal along the Atlantic coast of the United States and southern Canada, and are **diurnal** only in certain parts of the Gulf of Mexico (**Fig. 12.6**). However, pure semidiurnal and diurnal tides are rare. All tides have some mixed character, but the classifications are useful to mariners and to illustrate the complex tidal motions. In some locations, tides may change in character during the tidal month. For example, they may be semidiurnal for one part of the month and mixed for another part.

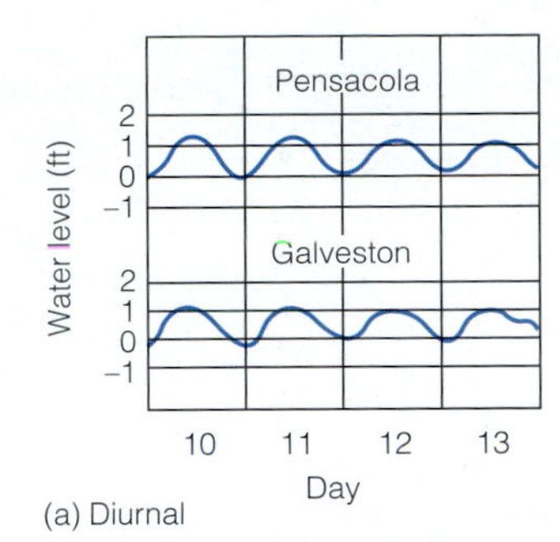

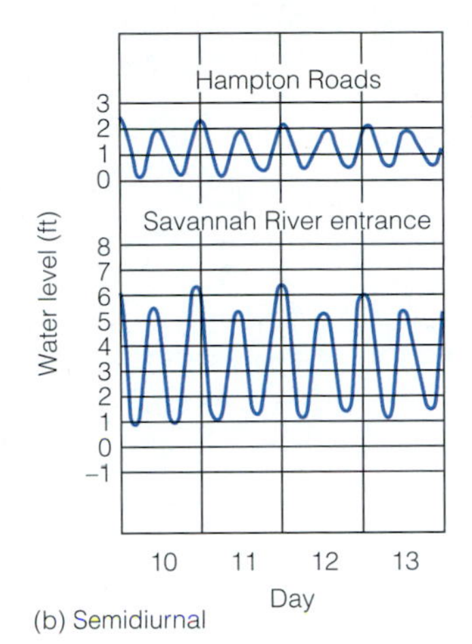

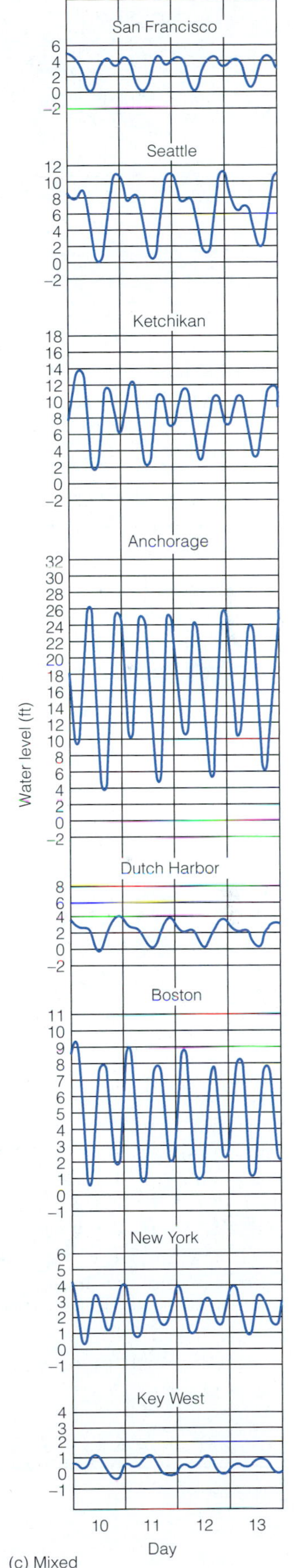

FIGURE 12.5 Selected tide records. (a) Diurnal tides occur at Pensacola, Florida, and Galveston, Texas. (b) Semidiurnal tides occur at Hampton Roads, Virginia, and the Savannah River entrance, Georgia. (c) Mixed tides occur at most locations, including San Francisco, California; Seattle, Washington; Ketchikan, Anchorage, and Dutch Harbor, Alaska; Boston, Massachusetts; New York, New York; and Key West, Florida. Note that even the Hampton Roads and Savannah River tides are not pure semidiurnal tides, but they are nearly so. Tidal heights are in feet because tide tables issued by the U.S. government still use these units.

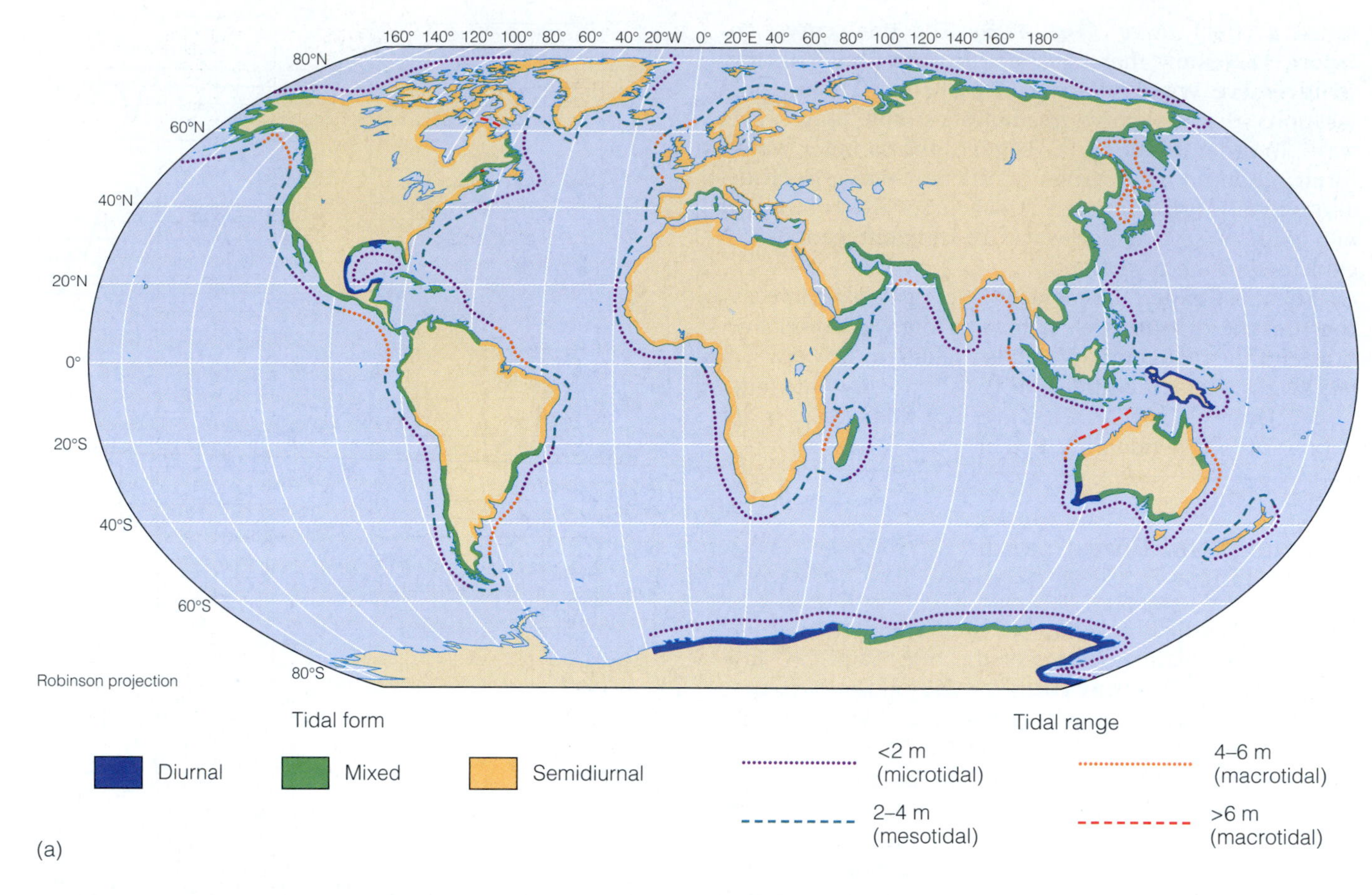

(a)

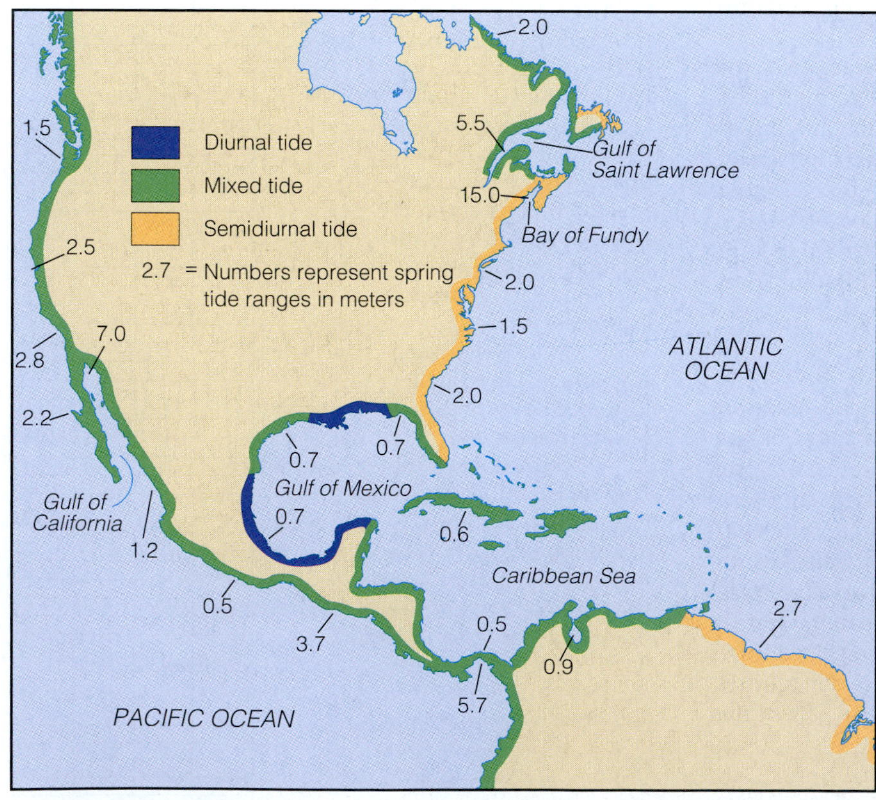

(b)

FIGURE 12.6 Although the characteristics of tides may change during the month at a given location, all locations have tides that can be characterized as primarily diurnal, semidiurnal, or mixed. (a) Global distribution of tides. (b) North American tides. Most of the coasts on the Atlantic Ocean have semidiurnal tides, and most coasts on the Pacific Ocean have mixed tides. Diurnal tides are relatively rare. This pattern is generally true for the North American continent, except that much of the Gulf of Mexico has diurnal tides, and parts of the Atlantic coast of Canada have mixed tides.

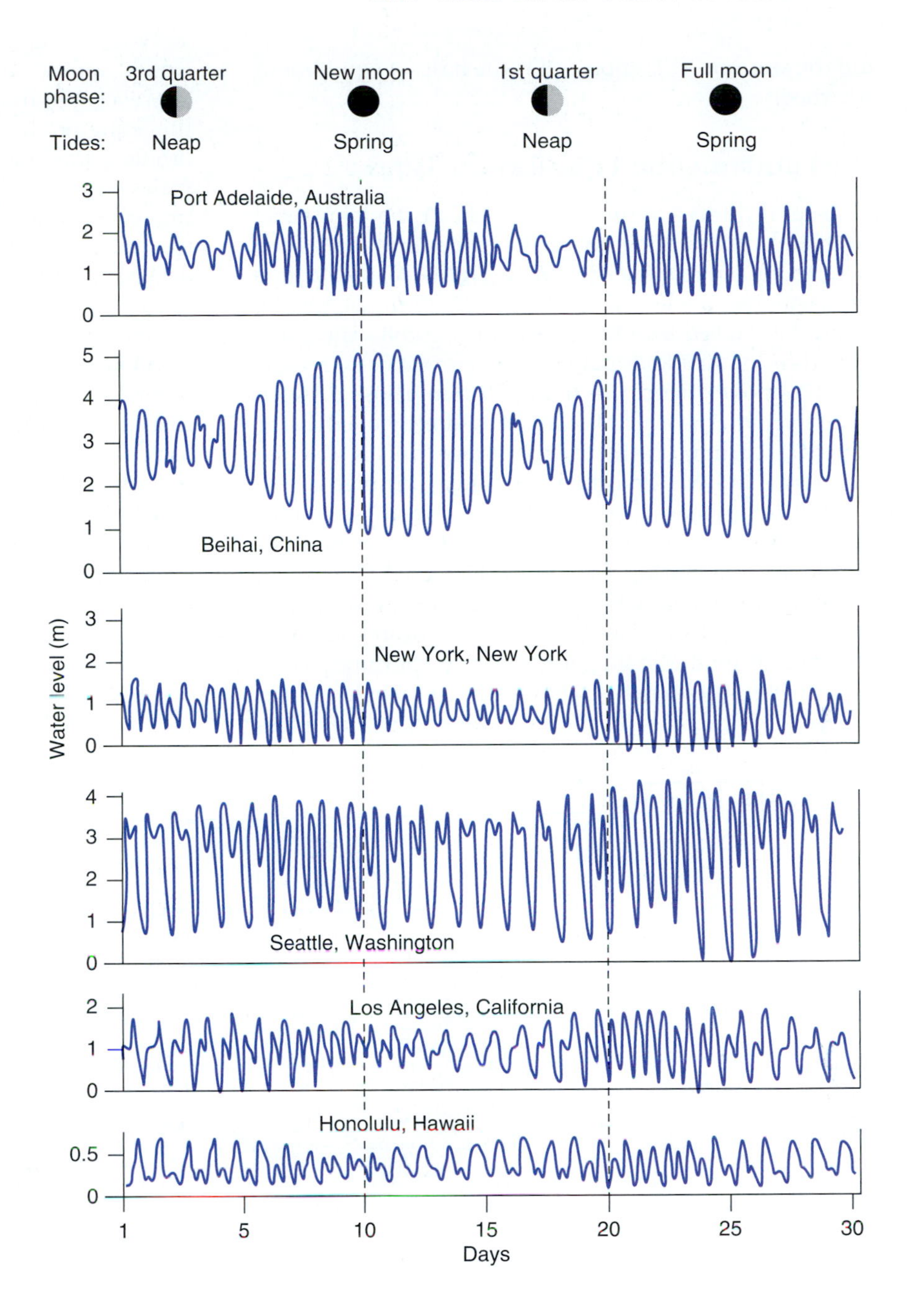

FIGURE 12.7 Monthly tide records show the twice-monthly occurrence of spring tides associated with new moon and full moon, and the twice-monthly occurrence of neap tides associated with the quarter moons. This pattern occurs at all of the locations shown, but it is more pronounced in some areas, such as Port Adelaide, Beihai, and New York.

Spring and Neap Tides

If we look at tide records for various locations over a period of a month, we see that the daily tidal range varies during the month (**Fig. 12.7**). This variation occurs at all locations, regardless of whether tides are diurnal, semidiurnal, or mixed. Although the variations are often complex, the daily tidal range generally reaches a maximum twice during each **lunar month** ($29^1/_2$ days). The tidal range thus oscillates back and forth twice each lunar month. Tides with the largest tidal range during the month are **spring tides**. Tides with the smallest tidal range are **neap tides**. Two sets of spring tides and two sets of neap tides occur each lunar month.

If we look at tide records for an entire year or for several years, we find that the tidal range varies during the year and between years. However, these variations are much less than the variation between spring and neap tidal ranges during the lunar month.

12.3 TIDES ON AN OCEAN-COVERED EARTH

We can understand many features of tides and their variability by examining the theoretical effect of the sun and moon on a hypothetical planet Earth that is entirely covered by deep oceans and has no **friction** between water and the seafloor. This approach is the basis of equilibrium tide theory.

The Fundamental Equilibrium Tides

Earth spins on its axis once every 24 h. At the same time, the moon orbits the Earth in the same direction as the Earth's spin, but much more slowly (**Fig. 12.8**). By the time the Earth has made one complete rotation (24 h), the moon has moved forward a little in its orbit. The Earth must turn for an additional 49 min before it "catches up" with the moon. Hence, 24 h and 49 min elapse between successive times at which the moon is directly overhead at a specific location on the Earth's surface. This is why the moon rises and sets almost an hour later each night.

On a hypothetical planet covered by oceans and without friction, we can consider what would be the effect of the moon alone during a solar day (**Fig. 12.8**). The moon would create upward bulges on the oceans at points aligned directly toward and directly away from the moon. As the Earth spun, the tidal forces would continue to pull water toward these points. The tide bulges would remain aligned with the moon as the Earth rotated, so the bulges would appear to migrate around the Earth each day. However, the moon does not always orbit directly over the equator. Instead, the angle between the moon's orbital plane and the equator, called the **declination**, varies with time during the lunar month and on longer timescales (**Fig. 12.8a**). For now, we need only consider the much simplified case of the moon at its maximum declination to see how diurnal, semidiurnal, and mixed tides might result (**Fig. 12.8b**).

Imagine an observer standing at a point along latitude C in **Figure 12.8b** and rotating with the Earth. When the observer is at point C_1 there is a high tide. As the Earth rotates, the observer passes through the low tide at the back of the Earth (C_2) after the Earth has rotated a little more than 90°, or a little more than 6 h later. After 12 h and 24½ min, the observer passes a second high tide (C_3), but it is not as high as the original high tide. C_3 is further from AM than C_1 is from TM. Thus, as the bulge passes the longitude of this location (C_3), the highest point of the bulge is farther away from the observer than it was when the bulge passed the longitude of the observer's original position (C_1). The observer rotates farther with the Earth, passing a second low tide after about 18 h (C_4),

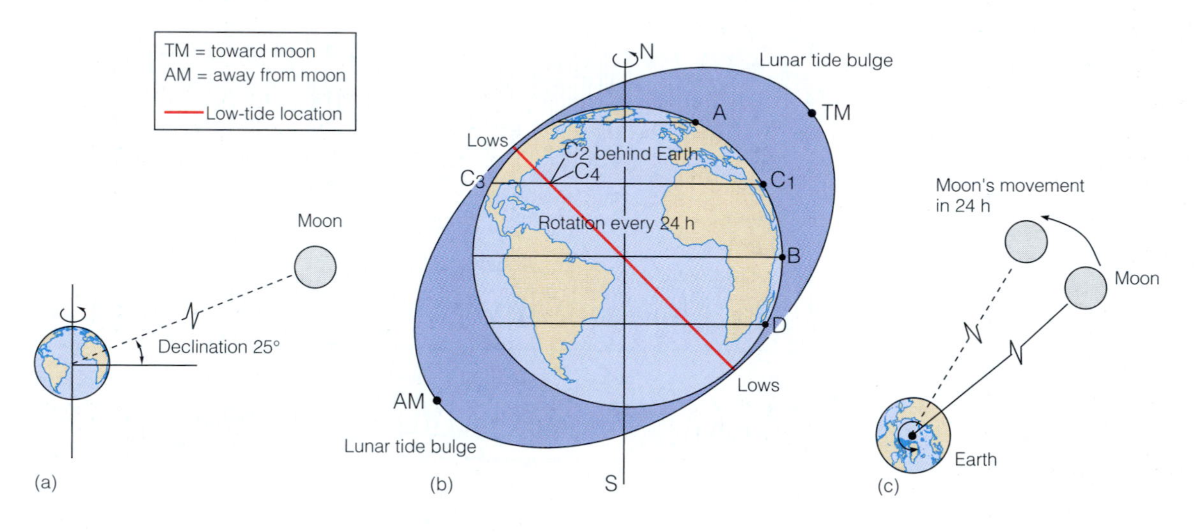

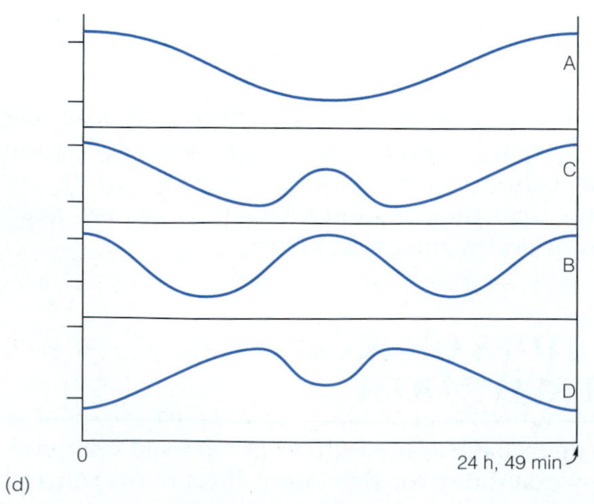

FIGURE 12.8 A simplified system with the moon and an ocean-covered Earth illustrates the principal feature of the tides. (a) The angle of the moon's orbit with respect to the equator (the declination) oscillates regularly. The maximum declination is about 25°. (b) The tide bulges caused by the moon's tidal forces are shown here with the declination and size of the bulges greatly exaggerated, and the effects of the continents ignored. (c) Looking down from above the North Pole at the Earth and moon, we would see that the moon moves through a small segment of its orbit each time the Earth rotates. Hence, the moon is "overhead" at any point on Earth a little later each day. (d) In the simplified system depicted in (b), an observer at latitude A would see a diurnal tide, an observer at B would see a semidiurnal tide, and observers anywhere else on Earth, such as C and D, would see mixed tides.

and another high tide after 24 h and 49 min (C_1). The extra 49 min are due to the moon's progression in its orbit during the day (**Fig. 12.8c**).

We can follow the tidal patterns that would be observed at points A, B, and D in a similar way. A diurnal tide is observed at A, a semidiurnal tide at B, and mixed tides at C and D (**Fig. 12.8d**). Note that, in this simplified model of tides, pure semidiurnal tides occur only at the equator. Note also that, in mixed tides, low tides do not occur exactly midway between high tides.

In the simple model shown in **Figure 12.8**, the low tide would be the same height at all locations. However, the tide records in **Figures 12.5 and 12.7** show that this is not true. The reason is that the simplified system in **Figure 12.8** does not include solar tides or the effects of continents that create much greater complexity in the tides.

The Origin of Spring and Neap Tides

The sun exerts tidal forces on the Earth that are about half as strong as the lunar tidal forces. We can see how the solar and lunar tides interact by again considering a simplified ocean-covered Earth. Tide bulges are created by both the sun and the moon (**Fig. 12.9**). The tidal height at any point on our model Earth is the sum of the tidal heights of the lunar and solar tides. The moon's orbit around the Earth and the Earth's orbit around the sun are not quite in the same plane, but they are nearly so. We can ignore this small angle for now.

Figure 12.9a shows the relative positions of sun, moon, and Earth at a full moon. Because the entire disk of the moon is sunlit as seen from the Earth, the Earth must be between the sun and the moon. Lunar eclipses occur when the three bodies are lined up precisely. At full moon, we can see that the high-tide bulges of the solar and lunar tides are located at the same point on the Earth. The lunar high-tide bulge directly under the moon and the solar high-tide bulge at the opposite side of the Earth from the sun combine in the same location, and vice versa. Thus, at full moon the solar and lunar high tides are added together and the solar and lunar low tides are added together, producing the maximum tidal range with the highest high tides and also the lowest low tides during the lunar month. These are spring tides.

Each day, the moon's movement in its orbit is the equivalent of the angle through which the Earth rotates in about 49 minutes. Approximately 7½ (7.38) days after full moon, the moon, Earth, and sun are aligned at right angles (**Fig. 12.9b**). The left half of the moon (as seen from the Northern Hemisphere) is now lit by the sun. The other part of the moon that is lit is hidden from an observer on the Earth. This is the moon's third quarter (The lunar month is counted from a new moon, while Figure 12.9 starts from a full moon). At third-quarter moon, the lunar tide bulge coincides with the low-tide region of the solar tide, and vice versa. Because the tides are additive, the low tide of the solar tide partially offsets and reduces the height of the lunar high tide, and the high tide of the solar tide partially offsets the lunar low tide. The tidal range is therefore smaller than at full moon. These tides are the neap tides.

Fifteen days after full moon, the moon is between the Earth and the sun (**Fig. 12.9c**). Because only the back of the moon is lit as seen from the Earth, there is a new moon. At new moon, the solar and lunar tide bulges again coincide and there is a second set of spring tides for the month. These tides are equal in height and range to the spring tides that occurred 15 days previously at full moon. Similarly, a little more than 22 days (22.14) after full moon, there is a first-quarter moon. The right side of the moon is lit for Northern Hemisphere observers (**Fig. 12.9d**). The solar and lunar tides again offset each other, and there is a second set of neap tides.

After about 29½ (29.53) days, the moon is full again. We have a new set of spring tides and begin a new lunar month. From this simple model, we can see why we have two sets of spring tides and two sets of neap tides in each 29½-day lunar month (**Fig. 12.8**).

CRITICAL THINKING QUESTIONS

12.1 What might happen if the mass of the moon doubled? How would the moon's orbit be affected? How would ocean tides on the Earth be affected?

12.2 Why is the tide wave a shallow-water wave? How would it be refracted as it approached a large island with a wide continental shelf?

12.3 Why is it necessary to build a dam across an estuary to get the greatest possible amount of energy from tidal power? Why couldn't we just put a number of turbines side by side across the entrance?

12.4 There is a proposal to build a sea-level canal with no locks to connect the Atlantic and Pacific Oceans across the Isthmus of Panama near the existing Panama Canal. Imagine that this canal was built instead across Mexico from the Gulf of Mexico and that it was several hundred meters wide and 30 m deep.
- (a) Would it affect the nature of the tides in the Gulf of Mexico? If so, how?
- (b) In the canal itself, would there be tidal-height variations of water level?
- (c) Would there be tidal currents? If so, hypothesize what characteristics the tides and tidal currents would have.

12.5 Tidal bores are formed when a flood tide moves up an estuary or inlet faster than the speed that a shallow-water wave of tidal wavelength can sustain. When the tide ebbs in such inlets, is there a tidal bore that flows seaward? Explain why such an ebb tide bore does or does not exist.

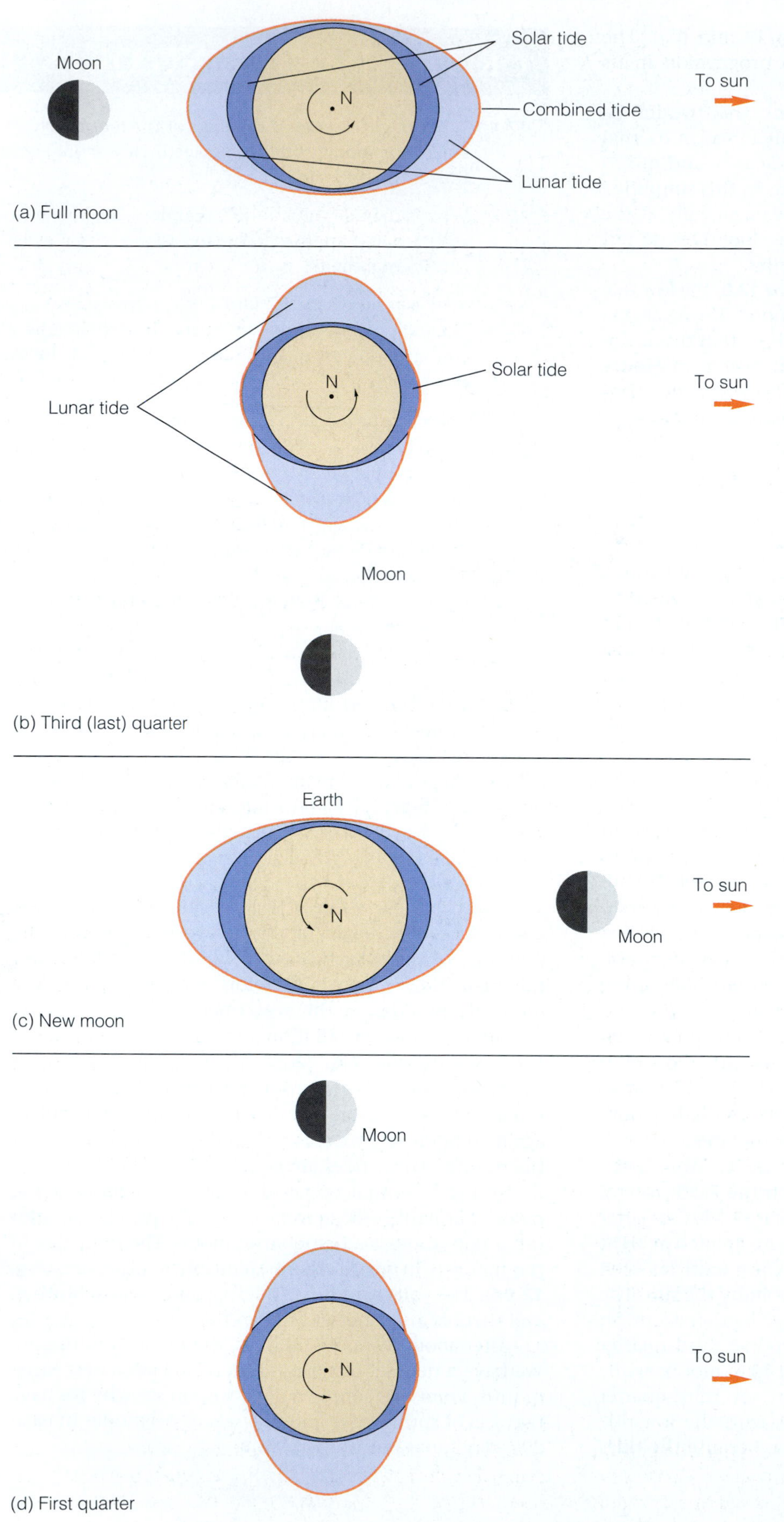

FIGURE 12.9 A simplified depiction of the Earth–moon–sun system shows how the solar and lunar tides interact to create spring and neap tides. (a) At full moon, spring tides with a greater tidal range (higher high and lower low tides) occur because the tide bulges caused by the moon and the sun (both greatly exaggerated here) are aligned so that their maxima are at the same locations. (b) At third-quarter moon, neap tides occur because the moon's bulge and the sun's bulge are aligned at 90° to each other and the maxima of the moon's tide coincide with the minima of the sun's tide. (c) At new moon, spring tides occur again as the moon's and the sun's bulges are again aligned with their maxima at the same locations. (d) At first-quarter moon, neap tides occur again as the moon's and the sun's bulges are again aligned at 90° to each other and the maxima of the moon's tide coincide with the minima of the sun's tide. The observed spring and neap sequences on the Earth are more complicated for various reasons, including that the orbits of the moon and sun are not exactly in the same plane as this simplified figure depicts. These orbits are inclined to each other, and the inclination changes with time.

TABLE 12.1 Selected Tidal Components

Tidal Component	*Period (h)*	*Relative Amplitude*	*Description*
SEMIDIURNAL			
Principal lunar	12.42	100.0	Rotation of Earth relative to moon
Principal solar	12.00	46.6	Rotation of Earth relative to sun
Larger lunar ecliptic	12.66	19.2	Variation in moon–Earth distance
Lunisolar semidiurnal	11.97	12.7	Changes in declination of sun and moon
DIURNAL			
Lunisolar diurnal	23.93	58.4	Changes in declination of sun and moon
Principal lunar diurnal	25.82	41.5	Rotation of Earth relative to moon
Principal solar diurnal	24.07	19.4	Rotation of Earth relative to sun
BIWEEKLY AND MONTHLY			
Lunar fortnightly	327.9	17.2	Moon's orbit declination variation from zero to maximum and back to zero
Lunar monthly	661.3	9.1	Time for moon–Earth distance to change from minimum to maximum and back
SEMIANNUAL AND ANNUAL			
Solar semiannual	4382.4 (182.6 days)	8.0	Time for sun's declination to change from zero to maximum and back to zero
Anomalistic year	8766.2 (365.2 days)	1.3	Time for Earth–sun distance to change from minimum to maximum and back

Other Tidal Variations

From the simple model that we have been discussing, we have seen that the height of the tide due to the moon varies on a cycle that is 24 h 49 min long. High tides actually occur every 12 h 24½ min because there are two bulges. Similarly, the solar tide varies on a 24-h cycle, and high tides occur every 12 h. The motions of the moon and the Earth in their orbits are much more complex than the simple model suggests, and tidal variations occur on many other timescales. These additional variations are generally smaller than the daily or monthly variations, but they must be taken into account when precise tidal calculations are made. They are due to periodic changes in the distances between the Earth and the sun and between the Earth and the moon, and the various declinations. Declinations are the angles between the Earth's plane of orbit around the sun, the Earth's axis of spin, and the moon's orbital plane. **Table 12.1** lists some of the more important variations of Earth–moon–sun orbits, along with the periodicity of the **partial tides** that they cause.

12.4 TIDES IN THE EARTH'S OCEANS

Tides in the Earth's oceans behave somewhat differently from those on the hypothetical ocean-covered Earth. For example, on the model Earth, pure semidiurnal tides would occur only at the equator (**Fig. 12.8**), except when the moon's declination was zero, and then all points on the Earth would have pure semidiurnal tides. However, on the real Earth semidiurnal tides occur at places other than the equator (**Figs. 12.5, 12.6, 12.7**).

Four major interrelated factors alter the Earth's tides from the equilibrium model: the Earth's landmasses, the shallow depth of the oceans in relation to the **wavelength** of tides, the latitudinal variation of **orbital velocity** due to the Earth's spin around its axis, and the **Coriolis effect** (**CC12**). When these factors are included in calculations of tides, the calculated tides are called "dynamic tides."

Effects of Continents and Ocean Depth

In discussing the factors that affect tidal motions, it is convenient to envision tide waves moving across the Earth's surface from east to west, even though, in fact, it is the Earth that is spinning. The peaks of the tide waves remain fixed in their orientation toward and away from the moon or sun, moving only slowly as the moon, Earth, and sun move in their orbits.

The presence of landmasses prevents the tide wave from traveling around the world. The continents are generally oriented north–south and bisect the oceans. The equilibrium tide wave moves from east to west. When the tide wave encounters a continent, its energy is dissipated or reflected, and the wave must be "restarted" on the other side of the continent. Because continents, landmasses, and the ocean basins are complicated in shape, the tide wave is dispersed, **refracted**, and reflected in a complex and variable way within each ocean basin.

The average depth of the oceans is about 4 km, and the maximum depth is only 11 km. In contrast, the wavelength of the tide wave is one-half of the Earth's circumference, or about 20,000 km at the equator. Hence, the water depth is always considerably less than 0.05 times the wavelength of the tide wave (that is, the water depth is considerably less than $L/20$), and the tide wave acts as a **shallow-water wave** (Chap. 11). The speed of a shallow-water wave is controlled only by the water depth. In the average depth of the oceans (4 km), the tide **wave speed** is approximately 700 km·h^{-1}. The equilibrium model tide wave always travels across the Earth's surface aligned with the movements of the moon and sun. However, to travel across the Earth's surface and remain always exactly lined up with the moon (or sun), the tide wave would have to travel at the same speed as the Earth's spin. At the equator, the wave would have to travel at about 1620 km·h^{-1}. Because it is a shallow-water wave that can travel at only 700 km·h^{-1}, the tide wave at the equator must lag behind the moon as the moon moves across the Earth's surface.

The interaction between the moving tide wave that lags behind the moon's orbital movement and the tendency for a new tide wave to be formed ahead of the lagging wave is complex. However, the result is that the tide wave lags behind the moon's orbital movement, but not by as much as it would if it were not continuously re-created. The tide wave is said to be a forced wave because it is forced to move faster than an ideal shallow-water wave.

Because the tide wave is a shallow-water wave and its speed depends on the water depth, the tide wave is refracted in the same way that shallow-water wind waves are refracted as they enter shallow water. Dynamic-tide theory therefore must include the refraction patterns created by the passage of tide waves over **oceanic ridges**, **trenches**, shallow **continental shelves**, and other large features of the ocean basins.

Latitudinal Variation of the Earth's Spin Velocity

The tidal time lag changes with **latitude** because the orbital velocity (due to the Earth's spin) of points on the Earth's surface decreases with latitude (**Fig. 12.10**). At latitudes above about 65°, the shallow-water wave speed is equal to or greater than the orbital velocity. Hence, there is no tidal time lag at these latitudes, and the tide wave even tends to run ahead of the moon's orbital movement. Another complication in calculating the tide wave time lag is the fact that the Earth's axis is usually tilted in relation to the moon's orbit (and to the Earth's orbit with the sun). Therefore, even on a planet with no continents, the tide waves would not normally travel exactly east–west.

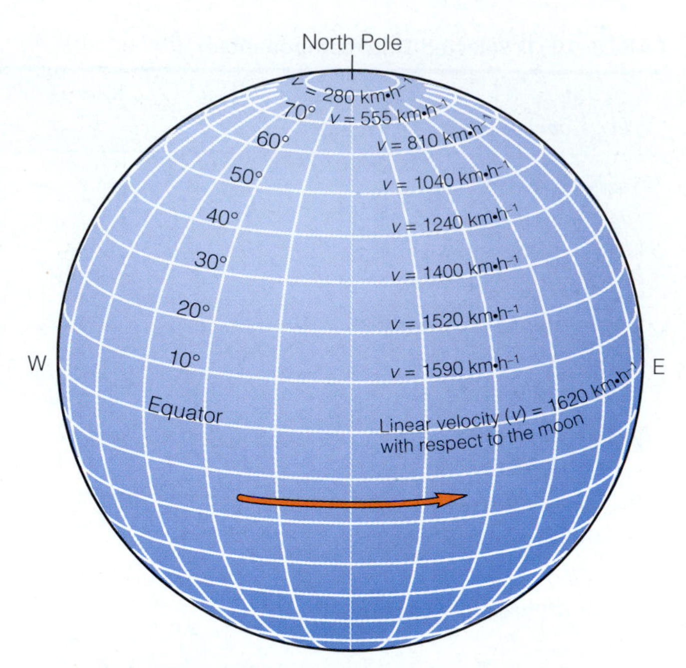

FIGURE 12.10 The tide wave must move across the Earth's surface as the Earth spins in relation to the moon and sun. If it were to do so exactly in time with the Earth's spin, it would have to move at different speeds across the Earth's surface at different latitudes. This is because the orbital velocity of points on the Earth's surface due to the Earth's spin is at a maximum at the equator and decreases with increasing latitude until it is zero at the poles.

Coriolis Effect and Amphidromic Systems

Tide waves are shallow-water waves in which water particles move in extremely flattened elliptical orbits. In fact, the orbits are so flattened that we can consider the water movements to be horizontal. Water moving within a tide wave is subject to the Coriolis effect (**CC12**). One important consequence of the Coriolis deflection of tide waves is that a unique form of **standing wave** can be created in an ocean basin of the correct dimensions. This type of standing wave is called an **amphidromic system,** in which the high- and low-tide points (the wave **crest** and **trough**, respectively) move around the basin in a rotary path—counterclockwise in the Northern Hemisphere and clockwise in the Southern Hemisphere.

Figure 12.11 shows how an amphidromic system is established in a wide Northern Hemisphere basin. In the Northern Hemisphere, a standing-wave crest (**antinode**) enters the basin at its east side. As water flows westward, it is deflected ***cum sole*** to the north, causing a sea surface elevation on the north side of the basin as the wave height at the east boundary decreases behind the crest.

Water now flows toward the south with the north-to-south pressure gradient created by the sea surface elevation on the north side of the basin. As water flows south, it is deflected *cum sole* to the west. The sea surface elevation continues to move around the basin counterclockwise in this way until it returns to the east side of the basin. If the standing-wave crest returns to the east side of the basin after exactly 12 h and 24½ min (or any multiple thereof), it will meet the crest of the next lunar tide wave and the oscillation is then said to be **tuned**.

One important characteristic of amphidromic systems is the amphidromic point (**node**) near the center of the basin, at which the tidal range is zero (**Fig. 12.11**). Amphidromic systems are established in all the ocean basins (**Fig. 12.12**). However, different amphidromic systems are set up for each tidal component (**Table 12.1**)

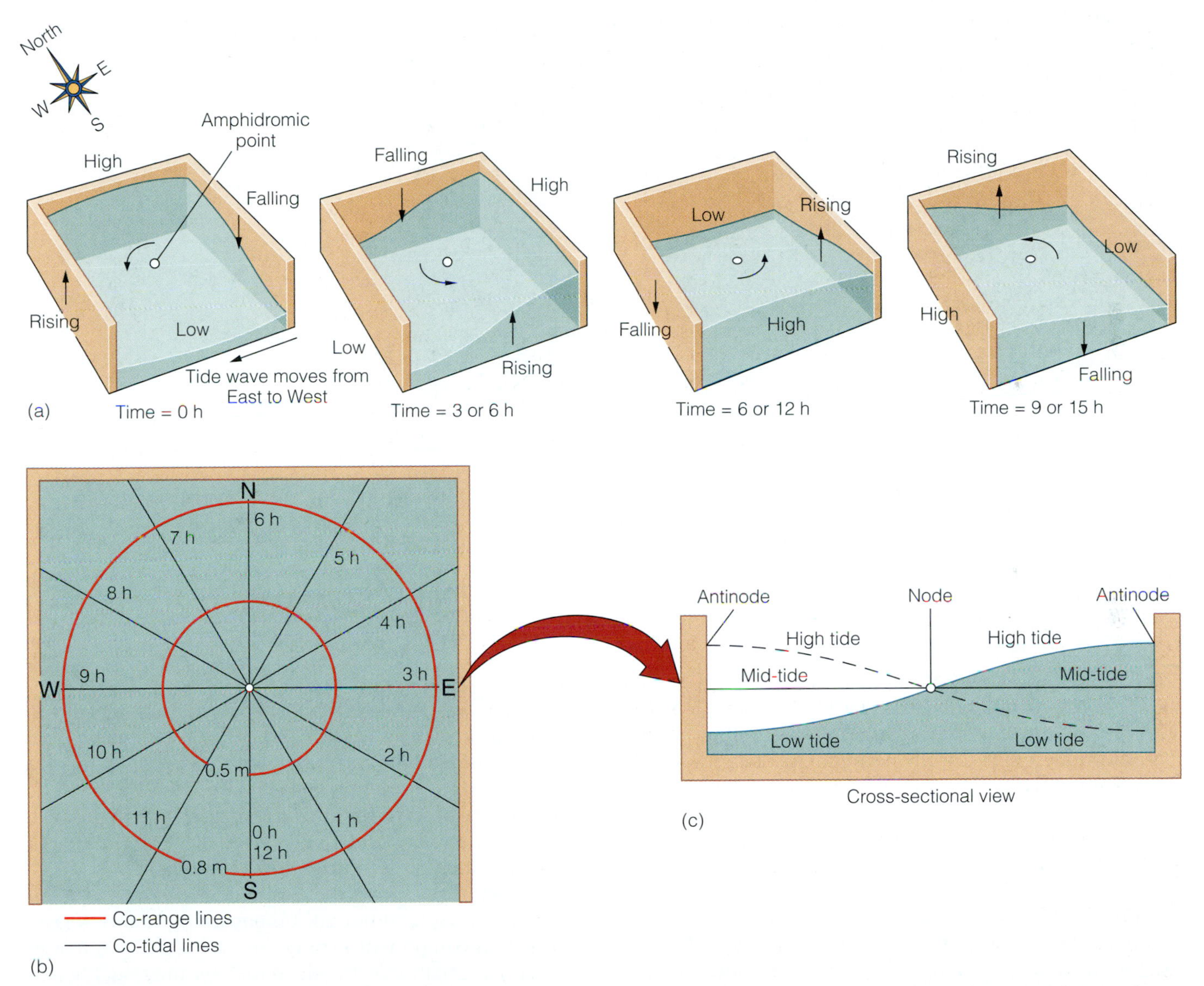

FIGURE 12.11 Development of an amphidromic standing-wave oscillation in response to the Coriolis effect in the Northern Hemisphere. (a) An amphidromic system is created in the principal solar semidiurnal or diurnal component of the tide when the tide wave enters a basin and is deflected by the Coriolis effect so that it rotates around the basin and returns to the entrance exactly 12 or 24 h later. (b) In an idealized Northern Hemisphere amphidromic system, the tidal range is the same for all places on any circle centered on the amphidromic point. The line along which the low tide or high tide is located progresses around the basin counterclockwise (clockwise in the Southern Hemisphere). In this plan view, high tide or low tide occurs along each of the radial lines at the indicated number of hours after occurring along the line labeled "0 h". (c) A cross-sectional view of the basin shows the up-and-down motion about the node upon which the rotation around the basin is superimposed.

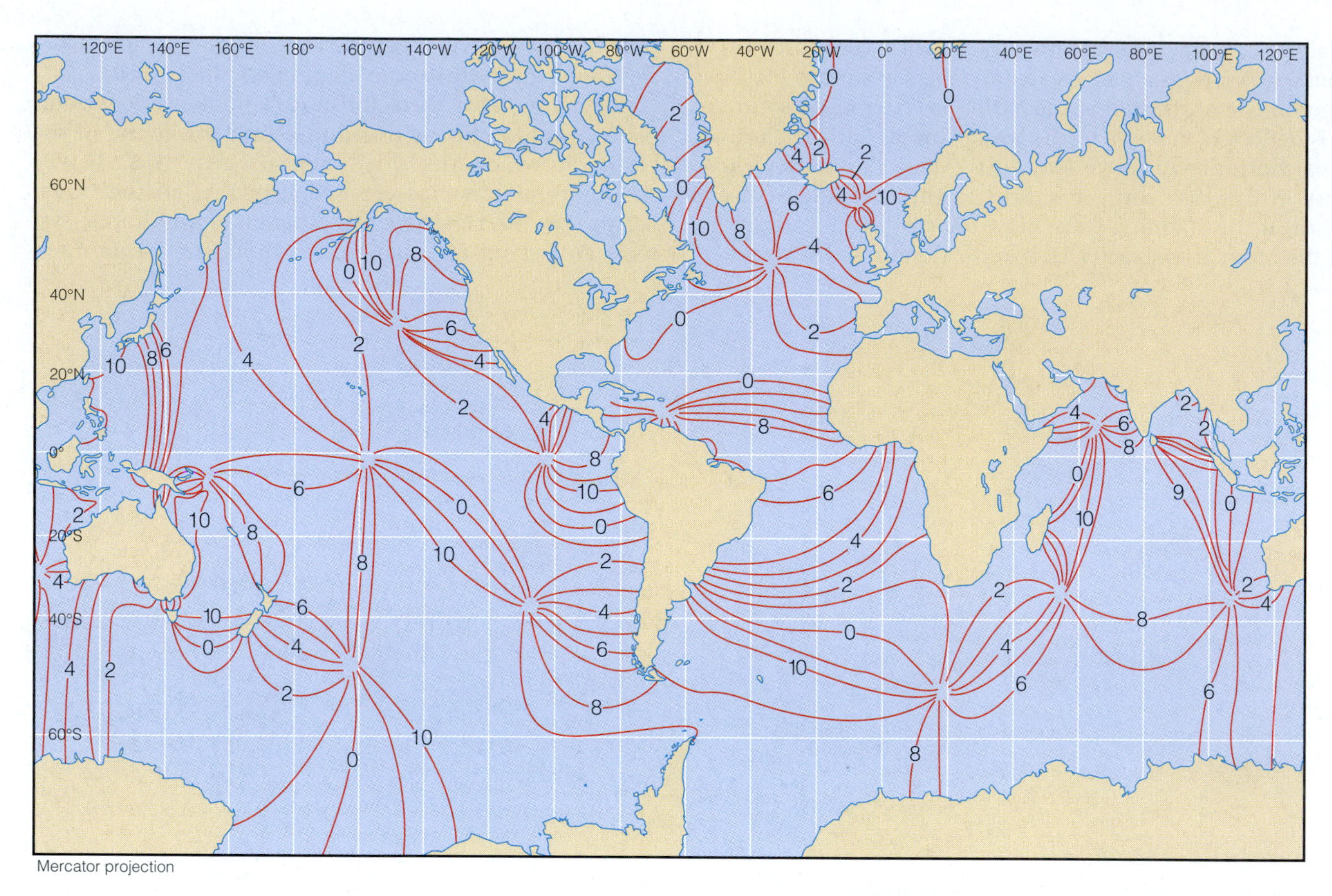

FIGURE 12.12 Amphidromic systems of the principal lunar semidiurnal tidal component in the Earth's oceans. The lines show the location of the tidal maximum. The numbers indicate the number of hours that elapse as the maximum tide (crest) travels from the line labeled 0 to the indicated line. Notice that the crest of the tide wave moves in rotary amphidromic systems in most of the ocean basins, and that the rotation is counterclockwise in the Northern Hemisphere and clockwise in the Southern Hemisphere. Because it is a shallow-water wave, the tide wave is also refracted as water depth changes, resulting in some of the complex bending of the wave in certain parts of the oceans.

because the components have different wavelengths. Hence, a location in an ocean basin that is an amphidromic point for one component of the tide will have a zero tidal range for that component, but will have tides generated by tidal components with other periods. In addition, some basins tune more easily with a particular tidal component. In parts of such basins, the tidal range due to that tidal component can be enhanced in relation to the ranges due to other components. This effect explains the presence of dominant diurnal and semidiurnal tides at many locations where they would not be predicted by equilibrium tide theory.

12.5 TIDES IN THE OPEN OCEANS

Having examined the many factors that control tides, we can now look at some characteristics of real tides in the ocean basins. Much of what we know about the behavior of the tides is based on tidal-height measurements made along the coasts and on islands. The details of tidal movements in the deep oceans generally are derived from these measurements and dynamic-tide theory, although a number of deep-sea tide gauges have been deployed to make direct observations during the past several decades. Satellites now also provide data that can help refine our understanding of deep-ocean tidal motions.

Figure 12.13 shows the progression and height of the principal lunar semidiurnal component of Atlantic Ocean tides. This component is only one of the many different partial tides that must be added to determine tidal height at any time and location.

The most striking feature of **Figure 12.13** is that the tide wave does not move from east to west with the moon, as equilibrium tide theory predicts. This is because the Atlantic Ocean is a relatively narrow basin in which only a small tide wave can be generated during a single pass of the moon. The small east–west tide wave soon encounters the American continent, where it is partially reflected and much of its energy is dissipated.

The only segment of the Earth where the tide wave can travel east to west around the world without encountering a landmass is near Antarctica. At this high latitude, the orbital velocity of the Earth's surface is low enough (**Fig. 12.10**) that the shallow-water tide wave can travel fast enough to keep up with the moon's orbital movement. The tide wave around Antarctica is therefore well developed (**Fig. 12.12**). It enters the Atlantic Ocean between the tip of South Africa and Antarctica and is partially deflected and dispersed into the South Atlantic Ocean. The tide wave moves northward through the South Atlantic Ocean as a progressive wave. As it travels north, it interacts with the weaker east–west wave formed in the Atlantic and is reflected and refracted in complex ways. It is also deflected by the Coriolis effect. Although the deflection is obscured by other factors in the southern part of the South Atlantic, it causes tides to be slightly higher on the South American coast north of Rio de Janeiro than on the opposite African coast (deflection to the left in the Southern Hemisphere). The Coriolis deflection is also partially responsible for tides being slightly higher on the coasts of Europe and North Africa than on the North American coast (deflection to the right in the Northern Hemisphere).

In the North Atlantic, the progressive wave traveling north from Antarctica is converted into the standing wave of an amphidromic system (**Fig. 12.13**). The high tide moves around the North Atlantic basin counterclockwise and arrives back where the next crest of the Antarctic progressive wave arrives almost exactly 12 h and 24½ min later. The North Atlantic Ocean basin is therefore well tuned to the semidiurnal tidal component, and this component dominates and produces semidiurnal tides in this region (**Fig. 12.6**).

Amphidromic systems similar to the system in the North Atlantic are also present in the North Sea and the English Channel. However, the Gulf of Mexico has a natural period of about 24 h. Therefore, the semidiurnal tidal component in the Gulf of Mexico is poorly developed, and the diurnal component is stronger, so diurnal tides dominate in some parts of the Gulf (**Fig. 12.6**). The Pacific Ocean basin is wider than the Atlantic Ocean basin. Therefore, the Pacific has a more developed east–west tide wave and greater complexity than the Atlantic Ocean. In both the Pacific Ocean and the Caribbean Sea, diurnal and semidiurnal waves are relatively well tuned and tides are generally mixed tides.

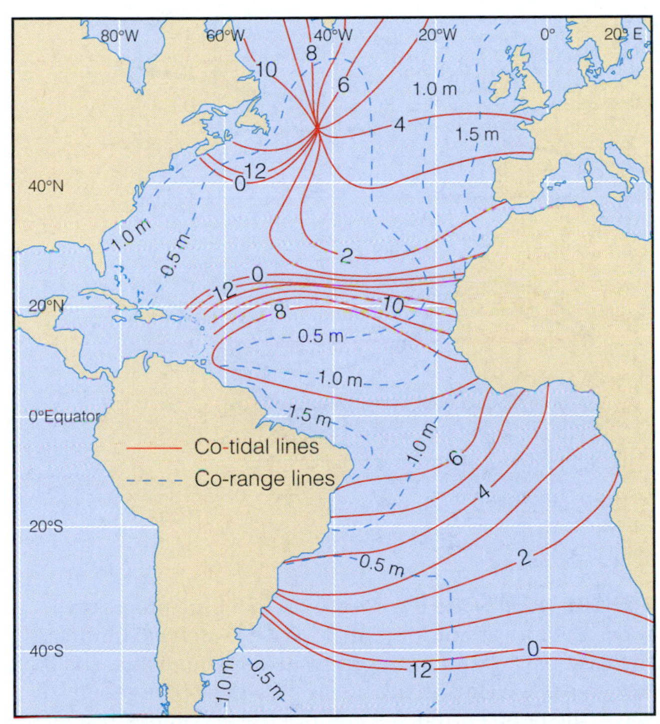

FIGURE 12.13 The principal lunar tidal component wave enters the Atlantic Ocean from the Indian Ocean around Antarctica and travels north into the North Atlantic, where it forms an amphidromic wave. Red lines show the location of the tidal maximum. Dashed blue lines are contours of tidal range. Note the higher tidal range on the west side of the South Atlantic compared to the east side. Also note that the tidal range increases in all directions from the center of the amphidromic system.

12.6 TIDAL CURRENTS

Orbits of water particles in tide waves are so flattened that water movements in the tide wave are essentially oscillating currents. The magnitude and periodicity of these currents are as important to mariners as the tidal range is, particularly in coastal and estuarine regions. Although tidal currents are relatively weak in the open oceans, their speed increases as the tide wave moves inshore and its energy is compressed into a shallower depth of water. Tidal-current speed also increases where the tide wave must move large quantities of water through a narrow opening into a large bay or estuary, such as the entrances to San Francisco Bay and New York Harbor.

Open-Ocean Tidal Currents

Tidal currents are generally weak and rotary in character in deep-water areas away from the coast in the large ocean basins (**Fig. 12.14**). In areas of the Northern Hemisphere with semidiurnal tides, the tidal-current direction rotates progressively, usually clockwise, and completes a 360° rotation in 12 h 24½ min (**Fig. 12.14a**). The current speed varies but is never zero.

In areas with mixed tides, the progression of tidal-current speed and direction is more complicated (**Fig. 12.14b**) because both variables represent a combination of two different tidal components. Note that the times of maximum and minimum tidal-current speed are apparently not related to the times of high and low tide.

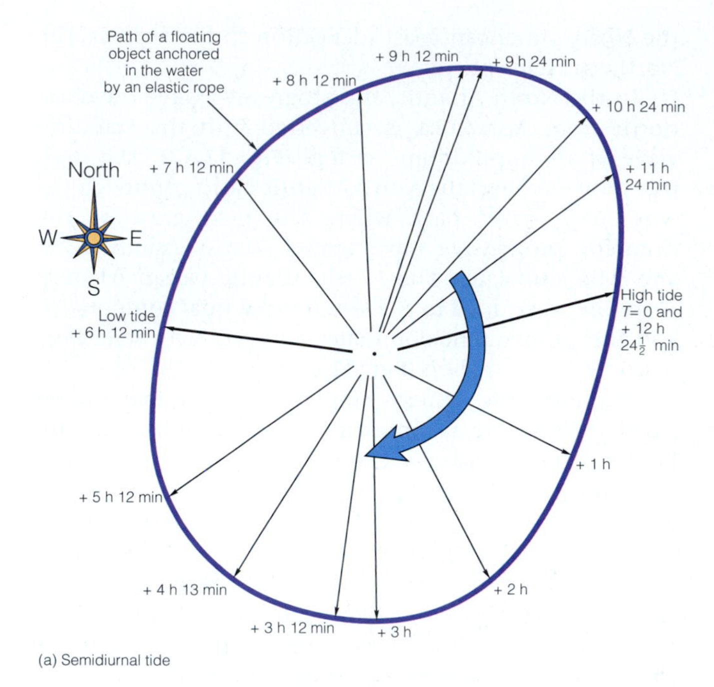

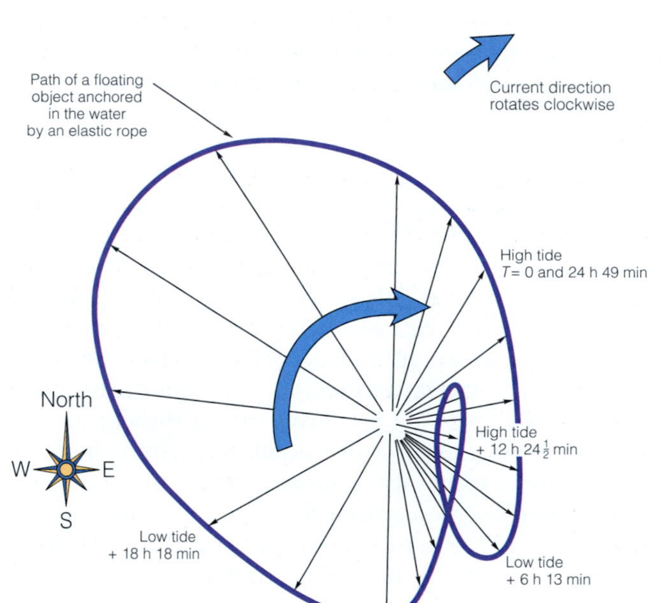

FIGURE 12.14 Changes in the direction and magnitude of open-ocean tidal currents in the Northern Hemisphere. The arrows indicate the direction of the current at approximately hourly intervals, and the length of the arrows is proportional to the current speed. The outer blue line describes how an object would move in the water if there were no currents or water movements other than tidal currents. (a) Where there is a simple semidiurnal tide, the current velocity varies little, but its direction rotates clockwise through 360° every 12 h 24½ min. (b) Where there is a mixed tide, the variations of current speed and direction can be more complex. In this case, the current direction rotates clockwise twice within 24 h 49 min, but the rate of rotation and the current speeds follow very different patterns within the two successive 12-h periods.

Temporal Variation of Tidal-Current Speeds

Many people mistakenly believe that the tidal currents will stop when the tide reaches its highest and lowest points. This is rarely true. In fact, if the tide were a pure progressive wave, the tidal current would be at its fastest at high or low water. We can see why by considering the motions of water particles in the tide wave (**Fig. 12.15a**).

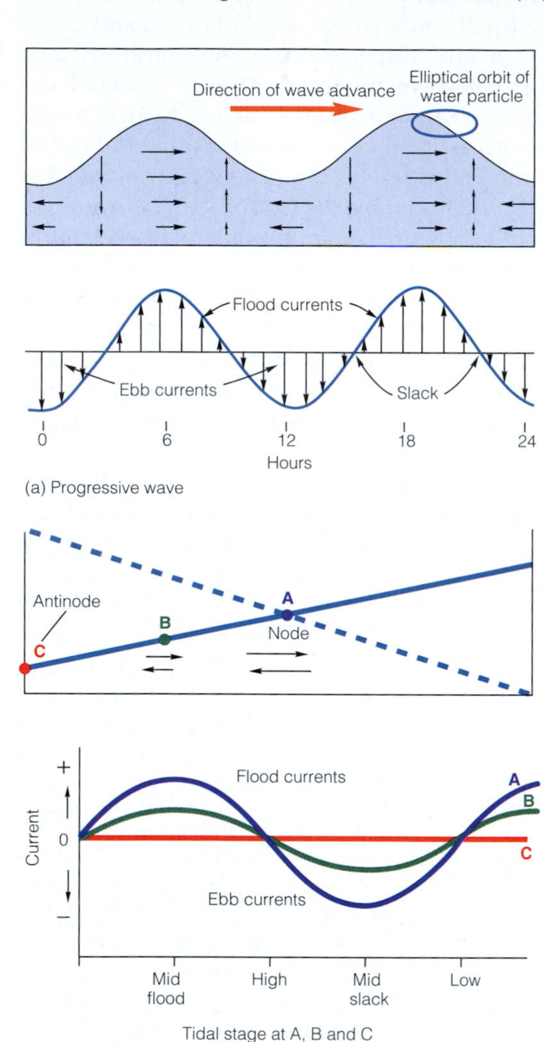

FIGURE 12.15 Tidal currents. (a) In a pure progressive wave, the forward motion in the wave is at a maximum as the crest passes, and in the opposite direction when the trough passes. Thus, if the tide wave were a pure progressive wave, the flood currents would reach a maximum at high tide and the ebb currents would reach a maximum at low tide. (b) In a standing wave, the horizontal currents are always zero at the antinodes, and they are reversing currents whose maximum velocity increases from zero at the antinode to a maximum at the node. The maximum current speeds are reached at mid flood and mid ebb. Because most tides have both progressive-wave and standing-wave components, there is no fixed relationship between the times of high and low tide and the times of maximum tidal current and slack water at different locations. Tidal currents are also complicated by seafloor friction.

However, the maximum tidal currents rarely coincide with high or low tide for several reasons. The most important is that all tides are complex combinations of many components. Some components may be progressive waves, some may be standing waves, and some may have both progressive-wave and standing-wave characteristics. In addition, tide waves are reflected and refracted, so the observed tide may be the result of several different waves moving in different directions.

In a standing wave, currents are at their maximum midway between high and low tide, when the sea surface is exactly level (**Fig. 12.15b**) Current speed varies not only with time, but also with location in the wave. It is highest at the node and is always zero at the antinodes, where the vertical water surface displacement, or tidal range, is greatest.

In most places, the tide is a complex mixture of progressive-wave and standing-wave components that vary from location to location. Hence, the relationship between the timing of **slack water** (or, conversely, maximum current) and high and low tide is different for each location. Curiously, at certain times, some locations have tidal currents but no tide (zero tidal range), and other locations have a large tidal range but no tidal currents. We can envision such locations to be the node and antinode, respectively, of a standing wave that is the dominant component of the tides, although the situation is generally more complicated because the tide is the sum of many different components of both solar and lunar tides.

Tide tables that list only times of high and low tide are useless for determining tidal-current velocities because there is no consistent relationship between the times of high and low water and the times of highest current speed and slack water. The only generalization that can be made about the tidal currents is that their maximum speed will increase or decrease as tidal range increases or decreases from day to day. Therefore, tidal currents must be measured in each location for which a forecast of the current velocity is important, just as tidal ranges must be measured. Tidal-current information from such measurements is subject to **harmonic analysis** that is similar to the analysis of tidal-height data described later in this chapter. From this analysis, tidal-current tables are produced.

Tidal Currents in Estuaries and Rivers

Tides extend far into many bays, estuaries, and rivers. Tides in such locations are affected by the extremely shallow water depths, freshwater flow, and friction with the seafloor. In very shallow water, the crest of the tide wave moves in significantly deeper water than the trough. Therefore, the high tide tends to catch up to the low tide in estuaries or rivers where the tide travels long distances through shallow water. As a result, river tides can be modified so that there is a long period between high and low tide, but a very short period between low and high (**Fig. 12.16**).

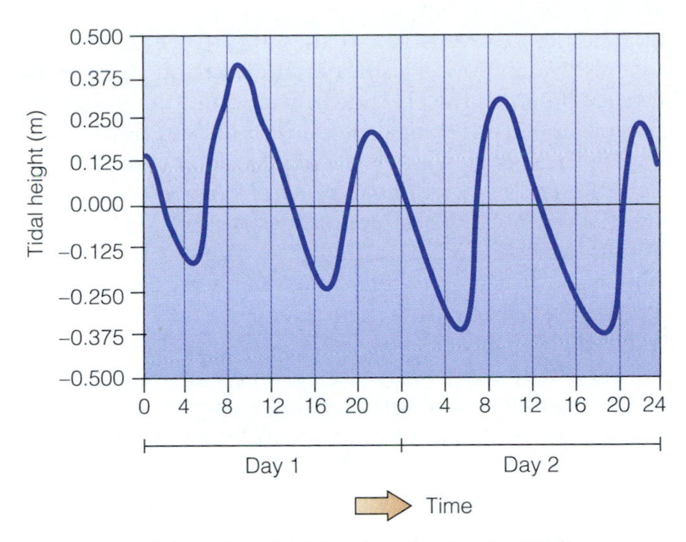

FIGURE 12.16 This tidal-height plot for the Hudson River near Albany, New York (about 220 km from New York Harbor), shows that the tide rises much faster (that is, the plot is steeper) than it falls.

In some areas where tidal ranges are large and the tide enters a channel or bay that narrows markedly or has a steeply sloping seafloor, tidal bores may occur. A tidal bore is created when the currents in the flooding tide are faster than the speed of a shallow-water wave in that depth. The leading edge of the tide wave must force its way upstream faster than the wave motion can accommodate. The wave therefore moves up the bay or estuary as a wall of water, much like a continuously breaking wave. Well-known tidal bores occur in the Bay of Fundy in Nova Scotia and in Turnagain Arm off Cook Inlet in Alaska (**Fig. 12.17**). Most tidal bores are less than 1 m high, but they can reach as much as 10 times that height. Some bores, notably one in China, are high enough that they attract surfers who are looking for the longest ride of their lives.

FIGURE 12.17 Tidal bore in Turnagain Arm, Alaska.

The tidal currents in estuaries and rivers are affected by freshwater flow. Freshwater flowing into a river or estuary prolongs the ebb tide in relation to the flood tide because more water must be transported out of the river during the ebb than enters during the flood to accommodate the freshwater discharge. River flow rates can substantially change the nature of tides and tidal currents in estuaries. Accordingly, tide tables and tidal-current tables for bays, estuaries, and rivers must be used only as a general guide, particularly when river flow rates are abnormally high or low.

Tidal currents within bays, rivers, and estuaries can behave as progressive waves, standing waves, or, in some cases, a combination of the two. In long bays with progressive tides, tidal currents can flow in opposite directions at the same time in different sections of the bay. Chesapeake Bay has this type of circulation (**Fig. 12.18**). Slack water from each tide migrates up the bay about 60 km each hour and takes about 10 h to travel the length of the bay. Successive slack tides enter the bay approximately every 6 h 12 min because each semidiurnal tide has slack water associated with both flood and ebb. This progressive wave is not modified significantly by an east–west wave, because the bay is too narrow for a significant wave to be generated directly by the tide-generating forces.

Tidal currents in Chesapeake Bay are somewhat more complex than those shown in **Figure 12.18** because the

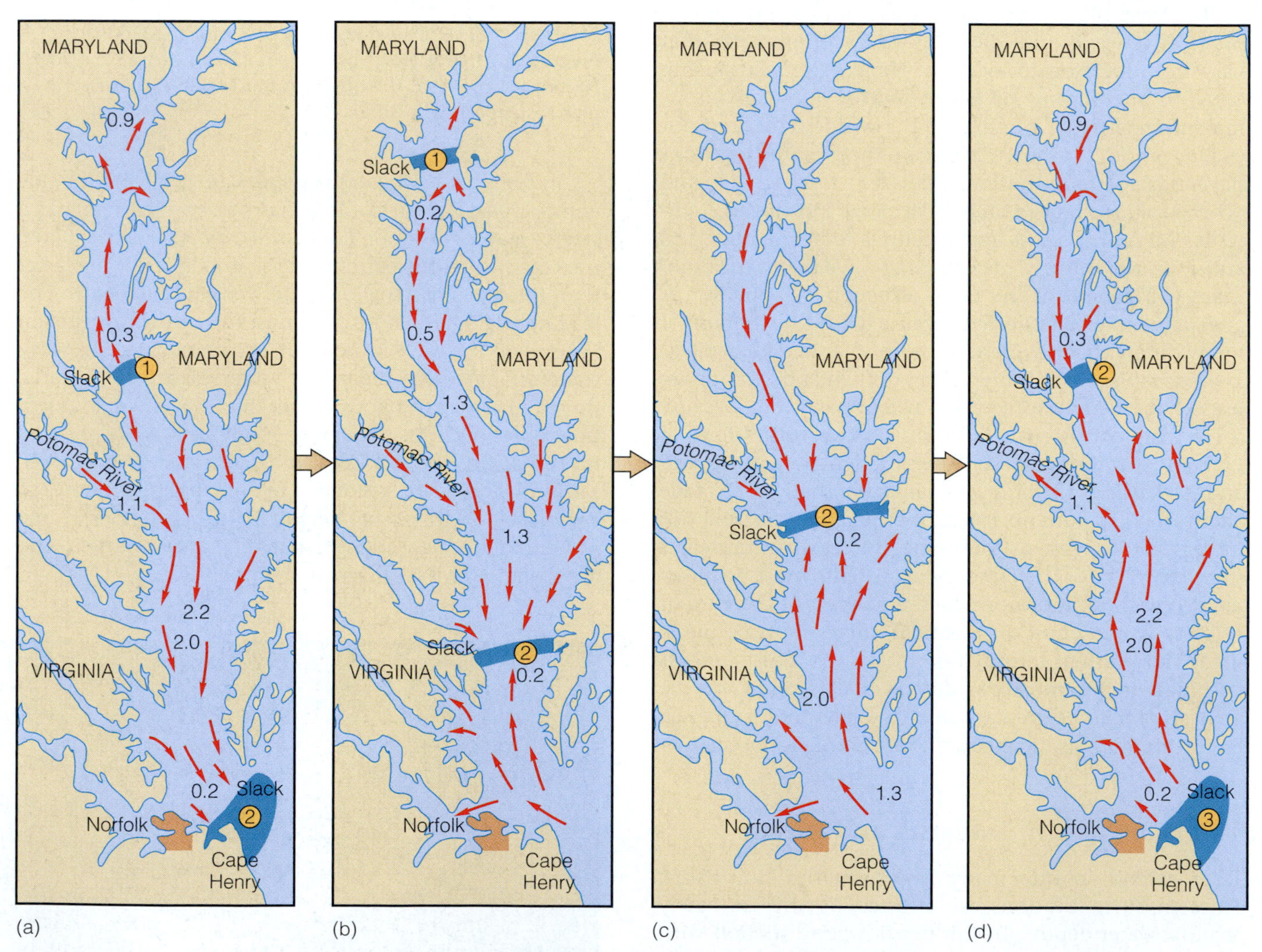

FIGURE 12.18 Chesapeake Bay has a primarily progressive-wave tide. The tide wave enters the estuary and moves progressively up the bay, reaching the north end about 10 h after entering the bay. Follow the successive slack-water occurrences numbered 1, 2, and 3 in these diagrams as they move north. (a) Slack water occurs at the mouth of the bay just as the flood begins and also at a location just above mid bay. (b) Two hours after the start of the flood, the two areas of slack water have migrated northward. (c) Four hours after the start of the flood, the northernmost slack has reached the north end of the bay and the second slack is almost in mid bay. (d) Six hours after the start of the flood, the slack that entered the bay 6 h earlier is now where its preceding slack had been at that time, and a new slack-water area occurs at the mouth of the bay as the ebb begins.

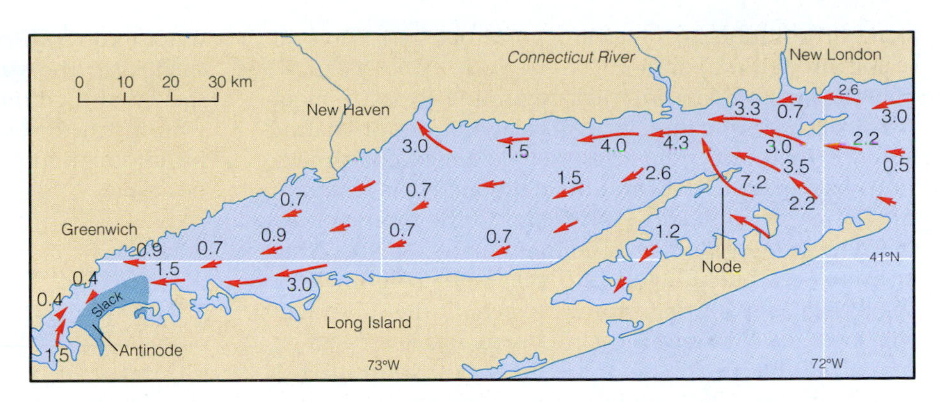

FIGURE 12.19 Long Island Sound has a predominantly standing-wave tide, with a node at the entrance to the sound and an antinode at the western end. There is little or no tidal current at any time during the tidal cycle at the western end where the antinode is located. In the remainder of the sound there are reversing tidal currents, with the maximum current speed increasing from the western end to the eastern opening to the ocean, where a node occurs.

bay does not have uniform depth. The tide wave is slowed more rapidly at the sides of the bay, where the water is shallower and friction with the channel sides is enhanced. In this and other bays with a deep central channel and shallower sides, the tidal current reverses sooner at the sides than in the deep channel. This effect can be useful to boaters who want to avoid the fastest opposing currents. It is also a mechanism for mixing water from the deep main channel with water from the shallow sides, and therefore is of interest in studies of the dilution of **contaminant** discharges.

Long Island Sound has a predominantly standing-wave tide (**Fig. 12.19**). Tidal currents first flow westward throughout the length of the sound, then slack water occurs throughout the sound, and subsequently the current reverses and flows eastward throughout the sound. The area of nearly permanent slack water at the west end of the sound is the standing wave antinode and has a large tidal range, but little or no tidal current. The east end of the sound is the node and has a small tidal range.

12.7 TIDE PREDICTION

As we have seen, the dynamic-tide theory must take into account many different variables to predict tidal height at any given location at a specific time. The best predictions made by applying dynamic-tide theory closely approximate the actual tides. However, the predictions are not accurate enough for use by mariners, particularly in complex estuaries and bays. Therefore, tide predictions are made from tide observations. Tidal variations are observed and recorded continuously at many locations throughout the globe. Reasonably accurate tide predictions can be made from a tide record covering only a few months, but much longer records produce more accurate predictions.

Once the tide record is available, computers may be used to perform harmonic analysis of the data to identify components of the complex wave. For example, harmonic analyses would distinguish the five component waves in **Figure 11.12b** from the complex wave record in **Figure 11.12a**. Harmonic analysis determines the wave heights and time lags (in relation to the moon's orbital movement or other motions) of each of the many different tidal components (**Table 12.1**). More than 390 different tidal components have been identified. Once the timing, periodicity, and **amplitude** of each component have been determined for a particular location, computing the magnitude of each component for any point in time in the future is relatively easy. Simple addition of the components gives the tidal height at the future time and location.

Tide predictions made in this way are used extensively. They are the basis for the tide predictions published daily in many newspapers and the printed tide tables offered in bait-and-tackle stores, dive shops, and marinas. Recently, satellite-based radar sensors accurate enough to measure tidal heights have allowed detailed mapping of the tides in the open-ocean areas where few tidal-height measurements have previously been made. The comprehensive tidal-height data sets collected through satellites are steadily accumulating and will eventually allow extremely accurate computer models of the tides. Tidal currents vary locally, and predictions must be based on several months of measurements at each specific location.

12.8 MARINE SPECIES AND THE RHYTHM OF THE TIDES

Tides play an important role in the life cycles of many marine species. The most obvious connection is with species that live in the intertidal zone, for which the movements of the tides mean that they and their **habitats** are exposed to the atmosphere for variable periods during the day. These species have developed strategies to avoid dehydration and/or being eaten by birds and other predators during the times when their habitat is exposed to the air. For example, most species that live in sandy or muddy intertidal **environments** bury themselves in the sand when the tide is out. Many species that live on rocky shores have shells to retreat into or leathery outer skins that, in addition to deterring predators, resist loss of water through evaporation.

Although we are only just beginning to learn about many of the subtle influences of tidal motions on marine species, there are numerous well-documented connections between the tides and the reproductive behavior of many species, especially with regard to the timing of the reproductive cycle. For example, many species that **spawn** eggs and **larvae** into the water column have spawning behavior that is timed to benefit the survival of the species. For example, many **coral** species spawn en masse at a specific phase of the lunar tide, most often 1 or 2 days after a full moon in the spring months, when the tidal range is at its greatest. Spawning at this time ensures the greatest possible dispersal of the eggs and larvae to potential new sites for colonization. Spawning strategies are discussed in more detail in Chapter 16.

There is one connection between tides and species' reproductive strategies that you may be able to see for yourself. This is the use of high spring tides by some species at a certain time of year to place fertilized eggs in the sand high up on a **beach** where they cannot be reached by most predators that live in the water. Perhaps the best-known examples are sea turtles that haul themselves high up the beach, usually at night when there is a high spring tide, dig a hole in the sand, lay and bury their eggs, and then leave the eggs to hatch some weeks later. However, many other species use a very similar strategy. Some of these species can be seen on beaches or mudflats around North America. For example, on the Atlantic coast, horseshoe crabs (*Limulus polyphemus*) spawn on the beaches between the high- and low-tide lines on nights when the spring high tide occurs. Female crabs crawl out of the water as the tide advances, dig a cavity about 15 cm deep in the sand, and deposit several thousand large, greenish eggs. The male, which is attached to the female's back, fertilizes the eggs, and the eggs are then buried by sand moved by the advancing tide. The eggs hatch 14 days later when the next high spring tide covers the sand.

On the Pacific coast, several fishes use a similar strategy. The most famous of these fish species is the grunion (*Leuresthes tenuis*). This small, silvery fish comes completely out of the water onto the southern California and Baja beaches to spawn just after each maximum nighttime spring tide during March through September. The eggs are protected from wave action for the next 14 days because they are laid above the lower high-tide lines that occur at high tide between one spring tide and the next. Although they mature in about 9 days, the eggs do not hatch until they are disturbed by water and the wave **turbulence** that occurs as the next spring high tide occurs 14 days after they were laid. The eggs hatch within minutes of being exposed to the water.

Although the grunion is the most famous fish that spawns on the beach at high tide, other species also use this strategy, but they do not come completely out of the water to do so. One example of such a species is another small fish, the **surf** smelt (*Hypomesus pretiosus*), which spawns at high tides on the beaches of Puget Sound. Ask your instructor or do some research at your local library for information on species that might spawn on beaches near you.

12.9 ENERGY FROM THE TIDES

The tides are a potential source of inexhaustible and clean energy. Tidal energy is dissipated as heat by friction between **water masses** and the seafloor. Some of the energy can be captured by electricity-generating turbines placed in locations where tidal currents are very fast. However, very few locations have maximum tidal currents that are fast enough to drive turbines efficiently. Even where tidal currents are swift, they must be enhanced by the tidal flow being channeled into the turbines through narrow openings. The best locations for capturing tidal energy are bays or estuaries with a large tidal range and swift tidal currents and where a dam can be built across the entrance. The dam is constructed with a number of narrow openings in which turbines are located. Water entering the bay is channeled through the turbines as the tide enters the bay, and again as the tide flows back out of the bay.

For a tidal power-generating plant to be feasible, the tidal range must exceed about 5 m, and it must be possible to build a dam that isolates an area into which such a tide flows from the open ocean. The most promising locations are few and widely distributed (**Fig. 12.20**). A large tidal power-generating plant has operated successfully at La Rance in France since the 1960s (**Fig. 12.21**). There has been continuing interest in constructing other plants, notably in the Bay of Fundy, Nova Scotia. Such projects have a number of problems, two of which are especially important.

First, a tidal power-generating plant cannot generate energy throughout the 24 h in a day. Power cannot be generated during the period of slack water when tidal currents are low. Judicious operation can retain water outside or behind the dam to some extent and thus lengthen the period during which power can be generated, but generating power uniformly over a 24-h period is apparently not possible.

The second problem is that building a dam across a bay substantially alters the **ecology** and other aspects of the bay. For example, the Bay of Fundy sustains rich and **diverse** marine **communities** and fisheries and has a famous tidal bore. Building a dam across the mouth of the bay would drastically alter or destroy both the fishery and the tidal bore. Fishers who live around the bay would lose both their local catch and access to the open ocean for their boats. Tourism would decline with the loss of the tidal bore.

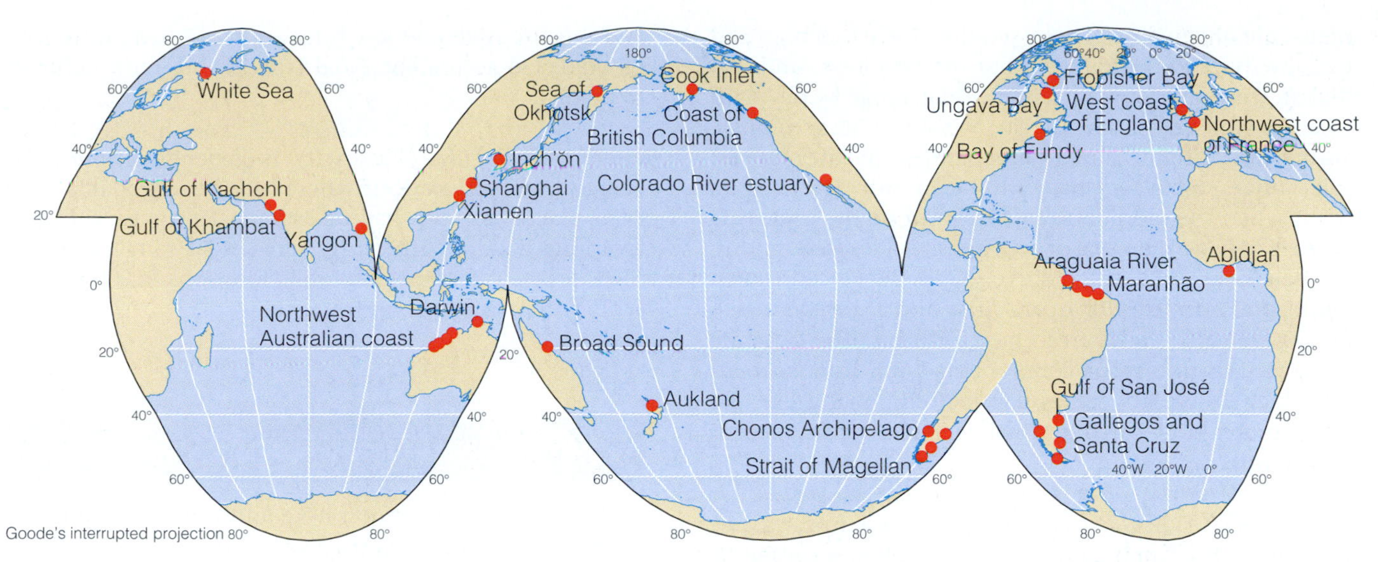

FIGURE 12.20 There are relatively few locations with a tidal range sufficiently large (more than about 5 m) that tidal power-generating plants are considered to be feasible.

On the positive side, a Bay of Fundy tidal power-generating plant would produce "clean" energy. It would help to lessen enhancement of the **greenhouse effect** and **acid rain** by reducing the need for power plants that burn **fossil fuel**. The benefits and losses associated with tidal power-generating stations will be difficult to balance. A decision to build or not build such a plant at the Bay of Fundy (or anywhere else) will be controversial and will generate opposing views from reasonable people.

FIGURE 12.21 La Rance tidal power-generating plant in France.

CHAPTER SUMMARY

Tide-Generating Force. Bodies orbiting together do so around their common center of mass. A centripetal force acting toward the other body is needed to keep each body in orbit. This force is supplied by the gravitational attraction between the bodies. The centripetal force needed to maintain the Earth in its common orbit with the moon (or sun) is the same at all points within the Earth. However, the gravitational force of the moon (or sun) is inversely proportional to the distance from the moon's (or sun's) center of mass. The gravitational force exerted by the moon (or sun) is therefore slightly higher on the side of the Earth facing the moon (or sun) than on the other side. Hence, there is a slight excess of gravitational force over centripetal force at the Earth's surface facing the moon (or sun), and a slight excess of centripetal force over gravitational force at the opposite side of the Earth. The slight imbalances are the tide-generating forces. Tide-generating forces exerted by the sun are only 46% as strong as those exerted by the moon because even though the sun is more massive than the moon, it is much farther away from the Earth than the moon is.

Characteristics of Tides. Tides resemble patterns of oscillating sea surface elevation caused by simple progressive waves or by the addition of two or more progressive waves. Because the moon's orbit is inclined to the equator, tides are usually predominantly one of three types. Diurnal tides have one high and one low tide per tidal day (24 h 49 min). Semidiurnal tides have two high and two low tides each tidal day, and the two lows (and two highs) are of equal height. Mixed tides have two high and two low tides each tidal day, but the two lows (and two highs) are not of equal height. Tidal range varies with location and from day to day within a lunar month. Spring tides (highest tidal range) and neap tides (lowest tidal range) occur twice each lunar month. Spring tides occur when the moon and sun are both on the same side of the Earth (new moon) or on opposite sides (full moon), so their tidal pulls are additive. Neap tides occur when moon and sun are 90° apart (first and third quarters). Tidal range also varies from month to month. Tides are composed of numerous components with different periods, each caused by a regular periodic change in the Earth–moon and Earth–sun orbits, such as distance or declination.

Tide Waves. Tide waves appear to progress from east to west as the Earth spins under the tide bulges. Continents interrupt the tides everywhere except around Antarctica and cause tide waves to be dispersed, refracted, and reflected in complex ways that affect tidal characteristics differently at different locations. Because its wavelength is so long, a tide wave is a shallow-water wave that is refracted by seafloor **topography.** The tide wave is a forced wave because it is too slow, even in deep oceans, to match the orbital velocity of the Earth's surface as it spins, except near the poles. Tide waves are deflected by the Coriolis effect. In ocean basins of suitable dimensions, this deflection can create tuned-oscillation standing waves called amphidromic systems. The crest of one tidal component enters the basin and passes counterclockwise (Northern Hemisphere) or clockwise (Southern Hemisphere) around the basin to arrive back at the entrance exactly when another crest arrives.

Tidal Currents. Orbits within tide waves are so compressed, particularly in coastal and estuarine regions, that vertical tidal motion is very small in relation to the horizontal motions. The horizontal motions are tidal currents. Open-ocean tidal currents are weak, and their direction progresses in a rotary pattern during the tidal cycle. In coastal and estuarine areas, tidal currents generally reverse direction 180° as flood and ebb currents during the tidal cycle. Tides in such areas are usually combinations of progressive and standing waves, and the relationship between high or low tide and times at which slack water or maximum currents occur is different at each location. Tidal currents can be very strong, tidal range particularly high, and the progression of the tide wave complicated in estuaries where standing waves occur for some tidal components.

Tide Predictions. Tides are too complex to be predicted without measurement at each location for which predictions are needed. Tidal-height measurements may now be made with satellite sensors. Tidal-height measurements recorded for several months or longer can be subjected to harmonic analysis. This analysis can allow the tidal range and times of high and low tides at the studied location to be predicted accurately. Tidal currents are more difficult to predict.

Energy from Tides. Electricity is generated from tidal energy in a few locations where turbines and dams have been placed across a bay or estuary. However, very few locations have a sufficiently large tidal range to make such projects feasible. Suitable locations have unique ecology that depends on those tides.

STUDY QUESTIONS

1. If the mass of the moon were four times what it is, the distance between the Earth and the moon would be twice what it is now. Is this a correct statement? Explain why or why not.
2. Describe the imbalance between centripetal force and the gravitational attraction force at different locations on the Earth's surface. How is it created?
3. Where on the Earth's surface, in relation to the moon's direction, is the tide-generating force due to the moon equal to zero?
4. Why does the moon exert a greater tide-generating force than the sun, even though the sun's gravitational attraction at the Earth's surface is almost 200 times stronger than the moon's?
5. How are diurnal, semidiurnal, and mixed tides different? Why is it important to some sailors to know which type of tide is present in a port that they visit?
6. Spring tides and neap tides occur twice each month but are not the same height each month. What variations of the Earth's orbit around the sun might explain seasonal variations in the tidal range of spring or neap tides?
7. Why is the tide wave a forced wave in some places and not in others? Where is it a forced wave, and where is it not a forced wave? Why?
8. Why are amphidromic systems not established in all ocean basins? Why do some tidal components in a particular basin form amphidromic systems, whereas other components do not?
9. Why are tidal currents in the deep oceans slower than those in coastal waters? Why are deep-ocean tidal currents generally rotary, whereas coastal tidal currents are reversing?
10. How are tide tables able to predict the future times and heights of high and low tides? These predictions are generally accurate, but weather conditions can cause the heights and times of high and low tide to be shifted somewhat from those predicted. How can this happen? List and explain more than one possible reason.

KEY TERMS

You should be able to recognize and understand the meaning of all terms that are in boldface type in the text. All of those terms are defined in the Glossary. The following are some less familiar key scientific terms that are used in this chapter and that are essential to know and be able to use in classroom discussions or on exams.

amphidromic system (p. 324)
amplitude (p. 331)
antinode (p. 324)
centripetal force (p. 313)
Coriolis effect (p. 323)
crest (p. 324)
cum sole (p. 324)
current (p. 312)
declination (p. 320)
diurnal tide (p. 317)
ebb (p. 312)
estuary (p. 312)
flood (p. 312)
harmonic analysis (p. 329)
high-tide line (p. 312)
intertidal zone (p. 312)
low-tide line (p. 312)
lunar month (p. 319)
mixed tide (p. 317)
neap tide (p. 319)
node (p. 325)
orbital velocity (p. 323)
partial tide (p. 323)
pressure gradient (p. 315)
progressive wave (p. 317)
refracted (p. 324)
semidiurnal tide (p. 317)
shallow-water wave (p. 324)
slack water (p. 329)
spring tide (p. 319)
standing wave (p. 324)
tidal range (p. 312)
trough (p. 324)
tuned (p. 325)

CRITICAL CONCEPTS REMINDER

CC12 **The Coriolis Effect** (pp. 313, 314, 323, 324). Water masses move freely over the Earth's surface while the solid Earth itself is constrained to move with the Earth's rotation. This causes moving water masses, including long wavelength waves that comprise the global tide wave motion, to appear to follow curving paths across the Earth's surface. The apparent deflection is called the Coriolis effect. Coriolis deflection can create a rotary motion of the tide wave, called amphidromic systems, within certain ocean basins. Amphidromic systems rotate counterclockwise in the Northern Hemisphere and clockwise in the Southern Hemisphere. To read **CC12** go to page 32CC.

CHAPTER 13

CHAPTER OUTLINE

CRITICAL CONCEPTS USED IN THIS CHAPTER

This beautiful rocky coastline is in the Point Lobos State Park just south of Monterey, California. In the distance, at the far side of the bay shown in the image, is the wide, white sand beach of Carmel, California.

Coasts

The sea is flowing ever,
The land retains it never.

—John Wolfgang von Goethe,
Hikmet Nameh—Book of Proverbs

This proverb captures the dynamic interdependence of the land and ocean, but perhaps the second line should read instead, "The land resists but defeats it never," especially during the next several decades. Despite the wide public discussion of the potential effects of climate changes wrought by humans through the greenhouse effect, relatively little attention has been paid to what is likely to be one of the most costly and disruptive changes that human civilization will face as a result of greenhouse warming: the relentless advance of the rising sea on our coasts.

Various scientific studies have predicted that thermal expansion of the water in the oceans and the melting of polar ice will accelerate the current rate of sea-level rise, which was several millimeters per year, or a total of 25 to 30 cm, over the last century. A 1995 Environmental Protection Agency study projected that this rate of rise might double over the next century. A sea-level rise of 50 cm to 1 m over the next century does not seem like much. However, coastal cities and beaches, and even some entire island nations, could be completely submerged and eroded away by wave action. Many inhabited areas near the present-day coastline that are not now impacted by the ocean would be exposed to storm wave damage and erosion. Groundwater aquifers could be polluted by salt water. Freeways, tunnels, harbor facilities, and coastal wetlands could be inundated by salt water. In addition, sea-level rise

could be much greater if greenhouse warming is accelerated by feedback effects, as many scientists suggest is quite possible.

Living with the inevitable sea-level rise in our future will require that we better understand the interactions between the sea and the land and that we recognize the inevitability that the land cannot retain the sea and resist its advances indefinitely. The devastation of New Orleans by floodwaters in 2005, when the city's levees were breached by the effects of hurricane Katrina, is an event that should raise awareness of this fact.

Most of the world's population lives within a few tens of kilometers of a **coast.** In the United States, 53% of the total population, or more than 150 million people, live in coastal counties. That total is expected to grow to 165 million within the next 10 years. The concentration of human populations on or near the coast reflects the importance of the **coastal zone** to civilization.

The term *coast* has various definitions, but it is generally considered to be the strip of land between the **coastline** (where water meets land) and the inland location where there is no longer any environmental influence of the ocean. We build houses, factories, piers, docks, and marinas in this zone, primarily because the ocean is valuable for transport, recreation, and disposal of wastes, particularly sewage. In a human lifetime, most coastlines seem fixed and permanent. In reality, they are in a continuous state of change and movement. This chapter examines different types of coasts and the processes that form and reshape them. It also examines human attempts to control and mold the coasts, and the futility of many such efforts.

13.1 PROCESSES THAT FORM AND MODIFY THE COASTLINE

All coasts change in character along their length, but their general character tends to be similar for thousands of kilometers. **Figure 13.2** shows various coast types of the United States and Canada. Each type is continuously changing in response to various processes that form and modify it over decades, centuries, and millennia.

Most coasts can be classified as either **erosional** or **depositional,** depending on whether their primary features were created by erosion of land or deposition of **sediments.** Erosional coasts develop where the **shore** is actively eroded by wave action or where rivers or **glaciers** caused erosion when sea level was lower than it is now. Depositional coasts develop where sediments accumulate either from a local source or after being transported to the area in rivers and glaciers or by ocean **currents** and waves (**Fig. 13.1**). Erosional coasts are often dominated by sea cliffs and rocky shores, whereas depositional coasts include **deltas, mangrove** swamps, **salt marshes, barrier islands,** and **beach-sand dunes.**

Coasts can also be classified as either primary or secondary. Primary coasts are shaped predominantly by terrestrial processes, including erosion or deposition by rivers, streams, glaciers, volcanism, and tectonic movements. Secondary coasts are shaped by marine erosion or deposition caused by wave action, sediment transport by currents, and marine organisms (e.g., those that form **reefs**). Many coasts have characteristics of both marine and terrestrial processes.

Formation of Coasts

New coasts are formed either when the relative levels of the ocean surface and coastal landmass change or when the edge of the landmass is added to or removed. Some processes that form coasts, such as volcanic eruptions and earthquakes, can occur instantly or over a very short period, but most other processes, such as sea-level change and **coral reef** growth, continue slowly over centuries. Many of these processes can occur on the same coast at the same time, but at different rates or frequencies.

Tectonic Processes

Chapter 4 describes movements of **lithospheric plates** that create major **topographic** features of the ocean floor. It also explains how these movements build mountain chains and **magmatic arc** and **sedimentary arc** islands at **convergent plate boundaries**, create new oceans at **continental divergent plate boundaries,** and create volcanic islands at **hot spots** and particularly active locations on **oceanic ridges.** Tectonic processes take millions of years to re-form the planet. For example, the conversion of Pangaea to the present configuration of continents required about 150 to 200 million years.

Plate tectonic movements are slow but take place continuously, modifying coasts at plate boundaries and hot spots. At hot spots, new coast is formed when volcanoes erupt. For example, several hundred hectares have been added to the island of Hawaii by an eruption of Kilauea Volcano that started in 1983 and still continues. Kilauea **lava** flows into the sea, hardens, and extends the coastline (**Fig. 13.2e**). Tens of thousands of years from now a new island, Loihi, will emerge south of Hawaii (Chap. 4).

New coast is also formed by volcanic eruptions in magmatic arcs at convergent plate boundaries. Indonesia and the Aleutian Islands are good examples (Chap. 4). At convergent plate boundaries or at **transform faults**, new coast is formed when earthquakes

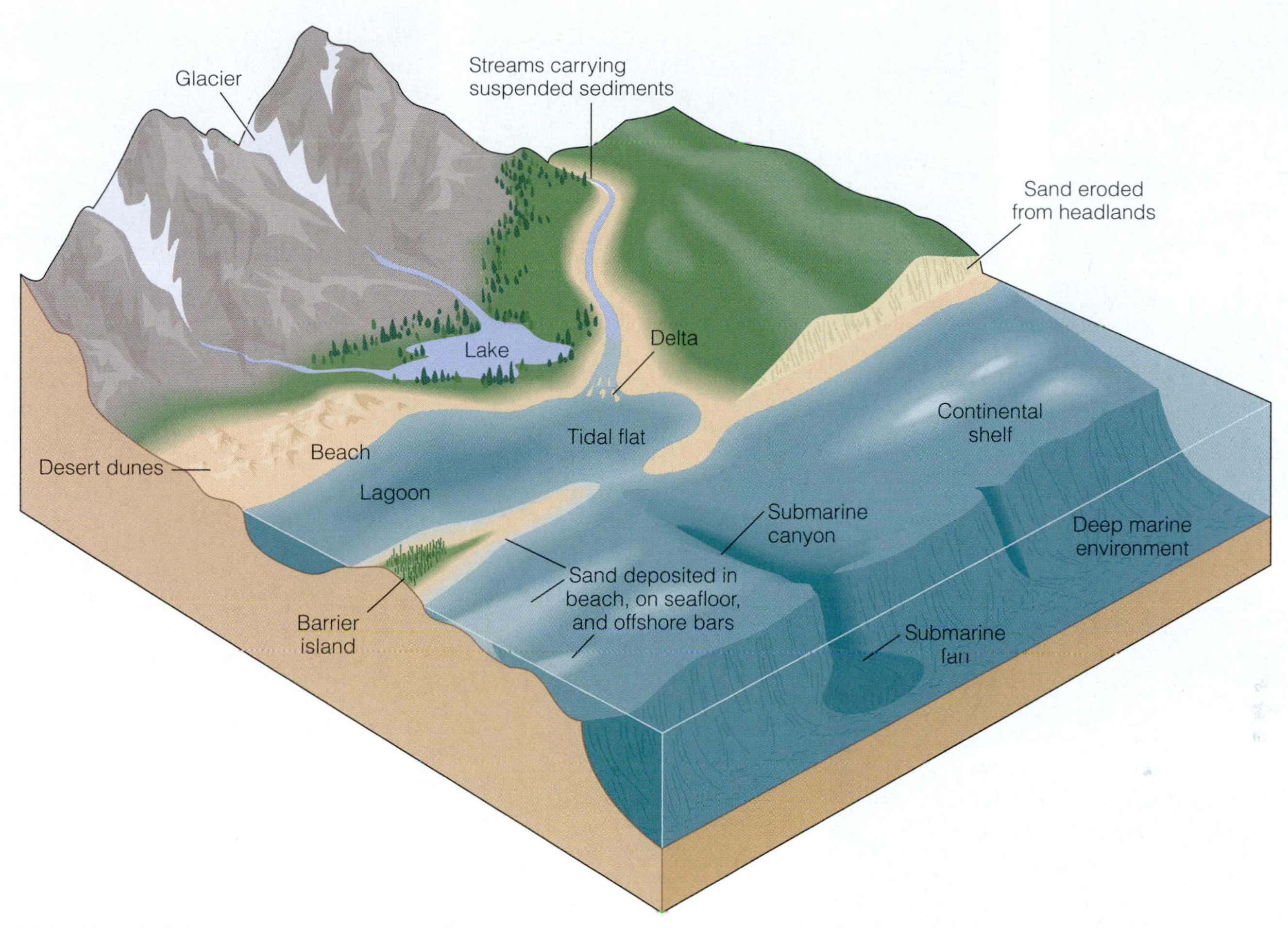

FIGURE 13.1 The major sources of sediments in the coastal ocean are erosion of the shoreline and transport of particulate matter by streams, rivers, and glaciers. Many river-borne particles are deposited in estuaries or lagoons along the coastal plain. Once in the coastal marine environment, sediment particles can be transported along the coastline and deposited on beaches, barrier islands, and offshore bars. Small sediment particles can be transported offshore by currents and deposited on the deep-ocean floor. Larger particles can be transported offshore by turbidity currents that sometimes travel down submarine canyons.

uplift a continent edge as the oceanic plate is **subducted** beneath it, or when earthquakes uplift ocean sediments at a sedimentary arc. Although earthquakes that uplift land to create new coast are infrequent, coastal erosion processes are slow. Therefore, new uplifted coast is formed faster at some convergent plate boundaries than it is modified by ocean processes. For example, the Loma Prieta earthquake in October 1989 raised the coastal mountains and coast near Santa Cruz, California, by as much as 1.5 to 2 m. A strong earthquake in Alaska in 1964 raised parts of the seafloor of Prince William Sound by as much as 8 m, creating a new strip of land from the former seafloor that was up to several hundred meters wide.

Coast can also be destroyed by tectonic processes. For example, coasts of what is now northern India were destroyed as India collided with Asia (Chap. 4). Earthquakes at convergent plate boundaries or at transform faults can cause sections of coastal land to move vertically downward, although such changes at these boundaries are often temporary because subsequent earthquakes may uplift this same section as the often complex subduction process continues.

Landslides

The simultaneous destruction of old coast and formation of new is particularly dramatic where volcanoes form islands, such as Hawaii, whose underwater flanks are much steeper than most other terrestrial margins. The steep flanks can become unstable as lava accumulates from continuing eruptions or when the island cools and sinks **isostatically** after moving off the hot spot. When this happens, a section of the island can break loose and slide down to the deep-ocean floor like a giant avalanche,

(o)
(n)
(a)
(b)
(c)
(d)
(e)
(f)

FIGURE 13.2 Coastlines of the United States and Canada (counterclockwise from top left). (a) Rocky coast (west coast of Vancouver Island, British Columbia, Canada). (b) Rocky coast (Monterey, California). (c) Fringing coral reef (Guam). (d) Island beach and cliffs (Polihale, Kauai, Hawaii). (e) Lava flows (south shore, Hawaii). (f) Beach and cliffs (Santa Barbara, southern California). (g) Barrier island beach (Galveston Island, Texas). (h) Mangrove swamp (Florida Everglades).(i) Barrier island beach and dunes (Pea Island, Cape Hatteras, North Carolina). (j) Barrier island beach (Virginia Beach, Virginia). (k) Sand dunes (Cape Cod, Massachusetts). (l) Rocky beach with high wave energy (New England). (m) High tidal range (Bay of Fundy National Park, New Brunswick, Canada). (n) Arctic coastal plain (Beaufort Sea, Alaska). (o) Fjord (Turnagain Arm, Alaska).

destroying the old coast and creating a new coast where the break occurs. Huge sections of the Hawaiian Islands have apparently broken off in this way in the past. As much as 10% to 20% of Oahu apparently instantly broke loose at one time. There is evidence that about 70 more such landslides have occurred around the Hawaiian Islands during the past 20 million years. The remains of these giant landslides are littered over vast areas of seafloor extending more than 200 km around the islands (**Fig. 13.3**).

Little is known about these monster landslides or the probability of another one occurring on Hawaii or other volcanic islands. However, in November 2000 a 20-km-long by 10-km-wide section of the southeast slope of the Kilauea Volcano on Hawaii slipped about 10 cm in only 36 h, millions of times faster than most tectonic motions. This occurrence may have been a forewarning of an imminent (in geological time) collapse of this section of the island. Such a slide not only could destroy a large section of the islands and their inhabitants, but also could cause a huge **tsunami** (Chap. 11), which might be several tens of meters high when it impacted the west coast of North America. Fortunately, such large slides apparently occur only at intervals of about 100,000 years or more in Hawaii and about once in every 10,000 years on average worldwide.

Although not as dramatic in size or impact, landslides smaller than those observed around Hawaii are important processes of coastline modification on most uplifted coasts. On eroding coastlines such as those found in southern California, many homes built years ago and tens of meters from the then existing cliff edge have been destroyed as the cliffs have progressively collapsed.

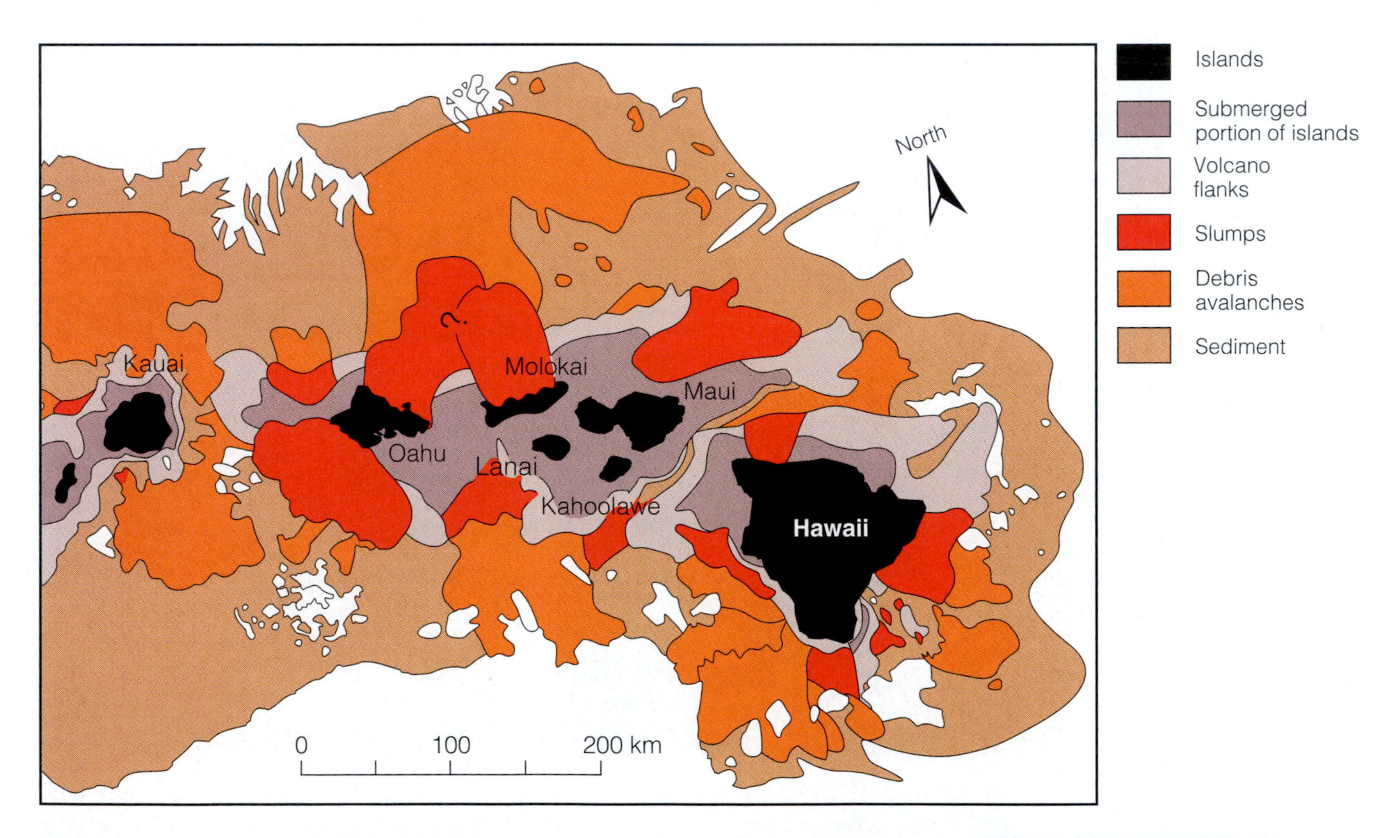

FIGURE 13.3 Large areas of the ocean floor surrounding the Hawaiian Islands are covered by slumps, fields of large debris, and sediments that originated from the islands. The data depicted on this map were obtained by the extensive use of precision sonar and three-dimensional sonar and the collection of many sediment samples. The slumps are areas where large blocks of the side of the volcano slipped downward. Slippage may occur both by a slow creeping motion and by periodic surges of several meters that cause large earthquakes. The debris fields and sediments are the result of catastrophic landslides in which huge blocks of the islands, up to 10 km or more in size, break loose instantaneously in massive avalanches that slide and flow down the steep volcano sides to the deep-sea floor. Although not recorded in recent history, these landslides must cause massive turbidity currents and, probably, massive tsunamis that may be hundreds of meters high when they reach the adjacent islands.

Isostatic and Eustatic Sea-Level Changes

If sea level rises, coasts are drowned and a new coastline is formed inland from the previous location. Similarly, if sea level falls, ocean floor is exposed and becomes the new coast. Sea level can change on a particular section of coast because the continent edge rises or sinks isostatically (Chap. 4, **CC2**). Sea level can also change **eustatically** if the volume of water in the oceans changes or the volume of the ocean basins themselves changes (Chap. 3, **CC2**). Eustatic changes take place uniformly throughout the world's oceans, whereas isostatic leveling occurs locally or regionally. At present, worldwide sea level is rising slowly because of eustatic processes (melting of glaciers and warming and expansion of ocean water), but sea level is not observed to be rising on all coasts. Some coasts are rising isostatically as fast as, or faster than, the rate of eustatic sea-level rise. The net result is that observed sea level is stable, or falling, on these coasts. Other coasts are sinking isostatically, and the observed sea level on these coasts is rising faster than the rate of eustatic sea-level rise.

Eustatic changes of sea level have made drastic changes in the location of coastlines on the continents. During approximately the past 18,000 years, sea level has risen about 120 m (**Fig. 8.17**). The history of sea-level change during this 18,000-year period can be determined by, for example, studies of the ages of **relict sediments** or buried sediments on the **continental shelf** (Chap. 8). Determining the history of sea level before 18,000 years ago is more difficult, but sea level seems to have oscillated many times during the present **spreading cycle**, from about 130 m or more below its present level, to about 40 to 50 m above the present level. Consider where the coastline would be if no isostatic changes had occurred during this oscillation of sea level. At the highest sea level, the Gulf of Mexico would extend across the central plains states of the United States as far north as southern Canada. At the lowest sea level, the Texas coastline would be about 150 km farther south in the present-day Gulf of Mexico and the Florida Peninsula would be about twice as wide.

During the past 4000 years, the eustatic rise in sea level has been slower than in the immediately preceding period. Therefore, most present-day coasts were formed several thousand years ago as sea level rose rapidly over what is now the continental shelf. Sea level may rise more rapidly in the future as a result of global **climate** changes caused by enhancement of the **greenhouse effect**. In any event, the relatively slow sea-level change of the past 4000 years cannot continue indefinitely. If sea level does rise more quickly, the types of coasts at various locations will change because the rate of sea-level rise greatly affects the formation and migration of coastal features such as barrier islands. Coastal changes may disrupt the Earth's **ecosystems**, thereby possibly causing more damage than even that caused by the flooding of coastal cities.

Because sea level may rise more rapidly in response to climate change induced by the enhanced greenhouse effect, oceanographers are currently mounting intensive studies of coasts. Critical questions that remain to be answered include how fast sea level is rising eustatically and whether it will continue to rise, how isostatic changes will enhance or mitigate eustatic changes in sea level on specific coasts, and how coasts will change if the rate of eustatic sea-level rise increases.

CRITICAL THINKING QUESTIONS

13.1 In the Hawaiian Islands the youngest island, Hawaii, is the biggest, has sandy beaches along the smallest fraction of its coastline, and has more black sand beaches than the other islands. Oahu, which is older than Hawaii, is smaller, has a greater fraction of its coastline occupied by beaches, and does not have black sand beaches. Kauai, which is older than both Oahu and Hawaii, is smaller than either of these other islands, has the largest fraction of its coastline occupied by beaches, and also does not have black sand beaches. Explain why this progression of island characteristics with age occurs. (These processes are described individually in several different chapters of this text.)

13.2 Sea level may soon begin to rise faster than has ever occurred in the history of the Earth. Hypothesize the possible effects of such a rapid sea-level rise on the number and health of coral reefs in the world's oceans and on the geographic extent and distribution of wetlands.

Glaciers

As glaciers flow, they scour out steep-sided valleys (**Fig. 13.4a**). Rocks and smaller particles that have been eroded from the valley walls and floor are carried by the glacier and deposited where the ice melts at the glacier's end. During the last 15,000 years, the Earth's climate warmed and the glaciers retreated, each one leaving one or more sedimentary deposits called "moraines" at the former location of the glacier's end. At the same time, sea level was rising as ocean waters warmed and expanded, and as more water entered the oceans from melting glaciers. The rising sea inundated many steep-sided valleys cut by glaciers and created deep, narrow **fjords,** many of which are partially closed off from the ocean by a submerged **sill,** which is usually a moraine (**Fig. 13.4b,c**).

Because fjords are long narrow inlets, they are generally well protected from erosion by ocean waves, and their shores are little altered from the original sides of the

(a)

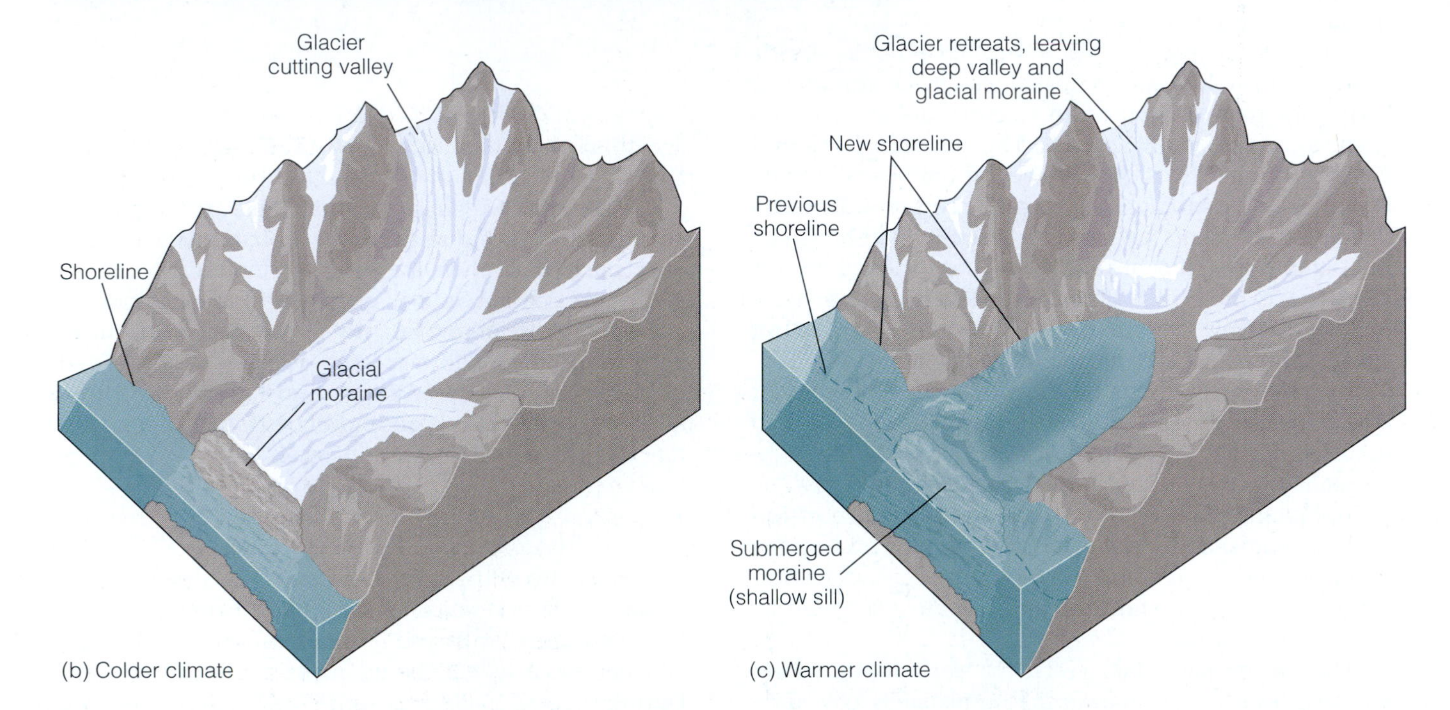

FIGURE 13.4 Glaciers and fjord formation. (a) The Franz Josef Glacier on the South Island of New Zealand has retreated substantially in recent decades. Note the very steep-sided valley that the glacier has left behind as it has receded. (b) When sea level was lower and glaciers more extensive, the glaciers cut deep, steep-sided valleys down to the ancient sea level. Rock fragments carried by the glaciers were deposited to form one or more glacial moraines at the foot of each glacier. (c) As the Earth's climate warmed and sea level rose, the glaciers retreated and their steep-sided valleys were filled by seawater to form fjords. The glacial moraines left behind by the glaciers formed shallow submerged sills at the mouths of many fjords.

glacial valley. Many high-**latitude** areas where glaciers cut through coastal mountain ranges have extensive fjord systems. Excellent examples are found on the South Island of New Zealand, in Scandinavia, and on the Pacific coast of Canada and Alaska.

River-Borne Sediments

New coasts are formed where large amounts of river-borne sediments are deposited. The extended delta of the Mississippi River (**Fig. 13.5**) and similar deltas elsewhere are examples of coasts formed by river-borne sediments. Deltas, discussed in more detail later in this chapter, are present at the mouths of relatively few rivers. Most of the world's rivers flow across a gradually sloping **coastal plain** before reaching the sea. The sediment load of the river is deposited in the river valley as the flow slows in this flatter area.

Only a few rivers other than the Mississippi carry such large sediment loads that their river valleys have filled enough for large quantities of sediment to be transported to the sea. Most rivers that flow across coastal plains, such

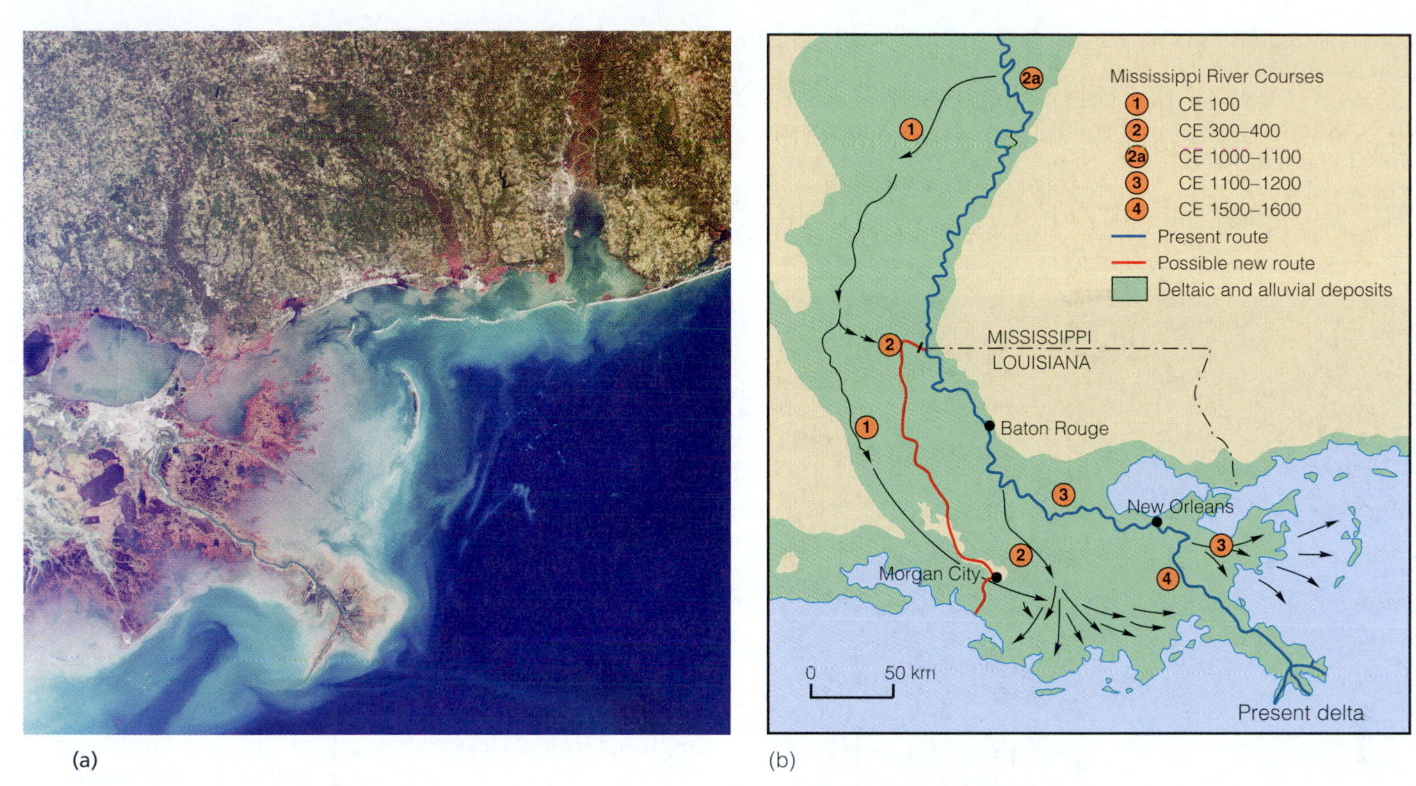

FIGURE 13.5 The Mississippi delta. (a) This satellite image shows the extensive delta with its many distributary channels. (b) The Mississippi River periodically changes its course through the coastal plain and delta, distributing its sediments across the entire area. Flood-control levees now prevent this natural process.

as those on the Atlantic coast of North America, carry considerably less sediment than the Mississippi. In addition, rivers emptying to the Atlantic Ocean have only recently (in geological time) begun to flow toward that ocean. In the region now drained by the rivers emptying to the Atlantic Ocean, rivers flowed away from the Atlantic Ocean until about 100 million years ago, when the newly formed **passive margin** of the Atlantic coast sank isostatically sufficiently far to reverse the slope (Chap. 4).

Rivers that flow across tectonically active coastal margins generally flow through steep coastal mountain ranges. Because they drain relatively small land areas and flow through steep valleys to the sea, many carry relatively little sediment, but most of it is transported to the oceans. The continental shelf is steeper and narrower on these active coastal margins, so sediment carried to the ocean can be transported to the deep-sea floor more easily than at passive margins.

Biological Processes

Reef-building **corals** cannot grow and build reefs unless they are underwater. However, reefs grow fastest in shallow waters where light intensity is high (Chap. 17), and some reef tops emerge above water at low **tides**. Although they are not truly land, these reefs constitute an important feature of the coast because ocean waves break on them and lose much of their energy. **Fringing reefs** and **barrier reefs** are present on many coasts in tropical and subtropical regions. A small drop in sea level or a small tectonic or isostatic uplift of a coastal margin

CRITICAL THINKING QUESTIONS

13.3 Describe and explain what happens to deltas when humans build levees to prevent frequent flooding. Many lives and huge expenditures needed to fix property damage have been saved by the construction of levees on the Mississippi and San Francisco Bay deltas. Deltas in other parts of the world, such as the delta on which most of the population of Bangladesh lives and farms, are still regularly flooded by hurricane storm surges or intense monsoon rains, often with great loss of life and damage to property.

(a) Discuss what you think should be done to manage the Mississippi River, San Francisco Bay, and Bangladesh deltas in the future.
(b) What information should be considered in making such decisions?
(c) How do you think these decisions should be made, and by whom?

can raise coral reefs above the sea surface, where the corals cannot survive.

When corals die, they leave behind their hard "skeletons." Many tropical and subtropical islands and coasts are characterized by rocks composed of old coral reefs. These coral rock shores are eroded to form a jagged surface, often called "ironshore," that makes walking difficult (**Fig. 13.6**). The Cayman Islands in the Caribbean Sea have excellent examples of reef-dominated shore. In fact, these islands are predominantly uplifted coral reef. The ironshore on part of one island is so jagged that the local community has been named "Hell."

FIGURE 13.6 Uplifted and eroded coral reef forms a jagged shore often called "ironshore." (a) Coastlines of coral islands, like this island in Palau, often have interspersed sandy beaches and eroding ironshore. (b) The jagged rocks of this ironshore segment in Fiji are razor-sharp.

Many coral reefs are located on islands or submerged pinnacles that are sinking isostatically (**Fig. 4.27**). Isostatic sinking and sea-level rise both tend to increase the depth of the water column over a reef. If the combined rate of these deepening processes exceeds the rate at which a coral reef can build upward, the top of the reef becomes progressively deeper and may eventually become too deep to sustain the **photosynthesis** on which the corals depend. Sea level has been rising for about the past 18,000 years, and many coral reefs appear to have been drowned in this way. If global climate change leads to an increase in the rate of sea-level rise, many more coral reefs may die, primarily those with lower maximum growth rates.

The maximum upward growth rate of coral reefs varies with latitude, depth, and water clarity. Upward growth rates are lower at higher latitudes, where water temperatures are cooler; and they are lower at deeper depths or in less clear waters, where the light levels are reduced. However, the maximum upward growth rate is about 1 to 10 mm per year at 10 m depth, and the current rate of sea-level rise is estimated to be about 3 mm per year.

Modification of the Coast

All shores and coasts are continuously, but slowly, modified by waves, tides, winds, and biological processes. The present form of coasts represents a balance between modification processes and formation processes. Older coasts and coasts with higher wind, wave, and tide energies are generally more extensively modified. The extent of modification also depends on the type of rock constituting the coast and, in some cases, on the types of vegetation on the coastal land and in the nearshore zone.

Waves

The breaking of waves on the shore is the principal coast-modifying process. On rocky coasts, breaking waves progressively erode the rock away. Soft sedimentary rocks are eroded much faster than harder volcanic rocks. In addition, erosion is faster on coasts that are exposed to greater wave action. Wave action is greater in areas of frequent storms or where the coastline is impacted by waves that travel far across the ocean.

As discussed in Chapter 11, wave energy is focused on headlands and spreads out along the interior shores of bays. Consequently, on an indented coastline, erosion occurs fastest at headlands, and the products of erosion (sand) accumulate within the intervening bays. Preferential erosion of headlands along a coastline tends to straighten the coast progressively until no headlands remain. This process is complete on many coasts, particularly where rocks are easily eroded. The New Jersey coast is an example (**Fig. 13.7**).

On rocky coasts, waves cut away rock between the **high-** and **low-tide lines**. As rock is cut away, the land becomes unstable and breaks away, leaving behind a cliff that may be nearly vertical (**Figs. 13.2d, 13.8**). The debris from the cliff temporarily alters the shape and nature of

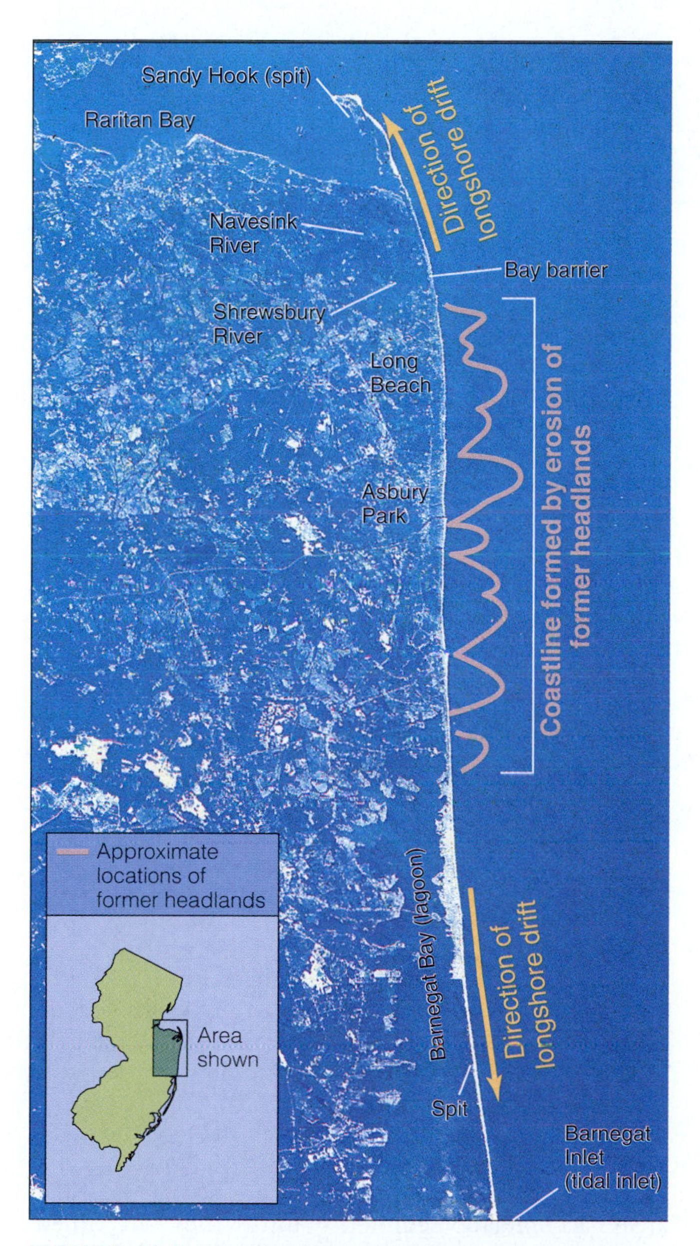

FIGURE 13.7 The New Jersey coastline has long straight beaches and barrier islands. Headlands that used to exist along the northern shore have been eroded away, providing sand for the formation and maintenance of the spits to the north and south.

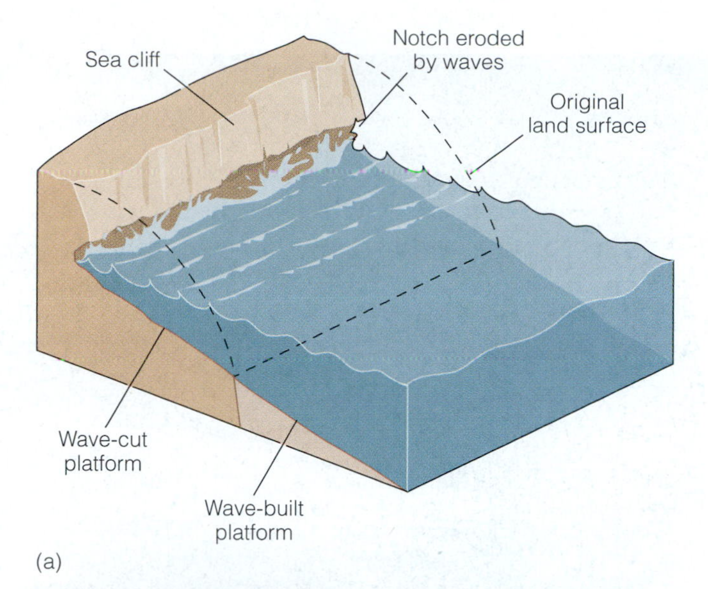

FIGURE 13.8 Wave erosion on steep coastlines can create cliffs. (a) As rock is eroded by waves, the land is undercut, becomes unstable, and slumps. The slumped material is eroded and transported away by the wave action. Thus, a gently sloping wave-cut platform is maintained at the base of the cliff. (b) Cliff and beach near Monterey, California.

FIGURE 13.9 Wave erosion, combined with boring and dissolution by marine organisms, undercut this rocky coastline on the island of Palau in the Pacific Ocean. Erosion is confined to a very narrow height range because the tidal range is nearly zero.

the beach where the cliff face has collapsed. However, these rocks fall into the wave-breaking zone, where they are eroded away relatively quickly. The beach is thus restored, and the waves renew their attack on the base of the cliff. In many locations, there is no beach at the base of the cliffs because wave energy is too high or wave erosion has not continued long enough for sand to accumulate (**Fig. 13.9**).

As waves cut into coastal cliffs and headlands, they encounter rocks of variable resistance to erosion. Rapid erosion of the less resistant rock often leads to the formation of sea caves at the base of a cliff (**Fig. 13.10a, b**). Sea caves that are cut into either side of a headland can continue to be eroded until they meet under what remains of the headland, resulting in the formation of a sea arch (**Fig. 13.10b,c**). As the headland erodes further, the arches collapse, and the remaining pinnacles of rock, called "stacks" (**Fig. 13.10b,d**), are eventually eroded away.

Coasts that have been substantially eroded usually have beaches. Off these coasts, waves transport and distribute particles (e.g., sand, silt, or pebbles) that make up the beach. These processes are discussed later in the chapter. Beaches help protect the coast from wave erosion.

CRITICAL THINKING QUESTIONS

13.4 Describe barrier islands and how they form and move. Explain why rising sea level is important for barrier island formation, and discuss what would happen to barrier islands if sea level were to fall.

13.5 Privately owned houses on barrier islands may be severely damaged by hurricanes.
(a) Should these be rebuilt?
(b) Many of the owners of such homes have no flood insurance. Should federal disaster funds be used to help them rebuild their homes?
(c) If it is decided that these homes should not be rebuilt, should the owners be compensated? If so, how and by whom?

Tides

The area along the coast that lies between the lowest point exposed at the lowest tide and the highest point reached by storm waves is the shore. The shore consists of the **foreshore**, which is the area between the low-tide line and the high-tide line, and the **backshore,** which is the area above the high-tide line that is affected by storm waves (**Fig. 13.11**). **Tidal range** determines the height range over which wave erosion occurs and, thus, the width of the shore. Where the tidal range is large, wave erosion energy is spread over a large vertical range. Generally, coasts with small tidal ranges are eroded faster because wave energy is concentrated in a narrow zone. The swift currents associated with large tidal ranges have little erosional effect because they are slow in relation to the speed of water in waves (**CC4**).

Tides are particularly important in shaping and maintaining **wetlands**. Tidal motions in shallow bays and **estuaries** transport and redistribute sediments. Such areas accumulate sediments until they are filled. Extensive mudflats form just below the high-tide line and are dissected by drainage channels. During the tidal cycle, the mudflats of tidal wetlands are alternately exposed and covered by shallow water.

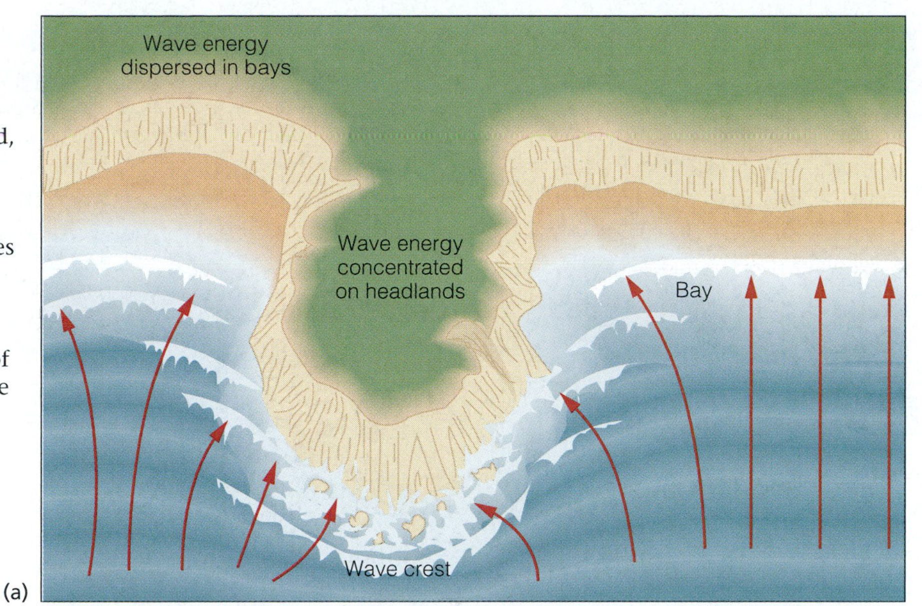

(a)

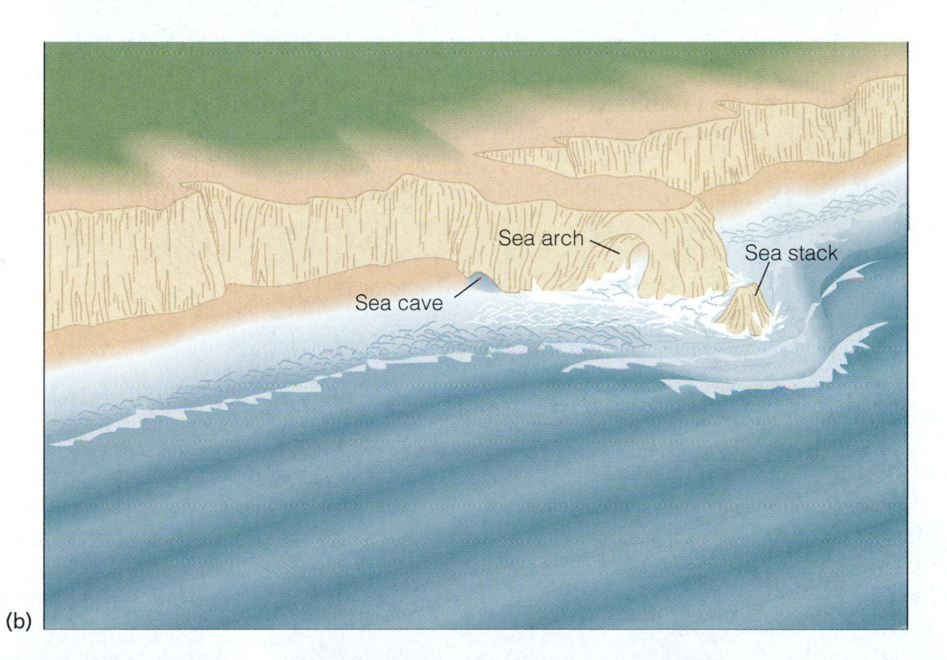

(b)

(c)

(d)

FIGURE 13.10 (a) Wave energy is concentrated on headlands by wave refraction. Because the seafloor topography is rough and varies as slumps occur from different parts of the headland, wave energy is distributed unevenly. In addition, the rocks of the headland differ in composition and susceptibility to erosion. As a result, the wave action causes the headlands to be eroded unevenly. (b) The uneven erosion of the headland can cause sea caves and sea arches to be formed. (c) This headland on the island of Molokai in Hawaii has been eroded by the intense wave action to form several sea arches and sea stacks. (d) There are many sea stacks on the Pacific coast of North America, including these on the Big Sur coast south of Monterey, California.

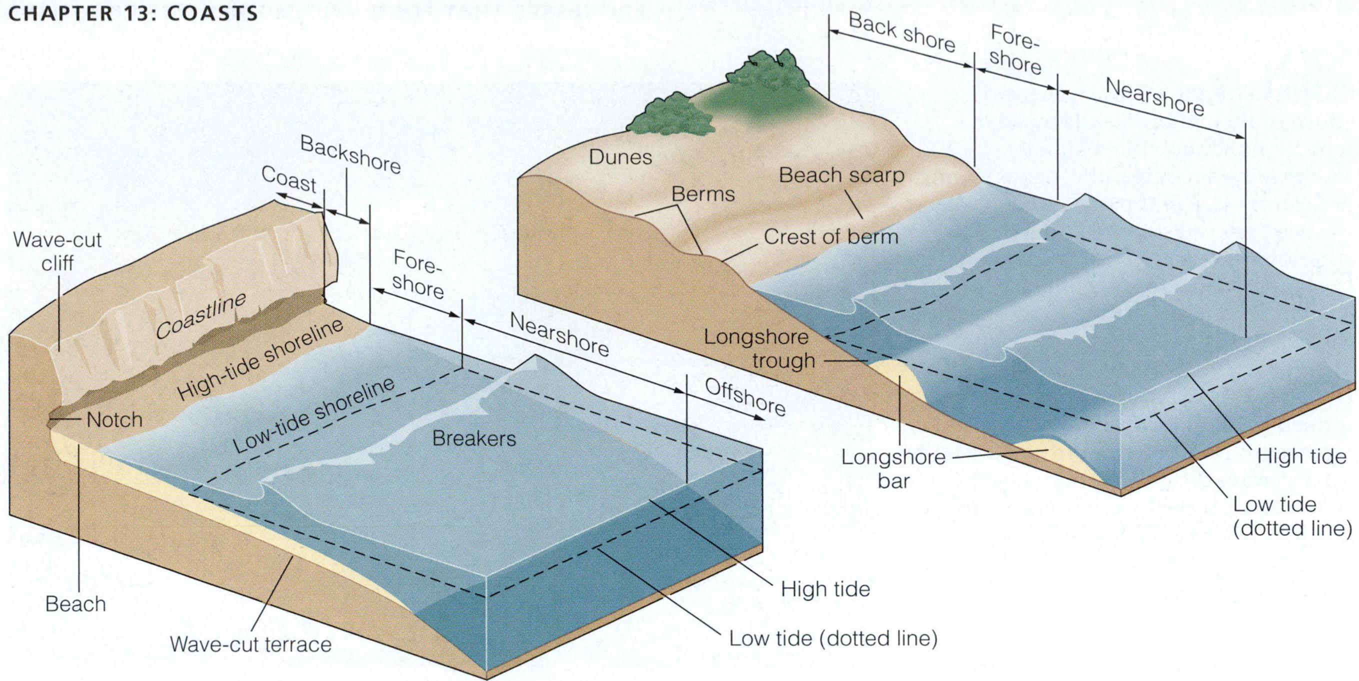

FIGURE 13.11 The coastline is separated into several zones that reflect different exposures to erosion and sediment movement due to the variation in waves and tides. The backshore is covered only in storms, but it is subject to seawater wind spray. The foreshore is the region between the high- and low-tide lines, and the nearshore is the region between the low-tide line and the depth at which wave action no longer affects the seafloor. Longshore bars and troughs may be present on the seafloor within the nearshore zone. The foreshore is often separated from the backshore by a berm. A scarp may be present in the foreshore at the height of the most recent high tide.

Winds and Weather

Onshore winds can carry sand from beaches and deposit it on the backshore above the highest point reached by waves. This process leads to the formation of the sand dunes (**Fig. 13.12**) that characterize many coastlines. Conversely, sand dunes and soils can be eroded by offshore winds and deposited in the oceans. Just as they are elsewhere, rocks, sand dunes, and soils of the coast are eroded by water in streams and rainfall, and by various processes, including alternate cycles of freezing and thawing and dissolution in acidic rainwater. Wind-driven particles and ocean spray may also erode rocks or soils.

FIGURE 13.12 Sand dunes are characteristic of many coasts. (a) Dunes usually separate the beach from a low marshy area, as they do on this Aruba coast. (b) Dunes can be 10 m or more high, like these in Monterey Bay, California. Note the patches of vegetation on the dunes.

Vegetation

The type and extent of vegetation on the coast affect the rate at which winds, streams, and storm waves erode the land. Grasses are particularly important in protecting sand dunes from erosion. Similarly, rooted plants that grow in the water, including **sea grasses** and mangroves (Chap. 17), help prevent erosion of mudflats by waves and currents. In contrast, tree roots and animal activities, such as burrowing, contribute to the continuous erosion of rocks and soils of land near the coast just as they do elsewhere. In addition, especially on rocky coastlines, many animals that live in the zone between high and low tides erode the rocks as they bore or chemically dissolve their way into the rock or probe into cracks in the rocks, either to find food or shelter or to be able to "hold on" to the rock against the power of the waves.

13.2 BEACHES

Beaches are endlessly fascinating. We love to sunbathe on them, play in the **surf**, search for shells and other things washed up by the tide, and simply walk along them enjoying their natural splendor. To many people, beaches are the best place to enjoy what is usually an unspoiled natural **environment**. Civilizations have dramatically altered much of the dry land adjacent to the oceans, but beaches have been substantially changed in only a few locations.

Although beaches appear to be resistant to change, in reality they are places of continuous movement and transformation. Beaches are formed by processes that occur in the zone between the seaward boundary of land vegetation (or the base of a cliff, if present) and the point where the seafloor reaches a depth at which sediment is no longer disturbed by waves, commonly 10 to 20 m (CC4). This zone includes the coast and shore and is known to geological oceanographers as the **littoral zone**. Unfortunately, the term *littoral zone* means something different to biological oceanographers: the zone between the high- and low-tide lines.

The Littoral Zone

The landward side of the beach is delineated by either a cliff or sand dunes. Cliffs vary in height and steepness. Sand dunes vary from a few centimeters to more than tens of meters high and can be hundreds of meters wide. The beach that stretches out to sea from the cliffs or dunes consists of three zones. First, the backshore zone is the low-slope area of upper beach that is normally dry, even at high tide (Fig. 13.11). Although it is generally flat and slopes only gradually downward toward the sea, the backshore may have well-defined areas where the downward slope steepens abruptly (Fig. 13.11). Such areas are called **berms**. The top of a berm is usually flat, and the seaward side slopes downward relatively steeply (Fig. 13.13a). Sometimes the berm appears as a slight mound or ridge (Fig. 13.13b). Berms generally continue along the length of the beach at a fixed height above the water. In many places, we see several berms on the same beach at different heights above the water.

Berms are created by storm waves. The highest berms are created by storm waves that reach farthest up the beach. Because the strongest storms occur in winter in most locations, the uppermost berm is often called the "winter berm," even though it may actually have been created by an infrequent great storm that occurred one or more years earlier. Berms are not permanent features. They are continuously destroyed and new ones are formed as **wave heights** and tidal ranges vary.

The second zone of the beach is the foreshore (Fig. 13.11). It is the area between the highest point reached by waves at high tide and the lowest point exposed at low tide. This area is commonly an almost flat and featureless slope and is often called the **low-tide terrace**. Foreshore slopes can vary greatly from beach to beach

CRITICAL THINKING QUESTIONS

13.6 Imagine that you are a member of the U.S. Congress representing a district that includes a barrier island community, such as Miami Beach or Galveston, that has just been hit by a hurricane causing such extensive property damage that many homes and hotels near the beach must be torn down. The existing seawall is clearly too low to fully protect the community from future hurricanes. You are aware that emissions of greenhouse gases during the past century will almost certainly result in climate warming, more rapidly rising sea level, and increased intensity and frequency of hurricanes. You are also aware that any reduction in the current rate of greenhouse gas emissions will be very expensive and probably detrimental to the economy. Furthermore, it will only slow, not stop, the warming trend, and it will have no effect unless other nations also reduce their use of fossil fuels, including those underdeveloped nations that need to increase such use if they are to improve their living standard. The national budget deficit is getting worse, and money for social programs is being reduced.

(a) On the basis of what you have learned about barrier islands, the effects of sea-level rise, and the current state of knowledge of the enhanced greenhouse effect, what would you do?

(b) Discuss your reasons for choosing the action or course of actions that you have chosen.

(c) Would you feel differently if you owned a home on the barrier island in question? If so, how and why?

(d) Would you feel differently if you represented a congressional district in Colorado? If so, how and why?

FIGURE 13.13 Berms and scarps. (a) There are two berms on this beach just south of San Francisco, California: one in the center of the photograph, where the waves of a recent storm have reached, and another much higher on the left. This upper berm must have been formed during a strong storm. (b) The berm on this beach on the east coast of the North Island of New Zealand forms a distinct ridge with a depression on its landward side. During major storms, the waves will pass far over this berm to erode the cliffs seen in the background. (c) This beach scarp on the California coast just south of San Francisco is about 1 m high. (d) The same scarp as in part (c), seen from a different angle.

and from month to month at the same beach, but they are generally greatest on beaches composed of coarse-grained material.

At the landward edge of the low-tide terrace (the high-tide line), there is often an abrupt vertical face of sand, usually a few centimeters high, before the beach slopes normally upward again as dry backshore. This feature is called a **scarp** (**Fig. 13.13c, d**). Like a berm, a scarp stretches along the beach, and there may be more than one. Scarps are caused by waves cutting away sand from the beach, and the location of the scarp represents the height that waves normally reach at high tide. Waves flatten the low-tide terrace when it is covered with water. Therefore, if two scarps are present, the one higher on the beach was formed before the lower one at a time when either the tide (**spring tide**), waves, or both were higher. The lower scarp would have been flattened and destroyed if waves had passed over it to cut the higher scarp.

The boundary between backshore and foreshore is defined as the location reached by waves at high tide. This location is not fixed, because the maximum tidal height changes from day to day (Chap. 12) and wave action varies. When waves are small, they do not reach far up the beach. However, storm waves can crash far beyond the highest point reached by smaller waves. In addition, when strong coastal winds blow, **Ekman transport** can raise or lower sea level on a coast by a meter or more (Chap. 10).

Seaward of the low-tide line, the beach continues to slope downward. However, if we were able to remove the water, we would see, at some distance offshore, a trough parallel to the beach. Seaward of the trough there is a rounded, elongated mound of sand, called a **longshore bar**, before the seafloor resumes its downward slope (**Fig. 13.11**). Sometimes there is more than one longshore bar, each located at a different depth and distance from shore. Longshore bars are created and maintained by wave-driven sand movement. Although we cannot see submerged longshore bars, we may see evidence of them. When high waves with a long **wave period** approach the shore, a line of surf generally forms tens or hundreds of meters offshore. The offshore surf line is caused by a longshore bar or reef where the seafloor is elevated and water is shallower than it is immediately inshore or offshore. Waves may break over this bar (Chap. 11), then re-form over the deeper water inshore and break again near the beach.

In some areas, longshore bars have been built up high enough to emerge above the normal water line and become barrier beaches or even barrier islands. Longshore bars and barrier islands are dynamic features that are continuously created, destroyed, and moved. Some consequences that often ensue when we ignore this basic truth are discussed later in this chapter.

Sources of Beach Materials

For most of us, the word *beach* conjures up visions of a long strip of land next to the water that is composed of sand grains. The grains and beach usually have a color so familiar that it is called "sandy." However, not all beaches are composed of the familiar sandy materials. Some are composed of rounded pebbles or cobbles (**Fig. 13.14a**), and the color of the sand on a beach can vary dramatically.

For example, volcanic island shores where the coast consists of recently solidified black **basaltic** rocks have black sand beaches (**Fig. 13.14b**). The island of Hawaii has several black sand beaches. Other beaches on Hawaii and Maui consist of green (see Chapter opener) or red sand, where the coasts consist of volcanic ash with high concentrations of iron or other elements that form colored minerals. In some cases, these odd-colored Hawaiian beaches are only a few hundred meters around a headland from beaches that have a more typical sandy color. The sand grains on such beaches obviously originate from erosion of local rocks. Erosion is continuous on all shores and is due primarily to the action of waves.

When rocks are **weathered,** some minerals slowly dissolve and others change in composition. After prolonged exposure and erosion, only the most resistant minerals remain. The most resistant rocks include granite, which is composed mostly of the solution-resistant silicate minerals, primarily quartz and feldspars. Beaches on many shores consist mainly of grains of these minerals that, with their impurities, have the common sandy color. Sand grains on such beaches have been subjected to long exposure to water and may have been eroded upland and brought to the beach by rivers.

Many low-lying coasts have no rivers that can carry large amounts of quartz and other mineral sand grains to the ocean and have little coastal erosion. Beaches in such areas are absent or poorly developed and may be composed of other materials. In many tropical or subtropical areas, particularly on low-lying coasts, beaches are composed primarily of calcium carbonate in the form of shell fragments from **foraminifera, mollusks, echinoderms,** and of platelets from **calcareous algae** such as *Penicillus* sp. and *Halimeda* sp.

FIGURE 13.14 Many beaches are not soft and sand-colored like those that are so popular for recreation. (a) This beach in New England is covered by large cobble-sized deposits, a sure sign that intense storm waves often reach this beach and transport all of the fine-grained material elsewhere. (b) The sand on this beach on the south shore of the island of Hawaii is black. The sand particles that make up the beach are eroded from nearby black volcanic rocks. Unfortunately, the beach shown here has now been destroyed by lava flowing from the Kilauea volcano.

(a)

(b)

If we look carefully at a handful of sand, we see that its grains consist of several materials. Sand is a mixture of particles from different sources. Sand composition may differ on two beaches that are close to each other, indicating that local river input, local erosion, or shell fragment washup contributes most of the material on at least one of the beaches. However, along most coasts, sand has much the same composition on all the beaches between any two river mouths, aside from some variations in **grain size.** On these coasts, river transport may be the dominant source of beach sands, and coastal erosion may be less important. However, for many locations the relative contribution of each source is not well known. If rivers supply most of the sand for beaches stretched out along many kilometers between river mouths, sand must move along the coast from the river mouths to form the beaches. This movement is achieved by **longshore drift.**

Longshore Drift

As waves travel through shallow water, some of their energy is transferred to sand grains, which sets them in motion (**CC4**). This interaction continuously forms and re-forms beaches and transports sand along the coast.

When a wave breaks, it rarely does so exactly parallel to the **shoreline.** Usually the wave approaches from a slight angle (Chap. 11), and the water that crashes on a beach does not move directly up the beach slope (**Fig. 13.15**). As the breaking wave moves up the beach (the **swash**), it **resuspends** sand grains from the beach and carries them with it. Thus, sand grains move up the beach, and a small distance downcoast, in the direction of the waves. When the water from the broken wave flows back down the beach (the **backwash**), sand grains flow with **gravity** directly down the beach slope (**Fig. 13.15**). Each swash and backwash cycle may move the grain only a centimeter or two along the beach. The net result is that sand grains are carried along the beach in the direction of the waves by a series of saw-toothed swash and backwash movements. Water from the wave is also transported along the beach, creating a **longshore current** in the zone landward of the breakers. Movement of the sand with this current is called **longshore drift** or **littoral drift.**

Although each wave moves sand grains only a very small distance along the beach, waves follow each other every few seconds in a continuous series. As a result, sand can be moved by longshore drift at speeds that range from the typical rate of a few meters per day to as fast as 1 km per day. Large quantities of sand are moved along the coast by longshore drift. Transport along the east and west coasts of the United States is estimated to range from several hundred to more than 5000 m^3 of sand per day, depending on factors such as the height and **frequency** of waves.

Longshore drift is directed downcoast, in the direction away from which the waves approach, and it can reverse if that direction changes. Waves on both the east and west coasts of the continental United States, particularly the larger and more energetic storm waves, come predominantly from the north. Therefore, longshore drift is generally to the south along both coasts.

If so much sand moves along the coast every day, where does it go and why is there any sand left on the beaches? As sand moves along the coast, it may eventually meet the head of a **submarine canyon**. Instead of continuing to move along the coast, it is funneled into

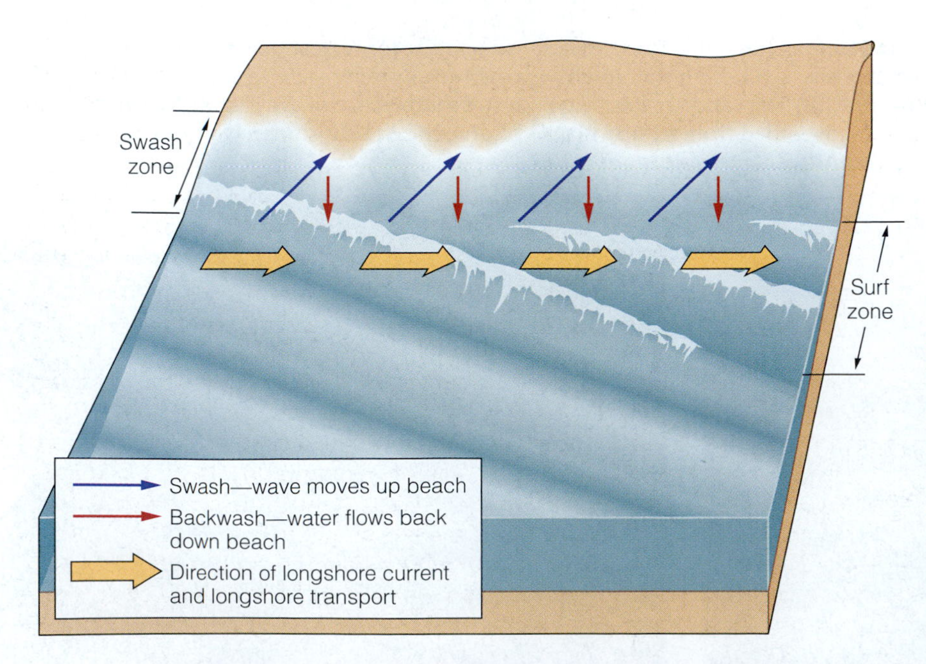

FIGURE 13.15 Water moves up a beach face at an angle with the wave direction but returns directly down the beach slope. Sand picked up by the incoming wave is washed up the beach and is returned seaward a small distance further down the beach in the direction of the waves. Successive waves move sand progressively along the beach—a process known as "longshore drift" or "littoral drift."

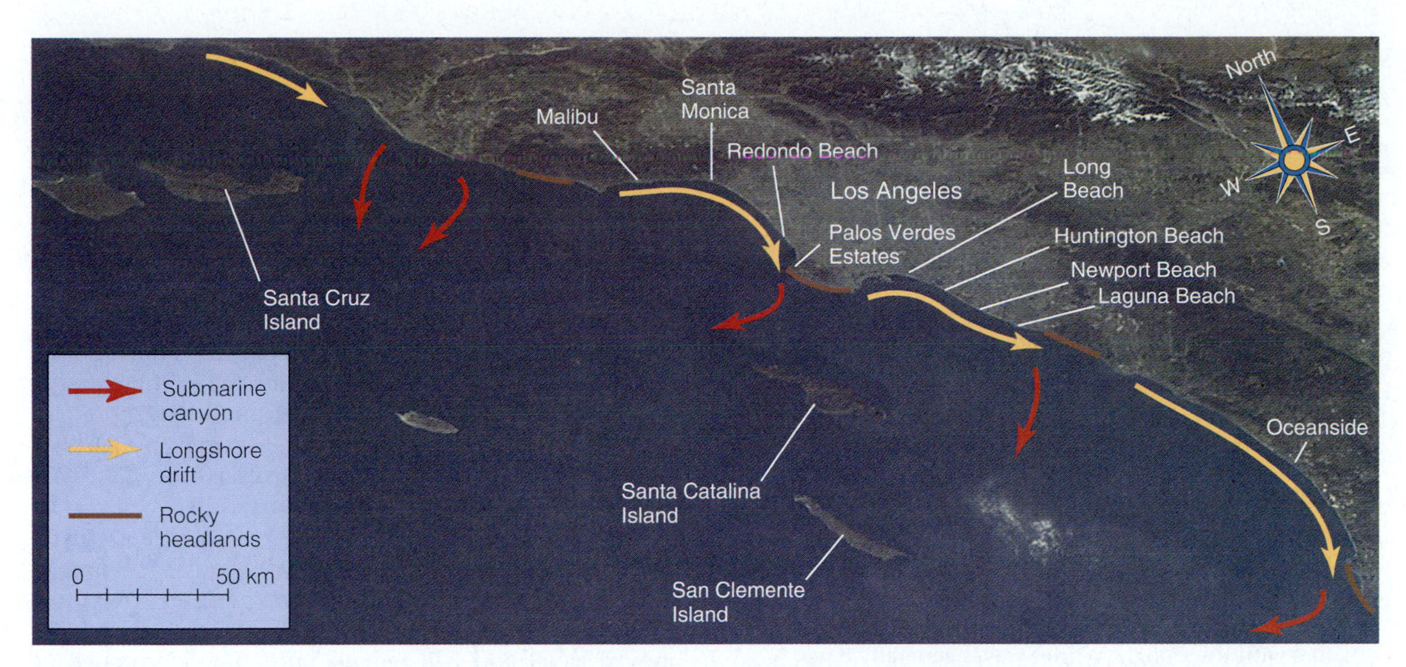

FIGURE 13.16 There are several north-to-south littoral drift cells in southern California. Each of these cells ends at the head of a submarine canyon where the sand is lost as it is transported down the canyon.

the canyon and flows down onto the deep-sea floor. As this sand is lost, new sand is brought to the beach by river flow and erosion of the shore. The beach is maintained by the balance between sand supplied from these sources and sand lost down the canyon. Thus, along the coast there is a series of separate cells within which beach sand is supplied and transported until it meets the head of a submarine canyon. Southern California has four well-defined coastal cells of this type (**Fig. 13.16**). Little or no sand passes south from one cell to the next. If you drive down the California coast, you can see the southern ends of these longshore drift cells where the sand beach ends abruptly. You see cliffs or cobble beaches as you continue south from these points until you reach an area sufficiently supplied with new sand to form the first beach of the next cell.

The future of California's beaches is in some doubt. Most rivers emptying into the Pacific Ocean have been dammed. Each dam acts as a trap for sand moving down the river. In addition, the dams and water withdrawals reduce the rivers' flow rate and velocity, thereby reducing their ability to erode rocks and to carry sand (**CC4**). If rivers, rather than coastal erosion, are the major historical sources of most sand on California's beaches, this reduction in river flows will lead to the progressive depletion of sand on the beaches. Beaches will become narrower, and waves will more easily reach the backshore, where communities such as Malibu are already susceptible to wave damage. Studies are being done to determine the extent to which California beaches depend on river-borne sand, but at this time the results are uncertain.

Longshore drift cells on many other coasts are not as well defined as those on the California coast. For example, on the North Atlantic coast of the United States, the cells are irregular and complex because of the many local barriers to longshore drift, such as headlands, deep river mouths, and rocky shores. In addition, this coast has a broad, flat continental shelf and few steep canyons through which sand can be transported offshore.

Wave Sorting of Beach Sands

Waves move sand if the speed of orbital motion is fast enough to resuspend the sand grains (**CC4**). Water in the wave moves toward the beach with the wave **crest** and then away from the beach when the **trough** passes over. Therefore, we would expect resuspended sand grains to move forward and back and return to their original location with each wave pass. However, the **orbital velocity** of the wave is distorted by its interaction with the seafloor. Water particles are farther from the seafloor as they flow toward the beach in the wave crest than they are as they flow back away from the beach in the trough. Because bottom **friction** increases nearer the seafloor, orbital velocity is lower in the wave as water moves offshore (trough) than it is when the wave moves onshore (crest). Although offshore (trough) flow is slower, it lasts longer than onshore (crest) flow. If this were not so, the wave orbit would not be complete (**Fig. 13.17**).

The responses of sand grains to the difference in orbital velocity between crest and trough and to the varying height and **wavelength** of waves reaching the beach

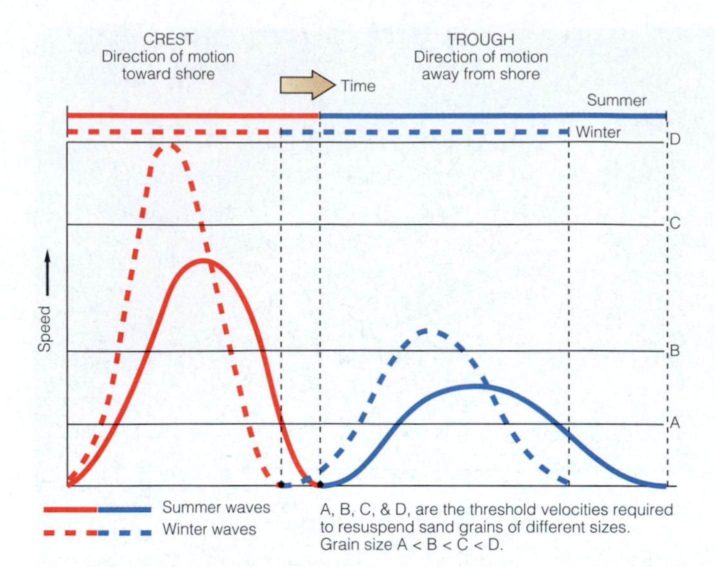

FIGURE 13.17 Water in a wave moves onshore with the crest at higher speeds and for a shorter period than it moves offshore with the trough. Winter waves generally have shorter wavelengths, greater wave heights, and thus, faster orbital speeds than summer waves. The text describes how variations in wave orbital velocities affect the transport of different sizes of sand grains (A, B, C, D) on a beach and how these processes sort sand grains by size.

are the factors that control the movement of sand grains across the littoral zone. Winter storm waves are generally higher than summer waves, and they have shorter wavelengths and periods. The orbits are larger (higher waves) and must be completed in less time (shorter period) in winter waves, so they have higher orbital velocity.

Consider the movements of sand grains of different sizes in response to winter and summer waves. In **Figure 13.17**, sand grains of size A are small enough to be resuspended easily by passage of both the crest and the trough of winter and summer waves. Even if there is time for such grains to be deposited temporarily between crest and trough, they are immediately resuspended and will not be deposited on the beach. These particles eventually are carried offshore by **rip currents**. Once in deeper water, they settle to the seafloor, where waves will no longer resuspend them, because the orbital velocity of waves in deeper water is reduced with depth below the surface (Chap. 11). Thus, fine-grained sand is winnowed from the beach, transported seaward, and deposited in deeper water below the influence of waves. The finest-grained sand is resuspended and transported offshore, where it is carried by ocean currents until deposited elsewhere (Chap. 8).

Sand grains of size B (**Fig. 13.17**) are resuspended under the crest of summer waves, but they cannot be resuspended by the slower orbital velocity under the trough. These particles are resuspended, moved shoreward with the crest, and redeposited as the velocity drops. They remain in place through the return flow (trough), which has lower velocity, and are resuspended and moved farther shoreward by the next crest. Thus, summer waves move sand grains of this size range from offshore onto the beach. When larger winter waves arrive, size-B grains are resuspended under both crest and trough and therefore are moved offshore. Because size-B grains are relatively large, they are not transported by ocean currents (whose speeds rarely approach the orbital velocities of waves), so they are deposited offshore just beyond the depth of wave influence. Hence, during winter, size-B particles are removed from the foreshore, transported offshore, and deposited to form a longshore bar. The longshore bar builds until it is just deep enough that the orbital velocity of passing waves is not quite fast enough to resuspend the sand grains.

Larger sand grains, of size C (**Fig. 13.17**), are moved up the beach by strong winter waves, but they are too large to be moved back offshore, even by the most intense storms. The largest grains (size D) cannot be moved at all and will remain until they are eroded or physically broken down into smaller grains that can be moved.

If wave action were the only process occurring, we would expect beaches to be composed of a mixture of sand grains of different sizes, lacking only those small enough to be continuously resuspended and removed to the deeper ocean. Beaches protected from strong wave action would be an unsorted mixture ranging from fine-grained sand to pebbles and boulders. Beaches with greater wave action would have no sand grains below a certain size. This grain size minimum would be larger for beaches subjected to strong wave action, but all grain sizes above the minimum would be present. However, we rarely find such a mixture. Each beach tends to be characterized by a narrow range of grain sizes, although the range varies from beach to beach (e.g., **Fig. 13.14**).

Why are virtually no large grains found on a fine-sand beach and no pebbles found on a sandy beach? The answer is that the range of grain sizes present depends on the sediment supply as well as on wave energy. The sand that makes up a beach is supplied mainly by river inflow or erosion of rocky headlands and is carried to the beach by longshore drift. Large particles, such as those of size D in **Figure 13.17**, cannot be transported, because they cannot be resuspended and moved by longshore drift, even by the most intense waves. Therefore, any boulders and other large particles present on a beach must have been eroded from the adjacent cliffs.

The size range of beach sand particles is thus limited because particles must be small enough to be resuspended and moved by longshore drift, but large enough not to be resuspended easily by the troughs of the waves. This explains why we normally see only a narrow range of grain sizes among particles on most beaches. Sedimentary deposits are said to be well **sorted** when they have only a narrow range of grain sizes (Chap. 8,

CC4). Beaches with extremely fine sand are those best protected from wave action (and any coarser material present is rapidly buried). Pebble or cobble beaches, such as that shown in **Figure 13.14a**, are regularly exposed to intense storm waves that winnow away smaller particles.

Seasonal Changes in Beach Profiles

If we look carefully, we can see that the sand of beaches with calm summers and stormy winters is finer-grained in summer than in winter. In winter, large and small sand grains are transported along the beach by longshore drift, but the finer grains are also moved offshore and deposited as longshore bars. In summer, these finer grains are moved back onto the foreshore, where they cover or mix with the coarser grains of the winter beach.

Seasonal movement of sand changes the beach profile. In winter, the foreshore moves back toward the cliffs or dunes as sand is winnowed by winter waves and deposited offshore. The foreshore is more gently sloped in winter than in summer, but the backshore is steeper (**Fig. 13.18**). When "quiet" conditions of summer return, fine-grained sand is returned to the beach from offshore. However, the summer waves do not reach as far as the winter waves, so a rounded ridge of sand may be formed at the highest point that these summer waves reach (**Fig. 13.18a**). This beach ridge, or crest, is a summer berm. When a summer storm temporarily increases wave energy on a steep summer beach, the beach below the berm can be cut away temporarily to form a scarp (**Fig. 13.13c, d**). In most cases, the scarp dries and is flattened by **slumping** and winds, or is destroyed by waves of subsequent storms.

Reconfiguration of the beach and winter removal of its sand to longshore bars is an important process that protects the coast from erosion by the huge waves generated by exceptionally large storms. Longshore bars are built by shorter-wavelength waves from the less intense storms that characterize most of the winter. Therefore, they are shallow enough to cause the exceptional huge waves to break. Wave energy is partially dissipated as the wave breaks over the longshore bar, and the weakened wave causes less damage when it hits the shore. Some of the consequences of ignoring the protective nature of longshore bars are discussed later in this chapter.

Beach Slope and Grain Size

Beach slope between high- and low-tide lines depends not only on wave size, but also on the grain size of beach materials. Pebble and cobble beaches can have slopes of 10° to 20°, whereas the finest sand beaches have slopes of 1° or less. The slope reflects the conditions needed for the equilibrium of particle transport up the beach in the swash and down the beach in the backwash.

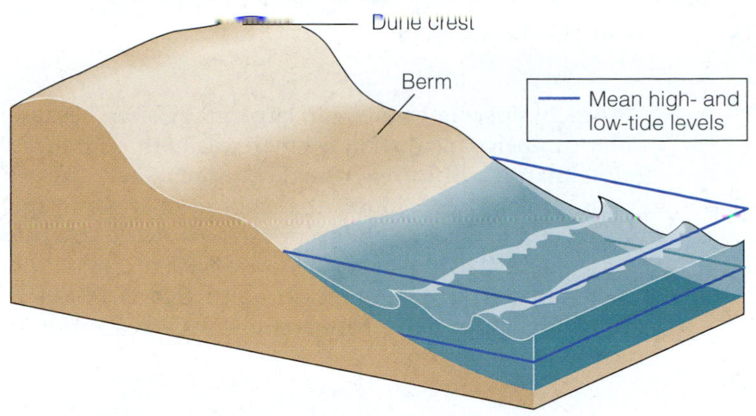

(a) Summer profile

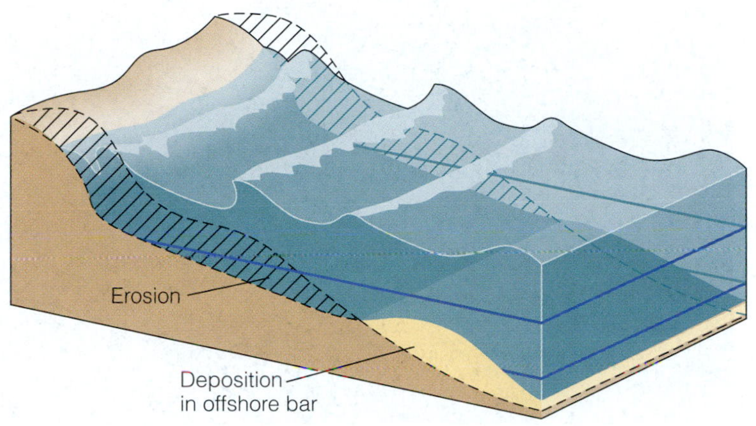

(b) Winter profile

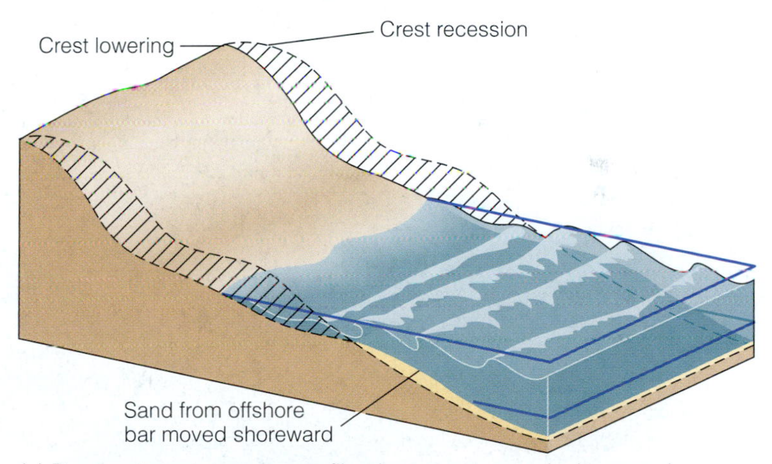

(c) Beach returns to summer profile when wave energy is decreased

FIGURE 13.18 Seasonal variation in beach profile. (a) During summer, when waves are relatively gentle, the beach profile is steep, and one or more summer berms form immediately above the high-tide line. (b) Winter or storm waves, especially when combined with storm surges, reach farther up the beach and erode sand from the berm, moving the sand seaward to form one or more offshore bars. The result is a narrower flat-beach foreshore that is cut back to dunes or a winter berm left from periodic extreme winter storms. (c) When storm waves cease, the gentler waves return sand from the offshore bar to the beach, rebuilding its summer profile. If storms are strong in a particular winter, the beach may recede inland, and dunes may be somewhat lowered.

Beach slope is greater for large grain sizes because water can percolate downward more easily through large grains than it can through smaller, more closely packed grains. This process reduces the amount of water in the backwash and thus its ability to carry particles seaward. Consequently, sediment moves onshore and builds the beach slope. The slope increases until the backwash is strong enough to move particles down the beach as fast as they are moved up.

13.3 BARRIER ISLANDS AND LAGOONS

Many coasts, including large stretches of the Atlantic Ocean, Gulf of Mexico, and Beaufort Sea (Alaskan Arctic) coasts of the United States, have low, elongated sandy islands offshore from the mainland coast (**Figs. 13.2, 13.19**). These islands are aligned approximately parallel

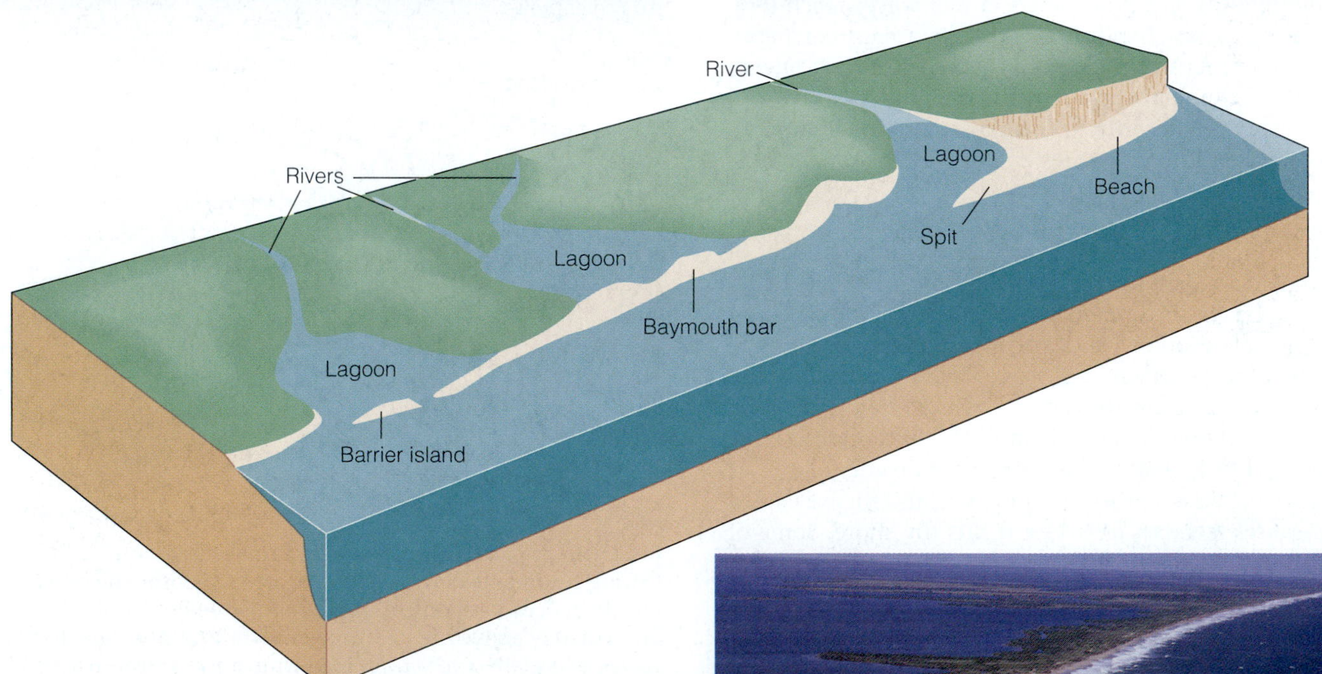

FIGURE 13.19 Spits, baymouth bars, barrier islands, and lagoons are all formed on coastlines where erosion supplies large amounts of sand to the longshore drift. (a) Homer Spit, Alaska. (b) A baymouth bar at the mouth of the Carmel River, California. (c) A barrier island at Canaveral National Seashore, Florida.

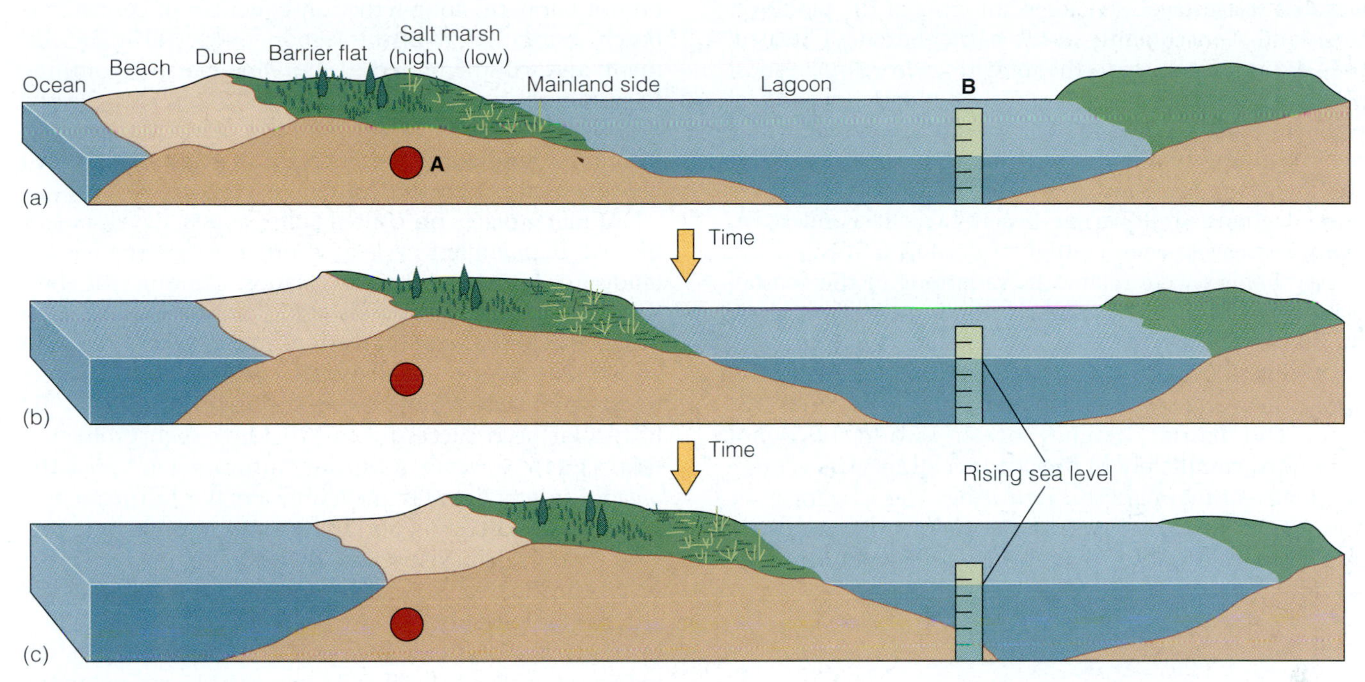

FIGURE 13.20 As sea level rises, barrier islands retreat toward the mainland as storm waves erode sand from the beach and transport it over the barrier island to accumulate on the landward side of the island. Note the sea-level rise shown at B, the inundation of the coast as sea level rises, and the landward progression of the island. Also note that the sand at point A remains in the same place during this process as the island migrates. This sand would eventually be eroded if sea level continued to rise and the barrier island continued to retreat. Vegetation stabilizes the barrier island dunes, but it is continuously eroded on the ocean side of the island and replaced by new growth on the growing landward side.

to the coast and are called "barrier islands." Landward of the barrier island, there is usually a calm shallow **lagoon** in which sea grasses, mangroves, or marsh grasses may grow. Barrier islands and their sheltered lagoons are important because they protect the coast from erosion caused by storm waves. In addition, the sheltered lagoons provide **habitat** for the juveniles of many marine **species**.

Formation of Barrier Islands

The mechanisms that lead to the formation and migration of barrier islands are not fully understood. Some barrier islands may be formed from **spits** created by longshore drift, or they may be coastal sand dunes behind which rising sea level has flooded. Others may originate as a longshore bar when particularly strong storms raise the sea level and transport sand from the beach to the bar. The enlarged bar emerges when the storm has abated and sea level has dropped. Almost all possible mechanisms depend on or are aided by rising sea level.

Many or most barrier islands are thought to have been formed by erosion of the flat, sediment-filled coastal plains of passive-margin coasts as sea level rose during the past 18,000 years. As sea level rose, the soft sediments and rock of the newly flooded coastal plain were rapidly eroded, providing an abundant supply of beach sands that entered the longshore drift. As it migrated along the coast past headlands, the abundant sand accumulated downcurrent of the headlands to form spits (**Figs. 13.7, 13.19**). Spits often grow completely across the mouths of bays and inlets between two headlands to become a **baymouth bar** (**Fig. 13.19**).

As sea level rose further, the spits and bars were breached in places, allowing the sea to inundate more of the coastal plain behind the accumulated beach sands. Where sections of beach were separated from the headlands on both sides, barrier islands were formed (**Fig. 13.19**). Once formed, barrier beaches (including islands, spits, and baymouth bars) continued to be fed with large amounts of sand from the easily eroded coast and began to retreat landward in concert with the rising sea level (**Fig. 13.20**). As sea level rose, storms piled sand onto and across the barrier islands, eroding their ocean shores but building the lagoon side by the accumulation of overwashed sand (**Fig. 13.20**).

As sea level rose rapidly between about 15,000 and 4000 years ago, the barrier islands retreated with the advancing sea (**Fig. 13.20**). However, barrier beaches do not retreat continuously. The abundant sand in the longshore drift system is accumulated on the seaward side of a barrier island until an exceptionally strong storm and elevated sea surface carry large amounts of the sand over the island. Consequently, as the barrier beaches retreated, they did not catch up to the retreating shoreline in most places. The rising sea continuously flooded more coastal plain through breaches in the barrier beach system, while the shoreline, lagoon, and barrier beach complex retreated onto the continent as sea level rose. The shorelines, lagoons, and barrier beach breaches underwent many changes in their configuration during this process. These changes were related to variations in the coastal-plain topography and in the distribution of storms along the coast.

About 4000 years ago, the rapid rise in sea level diminished, and sea level has risen only slowly since that time. The barrier beaches of the world have not responded completely to the virtual halt in sea-level rise, and many of them are still retreating. The retreat can be quite rapid. In some cases, barrier islands have retreated by a distance equal to their entire width within the past several decades.

Future of Barrier Islands

If sea level remains the same as it is today, barrier islands should retreat more slowly, but the retreat may be necessary for their continued existence. Abundant sand eroded from newly inundated coast is needed to sustain a barrier beach, and rising sea level is needed to prevent shallow lagoons behind the barrier island from slowly filling with sediment. If sea level remains stable, barrier islands may slowly disappear. If sea level falls significantly, barrier islands may no longer form. In contrast, rising sea level, often predicted as a consequence of enhanced greenhouse effect, would probably result in a renewed cycle of active barrier island formation and retreat.

Development on Barrier Islands

Barrier islands are particularly inviting places to build houses and resorts. In addition, the lagoons behind them are well suited to be harbors for small boats if the channel to the sea between islands is maintained at an adequate depth for navigation. Unfortunately, barrier beaches are dynamic features that undergo continuous change. Each time a major storm or **hurricane** hits one of the many heavily developed barrier islands, massive damage occurs and "undesirable" sand accumulates in the lagoon and often in channels. Large sums of money must be spent to rebuild and to move sand out of the lagoon and back to the depleted beach.

Although periodic destruction by storms will continue indefinitely, it may be too technologically difficult, too costly, or too environmentally undesirable to indefinitely "restore" the barrier islands as natural processes continue to cause their landward retreat. In addition, we do not know the long-term consequences of continuous beach restoration. Will the lagoon slowly fill with sediment, and will the protected shoreline marshes continue to sustain their vital **ecological** role (Chap. 15)? Clearly, if sea level rises more rapidly, the task of maintaining civilization's temporary presence on barrier islands will become increasingly difficult and expensive.

At many places on United States coasts, development has led to an endless cycle of efforts to arrest the natural landward retreat of barrier islands. Among the best known of the barrier islands are Miami Beach and Palm Beach in Florida, and Galveston and Corpus Christi in Texas. In Galveston, a **seawall** constructed to protect houses that were rebuilt after a strong hurricane in 1900 has so far been successful in providing such protection (**Fig. 13.21**). However, no strong hurricane has struck the area for several decades. Inevitably a major hurricane will again strike Galveston, and the seawall is unlikely to prevent massive damage because the shoreline there retreats several meters each year. In addition, the seawall has altered the longshore sand transport processes, resulting in the virtual disappearance of the formerly wide beach in front of it. The seawall and town will be exposed to the full fury of any future hurricane-driven **storm surges** and high waves.

Corpus Christi has similar problems. Instead of building a seawall, the approach in Corpus Christi was to try to replenish the vanishing beach. Unfortunately, the sand used for replenishment had a larger grain size than the natural beach sand. As a result, the beach stabilized with a much steeper slope than it previously had, and thus it became more dangerous for people entering the water. In addition, beach replenishment must be repeated periodically and continued indefinitely if the beach is to be maintained. Meanwhile, the barrier island adjacent to the area where the beach is replenished continues to retreat, further increasing the exposure of Corpus Christi to ocean waves.

13.4 BEACHES AND HUMAN STRUCTURES

The beachfront is a preferred location for constructing houses, condominiums, hotels, marinas, and other commercial establishments. Structures are built next to the beach despite the knowledge that beaches are constantly changing and subject to erosion and, hence, that the structures are vulnerable to damage by storm waves.

Most beaches migrate inland as coastal storm erosion transports sand offshore. The migration occurs more quickly in some places, particularly some barrier beaches, than in others. Nevertheless, coastal erosion continues inexorably on most coasts, carving away the base of coastal cliffs, removing sand by longshore drift, and pushing barrier islands back toward the mainland. In a

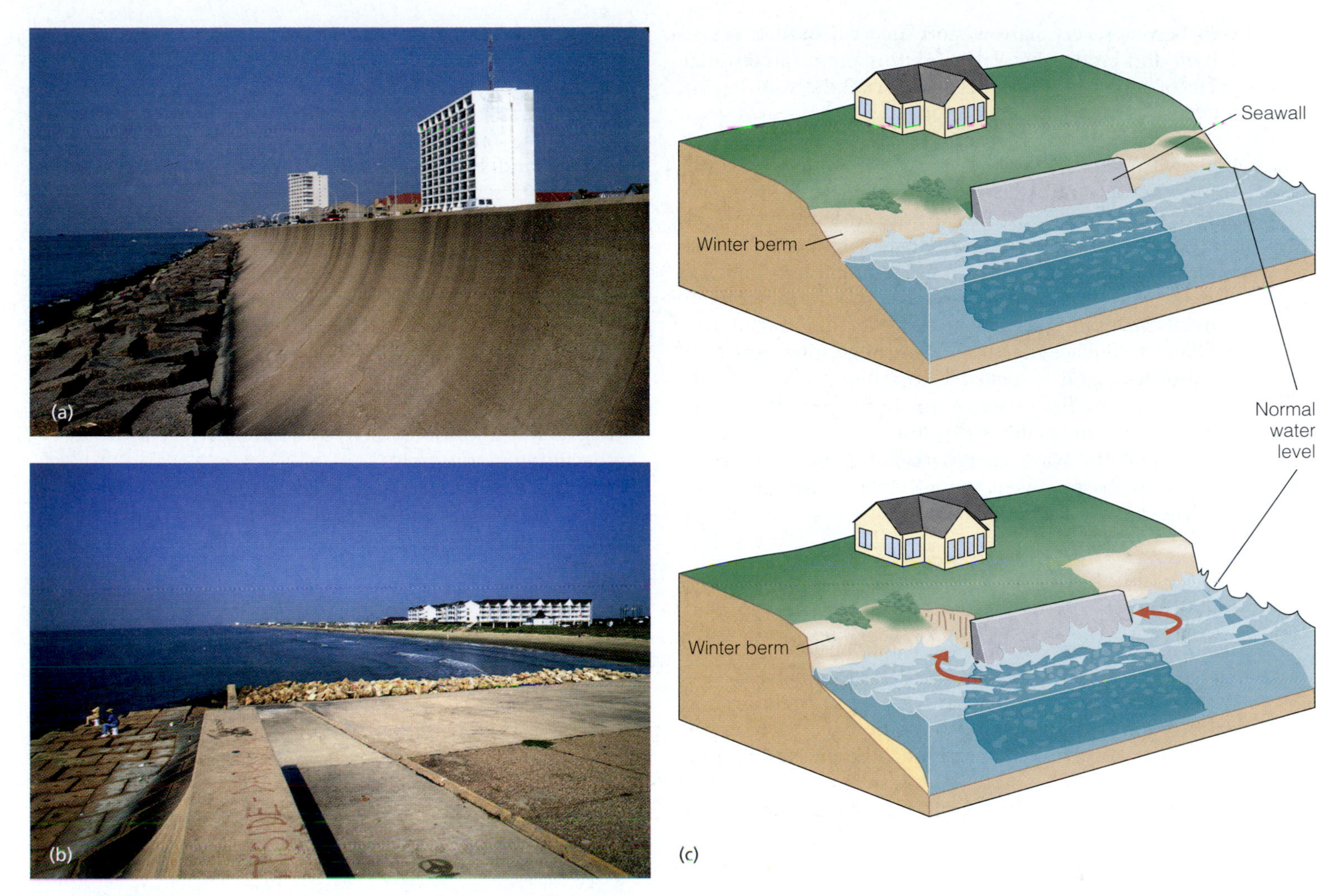

FIGURE 13.21 Seawalls are designed to protect beachfront property from erosion. (a) The Galveston seawall was constructed after the destructive hurricane of 1900. Since the wall was completed in 1962, the beach has eroded along much of its length, leaving the shoreline more exposed to attack by the highest waves of future hurricanes. (b) The barrier island has continued to retreat since the Galveston seawall was constructed. The beach just beyond the end of the wall has now retreated substantially farther shoreward than the wall's location as a result of the natural process of barrier island retreat and the consequent reduction in the amount of sand in the littoral drift system that occurred once the beach had been eroded away in front of the wall. (c) Seawalls are undermined from the sides and eventually destroyed as sea level rises.

few locations, beaches are built up, sometimes very rapidly, by coastal sand transport. Usually, however, beaches are built and extended seaward only where the land edge is uplifted by tectonic processes or where isostatic processes raise the land edge more quickly than erosion can occur. Accordingly, in most locations, maintenance of beachfront property requires a defense against the continual erosional loss of beach and the consequent increasing vulnerability of the property to storm waves.

When the beach in front of beachfront property loses a significant amount of sand, the traditional engineering response is to build seawalls or **groins** to restore beach sand and protect the property from wave damage. These structures interfere with the normal movement of sand in the beach system and often have unintended consequences.

Seawalls

When the beach in front of buildings or a highway has eroded or retreated so far that winter storm waves can reach and damage the structures, a seawall is often built to break up or reflect storm waves. The seawall is usually constructed parallel to the beach and behind the area of beach used by summer visitors (**Fig. 13.21**). It is made of large boulders, concrete, or steel, and either replaces or covers beach sand that normally would be mobile and eroded by the strongest winter storm waves. The wall usually works well for some years, but beach erosion continues on either end of the walled section. As the adjacent coastline retreats, the wall protrudes progressively farther seaward on the beach (**Fig. 13.21**) and is increasingly exposed to waves. Eventually the beach in front of the wall becomes very narrow, and the wall itself is undermined and eroded by waves. At this point, a continual cycle of rebuilding and destruction of the wall begins, and the beach essentially disappears.

A seawall has another unintended effect. Because it is built within the area of wave action, it prevents waves from eroding sand from what was previously the upper part of the beach. Consequently, the amount of sand carried offshore to form longshore bars is reduced. The seawall also prevents the beach from migrating landward and the waves from progressively eroding the coast to provide new sand. Thus, it reduces the amount of sand available to replace the sand lost through longshore drift. Because less sand is available, the beach becomes narrower and longshore bars are further depleted. The narrower beach and smaller longshore bars absorb less wave energy, and the wave energy reaching and eroding the seawall increases, even though storms are no more intense.

Groins and Jetties

Beaches progressively lose sand and become narrower when structures prevent the normal erosional retreat of the coastline or when the sand supply to the longshore drift system is reduced by other means, such as the damming of rivers. When the beach erodes and narrows in front of a valuable piece of real estate, groins are often built to try to restore it. A groin is a wall built perpendicular to the beach from the backshore out to beyond the **surf zone** (**Fig. 13.22a**). The groin's purpose is to block the longshore drift so that sand accumulates on the upcurrent side of the groin, widening the beach at that point. The problem with groins is that they further deplete sand supply to the beach on their downcurrent side, where severe erosion may ensue. A common solution to this problem is to build another groin downcurrent from the first. Eventually a long series of groins may be built along the entire length of the beach (**Fig. 13.22b**). The net effect is to alter the beach so that wave erosion is enhanced on the downcurrent sides of the groins while segments of "normal"-width beach are maintained on the upcurrent sides (**Fig. 13.22a**). In many areas, groins are a temporary solution at best because they do not stop shoreline retreat, but merely redistribute a declining sand supply.

Other structures built on the coast also cause problems by interfering with the longshore drift system. For example, rock **jetties** are built to protect inlets between barrier beaches (**Fig. 13.23**). Such inlets are dynamic features that normally appear, disappear, and move up and down the coast as sand is moved in the longshore drift system. Jetties stabilize an inlet by blocking waves that otherwise would erode its banks and by blocking longshore drift into the inlet that would cause **shoaling.** Beaches often become wider on the upcurrent sides of inlet jetties. In contrast, downcurrent beaches can be eroded and severely depleted of sand, and the inlet behind the island can be silted up as the island retreats inland.

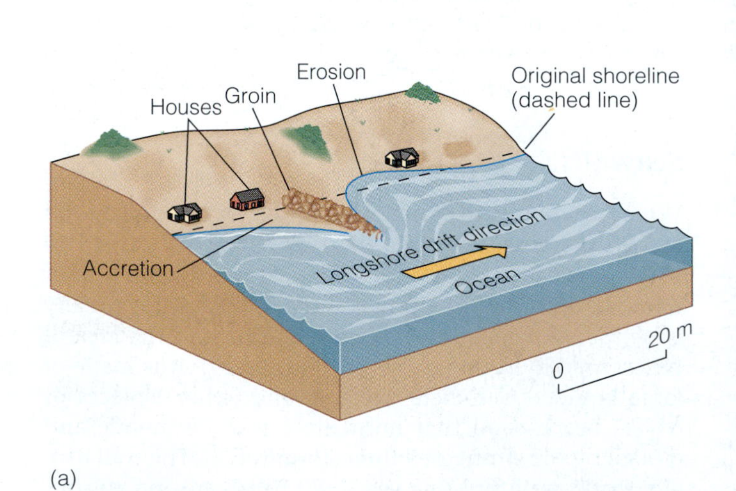

FIGURE 13.22 Groins designed to stop beach erosion modify the longshore sand transport and, consequently, the beach. (a) Sand accumulates on the upcurrent (longshore drift current) side of the groin and is eroded from the downcurrent side. (b) Often a line of groins is constructed along the length of a beach, which creates a saw-toothed beach shoreline like this one at Ship Bottom, New Jersey.

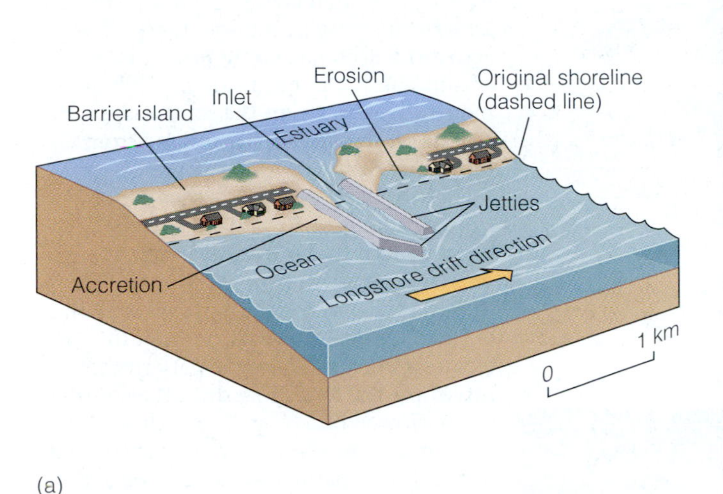

(a)

(b)

FIGURE 13.23 Jetties. (a) Rock jetties are often constructed on either side of an inlet between barrier islands to prevent silting of navigation channels and erosion of the sides of the inlet by waves from boats' wakes. These jetties obstruct the longshore sand transport, leading to the accumulation of sand and widening of the beach on the upcurrent side, and erosion and loss of the beach on the downstream side. (b) Mattituck Inlet, Long Island, New York. Can you tell which direction the longshore drift is?

Harbors

On coasts with few natural inlets, jetties or **breakwaters** commonly are built to provide a safe harbor for small boats. Two such harbors are shown in **Figures 13.24 and 13.25**. The dogleg **jetty** on the upcurrent side of the Santa Barbara harbor (**Fig. 13.24**) interrupts the longshore drift so that it carries sand around the jetty but not across the deep harbor entrance, which is too deep for waves to sustain the sand movement. Sand flows around the jetty and accumulates in the harbor, making parts of the harbor too shallow for navigation. The beach that is downcurrent of the harbor is depleted of sand and threatened by severe erosion. These problems are addressed by a very expensive, energy-intensive dredging project. Sand is continuously dredged from the harbor, pumped through a pipeline, and discharged downcurrent, where it re-joins the longshore drift.

The Santa Monica harbor (**Fig. 13.25**) was created by a simple breakwater that was built parallel to the beach beyond the surf zone, and was designed to shelter boats. Because this breakwater does not encroach on the surf zone, it seemingly should not interfere with the longshore drift. Unfortunately, this is not the case, because the breakwater prevents waves from reaching a segment of the beach behind it. Since only low-energy waves break on this segment, the longshore drift is reduced just as effectively as it would be by a groin passing through the surf zone. Soon after the harbor was constructed, the beach behind the breakwater became greatly enlarged, reducing the area of safe anchorage for boats. The only solution is an expensive dredging program.

13.5 CORAL REEFS AND ATOLLS

Reef-building corals are **communities** of microscopic animals that have photosynthetic **dinoflagellates**, called **zooxanthellae**, living **symbiotically** within their tissues. These corals cannot grow successfully without the zooxanthellae or where there is not enough light for the zooxanthellae to photosynthesize. Reef-building corals also require warm water (≥18°C) and are intolerant of low **salinity** and high concentrations of **suspended sediments**. Hence, coral reefs grow only in shallow tropical and subtropical waters that are not subject to low salinity and high **turbidity** caused by stream **runoff** or abundant **plankton**.

The three basic types of coral reefs are fringing reefs, barrier reefs, and **atolls** (**Fig. 4.27**). Fringing reefs grow in shallow waters along the shore and are best developed off coasts with arid climates and where river runoff is limited. Because fringing reefs are easily damaged by excess freshwater runoff or high concentrations of suspended sediment due to storm waves, they are distributed in a patchy fashion along most coasts where they occur. Human

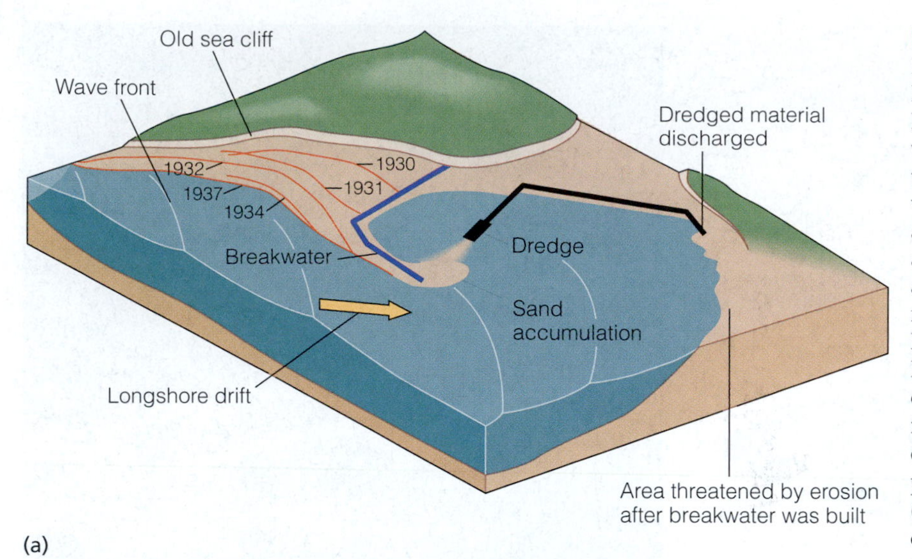

FIGURE 13.24 The jetties of the Santa Barbara harbor have interfered with the longshore drift transport system. (a) Sand has accumulated and progressively widened the beach on the north side of the main jetty (the direction from which waves most often come). Sand transported around this jetty cannot be transported across the deep harbor mouth and accumulates in a spit inside the harbor mouth. The beach on the south side of the harbor has had its supply of sand from longshore drift cut off. Continuous dredging is needed to keep the harbor from silting up. The dredged sand is deposited on the south-side beach to prevent this beach from eroding further. (b) This aerial photograph shows how the deeper channel in the harbor mouth creates a break in the wave pattern that interrupts the longshore drift. It also shows the narrowed beach on the south side of the harbor. The dark object just at the exit of the inner boat harbor is the dredge that continuously removes sand from the harbor and discharges it south of the harbor, back into the longshore drift system.

activities such as flood-control projects, dredging, sewage discharge, and coastal modification often result in elevated concentrations of suspended sediments and enhanced runoff of low-salinity water during storms. Consequently, fringing coral reefs are damaged by these activities in many locations, but it is often difficult to distinguish damage due to human influence from natural variations.

Barrier reefs and atolls are formed only where there has been a change of sea level on the adjacent coast. If sea level rises or the coast subsides in an area that has a fringing reef, the reef grows upward. The upward growth is faster at some distance offshore where conditions are optimal. In the shallowest water close to the shoreline, wave resuspension of sediments and the effects of runoff inhibit coral growth. In deeper offshore water, reduced light levels inhibit photosynthesis by zooxanthellae and thus coral growth. Therefore, the optimal zone for coral growth is a strip of ocean some distance offshore, but not extending into deep water. The width of this zone varies with factors such as the seafloor slope and the amount of runoff. As the land subsides, the reef grows upward in this zone faster than coral can grow in either shallower or deeper water (**Fig. 4.27**), and a barrier reef is formed. Once formed, the barrier reef continues to grow upward. The rate of upward growth is usually just sufficient to match the changing sea surface level. The top of the reef remains just below the water surface, because corals are damaged if exposed to air for prolonged periods. If the maximum possible rate of upward growth is less than the rate at which the land subsides, the reef is submerged and eventually dies.

Once a barrier reef is formed, it closes or partially closes a lagoon between the reef and the coastline. Corals

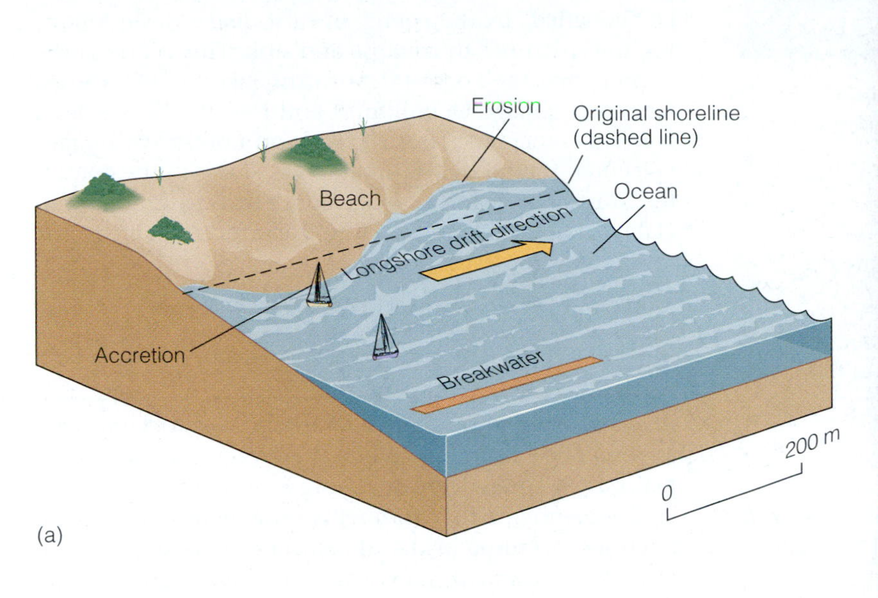

FIGURE 13.25 (a) Some boat harbors consist of a breakwater or jetty built parallel to the shore. The breakwater protects boats but also interferes with the longshore drift by blocking the waves from reaching the beach behind the breakwater. (b) This aerial image of Santa Monica shows the harbor wall parallel to original beach line, with a number of small boats and a pier behind it. The sand build up on the beach due to the breakwater blocking the wave energy is clearly visible. The pier is elevated on pilings, which allows the wave energy to pass through. As a result, the pier does not block the longshore drift.

grow in this lagoon, but they are limited by higher turbidity due to land runoff, by salinity variations, or by lack of **nutrients** as a result of the long **residence time** of the lagoon water (Chap. 14, **CC8**).

Barrier reefs are especially abundant on the Earth at present because sea level has been rising eustatically for approximately the past 18,000 years. The Great Barrier Reef of Australia, which is about 150 km wide in places and more than 2000 km long, is the best-known barrier reef.

Islands formed by volcanoes on oceanic **crust** are particularly good locations for the development of barrier reefs. Once a volcanic island has moved away from the oceanic ridge or hot spot on which it was formed, it cools and sinks isostatically (Chap. 4, **CC2**). Hence, even in times of falling sea level, barrier reefs can be formed around such islands, as long as the island sinks faster than the sea level falls. Sinking of volcanic islands leads to the formation of the third type of coral reef, the atoll (**Fig. 4.27c**). As the island sinks, a barrier reef develops from the fringing reef and continues to grow upward, remaining at or close to the sea surface as the island continues to sink. Eventually the island sinks entirely below the ocean surface, and its reef is left in the form of an atoll. The atoll continues to grow and maintain itself as the island sinks (or as sea level rises).

Low islands may form on an atoll, particularly on the side of the atoll from which wind and waves most often approach. Such islands are composed of debris from the reef and of calcareous algae that grow in very shallow water and can survive exposure at low tide. Islands can also be formed on an atoll by earthquakes that uplift or tilt the sinking volcano.

13.6 WETLANDS

On many shores protected from wave action, tidal wetlands develop. Tidal wetlands are flat, muddy areas covered by water during only part of the tidal cycle. Most tidal wetlands are characterized by an abundance of emergent plants: marsh grasses in salt marshes, and mangroves in mangrove swamps. Both plant types grow well in muds rich in organic matter that are inundated periodically by salt water. Grasses grow in salt marshes at all latitudes, but mangroves grow only in tropical and subtropical latitudes between about 30°N and 30°S. Where both are present, mangroves usually rapidly outgrow and eliminate marsh grasses.

Wetlands are created by the accumulation of sediments from a variety of sources. The most important are usually river-borne muds laden with organic **detritus**. Sediments accumulate in shallow protected waters, particularly at the edges of estuaries and on lagoon shores, until they extend above the low-tide line and a wetland forms.

CRITICAL THINKING QUESTIONS

13.7 Explain what happens to a beach if a solid jetty is built out through the surf zone. Describe how the situation might be different if the jetty were built on widely spaced pilings.

13.8 What would happen to the beach if a solid jetty were built at an angle of 45° instead of at right angles to the beachfront? Explain your answer.

13.9 Why are there few deltas in the United States? In the next few million years, where would you expect to see new deltas form in the United States? Explain why.

A wetland consists of a muddy, flat terrace between the low- and high-tide lines that is dissected by shallow channels. As the tide rises, water moves through the channels and spreads over the mudflats. The current velocity within the channels is fast in relation to the water's velocity as it spreads over the flats. As the velocity decreases over the mudflats, suspended particles carried up the channels are deposited (**CC4**) as wetland mud. Particles of organic detritus have low **density** and can be fragmented to very small sizes. Consequently, wetlands tend to accumulate muds that are rich in organic matter, with much of the detritus contributed by the local marsh grasses or mangroves.

The abundant supply of detritus makes wetlands very attractive places for marine animals to feed and for the juvenile stages of many marine species to spend part of their life cycle before migrating to the sea. Wetlands provide not only abundant food, but also safety from predators. Large marine predators are not able to enter these shallow areas, and the grasses, mangrove roots, and abundant leaf litter offer many excellent places to hide from other predators. Wetlands have such an abundance of vegetation and juvenile marine animals that many birds use wetlands as their principal, or only, feeding areas.

Wetlands are extremely valuable ecologically, but generally they are not considered to have great scenic beauty. Therefore, because their proximity to the water makes them very attractive for development, vast areas of wetlands in the United States and elsewhere have been filled or drained. The result has been the destruction of habitat that is important for many commercially valuable fishes and **shellfish** species and a number of species of ducks and other birds. Filled wetlands are not ideal places to build or farm. Because they are low-lying, they are susceptible to flooding and to storm surges or storm waves. Structures built on former wetlands also are susceptible to subsidence as the former wetland sediments upon which they are built are compacted.

Even more importantly, structures built on filled wetlands are very vulnerable to earthquake damage. The sediments underlying structures in filled wetlands, if not properly compacted during the construction process, can be "liquefied" by the energy of earthquake waves, allowing the structures to collapse and sink. This is one of the reasons that the Loma Prieta earthquake of 1989 caused far more damage to buildings and freeways in Oakland and San Francisco, about 100 km north of the earthquake epicenter, than it did in urban areas closer to the epicenter. Most of the badly damaged or destroyed structures in Oakland and San Francisco were constructed in areas that formerly were wetlands.

13.7 DELTAS

Rivers flowing toward the sea carry eroded rock particles as suspended sediment. When they reach the flat coastal plain, their speed decreases because of diminishing slope and widening of the channel. Consequently, larger size fractions of suspended sediment are deposited (**CC4**). Deposited sediment slowly fills the river valley. Many rivers flow through valleys cut by glaciers or rivers when sea level was lower during the most recent **ice age,** and the lower ends of these valleys have not yet filled with sediment. Because most rivers on the east coast of the United States are in this category, they deliver little sediment to the Atlantic Ocean. Some rivers, such as the Mississippi, carry so much sediment that they have completely filled their river valleys and the bulk of their sediment is now deposited beyond the river valley to form a delta that extends out to sea. The river flows in distributary channels through the deposited sediment on the delta, and in some places, such as the Mississippi delta, the channels pass through long extensions or lobes of these deposits (**Fig. 13.5**).

Deltas resemble wetlands in function. During normal river flow, relatively small quantities of suspended sediment are transported down the narrow river channels and deposited in the channel or carried out to sea. When the river floods, much larger quantities of sediment are brought downstream into the delta, but the river overflows its banks and inundates the flat land of the delta. The current speed of river water decreases as it leaves the channels and spreads across the delta, allowing suspended sediment to be deposited. Sediment distribution across the delta is aided by the river's occasional abandonment of channels and creation of new ones (**Fig. 13.5**). In this way, surface soils of the delta are periodically enriched by additions of organic matter and nutrients from the river.

In some deltas, the deposition rate has been high enough that large quantities of organic-rich muds have been continuously buried by newly deposited sediments over long periods of geological time. These buried sediments are compacted by the overlying layers, water is squeezed out, and in many such areas, oil and gas deposits have been formed as the organic matter has decomposed. Many oil and gas deposits are located in deltas or offshore in areas where deltas existed when sea level was lower.

Major deltas of the world include the Mississippi delta, the San Francisco Bay delta, the Copper River delta in Alaska, the Mackenzie delta in northwest Canada, the Nile delta in Egypt, the Niger delta in Nigeria, and the Ganges–Brahmaputra delta in Bangladesh and India.

Muddy delta soils are slowly compacted as water is squeezed out of them by the weight of overlying sediment. This process can be accelerated by the weight of structures built on the delta. In addition, the organic matter in the delta soils decomposes if it is not buried by additional river-borne sediments, and the soils themselves are further eroded by winds and rainfall. Therefore, periodic influxes of new sediment are needed to maintain the delta land surface slightly above sea level.

Because they are particularly rich in nutrients and organic matter, delta soils are among the best for agriculture. Consequently, many deltas, such as the Mississippi delta, are populated and intensively farmed. Unfortunately, deltas are also very easily flooded. Almost the entire population of Bangladesh lives on a delta that is regularly inundated by swollen rivers and hurricane storm surges (Chap. 11), often with great loss of life.

To control the frequent flooding of two major deltas in the United States—the Mississippi delta and the San Francisco Bay delta—extensive engineering projects have been carried out in which banks called **levees** have been built along each side of the river channels. Levees have worked extremely well by some measures because they have dramatically reduced flooding, but they also have unintended consequences. Without the periodic influx of nutrient-rich suspended sediment to the delta during floods, the fertility of the soils steadily declines. In addition, without the periodic inputs of new soil, the land slowly sinks as soil compacts and erodes. Sinking of the land surface is actually accelerated by agriculture through increased soil erosion and decomposition of soil organic matter. Worse still, withdrawal of freshwater from aquifers beneath the delta further accelerates sinking, particularly in the Mississippi delta.

As the land sinks in the Mississippi and San Francisco Bay deltas, more and higher levees are needed to prevent flooding. In the San Francisco Bay delta, most of the land is now meters below the normal river water level. Any breach of the levee allows the land to be covered completely with water, even when the river is not swollen by flood rainwater. For example, a 100-m-wide breach occurred in a levee in the San Francisco Bay delta in the summer of 2004. The breach flooded nearly 50 km^2 of farmland, destroying crops, homes, and other structures. It took about 200,000 tonnes of rock to seal the levee breach. The total cost of the crops and structures destroyed, repairing the levee, and pumping water out of the flooded land was nearly $100 million. Significant as this event was, it was just a foretaste of the devastation, chaos, and loss of life and property that was caused by the breaks in the protective levees of New Orleans that flooded most of that city to a depth of up to 10 m or more in 2005 following the passage of hurricane Katrina.

Higher and stronger levees could be built, but they would be very expensive and difficult to construct because they would need to be capable of restraining increasing **hydrostatic pressure** as the delta floor sinks and sea level rises. In addition, levees constrain the river to flow very swiftly through a narrow channel, so erosion of existing levees is a major problem and would remain so even if levees were enhanced.

What is the future of the Mississippi and San Francisco Bay deltas, particularly if global climate change causes sea level to rise more quickly? If we continue on our present course, these delta lands will continue to sink and levees will continue to be built higher until holding back the water becomes impossible. Meanwhile, soils will become impoverished and the land's agricultural value will decline. One alternative is to tear down the levees. Within a few years or decades, the deltas would return to their normal functioning, and valuable agricultural lands would again be available for the future. If we are to retain the highly productive deltas, the price we must pay is periodic flooding.

In Louisiana, where the levees on the Mississippi delta have been a major contributor to the loss of an average of about 65 km^2 of coastal land per year since 1930, a program has begun to establish breaches in the levees that will replenish delta sediments in key areas, converting these areas back to wetlands. So far, this program has focused mainly on areas that had become shallow coastal lagoons as a result of the levee diversions of sediments. Strong opposition to the program from fishers who used the lagoons to collect oysters had to be overcome. Even greater opposition is, and will continue to be, encountered when the proposed replenishments involve land that has been drained for agriculture. It remains to be seen whether this replenishment program and similar programs that may be proposed on other deltas will ultimately succeed.

CHAPTER SUMMARY

Processes That Form and Modify the Coastline. Coasts can be erosional or depositional. Primary coasts are formed by volcanoes, earthquake movements, landslides, deposition of river-borne sediment, and sea-level change. Sea level has risen about 130 m in the past 18,000 years and continues to rise. Secondary coasts are formed by erosion or deposition by waves, tidal and ocean currents, and the formation of coral reefs.

Beaches. The littoral zone extends between the seaward boundary of land vegetation and the depth where wave action does not disturb the sediments (about 10 to 20 m).

Beaches are part of the littoral zone and consist of a backshore, foreshore, and offshore. The backshore is normally dry at high tide and may be characterized by berms created by storm waves. The foreshore is the gently sloping area between the high- and low-tide lines, where scarps may form. The offshore extends seaward of the low-tide line and may have a longshore bar near its seaward boundary.

Beaches are a mixture of grains of varied composition supplied by erosion or by longshore drift. Longshore drift, caused by the incomplete **refraction** of waves, transports sand downcoast of the direction from which the waves approach. Sand is moved along the coast until it is transported down a submarine canyon. Some beaches may be in danger of losing sand because the supply of river-borne sand to the longshore drift system has been reduced by dams and other factors.

Particles that make up a beach are sorted by waves. Small particles are removed to deep water. Large particles are not moved by longshore drift and are present only if they are formed locally by the erosion of cliffs next to the beach. Grain size is larger on beaches with greater wave action. Quiet beaches are composed of fine-grained sand, whereas those with greater wave action are composed of coarse-grained sand or gravel.

Large winter waves move sand from the beach offshore to build a longshore bar. In summer, gentler waves return sand to the beach. In winter, a beach is narrower with a flat foreshore and steep backshore. In summer, the beach is broader, the foreshore is steeper and a summer berm or berms are formed. Beach slope increases as grain size of the beach materials increases.

Barrier Islands and Lagoons. Barrier islands protect the coast from wave erosion and enclose lagoons that are nursery areas for many marine species. They are formed and maintained by sand that is moved by longshore drift. Barrier island–lagoon systems are retreating landward as sea level rises. Development on barrier islands presents problems because of this landward migration. Dredging of the lagoons and beach replenishment are necessary to reverse migration that would destroy developed structures on the islands, but these measures cannot be successful indefinitely. Developments on barrier islands repeatedly sustain expensive damage when major storms occur.

Beaches and Human Structures. As beaches migrate inland or lose sand because river sources are reduced, seawalls and groins are built to retain the sand or protect coastal structures. Seawalls temporarily protect coastal structures from waves, but eventually they are undermined and destroyed as the shoreline recedes. Seawalls also reduce erosion that would otherwise supply sand to the longshore drift system, and consequently, they reduce beaches and longshore bars that would absorb wave energy. Groins are built across the littoral zone to retain sand on beaches by blocking longshore drift. Sand builds up on the upcurrent side of the groin and is lost from the downcurrent side. Groins are often only temporarily effective and can cause a loss of sand on downcurrent beaches that leads to the building of more groins. Jetties or breakwaters protect harbors from wave action but interfere with longshore drift and often cause sediment to accumulate in the harbor, making continuous dredging necessary.

Coral Reefs and Atolls. Reef-building corals grow only in warm water shallow enough that their zooxanthellae can photosynthesize. They grow best in waters with lower turbidity and little salinity variation. Fringing coral reefs grow in shallow water off shores where erosion and freshwater inputs are low. Barrier reefs and atolls form where coral growth is faster some distance offshore than immediately along the shore, and where sea level rises or the coast subsides. Most atolls are formed from barrier reefs as volcanoes sink isostatically below the sea surface.

Wetlands. Tidal wetlands are flat, muddy areas covered by water part of the time. They retain detritus from river flow and vegetation that grows in them. The detritus provides food and the grasses and shallow water provide shelter for juveniles of many marine species. Large areas of coastal wetlands have been filled for development.

Deltas. Deltas are formed after sediments fill river valleys and then accumulate at the mouth of the river. They are flat areas maintained at an elevation slightly above sea level by periodic river floods that bring sediment, detritus, and nutrient-containing particles to be deposited on the flooded delta. They are good agricultural land because of these periodic inputs, but they are often flooded. Although levees have been built to protect many deltas from flooding, these structures cut off the supply of new sediment, so the land slowly subsides, is eroded away, and loses its nutrients and detritus.

STUDY QUESTIONS

1. Describe the processes that modify a coastline once it has been formed. How and why are the coastlines of passive and active margins different?
2. Describe the features of a beach at the end of a summer in which no large storm waves arrived at the beach. How would these features change if a storm with very large waves hit the shoreline?
3. Why are the particles of sand on most beaches limited to a narrow range of sizes? Why can the size ranges be different on two beaches located near each other?
4. Where does beach sand come from, and how is sand moved along a coastline?
5. It has been proposed that the energy of waves can be harnessed by various devices placed seaward of the surf line. If many such devices extended along a

coastline and they collected a substantial fraction of the incoming wave energy, how would the beaches be changed?

6. Why is replacing sand lost from a barrier island beach, or building a seawall to protect beachfront property, only a temporary solution for beach erosion?
7. In many indented bays there is a wide, flat, sandy beach at the bay's center but the beach disappears near each headland. Another sandy beach is usually found in the next bay. Describe the processes responsible for this distribution of beach sand.
8. Why don't fringing coral reefs grow on all tropical coastlines? What is the relationship between rising sea level and the number and distribution of atolls and barrier reefs in the oceans?
9. Describe the differences between fringing coral reefs, barrier reefs, and atolls. The island of Hawaii has mostly fringing reefs. Why?
10. What are the important physical, biological, and geological features of a wetland? Many wetlands have had roads built across them, either on gravel or dirt roadbeds or on elevated concrete pilings. Which type of road is environmentally preferred and why? What could be done to lessen the impact of the lesser-preferred option, short of rebuilding it?

KEY TERMS

You should be able to recognize and understand the meaning of all terms that are in boldface type in the text. All of those terms are defined in the Glossary. The following are some less familiar key scientific terms that are used in this chapter and that are essential to know and be able to use in classroom discussions or on exams.

atoll (p. 363)
backshore (p. 348)
backwash (p. 354)
barrier island (p. 338)
barrier reef (p. 345)
baymouth bar (p. 359)
beach (p. 338)
berm (p. 351)
breakwater (p. 363)
coast (p. 338)
coastal plain (p. 344)
coastal zone (p. 338)
coastline (p. 338)
coral reef (p. 338)
delta (p. 338)
depositional (p. 338)
erosional (p. 338)
estuary (p. 348)
eustatically (p. 343)
fjord (p. 343)
foreshore (p. 348)
fringing reef (p. 345)
groin (p. 361)
high-tide line (p. 347)
hydrostatic pressure (p. 367)
isostatic leveling (p. 339)
jetty (pp. 362, 363)
lagoon (p. 359)
levee (p. 367)
littoral drift (p. 354)
littoral zone (p. 351)
longshore bar (p. 353)
longshore current (p. 354)
longshore drift (p. 354)
low-tide line (p. 347)
low-tide terrace (p. 351)
mangrove (p. 338)
orbital velocity (p. 355)
reef (p. 338)
resuspend (p. 354)
rip current (p. 356)
runoff (p. 363)
salt marsh (p. 338)
sand dune (p. 338)
scarp (p. 352)
sea grass (p. 351)
seawall (p. 360)
shore (p. 338)
shoreline (p. 354)
sorted (p. 356)
spit (p. 359)
storm surge (p. 360)
submarine canyon (p. 354)
surf zone (p. 362)
swash (p. 354)
symbiotically (p. 363)
tidal range (p. 348)
turbidity (p. 363)
weathered (p. 353)
wetland (p. 348)

CRITICAL CONCEPTS REMINDER

CC2 Isostasy, Eustasy, and Sea Level (pp. 343, 365). Earth's crust floats on the plastic asthenosphere. Sections of crust rise and fall isostatically as temperature changes alter their density or as their mass loading changes. This, in turn, causes isostatic changes of sea level. Eustatic changes of sea level occur globally when the volume of water in the oceans changes or when the volume of the ocean basins themselves change. Sea level changes create new coasts. Oceanic crust cools progressively after it is formed and sinks because its density rises. Thus, the hot spot volcanic islands slowly sink after they move away from the hot spot. To read **CC2** go to page 5CC.

CC4 Particle Size, Sinking, Deposition, and Resuspension (pp. 348, 351, 354, 355, 356–57, 366). Suspended particles in the ocean sink at rates primarily determined by particle size: large particles sink faster than small particles. Once deposited, particles can be resuspended if current speeds are high enough. Generally large particles are more difficult to resuspend, although some very fine particles may be cohesive and therefore, also difficult to resuspend. Sinking and resuspension rates are primary factors in determining the grain size characteristics of beach sands and sediments at any given location. To read **CC4** go to page 12CC.

CC8 Residence Time (p. 365). The residence time of seawater in a given segment of the oceans is the average length of time the water spends in that segment. In restricted arms of the sea, such as lagoons behind barrier islands and fringing reefs, residence time can be long, in which case the nutrients may be depleted, limiting the growth of corals and other marine species. To read **CC8** go to page 19CC.

CHAPTER 14

CHAPTER OUTLINE

CRITICAL CONCEPTS USED IN THIS CHAPTER

Although it takes many wonderful forms, life in the oceans is all made possible by a few simple physical and chemical processes. When we found it, this midring blue-ringed octopus (*Hapalochlaena* sp., Indonesia) was a drab, dark color and blended in almost totally with the background. After we approached, it changed color rapidly and displayed the blue rings as a warning. A very closely related species of blue-ringed octopus is venomous with a toxin in its saliva for which there is no antidote. This species is thought to be less toxic, but little is known about it. This individual octopus entertained us by swimming and stopping several times, all the while displaying the blue rings and rapidly varying its body color before it settled and assumed the shape and color to blend perfectly with the rock on which it had landed. The smaller image shows the same octopus a few seconds after the larger image was taken.

Foundations of Life in the Oceans

In 1843 Sir Edward Forbes, a British marine biologist, wrote,

> *The distribution of marine animals is determined by three great primary influences, and modified by several secondary or local ones. The primary influences are climate, sea-composition and depth, corresponding to the three great primary influences which determine the distribution of land animals, namely climate, mineral structure and elevation. . . . The secondary influences which modify the distribution of animals in the Aegean are many. First in importance ranks the character of the sea-bottom, which, though uniform in the lowest explored region, is very variable in all the others. Accordingly as rock, sand, mud, weedy or gravelly ground prevails, so will the numbers of the several genera and species vary.*
>
> —Edward Forbes, ***Report on the Mollusca and Radiata of the Aegean Sea, and on Their Distribution, Considered as Bearing on Geology.*** **Report of 13th Meeting of the British Association for the Advancement of Science, Cork, August 1843, published by J. Murray, London (p. 152)**

The conclusions Forbes made about the factors that determine the species and their abundance in any part of the oceans are as valid today as they were in 1843. Neither Forbes nor any of his contemporaries had any significant knowledge of the true foundations of life in the oceans, the many different kinds of

organisms of microscopic size that are dispersed in all ocean waters. Only much more recently has science uncovered knowledge of the organisms and of how physical and chemical parameters of their environment control their growth and distribution. This knowledge is far from complete. Indeed, the first edition of *Introduction to Ocean Sciences,* published in 1997, made no mention of archaea, which were then virtually unknown. Some species of archaea are now thought to be perhaps the most abundant organisms in the oceans and possibly on the planet.

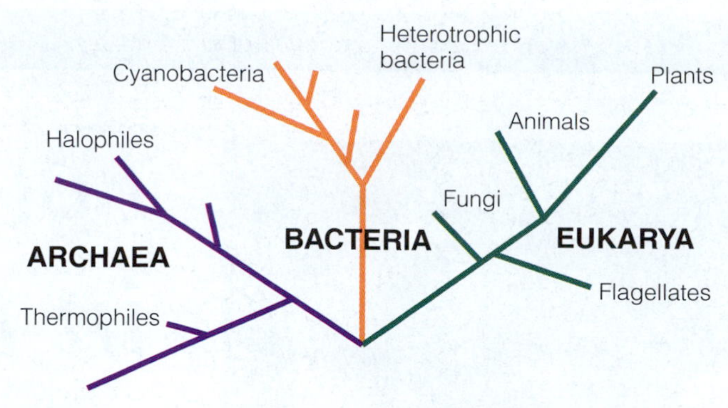

FIGURE 14.1 The tree of life. The prokaryotes—Bacteria and Archaea—are single-celled organisms with no membrane-bound (enclosed by a membrane) nucleus. The eukaryotes (Eukarya)—protists, fungi, plants, and animals—are single- or multi-celled organisms whose cells contain a membrane-bound nucleus.

In many ways the oceans are an ideal **environment** for life. Temperatures are much more uniform than on land, so most marine organisms do not have to contend with the temperature variations and extremes that terrestrial organisms experience. In addition, all elements essential to life are present as dissolved **ions** in seawater, albeit sometimes at very low concentrations, and they are readily available to marine plants and **algae.**

Living in the ocean also presents special problems, and the abundance of ocean life is limited by a variety of physical and chemical factors. This chapter examines the basic processes of life in the oceans, and the physical and chemical parameters that control the quantities and types of organisms in different regions of open oceans. The parameters that determine the distribution of life include the physical and chemical nature of seawater, the distribution of winds and **currents,** and even **plate tectonics.** Chapter 15 discusses how life in coastal and **estuarine** zones differs from open-ocean marine life as a result of the different physical and chemical environments.

14.1 HOW DO WE DESCRIBE LIFE?

Before we examine the many complex physical, chemical, and biological interactions that are the foundations of life in the oceans, it will help to look briefly at how biologists group and classify living organisms. All living things are arranged into formal groups according to their anatomy, physiology, and, more recently, **genetic** differences, through a system known as "taxonomy." Taxonomy is discussed more fully in Appendix 3. For now, we need to know that, at the highest level of organization, called the "tree of life," all **species** are classified into one of three domains (**Fig. 14.1**, **Table 14.1**): Bacteria, Archaea, and Eukarya.

Archaea were once considered part of the Bacteria domain but are now considered separately. Together, **bacteria** and archaea make up a group of species known as **prokaryotes.** All other living species belong to a group called **eukaryotes**, which includes four kingdoms: Protista (**protists**), Fungi (**fungi**), Plantae (plants), and Animalia (animals). As discussed in Appendix 3, species within each kingdom are classified into groups that are arranged in a hierarchy containing a number of levels. All species eventually are given a formal name that consists of **genus** and species, the two lowest levels in the hierarchy. For example, humans are genus *Homo*, species *sapiens*; we are all members of the species *Homo sapiens* (note the italics always used for genus and species names).

Although the formal taxonomic classification of species addresses the need to classify organisms according to their physiological and genetic differences, it does not always serve well to classify species with regard to their functions in the **ecosystem.** Therefore, many other functional grouping schemes are used for specific situations. For example, marine biologists may group organisms according to whether they live in the water column or on the seafloor, or whether they are able to swim against currents or primarily drift with the water.

14.2 PRODUCTION, CONSUMPTION, AND DECOMPOSITION

Life is based on the production of an enormous variety of organic compounds, each of which serves a different function in the cells of living organisms. Organic compounds are created when carbon atoms are combined in chains or rings. Chemical properties of a compound are determined by the number of carbon atoms and how they are arranged, and by the number, position, and elements in other groups of atoms (containing, for example,

TABLE 14.1 The Major Taxonomic Groupings of Living Organisms

PROKARYOTES: Generally microscopic, single-celled organisms that have no membrane-bound (enclosed by a membrane) nucleus or internal structure. May be photosynthetic, chemosynthetic, or heterotrophic.	
Two Domains	
Domain Bacteria	Microscopic and relatively simple single-celled organisms. Bacteria have no organelles (membrane-bound nucleus or internal structural features). However, most possess a carbohydrate-based cell wall.
Domain Archaea	Archaea are similar to bacteria but have different cell wall structures, different constituent organic compounds, and many different genes. Many archaea live in extreme environments, such as those found at hydrothermal vents, those where the temperature exceeds the boiling point of water (e.g., geysers), those where salinities are extremely high, or those that are strongly acidic or alkaline. Some archaea are single-celled, whereas others form filaments or aggregates.
EUKARYOTES: Single-celled or multi-celled organisms whose cells have a nucleus and other internal structure. Generally larger than prokaryotes.	
Domain Eukarya	
Kingdom Protista	A very diverse group of species, including all the eukaryotes except for the plants, fungi, and animals. The vast majority of protists are single-celled organisms, typically only 0.01 to 0.5 mm in size, yet generally larger than the prokaryotes. Examples are diatoms and dinoflagellates, foraminifera, radiolaria, marine algae, and seaweeds. Protists commonly can survive dry periods in the form of cysts; some are important parasites; a few forms are multicelled—for example, the brown and red algae.
Kingdom Fungi	Mostly single-celled, including many decomposers and parasites that infect animals and plants. Fungi release enzymes that break down organic matter that they absorb for food and energy. Examples are molds and mushrooms.
Kingdom Plantae	Multicelled photosynthetic autotrophs. Flowering plants, including sea grasses, mangroves, ferns, and mosses. Note that marine algae and seaweeds are *not* generally considered to be plants.
Kingdom Animalia	Multicelled heterotrophs, including all invertebrates and vertebrates.

oxygen, nitrogen, sulfur, or phosphorus) attached to the carbon chains or rings. Organic compounds made by living organisms can contain hundreds of carbon atoms and attached groups arranged in an enormous number of ways. Even the simplest living organisms consist of a bewildering array of organic compounds.

The most important use of organic molecules by living organisms is to provide energy for the **biochemical** processes that control their growth, movement, feeding, and reproduction. Energy is needed to combine carbon atoms into organic compounds. This energy can be released from organic compounds by decomposition of their molecules to carbon dioxide and water. All living organisms use a decomposition process called **respiration** to provide their needed energy.

There are two fundamental types of organisms. Those of the first type, **autotrophs,** create their own food from inorganic compounds by using an external source of energy. The process of converting inorganic compounds to organic matter is called **primary production.** Autotrophs use up some of the food they synthesize to fuel their life processes through respiration. Plants, algae, and some bacteria and archaea are autotrophs. Organisms of the second fundamental type, **heterotrophs,** cannot make organic compounds from inorganic compounds and must obtain organic matter as food. Animals and most bacteria and archaea are heterotrophs.

The distribution of life depends on the distribution and growth rate of autotrophs. Heterotrophs can live only where and when autotrophs supply enough food and **nutrients.** Autotrophs synthesize organic matter by one of two fundamentally different mechanisms: **photosynthesis** and **chemosynthesis.**

Photosynthesis

The dominant process of primary production in the oceans is photosynthesis (**CC14**). The basic raw materials needed for photosynthesis are carbon dioxide, water, and nutrients. An ample supply of carbon dioxide is dissolved in seawater in the form of carbonate and bicarbonate ions (Chap. 6). Of course, water is readily available. The source of energy used to combine carbon atoms is solar radiation. Therefore, light availability is an important determinant of where photosynthesis can occur and at what rate.

Nutrient elements, such as nitrogen, phosphorus, magnesium, and sulfur, are also necessary for photosynthesis. For example, nitrogen atoms are part of all protein molecules and of the molecules of **chlorophyll**, which is essential to the photosynthetic process (**CC14**). The following sections explain how the distribution of light and nutrients and the processes that affect their distribution control the distribution of life in most of the oceans.

Chemosynthesis

Some types of bacteria and archaea can convert carbon dioxide and water to organic matter by using energy from chemical reactions rather than from light. This process, called "chemosynthesis," uses energy obtained from the oxidation of hydrogen sulfide to sulfate, of metals from a reduced to an oxidized form, of hydrogen to water, or of methane to carbon dioxide and water.

Hydrogen sulfide, metal sulfides, hydrogen, and methane are all oxidized fairly rapidly by chemical processes in environments where oxygen is present. Therefore, the reduced compounds that fuel chemosynthesis are not present in the atmosphere or in most ocean waters, because these environments contain appreciable concentrations of oxygen. In reducing environments where chemosynthetic fuels are abundant, free oxygen is absent because it has been consumed by respiration or chemical reactions. Consequently, in the oceans chemosynthesis occurs primarily in limited environments where both reducing and oxygenated waters meet and mix. However, there are also transitional environments in which oxygen is depleted, but some chemosynthetic organisms can use the oxygen from nitrate, nitrite, and sulfate until these sources are also depleted.

Chemosynthesis is most common near **hydrothermal vents** on **oceanic ridges** and volcanoes, but it also occurs in water that is released to the oceans by vents or in seeps from ocean **sediment**. Some vents or seeps are in **subduction zones**, where water with dissolved methane and hydrogen sulfide is squeezed out of **subducted** ocean sediment. Others are in certain areas of the **continental shelf** where **groundwater** from the continents migrates through the sediment and accumulates methane or hydrogen sulfide. Chemosynthesis also occurs in the surface sediments of **salt marshes** and swamps, at the interface between oxygenated surface waters and oxygen-deficient bottom waters in **fjords**, and in other ocean areas where oxygen is depleted. Recent studies have found living microorganisms in many extreme environments, including deep within ocean sediments and even within oceanic **crust** that is millions of years old. So little is known about these organisms at the moment that it is not clear how they live. However, many may be chemosynthetic, obtaining their energy needs by chemical reactions such as the combination of hydrogen and carbon dioxide to form methane.

Secondary Production and Decomposers

All nonautotrophic marine organisms are heterotrophic and must obtain food to supply all or part of their needs for organic matter and energy. Heterotrophs convert organic compounds to carbon dioxide and water during respiration, thereby releasing energy that was originally captured from the sun by photosynthesis or from reduced chemicals by chemosynthesis. The released energy fuels the metabolic processes of the heterotrophs.

Heterotrophs include all animals and most species of bacteria and fungi. Animals eat plants, algae, other animals, and organic **detritus** formed by the partial decomposition of dead organisms. Animals that are **herbivores** eat only plants, **carnivores** eat only animals, **omnivores** eat both plants and animals, and **detritivores** eat detritus. The production of animal **biomass** by plant eaters from their food is called "secondary production."

Heterotrophs use food inefficiently (**CC15**). They **excrete** part of their food as solid waste organic particles, called **fecal pellets**, or as dissolved organic matter in liquid excretions. Humans are no exception to this rule.

Many species of microorganisms, including some bacteria, archaea, and fungi, are heterotrophs but are also called **decomposers**. Decomposers obtain energy from organic particles or dissolved organic compounds, which they convert to carbon dioxide and water. Recent studies suggest that some species of microorganisms may be capable of growing either autotrophically or heterotrophically, depending on their environment.

Photosynthesizers and chemosynthesizers, as well as food-eating and decomposing organisms, function together as a **community** within an ecosystem (**Fig. 14.2**). In an ecosystem, carbon dioxide, water, and nutrients are synthesized to organic matter, which is then processed through herbivores, carnivores, omnivores, detritivores, and decomposers. Each of the organisms in this **food chain** breaks down some organic matter and releases carbon dioxide, water, and nutrient elements back to solution. In addition to this well-known food chain, there is also a less well-known microbial loop. As much as 50% of the organic matter produced by primary producers is released to solution because of inefficiency by the primary producer, loss during sloppy feeding by **zooplankton**, or destruction of the cells by **viruses**. Much of this organic matter is rapidly taken up by bacteria and archaea, then possibly consumed by **protozoa** and by other organisms leading to the higher levels of the food chain.

Nutrients and other elements are recycled through ecosystems (**Fig. 14.2**). Recycling is an important factor in the distribution and abundance of ocean life. In parts of the oceans, life is limited because nutrients are not recycled fast enough to support photosynthesis.

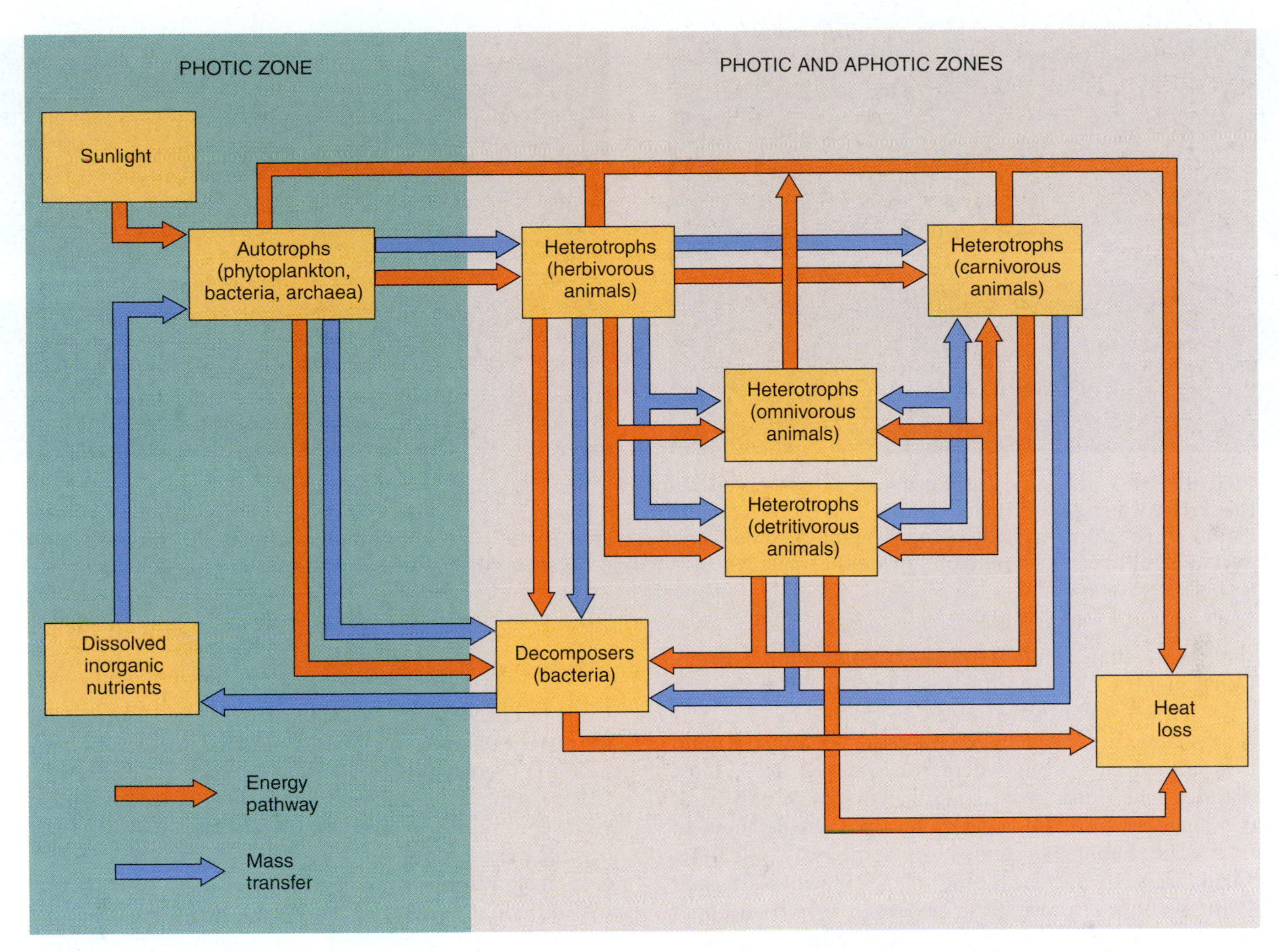

FIGURE 14.2 Simplified model of the energy and mass (organic matter and nutrient element) cycles in a marine ecosystem. Sunlight and dissolved nutrients are essential for autotrophs to perform the photosynthetic primary production of organic matter from carbon dioxide. Energy collected and mass produced by autotrophs is cycled through various heterotrophs, followed by decomposers. The mass is transferred back to solution by decomposers, and the energy is eventually released as heat. Primary production of organic matter takes place only in the photic zone (except for limited areas where chemosynthesis occurs). All other steps in the ecosystem take place partly in the photic zone and partly in the aphotic zone.

14.3 PRIMARY PRODUCTION AND LIGHT

Seawater absorbs sunlight, so no light at all reaches ocean depths greater than a few hundred meters, even in the clearest open-ocean waters (Chap. 7). In coastal waters with high concentrations of **suspended sediments**, **turbidity** is high and light is more effectively absorbed and **scattered**. Turbidity is generally highest in shallow coastal waters where bottom sediments are **resuspended** by waves and in estuaries fed by rivers with high suspended sediment loads. In such waters, light penetrates a few meters at most, and only centimeters in extreme cases.

Photosynthesis cannot take place without light. Light is absorbed by the water, and its intensity is reduced with depth. Light of sufficient intensity to support photosynthesis, usually thought to be about 1% of the intensity at the surface, rarely penetrates more than about 100 m, and photosynthesis is restricted to the water column above this depth. This upper layer is called the **photic zone**.

Terrestrial plants must obtain nutrients from soil or, in some cases, from raindrops. In contrast, marine organisms can extract nutrients from the surrounding seawater. As a result, they do not have to be attached or rooted to the seafloor. Furthermore, attached marine plants or algae can receive enough light for photosynthesis only if they are attached to the limited areas of seafloor that are

FIGURE 14.3 Many species of macroalgae have gas enclosures within their structure that enable them to float and stay in the surface waters where there is ample light for photosynthesis. (a) The round ball-like structures in this *Sargassum yendoi* (South Korea Sea) are gas-filled. (b) Masses of kelp (*Macrocystis* sp.) grow in this cove near Monterey, California. The plants are anchored to the seafloor, but you can see the upper ends of the kelp fronds covering much of the water surface, held up by their gas enclosures.

shallower than the depth of the photic zone. Consequently, most marine algae live in the water column and are not attached to the seafloor.

Because they need light and therefore must remain near the surface, even where the water is deep, most marine algae must avoid sinking. Pockets of gas or air contained within some species of algae enable them to float. The bubblelike lumps on some seaweeds (**Fig. 14.3a**) are such pockets. Some species that live attached to the seafloor also use these devices to keep their upper fronds in the light, including the giant **kelp** in California coastal waters (**Fig. 14.3b**).

Phytoplankton are among the major primary producers in the oceans and are abundant in productive ocean waters. Most phytoplankton species are microscopic and drift freely in the photic zone. For reasons discussed later in this chapter, the microscopic size of most phytoplankton species also aids them in taking up dissolved substances. Another major group of primary producers in the oceans, which includes certain species of bacteria and the related archaea, are even smaller single-celled organisms.

Phytoplankton can remain in the photic zone because of the relatively high **viscosity** of water (Chap. 7) and consequent very slow sinking rate of small particles (**CC4**). Most phytoplankton are small and sink very slowly, even if their **density** is higher than that of seawater. In addition, **turbulence** due to waves and currents counteracts the tendency of small phytoplankton to sink (**CC4**). Some species of **diatoms**, which are among the largest phytoplankton, store oils that increase their **buoyancy**, and many also have spines or aggregate in chains, which also reduces their tendency to sink. **Dinoflagellates**, another group of phytoplankton species, are able to counteract sinking because they have whiplike **flagella** that provide them with limited motility.

Photosynthesis, Light, and Depth

Until recently, it was thought that almost all ocean life ultimately depends on photosynthetic primary production for food. Other food sources, such as chemosynthetic primary production (especially in hydrothermal vent communities) and the dissolved or particulate organic matter supplied by rivers, were previously considered insignificant in comparison. Recent research in hydrothermal vent areas and studies in estuarine and coastal waters suggest these sources may be much more important than previously thought. Because the relative importance of these other sources is not yet well known, we will focus almost exclusively on photosynthetic primary production in this chapter.

Because photosynthesis requires light energy, most of the primary production in the oceans takes place in the upper layer of the water column, down to the depth that sunlight penetrates. Light intensity at any given depth is determined by the intensity of sunlight reaching the sea surface, the **angle of incidence** of the sun's light to the sea surface, and the water turbidity. The intensity of sunlight reaching the sea surface varies with the seasons, throughout the day as the angle of incidence of the sun changes, and as cloud cover varies. Light intensity at a given depth changes continuously. Fortunately, at any given location and depth, the variation of daily mean light intensity within a season is relatively small in comparison with the variation between seasons. Hence, the daily average intensity is a reasonable measure of the amount of light available for photosynthesis at any location and depth.

Light intensity decreases rapidly with depth (**Fig. 7.10**). In the clearest ocean waters, light intensity is reduced to 1% of the surface intensity at a depth of about

100 m. In coastal-ocean waters, the corresponding depth is variable, but it is often about 5 to 10 m. The rate at which photosynthesis can occur is reduced as light intensity decreases (**Fig. 14.4**). In contrast, the rate of respiration remains virtually constant at all depths because the energy needed to fuel the life processes of a plant cell does not change significantly. At relatively shallow depths, the rate of photosynthesis exceeds the rate of respiration, and photosynthetic autotrophs can grow larger and reproduce. At depths where the respiration rate exceeds the photosynthetic rate, autotrophs must metabolize more organic matter than they create by photosynthesis in order to survive. At least some species of photosynthetic autotrophs can survive for a limited time at these depths, but they cannot grow or reproduce. Although many species die if they do not return to conditions in which photosynthesis exceeds respiration, some species are capable of entering a resting phase, much as terrestrial plant seeds do, and can survive prolonged periods below the photic zone.

The depth at which photosynthesis equals respiration is called the **compensation depth**. The water column between the ocean surface and the compensation depth is the photic zone, and depths below this region constitute the **aphotic zone**. The compensation depth identifies the depth range within which phytoplankton and other photosynthesizers can produce more organic carbon than they consume. The compensation depth lies approximately where light intensity is reduced to 1% of that at the surface. Hence, the photic zone is rarely deeper than 100 m and is much shallower in turbid coastal waters (**Fig. 7.10**). Except in tropical regions, where the day length does not vary much, the compensation depth generally increases during local summer because the angle of incidence of the sun is smaller and the days are longer. However, in some regions where phytoplankton reproduce more rapidly, the additional phytoplankton cells and increased numbers of zooplankton that eat them contribute additional "particles" that absorb and scatter light. Consequently, in these regions the **plankton** reduce light penetration and the compensation depth, offsetting the effect of greater light intensity in summer.

Primary Production and the Ozone Hole

Although the rate of photosynthesis usually decreases with increasing depth (**Fig. 14.4**), in many areas it is often lower in water immediately below the surface than at a depth of several meters. One reason for the lower photosynthetic rate near the surface is that high light intensity, especially ultraviolet light, appears to interfere with photosynthesis. Because ultraviolet light is absorbed rapidly in the first few meters of seawater (Chap. 7) the maximum photosynthesis often occurs slightly below the surface.

Chapter 9 discussed depletion of the Earth's **ozone layer** by synthetic chemicals that may cause an increase in ultraviolet light intensity at the Earth's surface. Higher ultraviolet light intensity may significantly increase near-surface inhibition of photosynthesis, reduce the total food supply, and adversely affect ocean ecosystems.

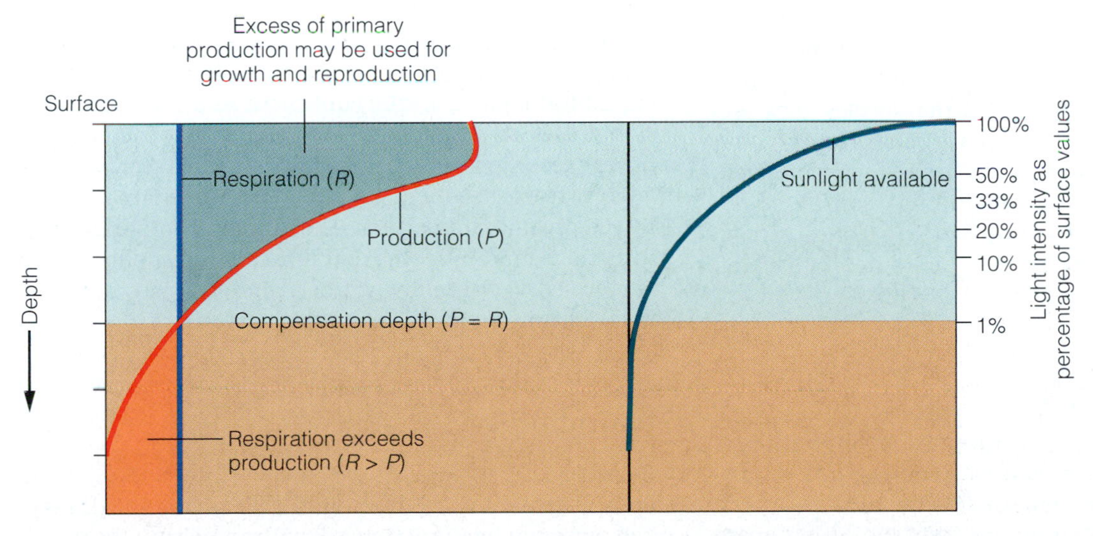

FIGURE 14.4 Photosynthesis depends on the availability of light energy. Therefore, the rate of primary production by phytoplankton decreases with decreasing light intensity and, thus, with depth. The rate of metabolic energy use and, thus, respiration by phytoplankton remains almost constant regardless of depth. The depth at which the rate of phytoplankton primary production equals the rate of phytoplankton respiration is called the "compensation depth." Below this depth, phytoplankton use more organic matter for respiration than they can produce by photosynthesis. The water column above the compensation depth is the photic zone, and everything below this depth is in the aphotic zone.

14.4 PRIMARY PRODUCTION AND NUTRIENTS

Photosynthetic organisms require many elements for growth. Carbon, hydrogen, and oxygen are obtained from dissolved carbon dioxide and water. Because carbon dioxide is abundant in seawater (Chap. 6), carbon is plentiful in the ocean. Several other elements are required, each in different amounts. Some, such as magnesium and sulfur, are present in high concentrations in seawater. As with carbon dioxide and water, they are always available in plentiful amounts. Other elements, such as nitrogen, phosphorus, iron, zinc, cobalt, and, for some species, silicon, are needed sometimes in substantial quantities but are present in low concentrations in seawater. In some areas, uptake by organisms or by processes such as **adsorption** on **lithogenous** particles can remove enough of these elements from solution that the remaining concentrations are too low to meet the needs of some species of autotrophs. Growth of phytoplankton, marine algae, and other species can be limited by the unavailability of a single essential element. In this condition, the system is said to be **nutrient-limited.**

Where growth is slowed or prevented because the concentration of a particular element is too low, that element is called a **limiting nutrient**. The most important limiting nutrients are nitrogen as dissolved inorganic ions nitrate (NO_3^-), nitrite (NO_2^-), and ammonia (ammonium ion [NH_4^+] and NH_3); phosphorus (phosphate ion, PO_4^{3-}); iron; and silica (silicate, H_4SiO_4). When inorganic nitrogen and phosphorus ions are at very low concentrations, phytoplankton will often use organic nitrogen and/or phosphorus compounds. Although other nutrient elements, such as cobalt and zinc, are present only as trace elements, biological requirements for trace elements are so small that these elements generally do not limit photosynthetic growth. Some marine species that are otherwise autotrophs may need certain organic compounds that they are unable to synthesize for themselves. In certain circumstances, including some newly **upwelled** water in **coastal upwelling** zones (Chap. 15), **primary productivity** may be limited by lack of such organic nutrients.

Nutrient Uptake

Phytoplankton and other autotrophs obtain nutrients from seawater across their porous outer membrane by **diffusion**. Diffusion is a random-motion process, so at lower concentrations, the nutrient ions are farther apart and the probability that an ion can diffuse across the cell outer membrane is lower. More nutrient ions will be in contact with a cell that has a larger surface area. Hence, autotrophs improve their chances of capturing nutrient ions if they have a large surface area.

Because phytoplankton and other autotrophs are composed of similar organic compounds, their nutrient needs are approximately proportional to their volume. The ratio of surface area to volume generally is greater for smaller than for larger organisms. We can visualize why if we consider an apple cut into two halves. The apple's total volume does not change when it is cut, but the total surface area is increased by the area of the newly exposed interior surface of each apple half. If the apple is cut again, the total volume still remains the same, but the total surface area again increases. Thus, phytoplankton compete more effectively for nutrients at low concentrations if they are small (their surface area is large in relation to their volume). This is one reason that most marine autotrophs are small, particularly where nutrient concentrations are low.

Small size is not the only adaptation that allows phytoplankton to compete for a limited supply of nutrients. When phytoplankton grow in seawater with relatively high nutrient concentrations, they take up more nutrients than are immediately needed. The excess nutrients are stored and can be used as a reserve to support growth and reproduction after nutrients have been depleted to concentrations that would otherwise limit growth. If concentrations remain growth-limiting for more than a few days, even the stored nutrients will be used up and primary productivity will slow. However, photosynthesis is not completely halted, because some nutrients are continuously made available by recycling (**Fig. 14.2**).

Nutrient Recycling

Nutrients are recycled to solution when organic matter is decomposed during animal and decomposer respiration. Animals excrete nutrients in fecal material and in dissolved form, either in the equivalent of urine or by diffusion losses through external membranes. Bacteria and other decomposers release nutrients as they extract energy from dead organisms, fecal material, and dissolved organic compounds. The four most important limiting nutrients are recycled at different rates: phosphorus is recycled very rapidly, nitrogen more slowly, and iron and silica more slowly yet.

Phosphorus

Phosphorus recycling is relatively simple (**Fig. 14.5**). Phosphate is returned to solution rapidly after the death of any organism by the action of **enzymes** within the organisms that break phosphorus away from organic molecules with which it is combined. Phosphorus is converted directly to phosphate and released to solution. Enzyme-mediated release of phosphate also occurs within living zooplankton and other animals as they digest their food.

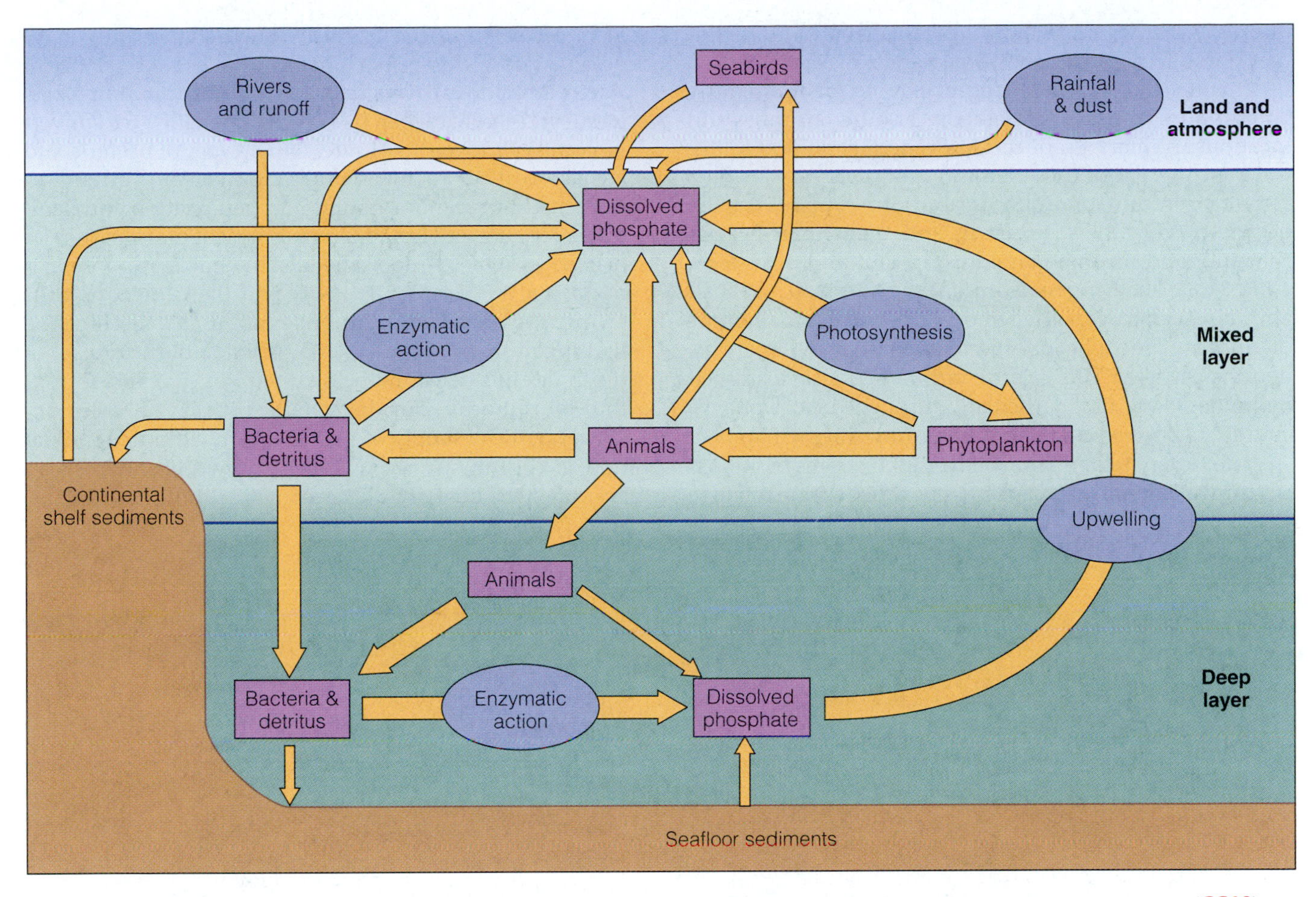

FIGURE 14.5 Simplified representation of the phosphorus cycle in the oceans. The boxes represent compartments (**CC10**) or components of the system among which the phosphorus is distributed. The ovals represent processes that transport or convert the phosphorus between these compartments. Thicker arrows show more important pathways. Phosphorus is recycled rapidly between solution and organisms in the photic zone. Thus, although phosphorus is progressively transported below the thermocline, the low concentrations present in some ocean surface waters are still usually sufficient to support phytoplankton growth.

Certain zooplankton species are known to excrete as much as 60% of the phosphorus taken in with their food. More than half of what remains is disposed in fecal pellets, from which it is rapidly released by continued enzyme activity as the fecal material is attacked by decomposers.

Because it is recycled rapidly, phosphorus is usually not a limiting nutrient in the oceans. In contrast, phosphorus is often the limiting nutrient in lakes. As a result, phosphate-free detergents are required by law in many areas, primarily to prevent **eutrophication** in lakes and rivers.

Nitrogen

Nitrogen, which occurs in living tissue mainly in **amino acids,** is released during the decomposition of organic matter, primarily as the ammonium ion (**Fig. 14.6**). Animals also release nitrogen in their liquid excretions as **soluble** organic compounds, such as urea and uric acid, which are broken down by decomposers to the ammonium ion. Decomposition of particulate organic matter to release ammonium is much slower than processes that release phosphorus. Consequently, nitrogen often becomes the limiting nutrient in the marine environment.

The ammonium ion is used as a source of energy by specialized bacteria that oxidize ammonium to nitrite. In turn, nitrite is oxidized to nitrate by yet another group of bacteria. Conversions from ammonium to nitrite and nitrate take place relatively slowly. Nevertheless, in waters below the photic zone this process can be completed, and nitrate is the principal form of dissolved nitrogen other than molecular nitrogen. Conversion to nitrate may also be completed in temperate **latitudes** during winter, when reduced light availability limits primary productivity. Such seasonal variations are discussed in Chapter 15. Because nitrogen may be predominantly ammonium, nitrite, or nitrate ions at different times and locations, primary producers are generally able to utilize nitrogen in each of these forms.

Nitrogen and phosphorus are both supplied to the oceans by river flow, **erosion,** and **weathering** from the land, and, in small amounts, in rain and dust (**Figs. 14.5, 14.6**). The **biogeochemical cycle** (Chap. 6) is balanced in each case by the removal of equivalent amounts of nitrogen and phosphorus to the ocean sediment in detritus particles. This balance may be seriously disturbed by human activities, particularly through the use of agricultural fertilizers and the discharge of sewage (Chap. 19).

The supply to and removal from the oceans of biologically available nitrogen are complicated by the exchange of molecular nitrogen gas between the atmosphere and ocean waters. Molecular nitrogen dissolved in ocean waters is biologically available only to nitrogen-fixing bacteria (and probably archaea) that are able to convert molecular nitrogen to nitrate. Other types of bacteria and archaea, called "denitrifying," can convert nitrate to molecular nitrogen. Although nitrogen fixation and denitrification are very limited, nitrogen fixation in particular may be extremely important in some nutrient-poor environments, such as **coral reefs** and the centers of **subtropical gyres.** Phosphorus does not usually become the limiting nutrient because a small fraction of the phosphorus in detritus does sink below the photic zone before all the phosphorus can be released back to solution. However, if the supply of nitrogen through nitrogen fixation is continuous, even if very slow, phosphorus can eventually be depleted and become the limiting nutrient.

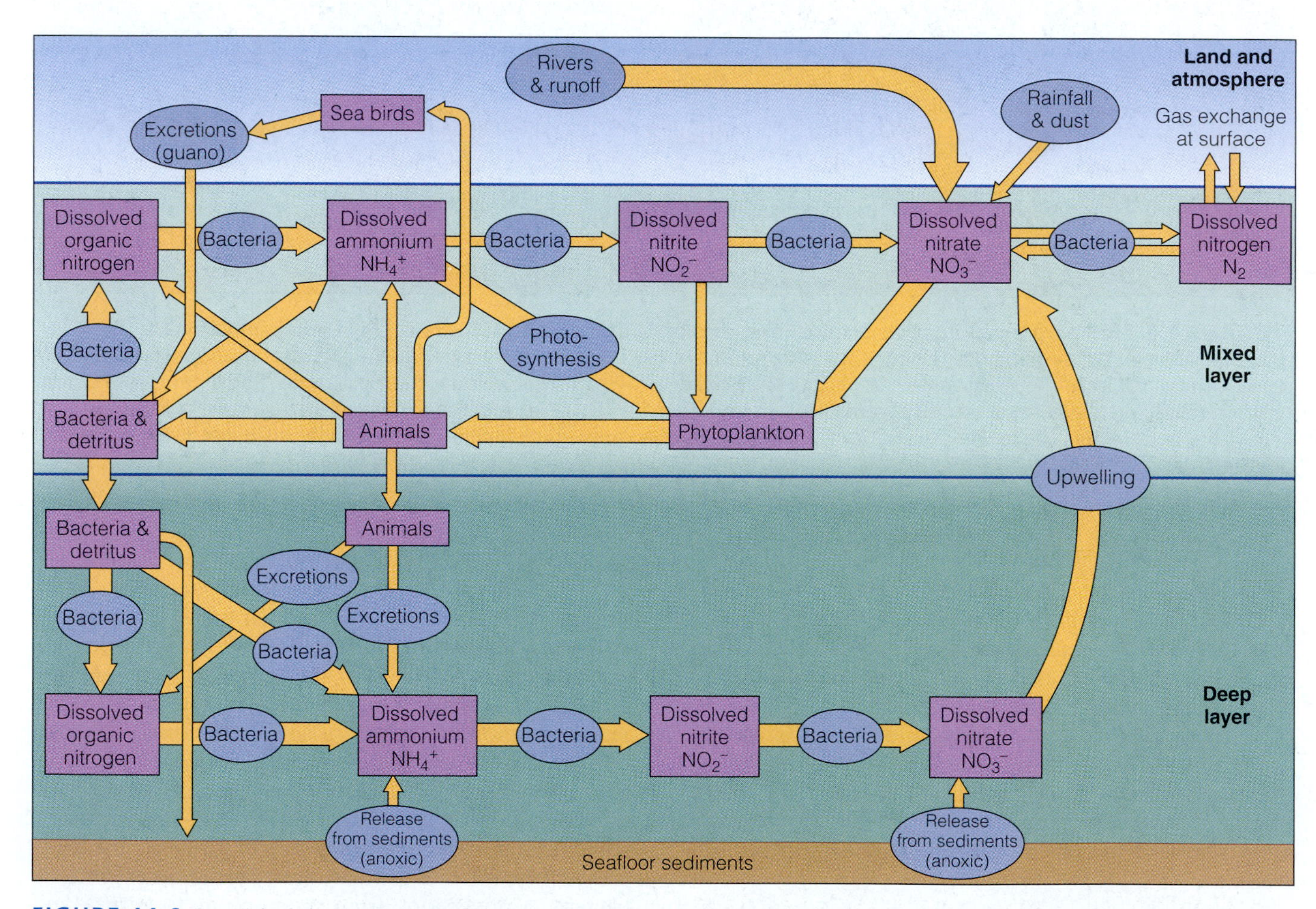

FIGURE 14.6 Simplified representation of the nitrogen cycle in the oceans. The boxes represent compartments (**CC10**) or components of the system among which the nitrogen is distributed. The ovals represent processes that transport the nitrogen or convert its chemical form between these compartments. Thicker arrows show more important pathways. Unlike phosphorus, nitrogen is not recycled rapidly in the photic zone, and it tends to be transported below the thermocline. Thus, nitrogen is often the limiting nutrient. Nitrogen can be recycled from shallow seafloor sediments (as depicted for phosphorus in Figure 14.5), but these pathways have been omitted in this figure to reduce its complexity.

Silica

Silicon is essential for the construction of **hard parts** (the shells or skeletons) of a variety of marine organisms, of which diatoms (**Fig. 8.5**) are the most important. Diatoms are larger than many other types of phytoplankton and are covered by an external silica **frustule.**

Because silica is not utilized or released to solution by decomposers, silica hard parts must dissolve chemically in seawater before silicate ions are again biologically available. Silica dissolves very slowly (**Fig. 8.9**). Consequently, once silicate is depleted in the photic zone, silicate limitation inhibits the growth and reproduction of organisms with silica hard parts. Because phytoplankton that do not have silica hard parts are not affected, silicate depletion can lead to a change in the dominant species, but it does not limit primary productivity. Silicate is supplied to the oceans in large quantities by rivers and is slowly dissolved from ocean sediment. Therefore, silicate limitation is generally only a temporary situation that develops in locations where diatoms undergo an explosive population growth called a **bloom.**

Iron

Iron is essential to certain reactions within the process of photosynthesis. Iron is supplied to the surface waters of the oceans from the land, primarily in **runoff,** but also through the transport and **deposition** of airborne dust. Iron is also continuously transported below the photic zone in detritus. As a result, iron can become the limiting nutrient in areas of the oceans where inputs from the continents are too small to replace the iron that is lost from the surface layers by sinking of detritus. Iron is now thought to be the nutrient that limits primary production in large areas of the open oceans where river and dust inputs of iron are low. These areas include the sub-Arctic Pacific Ocean, large areas of the equatorial Pacific Ocean, and much of the Southern Ocean.

Nutrient Transport and Supply

Substantial quantities of nutrients are supplied to the oceans by rivers, and smaller quantities are supplied in rainfall. In the open ocean remote from river influences, nutrients in the photic zone are depleted rapidly by growing phytoplankton and other autotrophs. Therefore, if primary production is to continue, nutrients must be resupplied. Some nutrients are continuously recycled and resupplied directly to the photic zone by decomposition. However, there must also be nutrient resupply by upwelling from below the photic zone because a proportion of the nutrients taken up by primary producers is released back to solution in waters below this zone.

Nutrients are transported from the photic zone to the aphotic zone by two major mechanisms. Almost all phytoplankton are eaten by **grazers** (herbivores or omnivores). Although these grazing animals are mainly small zooplankton, most are much larger than phytoplankton. Much of these animals' undigested waste material is excreted as fecal pellets. Fecal pellets from even the smallest zooplankton are packages of partially digested remains of numerous phytoplankton cells. Consequently, fecal pellets are larger than phytoplankton and sink much faster (**CC4**). As they sink, fecal pellets provide food for other animals and are progressively decomposed by bacteria and fungi. Much of the decomposition and release of nutrients occurs below the photic zone, including on the ocean floor.

The second major mechanism of nutrient removal from the photic zone is the vertical migration of zooplankton and other animals. Phytoplankton generally remain in the photic zone unless carried out of it by turbulence or **downwelling.** However, many species of zooplankton and other animals migrate vertically between the photic and aphotic zones each day, rising into the photic zone at night to feed, and then sinking or swimming down to waters below the photic zone during the day. This **diurnal** migration is probably a defensive mechanism that prevents many potential predators from feeding on the zooplankton when it is light enough to see them easily. Zooplankton and other animals that practice diurnal migration continue to digest food and excrete fecal pellets and liquid wastes during the day when they are below the photic zone. Consequently, much of the organic matter that they consume is transported below the photic zone, where it is released, with its nutrients, to the decomposer community.

Vertical Distribution of Nutrients

The photic zone is usually less than 100 m deep in the open oceans and normally restricted to water above the permanent **thermocline** (Chap. 10). Phytoplankton and nutrients are distributed throughout the **mixed layer** above the thermocline by turbulent motions of wind and wave mixing (Chap. 10). Where light is sufficient to support primary production in the open oceans away from terrestrial sources of nutrients, nutrients are continuously transported below the thermocline by the mechanisms already described. Below the thermocline, these nutrients are released by decomposers, but the water into which they are released does not mix with the mixed-layer or photic-zone water above, because vertical mixing is inhibited by the density difference across the thermocline. Consequently, the mixed layer becomes nutrient-depleted, while deeper water becomes more nutrient-rich (**Fig. 14.7**).

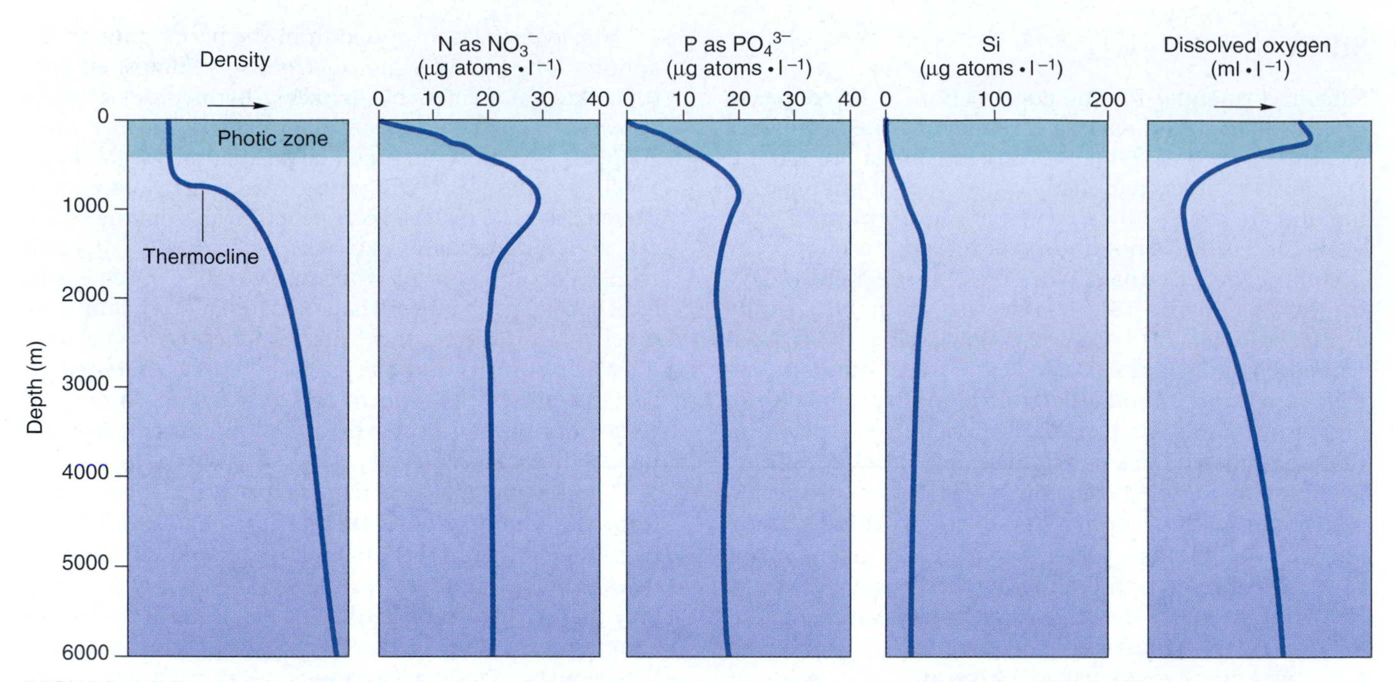

FIGURE 14.7 General representation of the vertical profiles of water density, nitrate, phosphate, silicate, and dissolved oxygen concentrations in the center of the Atlantic Ocean basin. Dissolved oxygen concentration is high above the thermocline, reaches a minimum just under the thermocline, and then increases to somewhat higher levels in the deep water masses formed by cooling of surface water at high latitudes. Nutrients (nitrate, phosphate, silicate) are generally depleted in the photic zone, increase rapidly in concentration until they reach a maximum at about the same depth as the dissolved oxygen minimum, then remain at relatively constant high levels in the deep water masses.

Organic matter is decomposed continuously at all depths in the oceans, although the rate of decomposition varies with depth. Hence, water that leaves the surface to form deep **water masses** progressively accumulates recycled dissolved nutrients. Nutrient concentrations are usually highest in water immediately below the thermocline layer. The primary reason for this nutrient maximum is that this is the "oldest" deep water because it is formed by slow upward mixing of bottom waters that were formed near the poles (Chap. 10). The effect of water-mass "age" in determining nutrient concentrations is illustrated by the nutrient concentration differences between the deep waters of the Atlantic Ocean and the much older deep waters of the Pacific Ocean (**Fig. 14.8**).

Several additional factors cause concentrations of nutrients to be higher in waters just below the bottom of the thermocline than in bottom waters. First, bacterial decomposition processes are slowed at the low temperatures and high pressures near the deep-ocean floor. Second, the most easily oxidizable organic matter is decomposed long before it reaches the ocean floor. Third, animals that migrate vertically generally do not descend far below the permanent thermocline. However, these influences on the vertical distribution of nutrients, even taken together, are less important than the age of the water and the length of time during which recycled dissolved nutrients have accumulated in it.

The permanent thermocline is a persistent and widely distributed feature. Primary productivity above the thermocline is nutrient-limited except in areas where nutrients are transported into the mixed-layer waters by inputs from the continents or by upwelling of nutrient-rich water from below the thermocline. Major zones of high primary productivity and highly productive fisheries are found in coastal upwelling regions, off the mouths of rivers (which discharge nutrients), and in shallow areas where nutrients in sediments are returned to solution by decomposition and then released directly into the photic zone.

14.5 FOOD WEBS

Heterotrophic species from the smallest microbe to the largest whale depend on autotrophs to produce organic matter for food. Herbivores eat primary producers directly, whereas carnivores eat herbivores and other carnivores.

In the simplified food chain in **Figure 14.9**, photosynthetic organisms are eaten by herbivorous zooplank-

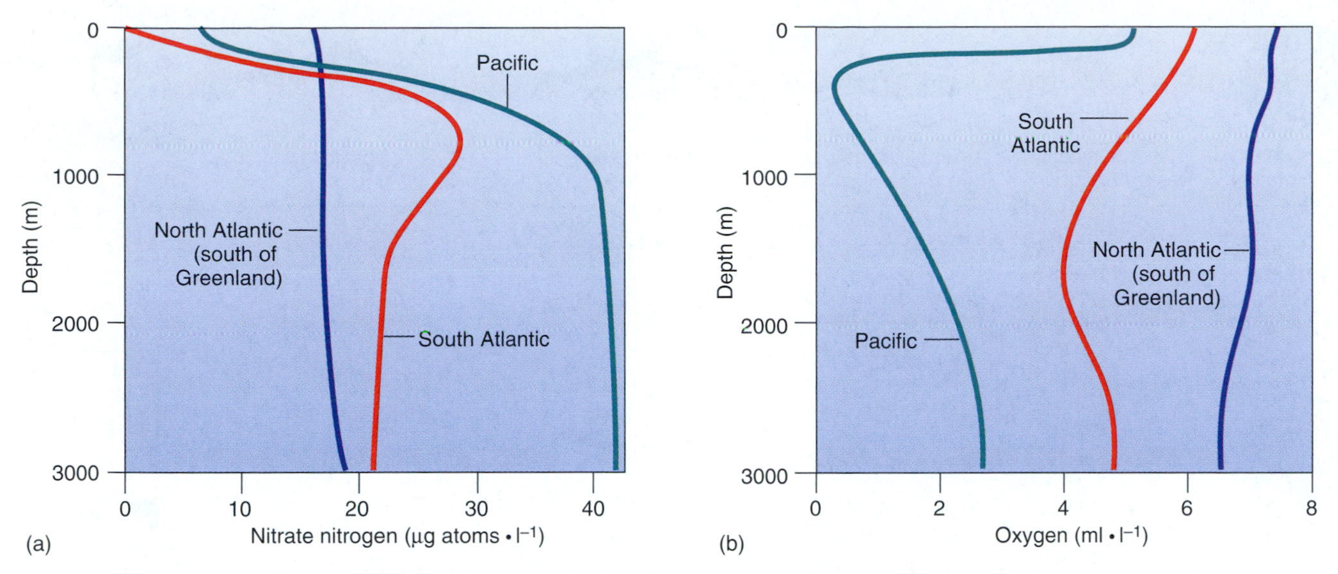

FIGURE 14.8 The vertical profiles of nitrate nitrogen and dissolved oxygen in the North Atlantic (south of Greenland), South Atlantic, and North Pacific Oceans are very different. In the North Atlantic, where cooled surface water is sinking to form North Atlantic Deep Water (NADW), the concentrations vary little with depth. During its passage southward to the South Atlantic, this water mass loses oxygen and gains dissolved nitrate as respiration by animals and decomposers uses oxygen and releases dissolved nitrate. This process continues as the water mass moves around Antarctica and northward into the deep water of the Pacific Ocean. Thus, the ratio of nitrate concentration to oxygen concentration is an indication of the "age" of the water mass since it left the surface layer. Higher ratios indicate "older" water.

ton, which are eaten by carnivores, which are eaten by larger carnivores, and so on. Each step in this food chain is a **trophic level**. As organisms at one trophic level are consumed by those at the next-higher trophic level, the ingested food biomass is not used entirely to create biomass of the consumer species at the higher trophic level. An average of only about 10% (varying from 1% to 40%) of food consumed at each level is used for growth. The remaining 90% is used during respiration to provide the consumer with energy or is excreted as waste. Clearly, the transfer of food energy in food chains is inefficient (**CC15**).

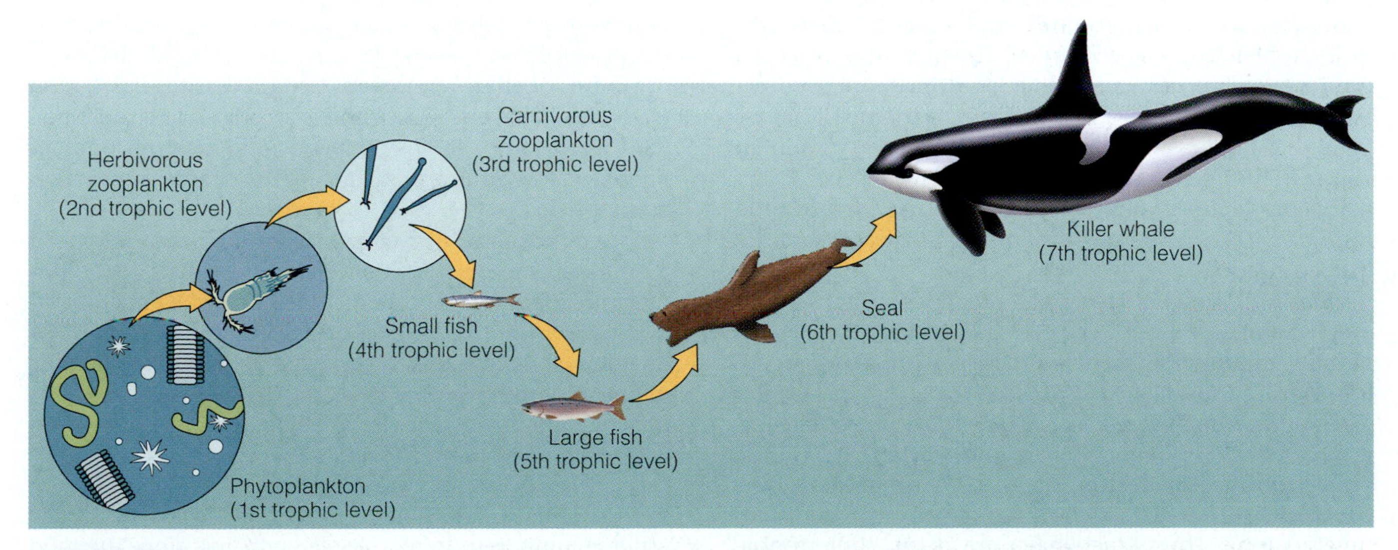

FIGURE 14.9 A simple food chain. Phytoplankton are eaten by herbivorous zooplankton, which are eaten by carnivorous zooplankton, which are eaten by small fishes, which are eaten by large fishes, which are eaten by seals, which are eaten by killer whales. Each step in this chain is one trophic level higher.

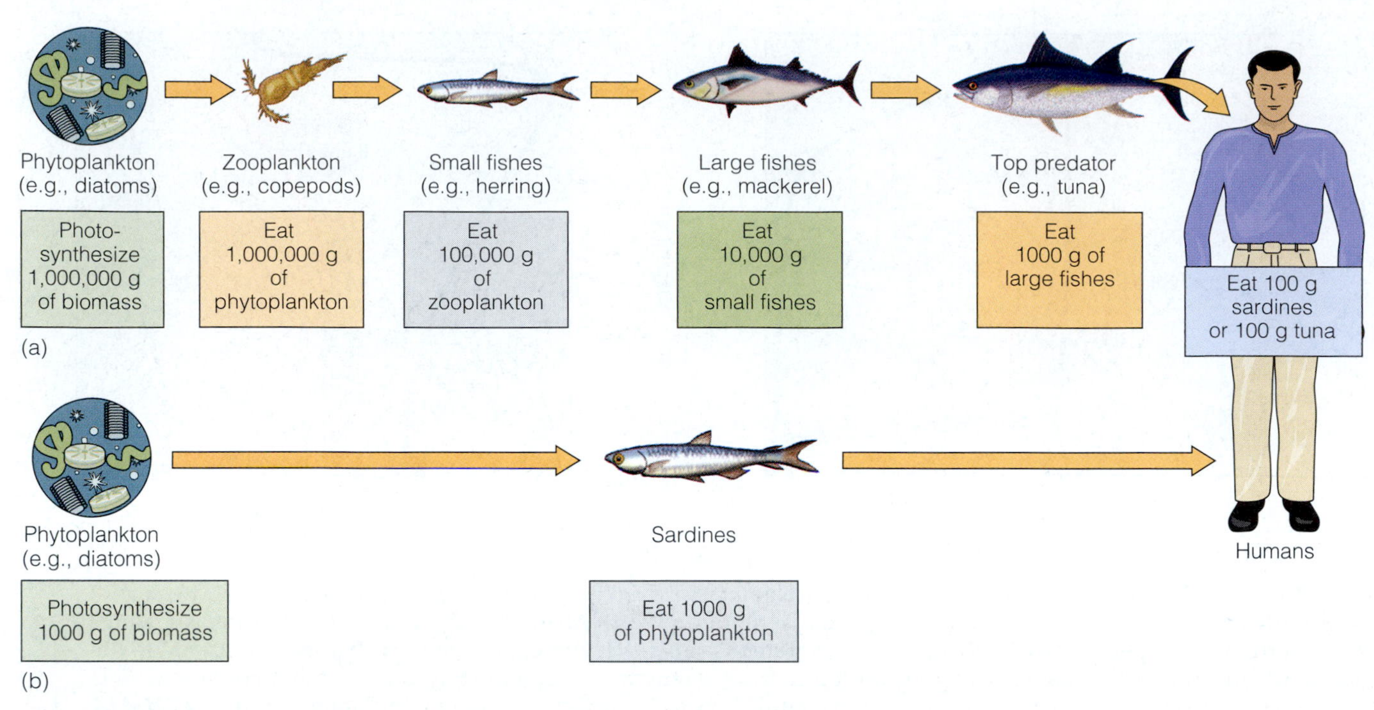

FIGURE 14.10 The food chains leading to (a) tuna and humans, and (b) sardines and humans, assuming that the trophic efficiency at each trophic level is 10%. Phytoplankton must photosynthesize 1 kg of organic matter to feed the human 100 g of sardines, but must photosynthesize 1000 kg (1 tonne) of organic matter to feed the human 100 g of tuna. However, because sardine species feed on both phytoplankton and zooplankton, their overall trophic efficiency is somewhat less than depicted. In addition, some tuna feed on both small and large fishes, so their overall trophic efficiency is somewhat better than depicted.

The consequences of food chain inefficiency are illustrated by the food chains in **Figure 14.10**. A good-sized tuna sandwich may contain about 100 g of tuna. For each 100 g of tuna produced, a tuna must eat about 1000 g of a large fish (e.g., mackerel). In turn, mackerel must eat 10 times their weight of smaller fishes (e.g., **herrings**), herrings must eat 10 times their weight of zooplankton (e.g., copepods), and copepods must eat 10 times their weight of phytoplankton (e.g., diatoms). The next time you eat a tuna sandwich, remember that phytoplankton synthesized about 1 million g (1 metric ton, or tonne) of phytoplankton biomass to produce your 100 g of tuna. In contrast, 100 g of sardines represent only 10 kg of primary production (**Fig. 14.10b**). We use ocean food resources 100 times more efficiently when we eat sardines instead of tuna.

Marine animals usually eat organisms that are not much smaller than themselves. Therefore, most large marine animals are high-trophic-level predators at the top of long food chains. **Baleen** whales are spectacular exceptions. The blue whale, the largest known ocean animal, can be 30 m long and weigh more than 100 tonnes. This magnificent creature eats only shrimplike **crustaceans** called **krill**, each of which is only a few centimeters long. Thus, blue whales are at the same trophic level as herrings (**Figs. 14.10a, 14.11**).

Food relationships in the ocean are far more complex than the simplified food chains depicted in **Figure 14.10** because they almost always include opportunistic carnivores that eat animals from several different trophic levels. The killer whale in the Antarctic marine ecosystem is a good example (**Fig. 14.11**). Complex food relationships, such as that shown in **Figure 14.11**, are called **food webs**. Food webs are further complicated by omnivorous species that will eat almost any plant or animal and sometimes even detritus. Detritus feeders eat particulate organic matter produced as waste products and/or dead tissues of organisms of all trophic levels. Certain detritus-based food webs are somewhat independent of phytoplankton-based food webs, although they depend ultimately on primary producers to synthesize organic matter that becomes their detrital food.

There are also, as yet poorly understood, microbial food loops. By some estimates, about one-half of the primary production in the oceans is performed by microscopically small photosynthetic eukaryotes and bacteria. The microbial loop also uses dissolved and particulate organic matter. Heterotrophic bacteria and archaea can absorb this material and utilize it to grow. These organisms may then be consumed by protozoa and by other organisms that are themselves consumed by organisms in the higher levels of the food chain. Free-living heterotrophic bacteria in the water column help to recycle elements lost from the food web back into the food web through the microbial loop, or by releasing them as they consume organic matter for energy and growth.

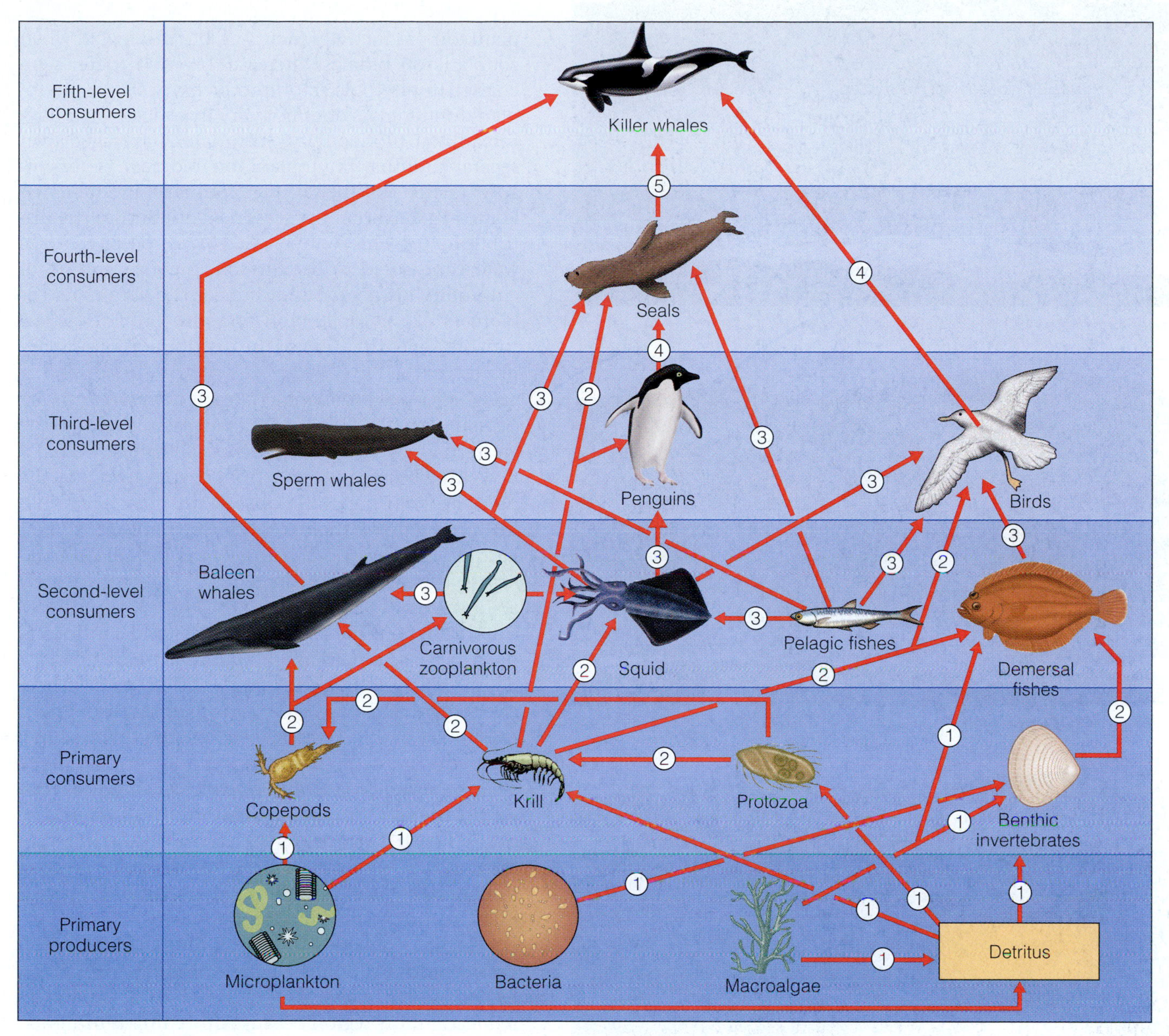

FIGURE 14.11 Simplified representation of the pelagic food web of the Southern Ocean. The number beside each arrow represents the trophic level at which the consuming organism is feeding. Note that many species feed at more than one level. For simplicity in this diagram, a number of pathways have been left out. For example, pelagic fishes feed on various prey including phytoplankton, zooplankton, and krill.

14.6 GEOGRAPHIC VARIATION IN PRIMARY PRODUCTION

All marine animals ultimately depend on photosynthetic or chemosynthetic primary producers, which include phytoplankton, some bacteria, and some archaea, for food. Consequently, zooplankton, fishes, and other **pelagic** animals are most abundant in areas of high primary productivity. Most **benthos** rely on the rain of detritus from above. Exceptions are some species that live in limited areas where the seafloor is shallower than the photic-zone depth and in hydrothermal vent and other chemosynthetic communities. The quantity of detritus depends on the abundance of organisms in the overlying water, so **benthic** communities are also more abundant in areas of high primary productivity.

Phytoplankton biomass (**standing stock**) is determined by phytoplankton growth and reproduction rate versus consumption rate. If the productivity is high and the rate of grazing of phytoplankton by zooplankton is low, phytoplankton biomass increases, and vice versa.

Zooplankton populations and biomass adjust to changes in food supply, but the changes lag days or weeks behind changes in phytoplankton biomass. As a result, phytoplankton biomass initially increases when phyto-

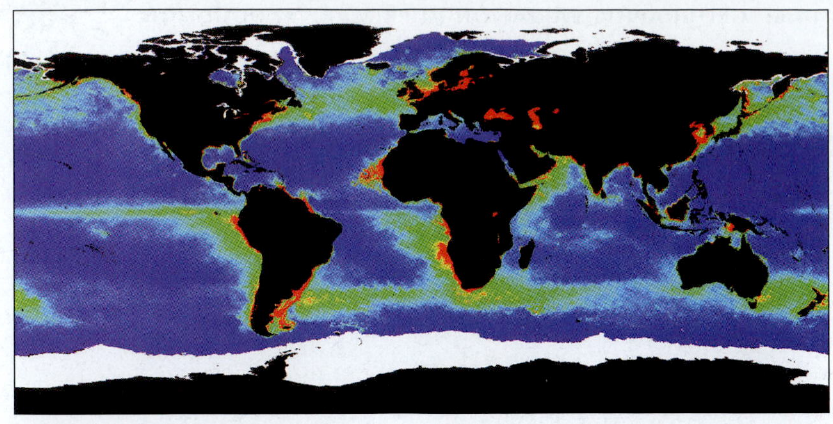

(a) September–November 1998

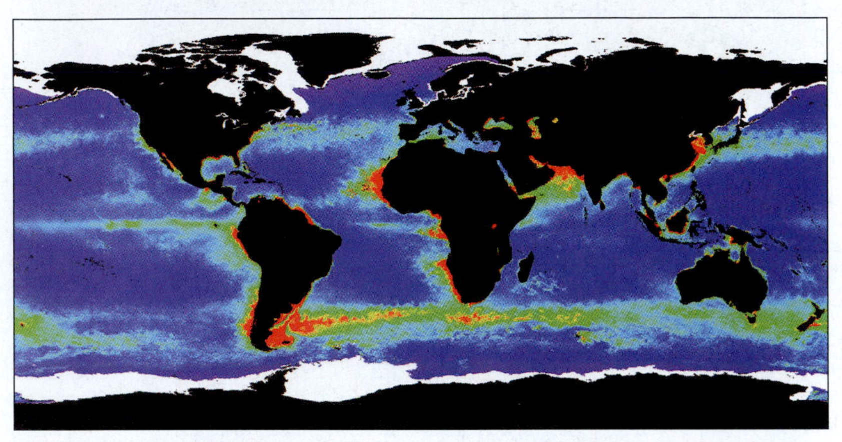

(b) December 1998–February 1999

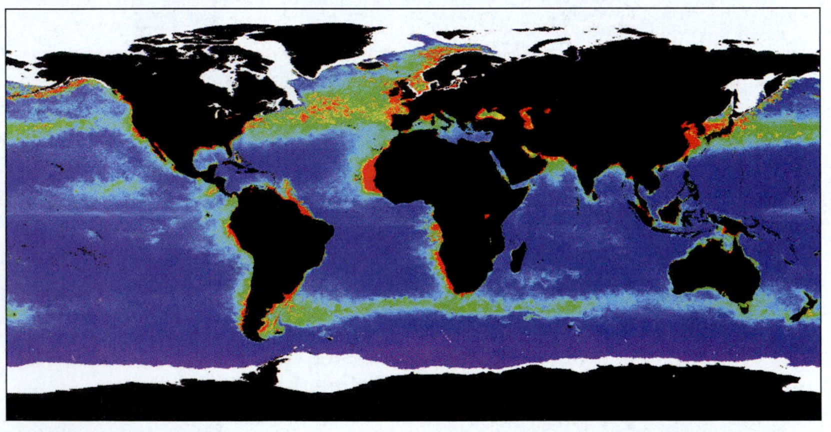

(c) March–May 1999

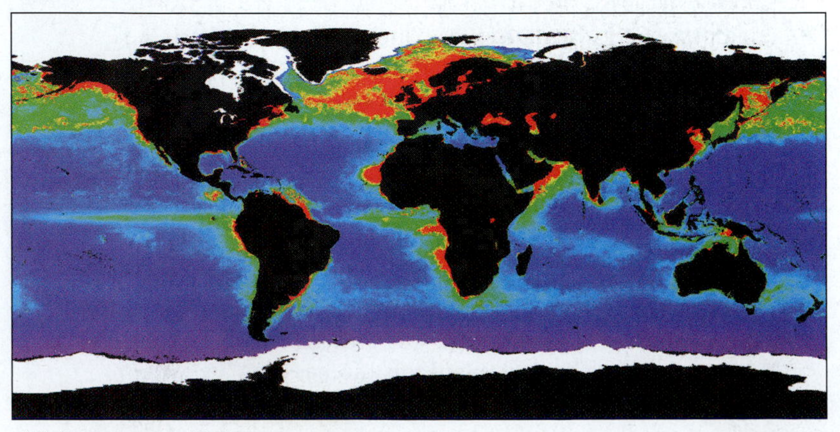

(d) June–August 1999

plankton productivity increases, but it may fall when zooplankton biomass increases, even if there is no change in phytoplankton productivity. Thus, phytoplankton biomass does not vary in concert with phytoplankton productivity. Fortunately, averaged over several months, phytoplankton biomass is reasonably related to the average primary productivity. **Figure 14.12** shows the seasonal variation in the distribution of chlorophyll (**CC14**) in ocean surface waters measured by satellites. The concentration of chlorophyll is a good indicator of the abundance or biomass of phytoplankton and also provides a reasonable approximation of the distribution of primary productivity.

The most productive parts of the ocean are coastal regions (**Fig. 14.12**), particularly along the western margins of continents, where coastal upwelling brings nutrients from deep waters into the photic zone (Chap. 15). Most of the open ocean has low productivity, the exceptions being certain high-latitude regions and the equatorial upwelling band across the eastern Pacific and, to a lesser degree, across other oceans (**Fig. 14.12**).

Throughout most of the tropical and subtropical open oceans, a permanent thermocline begins at a depth of about 100 to 200 m. Light intensity is relatively high, and the photic zone extends throughout most or all of the mixed layer above the thermocline. Nutrients are depleted in the photic zone, and the steep thermocline inhibits vertical mixing that would be necessary to resupply nutrients from the nutrient-rich water below the thermocline. High productivity in tropical and subtropical open oceans is limited to areas of upwelling. The band of high productivity across the equator, particularly in the eastern Pacific, coincides with upwelling at the tropical **convergence** along the equator (**Fig. 10.2**). During **El Niño,** primary productivity is dramatically reduced in this region because upwelling is inhibited (Chaps. 9, 10; **Fig. 9.18**).

Upwelling at the Antarctic Divergence (**Fig. 10.2**) is responsible for high productivity in the circumpolar

FIGURE 14.12 Seasonal variations in the distribution of chlorophyll in the surface waters of the world oceans. Chlorophyll concentration is a reasonable measure of primary production. The data were obtained by a satellite sensor that measures ocean surface color within several wavelength bands. Red indicates the highest chlorophyll concentrations. White indicates areas where there were no data, because of cloud or ice cover or the limited orbital coverage of the satellite near the poles. (a) Northern fall. (b) Northern winter. (c) Northern spring. (d) Northern summer. Note the high productivity of many coastal regions and of the high-latitude seas in the Northern Hemisphere.

ocean around Antarctica. The Northern Hemisphere has no comparable **divergence**, primarily because of the presence of continents. At high latitudes in the North Atlantic and North Pacific Oceans, productivity is high because cooling of surface waters, strong **westerly** winds, and **extratropical cyclones** effectively mix surface and subsurface waters during winter, when light levels are low. Nutrients supplied by winter mixing support high productivity when light intensity increases in spring.

Productivity is lowest in the interior of subtropical gyres in each ocean. These regions are remote from nutrient inputs in runoff, have low rainfall (which can carry small amounts of nutrients), and are characterized by downwelling and deep thermoclines (**Fig. 10.14**). Light is plentiful, but lack of nutrients prevents phytoplankton growth. Ocean waters are a brilliant blue in these regions because of the lack of suspended particles. However, areas of the Sargasso Sea (the interior of the North Atlantic subtropical gyre) are covered by vast rafts of *Sargassum* seaweed, despite the lack of nutrients (Chap. 17).

14.7 DISSOLVED OXYGEN AND CARBON DIOXIDE

Oxygen is released to solution during photosynthesis and consumed during respiration. The distribution of dissolved oxygen is controlled by these processes and by exchanges between atmosphere and ocean.

Oxygen is exchanged continuously between atmosphere and ocean. Hence, dissolved oxygen concentrations in the upper few meters of ocean water almost always equal the **saturation solubility** for the water temperature (**Fig. 6.6**). However, oxygen is produced by photosynthesis much faster than it is consumed by respiration in this layer, particularly when primary productivity is intense. As a result, oxygen concentration increases and the water becomes **supersaturated.** The depth and concentration of the oxygen maximum in the photic zone depend on the depth of maximum primary productivity and the depth and intensity of wind mixing that brings water to the surface, where excess oxygen can be released to the atmosphere. By many estimates, marine algae, not land plants, contribute most of the oxygen to our atmosphere.

As they release oxygen to ocean waters and atmosphere, phytoplankton also remove dissolved carbon dioxide. Carbon dioxide is highly soluble in seawater as it reacts to form carbonate and bicarbonate ions (Chap. 6). The carbon dioxide concentration in seawater is so high that its removal by photosynthesis does not produce a marked minimum concentration of carbon dioxide at the oxygen maximum. Nevertheless, much of the carbon dioxide that primary producers remove from ocean waters is carried down into the deep ocean, where it is removed from contact with the atmosphere for hundreds of years.

It has been estimated that almost half of all carbon dioxide released to the atmosphere by human activity since the Industrial Revolution has entered the oceans. Substantial research is now focused on identifying the fate of this carbon dioxide. Predictions of future **climate** change depend on the rate at which carbon dioxide is transported into the deep ocean. Global climate models are limited by a lack of understanding of this rate and of various other critical ocean processes. For example, we do not know whether and to what degree increased carbon dioxide concentrations might increase ocean primary productivity, or how much of the additional organic matter that might be produced by such an increase would be transported below the thermocline. Proposals have been made to lessen or avoid global climate changes due to the increasing carbon dioxide concentrations in the atmosphere by injecting some of the carbon dioxide into the deep. However, the likely effectiveness and possible adverse side effects of implementing such proposals cannot be properly assessed, given the present state of knowledge of the biogeochemical cycles of the oceans, and much current research is focused on this area.

Organic matter transported below the thermocline as detritus or through animal migration is decomposed, releasing nutrients and carbon dioxide. At the same time, the processes that decompose this organic matter consume dissolved oxygen. The continuous decomposition of organic matter progressively reduces the dissolved oxygen concentration of deep water. An oxygen concentration minimum below the bottom of the thermocline coincides with the nutrient maximum (**Fig. 14.7**).

At present, oxygen is fully depleted only in limited areas of the oceans, mainly where the **residence time** of subthermocline water is extremely long and/or where mixed-layer primary productivity is exceptionally high. Such conditions are present, for example, in the Baltic Sea, where water residence time is long and primary productivity high because of large **anthropogenic** nutrient inputs. Oxygen depletion also occurs in many fjords, where water residence time is very long. Once dissolved oxygen has been completely depleted, the water is **anoxic.** Most marine species cannot survive in anoxic water. However, certain bacteria in anoxic water obtain energy by reducing molecules that contain oxygen. First, bacteria that reduce nitrate (NO_3^-) to ammonium (NH_4^+) thrive, and then, when all nitrogen compounds are reduced, they are replaced by other species that reduce sulfate (SO_4^{2-}) to sulfide (S^{2-}). Sulfide is highly toxic and can be released from anoxic bottom waters into the overlying photic zone if vertical mixing is temporarily enhanced. In some cases, water containing hydrogen sulfide reaches the surface, and its foul smell is released to the atmosphere. Anoxic conditions are becoming more common and widespread in some coastal-ocean regions and estuaries because of inputs of nutrients from sewage and agricultural chemicals.

Anoxic conditions have been more widespread in the oceans in the past. During anoxia, detritus that falls into the anoxic layer is no longer subject to decomposition by **aerobic** respiration, because there is no oxygen. In such circumstances, organic matter can accumulate in large quantities in sediments. Oil and gas deposits are probably the result of **diagenetic** changes to such sediments over the millennia.

14.8 ORGANIC CARBON

Most organic matter in the oceans is nonliving and exists as both dissolved compounds and particulate matter (detritus). The reason most organic matter is nonliving is that organic matter created by phytoplankton and other marine algae undergoes a series of transfers between organisms, and conversions among physical and chemical forms, before finally being converted back to dissolved inorganic constituents by decomposers.

Under normal conditions, about 10% of the organic carbon that phytoplankton create by photosynthesis is excreted to seawater as dissolved substances. When phytoplankton are stressed by low nutrient or low light levels, nonoptimal temperatures, or other unfavorable factors, they may excrete 50% or more of the organic matter they create by photosynthesis. Although most phytoplankton are eaten before they die, those that are not eaten excrete substantial amounts of dissolved and particulate organic material as they reach the end of the life cycle. In addition, dissolved and particulate organic matter are released when phytoplankton cells are ruptured by viruses and during inefficient feeding by herbivores.

The distinction between dissolved and particulate matter is arbitrary. Oceanographers usually consider compounds to be dissolved if they pass through a filter with a pore size of 0.5 μm (human hair is about 100 μm in diameter), although filters used to separate dissolved and organic matter often have pore sizes somewhat larger or smaller than this value. Material that collects on such a filter is called "particulate." Much of the "dissolved" material that passes through a 0.5-μm filter is not truly dissolved, because it consists of **colloidal**-sized (**CC4**) inorganic or organic particles in addition to living organisms (such as microbacteria and archaea) and viruses. Nevertheless, the distinction between dissolved and particulate is useful. "Particles" less than 0.5 μm in diameter are believed to be too small to be eaten by most species in food chains that lead to higher animals, such as fishes.

Both particulate and dissolved organic matter are produced continuously wherever organisms are present in the oceans. Most organic matter is released in the photic zone, where marine organisms are concentrated. Many organic particles, such as fecal pellets and phytoplankton or zooplankton fragments, are relatively large (ranging from about 1 μm to more than 1 mm in diameter) and sink relatively fast (Chap. 8, **CC4**).

Marine bacteria and fungi utilize organic particles and dissolved organic matter that is present throughout the ocean depths. These decomposers rapidly break down the more easily oxidized organic compounds in solution and in detritus. As particles are decomposed, smaller particles and more dissolved organic matter are released. Decomposition continues until all organic matter is converted to carbon dioxide and water, which may take many years or even centuries. There are three reasons for the long-term persistence of some dissolved and particulate organic matter. First, many organic compounds are extremely resistant to oxidation. Second, the rate of bacterial decomposition is very slow in the cold temperatures and high pressures of the deep oceans. Finally, dissolved organic molecules are so well dispersed throughout seawater that they may be present for a long time before they are encountered by a decomposer.

The mass of dissolved and particulate matter in the oceans far exceeds the mass of living organisms because dissolved and particulate organic matter is produced continuously but decomposed only very slowly. In the photic zone, dissolved and particulate nonliving organic matter usually makes up more than 95% of the total organic carbon. An unknown proportion of the "dissolved" organic matter is, in fact, living microorganisms (including microbacteria and archaea) and viruses. Although the number of these microorganisms is extremely large, they are very small and their total biomass is unlikely to be more than a small fraction of the nonliving organic matter. **Table 14.2** gives approximate relative abundances of

TABLE 14.2 Approximate Relative Abundances of Organic Carbon in Living and Nonliving Forms in the Oceans

Organic Matter Form	*Percentage of Total Organic Carbon in Oceans*
Dissolved organic matter	94.9
Nonliving particulate organic matter	5
Phytoplankton	0.1
Zooplankton	0.01
Fishes	0.001

different forms of organic matter in the oceans. The declines in total mass of carbon from phytoplankton to zooplankton and from zooplankton to fishes reflect the low efficiency of food transfer between trophic levels (**CC15**).

Relatively little is known about the composition of dissolved or particulate organic matter. These materials are extremely difficult to study because they consist of hundreds of thousands or millions of chemical compounds, each present in extremely small concentrations. However, recent studies have revealed that much of the "dissolved" organic matter that passes through a 0.5-μm filter may consist of living organisms. Numerous bacteria and archaea have a diameter less than 0.5 μm. Such microorganisms apparently can both produce and consume the nonliving "dissolved" organic matter. In addition, even smaller viruses are present that, until recently, were counted as part of the dissolved organic matter. Extremely sophisticated techniques are needed to isolate and identify archaea and viruses, but apparently they exist in vast numbers in all ocean water. Concentrations of billions of viruses in every liter of water are not unusual. Although we do not know what role viruses play in ocean ecosystems, mounting evidence suggests that they cause diseases in marine organisms spanning the entire range from the smallest bacteria to the largest whales.

The concentration of particulate organic carbon is generally higher in the upper layers of the oceans than in deep waters. Within the surface layer, the concentration is lowest in areas where phytoplankton productivity is high. This seemingly paradoxical situation is created by the abundant populations of zooplankton and other **filter-feeding** organisms that thrive in productive waters because of the continuously abundant food supply. Particulate organic matter concentrations are high in deep waters below such productive areas because large numbers of fecal particles fall from the abundant filter-feeding organisms.

Organic particles generally account for about one-quarter of the suspended particles in ocean waters. In most areas, the majority of suspended particles are fragments of plant and animal hard parts with lesser amounts of lithogenous particles (Chap. 8).

CRITICAL THINKING QUESTIONS

14.1 Very high productivity is found in the circumpolar ocean around Antarctica. This is also the region most often under the hole in the ozone.
(a) Does the high productivity indicate that the ozone hole does not have a significant adverse effect on the ocean ecosystem? Explain why or why not.
(b) What other information did you use to reach your answer?
(c) What other studies or data would you like to have to make sure that your answer is correct?

14.2 The enhanced greenhouse effect is expected to cause climate changes that will lead to warming of the mixed layer of the oceans by several degrees.
(a) Hypothesize how this change might affect winds, the characteristics of the pycnocline, and ocean currents.
(b) Would these changes affect the amount of nutrients available to phytoplankton? If so, how?

14.3 A number of years ago it was suggested that nuclear reactors should be placed on the seafloor in tropical deep-water areas and other areas of the oceans where the primary production in the mixed layer is nutrient-limited. The intention was that the heat generated by the reactor would cause artificial convection-driven upwelling, increase the supply of nutrients to the mixed layer, and increase primary production.
(a) Would this proposal work? If not, why not?
(b) If so, can you suggest reasons (other than the potential for the release of radioactivity) to explain why the idea was not adopted?

14.9 BIOLOGICAL PROVINCES AND ZONES

The communities of organisms in different parts of the oceans are as distinct and different from one another as the species that live in tropical rainforests, deserts, and Arctic tundra. The oceans are separated into distinct biological zones delineated by the availability of light, nutrients, and food, and by the temperature and **salinity** characteristics of the water.

Differences between communities that live in or on the seafloor and those that inhabit the water column are so great that the two locations are considered to be separate environments. The seafloor is the benthic environment, and the water column is the pelagic environment. In contrast to pelagic organisms, benthic organisms do not have to swim or control their density and/or size to avoid sinking, which can be energetically advantageous, but they must compete with each other much more intensely for living space and food. Although the pelagic and benthic environments support different communities, many species live part of the life cycle in one environment, then change form (in some species as dramatically as the **metamorphosis** of caterpillars into butterflies) and live in the other.

Benthic Environment

The benthic environment is separated into zones by depth. The deepest zone, the **hadal zone** (**Fig. 14.13a**), which occupies less than 1% of the ocean floor, is limited to the seafloor of deep-ocean **trenches** at depths greater than 6000 m. The seafloor between 2000 and 6000 m is the **abyssal zone,** and corresponds roughly to the **abyssal plains.** The hadal and abyssal zones are generally covered with soft, fine-grained muds. Much of these

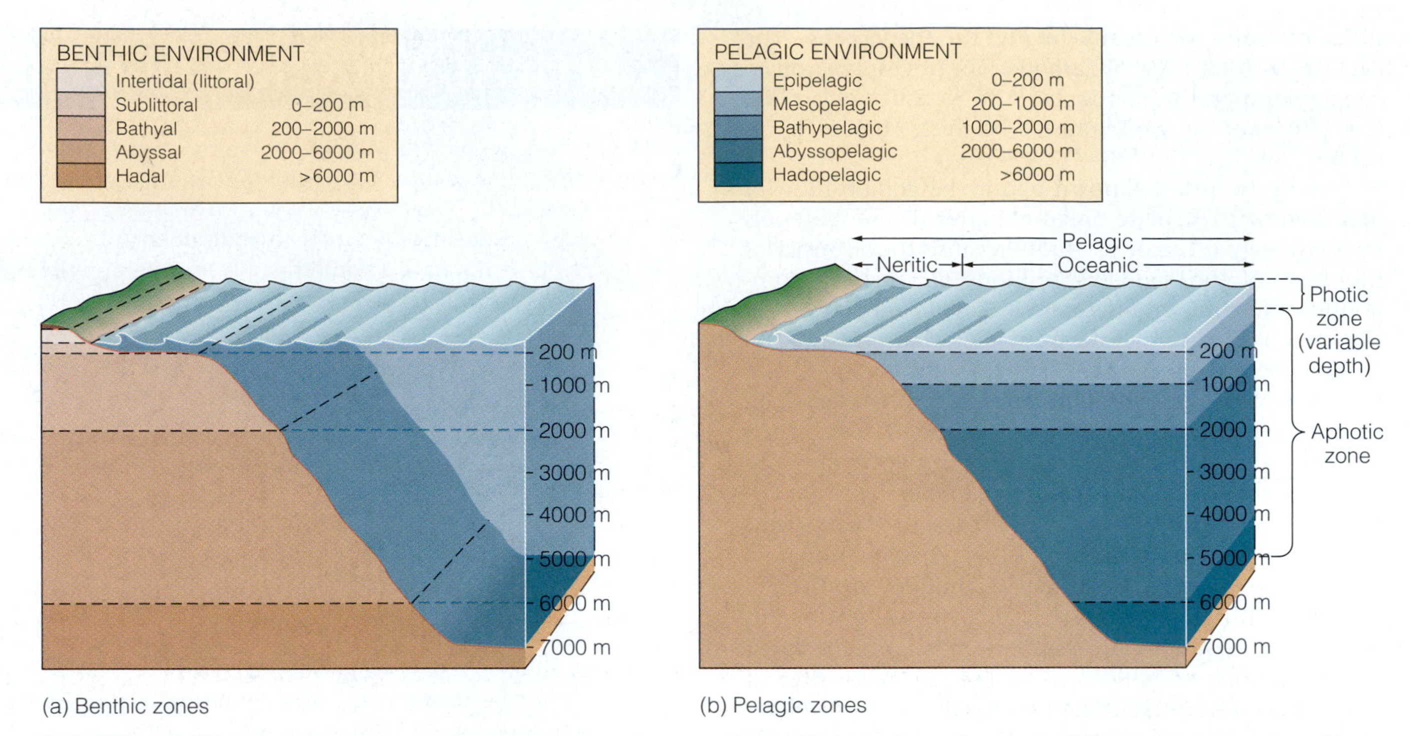

FIGURE 14.13 The oceans are separated into zones. The assemblage of species in each zone is different from that found in the other zones. (a) The benthic environment is separated into a number of zones, each of which spans a different depth range. The depths at which transitions occur between pelagic zones and between benthic zones are generally the same. (b) The pelagic environment (water column) is separated into a neritic province (water over the continental shelf) and an oceanic province. The oceanic province is divided into several zones that each span a different depth range.

zones is below the **carbonate compensation depth** (**CCD;** Chap. 8), and the surface muds generally contain little or no calcium carbonate.

Sediment characteristics are important to the species composition of the benthos. For example, some polychaete worms feed by passing sediment through the gut to digest bacteria and organic matter in the sediment, just as earthworms do. Such polychaetes are more successful in soft, fine-grained muds than in coarse-grained sediment, particularly if the muds contain a significant amount of detritus. Where sediments are coarse-grained with little organic matter, many benthic animals live on the sediment surface and obtain food from particles or other animals in the water column. Benthic animals are called **infauna** if they live much or all their life in sediments, and **epifauna** if they live on or attached to the seafloor.

The seafloor environment between 2000 m and the approximate depth of the continental **shelf break** (about 200 m) is the **bathyal zone** (**Fig. 14.13a**). Because sediments in the bathyal zone are more variable than those in the abyssal and hadal zones, the benthos is more variable in composition in the bathyal zone.

The hadal and abyssal zones and almost all of the bathyal zone are below the photic zone, so no photosynthetic organisms are present in the benthos. Instead, benthos in these zones are dependent on food transported to them from the overlying photic zone, primarily as detritus. All benthic communities in these zones thus belong to the detrital food web, with two exceptions. One is the unusual benthic communities around hydrothermal vents. These organisms are supported by a chemosynthesis-based food web (Chap. 17). A second, and minor, exception occurs in the shallowest parts of the bathyal zone in areas of the clearest waters, where light levels are very low but sufficient to support limited photosynthesis by species adapted to very low light levels. Such conditions are generally present only on some **seamounts** far from land.

The benthic environment between the continental shelf break (200 m depth) and the land is divided into three major zones. The **sublittoral zone** (subtidal zone) is the continental shelf floor that is permanently covered by water. It extends from the **low-tide line** to the depth of the continental shelf break (200 m). The **intertidal zone** (**littoral zone**) is the region between the low-tide line and the **high-tide line** and is covered with water during only a part of each tidal cycle. The **supralittoral zone,** often called the "splash zone," is the region above the high-tide line that is covered by water only when large storm waves or **tsunamis** reach the **coast** or during extremely high **tides** or **storm surges.**

Organisms that live in the supralittoral and intertidal zones must endure much more extreme conditions than other marine organisms. Such conditions include exposure to the temperature extremes of the atmosphere, loss of body fluids by evaporation while exposed to the atmosphere, salinity variations due to rainfall, predation by both marine and terrestrial species, and mechanical shock and turbulence created by breaking waves. In intertidal zones with sand or mud sediments, most organisms bury themselves to avoid extreme conditions. This is why **beaches** and mudflats appear to be unpopulated. Most sediment-covered **shores** support an abundance of infauna living a few centimeters beneath the sediment surface. Organisms that live in **rocky intertidal zones** adapt to their changeable environment in other ways (Chap. 17).

Pelagic Environment

The pelagic environment consists of the neritic province and the oceanic province. The neritic province is the water column between the sea surface and seafloor in water depths to about 200 m. Thus, it consists of the water overlying the continental shelf and shallow banks (**Fig. 14.13b**). The neritic province experiences greater variability in salinity, temperature, and suspended sediment concentrations than the oceanic province (Chap. 15). Further, the mixed layer reaches the seafloor in most of the neritic zone, at least for part of the year, which is important to nutrient cycles (Chap. 15).

The oceanic province comprises the entire water column of the open oceans beyond the continental shelf. It consists of several zones, each of which is a water layer distinct from the layer above or below because of differences in salinity, temperature, and light intensity. The **epipelagic zone** extends between the surface and 200 m, the approximate depth where light intensity becomes too low for photosynthesis. Note that this depth is not the same as the shallower (usually less than 100 m) compensation depth, where photosynthesis occurs just rapidly enough to match respiration (**Fig. 14.4**). The epipelagic zone is the only zone in the deep oceans in which food can be produced directly by photosynthesis. In all deeper zones, the original source of food is almost exclusively detritus falling from the epipelagic zone. Minor additional food sources include vertically migrating animals and chemosynthesis at hydrothermal vents (Chap. 17).

Below the epipelagic zone, between 200 and 1000 m is the mesopelagic zone (**Fig. 14.13b**). Although light intensity is too low to support photosynthesis, many organisms in this zone have photoreceptors or eyes adapted to detect very low light levels (Chap. 17). Some of these species migrate upward into the epipelagic zone at night to feed and return to the mesopelagic zone during the day. Below the mesopelagic zone are the bathypelagic zone (1000–2000 m), the abyssopelagic zone (2000–6000 m), and the hadopelagic zone (>6000 m). These are zones of perpetual darkness, high pressure, and low temperature, and their inhabitants feed on detritus or on each other (Chap. 17). The boundary between the bathypelagic and abyssopelagic zones is essentially the boundary between relatively young bottom water masses of the ocean basins and older overlying deep water masses (Chap. 10). The hadopelagic zone is restricted to the deep trenches, in which water movements are generally very slow and water residence time is long (**CC8**).

Latitudinal Zones

Most benthic and pelagic environments below the photic zone are relatively uniform at a given depth. Latitudinal variations in temperature and salinity are less in waters or sediment below the photic zone than in the epipelagic zone and in the pelagic and benthic environments of the continental shelf. Most marine organisms are adapted to live successfully within a narrow range of environmental conditions, particularly a narrow range of temperatures.

Variations in water temperature with latitude (**Figs. 9.13, 9.23**) act as effective barriers to latitudinal dispersal of many marine species. Hence, the marine **biota** of each biological zone varies with latitude (although the variation is small in the abyssal benthos). For example, coral reefs are present in warm-water areas in all oceans, but not at cold, high latitudes. The **fauna** and **flora** of the Arctic and Antarctic regions are also very different. Polar bears and most penguin species live only at high latitudes, but polar bears live only in the Arctic region, and most penguin species live only in the Antarctic.

In some cases, continents act as geographic barriers to the distribution of marine species. The tropical Atlantic Ocean is effectively separated from the tropical Pacific and Indian Oceans by the continents. To move between the Atlantic Ocean and either the Pacific Ocean or the Indian Ocean, marine organisms would have to pass around the southern tip of Africa or South America or through the Arctic Ocean. Many tropical species could not tolerate the low water temperatures they would encounter in such journeys. Therefore, almost all tropical Atlantic species are different from those in the tropical Pacific and Indian Oceans. In contrast, the tropical Pacific and Indian Oceans (the Indo-Pacific) have many species in common because they are connected at tropical latitudes.

Tropical Atlantic and Indo-Pacific species, although different, are more closely related than Arctic and Antarctic species. The reason is that the continents have drifted to their present positions in the latter stages of the present **spreading cycle** (Chap. 4). Previously, all the oceans were connected at tropical latitudes, and tropical species could move freely around the globe during this relatively recent period in evolutionary history.

14.10 PLANKTON

The term *plankton* includes all marine organisms and viruses that do not swim or are very weak swimmers and that do not live on or attached to the seafloor. Plankton generally do not settle to the seafloor, and they have very limited or no control of their horizontal movements, so they drift with the ocean currents. Phytoplankton are the autotrophs that photosynthesize more than 99% of the food used by marine animals. Zooplankton are planktonic herbivores, carnivores, or omnivores.

Plankton are often categorized by size. The largest, which are almost exclusively zooplankton, are the macroplankton (>2 mm). Most plankton are microplankton (20 µm–2 mm), nannoplankton (5–20 µm), or ultraplankton (2–5 µm), and these size ranges are dominated by phytoplankton. The smallest plankton are picoplankton (0.2–2 µm), thought to be predominantly bacteria and archaea, and femtoplankton (<0.2 µm), thought to be primarily viruses. Because even the finest mesh nets used by biologists are too coarse to collect the smaller species, they have not been well studied. Less is known about nannoplankton than about microplankton and macroplankton, and very little is known about the even smaller plankton or their **ecological** importance.

The better-known large marine plants and **macroalgae,** including kelp, seaweeds, and **sea grasses,** constitute only a tiny fraction of the photosynthetic species in the oceans. Macroalgae normally grow attached to the seafloor and are found on the beach only when they have been broken loose by storm waves or animals.

FIGURE 14.14 A typical plankton sample (photographed at 3× magnification) contains many species of both phytoplankton and zooplankton.

Phytoplankton

Phytoplankton are generally much smaller than 1 mm in diameter—too small to be seen by the naked eye. The oceans contain an abundance of phytoplankton dispersed throughout the surface waters at concentrations that may exceed 1 billion individuals per liter. There are only three known planktonic macroalgae species, all of the genus *Sargassum*. *Sargassum* grows in dense rafts, often many square kilometers in area, floating at the surface of the Sargasso Sea (Chap. 17).

There are tens of thousands of species of phytoplankton. A sampled phytoplankton assemblage always consists of many species (**Fig. 14.14**), but in any individual sample one species is often dominant and far outnumbers all others (as pine trees do in a pine forest).

Phytoplankton communities consist of different species in different climatic regions, and concentrations range from very low in some areas (equivalent to deserts) to very high in others (equivalent to rain forests). Unlike most land plant communities, phytoplankton communities can vary dramatically in composition and concentration within hours or days. Such variability occurs in part because phytoplankton are often concentrated in patches (tens to hundreds of meters across) that drift with ocean currents. Patches of phytoplankton can develop when rapid reproduction exceeds the rate of dispersal by mixing processes. Because phytoplankton can double their population within a day, or in even less time under favorable conditions, patches may develop quickly in calm seas. Phytoplankton are also concentrated by **Langmuir circulation** (Chap. 10). Smaller or less dense cells tend to concentrate at the surface convergence between Langmuir cells. Larger or higher-density cells, which tend to sink, may be concentrated below the surface where subsurface Langmuir cell currents converge (beneath surface divergences).

Individual phytoplankton species respond favorably to slightly different light intensity levels, temperatures, and nutrient concentrations. Under favorable conditions, one or more species can reproduce rapidly and become the dominant species. If conditions change and another phytoplankton species prospers in the changed conditions, the second species can become dominant within a few days because grazing zooplankton can rapidly remove the previously dominant species.

Diatoms. Diatoms (**Fig. 14.15a**) are among the most abundant types of phytoplankton in many of the more productive areas of the oceans. They are relatively large single cells (up to about 1 mm in diameter) with a hard, organically coated external **siliceous** casing called a "frustule," which is made of two halves much like a pillbox. The frustule is porous, allowing dissolved substances to diffuse through it and to be taken into or excreted from the cell. Because silica is denser than seawater, most diatoms contain a tiny droplet of a lighter-than-water natural oil to reduce their density and thus their sinking rate. Many diatoms have protruding, threadlike appendages and may link to form chains (**Fig. 14.15a**). These adaptations also reduce the cell's tendency to sink and may decrease predation by small zooplankton. The

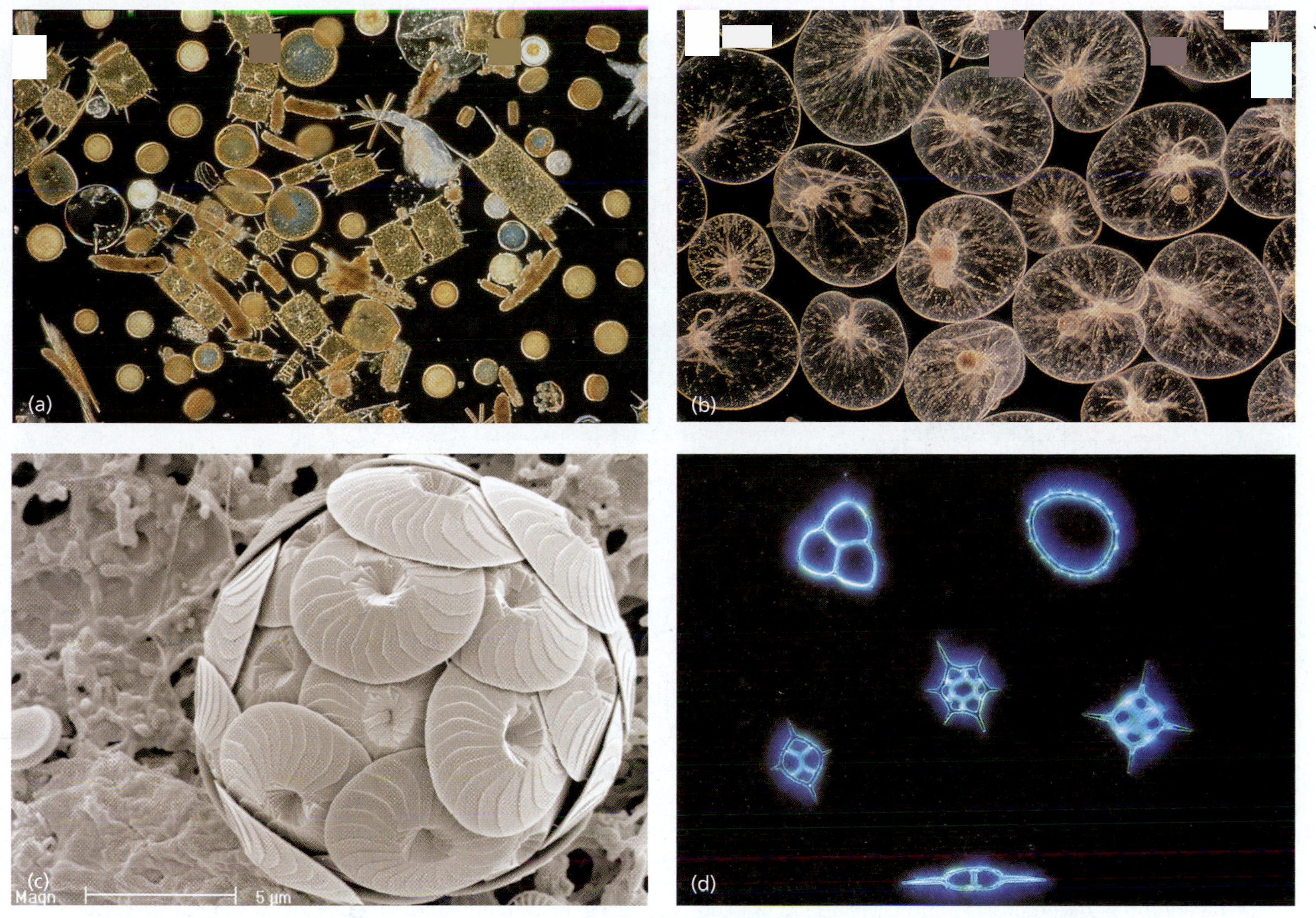

FIGURE 14.15 There are many varieties and species of phytoplankton (a) Various species of diatoms, photographed at 20× magnification. (b) Bioluminescent dinoflagellates (*Noctiluca* sp., approximately 10–20× magnification). (c) Coccolithophore (*Emiliania huxleyi*, approx. 5000× magnification). (d) Silicoflagellates (56× magnification).

threadlike appendages also increase the surface area over which nutrients can be taken up and light collected for photosynthesis.

Their oil droplet and relatively large size make diatoms a favored food source for many species of juvenile fishes and zooplankton. Consequently, food chains based on diatoms are generally shorter than those based on smaller classes of phytoplankton. Smaller phytoplankton must usually be eaten by small zooplankton before they can provide food for juvenile fishes and larger herbivorous and omnivorous zooplankton.

Diatoms reproduce asexually by cell division. At each division, which may follow the previous one by less than a day, the frustule separates, with one-half taken by each of the two new cells, and each cell then manufactures a new half. Consequently, one of the new cells is smaller than its parent (**Fig. 14.16**). After several such divisions, the now much smaller cells may reproduce sexually, restoring the cell size to its maximum.

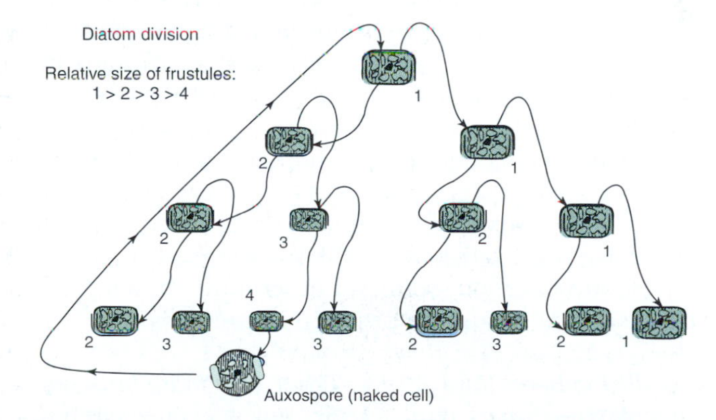

FIGURE 14.16 Most diatoms reproduce primarily by cell division. At each division, the two halves of the diatom frustule separate, and each half secretes a new half to its frustule. However, because one of the new halves is smaller than its parents, the mean size of the diatom is reduced with progressive divisions. Eventually, a small diatom reproduces sexually by producing an auxospore that at first has no frustule. After some period of growth, the auxospore secretes a new frustule.

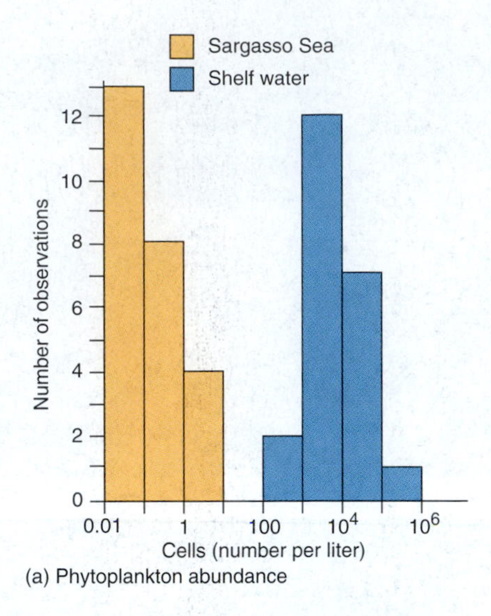

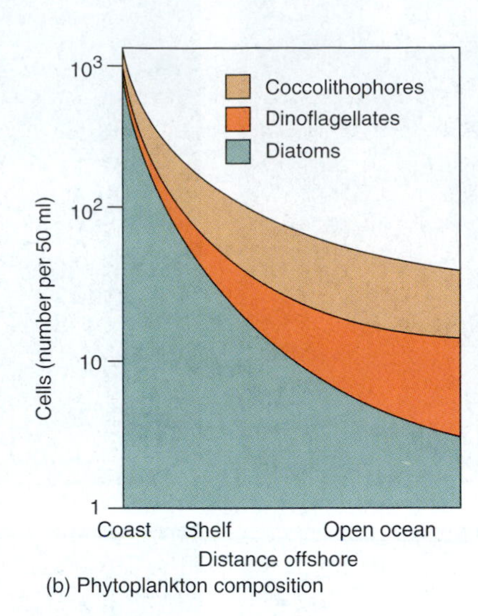

FIGURE 14.17 The composition of the phytoplankton community is different in coastal, offshore, and open-ocean waters. (a) Phytoplankton are much more abundant (note the logarithmic scale) in the surface layer waters of the continental shelf than they are in the Sargasso Sea, where nutrients are depleted. (b) In many areas, the relative proportion of diatoms generally decreases with distance offshore, reflecting the lower availability of dissolved silicate in the offshore waters. These data were obtained from the Caribbean Sea.

Dinoflagellates. Dinoflagellates range widely in size, but many are nannoplankton, smaller than most species of diatoms (**Fig. 14.15b**). Most species of dinoflagellates have two hairlike projections, called "flagella," that they use in whiplike motions to provide a limited propulsion ability. Some dinoflagellate species use this propulsion mechanism to migrate vertically and maintain a depth where light levels are optimal. Consequently, dinoflagellates of some species tend to concentrate around a single depth within the photic zone.

Many dinoflagellate species have an armored external cell wall made of cellulose, but others are "naked." Because cellulose is decomposed relatively easily by bacteria and other decomposers, dinoflagellates do not contribute significantly to deep-ocean bottom sediments. Dinoflagellates are not always autotrophs. Some species are able to use dissolved or particulate organic matter as food, and some are predators. Indeed, many species can live and grow both autotrophically and heterotrophically.

Dinoflagellates are more abundant than diatoms in the open oceans far from land because the silica needed to construct diatom frustules is in short supply. However, phytoplankton biomass is substantially lower in most such open-ocean areas (**Fig. 14.17a**). Silica can also be scarce in coastal waters at certain times of year. These temporary silica-deficient conditions can lead to explosive blooms of dinoflagellates if other nutrient, temperature, and light conditions permit (Chap. 15).

Coccolithophores and Other Types of Phytoplankton. **Coccolithophores** (**Fig. 14.15c**) are generally nannoplankton, and they are smaller and less abundant than diatoms and most dinoflagellates. Coccolithophores are single-celled **flagellates** whose external cell surface is covered by a mosaic of tiny **calcareous** plates (**Fig. 14.15c**). In certain areas, these plates are a major component of seafloor sediments (Chap. 8). Coccolithophores make up a greater fraction of the phytoplankton biomass in relatively nonproductive temperate and tropical open-ocean waters than they do in coastal waters.

In most of the oceans, microplankton consist primarily of diatoms, dinoflagellates, and coccolithophores (**Fig. 14.15a–c**). However, there are several other types of phytoplankton, including silicoflagellates (which have an intricate silica shell; **Fig. 14.15d**), cryptomonads, chrysomonads, green algae, and **cyanobacteria** (**blue-green algae**).

Bacteria and Archaea. The diatoms, dinoflagellates, and coccolithophores that make up most of the phytoplankton biomass are now known to be outnumbered by phytoplankton species so tiny that they escaped detection until the 1990s. These poorly studied, tiny ultraplankton are dominated by species of bacteria and archaea. One group of photosynthetic bacteria, *Prochlorococcus,* absorbs blue light efficiently at low light intensities, so it can grow throughout the depth of the photic zone. *Prochlorococcus* may be the most abundant component of the phytoplankton, especially in the tropical and subtropical oceans, and it is estimated to contribute 30% to 80% of the total photosynthesis in areas of the oceans where nutrients are scarce. Until recently it was thought that archaea existed primarily in extreme environments, but it is now known that archaeal species are found throughout the ocean environment and that some are primary producers. The discovery of these bacteria and archaea, along with the discovery that viruses are even more abundant throughout the oceans, has revolutionized ocean sciences. Extensive research efforts are currently directed toward understanding these organisms and viruses and their role in ocean ecosystems.

Zooplankton. Zooplankton are animals that do not swim strongly enough to overcome currents and so drift with the ocean water. As is the case for phytoplankton, and for similar reasons, the distribution of zooplankton is patchy. Dense patches of zooplankton attract fishes and other predators. **Nekton,** marine animals that swim strongly enough to move independently of ocean currents, are distributed nonuniformly because they are attracted to food sources, and they often exhibit **schooling** behavior, which is discussed in Chapter 16. The nonuniform distribution of both plankton and nekton, together with the temporal variability of their populations, makes it difficult for biologists to obtain precise population estimates, even when large numbers of samples are taken.

Zooplankton consist of a bewildering array of species from many different groups of organisms. Many zooplankton species tolerate only narrow ranges of environmental conditions, especially temperature. Consequently, zooplankton species composition changes from one water mass to another and with depth. At any one location, many species are represented, including species that are bacteriovores (bacteria eaters), herbivores, carnivores, and omnivores.

Zooplankton belong to one of two categories based on the life history of the species. Species that live their entire life cycles as plankton are known as **holoplankton**. The **larvae** (juvenile stages) of species that later become free swimmers or benthic species are **meroplankton**. Meroplankton include many species of fishes, **sea stars**, crabs, oysters, clams, **barnacles**, and other **invertebrates**. Holoplankton are the dominant zooplankton in surface ocean waters, whereas meroplankton are more numerous in continental shelf and coastal waters. In tropical coastal waters, larvae of benthic species make up as much as 80% of all zooplankton.

Many zooplankton species tend to concentrate at the same depth and collectively migrate between the photic zone and aphotic zone each day, but the depth to which this diurnal migration takes place differs depending on the species. Zooplankton also tend to collect at density interfaces between water layers because these interfaces inhibit (but do not prevent) vertical migration and sinking and thus collect food particles. When zooplankton are present in large numbers within a thin layer below the surface, they scatter or reflect sound and are observed by echo sounders as a "deep scattering layer." This layer changes depth during the day as the zooplankton make their daily migration between the photic and aphotic zones.

There are too many important species of zooplankton to describe in this text, but the characteristics of the major groups of holoplankton and meroplankton are described briefly in the sections that follow.

Holoplankton. The most abundant holoplankton are **copepods** (Fig. 14.18a) and **euphausiids** (Fig. 14.18b), which constitute 60% to 70% of all zooplankton in most locations. Copepods and euphausiids are both crustaceans, a class of invertebrates that includes crabs and lobsters. In the open oceans, most copepod species are herbivorous, whereas many coastal forms are omnivores. Copepods eat about half their body weight in phytoplankton or other food each day. They are abundant throughout the oceans and can double their population within a few weeks. Euphausiids are generally larger and reproduce more slowly than copepods. Euphausiid population doubling times are typically several months. Many euphausiids are omnivorous, eating smaller zooplankton, as well as their major food, phytoplankton.

Euphausiids called "krill" are especially abundant in waters around Antarctica, and they constitute the principal food source of the abundant marine animals there. Baleen whales, including the blue, humpback, sei, and finback, feed directly on krill. These baleen whales gulp large volumes of water, then squeeze it out through netlike baleen plates suspended from the roofs of their mouths. Krill collect on the baleen and are removed by the tongue and ingested.

Two groups of single-celled **amoeba**-like microplankton have hard parts that become important components of sediment in some areas: **foraminifera** (Fig. 14.18c), with shells composed of calcium carbonate, and **radiolaria** (Fig. 14.18d), with shells composed of silica. Both are holoplankton and feed on diatoms, small protozoa, and bacteria, often capturing them on their many long, sticky, spikelike projections. Radiolaria and foraminifera are most abundant in warm waters. Individual species, especially of foraminifera, are very sensitive to small changes in water temperature and salinity. Because of the sensitivity to temperature, the species compositions of radiolaria and foraminifera in sediments are important indicators of past climates. **Isotopic** compositions of the

CRITICAL THINKING QUESTIONS

14.4 If you measured the dissolved oxygen concentration carefully in the photic zone during a spring bloom of phytoplankton, describe how you would expect it to change over 24 h and why. Would the changes in concentration during the 24 h be different at different depths within the photic zone? If so, describe how and why.

14.5 Pelagic food chains that lead to large predatory fishes in the deep oceans far from land tend to be much longer than in coastal regions. What might account for this difference?

14.6 What is a detritus-based food web? Would you expect detritus-based food chains that lead to large fishes that feed on benthic animals to be longer than the pelagic food chains in the overlying water that lead to large carnivorous fishes? Why? Would your answer be different for coastal and deep-ocean regions? Why?

14.7 Many fewer species of reptiles and mammals live in the oceans than on land, and many more species of fishes and invertebrates live in the oceans than live on land and in freshwater. What do you think are the reasons for these differences?

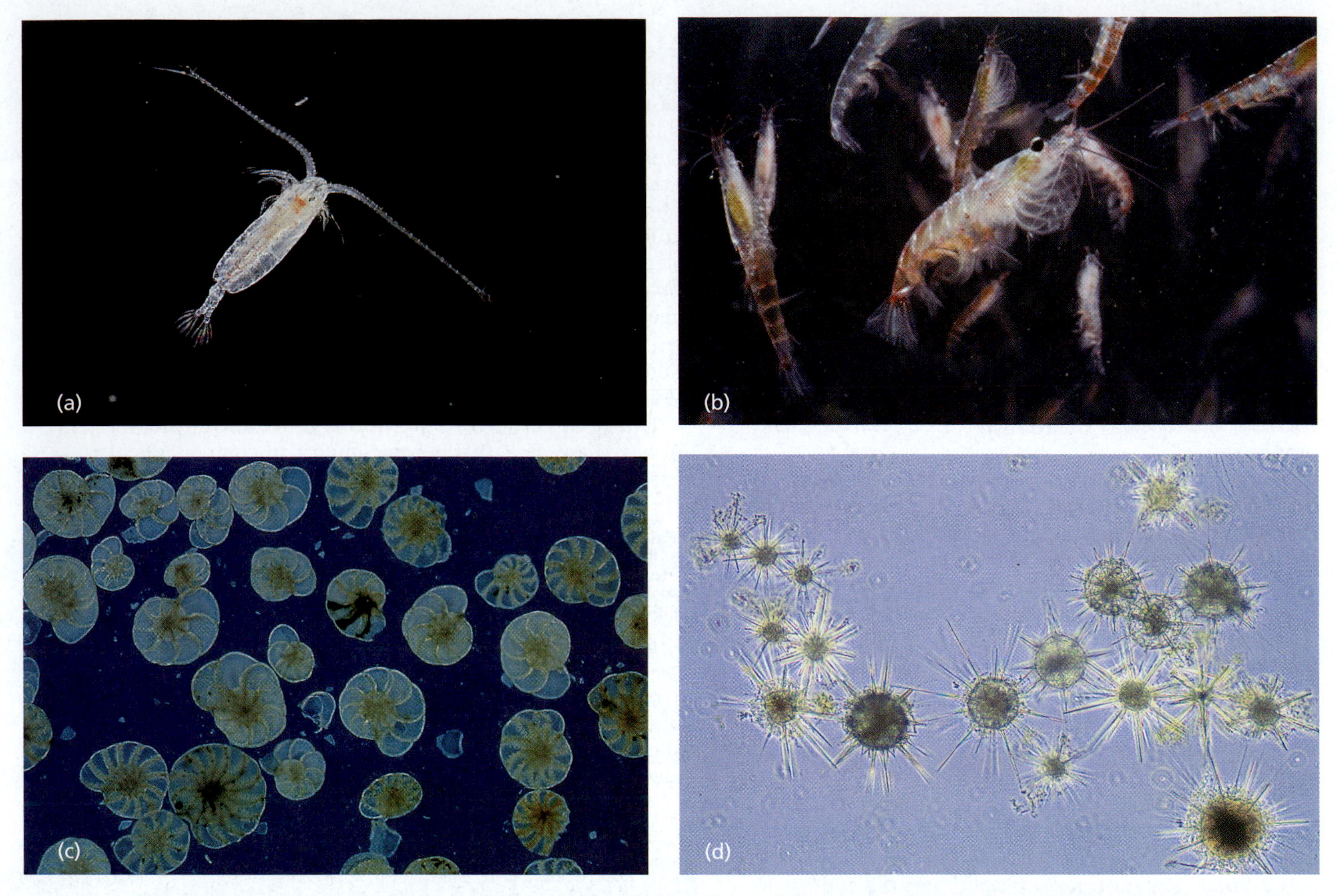

FIGURE 14.18 Typical holoplankton. (a) Copepods such as this *Calanus helgolandicus* are extremely abundant in some regions of high primary productivity. (b) Euphausiids (*Euphausia superba,* Southern Ocean). (c) Foraminifera (*Rheinbero* sp., Atlantic Ocean, photographed at 15× magnification). (d) Radiolaria (classes Acantharia and Radiolaria, Red Sea, 35× magnification).

shells of these organisms also provide a record of the temperature at the time they lived (Chap. 8).

The **pteropods** (**Fig. 8.7b**), another group of holoplankton, are also important in marine sediments. Pteropods are **mollusks** and are related to slugs and snails. In pteropods, the "foot" on which slugs or snails crawl is modified into a delicate transparent wing that undulates and propels the organism like a fin. This modified foot enables pteropods to migrate vertically hundreds of meters each day. Some pteropod species are carnivorous and do not have a shell. Others are herbivorous and have a calcareous shell that contributes to sediment, especially in tropical regions (Chap. 8), where they often occur in dense swarms.

Gelatinous Holoplankton. Many holoplankton differ from other holoplankton species because they have gelatinous bodies and are apparently not part of the food webs that lead to fishes and other marine animals exploited by humans. Although these species consume large amounts of other zooplankton, their gelatinous bodies provide little or no food for species at higher trophic levels.

The most familiar group of gelatinous holoplankton, the jellyfish, are **cnidarians** (phylum Cnidaria, or Coelenterata), a phylum which also include **corals** and **anemones**. All cnidarians have stinging cells, called "cnidocysts," that they use to inject toxins into their prey. In some species, the toxins are extremely strong and can paralyze or kill large fishes or even people.

Some jellyfish species (**Fig. 14.19a–c**) are very large in comparison with most other holoplankton, perhaps because their food value is low and they have relatively few predators. Some, such as the moon jelly *Aurelia* (**Fig. 14.19a**) and *Cyanea*, are holoplankton. Others are meroplankton that spend part of their lives in the plankton, then settle to the seafloor, where they attach with their stinging tentacles extended upward. In the benthos they resemble their close relatives: anemones and corals.

Among the most unusual jellyfish are species, such as the Portuguese man-of-war (**Fig. 14.19c**), that appear to be a single organism but in fact are a colony. In colonial forms, many individuals of the same species form a cooperative group that appears to be a single organism. Each

FIGURE 14.19 Typical gelatinous holoplankton. (a) Jellyfish *Aurelia* sp. This one was in a unique marine lake called Jellyfish Lake in Palau. (b) This upside-down jellyfish (*Cassiopeia andromeda,* Papua New Guinea, about 10–15 cm across) generally lies on its "back" on the shallow seafloor to provide light to the symbiotic algae that live in its tentacles. (c) The Portuguese man-of-war (*Physalia physalis*) is a colonial jellyfish whose long tentacles bear stinging cells that can severely injure a swimmer. (d) Ctenophores, like this *Mnemiopsis leidyi,* propel themselves by pulsating the columns of bright hairlike cilia visible through the body. (e) Each individual of this colonial salp species, photographed after it had floated into shallow water and settled on the sand in Indonesia, may be 10 cm or more across. The colonies, which normally drift through open-ocean surface waters, may exceed several meters in length.

colony member has a specialized task, such as protecting the colony, gathering food, digesting food, or reproducing. In the Portuguese man-of-war, one colony member is filled with gas to provide flotation and a "sail" that can partially control the colony's drift.

Widely occurring gelatinous plankton that are not jellyfish include **ctenophores** and **salps**. Ctenophores are transparent, **bioluminescent** organisms, some of which have long, trailing tentacles (**Fig. 14.19d**). Ctenophores propel themselves through the water by beating eight columns of hairlike cilia that are usually visible through the ctenophore's body (**Fig. 14.19d**) and that give these species their common name, "comb jellies." Small, rounded species of ctenophores are often called "sea walnuts" or "sea gooseberries."

Salps (**Fig. 14.19e**) are the holoplankton species of **tunicates**, most of which are benthos. Tunicates are among the most advanced invertebrates. They are **chordates** (phylum Chordata), a phylum which also includes the **vertebrates** (fishes and mammals). This close relationship is difficult to envision from the simple form of the adult salps or other tunicates, but tunicate larvae closely resemble vertebrate larvae. Adult tunicates have a simple baglike form with two openings. Water is pumped into one opening (**incurrent** opening) and out the other (**excurrent** opening). Food particles are removed from the water by a mucous layer spread over the interior of the tunicate's body.

Many salps are hollow and barrel-shaped, with incurrent and excurrent openings at opposite ends (**Fig. 14.19e**). They propel themselves slowly by pumping water through their bodies. Several salp species are bioluminescent. I witnessed a "magic moment" one moonless night on a research vessel sailing through an exception-

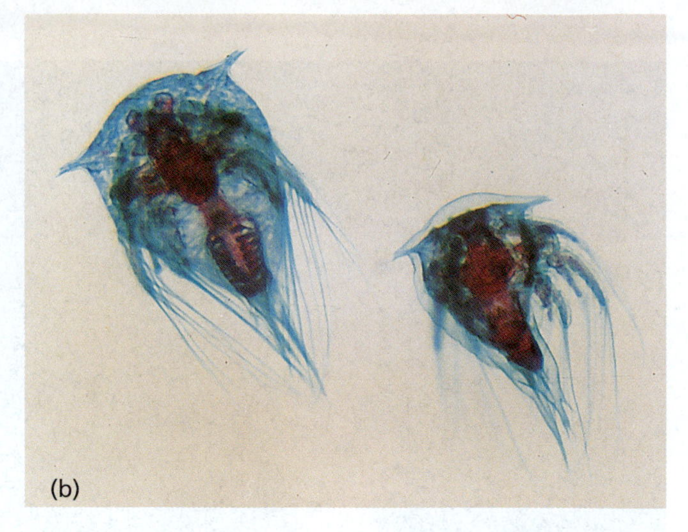

FIGURE 14.20 Typical meroplankton. (a) A porcellanid crab larva from the Gulf Stream. (b) Barnacle larvae (called "nauplii"), photographed at 50× magnification.

ally dense patch of salps in the tropical Atlantic Ocean. For miles, the ship's wake was a brilliant, sparkling light show, as a continuous stream of salps was disturbed by the wake and emitted pulses of light.

Meroplankton. Meroplankton are the eggs, larvae, and juveniles of species that spend their adult lives as benthos or nekton. Larvae or juvenile forms of the majority of benthic species, including clams, oysters, crabs, snails, lobsters, sea stars, **sea urchins**, corals, and **sea cucumbers,** spend their first few weeks of life as meroplankton. Many fish species also release eggs to the water column. These eggs, the larvae that hatch from them, and juvenile fishes that emerge from the larvae are meroplankton until the fishes become big enough to swim actively against currents as nekton. The release of eggs and larvae as meroplankton is an effective means of distributing species over wide areas (Chap. 16).

Many species of meroplankton larvae do not remotely resemble their adult forms (**Fig. 14.20**). In the past, numerous meroplankton have been named and classified as separate species, even though their adult forms were already well known. Modern scientific methods, particularly genetic studies, now enable us to match meroplankton with their adult forms.

14.11 NEKTON

Nekton are animals that swim strongly enough to move independently of ocean currents. Most nekton are vertebrates, which are organisms that have an internal skeleton, spinal cord, and brain. Marine vertebrates include fishes, **reptiles,** birds, and **marine mammals.** Some invertebrates, such as squid, are also nekton. The nekton are dominated by the varied and abundant fish species. There are three types of fishes: primitive jawless fishes including hagfishes and eels; fishes that have cartilage skeletons, including sharks and rays; and bony fishes that have bone (largely calcium carbonate and calcium phosphate) skeletons. More than 95% of all living species of fishes are bony fishes.

Bony Fishes

Bony nektonic fishes have a wide variety of body forms and range in length from just a few centimeters to more than a meter. Such pelagic giants as tuna or mackerel (**Fig. 14.21a**) can exceed 3 m. Species of fishes that live on or near the seafloor are called "demersal" fishes, and those living predominantly in the water column are called "pelagic" fishes.

The most abundant fishes in coastal waters are various species of silver-bodied schooling fishes including herrings (**Fig. 14.21b**), anchovies, pilchards, sardines, and menhaden. Because these fishes generally do not exceed a few tens of centimeters in length and have oily tissues, they are generally not valued for human consumption. Nevertheless, their numbers are so vast that the world's most productive commercial fisheries target these species. Much of the catch is processed and used as animal feed, fish oils for paints and diet supplements, soaps, and lubricants. These species have each been **overfished,** and their historical populations have been decimated in many coastal regions (Chap. 19). Various species of cod (**Fig. 14.21c**) are extremely abundant in mid and high latitudes, especially in the Northern Hemisphere, and they are commercially very valuable. Cod live most of their lives on or near the ocean floor, where they feed on invertebrates and small fishes.

The body shapes of fishes are determined primarily by their swimming habits (Chap. 16). Species that are members of the flatfish group have bodies modified to enable them to lie flat on the seafloor, where they remain well camouflaged (**Fig. 14.21d**). Juvenile flatfishes, like other fishes, have eyes on each side of their body. As they mature, one eye migrates across the body so that both eyes are on the same side of the head (**Fig. 14.21d**). If the eye migrates to the left side of the body, the species is usually called a "flounder." If it migrates to the right side, the species is usually called a "sole." However, some species, such as the starry flounder, have both left- and right-eyed forms.

Many fishes contain **swim bladders** filled with a gas that is usually a mixture of oxygen and carbon dioxide.

FIGURE 14.21 Representative species of fishes. (a) The dogtooth tuna (*Gymnosarda unicolor,* Papua New Guinea) is an oceanic predator. (b) A school of herring at the edge of a kelp forest in Monterey Bay, California. (c) The Atlantic cod (*Gadus morrhua*) was once extremely abundant, but its populations have now been substantially reduced by overfishing. (d) The flowery flounder (*Bothus mancus,* Solomon Islands) spends much of its life lying camouflaged on the seafloor.

Swim bladders are used to maintain buoyancy, and, at least in some species, they appear to be used in the fish's "hearing" mechanism. Because gas expands or contracts as pressure changes with depth, fishes with swim bladders must add or remove gas from their swim bladders as they move vertically. Some fishes absorb the gas back into the bloodstream, and others, such as herrings, can release the gas as bubbles. Many fishes bloat and die if brought rapidly to the surface because they cannot purge their swim bladders fast enough. Fish species that migrate vertically each day often have no swim bladders, but maintain buoyancy through high concentrations of lighter-than-water oils.

Sharks and Rays

Sharks and rays are more primitive than other fishes, and many species have existed essentially unchanged for millions of years. Unlike bony fishes, sharks and rays have skeletons made of cartilage. Sharks and rays have reproductive cycles that require eggs to be fertilized inside the female. Most other fish species reproduce by the simultaneous release of eggs and sperm by female and male. In many **cartilaginous** fish species, the fertilized eggs are hatched and develop into the adult form within the mother (or in some species the father) before being released in a live birth.

The common perception of sharks is that they are fearsome predators that will hunt and kill anything, including people (Fig. 14.22a). Although many species are able to hunt and kill large fishes and marine mammals, only a few species are known to attack people. Because humans are not the natural prey of sharks, most shark attacks on humans are thought to be cases of mistaken identity. The fact that most carnivorous sharks are opportunistic hunters, honing in on weak and dying prey, belies their reputation.

Species such as the reef whitetip, gray reef, and hammerhead (Fig. 14.22b–d) are regularly attracted to dead fishes set as bait by **scuba** divers. I have witnessed several occasions when these sharks ate such bait while swimming frantically among surrounding divers. Even when agitated, these sharks do not attack the divers and can almost always be chased away easily on the extremely rare occasions when they do attack. Other shark species, such as the tiger and great white, are more dangerous, and cautious divers will leave the water when they appear.

Although many sharks eat fishes and marine mammals, a number of shark species feed only on plankton. Plankton-eating sharks include basking sharks that grow to 12 m in length and whale sharks (Fig. 14.22e) that grow to more than 15 m. The beautiful and graceful

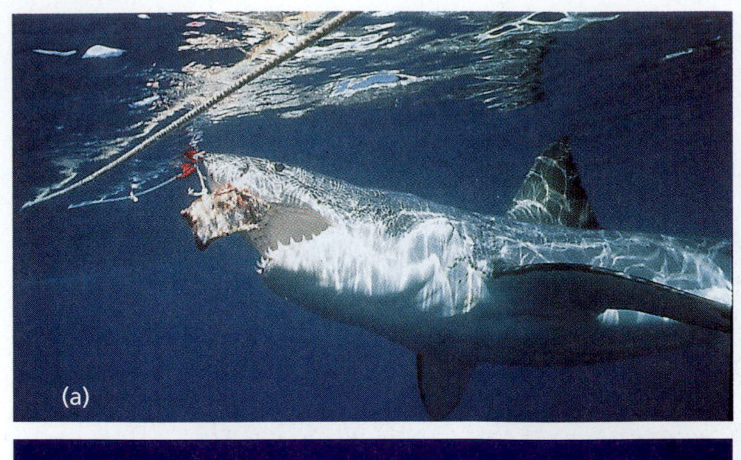

FIGURE 14.22 Representative sharks and rays. (a) Great white shark (*Carcharodon carcharias,* South Australia), attracted to a diver's bait. (b) Reef whitetip shark (*Triaenodon obesus,* Papua New Guinea) with a sharksucker, or remora (*Echeneis naucrates*) attached. (c) Gray reef shark (*Carcharhinus amblyrhynchos,* Fiji), which is called a "bronze whaler" in Fiji. (d) Scalloped hammerhead shark (*Sphyrna lewini,* Papua New Guinea). (e) The whale shark (*Rhincodon typus,* Papua New Guinea) is the largest known shark species. This one was quite small, only about 8 to 10 m long. (f) The graceful manta ray (*Manta birostris,* Philippines). (g) Up close and personal with a nurse shark (*Nebrius concolor,* Philippines). (h) Blue-spotted rays (*Taeniura lymma,* Red Sea) have a dangerous barb on the tail. (i) The torpedo ray (family Torpedinidae, Red Sea) can inflict a powerful electric shock. The whale shark and manta ray are plankton feeders, the nurse shark eats benthic invertebrates, and the other sharks are scavengers and carnivores.

manta ray is also a plankton eater (**Fig. 14.22f**). Plankton-eating sharks and rays swim with mouths wide open. Large volumes of water pass into the mouth and are filtered to capture the plankton before flowing out of the **gills.** The nurse shark (**Fig. 14.22g**) and many species of rays (**Fig. 14.22h,i**) feed only on mollusks and other benthic animals.

Many rays, such as the blue-spotted sting ray (**Fig. 14.22h**), have a barbed stinger embedded in the tail. They use the stinger as a defensive weapon against natural predators, such as sharks, and occasionally against a human foot. Torpedo rays (**Fig. 14.22i**) deliver a strong electric shock that can be lethal to humans, as much as 200 volts, from an organ in the head.

Squid and Their Relatives

Squid (**Fig. 14.23a**), nektonic mollusks of the class Cephalopoda, are extremely abundant in the oceans. Squid normally live in schools that can contain huge numbers of individuals. They are extremely fast swimmers (using their fins) and voracious predators of small fishes. They can also jet-propel themselves by ingesting water and forcibly squirting it out through a special cavity. They can even propel themselves several meters into the air. Most squid species live below the photic zone during the day and migrate to the mixed layer at night to feed. The largest squid species, the giant squid, can grow to 16 m or more in length (6 m of body and 10 m of tentacles). It

FIGURE 14.23 Cephalopod mollusks. (a) A common reef squid (*Sepioteuthis lessoniana,* Indonesia) photographed at night. (b,c) Broadclub cuttlefish (*Sepia latimanus,* Indonesia and Papua New Guinea). (d) A crinoid cuttlefish (*Sepia* sp., Papua New Guinea). (e) *Nautilus* (*Nautilus pompilius,* Papua New Guinea), captured at a depth of several hundred meters in a trap similar to a lobster pot, which was baited with fish. The *Nautilus* was released to swim back to its deep-water home after this photograph was taken. (f) This sectioned *Nautilus* shell shows the internal chambers that are maintained at a near vacuum to provide buoyancy. Because these chambers are incompressible, *Nautilus* can safely migrate great vertical distances.

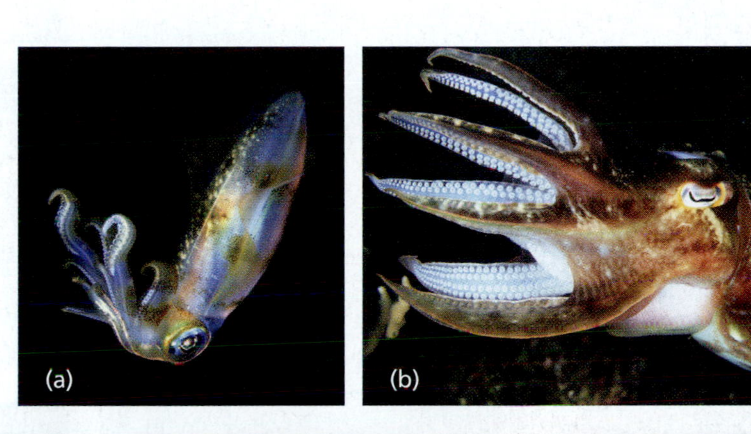

FIGURE 14.24 Cetaceans and sea cows. (a) The relative sizes of various toothed and baleen whales, compared to humans and the largest known dinosaur, *Apatosaurus*. (b) Humpback whale (*Megaptera novaengliae,* Hawaii). (c) Killer whale, often called "orca" (*Orcinus orca*). (d) A manatee mother (*Trichechus manatus*) with her two calves in the Crystal River, Florida.

lives in the deep ocean and is rarely caught or seen, but it is known to be a favorite food of sperm whales.

Cuttlefish (**Fig. 14.23b–d**) are **cephalopods** that have an unusual internal single "bone" or shell (the "cuttlebone" that is typically hung in parakeet cages). *Nautilus* species (**Fig. 14.23e**) have a calcium carbonate shell with a series of internal chambers (**Fig. 14.23f**). They are an ancient form of cephalopod that has changed little over millions of years. All cephalopods are believed to have had external shells at one time. *Nautilus* species, cuttlefish, and squid represent steps in cephalopod evolution in which the external shell, an excellent means of defense, has been discarded in favor of greater swimming speed, which is also an excellent means of defense, but more valuable for hunting prey.

Marine Mammals

Marine mammals, which include seals and sea lions, dolphins and other whales, and a number of other less familiar animals, are warm-blooded and breathe air. Their young are born live and nursed by their mothers, just as terrestrial mammals are.

FIGURE 14.25 Pinnipeds and otters. (a) Harbor seals (*Phoca vitulina*) haul out on the rocks near Monterey, California. (b) A scientist measures the length of an adult male elephant seal (*Mirounga angustirostris*) at the Año Nuevo State Reserve, California. This can be a dangerous job for the scientist because the elephant seal may weigh well over a tonne, can move surprisingly fast, and can be very aggressive when disturbed. (c) California sea lion (*Zalophus californianus*) populations are growing very quickly and becoming a nuisance in many harbors, where they aggressively compete for space when they haul out on any convenient platform, especially boat docks like this one in Monterey, California. (d) Walrus (*Odobenus rosmarus*) apparently use their tusks as the equivalent of sled runners that enable them to glide along the seafloor as they feed by sucking or slurping up sediments with their whiskered mouths. (e) Sea otters (*Enhydra lutris*, California) sometimes eat shellfish by resting the shellfish on their chests and pounding them open with a stone while floating on their backs.

The **cetaceans,** which include dolphins, porpoises, and other whales, live their entire lives in the water. The largest whales—the blue, finback, right, sei, and humpback (**Fig. 14.24a,b**)—are baleen whales that feed on plankton. The smaller gray whale, also a baleen whale, is unique because it feeds mainly on small crustaceans, mollusks, and worms that it stirs up from muddy sediments with its snout. Sperm whales (**Fig. 14.24a**) feed primarily on squid, whereas most small whales, including the pilot and beluga whales, dolphins, and porpoises, eat mostly fishes. However, the killer whale (**Fig. 14.24a,c**) lives up to its name by hunting and eating fishes, seals, sea lions, and even other whales. Killer whales often hunt in packs like wolves. On several occasions, pods of killer whales have been filmed systematically harassing a California gray whale mother and calf until they were separated. If they successfully separate the mother and calf, the killer whales, as a pack, set upon the calf, which they then kill and eat.

Manatees (**Fig. 14.24d**) and dugongs, commonly called "sea cows," are herbivorous sirenians that graze on vegetation in the shallow tropical coastal waters where they live. Sea cows are considered to be the source of the mermaid (or siren) legend.

Seals (**Fig. 14.25a,b**), sea lions (**Fig. 14.25c**), and walrus (**Fig. 14.25d**) are **pinnipeds**. Pinnipeds live and feed in the oceans for most of their lives, but they must haul themselves out of the water to breed and bear young and to rest and conserve body heat. The elephant seal (**Fig. 14.25b**), whose bizarre-looking adult males weigh 2 tonnes, is the largest pinniped. Most pinnipeds eat fishes, but walrus (**Fig. 14.25d**) eat clams and other **shellfish** from the sediment, and Antarctic leopard seals will eat invertebrates, seabirds, and other mammals.

The sea otter (**Fig. 14.25e**) is a unique shellfish-eating mammal that, unlike all other marine mammals, lacks an insulating layer of **blubber**. Instead, sea otters retain body heat with soft, thick fur that they must continuously groom if it is to retain its insulating properties.

Many marine mammals were once hunted intensively for their fur, meat, and oil. Because they have long life cycles, hunting quickly reduced some species to critically low population levels. Although certain species are still hunted, the hunting of marine mammals was stopped by most nations several decades ago. Most marine mammal populations that were decimated by hunting are now recovering slowly. However, some species are still threatened by the destruction of **habitat** (particularly pinniped breeding areas), by **pollution**, and by some fishing methods, such as the use of **drift nets** that kill mammals as well as targeted fishes (Chap. 19).

Reptiles

Only a small number of reptiles, which are air-breathing animals, live in the oceans. The best known are sea turtles, of which four large species are widely distributed in the oceans: the green (**Fig. 14.26a**), hawksbill (**Fig. 14.26b**), leatherback, and loggerhead. Green turtles are herbivorous and graze on sea grasses that grow in shallow **lagoons**. One common species of sea grass is known as "turtle grass" because it is a favorite food of these turtles. The hawksbill lives in tropical waters and eats mostly **sponges**. The loggerhead also eats sponges, but it adds crabs and mollusks to its diet. The biggest turtle, the leatherback, can weigh as much as 650 kg and is among the few large animals in the ocean that eat jellyfish.

Turtles have been hunted for centuries. They are easy to catch and, if placed on their backs out of water, live for many weeks but cannot escape. For this reason, turtles were an ideal source of fresh meat on sailing vessels before refrigeration was developed. Turtles were common just 200 years ago, but all turtle species have been decimated by humans, and most are threatened or endangered by extinction. The Kemp's ridley turtle, a relatively small species that lives only in the Gulf of Mexico and South Atlantic, is especially in danger of extinction.

Turtles lay their eggs in beach sands above the high-tide line. The eggs hatch in several weeks, and the young turtles immediately head to the sea. Turtle eggs are considered a delicacy in many areas, and the fact that they are easily dug out of the sand by human and other predators has contributed to the decline of turtle populations. In addition, turtles are easy for predators to catch when they climb onto the beach to lay their eggs and when they emerge as hatchlings.

Turtles use relatively few beaches throughout the world for breeding. They almost always return to lay eggs on the same beach where they hatched and usually will not lay eggs on any other beach. Because they are reluctant to lay eggs where human development has altered a beach, they have abandoned many traditional breeding beaches. The reduction in the number of breeding beaches that are suitably protected has seriously affected turtle populations.

The Pacific and Indian Oceans have about 50 species of sea snakes (**Fig. 14.26c,d**). Most are highly venomous, and their bite is generally lethal to people. Fortunately, sea snakes have small mouths and small fangs that cannot easily break human skin, unless they catch a finger or other small appendage. Sea snakes are also relatively shy and do not attack unless threatened. They live near the shore so that they can slither out of the water onto rocks to sun themselves. They eat small fishes, can dive as deep as 100 m, and can stay submerged for more than 2 h before they must come up for air.

In certain areas of Asia and the Pacific Islands, saltwater species of crocodiles feed on fishes in coastal waters. American alligators are also known to enter coastal-ocean waters, but normally they are confined to freshwater. The only marine species of lizard is the Galápagos marine iguana (**Fig. 14.26e,f**), which lives on land but frequently enters shallow coastal waters to graze on marine algae.

Birds

Penguins (**Fig. 14.27a**) are flightless birds that live in the oceans and feed on fishes. They leave the water only to lay eggs and brood their young. Most species are restricted to the cold waters near Antarctica, and none are present in the Northern Hemisphere.

Many species of seabirds feed exclusively on ocean fishes. Some, including most seagulls (**Fig. 14.27b**), feed at the ocean surface, but many, like the pelican (**Fig. 14.27c**), dive into the water to catch their prey. Others, such as the cormorant (**Fig. 14.27d**), are excellent swimmers that can dive tens of meters deep and stay submerged for as much as several minutes in pursuit of prey.

14.12 BENTHOS

The challenges and opportunities for marine animals in the benthic environment are different from those in the pelagic environment. The physical nature of the seafloor can change dramatically over relatively short distances, especially in coastal areas. The seafloor can be covered with soft mud into which organisms can easily burrow to live, hide, or feed, or it can be solid rock that provides a stable surface for organisms to attach to or bore into.

FIGURE 14.26 There are only a small number of marine reptile species. (a) Green sea turtle (*Chelonia mydas,* Hawaii). (b) Hawksbill turtle (*Eretmochelys imbricata,* Fiji). (c,d) Banded sea snake (*Laticauda colubrine,* Indonesia). (e,f) Marine iguana (*Amblyrhynchus cristatus*), which lives only in the Galápagos Islands.

Benthic communities have evolved to take advantage of each different type of seafloor between these two extremes. Consequently, benthic communities are more varied than pelagic communities, particularly in the **coastal zone**, where seafloor character varies the most. Deep-ocean benthos must obtain food from particulate organic matter that rains down from above or by preying on other organisms.

Chapter 16 discusses the constraints and strategies of life on the seafloor and the approaches that various species use to meet the challenges. Chapter 17 describes benthos in some specific regions of the seafloor, including those of the deep-ocean seafloor, the unique hydrothermal vent communities of the deep sea, and the remarkably different coastal benthos communities of the coral reef, kelp forest, and rocky intertidal zone.

FIGURE 14.27 Swimming seabirds that feed in the marine environment. (a) Adélie penguins (*Pygoscelis adeliae,* Antarctica). (b) Herring gulls (*Larus argentatus,* California), one of the many species of seagulls. (c) Brown pelican (*Pelecanus occidentalis,* Bonaire). (d) Double-crested cormorants (*Phalacrocorax auritus,* Monterey, California), which are excellent underwater swimmers.

CHAPTER SUMMARY

Production, Consumption, Decomposition. Almost all organic matter on which marine life depends for food is synthesized from carbon dioxide and nutrients by photosynthesis or chemosynthesis—a process called "primary production." Organisms are autotrophs if they synthesize their own food, and heterotrophs if they do not. Autotrophs and heterotrophs convert some organic matter to carbon dioxide and water in respiration. Heterotrophs can be herbivores, carnivores, omnivores, detritivores, or decomposers.

Primary Production and Light. Light is needed to provide energy for photosynthesis. Light is absorbed by seawater and scattered by suspended particles. It penetrates to a depth of a few hundred meters in clear waters and much less in high-turbidity waters. The photic zone is the upper layer of the oceans where light is sufficient to support photosynthesis. Most primary production is carried out by phytoplankton, which are floating microscopic algae. Benthic algae are present only where the seafloor is within the photic zone. As light decreases with depth, photosynthesis is slowed but respiration remains unchanged. The depth at which photosynthesis equals respiration is the compensation depth. This depth is shallower in winter and in high-turbidity water.

Primary Production and Nutrients. Phytoplankton require nitrogen, phosphorus, iron, and other nutrients. If the concentration of a particular nutrient is too low to support growth, it is a limiting nutrient. Nutrients are taken up through the cell membrane. The small size of phytoplankton provides a large surface area (per volume) for this uptake and to retard sinking. Phosphorus is recycled rapidly to solution in liquid excretions of animals and during the decay of dead organisms. Nitrogen is recycled more slowly through several chemical forms and is more often the limiting nutrient in the ocean. In some areas where river runoff and atmospheric dustfall are both low, iron can become the limiting nutrient. Silica is an important nutrient for diatoms and other plankton that have silica-based hard parts and is recycled very slowly. Nutrients are carried below the mixed layer by the sinking of dead organisms and fecal pellets. Nutrients are provided to the photic zone primarily by rivers, recycling, and upwelling. They are relatively abundant in waters below the permanent thermocline.

Food Webs. Phytoplankton are eaten by herbivores or omnivores, which are eaten by small carnivores, which in turn are eaten by larger carnivores. Each organism in this food chain is at a higher trophic level. The conversion efficiency of food to biomass is about 10% between each trophic level. Many carnivores and omnivores feed on organisms from more than one trophic level, forming a more complex food web.

Geographic Variation in Primary Production. The highest primary productivity is in upwelling regions, most of which are in coastal waters along the western margins of continents. Upwelling also occurs around Antarctica and across parts of the equatorial regions. Primary productivity is lowest in the centers of the subtropical gyres.

Dissolved Oxygen and Carbon Dioxide. Dissolved oxygen is released during photosynthesis and consumed during respiration. Gases are exchanged between ocean and atmosphere across the sea surface. Oxygen and carbon dioxide are saturated in surface waters. Oxygen can be supersaturated in shallow parts of the photic zone where photosynthesis exceeds respiration. Below the thermocline, oxygen is consumed in respiration and decomposition, and its concentration is below saturation. In some areas where the photic zone has high primary productivity and bottom waters have long residence time, bottom water is anoxic and may contain toxic hydrogen sulfide.

Organic Carbon. Organic carbon is present in dissolved and suspended particulate forms and in phytoplankton and animal tissues. More than 95% is nonliving dissolved organic matter, and most of the rest is particulate detritus. Most of the dissolved and particulate organic matter is difficult to decompose and has limited nutritional value. Organic particles are more abundant in the photic zone and in areas of high primary productivity.

Biological Provinces and Zones. The benthic and pelagic environments support fundamentally different biological communities. The benthic environment is divided into the hadal zone, which is the deep seafloor below 6000 m, the abyssal zone between 6000 and 2000 m, the bathyal zone between 2000 m and the continental shelf break, the sublittoral zone (water-covered continental shelf), and the intertidal zone. Benthic organisms in the hadal, abyssal, and most of the bathyal and sublittoral zones depend on detritus for food, with the exception of isolated chemosynthetic communities. The pelagic environment is separated into the neritic province in water less than 200 m deep, and the oceanic province in water deeper than 200 m. The oceanic province consists of the epipelagic zone, in which photosynthesis can occur; and the mesopelagic, bathypelagic, abyssopelagic, and hadopelagic zones, which are all below the photic zone. Communities of each of the benthic and pelagic zones are different, but there is some overlap.

In the sublittoral zone, the neritic province, and the epipelagic zone, latitudinal variations of temperature and salinity cause latitudinal zonation in biological communities. For example, tropical and polar regions support different species.

Plankton. Plankton are organisms and viruses that drift freely with currents and include phytoplankton (which are microscopic photosynthetic algae) and zooplankton (which are herbivorous, carnivorous, or omnivorous animals). Phytoplankton are abundant in productive photic-zone waters, exceeding 1 billion individuals per liter, and have patchy distributions. Diatoms are relatively large phytoplankton with silica frustules. They are important in food chains that lead to commercially valuable species. Dinoflagellates are smaller than diatoms, and most lack hard parts. They dominate where silica is depleted, particularly in the open oceans, are not always autotrophic, and often bloom explosively. Generally less abundant types of phytoplankton include coccolithophores that are covered in tiny calcareous plates, and silicoflagellates that have a silica shell. Microscopically small ultraplankton, which include bacteria, archaea, and viruses, are extremely abundant but are not yet well characterized.

Zooplankton are either holoplankton that live their entire life cycles as plankton, or meroplankton that spend only their larval stages as zooplankton. Many zooplankton migrate to the surface layer at night to feed and return below the photic zone by day. Holoplankton include copepods, euphausiids, and other crustaceans, foraminifera, radiolaria, and pteropods, as well as gelatinous forms such as jellyfish, ctenophores, and salps. Meroplankton include eggs, larvae, and juveniles of many invertebrates and fishes.

Nekton. Nekton are organisms that swim actively. They include many fish species. Sharks and rays are fishes that have cartilaginous skeletons. Many fishes have gas- or oil-filled swim bladders to maintain buoyancy. The nekton also include squid, which are extremely abundant, feed mostly near the surface at night, and migrate below the photic zone by day. Marine mammals, including dolphins, porpoises, other whales, seals, and sea lions, are also nekton. Sea turtles and sea snakes are nektonic reptiles.

Benthos. Benthic organisms comprise a profusion of species adapted to live on or in the sediment or on hard substrates such as rocks and coral reefs.

STUDY QUESTIONS

1. Why is it necessary to have primary producers, animals, and decomposers in a marine ecosystem?
2. Dissolved oxygen concentration varies substantially with depth. Describe and explain the distribution of dissolved oxygen in the deep ocean.
3. Why is nitrogen generally the limiting nutrient in the oceans? List the sources of biologically available nitrogen in the photic zone.
4. Of all the organic carbon in the oceans, 99% is in nonliving particulate or dissolved form. Why?
5. Why are most of the world's major fisheries in coastal waters? Why are many major fisheries off the west coasts of the continents? Why are the waters around Antarctica highly productive?
6. List and describe the principal differences between the pelagic and benthic environments. Why are the boundaries between zones of the benthic and pelagic environments defined by depth ranges?
7. At what depths in the pelagic and benthic environments would you expect to find substantial variation in the species composition of the fauna and flora between tropical, mid, and high latitudes? Why?
8. Describe the major types of plankton.
9. Why are phytoplankton small?
10. List the major categories of organisms that constitute the nekton. What are the principal distinguishing characteristics of each category?

KEY TERMS

You should be able to recognize and understand the meaning of all terms that are in boldface type in the text. All of those terms are defined in the Glossary. The following are some less familiar or often misused key scientific terms that are used in this chapter and that are essential to know and be able to use in classroom discussions or on exams.

abyssal zone (p. 389)
adsorption (p. 378)
aerobic (p. 388)
algae (p. 372)
anemones (p. 396)
angle of incidence (p. 376)
anoxic (p. 387)
anthropogenic (p. 387)
aphotic zone (p. 377)
archaea (p. 372)
autotroph (p. 373)
bacteria (p. 372)
baleen (p. 384)
barnacles (p. 395)
bathyal zone (p. 390)
benthic (p. 385)

benthos (p. 385)
bioluminescent (p. 397)
biomass (p. 374)
bloom (p. 381)
calcareous (p. 394)
carnivore (p. 374)
cartilaginous (p. 399)
cephalopods (p. 402)
cetaceans (p. 403)
chemosynthesis (p. 373)
chlorophyll (p. 374)
chordates (p. 397)
cnidarians (p. 396)
coccolithophores (p. 394)
colloidal (p. 388)
community (p. 374)
compensation depth (p. 377)
copepods (p. 395)
corals (p. 396)
crustaceans (p. 384)
ctenophores (p. 397)
decomposer (p. 374)
detritivore (p. 374)
detritus (p. 374)
diagenetic (p. 388)
diatoms (pp. 376, 392)
diffusion (p. 378)
dinoflagellates (pp. 376, 394)
diurnal (p. 381)
ecosystem (p. 372)
enzymes (p. 378)
epifauna (p. 390)
epipelagic zone (p. 391)
eukaryotes (p. 372)
euphausiids (p. 395)
eutrophication (p. 379)
excrete (p. 374)
fauna (p. 391)
fecal pellets (p. 374)
filter feeding (p. 389)
flagella (flagellates) (pp. 376, 394)
flora (p. 391)
food chain (web) (pp. 374, 384)
foraminifera (p. 395)
frustule (p. 381)
fungi (p. 372)
genus (p. 372)
gills (p. 401)
grazers (p. 381)
hadal zone (p. 389)
hard parts (p. 381)
herbivore (p. 374)
herring (p. 384)
heterotroph (p. 373)
holoplankton (p. 395)
infauna (p. 390)
intertidal zone (p. 390)
invertebrates (p. 395)
kelp (p. 376)
krill (p. 384)
limiting nutrient (p. 378)
littoral zone (p. 390)
macroalgae (p. 392)
marine mammals (p. 398)
meroplankton (p. 395)
metamorphosis (p. 389)
mixed layer (p. 381)
mollusks (p. 396)
nekton (p. 395)
nutrient (p. 373)
omnivore (p. 374)
overfishing (p. 398)
pelagic (p. 385)
photic zone (p. 375)
photosynthesis (p. 373)
phytoplankton (p. 376)
pinnipeds (p. 403)
plankton (p. 377)
primary production (pp. 373, 378)
prokaryotes (p. 372)
protist (p. 372)
pteropods (p. 396)
radiolaria (p. 395)
reptiles (p. 398)
respiration (p. 373)
rocky intertidal zone (p. 391)
salps (p. 397)
saturation solubility (p. 387)
sea cucumbers (p. 398)
sea grasses (p. 392)
sea stars (p. 395)
sea urchins (p. 398)
shellfish (p. 404)
siliceous (p. 392)
species (p. 372)
sponges (p. 404)
standing stock (p. 385)
sublittoral zone (p. 390)
supralittoral zone (p. 390)
swim bladder (p. 398)
trophic level (p. 383)
tunicates (p. 397)
vertebrates (p. 397)
zooplankton (p. 374)

CRITICAL CONCEPTS REMINDER

CC4 **Particle Size, Sinking, Deposition, and Resuspension** (pp. 376, 381, 388). Suspended particles in ocean water, including plankton, sink at rates primarily determined by particle size: large particles or plankton sink faster than small particles. Larger plankton species have various adaptations to reduce their sinking rate. Small organic particles can be aggregated into fecal pellets that are larger than the individual particles and, thus, sink faster. To read **CC4** go to page 12CC.

CC8 **Residence Time** (p. 391). The residence time of seawater in a given segment of the oceans is the average length of time the water spends in the segment. In the deep ocean trenches water column water mass residence time can be long and this affects the character of the biota that inhabits this zone. To read **CC8** go to page 19CC.

CC10 **Modeling** (pp. 379, 380). Complex environmental systems including the cycling of nutrients among ocean water, living matter, suspended particles and sediments can best be studied by using conceptual and mathematical models. Many oceanographic and climate models are extremely complex and require the use of the fastest supercomputers. To read **CC10** go to page 26CC.

CC14 **Photosynthesis, Light, and Nutrients** (pp. 373, 374, 386). Photosynthesis and chemosynthesis are the two processes by which simple chemical compounds are made into the organic compounds of living organisms. Photosynthesis depends on the availability of carbon dioxide, light, and certain dissolved nutrient elements including nitrogen, phosphorus, and iron. Chemosynthesis does not use light energy and instead depends on the availability of chemical energy from reduced compounds which occur only in limited environments where oxygen is depleted. To read **CC14** go to page 46CC.

CC15 **Food Chain Efficiency** (pp. 374, 383, 389). All organisms use some of their food as an energy source in respiration and for reproduction, and also lose some in excretions including wastes. On average, at each level in a food chain, only about 10% of food consumed is converted to growth and biomass of the consumer species. To read **CC15** go to page 49CC.

CHAPTER 15

CHAPTER OUTLINE

CRITICAL CONCEPTS USED IN THIS CHAPTER

The oceans and the land are intimately related where they meet. The vegetation, rocks, beach, waves, clouds, kelp seen floating on the water surface, and all the unseen plants and animals living within the coastal waters in this California bay are the result of processes that take place in the coastal oceans and estuaries.

Coastal Oceans and Estuaries

When they went ashore the animals that took up a land life carried with them a part of the sea in their bodies, a heritage which they passed on to their children and which even today links each land animal with its origin in the ancient sea. Fish, amphibian, reptiles, warm-blooded bird and mammal—each of us carries in our veins a salty stream in which the elements sodium, potassium, and calcium are combined in almost the same proportions as in sea water. This is our inheritance from the day, untold millions of years ago, when a remote ancestor, having progressed from the one-celled stage, first developed a circulatory system in which the fluid was merely the water of the sea. In the same way, our lime-hardened skeletons are a heritage from the calcium-rich ocean of Cambrian time. Even the protoplasm that streams within each cell of our bodies has the chemical structure impressed upon all living matter when the first simple creatures were brought forth in the ancient sea. And as life itself began in the sea, so each of us begins his individual life in a miniature ocean within his mother's womb, and in the stages of embryonic development repeats the steps by which his race evolved, from gill-breathing inhabitants of a water world to creatures able to live on land.

—Rachel Carson, ***The Sea Around Us,*** **published in 1951 by Oxford University Press, New York (pp. 13–14)**

Why is this quote relevant to a chapter on the coastal oceans and estuaries? For many reasons! This is the zone where ocean life is concentrated and where a number of species live part of their lives in the oceans and part on the land. This is the region where many of the chemicals that support ocean life are discharged

to the sea. This is a region where many ocean species find habitat, shelter, and food for their young. This is the region where aquatic habitats vary the most, both in the salinity and chemical composition of seawater and in the composition and character of the seafloor. This is the only region of the oceans where rooted aquatic plants may grow. This is the part of the oceans next to which the majority of human populations and our industry have chosen to live and upon which much of our lives relies. What goes on beneath the surface of our estuaries and oceans is hidden from view. Rachel Carson's observation should remind us that, though we may not see all of its parts and connections, we are just one species that lives within the complex ecology of this living ocean planet.

In the coastal oceans, **salinity,** temperature, **density, turbidity,** chemical composition, and water movements are affected by freshwater discharges from rivers, and by the physical presence of land and shallow seafloor. These influences vary in **spatial** extent as seasonal changes in rainfall affect **runoff** and river input of freshwater and as **weather** and seasonal changes in winds affect wind-driven **currents**. Hence, the boundaries of the **coastal zone** are not defined precisely. On **passive margins** where the **continental shelf** is wide, the coastal zone is generally the area between the **coastline** and the continental **shelf break.** Where large river discharges occur on tectonically active margins with narrow continental shelves, the coastal zone may extend beyond the continental shelf.

The continental shelf comprises approximately 8% of the total area of the oceans. The coastal zone, which includes only limited areas beyond the shelf, comprises less than 10%. Nevertheless, the coastal ocean is disproportionately important because more than 99% of the total world fishery catch is taken from coastal waters. The coastal ocean is also important for ports and shipping; recreational activities; mining of sand, gravel, and other minerals; oil exploration and extraction; and waste disposal. Many of these uses are discussed in Chapters 18 and 19. This chapter examines the factors that make the coastal zone and its **ecosystem** different from the open ocean. It also describes the characteristics of **estuaries** and **lagoons** and their relationships to ocean ecosystems.

15.1 SPECIAL CHARACTERISTICS OF THE COASTAL OCEANS

Because the coastal oceans are shallow, water movements are affected by the seafloor and the coastline. In addition, rivers discharge freshwater into coastal waters. As a result,

- The salinity and temperature of coastal waters are more variable than those of open-ocean waters.
- Coastal currents are generally independent of the open-ocean **gyre** currents.
- The **mixed layer** has more sources of **nutrients** in coastal waters than in the open ocean.
- **Benthos** of the coastal oceans are more **diverse,** abundant, and commercially valuable than open-ocean benthos.

These factors make the coastal oceans different and generally more variable than the open oceans. Some of the ways in which these factors affect the coastal oceans are discussed in this section.

Salinity

Freshwater from rivers reduces the salinity of coastal waters. Where river discharges are large, low-salinity water spreads as a surface layer over higher-salinity ocean water unless vertical mixing is intense. Consequently, near mouths of major rivers, a **halocline** often is present a few meters below the surface. The area affected by the halocline and the halocline's strength depend on the rate of discharge of low-salinity water from the river and on the extent of wind-, wave-, and **tide**-induced vertical mixing of the water column (**Fig. 15.1**). Haloclines inhibit **upwelling** and vertical mixing.

Low-salinity water discharged by rivers is mixed with ocean water and transported away from the river mouth by coastal currents. Chapter 10 explained that coastal currents generally flow parallel to the coastline and are often separated from the open-ocean gyre currents by sharply defined **fronts**. Thus, mixing of coastal water with water from the open oceans is slow, and **residence times** of water in the coastal ocean can be long.

The residence time of freshwater in the coastal **water mass** can be calculated if the volume input of freshwater from rivers, the average salinity in the coastal waters, and the volume of the coastal water mass are known (**CC8**). For example, the residence time of freshwater in the coastal oceans of the northeastern United States from Cape Cod, Massachusetts, to Cape Hatteras, North Carolina, is about 2 1/2 years. The long residence time has implications for the use of this area for ocean waste disposal. Although the volume of coastal water is large, the quantities of **contaminants** released into the northeastern United States coastal ocean in discharges from industries and cities are also very large. For contaminants not removed by **sedimentation** or decomposition (Chap. 19), an amount equivalent to 2 1/2 years of discharges is retained within this region.

Salinity varies seasonally with river flow in many coastal areas. In mid and low **latitudes,** salinity is usually lowest in the local rainy season. At high latitudes, salinity is always lowest in late spring or summer when snowmelt swells rivers and seasonal sea ice melts.

Salinity extremes occur in **marginal seas,** lagoons, or other bays with restricted connections to the oceans. In locations such as the eastern Mediterranean and the Red Sea, evaporation is high and rainfall low, so surface

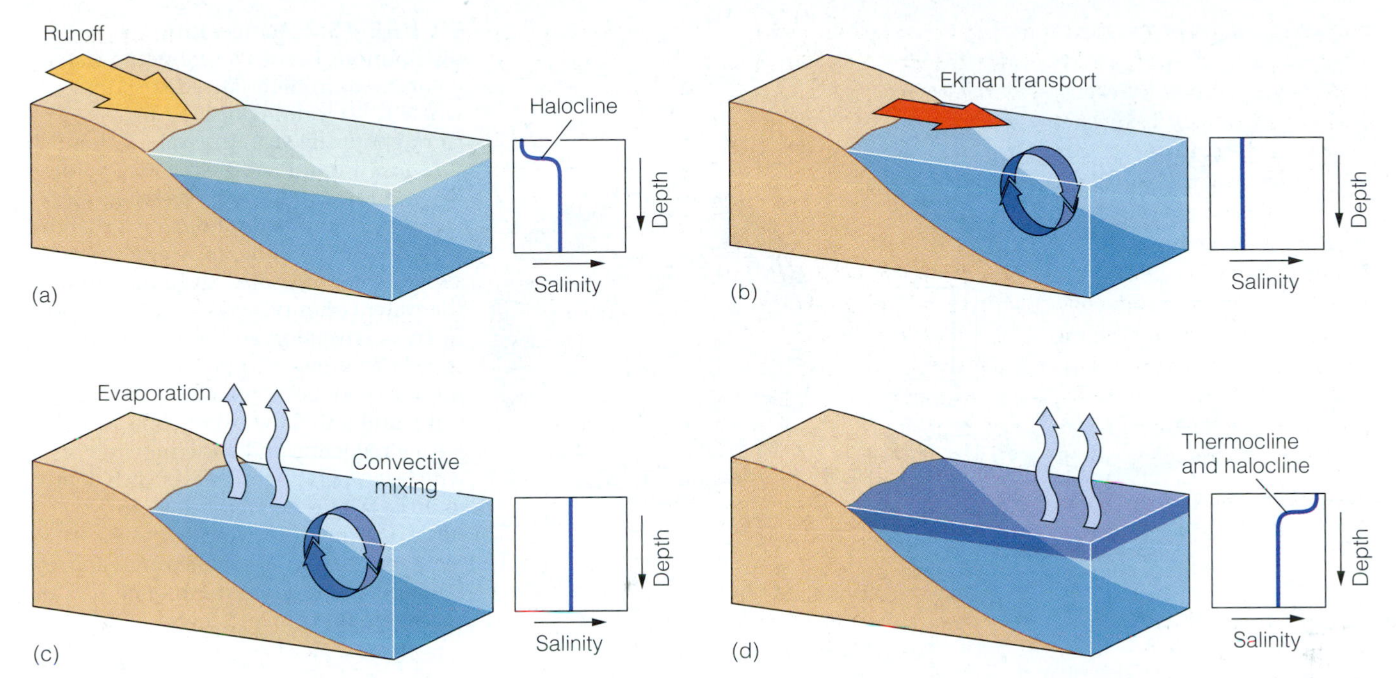

FIGURE 15.1 Salinity distributions in the coastal-ocean water column. (a) Near major rivers, discharges of low-salinity river water over higher-salinity ocean water create a halocline. Where winds transport surface waters offshore, the layer of low-salinity water may spread farther and be thinner. Lower-volume river discharges produce thinner, low-salinity layers that do not extend far from the river mouth. (b) Where winds are strong and river discharge is relatively small, the low-salinity river water may be completely mixed with ocean water, and as a result, there is no halocline. (c) In areas of strong evaporation and low rainfall, warm, high-salinity surface water is produced that may be dense enough to sink all the way to the seafloor. (d) Conversely, this warm, high-salinity water may be warm enough that its density is lower than the lower-salinity but cooler offshore water. In this case, the warm, high-salinity water forms a higher-salinity surface layer overlying a halocline in which salinity decreases with depth.

salinity is high—in some areas exceeding 40. In these regions, the mixed layer is usually deep because vertical circulation is enhanced by the formation of surface waters with relatively high density due to high salinity. The mixed layer often extends to the seafloor and consists of warm, high-salinity water. The vertical circulation helps to recycle nutrients from relatively deep waters into the **photic zone.** Consequently, many of these regions have moderately high productivity, despite their limited inputs of nutrients in runoff from land. However, in some parts of such regions, lower-salinity surface water from the open ocean may flow as a surface layer over the high-salinity water created by evaporation in another part of the basin (**Fig. 15.1c**). In some basins with high evaporation and low rainfall, higher-salinity but warmer surface water has lower density than the lower-salinity but colder open-ocean water. Thus, a warm, high-salinity surface layer is formed (**Fig. 15.1d**).

Salinity is lowest where river outflow is large and evaporation is low. Such conditions generally occur in mid- and high-latitude locations, including many **fjords** and the Baltic Sea. The huge **monsoon** flows of the Ganges and Brahmaputra rivers lower the surface salinity of the entire Bay of Bengal (**Fig. 9.15**).

Temperature

Seasonal changes in solar intensity and air temperature cause larger temperature changes in shallow coastal waters than in the open ocean. Temperature variations are particularly large in enclosed marginal seas, fjords, and other embayments that have restricted exchange with the open ocean. In tropical regions, coastal water temperature is uniformly high year-round, and the water column is often **isothermal** to the seafloor, particularly in shallow lagoons (**Fig. 15.2c**). In deeper coastal waters, the depth of the mixed (or isothermal) layer is determined by physical processes, including wind-, tide-, and wave-induced mixing.

In high latitudes, coastal water temperatures vary little between seasons (**Fig. 15.2a**). Sea ice in these regions melts during summer and re-forms during winter. This process maintains the water temperature at the freezing point of seawater, usually above –2°C (Chap. 7, **CC5**).

Seasonal temperature changes in coastal waters are most evident in mid latitudes (**Fig. 15.2b**). In these regions, the surface water layer becomes warmer and less dense as solar heating increases in spring, reaching a temperature maximum in summer. In addition, in many

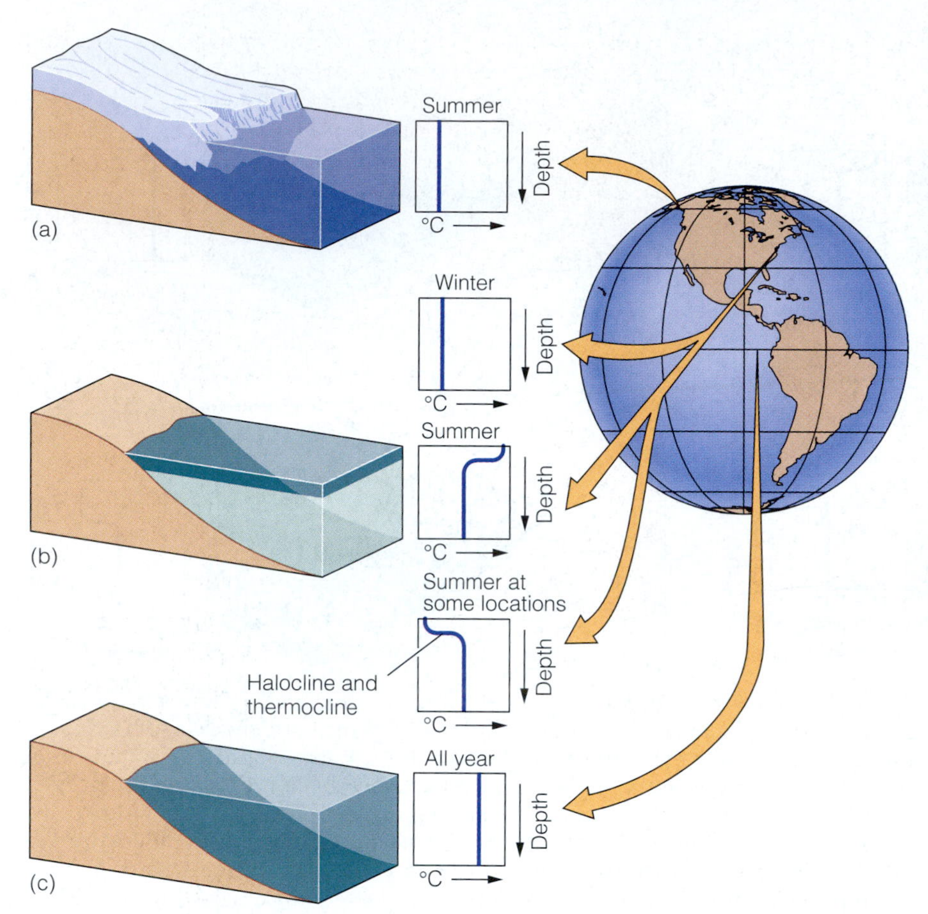

FIGURE 15.2 Temperature distributions in the coastal-ocean water column. (a) In high-latitude regions where ice is forming or melting throughout the year, the water column is uniformly at or near the freezing point. (b) In mid latitudes, surface waters are warmed in summer, producing a shallow seasonal thermocline. In winter the surface water is cooled and sinks, and the water column is generally well mixed by this convection and by winds. In some areas, low-salinity runoff may produce cool but low-salinity surface water and a shallow pycnocline that is both a halocline and a thermocline in which temperature increases with depth. (c) In low-latitude regions where freshwater inputs are limited, the temperature of the water column above the permanent ocean thermocline depth is usually uniform and high.

locations, mixing by wind and waves decreases during summer. As a result, a shallow **thermocline** is formed and the mixed-layer depth is reduced to about 15 m (**Fig. 15.2b**). This seasonal thermocline breaks down in winter when surface water is cooled and **convection** and wind mixing increase. A uniformly cold mixed layer is produced that extends to the top of the permanent thermocline (about 100 m) or the seafloor, if it is shallower. Because most coastal waters are less than 100 m deep, the mixed layer extends to the seafloor in most of the temperate coastal region during winter.

Coastal-ocean water temperature in mid latitudes is usually lower in summer and higher in winter than air temperatures over the adjacent coastal land. This difference affects the **climate** of coastal regions (Chap. 9) by moderating temperature variations.

Waves and Tides

Tidal-current speed increases as the tide wave enters shallow coastal waters and its energy is compressed into a shallower depth (Chap. 12). Consequently, tides are an important contributor to vertical and horizontal mixing in the coastal zone.

As tides **ebb** and **flood,** they cause **turbulence** that enhances vertical mixing. Tides also mix water across the shelf. Near-bottom tidal currents are slowed by **friction** with the seafloor. In addition, tidal currents may be different above and below a **pycnocline** when one is present. Thus, water in different layers moves different distances across the shelf during a tidal cycle. Turbulent mixing between layers transfers water from one layer to another.

Wind-driven waves create turbulence and vertical mixing. In the coastal zone, vertical mixing is enhanced where waves move into shallow waters and break. **Langmuir circulation** and **internal waves** (Chap. 10) also contribute to vertical mixing.

Turbidity and the Photic Zone

The photic zone extends to the seafloor in many areas of the coastal oceans. In these areas, **kelp** and other **macroalgae** (**Fig. 15.3**) can live and **photosynthesize** on the seafloor. However, the photosynthetic **community** of the shallow seafloor is often dominated by **species** of single-celled **algae** similar to **phytoplankton**. These algae **encrust** rocks, dead **coral**, and other organisms that live on the seafloor. A number of such species live inside the tissues of corals, clams, and other **benthic** animals in a **symbiotic** relationship (Chap. 16).

The photic zone is generally shallower and more variable in depth in the coastal oceans than in open-ocean waters because the concentration of suspended particles, and hence turbidity, is higher. Turbidity is higher primarily because of the proximity of river discharges that can carry large quantities of **suspended sediment**. The finest particles are continuously **resuspended** and transported in the coastal zone by waves and currents, so their concentration remains high. Once they have been transported to the lower-energy deep oceans, these particles sink below the photic zone.

The concentration of suspended sediment in coastal waters is variable because it depends on such factors as location in relation to river discharges, river discharge rates, intensity of wave resuspension activity, and the speed of coastal currents. Each of these factors varies spatially and temporally. Similarly, phytoplankton concentrations are highly variable in the coastal zone. As a result, the photic-zone depth can vary substantially over short distances and time periods. Populations of benthic algae, particularly in the relatively deep waters of the mid-shelf region, vary accordingly. High turbidity also affects the health and distribution of **coral reefs** by reducing the light available for photosynthesis by their symbiotic algae, the **zooxanthellae**.

Currents

Because coastal currents (Chap. 10) are dominated by local winds, they vary seasonally and on shorter timescales in response to weather systems. Coastal currents are generally steered by their interaction with the seafloor, so they flow parallel to the coastline. However, interactions between currents and the seafloor or **shoreline** can also cause permanent or semipermanent **eddies** to form.

Nutrients

The supply of nutrients to the coastal photic zone is generally greater and more variable than the supply to open-ocean photic zones. The sources of, and variations in,

FIGURE 15.3 Benthic algae. (a) Various species of green, brown, and red macroalgae in a tide pool on the California coast near San Francisco. (b) A red calcareous macroalga (phylum Rhodophyta, Philippines). (c) The upper fronds of kelp (*Macrocystis* sp.) floating on the water surface in Monterey Bay, California. (d) A coral reef green macroalga called "grapeweed" (*Caulerpa racemosa,* Papua New Guinea). (e) An encrusting red alga (phylum Rhodophyta, Hawaii) covering a platelike hard coral.

nutrient supply are important to the biological characteristics of coastal waters and are discussed in detail in the next section.

15.2 NUTRIENT SUPPLY TO THE COASTAL PHOTIC ZONE

The most productive areas of the oceans are in coastal regions (**Fig. 14.12**). With only a few exceptions, coastal waters have higher **primary productivity** than the adjacent open-ocean waters because a variety of mechanisms supply or resupply nutrients to the photic zone in coastal regions. The most important mechanisms are river inputs and coastal upwelling.

Some rivers discharge large quantities of nutrients into coastal waters, where currents carry water parallel to the coastline so that it mixes only slowly with offshore waters. Consequently, nutrients discharged by rivers tend to be distributed along the **coast** in the direction of the prevailing coastal current. The influence of river-borne nutrients and coastal-current transport can be seen in the high primary productivity near the Amazon River mouth in Brazil (**Fig. 15.4**).

River-borne nutrients are important in many areas, but in other areas, much larger quantities of nutrients are supplied to the coastal photic zone by wind-driven coastal upwelling (Chap. 10). In addition, various other physical mechanisms mix nutrients into photic-zone waters of the continental shelf.

Wind-Driven Coastal Upwelling

Coastal upwelling occurs when wind-driven **Ekman transport** (**Figs. 10.3, 10.5**) moves the surface water layer offshore. Coastal upwelling brings offshore water from depths of 100 m or more to the surface to replace the mixed-layer water that is transported offshore. If the water raised to the surface has high nutrient concentrations and replaces nutrient-depleted surface water, and if the upwelling occurs where photosynthesis is not **light-limited,** the nutrients enhance primary productivity. In coastal-ocean regions where these conditions are met and where winds that cause the upwelling are persistent, primary productivity is high and remains high while upwelling persists. The term "**coastal upwelling region**" is usually applied to such areas. Ekman transport also causes upwelling in other areas where the upwelled water originates from above the permanent thermocline. This water typically is poor in nutrients, and the upwelling has little effect on productivity.

Nutrient-rich water from below the permanent thermocline generally does not penetrate far onto the conti-

FIGURE 15.4 This false color–enhanced Coastal Zone Color Scanner satellite image of the Atlantic coast of South America depicts the elevated (red colors) chlorophyll concentrations associated with the nutrient inputs in the outflow of the Amazon and Orinoco Rivers. The river outflow is transported north along the coast by the coastal currents.

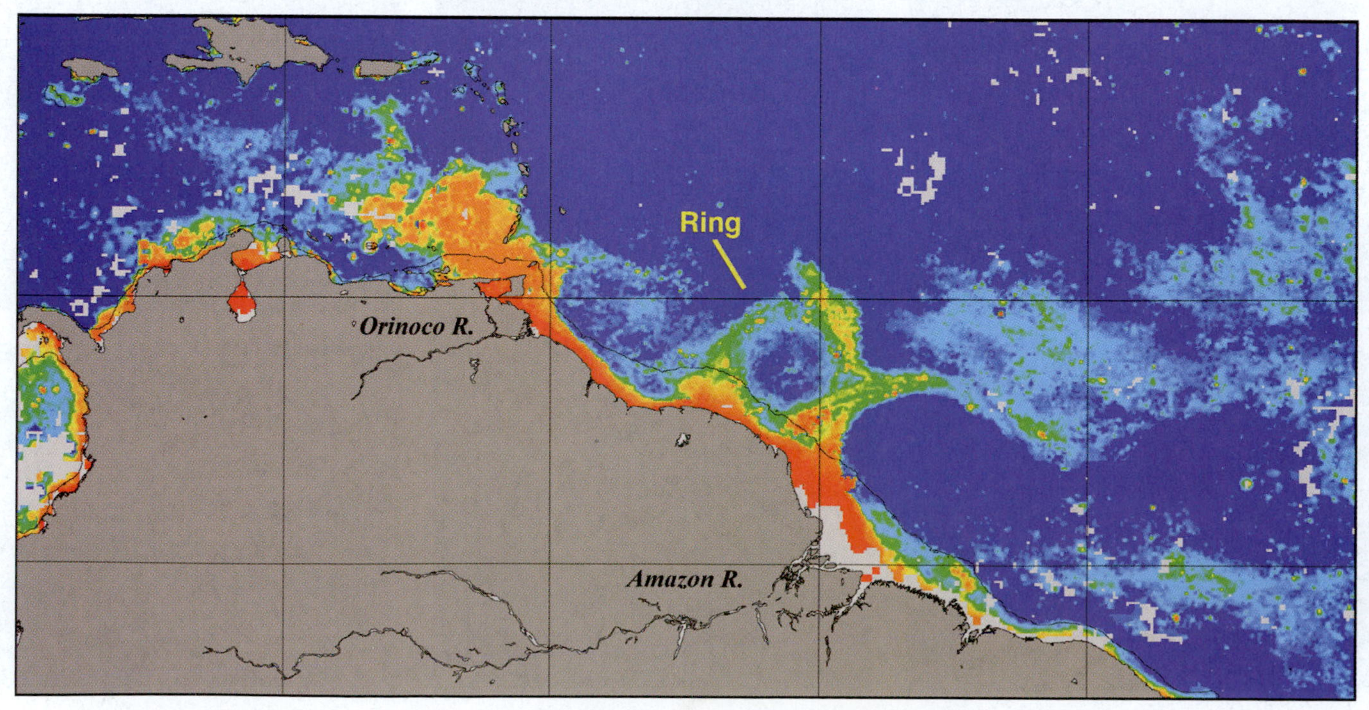

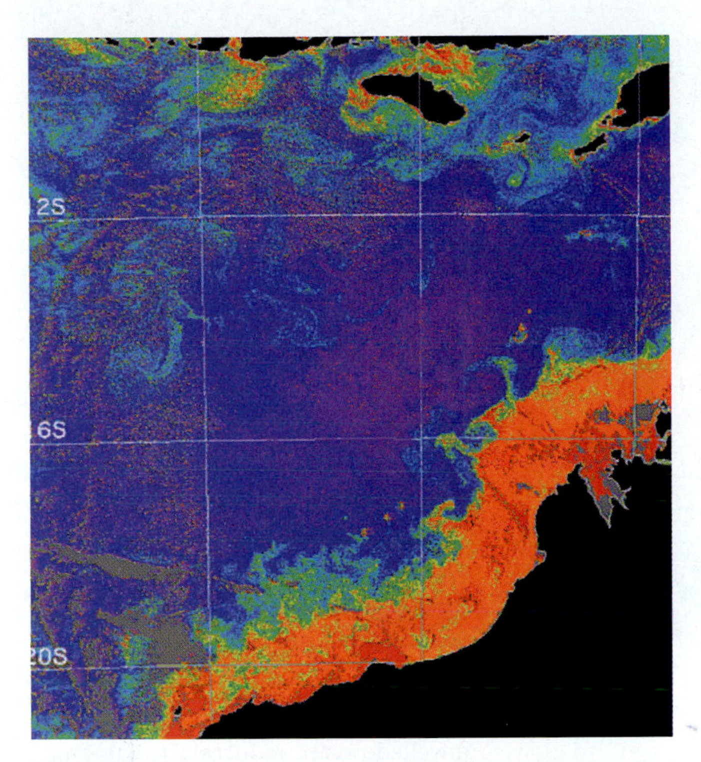

FIGURE 15.5 Surface water chlorophyll concentrations in the Indian Ocean region between Indonesia and northwestern Australia (June 1981). High concentrations are shown in red and are associated with high rates of primary production. The high productivity off northwestern Australia occurs because winds cause offshore Ekman transport and coastal upwelling. Note the complex local variations along the coast. The productivity is much lower near Indonesia, where there is no major coastal upwelling.

nental shelf off mid-latitude east coasts because of the swiftly flowing, deep **western boundary currents** of the ocean gyres (**Fig. 10.10**). In contrast, nutrient-rich water does penetrate onto the continental shelf off the west coasts of the continents. Hence, high-productivity coastal upwelling regions are concentrated off the west coasts of the continents, notably off California, Peru, and the western Africa coast both north and south of the equator (**Fig. 14.12**).

The intensity of upwelling varies locally in coastal upwelling regions (**Fig. 15.5**). It is especially strong and persistent off coasts where prevailing winds produce offshore Ekman transport in regions with dry or desert climates, such as California and Peru. In some other coastal regions, winds cause offshore Ekman transport, but river inputs of freshwater are substantial. In these areas, the low-salinity water discharged by the rivers spreads to form a low-density, relatively shallow surface layer. Because this layer "slides" relatively easily over the higher-density water below it, offshore Ekman transport tends to spread it farther offshore to be replaced by additional low-salinity water from the river. The low-salinity, low-density surface layer inhibits upwelling. If Ekman transport is strong and river flow rate relatively low, upwelling does occur, but the upwelled water may be low-nutrient shelf water from below the shallow halocline rather than high-nutrient water from below the deeper permanent thermocline. Upwelling of nutrient-rich water from below the permanent thermocline can occur only if the winds and offshore Ekman transport are persistent and strong, and if river input remains relatively low.

For coastal upwelling to occur, winds must blow in the appropriate direction long enough to initiate Ekman transport, which then must move sufficient surface water offshore to bring nutrient-rich waters to the surface. This process may take many hours or several days, depending on the wind strength and direction and the depth of the thermocline. In locations where winds are variable in strength and direction, upwelling may be sporadic and highly variable from day to day and from year to year. Coastal upwelling is generally most consistent in the **trade wind** region because trade winds are less variable than winds at other latitudes (Chap. 9).

In many regions, such as off the California coast, climatic winds and coastal currents change direction seasonally (**Fig. 10.17**) and upwelling is also seasonal. For example, upwelling occurs during summer and fall off the California coast, when winds blow from the north. Seasonal upwelling may occur each year, but the timing, intensity, and persistence of upwelling vary from year to year. These variations are important to major fisheries that depend on the upwelling.

Coastal upwelling often extends farther offshore and is stronger and more persistent near capes (**Fig. 15.6**). Fishers have long known about this phenomenon because capes have especially abundant fish, seabird, and often **marine mammal** populations. The reason is still not totally understood, but the phenomenon appears to be the result of an interaction of coastal currents with the undersea **topographic** ridge that usually extends onto the continental shelf as an extension of the cape. The coastal current flows parallel to the coastline, and as it meets the ridge, it is deflected first offshore and then back toward the **shore** as it passes up and over the ridge. This deflection sets up a complex eddy circulation similar to the Gulf Stream rings discussed in Chapter 10. Upwelling is especially intense immediately downstream of the ridge, but upwelled water is also entrained in eddies that form farther offshore.

Biology of Coastal Upwelling Zones

In most highly productive coastal upwelling regions, upwelling persists for several months or more each year. During this period, continuous, although variable,

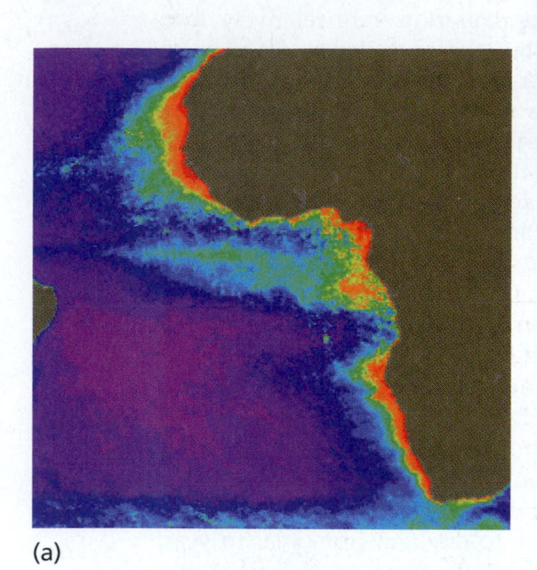
(a)

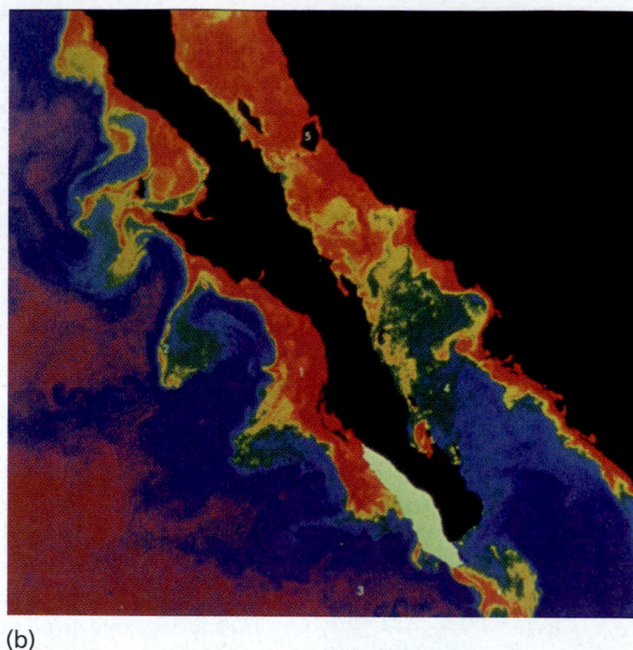
(b)

FIGURE 15.6 These false color–enhanced Coastal Zone Color Scanner satellite images show the elevated (red colors) chlorophyll concentrations associated with the interaction of coastal currents and capes. (a) Upwelling is especially strong and extends farther offshore at each of the several capes on the western Africa coast. (b) The complex gyres that form at capes and that support enhanced primary productivity are clearly seen in this image of the coastal region off Baja California.

onshore–offshore transport processes are established (**Fig. 15.7**). Surface layer water flows offshore and is replaced by cold, nutrient-rich water that flows inshore near the seafloor. As the upwelled cold water is transported offshore, it is progressively warmed by solar heating and mixed with warmer offshore surface water.

Although the newly upwelled water near the coast has high nutrient concentrations and adequate light, phytoplankton growth does not begin immediately. There may be several reasons for the time lag, including the following:

- Newly upwelled water contains few viable phytoplankton or other primary producer cells.
- Dissolved **micronutrient** organic compounds may be lacking in newly upwelled water.
- Dissolved trace metals in the newly upwelled water may be in a more toxic ionic form because the concentration of dissolved organic compounds is low.
- Dissolved nutrient trace metals, such as iron, may be in an ionic form unavailable to some species of primary producers, again because the concentration of dissolved organic compounds is low.

The toxicity and biological availability of metals in solution are known to be changed by chelation, a process in which the metal atom or **ion** is surrounded by an organic molecule or molecules and stable bonds are formed with part of the organic molecule structure. Generally, toxicity is decreased by chelation, and the metal may become more readily available for uptake by living organisms. In laboratory experiments, the addition of a simple organic compound known to be a chelating agent to newly upwelled water reduces the lag time. Therefore, either toxicity or unavailability of metals must account for at least some of the lag time. In newly upwelled water, a variety of small **flagellate** phytoplankton species are apparently less affected by the factors that create the lag time. These species are the first to grow and reproduce actively in the newly upwelled water. In doing so, they are thought to synthesize a variety of organic compounds that they then, at least in part, release to solution. These compounds must include chelating agents and/or the micronutrients needed by other phytoplankton species, which then allows those species to grow.

As upwelled water moves offshore, larger phytoplankton, including chain-forming **diatoms**, reproduce rapidly. In the mid-shelf region, these **blooms** peak, and phytoplankton **biomass** reaches more than 100 times that present in nutrient-poor surface waters farther offshore. The blooms consume and thus deplete nutrients (nitrogen, phosphorus, and silica) in the previously upwelled water. As the nutrient-depleted water is transported farther offshore and mixed with offshore surface waters, the phytoplankton, which are now **nutrient-limited** and no longer actively growing, are reduced by **grazers**. Consequently, the larger phytoplankton are replaced by smaller numbers of flagellate species that are able to grow at low nutrient concentrations.

Zooplankton and **nekton** populations within the coastal upwelling ecosystem are distributed to take advantage of the local food supply. Few zooplankton or nekton are present in newly upwelled water, where primary productivity is still low. The mid-shelf area where concentrations of diatoms are highest has large populations of **herbivorous** fishes and zooplankton. Farther offshore, larger **carnivorous** zooplankton and carnivorous fishes feed on the herbivores brought to them by the offshore flow. These **food webs** are short in comparison with open-ocean food webs (**CC15**). Hence, coastal upwelling fisheries use photosynthetically produced organic matter more efficiently than open-ocean fisheries do (**CC16**).

Fecal pellets and **detritus** produced by organisms in the surface water layer of the upwelling zone fall below the surface layer into the near-bottom water or to the seafloor, where they are decomposed. Nutrients released by this decomposition are added to the water that moves inshore to be upwelled. Thus, upwelling-zone circulation ensures that nutrients are rapidly recycled.

Coastal upwelling circulation is also important to the life cycles of species in the upwelling-zone ecosystem, particularly **plankton,** because they are continuously transported offshore in the mixed layer. If they are carried offshore beyond their optimal location, members of these species can return by sinking through the water column so that they are carried onshore by the upwelling circulation. Many phytoplankton species form resting phases called **spores** or **cysts,** and zooplankton species lay eggs in offshore waters, where they sink into cold subthermocline water that moves onshore. Once upwelled into an area where nutrients or food is available, the phytoplankton revive and zooplankton eggs hatch.

Nekton use a similar strategy to distribute their eggs and **larvae** to favorable locations. Many nekton species migrate inshore or to estuaries to **spawn,** but others may spawn offshore. Their eggs are carried in the near-bottom waters toward shore, where they are upwelled and then hatch. The larvae may change form and feeding strategy

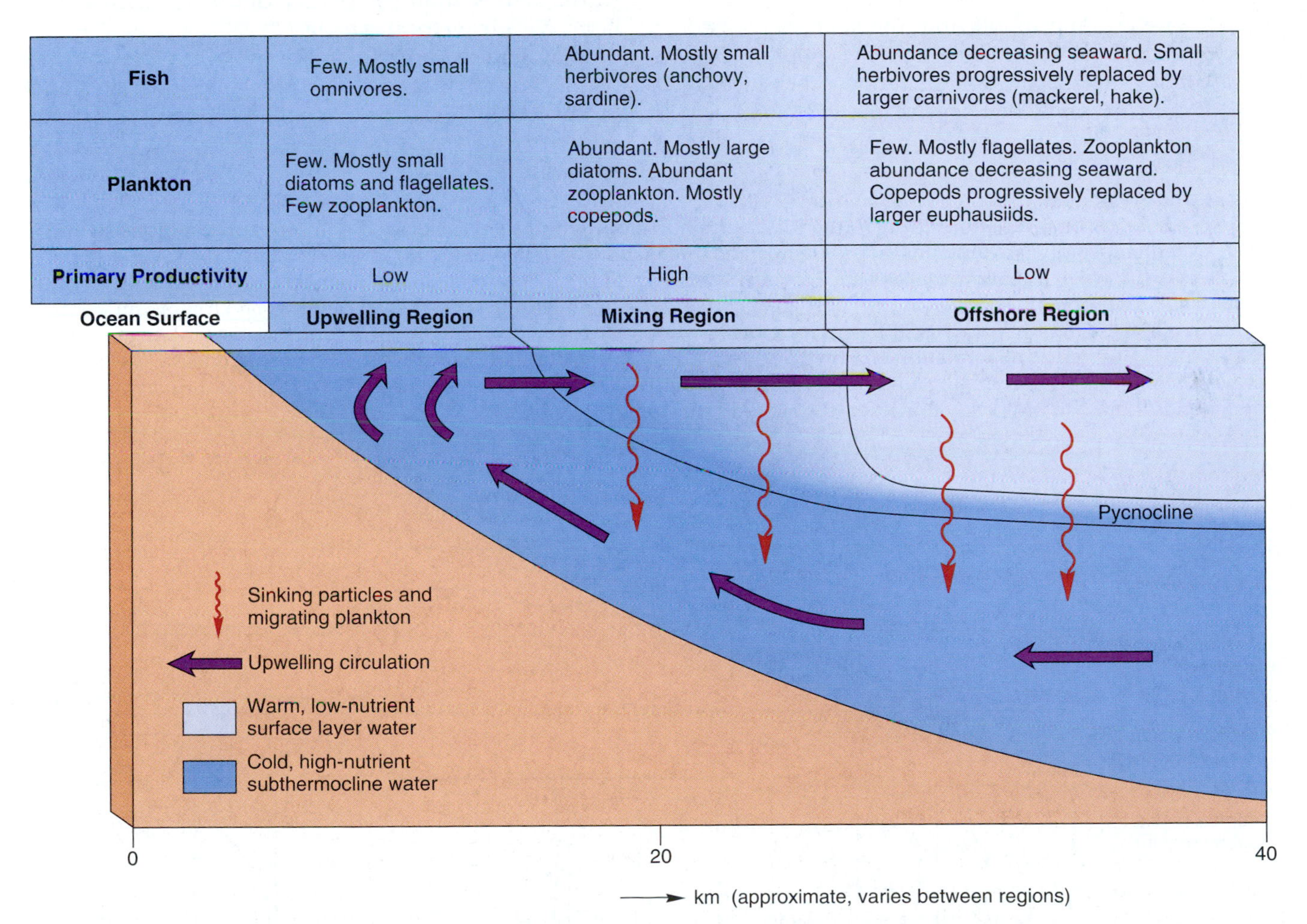

Fish
Few. Mostly small omnivores.
Abundant. Mostly small herbivores (anchovy, sardine).
Abundance decreasing seaward. Small herbivores progressively replaced by larger carnivores (mackerel, hake).
Plankton
Few. Mostly small diatoms and flagellates. Few zooplankton.
Abundant. Mostly large diatoms. Abundant zooplankton. Mostly copepods.
Few. Mostly flagellates. Zooplankton abundance decreasing seaward. Copepods progressively replaced by larger euphausiids.
Primary Productivity
Low
High
Low
Ocean Surface
Upwelling Region
Mixing Region
Offshore Region
Pycnocline
Sinking particles and migrating plankton
Upwelling circulation
Warm, low-nutrient surface layer water
Cold, high-nutrient subthermocline water
0
20
40
km (approximate, varies between regions)

FIGURE 15.7 In coastal upwelling regions, the distribution of water properties and biota is related to the upwelling circulation. Upwelled water, high in nutrients, supports only low levels of primary production at first, but as it moves offshore as the surface layer, it supports a diatom bloom that depletes the nutrients, after which flagellates become dominant. Zooplankton and fish species are distributed to take advantage of this distribution of food. Many species of plankton use the upwelling circulation by changing their depth to be transported either onshore or offshore. Many fish and invertebrate species also use the circulation by placing their eggs and larvae at the right place and depth so that they are transported to a location where the appropriate food source is available at the different stages of their life cycle.

several times to exploit different food sources as they are transported offshore as plankton. A final change to the adult nektonic form occurs when an appropriate location in the upwelling circulation is reached.

Many upwelling-zone species rely on the upwelling circulation to deliver their eggs and larvae to locations where appropriate food is available at specific times during their life cycles. Consequently, in years when upwelling occurs at a different time or differs in intensity and duration, some species may be advantaged or disadvantaged in relation to others. The result is considerable year-to-year variation in the breeding success of many species. Many fisheries may have collapsed because they were **overfished** during one or more years when environmental factors substantially reduced reproduction.

Other Mechanisms of Nutrient Supply

Coastal upwelling and runoff from the land are the two most important mechanisms of nutrient supply to the coastal photic zone. However, there are other mechanisms of nutrient supply that depend on waves and tides, the interaction of ocean currents with the seafloor, and thermal convection.

The mixed-layer depth is partially determined by the intensity and duration of winds. Winds cause vertical mixing due to Langmuir circulation (Chap. 10) and wave action. In regions where a seasonal thermocline is formed, intense storms in summer or fall generate waves with long **wave periods** that disrupt the thermocline. This disruption mixes water from above and below the thermocline, weakens the thermocline, and transports some nutrient-rich waters upward into the mixed layer. Nutrients are also returned to the mixed layer by convection that occurs when the surface layer water cools in fall–winter and the seasonal thermocline disappears (**Fig. 15.8a**). Where the seafloor is shallower than the permanent thermocline depth, detritus falls through the photic zone and accumulates on the shallow seafloor. As this detritus is decomposed, nutrients are released into near-bottom waters, from where they can be returned to the photic zone by vertical mixing.

Long-**wavelength** internal waves that form on the thermocline are slowed as they reach shallow water, and they eventually break on the outer portion of the conti-

FIGURE 15.8 Seasonal variations in solar intensity, nutrient concentrations, phytoplankton and zooplankton biomass, and surface water temperature are different in mid (temperate), high, and tropical latitudes. (a) In mid latitudes, there is a distinct seasonal cycle in which a shallow thermocline forms in summer, a spring bloom of plankton depletes the nutrients, and a smaller fall bloom sometimes occurs as nutrients are recycled. (b) In high latitudes, there is an intense plankton bloom in the short summer when light intensity is sufficient for growth. Nutrients are not depleted, because strong vertical mixing occurs and this mixing prevents the formation of a thermocline. (c) In tropical latitudes, a strong permanent thermocline exists. This thermocline inhibits vertical mixing, and nutrients in the surface layer are permanently depleted. As a result, primary production is limited to uniformly low levels throughout the year.

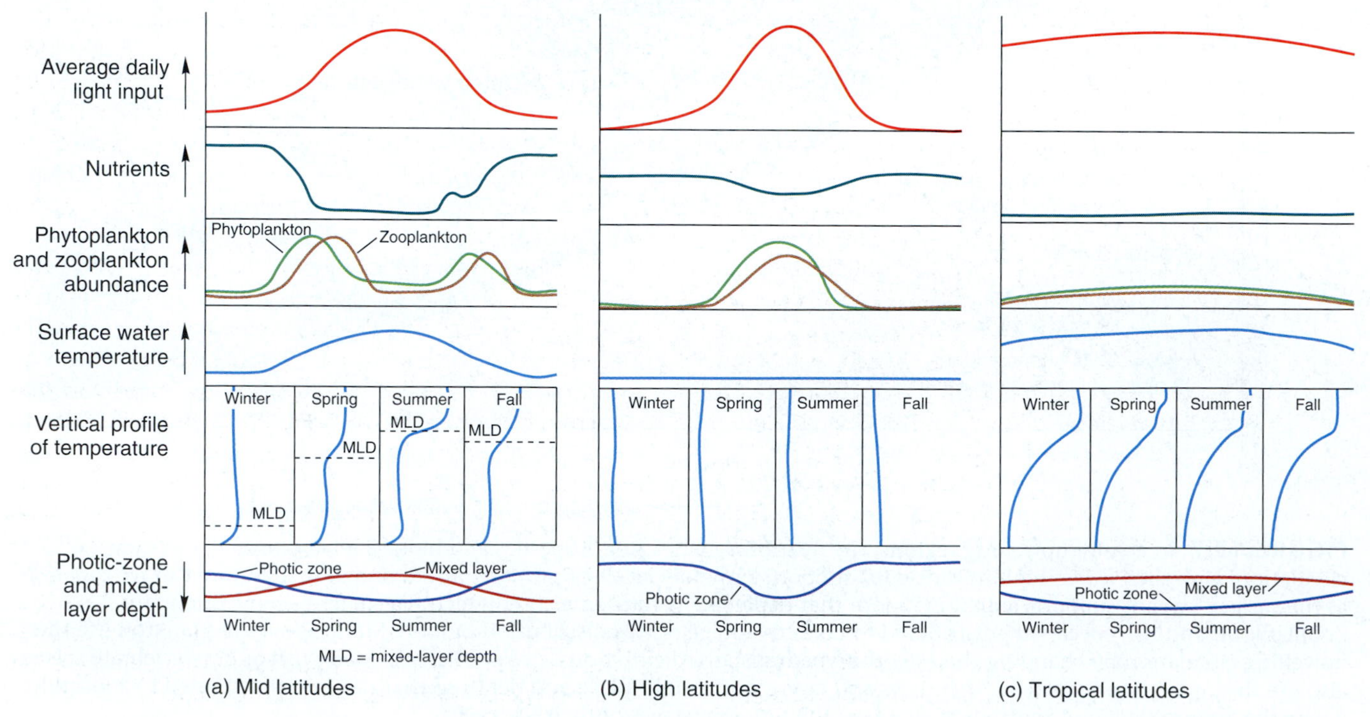

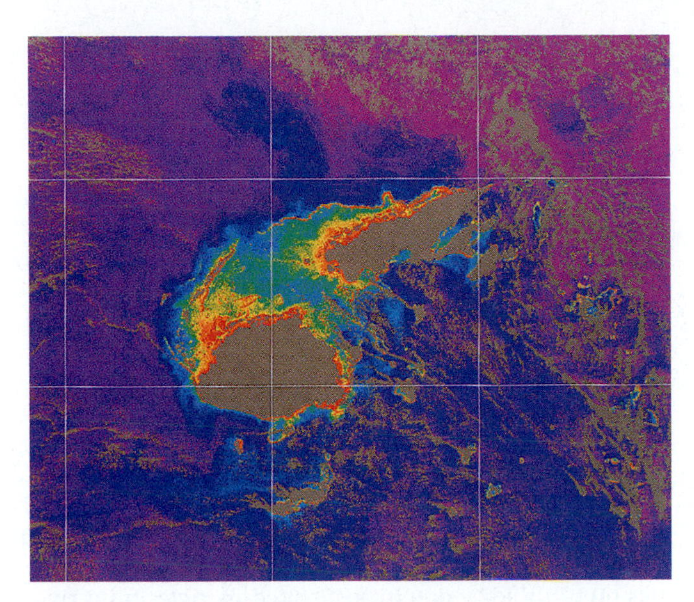

FIGURE 15.9 Elevated chlorophyll concentrations are found around islands, including the Fiji Islands, which are depicted in this Coastal Zone Color Scanner image. Primary productivity is enhanced by the vertical mixing that is caused by the interaction of ocean currents with the island topography, which brings nutrients to the surface layer, and by nutrients in freshwater runoff from the islands.

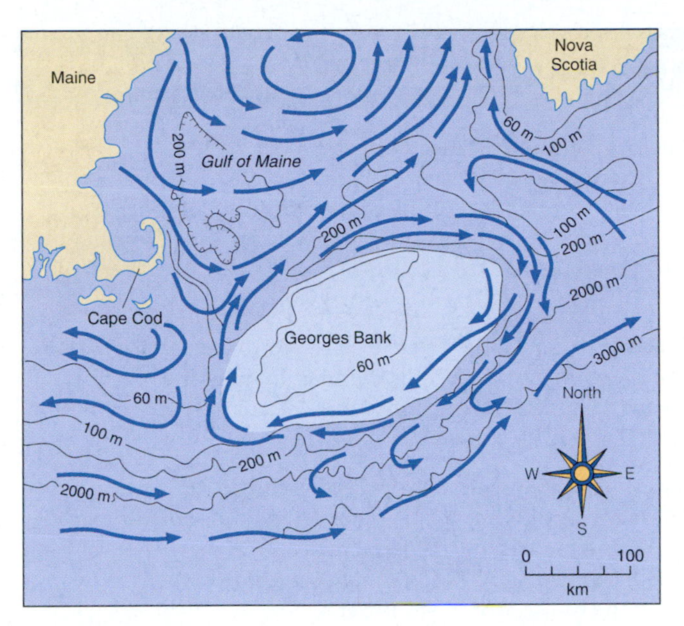

FIGURE 15.10 The water circulation around Georges Bank, New England, is characterized by two permanent gyres. One is a clockwise-rotating gyre that runs around the edge of Georges Bank. The other is a counterclockwise gyre in the Gulf of Maine. The circulation around the bank causes upwelling of nutrients and retains plankton in the water over the bank so that they are able to reproduce to high population densities. These plankton provide food for large fish populations.

nental shelf (Chap. 11). In some areas, such as the outer continental shelf next to Long Island and New Jersey, these waves may transport nutrient-rich subthermocline water into the mixed layer.

Tides in certain regions generate swift currents, particularly in bays, estuaries, and gulfs. Turbulence created by these currents can prevent a seasonal thermocline from forming. Thus, nutrients are recycled more effectively in such locations.

Where currents encounter seafloor topographic features, such as ridges, plateaus, **seamounts,** banks, or islands, water must flow over or around the feature, and nutrient-rich deep water can be mixed into the photic zone by complex eddies and turbulence. The increase in productivity caused by the interaction of ocean currents with islands can be seen in the increased **chlorophyll** concentrations around some such islands (**Fig. 15.9**).

When surface mixed-layer water is cooled, it becomes dense and sinks while convective motions replace it with nutrient-rich water from below. Convective mixing is extremely important in the high-latitude oceans (**Fig. 15.8b**). In these regions, surface waters are cooled continuously except in midsummer, so there is no thermocline, and convective mixing is nearly continuous. When light intensity is sufficient, these regions sustain intense phytoplankton blooms that are rarely slowed by nutrient depletion.

Georges Bank off the New England coast is an extremely productive fishery, primarily because of a plentiful supply of nutrients. The bank is a huge shallow **shoal** in the mouth of the Gulf of Maine (**Fig. 15.10**). Cold, nutrient-rich water flows out of the Gulf of Maine and intersects the north side of the bank, where it is deflected. Some water flows over the bank, but most forms a huge eddy, and a large fraction (approximately 10% to 30%) flows all the way around the bank. Wind waves, tidal currents, and turbulence and upwelling caused by deflection of the water mass into and around the bank continuously inject nutrients into the eddy and the water on the bank.

Productivity is especially high because the eddy that circulates around the bank enables phytoplankton to remain in the water on the bank, where nutrient supplies are high. The residence time of water and plankton on the bank is about 2 to 3 months, so phytoplankton may go through many cell divisions before being swept off the bank. The phytoplankton provide abundant food for zooplankton, which also remain on the bank long enough to feed and reproduce, ensuring that their young have ample food. Fishes, in turn, exploit the abundant phytoplankton and zooplankton. Unfortunately, Georges Bank fisheries have declined substantially because of overfishing especially prior to 1994 when some fishing restrictions were put in place. The populations of commercially valuable species of fishes have yet to recover.

CRITICAL THINKING QUESTIONS

15.1 Some areas of the coastal oceans have lower oxygen concentrations and a greater probability of developing periodic anoxia in the lower part of the water column than other areas. List and explain the factors that make some coastal areas more susceptible to anoxia than others.

15.2 Describe what would happen to the low-density organic particles in sewage if they were discharged to the surface layer at the center of a partially mixed estuary. Would the fate of these particles be different if they were discharged to the lower layer? If so, how?

15.3 It has been suggested that, if the Earth's climate does warm by several degrees, the seasonal cycle of primary production may be disrupted more in polar coastal regions than anywhere else on the Earth. Do you agree or disagree with this prediction? Explain the reasons for your answer.

15.4 Is it likely that a global climate change in which the Earth's atmosphere warms by several degrees will alter the frequency or intensity of dinoflagellate blooms? Why? Is it likely that the frequency and geographic extent of periodic anoxia in coastal waters will change? If so, how and why?

15.3 SEASONAL CYCLES

Substantial seasonal variations in primary productivity occur in mid and high latitudes, but not in the tropics and subtropics (**Figs. 14.12, 15.8**). Seasonal variations are complex and controlled by the availability of light and nutrients and the depth of the mixed layer, each of which varies with latitude, location, and year-to-year climatic variations. As a result, seasonal cycles in high latitudes, mid latitudes, and tropical latitudes have different general characteristics that are representative of each of these broad regions, but there are no sharply defined latitudinal boundaries between areas with these characteristics, and characteristics are substantially modified by local conditions. Seasonal cycles are generally similar in coastal and open-ocean waters within each latitudinal region. However, seasonal cycles are especially important in coastal and estuarine ecosystems.

Polar and Subpolar Regions

Figure 15.8b is a simplified representation of the annual cycles of light intensity, nutrient concentrations, phytoplankton biomass, zooplankton biomass, and water temperature in subpolar seas remote from major freshwater inputs. In these regions, the water column remains well mixed year-round. Frequent storms and low light intensity (limited solar heating) prevent the formation of a seasonal thermocline, and surface cooling produces year-round convective circulation, so there is no permanent thermocline. Nutrients are plentiful because they are supplied continuously by this convective mechanism, but light intensity is sufficient to sustain phytoplankton growth only during a short period in summer, when an explosive bloom of mostly diatoms occurs. Because nutrients are continuously resupplied and do not become **limiting**, the bloom continues unchecked until light intensity declines at the end of summer.

In polar regions, subpolar regions near land, and areas of seasonal sea ice, this seasonal cycle is modified by freshwater inputs. During spring and summer, substantial quantities of freshwater mix with surface waters. The freshwater comes from river runoff fed by snowmelt and from the melting of seasonal sea ice or **glacial** ice. It is mixed with surface ocean water to form a cold but low-salinity surface layer separated from the water below by a halocline. In many areas, the surface layer has high turbidity due to suspended particles from river or glacial ice inputs. High turbidity and low sun angle restrict the photic zone to a very shallow depth. Because the halocline restricts the vertical movement of nutrients and the photic zone does not extend below this layer, **primary production** quickly becomes nutrient-limited after an initial bloom that occurs in early summer when light intensity first increases. Frequent and intense storms in many subpolar regions vigorously mix subpolar seas, break down the halocline, and resupply nutrients to the surface layer. Hence, primary production is nutrient-limited primarily in protected inshore regions such as fjords.

Productivity is also nutrient-limited throughout most of the Arctic Ocean that is ice-free in summer (**Fig. 10.28**). With the exception of a few areas of upwelling, these ice-free waters have a steep, almost permanent halocline a few meters below the surface for several reasons. First, freshwater input from rivers and from the melting of the permanent ice pack is large in volume and continuous during the short summer. In addition, salt is removed from the surface layer during the formation of seasonal sea ice by **ice exclusion** (Chap. 9), which creates high-salinity **brines** that sink below the surface layer. Water exchange between the Arctic Ocean and the Pacific and Atlantic Oceans is limited, so the low salinity Arctic Ocean surface water is transported to lower latitudes very slowly. Furthermore, because the Arctic coast is close to the polar atmospheric **downwelling** area (Chap. 9), there are relatively few storms in summer. Storms that do occur have only a limited **fetch** in which to build wind waves because the ice-free Arctic Ocean is restricted to a narrow zone (a few kilometers to about 200 km wide) between the edge of the permanent, floating polar ice shelf and the coast (**Fig. 10.28**).

In the Southern Hemisphere, the oceans around Antarctica do not have a halocline like the Arctic Ocean halocline, because Antarctica is a desert and the continental ice sheet melts very little during summer. The very

FIGURE 15.11 Reef-building corals. (a) A typical reef-building hard coral colony (probably *Diploastrea heliopora*, Papua New Guinea). (b) Another typical hard coral (possibly *Pocillopora* sp., Papua New Guinea). (c, d) Individual polyps of *D. heliopora* (Papua New Guinea) when not open during the day (c) and photographed at night (d), when the polyps have extended their soft tentacles to feed on zooplankton or particles.

limited freshwater input from the melting of sea ice and from runoff is readily and rapidly mixed and transported into the much larger volume of the Southern Ocean. Because persistent haloclines do not form in the oceans around Antarctica, the area has higher annual productivity than the Arctic Ocean ice-free regions have.

Tropical Regions

In the tropical and subtropical oceans, there is generally little seasonality of plankton growth because of the general uniformity of light intensity year-round and a very steep permanent thermocline that begins below a relatively shallow mixed layer (**Fig. 15.8c**). Hence, light intensity is always high and the photic zone is deep. However, nutrients transported below the relatively shallow mixed layer are removed from the photic zone. Consequently, everywhere but in upwelling regions, primary productivity is low because of nutrient limitation.

Although productivity is nutrient-limited throughout the tropical oceans, most coastal areas are characterized by extremely rich and abundant coral reef communities, except where turbidity is high or salinity is variable because of runoff. Despite the extremely low nutrient concentrations in the water column above, the coral communities thrive as a result of a unique symbiotic relationship between algae and **reef**-building corals.

Reef-building corals consist of millions of individual tiny animals whose **hard parts** are cemented together to form the coral mass (**Fig. 15.11**). Each coral animal, or **polyp**, contains within its tissues a large number of algae known as zooxanthellae. The relationship between coral and zooxanthellae is complex and not fully understood, but one of the principal effects of, or reasons for, this partnership is thought to be more efficient utilization of the small amounts of nutrients available in tropical waters. The zooxanthellae and coral polyps continuously and rapidly recycle nutrients between themselves, thus retaining these nutrients in the photic zone.

Reef-building corals can grow only in the photic zone, where their zooxanthellae can photosynthesize. Hence, these corals are present only in relatively shallow water. In high-turbidity regions, the depth of the photic zone is reduced and coral reef communities do not develop. Coral polyps are adversely affected by high concentrations of suspended particulates and die if high concentrations are sustained. Increased turbidity due to human activities has destroyed or damaged many coral reef communities (Chap. 19).

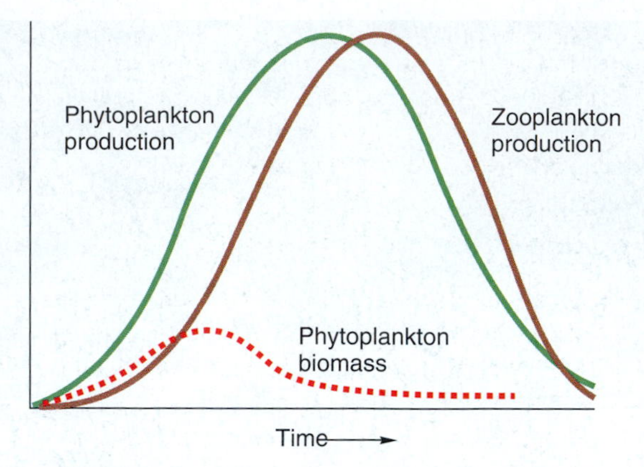

FIGURE 15.12 During a phytoplankton bloom, such as the spring bloom in mid latitudes, the production rate of new biomass by phytoplankton rises rapidly and is followed, with a time lag of a few days, by a similar increase in zooplankton production. The phytoplankton biomass (standing stock) increases somewhat during the early part of the bloom, but it is rapidly reduced by zooplankton feeding, even though the phytoplankton production rate continues to increase.

Mid Latitudes

In mid latitudes during winter, cooling of surface waters and mixing by wind waves eliminate the seasonal thermocline and mix water that has high nutrient concentrations from below the thermocline into the mixed layer (**Fig. 15.8a**). Phytoplankton are distributed almost randomly throughout the mixed layer because of their limited swimming capabilities. Because the mixed layer is much deeper than the photic zone, phytoplankton are within the photic zone only a small percentage of the time. Therefore, during winter, phytoplankton are light-limited, populations are low, and many species are in a hibernation-like resting phase.

In spring, light intensity increases and the photic zone becomes deeper. At the same time, storms are reduced and surface waters begin to warm, the mixed-layer depth is reduced, and a seasonal thermocline begins to form. Phytoplankton are now in the photic zone for an increased percentage of time, and nutrients remain plentiful. The phytoplankton begin to photosynthesize, grow, and reproduce rapidly.

The **standing stock** (biomass) of phytoplankton increases rapidly during the spring bloom period, but the zooplankton population, which now has an increasing food supply, also begins to feed, grow, and reproduce rapidly. Consequently, the majority of new phytoplankton cells are eaten by zooplankton, and the rapid reproduction of phytoplankton does not lead to sustained high phytoplankton concentrations (**Fig. 15.12**). The standing stock of phytoplankton represents only a small percentage of the total production. The remaining fraction is lost to consumption by herbivores or to sinking.

As spring continues, phytoplankton rapidly deplete nutrients in the mixed layer. In addition, the mixed layer becomes shallower (about 10 to 15 m) as the seasonal thermocline forms. At this time, the phytoplankton have adequate light but are nutrient-limited. Consequently, their growth slows and the spring bloom collapses as zooplankton continue to feed on them. Soon the zooplankton population also declines as its food supply decreases, and carnivores continue feeding on zooplankton.

During summer, primary productivity continues at a low level, supported by nutrients recycled by the consumers and **decomposers** in the mixed layer. However, most of these nutrients are recycled below the seasonal thermocline, where light levels are too low to support significant phytoplankton growth.

In fall, cooling and winter storms again weaken the seasonal thermocline, and nutrients are mixed into the photic zone from waters below. The addition of nutrients often supports a fall bloom of phytoplankton. However, the photic zone is now shallower because of the declining light intensity. Variations in timing, location, and year-to-year climate are particularly important in determining the magnitude of the fall bloom. If nutrients are mixed into the surface layer before the light intensity declines too much and before the mixed layer becomes too deep, a fall bloom occurs. Fall blooms are generally smaller and of shorter duration than spring blooms. If the nutrients are not mixed into the surface layer until late fall, when light intensity is already very low, no fall bloom occurs. In addition, no fall bloom occurs if intense cooling or strong storms eliminate the seasonal thermocline early and create a continuously mixed, very deep mixed layer.

In early winter, the enhanced mixing by storm winds, surface water cooling, and declining light levels combine to return the system to its winter state, in which the phytoplankton are light-limited.

Each year, at each location, the seasonal phytoplankton cycle is different from that of the previous years because of variations in weather. In addition, the details of the cycle vary in response to other influences, such as the composition of planktonic species and the concentrations of nutrients other than nitrogen and phosphorus (particularly the concentration of silicate). The subtle and complex interactions between ocean physics, chemistry, and biology that control the seasonal plankton cycle are further complicated in some coastal regions by other processes, such as the introduction of suspended sediment and nutrients by rivers and coastal upwelling.

Phytoplankton Species Succession

Many herbivores are selective feeders that require or prefer certain characteristics in the phytoplankton they eat. Most commonly, the requirement or preference is that the food species be within a range of sizes suited to the animals' feeding methods. Larger zooplankton and juvenile fishes generally require large phytoplankton cells that are easy to capture and provide large amounts of food. Smaller zooplankton specialize in eating smaller phytoplankton. Most coastal **food chains** that lead to commercially valuable fish and **shellfish** species are short and based primarily on zooplankton and juvenile fishes that eat larger phytoplankton species. Food chains based on the smaller phytoplankton species are generally longer and less efficient (**CC15**), and support fish and shellfish species of generally lesser commercial value. Microbial food chains may be short, but they are thought to be primarily self-contained, with much of the organic matter produced being recycled by microbial species rather than consumed in food chains leading to commercially valuable species.

Many factors influence phytoplankton species composition, including temperature, salinity, light intensity, and nutrient concentrations. Because these factors vary seasonally, phytoplankton species composition also varies during the year in most locations. Many different species of phytoplankton are always present at any location, but one or, at most, a few species generally dominate and constitute the majority of phytoplankton. As environmental factors change, one dominant species is replaced by another in a progression called **species succession**.

Phytoplankton species succession can be highly complex and can vary, often greatly, from year to year. One species may become dominant at approximately the same time of year for many years, but may not become dominant at all in other years.

In high latitudes and in regions of persistent upwelling, nutrients are present in relatively high concentrations throughout the phytoplankton growth period. High nutrient concentrations favor larger phytoplankton species, which are predominantly diatoms. Diatoms dominate the phytoplankton community throughout the active growth period in these regions, which consequently support short, efficient food webs (**CC15**).

In mid latitudes, nutrient concentrations vary during the summer growing period. In spring, nutrient concentrations are high, and, consequently, diatoms dominate. The spring bloom of diatoms removes most nutrients from the water column, causing the bloom to collapse and to be replaced by flagellates (**Fig. 15.13**). During summer, after the spring diatom bloom, flagellates (sometimes a succession of different species) dominate the phytoplankton community. The flagellates can obtain their nutrients from the very low concentrations that are sustained by nutrient recycling in the photic zone. In fall, storms return enough nutrients to the photic zone to support fall diatom blooms in some years and areas.

As a result of the seasonal availability of diatoms, many species of zooplankton and juvenile fishes that pre-

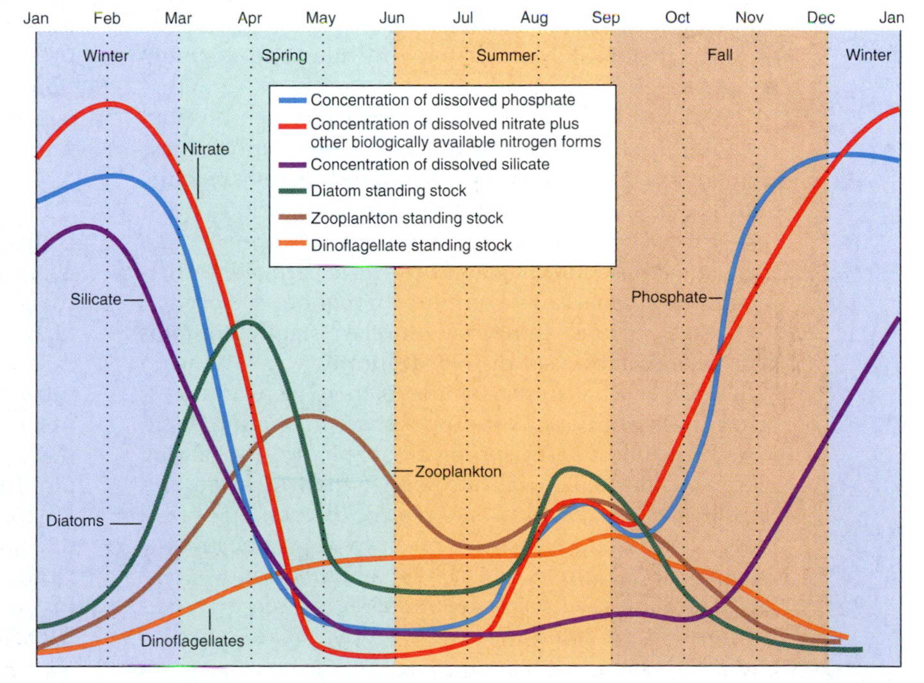

FIGURE 15.13 The seasonal variations of nitrogen, phosphorus, silicon, diatoms, flagellates, and zooplankton in the mid-latitude coastal oceans usually follow a similar pattern each year. The spring bloom of diatoms rapidly depletes the nutrients, and the diatoms are superseded by dinoflagellates. Because the dinoflagellates are not suitable food for many zooplankton species, the zooplankton population also declines. By utilizing nutrients recycled within the photic zone, flagellates and some diatoms continue primary production at relatively low levels throughout the summer. A fall bloom of diatoms often occurs when storms and cooling weaken the thermocline and return some nutrients to the surface layer by mixing. In late fall, the light intensity becomes too low to support substantial primary production. Thus, decomposition, cooling, and wind mixing enable the return of nutrients to the surface layer before the following spring.

fer or require diatoms as food have life cycles that are attuned to the seasons. For example, some fish and **invertebrate** species spawn in spring, and their eggs hatch into larvae that feed voraciously on diatoms. The juvenile fishes reach a sufficient size by late spring, when the diatom bloom collapses, to be able to feed on zooplankton. Some species spawn twice during the year to take advantage of both spring and fall blooms.

Many mid-latitude fish and invertebrate species have life cycles that depend on the availability of abundant diatoms as food for their juveniles during only a few days or weeks at a specific time of year. However, the timing, intensity, and dominant species of seasonal diatom blooms vary from year to year. As a result, the survival rate of the larval stages of many fish and invertebrate species also varies from year to year. The number of juveniles that survive their first critical year of life is called the "year class strength." Year class strength may be high if the species spawns at the optimal time and locations in relation to the seasonal phytoplankton cycle. If year-to-year variability causes spawning and the phytoplankton seasonal succession to be misaligned in time or location in any specific year, sufficient food is not available when needed and year class strength is low.

Phytoplankton species succession is further complicated by variations in the availability of specific nutrients, particularly silicate and nitrogen compounds. During the spring diatom bloom in mid latitudes, silicate may be depleted before other nutrients (**Fig. 15.13**), causing diatoms to be replaced by flagellates that do not require silica. Silica is recycled extremely slowly into solution once it has been incorporated in diatom **frustules**. As a result, silica is not returned to the photic zone until the fall, when nutrient-rich waters from below the seasonal thermocline are mixed back into the mixed layer. These deeper waters have relatively high concentrations of silica dissolved from diatom frustules and **terrigenous** particles.

All phytoplankton cells require and take up phosphorus and nitrogen in approximately the same ratio as these elements occur in seawater. Similarly, diatoms require and take up silica in a reasonably constant ratio to other nutrients. The proportions of silicate, phosphate, and nitrogen compounds in water below the permanent thermocline are remarkably constant throughout the oceans. In coastal regions, terrestrial runoff normally contains high concentrations of dissolved silicate.

Consider what might happen if there were an excess of nitrogen compounds and phosphate in relation to silicate at the onset of the spring diatom bloom. The diatom bloom would deplete silicate sooner than nitrogen and phosphorus. Flagellates would replace diatoms and could bloom explosively because nitrogen and phosphorus would still be ample. This bloom might produce very large standing stocks of flagellates if, as is often true, herbivores that are able to graze the tiny flagellate cells efficiently were not abundant and could not reproduce quickly. Flagellate blooms do occur periodically, although infrequently, in limited regions where runoff has high nitrogen and phosphorus concentrations but low silicate concentrations. It appears that blooms are becoming more frequent and widespread in regions near centers of human population, particularly where coastal waters have long residence times. It is likely, but not yet proven, that some of these blooms are caused by the huge volumes of human sewage wastes discharged to the oceans (Chap. 19). Untreated or treated sewage has high concentrations of nitrogen compounds and phosphate but relatively low silicate concentrations.

15.4 ALGAL BLOOMS

Intense **dinoflagellate** blooms occur naturally, but infrequently, in many parts of the coastal oceans, particularly in subtropical to temperate climates. The frequency of such blooms appears to be increasing, and they are occurring in areas where they have not been seen before.

Dinoflagellate blooms can be caused by a variety of mechanisms. For example, a large increase in freshwater runoff can form a shallow surface layer of low-salinity water in a previously well-mixed water column. Phytoplankton that were previously light-limited because they were frequently mixed below the photic zone may be restricted to this shallow surface layer where they are always in the photic zone. This situation may trigger a bloom. In some instances, it is thought that runoff containing dissolved inorganic or organic compounds may trigger blooms because these substances react with toxic substances, such as copper or mercury, and reduce their toxicity to the phytoplankton. Many blooms are caused by such natural events, but increasing nutrient fertilization due to human activity (Chap. 19) is generally believed to contribute to the increasing frequency and severity of blooms.

Because the nitrogen compound concentration in discharges of sewage and of agricultural and industrial waste is high in relation to the silicate concentration, such discharges are generally believed to be responsible for the higher frequency of dinoflagellate, as compared to diatom, blooms. However, several **cyanobacteria** (also called "blue-green algae") convert dissolved molecular nitrogen, a nitrogen form that most phytoplankton cannot use, to usable nitrate. In some areas, cyanobacteria may provide the nitrogen needed to initiate or sustain dinoflagellate blooms.

Dinoflagellate blooms can appear quickly but often disappear within a few days as nutrients become limiting. However, under appropriate conditions, blooms can last for many weeks if the source of nutrients is sustained. Dinoflagellates reproduce very rapidly. In optimal conditions, dinoflagellate populations can double in as little as a few hours. Accordingly, blooms can grow within a few days to concentrations of dinoflagellate cells a million or more times greater than normal. Dinoflagellates can also concentrate themselves into blooms by using their weak swimming ability to congregate at depths where light levels are optimal.

The dense concentration of phytoplankton in blooms colors the surface water yellow, green, brown, or reddish, depending on the species. Many dinoflagellates are reddish in color, so their blooms are often called "red tides."

Dinoflagellate blooms often have adverse effects on other marine organisms and, occasionally, on human health. These effects are caused by toxic substances synthesized by certain phytoplankton species, particularly when they are stressed in bloom situations, or by the depletion of oxygen caused by the decaying organic matter that remains when the bloom collapses.

Dinoflagellate and Other Phytoplankton Toxins

Like many terrestrial plants, certain phytoplankton species produce toxic substances as a defense against herbivores. Some, but not all, species of dinoflagellates produce particularly virulent toxins, including saxitoxin, brevetoxin, okadaic acid, and domoic acid. These substances are complex organic compounds and often mixtures of several compounds, some of which are many times more toxic than, for example, strychnine. The toxins are selective: some are toxic to **vertebrates** only, whereas others also affect some invertebrates, particularly **crustaceans** such as shrimp and crabs. Toxins produced by dinoflagellate blooms often kill large numbers of fish.

In addition to fish kills, dinoflagellate toxins can cause serious human health problems because the toxicity is selective for vertebrates. Most **filter-feeding** organisms (organisms that strain or sift food particles from the water), including clams, mussels, and oysters, are unaffected by dinoflagellate toxins but concentrate the toxins in their body tissues. If these shellfish are then eaten by vertebrates, the toxins can still produce their deadly effects in the vertebrates. Unfortunately, humans are one of the vertebrate species that feed on shellfish.

Human poisoning by dinoflagellate toxins concentrated in shellfish is well known. At least four major pathologies are caused by shellfish-borne toxins. The most widespread is paralytic shellfish poisoning (PSP). The toxins that cause PSP are strong nerve poisons that can cause permanent nerve injuries or, if the concentrations are high enough, paralysis and death. There are an estimated 1600 annual cases of PSP worldwide, and an estimated 300 fatalities among these cases. Neurotoxic shellfish poisoning (NSP) is less common and less serious than PSP, but it produces symptoms that are often mistaken for the more common types of **bacterial** food poisoning. Diarrhetic shellfish poisoning (DSP) is also less serious but probably more widespread than NSP, and it causes severe diarrhea. Amnesic shellfish poisoning (ASP) is a more recently recognized pathology in which short-term memory is destroyed, perhaps permanently.

Winds can blow ashore sea spray containing dinoflagellate cells or their detritus from nearshore blooms. When this happens, the toxins cause the local human population to suffer from allergylike reactions, such as irritated eyes, running noses, coughs, and sneezes.

Most dinoflagellate toxins are stable compounds that remain toxic even after prolonged cooking of contaminated shellfish. Once the bloom has disappeared, some shellfish lose their toxicity within days, but other species may remain toxic for months. The only certain way to avoid human poisoning by these toxins is to identify affected shellfish and prevent them from reaching the marketplace. Many countries, including the United States, have extensive seafood monitoring programs to ensure toxin-free shellfish. Unfortunately, these monitoring programs are extremely expensive and difficult to implement. As an alternative, in areas where dinoflagellate blooms occur with reasonable frequency, shellfish areas are closed to harvesting during the summer and fall, when blooms are most likely. Such closures lead to substantial loss of income to fishers. Areas closed to shellfishing because of dinoflagellate blooms are expanding each year, perhaps indicating an increase in the frequency and geographic extent of such blooms.

At one time, only dinoflagellates were thought to produce the powerful toxins that affect people and other vertebrates. However, blooms of at least two different diatom species have now been observed to produce domoic acid. An event that occurred in 1987 in Prince Edward Island, Canada, left several victims afflicted with ASP. The victims suffered short-term memory loss that lasted at least 5 years. A second event, in 1991 off the west coast of California, Oregon, and Washington, did not affect people. In this case, pelicans were poisoned by domoic acid produced by a diatom species that had been consumed by anchovies, on which the pelicans were feeding. Shellfisheries throughout this huge area were closed to harvesting for many weeks. According to a recent hypothesis, the toxins may be produced by bacteria or **viruses** living in the phytoplankton cells, rather than by the phytoplankton themselves.

Humans and pelicans are not the only vertebrates poisoned by phytoplankton toxins. For example, in 1986, hundreds of bottlenose dolphins died off the coasts of New Jersey and Maryland after consuming menhaden contaminated with a dinoflagellate toxin, and more died in the same way off the Carolina coast in 1987. In these cases, the menhaden were either less susceptible to the toxin than the dolphins, or the dolphins accumulated more of the toxin as they fed on large quantities of menhaden and were unable to destroy or **excrete** the toxin.

Oxygen Depletion

Below the permanent thermocline, oxygen concentrations are reduced by oxygen consumption during **respiration** and decomposition of detritus transported through the thermocline from the photic zone (Chap. 14). The same processes consume oxygen below seasonal thermoclines, but the reduction in oxygen concentration is usually small for two reasons. First, the residence time of water in the layer below a seasonal thermocline (but above any permanent thermocline) is at most a few months. When the seasonal thermocline breaks down in the fall, oxygen is returned to this water when it becomes part of the mixed layer in contact with the atmosphere. In addition, in locations where a seasonal thermocline forms, the total production of phytoplankton during the summer is limited by nutrient availability in the shallow mixed layer above the seasonal thermocline. Hence, only a limited quantity of organic matter is transported through the thermocline to be decomposed. Areas with high productivity are generally areas of upwelling or other physical processes that prevent the formation of a steep thermocline. Vertical mixing in these areas continuously transports oxygen-deficient water to the surface, where it regains oxygen from the atmosphere.

Severe oxygen depletion does not generally occur below seasonal thermoclines, but it may occur in specific circumstances when the cumulative **oxygen demand** in the subthermocline layer is very large. The cumulative oxygen demand is the total amount of oxygen consumed before the water mass recontacts the atmosphere or before additional oxygen is supplied by mixing with water that has a higher oxygen concentration. Two circumstances tend to produce high cumulative oxygen demand: high productivity in the overlying water, and long residence time of water below the thermocline. If oxygen concentrations become too low, many animal species cannot survive for more than a short period of time. This low oxygen condition is called **hypoxia**. If the oxygen concentration drops to zero—a condition called **anoxia**—animals cannot survive at all.

Residence times of water below seasonal thermoclines on the continental shelf are generally short, but residence times of water below permanent thermoclines in marginal seas or ocean inlets and bays may be very long (often many decades). Consequently, despite the relatively low productivity of marginal seas that have steep thermoclines, the cumulative oxygen demand is sufficient in certain locations to deplete oxygen. The result can be anoxia and the formation and accumulation of toxic sulfide in bottom waters. Examples of such locations are many Norwegian and other fjords and the Baltic Sea, where anoxic conditions have been present continuously in the deep waters of the central region for at least 100 years. The geographic area of anoxic bottom waters has grown during that time in response to increasing inputs of nutrients and oxygen-demanding organic matter by humans (Chap. 19).

In some marginal seas or other coastal areas, normal conditions result in periodic (often seasonal) or permanent hypoxia. In some of these regions, increased phytoplankton growth and blooms can occur as a result of nutrient inputs from sewage, agriculture, and other human sources. Blooms produce large amounts of detritus, and as a result, the naturally low oxygen concentrations below the thermocline are reduced further. This may lead to occurrences of anoxia, causing most benthic **biota** and some deep-water nekton to be either killed or deprived of these areas as **habitat.**

Anoxia may occur in any particular year in a coastal region if primary productivity is high and/or residence time is long. For example, on the continental shelf of New York and New Jersey in 1976, wind mixing was limited and an intense bloom of phytoplankton developed. The combination of long residence time and high productivity caused a widespread anoxia that extended for hundreds of miles along the coast in a band tens of miles wide across the shelf (**Fig. 15.14**). The anoxia killed enormous numbers of many marine species, including clams and other shellfish valued at more than $500 million.

FIGURE 15.14 During the summer of 1976, eutrophication and unusual climate conditions combined to produce anoxia or conditions in which oxygen concentrations were too low to support animal life (values of less than 2 in this figure) throughout a large area of the New Jersey continental shelf. This event caused the deaths of massive numbers of clams and other benthic fauna.

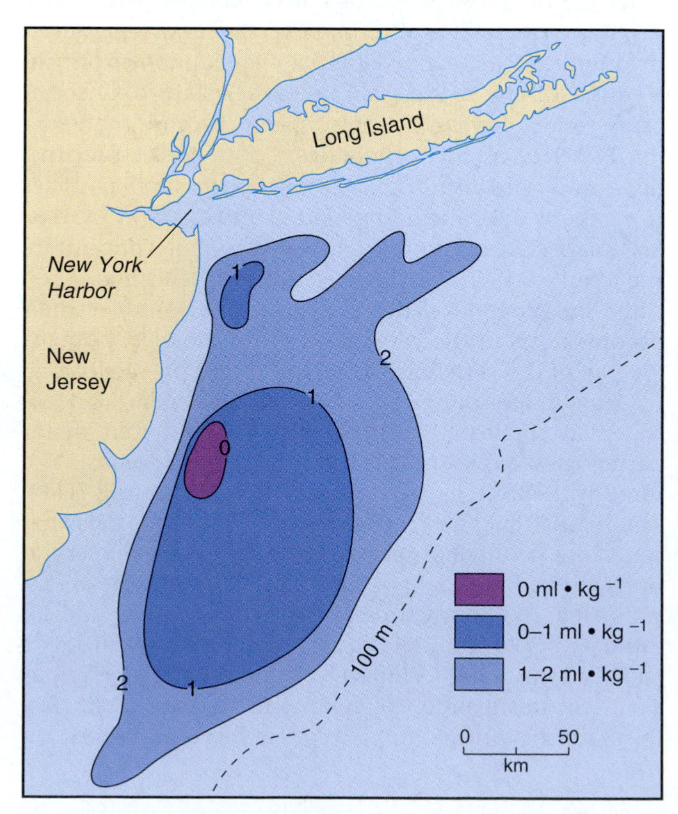

The 1976 anoxia was not caused directly by excess nutrients discharged to the oceans from New York and New Jersey, because natural conditions were responsible for the unusually long residence time of that year. However, nutrient inputs to this region in sewage increase productivity and, thus, the probability of blooms. As nutrient inputs increase, hypoxia and anoxia will occur at shorter residence times and, hence, probably more frequently.

Hypoxia or anoxia has been found in many parts of the coastal zone. For example, in the Adriatic Sea, the Saronikos Gulf in Greece, the Sea of Japan, and many bays and harbors, periodic massive algal blooms exceed the consumption capacity of the zooplankton population and cause anoxia in the bottom waters. The blooms have been so bad in several years in the Adriatic Sea that the rotting algae have fouled **beaches** and the air, severely affecting tourism in Italy and Croatia. In the United States, hypoxia or anoxia has occurred in the North Atlantic coastal zone, the Gulf of Mexico, Chesapeake Bay, the Oregon coastal region and a number of other bays and estuaries.

Overfertilization of natural waters with nutrients that cause blooms and sometimes hypoxia or anoxia is called **eutrophication**. It has been well known in lakes for decades, but has only recently become the subject of intense study in the coastal oceans.

15.5 FISHERIES

Because they ultimately depend on primary production for their food, fishes are most abundant in areas where primary productivity is high (**Fig. 14.12, Fig 15.15a**). Hence, fish population density and productivity are

FIGURE 15.15 (a) Primary productivity, measured as the total amount of organic carbon produced annually in each square meter of water column, is much higher in upwelling areas than in coastal (nonupwelling) or oceanic areas. (b) The average trophic efficiency is high in upwelling regions, intermediate in coastal regions, and lowest in oceanic regions. (c) Most of the ocean is oceanic in character, whereas upwelling areas constitute only a very small percentage of the total area. (d) Because the area occupied by oceanic regions is so large, the oceanic regions are responsible for most of the worldwide primary production. (e) The total production of fish biomass in oceanic areas is small, despite the high primary production, because this primary production is performed mostly by dinoflagellates and because the average trophic efficiency is low. Dinoflagellates must pass through more trophic levels than the diatoms that dominate in coastal and upwelling regions before they are used to build fish tissue. Thus, more dinoflagellate biomass than diatom biomass is needed to sustain the same fish biomass. Fish production is high in coastal regions because of the relatively high primary productivity and trophic efficiency. Despite their very small area, upwelling regions are responsible for a large proportion of the world's fish production because of their very high primary productivity and high trophic efficiency.

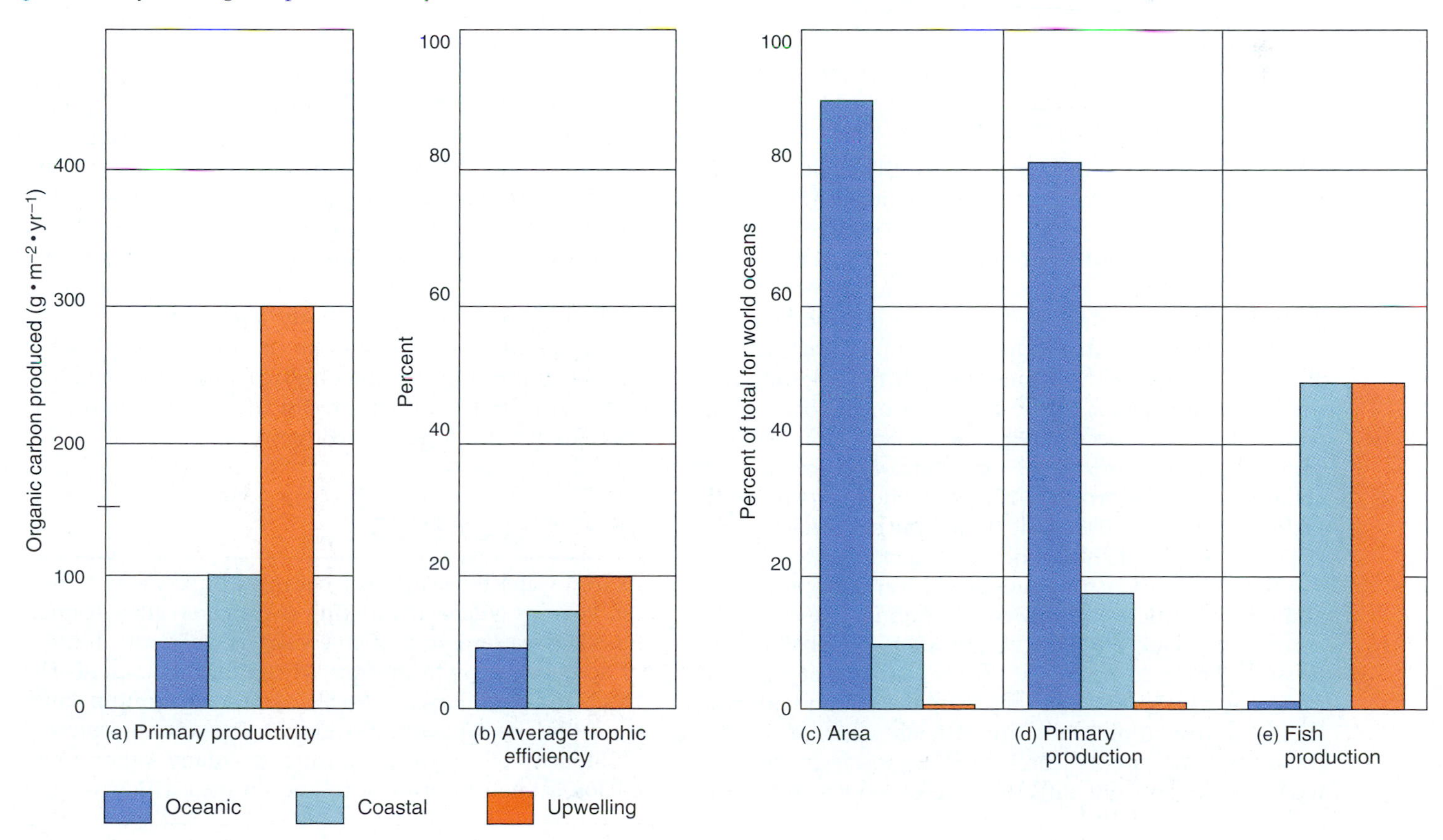

CRITICAL THINKING QUESTIONS

15.5 Many marine species are overfished. The current method of managing fisheries is to restrict the total catch of individual species to below the maximum sustainable yield of that species. However, this approach does not prevent adverse effects on other species that may depend on the targeted species for food. For example, it is hypothesized that Pribilof seal populations in the Bering Sea are declining because the fish species that they consume are being heavily fished, reducing the seals' food supply. A new approach to fishery management is developing in which ocean ecosystems are managed as a whole. For example, the growing fishery for krill in the Southern Ocean is managed by an ecosystem-based international treaty. Krill are the food supply on which a number of whale and seal species are dependent. Populations of these whale and seal species were drastically reduced by hunting in the first half of the twentieth century and have not yet recovered. Populations of penguin species that also rely on krill have increased to take advantage of the food previously eaten by whales and seals. If the krill fishery is to produce close to its maximum sustainable yield and contribute significantly to the solution of world hunger and protein deficiency, the whale and seal populations will never recover to their previous population levels, and it may be necessary to reduce penguin populations to protect whales and seals from renewed decline.

(a) Do you think that marine ecosystems and fishing should be managed in this way? Explain your reasons.

(b) What alternatives would you propose to solve the world hunger problem?

higher in the coastal oceans, especially in upwelling regions, than in open-ocean areas where there is no upwelling.

The average **trophic efficiency** (**CC15**) is higher in coastal food webs than in open-ocean food webs and is highest in food webs of upwelling areas (**Fig. 15.15b**). As a result, the total fish production of upwelling areas is about half of the world's total (**Fig. 15.15e**), despite the very small area (0.1% of the total area of the oceans) in which upwelling occurs (**Fig. 15.15c**). Coastal regions constitute about 10% of the ocean area and account for about 50% of the world's fish production. Open-ocean regions constitute 90% of the oceans but sustain less than 1% of the world's fish production (**Fig. 15.15e**).

The most successful fisheries are **herring,** anchovy, and sardine fisheries in upwelling regions. These species constitute about 25% by weight of the global catch. Because they feed primarily on zooplankton at the second **trophic level,** they are extremely abundant and their harvest represents an efficient use of ocean resources. Much of the catch of these species is used to feed animals and not directly to feed people, so we are using these resources at only about 10% of their potential efficiency. About 40% of the global fishery catch of all species is used as animal food.

Because of the distribution of fish biomass production, the world's major fisheries are almost all located in the coastal zone, especially in upwelling areas. Most of these major fisheries are being exploited at or near their **maximum sustainable yield** (**CC16**), and many are overfished. Probably the total world fishery catch cannot be increased substantially from its present level without causing detrimental effects on the ocean ecosystem, and particularly on higher carnivores (including the largest fishes, marine mammals, and seabirds) that rely on fish stocks for their food.

Natural variations in fish stocks are caused by complex interactions of ocean physics, chemistry, and biology. For example, many fish species release their eggs into the plankton community that drifts with the ocean currents. The eggs hatch into larvae that must have the right type of food. Anchovy larvae, which eat only phytoplankton that are at least 40 mm in diameter throughout the first few days after hatching, typify this requirement. Larval survival varies from year to year because the availability of such phytoplankton varies, as do many other factors, including the concentrations of species that compete for food with or prey on the larvae. Survival and success are no less complicated for the larval fish when it becomes a juvenile, because it has to find food and is subject to predation, disease, **parasitism**, and the effects of **pollution.** Each of these factors, in turn, can be variable from year to year, at least in part because of year-to-year climatic variations.

Life cycles and their interactions with physical, chemical, and biological variables are so complex that studies must be conducted over many years to obtain even a limited understanding of the population variations in a single fish species. Even after such extensive studies, it is impossible to predict the future of the fish stock with any certainty. Many of the influences on fish stock size and age composition are nonlinear. Consequently, fish stocks of many species are inherently **chaotic** (**CC11**) and appear to fluctuate in a random or unpredictable way. It may be impossible to predict future fish stocks accurately, just as it is impossible to forecast the weather accurately more than a day or so in advance.

15.6 ESTUARIES

As freshwater flows into the oceans, it mixes with seawater. Regions where this mixing occurs are called "estuaries." The seaward limit of an estuary is where the dilution of ocean water by freshwater is insignificantly small. The landward boundary, or "head," is the maximum landward limit of saltwater movement. Estuaries are present at mouths of major rivers and in many other semi-enclosed inlets or arms of the ocean into which streams and rivers flow. If freshwater discharge is very large, the seaward boundary of the estuary can be many kilometers offshore. The inland boundary can be tens or even hundreds of kilometers inland. The boundaries are not fixed. They can move seaward if freshwater flow rate increases and farther inland if it decreases.

The fact that most major cities are located on estuaries reflects the historical importance of estuaries as harbors and ports. Estuaries are also important to marine ecosystems because many species of fishes and invertebrates spend part of their life cycle in the oceans and part in an estuary or river, and many major fisheries and shellfisheries are in estuaries.

Estuaries differ in characteristics that include their length, width, depth, **tidal range,** freshwater flow rate, shape, and coastal character. These factors affect ocean and river water movements and mixing processes in the estuary. Consequently, there is no simple description of a typical estuary, and no single classification system can capture the many variations. Every estuary is different from all others and has its own unique behavior. However, estuaries are classified into general groups according to the geological processes that formed their embayment or, alternatively, according to their water circulation and mixing characteristics.

Geological Origin of Estuaries

Almost all present-day estuaries were formed as sea level rose in the past 18,000 years from a low point approximately 130 m below current sea level (**Fig. 8.17e**). As the most recent **ice age** ended and glacial ice melted, sea level rose. River and glacial valleys were flooded and coasts were modified by **erosion**, forming **barrier islands** in some locations (Chap. 13).

Estuaries are filled progressively with **sediment** supplied by the river. If sea level were to remain stable long enough, most estuaries would eventually become filled with sediment and **deltas** would form (Chap. 13). The inexorable process of sediment filling can be seen at locations such as the former Greek port of Ephesus, now in Turkey. This city, a bustling seaport less than 3000 years ago, is today many kilometers inland, behind a low **coastal plain** formed by the accumulation of sediment eroded from surrounding mountains (**Fig. 15.16**).

In the past, during periods when sea level was falling, there were few estuaries. As sea level falls, the coastal landform consists of newly exposed continental shelf. This new coastal plain is relatively featureless because the topographic lows in the shelf floor are filled with sediment transported by tides, waves, and currents. Estuaries will virtually disappear again when sea level falls in the future.

Rising sea level has created estuaries that differ according to the character of land that has been inundated. Four types are recognized: **coastal-plain estuaries, bar-built estuaries, tectonic estuaries,** and fjords. Coastal-plain estuaries were formed as sea level rose to flood river valleys (**Fig. 15.17a**). These estuaries, often called "drowned river valleys," are especially abundant on passive margins, such as the east coast of the United States. Chesapeake Bay, Delaware Bay, and the New York Harbor are examples of coastal-plain estuaries. The Mississippi

FIGURE 15.16 The ancient Greek city of Ephesus, now in Turkey. In ancient Greek times, Ephesus was a port city. The waterfront was just at the foot of the hill behind the temple in the center of this photograph. The ruins of the port are now several kilometers inland. In the background, you can see the wide flat coastal plain. This was once a large bay, but since ancient times it has filled with sediments.

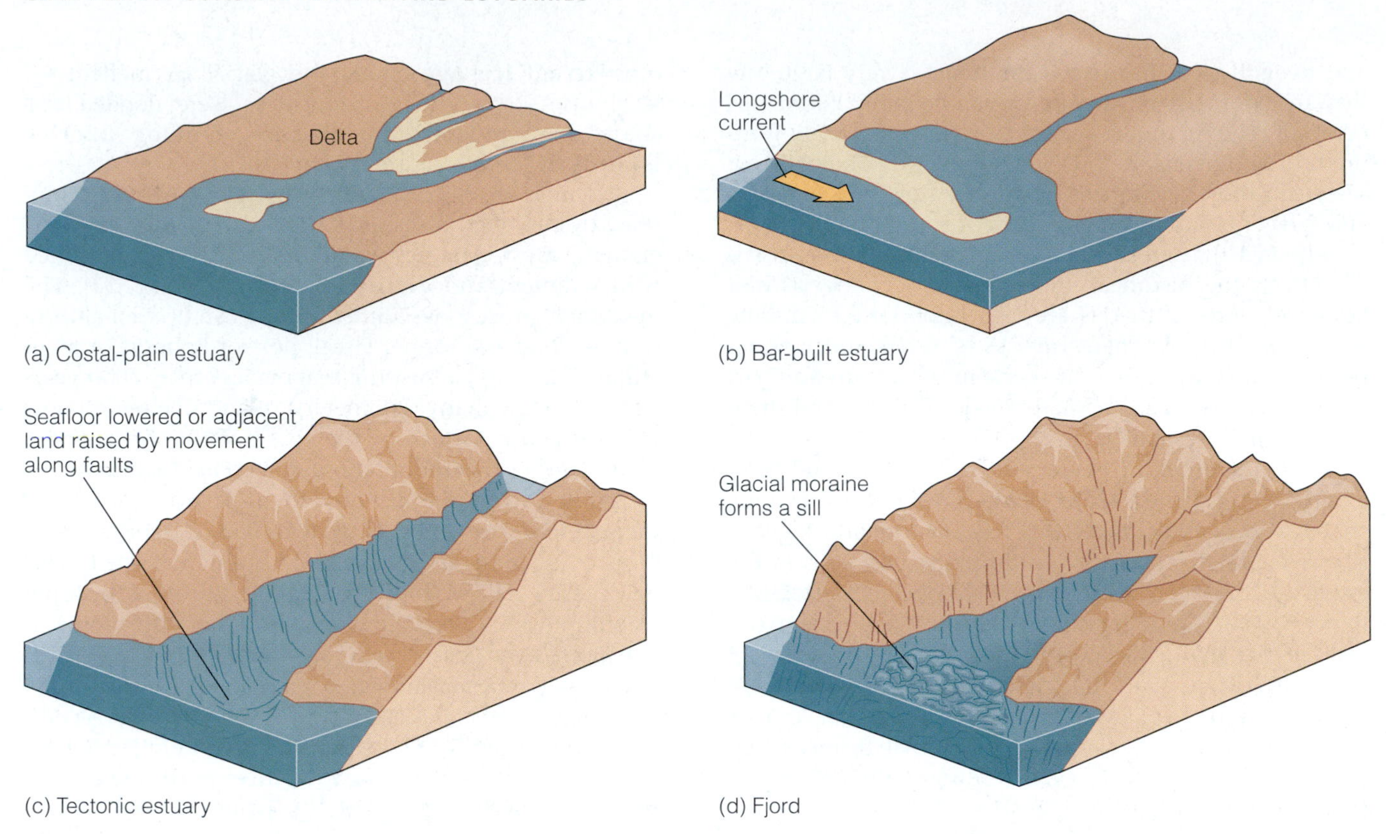

FIGURE 15.17 The rise of sea level during the past 18,000 years has been a major factor in the creation of all present-day estuaries. Estuaries have a variety of geological settings and origins. (a) The most common estuary is a coastal-plain estuary (also sometimes called a "drowned river valley") that forms as rising sea level floods the mouth of a river. (b) Some estuaries form where longshore drift builds a bar, spit, or barrier island that isolates a bay from free exchange with the ocean. (c) Tectonic estuaries are formed where a block of the Earth's crust is lowered by tectonic processes at earthquake faults. (d) Fjords are created when rising sea level floods the steep-sided valleys that are cut by glaciers and exposed after the glaciers retreat.

River delta is an example of a former coastal-plain estuary that has been filled with river-borne sediment.

Bar-built estuaries are formed when a sandbar is constructed parallel to the coastline by wave action and **longshore drift** (Chap. 13), and the bar separates the ocean from a shallow lagoon (**Fig. 15.17b**). Lagoons behind barrier islands are bar-built estuaries. Examples include Albemarle Sound and Pamlico Sound in North Carolina. Some estuaries have mixed characteristics. For example, the Hudson River estuary that passes through the New York Harbor and Raritan Bay is primarily a drowned river valley, but the Sandy Hook **spit** gives Raritan Bay certain bar-built estuary characteristics (**Fig. 15.18**).

FIGURE 15.18 This satellite image of the Hudson River basin–New York Harbor–Raritan Bay estuary shows the complex character of this estuary. The Hudson River valley is a drowned river and glacier-cut valley and so can be considered a coastal-plain estuary. Parts of the river are steep-sided and deep enough that they may even be considered a fjord. Sandy Hook Bay and Jamaica Bay are bar-built. Raritan Bay and New York Harbor could be considered either coastal-plain estuary segments or bar-built, or both.

Tectonic estuaries are formed when a section of land drops or tilts below sea level as a result of vertical movement along a **fault** (**Fig. 15.17c**). Such estuaries are most often present on coasts along **subduction zone** or **transform fault** plate boundaries. The best-known tectonic estuary in North America is San Francisco Bay (**Fig. 15.22**), which is on a **transform plate boundary.**

Fjords are estuaries in drowned valleys that were cut by glaciers when sea level was lower (**Fig. 15.17d**). These estuaries are generally steep-sided, both above and below sea level, and are often deep. Many fjords have a shallow **sill** near the mouth that was formed by sediment deposited at the lower end of the glacier when it flowed through the valley. Fjords are common on the Norwegian coast, the southwest coast of New Zealand, and the Pacific coast of Canada and southern Alaska. Puget Sound in Washington State is also a fjord.

Estuarine Circulation

In estuaries, less dense river water flows over seawater, causing vertical mixing between the two layers. The movements and mixing of freshwater and seawater are affected by many factors, including tidal currents and mixing, wind-driven wave mixing, shape and depth of the estuary, rate of freshwater discharge, friction between the moving freshwater and seawater layers and between the water and seafloor, and the **Coriolis effect** (**CC12**). Because some or all of these factors vary among estuaries between seasons within individual estuaries, and from one part of an estuary to another, estuaries have a wide variety of circulation patterns.

Estuarine circulation is extremely important because many major cities are located on estuaries. These cities discharge large quantities of wastes, particularly sewage treatment plant **effluents** and storm water runoff, into the estuaries (Chap. 19). Estuarine circulation patterns determine the fate of these contaminants.

In many estuaries, circulation is altered by piers and other port structures, dredging of navigation channels, filling of **wetlands**, and construction of **levees** and other coastal structures. These alterations can affect life cycles of marine species that inhabit or transit the estuary (Chap. 19).

Because circulation within each estuary is complex, varies temporally, and is unique to that estuary, detailed, multiyear physical oceanography studies of each estuary are necessary to understand its circulation. Such studies are difficult and expensive, particularly in large, complex estuaries such as Chesapeake Bay or San Francisco Bay.

Types of Estuarine Circulation

Estuaries can be classified according to the major characteristics of their circulation. The major types are **salt wedge estuaries, partially mixed estuaries, well-mixed estuaries,** and fjord estuaries.

Salt Wedge Estuaries. In a salt wedge estuary, freshwater flows down the estuary as a surface layer separated by a steep density interface (halocline) from seawater flowing up the estuary as a lower layer that forms a wedge-shaped intrusion (**Figs. 15.19a, 15.20a**). Vertical mixing across the steep density gradient is slow in salt wedge estuaries, but the velocity difference between seaward-flowing river water and the underlying seawater creates friction between these layers. The friction causes turbulence and internal waves at the interface. The internal waves grow, and when they break, small quantities of high-salinity water are injected and mixed into the upper layer, which progressively increases in salinity toward the estuary mouth (**Figs. 15.19a, 15.20a**). Almost no freshwater is transferred into the lower seawater layer.

Salt wedge estuaries are most likely to be present where freshwater input is relatively large and allows a thick, freshwater surface layer to develop. Estuaries that are narrow in relation to their depth also favor the development of a thick freshwater layer. Other conditions that favor salt wedge estuaries are small tidal range, and hence limited tidal currents and tidal mixing, and limited vertical mixing due to wind-induced motions (e.g., waves, Langmuir circulation, and Ekman transport). To understand how these factors favor the two-layer salt wedge configuration, compare the estuary with a blender filled with water and cooking oil. If the blender runs at high speed, the cooking oil is quickly mixed with and dispersed in the higher-density water. If the blender runs more slowly, the oil will tend to remain in a distinct upper layer, particularly if there is enough oil to create a thick oil layer. A thin oil layer will mix and disperse at a lower blender speed than a thick oil layer.

Partially Mixed Estuaries. Vertical mixing between freshwater and seawater is greater in partially mixed estuaries than in salt wedge estuaries, and therefore the density gradient between the two layers is much less pronounced (**Fig. 15.19b**). The most important process that causes vertical mixing across an estuarine halocline is the friction and resulting turbulence created by reversing tidal currents. Hence, partially mixed estuaries are most common where tidal currents are relatively fast, river flow rate is moderate, and river current speed does not greatly exceed tidal current speed. As in a blender running at moderate speed with roughly equal volumes of water and oil, some mixing occurs at the interface, but two distinct layers persist.

In partially mixed estuaries, freshwater and seawater layers are separated by a relatively weak halocline. Seawater moves landward up the estuary as a bottom layer and is diluted progressively with freshwater from above (**Figs. 15.19b, 15.20b**). Both layers move up and down the estuary with each tidal cycle. The distance that water moves in the estuary with each ebb and flood is the tidal excursion. The tidal excursion is usually much greater than the landward or seaward net movement that

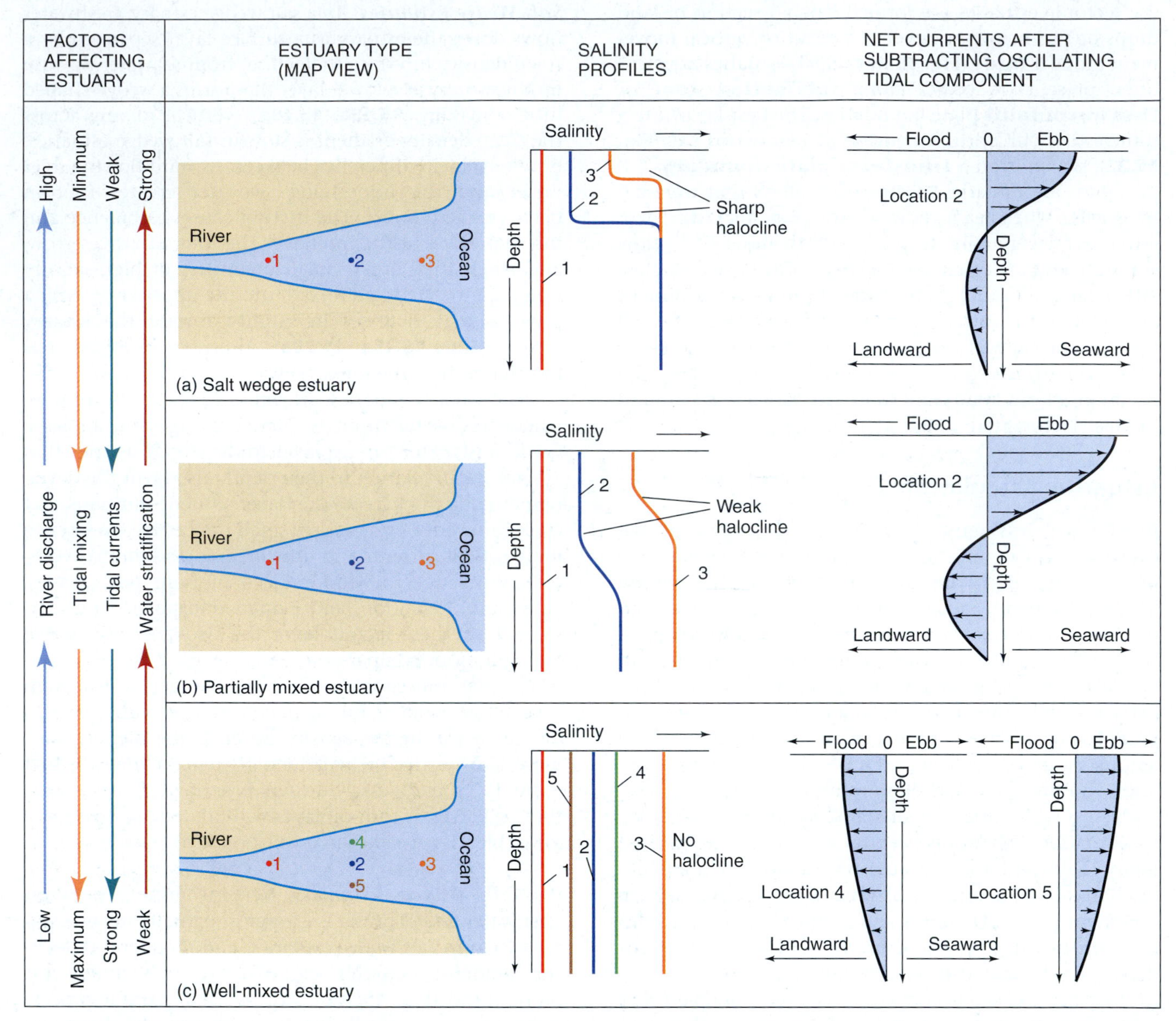

FIGURE 15.19 One way to classify estuaries is on the basis of their circulation characteristics and the factors that influence their circulation. (a) Salt wedge estuaries are river-dominated and are strongly stratified with a sharp halocline. (b) Partially mixed estuaries have a weak halocline because the surface and bottom layers are partially mixed by turbulence induced by tidal currents. (c) In well-mixed estuaries, the water column is thoroughly mixed from surface to seafloor, and there is no halocline. However, salinity does increase progressively with distance from the head of the estuary to the sea and across the estuary as a result of the Coriolis effect.

occurs as a result of the **residual currents** of the estuarine circulation (currents caused by the river flow and landward flow of seawater). **Figure 15.19** shows only these residual currents. Residual currents are often difficult to measure because they are masked by the faster reversing tidal currents that must be averaged out to reveal the residual current.

As seawater moves landward and freshwater moves seaward in an estuary, the moving water is deflected by the Coriolis effect. Consequently, if we look at an estuary from the sea in the Northern Hemisphere, seawater moving up the estuary tends to be deflected to our right, and freshwater moving down the estuary tends to be deflected to our left. In salt wedge estuaries, this deflection causes the halocline to be tilted slightly upward toward the right bank (**Fig. 15.20a**). In partially mixed estuaries, the Coriolis deflection affects not only the mean flow, but also tidal currents. As tidal currents flow into the estuary, they are deflected to the right side (as viewed from the sea); as they ebb, they are deflected to the left side. Consequently, in a partially mixed estuary, movement of seawater landward in the lower layer is concentrated on the right side, and the lower-salinity estuarine outflow is concentrated on the left side. The halocline is strongly inclined upward to the right (as viewed from the sea) in many partially mixed estuaries. The incline is in the opposite direction in the Southern Hemisphere.

Well-Mixed Estuaries. In estuaries with swift tidal currents, mixing is very strong (like the blender at high speed), the water column is completely mixed, and no halocline is present (**Fig. 15.19c**). In these well-mixed estuaries, seawater moving landward is continuously mixed with freshwater moving seaward. Salinity decreases progressively from the ocean toward the head of the estuary. Although the salinity is uniform from surface to bottom, it varies during the tidal cycle, increasing as the tide floods and decreasing as it ebbs.

In large well-mixed estuaries, the Coriolis effect tends to separate the landward and seaward estuarine flows, so they are concentrated on opposite sides of the estuary. This separation is especially important in wide and shallow estuaries. Many such estuaries have a small residual landward movement of water along the right side (as viewed from the sea in the Northern Hemisphere), and a small residual seaward current on the left side. This is made possible by a residual current that flows from one side of the estuary to the other (**Fig. 15.20c**). This circulation causes salinity to be higher on the right side of the estuary and to gradually decrease across the estuary to the left side (**Fig. 15.19c**).

Well-mixed estuaries tend to be wide and relatively shallow. They usually have limited freshwater inputs and strong tidal currents. These factors and others that determine the circulation characteristics change with location in the estuary and sometimes with time. Consequently, an estuary's circulation can vary from well mixed, to partially mixed, and to salt wedge at different times and at different locations within the estuary. San Francisco Bay is a good example (**Fig. 15.22**). The central portion of the bay (closest to the ocean) has relatively strong tidal currents, is wide, and is generally partially mixed. The upper part of the estuary, which is closer to the Sacramento and San Joaquin rivers, is narrower and generally has salt wedge characteristics because tidal currents are somewhat reduced and freshwater inputs are large enough to form a thick, low-salinity surface layer. The southern part of San Francisco Bay is shallow, has very little freshwater input, and is generally well mixed. When river flow rates are low, seasonally or in drought years, the northern part of the estuary becomes partially mixed. During high spring river runoff, the southern and central parts of the bay can become partially mixed or even salt wedge estuaries.

Fjord Estuaries. Fjords are generally narrow and usually much deeper than most other estuaries. Most fjords are deep enough that vertical mixing does not reach the bottom waters, even if tidal currents are strong. Consequently, almost all fjords have a halocline separating high-salinity bottom water from lower-salinity surface water, and fjords are almost never well-mixed.

Circulation in many fjords is complicated by the presence of a shallow sill at the seaward end of the estuary that prevents the free exchange of deep water between ocean and fjord (**Fig. 15.20d**). Therefore, estuarine circulation is established only above the level of the sill. The fjord's interior above the sill depth behaves as a salt wedge or partially mixed estuary. At the sill, vertical mixing is enhanced by turbulence caused by the flow of water over the sill (**Fig. 15.20d**). The deep water below the sill depth within the fjord is stagnant and is not involved in the estuarine circulation. In many fjords, the deep water becomes anoxic because of the decomposition of organic particles that have settled from above. Periodically this deep water may be displaced by high-salinity ocean water that floods over the sill in response to some unusual set of water movements. In such circumstances, hydrogen sulfide in the fjord bottom water is displaced into surface waters of the fjord and ocean, where it may cause extensive fish mortality.

Inverse Estuaries. Shallow estuaries in arid regions can sustain sufficiently high rates of evaporation that salinity is higher than that of ocean water outside the estuary. In this situation, estuarine circulation is inverted. High-salinity estuarine water flows seaward in a bottom layer, and ocean water flows landward as a relatively lower-salinity (and less dense) surface layer. Inverse estuaries generally occur at about 30°N or 30°S, in the region of atmospheric subtropical highs, where evaporation is strong and rainfall typically is low. Some well-known examples of inverse estuaries are the Red Sea, San Diego Bay in California, and Laguna Madre in Texas. In addition, the Mediterranean Sea behaves as though it were an exceptionally large inverse estuary.

Particle and Contaminant Transport in Estuaries

Most major cities have developed next to rivers and estuaries, partly because of their convenience for the disposal of sewage and other wastes. The common belief is that wastes discharged to a river or estuary are simply swept out to the ocean, where they are diluted in the vast volume of ocean water. However, a simple explanation of water movements and suspended-particle transport within estuaries reveals that this is not what happens.

Freshwater input at the head of the estuary creates the estuarine circulation. To sustain this circulation, the amount of seawater that flows through the estuary must greatly exceed the volume of freshwater discharged during the same period of time. To understand this process, we must first observe that freshwater entering the head of most estuaries has a salinity close to zero, whereas water

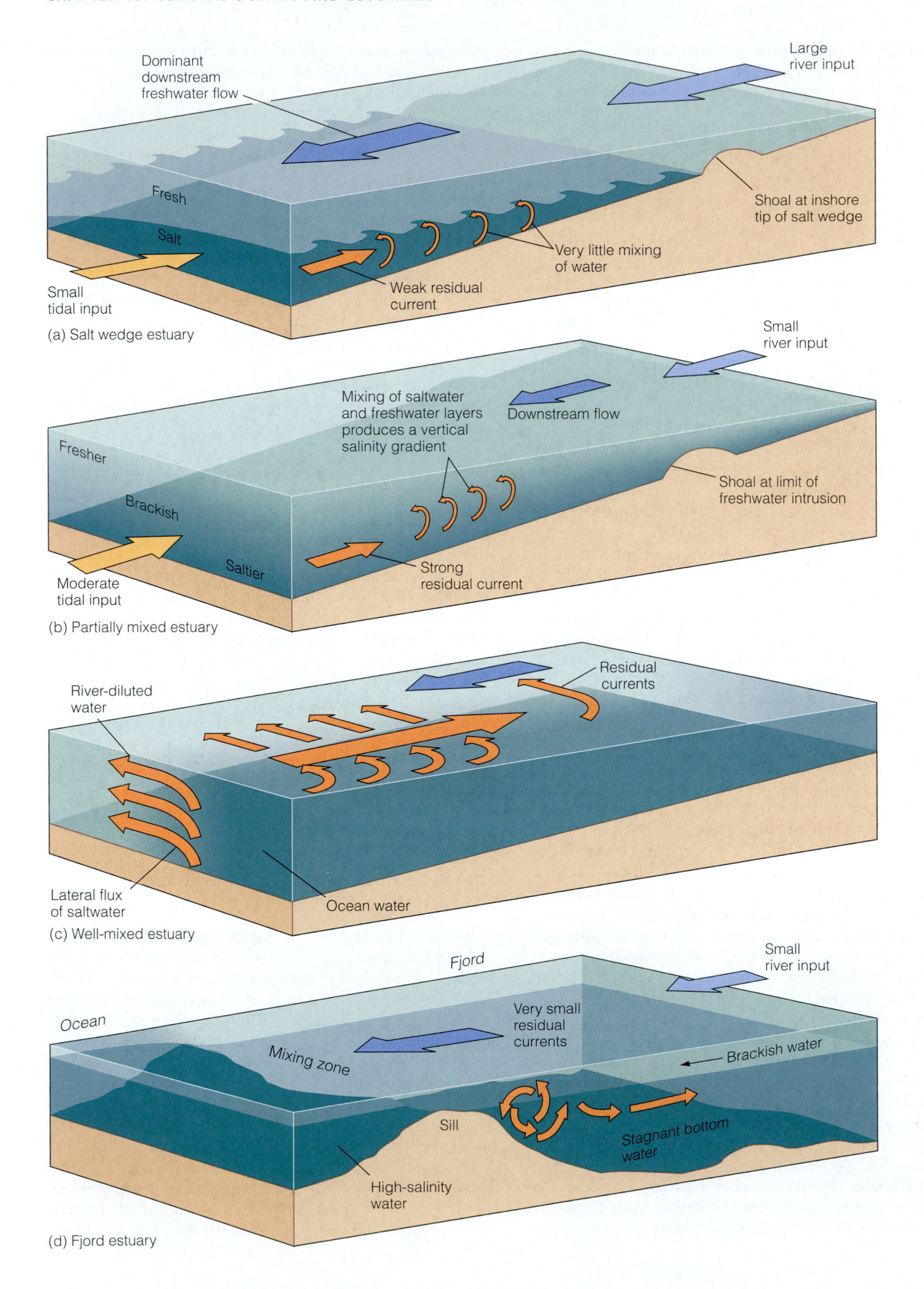
Dominant downstream freshwater flow
Large river input
Fresh
Salt
Shoal at inshore tip of salt wedge
Very little mixing of water
Weak residual current
Small tidal input
(a) Salt wedge estuary
Small river input
Mixing of saltwater and freshwater layers produces a vertical salinity gradient
Downstream flow
Fresher
Brackish
Saltier
Shoal at limit of freshwater intrusion
Strong residual current
Moderate tidal input
(b) Partially mixed estuary
Residual currents
River-diluted water
Lateral flux of saltwater
Ocean water
(c) Well-mixed estuary
Fjord
Small river input
Ocean
Very small residual currents
Mixing zone
Brackish water
Sill
Stagnant bottom water
High-salinity water
(d) Fjord estuary

FIGURE 15.20 *(opposite)* Types of estuaries. (a) Salt wedge estuaries are characterized by a net (after tidal motions are averaged out) landward current in the lower layer and a net seaward current in the upper layer. There is little vertical mixing, but breaking internal waves on the halocline mix some seawater up into the surface layer. (b) Partially mixed estuaries are characterized by relatively strong net landward currents near the seafloor and relatively strong net seaward currents in the surface layer. (c) Well-mixed estuaries have no vertical salinity variations, but they do have variations in salinity both across and down the length of the estuary. The net estuarine currents are landward at all depths on the right side of the estuary in the Northern Hemisphere (viewed from the ocean), and seaward at all depths on the left side. These currents are on the reverse sides in the Southern Hemisphere. (d) Estuaries in fjords have little current in the deep water below the depth of any sill. In the upper layers above the sill depth, there is generally a salt wedge or sometimes a partially mixed estuary circulation pattern.

discharged from the estuary mouth into the ocean has a salinity almost equal to that of the seawater. Freshwater mixes with seawater, so the salinity of the seaward-moving water progressively increases along the length of the estuary (**Figs. 15.19, 15.21**).

Figure 15.21 depicts the relative residual flow rate (the volume of water that passes a particular point in the estuary per day) in various parts of an estuary. From the figure, we can see that the residual flow rate of seawater moving landward and the residual flow rate of estuarine water flowing seaward at the mouth of an estuary are both many times greater than the river flow rate. In addition, the seaward and landward flow rates are highest near the mouth of the estuary and decrease toward the head of the estuary. The residual flow rate is equal to the residual current speed multiplied by the cross-sectional areas of the estuary. Consequently, flow rate does not always translate directly into current speed, because the cross-sectional area of an estuary changes, sometimes in complex ways, along its length. However, the large differences in flow rate along an estuary generally cause both landward and seaward residual currents to be higher near the mouth of the estuary than near the head.

Rivers carry large quantities of suspended sediment particles, the largest of which are quickly deposited when the river reaches the flat coastal plain or, if there is no coastal plain, when the river enters the head of an estuary. The finest particles can remain in suspension as they are transported to the oceans, but the increased salinity in the estuary causes most of them to clump. Because the clumps are larger, they are then deposited.

Particles of intermediate size, which fall slowly through the water column, are usually deposited within the estuary. In well-mixed estuaries, these particles are deposited primarily on the left side (as viewed from the sea in the Northern Hemisphere). In other estuaries, seaward-moving particles fall through the water column into the landward-moving bottom layer, where they are transported back toward the head of the estuary. These particles often are deposited near the estuary's head where tidal and residual currents are low, often accumulating as a shoal (**Fig. 15.20a,b**). However, when elevated river flows extend the head of the estuary seaward, the particles may be resuspended by the swifter river currents. In many estuaries, these recycled particles are eventually deposited in wetlands (Chap. 13).

Sediments also can be transported from the ocean into estuaries, particularly in estuaries where tidal currents are strong enough to resuspend sediments that accumulate on the nearshore continental shelf.

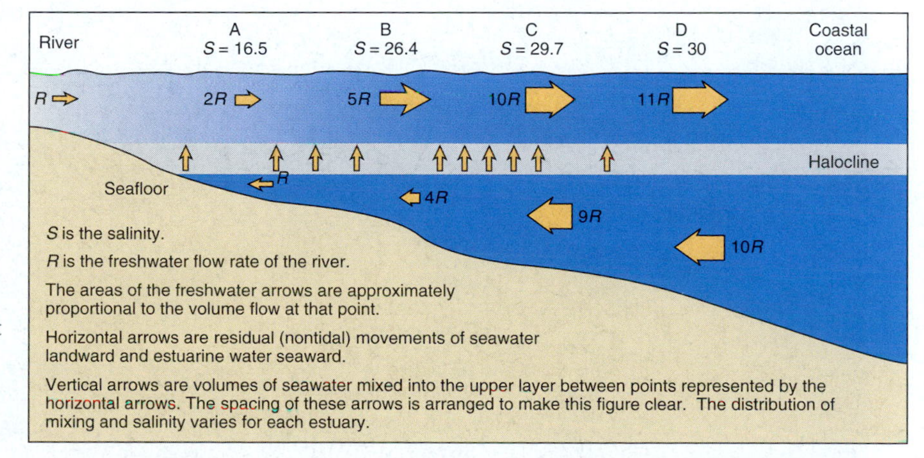

FIGURE 15.21 Seaward-moving surface water increases in salinity from nearly zero at the head of the estuary to 30 where it enters the ocean. Ocean water entering the estuary has a salinity of 33. Where seaward-flowing water reaches a salinity of 16.5 (half that of ocean water), it must consist of one-half river water and one-half ocean water. Seawater must flow landward to point A at a rate (volume per unit time) equal to the river flow rate, and seaward transport of low-salinity estuarine water must equal twice the river flow rate. Similarly, the landward flow rate of seawater at point B must equal 4 times the river flow rate, and the seaward flow rate of low-salinity estuarine water must be 5 times the river flow rate.

Substantial quantities of suspended sediment can be transported from the ocean into well-mixed estuaries, where they accumulate along the right side (as viewed from the sea in the Northern Hemisphere). Smaller quantities of suspended sediment are carried from the ocean into partially mixed and salt wedge estuaries. Even particles carried into the ocean by the seaward-moving surface layer may return to the estuary if they sink and are transported landward in the seawater layer.

The fate of particles in an estuary depends on the particle size and where the particles are introduced to the estuary. Floating and extremely small particles introduced to the surface layer (or to the left side of a well-mixed estuary) are transported to the oceans, although many will clump at higher salinities. Small or low-density particles in bottom waters are transported landward until the current slows and they are deposited. Such particles may accumulate in the estuarine sediments or in the sediments of adjacent wetlands, and they may be resuspended and carried seaward when river flow rates are high. Particles of intermediate size in the surface layers of salt wedge or partially **stratified** estuaries are carried seaward until they sink below the halocline, then are transported landward until they settle out in the upper parts of the estuary where current speeds are lower. Large particles tend to accumulate near the point of introduction.

The important point of the preceding discussion is that suspended particles tend to be retained within the estuary by the estuarine circulation and are not simply flushed into the sea. Because most toxic contaminants

FIGURE 15.22 This image of San Francisco Bay, taken by the Landsat satellite, shows the high turbidity in the upper estuary, where there is a salt wedge circulation, and the location of the Alcatraz dredged-material dump site near the estuary mouth, where the circulation is partially mixed.

attach to particles, particularly the organic particles in treated sewage effluents, contaminants tend to accumulate in the estuary and are not flushed out to the ocean.

A lack of understanding of estuarine circulation and its effect on particles and contaminant fates is one reason that many estuarine ecosystems have been and continue to be severely impacted by human activities. One of many examples of the misuse of estuaries for waste disposal is in San Francisco Bay, where navigation channels, marinas, and ports must be dredged to maintain desired depths. Most of the dredged sediments were, for decades, and some still are, dumped near the bay's mouth at a site between Alcatraz Island and San Francisco (**Fig. 15.22**). The swift tidal currents were expected to flush the material out to the ocean, where it would be dispersed. However, the dump site is near the right bank (viewed from the sea) of a partially mixed estuary, and the dumped dredged sediments instead fall to the floor of the bay and disperse in the lower part of the water column.

With an understanding of estuarine circulation, we realize that much of the dredged material dumped at the Alcatraz dump site, with its toxic contaminant load, is not transported to the ocean. It is instead transported back within San Francisco Bay by estuarine circulation and eventually is deposited in low-energy areas, including the same channels and harbors from which it was originally dredged. Perhaps more importantly, continual dredging and disposal disturbs the particle-associated toxic contaminants, which were relatively safe when buried in sediments, and resuspends them in the estuarine circulation, where they are more directly in contact with the biota (Chap. 19). Fortunately, the most contaminated dredged material is no longer dumped at this dump site.

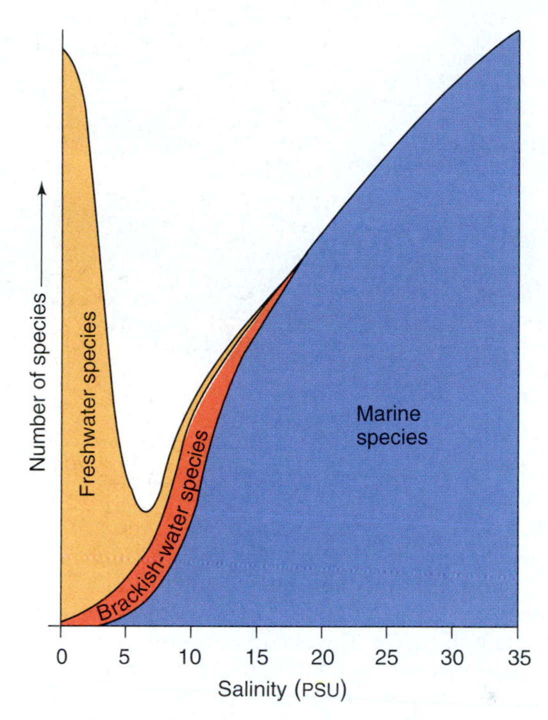

FIGURE 15.23 The number of species of marine, estuarine, and freshwater origin varies within the salinity gradient of a typical estuary. The fact that the total number of species is much lower in the brackish-water zone than in the freshwater or the marine zones reflects the stressful environmental conditions, particularly variable salinity, in the brackish-water region.

Estuarine Biology

Estuaries are difficult places for organisms to inhabit because they are subject to rapid and irregular changes in environmental factors such as salinity, temperature, and current speed. The stress induced by salinity changes is particularly challenging because organisms must be able to cope with widely varying **osmotic pressures** (Chap. 16).

Because estuaries are high-stress **environments**, they support fewer species than the adjacent ocean or freshwater environments support (**Fig. 15.23**). However, there is an abundant supply of nutrients and sunlight in most estuaries. As a result, estuarine biomass is typically much greater than biomass in freshwater and ocean environments, with the exception of the most productive coastal upwelling areas.

Although estuaries typically have relatively high turbidity, they are also generally shallow, and in many, light penetrates to the bottom. In such estuaries, primary production is dominated by benthic **microalgae**, particularly diatoms, that form mats that carpet the sediments. Macroalgae can grow on the estuary floor in areas where current speeds are low. In addition, rooted aquatic plants are abundant in the tidal **salt marshes** and **mangrove** forests along the edges of many estuaries. In contrast, because phytoplankton are continually being swept seaward with the estuarine surface layer flow, their populations are relatively limited in estuaries that have short residence times.

Nutrients are generally abundant in estuaries because they are supplied in substantial quantities in river water that has **leached** them from soils and rocks of the drainage basin. In addition, once in the estuary, nutrients tend to be retained and recycled within the estuarine circulation. Nutrient-containing detritus, which flows seaward in the surface layer, sinks below the halocline and then returns with the landward-flowing ocean water. This circulation tends to concentrate detritus in the region near the landward limit of seawater intrusion (**Fig. 15.24**), which has low residual currents. This region has high turbidity as a result of the accumulated detritus and other particles. The nutrient-rich detritus supports a varied community of fish and invertebrate larvae and juveniles.

In many estuaries, much of the detritus that accumulates in the upper part is derived from algal mats from the adjacent wetlands (Chap. 13). Where residual currents are

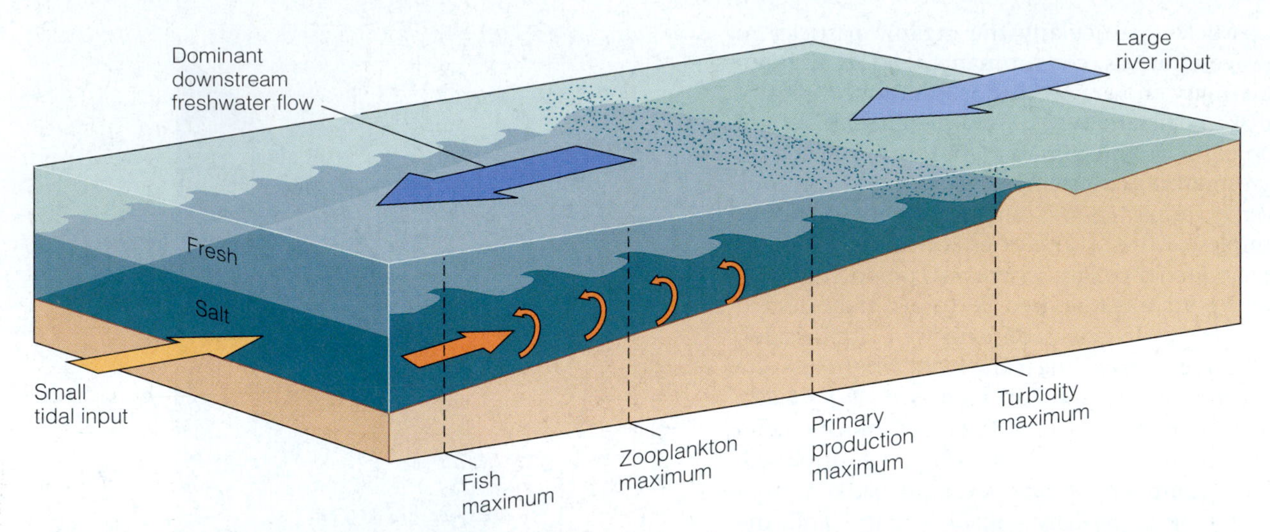

FIGURE 15.24 In many estuaries, an accumulation of particulate matter creates high turbidity near the landward extent of seawater intrusion. Particles transported seaward in the upper layer may sink to the lower layer before leaving the estuary and then return to the high-turbidity area in the landward-flowing lower layer. Many planktonic estuarine organisms use this pathway to stay within the estuary or to reach an appropriate point in the estuary at the right time in their life cycle. The maximum primary productivity in an estuary is usually just seaward of the high-turbidity zone. In this region, more light can penetrate. In addition, nutrients are available because they are regenerated by the decomposition of suspended detritus particles transported up the estuary in the lower layer and are then mixed into the surface water at the turbidity maximum. Maximum zooplankton populations and growth occur downstream of this area to take advantage of the phytoplankton drifting seaward in the upper layer.

slow or the estuary is long, a food chain from phytoplankton to zooplankton to estuarine fishes is established that is analogous to the food chain in the coastal upwelling ecosystem described previously in this chapter. Zones of high phytoplankton abundance, high zooplankton abundance, and high fish abundance form a seaward progression from the head of the estuary (**Fig. 15.24**). Some planktonic organisms and larval fishes sink into the seawater layer as they are transported seaward away from the region of abundant food supply. They are then transported back into the food-rich region.

Because they provide abundant food and substantial protection, estuaries are ideal habitats for the larvae and juvenile stages of marine animals. Particularly in wetlands, there are many shallow-water hiding places that are accessible to small swimming organisms but not to their larger predators. In addition, predators are less common in estuaries than in less stressful freshwater or ocean environments. Many marine fishes and invertebrates use estuaries as nursery grounds. Some species briefly visit the estuary to spawn, and others release vast numbers of eggs in the adjacent ocean so that they are transported into the estuaries by the estuarine circulation. Along the southeast coast of the United States, more than half of the commercially important marine fish species are known to use estuarine wetlands as nursery areas or for breeding. Because of the abundant supply of detritus, estuaries also support huge populations of commercially important shellfish, including clams, oysters, mussels, and crabs.

Anadromous and Catadromous Species

Estuaries are important to two categories of migratory fishes that live most of their lives either in the oceans or in freshwater. **Anadromous** fishes, such as salmon and striped bass, live most of their lives in the ocean but return to freshwater to spawn. The stress involved in the return to freshwater and the energy lost in spawning are such that the adults die soon after they spawn and do not return to the ocean. Their offspring develop in freshwater and migrate to the estuary, which provides abundant food and relative safety from predators. Eventually, they migrate out to sea, where they live several years before returning to the rivers to spawn.

Catadromous fishes use a strategy opposite that of anadromous fishes. Catadromous species live most of their lives in freshwater and then journey to the oceans to spawn and die. Freshwater eels are the best-known and most abundant catadromous species. Sexually mature eels in rivers on both sides of the North Atlantic Ocean make a one-way journey to spawn in the southwest part of the Sargasso Sea. Their offspring return to the rivers, riding part of the way on the Gulf Stream. North Pacific eels make a similar journey to the equivalent southwest area of the North Pacific Gyre, and their offspring use the Kuroshio (Japan) Current for the ride home.

Anadromous and catadromous species do not generally spend much time in the estuary during their reproductive migration, but being there is especially stressful because they must make the osmotic-pressure adjustment from seawater to freshwater, or vice versa, very quickly. Striped bass, for example, rest in the outer part of the estuary to adjust to lower salinity before moving farther upstream. Many estuaries also have relatively high concentrations of toxic contaminants that are likely to cause additional stress on anadromous and catadromous species as they transit the estuary. Such contaminants may account, in part, for the decline of these species in some areas.

CHAPTER SUMMARY

Special Characteristics of the Coastal Oceans. Salinity and temperature are more variable and turbidity is generally higher in coastal waters than in the open ocean. River discharges often lower salinity and increase turbidity near river mouths, and a shallow halocline is often present in this region. Coastal currents are driven by local winds and are generally independent of ocean gyre currents, often flowing in the opposite direction. Benthos of the coastal oceans are more diverse and abundant than open-ocean benthos.

Nutrient Supply to the Coastal Photic Zone. The mixed layer in many coastal regions receives nutrients from river runoff, coastal upwelling, decomposition of organic matter on the shallow seafloor, eddies formed around shallow seafloor topographic features, and breaking internal waves. Nutrients tend to be retained in coastal waters because residence times of coastal water masses are often long. Coastal upwelling is most prevalent off the west coasts of continents where Ekman transport moves the surface layer offshore. Newly upwelled water may support only limited primary productivity until needed organic compounds are synthesized by flagellates. As upwelled water moves offshore, diatoms bloom and are consumed by zooplankton, and eventually nutrients are depleted and primary productivity declines.

Nutrients may be recycled as they are transported into subthermocline waters in fecal pellets, released there by decomposers, and transported back inshore and upwelled. The eggs or spores of many species are carried inshore in subthermocline water, are then upwelled, **metamorphose** or hatch, and move back offshore where phytoplankton food supply increases. There they mature and produce eggs or spores that sink and reenter the circulation.

Seasonal Cycles. In subpolar regions, primary productivity is very low, except during a short summer when light intensity is sufficient for photosynthesis. Nutrients are plentiful in areas where strong vertical mixing prevents a pycnocline from forming. In polar and some subpolar regions, vertical mixing is restricted by a halocline formed by freshwater runoff and sea-ice melt; hence, nutrients are limited and productivity is low.

In tropical regions, there is little seasonality, and nutrients are always limiting, except in upwelling areas. However, coral reef ecosystems are highly productive because of nutrient recycling, primarily between corals and their symbiotic zooxanthellae.

In mid latitudes, the winter water column is well mixed, and primary production is light-limited. In spring, as light intensity increases, a plankton bloom occurs and continues until nutrients are depleted. Nutrient depletion is often accelerated by the formation of a shallow seasonal thermocline. Productivity remains low until fall, when storms provide additional nutrients by vertical mixing and a limited phytoplankton bloom may occur. Nutrients are returned to the photic zone by winter mixing.

There is a succession of dominant phytoplankton species during seasonal cycles of mid and high latitudes. Generally, diatoms dominate when nutrients are abundant and are replaced by flagellates as nutrients decline or as silica is depleted. Species succession is similar each year, but it varies in timing and dominant species because of differences in climate and the physical and chemical characteristics of the ocean water. Juveniles of many animal species rely on particular phytoplankton species or types at particular times and locations. When the species succession is altered and this food is unavailable, these species' larval survival can be drastically reduced.

Algal Blooms. Intense blooms of dinoflagellates occur infrequently in coastal water, primarily at subtropical and temperate latitudes. These episodic blooms can be natural. However, their frequency appears to be increasing off many coasts, and evidence suggests that sewage-derived nutrients may favor the dominance of dinoflagellates over diatoms and contribute to some blooms. Certain dinoflagellates and diatoms produce substances toxic to vertebrates. Hence, blooms can kill fishes. They can also adversely affect human health, because the toxins do not harm shellfish but are toxic to people who eat the shellfish. The most prevalent dinoflagellate toxin causes paralytic shellfish poisoning in humans. Many valuable shellfishing areas are closed to harvesting during months when dinoflagellate blooms may occur.

When phytoplankton blooms collapse as a result of nutrient depletion, their decomposing remains can deplete oxygen in bottom waters, particularly where residence times are long and bottom waters are separated from surface waters by a seasonal thermocline. Anoxia can decimate benthic communities. Some evidence suggests that nutrients in sewage may increase the frequency and geographic extent of hypoxia and anoxia in coastal waters adjacent to major cities.

Fisheries. Fishes are most abundant in areas of high primary productivity, especially in coastal waters, where trophic efficiency is higher because food webs are shorter. Coastal upwelling regions account for about half of the world's fishery production. Natural variations in fish stocks and reproductive success are so great that maximum sustainable yield cannot be established accurately. Many fisheries are currently exploited at or above their maximum sustainable yield, and a number have collapsed.

Estuaries. Estuaries are regions where seawater mixes with freshwater from rivers. Most present-day estuaries were formed as sea level rose during the past 18,000 years. Coastal-plain estuaries are drowned river valleys, and are abundant on passive margins. Bar-built estuaries are located landward of sandbars or barrier islands. Tectonic estuaries are formed by uplift or subsidence of the coast at a fault. Fjords are drowned valleys previously cut by glaciers.

In salt wedge estuaries, seawater forms a lower layer separated by a sharp halocline from a low-salinity surface layer. Seawater progressively mixes into the surface layer as this layer moves seaward, raising its salinity. In partially mixed estuaries, the halocline is much broader, usually because of greater vertical mixing by tidal currents. In well-mixed estuaries, the water column has vertically uniform salinity, but salinity progressively increases seaward. In all but inverse estuaries, nontidal currents due to the estuarine circulation flow landward in the bottom part of the water column and seaward in the surface layers. The flow rates of seawater into the estuary and of lower-salinity water into the ocean are many times greater than the flow rate of freshwater into the estuary. In the Northern Hemisphere, the Coriolis effect deflects flow into the estuary toward the right side as viewed from the ocean and deflects the outflow to the left side. Fjords have estuarine circulation only at depths above the top of their sill. Water below the sill depth is stagnant and often anoxic. Occasional flushing of anoxic bottom water enriched in hydrogen sulfide can cause fish kills.

Particles and associated toxic contaminants tend to be trapped in estuaries. They can move seaward in the upper layer, but they sink to the lower layer that moves landward in estuarine circulation.

Estuaries are stressful environments for organisms because of variable temperature, salinity, turbidity, currents, and other environmental factors. However, they provide food and shelter from ocean predators, particularly in bordering wetlands. The estuarine circulation enables weakly swimming or planktonic species or their eggs and larvae to remain within the estuary. Anadromous fishes live in the ocean and transit estuaries to spawn in rivers. Catadromous fishes live in rivers and transit estuaries to spawn in the oceans.

STUDY QUESTIONS

1. What characteristics of the coastal oceans make them different from the deep oceans?
2. How does coastal upwelling occur? How could you tell whether upwelling is likely in waters off a particular coast without making any direct observations of the oceans?
3. Describe the circulation in coastal upwelling zones. How do organisms take advantage of this circulation?
4. What are the sources of nutrients to the coastal photic zone? Discuss their distribution in the oceans.
5. Why are there seasonal cycles in populations of marine organisms in mid latitudes and polar regions? Why is there little or no seasonality in tropical regions?
6. Why does primary production slow during midsummer in mid latitudes, and why is there sometimes a fall bloom?
7. Why is there a succession of dominant phytoplankton species during a typical seasonal cycle? Why are diatoms usually replaced by flagellates during this succession?
8. What are the relative magnitudes of fish biomass and production in the deep oceans and coastal regions? Why are these not proportional to the relative areas of these regions?
9. What are the major types of estuarine circulation? Why do they occur in different estuaries?
10. If you were in Australia and operating a boat that had only a weak engine, on which side of an estuary would you choose to travel out to the ocean, and on which side would you choose to return? Why?
11. Why are estuaries stressful environments for marine organisms?

KEY TERMS

You should be able to recognize and understand the meaning of all terms that are in boldface type in the text. All of those terms are defined in the Glossary. The following are some less familiar key scientific terms that are used in this chapter and that are essential to know and be able to use in classroom discussions or on exams.

algae (p. 414)
anadromous (p. 440)
anoxia (p. 428)
bar-built estuary (p. 431)
barrier island (p. 431)
benthic (p. 415)
benthos (p. 412)
biomass (p. 418)
bloom (p. 418)
carnivorous (p. 419)
catadromous (p. 440)
chlorophyll (p. 421)
coast (p. 416)
coastal plain (p. 431)
coastal-plain estuary (p. 431)

coastal upwelling (p. 416)
coastal zone (p. 412)
coastline (p. 412)
community (p. 414)
contaminant (p. 412)
continental shelf (p. 412)
coral (p. 414)
coral reef (p. 415)
cyanobacteria (p. 426)
delta (p. 431)
detritus (p. 419)
diatoms (p. 418)
dinoflagellates (p. 426)
ebb (p. 414)
ecosystem (p. 412)
eddy (p. 415)
effluent (p. 433)
Ekman transport (p. 416)
estuary (p. 412)
eutrophication (p. 429)
fjord (p. 413)
flagellate (p. 418)
flood (p. 414)
food chain (web) (pp. 419, 425)
habitat (p. 428)
halocline (p. 412)
hard parts (p. 423)
herbivorous (p. 418)
hypoxia (p. 428)
internal wave (p. 414)
invertebrates (p. 426)
isothermal (p. 413)
kelp (p. 413)
lagoon (p. 412)
larva (p. 419)
longshore drift (p. 432)
macroalgae (p. 413)
mangrove (p. 439)
marginal sea (p. 412)
maximum sustainable yield (p. 430)
microalgae (p. 439)
micronutrient (p. 418)
mixed layer (p. 412)
nekton (p. 418)
nutrients (p. 412)
nutrient-limited (p. 418)
osmotic pressure (p. 439)
overfished (p. 420)
oxygen demand (p. 428)
partially mixed estuary (p. 433)
photic zone (p. 413)
photosynthesis (p. 414)
phytoplankton (p. 414)
plankton (p. 419)
polyp (p. 423)
primary production (p. 422)
primary productivity (p. 416)
pycnocline (p. 414)
residence time (p. 412)
residual current (p. 434)
respiration (p. 428)
resuspended (p. 415)
runoff (p. 412)
salt marsh (p. 439)
salt wedge estuary (p. 433)
shelf break (p. 412)
shellfish (p. 425)
shoal (p. 421)
shore (shoreline) (pp. 415, 417)
sill (p. 433)
spawn (p. 419)
species succession (p. 425)
spit (p. 432)
standing stock (p. 424)
stratified (p. 438)
suspended sediment (p. 415)
symbiotic (p. 415)
tectonic estuary (p. 431)
thermocline (p. 414)
trophic level (p. 430)
turbidity (p. 412)
turbulence (p. 414)
upwelling (p. 412)
vertebrates (p. 427)
water mass (p. 412)
well-mixed estuary (p. 433)
wetland (p. 433)
zooplankton (p. 418)

CRITICAL CONCEPTS REMINDER

CC5 Transfer and Storage of Heat by Water (p. 413). Water's high heat capacity allows large amounts of heat to be stored in the oceans and released to the atmosphere without much change in the ocean water temperature. Water's high latent heat of fusion allows ice to act as a heat buffer, which keeps the ocean surface water layer temperatures in high latitudes relatively uniform and near the freezing point. To read **CC5** go to page 15CC.

CC8 Residence Time (p. 412). The residence time of seawater in a given segment of the oceans is the average length of time the water spends in that segment. The residence times of some coastal water masses are long and therefore some contaminants discharged to the coastal ocean can accumulate to higher levels in these regions than in areas with shorter residence times. To read **CC8** go to page 19CC.

CC11 Chaos (p. 430). The nonlinear nature of many environmental interactions, including some of those that control annual fluctuations in fish stocks, mean that fish stocks change in sometimes unpredictable ways. To read **CC11** go to page 28CC.

CC12 The Coriolis Effect (p. 431). Water masses move freely over the Earth and ocean surface, while objects on the Earth's surface, including the solid Earth itself, are constrained to move with the Earth in its rotation. This causes moving water masses to be deflected as they flow. The apparent deflection is at its maximum at the poles, is reduced at lower latitudes, and becomes zero at the equator. In Northern Hemisphere estuaries, the Coriolis effect tends to concentrate the lower salinity water that flows down the estuary toward the left side of the estuary as viewed from the ocean, while the high salinity water flowing up the estuary tends to be concentrated toward the right side (In the Southern Hemisphere these directions of concentration are reversed). To read **CC12** go to page 32CC.

CC15 Food Chain Efficiency (pp. 417, 423). All organisms use some of their food as an energy source in respiration and for reproduction. They also lose some of their food in excretions (including wastes). On average, at each level in a food chain, only about 10% of food consumed is converted to growth and biomass of the consumer species. To read **CC15** go to page 49CC.

CC16 Maximum Sustainable Yield (pp. 418, 423, 427). The maximum sustainable yield is the maximum biomass of a fish species that can be depleted annually by fishing but that can still be replaced by reproduction. This yield changes unpredictably from year to year in response to the climate and other factors. The populations of many fish species worldwide have declined drastically when they have been overfished (beyond their maximum sustainable yield) in one or more years when that yield was lower than the average annual yield on which most fisheries management is based. To read **CC16** go to page 51CC.

CHAPTER 16

CHAPTER OUTLINE

CRITICAL CONCEPTS USED IN THIS CHAPTER

Each of these images show the flamboyant cuttlefish (*Metasepia pfefferi*), which can grow to about 10 cm long. The four larger images (from top left to bottom right) illustrate the mastery that this species has with changing its colors. These four photos, all of the same cuttlefish, were taken in succession over a period of less than one minute. Unfortunately, the full startling range of its color cannot be captured on film (or video), it can only be seen in person. The smaller image in the upper right corner is the cuttlefish we encountered in the Chapter Opening story to the right. The smaller image in the bottom left corner is a young cuttlefish, which is only about one centimeter long. It is no more than a few days old, but is already capable of the same color changing artistry that the adults of its species possess. In fact, the cuttlefish can alter their colors even before hatching from their transparent egg cases, as seen later in this chapter in Figure 16.27j.

Marine Ecology

It started as an unremarkable dive. Little did we know that, even after having done more than 1000 dives, we were about to experience one of the most exciting encounters of our diving lives. Sharks or whales? No, this creature was much smaller, but much more unusual and interesting. What we came across was a flamboyant cuttlefish, a relatively small cuttlefish species, this one an adult about 10 cm long. We had heard about this species and had always wanted to see one. Now here it was, and it was out hunting, totally oblivious to our presence, allowing us to approach within less than a meter.

What was so special about this species? First, it hunts small prey such as shrimp that hide just buried in the sand. To do so, the species has learned to "walk" along the seafloor using two of its eight tentacles, giving it a very unusual appearance that reminded me of a fighting machine from the Star Wars movies. Second, like other cuttlefish, it hunts by extending another of its tentacles slowly forward beyond the length of its body and then, in a lightning motion, too fast for the human eye to follow, strikes out farther to capture its prey. In this same motion, the feeding tentacle and captured prey are retracted out of sight in the cuttlefish's mouth.

The cuttlefish's hunting technique is simply amazing to watch, but even that does not tell the whole story, for while it was hunting, this animal was constantly changing its color patterns. All cuttlefish can rapidly change color using chromatophores distributed on their bodies, but the flamboyant cuttlefish is, as its name implies, a master of this art. As it hunted, this animal exhibited bands of bright colors that flowed along its body, accompanied by pinpoint rows of white dots that also flowed along its body in a manner that can only be described as exactly like the flowing news ticker in Times Square.

A special encounter such as this, our first of what have now been many encounters with flamboyant cuttlefish, is rare. In this encounter,

we watched the animal hunt for 20 to 30 minutes, until we reluctantly had to return to the surface. In all of our many subsequent encounters with flamboyant cuttlefish, we have never seen one perform such a spectacular display—or is it just that a first encounter with any of the multitude of wondrous species in the oceans is always the most memorable?

The oceans are populated by uncounted millions of **species,** most of which have not yet been identified. Marine **ecology** is the study of relationships between species and between species and their **environment.**

Because each species has a unique relationship to the ocean environment and to other species, marine ecology is a complex discipline. This chapter describes the environmental factors that govern the distribution of marine organisms, and the adaptations that have evolved in marine species to meet the challenges of living in the ocean.

16.1 ECOLOGICAL REQUIREMENTS

A fundamental imperative of life is survival of the species. Hence, all species must meet two requirements. First, an adequate food supply must be ensured for each life stage. Second, enough individuals of the species must survive until they reproduce successfully. The principal challenges are

- Where to live
- What food to eat and how to obtain it
- How to avoid predators
- Where, how, and when to reproduce

In meeting these challenges, marine species have adopted an extraordinary variety of life cycles to take advantage of the various attributes of the marine environment.

16.2 HABITATS

The major **habitats** available to marine organisms are in the water column, on the seafloor surface, and in seafloor **sediment** (Chap. 14). For **photosynthetic** organisms (**CC14**), only the first two habitats are suitable, and only within the **photic zone.** Consumer and **decomposer** species are present in each of the three habitats: **pelagic** species in the water column, **benthic epifauna** and epiflora on the seafloor, and benthic **infauna** within the sediment.

The three habitats offer different advantages and challenges. Each species has selected its own unique balance between those opportunities and challenges. Many species take advantage of different balances at successive stages of their life cycle by changing their habitat between stages.

In shallow water, photosynthesis may take place at the seafloor if it is within the photic zone, and **chemosynthetic communities** are present in restricted areas of the seafloor below the photic zone. In both locations, **primary production** by benthic organisms creates some food. This primary production can be substantial locally, particularly in coastal waters and at **hydrothermal vents,** but in most of the ocean the food supply is produced by photosynthesis by **phytoplankton,** and certain microscopic **prokaryotes** (Chap. 14).

Pelagic Habitat

Phytoplankton and other photosynthetic microorganisms in the pelagic environment are effectively **grazed** by **zooplankton** and single-celled **protists** that, in turn, are eaten by **carnivores** or **omnivores**. During this process, most of the **biomass** produced by photosynthesis is used by photic-zone pelagic consumers to fuel their energy needs (Chap. 14). Consequently, only a small fraction of the biomass produced by photosynthesis eventually sinks or is transported below the photic zone, and hence food is scarce in the environments below the photic zone.

Most pelagic organisms must actively seek food and avoid or fight off predators at the same time. Whatever strategy an organism uses for these activities, it uses up energy that must be obtained from food. Hence, one penalty for living in the pelagic habitat is that more food must be consumed than is needed for growth and reproduction. The need to avoid sinking also carries an energy penalty for pelagic organisms. **Buoyancy** can be maintained by swimming, which uses energy directly, or by including in the body pockets of gas or low-**density** oil, each of which requires energy to produce.

All pelagic organisms must evade predators, compete for food, and counteract the tendency to sink. Below the photic zone, predators are less abundant and easier to avoid because they cannot hunt visually in the darkness. Offsetting this advantage is the scarcity of food below the photic zone. Many species, especially of zooplankton and squid, migrate between the photic zone and the **aphotic zone** to take advantage of this trade-off between the zones. These organisms feed in the photic zone at night and then descend to the darkness below in the day. Nevertheless, even this behavior has an energy cost, because energy must be expended to move vertically between layers.

Benthic Epifaunal Habitat

Organisms that live on the seafloor do not need to expend energy to control buoyancy and generally require less energy to find and capture food than pelagic organisms require. Hunting is reduced to a two-dimensional problem, and because many benthic prey are sedentary or nearly so, hunting movement can be slow and energy-efficient. Even more energy is saved by **benthos** that remain sedentary and wait for prey to walk, swim, or drift by.

The energy savings of living in the benthic epifaunal environment may be offset by the scarcity of food and the difficulty of evading predators. Organisms that move slowly and sit on the sediment may be easy to find and capture. Hence, many benthic epifaunal species have developed defensive mechanisms such as poisons, spines, and shells or tubes into which they withdraw. Such mechanisms require energy that otherwise could be used for growth and reproduction. Where food is readily available, epifauna may be abundant and the competition for available space may be a significant disadvantage. Finding living space is a problem, particularly for epifauna that attach to rocks, because most of the seafloor is covered by sediment and rocky substrates are very limited.

Although the benthic epifaunal environment in shallow water is similar to that in deeper water, benthic plants enhance the food supply in many shallow areas. In some, such as **kelp** forests (**Fig. 14.3**), the kelp provides abundant hiding places for predator avoidance. **Coral reefs** sustain exceptionally **diverse** (**CC17**) and abundant **encrusting** algae (plant) communities.

Benthic Infaunal Habitat

Even though food is scarce in most of the benthic infaunal environment, some organic **detritus** accumulates continuously in sediments. Many animals, such as worms and certain **crustaceans**, feed by sifting sediment grains to obtain organic particles or by digesting organic matter as sediment is passed through the gut. Other benthic infaunal species live in the sediment primarily for protection from predators and feed on particles in the water above or prey on passing animals.

The relatively poor supply of food available to benthic infauna is offset by the energy savings of relatively sedentary lifestyles, by the lack of need for buoyancy control, and by the reduced need for defenses against predators. In **estuarine** and other coastal environments, much of the **salinity** and temperature variation that can occur in overlying waters does not occur in the sediment, so infauna are not subject to these environmental stresses. Intertidal benthic infauna also avoid exposure to air during low **tides**.

Benthic infauna must expend energy to dig into or move through sediments or rock. They must also cope with an environment that varies because of **biochemical** and chemical processes within the substrate. The most important of these processes consume dissolved oxygen and produce toxic hydrogen sulfide, so most infaunal species can live only in the oxygenated surface layer of sediment. In areas where detritus inputs are high and water movements slow, the oxygenated layer is thin or absent, but elsewhere it ranges from a few millimeters to several meters in thickness. Where sulfide is present, the **biota** consists primarily of **bacteria** and **fungi** adapted to the sulfide environment and animals that build burrows or tubes through which they can obtain oxygenated water from above.

Other Ocean Habitats

Special challenges and opportunities are also found in other ocean habitats, including the **surface microlayer**, the **intertidal zone**, and hydrothermal vents. The surface microlayer, which is only a few molecules thick, concentrates a variety of natural and **contaminant** organic compounds. A few species attach themselves to the surface microlayer to keep from sinking. **Surface tension** tends to prevent small particles within the microlayer from sinking, even if their density is higher than that of seawater. Many species distribute their eggs and **larvae** by placing them in the surface microlayer, where **currents** distribute them until they grow large enough to feed and swim independently. Placement in the surface microlayer reduces the need to use energy to provide the eggs and larvae with oils or other means of buoyancy control.

Although the intertidal zone and hydrothermal vents are sites of very high **primary productivity** and food is abundant, each of these habitats imposes substantial offsetting challenges due to variations in environmental conditions. These two unique environments are discussed in Chapter 17.

16.3 FEEDING

With the exception of organisms that obtain their energy needs through photosynthesis or chemosynthesis, all marine species must obtain energy from organic compounds that they obtain as food. The food can consist of another living organism, nonliving organic particles, or dissolved organic compounds.

Because concentrations of dissolved organic compounds in seawater are extremely low, few marine species are known to rely on dissolved organic matter as their principal source of sustenance. Nevertheless, this source of food may be important to some species, including bacteria and other decomposers, and a few higher animals. To use dissolved organic matter, an organism must take up individual dissolved molecules through its membrane surface. This process is very slow if dissolved compounds are transported to the membrane by **diffusion,** but it can be enhanced if water flows over the membrane.

In terrestrial **ecosystems,** animals feed in three basic ways: by grazing the abundant macroscopic (large) plants that cover much of the Earth's surface, by consuming detrital organic particles in leaf litter and soils, and by hunting and eating other animals. Similarly, in the oceans many species graze on the seafloor, eat detritus in the sediment, and hunt and eat other animals. However, because most plant material and detritus in the oceans consists of small particles dispersed in the water column, many marine species are **suspension feeders** that feed on particles.

Suspension Feeding

Suspension feeders must be able to gather efficiently the phytoplankton, zooplankton, and detritus particles that constitute their food. They do so in several different ways. Many suspension feeders are **filter feeders** that have a weblike or matlike structure to capture particles as water flows through it. This simple method works efficiently for suspension feeders that filter out and eat large particles, but filter feeding becomes more difficult as the size of the filtered particles becomes smaller. Zooplankton that eat small phytoplankton and detritus particles would need a filter mesh so small that **viscosity** would restrict the flow of water through the filtering apparatus. Consequently, these zooplankton generally have evolved hairlike appendages, called "setae," on their mouths. As water passes through the mouth, particles are captured by the setae.

As an alternative or sometimes an addition to filter feeding, many suspension feeders secrete mucus to which particles adhere as water flows past. The organism then ingests the mucus and the food that it contains. Many other suspension feeders capture particles that come in contact with appendages designed to grab or grasp a particle and transfer it to the mouth. A number of zooplankton species feed on relatively large prey by capturing them in this way with setae on their appendages. When this mechanism is used to capture live prey, it can be considered a hunting method and the organism can be considered a predator.

Many suspension feeders ingest particles selectively, most often by size. Particles smaller than a certain size are not captured by the filtering or other feeding apparatus, and particles larger than a certain size cannot be captured or passed into the mouth. Different species have evolved to collect different size ranges of particles. Certain zooplankton species that are adapted for capturing large phytoplankton, such as **diatoms,** tend to be most abundant where and when diatoms are the dominant phytoplankton. Other zooplankton species are better adapted to capture smaller phytoplankton, such as **flagellates,** and therefore are most abundant when flagellates dominate. Because different species of carnivores feed on different **herbivorous** zooplankton species, the size of food particles available to suspension feeders can affect the entire **food web.**

Many suspension feeders selectively ingest only certain species within a selected size range of particles. For example, herbivores consume only living phytoplankton cells and reject detritus particles of similar size. Such selective feeding requires the feeding organism to expend energy to sort its food supply. The benefit gained is that the selected food has higher nutritional value than nonselected species or detritus has. In contrast, many suspension feeders are nonselective omnivores that ingest plant, animal, and detritus particles of appropriate size without discrimination.

Suspension-feeding species inhabit both pelagic and benthic environments. Pelagic suspension feeders include zooplankton and **nekton.** Some **planktonic** suspension feeders simply drift through the water, capturing food particles brought to them by **turbulence** and diffusion. The chance of encountering a food particle in this way is low because of the low concentrations and slow movement of the particles in the water, so filter feeders generally have evolved methods of ensuring a flow of water through their feeding apparatus. The three basic methods are

- Actively pumping water through the filtering apparatus
- Moving the filtering apparatus through the water
- Keeping the filtering apparatus stationary and allowing ocean currents to move water through it

Marine organisms have developed numerous physiological variations to apply one or combinations of these methods.

Pelagic Suspension Feeders

Many pelagic suspension feeders have evolved mechanisms to move or pump water past or through their feeding apparatus as they drift through water. Crustaceans such as **copepods** and **euphausiids** (**Fig. 14.18a,b**) have long, slender forelimbs or appendages surrounding their mouths (**Fig. 16.1a**), with which they grasp or direct suspended food particles toward the mouth. In contrast, the equivalent mouthparts of most shrimp, also crustaceans, are designed to cut or crush, reflecting the shrimp's hunting or **scavenging** feeding habits (**Fig. 16.1b,c**). Suspension-feeding crustaceans and many other zooplankton can swim weakly through the water to increase their chances of encountering food particles.

Pteropods, which are **mollusks** related to slugs and snails, have adapted to suspension feeding by evolution of their foot (the foot on which a snail crawls) to be a membrane that extends from the body (**Fig. 16.2**). The pteropod uses this sail-like membrane as a paddle to propel itself slowly through the water. This action moves water past the mucus-covered membrane, which captures suspended food particles that brush against it.

FIGURE 16.1 Forelimb use in crustaceans. (a) A deep-sea euphausiid, or krill. (b) Banded coral shrimp (*Stenopus hispidus,* Papua New Guinea). (c) Harlequin shrimp (*Hymenocera elegans,* Papua New Guinea). The forelimbs of the krill are adapted for filter feeding. Compare these with the forelimbs of the coral shrimp, which eats larger food particles than the krill does, and the massive forelimbs of the harlequin shrimp, which feeds on tough sea star bodies.

Salps, a type of **tunicate**, evolved in aquatic environments to take advantage of suspended particulate food. Many salps have a simple body resembling an elongated barrel (**Fig. 14.19e**) with an opening at either end. Within the barrel, the salp is coated with a continuously moving mucous sheath, which captures food particles from water pumped continuously through the salp's body by sequenced contractions of bands of muscles. There are also colonial salps, which are composed of individual members joined to form the outer wall of a saclike structure. Each individual's **incurrent** opening faces outward, and each **excurrent** opening discharges into the common open space inside the sac (**Fig. 16.3**). This arrangement enables the colony to "swim" as water is forced out of the relatively narrow sac opening by the combined pumping efforts of its members. As a result, each individual is more likely to encounter food particles than it would if it operated alone.

Most jellyfish and **ctenophores** (**Fig. 14.19d**) feed on suspended particles by extending long trailing tentacles that, in some species, have poisonous stinging cells and, in others, are covered with a sticky substance to capture food. These organisms are generally carnivores that feed selectively on zooplankton or small fishes, and many propel themselves by contracting rhythmically while trailing their tentacles through the water in order to increase their chances of encountering food. The Portuguese man-of-war achieves a similar increase in mobility and feeding efficiency by allowing winds to blow the colony across the ocean surface.

Although many jellyfish and ctenophores are suspension feeders, larger species of these organisms are clearly hunters that selectively seek, kill, and consume small

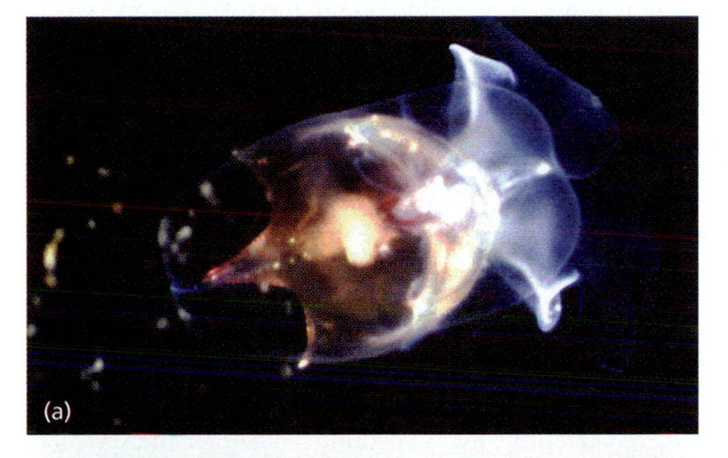

FIGURE 16.2 Some adaptations of the foot in gastropods. (a) Typical pteropod of the order Thecosomata (Papua New Guinea). (b) Marine gastropod, a harp shell (*Harpa major,* Indonesia). The transparent part of the pteropod is its foot, which is very different from that of other gastropods, such as the harp shell. The harp shell's foot is extended from its shell flat onto the seafloor and allows it to "walk."

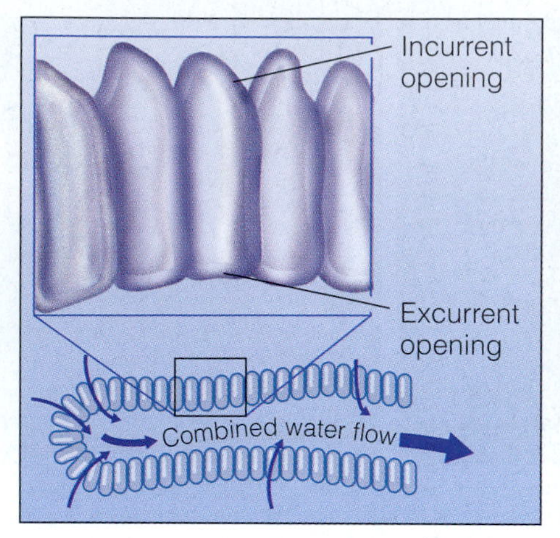

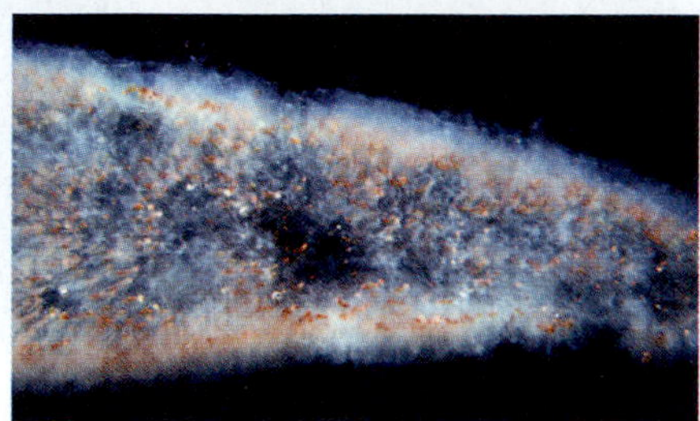

FIGURE 16.3 Salps. Each individual of a salp colony draws water into its body and filters it for food. The combined water flow from all individuals is passed out of the open end of the saclike colony, which slowly propels the colony through the water. If you look carefully in the inset photograph, you will see the individual members of the salp colony. Note that they resemble their relatives, the tunicates shown in Figure 16.8.

fishes and other nekton. The distinction between suspension feeding and hunting is not precise. Some species are clearly suspension feeders and some are clearly hunters, but other species use elements of both approaches and feed on both suspended particles (plankton and/or detritus) and small nekton.

Benthic Suspension Feeders

Like other benthos, benthic suspension feeders benefit from the low energy needs of their sedentary lifestyle and the lack of a need to control buoyancy. They have the additional advantage that ocean currents bring food to them, which reduces the energy needed to hunt for food. Consequently, many benthic epifaunal and infaunal species are suspension feeders. Suspension feeders are especially abundant in coastal regions where suspended particulate food is plentiful.

In soft sediments that cover much of the seafloor, certain infaunal suspension feeders pump water into their feeding apparatus through tubes that they extend upward into the water (**Fig. 16.4**). This feeding method is particularly advantageous in intertidal mudflats because a buried mollusk can withdraw its siphon and close its shell for protection when the sediment is exposed at low tide. On mudflats or muddy sand **beaches** where cockles are abundant, buried cockles can be detected as the tide **ebbs** because they squirt a small fountain of water into the air as they close abruptly when the water recedes.

The sea pen is a benthic infaunal suspension feeder that extends a beautiful and intricate fanlike structure into the water column to feed (**Fig. 16.5a,b**). Suspended particles are captured as they drift with the current into the fan. The fan can be folded up and withdrawn into the sediment where the main body of the organism is buried. Many sea pens withdraw by day and open to feed only at night, when zooplankton are more abundant because many species migrate from below the photic zone and nocturnal zooplankton emerge from their daytime hiding places. Other types of **invertebrates**, including some species of **tube worms, sea cucumbers,** and **anemones**, live in soft sediment and extend feeding tentacles or webs into the water (**Fig. 16.5**).

Epifaunal suspension feeders are especially abundant and diverse in **continental shelf** regions, where there is a solid substrate to which they can attach without being covered by sediment. In temperate and high **latitudes** these areas are limited to a few rocky outcrops on the seafloor, but in tropical regions coral **reefs** provide extensive areas of suitable substrate. Suspension-feeding epi-

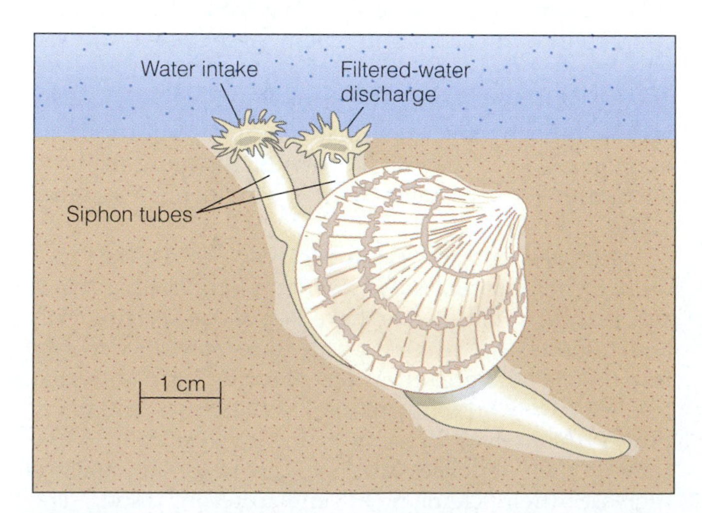

FIGURE 16.4 Some bivalve mollusk species, such as this cockle (family Cardiidae), live buried within the sediments but extend two tubes up into the water column just above the surface. They pump water in from one tube and out the other, and they filter food particles as the water passes through their bodies.

FIGURE 16.5 Examples of sand-dwelling benthic infaunal suspension feeders. (a) A sea pen (order Pennatulacea, Papua New Guinea) with its feeding polyps oriented into the current. (b) The reverse side of the same sea pen shown in (a). (c) This sand anemone (Papua New Guinea) feeds on large "suspended" matter such as small fishes. If it is threatened or when it captures prey, it can withdraw these tentacles almost instantly and completely into a hole in the sand, where the remainder of its body is hidden. (d) A tube worm (family Sabellidae, Papua New Guinea) with its netlike feeding apparatus extended into the water column.

faunal invertebrates include species of mollusks, crustaceans, tunicates, **corals**, anemones, sea cucumbers, **sea stars,** and worms.

The most familiar epifaunal suspension-feeding mollusk is the mussel (**Fig. 16.6a**), which attaches itself to any available substrate and partially opens its shell to pump water through its body and feed. Other bivalve mollusks, such as the spiny oyster (**Fig. 16.6b**), use a similar method in coral reef communities, where they are firmly attached to the reef. Some scallop species in temperate latitudes do not attach to the seafloor, but instead live on coarse sand or gravel bottoms, from which they can "swim" up into the water column for short distances by snapping their shells shut to create propulsive jets of water.

Barnacles are suspension-feeding epifaunal crustaceans (**Fig. 16.6c**) that look nothing like their close relatives shrimp and crabs, because they have evolved an unusual feeding method. Barnacles, in essence, lie on their backs, strongly attached to rocks or other hard substrates such as ships' hulls, and are protected by hard plates and shells. They open the plates and extend their much altered legs as a weblike structure that sweeps through the water to grasp and capture suspended particles (**Fig. 16.6d**).

Most of us think of coral reefs as large, strangely shaped, hard, rocklike structures (**Fig. 16.7a,b**). However, reef-building corals are actually extremely small suspension-feeding epifaunal organisms that grow in colonies, each of which may contain millions or even billions of individuals, called **polyps.** Some of the many species of corals have **hard parts** that are left behind when individuals die and thus serve as a base on which other polyps can grow.

FIGURE 16.6 Examples of suspension-feeding invertebrates. (a) California mussels (*Mytilus californianus,* Monterey, California) grow in profusion in shallow waters. Other species of mussels are common in other areas. (b) A Pacific spiny oyster (*Spondylus varians,* Papua New Guinea) attaches to a coral reef wall and opens its shell so that it can feed by pumping water through its body. (c) A colony of barnacles in Monterey, California. The colony contains more than one species, in this case probably primarily *Balanus glandula* and *Chthamalus* sp. (d) Barnacles, including these coral barnacles (order Pyrgomatidae, Indonesia), use fanlike modified feet to sweep through the water and collect food particles.

The living coral individuals, or polyps, feed by extending tentacles that capture food particles, primarily zooplankton, and draw them down into the mouth (**Fig. 16.7c**). Most hard corals extend their tentacles to feed only at night, but many species of **soft corals,** which are found mostly in the Pacific and Indian Oceans, feed both day and night. Many of these spectacular soft corals (**Fig. 16.7d,e**) come out to feed only when there is a strong current to bring abundant food supplies to the colony. Many corals, both soft and hard, produce colonies that extend upward or outward into the water from their attachment point (**Fig. 16.7b,d,f**). This spreading maximizes the volume of water that passes over the colony and, thus, the amount of particulate food that passes within reach. Many species of zooanthids and anemones, close relatives of corals, are also suspension feeders that eat plankton and detritus (**Fig. 16.7g,h**), but larger anemones (**Fig. 16.7i**) are primarily carnivorous hunters that eat fishes and invertebrates.

There are many species of suspension-feeding epifaunal tunicates, particularly on coral reefs. They feed in the same way as salps but are attached to the substrate and have both their incurrent and excurrent openings on their upper bodies. Tunicate species include a variety of single and colonial forms (**Fig. 16.8**).

Certain species of sea cucumbers, sea stars, and worms suspension-feed by extending intricate tentacles or weblike structures into the water column (**Fig. 16.9**). The elaborate basket star (**Fig. 16.9b**) comes out to feed only at night and spreads its intricate arms, which can extend more than a

FIGURE 16.7 Polypoid species (class Anthozoa) take many different forms. (a) A hard coral known as brain coral (*Diploria* sp., Papua New Guinea). (b) A hard coral often called staghorn coral (*Acropora* sp., Papua New Guinea). (c) Close-up showing polyps of a hard coral species (*Tubastrea* sp., Fiji) extended to feed at night. (d) Colony of a soft coral (*Dendronephthya* sp., Fiji). *Dendronephthya* occur in many different colors. (e) Close-up of a soft coral colony (*Xenia* sp., Papua New Guinea) showing individual polyps at different stages of feeding. The eight fingers surrounding each polyp's mouth open wide to capture suspended particles and then close like a fist to move the particles to the mouth. (f) Typical gorgonian sea fan (*Melithaea* sp., Papua New Guinea). These fans can consist of millions of individual polyps and be several meters across. (g) A small suspension-feeding colonial anemone (*Nemanthus annamensis,* Fiji). (h) A colonial zooanthid (*Protopalythoa* sp., Papua New Guinea). (i) A "magnificent sea anemone" (*Heteractis magnifica,* Fiji), here providing a home for a pink anemonefish (*Amphiprion perideraion*).

FIGURE 16.8 Tunicates (also called "ascidians") are among the most advanced invertebrates, although they appear very simple in structure. (a) Lightbulb tunicate (*Clavelina* sp., Philippines). (b) The colonial tunicate species *Didemnum molle* (Papua New Guinea). Each of the rounded forms is a colony. (c) There are two species of tunicates in the photograph. The three tall, blue and blue-green individuals are *Rhopalaea* sp., and the three smaller individuals with darker bodies and yellow markings around their incurrent and excurrent openings are *Clavelina robusta* (Indonesia). (d) A very different form of colonial tunicate that covers and encrusts the reef (*Botryllus* sp. Indonesia).

meter, across the current to capture food efficiently from huge volumes of water. During the day, the basket star coils into a tiny ball and hides in crevices in the reef. Fan worms (**Fig. 16.9c**) and Christmas tree worms (**Fig. 16.9d**) live in tubes that are usually drilled into, or built with the surrounding growth of, a hard coral colony. They feed by day but have an amazing ability to sense movement or shadows of moving objects and withdraw into their tubes instantaneously as divers or predators approach.

Surface Grazing

Although shallow seafloor areas with macroscopic plants that can be grazed by herbivores are rare, much of the seafloor provides sufficient, and in some areas abundant, food for grazers. This food supply varies in composition from location to location, but on most of the ocean floor it consists primarily of detritus and the bacteria and other decomposers associated with the detritus. The concentration of these foods generally decreases with depth and is higher where the seafloor is below a region of high pelagic primary productivity.

Note that, in marine ecosystems, the term *grazing* includes feeding on detritus and animals and is not restricted to plant eating, its common terrestrial usage. On seafloor within the photic zone, detritus and associated decomposer biomass are supplemented by other sources of grazer food, including benthic **microalgae** and **macroalgae.** Benthic microalgae, particularly diatoms, are abundant on shallow seafloor, where light

FIGURE 16.9 Examples of suspension feeders that employ weblike structures to capture food particles. (a) A creeping sea cucumber (*Cucumaria* sp., Philippines) with its suspension-feeding apparatus extended. (b) A basket star (family Gorgonocephalidae, Papua New Guinea) with its arms extended into the current to feed at night. (c) A sabellid spiral fan worm (*Protula magnifica*, Indonesia). (d) A Christmas tree worm (*Spirobranchus giganteus*, Papua New Guinea).

and **nutrients** are abundant and where waves or other water motions rarely **resuspend** sediments to cause abrasive scour.

Many marine animals that live on the seafloor, especially species that live on hard substrates, are colonial forms whose individuals are small and immobile. The colonies can be grazed by other animals without significant harm to the colony because grazing removes only a limited proportion of the colony's individuals, which can be replaced relatively quickly by reproduction. Many colonial animal species thus provide a renewable food supply for grazers in much the same way that plants do for terrestrial grazers.

Numerous adaptations have evolved for surface grazing. Many species of surface grazers obtain food by sifting the upper layer of sediment and therefore can also be considered **deposit feeders.** Deposit feeders, which are species that obtain food from within sediments, are described later in this chapter. Species that surface-graze may also suspension-feed or hunt larger prey.

Some surface grazers, such as **sea urchins** (**Fig. 16.10a**), have evolved such that their mouth is on their underside. These organisms crawl over sand and rocks, scraping off encrusting benthic **algae** ingesting detritus as they move. Tiny allied cowries crawl across their favorite food, the **sea fan,** and ingest the individual polyps (**Fig. 16.10b**). Other surface grazers include **nudibranchs,** or sea slugs (**Fig. 16.10c,d**), that crawl across the substrate (including corals and other immobile animals that constitute or cover the hard surface of the sub-

FIGURE 16.10 Examples of surface grazers. (a) Long-spined sea urchins (*Diadema savignyi,* Papua New Guinea), browsing across the seafloor. (b) An allied, or egg, cowrie (*Primovula* sp. Indonesia) on a gorgonian sea fan. Notice that the gorgonian's polyps have been eaten on the areas of the sea fan along which the cowrie has moved. (c) A flabellina nudibranch (*Flabellina rubrolineata,* Indonesia). These shell-less gastropod mollusks advertise their toxic nature with their flamboyant coloration and body shapes. (d) This spectacular nudibranch photographed in Papua New Guinea may be a previously unknown species.

strate) to eat algae, detritus, or animals such as corals, **sponges**, **hydroids,** and tunicates.

Surface grazers do not necessarily live on the sediment surface. For example, some clams live buried in the sediment and extend long siphon tubes to the water above. They use the siphon tubes to select food particles and "vacuum" them off the sediment surface (**Fig. 16.11a**).

In some shallow areas, large concentrations of macroalgae (**Fig. 15.3**) are extensively grazed by many animal species. In extremely shallow water, the seafloor may be covered with the **seagrass** *Thalassia* (**Fig. 16.11b**) or with other rooted plants, such as the marsh grass *Spartina* (**Fig. 16.11c**). These plants are not heavily grazed, because they are difficult to digest. The plants must be broken down by bacteria and fungi to detritus before they become a desirable and widely used food source. Some notable species, such as the manatee (**Fig. 14.24d**) and the green sea turtle (**Fig. 16.11d**), are adapted to feed directly on seagrasses. *Thalassia*, the favorite food of green sea turtles, is commonly called "turtle grass."

Deposit Feeding

Organic detritus that falls to the sediment surface is used mostly by decomposers and surface-grazing benthic **fauna,** but much detritus is buried by **bioturbation** and by accumulating sediment, especially in areas of high **sedimentation rate** or **anoxic** bottom water.

Organic matter buried in sediments serves as food for decomposers and deposit-feeding animals (**Fig. 16.12**). As

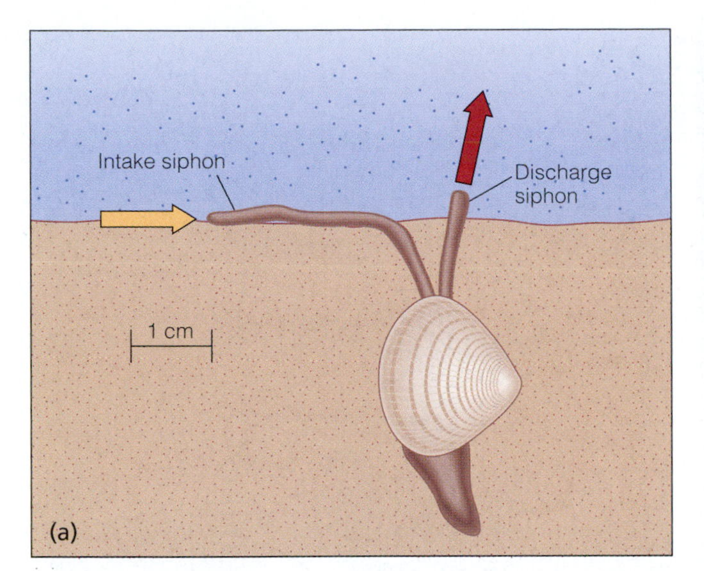

FIGURE 16.11 A benthic infaunal surface grazer, and seagrass communities. (a) The clam *Tellina* sp. feeds on the sediment surface. (b) Seagrass, especially turtle grass (*Thalassia* sp.), is a favorite food of the green sea turtle. Turtle grass beds are excellent places for animals such as this seagrass filefish (*Acreichthys tomentosus,* Papua New Guinea) to hide. (c) Rooted marsh grasses, primarily *Spartina* sp. (Aruba, Caribbean Sea), form dense beds in many coastal wetlands where the water is brackish. (d) Green sea turtle (*Chelonia mydas,* Hawaii).

this food is depleted, the concentration of organic matter generally decreases with depth below the sediment surface, which is one reason that deposit feeders are concentrated in the upper layers of sediment.

Deposit-feeding infauna generally eat their way through sediments, digesting organic particles or the organic matter that coats some particles in much the same way that earthworms eat their way through soil. Much of the detrital organic matter that survives decomposition to be buried in sediment is refractory (difficult to decompose) and has low food value. Consequently, many deposit-feeding infauna obtain most of their food by consuming bacteria and fungi, many species of which live within the sediment. Other deposit-feeding infauna have evolved means of decomposing refractory detritus.

Because detritus particles are generally small and have lower density than inorganic sediment particles, detritus tends to settle and concentrate in low-energy areas where fine-grained sediments also accumulate. Consequently, fine mud deposits tend to have higher organic matter concentrations than coarser sediments have, and they are more favorable habitats for deposit feeders. However, to live and feed in sediment, deposit feeders must have a supply of oxygen for **respiration.** In muddy sediment where organic matter is abundant, respiration by animals and decomposers quickly uses up

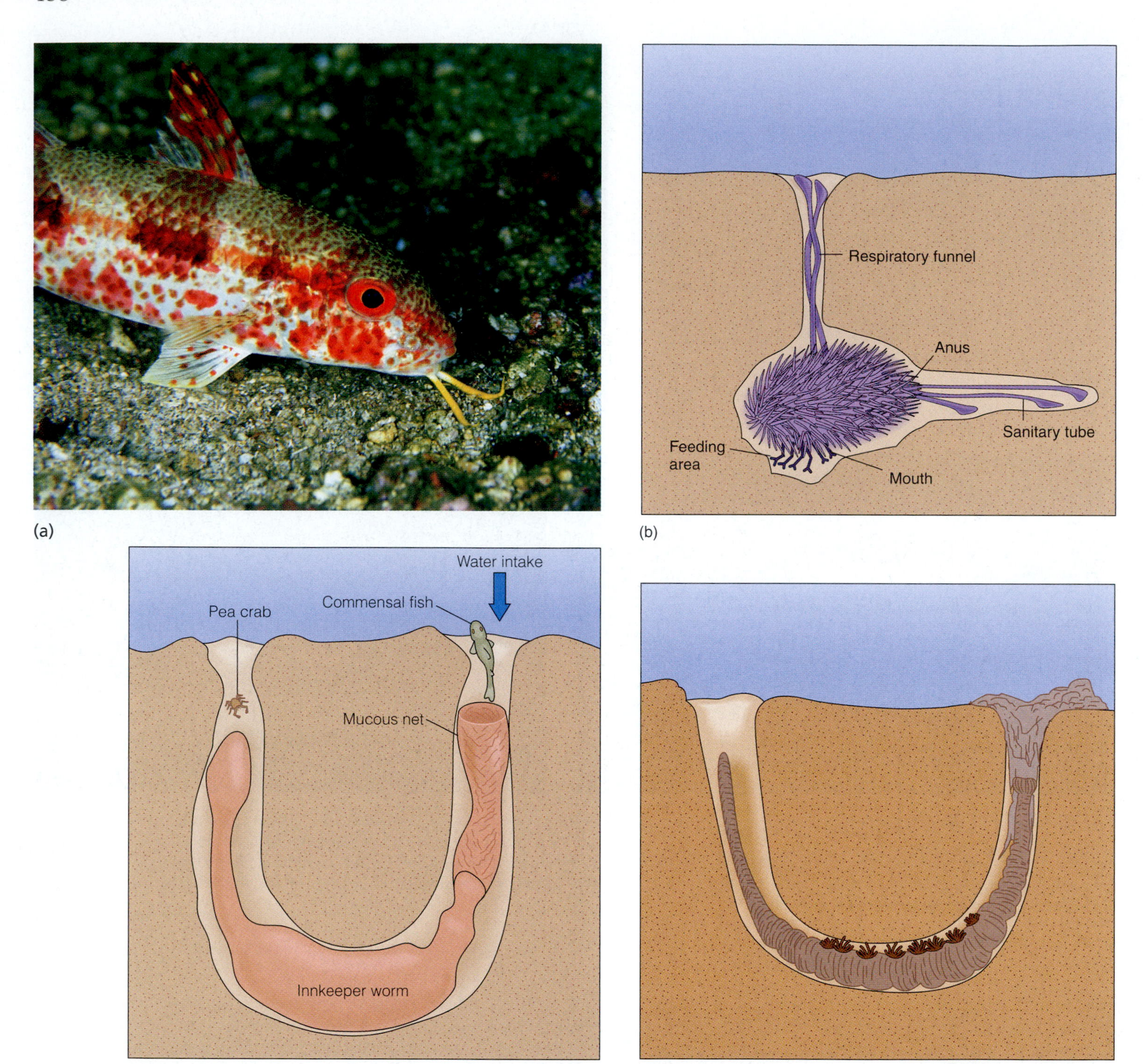

FIGURE 16.12 Examples of deposit feeders. (a) Goatfishes such as this blackstriped goatfish (*Upeneus tragula,* Indonesia) use their barbells (the two yellow "whiskers" protruding from just under the mouth) to sense and smell out small invertebrates under the surface of the sediments. The goatfish digs into the sediments with the barbells to capture and eat these invertebrates. Other fishes, especially some wrasses, often follow the goatfish to capture a free meal from among the invertebrates that the goatfish exposes as it digs. (b) The heart urchin (*Echinocardium* sp.) feeds on detritus buried in the sediments and respires through long tubes extended to the surface of the sediment. (c) The innkeeper worm (*Urechis caupo*) builds a *U*-shaped tunnel through which it draws oxygen-containing water for respiration. The tunnel forms a perfect home for several associate species, sometimes including several gobies and two species of small crabs. Although it lives in the surrounding sediment, a small clam species also uses the burrow by extending its siphon tubes into the burrow to obtain oxygenated water. (d) Lugworms (*Arenicola brasiliensis*) use a similar *U*-shaped burrow.

the oxygen dissolved in **pore waters.** Oxygen is not readily replenished from the water column above, because downward diffusion of oxygenated water through the sediment is slow.

Deposit feeders either are restricted to the near-surface oxygenated sediment layer or have evolved mechanisms to acquire oxygen from the water column above. For example, the heart urchin has adapted some of its tube feet to create a periscope-like breathing apparatus that it extends upward to the sediment surface (**Fig. 16.12b**). Deposit-feeding clams extend their siphons to the sediment surface for the same purpose. Many other infaunal deposit feeders form and live in a *U*-shaped or similar burrow and draw oxygenated water through the burrow as they feed. The innkeeper worm (**Fig. 16.12c**), lugworm (**Fig. 16.12d**), and many other species use variations of this method.

Muddy sediments tend to be dominated by **annelid** worm populations, whereas sandy substrates are more likely to be dominated by various species of deposit- or filter-feeding clams. Many deposit feeders pass sediment through the gut and discharge feces containing large volumes of processed sediment through the excurrent end of their burrow or breathing tube. In this way, food-depleted material is transported to the sediment surface, where it will not be reingested. Some species, particularly annelid worms, produce tightly packaged **fecal pellets** that accumulate on the sediment surface and protect it from **erosion.** Other species, especially certain clams, **excrete** their fecal material as a cloud of loose particles.

Clams may discharge their wastes as a cloud of particles for reasons other than to get rid of food-depleted sediment. This form of waste disposal may protect the clams' environment from invasion by other species. By working through the upper layer of sediment and continuously dispersing this material at the surface, clams prevent the silty sand in which they live from compacting and becoming firm enough for epifauna to attach to or lie on without being buried. In addition, the clouds of fine inorganic particles that the clams continuously create tend to clog the feeding apparatus of filter-feeding organisms with non-nutritious material.

16.4 HUNTING AND DEFENSE

All marine species must ensure that enough individuals survive predation to produce the next generation. Carnivorous species must also adopt a successful hunting method that can counter the defensive methods of their prey and provide adequate food. Although there are only a few basic approaches to either offense or defense, combining these approaches makes possible an incredible variety of offensive and defensive strategies. The situation is analogous to the game of chess, in which a few simple permitted moves of the chess pieces can be combined into an almost infinite variety of offensive or defensive strategies. Each species has its own unique combination of defensive and offensive strategies. In a chess game, the most successful strategy is usually one in which individual moves are combined in such a way that each contributes to both defense and offense. Similarly, marine species often use the same approach for both purposes.

Basic offensive and defensive approaches used by marine organisms are summarized in **Table 16.1.** The following sections describe and illustrate how these approaches are used by various species, and how fish species have evolved to optimize variations of them. The unusual sensing mechanisms that some marine species use to locate prey are discussed in a subsequent section.

TABLE 16.1 Some General Approaches to Hunting and Defense

Behavior	*Offensive Uses*	*Defensive Uses*
Speed	Chase down prey	Escape predators
Lures	Attract prey	Confuse predators
Camouflage and mimicry	Ambush	Escape detection Confuse or deter predators
Concealment	Ambush	Escape detection Evade capture
Spines and armor	Overcome armor or other defenses	Deter predators
Poisons	Kill or disable prey	Reduce predator population Reduce competitor population Deter predators
Group cooperation	Overwhelm and confuse defenders	Confuse predators Minimize predation

Speed

The most familiar approach to hunting and defense in terrestrial ecosystems is the use of speed. The predator chases the prey, and the prey tries to outrun the predator. For example, lions and tigers chase antelopes, but antelopes often can escape by outrunning their pursuers. Speed is used in much the same way in the marine environment.

Predators that hunt by using their swimming speed include a variety of sharks, bony fishes, **marine mammals,** and squid. The prey are generally other species of fishes, marine mammals, or squid. In some cases the hunt is similar to the lion–antelope chase, but it occurs in three dimensions. Prey outnumber the predator and live uneasily in the predator's presence, always carefully monitoring its movements and maintaining a respectful distance. Suddenly the predator selects a potential victim and begins the chase. Unwary prey or prey that are weak or injured fail to elude the predator and are consumed. Stronger individuals and those that are more successful in avoiding predators preferentially survive to reproduce and pass on their more successful genes to future generations. The predatory species is subject to a similar natural selection process because stronger and more skillful predators outcompete weaker and less successful members of their species for the available food resources.

Only the largest predators and prey can invest the considerable energy required to overcome water resistance and swim long distances at high speed. These species must obtain large quantities of food to replace energy reserves used in the chase. Consequently, most ocean predators use speed in ways that ensure a quick kill, and prey species seek ways to escape their predators quickly.

Predators may use a short burst of speed as the final component of a hunting strategy in which stealth or other approaches are used initially to get within striking distance of the prey. A variety of fish species, including lizardfishes (**Fig. 16.13a**), frogfishes (**Fig. 16.13b**), and hawkfishes (**Fig. 16.13c**), lie quietly in wait on the seafloor until their prey passes nearby. Then, a quick burst of speed is sufficient to capture the prey. Lizardfishes rely on their prey's mistake in approaching within half a meter or so, a distance across which a lizardfish can make a very fast attack. Frogfishes are among the fastest-moving marine animals, but they can maintain their speed for only a few tens of centimeters and must be very close to their prey for a successful hunt. To ensure that its prey approaches close enough for such attacks, a frogfish uses two other approaches in its hunting strategy: camouflage and a lure (discussed below).

Speed is also used in a variety of ways as a component of defensive strategies. **Gobies** use short bursts of speed to retreat to their home, which is generally a hole in the sand or reef where they are safe from predators (**Fig. 16.13d**). Some invertebrates, such as the fan worm (**Fig. 16.9c**), Christmas tree worm (**Fig. 16.9d**), and some sand anemones (**Fig. 16.5c**), use a high-speed, almost instantaneous, retreat into a protected tube or burrow to avoid predation. Many fish species change direction quickly to evade an onrushing predator. For example, butterflyfishes (**Fig. 16.13e**) can evade predators by swimming quickly in and out of the tortuous nooks and crannies of a reef or by simply making tight turns.

Lures

Lures are used by ocean predators to attract fishes and other prey just as they are used in sportfishing. The lure is made to look like a tasty morsel of food to attract the prey species to the concealed predator (or fishhook). The best-known practitioners of this technique are frogfishes (**Fig. 16.13b**). Lying still and camouflaged, a frogfish wiggles a fleshy knob on the end of a stalklike projection extending from above the mouth. When another fish approaches, this lure appears to be a small fish, shrimp, or other food morsel. By the time an unwary approaching fish is close enough to realize its mistake, it is within striking distance of the frogfish's lightning-quick, short-range attack. Because the predator uses little energy while lying in wait, it requires successful hunts only infrequently. In addition, the lure can attract more prey than would normally swim by. The deep-sea anglerfish (**Fig. 16.14**) uses the lure-based ambush strategy to perfection. Its **bioluminescent** lure appears as a tantalizing tiny point of light to prey that live in perpetual dark, although many potential prey species in this zone are sightless.

Ironically, lures can also be used for defensive purposes. They are used by many reef fish species, especially butterflyfishes, which have a dark spot near the tail (**Fig. 16.13e**) that resembles an eye. The real eye is often camouflaged in a vertical black line. When a predator makes its initial move toward a butterflyfish, it aims its attack toward the front of the fish to block its expected escape route. If the predator is lured into orienting its attack to the false eye location, the butterflyfish gains valuable moments in which to swim off in a direction other than that expected by the predator. The false eye may also be

CRITICAL THINKING QUESTIONS

16.1 Most marine species are adapted to eat only food of a certain type, such as suspended particles, organic coatings on sediment particles, phytoplankton cells of a certain size range, zooplankton, or small fishes. Discuss the possible reasons why most species do not develop (or do not retain) the ability to feed on many different types of food.

16.2 Nudibranchs are snails that have no shell. They generally live on reefs and feed on algae, sponges, or corals. There are many holes in reefs that make excellent hiding places. Why do nudibranchs use poisons and bright coloration to warn off predators rather than just using the many available hiding places in the reef?

FIGURE 16.13 Some fishes that employ speed for offense or defense. (a) A twospot lizardfish (*Synodus binotatus,* Indonesia) lying in wait for passing prey. (b) A striated frogfish (*Antennarius striatus,* Indonesia). The frogfish wiggles the lure, called an "esca," on the end of its illicium, a modified first ray of the dorsal fin located just above the eye. When wiggled, the esca looks like a small polychaete worm to the frogfish's unsuspecting prey. (c) A Falco, or dwarf, hawkfish (*Cirrhitichthys falco,* Indonesia) that has just been rewarded for its time spent lying in wait by capturing a favorite food item, a shrimp. (d) A fire goby or dartfish (*Nemateleotris magnifica,* Papua New Guinea) hovering above its hole in the reef, into which it is ready to dart and take refuge. (e) Bennett's butterflyfish (*Chaetodon bennetti,* Papua New Guinea). Note the large false eyespot near its tail.

FIGURE 16.14 This deep-sea anglerfish (*Chaunax pictus*) is bloated from the pressure reduction as it was brought to the surface. Look carefully and you can see the angler's bioluminescent lure, a small black dot on the end of a stalk just between its eyes. The stalk is now lying limp on the fish's body. The inset shows how the lure is normally deployed.

larger than the real eye to make the prey seem much bigger and thus more capable of retaliatory defense. You may have seen some of the many butterflies and moths that have false eyes for the same reason.

Camouflage and Mimicry

Camouflage (the art of making an object difficult or impossible to see against its background) and mimicry (the art of making an individual look like a completely different species or object) are practiced by many marine species. The principal use of camouflage is probably to enable an organism to escape detection by its predators, but it is also used by predators to enable them to get close to their prey without detection. In many cases, camouflage is extremely effective (**Fig. 16.15**). Countless times when my diving partner has pointed out an organism on a reef, I have needed many seconds of careful visual examination before I could suddenly see through the effective camouflage to discern the organism.

Marine animals take many different approaches to camouflage, but the predominant basic approach is to make texture, color, and pattern appear the same as those of the background. Some animals, such as octopi and cuttlefish, can change their colors almost instantly to blend into their background. These species can also change their body texture to blend into a reef. Octopi, cuttlefish, and many other species change color by controlling the chemistry of special cells in their outer body surface called "chromatophores." However, this is not the only way to match the color of a background. For example, some species of frogfishes (**Fig. 16.13b**) have a variety of colors and color patterns, and each individual finds itself a permanent or semipermanent home where it is surrounded by similarly colored sponges or sponge-encrusted rocks. The frogfish's body is lumpy and irregular, so its shape and texture do not instantly reveal its presence. In addition, the frogfish has small eyes surrounded by confusing decoration, again for camouflage. Scorpionfishes and many seahorses (**Fig. 16.15d,h,i**) use a very similar camouflage technique: a mixture of drab colors that blend into their background on a reef or rubble-covered bottom, and frilly appendages that disguise their body outline.

The differences in frogfish and scorpionfish camouflage reflect their different preferred habitats and habits. Frogfishes live in or on a reef and rarely move from one preferred location. Scorpionfishes live primarily on a coral rubble seafloor and move periodically from place to place to hunt. Scorpionfishes therefore need a more generalized camouflage than frogfishes do, to blend with the variety of backgrounds they encounter. Like frogfishes and scorpionfishes, the ghost pipefish (**Fig. 16.15e**) is a master of both color and body form camouflage. Some species live among

FIGURE 16.15 *(opposite)* Examples of camouflage. (a,b) Crinoid shrimp (*Periclimenes amboinensis,* Papua New Guinea and Indonesia) on their host crinoid. These two photographs show how the shrimp are camouflaged to hide in crinoids of different colors. (c) A peacock sole partially buried in the sand (*Pardachirus pavoninus,* Indonesia). (d) This raggy scorpionfish (*Scorpaenopsis venosa,* Indonesia) is well camouflaged against the sand and rubble seafloor, but it even enhances its natural camouflage by allowing algae to grow on its body. (e) The algae ghost pipefish (*Solenostomus paradoxus,* Papua New Guinea) is well camouflaged to look just like a frond of this green calcareous alga. The ghost pipefish even spends most of its time swimming slowly in a vertical position to look even more like the alga. The pipefish is on the left-hand side of the clump of algae in this photograph. (f) There are many species of decorator crabs, each of which covers itself in sponges and other epifauna and algae. This one is a spider crab of the family Majidae (Papua New Guinea). (g) A commensal spider crab (*Xenocarcinus* sp., Palau) on a gorgonian coral. The crab is about 1 cm in length. (h) A pygmy (or gorgonian) seahorse (*Hippocampus bargibanti,* Indonesia) blends in with its host sea fan with coloration and body "bumps" that resemble the polyps of the fan when they are closed. (i) The spotted seahorse (*Hippocampus kuda*) is found in many colors, often to blend in with its background. This individual in Papua New Guinea, with its color and coating of algae and organic detritus, blends perfectly with the dead leaves and detritus of its home near the mouth of a stream. (j) The coral shrimp (*Dasycaris zanzibarica,* Indonesia) lives on a whip coral and is camouflaged by its body protuberances and perfect color to match the whip. (k) Like the coral shrimp, this tiny spider crab (*Xenocarcinus tuberculatus,* Indonesia) is camouflaged on its whip coral. (l) A tiny (about 3 to 5 mm long) shrimp (possibly *Allopontonia* sp., Indonesia) is almost transparent, but it has flecks of color to match its sea star host. It can be seen in this photograph only because the lighting was just right to leave it a little shadowed.

(a)
(b)
(c)
(d)
(e)
(f)
(g)
(h)
(i)
(j)
(k)
(l)

algae or turtle grass (*Thalassia*) and even swim in a vertical orientation to parallel the algae or turtle grass blades.

Tiny shrimp, crabs, and other invertebrates that live on **crinoid**s and soft corals make very effective use of camouflage (**Fig. 16.15a,b,g,j,k,l**). Many of these tiny and beautiful invertebrates spend their entire life cycle on a single host crinoid or soft coral that they perfectly mimic. These hosts have a wide variety of colors and color patterns, and the guest species often matches the pattern. Many of the tiny invertebrate species that inhabit crinoids and soft corals have not been well studied, so we do not know how each of these species achieves its perfect camouflage. Some species of shrimp will change color within several days to match a new host perfectly when they are moved from one crinoid or soft coral to a differently colored individual or colony of the same species. The shrimp achieves this color change by using its chromatophores, or by eating small parts of its host and incorporating the host's colored pigments into its own body, or by a combination of these techniques.

Sometimes even the perfect color-matching capabilities of crinoid and soft coral inhabitants appear to provide insufficient camouflage. Many invertebrates, particularly several species of tiny crabs, also have modified body shapes to match the structure of their host. In some cases, the crab's or other lodger's body is covered with spines and protuberances that mimic the appearance of the host or its individual polyps. In other cases, the lodger plucks off some of its host's polyps and places them on its shell or appendages (**Fig. 16.15g**). The technique of placing other organisms on one's own body for camouflage is practiced by many invertebrate species. For example, the decorator crab (**Fig. 16.15f**) covers its entire body with algae, sponges, and other invertebrates, beneath which the crab is not easily seen, particularly if it stays motionless when a predator nears. The polyps, sponges, and other species used for camouflage can live and reproduce while attached to the camouflaged crab. As will be discussed later in the chapter, the association may even be beneficial to both species.

Pelagic fishes and marine mammals that swim in the water column away from the seafloor take advantage of somewhat different camouflage. Many of these species have **countershading** that minimizes the contrast of their bodies against the background as viewed by predators from above or below. Species that rely on countershade camouflage have a white or silver and highly reflective underside (**Fig. 16.16a,b**) that efficiently reflects the ambient light downward and reduces the contrast between the fish's (or mammal's) body and the water surface as seen by a predator from below. The upper side of the countershaded species is dull, nonreflective, and often mottled gray or blue-gray to reduce upward reflection of light and soften the outline by providing variations in the light reflected. A predator looking downward must distinguish the dull and confusing shape of the countershaded prey seen against the murky, confused background of the seafloor or deep water below.

Some species conceal their identity by mimicking another species. Again, this approach can be used in either hunting or defense. The false cleaner wrasse is a small carnivorous fish that mimics a species of cleaner wrasse (**Fig. 16.16c**). Cleaner wrasses establish stations where they wait to eat **parasites** off larger fishes, and they advertise their services by their bright colors and a jerky dance that is similar for all cleaner species. The cleaner benefits from the food supply of parasites, and the cleaned fish benefits by being rid of the parasites. The false cleaner mimics not only the cleaner's coloration and body shape, but also its advertising dance. The unsuspecting parasitized fish moves in and lowers its defensive guard to the false cleaner, which simply takes a bite out of the fish. Often the fish is startled but stays and allows the false cleaner a second bite before realizing its mistake and scurrying away in complete confusion.

The black-saddled mimic filefish (also known as the mimic leatherjacket) uses mimicry for defensive purposes. The filefish (**Fig. 16.16d**) is virtually indistinguishable in size, shape, and markings from a pufferfish, the black-saddled toby (**Fig. 16.16e**). Because the puffer is poisonous and predators are unable to distinguish the two species, they ignore both the puffer and its mimic. The juvenile mimic surgeonfish (**Fig 16.16f**) mimics both the looks and the behavior of the pearlscale angelfish (**Fig 16.16g**), and it is about the same size as the angelfish, but the larger adult mimic surgeonfish (**Fig 16.16f**) looks completely different from the juvenile. Angelfishes have sharp spines on their cheeks to deter predators, so the predators do not attack the juvenile mimic surgeonfish even though surgeonfishes do not have such a spine.

Concealment

Although many marine predators use senses other than vision to locate their prey, most marine predators in the photic zone hunt primarily by sight. Consequently, concealment is an excellent technique that is widely used for defensive purposes and also by certain predators to ambush their prey. Camouflage and concealment are often used together to prevent detection. For example, flounders and soles (**Fig. 16.15c**) are not only camouflaged, but also may partially bury themselves in sand to conceal their presence further.

Concealment is a very effective strategy for many invertebrates and fishes that bury themselves in sand or mud; hide in cracks, crevices, or caves of a reef or rubble-covered seafloor; or build and live in holes in a reef or rocks. Most species that use concealment for defense against visual hunters remain concealed by day and come out to feed at night. Other species emerge during the day but seldom stray far from their hiding places. These species use concealment places in two ways: to hide in to escape detection when predators approach, and as a refuge into which the usually much larger would-be predator cannot follow. For example, fire gobies (**Fig. 16.13d**) live in holes excavated in coarse sand or rubble areas

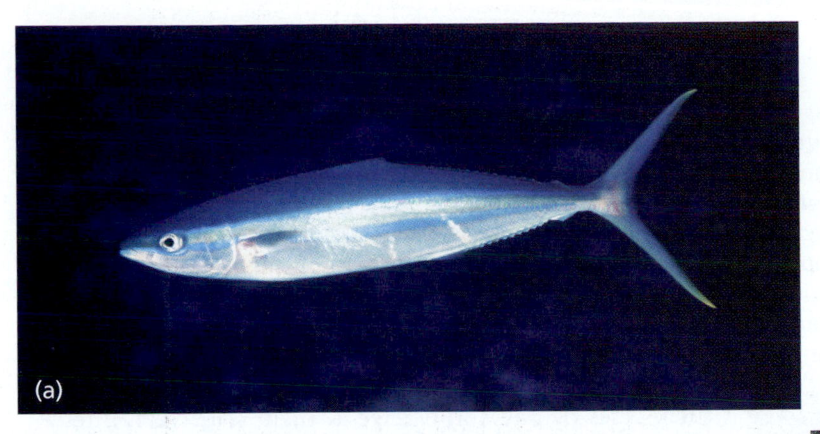

FIGURE 16.16 Examples of countershading and mimicry. (a) This trevally, called a "rainbow runner" (*Elagatis bipinnulata,* Solomon Islands) lives mostly in the open water column. Note the light-colored underside and nonreflective upper surface that provide countershading. (b) Another trevally species, the bluefin trevally (*Caranx melampygus,* Solomon Islands), is also countershaded with a light underside and darker upper surface, but this species spends much of its life close to the reef and seafloor. Perhaps the spotting on the trevally's upper side helps it to visually blend in better with the seafloor when seen from above. (c) A cleaner wrasse, similar to *Labroides phthirophagus,* Hawaii shown in the inset, is removing parasites from a Celebes sweetlips (*Plectorhinchus celebicus,* Papua New Guinea). (d) Blacksaddle mimic filefish or leatherjacket (*Paraluteres prionurus,* Philippines). (e) Black-saddled toby (*Canthigaster valentini,* Papua New Guinea). Can you tell the difference between this puffer (toby) and the leatherjacket in part (d)? The puffer lacks a first dorsal fin, and it has shorter dorsal and anal fins than the leatherjacket has. (f) A juvenile mimic surgeonfish (*Acanthurus pyroferus,* Indonesia). The inset shows an adult (Papua New Guinea). (g) A pearlscale angelfish (*Centropyge vrolikii,* Papua New Guinea).

around coral reefs. Some will dive into these holes, of which they may have several, when a predator is in sight. Others will stay out until the predator reaches a distance of a meter or two and then dive into their holes at lightning speed. Photographing these gobies, and many other reef fishes that have similar habits, requires the patience to lie quietly in wait until the fish re-emerges.

Nocturnal species that conceal themselves by day and hunt at night are often a deep red color and have big eyes (**Fig. 16.17a,b,e**) that enable them to hunt when light levels are very low. The red color aids concealment in caves or at night because red light is strongly absorbed by seawater. Consequently, these organisms appear dark gray or black unless seen near the surface in daylight or when illuminated with artificial light. In addition, many marine species have eyes that are not sensitive to red **wavelengths.**

A wide variety of fishes, shrimp, crabs, sea urchins, octopi, and many other invertebrate species conceal themselves by day and emerge at night (**Fig. 16.17**). In contrast, many species that hunt or feed during the day conceal themselves at night (**Fig. 16.18a**). Diving at night on a reef or rocky area reveals a community very different from the one seen during the day. Many invertebrates,

FIGURE 16.17 Some nocturnal species. (a) A nocturnal shrimp (Fiji). Note the large eyes. (b) This fish, called a crescent-tailed bigeye (*Priacanthus hamrur,* Fiji), spends the daytime mostly in dark caves, emerging at night to feed. Note the large eyes and predominantly dark red coloration. (c) Most crabs are nocturnal, including this coral crab (probably *Cancer* sp., Fiji). (d) Fire urchin (*Asthenosoma intermedium,* Papua New Guinea). This species is normally seen only at night. (e) Many octopi, such as this one (*Octopus luteus,* Indonesia), hunt mostly at night. Note again the red color.

FIGURE 16.18 Examples of nocturnal defense and predators, and of spines and armor in fishes. (a) This black-headed parrotfish (*Scarus gibbus,* Fiji) is sleeping at night. Like most parrotfish species, it excretes a transparent mucous cocoon (visible in this photograph primarily because of the particles deposited on its surface) in order that potential predators cannot chemically sense, or "smell," it. (b) A mud snail (*Nassarius papillosus,* Hawaii) that has emerged from the sand to hunt at night. (c) This cone shell (*Conus geographicus,* Red Sea) is hunting at night across the reef. (d) This orbicular burrfish (*Cyclichthys orbicularis,* Philippines) has already partially inflated to respond to a threat from a predator. (e) When fully inflated, this same burrfish is too large for most predators to grasp. It would even be difficult to take a bite out of, because of its round shape and sharp spines. (f) Boxfishes, or trunkfishes, such as this male spotted boxfish (*Ostracion meleagris,* Papua New Guinea), have a hard, box-shaped exterior to make it difficult for predators to grasp or bite.

including certain sea pens (**Fig. 16.5a,b**), shrimp, crabs, and mollusks (**Fig. 16.18b**), emerge only at night and bury themselves in the soft seafloor by day. Nocturnal predators use the cover of night as a form of concealment to ambush their prey.

Concealment would seem to be impossible for species that live in the water column. Nevertheless, squid and octopi are able to conceal themselves by speeding off behind an opaque cloud of ink that they release when threatened.

Spines and Armor

Numerous invertebrates have evolved thickened shells or long spines to prevent predators from reaching their vital soft parts. Many mollusks have external shells that protect them after they retreat inside. Bivalve mollusks, including clams and scallops, have two hinged valves that they can close tightly together. Gastropod mollusks have a single shell (**Fig. 16.18c**) into which the animal's soft parts can be withdrawn. The entrance to the shell is closed off with a tough door called an "operculum." Some fishes, such as puffers and burrfishes (**Fig. 16.18d,e**), have developed special jaws to crush mollusk shells. Crabs have a large armored crushing claw for the same purpose. In turn, many mollusks have developed extremely thick and strong shells or armored spines. Spines increase the size of the shell and make it more difficult for a crab's claw or fish's jaw to grasp and crush it. Many mollusks conceal themselves during the day and hunt only at night to avoid predators, but some predators, particularly crabs, are also nocturnal. There is a perpetual contest between predator and prey as each evolves better adaptations for survival.

Many sea urchins (**Fig. 16.17d**) have evolved long sharp spines that cover their entire upper body. The spines are often tipped with reverse barbs or venom injectors. A direct approach by a predator to one of these sea urchins is likely to result in severe stab wounds or even paralysis and death. However, some triggerfishes will methodically pick off one spine at a time until they can attack the underlying external shell of its then helpless inhabitant. Other triggerfishes use a jet of water from the mouth to blow the sea urchin over, exposing the vulnerable underside that has no spines—an elegant means of circumventing the urchin's defenses. In addition to their formidable spines and, in some species, highly toxic venom, many sea urchins have become nocturnal to avoid daytime predators.

Crustaceans, including shrimp, crabs, and lobsters, have developed hard external shells that they must periodically shed and regrow as the soft animal within becomes larger. Worms, such as the fan worm and Christmas tree worm (**Fig. 16.9c,d**), use rocks, coral, or sand to build armored burrows into which they can withdraw. Sea cucumbers do not have hard shell armor, but they have developed thick leathery outer skins containing calcium carbonate that, although pliable, are almost impossible to bite through.

Certain fish species have also developed armor and spines. Boxfishes (**Fig. 16.18f**) have a hard, boxlike body that allows only slow and awkward swimming but makes them large in cross section to prevent a predator's jaw from grasping and crushing them. The armor also prevents the predator from taking small bites off a boxfish's body. When attacked, burrfishes (**Fig. 16.18d,e**) gulp large amounts of water and swell up, extending spines that normally lie flat on their body. Burrfishes reportedly may even inflate when already caught in a predator's jaws and simply wait for the predator to weaken or die from hunger before deflating and releasing themselves.

Poisons

A variety of invertebrate species, many marine algae, and certain fishes produce or concentrate from their food a wide spectrum of toxic substances. These substances are sometimes used to kill parasites and larval stages of other organisms that may settle on the toxin-producing species, but their main use is to discourage predators. Most of the myriad toxic substances synthesized by marine species have not been identified, and many thousands of these compounds are likely to be found. Because most toxic substances discourage predators, including bacteria and **viruses,** by interfering with their biochemistry, these substances hold the potential to be used as, or to lead to the development of, drugs that can treat human illness. Several drugs derived from marine toxins are already in routine use or are undergoing clinical trials.

Many species that produce toxins to make themselves inedible, notably nudibranchs (**Fig. 16.10c,d**), are brightly colored. The conspicuous color schemes of nudibranchs and similarly toxic tunicates, sponges, and other invertebrates warn predators of the poison defense. In response, many predators have evolved mechanisms to detoxify the poison produced by their prey species.

Numerous species inject toxins into the predator to repel its attack. Sea urchins inject such venom with the tips of their spines. Stonefishes and scorpionfishes (**Fig. 16.15d**) inject deadly venom with sharp spines along their dorsal (upper) fin. Anemones, corals, hydroids, and other invertebrates all inject venom by firing little poisonous darts called "nematocysts" from their tentacles.

Toxins are used for hunting as well as defense. For example, anemones (**Fig. 16.19a**) use toxin-laden nematocysts to attack and stun or immobilize invertebrates and small fishes that blunder into their trap. Once stunned, the prey is trapped by mucus-laden anemone tentacles that fold over and pass the prey toward the mouth or open stomach at the anemone's center. Anemonefishes (often called "clownfishes" or "clown anemonefishes") live in or near the anemone (**Fig. 16.19b–f**), retreating into its protective toxic folds and tentacles when threatened. The anemonefishes are stung by the anemone, but they build immunity by deliberately exposing themselves to the toxin.

FIGURE 16.19 Anemonefishes, also called "clownfishes" or "clown anemonefishes," live in family groups in association with a number of different species of large carnivorous anemones, such as the Haddon's sea anemone (*Stichodactyla haddoni,* Indonesia) shown in (a). There are a number of anemonefish species, including the examples shown here. (b) Tomato anemonefish (*Amphiprion frenatus,* Papua New Guinea). (c) Clown anemonefish (*Amphiprion percula,* Papua New Guinea). (d) Pink anemonefish (*Amphiprion perideraion,* Papua New Guinea). (e) Spinecheek anemonefish (*Premnas biaculeatus,* Papua New Guinea). (f) Clark's anemonefish (*Amphiprion clarkii,* Papua New Guinea). All anemonefish species are damselfishes of the subfamily Amphiprioninae, except the spinecheek, which belongs to another subfamily.

Other notable species that use toxins to immobilize their prey include cone shells (**Fig 16.18c**), which inject their venom through a long, thin tube extended from their body; and blue-ringed octopi (**Chapter 14 opener**), which transmit their toxin in mucus as they bite their prey. A number of human deaths have been caused by cone shell toxin injected when people inadvertently stepped on a cone shell living just buried in the sand in shallow water near the beach. One species of blue-ringed octopus, which lives in tide pools and on shallow reefs around Australia, is not aggressive but will bite if threatened or stepped on. The venom injected by this species is especially toxic, and there is no known antidote. A bite often results in death.

Numerous species excrete poisons into the surrounding water to discourage their competitors for food resources and living space or to discourage potential predators. Perhaps the most well-known is a **dinoflagellate** species, *Pfiesteria piscicida,* which inhabits estuaries along the east **coast** of the United States and which has been reported to produce a toxin that kills fishes. The toxin thought to be produced by *Pfiesteria* may also cause a persistent but nonfatal illness in humans who swim in, or breath aerosols from, *Pfiesteria*-infested waters. The details of *Pfiesteria*'s life cycle and of its role in fish kills and human health problems are still somewhat controversial, and much research remains to be done.

Group Cooperation

Both hunting and defense can be aided by cooperative approaches involving several or many individuals of the same species. The Portuguese man-of-war is a good example (see chap. 14). Group cooperation is a common adaptation, but two special categories are of interest: colonial forms and **schooling** in pelagic animals.

Colonial species abound in the oceans and include many sponges and **cnidarians.** In some colonial forms, each individual feeds and reproduces separately, but the individuals cooperate to enhance each other's and the species' success. For example, individual polyps of sea fans and corals, by growing attached to each other, extend the colonies' reach into the water column. This extension gives each polyp a better chance of encountering food and reduces competition by other species that live on sediment or rock surfaces.

Pelagic species, including many fishes, squid, and marine mammals, congregate in schools. Similarly, phytoplankton and zooplankton often have patchy (clustered) distributions. Schooling and clustering behavior affords advantages. For example, a group of fish can often overcome the defenses of a single individual of an otherwise superior species. Damselfishes lay their eggs in clusters on a reef surface and then defend them aggressively. Wrasses and other fishes arrive in groups to feed briefly but voraciously on the eggs as the damselfishes frantically try to chase them off but fail because there are too many attackers.

Fish schools have no leader, but each fish precisely matches the seemingly random twists and turns of the school, maintaining a precise distance from each of its neighbors. Movements of the school may confuse predators by making the school appear to be a single large organism. In addition, predators may have difficulty singling out a victim from the moving school. Schooling may have benefits in reproduction as well. Finding a mate is easier in a school, and a high rate of fertilization is ensured if eggs and sperm are released to the water column simultaneously by many members of the school. Mass **spawning** can in fact be considered a form of "schooling" of the fertilized eggs and larvae.

When prey is concentrated in schools, the predator spends much of its time searching for the school. Once it encounters a school, it cannot consume the entire school, so most individuals will survive, even if the predator gorges itself. Predators are able to consume less food through periodic gorging than they could by continuous steady feeding. Schooling of the prey and mass spawning thus reduce the efficiency of predator feeding, and consequently more of the prey species survive. This outcome may be the principal advantage of schooling.

16.5 SELECTED ADAPTATIONS IN FISHES

Each marine species is adapted in different ways that enable it to survive. Fishes are ideal subjects to illustrate adaptations because they can be viewed at the fish counter of the local supermarket or in aquariums. Fishes are adapted in many different ways to respond to such challenges as swimming in a manner that best supports their hunting and defensive strategies, the high **osmotic pressure** of seawater, the variability of osmotic pressure in estuarine waters, and control of their buoyancy.

Swimming Adaptations

Water is much denser than air, so it is much more difficult to travel through water than through air. The next time you go swimming, try running through knee-deep water and you will understand just how much more difficult it is to move through water. Fishes must overcome water resistance to swim. Therefore, the body shape of a fish must be optimized to facilitate its specific swimming habits.

While swimming, fishes must overcome three types of resistance or drag: surface drag, form drag, and turbulent drag. Surface drag is the **friction** between a fish's body surface and the water. Form drag comes about because water must be pushed out of the way, and turbulent drag is related to the smoothness of water flow past the swimming object.

Surface drag increases as the surface area in contact with the water increases. Consequently, surface drag is

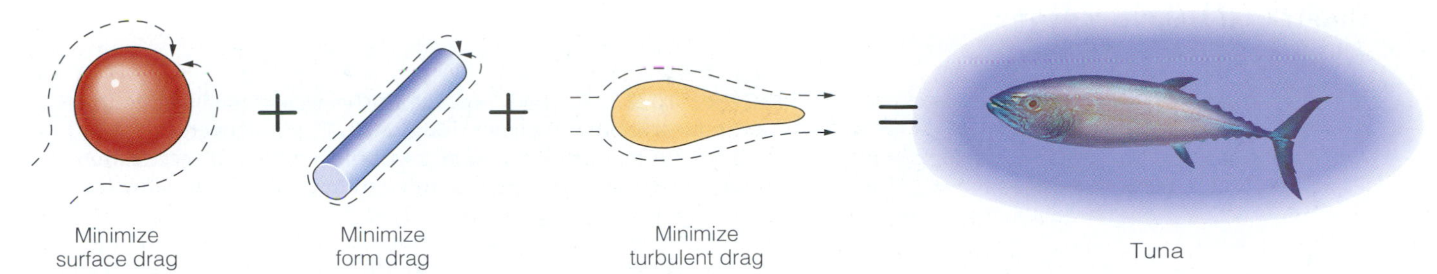

FIGURE 16.20 Three types of drag slow an object as it moves through water: surface drag, form drag, and turbulence drag. For each type, the magnitude of the drag varies with the shape of the moving object. This figure shows the optimum shape required to minimize each type of drag. Body shapes of most fish species are a compromise between the need to minimize drag and other needs of the animal, such as acceleration and turning speed. Fishes that swim constantly at high speed, like the tuna in this figure (*Thunnus* sp., Fiji), have a body shape that provides the best compromise for reducing the sum of the three types of drag: a rounded body that is narrow at the tail and thicker near the front. Nuclear submarines also are built in this general shape.

minimized if the swimming object is spherical because a sphere has the smallest surface area per unit volume of any solid object. Form drag increases in proportion to cross-sectional area (**Fig. 16.20**). The perfect shape to minimize form drag is needlelike, but this shape has a high ratio of surface area to volume and is subject to increased surface drag. A few fish species are needle-shaped (**Fig. 16.21a**), and others have an extremely thin platelike body (**Fig. 16.21b,c**) to minimize form drag. Most fish shapes are a compromise: rounded or oval in cross section, but elongated in the swimming direction. This shape tends to minimize total (form plus surface) drag (**Fig. 16.20**).

The third form of drag, turbulent drag, dictates the final refinement in fish body shape. The turbulent flow of water around the fish is reduced if the front is rounded and blunt and the rear tapers to form a teardrop shape (**Fig. 16.20**). This shape is similar to the cross section of an airplane wing, or the form of a blimp or submarine, all of which are designed to minimize drag.

The fastest and most continuously swimming fishes, such as tuna (**Fig. 16.20**), have generally a teardrop shape to minimize total drag and thus minimize energy used in swimming. Eyes that are flush and smoothly contoured against the body and a slimy coating are other adaptations that reduce surface drag. Fishes that do not swim continuously and fishes that are adept at fast turns rather than high speed have less need to reduce drag than fast swimmers have. Thus, these fishes often have body shapes that are greatly modified from the teardrop form

FIGURE 16.21 Some fish shapes that minimize form drag. (a) The cornetfish (*Fistularia commersonii*, Red Sea) has a very long, needle-shaped body to minimize form drag. (b) Side view of Klein's butterflyfish (*Chaetodon kleinii*, Papua New Guinea). (c) Front view of the same species of butterflyfish. Note the lateral compression of the body (narrower across than vertically), which reduces form drag and enables the fish to execute turns quickly.

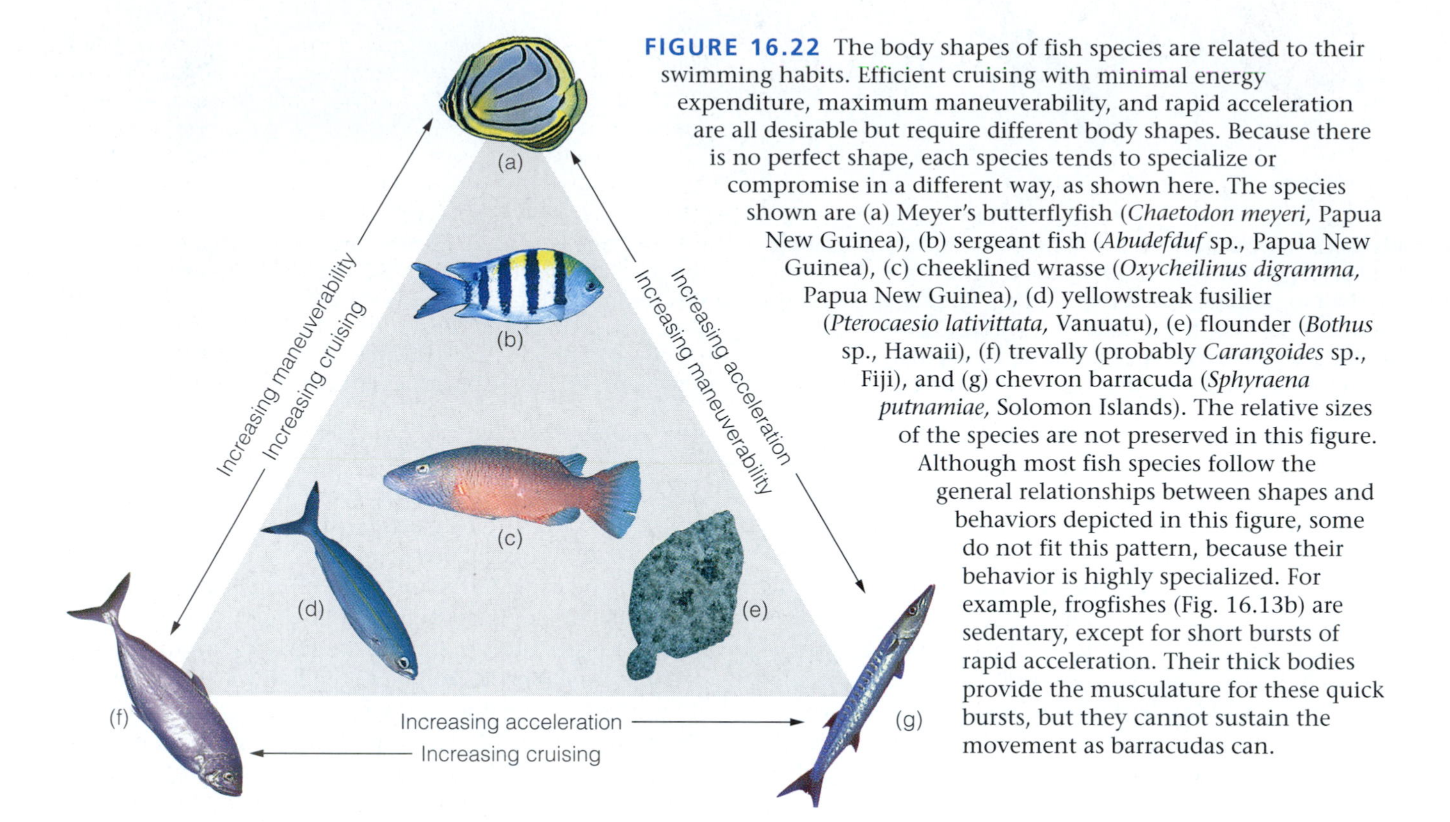

FIGURE 16.22 The body shapes of fish species are related to their swimming habits. Efficient cruising with minimal energy expenditure, maximum maneuverability, and rapid acceleration are all desirable but require different body shapes. Because there is no perfect shape, each species tends to specialize or compromise in a different way, as shown here. The species shown are (a) Meyer's butterflyfish (*Chaetodon meyeri,* Papua New Guinea), (b) sergeant fish (*Abudefduf* sp., Papua New Guinea), (c) cheeklined wrasse (*Oxycheilinus digramma,* Papua New Guinea), (d) yellowstreak fusilier (*Pterocaesio lativittata,* Vanuatu), (e) flounder (*Bothus* sp., Hawaii), (f) trevally (probably *Carangoides* sp., Fiji), and (g) chevron barracuda (*Sphyraena putnamiae,* Solomon Islands). The relative sizes of the species are not preserved in this figure. Although most fish species follow the general relationships between shapes and behaviors depicted in this figure, some do not fit this pattern, because their behavior is highly specialized. For example, frogfishes (Fig. 16.13b) are sedentary, except for short bursts of rapid acceleration. Their thick bodies provide the musculature for these quick bursts, but they cannot sustain the movement as barracudas can.

to conform with their own special habits. Fishes that specialize in short bursts of speed with quick accelerations generally are somewhat thickened in the middle (**Fig. 16.22**) by the heavy musculature necessary for such maneuvers (similar to the difference between long-distance runners and weight lifters). Fishes that specialize in quick turns generally are somewhat flattened (**Fig. 16.21b,c**) so that the flat sides of their body can be used in turning in much the same way that a boat's rudder is used. Fishes that hunt mostly by stealth or feed on plankton and do not need to swim quickly often have bizarre body shapes (**Fig. 16.15e**). Fin shapes are also modified to accommodate different swimming habits.

Adaptations of Fins

Fish fins are extremely important in swimming and executing turns, just as the tail fin, rudder, short rear wings, and wing flaps of an airplane are important in controlling course and stability. Fishes generally have a pair of pectoral fins (one on each side of the body behind the head), two dorsal fins along the center of the back, and an anal fin beneath the rear half (**Fig. 16.23a**). They also have a caudal fin at the posterior, and a pair of pelvic fins on either side of the lower body forward of the anal fin. These fins vary greatly among species in size, shape, and location, and some fins may be absent in certain species.

The pairs of pelvic and pectoral fins are used primarily to execute maneuvers, including turns and stops, and usually can be folded flat against the body when not in use. The vertical dorsal and anal fins serve primarily as stabilizers during swimming and, in some species, can be folded against the body when not needed. The caudal fin is the primary provider of propulsion in most species. Most fishes also alternately contract and relax muscles along their body to create a wavelike motion that travels along the body and produces a forward thrust (**Fig. 16.23b**). The caudal fin that provides the final thrust is flared out vertically to provide a large surface area and, consequently, strong thrust as it is moved from side to side. You can experience such a thrust increase if you use swim fins.

Increasing the caudal fin surface area increases thrust but also increases surface drag. Therefore, the caudal fin is modified to reflect the swimming habits of the species. These modifications are shown by the **aspect ratio** of the fin, which is defined as

$$\frac{(\text{fin height})^2}{(\text{fin area})}$$

Five major types of caudal fins are distinguished by different ranges of aspect ratio and fin shape (**Fig. 16.24**). Rounded fins (aspect ratio 1; **Fig. 16.24a**) are useful for maneuvering and quick acceleration by species such as butterflyfishes (**Figs. 16.13e; 16.21b,c; 16.22a**). Truncate fins (aspect ratio 3; **Fig. 16.24b**) and forked fins (aspect ratio 5; **Fig. 16.24c**) reduce drag in comparison with rounded fins, but they still provide substantial maneuvering assistance. Truncate and forked fins are used by many species that swim reasonably fast but also maneuver relatively quickly (**Fig. 16.22**). Lunate caudal fins (aspect ratio up to 10; **Fig. 16.24d**) are typical on fast and continuously swimming species, such as trevally (**Fig. 16.22f**),

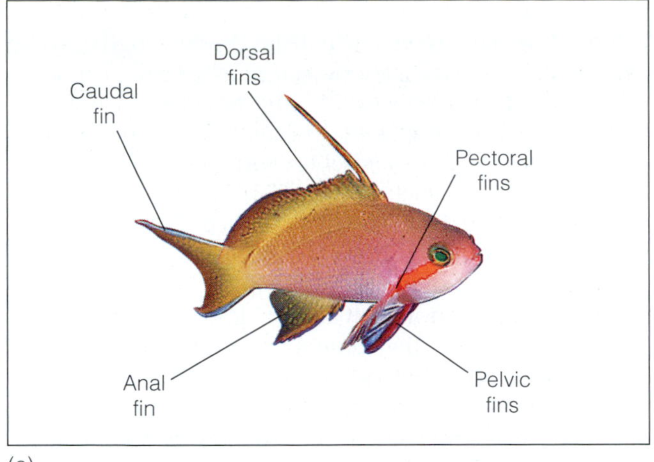

(a)

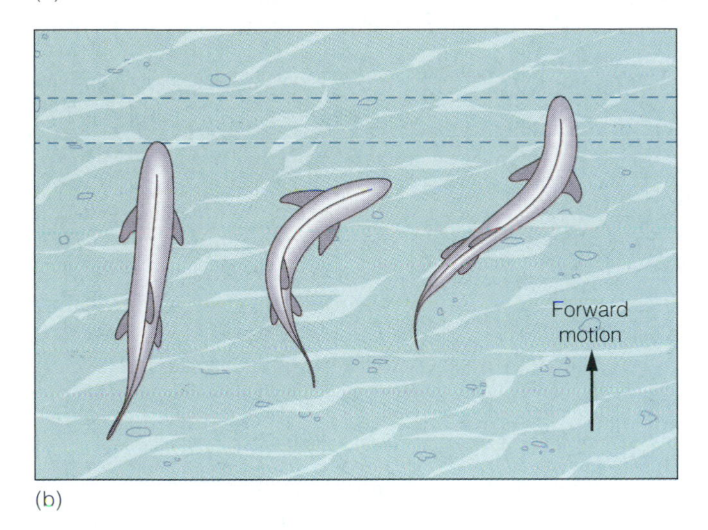

(b)

(c)

FIGURE 16.23 Fish fins and swimming motion. (a) Fishes have a dorsal fin or fins along the back, pectoral fins on either side of the body, a pair of pelvic fins under the forward half of the body, an anal fin under the rear half of the body, and a caudal fin at the tail. Each of these types of fins may be much modified for special purposes in different fish species. The fish shown here is a threadfin, or red-cheeked fairy basslet (*Pseudanthias huchtii,* Papua New Guinea). (b) Most fish species swim by using a series of undulating body motions. (c) Triggerfishes and their relatives, such as this barred filefish (also called an orange-fin filefish; *Cantherhines dumerilii,* Hawaii) swim not by undulating their bodies, but by back-and-forth motions of their specially adapted soft dorsal and anal fins. This movement gives them great maneuverability, but it is less efficient in providing acceleration than is the undulating body motion method of swimming.

tuna (**Figs. 14.21a, 16.20**), marlin, and swordfish. Lunate caudal fins in these species are rigid and have little use in maneuvering, although they are very efficient in forward propulsion. Fast-swimming predatory species with lunate caudal fins can outrun other species with truncate, forked, or rounded caudal fins, but the prey species have an excellent chance of avoiding the predator by using their greater maneuverability.

Heterocercal caudal fins are asymmetrical, the upper lobe being longer and taller than the lower lobe (**Fig. 16.24e**). This type of caudal fin is used primarily by sharks (**Fig. 14.22**). It is similar to the lunate caudal fin in that it provides very efficient forward thrust but little help in maneuvering. However, its asymmetrical shape also provides upward lift, which is important to sharks because they have no **swim bladder** and tend to sink if

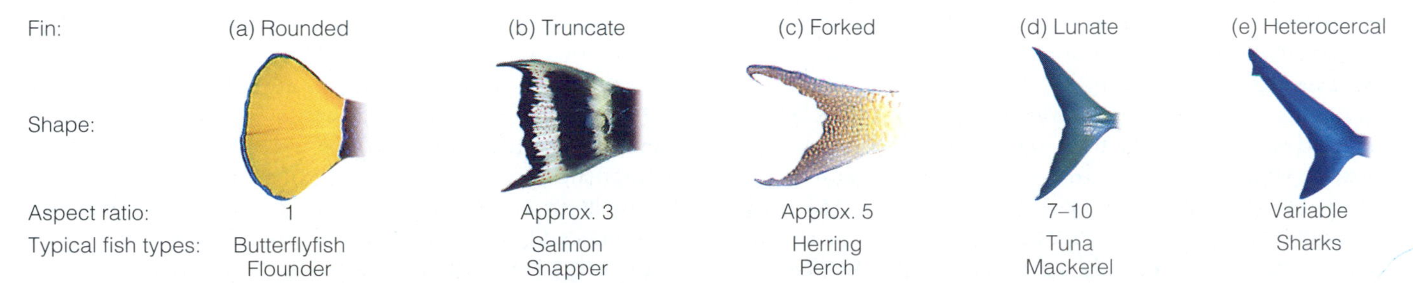

FIGURE 16.24 The shape of the caudal fin in fish species is related to their swimming behavior. There are five basic types of caudal fins, characterized by different aspect ratios, although caudal fins of many species have characteristics intermediate between these categories. The caudal fins shown here belong to (a) a blue-girdled angelfish (*Pomacanthus navarchus*), (b) a redbreasted Maori wrasse (*Cheilinus fasciatus*), (c) a male black-spot angelfish (*Genicanthus melanospilos*), (d) a trevally (*Carangoides* sp.)—all photographed in Papua New Guinea—and (e) a gray reef shark (*Carcharhinus amblyrhynchos*), photographed in Palau. Their relative sizes are not preserved in this figure.

they stop swimming. Lift is also provided by the sharks' pectoral fins.

Unlike those of most other fishes, sharks' pectoral fins are large, flat, and relatively inflexible (**Fig. 14.22**), providing lift like aircraft wings. The pectorals are relatively far forward on the shark's body (**Fig. 14.22**) and lift the front of the shark, while the caudal fin lifts the tail. Although sharks are powerful swimmers, their fins are designed poorly for maneuvering, and they are not adept at capturing prey that can anticipate their charge and perform evasive maneuvers. Unfortunately, this knowledge is of little use to swimmers who may be attacked by sharks, because human body shapes do not allow for quick maneuvers in the water.

Although most fish species use their fins for swimming, numerous species have fins adapted to perform highly specialized functions. A number of species have dorsal fins modified to act as defensive, and perhaps offensive, weapons. In these species, the individual rays or spines of the dorsal fins, or parts of the dorsal fins, are needle-sharp and may contain a venom that is injected into any predator that challenges the fish. Species with this type of dorsal fin include stonefishes and scorpionfishes (**Fig. 16.15d**), which lie on a reef or the seafloor and use their dorsal defenses to protect against attacks from above. They also include lionfishes (**Fig. 16.25a**), which swim close to a reef, turning their backs toward any approaching predator and their more exposed undersides toward the reef. In some species, such as frogfishes (**Fig. 16.13b**) and anglerfishes (**Fig. 16.14**), the forwardmost ray or rays of the dorsal fin are adapted to become lures that can be dangled in front of the fish's mouth to entice prey.

In triggerfishes (**Fig. 16.25b**), the forward dorsal fin consists primarily of a single strong, rigid spine that normally lies flush against the fish's body. This spine can be extended from the body to become a fearsome weapon that gives the triggerfish its name. Certain triggerfish species may also use this spine defensively. When chased, they swim headfirst into holes in the reef and extend their trigger to lock themselves in. Because of the way the trigger is hinged, almost no amount of tugging by a predator can pull the triggerfish out. Once the predator leaves, the trigger can be relaxed, allowing the triggerfish to back out of its refuge. Triggerfishes also use modified fins for swimming. The anal and dorsal fins are enlarged and undulate back and forth in wavelike motions that replace the body undulations used by other fishes. These fins enable triggerfishes to hover, turn, and swim slowly forward or backward to enter holes in the reef just wide enough for them to fit through.

Many species that live or rest frequently on the seafloor have pectoral and sometimes caudal fins that are elongated and have strong rigid spines on which the fish can rest. Sandperches (**Fig. 16.25c**) have this type of adaptation. In certain deep-sea species that inhabit areas where currents are generally weak, both the pectoral fins and the lower lobe of the caudal fins are elongated to an extent that can exceed the fish's body length. These fishes "walk" on the seafloor as if perched on a tripod.

Like triggerfishes, wrasses (**Fig. 16.25d**) do not normally swim by using body undulations for propulsion. Instead, they propel themselves with their pectoral fins, which they stroke back and forth in much the same way that an oar is used. The fin is moved backward while spread vertically to push the water back and the fish forward. Then it is rotated and moved forward while in a horizontal orientation that minimizes drag. Wrasses can also swim by using their caudal fin and body undulation in the same way that other fishes do, but the oarlike propulsion created by using the pectoral fins provides better control of movements at slow speeds. Such control is ideally suited to the wrasses' feeding habit of picking small crustaceans, algae, and individual coral polyps from cracks and crevices in reefs.

Many fishes spend most of their lives concealed in holes in a reef and do very little swimming. These fishes generally have greatly reduced fins, and their dorsal, caudal, and anal fins are often fused into one continuous fin extending around the fish. In extreme cases, the dorsal and anal fins may be missing entirely. Such fishes usually have an elongated body and swim by using sinuous body undulations. Swimming without the use of dorsal and anal fins is slow and inefficient, but it is perfectly suited to the lifestyles of these fishes. Snakelike flexibility and the lack of protruding fins enable them to swim easily through the narrow, tortuous passages of holes in the reefs where they live. Species adapted in this way include gobies (**Fig. 16.25e**) and moray eels (**Fig. 16.25f**).

Flyingfishes have some of the most bizarre fin adaptations. These warm-water fishes have elongated pectoral (and, in some species, also pelvic) fins that can be spread out from the sides to resemble bird wings. When they sense danger, flyingfishes swim upward with a rapid burst of speed that carries them through the water surface and into the air. Once in the air, they spread their fins and, using these "wings," sail a few tens of centimeters above the waves. They can glide for as long as 30 s, and some species can prolong their glide by flailing an elongated lower lobe of the caudal fin at the sea surface as they descend close to the water. Although flying is a very effective strategy to escape from some predators, flyingfishes expose themselves to predation by seabirds.

A number of species, such as clingfishes and remoras (**Fig. 16.25g**), use fins as suction devices with which they can hold onto the seafloor or another organism. Clingfishes use modified pelvic fins to cling tenaciously to rocks in coastal areas where wave action is intense. Remoras use a modified dorsal fin as a suction cup to attach the top of their head to sharks, manta rays, other large ocean animals, and even boats and **scuba** divers. The remora's sucker looks so little like a fin that careful research was needed to identify its origin. Although remoras must swim rapidly to catch and attach to a host, they are then transported without having to expend energy. Remoras can detach themselves to feed on any

FIGURE 16.25 Examples of fin adaptations. (a) The zebra lionfish (*Dendrochirus zebra,* Papua New Guinea) has long poisonous dorsal-fin spines to protect itself. (b) The clown triggerfish (*Balistoides conspicillum,* Philippines) has a strong, sharp spine that it can raise when threatened. Like many other reef fishes, clown triggers are territorial and will chase off any intruder of their own species. (c) The speckled sandperch (*Parapercis hexophtalma,* Fiji) uses its pectoral fins for support as it sits and waits for its prey. (d) Wrasses, like this moon wrasse (*Thalassoma lunare,* Papua New Guinea), use their pectoral fins like oars to propel themselves. (e) Many fishes, especially those that live in holes in the reef or seafloor, such as this Old Glory goby (*Amblygobius rainfordi,* Papua New Guinea) have long, fused dorsal and anal fins. (f) This darkspotted moray eel (*Gymnothorax fimbriatus,* Papua New Guinea), photographed while hunting at night, showed aggression when disturbed by the diver's lights. Note the sinuous snakelike appearance. (g) This remora, or sharksucker (*Echeneis naucrates,* Papua New Guinea), was attached to a scuba diver's leg, where it stayed for some time before leaving to try to find another ride.

nearby available food, particularly scraps of their host's meal if they are riding on a shark or other predator.

Ghost pipefishes (**Fig. 16.15e**) and seahorses (**Fig. 16.15h,i**) have perhaps the most extreme adaptation of body form and fins. Some species swim in a vertical head-down or head-up position and propel themselves slowly by rapid back-and-forth oscillations of their small dorsal fins. Because the fish is oriented vertically, the thrust from these fins is oriented perpendicular to the fish's body. Ghost pipefishes thus swim sideways, although they swim in a normal horizontal position when they need to swim rapidly to avoid danger.

Osmoregulation

The relative proportions of dissolved chemicals in body fluids of fishes, other **vertebrates**, and invertebrates are remarkably similar to their relative proportions in seawater. In most invertebrates, the salinity of internal fluids is also the same as the external seawater salinity. However, the internal fluids of fishes are less saline than seawater. The reason is that bony fishes are thought to have evolved in freshwater. Because **osmosis** causes water to diffuse across cell membranes from lower salinity to higher salinity, fishes must be able to counteract osmosis.

Osmosis of water molecules across a semipermeable membrane (such as the cell surface) from lower to higher salinity is easy to understand. Because the "concentration" of water molecules is higher in the lower-salinity fluid, more water molecules are in contact with that side of the membrane. Thus, more water molecules diffuse through the membrane toward the higher-salinity fluid than diffuse in the opposite direction. In contrast, dissolved salt molecules would be more likely to diffuse from high to low salinity because of their higher concentrations, but **ions** of dissolved salts are generally much larger than water molecules and hence less likely to pass through the openings in a semipermeable membrane.

One way to counteract osmosis is to increase the pressure on the high-salinity side of the membrane, which forces more water molecules through the membrane to the lower-salinity fluid. The pressure needed to balance water migration across a membrane is called "osmotic pressure." Osmotic pressure increases as the difference in salinity between the fluids on either side of the membrane increases.

Marine fishes must have a mechanism for counteracting osmosis or they would continuously lose water to their surroundings and dehydrate. Fishes cannot maintain a pressure difference across their external membrane, because they would be able to do so only if they had an impermeable body surface, like the pressure hull of a submarine. With an impermeable body surface, feeding and excretion of waste products would be extremely difficult because they would have to take place through the equivalent of a submarine airlock. In addition, all aquatic organisms must exchange oxygen, carbon dioxide, and nutrients with seawater through a porous membrane. Consequently, fishes have adapted methods called **osmoregulation** to counteract osmosis in seawater.

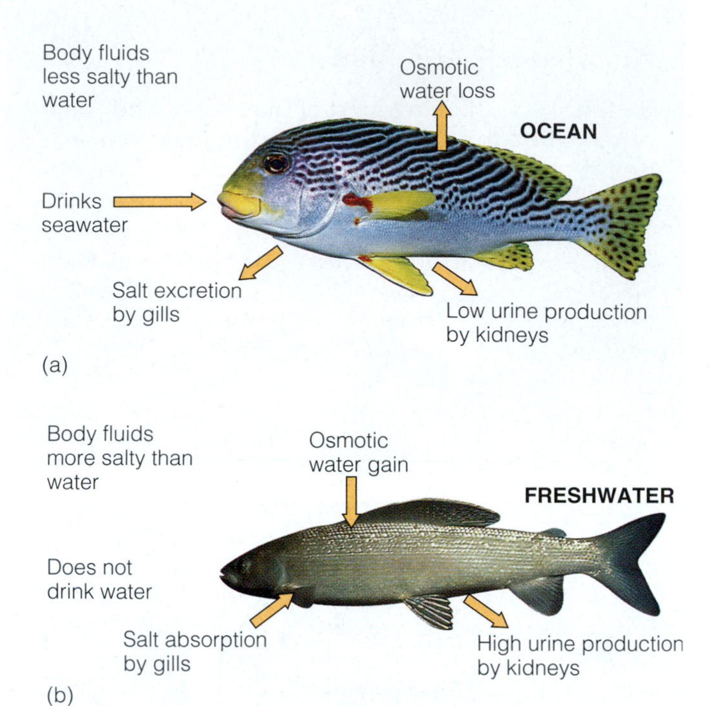

FIGURE 16.26 Osmoregulation. (a) Fishes that live in seawater, such as the Goldman's sweetlips (*Plectorhinchus goldmanni*, Vanuatu) illustrated here, lose water continuously by osmosis across their body surfaces and must drink seawater and excrete salt through specially adapted cells to replace the lost water. (b) Fishes that live in freshwater, such as the Arctic grayling (*Thymallus arcticus*, Prudhoe Bay, Alaska), absorb water continuously by osmosis and get rid of the excess by producing copious amounts of very dilute urine.

Fishes that live in seawater osmoregulate by drinking seawater to replace water lost from their internal fluids by osmosis. The excess salt ions ingested with the seawater are excreted through specially adapted cells in the **gills** (**Fig. 16.26a**). Freshwater fishes osmoregulate by drinking almost no water and excreting large volumes of very dilute urine to discharge the water that enters their bodies by osmosis. They also must take up dissolved salts through their gills (**Fig. 16.26b**). Certain estuarine fish species or species that live in environments of variable salinity, such as **tide pools,** must be able to osmoregulate in both directions. Because few fish species have such ability, most estuarine fishes migrate within the estuary as river and tidal flows vary to remain at the same approximate external salinity.

Two special categories of fishes have evolved life cycles that require them to cross the salinity gradient between freshwater and seawater twice during their life cycles. These are the **anadromous** and **catadromous** fishes (Chap. 15).

Swim Bladders and Buoyancy

Some fish species that live most of the time on the ocean floor can afford to be negatively buoyant (dense enough to sink) because they expend relatively little energy swimming against **gravity** during their limited excursions above the seafloor. Examples include most species of frogfishes (**Fig. 16.13b**) and scorpionfishes (**Fig. 16.15d**). Most fishes live in the water column and cannot afford to expend energy by swimming continuously to counteract gravity and maintain their depth. Consequently, most pelagic fishes (and other vertebrates and some invertebrates) must find a way to adjust their buoyancy to be approximately equal to that of seawater. The two primary ways of doing this are to synthesize and retain low-density oils, or to develop a gas-filled bladder.

Many fish species that live in surface layers and mid depths achieve neutral buoyancy by filling an internal swim bladder with gas. The amount of gas within the swim bladder must be adjusted as the fish changes depth. Otherwise, the gas will expand or contract and change the buoyancy of the fish as it ascends or descends. Some fishes with swim bladders make only limited and slow vertical excursions because the exchange of gas between blood and swim bladders is slow. Such species often die if caught and brought rapidly to the surface, because the swim bladder expands faster than the fish can evacuate the gas (**Fig. 16.14**). Other species have a special duct that connects the swim bladder with the esophagus, and they can ascend rapidly by "burping" to release excess gas from the swim bladder. Moray eels (**Fig. 16.25f**) have this duct, allowing them to change depth rapidly while hunting prey.

In shallow-water fishes, the gases in the swim bladder are similar to the atmosphere in composition: about 20% oxygen and 80% nitrogen. Fishes that live at greater depths have higher oxygen concentrations in their swim bladders, in some species up to 90% oxygen. The reason may be that they would be affected adversely by high nitrogen concentrations in their bloodstream in the same way that scuba divers suffer nitrogen narcosis (an affliction similar to the effects of drinking too much alcohol) if they breathe compressed air at pressures higher than that present at 30 to 40 m depth.

At a depth of 7000 m, the pressure is so great and gases so compressed that their density is approximately the same as that of fats. Consequently, many deep-water fishes have swim bladders filled with oil or fat instead of gases. These fishes do not need to adjust the amount of gas in the swim bladder as they change depth.

The largest and most active swimmers, such as mackerel and tuna (**Figs. 14.21a, 16.20**), have no swim bladder. These fishes can afford the relatively small energy penalty required to maintain their depth against their negative buoyancy because they expend much greater amounts of energy in swimming. Many of these species, especially sharks, have fins and body shapes designed to counteract their negative buoyancy as they swim. Sharks also have large livers with high concentrations of lighter-than-water oils to provide some compensation for their negative buoyancy.

16.6 REPRODUCTION

To reproduce successfully, each species must ensure that enough of its offspring survive to reproductive age and that these progeny in turn produce enough of their own offspring to continue the cycle. Reproductive cycles are poorly documented or poorly understood for all but a few marine species, but the basic elements of their diverse strategies are known.

Separate-Sex Reproduction

The majority of marine species reproduce by sexual interaction between male and female. Sexual reproduction increases **genetic** diversity (**CC17**), which improves the species' ability to survive and adapt to environmental change.

In sexual reproduction, sperm is transferred from male to female by a copulatory organ or directly from the male sexual organ to eggs laid previously by the female. Alternatively, sperm is either discharged into the water, where it fertilizes eggs retained by the female, or eggs released into the water, where fertilization occurs.

Direct transfer of sperm from male to female requires relatively few eggs and sperm to ensure sufficient fertilizations. Hence, this strategy minimizes the energy needed to produce eggs or sperm. However, it requires that males and females locate each other for breeding, which for many species is a major problem. Many fishes and invertebrates find mates in mating congregations that assemble at a specific time of year and place. Mating congregations enable fishes to select mates that have desirable characteristics.

The mate location problem has been solved in an unusual way by a number of species, including some barnacles and fishes that live at depth, in the dark. In these species, one sex is relatively small and resides either near or directly on the other. The smaller sex is usually the male because producing sperm requires less energy, and hence less body weight, than producing eggs does. The deep-sea anglerfish (**Fig. 16.14**) is a good example of a species that uses this strategy.

In pelagic reproduction, sperm and often eggs are released into the water (**Fig. 16.27a,b,c,l**). Sperm are attracted to eggs or females of their own species probably by chemotaxis ("tasting" or "smelling" of chemical clues released into the water by eggs or females). If sperm are released far from the eggs or females, or at a different time than the eggs, the probability of fertilization is very small. Consequently, pelagic spawning often occurs synchronously among all males and females of a given

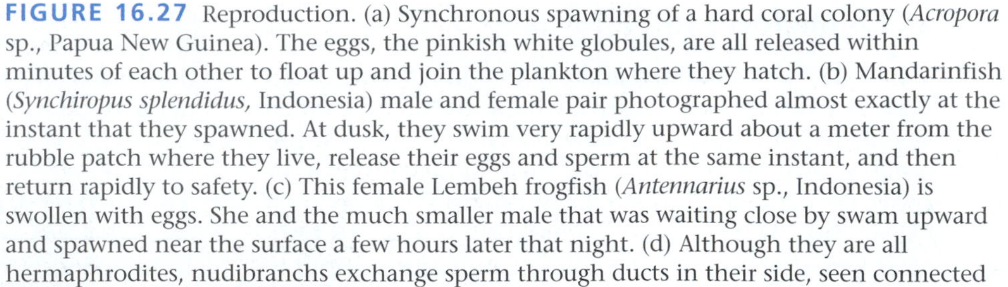

FIGURE 16.27 Reproduction. (a) Synchronous spawning of a hard coral colony (*Acropora* sp., Papua New Guinea). The eggs, the pinkish white globules, are all released within minutes of each other to float up and join the plankton where they hatch. (b) Mandarinfish (*Synchiropus splendidus,* Indonesia) male and female pair photographed almost exactly at the instant that they spawned. At dusk, they swim very rapidly upward about a meter from the rubble patch where they live, release their eggs and sperm at the same instant, and then return rapidly to safety. (c) This female Lembeh frogfish (*Antennarius* sp., Indonesia) is swollen with eggs. She and the much smaller male that was waiting close by swam upward and spawned near the surface a few hours later that night. (d) Although they are all hermaphrodites, nudibranchs exchange sperm through ducts in their side, seen connected in this image (*Hypselodoris bullocki,* Indonesia). Each individual retains and lays its own fertilized eggs. (e) On rare occasions, three nudibranchs may exchange eggs and sperm simultaneously (*Nembrotha rutilans,* Indonesia). (f) Nudibranch laying eggs (*Kentrodoris rubescens,* Papua New Guinea). (g) Nudibranch eggs are laid in a characteristic spiral ribbon pattern encased in a protective gel (species unknown, Hawaii). (h) Snails, although related to nudibranchs, nearly all reproduce sexually, but they also lay eggs encased in a protective gel (*Epitonium billeeanum,* Papua New Guinea). (i) Common reef squid eggs (*Sepioteuthis lessoniana,* Papua New Guinea) laid in rubbery gel casings in a seagrass bed. (j) Flamboyant cuttlefish eggs (*Metasepia pfefferi,* Indonesia) laid under a coconut shell half. The individual near the center of the image is close to

hatching, and the unhatched cuttlefish can be seen through the transparent egg case. (k) Many species of damselfishes, including this Indo-Pacific sergeant (*Abudefduf vaigiensis,* Vanuatu), lay eggs on the reef surface and then stay to defend the eggs aggressively from potential predators until they hatch. The eggs are the tiny reddish brown dots covering the reef surface behind the sergeant. (l) Sea cucumbers that normally lie flat on the substrate (see inset) rear up as high as they can before releasing clouds of eggs or sperm. Fertilization takes place in the water column, where the sperm are attracted to the eggs by chemical sensing; but the fertilization rate is low, and large numbers of eggs and sperm must be produced (*Thelenota rubralineata,* Papua New Guinea). (m) Most of this burrowing sponge (*Oceanapia sagittaria,* Indonesia) lies buried under the sediments. It reproduces vegetatively by growing the round ball-like structure in the image, which is released to the currents.

(h)

(i)

(j)

(k)

(l)

(m)

species in a specific region. Often this synchronous spawning occurs at a specific time of year or on a specific day. The timing is usually coordinated with a particular phase of the tides, such as high **spring tides** at full or new moon. Corals (**Fig. 16.27a**), many other benthic invertebrates, and many fishes spawn synchronously. In some fish species, a male and female pair will spin around each other or swim in contact with each other (**Fig. 16.27b**), simultaneously releasing and mixing a cloud of sperm and eggs. Other species form spawning congregations in which vast clouds of sperm and eggs are released simultaneously. Often the individuals swim in a school or swarm that spirals inward and upward toward the surface as sperm and eggs are released. This movement ensures that sperm and eggs are concentrated in an intensely mixed, dense cloud; keeps the eggs away from benthic predators; and probably reduces loss to pelagic predators.

Hermaphroditism

In many species, hermaphroditic life cycles have solved the problem of finding or ensuring the presence of a mate. Hermaphrodites are individuals that have both the sexual organ (gonad) necessary to produce sperm and the sexual organ necessary to produce eggs. In certain species, including most nudibranchs, the individual may perform either the male or female function, or both, at any time in its sexually mature stage (**Fig. 16.27d,e**). The common acorn barnacle is also a hermaphrodite in which each individual has a penis that can be used to inject sperm into any other barnacle because each barnacle also has ovaries. Thus, although adult barnacles are permanently attached to a surface, a suitable mate is always present in any location where barnacles are well established (**Fig. 16.6c**).

Sequential hermaphroditism is the ability of an individual to change from female to male or from male to female at an appropriate time in its adult life. The female-to-male sex change confers an advantage to species in which reproductive success is aided by the presence of a strong, experienced male during reproduction. For example, damselfishes (**Fig. 16.27k**), including anemonefishes (**Fig. 16.19**), lay their eggs on the reef surface (under the edge of the anemone in the case of anemonefishes). Reproductive success in these species depends on the successful defense of fertilized eggs against predators. For a given anemone, an anemonefish family may consist of a large aggressive male, a somewhat smaller female, and usually 4 to 10 smaller subadults or juveniles. The male defends the eggs from predation and defends his anemone from predation by butterflyfishes. If the male dies, the female grows and changes sex, and one juvenile grows to become a sexually mature female.

A male-to-female sex change is the preferred strategy of many species in which the female produces more eggs as it becomes larger. This strategy is common in invertebrates. For example, the common eastern oyster (*Crassostrea virginica*) is a male for several years as it grows, and then it transforms into a female. In transitional individuals, both sperm and eggs may be produced and the individual may even fertilize itself.

Asexual Reproduction

There are several different methods of asexual reproduction, including binary fission, fragmentation, and vegetative reproduction. In another form of asexual reproduction, unfertilized eggs develop into adults, but this reproductive strategy is rare in marine life.

In binary fission, a single-celled organism divides into two offspring. Diatoms reproduce by binary fission (**Fig. 14.16**). Fragmentation is a reproductive process of multicelled organisms that is similar to binary fission. For example, fragments of certain macroalgae and of some invertebrates, including various worms and sea stars, can break off and develop into new individuals. In some cases, fragmentation occurs through the production of special buds that are designed to break off the main individual. This budding process is common in macroalgae and sponges.

In vegetative reproduction, a single individual divides into many individuals that may or may not be physically connected. Vegetative reproduction is the most important marine asexual reproductive mechanism and occurs in many marine algae and invertebrate species, particularly colonial forms, including corals, sponges (**Fig. 16.27m**), and anemones. In benthic invertebrates, a single larva may settle on a substrate, **metamorphose**, and then reproduce vegetatively to cover a broad area of substrate.

Asexual reproduction produces many individuals that are genetically identical **clones.** Therefore, colonies of vegetatively reproduced anemones or encrusting sponges, for example, are all the same sex and color. Cloned colonies can occupy a substrate very densely because the cloned individuals do not have aggressive territorial responses toward each other as they would toward other individuals of their species. This lack of aggression may be important in colonial species, such as the Portuguese man-of-war, in which millions of cloned individuals must cooperate, each performing only one of a variety of separate and different functions needed to feed, defend, move, and reproduce the colony.

Asexual reproduction allows the successful colonization of suitable benthic microhabitats encountered by a single settling larva and avoids the difficulties of ensuring that sperm and eggs are united successfully. However, asexually generated populations have little genetic diversity, and hence the colony, and the species, may be poorly adapted to survive any unfavorable changes in its environment.

Egg Laying

Most marine fish and invertebrate species are **oviparous** and lay large numbers of eggs that hatch on the seafloor (**Fig. 16.27f–k**) or in the water (**Fig. 16.27a,b,c,l**) to become larvae that are **meroplankton.** In most cases, the vast majority of the eggs and larvae (sometimes more than 99.999%) are consumed by carnivores before they reach adulthood. For benthic species, many additional larvae may die because they settle on unsuitable substrates. Therefore, species that have pelagic eggs and/or larvae generally must produce very large numbers of fertilized eggs to ensure survival of the species. Species that spawn eggs to the water for fertilization must produce especially large numbers of eggs because many will not even be fertilized.

Production of very large numbers of pelagic eggs ensures the wide dispersal of larvae, which enables them to take advantage of the dispersed phytoplankton food resource. For benthic species, the dispersal of large numbers of larvae also facilitates the colonization of suitable substrate where this substrate is found only in isolated areas. Because this strategy carries a substantial energy cost, many species of oviparous fishes, such as anemonefishes (**Fig. 16.19**) and other damselfishes (**Fig. 16.27k**), lay many fewer eggs on the substrate and protect the eggs from predators until they hatch. Other species, including many sharks, skates, and rays, octopi and squid (**Fig. 16.27i,j**) lay only a very few eggs that are protected from predators in tough envelopes or encased in a gel (**Fig. 16.27f–h**). Finally, **ovoviviparous** fishes retain fertilized eggs within their reproductive tracts until they hatch. Seahorses (**Fig. 16.15h,i**) and some pipefishes (**Fig. 16.15e**) have an unusual **incubation** mode whereby newly fertilized eggs are deposited in a pouch on the male's abdomen, where they remain until they hatch. Some cardinalfishes incubate eggs in the male's mouth.

Although it would seem to be energy-efficient to produce fewer eggs and protect them until they hatch, energy must be expended to protect the eggs, and this strategy does little to reduce mortality during the larval stage. Larval mortality can be reduced if larvae can evade some predators, but to do so, the larvae must be relatively large. Remember, if larvae are small, they cannot swim effectively, because of the enhanced effect of viscosity (Chap. 7). If larvae must be large when hatched, eggs must have large yolks to provide the energy for growth. This requirement offsets any energy gained by producing fewer eggs.

Viviparous species are at the opposite extreme from the prolific pelagic spawners in the trade-off between egg numbers and protection. In viviparous animals, including mammals, offspring are nurtured inside the mother's body until they become fully developed and assume adult or nearly adult form. Some shark and ray species are viviparous. Only one or two offspring are produced at a time because of the large amount of energy needed to nurture the offspring during its prebirth development, but the large live-born offspring have a high rate of survival to reproductive age.

CRITICAL THINKING QUESTIONS

16.3 Discuss the advantages and disadvantages of a spawning strategy in which large numbers of eggs and sperm are released simultaneously to the water column.
- (a) Explain why you might expect this strategy to be more common in coral reef communities than in the polar regions.
- (b) Are there any terrestrial species that employ a similar strategy? If so, what are they, and how does their strategy differ from the similar strategies of ocean species?

16.4 Describe the advantages and disadvantages that are experienced by species that live in association with other species.
- (a) Explain why species associations are so common in the oceans.
- (b) Do species live in associations in the terrestrial environment? If so, describe some of these associations.

Timing

The timing of reproduction during the life cycle and during the year can be important. Spawning in some species takes place over a period of days or weeks at the same time each year. In other species, individuals may spawn several times a year; and in some species, spawning is almost continuous because at any given time, some individuals in the population are spawning. In mid and low latitudes, most pelagic spawning species spawn only once a year, in spring. This pattern synchronizes the production of larvae with the availability of abundant food supplies from the spring phytoplankton **bloom.**

In many species, the time of spawning may be determined by variations in light intensity and temperature, which also influence the timing of the phytoplankton bloom. Dramatic year-to-year variations in the reproductive success of many species can and do occur. They are due in part to annual variations in factors controlling the timing of the phytoplankton bloom and of spawning. In some years, spawning may occur at the wrong time or place, in which case larvae miss the phytoplankton bloom and their population incurs massive starvation losses. Omnivores suffer in turn, and their populations decline.

In tropical latitudes and certain **upwelling** areas where plankton food supply varies relatively little during the year, pelagic spawners have a greater tendency to spawn year-round. However, this trend is not universal, because even in these areas, monthly, biannual, or annual synchronous spawning still provides many advantages for larval survival that are not afforded by continuous spawning.

In certain species, spawning is timed to provide the maximum protection of eggs from predators. The best-known users of this method are the small fish called grunion that spawn on southern and Baja California beaches, and the horseshoe crab that spawns along the Atlantic coast. These species deposit eggs to incubate in beach sands above the **high-tide line.** Thus, eggs are protected from the many predators that abound in coastal waters and sediments. Eggs must be placed in sand high enough up the beach to be free from wave action or they would soon be washed out and devoured.

Grunions cannot climb above the high-tide line, and both grunions and horseshoe crabs have pelagic larvae that cannot crawl across the beach to enter the water after they hatch. How do grunions and horseshoe crabs place their eggs in sand above the high-tide line and yet ensure that the sand is covered with water when the larvae hatch? Each species times its spawning to occur at or a little after some of the highest spring tides of the year. In the grunion's case, the high spring tides that occur on the Pacific coast during summer are perfect because the highest tide of the day (see the discussion of **mixed tides** in Chapter 12) occurs at night. At high tide in the darkness, when they are safe from air attacks by birds, grunions can move into very shallow water high up on the beach. There they lay eggs during a frenzied spawning that lasts about 30 minutes, until the tide recedes. After the eggs are laid, **tidal ranges** diminish as spring tides progress to **neap tides,** and then build back toward the next set of spring tides. Two weeks after the eggs are laid, the next set of spring tides occurs (Chap. 12). By this time, the eggs have matured. When the overlying sand is resuspended by waves of the high spring tides, eggs are released to the water and the larvae hatch, becoming meroplankton.

The factors that determine the timing of spawning within a given year are relatively simple to understand. It is more difficult to understand why some species reproduce only once and then die while others reproduce repeatedly, and why some species mature rapidly while closely related species may take much longer to reach reproductive age. The average mortality rate of larval and adult stages, energy requirements of spawning, and year-to-year variability in larval mortality rate are believed to be important and interrelated explanatory factors.

In species with high adult mortality and relatively low larval mortality, natural selection will favor early and one-time reproduction. Because adult mortality is high, individuals that mature late will be selectively removed from the population because they are more likely to die before they produce offspring. Individuals that spawn more than once will also be selectively removed because they are likely to die before reaching a second or subsequent spawning cycle. Thus, for many species it is a successful strategy to reach maturity quickly, spawn only once, and then die. Salmon and eels are prime examples of species that use this strategy.

If egg or larval mortality is high in relation to adult mortality, natural selection will favor individuals that mature late and spawn more than once. Individuals that mature early and divert a large fraction of their food energy resources to reproduction, but relatively little to growth and predator avoidance, will be more likely to die early. An individual that matures late, devotes most of its energy to growth and survival, and diverts only a small amount each reproductive year or cycle to reproduction will produce more eggs in its lifetime. The late maturer is larger when reaching reproductive age, and thus it is capable of producing more eggs than a younger-maturing individual could, while expending the same amount of energy as a percentage of body weight. If this late maturer invests a relatively small proportion of its food energy in reproduction during each year, it can survive to spawn repeatedly, further increasing its lifetime egg production.

Selection for late maturity and multiple spawning cycles is further enhanced by great year-to-year variability in the survival of eggs and larvae. Survival through multiple spawning cycles enables a species collectively to outlast excessive variations in egg production or survival in successive spawnings. Because adults of several ages contribute to each year's total egg production, the effect of a small or missing year class (resulting, for example, from a poor year for larval survival) is minimized. Pelagic eggs and larvae are subject to intense predation and to ocean current and **climate** variations. Many fishes and benthic invertebrates with pelagic eggs and larvae have the relatively late-maturity and multiple-cycle reproduction that is associated with high and variable mortality of eggs and larvae.

Even closely related species can have very different reproductive timing. For example, the common mussel (*Mytilus edulis*) and California mussel (*Mytilus californianus*) are closely related and coexist in some areas. Because *M. edulis* has high mortality due to predation, it is outcompeted by *M. californianus* for living space in places where the larvae of both species settle. *M. edulis* invests a very large amount of energy in reproduction in comparison with *M. californianus* and spawns once a year in winter, whereas *M. californianus* spawns throughout the year.

Migration

Many marine animals migrate between different regions during their life cycles. The migrations can cross tens of thousands of kilometers, but typically they are much shorter in distance. Generally, one area on the species' migration route is its reproductive site, where eggs are fertilized, released, or deposited. The other end of the migration route (or other points on a complex migration pathway) is the main feeding ground for adults and/or juveniles. Hence, for most species, migration can be viewed as a strategy for ensuring appropriate habitat and food supply for each life stage of the species.

Migration may place larvae where appropriate food supplies are abundant and adults in another region where their different food supply is plentiful. However, migration is often not this simple, because adequate food supplies for each stage often are present throughout the migration route or at only a single location.

Why then do species not avoid the energy expenditure imposed by migration? There may be several answers. For example, the species might deplete available food in a specific region if it did not migrate to allow recovery of the food species populations. Migration may also place eggs, larvae, and/or the adult populations in locations where they are less subject to predation. Alternatively, migrations may enable species that are usually scattered widely across large ocean areas to congregate in one or more breeding areas, thus improving the chances of finding a mate and enhancing genetic mixing and diversity. Finally, adult migration to a spawning ground may ensure that pelagic eggs and drifting or weakly swimming larvae are carried by ocean currents to locations where food suitable for young adult stages is abundant. After drifting to such feeding grounds, larvae can metamorphose to the adult stage. Each of these possible advantages of migration and probably others are important for some species.

The North Sea **herring** has a migration–reproduction pattern typical of many abundant coastal pelagic fishes. Adults feed for much of the year in areas of rich plankton production off Norway (**Fig. 16.28**). In spring, herring migrate across the North Sea to spawning grounds near the Scottish coast. There they breed and lay eggs, which attach to stones or gravel on the seafloor. After they hatch, the larvae are transported eastward by currents to the adult feeding grounds, where they assume adult form.

Anadromous and catadromous fish species have perhaps the most amazing migrations. For example, the catadromous Atlantic eel, *Anguilla*, spends its approximately decade-long adult life in freshwater rivers of North America, Europe, and the Mediterranean Sea. At the end of this period, the eels undergo changes that include exchanging their dull gray color for a silver hue and growing enlarged eyes typical of pelagic fish species that live below the photic zone. The eels then migrate down their home rivers and thousands of kilometers across the oceans to the southeastern part of the Sargasso Sea (**Fig. 16.29a**). There, at locations not yet precisely known, the eels breed and then presumably die. Their

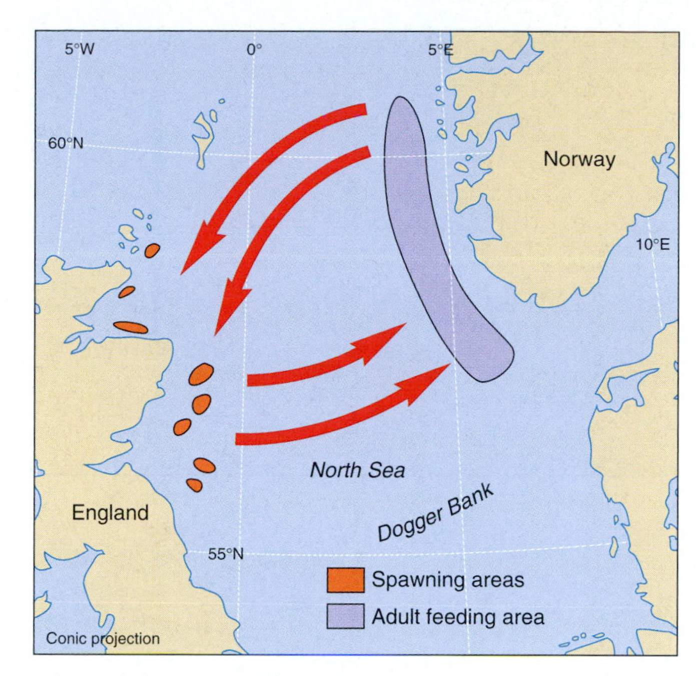

FIGURE 16.28 North Sea herring migrate from feeding areas off the coast of Norway to a number of localized areas off the Scottish coast to spawn.

leaf-shaped larvae (called leptocephali) are a few millimeters long when they begin their journey to the streams that will become their adult homes.

The American population drifts north with the Gulf Stream for 1 or 2 years before somehow sensing the proximity of the rivers of their ancestors, at which point the eels swim upstream to reach their new freshwater homes. European eels take 2 to 3 years to drift to their European homes. Mediterranean eels take an additional year to drift north with the Gulf Stream and then back to the south with the **eastern boundary current** of the North Atlantic Gyre until they can enter the Mediterranean. Although all Atlantic eels are considered the same species, the American, European, and Mediterranean eels are genetically distinct. These distinct populations may breed in slightly different areas of the Sargasso Sea so that larvae can join the Gulf Stream at the best point for the most efficient ride home.

In the Pacific Ocean, a similar species of catadromous eel that has adult populations in North America and Asia migrates to a spawning region in the southwestern part of the North Pacific Gyre that corresponds to the southwestern corner of the Sargasso Sea. Many details of the eel's migration, breeding behavior, and spawning locations are still unknown. In addition, as with all migrations, the chemical, magnetic, visual, or other clues that eels use to guide their migration to appropriate locations are unknown and subject to much research.

Pacific salmon are anadromous species that use a migration pattern almost opposite that of the eel (**Fig.

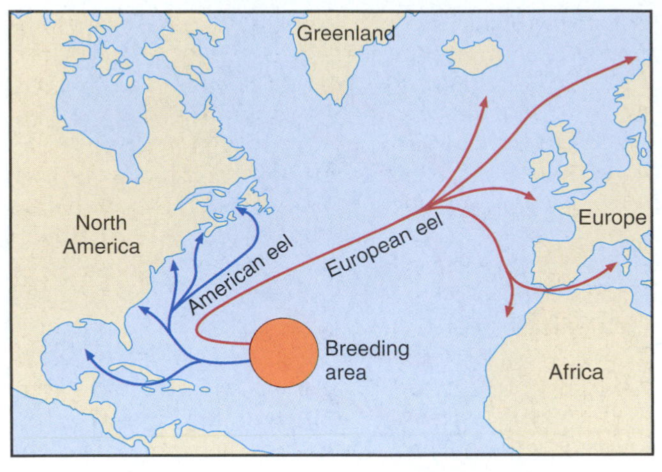

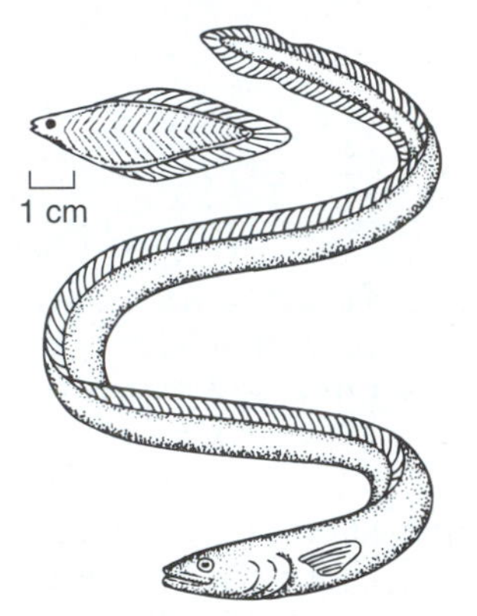

(a)

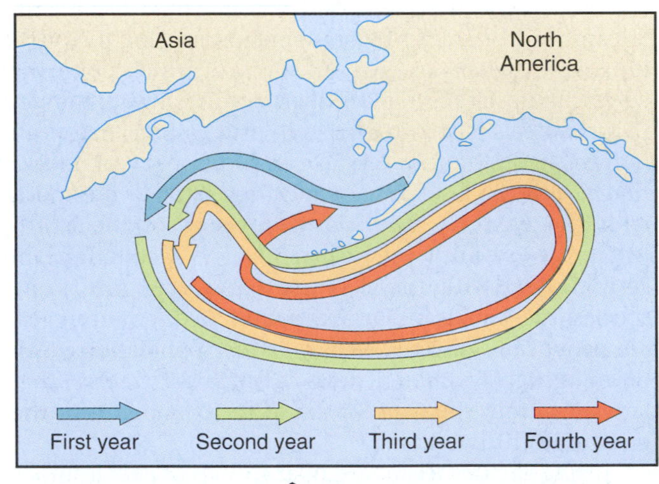

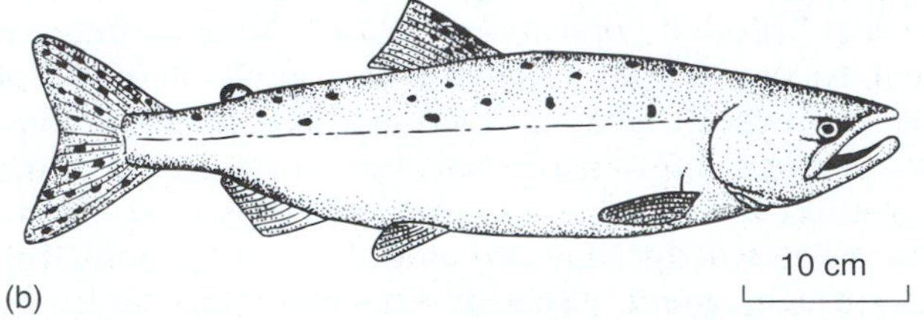

(b)

FIGURE 16.29 Migration routes. (a) Adult Atlantic eels migrate from their river homes in Europe and North America to the southwestern corner of the Sargasso Sea, where they spawn and die. Their larval offspring travel back to North America and Europe via the North Atlantic subtropical gyre currents and enter rivers (generally the home rivers of their parents), where they grow to maturity. (b) There are several species of North Pacific salmon, each of which has a somewhat different migration pattern. This figure shows a typical pattern in which the juvenile salmon migrate from their home streams to sea to spend 3 or more years migrating and feeding in the North Pacific Ocean between Alaska and Russia. The adults then return to rivers (usually to their home streams) to spawn and die.

16.29b). Salmon start life as eggs laid by the parent in gravel beds of rivers along the North Pacific coast. The eggs hatch 1 to 4 months later, and the resulting plankton-eating larvae grow quickly. During either their first or second year, juvenile salmon swim down the river to the ocean. By the time they reach the ocean, they are small but voracious predatory fishes. The salmon grow into adults and migrate across the Pacific Ocean to their selected feeding grounds, which range throughout the North Pacific Ocean. After remaining in the ocean up to about 5 years, they return to their home streams to breed and then die.

Salmon usually lay only several thousand eggs, far fewer than many other fish species. However, the eggs and larvae have relatively low mortality because they are protected from the abundant ocean predators through the spawning migration. Hence, the features of the salmon reproductive cycle are comparatively low egg and larval mortality, a single relatively early reproductive cycle followed by death, enormous energy costs of spawning migration, and production of a relatively small number of eggs in a favorable habitat. These features constitute a stable and successful reproductive strategy, as described earlier.

Almost all salmon return to the stream in which they hatched. They are thought to achieve this feat by "smelling" the distinctive chemical compositions of their home streams. However, contrary to popular misconception, not all salmon successfully migrate back to their home streams. As many as 10% to 20% are known to lose their way and migrate to streams other than their original home. Such "mistakes" are probably important to the maintenance of genetic diversity in the species and to the recolonization of streams where catastrophic events may have destroyed the spawning population.

The longest known ocean migrations are those of sea turtles and marine mammals, particularly the California gray whale. Sea turtles range far and wide throughout the oceans and return with great reliability to the beaches where they hatched.

All **baleen** whales, which, with the exception of the gray whale, are filter feeders, migrate seasonally. They feed in plankton-rich, high-latitude waters during spring and summer, and they return to warm tropical waters to breed in winter. California gray whales are unique among baleen whales because they feed by sifting sediment to eat small sediment-dwelling crustaceans called **amphipods** and they undertake one of the longest known seasonal migrations. In summer, they feed in the Bering and Chukchi Seas (**Fig. 16.30a**). When winter approaches, they migrate south along the North American coast, generally staying within sight of land. Their migration takes them some 11,000 km south to the shallow coastal **lagoons** of Baja California, where food is extremely limited and where adults must live off the fat reserves built up during summer. In the warm lagoon water, gray whales breed and females give birth 1 year later. Although young whales weigh about 2 tonnes, they have relatively thin layers of **blubber,** the fat that protects these warm-blooded mammals from losing body heat. If they were born in colder waters where the amphipod food supply is abundant, their mortality rate would be very high. The annual 22,000-km migration is therefore undertaken to ensure both adequate food for adults in summer and the high survival rate of offspring necessary to this species, which, like all other mammals, produces few young during its lifetime. Other whales, such as the humpback, also make long annual migrations (**Fig. 16.30b**).

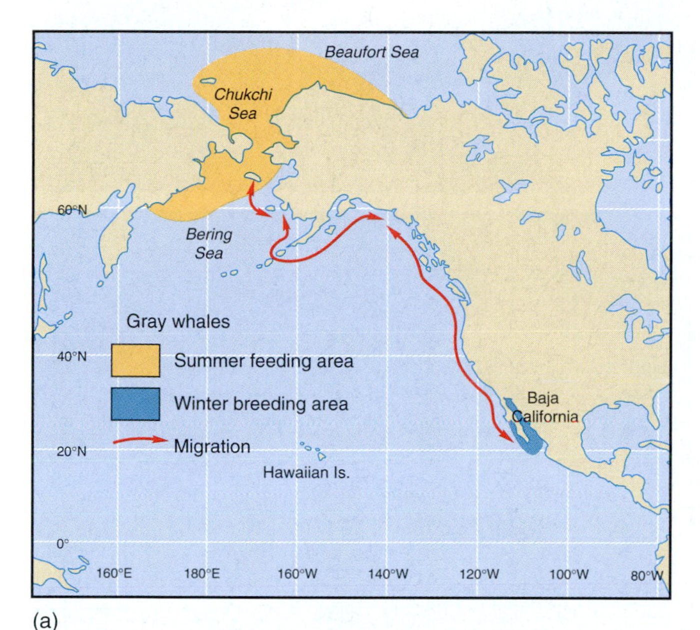

(a)

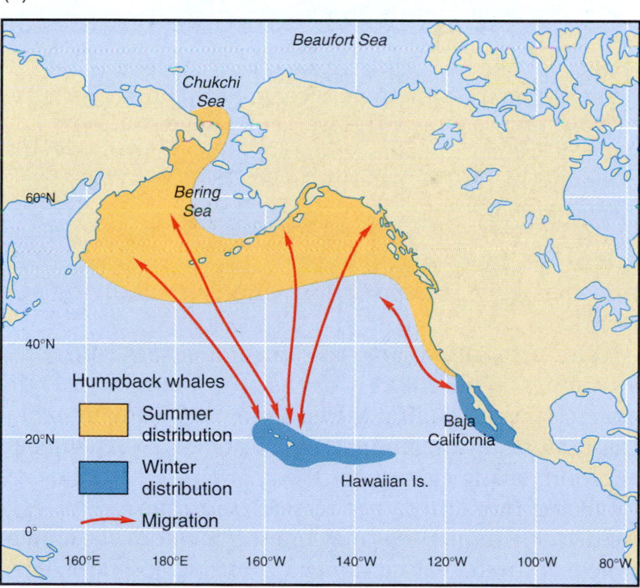

(b)

FIGURE 16.30 Some whale migrations. (a) California gray whales migrate annually between feeding areas in the Beaufort and Chukchi seas and breeding areas off the coast of Baja California. (b) Humpback whales feed in the far North Pacific Ocean and Bering Sea and migrate to breeding areas around Hawaii and Baja California.

16.7 ASSOCIATIONS

Until relatively recently, competition was viewed as the dominant force shaping marine ecosystems. According to this view, ecosystems sustain a number of species, each of which competes with all others for food and living space. However, we now know that many species enter into associations with other species as an essential part of their life cycles and that these associations are critically important to the functioning of marine ecosystems.

Associations between species, called **symbiosis,** can take a bewildering variety of forms but are of three basic types: **parasitism,** whereby one species benefits from the association and the other is disadvantaged; mutualism, whereby both species (or all species if more than two species are involved) benefit from the association; and commensalism, whereby one species benefits and the other does not benefit but suffers no disadvantage from the association. It is often difficult to determine what, if any, advantage or disadvantage the partners receive in a given association.

A parasite lives in or on, or frequently visits, another organism (the host) and feeds on the host's tissues or steals the host's food, thereby disadvantaging, but usually not destroying, the host. A tremendous variety of parasitic species are present in the oceans, including marine worms, crustaceans, and snails. The majority are small and live inside their host's body. Some, however, such as the fish and sea star parasites shown in **Figure 16.31**, are partly or fully external to their hosts. Fish parasites are abundant. As scuba divers can observe on a coral reef, several species of small fishes and shrimp make their living by eating parasites off larger fishes' bodies (**Fig.**

FIGURE 16.31 Parasites of fishes. (a) This whip goby (*Bryaninops yongei,* Indonesia) has a large parasite, a copepod crustacean, on its body. (b) A large black isopod parasite (a crustacean of the family Cymothoidae commonly known as a "fish doctor") on a lemon damsel (*Pomacentrus moluccensis,* Papua New Guinea). There are two parasites. Look carefully and you might be able to see the much smaller male just under the large female. (c) Fish doctor parasites are common on many species of fishes, including this flasher scorpionfish (*Scorpaenopsis macrochir,* Indonesia). (d) This parasitic snail (*Thyca crystallina,* Indonesia) is apparently found only on this one sea star species (*Linckia laevigata*).

16.16c). The larger fishes seek out the cleaner, which often establishes a cleaning station to which its customers return for periodic service.

Because they live either in or on another organism, many parasitic species do not need sensory, locomotory, or skeletal organs. Consequently, many have become degenerate and lack such organs. Many species are little more than a digestive and reproductive system. The reproductive system is generally large because the parasite must normally produce enormous numbers of larvae so that a few of these larvae can encounter suitable hosts. Host location is a major problem for parasites, many of which are parasitic on only one host species. Accordingly, parasite life cycles are often very complex, with one or more intermediate hosts (**Fig. 16.32**).

Commensal or mutual associations can provide various benefits to the cooperating species. These include ready availability of food, avoidance of predators or parasites, easy transportation, and suitable living surfaces.

The association between anemonefishes and their anemones (**Fig. 16.19**) has major benefits for the anemonefishes. Hiding in the anemone's stinging tentacles protects an anemonefish from potential predators. In addition, laying eggs under the anemone where they are protected from predators reduces the numbers of eggs needed. Anemonefishes may use the anemone for food in times of need, eating mucus, food scraps, and even anemone tentacles. They may also benefit by having external parasites stung to death and removed by the anemone.

Because many anemones are found without anemonefishes but anemonefishes are never found without an anemone, the benefits of the association to the anemone are less obvious. The anemone does benefit by being cleared of dead tissue, mucus, food wastes, and

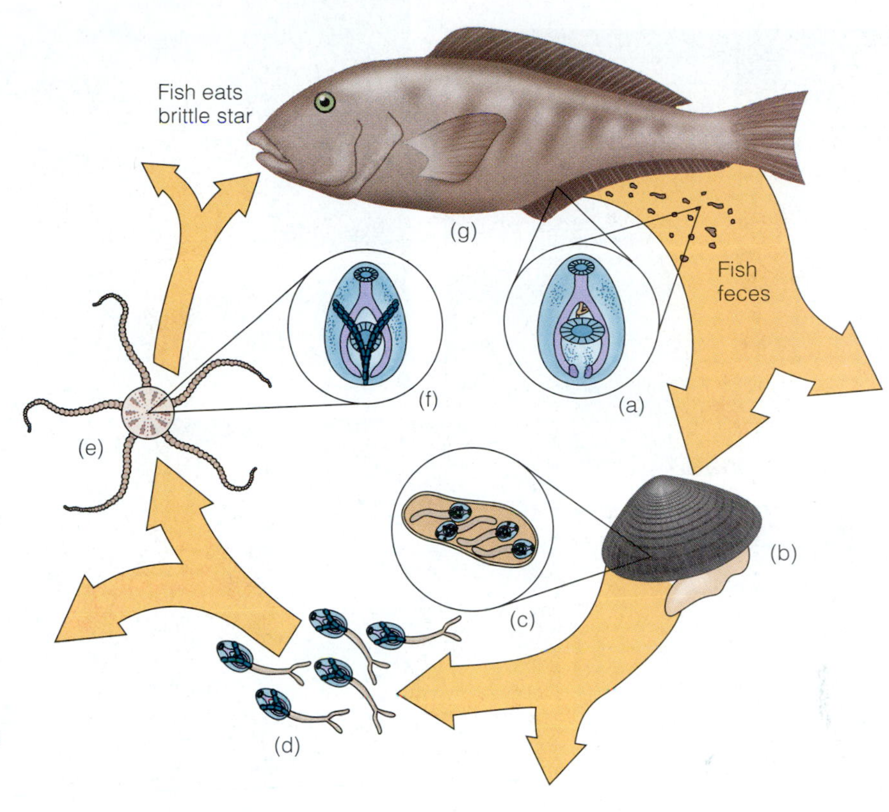

FIGURE 16.32 The life cycles of parasites are often complex. The adult fish fluke lives in the gut of its fish host (a). It produces very large numbers of eggs that are discharged with the host's feces and hatch into a juvenile stage called a "miracidium." Although most of the miracidia do not survive, a very small number enter a bivalve mollusk (b), which acts as an intermediate host for the parasite. While in the bivalve, the miracidium develops into a larval form (c), which in turn develops a large number of another life stage, called a "cercaria" (d). The cercariae are released to the water, and a very small number of these cercariae eventually enter a second intermediate host, a brittle star (e). Inside the brittle star, the cercariae develop into yet another life stage, the metacercaria (f). If this intermediate host is eaten by a fish (g), the metacercaria develops into an adult fish fluke parasite. At each stage in its life cycle, only an extremely small fraction of the individuals survive to reach the next host. This is true because many parasites, such as the fish fluke, can use only a single species or only a very small number of species as host for one or more of their life stages.

parasites by the anemonefish. The anemonefish also protect the anemone from its few predators. Without the anemonefish, the anemone could be consumed by certain butterflyfishes and might have to defend its delicate tentacles from such predators by closing up if the anemonefish were not present. Because the anemone cannot feed when closed, the freedom to remain open when predators are nearby is undoubtedly a benefit of the anemonefish association. In some cases, the anemone may benefit by being fed by the anemonefish, which, on occasion, have been seen to carry food to the anemone.

Many other associations also involve a mobile species that uses a nonmobile or less mobile host for protection and food. Examples are associations in which a variety of crabs (**16.33a–e**), small fishes (**Fig 16.33f,g**), shrimp (**Figs. 16.33h–q, 16.34e**), scale worms (**Fig 16.33r**), and mollusks (**Fig 16.33s**) live on crinoids (**Fig. 16.15a,b**), corals (**Fig. 16.15g,h,j,k**), sea stars (**Fig.16.15l**), and sea fans (**Fig. 16.10b**). In most of these associations, the small lodger is well camouflaged or hidden within the host. The lodger eats mucus, dead tissue, parasites, the host's body parts or food supply, or any combination thereof. Generally, the host benefits by being cleaned of mucus, dead tissue, and/or parasites.

Even if it is eaten by its partner species, the host may gain. Many species that feed on their host are present as only a single pair on their much larger host. This pair nourishes itself from the host's tissues but eats the host only at a rate that can be replaced by normal growth. The resident pair aggressively protect their host against colonization by others of their species, parasites, and other species that would also consume the host. Thus, although it loses some tissue to the associate species, the host is better protected.

Many associations involve a normally nonmobile species that lives on a mobile host. Examples are the associations formed by the decorator crab (**Fig. 16.15f**) and the anemone crab (**Fig. 16.34a**). The decorator crab gains camouflage and protection from its covering of algae, hydroids, and sponges, which are unpalatable and rarely attacked by predators. The immobile associates may gain by being transported through the water, which increases their chances of obtaining food or dissolved nutrients, and by using the host as a substrate on which to grow. In similar associations, the sponge may be an indifferent partner removed from its home on the reef and carried on the crab's back by a pair of specially adapted legs. The anemone **hermit crab** is well protected by the adopted mollusk shell in which it lives, but it is further protected by the stinging anemones that it attaches to the shell (**Fig. 16.34a**). The anemone gains by being transported through areas where food can be obtained, including scraps of food released as

FIGURE 16.33 Examples of the many associations between species. (a) The arrowhead crab (*Huenia heraldica,* Papua New Guinea) shown here not only blends into the algae (*Halimeda* sp.) that it lives on, but usually carries pieces of the algae around with it, as this one is doing, to further enhance its camouflage. (b) Many small crabs live in or on a larger invertebrate for camouflage, to hide and sometimes to steal the host's food or feed on the host itself. Many species, such as the one shown here on a soft coral, are unidentified or poorly studied and difficult to identify (Indonesia). (c) This tiny spider crab (*Xenocarcinus tuberculatus,* Indonesia) lives on a coral whip, usually just one pair per whip. (d) The porcelain crab (probably *Porcellanella triloba,* Indonesia) lives on a sea pen, even when the sea pen withdraws into the sand for protection. (e) Another tiny porcelain crab (probably *Porcellanella* sp., Indonesia) lives on a whip coral. (f) The weed cardinalfish (*Foa brachygramma,* Papua New Guinea) spends its life hidden in weeds or, as in this case, in a crinoid where it cannot be seen or reached by predators. (g) The crinoid clingfish (*Discotrema crinophila,* Papua New Guinea) lives and hides in a crinoid and usually has a body color that matches the crinoid. (h) Emperor shrimp (*Periclimenes*

imperator, Papua New Guinea) like the pair shown here live on a variety of hosts, including nudibranchs and sea cucumbers. (i) Sea star shrimp (*Periclimenes soror,* Indonesia), live on sea stars and match their body color to blend in with the host's body. (j) This sea star commensal shrimp (possibly *Allopontonia* sp., Indonesia) is partially transparent, and with its spots it blends in so well with its background when it is on some parts of the sea star's body that we had to wait for the shrimp to move before getting a photograph in which the shrimp is more visible. (k) A pair of Coleman's shrimp (*Periclimenes colemani,* Papua New Guinea) living protected on their venomous fire urchin host (*Asthenosoma varium*). (l) Many species of shrimp, like this one (*Periclimenes* cf. *venustus,* Indonesia), live in anemones. (m) Many small commensal shrimp species, including this one (*Periclimenes* cf. *tosaensis,* Indonesia), have mostly transparent bodies, so they can blend in with several different hosts. This individual is living in a bubble coral, and the eggs that it's carrying can be seen as a pink egg mass through its transparent body. (n) This species of coral shrimp (*Vir philippinensis,* Indonesia) lives only in bubble corals. (o) A humpbacked shrimp (*Hippolyte commensalis,* Indonesia) can hardly be seen when it withdraws into its soft coral host. (p) This tiny urchin shrimp (*Gnathophylloides mineri,* Indonesia) is also virtually impossible to see when it hunkers down among the spines of its urchin host where it normally hides. (q) This saw blade shrimp (*Tozeuma armatum,* Indonesia) has an unusually elongated body so that it can lie flat against the strands of black coral on which it lives. (r) Scale worms, including this species (*Gastrolepidia clavigera,* Vanuatu), have flattened bodies and coloration to match their host, usually a sea cucumber, as seen here. (s) This egg cowrie (*Pseudosimnia* sp., Red Sea) lives on and probably eats soft corals such as the one shown here (*Dendronephthya* sp.).

FIGURE 16.34 Additional examples of associations. (a) This anemone hermit crab (*Dardanus* sp., Papua New Guinea) has found a strong gastropod mollusk shell to live in and has attached several anemones on the shell to further discourage potential predators. (b) This crab (*Dorippe frascone,* Indonesia) has a pair of legs that are modified to carry an urchin, in this case *Astropyga radiata,* on its back. The urchin's toxic spines protect the crab from predators. (c) This tiger pistol shrimp (*Alpheus bellulus,* Indonesia) lives in a single burrow with the yellow shrimp goby (*Cryptocentrus cinctus*). The shrimp digs and maintains the burrow. The goby warns the shrimp, which is blind, when a predator approaches, and they both disappear into the burrow. (d) A giant clam (*Tridacna gigas,* Papua New Guinea). (e) This small pistol shrimp (*Synalpheus* sp., Papua New Guinea) lives in its host crinoid.

the hermit crab feeds by tearing apart its prey. This association, like many others, is highly specific. The association usually involves the same species of anemone, hermit crab, and mollusk shell. Some species of decorator crabs also cover themselves with sponges (**Fig 16.15f**) or carry an urchin on their backs (**Fig 16.34b**) to afford them protection.

The remora (**Fig. 16.25g**) is an example of a species that is mobile but hitches a ride on a larger swimming animal. In some cases, remoras gain by stealing food from their host. However, the major advantage to the remoras is probably transportation, since they often ride on manta rays, which feed on plankton. It is not clear whether sharks or manta rays gain anything from a remora's presence, but these hosts apparently are not harmed by carrying the passenger.

The association between gobies and shrimp is clearly beneficial to both partners. A goby and a shrimp live together in a common sand burrow, where they can retreat to safety from predators. The shrimp is blind, and the fish is incapable of digging its own burrow. The shrimp therefore digs and maintains the common burrow, but it is vulnerable to predators when it pushes excavated sand out of the hole. The goby repays the shrimp for its digging by sitting at the burrow entrance watching for predators (**Fig. 16.34c**). With one of its long, sensitive antennae, the shrimp maintains contact with the goby's caudal fin. If a predator approaches, the goby wiggles its fin to warn the shrimp, which immediately withdraws into the burrow. If danger increases, the goby itself darts into the burrow.

The few associations described here are all relatively easy to observe and study, but most associations are much more subtle. For example, coordinated feeding by two or more fish species that benefits both species does occur but may be infrequent and difficult to observe. In addition, many associations are difficult to study because one species is well hidden inside the other or because one or both partners are microscopically small. For example, the giant clam, *Tridacna* (**Fig. 16.34d**), has a mutually beneficial association with **zooxanthellae.** The algae live inside the soft **mantle** tissues of the clam and give this part of the clam its often brilliant coloration. The algae benefit from a secure location within the clam's tissues. There the algal cells can obtain needed sunlight when the clam is open, and they are protected from settling of other benthic organisms that might otherwise overgrow them. In addition, the clam closes when threatened, protecting both itself and its algae. The clam benefits from the association by using the algae's by-products as a supplemental food supply. Thus, the algae are essentially a carefully cultivated garden that supplements the clam's food so that it does not have to filter extremely large volumes of water. This feeding arrangement is very energy-efficient and probably accounts for the giant clam's ability to grow to its huge 1- to 2-m size. Hard corals have a similar association with zooxanthellae (Chap. 17).

16.8 COMMUNICATION AND NAVIGATION

The majority of species must be able to communicate at least on a rudimentary level—for example, to provide an indication of sex and attract a mate. Most predators cannot rely on finding their prey by chance but must actively seek them out. Many marine species must navigate from place to place in migrations or to find suitable habitat, food, and mates. Humans and most other terrestrial animals use their five senses—vision, smell, hearing, touch, and taste—to perform these functions, with vision being the predominant sense. Light is absorbed and **scattered** in seawater, and most of the ocean is dark. In addition, when light is present, its path is distorted by **refraction.** Therefore, marine species rely much less on vision than most terrestrial species do.

In the marine environment, hearing, chemical sensing that is equivalent to taste or smell, and touch have assumed much greater importance. Marine organisms have also developed other senses, including the ability to sense electrical fields and probably the Earth's magnetic field. Our knowledge of these unusual and different senses and even of how marine organisms use their greatly enhanced hearing, chemical sensing, and touch is still extremely limited.

Although vision is much less useful in the oceans than on land, it is still widely used by many marine species, and many adaptations of vision are found in species that live in the ocean environment. The most important adaptations are greatly enlarged eyes or multiple-lens eyes that can see in very low light levels and distinguish subtle changes in light intensity (**Fig. 16.17a,b**). Most marine species do not see sharply focused images, and many, particularly some marine invertebrates, have bizarre-looking eyes (e.g., **Figs. 16.17a, 16.34a**). Some species have light-sensing organs that cannot be readily identified as eyes. For example, the giant clam (**Fig. 16.34d**) and many other benthic invertebrates are able to sense very small variations in light intensity, which enables them to withdraw into their shells, tubes, or other protective environments when predators approach.

Chemical sensing is very important in the marine environment, but very little is known about it. For example, anadromous and catadromous species are thought to use a well-developed chemical sensing capability to detect and navigate to their home streams. They apparently can sense small differences in the composition of chemicals present in very low concentrations in these home streams. Sharks also appear to have an extremely keen sense for chemicals. They are attracted to blood even when they are kilometers away from its source and the concentration of blood chemicals is exceedingly small. Adventurous scuba divers deliberately attract sharks by dumping blood and shredded fish into the water.

Sound travels through seawater with relatively little loss of intensity. Consequently, sound sensing, or hearing, is well developed in many marine species, and sound is used extensively for communication. Many species, particularly **cetaceans** such as whales and dolphins, use sound not only for communication, but also to locate objects, including prey. Most cetaceans can generate sound pulses that bounce off objects in their path. They focus and process the resulting echoes, and thus are able to "see" the objects. This **echolocation** ability is extremely sophisticated and capable of high sensitivity and precision in at least some species. For example, the dolphin echolocates by emitting low-**frequency** clicks to scan objects at distances up to hundreds of meters. As it nears the object, the dolphin uses higher-frequency clicks to get a more detailed "picture." With their echolocation abilities, dolphins can identify specific objects that differ only slightly in shape, size, thickness, or material composition. Dolphins and other cetaceans can produce and sense an extraordinary range of sound frequencies far beyond human hearing at both high and low frequencies.

We do not know exactly how cetaceans produce sound pulses or how they focus and sense returning echoes. However, some species are believed to create the sounds by moving air within hollow bony structures in the head. The bulbous brow of most cetaceans (**Fig. 14.24**) contains oil- or fat-filled structures believed to be responsible for focusing sound pulses when they are generated and perhaps when echoes are received.

In addition to cetaceans, many other marine species produce sounds thought to be used primarily for communication. For example, certain fishes are known to generate sound by using their swim bladders, and they may sense other fishes' sounds or echoes in the same way. Scuba divers, if they listen carefully, can hear a continuous cacophony of noises in the oceans that sounds like the static on a radio. Some species use sound for more sinister purposes. The pistol (or snapping) shrimp (**Fig. 16.34e**) has one very large, overdeveloped claw, which, when snapped closed, creates a loud noise that can be heard across a room if the shrimp is in an aquarium. The concussion of the shrimp's snapping sounds stuns its invertebrate or small fish prey, sometimes into unconsciousness. The stunned prey is then easy game for the shrimp. Other species, notably killer whales, also may generate and use intense sound pulses to stun or even to kill their prey.

Fishes can sense small changes in pressure through sensory organs located in a row along each side of their bodies that form what is called the "lateral line." Lateral-line sensors are apparently used by schooling fish to sense motions of their neighbors. Together with visual clues, these sensors may explain how schooling fish can turn in unison. Sharks also may be able to sense pressure variations, since they are attracted from long distances by the thrashings of a sick or dying fish. The low-frequency pressure variations caused by such a fish are transmitted through the water as very low-frequency sound waves, and the shark's ability to sense such pressure fluctuations may simply be an extension of its "hearing."

Evidence is steadily accumulating that many marine species are able to sense magnetic fields, including the Earth's magnetic field. This ability may be vital for migrating species such as sea turtles. In addition, some marine species are able to generate and sense electrical fields.

CHAPTER SUMMARY

Ecological Requirements. The fundamental needs of all species are a place to live, food, safety from predators, and successful reproduction.

Habitat. The water column, seafloor, and sediments are three fundamentally different habitats. Pelagic organisms must avoid sinking, and most pelagic animals must actively seek food. Benthic epifauna save energy because they do not have to control buoyancy, but they have difficulty avoiding predators and, in most areas, must rely on detrital food that rains down from above or on predation. Benthic infauna have some protection from predators and do not require buoyancy control, but because food in sediments is limited, they must expend energy to move or dig through the sediment, and they must obtain oxygen that is depleted in most sediments below the upper few centimeters. The surface microlayer, intertidal zone, and hydrothermal vents are special habitats that present unique problems and opportunities.

Feeding. Suspension feeders eat particles suspended in the water column, including living phytoplankton and zooplankton, and detritus. Many suspension feeders are filter feeders that strain water through a meshlike structure to capture food. Other suspension feeders use mucus to capture food particles, or they are able to grab or grasp particles with armlike appendages. Many suspension feeders increase their feeding efficiency by pumping water through their collection apparatus, moving the apparatus through the water, or placing it in currents. Many suspension feeders are present among both pelagic fauna and benthic epifauna, and a few benthic infaunal species suspension-feed using a feeding apparatus that they extend above the sediment surface or by pumping water through their tubes or burrows.

Surface grazers consume algae, small sedentary animals, and detritus from the seafloor. Food is abundant only where the seafloor is within the photic zone, where the substrate may be covered with benthic microalgae, macroalgae, and sponges, tunicates, and other animals. In the deep sea, surface grazers eat primarily detritus or bacteria. Surface grazers may feed by using specially adapted mouths on the underside of the body that can either rasp off food organisms or grasp food as they sift through the surface sediment. Others suck off and sift surface sediment to obtain food particles. Some surface grazers also suspension-feed or hunt.

Most deposit feeders move through the sediment, taking sediment into the gut, where they digest organic matter from detritus or coatings. In sediments with anoxic pore waters, deposit feeders must obtain oxygen from the water above or remain in the oxygenated upper sediment layer.

Hunting and Defense. Speed is used by both hunter and hunted; the hunter chases, and the potential prey seeks to escape. Prolonged movement at high speed in the oceans is energy-intensive because of water resistance. Consequently, all but a few of the largest ocean animals use speed in short bursts rather than in prolonged chases. Other strategies are used to get close to the prey, and then a short burst of speed over the remaining small distance is used to capture the prey. In defense, speed generally is used for immediate escape, followed by other strategies to ensure ultimate safety.

Many marine hunters use lures to attract prey. The lure is typically an appendage on or near the mouth that resembles a bite-sized meal to the prey. Some species use false markings such as eyespots to lure predators into attacking in the wrong direction or at the wrong place as the potential prey speeds off in the opposite direction.

Hunters use camouflage to lie in wait unseen by their prey, or they mimic species that clean other species so that they can approach unsuspecting prey. Many species use camouflage to conceal themselves from predators, or they mimic fiercer or poisonous species to deter predators. Concealment in sediment, within other species, or in holes in reefs and rocks is a much used defensive strategy. Many species conceal themselves by day and emerge to hunt only at night. Hunters also conceal themselves to ambush prey.

Spines and armored bodies are used by many marine species as defensive mechanisms, and in some cases as weapons. Hunters use crushing claws and jaws to overcome the armor of prey species. Poisons are used by hunters that inject venom into their victims to stun or kill them, and by potential prey species that inject venom into attacking species or make themselves toxic to eat. Many species that are poisonous to predators are brightly colored to advertise their toxicity.

Schooling or cooperation can be used to overcome defenses of a confused prey species, and to confuse predators or frighten them off by creating the appearance of a target much larger than any one individual.

Selected Adaptations in Fishes. Fish body shapes differ by species to reflect swimming habits. Species that swim continuously have teardrop shapes—a compromise that minimizes total drag, including surface, form, and turbulent drag. Fishes that make fast turns are flattened so that the flat sides can act like a rudder. Fishes that specialize in a quick burst of speed are thickened in the middle by the needed musculature.

Fin shapes also vary with swimming habits. The caudal fin is rounded in fast-maneuvering and -accelerating species, truncate and forked in faster-swimming species, and lunate in the fast continuous swimmers. Heterocercal caudal fins and pectoral fins in sharks provide lift to maintain buoyancy. Fins are also adapted for special purposes, such as "walking," defense and venom injection, attachment to substrates or hosts, and even gliding through the air.

Unlike invertebrates, bony fishes have internal body fluids that have lower salinity than seawater, and therefore they must prevent the continual loss of water by osmosis. Marine fishes osmoregulate by drinking seawater and excreting salts through a special gland. Freshwater fishes gain water by osmosis and osmoregulate by producing large volumes of urine. Some species that live where salinity is variable must be able to osmoregulate in both ways.

Most fish species maintain buoyancy by producing and storing low-density oils or by filling a swim bladder with gas. Gas must be released from or added to the swim bladder as the fish changes depth. Oils are favored in abyssal species because gas compressed at depths below about 7000 m is more dense than oil. Some species have no swim bladders, must swim continuously to avoid sinking, and have fins and body adapted to provide lift.

Reproduction. Many species reproduce sexually, either by the direct transfer of sperm from male to female or, more often, by the release of sperm, or of eggs and sperm, to the water column for fertilization. Mating congregations and synchronous spawning by all males and females of a species maximize the probability of egg fertilization, improve egg and larval survival, and maximize genetic diversity.

Because mates may be difficult to locate, many species are hermaphroditic, and some can even function as both sexes at the same time. Asexual reproduction is common in algae and in colonial invertebrates.

To ensure egg fertilization and survival, either very large numbers of eggs must be produced and released or smaller numbers of eggs must be defended against predators. Many fishes and invertebrates release large numbers of eggs to the water column, where they hatch to become meroplankton. Other species lay eggs on the substrate and protect them until they hatch, and still others retain fertilized eggs in or on their bodies until they hatch. Spawning often occurs at times and locations that match the availability of suitable food for larvae. Species with high adult mortality and relatively low larval mortality tend to reproduce early and only once. Species with high egg and larval mortality tend to mature late and spawn more than once.

Many species migrate during their life cycle. Most often they move from adult or seasonal feeding areas to mating or spawning areas. Anadromous and catadromous fishes migrate between freshwater and the ocean.

Associations. Symbiosis between species is very common in the marine environment. In some associations, one species benefits and the other is disadvantaged (parasitism). In other associations, both species benefit (mutualism), or one species gains and the other is neither advantaged nor disadvantaged (commensalism). Benefits gained from associations include food, living space, camouflage, protection from predators, removal of parasites, and transportation.

Communication and Navigation. Many marine species have and use vision, but other senses are often more important. These senses are used to locate prey, mates, and predators and for navigation. Sensing of dissolved chemicals, akin to smell or taste, is especially well developed in many species. Certain species generate sounds that are used for echolocation, to stun prey, or to communicate. Many species can sense small changes in pressure, electrical fields, and probably the Earth's magnetic field.

STUDY QUESTIONS

1. What are the four fundamental challenges that any species must meet to survive?
2. What are the advantages and disadvantages of the benthic infaunal habitat in comparison with the benthic epifaunal and pelagic habitats? Describe how these factors change with depth and substrate type.
3. How do suspension feeders obtain food, and why are there so many suspension feeders in the oceans?
4. Why do most deposit feeders live in only the upper few tens of centimeters of the sediment?
5. Why are some pelagic predators not able to swim fast? What other capabilities do they use to capture prey?
6. Why do many fish species congregate in schools?
7. What are the physical characteristics of a fish species that swims continuously at high speeds, and of a species that waits on the seafloor for prey to swim by?
8. What are the three ways in which different fish species maintain their buoyancy? Which of these is more common in abyssal species, and why?
9. For what reasons do many species migrate at some stage in their life cycles?
10. What aspects of their environment might marine species be able to detect with senses other than sight?

KEY TERMS

You should be able to recognize and understand the meaning of all terms that are in boldface type in the text. All of those terms are defined in the Glossary. The following are some less familiar key scientific terms that are used in this chapter and that are essential to know and be able to use in classroom discussions or on exams.

anadromous (p. 476)
anoxic (p. 456)
benthic (benthos) (pp. 446, 447)
biomass (p. 446)
carnivores (p. 446)
catadromous (p. 476)
cnidarians (p. 470)
countershading (p. 464)
decomposer (p. 446)
deposit feeder (p. 455)
detritus (p. 447)
epifauna (p. 446)
filter feeder (p. 448)
grazed (grazer) (p. 446)
habitat (p. 446)
herbivores (p. 448)
infauna (p. 446)
intertidal zone (p. 447)
invertebrates (p. 450)
meroplankton (p. 481)
metamorphose (p. 480)
nekton (p. 448)
omnivores (p. 446)
osmosis (osmotic pressure, osmoregulation) (pp. 470, 476)
oviparous (p. 481)
ovoviviparous (p. 481)
polyp (p. 451)
protists (p. 446)
suspension feeder (p. 448)
swim bladder (p. 473)
symbiosis (p. 485)
vertebrates (p. 476)
viviparous (p. 481)
zooxanthellae (p. 491)

CRITICAL CONCEPTS REMINDER

CC14 Photosynthesis, Light, and Nutrients (p. 446). Photosynthesis and chemosynthesis are two processes by which simple chemical compounds are made into the organic compounds of living organisms and upon which all species are ultimately dependent. Photosynthesis depends on the availability of light and can only take place in a shallow upper layer of water or on the shallow seafloor. Chemosynthesis does not use light energy but instead uses chemical energy from reduced compounds. Therefore, chemosynthesis can occur in all ocean environments in which oxygen is depleted, but these environments are very limited in extent in the present day oceans. To read **CC14** go to page 46CC.

CC17 Species Diversity and Biodiversity (pp. 447, 477). Biodiversity is an expression of the range of genetic diversity; species diversity; diversity in ecological niches and types of communities of organisms (ecosystem diversity); and diversity of feeding, reproduction and predator avoidance strategies (physiological diversity), within the ecosystem of the specified region. Species diversity is a more precisely-defined term and is a measure of the species richness (number of species) and species evenness (extent to which the community has balanced populations with no dominant species). High diversity and biodiversity are generally associated with ecosystems that are resistant to change. To read **CC17** go to page 53CC.

CHAPTER 17

CHAPTER OUTLINE

CRITICAL CONCEPTS USED IN THIS CHAPTER

Coral reef ecosystems have a species diversity that rivals, and perhaps exceeds, the species diversity of tropical rain forests. In this photograph of Deacon's Reef in Papua New Guinea, several species of fishes are easily seen. However, the majority of species that live on this reef live in or on the reef itself, or live hidden in the seafloor substrate. The beautiful red-and-white coral "tree" is a colony of the soft coral *Dendronephthya* sp. Less obvious are many species of hard corals, crinoids, sponges, tunicates, hydroids, and algae. Even this profusion of species is only a tiny fraction of the species that live on this reef. Hidden from view in or among the corals and other large animals, there are literally thousands or tens of thousands of species of algae, shrimp, crabs, sea stars, brittle stars, urchins, fishes, barnacles, worms, snails and other mollusks, and other less familiar types of animals.

Ocean Ecosystems

In 1985, the shores of the Big Island were suddenly littered with large numbers of dead and dying fantail filefish. What was happening? An environmental disaster caused by human pollution? This is often the assumption made when such extraordinary events occur, but in this case, the event was quite natural. We had dived many times along this coastline and had never seen this species before. In 1985, however, almost whenever we entered the water there were many fantail filefish to be seen. Not only were there many of this species, but a large percentage of the individuals that were present appeared to be emaciated, as though these animals were seriously short of food. Indeed, many were apparently seeking food and even aggressively swam directly up to us and attempted to bite at our cameras, dive fins, and elsewhere. After we left the islands on this trip, we heard that the "plague" of fantail filefish on the beaches had continued for some weeks and had even grown worse after we left. When we returned to Hawaii a year later, there was, again, not a single fantail filefish to be seen on any of our dives.

How do we know the event was natural? Fortunately, historical records relate that this event occurred infrequently and periodically in the past. In fact, the natives of old Hawaii were reported to have collected the dead filefish when they appeared and to have used their dried bodies as fuel. We also now know from scientific surveys that the fantail filefish is abundant in Hawaiian waters, but, except for periodic very unusual years such as 1985, they are found in deeper waters on the outer edge of the reef beyond recreational scuba diving depths.

What causes large numbers of the fantail filefish to move to shallower inshore waters and starve in some years but not in most others? We simply do not know. There could be subtle changes in the temperature and salinity or in the chemical composition of the

waters that trigger these events. Equally, there could be periodic cycles of reproduction of the fantail filefish itself or periodic cycles of its normal food source. Because so many complex and interacting factors may be involved, the scientific studies that would be needed to unravel this mystery would be lengthy, extensive, and very expensive, and they would not be guaranteed to be successful. Ocean sciences are, at present, not well enough developed for us to understand, in anything more than a very general way, how physical, chemical, and geological characteristics and variations interact to determine the characteristics of various marine ecosystems. What we do know is that marine ecosystems are in a constant state of variation and change—some cyclical, some irreversible—in response to changes in the physical environment both natural and induced by human influence.

The term **ecosystem** encompasses both living organisms and the physical **environment** of a particular volume of space. Within the ecosystem, the components and characteristics of the living and nonliving environment are interdependent, each influencing the other. Hence, conceptually the global environment can be broken down into separate ecosystems within which the **species** that are present, the physical environment, and the relationships among the species and with their environment are distinct and different from those in adjacent areas. For example, we can define each geographically distinct **coral reef** as a separate ecosystem, but all coral reef ecosystems have common characteristics in terms of the types of organisms they sustain, their physical environment, and processes driven by biological–physical interactions. These common characteristics are different from those of other ecosystems, such as **mangrove** swamps, the **rocky intertidal zone**, and the open-ocean **photic zone.**

This chapter briefly reviews several ocean ecosystems, emphasizing the relationship between their individual physical environmental conditions and their biological **communities**. However, the classification of separate ecosystems in the oceans is not precise. In reality, all such ecosystems are linked and interdependent.

17.1 COMMUNITIES AND NICHES

The world ocean can be considered a single ecosystem, but it is too large and complex to be studied as a whole. Consequently, marine **ecologists** often separate the oceans into ecosystems that have common biological and physical characteristics. Because even the simplest of these is complex, marine ecologists study the relationships of species with each other and their environment at a number of different levels, ranging from ocean-sized ecosystems to individual species. These levels are generally not well defined or distinct from each other.

Within an ecosystem, species are distributed in a nonuniform way because of small-scale variations in the physical environment and competition and other interactions among species. Species are often clustered in their communities. The term *community* is inexact and can refer either to all organisms that coexist in a specific location or to groups of species that are found together in many different locations. Thus, a community can be all the species within an entire coral reef, or a specific species of **sea fan** and all species that commonly live in or on it wherever this sea fan is present.

The distribution of species and their relationships with each other and the physical environment can be studied at the species level. In such studies, a very useful concept is the biological **niche.** Each species has evolved to take advantage of certain characteristics of the physical environment. Hence, each species has a range of environmental variables within which it can survive. These variables include **salinity,** temperature, **suspended sediment** concentration, substrate **grain size**, light intensity, and **nutrient** and other chemical concentrations. The combination of the ranges of each of these environmental variables within which the species can survive is the species' **survival niche**. **Figure 17.1a** shows how such a niche can be defined by two environmental parameters. In practice, niches are defined by many more parameters and cannot be fully represented in a two-dimensional diagram.

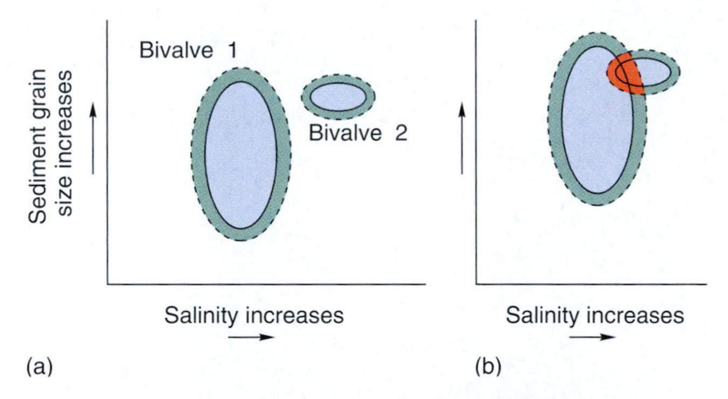

FIGURE 17.1 Ecological niches can be defined in several ways and by many parameters. These diagrams illustrate the concepts using just two such parameters: salinity and sediment grain size. (a) The fundamental niches of two bivalves are defined by the range of values of the two parameters defined by the blue areas. The survival niches are defined by a wider range of environmental conditions and include both the blue and green areas within each dotted line. Because the niches of these two species do not overlap, they would not occur in the same environment. (b) If the niches of the two species overlap, as they do in the red area, they may both occur in environments within this range of the two parameters, but they may also compete such that one species may be excluded from this part of its niche.

Within the survival niche of a particular species is a somewhat smaller range of environmental variables within which the species can both survive and successfully reproduce (**Fig. 17.1a**). This range is often referred to as the organism's **fundamental niche.** Species rarely, if ever, occupy every area within the marine environment where environmental parameters are within their fundamental niche. Two factors restrict the distribution of a species within its fundamental niche. First, each species has a smaller niche within its fundamental niche in which it finds optimal conditions. Second, species with overlapping niches may compete, and one species may be excluded from parts of its fundamental niche (**Fig. 17.1b**).

The interaction of species with each other and with the environment is wonderfully complex and may never be fully understood for all species in the oceans. Thus, descriptions of ecosystems in the discussion that follows represent only a cursory view of the factors that contribute to the amazing profusion and **diversity** of species in the oceans.

17.2 CORAL REEFS

Although some types of **corals** are present in all parts of the oceans, including Arctic seas, species of corals that build coral reefs, called **hermatypic corals,** grow only in areas where the water temperature never falls below about 18°C. Coral reefs are thus restricted to a broad band of mostly tropical waters between about 30°N and 30°S (**Fig. 17.2**). The range of coral reef occurrence extends into somewhat higher **latitudes** on the western side of each ocean because warm **western boundary currents** flow poleward.

Environmental Requirements for Coral Reef Formation

Reef-building corals require an appropriate substrate on which to attach, and they have a **symbiotic** relationship with **zooxanthellae,** a nonmotile form of **dinoflagellate** that lives within the coral's tissues. Because zooxanthellae need light to **photosynthesize** (**CC14**), living coral reefs are present only in waters where the seafloor is within the photic zone. In clear waters, corals can grow to depths of about 150 m, but in high-**turbidity** waters coral growth is reduced or prevented by two mechanisms: First, the higher turbidity reduces light penetration and limits the depth at which zooxanthellae can photosynthesize and, therefore, also limits the depth at which corals can grow. Second, large quantities of suspended sediment that cause high turbidity may smother the corals. Corals can clear away a certain amount of sedimented material, but they must use energy to do so, and they lose additional energy because they must also stop feeding. When these energy costs are too high, the corals cannot survive. Therefore, coral reefs do not grow in coastal areas near river mouths or near other sources of large amounts of suspended matter, such as dredging

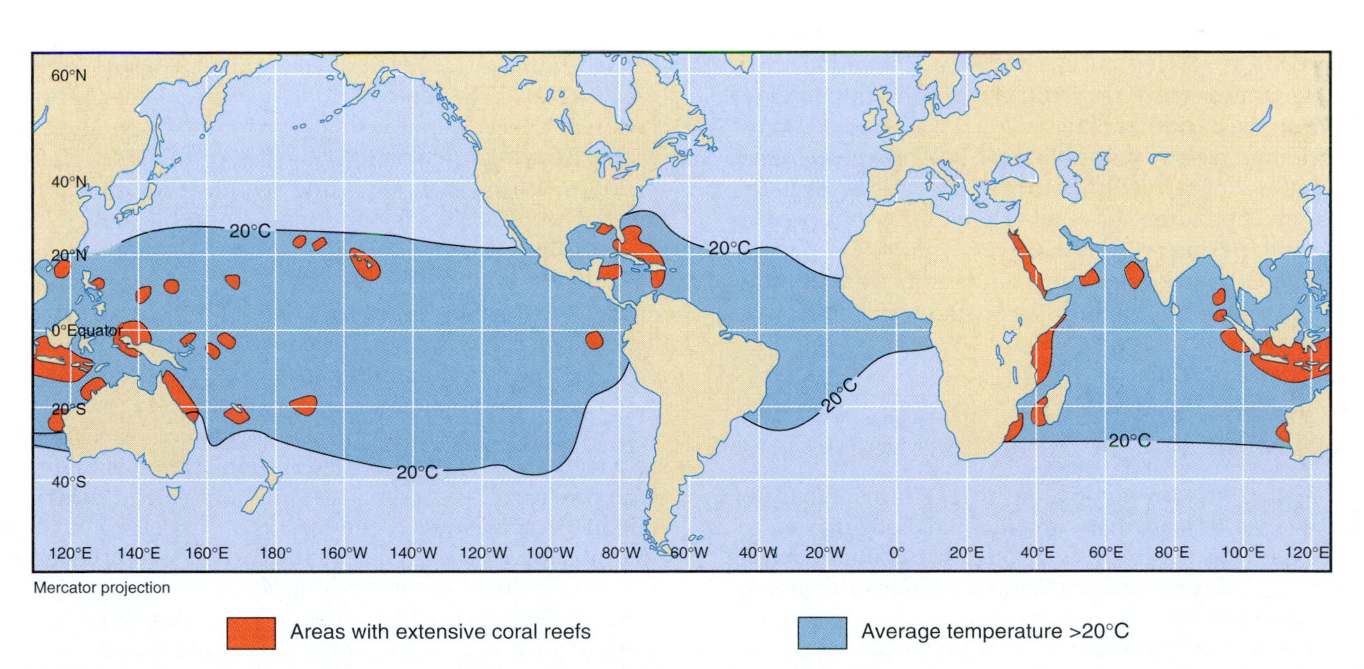

FIGURE 17.2 The area of the oceans with average sea surface temperatures above 20°C extends to about 30° north and south of the equator in the Pacific and Indian oceans, and somewhat less in the Atlantic Ocean. The temperature never falls below 18°C in most of this area. Hermatypic corals grow only in these areas and where other environmental conditions are suitable within these areas. Although many small tropical islands have extensive coral reefs, they are too small to show in this figure.

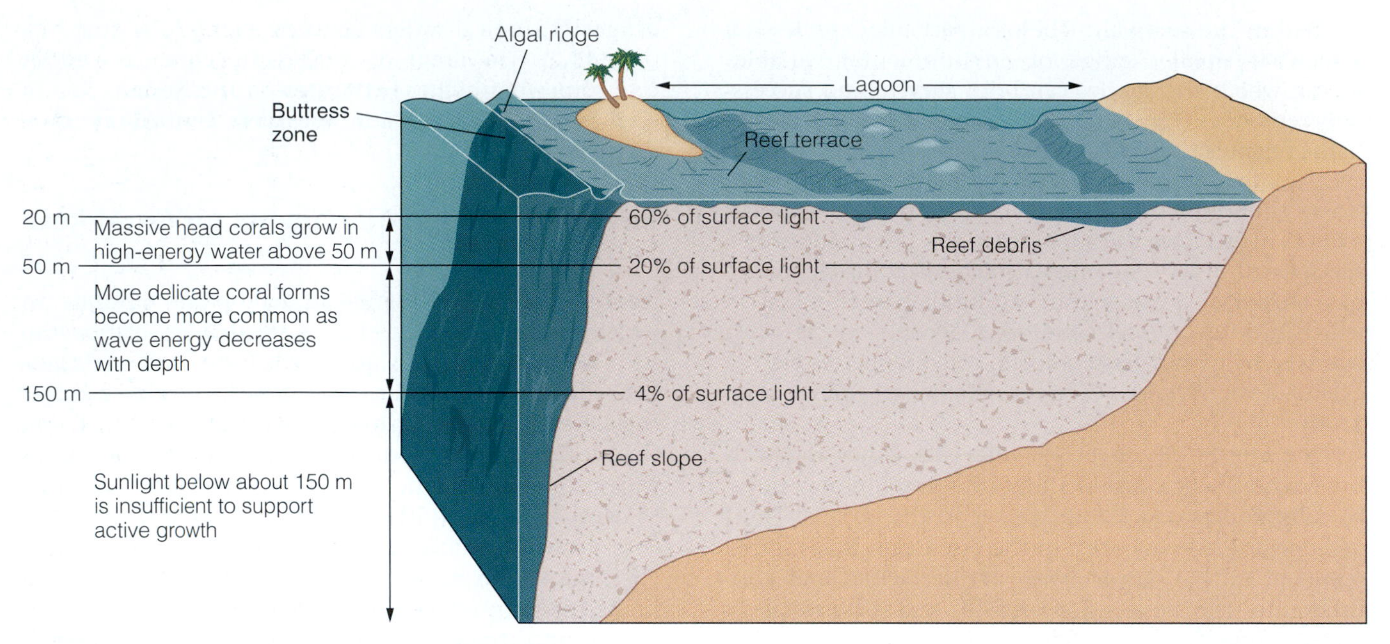

FIGURE 17.3 Structure of a typical coral reef.

projects. Corals also grow poorly in water of low or variable salinity.

The physical conditions necessary for coral growth (clear, warm, shallow waters with relatively invariable normal ocean salinity) are present primarily between about 23.5°N and 23.5°S. This is a region that we might expect to be biological desert because surface waters are isolated by a steep permanent **thermocline** throughout most of the tropical oceans. In contrast, instead of biological deserts, coral reefs are areas of very high productivity with an amazing diversity (**CC17**) of fish and **invertebrate** species. In what may seem to be a paradox, waters surrounding reefs are clear blue and have extremely small populations of **phytoplankton**. The waters over the reef itself have somewhat higher phytoplankton populations and **primary productivity,** but these populations are extremely small in comparison with those in **upwelling**-zone ecosystems and cannot account for the high productivity of coral reefs.

Factors Affecting Coral Reef Productivity

The reasons for the anomalously high productivity of coral reefs are somewhat complex and not fully understood, but they appear to be related to a combination of physical conditions and the unique relationships among the reef organisms.

First, hermatypic corals and their associated zooxanthellae act together in a mutualistic (mutually beneficial) relationship to create a very effective mechanism for collecting, concentrating, and rapidly recycling nutrients. Zooxanthellae live embedded in the coral's tissues and use solar energy that penetrates the coral's transparent tissues to produce food by chemically recombining the coral's waste products. Carbon dioxide and nutrients, which are released by the coral through its digestive processes, are transferred directly to zooxanthellae and converted by photosynthesis to plant organic matter. In turn, as much as 60% of the organic matter created by the zooxanthellae through photosynthesis is released through the plant cell wall directly into the coral tissue, providing food for the coral. In this way, nutrients are continually recycled, and food is continually produced by the **algae** and consumed by the coral. The coral–zooxanthellae association ensures very little loss of either nutrients or food to the surrounding water. Corals feed on **zooplankton** to supplement the food supplied by their internal zooxanthellae. Thus, the small amounts of nutrients that are lost from the coral–zooxanthellae partnership are continuously replaced. This efficient nutrient retention and recycling mechanism is believed to be the main reason for the high productivity of coral reefs.

The second reason for their high productivity is that most reefs are built on the sides of submarine mountains or on the fringes of landmasses with very narrow **continental shelves.** Indeed, the growth of corals off **coasts** where sea level has risen during the past several thousand years has created very steep drop-offs at the outer edges of many reefs (**Figs. 4.27, 17.3**). Ocean **currents** that flow past such steep continental shelves and islands form **eddies.** The eddies create vertical water movements that can bring nutrient-rich deep water, at least episodically, into the photic zone of the reef.

The third reason for the high productivity of coral reefs is that many reef ecosystems are partially closed systems, within which most nutrients not retained in the coral–zooxanthellae association are continually recycled. Most of the reef floor is shallow and within the photic zone. Hence, organic **detritus** created on the reef settles largely on the reef floor, where it is consumed by **decomposers** that release nutrients to be recycled. In addition, most fishes are permanent residents of the reef. Nutrients in their urine and feces are released almost entirely back to waters of the reef, where they are rapidly recycled and taken up again by zooxanthellae or phytoplankton.

Primary Producers in the Coral Reef Community

Although zooxanthellae are important primary producers in coral reef ecosystems, they account for only a small proportion (generally less than 5%) of the **biomass** of photosynthesizers in these ecosystems. Reef ecosystems are dominated by **benthic** or **encrusting microalgae** and a variety of attached **macroalgae** (**Fig. 15.3**). The biomass of these plants generally exceeds the animal biomass in the reef ecosystem by as much as three times. In fact, the **hard parts** of **calcareous** algae are responsible for building much of the reef structure.

Algae attached to the solid substrate of the reef are favored over phytoplankton in coral ecosystems because nutrients are recycled, made available, and rapidly reassimilated at these surfaces. Attached algae remove most of the recycled nutrients before they can **diffuse** into the water column and become available to phytoplankton. In addition, attached algae remain in the reef ecosystem, where nutrients are available, whereas phytoplankton may be transported away from the reef into nutrient-depleted adjacent deep water.

Coral Reef Niches and Topography

Coral reef ecosystems contain a bewildering variety of species. These species are distributed in niches that are defined by current and wave action, and by variations in salinity, water depth, turbidity, temperature, and other factors.

Figure 17.3 shows the general **topographic** features of a typical coral reef. In shallow waters of the **lagoon,** currents are generally weak and there is little wave action. Hence, **sediments** tend to accumulate in the lagoon and are removed primarily during major storms. Lagoon sediments sustain a wide variety of invertebrates that feed on suspended or deposited detritus. These detritus feeders include **suspension feeders** such as sea pens (**Fig. 16.5a**), and a variety of bivalve **mollusks** that live in the sediment and extend their feeding apparatus into the water column. Also present are many species of **deposit feeders,** including mollusks and worms that live in and sift through the sediment, **sea cucumbers** (**Fig. 17.4a**), urchins (**Fig 17.4c**), sand dollars (**Fig 17.4d**), and other species that feed on benthic algae growing on the sediment surface, and predators such as goatfishes (**Fig. 16.12a**) and **sea stars** (**Fig. 17.4b**) that hunt mollusks and other animals living in or on the sediment.

In the lagoon, the growth of most corals is inhibited by the blanket of sediment that covers the seafloor. However, certain corals, such as *Acropora* (**Fig. 17.4e,f**), that grow up into the water column can prosper if **sedimentation rates** are sufficiently low and sediments are rarely **resuspended** by waves. Once established, these corals provide a sediment-free substrate for other species. Consequently, in quiet lagoons with little suspended sediment input and where salinity is not altered by freshwater input, irregularly shaped mounds of coral and associated species develop. These are called "patch reefs."

At the lagoon's outer edge is a **reef flat**, or **reef terrace,** that is relatively free of sediments because they are swept off the terrace by waves. It is an ideal location for coral growth. Coral grows upward until the reef terrace is only a few centimeters below the **low-tide line.** Further growth is inhibited because corals cannot survive for long periods out of water, although the surface of the reef terrace may be completely exposed to the atmosphere for short periods during low **spring tides** without killing the corals. Reef terrace corals are generally encrusting corals because the water is too shallow and the wave energy too high for corals that grow in other forms (e.g., **Figs. 16.7a,b, 17.4e,f**). The surface of the reef terrace is not smooth. It has many grooves and holes created primarily by invertebrates that eat or drill into the coral to obtain food or to create safe areas to shelter from larger predators.

On some reefs, a low island is formed by sediment accumulated during storms at the landward side of the

CRITICAL THINKING QUESTIONS

17.1 Sea otters are increasing in numbers in California coastal waters, but one of their favorite foods, abalone, is heavily harvested, so they may rely more on urchins for their food supply. Describe what you think will happen to the kelp forests over the next several decades as a result of these changes, and how these changes will affect other species in the ecosystem.

17.2 Describe the feeding, hunting, defensive, and reproductive characteristics that are desirable for carnivorous species that live in the *Sargassum* community. Compare and contrast these characteristics with the desirable characteristics for a carnivorous benthic species that lives in the deep oceans. Explain each of the differences.

FIGURE 17.4 Inhabitants of the coral reef lagoon and outer reef flat. Many invertebrates—including (a) sea cucumbers (family Synaptidae, Indonesia); (b) sea stars such as this rhinoceros, or horned, sea star (*Protoreaster nodosus,* Papua New Guinea); (c) urchins (*Astropyga radiata,* Indonesia); and (d) the urchins' close relatives the sand dollars (*Clypeaster* sp., Indonesia)— feed on algae, detritus, and microorganisms on sandy seafloors. (e) Spiked forms of hard corals, such as *Acropora* sp. (Papua New Guinea), grow in patches on sandy lagoon floors. (f) Elkhorn coral (*Acropora palmata,* Belize) grows in many areas on the outer reef flat that experience low wave energy.

reef terrace (**Fig. 17.5**). This island may be well enough established to support palm trees. The seaward edge of the reef usually has an irregular ridge, parts of which are shallow enough to emerge from the water, especially at low **tide.** The ridge is formed by intense wave action that periodically smashes against the reef's outer edge and dislodges large chunks of the **limestone** substrate of the reef. The chunks are cemented back onto the reef by the **calcareous** algae that live in abundance in this region. Calcareous algae are abundant because wave energy is too intense for corals to grow effectively. They cement themselves to the reef surface with their calcium carbonate hard parts. The cemented algae can withstand intense wave action and can quickly colonize any new surface created by storm wave damage. These algae benefit from a continuous supply of very low concentrations of nutrients brought to the reef edge by currents and waves. Because calcareous algae are abundant and help to maintain this ridge, it is called the **algal ridge.**

On the few reefs where wave energy is very low, no algal ridge or reef terrace is present. The outer reef may consist primarily of relatively robust massive corals, such as elkhorn coral (**Fig. 17.4f**), or the reef terrace may simply end at the reef edge. Stands of elkhorn coral characterize the seaward edge of several sheltered Caribbean reefs. Some reefs within tectonically active island chains, such as Palau in the Pacific Ocean, are protected from wave action by a **barrier reef** that surrounds groups of several or many islands. Inside some of these barrier reefs, the sheltered island **shore** has a **fringing reef** with a reef terrace that simply ends abruptly at the reef wall, which plunges vertically or even undercuts to depths exceeding several hundred meters (**Fig. 17.3**).

Seaward of the algal ridge (or reef terrace if no algal ridge is present), the downward slope of a reef can range from very gradual to nearly vertical. The upper part of the outer slope (**Fig. 17.3**), called the "buttress zone," is subject to strong scouring action of waves down to about 20 m. It is usually cut across at intervals by deeper channels separated by relatively high ridges, sometimes called a "tongue and groove" formation. The channels or grooves are cut by waves and may also be called "surge channels." Sand and debris created by wave action on the reef are transported seaward through these grooves. Coral growth is limited in the grooves, but many invertebrates and fishes feed on detritus transported along them. Coral growth is more extensive on the relatively sediment-free ridges. However, because wave energy is high, these ridges sustain primarily robust massive varieties of coral, such as brain coral (**Fig. 16.7a**), and encrusting corals and algae.

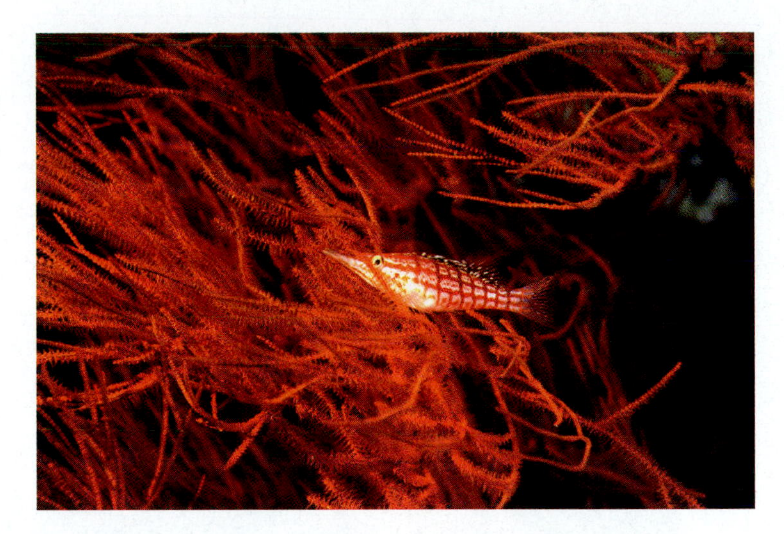

FIGURE 17.6 Black coral colony (probably *Cirrhipathes* sp.) with a longnose hawkfish (*Oxycirrhites typus,* Papua New Guinea) hiding among its branches. The hard parts of black corals are a deep black color and are often used to make jewelry, but the polyps may be a variety of colors in the living coral.

FIGURE 17.5 Islands often form behind the outer reef flat as wave-driven debris accumulates in this area. The island can become high and stable enough to sustain vegetation, such as the palm trees and other plants on this island on a fringing reef in Fiji.

Farther seaward, on the outer slope between about 20 and 50 m, is a transition from robust massive corals and encrusting species to still strong but less robust forms, such as *Acropora* (**Fig. 17.4e,f**), and then to more delicate varieties of corals, including black coral (**Fig. 17.6**), sea fans (**Fig. 16.7f**), and **soft corals** (**Fig. 16.7d,e**). The depth at which these transitions occur depends on the intensity of wave action. Delicate corals are present at shallower depths on **leeward** sides of coral-fringed islands than on **windward** sides.

Between about 50 and 150 m, the reef is dominated by delicate coral species that grow outward in slender fingers or arms. Such growth enables hermatypic corals to extend beyond the shadows of their neighbors in the never-ending competition to obtain light for their **zooxanthellae**. It also enables nonhermatypic corals, including the delicate and beautiful soft corals (**Figs. 16.7d,e**), to extend into the water that flows along the reef face to capture suspended food. Soft corals are most abundant in areas where currents are strongest and thus expose them to the greatest possible food supply.

Just as the types of coral that grow on the reef's seaward side are determined primarily by depth and the intensity of wave action, species of invertebrates and fishes that inhabit or feed on the corals change with these factors and in response to changes in coral species. The variations are too complicated to review here but can provide fascinating study for **scuba** divers when no big animals such as sharks, manta rays, or turtles are present to capture their interest.

17.3 KELP FORESTS

The name **kelp** is given to a number of species of brown macroalgae that are attached to the seafloor at depths as great as 30 to 40 m, where the water is clear enough to allow light to penetrate to that depth. Once established, kelp plants grow toward and eventually reach the ocean surface. Where conditions are suitable, kelp grows so densely that it forms forests (**Fig. 17.7**).

Extensive forests of one such kelp, *Macrocystis*, are present off the west coast of North America. Kelp fronds are **buoyant.** *Macrocystis* fronds, for example, have gas-

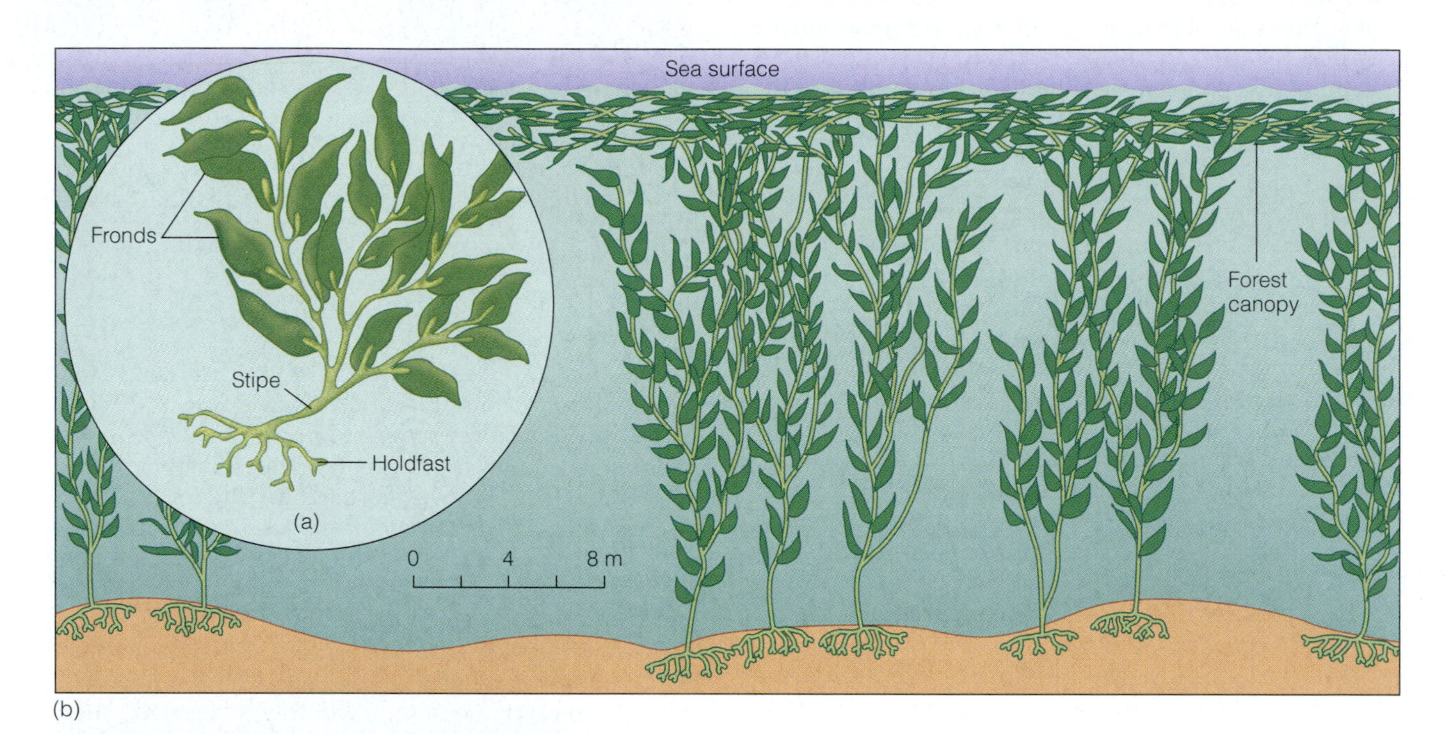

FIGURE 17.7 Kelp forest. (a) Kelp, such as this *Macrocystis,* is attached to the seafloor by holdfasts and has long flexible stems, called "stipes," that support the fronds in which photosynthesis occurs. (b) In some areas of abundant kelp, the ocean surface is covered by a canopy of kelp fronds, and the kelp growing upward from the seafloor forms an underwater forest. (c) The sunlight filters through the canopy and kelp forest in this underwater photograph from about 8 to 10 m depth in Monterey, California.

containing sacs or bladders distributed along them (**Fig. 14.3b**). Once a frond reaches the surface, it continues to grow, but the new growth floats on the surface to form dense mats or a canopy. These canopies can cover the entire ocean surface within the species' depth range along substantial stretches of coast.

CRITICAL THINKING QUESTIONS

17.3 Hydrothermal vent communities throughout the oceans are separated, sometimes by substantial distances. However, some of the same species are found at many of the vents, regardless of their location. In addition, chemosynthetic bacteria species that are identical or very closely related to those found at the hydrothermal vents on oceanic ridges have been found in hydrothermal vent environments at the bottom of Crater Lake, Oregon, and Lake Baikal, Russia.

(a) What are the possible reproductive strategies that could explain these observations?

(b) Hypothesize ways that chemosynthetic organisms could be transported between their present locations and the locations of new vents or the locations of new lakes similar to Crater Lake and Lake Baikal.

(c) What studies do you think would be needed to prove or disprove your hypotheses?

Kelp Community Environmental Characteristics

Kelp forests grow where the temperature and nutrient characteristics of the water column are very different from those required for coral reef formation. Kelp requires water cooler than 20°C and high nutrient concentrations. Although kelp is attached to the seafloor by holdfasts, it obtains its nutrients from the water column through the surface of its fronds. Kelp fronds grow as much as half a meter a day, and kelp forest primary productivity ranges from 500 to 1500 g of carbon per square meter per year. This rate exceeds the primary productivity in all but the most highly productive phytoplankton-based ecosystems and is approximately the same as **primary production** rates of terrestrial farms. Large quantities of nutrients are necessary to sustain such growth rates. Consequently, kelp forests are almost exclusively restricted to areas of intense upwelling. The fact that upwelled waters are cold probably accounts for kelp's temperature niche requirements.

Besides low temperatures and high nutrient concentrations, kelp needs a stable, generally rocky, seafloor to which it can attach. In addition, kelp can grow only in shallow waters where sufficient light penetrates to the seafloor to support the growth of newly settled plants.

Kelp Life Cycle and Communities

The kelp life cycle consists of two distinctly different stages. Although the large kelp plants that form kelp forests grow vegetatively (asexually), they also produce microscopic **spores** that germinate to form a life stage that reproduces sexually. The microscopic plants that result from this sexual reproduction become **plankton** and eventually settle to the seafloor. If they encounter suitable conditions, they grow asexually until they are full-sized and mature. The life cycle then begins again.

Kelp forests support amazingly diverse populations of invertebrates, fishes, and **marine mammals.** The kelp provides several benefits to species that live within its forests. First, kelp produces large quantities of detritus, which forms the base of the **food web.** Surprisingly, only a few animals, including only a few fish, snail, and **sea urchin** species, eat kelp itself, but kelp is easily torn apart, particularly at the ends where new growth occurs. Therefore, kelp releases large amounts of detritus, which is first modified by decomposers and only then consumed by animals. Second, the kelp canopy provides a hiding place and protection from predators, particularly seabirds. The fronds afford many escape routes that enable agile harbor seals and sea lions to evade their shark and killer whale predators, which are less agile and less able to maneuver through the kelp forest. In addition, kelp fronds and holdfasts provide substrate for encrusting organisms and their **grazers** or predators and many secluded places for small fishes and invertebrates to hide from predators.

The rocky seafloor and holdfasts of the kelp forest also sustain a surprisingly diverse community of nonhermatypic corals, **anemones,** crabs, shrimp, sea stars, **nudibranchs,** fishes, and other animals (**Fig. 17.8**). In many ways, kelp forest communities rival those of coral reefs in complexity and beauty.

Kelp, Sea Otters, and Sea Urchins

Two of the most important residents of the kelp community are sea otters (**Fig. 17.9a**) and sea urchins (**Fig. 17.9b**). Sea urchins eat kelp, and sea otters eat sea urchins. Because of this relationship, healthy and abundant kelp forests depend to a large extent on a healthy population of sea otters. In areas where otters are abundant and many sea urchins are eaten, sea urchin populations are low and dominated by small, young individuals. Such populations are too small to affect the kelp significantly (**Fig. 17.9c**). However, if otters are scarce, the sea urchin population multiplies and many large sea urchins are present to graze heavily on the kelp and reduce its

FIGURE 17.8 Members of the kelp holdfast community. (a) These small brown cup corals (*Paracyathus stearnsii,* California) cover many of the rocks at shallow depths. (b) Anemones, such as this rose anemone (*Tealia lineata,* California), are abundant on the kelp forest floor. (c) Many species of crabs, such as this northern kelp crab (*Pugettia producta,* California), feed on the abundant detritus and other animal life. (d) A kelp shrimp is camouflaged to look like the algae in which it lives and on which it feeds (Monterey, California). (e) Sea stars of many species live on the kelp forest floor, and sea pens are often found in the patchy areas of seafloor that are located in and around the forest (Monterey, California). (f) This horned nudibranch species (*Hermissenda crassicornis,* California) is common in the kelp community, together with many other nudibranch species. (g) Many rockfish species live and hunt in the kelp forest, such as these two: the quill-backed rockfish (*Sebastes maliger*) on the lower left, and the copper rockfish (*Sebastes caurinus,* California) swimming above, further to the right.

abundance (**Fig. 17.9d**). Large sea urchins, in particular, feed preferentially on the kelp holdfasts and may destroy the forest by cutting the kelp loose from the seafloor, even if they do not consume it.

The California coast was once almost completely fringed by kelp forests, but they were severely depleted by the early years of the twentieth century and have only recently begun to recover. The primary reason for the kelp forest decline is now understood to be hunting of sea otters during the eighteenth and nineteenth centuries, which drove the otters almost to extinction. Once the otters were removed, sea urchin populations increased and steadily overwhelmed and destroyed the kelp forests. Now sea otters are protected and their populations are slowly recovering. As a result, sea urchin populations are declining and kelp forests are gradually returning. The sea otter is called a "keystone predator" because, without it to prey on urchins, the urchin population growth produces large and fundamental changes in the ecosystem. Keystone predators are found in many ecosystems.

17.4 ROCKY INTERTIDAL COMMUNITIES

The rocky intertidal community that lives between the **high-tide line** and the low-tide line on rocky coasts is unique in two important respects. First, this community can be easily observed and studied by anyone who visits a rocky coast at low tide. Second, the community is normally arranged in well-defined depth zones that run parallel to the shore at different heights above the low-tide line. Each zone is populated by different species of plants

FIGURE 17.9 The sea otter–urchin–kelp relationship. (a) Sea otters (*Enhydra lutris*) were at one time abundant in the kelp forests of California, but their numbers plummeted during the nineteenth century because they were exploited for their pelts. Their populations are now recovering. (b) The sea otter's favorite and most important food is sea urchins, such as this red urchin (*Strongylocentrotus* sp., California). (c) At Amchitka Island, Alaska, where sea otters are abundant, sea urchins are rare in shallow water because they are preyed upon by the otters. As a result, there is little browsing of the kelp forests by urchins and the kelp is abundant. (d) In contrast, at nearby Shemya Island, where otters have become scarce, urchins are abundant and kelp is very sparse as a result of browsing by the urchins.

and animals. Species that live in each zone are determined by a combination of physical conditions of the environment and competition among species that have different tolerances to these conditions.

The most important characteristic of the rocky intertidal zone is the degree of exposure of the substrate to the atmosphere when the tide recedes. Unlike organisms that inhabit sandy **beaches** or mudflats, most rocky intertidal species cannot bury themselves in the rocky substrate to survive periods when they are exposed by the tide. Once exposed to the atmosphere, organisms are subjected to a number of stresses, such as

- Temperature variations and extremes that far exceed those in seawater
- Variable water exposure (spray or rain) and **humidity,** and the consequent variable tendency to lose body fluids by evaporation
- Variable salinity because the remaining pockets or pools of water may be subject to evaporation and/or dilution with rainwater or snow
- Variable oxygen concentrations and **pH** because physical and biological processes quickly alter these conditions in the small volume of seawater that remains in contact with the organism, and this water is not renewed until the water level rises again
- Predation by birds and terrestrial animals

Zonation of communities on a rocky intertidal shore is critically dependent on the frequency and duration of their exposure to the atmosphere and, therefore, to these stresses (**Fig. 17.10**).

The shore can be divided into four zones on the basis of atmospheric exposure time (**Fig. 17.11**). The highest zone, the **supralittoral zone,** is above mean higher high water (see Chapter 12) and thus essentially permanently out of the water, but it is frequently wet with seawater spray from breaking waves, at least during high tides. Below the supralittoral zone, the **high-tide zone** is covered with water for parts of the tidal cycle but remains exposed to the atmosphere most of the time. The **middle-tide zone** is usually covered by water, but it is exposed to the atmosphere during all or most low tides. Finally, the **low-tide zone** is water-covered almost permanently and exposed to the atmosphere only briefly during the lowest low tides. The boundaries between these zones are sometimes sharp, as revealed by abrupt changes in the biological communities that they support.

Even when covered by water, the zones of the rocky shore are subject to different physical conditions, particularly wave-induced **turbulence** and scour. Wave-induced turbulence decreases with depth but can still be substantial well below the low-tide zone.

Rocky coasts are present throughout the oceans, and each has its own unique communities with different species and species interactions. Consequently, only the general types of **flora** and **fauna** in rocky intertidal zones are described here. For most coastal areas, easy-to-read guides to shore and **tide pool** creatures can be found in local bookstores.

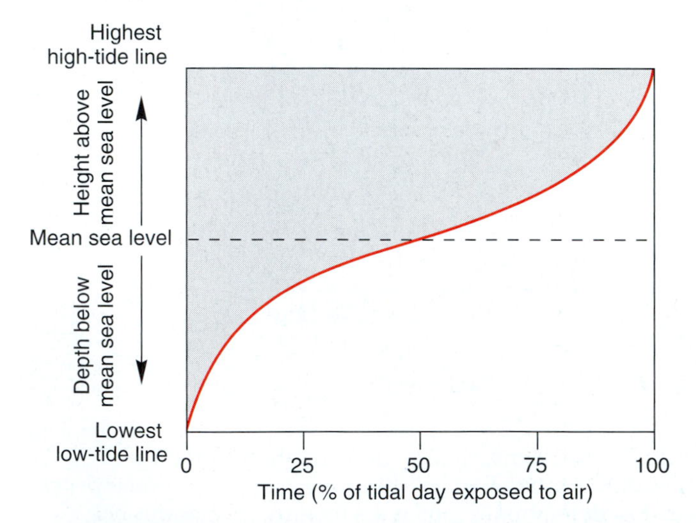

FIGURE 17.10 The length of time that a particular point on the foreshore (and the organisms that live at that point) are exposed to the air during the tidal cycle varies with height above the low-tide line.

Supralittoral Zone

The supralittoral zone is either permanently exposed to the atmosphere or covered by water only during occasional extreme high tides or **storm surges**. Hence, this zone can be considered land and not part of the ocean ecosystem. Nevertheless, the continuous or frequent spray of seawater that reaches this zone provides moisture and nutrients that support the growth of **lichens,** encrusting **blue-green algae**, and small tufts of various green algae. These plant communities are sparse but provide food for a variety of animals, including species of periwinkles (marine snails), other marine snails, **limpets, isopods,** and crabs (**Fig. 17.11**).

Periwinkles, other snails (**Fig. 17.12a**), and limpets are grazers that feed on algae encrusting the rocks. The periwinkles are often species of the **genus** *Littorina*. Many *Littorina* species are well adapted to life in the supralittoral zone. Some "breathe" air and may even drown if fully immersed in water for long periods. Supralittoral species of *Littorina* are **viviparous,** giving birth to live young. In contrast, the marine species of this genus that live completely submerged deposit eggs on the rocks or release them to the water. Periwinkles can completely withdraw into their shell and seal off the opening with an operculum, a rigid disk of hornlike material. This seal is watertight and provides protection from predators and especially from dehydration during long periods of exposure to the atmosphere.

FIGURE 17.11 The rocky intertidal shoreline can be separated into four zones based on length of time exposed to the atmosphere. Many species have ecological niches that restrict them to only one of these zones, although closely related species may occupy different zones. The flora and fauna in this figure are typical of a mid-latitude, cold-water, rocky shoreline. Generally, many more species occur in each zone than are shown here. Competition between species often determines the vertical limits of a particular species' distribution within the zone that it occupies.

FIGURE 17.12 Animals and algae of the rocky intertidal zone. (a) Group of black turban snails (*Tegula funebralis,* Monterey, California). (b) A limpet (order Archaeogastropoda, California). (c) Barnacles in the high intertidal zone, often called buckshot barnacles. These are probably primarily *Chthamalus* sp., with some interspersed *Balanus glandula.* (d) Lined chiton (probably *Tonicella lineata,* California). (e) Dense bed of California mussels (*Mytilus californianus,* Monterey, California). (f) These green algae are in the upper intertidal zone and have been out of water and in the sun for several hours. The algae survive this experience on a regular basis with no harm (south of San Francisco, California). (g) A hermit crab (*Pagurus* sp., California) living in its borrowed gastropod shell.

Limpets (**Fig. 17.12b**) are not able to withdraw entirely into their shell. Instead, they cling to the rocks with their bodies and pull their shells tightly down on top of them. If the shell edge fits well with the surrounding rock, the animal is sealed inside, where it is protected from predators and desiccation. Some limpets rasp the rock away to create a perfectly fitted platform for their shell. When immersed or wet, limpets may move over the rocks to graze, but they return to their prepared location during periods of prolonged exposure to the atmosphere. The limpet's bond to the rock is so strong that it cannot easily be turned over, and thus its soft body is protected from predators such as birds. The limpet's defensive action of clinging tightly to the rock is easy to observe. Limpets can be surprised and easily removed from their rock location by a quick sideways stroke of a chisel or blade. However, if the limpet is warned, perhaps by a light tap from a finger, it clamps down and is impossible to remove without breaking the shell.

Isopods often live in the supralittoral zone in great numbers. These animals are **scavengers** that feed on organic debris. They are rarely seen by most visitors to the shore because they remain in hiding places within the rocks by day and emerge to feed only at night. Various species of crabs also inhabit the supralittoral zone. Many are scavengers, but some are grazing **herbivores** or predatory **carnivores.**

Each zone of the rocky intertidal ecosystem varies in width according to factors that include the slope of the rocks and the **tidal range.** The width of the supralittoral zone is also affected by factors that determine the degree of wetness of the rocks, including the intensity of wave action, the average air temperature and humidity, and the location, roughness, and orientation of the shore (which control the degree of shading from the sun). For example, the supralittoral zone is wide in areas such as central California and Maine, where cool, damp, and foggy days are common.

High-Tide Zone

In the high-tide zone, organisms are exposed to the atmosphere for long periods and, like inhabitants of the supralittoral zone, must be able to withstand extremes of temperature and salinity and must be protected from dehydration. In addition, these organisms must be able to withstand severe wave-induced turbulence. To offset these disadvantages, a reliable supply of nutrients, plankton, and suspended organic particles becomes available each time the rocks are covered by ocean water. Consequently, the high-tide zone can support species that cannot tolerate the limited nutrient supply and prolonged dry periods of the supralittoral zone.

Rocks of the high-tide zone support many species of encrusting algae, particularly in the upper parts of the zone. In contrast to the supralittoral zone, this zone has several species of macroalgae. All these species have thick cell walls, which give them a leathery feel and protect them from excessive loss of water by evaporation when exposed to the atmosphere (**Fig. 17.12f**). They are firmly attached to the rocks by holdfasts, and their stipes (branches) are extremely flexible, enabling them to withstand wave turbulence. Although they are eaten by some species of snails, crabs, and other animals, these algae are tough and difficult to digest, and they contribute to the food web primarily in the form of detritus. The detritus is formed when parts or all of the algae are first broken loose from the rocks by waves or grazers and then broken down or modified by decomposers.

In many areas, the top of the high-tide zone is marked by a band dominated by small **barnacles** called "buckshot barnacles" (**Fig. 17.12c**). Several larger species of barnacles are present lower on the shore in the middle-tide zone. Because barnacles are suspension feeders, they cannot live above the highest high-tide line. However, they can survive in locations where they are able to feed for only a few hours on the few days of spring tides each month when they are immersed in water. When exposed to the atmosphere, the barnacle withdraws into its hard shell, which protects it from dehydration, predators, and wave-induced turbulence.

Because organisms of the high-tide zone, like species in the supralittoral zone, must withstand extended exposure to the atmosphere, the two zones sustain many similar species, including periwinkles and limpets. Toward the lower end of the high-tide zone, periwinkles and limpets become less abundant, whereas chitons and mussels become more abundant. This change marks the transition from the high-tide zone to the middle-tide zone. Chitons (**Fig. 17.12d**) feed and attach themselves to the rocks in the same way as limpets, but they have shells made up of eight separate connected plates. Mussels (**Fig. 17.12e**) are suspension feeders like barnacles, but they have a two-piece shell attached to the rocks by a network of strong threads called "byssal threads." These threads are formed by a liquid secreted from the mussel's foot that hardens in seawater. The threads are attached between the mussel shell and the rocks or, in dense mussel beds, between one shell and another. Periwinkles, barnacles, limpets, mussels, and other species living attached to rocks of the high- and middle-tide zones tend to have rounded shells, which can best withstand and dissipate the turbulent impacts of waves.

The lower limit of the zone inhabited by a rocky intertidal species is determined for many species by competition from other species. In contrast, the upper limit of the inhabited zone is normally determined by the tolerance limits of the species' fundamental niche.

Middle-Tide Zone

Macroalgae are generally less abundant or absent in the middle-tide zone because of competition by mussels and barnacles. In a middle-tide zone newly formed by vertical movements of the coast during earthquakes or by **lava** flows, or in a middle-tide zone partially denuded by extreme storms or **pollution** events, macroalgae quickly establish themselves and cover the rocks. However, as mussel and barnacle **larvae** settle and grow into new colonies, the macroalgae are steadily overcome and eventually disappear from the zone.

Within the middle-tide zone, the lower limit of mussel beds is determined by competition from predatory sea stars, which can grip and slowly pull open a mussel with the many tube feet on their undersides. On many shores, mussel beds terminate abruptly at their lower limit, almost as though the mussels were incapable of growing below that depth. However, this limit is simply the depth at which recolonization by mussels is less effective than predation by sea stars (whose fundamental niche does not extend as high up the middle-tide zone as the fundamental niche of the mussels does). Although some sea stars do prey on mussels within the mussel bed zone, they must withstand stresses associated with being at the limits of their fundamental niche (e.g., atmosphere exposure and turbulence), or they must migrate up and down the shore's zones with the tides. Hence, mussels are able to outcompete the predatory sea stars in this upper zone.

The mussel beds on many rocky intertidal shores are ideal **habitat** for a variety of algae and animals that include **hydroids**, worms, snails, clams and other mollusks, and crabs and other **crustaceans**. Acorn barnacles are interspersed in the mussel beds. These barnacles, and to a lesser extent the mussels, are eaten by snails that drill through the barnacle plates or mussel shell or force the mussel shell open to get to their prey. The middle-tide zone is also populated by numerous species of **hermit crabs** (**Fig. 17.12g**) and a number of species of anemones that can withstand periodic exposure to the atmosphere.

Low-Tide Zone

The low-tide zone sustains a variety of macroalgae and encrusting algae nourished by nutrients brought to them with each tide. In this shallow zone, algae have ample light, even when turbidity is relatively high. In addition, because they are exposed to the atmosphere for only short periods at low tides, they do not need the protection against desiccation that algae in higher zones require. Although the low-tide zone, unlike zones higher on the shore, is dominated by algae, it sustains numerous species of animals, many of which use the abundant mats of algae for shelter. Animals in the low-tide zone are similar to those of the kelp community (**Fig. 17.8**) and include anemones, **sponges**, sea urchins, nudibranchs, shrimp, sea stars, crabs, sea cucumbers, and fishes. These organisms are infrequently exposed to the atmosphere, so they include delicate forms that would be dehydrated or thermally shocked if exposed for more extended periods.

The principal physical hazard in the low-tide zone is wave-induced turbulence. Some species withstand this turbulence by attaching themselves to the rocks. Other species are active swimmers or simply use macroalgae or cracks and holes in the rock as protection from wave action. Because the low-tide zone has abundant macroalgae and a continuous supply of suspended detritus and plankton, the low-tide community comprises many species of **filter feeders**, detritus eaters, scavengers, grazers, and carnivores.

Tide Pools

On many coasts, the rocky shore is convoluted or pitted sufficiently that seawater remains in many depressions, even when the tide recedes. These depressions are called "tide pools." Depending on the size, location, and permanence of the tide pool, any of the species of any part of the rocky shore, from the high-tide zone to the low-tide zone, may be present.

Each tide pool is unique because evaporation, rainfall, solar heating, winter cooling, and other factors affect the physical properties (e.g., salinity, temperature, pH, and oxygen concentration) of the water in each tide pool differently. Tide pools high on the shore are isolated for long periods between tides and are subject to the greatest changes. Deep tide pools have a greater volume of seawater per unit area than shallow ones and are less affected by evaporation, rainfall, heating, and cooling. Consequently, small tide pools tend to support only microscopic algae and highly tolerant **copepods** and other microscopic animals. In contrast, large tide pools may contain many species of algae, sea urchins, anemones, crabs, shrimp, small fishes, and other animals that are tolerant to relatively small changes in salinity, temperature, and other physical characteristics of the tide pool water. The next time you visit a shore with tide pools, you can see which species are more tolerant by simply examining several different tide pools.

17.5 SARGASSO SEA

The Sargasso Sea is the region of the North Atlantic Ocean surrounded by the North Atlantic **subtropical gyre**. The permanent thermocline in the Sargasso Sea is steep, persistent, and deep (Chap. 10). Hence, upwelling does not occur, and the surface layer is depleted of nutrients. Furthermore, there is little exchange between Sargasso Sea surface waters and adjacent surface waters from which nutrients could be resupplied. Consequently, primary productivity is very low, and we would expect the Sargasso Sea to be an ocean "desert." The phytoplankton-based **food chain** in the Sargasso Sea is indeed very limited. However, parts of the Sargasso Sea surface are covered by dense mats of floating brown macroalgae (**Fig. 17.13a,b**), within which lives a diverse community of animals. These mats often cover areas of many square kilometers.

FIGURE 17.13 The *Sargassum* community. (a) *Sargassum* weed floats on the Sargasso Sea surface in rafts that can exceed hundreds of meters in diameter. (b) Photographed from below, it can be seen that the *Sargassum* weed rafts float freely and do not extend deep into the water column. (c) A Sargassum frogfish (sargassumfish, *Histrio histrio*) well camouflaged in a mass of *Sargassum* weed. (d) A slender-horned prawn (*Leander tenuicornis*) living among *Sargassum* weed.

Why do these vast quantities of algae grow in what should be a desert? The answer illustrates the infinite adaptability of nature. The brown algae in the Sargasso Sea are species of the genus *Sargassum*. These species grow extremely slowly and are thought to have lifetimes of decades or even centuries. The very small amounts of nutrients available in Sargasso Sea surface waters are adequate to sustain the very slow growth of *Sargassum*. However, the critical factors that enable *Sargassum* to populate the Sargasso Sea so heavily are very weak currents and very limited wind mixing. Because the surface currents are weak and flow in directions that are oriented partially toward the center of the subtropical **gyre**, they tend to retain the floating algae within the Sargasso Sea, and the mats of the long-lived algae tend to stay together. The algal population therefore can develop a large biomass despite its very slow growth and reproduction rate. In addition, because wind mixing is very limited, nutrients released into the water column by the consumers and decomposers that live in the *Sargassum* are not rapidly diluted, and are available to be recycled back into new *Sargassum* growth.

Within the *Sargassum* mats lives a community of many different species of fishes, crabs, snails, and other animals, including a frogfish called the sargassumfish (**Fig. 17.13c**) and a slender-horned prawn (**Fig. 17.13d**). Many of these species have a body that mimics the structure and color of the *Sargassum*, and they can cling to the branches of the algae unnoticed by predators or prey that swim by. In contrast to many other ocean ecosystems, the *Sargassum* community has relatively few grazers and a relatively small animal biomass in relation to the algal biomass. This composition reflects the evolution of the community to a stable condition. If there were large populations of grazers or other animals, the algal biomass would be rapidly reduced to the point where the animal populations could not be sustained. The animals that have survived in this community include many long-lived, slow-growing species with sedentary lifestyles. Each of these characteristics represents an adaptation to minimize food intake needs.

Although some species similar to those found in the Sargasso Sea are found within the interior of other subtropical gyres, for reasons as yet not fully understood, none of these other areas sustain the high biomass found in the Sargasso Sea.

17.6 POLAR REGIONS

The marine ecosystems of the north and south polar regions differ from one another physically and biologically because of the configuration of the landmasses and because they are separated by warm tropical waters through which most cold-adapted marine species cannot transit. Nonetheless, the two ecosystems have a number of environmental similarities, including extreme seasonal variation in light availability, generally low surface water temperatures, and seasonally variable sea-ice cover.

In both polar regions, cooling of surface ocean waters and vigorous wind mixing caused by storms formed at the polar **front** (Chap. 9) prevent the formation of a permanent thermocline and promote vertical mixing. Therefore, nutrients are generally abundant, especially in surface waters of the upwelling region between the Antarctic **Convergence** and the near-coastal east-wind drift current that flows around Antarctica. Primary productivity in this region apparently is limited by the availability of **micronutrients**, particularly iron, even when light is ample and nitrogen, phosphorus, and silica are abundant.

Special Characteristics of Arctic Marine Environments

In the Northern Hemisphere, nutrients are abundant in surface waters of the **marginal seas** that surround the Arctic Ocean, particularly the Bering, Norwegian, and North Seas. Nutrients are less abundant in seasonally ice-free surface waters of the Arctic Ocean itself, because freshwater **runoff** and **ice exclusion** lower surface salinity and establish a strong **halocline** that inhibits the vertical mixing of higher-salinity, nutrient-rich deep water into the surface layer. In addition, the permanent ice cover of most of the Arctic Ocean and its location in an atmospheric **downwelling** zone (zone of weak winds) minimize wind-induced vertical mixing.

In coastal regions of the Arctic Ocean, nutrients are generally available in relatively high concentrations during summer because they are supplied to some extent in freshwater runoff. In addition, nutrients are returned to the water column through decomposition during the darkness of winter, when these regions are covered by ice. Hence, nutrients are readily available during at least the early part of the short spring–summer ice-free season.

Common Characteristics of Arctic and Antarctic Marine Environments

In both the Arctic and the Antarctic, substantial populations of microscopic ice algae live, or form resting phases, within liquid pockets in the ice or on the underside of the ice. During spring and summer, ice algae can grow in or under the ice as the light increases in intensity and begins to penetrate the ice. As the ice melts, the resting spores of many species of phytoplankton are released into the water column, where they grow rapidly in the nutrient-rich and now well-illuminated open water. Both the Arctic and the Antarctic have a zone of maximum productivity that coincides with the edge of the floating ice. Each spring, this zone moves poleward with the melting ice edge.

Several important characteristics of polar ecosystems determine the species that are able to live there. First, because nutrients are generally available, phytoplankton usually grow quickly and are abundant when light is also available. Second, the extreme seasonal variation of light intensity and duration, and in some areas the extent of sea-ice cover, limit the period of ideal conditions for phytoplankton growth to only a few weeks or months in summer. Third, upwelling and wind mixing, which supply nutrients to surface waters and at the same time affect the **residence time** of phytoplankton in the nutrient-rich photic zone, are particularly variable because of eddies, turbulence induced by seafloor topography, and especially changes in **weather** and **climate**.

As a result of the unique physical characteristics of their environment, many animals that live in polar regions must migrate seasonally or be able to obtain all the food they need for the year during the short productive summer period. In addition, they must be capable of ensuring the survival of their species during years when climatic extremes may result in a partial, or even total, loss of their food supply. The evolutionary response to these constraints is for the animals to be long-lived, to reach sexual maturity late, and to bear only one or a few offspring each year for several years. This response ensures that the species will survive if one or more successive years are poor years in which none of the offspring survive. In addition, the animals must be able to store the food energy needed to survive the winter when food is unavailable or to migrate to warmer regions for the winter. The food energy is stored as fat. Many species that live in polar ecosystems, including certain fishes, marine mammals, and birds, have these characteristics.

Many of the characteristics shared by polar animals make them more susceptible than the marine species of lower latitudes to the adverse effects of certain **contaminants**. Because fat-**soluble** contaminants are concentrated in an animal's fatty tissues, they can accumulate for several years before polar animals reach reproductive age. This buildup maximizes the possibility of **carcinogenic**, **mutagenic**, and **teratogenic** effects in the species or its offspring (**CC18**). Fortunately, the Arctic and Antarctic marine ecosystems are substantially less contaminated than marine ecosystems at lower latitudes.

Antarctic Communities

The **pelagic** food web of the Southern Ocean is described in Chapter 14 (see **Fig. 14.11**). It consists of many species of whales, seals, and penguins that feed on the abundant **krill** and other zooplankton. Most of the penguin and seal species live year-round in Antarctic waters and haul themselves out on the continent or one of the nearby islands to breed and bear their offspring. Like many polar species, most of these mammal and bird species mature only after they are several years old, normally have one or two offspring per year for several years, and build up heavy layers of fat during summer, when food is abundant.

Although fat layers act as insulation against the Antarctic cold, particularly when the animal is on land, their most important function is to provide energy during the period when the animal is not feeding. For example, the majority of whale species in the Antarctic are **baleen** whales that visit the region only during summer, when food supplies are abundant. The huge store of fat that they build up during this time is used to supply them with energy during the remainder of the year, when most of these species migrate to breeding grounds in the tropical or subtropical ocean. During the migration and breeding season, the whales feed little or not at all.

Human hunting has dramatically reduced the populations of many seal and whale species in the Antarctic (**CC16**). These species are now protected, and most populations are showing signs of slow recovery.

Although fishes of the Antarctic ecosystem are much less well studied than the marine mammals and birds, we know that a high proportion of these species are present only in Antarctica and appear to be adapted to the cold and seasonally variable food supply in much the same way that Antarctic marine mammals are. Some species have evolved a unique blood chemistry that enables them to live at temperatures below freezing. The natural antifreeze of these fishes is the subject of considerable research because it could have commercial and medical applications.

CRITICAL THINKING QUESTIONS

17.4 If the Earth's climate warms and both polar ice caps melt away completely, which ecosystem is likely to be changed more: the Southern Ocean ecosystem or the Arctic Ocean ecosystem? Explain the reasons for your answer.

17.5 The temperature and salinity of the abyssopelagic environment vary little with latitude or from ocean to ocean. This environment is also uniformly dark and the pressure uniformly high. Does this mean that, if we were able to sample every species in a cubic kilometer of deep-ocean water, we could be certain that we had sampled a large majority of all abyssopelagic species that exist in the oceans? Why or why not?

17.6 Coral reefs and kelp forests both have high species diversity compared to some other ocean ecosystems. What are the possible reasons for this high diversity?

Arctic Communities

The biological populations of the Arctic region are similar in some ways to those of the Antarctic, particularly in their concentrations of marine mammals. However, species found in the Arctic are different from those found in the Antarctic. Many seal species live year-round in the Arctic and its adjacent seas and haul themselves out on land to breed, as other seal species do in the Antarctic. Most northern whale species migrate between their Arctic feeding grounds and tropical or subtropical breeding grounds. Penguins and leopard seals are not present in the Arctic. However, the Arctic is populated by polar bears and walrus (**Fig. 14.25d**), which are not present in Antarctica. Polar bears are voracious predators and superb swimmers, but they are land animals. However, because they range across the sea ice to hunt seals, they are primarily dependent on food from the marine environment, and so they are considered a part of the marine ecosystem.

Susceptibility to Climate Change

Global climate models all predict that climate changes will be amplified in the polar regions, and this conclusion is supported by historical data and by observations during the past several decades. The reasons for this special susceptibility are many and complex, but they include the positive **feedback** (**CC9**) due to reduced snow and ice cover. A covering of snow and ice is a good reflector of the sun's energy. Warming reduces snow and ice cover, which increases the amount of the sun's energy absorbed by land and ocean, causing further warming.

There is strong evidence that polar climate has warmed during the past 30 years, resulting in substantial reduction in the area of permanent sea ice in the Arctic

Ocean and causing ice sheets to be reduced in size in the Antarctic and **glaciers** to retreat in both polar regions. If sustained, this trend will not only cause sea level to rise worldwide, but will also have profound effects on polar biological communities. For example, polar bear populations may be devastated because the bears hunt mostly on floating summer sea ice, which is steadily reducing in area and retreating farther from the coast. Similarly, many species of seals and penguins may be affected by changing ice and snow cover in their traditional breeding grounds on the Antarctic coast. Furthermore, changes with unknown effects are likely to take place because a reduction in ice cover favors phytoplankton production over ice algae production in both polar regions.

17.7 BEYOND THE SUN'S LIGHT

Below the surface layers of the oceans, organisms must be adapted to extremely low light levels or no light at all, to uniformly low temperatures, to a detritus-based food supply that decreases steadily with depth, and to increasing pressure. These factors interact to give pelagic and benthic communities of the **bathyal zone** and **abyssal zone** characteristics that are very different from those of communities that live near the surface.

Relatively little is known about bathyal and abyssal communities because the vast volume of the ocean that they inhabit has been visited only for fleeting moments by research **submersibles**, which many denizens of the deep undoubtedly avoid. In addition, collecting samples from the deep oceans is very difficult and expensive. The creatures that live in the deep oceans range from rather familiar forms that resemble fishes and invertebrates of the shallower ocean to incredibly bizarre-looking creatures that would be well suited to science fiction movies. **Figure 17.14** shows just a small selection of the fishes that inhabit the **aphotic zone.**

Organisms living below the photic zone have four potential sources of food: particulate detritus that sinks slowly through the water column, carcasses of large animals that sink rapidly to the ocean floor because of their size, prey species that live in the aphotic zone, and prey species that live in the photic zone above. Of these, only prey species in the photic zone are abundant, so many deep-water species migrate to this zone to feed. All other deep-ocean biota must be adapted to a low and uncertain food supply.

Many species of crustaceans (such as shrimps, copepods, and **amphipods**), other invertebrates (such as squid), and numerous fish species live in the part of the water column immediately below the photic zone, between about 200 and 1000 m. Many of these species migrate vertically up into the photic zone at night to feed and then return to the depths during the day. Other species prey on species that live in their own depth zone. Vertical migrants include the unique *Nautilus* (**Fig. 14.23e,f**), which has a chambered shell to provide buoyancy. To avoid problems with pressure changes as the *Nautilus* migrates vertically, the internal buoyancy chambers contain gas at a very low pressure, close to a vacuum.

There is very little or no light below 200 m, and no red light penetrates to this depth. Many organisms of this zone are red-colored (**Fig. 16.14**), so they do not reflect any of the ambient light, which makes them difficult to see in the near darkness. However, many fishes that live at these depths have greatly enlarged eyes to enable them to hunt prey in the dim light.

The easiest way to hunt visually for prey in waters below about 200 m where light still penetrates is to look upward into the very dim light filtering down from above and search for the dark silhouette of the prey species. Consequently, many fish species of this zone have eyes that look directly upward. To counter this hunting strategy, many prey species have a series of light-producing organs called "photophores" arrayed along the underside of their body. By illuminating their underside with these photophores, they can reduce the sharpness of their silhouette and blend better into the dimly lit background above. Many species also have photophores arrayed on other parts of their body, presumably as devices to identify members of their own species or to attract prey species.

With increasing depth below 1000 m, fewer and fewer species are present that migrate vertically. Many organisms are brightly colored, although, in the absence of light, color does not advertise their presence and so is irrelevant. Eyes become less prominent and absent in many species, but some species that live in the absolute darkness of the deep oceans do have eyes, and it is likely that at least some potential prey emit light for recognition, mate attraction, or prey location.

Because of the low **density** of organisms, all fishes of mid and deep waters must be adapted to take advantage of any prey species they encounter. Therefore, many of these fishes have unhingeable jaws and soft expandable bodies, so they can swallow and digest prey species as large as, or larger than, themselves.

On the deep-ocean floor, the food supply comes from above as a rain of detrital particles and as occasional carcasses of large animals. Much of the detrital material has already been subjected to substantial decomposition during its slow descent through the water column. Hence, what remains is primarily material that is difficult to digest and that decomposes only very slowly. This material has little immediate food value for animals. This detritus is normally consumed by **bacteria** in the surface sediment, and it is the bacteria that become food for larger animals. Because the suspended particulate food supply near the deep seafloor is at very low concentrations and is of low nutritional value, there are few sus-

FIGURE 17.14 Mid-water fish species are often very strange-looking and have unusual adaptations, such as unhingeable jaws and hugely expandable stomachs, strange eyes, lures, and light-emitting photophores. Many of these adaptations aid these species to cope with the scarcity of food and difficulty of locating mates and prey.

pension feeders. Most animals of the deep seafloor are deposit feeders, including many species of sea cucumbers and **brittle stars.** The particulate food supply in some areas near the continents may be supplemented by detritus carried to the deep seafloor in **turbidity currents.**

Although we have little direct information about such events, it is known that bodies of large animals must sink rapidly to the seafloor with some frequency. When bait is lowered to the deep seafloor, a variety of fishes, sharks, crustaceans, and other invertebrate scavengers arrive to feed on the bait within as little as 30 minutes. Once the bait has been eaten, these animals disappear into the darkness. How these scavengers find the bait or their normal food is not known, but the speed at which they appear suggests that they must have extremely sensitive chemosensory (chemical-sensing) organs for this purpose. It has been observed that these scavengers can sense, but cannot find, bait if it is suspended even a meter or less above the seafloor, which demonstrates that they are completely adapted to feeding on material that lies on the seafloor.

The speed with which food is found by deep-ocean scavengers after it reaches the seafloor indicates fierce competition for food, because food does not decompose rapidly at such depths. Bacterial decomposition is inhibited by high pressures, so food items that fall to the deep-ocean floor would decompose only very slowly if they were not immediately consumed. Inhibition of bacterial decomposition by high pressure was first discovered when the research submersible *Alvin* sank unexpectedly in more than 1500 m of water. A lunch box containing an apple and a bologna sandwich was inside the submersible when it sank. A year later, when *Alvin* was recovered, these food items were wet but almost undecomposed. They did decompose within weeks after their return to the surface, despite being refrigerated. Inhibition of bacterial decomposition by high pressures has been confirmed by a number of experiments conducted since the *Alvin* discovery.

17.8 HYDROTHERMAL VENTS

Until 1977, it was thought that all areas of the abyssal oceans were biologically impoverished because of the limited availability of food that rains down from above. However, in 1977 the research submersible *Alvin* made a number of dives on the Galápagos Ridge (**Fig. 17.15**) that would change those ideas. The purpose of the dives was to study the geology and chemistry at this **oceanic ridge.** The researchers were looking for evidence that the high heat flow through the seafloor at the center of the ridge creates hydrothermal circulation. In hydrothermal circulation, seawater sinks through sediments or cracks in rocks of the seafloor and is heated, **convected** upward, and vented to the water column to be replaced by more seawater drawn through the rocks and sediments (**Fig. 17.16a**, Chap 8). The heat comes from the upwelling **magma** and cooling volcanic rocks beneath the seafloor.

The researchers found much more than they expected. In fact, their findings may represent one of the most surprising and profound scientific discoveries ever made. They found not only **hydrothermal vents,** but

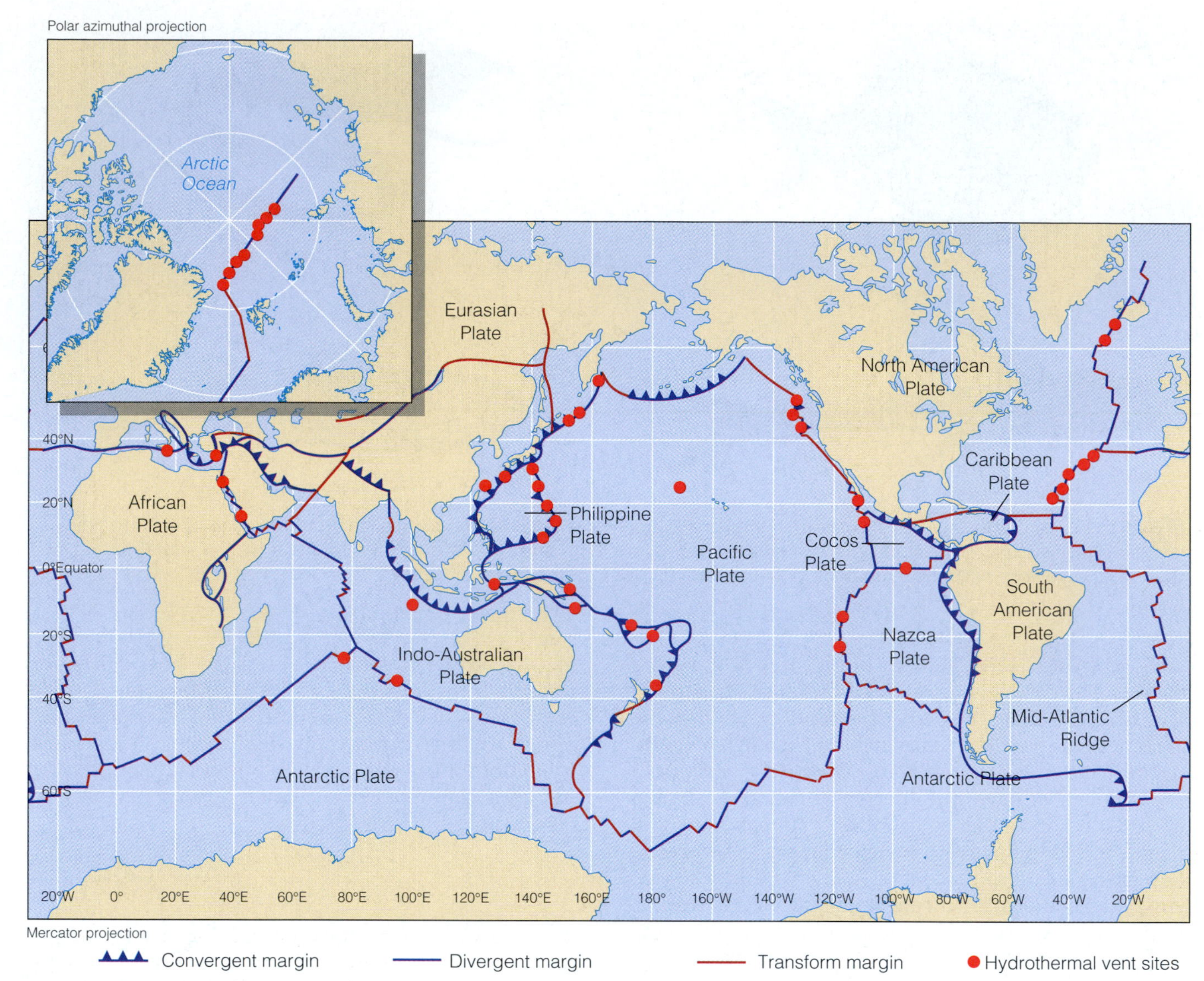

FIGURE 17.15 Hydrothermal vents are known to occur on the mid-ocean ridges of all oceans, including the locations shown here. Only a small fraction of the oceanic ridges have been visited or studied, and reports of newly found vents are frequent. Many more vents probably remain to be found.

also dense communities of marine organisms surrounding those vents. The communities included **tube worms,** clams, mussels, and many other invertebrates, most of which belonged to previously unknown species and many of which were very large in comparison with similar known species. The biomass in these vent communities is hundreds of thousands of times greater than that in any other community at comparable depths in the ocean.

However, even this finding of abundant oases of life in the "desert" of the abyss was to prove less surprising than the subsequent discovery that these communities do not depend on photosynthesis. In fact, they were found to be dependent on **chemosynthesis** for primary production of their food.

Hydrothermal Vent Environments

Initially, hydrothermal vents were thought to be rare and to occur only on fast-spreading oceanic ridges. We now know that hydrothermal vents and their associated biological communities are present in many locations, dispersed irregularly along the oceanic ridges in all oceans (**Fig. 17.15**). Vents have even been found on the ultraslow-spreading Gakkel Ridge in the Arctic Ocean, although very little is yet known about the biological communities at these vents.

Vents have also been found on the submerged volcanoes at island arc **subduction zones** such as the Mariana Arc. These **back-arc** volcano vents are especially interesting to scientists because many are at much

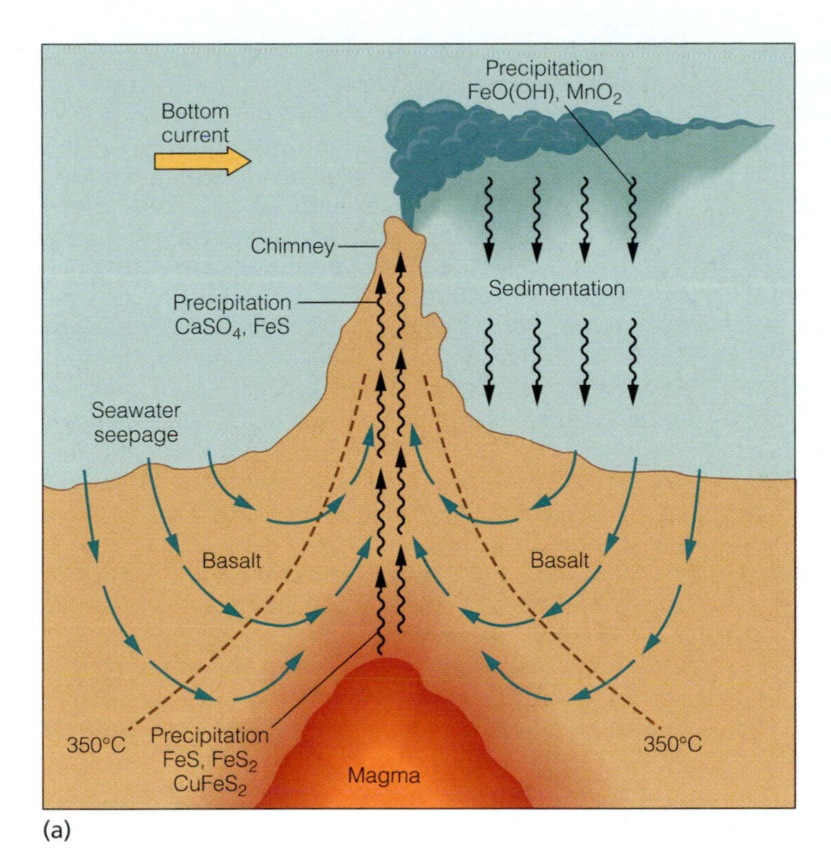

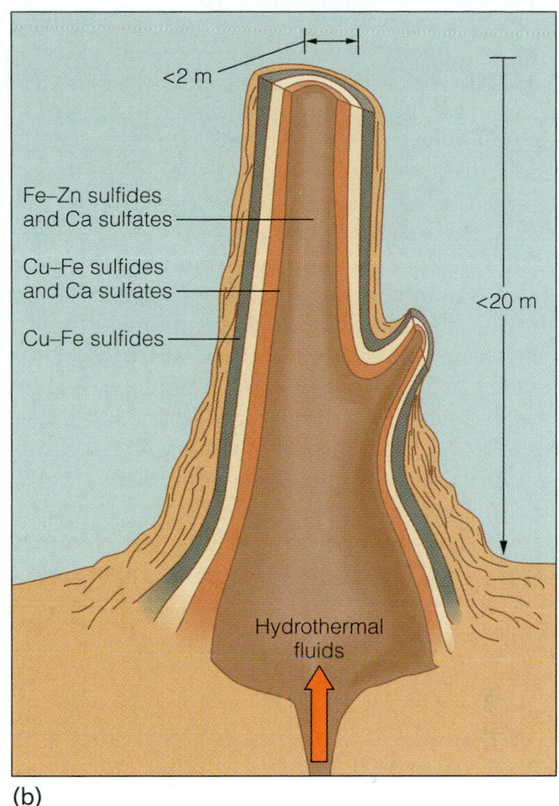

FIGURE 17.16 Black-smoker formation and composition. (a) Hydrothermal vents on the oceanic ridges are believed to be formed because heat from magma below the oceanic ridge sets up a convection circulation. In this circulation, water in the sediments is heated and rises at the vents and is replaced by water drawn through the rocks and sediment from the flanks of the ridge. (b) At black-smoker vents, the discharged water contains no oxygen but does contain high concentrations of iron, manganese, and other sulfides. As water is vented, the sulfides are oxidized by dissolved oxygen in seawater and sometimes are deposited around the vent mouth to form a chimneylike structure.

shallower depths than the oceanic ridge vents. The shallower depth makes them much easier to study, and the fluids they discharge are dispersed in the upper layers of the oceans, where they may have greater immediate effects on marine species that live in, or migrate periodically to, the photic zone. Some estimates now suggest that vents are abundant enough that a volume of ocean water equal to the entire volume of the world oceans may be processed through high-temperature vents about every 10 million years, which is a relatively short period in geological time.

Each vent differs from the others in terms of the temperatures and chemical characteristics of the water it discharges. Two types of vents are better known. Vents of the first type, which includes those first found on the Galápagos Ridge, discharge warm water (2–23°C) that usually contains substantial quantities of hydrogen sulfide but can also contain appreciable concentrations of dissolved oxygen and nitrate. Vents of the second type, known as "black smokers," discharge much hotter water (270–380°C) at much higher flow rates.

The water discharged by black smokers contains no oxygen or nitrate but has high concentrations of hydrogen sulfide and of certain metals, including iron (Fe) and manganese (Mn; **Fig. 17.16**). As the superhot water is discharged into cold, oxygenated seawater, it does not vaporize because of the high pressure. As it cools and its metal sulfides are oxidized by the dissolved oxygen in seawater, **hydrated** iron and manganese oxides precipitate to form a cloud of tiny black particles. This cloud gives the black smokers the appearance of a dirty smokestack and hence their name. As the black smoker continuously disgorges, precipitated metal sulfides accumulate around the vent exit and may help to construct a chimney at the vent outlet that can be up to 20 m high and several meters wide (**Fig. 17.16b**). Many of the metal sulfide ores mined on land today are believed to have originated in such hydrothermal vent environments and then to have been scraped off and added to the continents at subduction zones (Chap. 4).

In 2000, an entirely new type of hydrothermal vent was discovered on the Mid-Atlantic Ridge. Unlike previously known vents, these vents are not located along the ridge axis. Instead they lie on the flank of the ridge, where the underlying rock is 1.5 million years old. In this area the underlying rock is composed of peridotite, which must have been uplifted by earthquake activity, since it is formed much deeper in the oceanic **crust** than the **basaltic** rock that underlies other vent systems. Seawater interacts with peridotite to produce fluids that are rich in methane and hydrogen. These fluids are discharged at about 40°C to 75°C and have a higher pH than fluids discharged from the previously known ridge axis vents. When these fluids are discharged and mix with seawater, calcium carbonate and magnesium hydroxide are precipitated out and are deposited to form chimneys much like those at black smokers. However, some of these chimneys can be up to about 60 m high, much taller than black-smoker chimneys. The populations of larger animals, such as crabs, are much less abundant at these vents than are often found at ridge axis vents. However, the vents do support abundant microbial communities that are, as yet, not well characterized. Because it is common to find elevated concentrations of methane and hydrogen in deep-ocean waters, some researchers speculate that this new type of vent may be abundant in many parts of the global oceans.

Biological Communities Associated with Hydrothermal Vents

Although individual species vary, biological communities that surround ridge axis hydrothermal vents in the Atlantic, Pacific, and Indian Oceans are composed of generally similar species, and the communities form similar distinct zones around the vent.

At many of the Pacific vents, numerous giant tube worms and clams (**Fig. 17.17**) live closest to the vent. Other invertebrates, such as one or more species of limpets, shrimp, and scale worms, are present but less abundant. Farther from the vent are other plume worms, crabs, amphipods, other shrimp species, and several species of snails. Still farther from the vent are a wide variety of hydroids, species of worms, shrimp, anemones, and snails that are different from those closer to the vent. Also in this zone is a strange creature called a "dandelion" that is probably a colonial animal but is difficult to study because it disintegrates when brought to the surface. Many of these species are filter feeders, and others are predators. The biomass decreases progressively and rapidly with distance from the vent, and the more normal sparse fauna of the deep seafloor are present a few tens of meters from the vent.

More than 400 new species have been identified in the hydrothermal vent fauna, representing more than 20 new **families** and more than 90 new genera (plural of *genus*). The discovery of this bewildering array of new species is unique in the history of biological science, rivaled only by the findings of the *Challenger* expedition in the 1870s (Chap. 2).

The biomass of many Pacific hydrothermal vent communities is dominated by the giant tube worm, *Riftia pachyptila*, a very strange creature (**Fig. 17.17b**). This species has no mouth and no digestive system, but it can grow to several centimeters in thickness and more than 1 m in length. Like the clams that live in the same region near the vents, it has red flesh and blood. Both species get

FIGURE 17.17 Hydrothermal vent communities support many species that are not found elsewhere. (a) A typical biological community near a black smoker, like the one seen here, includes tube worms, crabs, mussels, and other invertebrates. (b) One of the most important members of most hydrothermal communities is the tube worm *Riftia pachyptila,* which can be a meter or more in length and which sustains itself by feeding on symbiotic chemosynthetic bacteria.

their red color from hemoglobin in their blood, the same oxygen-binding molecule that is present in human blood.

How does this giant tube worm feed without a mouth or digestive organs? How does it survive in the presence of hydrogen sulfide from the vents, which normally would poison the hemoglobin and kill the animal? How does it survive the great variations in temperature and other environmental parameters caused by the highly variable output rate of hot water from the vent? The answers to these and many other questions about hydrothermal vent communities are being sought as research on these unique environments continues.

The food source for the giant tube worm is apparently a population of chemosynthetic bacteria that it cultivates within its body in a symbiotic association similar to that between corals and zooxanthellae. The bacteria oxidize sulfide as an energy source to chemosynthesize organic matter (Chap. 14), and the tube worm assimilates either the waste products of this synthesis or the bacterial biomass itself, or both. This partnership is phenomenally successful, because the tube worms apparently grow very quickly. The tube worm also has a unique **enzyme** that is incorporated in its tissues, particularly in its surface tissues. The enzyme detoxifies hydrogen sulfide and protects the hemoglobin that carries oxygen needed for the worm to **respire.** However, this enzymatic protection is carefully adapted to allow a route by which the sulfide can be brought from outside the worm into the part of its body where the chemosynthetic bacteria reside.

Although it is certain that some hydrothermal vent species other than the tube worm have similar associations with chemosynthetic bacteria to provide a portion of their food, most species within the hydrothermal vent community are probably filter feeders. Hence, the bulk of their food must come from suspended particles. The source of these particles appears to be chemosynthetic bacteria and **archaea** that grow in profusion in the mouth of the vent and deep within the sediments and rocks of the seafloor. Clumps of the microbial biomass are broken loose periodically by the flow of water through the vent. The clumps fragment to form suspended particles that can be captured by the filter feeders. The concentration of these particles quickly declines with distance from the vent as the plume disperses and large particles are deposited.

Unanswered Questions about Hydrothermal Vents

Many questions about hydrothermal vents remain unanswered, not only about the abundance, geographic distribution, and physical/chemical characteristics of the vents, but also about the species that make up vent communities. For example, we do not know how the chemosynthetic bacteria or archaea can survive and grow at temperatures in excess of several hundred degrees. Also we do not know how these species are able to survive the trip across many kilometers of abyssal ocean to colonize new hydrothermal vents. Most vent species are adapted to higher temperatures and different water chemistry than are present in the abyssal ocean through which they would have to travel to colonize a new vent. Thus, vent species, or at least their eggs or larvae, must be able to survive a much greater range of environmental conditions than do most of the living organisms with which we are more familiar.

Observations have shown that new vents may be colonized within months or years. Most known vents are scattered along the ridges, some separated by substantial distances, and each may operate for a limited period, perhaps less than 20 years or so. Bottom currents that could carry eggs and larvae on some ridges may tend to follow the ridge, but new vents may be upcurrent of the old ones. Thus, the mechanisms by which new vents are colonized, which also likely differ for individual species, may be complex.

Many vent species produce extraordinarily large numbers of larvae. One possible explanation for the very large size of some vent species in comparison with similar species elsewhere in the oceans may be a need to produce very large numbers of larval offspring. It has been suggested that, in some species, these larvae may have an arrested development phase, so they could be transported by ocean currents for many decades or even centuries before encountering a new vent to colonize. Eggs and larvae of vent species may rise into shallower layers of the ocean, perhaps entrained in megaplumes of water heated slightly above ambient temperature that are known to occur in hydrothermal vent areas. In the shallower layers, larvae may be widely distributed before settling to the seafloor for a chance encounter with a new vent. There is also evidence that the partially decomposed carcasses of whales and other large mammals that fall to the ocean floor, or other slowly decomposing organic matter such as wood that reaches the deep seafloor, may be ideal sulfide-containing environments to support vent species during a "stopover" while being dispersed across the deep oceans. Any, all, or none of these mechanisms may be involved in new vent colonization.

Genetic analysis of vent species has just begun to reveal the rates and patterns of colonization of vents in different parts of the world ocean. Genetic and other studies of many more species from many more vents, exploration of the vast areas currently unsearched to fill in details of the distribution of hydrothermal vents worldwide, and studies to obtain much better knowledge of the current patterns, mixing, and dispersion in the deep oceans are all needed to enable us to unravel the mysteries of hydrothermal vent species life cycles. We are likely to have many more surprises as these investigations proceed.

Other Chemosynthetic Communities

Since the discovery of hydrothermal vents on the oceanic ridges, several chemosynthetic communities have been discovered in other locations in the deep sea, as well as in shallow-water **anoxic** environments such as marshes. White chemosynthetic bacterial mats have been found at the base of the **continental slope** off the west coast of Florida. These bacteria use hydrogen sulfide in water that seeps out from the limestone underlying sediments of the continental slope. In and around these mats is a diverse community of animals that may obtain some or all of their food from the chemosynthetic bacteria.

Chemosynthetic communities have also been found at oil and gas seeps at a depth of 600 to 700 m in the Gulf of Mexico south of Louisiana. These communities use either hydrogen sulfide and/or **hydrocarbons** as an energy source for their primary production. In the Juan de Fuca subduction zone and in other subduction zones, methane in **pore waters** squeezed out of the buried sediments provides the energy source for other chemosynthetic bacteria.

In addition to these ocean chemosynthetic communities, chemosynthetic communities that are generally dominated by archaea have been found to exist deep within the rocks of the Earth's **crust.** Studies of the organisms able to live and reproduce in such extreme environments have assumed an important role in the search for the origins of life on the Earth and in the search for life or evidence of life in the past on other planets and moons of the solar system.

CHAPTER SUMMARY

Communities and Niches. Species within an ecosystem are distributed nonuniformly, but they often cluster into communities with common characteristics. Each species has a survival niche that is defined by the ranges of environmental variables, such as salinity, temperature, turbidity, and nutrient concentrations, within which it can survive. Within its survival niche, each species has a fundamental niche within which it can survive and reproduce successfully. In some instances, a species may not occupy all of its fundamental niche, as a result of competition by other species.

Coral Reefs. Reef-building corals grow only on the seafloor in the photic zone of tropical waters between about 30°N and 30°S. These corals house photosynthetic zooxanthellae within their tissues. Reef-building corals obtain some food from their zooxanthellae. Productivity is higher in coral reefs than in other, nutrient-poor tropical marine areas because nutrients are recycled between coral and zooxanthellae, the reefs cause some upwelling, and most organisms in the reef community are residents and do not export nutrients. High turbidity adversely affects corals by reducing photosynthesis by zooxanthellae and requiring the corals to expend energy to clear away deposited particles.

Primary production in coral reef communities is performed by zooxanthellae and by benthic micro- and macroalgae, many of which are calcareous algae whose hard parts help to build the reef. A typical coral reef has a sheltered lagoon where coral growth is patchy because of variable salinity and turbidity and where many detritus, suspension, and deposit feeders are present. At the lagoon's seaward edge is a reef flat swept generally free of sediment by waves where coral growth is active and consists mainly of encrusting forms. Invertebrates live in holes and grooves cut in the reef flat.

Farther seaward, there is sometimes a low sandy island and usually an irregular ridge formed from broken coral thrown up periodically by waves and cemented by calcareous algae. Seaward of the ridge is the buttress zone, which may have a gradual to nearly vertical downward slope. It is scoured by waves to about the 20 m depth, characterized by massive robust corals, and often cut across by grooves in which there is little coral growth. Farther seaward, at depths below the reach of wave action, delicate forms of coral, including soft corals, grow in abundance.

Kelp Forests. Kelp forests grow where the seafloor is stable, preferably rocky, and within the photic zone, and where the water is cold and rich in nutrients. Kelp fronds grow as much as half a meter a day, and kelp primary productivity is very high. Kelp reproduce both vegetatively and by releasing spores to the water column. Kelp forests provide shelter and habitat for many species of fishes and invertebrates. Only a few of these species eat kelp itself, but kelp releases large quantities of detritus that enter the food chain when it is consumed by detritus feeders.

Sea urchins eat kelp, and sea otters eat sea urchins. In the eighteenth and nineteenth centuries, hunting of sea otters off California reduced sea otter populations and thus their predation on sea urchins. The increased sea urchin populations ate more kelp and destroyed the forest in many areas. Sea otters are now protected, and kelp is slowly returning.

Rocky Intertidal Communities. Species of the rocky intertidal zone are exposed to the atmosphere part of the time. They are subjected to variable conditions of temperature, water and air exposure, salinity, oxygen concentration, and pH, and they are vulnerable to birds and land predators. The rocky intertidal community is separated into four zones, distinguished by degree of exposure to air. The supralittoral zone is above high water and exposed permanently to air but is reached periodically by spray. It supports lichens, encrusting algae, grazers (including marine snails and limpets), and scavengers, primarily isopods. The high-tide zone is covered in water only during high tides and supports encrusting algae, tough attached macroalgae, filter-feeding barnacles, and periwinkle and limpet species different from those in the supralittoral zone. The middle-tide zone is covered and uncovered by water during most or all tidal cycles, has sparse macroalgae because of competition from mussels and barnacles, and supports a diverse community of invertebrates. The low-tide zone is uncovered only during the lowest tides and supports macroalgae and many species of invertebrates and fishes. The upper limit of each zone is generally determined by tolerance of the species to air exposure and other environmental factors, whereas the lower limit is generally determined by competition with other species.

Tide pools undergo substantial changes in temperature, salinity, and other factors because they are small and isolated from mixing with ocean water for part of the tidal cycle. Because small tide pools tend to have greater changes than large tide pools, they support fewer species, and these species are more tolerant.

Sargasso Sea. Extensive rafts of *Sargassum*, a macroalga, float on the surface of the Sargasso Sea, which is the interior of the North Atlantic Gyre. Nutrients are extremely limited in this region, and the large *Sargassum* biomass develops because it is very long-lived and currents tend to concentrate and retain it within the center of the gyre. A variety of small fish and invertebrate species live in the *Sargassum*, many of which are unique to this community. The animal biomass is very small in relation to the algal biomass, and there are few grazers because primary productivity is low and any food that is grazed is therefore replaced slowly.

Polar Regions. Polar regions have extreme seasonal variation in light availability and ice cover, and generally cold surface waters. River runoff or ice exclusion creates strong haloclines in places, but in all other polar waters, vertical mixing due to storms is intense and nutrients are abundant. Microscopic ice algae are important in both polar regions. They grow rapidly during the ice-melting season. Phytoplankton **bloom** during only a few weeks of summer when sufficient light is available. Many animal species are adapted to the short primary production period and the large year-to-year variability by having a long life span, maturing late, bearing only a few offspring each year, and storing food energy as fat to survive the winter.

Beyond the Sun's Light. Below the photic zone, food sources are limited to particulate detritus, carcasses that fall through the water column, and prey species. However, many species migrate vertically from the aphotic zone to the **mixed layer,** usually at night, to feed. Species that migrate vertically become less common with increasing depth. Many deep-sea animals are adapted to survive on very infrequent meals, and some are able to swallow prey larger than themselves. High pressures inhibit bacterial decomposition, which helps ensure that detrital food particles remain available.

Hydrothermal Vents. Hydrothermal vents located along oceanic ridge axes and on submerged volcanoes of volcanic island arcs discharge seawater that has percolated through the seafloor and been heated by magma or cooling magmatic rock. Warm water discharged from cooler vents contains hydrogen sulfide but also contains some oxygen and nitrate. The effluent from black smokers, which discharge the hottest water (270–380°C), has no oxygen and high concentrations of hydrogen sulfide and metal sulfides. Chemosynthetic bacteria use the sulfides to fuel primary production.

Communities at hydrothermal vents are composed of many species that are unique to these environments. At many Pacific vents, the community closest to the vent is dominated by giant tube worms and clams that feed on chemosynthetic bacteria that they cultivate within their bodies. Various species of other invertebrates occupy zones at different distances from the vent. Most are filter feeders that live on clumps of bacteria and archaea grown at the vent or beneath the seafloor and sloughed off to become suspended particles. Food availability declines rapidly with distance from the vent. Many questions remain about these communities, including how they cross large distances of abyssal ocean to colonize new vents.

A second type of vent has been discovered that lies on the oceanic ridge flank, discharges fluids rich in methane and hydrogen, and forms tall chimneys composed largely of calcium carbonate, as opposed to the metal sulfides that form the chimneys at other vents. Additionally, chemosynthetic communities that use methane or other hydrocarbons as their energy source are present in a few locations in the oceans.

STUDY QUESTIONS

1. Why don't species live in all locations where their fundamental-niche requirements are fulfilled?
2. Why do coral reefs sustain a very large variety of species, and why is the biomass of coral reef ecosystems greater than that of other ecosystems in tropical waters?
3. What factors determine the types of corals that live on different parts of a coral reef?
4. Describe the locations and environmental conditions that are appropriate for the development of kelp forests. Contrast these with the locations and environmental conditions appropriate for the development of coral reefs.
5. Why must rocky intertidal species be more stress-tolerant than many other marine species?
6. Why are rocky intertidal communities distributed in distinct bands of different species that follow depth contours?
7. Why does the Sargasso Sea sustain a community of photic-zone organisms that are different and distinct from those found elsewhere in the oceans?
8. What are the principal differences in physical characteristics between the Arctic Ocean and the ocean around Antarctica, and how do they affect primary production and species composition?
9. Why does the species composition of hydrothermal vent communities change rapidly with distance from the vents?

KEY TERMS

You should be able to recognize and understand the meaning of all terms that are in boldface type in the text. All of those terms are defined in the Glossary. The following are some less familiar key scientific terms that are used in this chapter and that are essential to know and be able to use in classroom discussions or on exams.

abyssal zone (p. 516)
amphipods (p. 516)
anemones (p. 505)
aphotic zone (p. 516)
archaea (p. 521)
barnacles (p. 511)
bathyal zone (p. 516)
biomass (p. 501)
brittle stars (p. 517)
chemosynthesis (p. 518)
copepods (p. 512)
crustaceans (p. 512)
decomposers (p. 501)
deposit feeders (p. 501)
detritus (p. 501)
filter feeders (p. 512)
fundamental niche (p. 499)
hermatypic corals (p. 499)
hermit crabs (p. 512)
hydroids (p. 512)
hydrothermal vent (p. 517)
invertebrates (p. 500)
isopods (p. 508)
kelp (p. 504)
krill (p. 515)
limpets (p. 508)
macroalgae (p. 501)
microalgae (p. 501)
middle-tide zone (p. 508)
mollusks (p. 501)
niche (p. 498)
nudibranchs (p. 505)
pelagic (p. 515)
photic zone (p. 498)
rocky intertidal zone (p. 498)
sea cucumbers (p. 501)
sea stars (p. 501)
sea urchins (p. 505)
soft corals (p. 503)
spores (p. 505)
supralittoral zone (p. 508)
survival niche (p. 498)
suspension feeders (p. 501)
symbiotic (p. 499)
tube worms (p. 518)
zooxanthellae (p. 499)

CRITICAL CONCEPTS REMINDER

CC9 The Global Greenhouse Effect (p. 515). Perhaps the greatest environmental challenge faced by humans is the prospect that major climate changes may be an inevitable result of our burning fossil fuels. The burning of fossil fuels releases carbon dioxide and other gases into the atmosphere where they accumulate and act like the glass of a greenhouse trapping more of the sun's heat. To read **CC9** go to page 24CC.

CC14 Photosynthesis, Light, and Nutrients (p. 499). Photosynthesis and chemosynthesis are two processes by which simple chemical compounds are made into the organic compounds of living organisms. Photosynthesis depends on the availability of carbon dioxide, light, and certain dissolved nutrient elements including nitrogen, phosphorus, and iron. Chemosynthesis does not use light energy, but instead depends on the availability of chemical energy from reduced compounds, which occur only in limited environments where oxygen is depleted. To read **CC14** go to page 46CC.

CC16 Maximum Sustainable Yield (p. 515). The maximum sustainable yield is the maximum biomass of a fish species that can be depleted annually by fishing but that can still be replaced by reproduction. This yield changes unpredictably from year to year in response to the climate and other factors. The populations of many fish species worldwide have declined drastically when they have been overfished (beyond their maximum sustainable yield) in one or more years when that yield was lower than the average annual yield on which most fisheries management is based. To read **CC16** go to page 51CC.

CC17 **Species Diversity and Biodiversity** (p. 500). Biodiversity is an expression of the range of genetic diversity; species diversity; diversity in ecological niches and types of communities of organisms (ecosystem diversity); and diversity of feeding, reproduction, and predator avoidance strategies (physiological diversity), within the ecosystem of the specified region. Species diversity is a more precisely-defined term and is a measure of the species richness (number of species) and species evenness (extent to which the community has balanced populations with no dominant species). High diversity and biodiversity are generally associated with ecosystems that are resistant to change. To read **CC17** go to page 53CC.

CC18 **Toxicity** (p. 515). Many dissolved constituents of seawater become toxic to marine life when the concentrations go above their natural amount. Some synthetic organic chemicals are especially significant because they are persistent and may be bioaccumulated or biomagnified. To read **CC18** go to page 54CC.

CHAPTER 18

CHAPTER OUTLINE

CRITICAL CONCEPTS USED IN THIS CHAPTER

The oceans provide humans with many valuable resources. These images illustrate both the wide ranging uses of the ocean as well as the importance of its resources to humankind. Clockwise from top right: Container ships docked in Port Klang, Malaysia; Fishermen clean their catch of cod from Stellwagen Banks just north of Cape Cod, Massachusetts; Some of the oil production platforms in the Ekofisk oil field in the North Sea off of Norway; An instructor teaches young Australians to surf on Manley Beach near Sydney, Australia; The USS *Nimitz* returns to her home port of Coronado, in San Diego Bay, California, after 8 months in the Persian Gulf; Buyers examine tuna for sale in the world's largest fish market in Tokyo, Japan.

Ocean Resources and the Impacts of Their Use

When I learned to scuba dive around the British Isles a little more than three decades ago, I remember that the first thing we did on most of our dives was to collect large crabs or lobsters for dinner. So plentiful were these crustaceans that, on more than one occasion, the group brought back more than we could eat and we fed the leftovers to the cats. How different it is today. Since then, these crabs and lobsters have become much rarer and are sold on the market for such high prices that they have become a luxury commodity. Rapidly rising human population, increasing demand for seafood, and overexploitation of the resources are, of course, to blame.

The story is the same in almost all of the world's fisheries. Rising demand drives up prices and fishing pressure until the resource is overfished and populations decline below the level needed to sustain the market demand.

Harris Stewart, a noted oceanographer with whom I worked many years ago, placed some of this in perspective:

> *Today man is a hunter of food in the sea, even as his ancestors were hunters of food on land. On land, however, he has become a farmer rather than a hunter. Not until he becomes a farmer of the seas, until he is as well versed in "aquaculture" as he now is in agriculture, will he begin to realize the great potential of the self-renewing food resource of the global seas.*
>
> —Harris B. Stewart, *Deep Challenge*, published in 1966 by Van Nostrand, Princeton, NJ

What this comment, although perceptive, misses is perhaps the key difference between farmers and hunters. Farmers live on and usually own their land, whereas hunters, especially ocean fishers, do not. The critical issue of who owns, and should care for and nurture, the resources of the oceans has yet to be fully resolved.

Seafood is one of the most valuable resources offered by the oceans, because it is high in protein, which is much needed in feeding the burgeoning human population of the world. If we cannot learn to use this most important resource wisely, then what will be the ocean kingdom's fate as we extend our exploitation of the mineral, energy, and other resources that have, as yet, been largely unexploited?

Throughout recorded history, the oceans have provided a variety of important resources for human populations, including natural resources and energy extracted from the oceans, and activities that depend on or take advantage of the ocean **environment.** As natural resource extraction and other ocean uses increase, the ocean waters and the seafloor, which historically have held little value as "property," have become ever more valuable. At the same time, many potential or real marine environmental issues or problems have emerged. In this chapter we examine the nature and value of ocean resources and their ownership, and we review the general nature of some of the conflicts or environmental issues that are associated with utilization of the resources. **Pollution** of the oceans is at the root of some of these environmental issues. Ocean pollution is discussed in more detail in Chapter 19.

18.1 WHAT ARE OCEAN RESOURCES?

Most people know that the oceans are fished for seafood and used for transport and recreation, and that oil and gas are extracted from the seafloor. However, it is not widely recognized how valuable these resources are. For example, in 2004 the U.S. Commission on Ocean Policy reported that, on an annual basis, U.S. ports handled more than $700 billion in goods, the commercial fishing industry was valued in excess of $28 billion, and $25 to $40 billion worth of offshore oil and gas was produced with royalties of about $5 billion paid to the U.S. Treasury. The same report estimated that cruise passengers accounted for $11 billion in spending annually, that the recreational saltwater fishing industry was worth about $20 billion, that the retail trade of aquarium fishes was worth about $3 billion, and that recreation based on the **coasts** and oceans supported more than 1.5 million jobs in the United States. There are also a number of other ocean resources or potential ocean resources that are not included in these figures. In the following sections, we briefly survey the wide range of ocean resources. For this purpose, we have chosen to group these resources in eight categories:

- Biological resources
- Transportation, trade, and military use
- Offshore oil and gas
- **Methane hydrates**
- Minerals and freshwater
- Recreation, aesthetics, and endangered **species**
- Energy
- Waste disposal

Before we look at the various categories of ocean resources, it is important to realize that the oceans and their resources are not owned by individuals or corporations, as is much of the land area of the Earth, together with its resources. Instead, ocean areas adjacent to coasts are owned by governments, and ocean areas remote from land are not owned at all. To understand why this is true and what it means, we must review a little about the history of national and international laws as they relate to the sea.

18.2 NATIONAL AND INTERNATIONAL LAW APPLIED TO THE OCEANS

Throughout most of history, the oceans have not been considered territory that could be claimed or owned. Indeed, until the 1600s, there was little thought or deliberation on the legal status of the oceans or the ownership of the associated resources. Individuals and nations were generally free to travel and fish anywhere in the oceans. In 1672, however, the British declared that they would exercise control over a **territorial sea** that extended 5.6 km (in British units, 1 league or 3 nautical miles) from the **coastline.** This distance corresponded roughly to the range of **shore**-based cannons. The 5.6-km territorial sea became commonly accepted as the standard for most nations and was formally accepted by the League of Nations in 1930. The remainder of the oceans outside the territorial seas, called the **high seas,** was considered to belong to no one, and in general, all nations were free to use and exploit high-seas resources.

The Truman Proclamation

The 5.6-km territorial sea remained almost universally accepted until after World War II, when President Harry Truman proclaimed in 1945 that "the exercise of jurisdiction over the natural resources of the subsoil and seabed of the continental shelf by the contiguous nation is reasonable and just" and that "the United States regards the natural resources of the subsoil and seabed of the continental shelf beneath the high seas, but contiguous to the coasts of the United States, subject to its jurisdiction and control." This proclamation, known as the Truman Proclamation on the Continental Shelf, was motivated by the "long-range worldwide need for new sources of petroleum and other minerals." The **continental shelf** was not defined in the proclamation. However, in a statement accompanying the proclamation, it was described as the land that is contiguous to the continent and covered by no more than 183 m (100 fathoms) of water. The 183-m depth was arbitrary because the continental shelf cannot be defined by a single depth (Chap. 4).

The Truman Proclamation caused a controversy and motivated a series of unilateral decisions by various coastal nations that claimed the resources under their own coastal oceans. However, there was no uniformity in the claims. Individual nations claimed jurisdiction over different widths of ocean off their coasts. For example, nations along the west coast of South America, where the continental shelf is very narrow (Chap. 4), claimed the resources offshore to 370 km (200 nautical miles). These nations went beyond the U.S. proclamation in another important way too: they extended their sovereignty over the entire 370-km zone, whereas the Truman Proclamation claimed jurisdiction only over the resources on or under the seafloor. Thus, these other nations were claiming ownership of the waters and fisheries in this zone, and the right to control any access to the zone by other nations' vessels.

"Law of the Sea" Conferences

In 1958 and again in 1960, the United Nations convened a conference on the "Law of the Sea." The conferences were intended to establish international uniformity in the ownership and access rights of nations to the resources of the oceans. The two conferences, attended by more than 80 nations, led to the adoption of several conventions. The conventions established a zone 22 km (12 nautical miles) wide within which nations had jurisdiction over fishery resources, and they affirmed that the "high seas" area beyond that zone was free for all nations to use for navigation, fishing, and overflight.

The conventions also attempted to define the right of coastal nations to own the seabed and sub-seabed minerals on the continental shelf. The continental shelf was defined as the area beyond the territorial sea out to a depth of 200 m or "beyond that limit, to where the depth of the superadjacent waters admits of the exploitation of natural resources." This definition was ambiguous. Coastal nations could claim the minerals out to a depth at which they could be exploited, and the size of this zone would expand as exploitation technologies improved. At the time, exploitation of seabed resources at depths greater than 200 m was considered unlikely. However, oil-drilling technology improved rapidly after 1960, and the United States was soon drilling in waters more than 200 m deep off California and more than 185 km (100 nautical miles) offshore in the Gulf of Mexico.

During the 1960s, the potential mineral resources of the deep oceans, particularly **manganese nodules** (Chap. 8), were first recognized. By the late 1960s, it was clear that more needed to be done to define and determine the rights of nations to use the oceans, particularly the rights to own and to exploit ocean mineral resources.

In 1967, the Maltese ambassador to the United Nations, Dr. Arvid Pardo, proposed that another Law of the Sea Conference be held. Pardo suggested a concept for a treaty that history may recognize as a critical turning point in the development of human civilization. His suggestion was that the ocean floor outside the zones of national jurisdiction should be reserved for peaceful uses, and that its resources should become the "common heritage of mankind." Pardo suggested that money generated from the exploitation of these resources be used for the benefit of less developed nations.

The common heritage principle subsequently has been applied to other global resources, including the atmosphere, Antarctica, and tropical rain forests. It is now an important basis for wide-ranging and growing international cooperation among nations. The common heritage principle was a major catalyst for the growing global understanding that the Earth's natural resources belong to all its peoples. The essence of the principle is that all nations have rights and responsibilities to protect the global environment and to use its resources wisely.

The Law of the Sea Treaty

In response to Pardo's call, the United Nations convened the Third United Nations Law of the Sea Conference in 1973. The conference met many times over the next 9 years. Finally, in 1982, the 151 participating nations adopted a new Convention on the Law of the Sea (commonly referred to as the "Law of the Sea Treaty") by a vote of 130 to 4, with 17 abstentions. The convention has now been signed by more than 150 nations. The United States was among the four nations that voted against adoption, even though many of the treaty provisions were originally written or supported by the U.S. delegation. The United States declared that it would not sign or ratify the treaty, because it did not agree with the ocean mining provisions, although it would follow many other individual provisions. Abstaining nations included the Soviet Union, Great Britain, and West Germany.

Since 1982, the Law of the Sea Treaty has been the basis for numerous laws written by many nations about their individual rights to the oceans and ocean resources. Many provisions of the treaty are now widely accepted, and the treaty was finally ratified by the required number of 60 nations and became fully effective in November 1994. Also in 1994, the United States signed the treaty, but the treaty must still be ratified by the U.S. Senate. The principal provisions of the Law of the Sea Treaty can be summarized as follows.

1. A territorial sea of 12 nautical miles (22.25 km) was established, within which each individual coastal nation has full sovereign rights to all resources and to controlling access by foreign nationals. (Some nations, including Peru, Ecuador, Somalia, and the Philippines still claim territorial seas that extend to 200 nautical miles, despite the treaty.)
2. An **exclusive economic zone (EEZ)** was established outside the territorial sea. The EEZ normally extends out to 200 nautical miles (370 km), but it can extend as far as 350 nautical miles (649 km) to the edge of the continental shelf where the shelf extends beyond 200 nautical miles. The edge of the continental shelf is defined by the treaty in geological terms, but in a way that is complicated and difficult to interpret in some areas. Within its EEZ, the coastal nation has jurisdiction over mineral resources, fishing, and pollution, and it may exercise control over access to the zone for scientific research.
3. Complicated procedures were established for drawing ocean boundaries of EEZs between nations that are closer to each other than 400 nautical miles (740 km), or whose coastlines meet in complex ways.
4. The right of free and "innocent" passage was guaranteed for all vessels outside the territorial seas and through straits used for international navigation that are within a territorial sea (straits narrower than 24 nautical miles, or 44.5 km).
5. Complicated rules were established for exploiting mineral resources from high-seas areas outside the EEZs. All such exploitation would be regulated by a new International Seabed Authority (ISA). Any nation or private concern wanting to extract minerals from the high-seas areas would have to obtain a permit from the ISA to mine a given site. In return for the permit, the nation or private concern would provide its mining technology to the "Enterprise," a mining organization set up within the ISA. The Enterprise would mine a separate site that would be paired with the permit site. Revenue from the permit site would go to the permittee, whereas revenue from the Enterprise site would be divided among the developing nations.

The treaty did not include precisely defined procedures and rules for technology transfer to the Enterprise or for the disbursement of Enterprise profits among the developing nations. Those and all other decisions of the ISA would require unanimous approval of all nations party to the treaty. Unanimous consent of all the nations is virtually impossible to achieve, especially when resources and profits are at stake. Therefore, this provision was felt by many to destroy any reasonable chance that the ISA could succeed. Nations that voted no or abstained when the Law of the Sea Treaty was approved included most of the nations that were interested in deep-ocean mining of manganese nodules. No mining has yet been done under ISA auspices, but this may be primarily because deep-ocean mining is not yet economical. In 2001, the ISA did grant exclusive 15-year contracts to seven national and industrial pioneering investor groups for exploration (but not production) at sites in the eastern equatorial Pacific Ocean.

Exclusive Economic Zones

The Law of the Sea Treaty has exerted considerable influence over the actions of the world's nations. Most significantly, the EEZ concept is now almost universally accepted. If the 200-nautical-mile EEZ definition is applied uniformly, the total area placed under the control of the various nations is approximately 128×10^6 km^2, or about 35.8% of the world ocean (**Fig. 18.1**). The United States has a larger EEZ than any other nation, not because of the lengthy Atlantic, Pacific, and Gulf of Mexico coastlines of the contiguous states, but because of the vast additional EEZ area contributed by Alaska, Hawaii, Puerto Rico, the U.S. Virgin Islands, and the various U.S. Pacific protectorates and islands, including American Samoa and Guam (**Fig. 18.1**). Indeed, the U.S. EEZ encompasses an area of approximately 7.5×10^6 km^2, which is larger than the nation's land area. The U.S. EEZ contains many of the world's most productive fisheries and probably a large proportion of the mineral wealth of the oceans.

Unfortunately, the treaty definition of what can be claimed as a nation's EEZ is controversial and open to interpretation. The rules of how this definition should be applied were not adopted until 1999, and they are so complex that nations were given until 2009 to submit their claims. Nations must provide extensive data and evidence to define where the seafloor descending from the landmass just touches down on the relatively flat ocean bottom. This location is almost impossible to define or measure precisely. In addition, nations may claim undersea mountains and ridges if they are "continental appendages." Again it is virtually impossible to define and demonstrate this connection. Indeed, Russia has already applied the rule to claim the ridge that lies across the center of the Arctic Ocean floor, which almost all geologists would agree is an **oceanic ridge** and not an appendage of any continent.

The EEZ areas associated with islands are often very large. A single tiny **atoll** remote from any other island can command an EEZ of more than 0.2×10^6 km^2, an area somewhat larger than the state of California. Ownership of such a large area is the real reason for many recent ter-

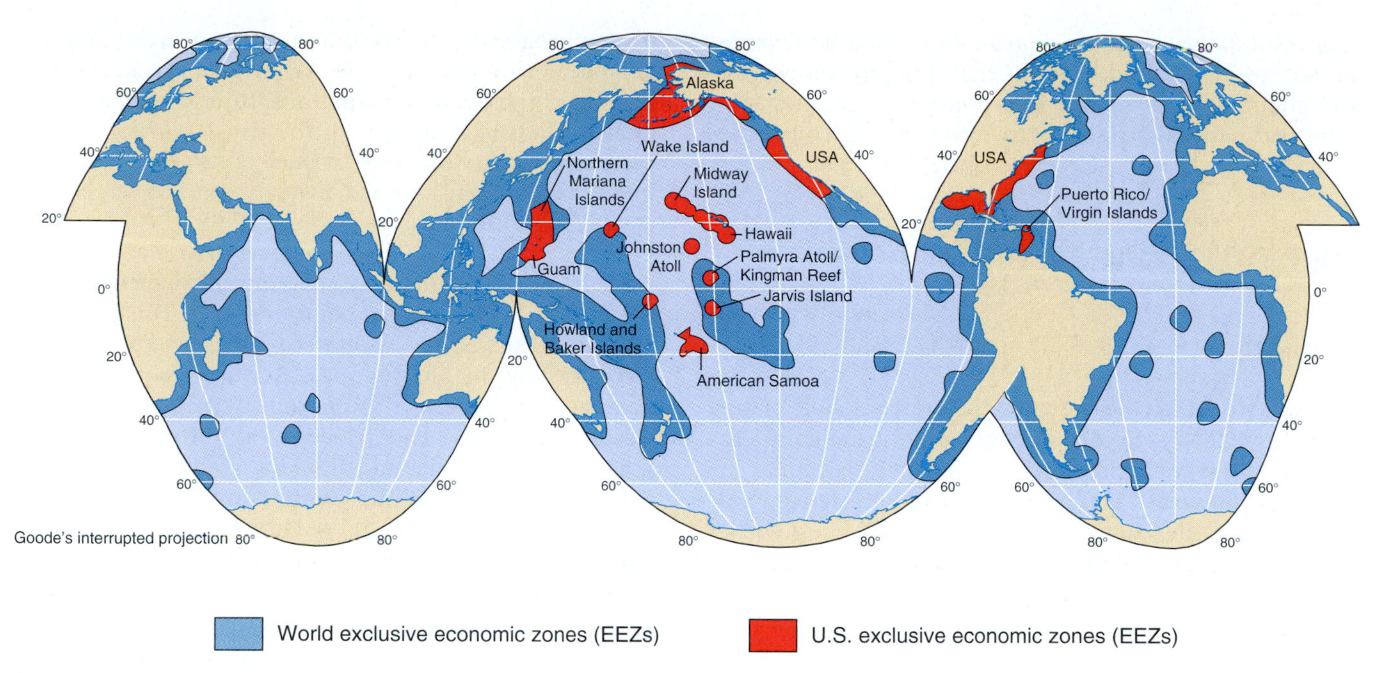

FIGURE 18.1 Exclusive economic zones (EEZs) of the United States and other countries. This map shows how large the U.S. EEZ is and how much of it is associated with Alaska, Hawaii, and various Pacific and Caribbean islands that are not among the current 50 states.

ritorial disputes over the ownership of islands that formerly were considered inconsequential. The war between Great Britain and Argentina over the Falkland Islands in 1982 was motivated by the resources of the islands' EEZ. A dispute still continues between Japan and Russia over the Kuril Islands north of Japan.

Probably the best example of an EEZ-motivated sovereignty dispute is the case of the South China Sea's Spratly Islands, which various nations claim to own. Although these tiny islands are inconsequential aside from the ocean resources that they command, various islands of the group are claimed by three or four nations (**Fig. 18.2**). Each of the claimant nations has stationed troops on one or more of the Spratly Islands, and tension often runs very high.

Sovereignty disputes over islands are not the only evidence of the importance of the EEZs. Many island nations, especially those in the Pacific, are extremely concerned about the possible reduction of their EEZs if sea level should continue to rise, as predicted. In some locations, such as the Seychelle Islands in the Indian Ocean, a sea-level rise of only a few tens of centimeters would submerge low-lying islands, each of which commands an enormous area of EEZ. Even if sea level does not rise, the smallest islands may disappear through **erosion**. Japan holds sovereignty over one tiny island that is remote from any other island and the mainland. The island rises to only a few tens of centimeters above sea level, and it has essentially no value other than the 0.25 million km^2 of EEZ that it commands. Because wave erosion could

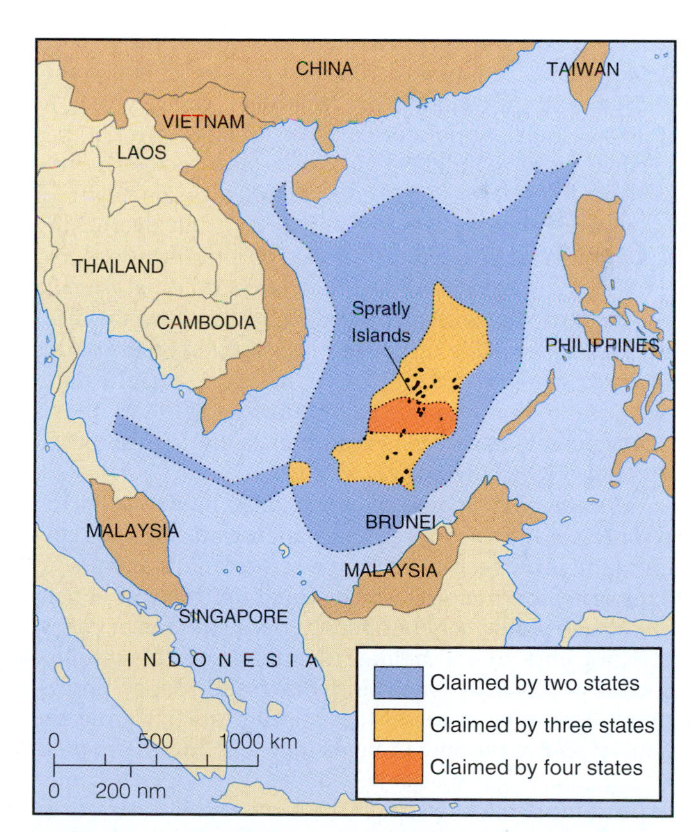

FIGURE 18.2 Boundary disputes in the South China Sea. Nations claiming ownership of one or more of the Spratly Islands include the Philippines, Vietnam, China, Taiwan, and Malaysia.

soon eliminate the island, Japan has committed large sums of money to build barriers that it hopes will protect and maintain the island. However, it is questionable whether such artificial means to preserve an EEZ are legal.

18.3 BIOLOGICAL RESOURCES

Fisheries

Fish and **shellfish** are probably the most valuable ocean resources (Chap. 15). Seafood has a very high protein content and is therefore of critical dietary importance. In many coastal areas, seafood is the basic subsistence food because no other significant source of food protein is available. Iceland and many Pacific Ocean island nations, including Japan, are good examples of such seafood-dependent areas. Fisheries are also a major part of the U.S. economy.

Until the past century, seafood resources were, for all practical purposes, inexhaustible because they were replaced by reproduction faster than they were consumed. However, human populations have burgeoned in the last 100 years and the demand for seafood has increased accordingly. Consequently, the seafood resources of many parts of the oceans have been exploited so intensively that many species are **overfished** and can no longer reproduce fast enough to replace their populations. Overfishing, which has resulted from a variety of technical and socioeconomic factors, poses perhaps the greatest threat to the health of ocean **ecosystems.** Most scientists feel that overfishing is a far greater threat than the oil spills or industrial and domestic sewage discharges that often dominate media coverage of the oceans. Indeed, the increasing need for the greatest possible utilization of ocean fishery resources to feed a hungry and growing population, and the damage done to these resources by unwise exploitation and management, have been important factors in the development of oceanography.

Historically, fisheries were, and many still are, resources not owned by any person, organization, or government. The fact that they were free for anyone to exploit led to a vicious cycle repeated in fishery after fishery. When a new fishery opens, a few fishing boat operators are able to catch large quantities of the resource species with little effort and make substantial profits. Other operators, aware of these profits, quickly enter the fishery. As the number of boats increases and the population of the target species is reduced, each operator must expend more effort to catch the same amount of fish. At the same time, the value of the catch may decline because the market is now flooded with this species. In response, each operator increases fishing efforts to catch more and recover profits lost in the price drop. Hunting and catching technologies are continuously improved, and the fishing pressure continues to rise. If unchecked, the continuous increase in fishing effort and effectiveness quickly leads to a collapse of the fish population as the **maximum sustainable yield** is exceeded (**CC16**). Many fishers then look for a new species to target, and the cycle begins anew.

In such instances, the fishery resource has declined, sometimes so precipitously that the commercial fishery has been essentially wiped out for decades. The California sardine fishery that supported the development of Cannery Row in Monterey, California (**Fig. 18.3a**), the menhaden fisheries off the middle Atlantic coast of the United States, the cod fisheries off the coast of New England and Canada, and the king crab fisheries of the Bering Sea (**Fig. 18.3b**) are prime examples of such decimated resources, although environmental factors may also have contributed to the declines. Many other fisheries have also collapsed or may be close to doing so (**Fig. 18.3c**). Perhaps the greatest known historical decline was a decline of 10 million tonnes in the anchovy catch off Peru in just one year, 1970. This decline constituted roughly 10% of the total world fishery catch at the time. A strong **El Niño** was certainly responsible for some of the decline, but many scientists believe that excessive fishing efforts contributed greatly to the problem. Most major fisheries in the world are today being fished close to, or above, their maximum sustainable yield (**CC16**).

Recently it has been recognized that the rapidly growing recreational fishing industry can also contribute to overfishing problems. For example, the recreational catch of striped bass on the east coast of the United States in 2001 was estimated to be about 19 million pounds, or three times the amount caught by commercial fishers.

Most traditional fisheries are in coastal waters near the consuming population. However, as coastal fisheries have been depleted, fishers have exploited resources from the deeper parts of the oceans and from distant coastal regions. The movement to far-flung ocean fisheries was accelerated by the development of refrigeration, which allowed catches to be stored and transported long distances in fresh or frozen condition. In addition, deep-ocean fisheries were traditionally free and open for anyone to exploit because no nation owned or controlled the resources. Unfortunately, the species that were the easiest targets of deep-ocean fishers were also those most vulnerable to overexploitation. The decimations of populations of many species of whales and seals are among the better-known examples of the overexploitation of ocean resources.

Overfishing is particularly serious in many developing countries surrounded by **coral reefs**. As modern medicine enables human populations in these countries to expand, the limited fishery resources of **reefs** near each village can no longer provide sufficient seafood for the growing population. Fishers are forced to exploit reefs in an ever larger area and must often resort to technological "improvements" in fishing techniques, such as dynamiting and spearfishing. Some of the fishes killed by dynamiting float to the surface for easy harvesting, but many of them sink and are lost to the fishers. Furthermore, dynamiting destroys the reef and **habitat** for the fishes that escape the blast or for future generations of fishes. Intensive spearfishing, particularly with **scuba,** can quickly remove breeding adults of a population and thus hinder reproductive replacement of the population. In these instances, technological advances in fishing have proven to be very destructive to the resources.

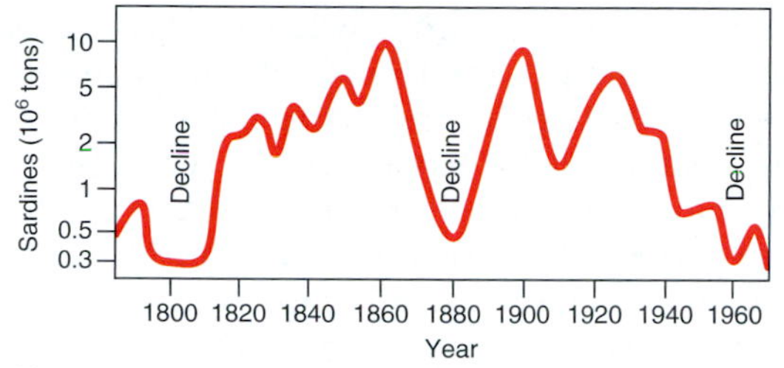

(a) California sardine

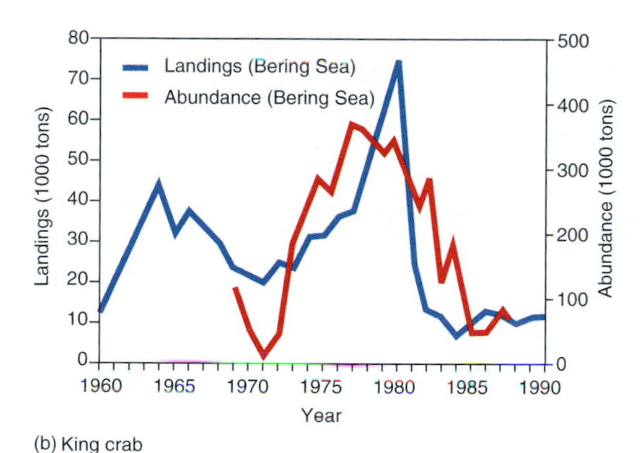

(b) King crab

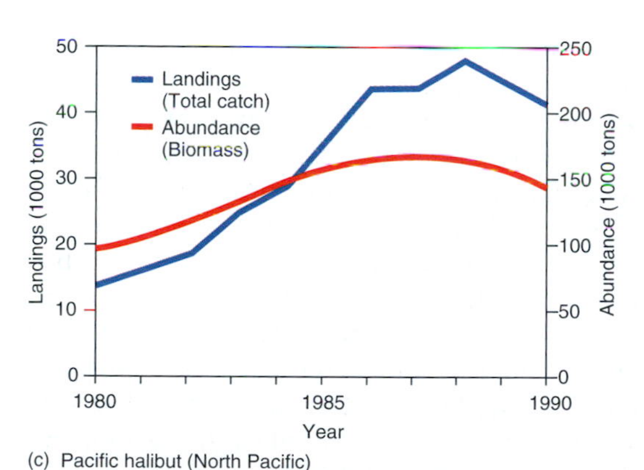

(c) Pacific halibut (North Pacific)

FIGURE 18.3 The effects of possible overfishing. (a) A history of changes in sardine stocks off California, as deduced from fish scales in sediments. Note that the dramatic decline that occurred in the 1940s, when fishing was intense, is not unique. Hence, the stock may have declined because of natural factors, possibly compounded by overfishing. (b) King crab abundance in Alaska declined dramatically in the early 1980s, when fishing pressure increased substantially. (c) The halibut fishery in the North Pacific Ocean may have been on the verge of collapse as a result of overfishing in 1990, but stringent catch limits have now apparently stabilized the resource.

Fishing threatens not only the targeted species, but also other species in the **food web.** For example, harvesting and drastic reductions in sea lion and elephant seal populations in California during the Gold Rush era (mid to late 1800s) led to a sharp reduction in the population of their natural predator, the great white shark. The **marine mammal** population became protected by law in 1972 and has recovered substantially since the early 1980s, and the great white shark population has also slowly increased. The precipitous decline of the now rare Pribilof fur seal in the Bering Sea was almost certainly caused largely by increased fishing pressure on its principal food species, including pollack.

Fishing can also have adverse effects on marine species even if those species are not the target of the fishing and not dependent on the target species for food. Many fishing techniques are not efficient in selecting the target species. For example, turtles are frequently caught in shrimp nets, and considerable effort has now been made to develop nets that, although less efficient in catching shrimp, do not catch and kill the endangered turtles. Until fishing technologies were modified, many dolphins were caught and killed in nets set to capture large **schools** of tuna. Dolphins, sharks, turtles, and many nontarget fish species also are caught in kilometer-long **drift nets** that form a barrier across the ocean **photic zone** and that capture, and usually result in the death of, anything large enough not to pass through the mesh. As in other fisheries, this incidental **bycatch** of "nonvaluable" species is simply discarded overboard. Such drift nets are now outlawed by most nations but, as with other illegal fishing methods, may continue to be used in areas far from surveillance by law enforcement officials. Enforcement of fishing regulations on the high seas is extremely difficult because no nation can afford to patrol the high seas adequately.

Many fisheries are now managed to avoid overfishing, but management is often ineffective because assessing the fishery stock size and its age composition is expensive and difficult. In addition, the maximum sustainable yield may vary dramatically from year to year because of the **chaotic** variations induced by natural factors (**CC16, CC11**). Consequently, managers have only a poor understanding of what maximum sustainable yield might be. Safe management requires that substantially less than the estimated maximum sustainable yield be caught each year to guard against errors inherent in the stock assessment data and against the inevitable years when stocks decline unexpectedly because of natural factors. However, managers are under pressure to allow the largest possible annual catch in order to provide adequate income to the owners and crews of fishing boats competing for the resource.

CRITICAL THINKING QUESTIONS

18.1 For what reasons do you think the Law of the Sea Treaty established a limit of 200 miles for exclusive economic zones (EEZs)? Why do you think the treaty allows some nations to establish limits that are farther offshore than 200 miles? Explain why you think such exceptions should or should not be allowed.

18.2 Who has the right to decide who can use the biological and mineral resources of the oceans in areas outside the EEZs of the nations? Who should benefit economically from this use? How should use of these resources be managed, and by whom? If a nation disagrees with a management decision and allows its citizens to exploit these resources without following the internationally agreed-upon rules, how should the situation be resolved?

18.3 Who do you think should finance the search for beneficial drugs and pollution-fighting organisms in the oceans? Who should benefit economically from the discovery of such commercially valuable compounds? Who should benefit from harvesting organisms to make the compounds? How should the harvesting be regulated, and by whom?

Setting the catch "too low" would mean lost income and possibly jobs for fishers, as well as unnecessary "wasting" of some of the resource value.

A number of new management concepts are now beginning to be applied to fishery resources to address this problem. These new approaches include assigning catch quotas to individuals, individual boats, or communities and granting exclusive limited fishing access to defined geographic areas. Each of these regulatory approaches embodies the principle that access to the fishery resources is a privilege and that access to these resources should no longer be open to any and all persons who choose to fish. Another conservation approach—establishment of fishing-free natural reserves where fish populations can reproduce freely—is also beginning to become more widely applied.

To this day, humans are primarily hunters and gatherers in the oceans, much as Stone Age people were on land. Human development from ocean hunter-gatherer to ocean farmer is long overdue. Mariculture (ocean farming) historically has been used on only a small scale in very few locations. However, it is developing rapidly, particularly in China, where it has been practiced for thousands of years, and in several other developing nations of the Pacific Ocean basin.

Overexploitation of resources is probably the most serious problem, but it has not been the only detrimental effect of human development on ocean fisheries during the last century. Human activities have also polluted many coastal waters, particularly **estuaries** and rivers, and have destroyed vast areas of coastal **wetlands** (Chap. 19). Rivers, estuaries, and wetlands are critically important to many fish and shellfish species, which use such areas as breeding, nursery, and feeding grounds.

In fact, fishing itself contributes to ocean contamination or pollution in many ways. Large quantities of fishing gear, including fishing line, fishing nets, and Styrofoam floats, are lost and have adverse effects on marine organisms, as discussed in Chapter 19. In addition, vessels intentionally or accidentally discharge oil and diesel fuel, oily bilge waters, sewage, and food and packaging wastes. Such discharges are regulated or banned by some nations, including the United States. However, as is true of fishing regulations, antidischarge laws are difficult to enforce on the high seas, and especially difficult in those areas that are not part of any nation's exclusive economic zone (EEZ), where many of these laws do not even apply. Fishing vessels also have hull paints that contain toxic chemicals to combat **fouling** by organisms such as **barnacles**, and these toxic chemicals can damage fishery resources. Antifouling paints are discussed in Chapter 19.

Other Biological Resources

Apart from their aesthetic value, marine species are an important pool of **genetic diversity.** Many marine species have developed unique **biochemical** methods of defending themselves against predators, **parasites,** and diseases, and of detoxifying or destroying toxic chemicals. Therefore, marine species are a major potential resource for the development of pharmaceutical drugs and pollution control methods. The search for beneficial drugs and pollution-fighting organisms in the oceans is extremely tedious and has barely begun. However, a substantial number of potentially valuable pharmaceutical products have already been isolated from marine species and are being tested for a variety of medical purposes. A small number of pharmaceuticals derived from marine organisms have already become approved for human use. These include compounds isolated from marine **sponges** such as the antiviral acyclovir and the HIV/AIDS drug azidothymidine (AZT). Many of the compounds that show promise have come from rare ocean animals or plants found only in limited areas of certain coral reefs or other threatened ocean ecosystems.

Coral reefs are like the tropical rain forests of the oceans, in that they are the most promising sources of pharmaceuticals because of their extremely high species diversity. They sustain large numbers of candidate species, any one of which may contain numerous chemical compounds potentially valuable as drugs. Some such naturally occurring compounds have been used to design similar molecules that have similar drug properties but that can be industrially produced from widely available raw materials. Unfortunately, other pharmaceutically active compounds isolated from marine species may not be readily synthesized or redesigned. If this proves to be the case, conservation measures will need to be developed and enforced to prevent impoverished villagers who live near a reef from using destructive harvesting techniques to supply the pharmaceutical industry.

An example of what could happen is provided by the industry that provides fishes for tropical aquariums. A small but growing number of fishes bred in aquariums are beginning to enter the market. However, most tropical fishes sold in the aquarium trade in the United States are captured from Philippine reefs. Some collectors in the Philippines and elsewhere capture fishes by releasing cyanide into cracks in the reef, even though this practice is illegal. When cyanide is released into the reef, many fishes and other species die, but some fishes are not killed immediately. They swim out of hiding to avoid the cyanide, but they are stunned by the chemical and, therefore, are easily captured. As many as 90% of the fishes that survive the initial cyanide collection die of the cyanide's effects during transport to the United States or after the buyer has placed the fish in a home aquarium. The ornamental (aquarium) fish market in the United States was estimated be worth about $3 billion in 2004. Only a very tiny fraction of this money ever reaches the collectors or is used to promote sustainable collection practices.

18.4 TRANSPORTATION, TRADE, AND MILITARY USE

Despite the rapid growth in air transport, surface vessels remain the principal and cheapest means of transporting cargo and people across the oceans. Large numbers of commercial, recreational, and military vessels enter or leave U.S. ports every day. The importance of the oceans for transportation is evidenced by the estimated $700 billion in shipped goods handled by U.S. ports in 2004, and the $28 billion passenger cruise industry. In addition to this value for commercial ocean transportation, the oceans have important military uses. The oceans are plied by many surface naval vessels, but submarines have become particularly important, especially to the U.S. Navy, as platforms to transport and deploy ballistic missiles. As a result, during the past 40 years intensive efforts have been made to improve ways to hide submarines and to find enemy submarines. Very extensive oceanographic studies, particularly studies of ocean surface and subsurface **currents** and of **acoustic** properties of the oceans, have been conducted to support these efforts.

The many benefits of our vessel use of the oceans for transportation, trade, military, and recreation do come at a cost to the ocean environment. Ships, particularly oil tankers, container ships, and naval vessels (especially aircraft carriers and submarines), have become progressively larger. The larger vessels require deeper ports and harbors and have increased the need for dredging of navigation channels in many bays, estuaries, and rivers. Dredging damages **benthos** at the dredging site. Dumping of the dredged material, which generally is done at a site in the estuary or coastal ocean not far from the dredging site, can also have serious impacts (Chap. 19).

The most obvious environmental effect of the millions of vessels that ply the oceans is the accidental or incidental release of petroleum products. Occasional huge spills of crude oil from tankers such as the *Exxon Valdez,* which ran aground in Prince William Sound, Alaska, in 1989, are well-known events. Although such incidents may seriously impact a small area, the amount of **petroleum hydrocarbons** released by tanker accidents is very small in comparison with the total of all the small releases that are inevitable as a result of boat operation. After all, most of the myriad vessels plying the oceans and estuaries of the world and using the thousands of ports and marinas use petroleum-fueled engines. Outboard motors, which are often poorly maintained, can be a particular problem.

Trash, garbage, and other solid wastes are still thrown overboard from most vessels on the high seas, although this practice is now illegal in the **coastal zones** of the United States and many other nations. Plastic and other floating debris discarded by vessels or thrown into streams and rivers litter **beaches** and oceans worldwide (**Fig. 18.4a**).

In many areas where vessel traffic is heavy, the seafloor is littered with cans, bottles, and other such debris. **Sediment** samples taken beneath the most heavily traveled trading routes of the last century also often contain substantial quantities of "clinker," the partially combusted pebble-sized residue of coal burning. Clinker was thrown overboard by coal-fired vessels that dominated marine trade and transport for several decades in the late nineteenth and early twentieth centuries.

The increase in vessel traffic has other less obvious environmental effects. Construction of marinas, docks, and other portside facilities (e.g., railheads, truck depots, and fueling facilities) can destroy wetlands or other valuable coastal environment (**Fig. 18.4b**), reducing the aesthetic value of the coast (**Fig. 18.4c**) and often severely affecting the coastal marine and associated terrestrial ecosystems. Such facilities contribute **contaminants**, such as petroleum products, that are spilled, washed off vehicles and roads, or deliberately discharged, both legally and illegally. Furthermore, as previously mentioned, ports, marinas, and navigation channels generally require periodic dredging to maintain adequate navigation depths.

An additional environmental effect of ocean vessel traffic, which has been identified and studied only recently, is the transport and introduction of species into areas where they do not occur naturally. These introduced or **nonindigenous species** are often carried in

FIGURE 18.4 Aesthetic degradation of the marine environment. (a) This littered beach is a turtle breeding beach at Playa Grande, Costa Rica. Note the variety of materials and the preponderance of plastics. (b) This marina at Antioch, California, was constructed on a former wetland in the San Francisco Bay delta. (c) The port of Richmond, California, occupies a large area of land, much of which was formerly wetland and might otherwise have a much higher scenic value.

the bilge waters or as fouling of ships' hulls, and they can cause major disruption of the ecosystem. For example, nonindigenous species, such as zebra mussels, Asian clams, shipworms, and aquatic weeds are among a number of serious threats, and hundreds of other nonindigenous species have now become established in various ocean and coastal ecosystems in the United States. More than 240 nonindigenous species are found in San Francisco Bay alone.

Vessels also cause environmental damage with their anchors, particularly on coral reefs. Some of the damage that anchors can cause is discussed when we consider recreational uses of the ocean later in this chapter.

18.5 OFFSHORE OIL AND GAS

Oil and gas are extracted from beneath the seafloor in many parts of the world. Most of the undiscovered oil and gas reserves are believed to lie beneath the continental shelves and **continental slopes**. The search for the sedimentary structures most likely to yield oil or gas under the oceans, and the development of technologies to drill for and produce oil and gas safely and efficiently, have been intense throughout the past several decades. Oil and gas have been produced from wells drilled in shallow water for many years. However, technological developments have steadily extended capabilities for drilling in deeper and deeper waters, and in areas where **weather**, waves, currents, and sometimes ice conditions are progressively more demanding. The search for oil and gas, and the need to identify and control the environmental impacts of this search, are a consistent and important focus of recent oceanographic studies.

In the EEZ of the United States, the most extensive known oil and gas deposits are in the Gulf of Mexico. However, substantial known reserves are present beneath the continental shelves of the northeastern and southeastern United States, California, and Alaska (**Fig. 18.5**). There may also be many undiscovered deposits, particularly off Alaska. The offshore petroleum industry is among the largest natural resource development industries on the globe. However, annual production of offshore oil and gas in U.S. waters is valued at only $25 to $40 billion a year, comparable to the value of the fishing industry but much smaller than the value of ocean transportation of goods and cruise passengers. Offshore oil and gas development elsewhere in the world, especially in the North Sea, is a much larger enterprise.

Oil and gas are used primarily as fuel for vehicles, industry, and heating, but they are also the basic raw materials for plastics, pharmaceutical and other chemicals, cosmetics, and asphalt. Although **fossil fuel** burning may be reduced to prevent further buildup of atmospheric carbon dioxide (**CC9**), petrochemicals will still be needed in the future.

Little or no petroleum is spilled during normal drilling or from producing wells. Drilling is usually done from offshore platforms supported by long legs anchored to the seafloor (**Fig. 18.6a,b**). Floating platforms (**Fig. 18.6a**) also are used, especially in deep water, and drilling in the nearshore Arctic is done from artificial gravel islands (**Fig. 18.6c**). A production platform may tap into 100 or more wells that have been drilled into the seafloor below. Directional drilling, a process in which the well pipe is drilled downward and then turned underground to drill at an angle or even horizontally away from the drill site, now allows a single well to drain oil from an area as much as 10 km in radius and to recover 20 times as much oil per drill site or platform as was possible just a few years ago. On an ocean production platform, the well pipes generally extend up through the platform's legs. On the platform, oil, gas, and water are separated, and oil and/or gas are usually transported ashore through a pipeline laid on, or buried in, the seabed.

The offshore oil industry has had a remarkable safety record, particularly in the United States. There have been very few major oil spills from offshore platforms or pipelines; the exceptions include the 1969 Santa Barbara oil spill (**Fig. 19.5a**) and the Mexican *IXTOC* platform explosion (**Fig. 19.5b**). The offshore oil industries of the United States and many other nations now equip each well and pipeline with blowout preventers and automatic shutoff valves designed to prevent the release of large quantities of oil if an accident occurs. The effectiveness of these safety precautions was well demonstrated in 2005 when hurricanes Katrina and Rita severely damaged or destroyed dozens of oil rigs in the Gulf of Mexico without causing any major oil spills. Spills and tank-cleaning discharges from oil tankers and discharges from land are much greater than spills and discharges from offshore platforms and pipelines. Tanker accidents and the effects of oil in the marine environment are discussed in Chapter 19.

During drilling, the drill pipe is lubricated with drilling muds that may contain toxic chemicals and that can cause **turbidity** and excess **sedimentation** if released. Drilling muds were once discharged into the water surrounding the platform, but they are now recycled, removed to landfills, or reinjected into the well. Oily water, called "produced water," is separated from the oil and gas on the production platform and is also now reinjected.

18.6 METHANE HYDRATES

Methane is released by the decomposition of organic matter. At low temperatures and high pressures, methane molecules can be trapped within the crystalline lattice of water ice crystals to form a combination called "methane hydrates" (on average, one molecule of methane for every five or six molecules of water ice). Methane hydrates were first observed in samples of cores drilled on land and the ocean floor several decades ago. These hydrates could be a potential source of methane, a clean-burning fossil fuel (it is completely combusted to carbon dioxide and water with virtually no chemical waste by-products or products of incomplete combustion such as are produced by the refining and burning of other fossil fuels). However, until recently, exploitation of this resource was considered unlikely for several reasons. Most importantly, methane hydrates are widely dispersed, usually occurring in the pore spaces of sediments and rocks.

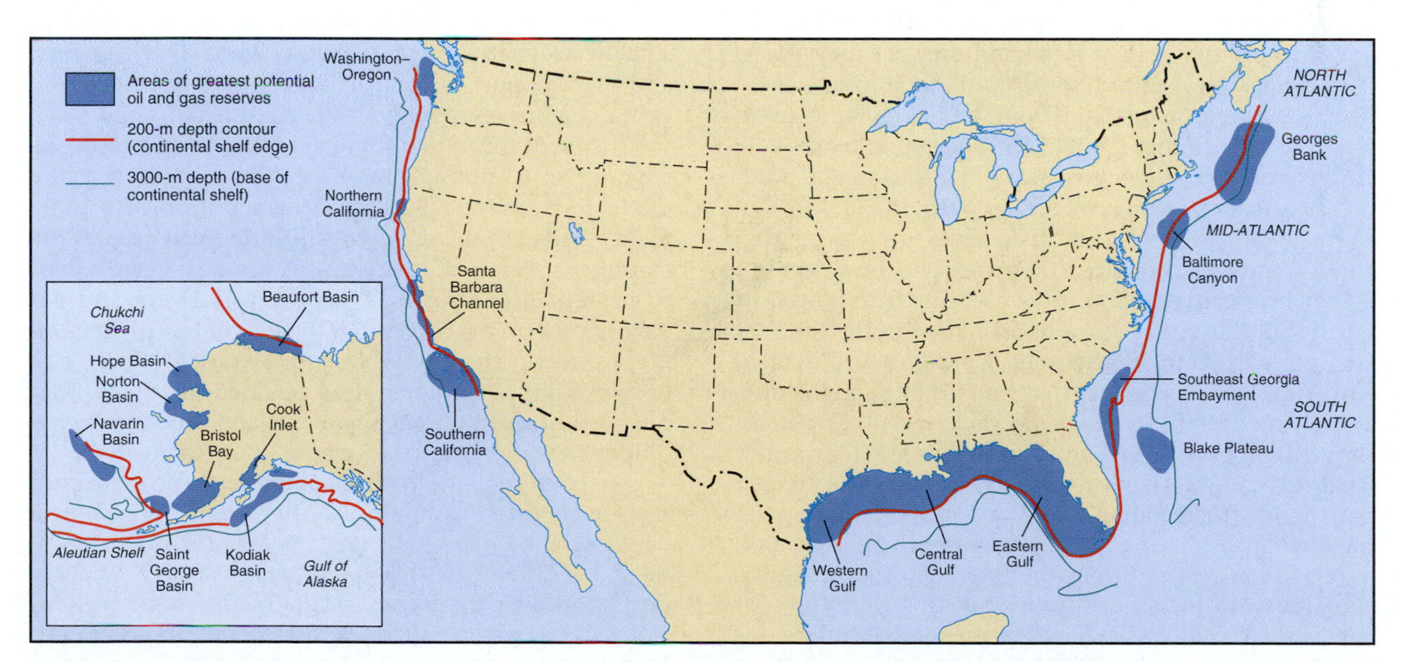

FIGURE 18.5 Offshore oil and gas deposits occur throughout much of the U.S. Atlantic, Pacific, and Arctic Oceans, Gulf of Mexico, and Bering Sea continental shelves. The areas currently thought to hold the greatest potential reserves are the deeper continental shelf and continental slope of the Gulf of Mexico and the Arctic Ocean continental shelf.

FIGURE 18.6 Structures of the offshore oil and gas industry. (a) These offshore exploration drilling rigs are under maintenance in Galveston, Texas. On the left, a typical jack-up drilling rig; on the right, a typical floating rig that rides on pontoons below the depth of most wave action. (b) Production oil rigs permanently emplaced in the seafloor of Cook Inlet, Alaska. (c) The Endicott development near Prudhoe Bay in the Alaskan Beaufort Sea has two artificial gravel islands on which a number of oil wells were drilled. The produced oil is separated from gas and water on the islands and then carried by a pipeline across a gravel causeway to the mainland. From there, it is carried south by the Trans-Alaska Pipeline to Valdez in Prince William Sound, Alaska, where it is loaded on tankers and transported to refineries.

Methane is rapidly released from the hydrates at normal atmospheric pressures and temperatures, but the deposits are generally too deep to mine and bring to the surface. It was not known until recently whether a means could be found to release methane from the hydrate in the sediment or rock so that it could be collected in a drill hole and recovered in the same way that petroleum and natural gas is recovered from oil and gas deposits.

In 2000, interest in methane hydrates was revived when some fishers dragged their trawl net at a depth of 800 m in a canyon about 50 km east of the mouth of Puget Sound in the northeastern Pacific Ocean. The fishers were startled to see their net rise to the surface filled with 1000 kg of icy chunks that were fizzing and melting. They hauled the "catch" aboard their vessel but quickly shoveled it back overboard because they had no idea what it could be. They tried to save samples in a freezer, but the low temperature alone was not enough to stop the methane from escaping the hydrate. The gas, expanding as it was released from the hydrate, even broke the containers in which they tried to store the hydrate. A year later, an ROV discovered what were described as "glaciers" of frozen methane hydrates forming outcrops on the seafloor about 50 km from where the fishers had found hydrates. Since that time, deposits of methane hydrates have been found on the seafloor or buried in shallow sediments in several other areas of the oceans, and they are now thought to occur in many parts of the world's oceans, especially on the continental slope.

The mechanism that forms methane hydrate deposits is poorly understood. Some hypotheses suggest that the methane is a product of the decomposition of organic matter buried in the sediments in **marginal seas** as the continents were pulled apart—the same source thought to be responsible for most of the world's oil and gas deposits (Chap. 4). Instead of being contained in nonporous rocks and converted to oil and natural gas, decomposing organic matter released methane that migrated through porous sediments and rocks until conditions were right for the formation of methane hydrates.

Methane hydrates are unstable except at high pressures and low temperatures, so they occur only below a depth of about 500 to 600 m in the oceans. However, these conditions are also present at relatively shallow depths beneath the frozen tundra of the Arctic regions. Not surprisingly, methane hydrate deposits have also been found in these environments. Recent estimates of the extent of methane hydrate deposits suggest that the total world resource may be more than 100 times the total volume of natural gas estimated to be recoverable from world oil and gas deposits. Thus, although most methane hydrate is widely dispersed and probably unrecoverable, if only 1% of the total were contained in concentrated deposits that could be recovered, the economic value would be huge. The methane would also be a very efficient and clean-burning fuel that could be readily adapted to most commercial and industrial energy uses.

As a result, large investments are being made in researching the occurrence and potential development of methane hydrate resources. In 2003, preliminary results of a drilling project in the Canadian Arctic suggested that simply lowering the pressure in a well hole can indeed cause methane to be released at depth. Several other projects are under way to determine whether this result can be replicated in methane hydrate deposits in different types of rock and sediment and to determine whether the methane can be released far enough away from the walls of a drilled well that it can be collected from a sufficiently large area around the well to be economic. In 2001, the Japanese began a 15-year program of exploratory drilling and research with the intention of recovering methane commercially in the future from methane hydrate deposits several hundred meters below the sediment surface in the Nankai trough 60 km off southeast Japan.

Although it appears that the engineering challenges associated with recovering continental slope methane hydrate resources can be met, little is yet known about the possible associated environmental impacts. For example, there is concern that the methane hydrate deposits may help cement the continental slope sediments in place and that mining these deposits could lead to sediment **slumps, turbidity currents,** and **tsunamis.** Methane is also a greenhouse gas (**CC9**) that could contribute to global **climate** change if released to the atmosphere through mining operations. Even if methane hydrates are not mined, they may be released if ocean waters warm significantly.

Methane released to the oceans and atmosphere might cause a significant positive **feedback** in a globally warming environment. Slumps, turbidity currents, and tsunamis may also result if the methane hydrates are melted as the oceans warm. Truly massive turbidity currents have occurred in the past (Chap. 8), and some evidence suggests that some of these may have been caused or enhanced by methane releases within the sediments that created a gas-rich layer in which **friction** was reduced. The reduced friction would allow the slumping sediments above to slide more easily.

CRITICAL THINKING QUESTIONS

18.4 Nearly 71% of the Earth's crust lies under its oceans. Available supplies of valuable minerals and oil and gas in the deposits beneath land are steadily becoming scarcer as they are depleted. Should minerals and oil and gas be removed from under the ocean? What factors need to be considered to make this decision?

18.5 A swarm of locusts threatens the crops of a large area of North Africa. Should the locusts be eradicated to protect the food? A growing population of fur seals threatens to consume a growing percentage of fish from a regional fishery where the human population depends heavily on seafood. If the fish populations are not to crash from overfishing, the catch must be reduced, creating high prices and market shortages, or the fur seal population must be controlled. Would you kill seals to keep the population from growing so large as to damage the fishery? Did you have different reactions to these two scenarios? If so, why?

18.7 MINERALS AND FRESHWATER

Ocean sediments contain vast quantities of mineral and material resources other than just oil and gas. They include sand and gravel (**Fig. 18.7a**), manganese nodules, **hydrothermal minerals, phosphorite nodules,** and heavy minerals such as gold that are often present in sediments of current or ancient river mouths (**Fig. 18.7b**). Sand and gravel are currently mined in large quantities from the shallow seafloor and used as construction materials in locations where no local land resources are available. At present, few efforts have been made to exploit other marine mineral resources. Cassiterite, a tin mineral, is dredged from shallow waters offshore from Thailand and Indonesia, and gold-bearing sands are dredged from shallow river mouth deposits offshore from Alaska, New Zealand, and the Philippines. Despite the very limited scope of current ocean mining activities, there is substantial interest in future development. Mineral deposits thought to be most likely exploitable include phosphorite-rich sands as potential sources of phosphorus for fertilizer, manganese nodules, and hydrothermal mineral deposits as potential sources of metals including iron, zinc, copper, nickel, cobalt, manganese, molybdenum, silver, gold, and platinum.

Because of the currently limited extent of ocean mining, we know relatively little about its potential environmental effects. However, sand and gravel mining causes increased turbidity that reduces light penetration in the area around the mining site. If a mining area is near contaminant sources and has relatively fine-grained sands, substantial quantities of toxic chemicals may be released in **suspended sediments** and to solution. Mining impacts are probably relatively small in coastal and estuarine locations that have naturally high suspended sediment loads, but they could be severe in low-turbidity waters, particularly near coral reefs, because **corals** and other organisms are adversely affected by high turbidity.

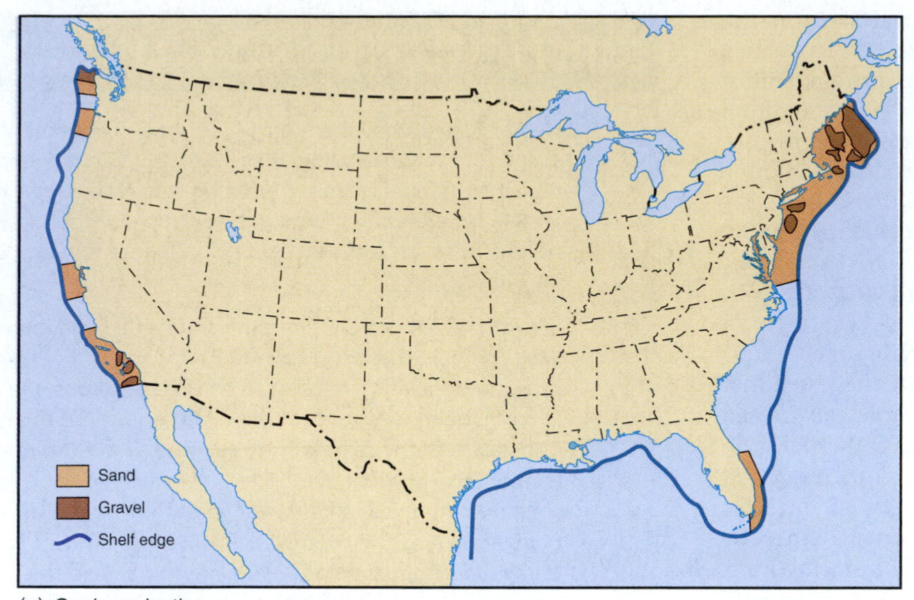

(a) Conic projection

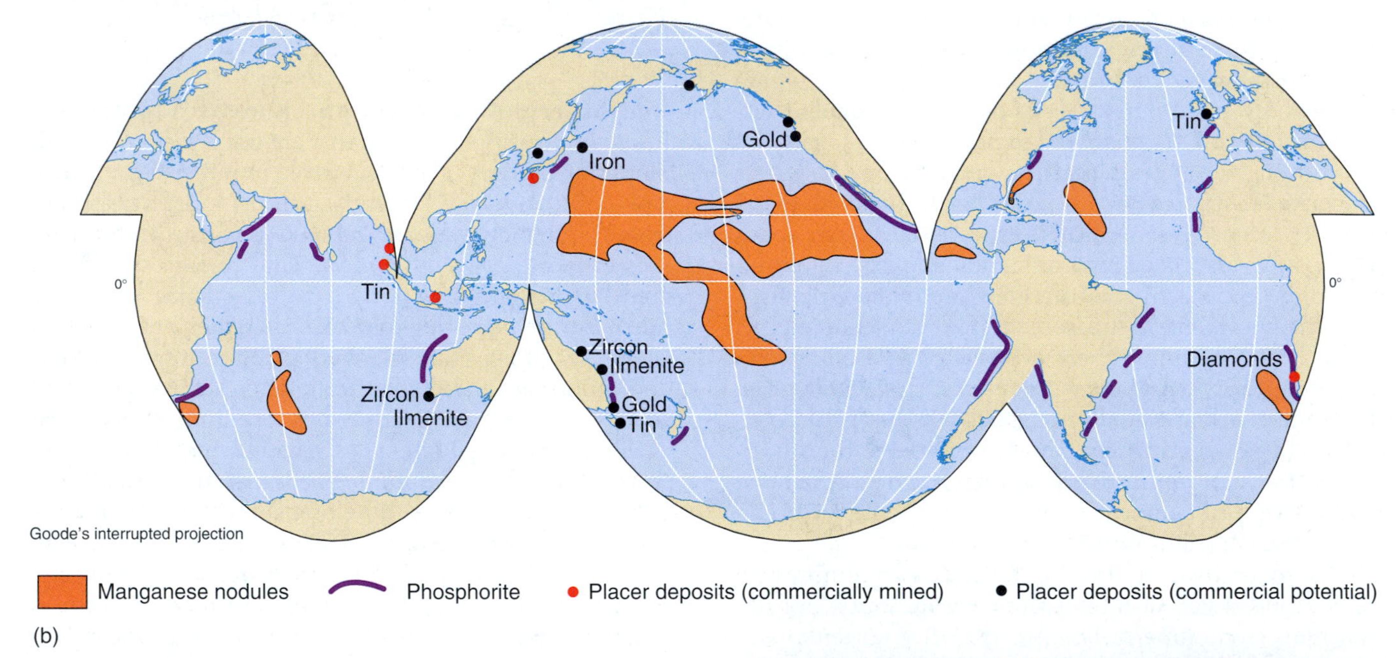

(b)

FIGURE 18.7 Some mineral resources of the oceans. (a) There are abundant sand and gravel resources on the continental shelves of the U.S. Atlantic and Pacific coasts in areas where fine-grained muddy sediments do not dominate. (b) The principal mineral resources of the seafloor are manganese nodules that occur in areas of abyssal seafloor, phosphorite nodules and deposits that occur primarily on the continental shelf, and placer deposits (minerals containing gold and other heavy metals) that accumulate at river mouths. There are a number of commercial mining operations in different parts of the world, as well as other areas where potentially commercial deposits occur. (c) This aerial photograph shows the levees that enclose some of the solar evaporation ponds used to produce common salt in South San Francisco Bay, California. The large white mounds are piles of salt collected from the ponds and awaiting processing, packaging, and sale.

Mining definitely has serious impacts on the **benthic community** and benthic habitat at the mining site itself, as a result of the physical disturbance of the seabed by the mining process.

Many mineral resources, particularly manganese nodules and hydrothermal minerals, are found primarily in the deep oceans far from land. The discovery of such mineral deposits has led to extensive oceanographic research to identify the processes that created them and to determine their distribution and abundance on the seafloor. At present, deep-ocean minerals are too expensive to mine in comparison to the dwindling, but still adequate, sources on land. However, the potential future value of those resources is considerable. If deep-ocean mining is developed, it may have significant environmental impacts if large quantities of fine-grained sediment are released into naturally clear waters of the open-ocean photic zone.

Until 1982, the mineral resources of the deep-ocean floor were not owned by any nation or individual and legally could be mined by anyone. During the 1960s and 1970s, a widespread fear arose in the international community that deep-ocean minerals might be exploited and depleted to the benefit of only one or two nations that commanded the technology to mine them. This fear was a principal driving force behind the negotiations that led to the Law of the Sea Treaty, described earlier in this chapter.

Table salt and freshwater are both produced from ocean waters. Salt is produced by evaporation in coastal **lagoons** (**Fig. 18.7c**). Although the salt industry has little direct impact on the marine environment, evaporation ponds are usually constructed in coastal wetlands that could otherwise serve as nursery areas for various marine species. Freshwater is produced by evaporation or by reverse-**osmosis** extraction units. The process generates high-**salinity brine** wastes that are discharged to the oceans. The high salinity may have adverse effects on the **biota,** especially if water temperature is also high and causes additional stress. Such situations occur in locations, such as the Persian Gulf (especially off Saudi Arabia), where freshwater production from seawater is practiced most intensively.

18.8 RECREATION, AESTHETICS, AND ENDANGERED SPECIES

Humans have probably always enjoyed living on the coast, not only for the ocean's food resources, but also for its aesthetic qualities and its moderating effects on climate (Chap. 9, **CC5**). It is the aesthetic qualities of oceans that have inspired poets and artists for millennia. However, only within the past generation or two have the oceans become popular for a variety of recreational activities, including sunbathing, swimming, surfing, sailing, luxury cruises, snorkeling, and scuba diving. With the development of such pastimes, a much wider spectrum of the human population now considers the ocean environment to be a special part of nature that should not be despoiled, as so much of the land has been. Underwater photography and video and the far-reaching impact of television have also introduced a large percentage of the human race to the beautiful, strange, and alien world of marine life.

As these recreational uses have become more popular, there has also been a growing recognition of the effects that such activities have on the marine environment. Oceanographic research is critical to the protection of marine ecosystems for, and from, recreational uses. For example, anchor damage by boats carrying recreational scuba divers has become so severe in some locations that anchoring is now illegal, and boats must tie up to permanent mooring buoys installed at carefully selected locations on the most heavily dived reefs. Scuba divers also can damage coral reefs by breaking coral with their fins or hands. Although incidental contact with the reef and some coral damage is inevitable, most divers are now careful to protect the reefs on which they dive. Damage to reefs by divers is a significant problem only on the most popular reefs, where many divers are in the water every day. Where divers are only periodic visitors, the reefs are able to recover from any minor damage they may sustain. Indeed, damage occurs naturally as a result of the feeding and other activities of large reef animals, such as parrotfishes, and as a result of the effects of periodic strong storms. Sometimes such damage is even beneficial to the health and species diversity of coral reef communities (**CC17**).

On a broader scale, we have finally recognized that rampant development of the planet has threatened the existence of many species. Although species destruction has probably been greater on land than in the ocean, certain species of whales, seals, penguins, turtles, and pelicans have been brought to the brink of extinction by human exploitation or pollution. Many other marine species are threatened, particularly species that live in coastal and estuarine environments where human impacts have been greatest. The protection of ocean species from destruction depends on our understanding of marine life and its interactions with the ocean environment. For example, Norway conducts an annual hunt to reduce the population of fur seals because fishers believe that there are too many seals and that the seals eat fishes that otherwise would be available for human consumption. More scientific data are needed to support the belief that the seals, if left to reproduce unharmed, would significantly reduce the available fish resource. However, if this is proven to be true, there will still remain a major debate as to whether we should kill seals in order to ensure that food is available for humans.

This is just one example of the ethical dilemmas that must be resolved if we are to manage fishery resources in a way that balances the needs of humans and marine animals. The protection of certain marine animals, notably marine mammals, has allowed their populations to recover, and continued protection, coupled with increasing human demand for seafood, will create many more situations in which fishers and marine animals are competing for a diminishing resource.

18.9 ENERGY

As fossil fuels have been depleted and nuclear energy has lost favor, the search has begun for alternate sources of energy that will be needed in the future. The search will probably accelerate as the environmental consequences of fossil fuel burning become better known. The oceans offer several potential sources of energy, including **tides**, waves, ocean currents, and thermal gradients. Full development of these energy resources, if it were possible, would allow substantial reduction in the use of fossil fuels and, consequently, in the release rate of carbon dioxide and air contaminants, such as sulfur dioxide, nitrogen oxides, and particulates. However, at present very little energy is generated from any ocean resource. Substantial engineering problems must be solved before ocean energy resources can contribute significantly to global energy demands. In addition, many questions must be answered about possible environmental impacts of ocean energy generation technologies.

The only ocean energy source currently exploited on a commercial scale is the tides (Chap. 12). The technology used is exemplified by the tidal power plant at La Rance, France (**Fig. 12.21**). Because this technology requires a dam across the mouth of a bay or other inlet in which the **tidal range** is very large, relatively few locations are suitable (**Fig. 12.20**). To generate power during most of the tidal cycle and adjust to fluctuating energy demands during the day, water must be stored behind the dam during part of the tidal cycle and restricted from moving into the inlet during the **flood.** Consequently, tidal currents may be considerably reduced, **residence time** increased, and sedimentation altered in a way that can be detrimental to benthic biota, including valuable shellfish and bottom-dwelling fishes. The dam also hinders the passage of organisms and vessels.

A number of technologies have been proposed for energy recovery from currents. The most likely technology would involve placement of huge fanlike turbines in a current, but many huge turbines placed side by side across the current would be needed to generate significant power. For example, there is a plan to deploy such turbines at the mouth of San Francisco Bay to capture the energy of the tidal currents.

The nature and scope of possible adverse environmental effects of the proposed technologies are unknown, but almost certainly they would include the effects of turbines on the transit of **pelagic** organisms and the effects of toxic paints or coatings that would be needed to prevent fouling of the turbines by marine organisms. Fouling could not be tolerated, because it would drastically reduce turbine efficiency and eventually cause turbine bearings to fail. Currents and **turbulence** undoubtedly would be altered in the immediate vicinity of the turbine, which could lead to increased sedimentation or erosion. However, the major concern is that modifying currents could have other far-reaching effects. For example, if enough turbines were deployed across the Gulf Stream to generate the power equivalent of several nuclear power plants, energy withdrawn from the current and the **eddy** motion caused by the turbines could conceivably alter the direction and speed of the Gulf Stream to some small extent. Even the remote possibility of such adverse consequences will probably prevent investment of the large sums of money needed to develop this technology.

The energy potentially available from ocean waves and currents is truly enormous, but such energy is widely dispersed and will not be easy to harness. Coastlines where wave energy is greatest, and therefore the potential for generating energy from waves is highest, are shown in **Figure 18.8a**.

Several design approaches have been developed to extract energy from waves, although none has yet been tested on a large scale. One relatively simple design uses the rise and fall of water at the **shoreline** to compress air within a boxlike structure or shaft, which is open to the ocean only below the waterline (**Fig. 18.8b**). The compressed air drives a turbine. A variant of this design uses

FIGURE 18.8 *(opposite)* Harnessing energy from ocean waves. (a) Waves could be harnessed for energy in areas where average wave energy is high. Such areas are primarily in high latitudes (where strong storms are more frequent), on west coasts of continents (because most storms move east to west with the westerlies), and on southern coasts of the Southern Hemisphere (because strong storms are frequent around Antarctica, and no continents limit the fetch or intercept the waves). (b) Various wave energy generators have been successfully used on some rocky coastlines. The design shown here has a chamber open to the sea just below mean sea level into which waves can force water. As water is forced in, the air in the chamber is compressed and drives a turbine. (c) One of several designs for wave energy generators that float on the ocean surface. This one is a "dam atoll." Most of its mass is below the depth of wave motion, such that it does not rise and fall as waves pass. Instead, water enters the generator at the surface, spirals downward, and drives a turbine. (d) A wave energy generator that uses a field of mechanical pumps on the shallow seafloor to pump seawater into a reservoir, where it can be released through a generator. (e) Floating OTEC power plants would consist of a long vertical pipe extending downward, and a floating platform to support the power generation equipment.

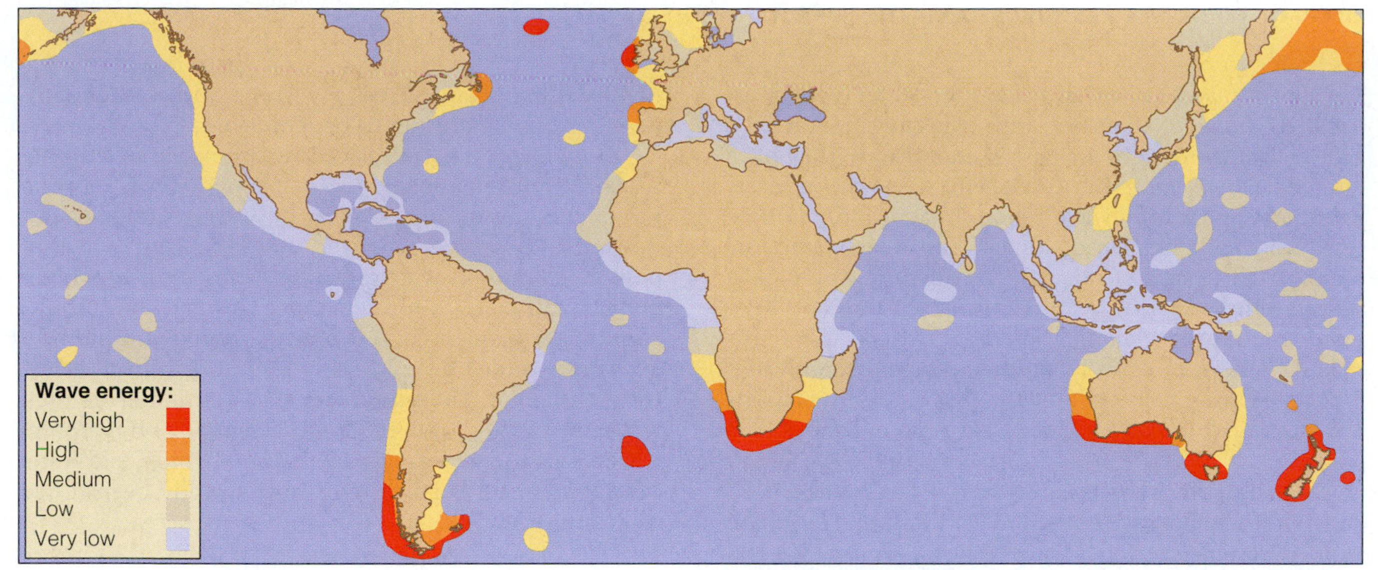
Wave energy:
Very high
High
Medium
Low
Very low

(a)

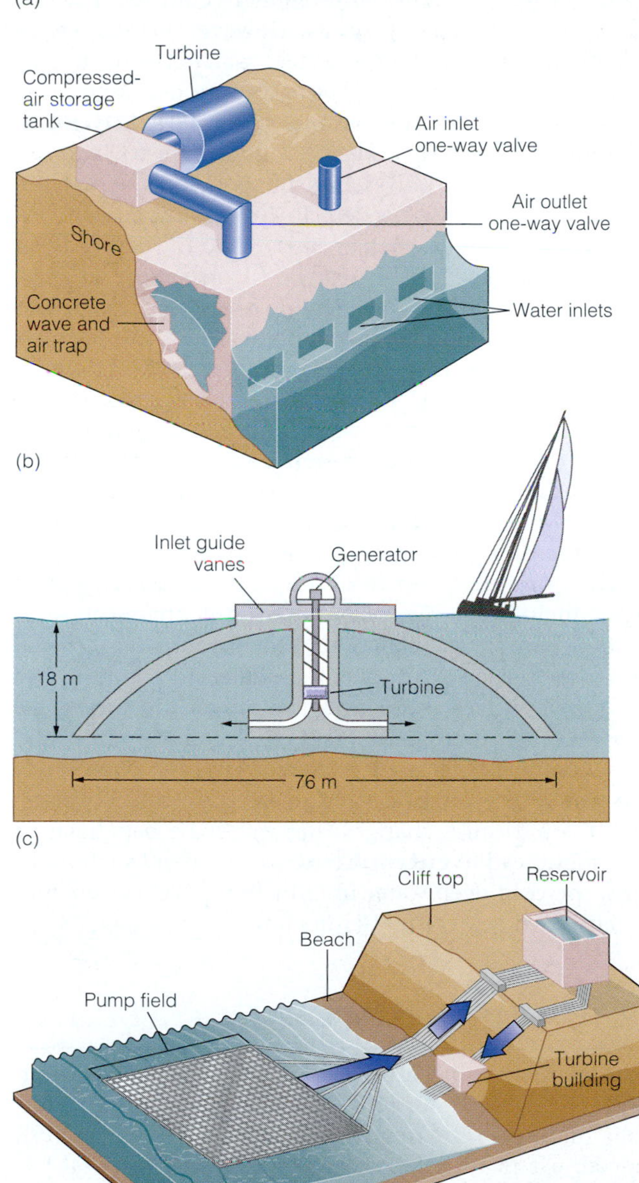
Turbine
Compressed-air storage tank
Air inlet one-way valve
Air outlet one-way valve
Shore
Concrete wave and air trap
Water inlets
Inlet guide vanes
Generator
18 m
Turbine
76 m
Cliff top
Reservoir
Beach
Pump field
Turbine building
Ocean
Pumped and released seawater

(b)

(c)

(d)

(e)

the upward and downward surge of water within an enclosed box or shaft to drive the turbine directly. Although these designs are simple and proven, they have limited value because they can be effective only on coasts with persistent strong wave action. In addition, multiple generators must be placed along a coast if large amounts of power are to be generated. Such structures would have adverse effects on the aesthetics of the shore and alter sediment transport along the coast. Generators of this type have been built to serve small coastal communities in Norway. Ironically, one of the first such installations was destroyed by severe storm waves soon after it was built.

Several wave power extraction devices have been designed to be deployed offshore. For example, the "dam atoll" is designed to focus waves toward a central generator shaft (**Fig. 18.8c**). A more recent design appears to overcome some of the disadvantages of earlier systems (**Fig. 18.8d**). In this design, a matrix of simple mechanical pumps would be placed just beyond the **surf zone,** where they would use wave energy to pump seawater up into a reservoir on the adjacent land. Power could then be produced, even when the wave energy was low, by release of the seawater from the reservoir through a turbine. The pumps would be submerged, minimizing the aesthetic impact. However, they would need to cover very large areas of seafloor and fill large reservoirs to produce as much power as a typical power plant, and the pumps would have to withstand wave impacts during even the largest storms. This system may soon be tested in northern California.

Presumably, some wave energy generator designs could succeed with appropriate development, but they would need to be deployed not far offshore in strings stretching along many kilometers of coast. They could be a navigation hazard and might interfere with movements of marine organisms, particularly marine mammals. In addition, they would probably require toxic antifouling paints or coatings. Even worse, a string of such wave generators along a coast would drastically reduce wave energy at the shore. The result would be reduction of **longshore drift**, change of sediment **grain size**, and change of both the sedimentation regime and the associated biology of the nearshore zone. On rocky coasts, the important **supralittoral zone** with its unique biota (Chap. 17) would be particularly affected.

The quantity of ocean thermal energy that potentially can be exploited is huge and may be easier to extract than wave or current energy is. Ocean thermal energy conversion (OTEC) systems exploit the temperature difference between deep and shallow waters in the oceans to drive a turbine and generate electricity. The process by which energy is extracted to run the turbines is analogous to a refrigerator running backward, or a power plant operating at unusually low temperatures. Conventional power plants use heat from burning fossil fuel or a nuclear reaction to vaporize water in a closed container. The resulting high-pressure steam drives a turbine and is then condensed by cooling water to be recycled.

In OTEC, water is replaced by ammonia or another suitable liquid that is vaporized at much lower temperatures. The ammonia is heated by warm surface waters flowing over heat exchanger tubes through which it passes. Ammonia evaporates, and the resulting high-pressure gas drives a turbine. Once through the turbine, ammonia is condensed as it passes through another heat exchanger cooled by cold water pumped up from below the permanent **thermocline**. The ammonia is then recycled. An OTEC power plant is essentially built around a wide pipe that reaches to depths from which cold water can be pumped. Such systems have already been tested successfully.

OTEC is very promising and is probably the form of ocean energy generation most likely to contribute significantly to world energy needs. However, OTEC can be used efficiently only where surface waters are warm and cold deep waters are accessible—conditions found year-round primarily in tropical and subtropical **latitudes**. Locations where deep water is found close to land are ideal because long power transmission lines across the seabed would not be needed. In North America, only a few coastal locations, such as Hawaii and the east coast of South Florida, are suitable. Pacific island nations are perfect locations, but the considerable investment needed to develop and build OTEC power plants is difficult for such nations to afford. This situation might change if floating OTEC platforms (**Fig. 18.8e**) can be developed that use the electricity they generate to produce hydrogen by electrolysis of water. The hydrogen could be either transported ashore in tankers to be used as a fuel or used to generate ammonia (for use in fertilizer and chemical manufacturing) from atmospheric nitrogen.

Potential environmental impacts of OTEC power plants include the release of antifouling chemicals, and aesthetic problems related to the onshore or offshore structures needed to support the plants. In addition, the biology of surface waters may be altered by **bacteria**, **viruses,** and other microorganisms brought up from the deep ocean and released in surface waters. **Nutrients** also would be brought up and released to surface waters, which could cause changes that might be beneficial or that might lead to **eutrophication.** To avoid such problems, the cold deep water used in the OTEC cycle could be reinjected below the thermocline.

18.10 WASTE DISPOSAL

The oceans have been used for waste disposal for thousands of years and, for most of that time, without significant harm to the marine environment (Chap. 19). Indeed, use of the oceans for sewage waste disposal has historically proven to be one of the most effective advancements ever made in human health protection.

Recent oceanographic research has documented a variety of major problems caused by the disposal of certain wastes in parts of the oceans. Although the research has led to mitigation of many of the worst impacts, much remains to be learned. The oceans will undoubtedly continue to be used for waste disposal, and this may be the most environmentally sound management approach for some wastes. However, a much better understanding of the oceans is necessary to determine which wastes can be disposed of in this way, in what quantities, where, and how, without adversely affecting the environment or human health.

The oceans have a very great capacity to assimilate, safely and completely, large quantities of natural wastes if these materials are widely dispersed or released slowly enough that they are thoroughly mixed into the huge volume of the open oceans. Hence, the oceans suffered little from waste disposal practices until the past century, when human populations and modern industry began to grow explosively. Problems arose when cities and industries grew and became concentrated in large urban areas. This concentration caused the rate of disposal in many coastal and estuarine areas to exceed the rate at which the wastes could be dispersed and assimilated. Additional problems arose with the development of synthetic chemicals and materials, because the ocean ecosystem had no mechanisms to destroy or neutralize many of these substances.

Most of the wastes currently disposed in the oceans are liquids or slurries, and they are disposed of by being discharged through pipelines, called **outfalls,** into rivers, estuaries, or the coastal ocean. At one time, quantities of a variety of solid wastes, including chemicals, low-level **radioactive** wastes, construction debris, and trash, were transported to sea on vessels and dumped with the assumption that they would simply fall to the ocean floor, most of which was thought to be lifeless, and not cause any harm. Growing understanding of ocean ecosystems and the effects of ocean dumping have now led to the elimination of almost all dumping of solid wastes in the oceans. Dredged material is now the only waste material dumped at sea from vessels in large quantities.

Waste disposal and its effects in the oceans are discussed in detail in Chapter 19.

CHAPTER SUMMARY

Law of the Sea. For most of recorded history, the resources of the oceans were not owned and could be exploited by anyone. In 1672, the British claimed ownership of a territorial sea that reached 5.6 km offshore. Such claims became common practice until 1945, when President Truman claimed for the United States ownership of the resources of the seafloor offshore to where the ocean reached a depth of 183 m. Many similar claims by other nations ensued. The resulting confusion was resolved by United Nations "Law of the Sea" conferences in 1958 and 1960, and passage in 1982 of a comprehensive Law of the Sea Treaty after more than a decade of negotiations.

The Law of the Sea Treaty grants to a coastal state ownership of all fisheries and mineral resources within an EEZ that is 200 nautical miles wide. The most controversial treaty provisions apply the principle that minerals of the deep-ocean floor outside of EEZs are the "common heritage of mankind." The United States and several other nations did not sign the treaty because of these provisions. In 1994, enough nations ratified the treaty for it to become effective, and the United States signed it.

The seabed and fishery resources of an EEZ can be very valuable. Consequently, sovereign nations now consider otherwise inconsequential islands valuable. The result has been a variety of territorial disputes, and even wars, over such islands.

Value of Ocean Resources. The oceans provide abundant resources, including fisheries and other biological resources; transportation, trade, and military use; offshore oil and gas; methane hydrates; minerals and freshwater; recreation, aesthetics, and endangered species; energy; and waste disposal.

Biological Resources. Most of the world's major fisheries are overfished. The principal reason is that most fisheries are an unowned resource open to anyone who wants to exploit them. Many fishing techniques collect and kill nontarget species that are often discarded. Fishing line, nets, Styrofoam floats, and other items are discarded or lost and cause beach pollution. Aside from fisheries and the aesthetic value of marine species, ocean biological resources include ornamental species used in aquariums, and pharmaceuticals extracted or developed from marine species.

Transportation, Trade, and Military Use. Most goods traded by humans are transported in surface vessels. The oceans are also extensively used by the world's navies, both surface vessels and submarines. Research to support naval uses, especially hiding and hunting submarines, has been critical to the development of oceanography. Recreation on cruise liners and small craft is increasing rapidly. Occasional spills, especially from oil tankers, releases of oil from boat motors and ship engines, dumping of trash by some vessels on the high seas, construction of ports and portside facilities, discharge of bilge water, antifouling paints, anchors, and dredging to maintain navigation channels all have environmental effects, often deleterious.

Offshore Oil and Gas. Most undiscovered oil is beneath the continental shelves and slopes. Offshore drilling and production platforms or islands are used throughout the world. These facilities have, on a few occasions, accidentally spilled large amounts of oil, and some of them discharge drilling muds and oily water.

Methane Hydrates. Large accumulations of methane hydrates have recently been discovered in ocean sediments, especially on the continental slope. Because methane burns cleanly, methane hydrates could provide a large and desirable source of energy. Studies are now being conducted to develop technological means for extracting the methane economically. Environmental impacts are uncertain but could include accidental release of methane (a greenhouse gas) and disturbance of continental slope sediments that could cause turbidity currents and tsunamis.

Minerals and Freshwater. Ocean mining for sand and minerals, although limited, is increasing. Mining alters habitat at the mine site and may lead to the discharge of tailings and other wastes. Coastal wetlands in some areas have been altered to construct evaporation ponds for salt production. Freshwater production from seawater generally discharges high-salinity brines.

Recreation, Aesthetics, and Endangered Species. Human populations have historically been concentrated on the coast for its aesthetic values and moderation of climate. These values are becoming increasingly important as human use of the oceans for many forms of recreation increases. Humans and their recreational activities have many effects on the marine environment that are becoming better studied and understood. The protection of ocean ecosystems and species has emerged as an important goal, but difficult conflicts often occur between the need for protection and the need for resource uses.

Energy. Winds, waves, tides, currents, and the temperature difference between surface and deep waters are all potential sources of energy. So far, only tidal energy is commercially developed. Facilities to generate energy from these sources may alter current, wave, and habitat characteristics and cause contamination from antifouling paints. OTEC may be the most promising technology, particularly for tropical island communities, if floating facilities can be developed.

Waste Disposal. The oceans are capable of assimilating large quantities of some wastes without any significant negative impacts. Wastes have been disposed of in the oceans for thousands of years, but increasing amounts have caused significant negative impacts on the ocean environment, especially when dumped or discharged in locations or by means that allow them to accumulate locally. Except for dredged material, most wastes now disposed of are liquids or slurries, such as treated sewage.

STUDY QUESTIONS

1. Why don't we generate a significant amount of energy from tides, waves, and currents?
2. List the human uses and resources of the oceans.
3. Why is an international treaty that establishes a special set of laws for the oceans important?
4. Discuss how you would apply the "common heritage of mankind" principle to the protection and hunting of terrestrial wild animals whose habitats cross national boundaries.

KEY TERMS

You should be able to recognize and understand the meaning of all terms that are in boldface type in the text. All of those terms are defined in the Glossary. The following are some less familiar key scientific terms that are used in this chapter and that are essential to know and be able to use in classroom discussions or on exams.

bycatch (p. 533)
drift net (p. 533)
eutrophication (p. 544)
exclusive economic zone (EEZ) (p. 530)
fouling (p. 534)
high seas (p. 528)
hydrothermal minerals (p. 539)
manganese nodules (p. 529)
maximum sustainable yield (p. 532)
nonindigenous species (p. 535)
outfall (p. 545)
overfished (p. 532)
phosphorite nodules (p. 539)
territorial sea (p. 528)

CRITICAL CONCEPTS REMINDER

CC5 **Transfer and Storage of Heat by Water** (p. 541). Water's high heat capacity allows large amounts of heat to be stored in the oceans and released to the atmosphere without much change of ocean water temperature. Water's high latent heat of vaporization allows large amounts of heat to be transferred to the atmosphere in water vapor and then transported elsewhere. Water's high latent heat of fusion allows ice to act as a heat buffer reducing climate extremes in high latitude regions. To read **CC5** go to page 15CC.

CC9 **The Global Greenhouse Effect** (pp. 536, 539). Perhaps the greatest environmental challenge faced by humans is the prospect that major climate changes may be an inevitable result of our burning fossil fuels. The burning of fossil fuels releases carbon dioxide and other gases into the atmosphere where they accumulate and act like the glass of a greenhouse, retaining more of the sun's heat. To read **CC9** go to page 22CC.

CC11 **Chaos** (p. 533). The nonlinear nature of many environmental interactions, including some of those that control annual fluctuations in fish stocks, mean that fish stocks change in sometimes unpredictable ways. To read **CC11** go to page 28CC.

CC16 **Maximum Sustainable Yield** (pp. 532, 533). The maximum sustainable yield is the maximum biomass of a fish species that can be depleted annually by fishing but that can still be replaced by reproduction. This yield changes unpredictably from year to year in response to the climate and other factors. The populations of many fish species worldwide have declined drastically when they have been overfished (beyond their maximum sustainable yield) in one or more years when the yield was lower than the average annual yield on which most fisheries management are based. To read **CC16** go to page 51CC.

CC17 **Species Diversity and Biodiversity** (p. 541). Biodiversity is an expression of the range of genetic diversity; species diversity; diversity in ecological niches and types of communities of organisms (ecosystem diversity); diversity of feeding, reproduction, and predator avoidance strategies (physiological diversity), within the ecosystem of the specified region. Species diversity is a more precisely defined term and is a measure of the species richness (number of species) and species evenness (extent to which the community has balanced populations with no dominant species). High diversity and biodiversity are generally associated with ecosystems that are resistant to change. To read **CC17** go to page 53CC.

CHAPTER 19

CHAPTER OUTLINE

CRITICAL CONCEPTS USED IN THIS CHAPTER

Ocean pollution comes from many sources and takes many forms, but one species is the ultimate origin of all pollution.

Pollution

As a child, I was taught that the oceans were vast and mostly untouched by humans, and that not only was it acceptable to dispose of waste materials in the oceans, but also this practice could cause little harm. The prevailing belief of the day was illustrated by the oft-repeated saying "The solution to pollution is dilution," and the oceans were the ultimate source of dilution. Today, we all "know" that the oceans are polluted, that this pollution has caused serious damage, and that the discharge or dumping of any wastes into the oceans is unacceptable. I have spent my entire professional life studying waste disposal and pollution, especially pollution in ocean environments, and it is clear to me (and to the majority of my scientific colleagues) that the system of beliefs in place when I was a child was wrong. However, it is just as clear to me that the system of beliefs that many accept as "facts" today is equally wrong. In fact, there are elements of truth and untruth in both sets of beliefs, as I hope you will be able to see for yourself as you read, critically assess, and discuss the information contained in this chapter.

The current belief that the oceans cannot be used for disposal of any of the many waste products of human society is based on the "precautionary principle." This principle, first proposed in Germany in 1986, argues that no wastes should be discharged into the sea, because we cannot reliably predict the effects of the addition of these wastes to the oceans. Consider what you have learned in other chapters in this text about how difficult it is to predict the weather and future climate, and how difficult it is to reach a scientific consensus on even a very well-supported theory such as plate tectonics. Ask yourself what other human activities we would have to ban if we were to apply the same precautionary principle to terrestrial environments and the atmosphere.

Chapters 2 and 18 reviewed the historical importance of the ocean and the growing use of ocean resources for fisheries; transportation; trade; extraction of offshore oil, gas, and other minerals; pharmaceuticals; energy; and recreational and aesthetic opportunities. Chapter 18 also reviewed some of the ways, or potential ways, in which various uses of ocean resources can have deleterious effects on ocean **ecosystems** and on other uses of the ocean. Such deleterious effects are encompassed by the term **pollution,** which can be characterized as "the addition of substances to, or alteration of, the ocean ecosystem in a manner that is deleterious to the ocean ecosystem or its resources." This definition, or a similar one, is generally accepted by many national and international organizations that have responsibilities for managing and protecting the oceans and ocean resources. For example, the formal definition accepted by the premier world scientific body in this area, the United Nations Joint Group of Experts on the Scientific Aspects of Marine Environmental Protection (GESAMP), is as follows:

> Pollution means the introduction by man, directly or indirectly, of substances or energy into the marine environment (including estuaries) resulting in such deleterious effects as harm to living resources, hazards to human health, hindrance to maritime activities including fishing, impairment of quality for use of sea water and reduction of amenities.

Notice that, according to the generally accepted official definition, the term *pollution* includes not only the discharge of harmful wastes, but also activities such as **overfishing,** construction of structures, and other human activities that adversely affect ocean ecosystems by changing **currents,** distributions of dissolved substances or heat, or **sediment** transport patterns. The definition also includes adverse impacts of one ocean use on another, such as the degradation of recreational and aesthetic value caused by trash on ocean **beaches** or offshore oil rigs "spoiling" the ocean view. Notice also that there must be harmful effects before any activity can be considered pollution. By contrast, the popular definition of *pollution* continues to be restricted to the addition of substances (chemicals, materials, or organisms) to the **environment,** and the term often is misused to include any such contamination, whether "deleterious" or not.

19.1 POLLUTION VERSUS CONTAMINATION

All too often, any human activity that releases wastes or introduces particulate or dissolved substances to the ocean, either by accident or incidental to other activities, is mistakenly reported as "pollution." In many cases, however, such releases are benign or even beneficial to the ocean ecosystem and ocean resources. In such cases, the material released is a **contaminant,** not a pollutant. Only when the ocean ecosystem or ocean resources are damaged should the activity be called "pollution." Hence, human activities may contaminate the oceans without polluting them.

For many years, the oceans were considered so vast that human populations could safely discharge all their wastes there and carelessly exploit ocean resources such as fisheries and mineral deposits without causing adverse effects. In recent years, the recognition that the oceans can be harmed by human exploitation and waste disposal has led to another viewpoint: that ocean ecosystems are so fragile that they must be protected from any human influence that may change them, and that no inputs of contaminants can be permitted. Just as the historical view of the oceans as limitless was wrong, this new view, although idealistic, is also incorrect. In some cases, the new view may have led to political decisions that, although they may possibly have reduced ocean pollution, also increased human health risks and terrestrial pollution. The oceans are naturally changeable, and humans have caused many changes in the oceans, just as we have in the terrestrial environment. If human civilization is to continue, further changes in both the terrestrial and ocean ecosystems are inevitable and necessary. We must learn to view the planet as a whole and recognize that we must use its resources wisely. We must accept contamination where necessary or desirable, and avoid pollution wherever possible, not only in the oceans but also on land and in the atmosphere.

Assimilative Capacity

The oceans receive millions of tonnes per year of many dissolved elements and organic substances from rivers, dust, and rain (Chaps. 6 and 8) and have done so since long before humans appeared on the Earth. These substances include many elements, such as copper, zinc, arsenic, and mercury, that are toxic to humans and other **species**. The oceans also receive organic matter from soils, plants, and animal wastes, the composition of which is substantially the same as that of human fecal and urinary wastes.

The amounts of these substances introduced naturally must have varied substantially as **plate tectonics** modified and moved the continents and as the Earth's **climate** changed. Hence, ocean ecosystems must be able to accommodate or adjust to a range of input rates of these substances. The maximum rate at which the oceans can accommodate such inputs without adverse effects is called the **assimilative capacity.** Chapters 6 and 8 discussed some of the chemical and biological processes that remove substances from ocean waters to balance inputs and prevent concentrations from rising continuously. The assimilative capacity is exceeded if the input rate increases so rapidly or by so much that removal processes are overwhelmed and the concentration rises to a level at which toxic or other adverse effects occur in the ecosystem.

Besides naturally occurring compounds, the oceans have an assimilative capacity, albeit sometimes small, for synthetic organic chemicals produced by human civilization. Although these compounds are new to the oceans, all are broken down by **decomposers** and chemical processes into other compounds and eventually inorganic compounds. Some are broken down quickly, whereas others, including **DDT** (dichloro-diphenyl-trichloroethane) and **PCBs** (polychlorinated biphenyls), are broken down very slowly.

Although assimilative capacity is a useful concept, it is very difficult to apply. Each element or substance has its own unique **residence time,** natural concentration, concentration at which it becomes toxic, variable toxicity to different species, and chemical and biological decomposition and removal processes. Each of these factors must be understood before the ocean's assimilative capacity for a single substance can be estimated. The oceans are not instantly and uniformly mixed. Consequently, the assimilative capacity can be exceeded for part of the oceans if inputs to a specific region exceed the rate at which they can be removed by chemical and biological processes and by mixing with the rest of the oceans. Hence, assimilative capacity and residence time are linked.

CC8 describes how residence times can be determined for individual substances. In a geographically distinct region, an increased input rate of a contaminating substance will cause the substance's concentration in the water (and thus, generally, in the sediments and **biota** of the region) to increase. The increase in concentration is greater if the residence time is longer. Thus, for example, organic material in the large volumes of sewage of a major city can far exceed the assimilative capacity of a river or bay if the water body has a relatively long residence time. The discharged organic material can be decomposed and deplete dissolved oxygen faster than the oxygen can be replaced by mixing with oxygenated ocean or river water or resupplied from the atmosphere. As a result, the water becomes **anoxic**, and toxic hydrogen sulfide is generated by **bacterial** action (Chap. 14). Parts of San Francisco Bay (**Fig. 19.1**), the New York Harbor, and many other **estuaries** developed chronic sewage-induced anoxia in the 1950s and 1960s.

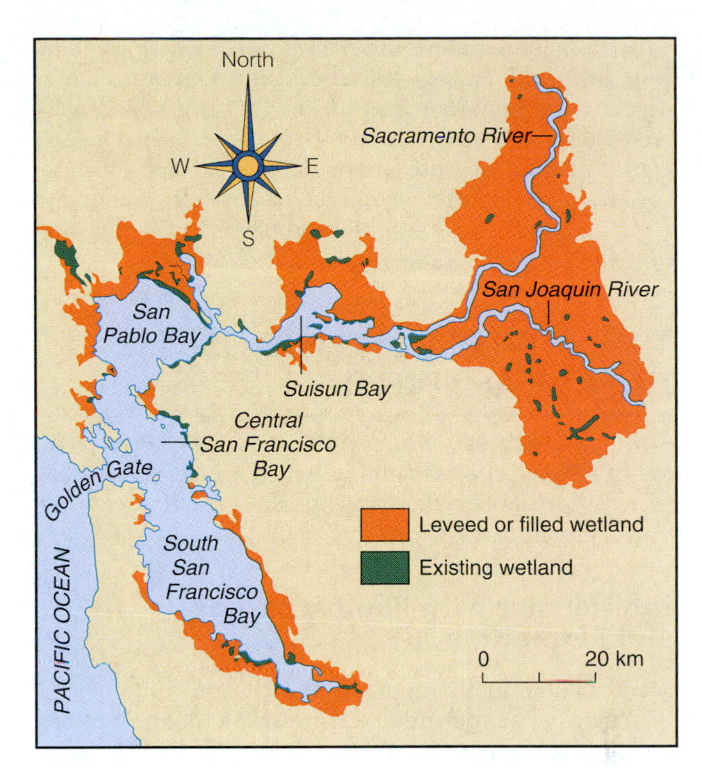

FIGURE 19.1 This map of San Francisco Bay and its delta shows the areas that were formerly wetland or part of the Bay but that have been filled or confined by levees in the past 150 years. The area of the Bay is now approximately one-half of what it was before 1850, and the area of remaining wetland is only a very small fraction of what it was then.

The anoxia that affected many estuaries in the United States and elsewhere only several decades ago has now been alleviated in most instances. Today, all sewage is treated to remove most organic matter before being discharged. In addition, some, but not all, treated sewage has been diverted to **outfalls** discharging directly to coastal oceans, where the residence time is much shorter and the assimilative capacity much greater. Unfortunately, even the reduced-contaminant inputs exceed the assimilative capacity in some estuaries. For example, the assimilative capacity for certain toxic metals, such as copper and silver, is now exceeded in South San Francisco Bay, and efforts are under way to reduce concentrations of these toxic elements in remaining discharges. This process is extremely expensive and difficult because concentrations of the metals in treated sewage must be reduced to levels that, in some cases, are lower than those in natural stream **runoff** or domestic water supplies.

Because of its very low inputs of freshwater and its semi-enclosed nature, South San Francisco Bay (**Fig. 19.1**) has a long residence time and a low assimilative capacity. As the local population continues to grow and the volume of sewage increases, even more stringent controls will be needed to avoid exceeding the assimilative capacity. An alternative for this region, which would almost certainly be environmentally preferable and, in the long term, less costly, would be to divert all treated discharges to the nearby Pacific Ocean. As long as sewage remains properly treated and outfalls are properly designed and located, there is little chance that the much greater assimilative capacity of the coastal Pacific Ocean will be exceeded, even with future increases in human population. However, the local population remains opposed to allowing any more discharges to the ocean, no matter how small the impacts on the ocean environment.

The critical lesson is that a discharge in one location can produce contamination but no pollution, whereas a discharge of identical volume and composition in a region with a longer residence time can produce a serious pollution problem. Hence, residence time and assimilative capacity are important parameters that must be considered in evaluating any activity that releases or may release contaminants. Because these parameters are location-specific, a release of a given type or amount of contaminant that has caused pollution problems in one location may not cause the same problems at other locations.

19.2 ADVERSE EFFECTS OF HUMAN ACTIVITIES

Contamination and other human activities cause a variety of adverse effects, some of which are described here. In many instances, one activity can lead to more than one of the impacts discussed.

Interference with Photosynthesis and Respiration

Many contaminant inputs to the oceans contain large quantities of **suspended sediment** and light-absorbing organic matter. If retained in the water column, high concentrations of these materials block the penetration of sunlight and reduce the amount of light available for **photosynthesis** (CC14).

Human-caused discharges of suspended particles to the oceans affect primarily coastal regions (Fig. 19.2), where much of the ocean's **primary production** occurs. **Benthic algae,** both **microalgae** and **macroalgae,** are important primary producers in many coastal and estuarine areas. Excess suspended sediment reduces light penetration and, thus, primary production by benthic algae.

Adverse effects on photosynthesis can be caused by discharges of large quantities of **nutrients**. Alterations in relative concentrations of nutrients can cause changes in **phytoplankton** species composition (Chap. 15). The result can be adverse effects on the **food web** if the newly advantaged phytoplankton species are unsuitable food for the **zooplankton** population.

Excessive nutrient discharges can increase **primary productivity** so dramatically that it exceeds the **grazing** capacity of consumers. In such cases, phytoplankton reproduce and grow rapidly in a **bloom** until the **limiting nutrient** is depleted. The bloom then collapses, and dead phytoplankton cells sink and decompose rapidly, removing some or all of the oxygen from the water. Reduction of oxygen concentration can adversely affect the **respiration** of fishes and **invertebrates.**

Oxygen depletion due to the presence of excessive nutrients or organic matter is called **eutrophication** and has long been a menace in lakes, rivers, and estuaries where nutrient inputs are large and residence times long. In many such areas, the nutrient and organic matter inputs are from sewage discharges. Sewage treatment has dramatically reduced eutrophication in U.S. lakes and rivers since the early 1970s. However, nutrients remaining in the **effluent** after sewage treatment, when combined with other nutrient inputs, are now causing increased eutrophication problems (**hypoxia** and anoxia) in coastal oceans near estuaries. Hypoxia and anoxia as a result of excessive nutrient inputs, and the increasing severity of these problems, were labeled in a 2000 report by the National Research Council of the National Academy of Sciences as the most pervasive and troubling pollution problem facing U.S. coastal waters.

Oxygen concentrations can also drop if photosynthesis is reduced. For example, increased concentrations of

FIGURE 19.2 There are many sources of contaminant discharges to the oceans. (a) This pipeline near Cape May, New Jersey, discharges treated sewage not far off the beach. (b) Sugar cane fields are often burned after harvest to clear the land for new planting. This practice increases erosion, and large amounts of suspended sediment are discharged in runoff. These fields are on the north shore of Hawaii. You can see the smoke from one of the fires and the plume of suspended sediment that spreads in the nearshore zone where there are reef-building coral formations that can be destroyed by excess sedimentation.

suspended sediment, copper, or other substances, such as **herbicides,** can inhibit growth or photosynthesis without significantly affecting respiration in a marine ecosystem.

Human activities adversely affect oxygen concentrations in at least two special situations. First, when rivers are constrained within **levees,** the surface area is diminished but the river depth and currents increase. Often the **photic-zone** depth is naturally shallow and may be further reduced by higher **turbidity** caused by faster river currents that **resuspend** particulate matter. Because both the surface area and the depth of the photic zone are reduced, the total volume of the photic zone is smaller. Furthermore, the reduced surface area lowers the rate of oxygen resupply from the atmosphere.

In the second special case, oxygen concentrations are reduced because power plant or other industrial effluents have a higher-than-ambient temperature. Because oxygen has a lower **saturation solubility** at higher temperature, oxygen concentrations in the heated effluents are generally lower than ambient concentrations. This disparity is seldom an immediate problem, because the saturated oxygen concentration is sufficient to support respiration at all but the highest ocean water temperatures (above about 30°C). However, the low concentration of dissolved oxygen reduces the assimilative capacity for **oxygen-demanding** substances. Because tropical waters have high ambient temperatures and low ambient oxygen concentrations, oxygen depletion and sulfide production can be caused by smaller inputs of nutrients or organic matter in the tropics than in other **latitudes.** Hence, water bodies in tropical regions have a lower assimilative capacity for organic wastes than do water bodies with similar residence times at higher latitudes.

Habitat Alteration

Each species that lives in the oceans has its own requirements for the physical and chemical characteristics that constitute its **habitat** (Chaps. 16 and 17). Benthic species are particularly dependent on the nature of sediments in or on which they live. **Pelagic** species require or prefer certain ranges of temperature, **salinity,** turbidity, and current or wave regime. Many pelagic species depend on the benthic environment for food or shelter during part of their life cycle, whereas many others depend on shallow-water environments of **mangrove** swamps, coastal **wetlands,** rivers, and estuaries during their juvenile phases.

Human activities have caused the destruction of vast areas of coastal wetlands and have caused adverse habitat alteration in other areas. For example, rivers have been dammed, preventing **anadromous** fishes, such as salmon, and **catadromous** fishes, such as eels, from migrating to the upper reaches of the rivers on which they depend. Levees, marinas, and ports have substantially altered current patterns, and thus suspended sediment transport routes, in many estuaries. Vast quantities of freshwater have been removed from many rivers, thus changing the salinity and other chemical characteristics of estuaries. Coastal structures have interfered with sand transport along many **coasts.** Large volumes of wastes have been discharged directly to the oceans or through rivers, and many parts of the seafloor have been damaged by dredging and dynamite blasting, and by anchors and fishing gear. All of these activities cause changes in **sedimentation rate,** sediment **grain size,** and sediment chemistry in affected regions. Sandy seafloor can be turned into mud and vice versa, and stable rocky or **reef** bottoms can be covered with sediment or broken up and **eroded** away.

Whenever marine habitat is altered, the species composition changes. Some species are disadvantaged and others benefit. Most often, the alteration causes at least a temporary reductions in species **diversity** (**CC17**) and dominance by species that are less sensitive to changing habitat. Opportunistic species not normally a major part of the biota commonly move into and dominate an environment when it has been altered by human activity, especially if the disturbance is ongoing, such as waste disposal or dredging. Although in some instances opportunistic species can enhance the **biomass** of the natural food web, more often they are less desirable or worthless in the natural food web.

Community Structure Alteration

The many species that make up a marine **community** depend on each other for food and in many other ways (Chaps. 16 and 17). The balance among species is determined by centuries of competitive and cooperative interaction that has enabled the community to reach a relatively stable state. If the balance is disturbed, the community structure can become unstable and change unpredictably.

The greatest direct human disturbances of community structure are caused by preferential exploitation of one or more species in a fishery or by the introduction, accidentally or otherwise, of **nonindigenous species** (discussed later in this chapter). Human activities may also introduce nonindigenous disease-causing organisms that affect some species important to the ecosystem. Other human influences, such as substrate alteration and introduction of toxic substances, can also advantage or disadvantage certain species and cause community structure to be altered.

Marine communities are periodically subjected to habitat disturbances due to natural events, such as earthquakes and climate changes. Some additional human disturbance can be tolerated and accommodated by the ocean ecosystem. However, in some cases, human-induced disturbance may be more rapid than natural disturbances. Furthermore, human disturbances are often continuous or increase progressively, and their scale may be unprecedented in some coastal and estuarine areas.

Contamination of Food Resources

Aquatic species can obtain elements and compounds from food and directly from solution. Therefore, toxic substances introduced in dissolved form or associated with organic particles can be assimilated by most marine organisms and passed through the food web. Many marine species are tolerant of relatively high concentrations of toxic elements or compounds in their environment or food, probably because their body surface and respiratory tissue are continuously exposed to seawater. Rather than building defenses against the absorption of toxic substances from their environment, many marine species simply take up such compounds and store them in some organ where they cannot interfere with essential **biochemical** processes. This method of detoxification has limits, but it enables many marine species to tolerate high concentrations of some toxic substances.

Fish and **shellfish** with high concentrations of stored toxic substances may suffer no adverse effects but may still pose a significant risk to human health. Shellfish, such as oysters, are particularly adept at concentrating trace metals, and many synthetic organic compounds, such as DDT and PCBs, are concentrated in fatty tissues of most marine animals.

High concentrations of metals and synthetic organic compounds have been found in the biota from many locations where human activities release such contaminants. In these locations, the fishery or shellfishery is closed, or people are advised to eat only limited amounts of seafood or a specific seafood species. Thus, the value of the fishery resource is diminished or lost. Fortunately, in only one recorded instance have people died from ingesting seafood contaminated with industrial toxic substances discharged into the oceans. That incident, in Minamata, Japan, is discussed later in this chapter.

Contamination of fish and especially shellfish with microorganisms that are human **pathogens** is a serious problem. Because some seafood is eaten raw or only lightly cooked, any microorganisms present will be passed on to the consumer. Seafood may be contaminated during handling and processing, although refrigeration and hygienic food-handling techniques have greatly reduced this problem in most developed countries. Most microbiological contamination of seafood now comes from the harvested waters. The problem areas are generally contaminated by discharges of raw or treated sewage or by animal feces carried in street and land runoff. The contamination is concentrated in estuaries and the **coastal zone** because human pathogens are progressively diluted and destroyed by marine microbes as they are transported to the open oceans. Unfortunately, shellfish beds are located mostly in coastal and estuarine zones.

Some pathogen-contaminated shellfish can be collected and cleansed by being kept in pathogen-free, constantly running and renewed seawater, but this procedure is expensive and requires a source of reliably pathogen-free seawater. Consequently, the only practicable way to prevent the spread of disease by contaminated shellfish is to prohibit harvesting in contaminated areas. At present, many potentially valuable shellfishing areas of the coastal oceans and estuaries in the contiguous United States are closed to shellfishing. The total area closed is increasing, despite immense expenditures for sewage treatment and control of other contaminant sources since 1972, when the United States passed the Clean Water Act requiring secondary treatment of almost all sewage.

Toxin-producing phytoplankton blooms are also a growing contamination problem. These blooms, the toxins they produce, and their effect on marine ecosystems and seafood values are discussed in Chapter 15.

Beach Closures and Aesthetic Losses

Many people look to the coastal oceans as places of aesthetic beauty where they can renew their contact with the natural world through various recreational activities. As a result, one of the largest industries on the planet has developed around coastal recreation. However, the value of the coast for such activities is compromised and diminished in many locations by the presence of human structures that mar the natural beauty (**Fig. 19.3**) and interfere with natural processes.

Many coasts that otherwise would be areas of high recreational value are sites of human industrial and residential developments that bar the public from reaching the **shore.** Such developments often cause major changes in the coastal form and function through direct modifications of the **shoreline,** such as **bulkheads,** piers, and **groins,** and through alteration of the coastal current, wave, and sediment transport patterns. In addition, vast areas of biologically important coastal wetlands have been drained and filled to accommodate development. Today there is a growing realization that many of these developments were ill-planned. Public opinion now favors protection of the coast from unreasonable degradation by development, although the legal system lags behind this imperative.

Outfalls discharge treated sewage, industrial waste, and storm-water runoff to rivers, estuaries, and oceans. Rivers also receive various materials that have been carelessly or deliberately dumped by humans. Vessels often discharge wastes directly into the ocean. Many beaches are periodically closed to swimming because the water is contaminated with pathogens from improperly treated sewage. In addition, floating debris mars beaches and, in extreme cases, leads to the temporary closure of beaches for recreational purposes. For example, medical wastes, including syringes apparently dumped in nearby rivers or from vessels, have periodically washed up on New York and New Jersey beaches, prompting closure of the beaches for fear that the materials could carry pathogens, including the HIV **virus.**

FIGURE 19.3 Coastal industries tend to reduce the aesthetic value of the shoreline. (a) The Moss Landing Power Plant, a very prominent structure located on the coast in the middle of Monterey Bay, California. (b) A paper mill at Port Angeles, Washington.

19.3 TOXICITY

Toxic substances are substances that have adverse effects on organisms, including humans. The term *toxic chemical* is now among the most emotion-evoking in our society. In fact, there is a widespread belief, fostered by the media and many environmental interest groups, that the release of any quantity of toxic chemicals to the environment, particularly the oceans, is harmful and must be stopped. This belief is unfounded (**CC18**). Many toxic chemicals are naturally occurring elements or compounds that have been present in the oceans throughout geological time. Some additional quantities, in many cases substantial quantities, of these toxic chemicals and even synthetic chemicals that do not occur naturally can be safely accommodated by ocean ecosystems. However, if they are to cause no harm, wastes must be disposed of in ways that ensure that the ocean's assimilative capacity is not exceeded either globally or locally. Marine pollution problems occur when toxic chemical discharges exceed the assimilative capacity. Therefore, determining this capacity for each local region into which toxic substances are discharged is an important scientific task.

A second widespread belief is that the ultimate solution to contamination by toxic chemicals is to recycle everything. Recycling is a highly desirable and rapidly growing practice that continues to reduce quantities of toxic chemicals released to the environment. However, recycling can never be 100% effective. Toxic chemicals are present in small quantities in almost every natural and synthetic product in our society, including human waste **excretions.** For many of society's waste products, the energy costs and associated adverse environmental effects of recycling and toxic-chemical removal may far outweigh the environmental costs of properly managed disposal of the waste. In addition, storing all toxic chemical–containing wastes in "secure" landfills is not feasible. In most instances, using valuable land to dispose of large-volume wastes that have low concentrations of toxic chemicals is inappropriate. Perhaps more importantly, no landfill can ever be secure on a geological timescale. Wastes containing high levels of long-lived toxic substances that are buried in landfills will eventually release their toxic substances to freshwater and **groundwater** and from there to the oceans. The Earth's freshwater resources are severely limited, and any toxic contamination of the freshwater poses a threat to human health and terrestrial ecosystems. Consequently, many environmental scientists believe that properly managed ocean disposal of wastes may be the environmentally preferable management option for wastes that cannot be recycled and that have low concentrations of toxic chemicals.

The fate and effects of toxic chemicals in the oceans are determined by how they are transported by currents, how they are removed and incorporated in sediments, and how they are taken up by the marine biota. Several chapters in this text describe the movements of ocean water and suspended sediment and the processes of marine **sedimentation** that determine the distribution and ultimate fate of dissolved and particulate constituents in the oceans.

Effects of Toxic Substances on Marine Organisms

CC18 describes the general principles that determine the effects of toxic chemicals on living organisms. The essential point is that most toxic substances are toxic to a specific organism only when their concentration in food or in solution in the surrounding water exceeds a certain level, above which the concentration of the chemical within the organism is high enough to interfere with one

CRITICAL THINKING QUESTIONS

19.1 What is the difference between naturally occurring toxic substances and synthetic toxic substances that suggests that they might be considered differently by policy makers and managers concerned with marine pollution?

(a) Do you think discharges of these two types of substances should be subject to different rules? Why or why not?
(b) Do you think the production of all synthetic toxic substances (without any exception) should be banned?
(c) If only some such substances are to be banned from being produced while others are allowed to be produced but their discharges regulated, what factors would you consider in deciding which of these two policies should be applied to a specific synthetic toxic substance?

19.2 In this chapter, it is speculated that the use of oceans for sewage waste disposal was one of the most effective advancements in human health protection in history. Do you think this statement is correct? Why or why not?

19.3 DDT has been banned from production and use in many countries, primarily in mid latitudes, where it was previously used on crops for insect control. However, DDT is still produced and used in large amounts in some nations, particularly developing nations in tropical regions where malaria is rampant. Although effective substitutes for DDT exist, they are not used in these countries, because they cost too much. What should be done about this situation, and how?

or more biochemical processes critical to the organism's life cycle. There is generally a concentration below which the chemical has no adverse effects.

The only exception to this rule may be for compounds that are **carcinogens** (cancer-causing), **mutagens** (causing **genetic** changes in the offspring by altering the parents' **DNA**), or **teratogens** (causing abnormal development of the **embryo**). There are conflicting views about whether a threshold concentration exists for such compounds. If there is no threshold concentration, any concentration of the compound will increase the incidence of disease. However, below a certain concentration, the number of individuals affected by a carcinogenic, mutagenic, or teratogenic chemical will be much smaller than the number similarly affected by naturally occurring compounds and natural **radioactivity,** which also have such effects. Hence, even for these compounds, there is a concentration below which they have no significant adverse effects.

Evaluating Toxicity

Assessing the toxicity of compounds to marine organisms is very difficult. Two general approaches are used. First, the distribution of species in an ocean region affected by toxic chemical inputs can be compared with the distribution in similar but unaffected regions or in the affected region before the **anthropogenic** inputs occurred. Second, **bioassays** can be performed in the laboratory to test the responses of organisms to various concentrations of the toxic chemical (**CC18**). The laboratory results must be extrapolated to what might occur in the environment. Both approaches are difficult and prone to errors and uncertainties.

The field approach to toxicity evaluation requires a very detailed survey of the affected ecosystem and of control sites. Researchers must characterize the population levels and health of many species to be certain that the most susceptible species are included. Moreover, changes observed in the ecosystem may be caused by natural factors, such as climate variations. Hence, even if a significant difference in the ecosystem is observed between test and control sites, researchers usually cannot eliminate the possibility that it is natural and unrelated to contaminant toxicity. In addition, most contaminated locations receive inputs that contain a variety of toxic chemicals, all of which vary in concentration with time. Assessing which contaminant might be responsible for an observed change is difficult.

At present, we do not have a detailed understanding of the effects of toxic substances in the oceans, and we cannot predict with certainty the consequences of specific concentrations. Management of ocean uses that involve the release of toxic chemicals to the marine environment is therefore difficult and controversial. However, our understanding of the effects of toxic chemicals in terrestrial, freshwater, and groundwater ecosystems, and on human health, is not significantly better than our understanding of their effects in the marine environment. There is general agreement that we must exercise caution and, to the extent practicable, limit the release of toxic chemicals to the environment. Nevertheless, wastes containing low concentrations of toxic chemicals will always require disposal. If properly managed, ocean disposal of these wastes may be both safe and environmentally preferable to other disposal methods.

Bioaccumulation and Biomagnification

Two processes cause concentrations of toxic substances in marine organisms to become elevated in relation to concentrations in their food or environment (**CC18**): **bioaccumulation** and **biomagnification**.

Uptake and excretion of toxic substances by marine organisms are typically complex processes that involve transfers of the toxic substance among several different tissues of the organism, as well as to and from the organism's food and the surrounding water. Hence, some toxic substances may be taken up quickly when the environmental concentration increases, but released much more slowly when the environmental concentration decreases. The complexity of these bioaccumulation equilibria generally increases as the organism becomes more complex.

Higher animals tend to store toxic substances in tissues where they are least harmful. Toxic substances are excreted only very slowly from these tissues after the exposure is reduced. In some cases, the loss is so slow that the concentration of the toxic substance tends to increase progressively during the individual organism's lifetime. The cumulative buildup of metals, such as lead, mercury, and arsenic, in human beings during their lifetime is an example. In such situations, short-term laboratory bioassay tests cannot accurately reflect the effects of long-term exposure to toxic substances or wastes.

Biomagnification of DDT and its decomposition products was responsible for the decline of pelican, sea lion, and elephant seal populations off California and throughout the eastern Pacific Ocean during the 1950s and 1960s. Almost all uses of DDT were banned in the United States in 1971. Since then, each of the affected species has recovered steadily in the eastern North Pacific Ocean. However, DDT continues to be used elsewhere in the world, particularly in tropical regions. Both DDT and its toxic but longer-lasting decomposition products are still found at high concentrations in marine species at higher **trophic levels** in all parts of the oceans.

Toxic trace metals are not biomagnified in marine food webs, except when organically combined, such as in the methylated form of mercury. Biomagnification in ocean food webs appears to be limited to compounds that are highly **soluble** in fatty tissues and have relatively low solubility in water. These include mainly synthetic organic compounds, of which DDT and PCBs, and perhaps a few synthetic or naturally produced metal–organic compounds, such as methylmercury and tributyltin, are the primary ones.

Synthetic and Naturally Occurring Toxins

Two distinct classes of toxic substances are released to the marine environment by human activities: naturally occurring substances, including trace metals (e.g., copper, lead, cadmium) and **petroleum hydrocarbons;** and synthetic chemicals produced only by human industries (e.g., DDT and PCBs). These two classes should be viewed differently. Marine organisms are adapted to generally low but variable concentrations of natural toxic substances in their environment. For example, lead and other toxic metals are either safely stored within marine organisms' tissues where they do not affect critical biochemical processes or safely excreted back to the environment. Petroleum hydrocarbons are metabolized to harmless substances by many fish species and used as food by many marine decomposers. Adverse effects occur only when these natural detoxification processes are overwhelmed by high concentrations of such compounds. This is fortunate, because total elimination of the releases of such compounds by all human activities is impossible.

Synthetic chemicals, such as DDT and PCBs, are not present naturally, and marine organisms may not have mechanisms to detoxify them. Consequently, these compounds have more unpredictable fates and effects in the marine environment. If they are persistent and not easily broken down by the chemical processes or metabolic processes of marine organisms, they tend to accumulate, persist, and cause adverse effects. Unlike contaminants that occur naturally, a synthetic chemical can be eliminated completely from our society and therefore from any further introduction to the environment. Synthetic chemicals have now been designed and developed that adequately fulfill the purposes for which DDT and PCBs were intended but that, in contrast, are readily and rapidly decomposed in the environment. Unfortunately, developing, testing, and manufacturing new, effective, and rapidly **biodegradable** chemicals is very costly. The widespread adoption of such products is impeded primarily by economic factors.

CRITICAL THINKING QUESTIONS

19.4 On an annual average, petroleum hydrocarbons from street runoff contribute more than twice the volume of oil spilled in the oceans from tanker accidents.

(a) What are the sources of petroleum hydrocarbons in street runoff?
(b) What would you suggest politicians do to reduce the amount of petroleum hydrocarbons released to bays and estuaries by this route?
(c) Should politicians require all storm-water runoff to be treated to remove the hydrocarbons?
(d) Should politicians attempt to control the release of hydrocarbons to the environment so that they do not get into the storm water? How could this be done?
(e) What can you personally do to reduce the amount of hydrocarbons that you release to the environment?

19.5 Describe the principal sources of petroleum contamination in the oceans. Discuss what would probably cause more pollution of the oceans: production of oil from drilling platforms in U.S. coastal waters and transport of the oil ashore in seafloor pipelines, or purchase of oil from foreign suppliers and transport to the United States in oil tankers.

FIGURE 19.4 Oil spills provide spectacular images that ensure media coverage. (a) The *Exxon Valdez* tanker 48 h after it ran aground on Bligh Reef in Prince William Sound, Alaska, in 1989. The tanker is still leaking some oil, and it is surrounded by a boom that has been placed on the water surface in an attempt to contain this oil so that it can be collected. (b) The *Amoco Cadiz,* Brittany, France, 1978. (c) A rescuer holds a badly oiled seabird (a guillemot or common murre, Uria aalgea) after the 1996 *Sea Empress* tanker ran aground on the coast of Wales in the United Kingdom. Rarely do such heavily oiled birds survive even when cleaned up. (d) A skimmer barge is sucking up oil collected by a floating boom placed around a patch of oil after the *Exxon Valdez* spill. Although this technique is expensive and is employed extensively in most spills, only a fraction of the spilled oil is normally recovered this way. (e) High-pressure water hoses are used to wash oil off a Naked Island beach in Prince William Sound one week after the *Exxon Valdez* oil spill. (f) Oiled rocks are collected from a beach and cleaned by hand before being replaced on the beach at Muxia, northwest Spain, after the tanker *Prestige* sank and released its cargo in 2003.

19.4 WASTE DISPOSAL

For centuries, the oceans were viewed as a limitless sink for waste disposal. Consequently, during the industrial era the oceans were used to dispose of sewage, dredged materials, construction dirt and debris, trash and garbage, chemical wastes, radioactive wastes, fish-processing wastes, used machinery and boats, and almost anything else that people needed to discard. Wastes have been dumped from vessels or discharged through outfalls into rivers, estuaries, and the ocean. Since the 1970s, indiscriminate use of the oceans for waste disposal has been recognized to cause adverse environmental impacts, some of them severe. As a result, ocean waste disposal is now viewed in the United States and elsewhere as being universally unacceptable. With the exception of dredged material, ocean dumping from vessels is now illegal in the United States and most other developed nations. In addition, stringent rules are in place that require sewage and industrial wastes to be treated before they are discharged through outfalls.

Contrary to popular belief, many marine scientists believe that properly managed disposal of certain wastes in the oceans is not only acceptable, but may be environmentally preferable to any possible alternative disposal or recycling technology for these wastes. Properly managed disposal of certain organic and nutrient-containing wastes, such as sewage and fish-processing wastes, might even have beneficial effects. However, because all ocean disposal is considered unacceptable by the media and the public, little or no research is being done to define how, where, and what types of wastes can and should be safely disposed of in the oceans. This situation may eventually prove detrimental to the global environment. Many toxic wastes, particularly wastes containing toxic trace metals, might be safely dispersed in and assimilated by the oceans and/or removed by natural processes to ocean sediments, but instead they are buried in "secure" landfills from which they will eventually be released.

19.5 PETROLEUM

Almost every year, somewhere in the world, one or more oil tankers accidentally spill some or all of their cargo into the oceans. Tanker accidents and oil well blowouts (**Figs. 19.4, 19.5**) generate enormous public and media attention. These dramatic events provide gripping visual images of stranded and damaged ships (**Fig. 19.4a,b**); of thick black oil floating on the water and washing up on beaches; of oiled and dying or dead seabirds (**Fig. 19.4c**), otters, and other animals; and of frantic efforts to clean up the spilled oil (**Fig. 19.4d–f**). Consequently, tanker spills and oil platform blowouts are widely believed to be the major source of oil contamination and the most damaging form of ocean pollution. Neither of these beliefs is correct.

Sources of Petroleum Contamination

Sources of oil in the oceans include natural seeps from the seafloor and from terrestrial watersheds that drain to the oceans, large and small spills, and incidental releases by vessels and offshore platforms. In 1985, natural releases of petroleum were estimated to be five times greater than releases from offshore oil platforms, and tanker accidents contributed about twice as much as natural sources (**Fig. 19.6a**).

Even in 1985, the total input of oil from routine operations of tankers and other vessels and the total input from other sources, such as street runoff (oil from motor

FIGURE 19.5 Major oil spills from the offshore oil industry are rare but dramatic events. (a) The 1969 Santa Barbara, California, oil spill. (b) This spectacular blowout and fire at the *IXTOC* oil rig in the Gulf of Mexico in 1979 resulted in a continuous release of oil that lasted for several months before the well was finally closed.

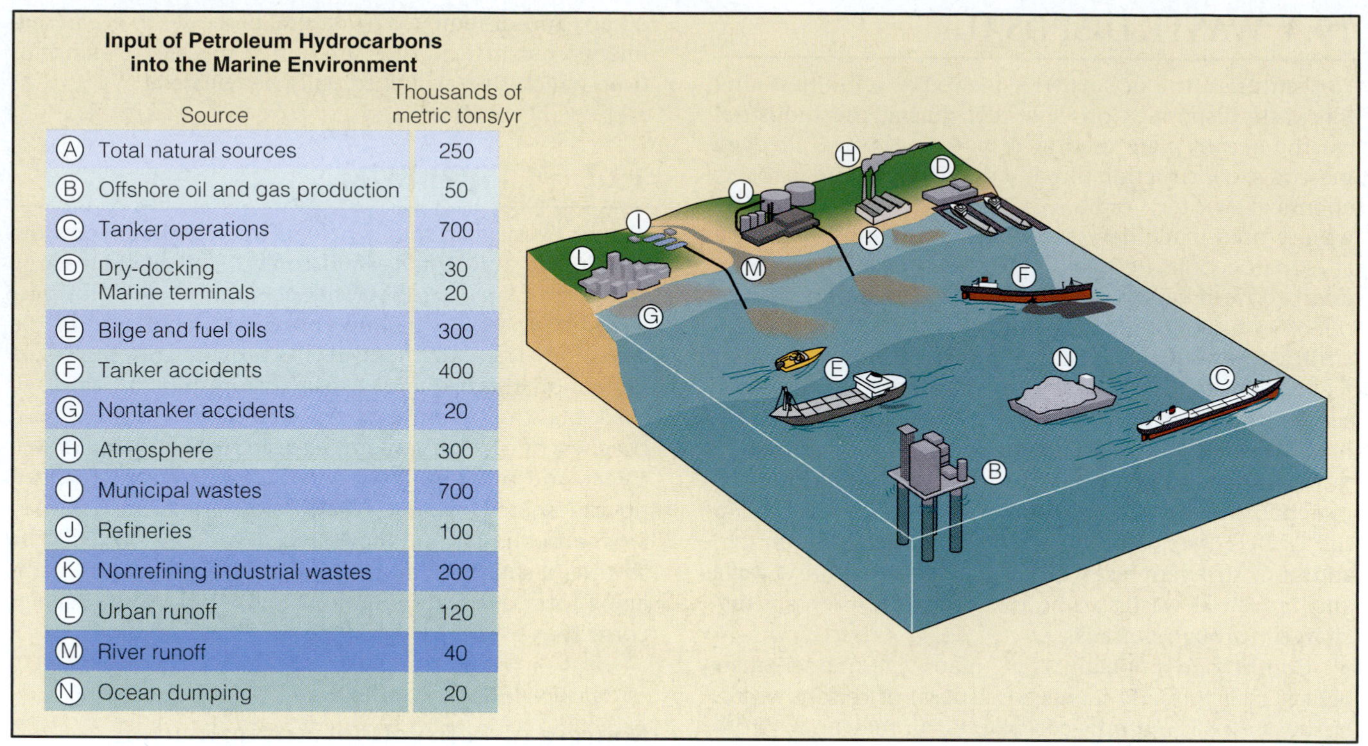

Input of Petroleum Hydrocarbons into the Marine Environment

Source	Thousands of metric tons/yr
(A) Total natural sources	250
(B) Offshore oil and gas production	50
(C) Tanker operations	700
(D) Dry-docking Marine terminals	30 20
(E) Bilge and fuel oils	300
(F) Tanker accidents	400
(G) Nontanker accidents	20
(H) Atmosphere	300
(I) Municipal wastes	700
(J) Refineries	100
(K) Nonrefining industrial wastes	200
(L) Urban runoff	120
(M) River runoff	40
(N) Ocean dumping	20

(a)

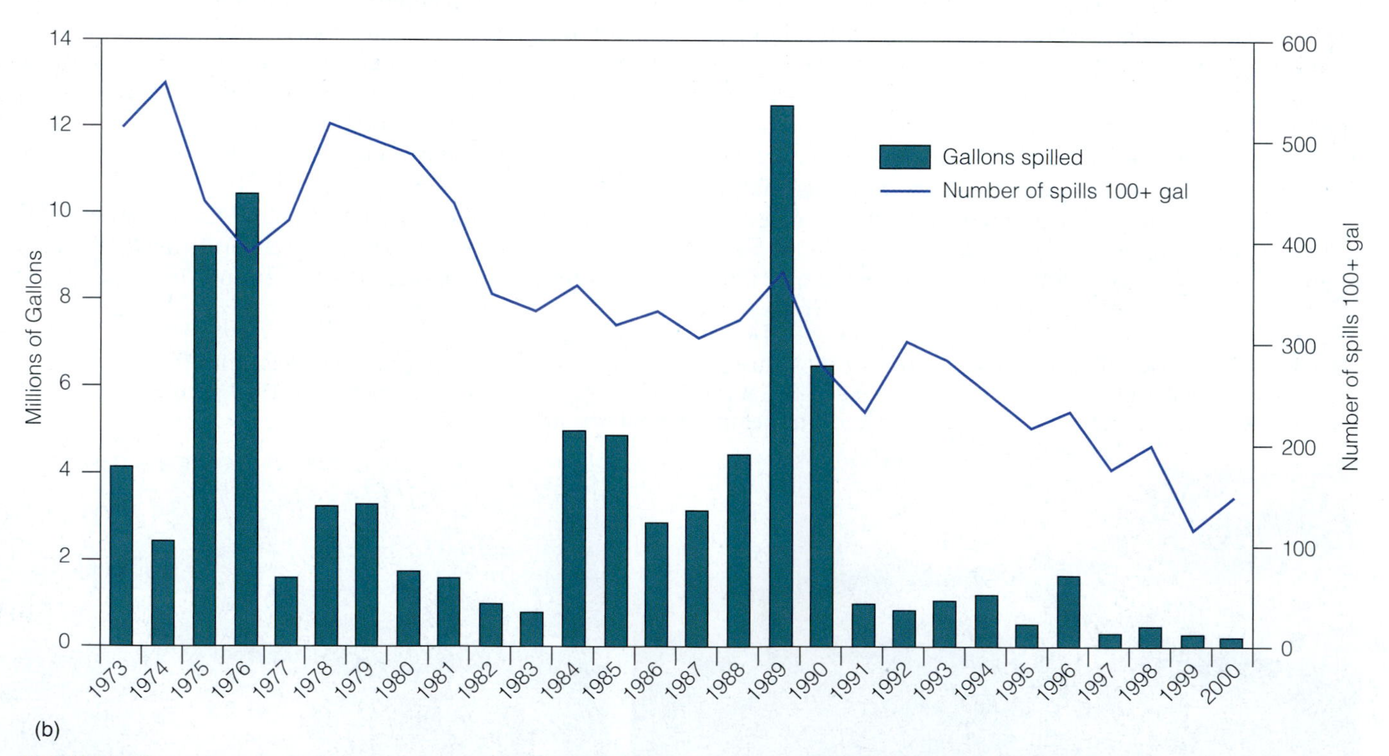

(b)

FIGURE 19.6 Sources of oil inputs to the oceans. (a) There are many sources of oil pollution in the marine environment. The greatest contributors are discharges of treated municipal sewage and of oily ballast water by tankers. Offshore oil production facilities and ocean dumping, often blamed for much of the pollution, are in reality only minor sources. (b) Both the number of spills and the volume of oil spilled annually in U.S. waters have decreased in recent decades, especially after a tough new law was passed in 1990 in response to the 1989 *Exxon Valdez* spill.

vehicles) and oil discharged into sewers and rivers, each exceeded twice the volume of oil spilled in tanker accidents (**Fig. 19.6a**). Since the *Exxon Valdez* oil spill in Alaska in 1989, extensive efforts to reduce accidents and incidental oil releases from vessels have substantially reduced the input from these sources, particularly in the United States (**Fig. 19.6b**). Unfortunately, it is difficult to compare newer estimates with the historical data because each data-gathering agency and each study have used a different classification scheme to distinguish sources, often combining specific inputs into different groupings. However, comparing data from **Figure 19.6** with data from 2000 included in a National Academy of Sciences report, it can be concluded that spills and operational inputs of tankers and offshore oil production have been substantially reduced. In contrast, there has been relatively little success in reducing the inputs of oil through urban and river runoff or in incidental discharges from recreational and other vessels.

Fate of Petroleum in the Oceans

When oil is spilled into the ocean, it spreads on the water surface to form a slick. The oil's fate then depends on several factors, including the oil composition, air and sea temperatures, concentrations of suspended sediment, presence of breaking waves, and whether the oil reaches a shore (**Fig. 19.7**).

Oil contains many different chemical compounds called **hydrocarbons**. Individual hydrocarbons differ widely in volatility, solubility, toxicity, and chemical properties. Crude oil composition depends on its source, and refined oil products are very different in composition from crude oil. Generally, refined oil products, such as gasoline and diesel fuel, contain a greater proportion of low-molecular-weight, more-volatile hydrocarbons, and these refined products are more toxic than crude oil.

After a spill, the volatile components of the oil evaporate into the atmosphere or are dissolved in the water.

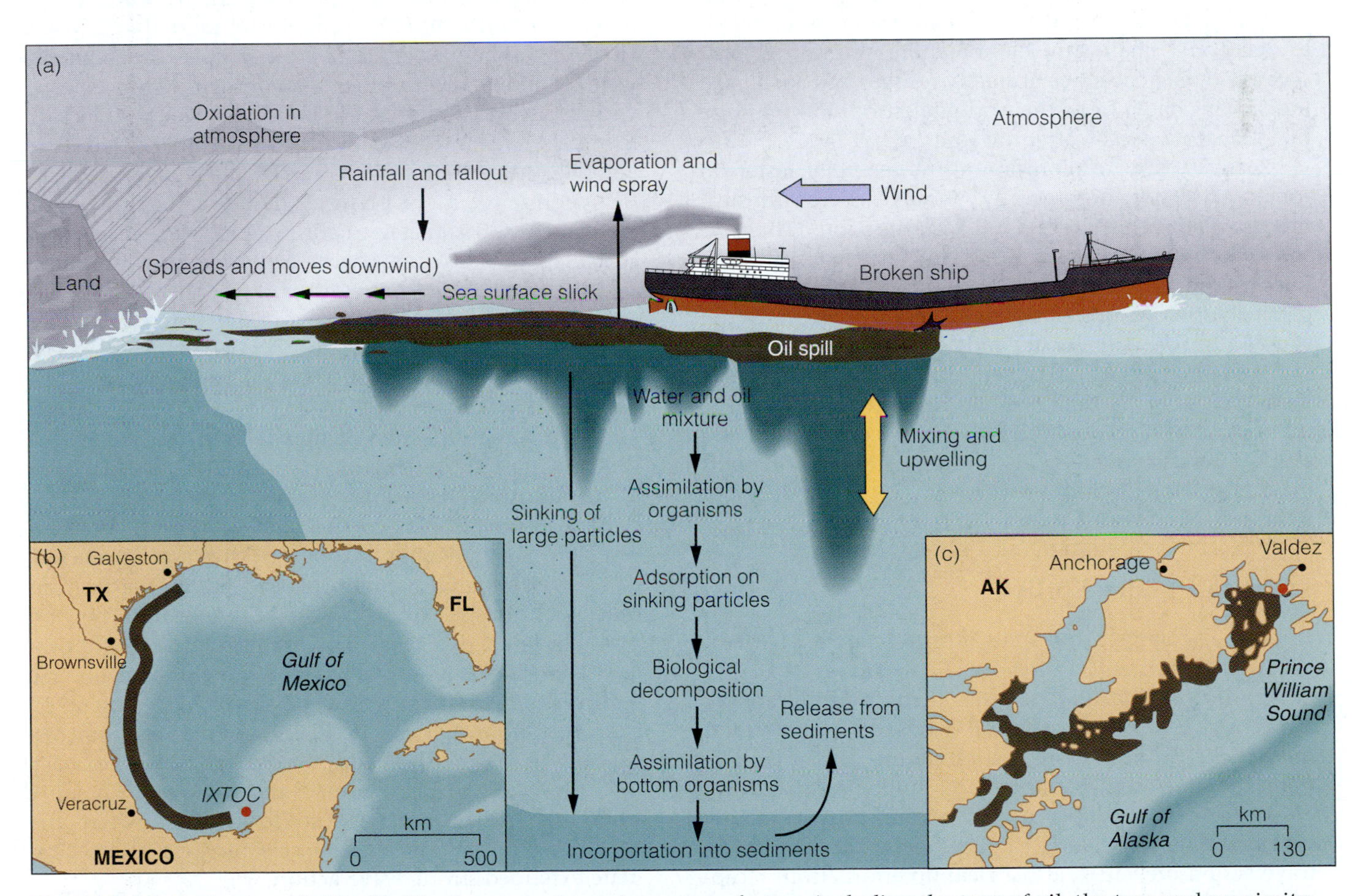

FIGURE 19.7 The fate of oil spilled in the ocean depends on many factors, including the type of oil, the type and proximity of the coastline, and the extent of wave energy. (a) Some oil evaporates quickly, but most coagulates to form tar balls or is adsorbed by particles and sinks to the sediments. The oil is eventually decomposed, primarily by bacteria. Oil spills generally do not cause lasting environmental damage, unless the oil reaches an ecologically sensitive shoreline, especially in high latitudes where oil degrades slowly because of the low temperature. (b) Areas affected by oil after the 1979 blowout of the *IXTOC* well, which took several months to cap. (c) Areas affected by oil after the 1989 *Exxon Valdez* accident.

These volatile compounds are largely evaporated from the slick within a day or two. As volatile components are removed, the oil becomes more **viscous** and, unless the seas are rough, begins to aggregate into lumps. In rough seas, oil may be mixed with air and water to form a gummy suspension that resembles and is often called "chocolate mousse." The oil dissolves or forms lumps, called "tar balls," composed primarily of high-molecular-weight, less-volatile hydrocarbons, or the oil is attached to sediment particles and deposited. If the slick does not encounter a shore, it is eventually completely dissipated. If the slick does reach a shore, oil clings to any substrate as an oily film. Once ashore, the oil film persists until it is washed off, buried in sediments by continuous strong wave action, or slowly decomposed by bacteria.

Because oil is a naturally occurring material, many decomposers are able to use hydrocarbons as food. Consequently, oil spills are eventually "cleaned up" naturally. The severe damage to birds and mammals and to **intertidal epifauna** and **infauna** that provides gripping television coverage after a spill is generally limited to relatively small stretches of shore. Even the most severely damaged shores normally recover and are almost indistinguishable from their original condition within a few years to a decade after the spill.

Recovery and recolonization are generally faster on rocky or other high-energy shores where physical processes limit the extent of oil accumulation during the spill and maximize its removal and dispersion into the open ocean in the postspill period. Low-energy shores, particularly wetlands, recover more slowly because oil can accumulate more easily and sediments into which it is mixed often have low concentrations of oxygen, an element that some bacteria need in order to decompose the hydrocarbons. Because both bacterial decomposition and evaporation are reduced at low temperatures, oil spills may persist longer and recovery may be slower in high-latitude environments than in warmer regions.

Effects of Major Spills

One of the largest oil spills from a tanker accident occurred in Brittany, France, in 1978 (**Table 19.1**). The tanker *Amoco Cadiz* (**Fig. 19.4b**) spilled its entire cargo of more than 200,000 tonnes of oil over several days after it hit a rock 13 km offshore and broke up. This was more than six times the amount of oil spilled in the *Exxon Valdez* accident in Alaska in 1989. Strong currents rapidly spread the oil slick along the Brittany coast, and strong wave action prevented at-sea cleanup efforts from recapturing much of the oil. Within a few days, 300 km of shore was affected by oil. Oil entered a number of low-wave-energy and low-current-energy estuaries and other embayments, where it accumulated in large quantities.

The biological impacts of the *Amoco Cadiz* spill were immediate, dramatic, and severe. More than 7000 seabirds, mostly diving birds such as cormorants, were oiled and died. The **plankton** biomass was substantially reduced for at least 2 months after the accident, and mortalities of benthic organisms, including **sea urchins,** clams, and **amphipods,** were massive. The major commercial species of the area, oysters, survived but were heavily contaminated and thus unfit for human consumption for many months. The rooted vegetation in coastal **salt marshes** and the **fauna** of intertidal mudflats were severely damaged. In contrast, only a small number of fishes were reported killed within the immediate vicinity of the wreck (about 10 km). Commercial flatfishes, including plaice and sole, showed no significant changes in population, although their average size in the spill year was somewhat below average, presumably because of the reduction in biomass of the juveniles' plankton food.

In the weeks and months after the *Amoco Cadiz* accident, extensive cleanup and oil removal efforts were made, especially in the low-energy estuarine environments. Visible signs of oil persisted in the water for as long as 6 months and in sediments for more than 3 years. However, all but very limited areas, such as the mudflats, were repopulated with most or all of their original species within several years. Some mudflats were subject to increased erosion caused by the loss of their vegetation, and oil contamination still could be detected in the mudflat sediments more than a decade after the spill. Nevertheless, this massive spill, which occurred in a particularly vulnerable area with extensive and important tidal wetlands and estuaries, caused severe ecosystem disruption in only a limited area, and the system recovered almost totally within a decade. This experience has been repeated in many other spills, including the *Exxon Valdez* accident (**Fig. 19.4a**), the huge *IXTOC* platform spill (**Fig. 19.5b**), and the massive deliberate spill by Iraq in the Persian Gulf during the Gulf War in 1991.

Cleanup activities after a spill may be essential to remove as much oil as possible and speed natural recovery. However, if cleanup is too aggressive and protracted, the environmental benefit of the additional actions quickly diminishes, and extending the "cleanup" beyond a certain point can cause more damage than would otherwise occur. Generally, it is beneficial to skim up and remove as much floating oil as possible and to mop up oil from the shore that can be easily removed without disturbing the sediment. In addition, oil can be washed back into the water to be skimmed and removed, but only from high-energy beaches and rocky areas where the high-pressure water jets used for cleaning essentially simulate extended strong wave action.

Other cleanup efforts, including aggressive removal of oiled sediment from low-energy environments and extended efforts to remove oil from below the surface of coarse sand or gravel beaches, are costly, provide little or no environmental benefit, and in some instances, cause additional **ecological** damage and retard recovery. In isolated wild areas, such as Alaska's Prince William

TABLE 19.1 Large Oil Tanker Spills

Tanker	*Date*	*Location*[a]	*Amount of Oil Spilled (tonnes)*[b]
Atlantic Empress	1999	Off Tobago	287,000
ATB Summer	1991	700 n.m. off Angola	260,000
Castillo de Bellver	1983	70 n.m. off Cape Town, South Africa	257,000
Amoco Cadiz	1978	Brittany, France	223,000
Haven	1991	Genoa, Italy	144,000
Odyssey	1988	700 n.m. off Nova Scotia, Canada	132,000
Torrey Canyon	1967	Scilly Isles, United Kingdom	119,000
Sea Star	1972	Gulf of Oman	115,000
Irenes Serenade	1980	Navarino Bay, Greece	100,000
Urquiola	1976	La Coruña, Spain	100,000
Hawaiian Patriot	1977	320 n.m. west of Hawaii	95,000
Independenta	1979	Istanbul, Turkey	95,000
Jakob Maersk	1995	Leixões, Portugal	88,000
Braer	1993	Shetland Isles, United Kingdom	85,000
Khark 5	1989	120 n.m. off Morocco, Atlantic	80,000
Prestige	2002	Off the Spanish coast	77,000
Aegean Sea	1992	La Coruña, Spain	73,000
Katina P.	1992	Off Maputo, Mozambique	72,000
Sea Empress	1996	Milford Haven, United Kingdom	72,000
Nova	1985	75 n.m. off Khārk Island, Persian Gulf	70,000
Sinclair Petrolore	1960	Off Brazil	60,000
Epic Colocontris	1975	60 n.m. northwest of Puerto Rico	60,000
Corinthos	1975	Marcus Hook, Philadelphia, USA	53,000
Assimi	1983	60 n.m. off Masqat, Oman	52,000
Metula	1974	Strait of Magellan, Chile	50,000
Andros Patria	1978	Off Cape Finisterre, Spain	50,000
World Glory	1968	90 n.m. off Durban, South Africa	48,000
Pericles GC	1983	200 n.m. off Doha, Qatar	46,000
British Ambassador	1975	Pacific Ocean	44,000
Ennerdale	1970	Off Port Victoria, Seychelles	41,000
Mendoil II	1968	340 n.m. off Washington State, USA	40,000
Wafra	1971	Off Cape Agulhas, South Africa	40,000
Juan A. Lavalleja	1980	Arzew, Algeria	40,000
Trader	1972	Off southwestern coast of Greece	37,000
Exxon Valdez	1989	Prince William Sound, Alaska, USA	37,000
Thanassis A.	1994	200 n.m. off Manila, Philippines	37,000
Burmah Agate	1979	Galveston, Texas, USA	36,000
Napier	1973	Off Guamblin Island, Chile	35,000

Note: For incidents in which some of the spilled oil burned, not all of the amount listed here was released into the ocean.
[a]n.m. = nautical mile(s).
[b]All are estimated values. Reports have varied for some spills.

CRITICAL THINKING QUESTIONS

19.6 Because of environmental concerns, offshore oil exploration is currently banned in large areas of the Pacific and Atlantic continental shelves of the United States, despite the successful safety record of the oil industry operating under U.S. laws and regulations. Some oil companies are drilling instead in other areas, such as the ice-filled waters near the Russian island of Sakhalin, where the safety and environmental rules are weaker and sometimes totally unenforced, and from where the oil must be transported long distances to the locations where it is used.

(a) Discuss the implications that drilling operations like the one conducted off Sakhalin Island might have for the ocean environment as a whole.
(b) Should the drilling bans in U.S. waters be continued? Describe all the reasons for your answer.
(c) If the ultimate political decision were to increase the drilling and oil activity in U.S. waters, what studies should be done or actions taken before such drilling took place?
(d) Have such studies been done already?

Sound, the extended presence of people and their cleanup activities on beaches can have adverse effects on shorebirds, terrestrial wildlife, and **marine mammals.** In addition, the use of chemicals to disperse and dissolve oil can be damaging because such chemicals can be more toxic and persistent than the oil.

Unfortunately, public pressure often requires that everything possible be done to clean up all the oil after a spill. As a result, spill cleanups often continue beyond the point at which many technical specialists believe they should be ended. Large sums of money are wasted in cleanup that yields no environmental gain or is even detrimental. Nature is very efficient at cleaning up oil spills within a few years. Our role should be to remove as much oil as possible quickly, and then let nature take its course. However, in some instances, naturally occurring hydrocarbon-degrading bacteria, or nutrients that encourage the growth of such bacteria, might be beneficially added to the oiled ecosystem to help nature heal itself.

Whether we should attempt to rescue, clean, and rehabilitate oiled birds and marine mammals in an oil spill area is debatable. For example, efforts to clean and rehabilitate sea otters after the *Exxon Valdez* spill are estimated to have cost more than $80,000 for each animal that was captured and eventually returned to the ocean (a total of only 197 animals). Estimates are that only about one-half of those animals survived a year after reintroduction, and many that did survive would probably have survived without cleaning because they were so lightly oiled. Mounting an animal cleanup program after a spill is emotionally satisfying, provides good material for the media, and provides those responsible for the spill with the ability to claim that they are making every effort possible to minimize the damage. However, almost all such efforts have proven ineffective.

Environmental scientists generally accept that even major oil spills do not cause lasting and widespread destruction of ocean ecosystems. However, concerns remain that major oil spills may have long-term adverse effects on some fish and other species because of the **chronic toxicity** of some of the more persistent hydrocarbons, particularly **polyaromatic hydrocarbons (PAHs)**. For example, the unusually low numbers of salmon, **herring,** and other species returning to Prince William Sound in some (but not all) years after the *Exxon Valdez* spill raised suspicions that the low returns were in some way chronic effects of the spill. These returns are highly variable from year to year because of natural factors, such as climatic variations and disease outbreaks. Consequently, determining whether variations, such as those in Prince William Sound since the 1989 spill, are natural or related to the spill or to other anthropogenic factors is exceedingly difficult.

The difficulty is partly related to the lack of studies of year-to-year variability before the spill. After many years of intensive research following the *Exxon Valdez* spill, there is a consensus in the scientific community that, although some lingering long-term effects could be identified, many of these were caused by the aggressive cleanup efforts after the spill and that the sum of the effects that still remained a decade or more after the spill was far less significant than natural changes in the ecosystem (possibly related to the climate change associated with the Pacific Decadal Oscillation; Chap. 9).

Chronic Inputs

Many hydrocarbons are extremely toxic to some species, even in small concentrations. PAHs especially have a range of toxic effects, including teratogenicity and carcinogenicity (**CC18**). Consequently, the chronic long-term impacts of oil contamination from sources other than spills are likely to be more serious and widespread than those of the much more dramatic spills. Nonspill sources generally are highly concentrated in estuaries and the coastal zone, particularly near ports, harbors, and major cities. Hence, many scientists believe that greater research emphasis should be placed on the chronic effects of these other sources of oil in environmentally important coastal areas than on the massive research programs that follow major spills.

Lessons Learned

Several lessons can be learned from a careful review of oil pollution studies. Most importantly, the extraordinary public attention to and concern about major spills is misplaced. Recovery after such spills is relatively rapid, and their long-term effects are almost certainly less than effects of the much more widespread chronic oil contamination from other sources.

Second, although cleanup of oil from the area of a spill helps to speed recovery, cleanup must be limited and done carefully, because overly aggressive cleanup can cause more damage than the spill itself.

Third, public opposition to offshore oil drilling in U.S. waters will result in greater oil pollution because offshore production and transport by pipeline is much safer than tanker transport. Although the United States must ultimately reduce its use of oil, it cannot do so immediately. If we do not produce oil in the United States, we must continue to import it in tankers and incur a greater risk of oil spills.

Finally, ecosystems are naturally variable, so after an oil spill or other major disturbance, appropriate actions may restore the ecosystem to a natural state, but the balance of species in this new state will be different from the preexisting natural state because ecosystems are in a constant process of change.

19.6 SEWAGE

The most fundamental and unavoidable wastes produced by human populations are human feces and urine. Almost all cities, towns, and villages in the United States and all the world's major cities have extensive sewer systems to collect these materials and transport this sewage away from our living environment. The development of sewers was probably the single most effective advancement for human health protection in history. In the Middle Ages, sewage was simply thrown into the streets in many cities. Apart from the obvious aesthetic problems caused by human excrement in the streets, this practice almost certainly fostered the devastating plagues of disease that decimated urban populations of entire continents during this period. Hence, for many decades, the practice of collecting sewage and discharging it into a river or ocean to be diluted and washed away was viewed as essential and beneficial. This view has changed almost totally since the 1970s. Sewage disposal to rivers and oceans is now seen by many people as a pollution problem that must be completely stopped.

Nature of Sewage Wastes

Sewage is a natural material that consists primarily of the water used to carry solid fecal material through sewers. The water contains small concentrations of dissolved nutrients, trace metals, and organic compounds that originate either in the water supply or in human excretions. In addition, sewage contains solid organic matter that has a wide range of particle sizes. This solid matter, which is composed partly of paper fibers, also includes low concentrations of nutrients and trace metals and a variety of bacteria, **parasites,** and viruses. Many of these organisms are benign to humans and aquatic species, but others are pathogenic to humans and some are thought to be pathogenic to some aquatic species.

Although aquatic and ocean ecosystems are capable of assimilating sewage with no harmful effects, sewage discharges have caused pollution in many lakes, rivers, estuaries, and coastal ocean areas. Pollution has resulted from one or more of three problems. First, pathogenic organisms in sewage can reinfect humans if sewage is discharged into waters where people swim or otherwise contact the water, or if it is discharged into waters where harvested shellfish concentrate the pathogens. Second, organic matter and nutrients in sewage can produce blooms, hypoxia, and anoxia in discharge waters if the assimilative capacity for these wastes is exceeded. Finally, sewage is often contaminated with toxic compounds that enter the sewers from industrial discharges, street runoff, or household chemicals.

CRITICAL THINKING QUESTIONS

19.7 Describe the principal adverse effects of sewage discharged to the oceans and how they can be reduced or eliminated. What actions have been taken to reduce or eliminate sewage pollution of the oceans in the United States? Have these actions been effective?

19.8 Billions of dollars have been spent on the construction of secondary sewage treatment plants in all cities and towns regardless of the locations of their discharges.

(a) Was this uniform approach justified scientifically? If so, why? If not, what other approaches to sewage treatment and disposal might have been taken?
(b) What are the characteristics of sewage that you would wish to modify by alternative treatment approaches before discharging it to the oceans?
(c) What are the factors that you would consider in deciding where you might not require secondary treatment or where you might allow only primary treatment?
(d) How would you determine the environmental benefits or detriments of alternative treatments?

19.9 Numerous sources release trace metals, toxic organic substances, nutrients, and pathogenic microorganisms to the environment. These releases are eventually discharged to the oceans in sewage, urban storm-drain runoff, and runoff from agricultural land.

(a) List as many of these sources as possible.
(b) What substances do the various sources contain?

Only a few decades ago, many rivers, estuaries, and coastal areas near major cities were severely impacted by sewage pollution. Many waterways were simply overwhelmed by large quantities of organic matter. As a result, they became anoxic, causing biota, including fishes, to die, and creating significant aesthetic problems (odors and floating material). As environmental consciousness arose and as city populations began to look toward the water for recreation, two approaches to solving the sewage pollution problem were adopted. First, pipelines called "outfalls" were built to carry sewage offshore into the ocean, where dilution would be greater and where pathogens would die before reaching beaches. Second, sewage treatment plants were designed and developed to reduce oxygen demand and floatable material in the sewage before discharge.

Sewage Treatment

The many different technological approaches to sewage treatment can all be characterized by one of three treatment levels (**Fig. 19.8b**):

- *Primary treatment* usually involves removing floating material, grinding up solids, and removing some particulate material by allowing it to settle out to form **sewage sludge**, which is not discharged but still must be disposed of. Some, but not all, pathogens die or are removed to the sewage sludge.
- *Secondary treatment* usually includes a primary treatment step, followed by a secondary step in which bacteria that use sewage organic matter for food are encouraged to grow in the sewage (**Fig. 19.8b**). Most of these bacteria are removed in the secondary sewage sludge. Secondary treatment generally removes more than 90% of the BOD (which stands for "biochemical oxygen demand" and is a measure of oxygen-demanding organic matter) and more than 90% of the bacteria found in raw sewage. Some nutrients, trace metals, and toxic organic compounds are also removed to the sludge, but secondary treatment is not designed for this purpose and a high proportion of each of these compounds remains in the treated liquid effluent that is discharged.
- *Tertiary treatment* processes are generally designed to remove nutrients, such as nitrogen and phosphate, from the effluent that remains after secondary treatment.

In the United States, all communities are required to perform secondary treatment, although in a very few exceptional cases this requirement is not met or is legally waived. Only a few communities are required to use tertiary treatment, which is very costly and often technologically unreliable.

Sewage treatment and diversion of treated sewage to the open ocean have dramatically improved the water quality in many rivers and estuaries. For example, salmon have returned to the Thames River in England after it had been anoxic and little more than an open sewer for many decades before treatment was started. Because some parts of San Francisco Bay were anoxic in the 1960s, the shore was covered in rotting debris and slime, and odors drove people away from the Bay. San Francisco Bay is now dramatically improved and no longer anoxic. Many other bays and estuaries have undergone similar improvements, but the story is not all good.

The United States has spent tens or hundreds of billions of dollars to build and operate secondary-treatment plants in almost all communities. In many locations, expenditures on sewage treatment have bought major environmental improvements. In others, where sewage inputs are small or where residence time at the discharge site is short, secondary treatment has bought little or no environmental benefit because assimilative capacity was not exceeded before treatment was started.

Many pathogens and toxic compounds are partially deposited in the sewage sludge, which is disposed of separately from the effluent discharge. However, secondary treatment was not designed to remove pathogens and toxic compounds, so a large proportion of many of these contaminants remains in the treated effluent and is discharged. Consequently, pollution by pathogens or toxic compounds in treated sewage effluents still remains a problem, particularly where residence times of the receiving waters are long and where commercially valuable shellfish beds are near discharges. For example, even though the Bay is no longer anoxic, shellfish beds remain closed and several metals exceed water quality criteria in South San Francisco Bay as a result of the inputs of treated sewage.

Sewage sludge must be incinerated, deposited in landfills, or spread on agricultural land. Toxic substances and pathogens in the sewage sludge cause environmental impacts with each of these options.

Ocean disposal of treated sewage still causes significant environmental problems in some areas. Problems include nutrient-induced eutrophication of the coastal ocean, contamination by toxics, contamination of beaches and shellfish by pathogens, and damage to benthic ecosystems as a result of high organic loadings.

Eutrophication

Eutrophication of the coastal ocean is discussed in Chapter 15. The primary cause of this eutrophication is believed to be nitrogen compounds discharged in treated sewage and agricultural runoff. Tertiary sewage treatment is successful in reducing phosphate concentrations in effluents, but potential technologies for nitrogen removal are more complex and costly. Consequently, the growing

FIGURE 19.8 Sewage treatment. (a) Sewage treatment plants are often located on the water's edge. This large secondary-treatment plant in Oakland is located on San Francisco Bay at the foot of the Bay Bridge that connects San Francisco and Oakland. (b) Sewage treatment is designed to remove suspended solids and organic matter from the waste stream before it is discharged to rivers, estuaries, or the ocean. Most areas of the United States are serviced by secondary-treatment plants that remove most of the suspended solids and organic matter but are not very effective at removing nitrogen, phosphorus, and toxic trace metals. Tertiary treatment, which removes nutrients, and disinfection, which kills most remaining pathogens, are employed in very few locations because they are technologically complex and costly. All sewage treatment processes produce a semisolid sludge that must be disposed of on land or by incineration. In some locations, this sludge was disposed of for many years by being dumped into the ocean, but in the United States this practice was stopped in 1992.

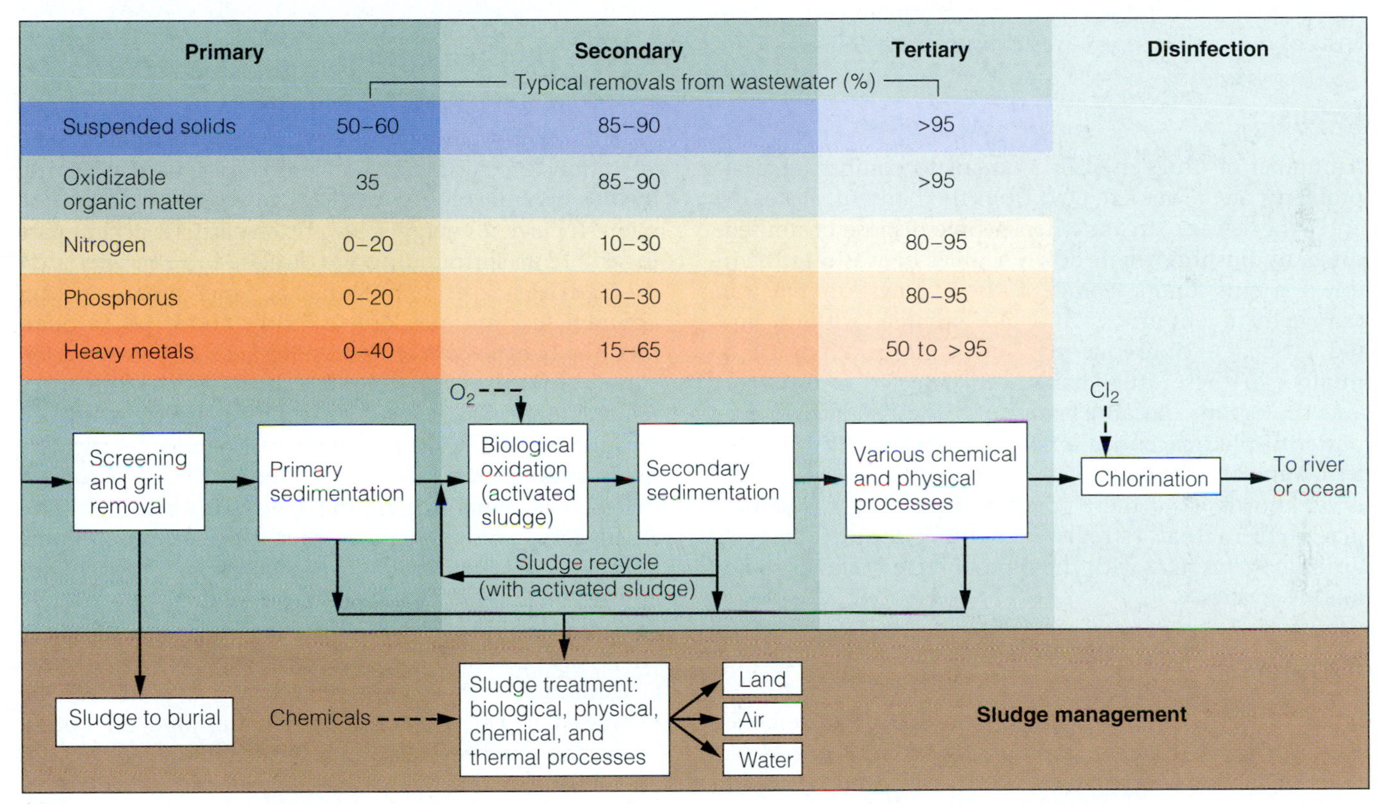

Typical removals from wastewater (%)	Primary	Secondary	Tertiary
Suspended solids	50–60	85–90	>95
Oxidizable organic matter	35	85–90	>95
Nitrogen	0–20	10–30	80–95
Phosphorus	0–20	10–30	80–95
Heavy metals	0–40	15–65	50 to >95

(b)

incidence and widespread recurrence of blooms and anoxia caused by eutrophication of the coastal oceans are growing pollution problems related to sewage effluent, for which, at present, sewage treatment has no remedy.

Toxic Substances

Although some toxic compounds are removed in sewage sludge, a considerable proportion of many toxic substances remains in the treated effluent. Once in the environment, toxic substances are taken up by organisms and sediment particles (especially fine-grained particles). In discharge locations where receiving waters have long residence times or low-energy sedimentation regimes, the continuous introduction of toxic substances can cause a buildup of concentrations in sediment or waters. In most locations where toxic substances accumulate from sewage discharges, other sources (e.g., runoff and industrial discharges) also contribute toxic substances, and it is difficult to assess the relative importance of sewage and the other sources.

Although toxic substances cannot be eliminated completely from sewage, they are unlikely to cause problems in the environment if their concentrations in sewage reflect only the background levels of these compounds in human food and water supplies. Consequently, major efforts are under way to prevent toxic substances from entering sewer systems. In the United States, a vigorous and extensive pretreatment program requires removal of toxic substances from industrial wastes before they are discharged to sewers. This program has drastically reduced inputs of toxic substances to sewage plants from industrial sources. However, only limited efforts have been made to reduce inputs of toxic substances from cleaning products, other household sources, and illegal disposal of chemicals. The principal source of toxic substances in most major cities' sewage is therefore the home. In some areas, sewers are connected to storm drains through which urban runoff may contribute substantial quantities of toxic substances. Frequently, toxic industrial wastes are dumped illegally into storm drains and sewers to save treatment costs or the costs of disposal in secure waste disposal landfills.

Drugs

A number of drugs used by humans to combat medical problems are either excreted from the body in wastes or enter the sewage stream when people dispose of unused drugs by flushing them down a toilet or washing them down a sink drain. Some of these drugs survive the sewage treatment process and are discharged with the liquid effluents. Consequently, low concentrations of a number of these drugs have been detected in estuaries and the oceans. Because many of these chemicals have powerful pharmaceutical effects on humans, it is likely that some also affect at least some marine species. Little is yet known about these possible effects. There is particular concern that estrogen and other fertility control drugs may interfere with the reproductive cycles of some marine species.

Pathogens

Sewage treatment is not designed to destroy or remove pathogens, but it does substantially reduce concentrations of some bacteria. Studies of the fate and effects of pathogens in sewage effluents discharged to the oceans are extremely difficult because of the low concentrations of pathogens in the environment and difficulties inherent in their isolation and measurement. Consequently, little is known about the fate of most pathogens in the oceans. Beach water quality and the safety of shellfish from pathogens are monitored by measuring concentrations of indicator bacteria, usually *Escherichia coli* (*E. coli*) or, more recently, *Enterococcus*. These organisms are present in all mammalian feces, not just human sewage, but they pose no significant human health risk themselves. Indicators are not perfect, because they are not specific to human feces and because pathogenic organisms are transported in different ways and die at different rates in the environment.

Bacterial monitoring for indicator organisms has reduced the incidence of diseases carried by shellfish or contracted through water contact. However, bacterial pollution still necessitates the periodic closing of beaches, and vast areas of coastal and estuarine zones in the United States and elsewhere in the world are closed to shellfish harvesting. Many pathogenic bacteria and viruses have been found to survive in seawater for much longer than was believed to be the case just a few years ago, and more human health problems may be due to sewage in the marine environment than was previously thought. In particular, the incidence of temporarily debilitating intestinal upsets caused by sewage-carried bacteria may be much higher than has been recognized. Such relatively mild upsets are poorly monitored and reported by health authorities.

Effects on the Benthos

The benthic ecosystem surrounding some ocean sewage outfalls is substantially altered by the discharge. The alterations are related to and probably caused primarily by the organic sewage particles deposited in the sediment. If current energy is low, these particles accumulate in sediments surrounding the outfall, and the effects on the **benthos** depend on the rate at which organic matter accumulates. At rates only slightly above background rate, the benthos are enhanced because the greater food supply increases the biomass and species diversity remains high (**Fig. 19.9a**). At higher rates, some species are disadvantaged by the higher suspended particulate loads and others are advantaged. Hence, biomass increases or remains high, but species diversity declines. As loading rates increase further, the benthos become dominated by a few tolerant species, and biomass eventually declines (**Fig. 19.9a**). Because loading rate usually decreases progressively with distance from the outfall, the effects also decrease as this distance increases.

Although it is somewhat arbitrary, a point can be identified in the gradient of effects around an outfall where the more severely impacted sediments are considered degraded and the less severely impacted sediments are not. The area of degraded sediments surrounding an outfall varies with the rate of discharge of organic particulate matter (**Fig. 19.9b**). For outfalls in similar current regimes, a small volume of discharge may cause no degraded sediments, whereas a higher volume of discharge may cause a significant area of degraded sediments. Hence, if sewage is discharged through many small outfalls that are designed to maximize the dispersion of organic matter, the benthos will be moderately enhanced. However, if the same treated sewage is discharged through a few large outfalls, significant areas of ocean floor may have degraded benthos.

These effects are analogous to those of fertilizer applied to a lawn. If the fertilizer is spread, the lawn will flourish. If it is dumped in one pile, it will burn and kill

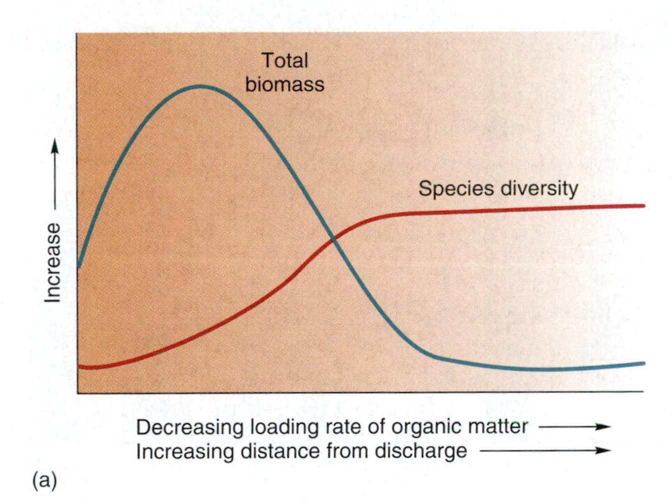

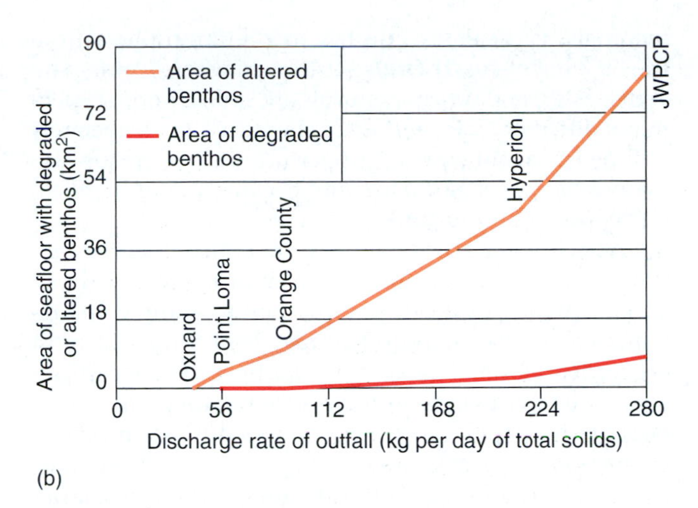

FIGURE 19.9 Degradation of the benthic environment in areas affected by sewage discharges to the marine environment. (a) Species diversity and biomass vary across a gradient of increasing sewage pollution. In the area on the right side where the loading rate of sewage is low, species diversity is not decreased and the total biomass is increased. Increasing biomass without decreasing the species diversity could be considered an enhancement of the environment. (b) The area of sediments with altered and degraded benthic infauna increases as the rate of sewage discharge from outfalls increases in similar hydrographic regimes. The large outfalls off the California coast—JWPCP (the Joint Water Pollution Control Plant in Los Angeles County) and Hyperion—adversely affect very large areas, whereas the smallest outfalls (Oxnard and Point Loma) do not cause any area to be degraded.

the grass close to and under it. Unfortunately, in the United States and elsewhere, environmental regulators have pursued a strategy of permitting only a small number of large outfalls in the belief that this restriction would limit any adverse impacts to only a few areas of the ocean.

For many years, large volumes of sewage sludge were dumped through an ocean outfall offshore of Los Angeles and from barges into the oceans offshore of New York, Philadelphia, and the United Kingdom. This practice was highly controversial. The media and public incorrectly blamed it for a wide variety of pollution problems, including the 1976 nutrient-induced anoxia off the New Jersey coast (Chap. 15) and the washup of floating medical wastes, including syringes, on New York beaches. Sewage sludge dumping contributed to these problems, at most, in a very minor way.

Lessons Learned

Public fears have now caused sewage sludge dumping in the ocean to be prohibited. However, many scientists believe that the high cost and environmental and human health impacts associated with land-based disposal alternatives far outweigh any environmental gains realized by terminating ocean dumping.

Sewage sludge dumping has caused or contributed to some pollution problems. These problems were related to toxic chemicals and pathogenic organisms in the sewage sludge and organic overloading of the seafloor, particularly at the Los Angeles outfall. If toxic substances were prevented from entering sewers and pathogenic organisms in sewage sludge were killed by modified treatment processes, sewage sludge could be disposed of safely in the ocean and might even have beneficial effects because of its food value. However, contrary to the historical practice of restricting dumping to small sites, disposal would have to be spread over a sufficiently large area to prevent overloading of sediments or the water column with oxygen-demanding organic matter.

Sewage treatment, although expensive, has dramatically reduced the worst impacts of historical sewage disposal in the marine environment. The remaining impacts can be reduced only by the adoption of new and different approaches to sewage treatment technologies that remove toxic substances and nutrients and kill all pathogenic organisms. Removal of all organic matter from sewage effluent is not necessary for disposal in the oceans, because, if adequately dispersed, such material will be assimilated without adverse effects and may in some instances even be beneficial.

19.7 URBAN AND AGRICULTURAL RUNOFF

Many chemicals are used in urban and agricultural communities. Fertilizers and pesticides are applied in large quantities to fields and gardens. Oil and rubber dust (containing toxic contaminants such as cadmium) are left on road surfaces by vehicles. Hydrocarbons and other chemicals are released from road-paving and other building materials. Particulates injected to the atmosphere from the burning of **fossil fuels** settle everywhere. Paints, solvents, acids, and other chemicals are spilled or released in many different ways, and toxic chemicals are deliberately dumped to avoid costs of proper disposal. There are also many other ways for toxic contaminants and nutrients to be deposited on the land.

When it rains, contaminants are either absorbed into the soil or washed off into storm drains, streams, rivers, and eventually estuaries and the ocean. Many contaminants are carried by runoff in dissolved form, but most are carried on small organic-rich particles. These particles are carried by streams and rivers to estuaries and coastal ocean, where they can become trapped in the estuarine circulation and sediments (Chap. 15). Contaminants from urban and agricultural runoff reach estuaries through numerous drainage channels spread throughout the watershed. Consequently, their sources are often referred to as **nonpoint sources.** Nonpoint source inputs are highly variable. Such inputs are greatest during the first few hours of a rainfall, especially if it follows a prolonged dry period. Thus, contaminants that are carried into the estuarine and marine environment generally are diluted by large quantities of freshwater. These factors make the identification and control of nonpoint sources of contaminants extremely difficult and make the cost of possible treatment in most instances prohibitively high.

Urban and agricultural runoff contributes a major proportion of the nutrients and toxic contaminants that enter many estuaries and coastal embayments (**Fig. 19.10**). Consequently, where such substances cause pollution problems, further controls on industrial and sewage inputs will have only a minor effect. The difficult task of controlling nonpoint source inputs must be addressed successfully if estuarine and coastal ecosystems are to be fully protected and restored.

Pollution problems caused by urban and agricultural runoff are many and varied. Three examples illustrate the range of such problems. First, in Chesapeake Bay, bottom waters in the center of the bay are relatively stagnant and prone to anoxia. What was once a small area of periodic anoxia has expanded to a large area of nearly permanent anoxia that has destroyed historical shellfish beds and fisheries. The cause is nutrient-induced eutrophication. Although sewage discharges contribute some nutrients, the predominant sources are fertilizer and animal waste in runoff from surrounding agricultural land, as well as atmospheric inputs. Extensive efforts have been made to reduce these nonpoint source inputs—for example, by reducing fertilizer use and avoiding its application near streams. Such efforts have resulted in some improvement, but the problem remains a difficult one.

The second example is the "dead zone," an area of the nearshore **continental shelf** of the Gulf of Mexico in which the bottom waters are seasonally hypoxic, that stretches from the mouth of the Mississippi River along the Louisiana and Texas coast. In midsummer, the

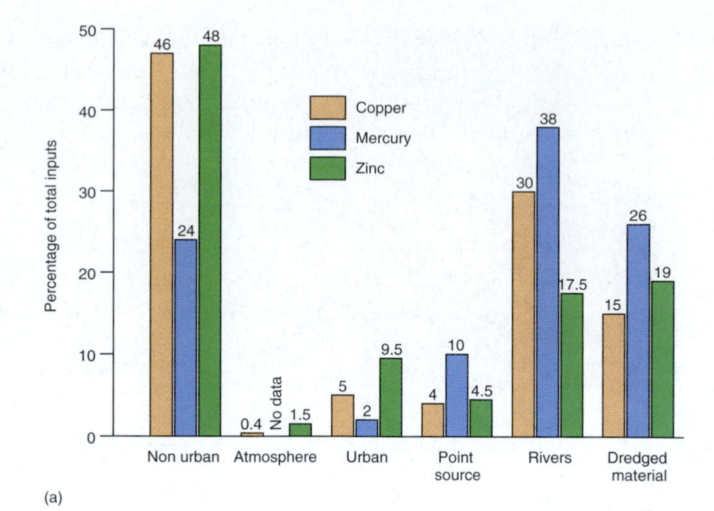

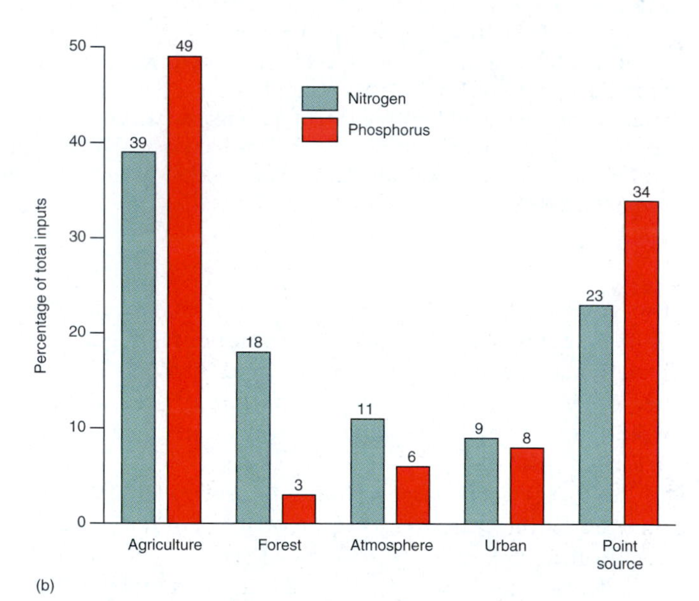

FIGURE 19.10 Sources of contaminants in estuaries. (a) Estimated inputs of copper, mercury, and zinc to San Francisco Bay from various sources. (b) Estimated inputs of nutrients to Chesapeake Bay. Atmospheric inputs are dust and rain. Inputs labeled "nonurban" and "urban" for San Francisco Bay and "agriculture," "forest," and "urban" for Chesapeake Bay are all nonpoint sources. These inputs enter the estuary in runoff through myriad streams, and they are difficult to estimate. Available estimates of some of these inputs range over an order of magnitude, and the numbers shown are averages of these estimates. River inputs to San Francisco Bay are partly natural and partly from agricultural and urban areas upstream. Dredged-material inputs are from dredging throughout San Francisco Bay, much of which is dumped at a site near Alcatraz Island. Agricultural inputs to Chesapeake Bay include primarily fertilizer used on fields and animal excretions. Point source input totals include all sewage treatment plant and industrial releases. The inputs from nonurban, urban, agricultural, forest, dredged-material, and atmospheric sources are the sum of both natural concentrations and contaminants.

hypoxic zone covers 8000 to 10,000 km^2 of ocean bottom—an area equal in size to the state of New Jersey. This area is located in the middle of one of the most important commercial and recreational fisheries in the United States. The Mississippi River basin drains about 41% of the Lower 48 states of the United States—a total of over 3.2 million km^2. The drainage basin encompasses all or part of 30 states, a population of about 70 million people, and extensive agriculture. The Mississippi River discharge contains a high nutrient level, some of which is natural but much of which comes from runoff of fertilizers applied to farmlands and treated sewage. In the last four decades of the twentieth century, the discharge of nitrogen by the Mississippi River basin tripled. It is this excess nutrient load that is believed to have created the dead zone, which was first documented in 1972. Efforts to reduce the nutrient loads of the Mississippi River are under way, but, as in the Chesapeake Bay region, these efforts are difficult and will take many years if the dead zone is to be completely eliminated.

The third example is in northern California, where agricultural runoff from irrigated fields passes through a series of drainage canals and ponds before eventually discharging to San Francisco Bay. Soils in this region contain naturally high concentrations of selenium. Irrigation waters dissolve selenium and carry it toward San Francisco Bay. Selenium accumulates in the freshwater ecosystems along the way, especially in a lake known as Kesterson Reservoir. There selenium is concentrated by rooted plants and consumed by ducks and other waterbirds that use the reservoir as a feeding stop on their annual north–south migration. Many birds have suffered severe selenium poisoning and have even died as a result of feeding at Kesterson. Fortunately, as selenium is carried farther into the marine environment, it is altered to a less toxic form, and it appears to have had less severe consequences in San Francisco Bay. Costly efforts have been made to reduce irrigation and remove selenium from excess water that drains from the fields. The selenium problem, which was especially severe in the mid 1980s has lessened but is still a concern.

19.8 INDUSTRIAL EFFLUENTS

Most industries produce solid or liquid wastes that differ widely in composition among industries and even among factories within an industry. Most of these wastes contain various toxic chemicals. A few decades ago, most industrial wastes were discharged to sewers or a local waterway or dumped in a nearby landfill.

The United States and many other industrial nations have had laws for several decades to control contaminant discharges to aquatic environments. Enforcement of most of these laws has focused on industrial discharges and sewage treatment. Industries have been forced to reduce drastically the concentrations of toxic contaminants in their liquid effluents before discharge. Most industries have changed technologies to reduce the amount of wastes they generate and avoid expensive treatment of effluents.

As a result of decades-long efforts to reduce toxic substances, nutrients, and other contaminants in industrial effluents, industry is now only a minor contributor to the contamination of estuaries and coastal oceans in most areas. This is true even in estuaries such as San Francisco Bay, with its large concentration of oil-refining, petrochemical-manufacturing, and ship-building and -repairing industries (**Fig. 19.10a**). Problems remain with some older industrial plants, with industrial discharges that are poorly located where residence time is long, and with the enforcement of discharge permits at a few plants whose unscrupulous owners or operators find ways to discharge wastes illegally. Nevertheless, the general public mistakenly still believes that industry is the principal polluter of the marine environment.

There are several reasons for this misunderstanding. The most important is probably that it is easier for media, politicians, and the public to blame industry for pollution problems than to face the difficult problems of cleaning up individual actions, homes, public utilities, and farms. Another reason is that the only demonstrated and confirmed incident in which the discharge of toxic chemicals to the oceans caused human deaths involved industrial effluents.

CRITICAL THINKING QUESTIONS

19.10 After you have answered Critical Thinking Question 19.9 on page 565, take a look at the labels on all of the containers of cleaning fluids and powders, cosmetics, shampoos, paints, and other materials around your house.

(a) What additional sources can you now list?
(b) Where do these products and their chemical constituents go after you have used them?
(c) Make two lists: one of the products whose containers tell you enough about their composition for you to decide what contaminants they might release to the environment, and another of those that do not. Do you know what is in the products that do not tell you on their label what they contain?
(d) Why do some products have information on their labels that divulge their composition, while other products do not? Should this situation be changed? If so, how?

19.11 Discuss why it is more difficult to control contamination of the oceans from urban and agricultural inputs than from industrial inputs.

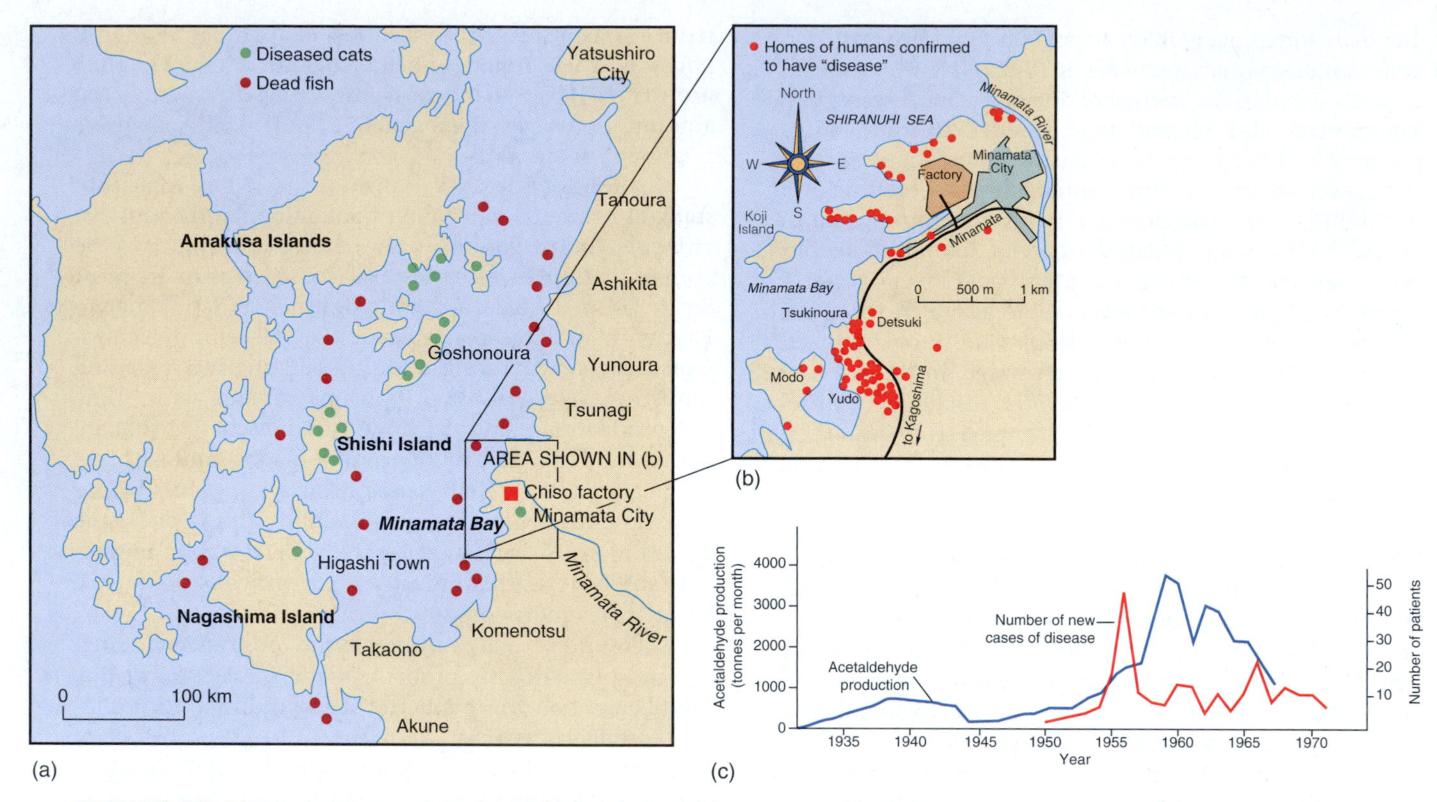

FIGURE 19.11 Methylmercury, discharged by a chemical factory at Minamata, Japan, and concentrated in fish and shellfish tissues, caused a major pollution problem in the 1950s. (a) Map of the locations around Minamata where diseased cats and dead fishes were reported many kilometers away from the discharge point. (b) Detailed map of the Minamata city area identifying the locations where humans lived who were confirmed to have contracted the "disease." (c) Forty-six people died of mercury poisoning, and many more suffered lasting effects. Even after the early poisonings, the Chiso Chemical Corporation continued to increase its production of acetaldehyde and its waste by-product, methylmercury. Poisonings continued for years after the release of mercury from the plant was eventually curtailed.

For many years, Chiso Chemical Corporation in Minamata, Japan, generated an organic form of mercury, methylmercury, as a by-product of its manufacture of acetaldehyde using mercury as a catalyst. A small portion of the methylmercury was discharged continuously to Minamata Bay in wastewater. Mercury, a toxic metal, accumulated in water, sediments, and fishes. Because methylmercury is more fat-soluble than elemental mercury and is biomagnified (**CC18**) in marine food webs, its concentration in fishes rose to very high levels.

The discharge of methylmercury to Minamata Bay started in 1952. By 1953, many cats were becoming ill, behaving erratically, and dying. Dead fishes were periodically found floating throughout the region (**Fig. 19.11a**). At the same time, some of the human population contracted a puzzling "disease" that produced numbness, disturbances in vision and hearing, and loss of the control of motor functions, similar to drunkenness. The "disease" quickly assumed epidemic proportions, and between 1953 and 1962 more than 40 people died and as many as 2000 other victims suffered what has now proven to be a persistent disability (**Fig. 19.11b**). In 1957, it was found that "Minamata disease" could be induced in cats by feeding them fishes taken from near the Chiso plant outfall. Although fishing was quickly banned, the discharge of methylmercury was not stopped until 1968, because Chiso denied that methylmercury caused the disease until it was overwhelmed by the weight of scientific evidence.

Minamata is apparently the only documented case of human deaths caused by industrial inputs to the marine environment, but many cases of deaths of fishes and other marine organisms caused by industrial effluents have been documented. For example, the Montrose Chemical Corporation in Los Angeles produced about two-thirds of the world's DDT in 1970. Starting in 1953, the plant discharged effluent containing DDT through a sewer to Santa Monica Bay. By 1970, this and other sources of DDT, which is fat-soluble and biomagnified, had caused the total collapse of many species at the top of the food web. It particularly affected pelicans, whose eggshells were so thinned by the DDT that they broke before hatching, and sea lions, which produced large numbers of stillborn offspring. DDT discharge was stopped, and the use of DDT was banned in the United States and Europe early in the 1970s.

Pelican, sea lion, and other affected populations have slowly recovered in the decades since the DDT ban. However, DDT and its persistent and toxic decomposition products can still be found in elevated concentrations in sediments and the biota of the southern California coastal zone. Unfortunately, despite its ban in the United States and many other countries, estimates suggest that more DDT is produced and used today than in 1970. Most of it is used for mosquito and malaria control in developing tropical countries that cannot afford more costly and less effective alternatives to DDT, especially since the resurgence of malaria in the 1990s.

19.9 DREDGED MATERIAL

With the exception of a few deep-water ports, most ports and harbors do not have sufficient natural depth for vessels to navigate safely and berth. Consequently, most ports and harbors must be dredged to increase navigable depth. Once sediments have been removed to deepen a channel or basin, suspended sediments tend to settle in the new **topographic** depression. Therefore, all dredged channels and harbors must be redredged periodically to remove newly accumulated sediment. In some high-energy regimes where the new channels are regularly swept by currents, or areas where very little suspended sediment is transported to refill the channel, dredging may be necessary only once every several years. However, other channels must be dredged more often, and some require almost continuous dredging.

Dredging is done either by scooping up buckets of sediment or by sucking sediment from the seafloor through a pipe lowered from a surface vessel. In most cases, the dredged material is loaded onto barges and transported to another location before being dumped into an estuary or the ocean.

Dredging destroys the benthos, causes increased turbidity, and, if the dredged material is contaminated, releases toxic substances to the water and suspended sediment at both dredging and dumping sites. These effects may be serious at some dredging sites, but they are generally more damaging at the disposal site. The United States has more than 150 aquatic disposal sites for dredged material. Most are in bays or coastal waters very close to the mouth of the estuary from which the material is dredged.

When dredged material is dumped, it descends quickly until it strikes the seafloor. Upon impact, it spreads rapidly across the seafloor as a suspended sediment cloud. Coarse-grained materials settle out quickly. Some smaller particles are partially buried with the coarse-grained material, and others are carried off by currents and deposited elsewhere where current speed is lower. Most or all benthos in the area buried by the dredged material are killed. The area is recolonized over the next few weeks or months, unless, of course, the dredged material is badly contaminated with toxic substances or dumping is frequent or continuous.

Because most dredged material consists of sediment taken from channels close to industrial and urban areas, it is usually contaminated with toxic substances from urban runoff, industrial discharges, sewage, and spills. Dumping at most dredged-material dump sites is continuous or is done several times each year. Consequently, the bottom at almost all such dump sites is covered with contaminated sediment, and the benthos is severely damaged on a continuing basis.

A proportion of the fine-grained contaminated dredged material is lost to the suspended sediment during dumping. Because the contaminants associated with these particles will be at least partially **bioavailable**, dredged material can contribute substantial quantities of toxic substances to the marine environment. For example, until several years ago the amount of several toxic contaminants dumped annually at the dredged-material dump site near Alcatraz Island in San Francisco Bay (**Fig. 19.12**) exceeded the combined total annual input of these toxic substances to San Francisco Bay from more than 100 industrial and treated sewage discharges (**Fig. 19.10a**). Some of the dredged material previously dumped at the Alcatraz site has now been diverted to disposal on land or at a deep-ocean dump site.

At some dump sites, very heavily contaminated dredged material is covered within a few days by a cap of clean, sandy dredged material several tens of centimeters deep. Once under this cap, contaminants are effectively removed from the biosphere, but it is not clear what fraction of the contaminants, especially those associated with fine-grained materials, escape before being capped or whether erosion caused by waves and currents eventually breaches the cap.

Most dredged-material dump sites are located in low-current areas to ensure that as much of the dumped material as possible is retained at the site. However, at some sites, such as the Alcatraz dump site in San Francisco Bay (**Fig. 19.12**), with its fast tidal currents, all but the largest dredged particles are swept away soon after their disposal. Alcatraz and other similar dump sites were originally selected near estuary mouths because it was asserted that the dumped material would be quickly swept to sea by the river flow. Unfortunately, the dredged material dumped at many such sites, including the Alcatraz dump site, is transported instead back into the estuary with the estuarine circulation (Chap. 15).

The practice of dredging and dumping the dredged material within estuaries has two important effects. First, the material is partially transported back to and deposited in dredging sites within the estuary, thus increasing the frequency and cost of dredging. Approximately 40% to

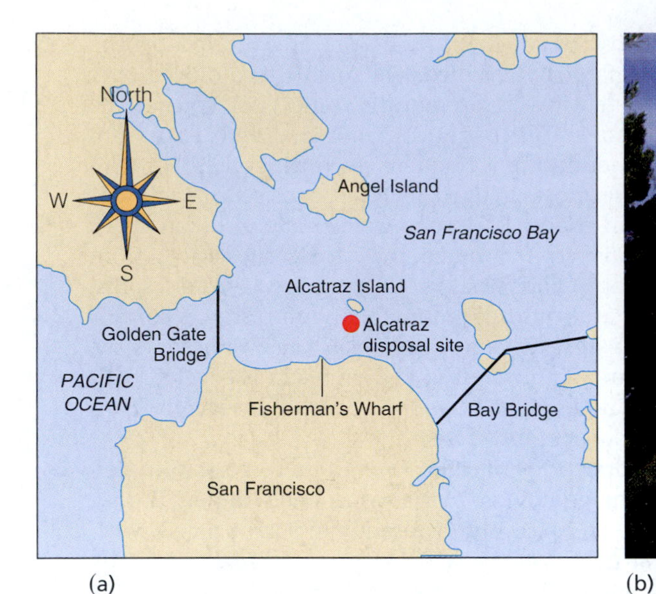

(a)

(b)

FIGURE 19.12 Dumping of dredged material in San Francisco Bay. (a) The Alcatraz dredged-material dump site in San Francisco Bay is located close to the estuary mouth and to the city of San Francisco. (b) The dump site, which is located between Alcatraz Island and the shoreline in this photograph, is only a few hundred meters from one of the famed tourist attractions of San Francisco, Fisherman's Wharf, and it is within easy view from much of the city.

50% of the dredged material dumped at the Alcatraz site returns to dredging sites and must be redredged. Second, contaminants that were discharged to estuarine sediments decades ago when discharges were largely uncontrolled are continuously redredged, dumped, and resuspended within the estuary, where they are in contact with the biota. This may be one reason why, in many estuaries, extensive contaminant source control efforts since the 1970s have not led to significant reductions in contamination of seafood or other species. The Alcatraz site has been used for dumping since the 1890s. Since that time, the once extremely valuable commercial fish and shellfish populations of San Francisco Bay have been largely destroyed. Dredged-material dumping has undoubtedly contributed to their decline.

Even if treatment techniques were available, treating estuarine sediments to remove historical contamination would not be possible, because the sediment volumes are too large. In addition, land disposal of the dredged material is acceptable only in a few locations, and then only if the material is not contaminated. However, if dredged material disposal sites were moved offshore beyond the biologically critical coastal zone, the contaminated material, if properly disposed of, would have less environmental impact on the limited benthos of the outer continental shelf and **continental slope.** In addition, if dredged material were dumped in the ocean, and if contaminant source control continued in the estuaries, contaminant concentrations in the estuaries would decrease and dredged material eventually would be clean enough to be used for beneficial purposes on land.

19.10 PLASTICS AND TRASH

Plastics do not completely decompose in the environment, but break down slowly into smaller particles that can be mistaken for food and eaten by marine organisms. Most organisms can neither digest the particles nor pass them through the gut, and plastic particles have been found to accumulate in the gut of some species until the gut is blocked and the animal dies of starvation. Larger fragments of plastics are ingested by large animals, especially turtles, which mistake plastic bags and other plastic debris for their favorite food, jellyfish. Plastic fishing line, six-pack rings, plastic fishing nets, and all kinds of other plastic debris act as drifting traps that entangle and kill turtles, birds, and marine mammals (**Fig. 19.13**).

It has been found that microscopically small fragments of plastic are ubiquitous in recent marine sediments and seawater. These fragments have been found in preserved plankton samples from the 1960s, and examination of preserved samples from more recent years shows that their concentration has increased significantly since then. Although adverse effects of plastic ingestion have been observed in larger animals that ingest relatively large particles of plastic, almost nothing is known about the effects of these microscopic fragments on zooplankton and benthic animals.

International law now prohibits the disposal of plastics in the oceans, but enforcement of this prohibition is virtually impossible and violations are common. Plastic also reaches the oceans from many sources other than direct disposal, including storm-drain runoff and helium-

FIGURE 19.13 Plastics can cause more than just aesthetic problems in the marine environment. (a) This northern fur seal pup is entangled by a section of lost or discarded gill net and would not have survived long without human intervention. (b) This seagull is entangled in a plastic six-pack ring and will not survive without help.

filled balloons released at sporting events and other festive occasions.

Metal, paper, and other trash items cause fewer problems in the ocean than plastics because they eventually corrode or decompose. Nevertheless, these discarded materials, as well as plastics, cause aesthetic problems when they wash up on beaches and shores or litter the seafloor where divers visit. Included in the floatable wastes are medical wastes, such as syringes, discarded illegally from vessels and through rivers and storm drains. These wastes have caused precautionary beach closures when they have washed ashore. Storm drains are thought to be the major source of most types of floatable wastes present in the oceans.

19.11 ANTIFOULING PAINTS

Unless protected, almost any surface introduced into the ocean is quickly covered, or **fouled,** with a wide variety of marine animals and algae. **Barnacles** have an especially strong natural adhesive that enables them to attach and hold on to surfaces even in swift currents that can sweep off other fouling organisms. Hence, barnacles dominate the fouling community on vessel hulls and on structures in strong currents.

When a normally smooth vessel hull or other moving surface is fouled, **friction** between hull and water increases and causes the vessel either to lose speed or to use more fuel to maintain speed. Consequently, vessel hulls, turbines of tidal power plants, and water intakes and drainpipes must be protected from fouling.

The predominant and most effective means of reducing fouling is to coat the substrate with paint containing a toxic substance that will kill any organism that settles. The toxic substance must be bioavailable to perform its function, so the paint must release its toxic substance slowly into solution. Consequently, vessels and other structures release these toxic substances to the marine environment, causing an especially serious problem in harbors where boats are concentrated and water residence times are long.

For many years, most antifouling paints contained copper as their active toxic ingredient. Copper is highly toxic at moderate concentrations in its **ionic** form. However, at lower concentrations it is nontoxic, and it is also readily **complexed** with organic matter, which reduces its toxicity. Thus, as copper is released from an antifouling paint, it has the desired antifouling action, but it is quickly diluted and complexed with natural organic matter as it **diffuses** away from the treated surface. Copper does accumulate in the water of harbors and marinas, but adverse impacts on biota other than those fouling the painted surface have been extremely rare.

Although copper-based antifouling paints are effective, they become less so as copper progressively dissolves. To reduce the frequency with which ships' hulls must be cleaned and repainted, a more effective antifouling paint was developed in which tributyltin replaced copper. Tributyltin is much more toxic than copper and is released more slowly from antifouling paints. Hence, each application of tributyltin-based paint is effective for a longer period than an application of copper-based paint.

Unfortunately, once released into the water, tributyltin is degraded only slowly to a less toxic chemical form. Consequently, it retains its toxicity and tends to concentrate in the water in areas where many boats have such paints. Tributyltin has caused serious marine pollution problems in some areas. Many ports and marinas where tributyltin paint was used extensively were essentially denuded of all animal life. Because tributyltin is particularly toxic to **mollusks,** it destroyed populations of oysters and other species over a wide area far from its source in some estuaries. In the United States, Europe, and some other nations, tributyltin-based antifouling paints are now strictly regulated, and their use is restricted to a few vessels with special needs. Fortunately, unlike DDT, tributyltin does not persist in the environment for decades. Recolonization and recovery have occurred rapidly in many areas that were previously affected by tributyltin.

19.12 RADIONUCLIDES

During the nuclear era, the oceans were contaminated with many **radionuclides.** They were introduced in fallout from atmospheric nuclear bomb tests, liquid waste discharges from nuclear facilities, and deliberate dumping of solid and liquid radioactive wastes at ocean dump sites. Some of the radionuclides do not occur naturally.

Once in the marine environment, radionuclides become involved in **biogeochemical cycles** and behave in the same way as the stable **isotopes** of their elements (if stable forms exist). Thus, each **radioisotope** has its own distinct behavior in the oceans. For example, tritium (hydrogen-3), a radioactive form of hydrogen released in nuclear explosions, quickly combines with oxygen to form water. The radioactive water molecules then join the hydrologic cycle and ocean circulation system, where they behave almost indistinguishably from other water molecules. An iodine radioisotope, iodine-131, enters solution in ocean water and is concentrated by marine biota, particularly certain species of algae. Plutonium does not occur naturally but is rapidly attached to particles when it enters the ocean environment. Most plutonium is carried with these particles to the sediment, where it remains.

Radioisotopes are easily measured, even at exceedingly small concentrations. They have proven invaluable as **tracers** to study geochemical, biological, and physical processes in the oceans (Chap. 10), but they can have adverse effects on marine ecosystems and on human health. The primary concern is that some anthropogenic radioisotopes are concentrated in seafood and that people who eat the seafood will have an increased risk of cancer.

With the exception of very limited areas surrounding nuclear bomb testing sites, the concentrations of anthropogenic radioisotopes in seafood are well below background levels of naturally occurring radioactive elements. Nevertheless, the prevalent, although not universally accepted, theory of carcinogenicity is that any increase in radioisotope intake will increase the risk of cancer. The assimilative capacity of the oceans for anthropogenic radioactive materials may therefore be considered essentially zero. In reality, a very small increase in radioactivity above natural background levels will not produce a measurable or significant increase in cancers. Hence, the oceans do have an assimilative capacity for radioactive materials, but it is small and not easily defined, because there is no generally accepted level of incremental risk.

Because the ocean's assimilative capacity for radioactive materials is so small, international agreements require that the release of such materials to the oceans be eliminated entirely. Almost all nations adhere to these agreements. The United Kingdom continues to discharge some liquid radioactive wastes from its plutonium reprocessing factory at Sellafield, but the discharges are being progressively reduced. In addition, after the fall of the Soviet Union, it was revealed that the Soviet Union had routinely dumped and discharged very large quantities of solid and liquid radioactive wastes directly into the Arctic Ocean and the Sea of Japan (**Fig. 19.14a**), and into rivers that empty into the Arctic Ocean. In fact, such ocean dumping continued until 2005 when Russia finally agreed to a global treaty that bans such practices.

The quantities of nuclear wastes dumped into the ocean by the Soviet Union are truly staggering (**Fig. 19.14b,c**). The dumped material includes a number of nuclear submarine reactors still containing their nuclear fuel. These fuel rods contain very large amounts of dangerous radionuclides, including strontium-90 and cesium-137, which could bioaccumulate and threaten human health if released to solution and dispersed. At present, very little radioactive material appears to have been released from sunken submarines and solid wastes. However, there is concern that these wastes, and radioactive wastes dumped on land and into rivers, may eventually be released to contaminate seafood in the Arctic Ocean and the Greenland and Bering seas, which are among the world's most important fisheries.

19.13 NONINDIGENOUS SPECIES

Marine species are transported around the world in ships' ballast water, attached to ships' hulls, and in other unintended ways. Species are also deliberately transported from one location to another to be used in aquariums or for aquaculture, and many of these are inadvertently or deliberately released to marine ecosystems in which they do not occur naturally. Once in their new environment, these species may die if they are unable to tolerate the new conditions or to avoid their new predators, or they may survive and reproduce with little effect on the rest of the ecosystem. However, some nonindigenous species not only survive but reproduce, spread rapidly, and outcompete native species. Estimates suggest that approximately 15% of introduced species cause severe harm to their new environment.

Many examples can be given of nonindigenous species that have severely damaged terrestrial and fresh-

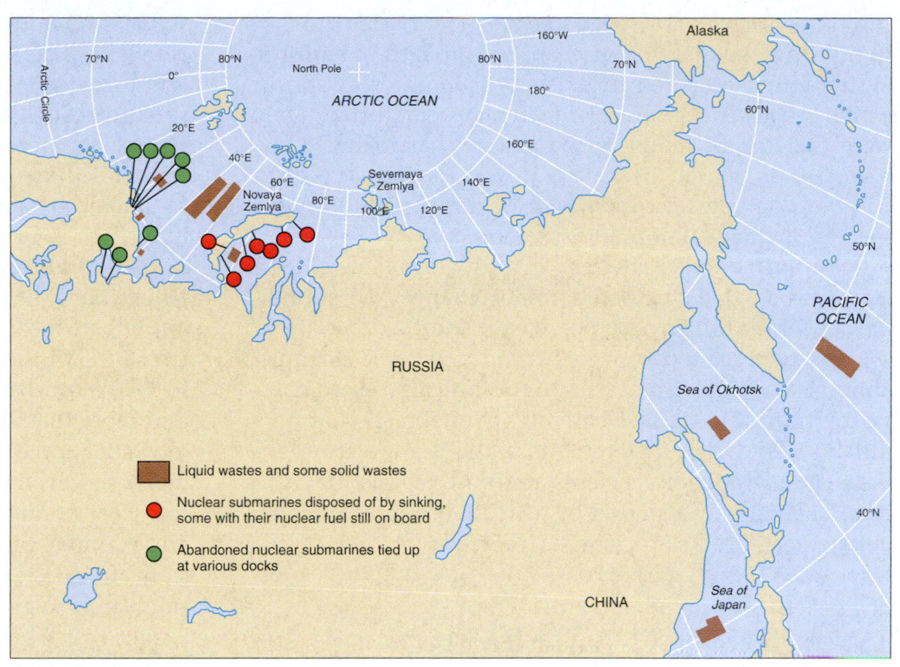

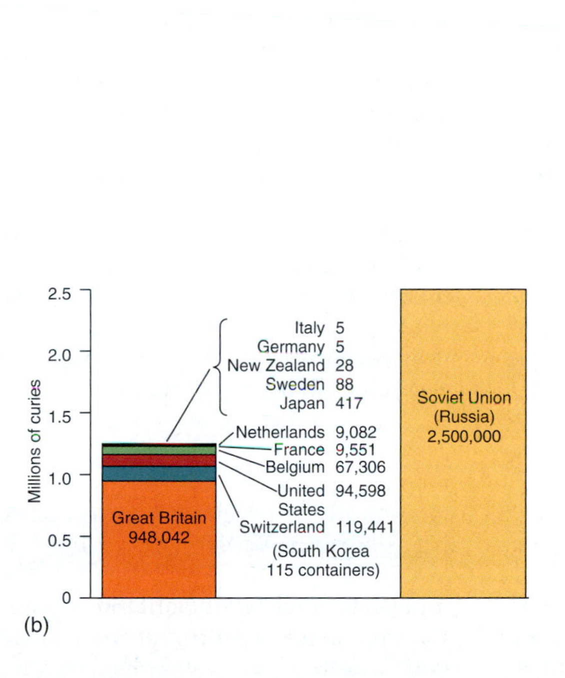

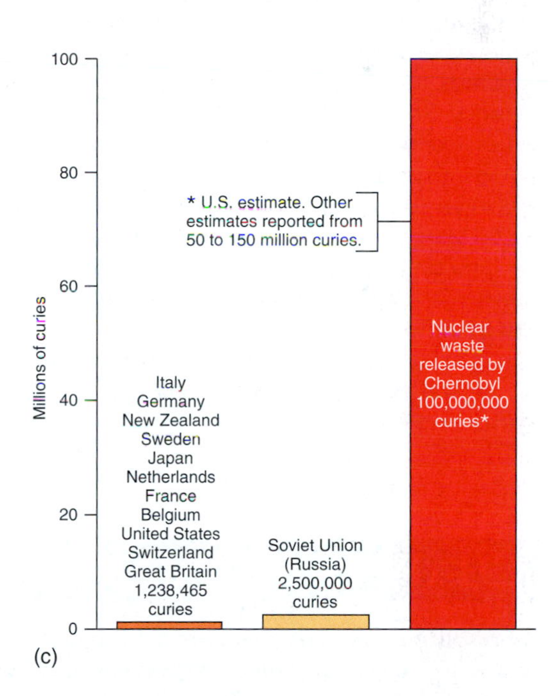

FIGURE 19.14 The former Soviet Union dumped large quantities of nuclear wastes in the oceans. (a) Damaged or decommissioned nuclear submarines were dumped and abandoned primarily in the Arctic and Pacific Oceans. (b) The quantities of radioactive wastes dumped in the oceans by the former Soviet Union far exceeded the total amounts dumped by all other nations. (c) Even though considerable amounts of nuclear wastes have been dumped in the oceans, these amounts are very small compared to the amount released to the environment as a result of the Chernobyl, Ukraine, nuclear power plant accident. Note that 1 curie is equal to the amount of radiation given off by 1 g of radium. The Three Mile Island, Pennsylvania, nuclear power plant accident in 1979 released only a few curies of radioactivity, an amount far too small to be visible if shown in these figures.

water ecosystems. Fewer examples of damage are known in the marine environment, probably for several reasons. First, marine biota are not as easily observed as freshwater and land biota, and, thus, adverse changes due to nonindigenous species may not be noticed. In addition, because the ecology of marine ecosystems is poorly known, declines of important species and other ecological changes may be attributed to other factors, even if caused by nonindigenous species. Open-ocean ecosystems are generally well connected with each other and more uniform than terrestrial or freshwater ecosystems. However, coastal marine ecosystems, especially estuaries, are more isolated from one another and therefore more vulnerable to the introduction of nonindigenous species.

San Francisco Bay is an excellent example of an estuary that has been severely damaged by nonindigenous species. It is the only major estuary on a long stretch of coast, and as a result, it is isolated from other estuaries. Therefore, the Bay had many unique plants and animals when European settlers arrived. The settlers deliberately introduced nonindigenous species, such as the eastern oyster (*Crassostrea virginica*), in an attempt to populate the bay with valuable food species. The oyster did not survive, but a variety of other species brought in with the oyster, either attached to its shell or in the seawater in which it was transported, did become established. They were the first of many introductions that included the striped bass, which became a major anadromous sport fish but competes for food and habitat with indigenous species, such as salmon and sturgeon.

Nonindigenous species introductions into San Francisco Bay have caused dramatic changes. Most original species in the Bay are now gone, and the majority of species present today were introduced. For example, almost half of the fish species in the critical wetland habitat are nonindigenous. Nearly all invertebrates that now inhabit shallow, nearshore parts of the Bay are also nonindigenous. Indigenous clams and other edible mollusks have been largely replaced by a small, economically valueless species of Asian clam.

19.14 HABITAT ALTERATION

Alteration or destruction of habitat is among the most damaging forms of pollution in estuaries and the coastal zone. The most serious habitat destruction is filling of wetlands to create dry land. An estimated 450,000 km^2 (almost 50%) of the historical wetlands of the United States, excluding those in Alaska, have been lost since the settlers arrived. In some areas, such as San Francisco Bay, less than 10% of historical wetland area remains (**Fig. 19.1**). Many tens of thousands of acres of wetland habitat for aquatic birds, juvenile fishes, and other species have been lost. Other forms of adverse habitat alteration or loss include beach erosion or loss due to coastal structures (Chap. 15), dredging of channels through mudflats, burial of seafloor with dredged material, and damage to **coral reefs** by anchors.

Less obvious habitat alteration is caused by human activities that lead to changes in salinity, temperature, and turbidity. Soil erosion has increased in many watersheds because trees and other vegetation have been removed. Increased suspended sediment loads have altered sediment grain size in many rivers, making the sediments unsuitable as **spawning** habitat for anadromous fishes. Excess turbidity has reduced light penetration and thus rendered both benthic and pelagic habitats less suitable, or unsuitable, for algae growth, with resultant effects throughout the food web. For example, more than 90% of the historical **sea grass** cover has been lost in Galveston Bay, more than 75% in Mississippi Sound, and more than 50% in Tampa Bay.

Reduction of freshwater flow into estuaries is a particularly serious form of habitat alteration. In many estuaries, species zonation is based on the salinity distribution. For example, plants rooted in freshwater are replaced by saltwater species in the lower parts of the estuary, where salinity is higher. If freshwater inputs are drastically reduced, as when rivers are dammed, salinity ranges can be shifted many kilometers up an estuary, dramatically altering the balance between freshwater-dominated and seawater-dominated habitat.

In some estuaries, such as San Francisco Bay, freshwater input reductions have caused the critical **brackish** water zone, where much estuarine productivity occurs, to migrate from a wide wetland-fringed part of the estuary into a deeper, narrower, faster-flowing section. Primary productivity has been lowered because of the smaller area of suitably brackish water and the high turbidity. Such loss of primary productivity may have contributed to dramatic declines in anadromous fish species in this and other estuaries. Reduction of freshwater inflows also tends to increase residence times within the estuary. Consequently, contaminants have longer residence times and reach higher concentrations.

CHAPTER SUMMARY

Assimilative Capacity. The oceans can safely assimilate wastes if the rate at which the waste is introduced does not exceed the assimilative capacity, which is the point at which toxic or other adverse effects occur. Assimilative capacity varies by waste and location because residence time determines the concentration of contaminants and each contaminant has a different concentration–effect relationship. Release of a specific type and amount of contaminant that causes pollution in one location may not do so elsewhere.

Adverse Effects of Human Activities. Increased turbidity can reduce primary productivity. Excessive nutrient inputs can alter phytoplankton species composition and can cause blooms, eutrophication, and hypoxic or anoxic bottom waters.

Habitat can be altered by development and by discharges of wastes. Many areas of mangrove swamps and coastal wetlands have been filled and developed. Damming of rivers has prevented or inhibited anadromous and catadromous fish migrations and has altered salinity regimes and circulation in estuaries. Dredging, anchoring, fishing, and other activities disturb and alter seafloor habitats.

Community structures within ocean ecosystems can be altered when human activities affect habitat or affect some species but not others. Selective harvesting of fish, invertebrate, and marine mammal species has caused major changes in communities in some areas.

Toxic substances released to the oceans can be concentrated in seafood and present a human health risk or result in economic loss if the resource is unfit to eat. Contamination of seafood, particularly shellfish, has caused many shellfisheries to be closed and sometimes has caused human disease outbreaks.

Coasts have been marred and altered in many locations by structures. Many beaches are periodically closed because of contamination with improperly treated sewage or medical wastes.

Toxicity. With the possible exception of carcinogens, teratogens, and mutagens, toxic substances are toxic only above a critical concentration that is different for each substance, species, and even each life stage of a particular species. The toxicity of a waste can be evaluated by comparing contaminated sites with control sites or by performing laboratory bioassays. Both approaches are difficult, and their results must be carefully interpreted. Most toxic substances are bioaccumulated by most species. Certain toxic substances are biomagnified, so their concentration increases at each trophic level.

Petroleum. The sources of most oil released to the oceans are storm-water runoff and discharges from the normal operations of oil tankers. Oil tanker accidents are a lesser source, and natural inputs from seeps are approximately equal to inputs from tanker accidents. Releases from offshore platforms are even smaller.

When oil is spilled, a slick forms and the volatile compounds evaporate to the atmosphere or dissolve, leaving lumps of tar or, if seas are rough, a gummy suspension called "chocolate mousse." If the slick reaches shore, oil coats any substrate but is eventually decomposed by bacteria.

Oil tanker accidents such as the *Exxon Valdez* spill damage the ecology of shores reached by the oil and kill seabirds and marine mammals in the spill area. Damaged areas recover naturally within a few years. Wetlands and other low-energy coastal habitats recover more slowly than high-energy rocky shores or beaches. Some spill cleanup efforts may have little or no benefit and may even retard natural recovery. Long-term effects may persist in some spill areas, but they are difficult to detect and document.

Sewage. Sewage is discharged to the marine environment to protect public health. Sewage is a natural material, composed mostly of water and particulate and dissolved organic matter contaminated with chemicals from household wastes and industry. Sewage also contains pathogens and can cause pollution problems because the pathogens can infect humans through water contact or seafood consumption. Organic matter and nutrients can cause blooms and anoxia, and toxic compounds can accumulate in discharge-area ecosystems. Sewage treatment removes floatables, particulate and some dissolved organic matter, toxic chemicals, and pathogens. Treatment has substantially reduced the incidence of anoxia in rivers and estuaries.

Particulate organic matter from sewage outfalls can accumulate in sediments around the outfall and cause alteration of benthic infaunal communities. If outfalls discharge large amounts of sewage, the benthic infauna is degraded in a zone around the outfall, but the effects decrease with distance. If the same amount of sewage were discharged from several small outfalls instead of a single large one, no areas would be degraded, and the benthic infauna around each outfall would be enhanced.

Urban and Agricultural Runoff. Runoff from streets and agricultural land contains toxic chemicals, nutrients, particles from paints, solvents, oil and other substances deposited by vehicles, combustion products, pesticides, fertilizers, and many other sources. In estuaries, these nonpoint source inputs are often much larger than sewage or industrial inputs.

Industrial Effluents. Industrial discharges of toxic substances and other contaminants have been considerably reduced in recent decades. However, violations of regulations and illegal discharges to sewers and streams are common. The industrial effluents of a chemical company in Minamata, Japan, in the 1950s caused the only documented case of human deaths due to toxic substances (methylmercury) discharged to the marine environment. DDT discharges have caused ecological damage in California and elsewhere.

Dredged Material. Navigation channels dredged to maintain navigation depths are often sites of accumulation of past and present contamination. Dredging damages the benthos at the dredging site and at the dump site, which is usually nearby in the estuary or coastal ocean. Toxic contaminants in dredged material are preferentially released to the suspended sediment during dumping and may be bioavailable. Most of the fine-grained, contaminant-rich dredged material dumped in estuaries or just outside estuary mouths is transported back into the estuary by the estuarine circulation and may accumulate in channels to be dredged again.

Plastics and Trash. Trash, particularly floating trash, is dumped into rivers and the ocean and can cause aesthetic problems when washed up on beaches or accumulated in areas visited by divers. Plastics are a problem because they degrade only very slowly, and they may strangle or choke birds and marine organisms or accumulate in the gut until the animals starve to death.

Antifouling Paints. To be effective, antifouling paints must be toxic and must be slowly released to solution. Tributyltin, used in some paints, is very persistent and highly toxic. It has accumulated in some estuaries to toxic levels, drastically reducing populations of commercially valuable shellfish and other invertebrates.

Radionuclides. The oceans have been contaminated by radionuclides from testing of nuclear weapons, liquid waste discharges, and dumping of radioactive wastes. Other than at a few nuclear bomb test sites, radionuclide concentrations in the oceans do not present a significant human health or ecological risk. However, the former Soviet Union disposed of nuclear submarine reactors and large quantities of radioactive wastes in the Arctic Ocean and Sea of Japan, and there is concern that these materials may eventually release radionuclides to the food web.

Nonindigenous Species. Species are carried in ships, in ballast water, and attached to hulls and are inadvertently introduced to ecosystems where they do not occur naturally. Species are also deliberately introduced. Some introduced species outcompete important species in their new habitat and may totally disrupt natural food webs. Estuaries are particularly severely damaged by nonindigenous species.

Habitat Alteration. Alteration or destruction of habitat, especially filling of wetlands, is among the most damaging forms of pollution, especially in estuaries and the coastal zone. Habitat can also be altered in many other ways, such as by dredging, by the construction of structures that affect circulation and erosion, and by the reduction of river flow rates that results from the withdrawal of freshwater for human uses.

STUDY QUESTIONS

1. What is the difference between contamination and pollution?
2. Explain assimilative capacity. How are assimilative capacity and residence time linked?
3. How can human activities alter primary production?
4. Which human activities cause alterations in marine habitats? How many can you list? In what way do each of these activities alter habitat?
5. How can we evaluate the toxicity of a substance to marine species? Why is it so difficult to decide what concentration is safe in the environment?
6. What is the difference between bioaccumulation and biomagnification? Why should we be more concerned about toxic substances that are biomagnified?
7. In what ways can fishing damage the marine ecosystem?
8. List the principal sources of contamination in the oceans. Summarize the characteristics of each type of waste or other source.
9. What are antifouling paints? What would be the desirable characteristics of an antifouling paint?

KEY TERMS

You should be able to recognize and understand the meaning of all terms that are in boldface type in the text. All of those terms are defined in the Glossary. The following are some less familiar key scientific terms that are used in this chapter and that are essential to know and be able to use in classroom discussions or on exams.

anoxic (p. 551)
anthropogenic (p. 556)
assimilative capacity (p. 550)
bioaccumulation (p. 556)
bioassay (p. 556)
bioavailable (p. 573)
biodegradable (p. 557)
biogeochemical cycle (p. 576)
biomagnification (p. 556)
carcinogens (p. 556)
catadromous (p. 553)
chronic toxicity (p. 564)
diversity (p. 553)
effluent (p. 552)
eutrophication (p. 552)
herbicides (p. 553)
hypoxia (p. 552)
nonindigenous species (p. 553)
nonpoint source (p. 570)
outfall (p. 551)
oxygen-demanding (p. 553)
polyaromatic hydrocarbons (PAHs) (p. 564)
residence time (p. 551)
sewage sludge (p. 566)

CRITICAL CONCEPTS REMINDER

CC8 **Residence Time** (p. 551). The residence time of seawater in a given segment of the oceans is the average length of time the water spends in that segment. The residence times of some coastal water masses are long, therefore some contaminants discharged to the coastal ocean can accumulate to higher levels in these long residence time regions than in areas with shorter residence times. To read **CC8** go to page 19CC.

CC14 **Photosynthesis, Light, and Nutrients** (p. 552). Photosynthesis is the primary process by which simple chemical compounds are made into the organic compounds of living organisms. Photosynthesis depends on the availability of carbon dioxide, light, and certain dissolved nutrient elements including nitrogen, phosphorus, and iron. Alterations of the availability of any of these elements by human activity can have adverse consequences in marine ecosystems. To read **CC14** go to page 46CC.

CC17 Species Diversity and Biodiversity (p. 553). Biodiversity is an expression of the range of genetic diversity; species diversity; diversity in ecological niches and types of communities of organisms (ecosystem diversity); and diversity of feeding, reproduction, and predator avoidance strategies (physiological diversity), within the ecosystem of the specified region. Species diversity is a more precisely defined term and is a measure of the species richness (number of species) and species evenness (extent to which the community has balanced populations with no dominant species). High diversity and biodiversity are generally associated with ecosystems that are resistant to change. Changes in diversity are often used as an indicator of the impacts of human activities. To read **CC17** go to page 53CC.

CC18 Toxicity (pp. 555, 556, 564, 572). Many dissolved constituents of seawater become toxic to marine life at levels above their natural concentrations in seawater. Some synthetic organic chemicals are especially significant because they are persistent and may be bioaccumulated or biomagnified. To read **CC18** go to page 54CC.

Units and Conversion Factors

SI Base Units

Base Quantity	Base Unit: Name	Base Unit: Symbol
Length	meter	m
Mass	kilogram	kg
Time	second	s
Electrical current	ampere	A
Thermodynamic temperature	kelvin	K
Amount of substance	mole[a]	mol
Luminous intensity	candela	cd

[a]1 mole is equal to the molecular weight in grams.

Some Examples of Derived SI Units

Derived Quantity	Name	Symbol
Area	square meter	m^2
Volume	cubic meter	m^3
Speed, velocity	meter per second	$m \cdot s^{-1}$
Acceleration	meter per second squared	$m \cdot s^{-2}$
Mass density	kilogram per cubic meter	$kg \cdot m^{-3}$
Specific volume	cubic meter per kilogram	$m^3 \cdot kg^{-1}$
Current density	ampere per square meter	$A \cdot m^{-2}$
Magnetic field strength	ampere per meter	$A \cdot m^{-1}$
Amount-of-substance concentration	mole per cubic meter	$mol \cdot m^{-3}$
Luminance	candela per square meter	$cd \cdot m^{-2}$
Mass fraction	kilogram per kilogram, which may be represented by the number 1	$kg \cdot kg^{-1} = 1$

Wherever possible, this text uses metric units of measurement. The metric system of measurement is used by all major countries but the United States, and it is the universal system used by all scientists. At present, an additional process is under way directed toward using only those metric units that are standard units approved in the International System of units (SI units). However, some of these standard units are unfamiliar even to many scientists, so they are not yet universally used. SI units are used in this text if they are generally in wide usage. In some instances where nonstandard units are much more widely used in the United States, the nonstandard units are used in this text. Over time, there will likely be a migration to the universal and exclusive use of SI units. The SI unit system has only seven base units from which all other units are derived (derived units are combinations of the base units). Tables listing the base units of the SI system, some derived SI units, and all of the units used in this text are included in this appendix. Some conversion factors to relate the SI units and other metric units to other units commonly used in the United States are also included.

SI Derived Units with Special Names and Symbols[a]

	SI Derived Unit			
Derived Quantity	*Name*	*Symbol*	*Expression in Terms of Other SI Units*	*Expression in Terms of SI Base Units*
Plane angle	radian	rad	—	$m \cdot m^{-1} = 1$
Frequency	hertz	Hz	—	s^{-1}
Force	newton	N	—	$m \cdot kg \cdot s^{-2}$
Pressure, stress	pascal	Pa	$N \cdot m^{-2}$	$m^{-1} \cdot kg \cdot s^{-2}$
Energy, work, quantity of heat	joule	J	$N \cdot m$	$m^2 \cdot kg \cdot s^{-2}$
Power, radiant flux	watt	W	$J \cdot s^{-1}$	$m^2 \cdot kg \cdot s^{-3}$
Celsius temperature	degree Celsius	°C	—	K
Luminous flux	lumen	lm	—	$m^2 \cdot m^{-2} \cdot cd = cd$
Illuminance	lux	lx	$lm \cdot m^{-2}$	$m^2 \cdot m^{-4} \cdot cd = m^{-2} \cdot cd$
Activity (of a radionuclide)	becquerel	Bq	—	s^{-1}

[a]For ease of understanding and convenience, a few SI derived units have been given special names and symbols. These are some that are relevant to this text.

The scientific community almost universally uses exponential notation for numbers. Exponential notation is explained in Chapter 1 and is used throughout this text. This appendix includes a table ("Exponential Notation and the Decimal System") for converting exponential numbers to their nonexponential equivalents. To multiply numbers expressed exponentially, add the exponents (superscript numbers). For example, $10^{-1} \times 10^3 = 10^2$ (one-tenth of 1000 is equal to 100). To divide, subtract the exponents. The "SI Base Units" table also lists the common terms used to identify certain exponent values. For example, 1000 is equivalent to 10^3, or 1×10^3, and 1000 can be expressed as "one thousand," while other units can be prefixed by "kilo-" to express one thousand of the units; for example, 1 kilogram is equal to 1000 grams.

Note that the naming system for large and small numbers is different in different parts of the world. In many non-English-speaking countries (and formerly in England), the term *billion* does not mean 10^9. Instead, in this alternate system billion is 10^{12} and trillion is 10^{18}. This is one reason why it is always better to use exponential notation for large or small numbers.

Exponential Notation and the Decimal System

Value	*Exponential Expression*	*Name*	*Prefix*
1,000,000,000,000	10^{12}	1 trillion	tera-
1,000,000,000	10^9	1 billion	giga-
1,000,000	10^6	1 million	mega-
1,000	10^3	1 thousand	kilo-
100	10^2	1 hundred	centa-
10	10^1	ten	deca-
1	10^0	one	uni-
0.1	10^{-1}	1 tenth	deci-
0.01	10^{-2}	1 hundredth	centi-
0.001	10^{-3}	1 thousandth	milli-
0.000,001	10^{-6}	1 millionth	micro-
0.000,000,001	10^{-9}	1 billionth	nano-
0.000,000,000,001	10^{-12}	1 trillionth	pico-

Units of Measurement Used in This Text (with Abbreviations)

Length/distance:	**meter (m) — *SI base unit*** millimeter (mm) centimeter (cm) kilometer (km) foot (ft) — *obsolete nonmetric unit* nautical mile (n.m.) — *equals 1.853 km (1.15 statute miles) or 1 minute (1/60 of a degree) of latitude*
Mass:	**kilogram (kg) — *SI base unit*** gram (g) milligram (mg) microgram (μg) tonne (t) — *metric ton, equal to* 10^3 *kg*
Time:	**second (s) — *SI base unit*** year (yr) day (day) hour (h) minute (min)
Area:	meter squared (m^2) — *derived SI unit* centimeter squared (cm^2) kilometer squared (km^2)
Volume:	meter cubed, cubic meters (m^3) — *derived SI unit* centimeter cubed (cm^3) kilometer cubed (km^3) liter (l) — *equals* 10^3 cm^3*, so* $1\ m^3 = 1000\ l$
Pressure:	pascal (Pa) — *derived SI unit (not used in this text)* kilogram per centimeter squared ($kg{\cdot}cm^{-2}$) — *equals* 9.56×10^6 *Pa* atmosphere (atm) — *equals* $1.03\ kg{\cdot}cm^{-2}$
Mass density (absolute density):	kilogram per cubic meter ($kg{\cdot}m^{-3}$) — *derived SI unit* gram per cubic centimeter ($g{\cdot}cm^{-3}$)
Speed/velocity:	meters per second ($m{\cdot}s^{-1}$) — *derived SI unit* centimeters per second ($cm{\cdot}s^{-1}$) kilometers per hour ($km{\cdot}h^{-1}$) — $1\ km{\cdot}h^{-1} = 0.28\ m{\cdot}s^{-1} = 28\ cm{\cdot}s^{-1}$ kilometers per day ($km{\cdot}day^{-1}$)
Acceleration:	meters per second per second ($m{\cdot}s^{-2}$)
Temperature:	degrees Celsius (°C) — *derived SI unit*
Energy and related units:	joule (J) — *derived SI unit* joule per meter squared ($J{\cdot}m^{-2}$) — *used as measure of wave energy* joule per gram ($J{\cdot}g^{-1}$) — *used as measure of latent heat* joule per gram per degree Celsius ($J{\cdot}g^{-1}{\cdot}°C^{-1}$) — *used as measure of heat capacity* calorie (cal) — *obsolete measure of energy* calories per gram ($cal{\cdot}g^{-1}$) — *obsolete, used as measure of latent heat* calories per gram per degree Celsius ($cal{\cdot}g^{-1}{\cdot}°C^{-1}$) — *obsolete, used as measure of heat capacity*
Concentration:	mole per cubic meter ($mol{\cdot}m^{-3}$) — *derived SI unit (not used in this text)* milligram per kilogram ($mg{\cdot}kg^{-1}$) microgram per kilogram ($\mu g{\cdot}kg^{-1}$)
Notes:	*Concentrations are expressed as mass of dissolved (or constituent) substance in one unit mass of the solution (or combined mixture).* *1 mole is defined as the molecular weight of a substance expressed in grams. To convert concentration in* $mg{\cdot}kg^{-1}$ *to* $mol{\cdot}m^{-3}$*, divide by the molecular weight of the dissolved substance (or constituent); then divide by the solution (or mixture) density in* $kg{\cdot}m^{-3}$ *and multiply by 1000.*
Miscellaneous:	grams per year ($g{\cdot}yr^{-1}$) — *mass transport rate* cubic meters per second ($m^3{\cdot}s^{-1}$) — *volume transport rate* practical salinity unit — *dimensionless ratio; no abbreviation, but sometimes listed as PSU* millions of years ago (MYA) — *millions of years before the present date* before the common era (BCE) — *number of years before the year 1 of the Christian calendar, equivalent to* BC the common era (CE) — *number of years after the year 1 of the Christian calendar, equivalent to* AD

Conversions between Basic Units of Length, Mass, and Time

Length:	1 kilometer = 1000 meters (m) 1 meter = 100 centimeters (cm) 1 centimeter = 10 millimeters (mm) 1 millimeter = 10^3 micrometers (μm) 1 micrometer = 10^3 nanometers (nm)
Mass:	1 tonne (metric ton) = 1,000 kilograms (kg) 1 kilogram = 1000 grams (g) 1 gram = 10^3 milligrams (mg) 1 milligram = 10^3 micrograms (μg) 1 microgram = 10^3 nanograms (ng) 1 nanogram = 10^3 picograms (pg)
Time:	1 hour = 3600 seconds (s)

Additional Conversions

Metric to Other Unit Conversions	*Other Unit to Metric Conversions*
LENGTH	
1 km = 0.62 statute miles	1 statute mile = 1.609 km
1 km = 0.54 nautical miles	1 nautical mile = 1.853 km
1 km = 1093 yards	1 yard = 91.4 cm
1 m = 39.4 inches	1 foot = 30.5 cm
1 m = 3.28 feet	1 inch = 2.54 cm
1 cm = 0.394 inches	
MASS	
1 kg = 2.2 pounds	1 pound = 0.45 kg
1 tonne (metric ton) = 2205 pounds	1 pound = 454 g
1 tonne (metric ton) = 1.10 U.S. tons	1 U.S. ton = 907 kg
1 g = 0.035 ounce	1 ounce = 28.4 g
AREA	
1 km^2 = 0.386 statute mile squared	1 statute mile squared = 2.59 km^2
1 km^2 = 247.1 acres	1 acre = 4046 m^2
1 m^2 = 10.7 feet squared	1 foot squared = 929 cm^2
1 cm^2 = 0.155 inch squared	1 inch squared = 6.45 cm^2
VOLUME	
1 m^3 = 35.3 feet cubed	1 foot cubed = 28.32 l (0.028 m^3)
1 m^3 = 264 U.S. gallons	1 U.S. gallon = 3.78 l (0.0037 m^3)
1 cm^3 = 0.061 inch cubed	1 inch cubed = 16.4 cm^3
1 l = 2.12 pints	1 pint = 0.47 l
PRESSURE	
1 $kg{\cdot}cm^{-2}$ = 14.2 pounds per inch squared	1 atmosphere (sea level) = 1.03 $kg{\cdot}cm^{-2}$
1 $kg{\cdot}cm^{-2}$ = 0.97 atmosphere	1 bar = 1.02 $kg{\cdot}cm^{-2}$
1 $kg{\cdot}cm^{-2}$ = 981 millibars	1 pascal = 0.000,01 $kg{\cdot}cm^{-2}$
1 $kg{\cdot}cm^{-2}$ = 0.981 bar	1 inch of mercury at 0°C = 0.034 $kg{\cdot}cm^{-2}$
1 $kg{\cdot}cm^{-2}$ = 98,068 pascals	1 mm of mercury at 0°C = 0.0014 $kg{\cdot}cm^{-2}$
1 $kg{\cdot}cm^{-2}$ = 29.0 inches of mercury at 0°C	
1 $kg{\cdot}cm^{-2}$ = 736 mm of mercury at 0°C	
SPEED	
1 $km{\cdot}h^{-1}$ = 0.62 mile per hour	1 mile per hour = 1.61 $km{\cdot}h^{-1}$
1 $km{\cdot}h^{-1}$ = 0.54 knot	1 knot = 1.85 $km{\cdot}h^{-1}$
1 $cm{\cdot}s^{-1}$ = 1.97 feet per minute	1 foot per minute = 0.51 $cm{\cdot}s^{-1}$
1 $cm{\cdot}s^{-1}$ = 0.033 foot per second	1 foot per second = 30.5 $cm{\cdot}s^{-1}$
TEMPERATURE	
0°C = 32°F	–40°F = –40°C
10°C = 50°F	0°F = –17.8°C
20°C = 68°F	32°F = 0°C
30°C = 86°F	40°F = 4.4°C
40°C = 104°F	50°F = 10°C
100°C = 212°F	70°F = 21.1°C
	90°F = 32.2°C
	100°F = 37.8°C
	212°F = 100°C

Dimensions of the Earth and Oceans

The Earth

Radius at equator	6,378 km
Radius at poles	6,357 km
Average radius	6,371 km
Circumference at equator	40,077 km

Percentage of Ocean Area Occupied by Ocean Provinces

Continental shelf and slope	15.3%
Continental rise	5.3%
Abyssal seafloor	41.8%
Volcanoes and volcanic ridges	3.1%
Oceanic ridges and rises	32.7%
Trenches	1.7%

Areas of Land and Oceans

Land (29.22% of total)	149,000,000 km^2
Oceans and seas (70.78%)	361,000,000 km^2
Pacific Ocean (marginal seas included)	180,000,000 km^2
Atlantic Ocean (marginal seas included)	107,000,000 km^2
Indian Ocean (marginal seas included)	74,000,000 km^2
Ice sheets and glaciers	15,600,000 km^2
Land plus continental shelf	177,400,000 km^2
Land in the Northern Hemisphere	100,200,000 km^2
Ocean in the Northern Hemisphere	154,800,000 km^2
Land in the Southern Hemisphere	48,700,000 km^2
Ocean in the Southern Hemisphere	206,300,000 km^2

Density and Mass of the Earth's Parts

	Average Thickness (Radius for core) (km)	*Mean Density ($g \cdot cm^{-1}$)*	*Total Mass ($\times 10^{24}$ g)*
Atmosphere	—	—	0.005
Oceans and seas	3.8	1.03	1.41
Ice sheets and glaciers	1.6	0.90	0.023
Continental crust (includes continental shelves)	35	2.8	17.39
Oceanic crust (excludes continental shelves)	8	2.9	7.71
Mantle	2881	4.53	4068
Core	3473	10.7	1881

Elevations and Depths

Average elevation of land	840 m
Average depth of oceans	3,800 m
Average depth of Pacific Ocean (marginal seas excluded)	3,940 m
Average depth of Atlantic Ocean (marginal seas excluded)	3,310 m
Average depth of Indian Ocean (marginal seas excluded)	3,840 m
Highest elevation of land (Mount Everest)	8,848 m
Greatest depth of oceans (Mariana Trench)	11,035 m

APPENDIX 3

Classification of Marine Organisms

Understanding life on the Earth, and how this life has evolved and may evolve in the future in response to human influences, requires that we understand the differences between the many forms of life on the Earth, and how and why the differences have developed. Fundamental to such studies is taxonomy, the classification of organisms to express their relationships to each other.

One fundamental concept of taxonomy is that organisms can be identified as belonging to a species. The definition of a species is generally accepted to be a population of organisms whose members interbreed under natural conditions and produce fertile offspring that are reproductively isolated from other such groups. A new definition based on genetics, the composition of the DNA, may be developed in the future. The current definition has generally worked well, although there are difficulties, for example, when two populations that are capable of interbreeding are geographically separated and do not interbreed. In some such cases, scientists disagree as to whether these separate populations are the same species or are different enough to be considered separate species. Populations of king salmon that breed in different rivers along the coasts of the United States and Canada are an example. Other difficulties arise because the members of a single species may look very different. For example, bulldogs, terriers, and poodles all belong to the same species. Conversely, organisms that appear to be very similar may belong to different species; red squirrels and gray squirrels are an example.

Millions of species on the Earth have been identified and studied, and certainly many more millions remain to be found, particularly in tropical rain forests and coral reefs. Species are arranged in a hierarchical classification similar to a human family tree. Taxonomists have established a series of levels for this hierarchy, from species at the bottom to kingdom, which was the top level until recent years, when a new top level—domain—was added.

This new top level became needed when archaea were discovered. Initially placed in a kingdom along with bacteria, archaea were subsequently found to be entirely different from bacteria. In fact, archaea are different enough that they do not fit into any of the five kingdoms that were previously recognized and that are still listed in many texts. Although the new top level (domain) is not yet universally accepted, the generally accepted classification now identifies three such domains:

Three Domains of Living Organisms

Domain I: Bacteria—Most of the Known Prokaryotes ***Although it is clear that the bacteria need to be separated into kingdoms, they are not well enough studied for this to have been done yet. At present they are separated into five phyla.***	
Phylum Proteobacteria	Nitrogen-fixing bacteria
Phylum Cyanobacteria	Photosynthetic bacteria (formerly blue-green algae)
Phylum Eubacteria	True gram-positive bacteria
Phylum Spirochaetes	Spiral bacteria
Phylum Chlamydiae	Intracellular parasites
Domain II: Archaea—Prokaryotes of Extreme Environments	
Kingdom Crenarchaeota	Thermophiles
Kingdom Euryarchaeota	Methanogens and halophiles
Kingdom Korarchaeota	Some hot-springs microbes
Domain III: Eukarya—Eukaryotic Cells	
Kingdom Protista	Single-celled organisms that have a nucleus and other internal structure in the cell (e.g., diatoms)
Kingdom Fungi	Multicelled organisms that are not able to photosynthesize (or chemosynthesize)
Kingdom Plantae (or Metaphyta)	Multicelled photosynthetic autotrophs (e.g., kelp)
Kingdom Animalia (or Metazoa)	Multicelled heterotrophs (invertebrates and vertebrates)

The taxonomic hierarchy generally has nine levels, although sometimes one or more levels are omitted, sometimes intermediate levels are added within the hierarchy, and there is, as yet, no universal agreement on the use of the top level, domains. The major levels in the hierarchy are as follows:

Domain
Kingdom
Phylum
Subphylum
Class
Order
Family
Genus
Species

The principal phyla and classes to which all common marine species belong and some examples of each type are listed at the end of this appendix.

Scientific names of organisms are important for many reasons, one of which is that common names are often confusing. For example, the common name "red snapper" is used in the Gulf of Mexico for the species *Lutjanus campechanus*. In the Pacific Ocean the same common name, "red snapper," is used for at least three different species (*Lutjanus bohar*, *Lutjanus malabaricus*, and *Lutjanus gibbus*) that are related but very different from one another. This difference does not matter to the local populations who eat these fishes (indeed, when you order red snapper in a restaurant, you may be served any one of many species that may not even be snappers). However, scientists must be able to distinguish them in the literature, and they do so by using scientific names.

The scientific names given to species always consist of two parts: a genus name followed by a species name. For example, mussels are classified as follows:

- Domain: Eukarya
- Kingdom: Animalia (Metazoa)
- Phylum: Mollusca
- Class: Bivalvia
- Order: Mytiloida
- Superfamily: Mytilacea (this is good example of an intermediate level used in the taxonomic hierarchy for only some orders)
- Family: Mytilidae
- Genus: *Mytilus*

The common blue mussel of the east coast of North America is *Mytilus edulis*, and a different species, *Mytilus californianus*, is present only on the west coast.

These and other scientific names may seem complicated, but they are either Latin or Greek words modified to explain something that the taxonomist considers important in identifying the organism. For example, "*Mytilus*" is derived from the Greek word for sea mussels, "*californianus*" expresses that this species is present only on the West Coast ("california"); and "*edulis*" means "edible," a suitable term for this species that is good to eat. Similarly, the genus name *Sargassum* is applied to species of algae present in the Sargasso Sea, and *Enterococcus* is the genus name for species of bacteria ("*-coccus*") that are present in vertebrate intestines ("*Entero-*").

Scientific names can be complicated and sometimes obscure, but in general, they follow an internationally accepted set of rules. Note that the convention is to use italics for scientific names and to capitalize the genus name only. In addition, shortening the genus name is acceptable where it is clear. For example, this text refers to mussels as *M. edulis* and *M. californianus* after the full names have been used once. There are other conventions as well. For example, "sp." after the genus name refers to a single species of the genus whose species name is not known, and "spp." after the genus name indicates that all (or many) of the species of that genus are being referred to (e.g., "*Mytilus* sp." refers to an unstated or unknown species of mussel, and "*Mytilus* spp." refers to all or any species of mussel). The "sp." and "spp." are often omitted, as is done for simplicity in many cases in this text. Sometimes the genus name is followed by abbreviations such as "cf." and "aff." The first of these abbreviations ("cf.") means that the genus listed is thought to be correct and that the species is very similar to the species identified by the name following the genus name and "cf." but there are enough differences to lead the investigator to believe that this may turn out to be a new, but very closely related, species or a subspecies. The second abbreviation ("aff.") means that the species is very similar to the species identified by the name following the genus name and "aff." but it is sufficiently different that it is most likely an undescribed species similar to the one named.

TAXONOMIC CLASSIFICATION

Phyla that have no members in the marine environment, as well as phyla in which all species are extinct, are omitted from the following list. The list includes general descriptions of members of each classification and, in some cases, examples of common species and a listing of figures in this text that show a photograph of a member of this classification. The classification presented is not accepted by all taxonomists. Classifications, including the distinctions between phyla, continually change as more is learned.

- **DOMAIN ARCHAEA**. Organisms that have no nucleus, predominantly single-celled. Occur primarily in extreme environments.
  - ***Kingdom Crenarchaeota.*** Thermophiles.
  - ***Kingdom Euryarchaeota.*** Methanogens and halophiles.
  - ***Kingdom Korarchaeota.*** Some hot-springs microbes.
- **DOMAIN BACTERIA.** Organisms that have no nucleus, predominantly single-celled.
  - ***Phylum Proteobacteria.*** Nitrogen-fixing bacteria.
  - ***Phylum Cyanobacteria.*** Photosynthetic bacteria (formerly blue-green algae).
  - ***Phylum Eubacteria.*** True gram-positive bacteria.
  - ***Phylum Spirochaetes.*** Spiral bacteria.
  - ***Phylum Chlamydiae.*** Intracellular parasites.
- **DOMAIN EUKARYA, KINGDOM PROTISTA.** Organisms with a nucleus confined by a membrane, predominantly single-celled.
  - ***Phylum Chrysophyta.*** Diatoms (Figs. 8.5, 14.15a), coccolithophores (Figs. 8.6, 14.15c), silicoflagellates (Fig. 14.15d).
  - ***Phylum Pyrrophyta.*** Autotrophic dinoflagellates (Fig. 14.15b), zooxanthellae.
  - ***Phylum Chlorophyta.*** Green algae (Fig. 15.3a,d; 17.12f).
  - ***Phylum Phaeophyta.*** Brown algae, Kelps (Figs. 14.3b; 15.3a,c; 17.7c), *Sargassum* (Figs. 14.3a, 17.13a–c), and others.
  - ***Phylum Rhodophyta.*** Red algae (Fig. 15.3a,b,e).
  - ***Phylum Zoomastigophora.*** Flagellated protozoa, including heterotrophic dinoflagellates.
  - ***Phylum Sarcodina.*** Amoebas and relatives, including foraminifera (Figs. 8.7a, 14.18c) and radiolaria (Figs. 8.8, 14.18d).
  - ***Phylum Ciliophora.*** Protozoa with cilia.
- **DOMAIN EUKARYA, KINGDOM FUNGI.** Fungi, lichens.
  - ***Phylum Mycophyta.*** Marine fungi are primarily benthic decomposers. Marine lichens are primarily intertidal.
- **DOMAIN EUKARYA, KINGDOM PLANTAE (METAPHYTA).** Autotrophic multicelled plants.
  - ***Phylum Tracheophyta.*** Plants that have roots, stems, and leaves and special cells that transport nutrients and water.
    - **Class Angiosperma.** Flowering plants with seeds contained in a closed seedpod. Turtle grass (*Thalassia*; Fig. 16.11b), marsh grasses (*Spartina*; Fig. 16.11c).
- **DOMAIN EUKARYA, KINGDOM ANIMALIA (METAZOA).** Multicelled heterotrophs (animals).
  - ***Phylum Placozoa.*** Amoeba-like multicelled animals.
  - ***Phylum Mesozoa.*** Wormlike parasites of cephalopods.
  - ***Phylum Porifera.*** Sponges (Fig. 16.27m).
  - ***Phylum Cnidaria (Coelenterata).*** Jellyfish and related organisms, all of which have stinging cells.
    - **Class Hydrozoa.** Polypoid colonial animals, most with a medusa-like stage. Includes Portuguese man-of-war (Fig. 14.19c).
    - **Class Scyphozoa.** Jellyfish with no (or reduced) polyp stage in life cycle. Medusa stage dominates (Fig. 14.19a,b).
    - **Class Anthozoa.** Zooanthids (Fig. 16.7h), anemones (Figs. 16.5c; 16.7g,i; 16.19a; 16.34a; 17.8b), sea pens (Fig. 16.5a,b, 17.8e), and corals (Figs. 15.11a–d; 16.7a–f; 16.27a; 17.4e,f; 17.6; 17.8a) that have only a polypoid body form.
  - ***Phylum Ctenophora.*** Comb jellies (Fig. 14.19d), “sea gooseberries.” Predatory, predominantly planktonic.
  - ***Phylum Platyhelminthes.*** Flatworms, flukes, tapeworms. Many are parasitic; many others are free-living predators.
  - ***Phylum Nemertea.*** Ribbon worms. Benthic and pelagic.
  - ***Phylum Gnathostomulida.*** Microscopic, wormlike, meiofaunal (small organisms that live in the spaces between sediment grains).
  - ***Phylum Gastrotricha.*** Microscopic, ciliated, meiofaunal.
  - ***Phylum Rotifera.*** Ciliated, less than about 2 mm long. Most species freshwater. Planktonic or epibenthic.
  - ***Phylum Kinorhyncha.*** Small, spiny, segmented worms; meiofaunal.
  - ***Phylum Acanthocephala.*** Spiny-headed worms. All are intestinal parasites of vertebrates.
  - ***Phylum Entoprocta.*** Small polyplike suspension feeders.
  - ***Phylum Nematoda.*** Roundworms. Most are 1 to 3 mm; marine species are infaunal.
  - ***Phylum Bryozoa.*** Small moss animals that form encrusting or branching colonies on seafloor.
  - ***Phylum Phoronida.*** Suspension-feeding tube worms. Infaunal; inhabit shallow and temperate sediments.
  - ***Phylum Brachiopoda.*** Lamp shells. Bivalve animals that superficially resemble clams. Mostly deep-water.

Phylum Mollusca. Soft-bodied animals with a mantle and muscular foot. Most species secrete a calcium carbonate shell.

Class Monoplacophora. Limpetlike shells, segmented bodies, abyssal only.

Class Polyplacophora. Chitons (Fig. 17.12d). Oval, flattened body covered by eight overlapping plates.

Class Aplacophora. Worm-shaped mollusks; soft-sediment infauna.

Class Gastropoda. Snails (Figs. 16.2b, 16.18b, 16.27h, 16.31d, 17.12a), limpets (Fig. 17.12b), abalones, pteropods (Figs. 8.7b, 16.2a), nudibranchs (Chapter 3 Opener, Figs. 16.10c,d; 16.27d–g; 17.8f), and many others (Figs. 16.10b, 16.18c, 16.33s).

Class Bivalvia. Clams (Fig. 16.34d), scallops, oysters (Fig. 16.6b), mussels (Figs. 16.6a, 17.12e). Bivalves (shell with two fitted segments). Mostly filter-feeding.

Class Scaphopoda. Tusk shells. Soft-sediment infauna.

Class Cephalopoda. Squid (Fig. 14.23a, 16.27i), octopi (Chapter 14 Opener, Fig. 16.17e), cuttlefish (Chapter 16 Opener, Fig. 14.23b–d, 16.27j), *Nautilus* (Fig. 14.23e,f). No external shell except in *Nautilus* spp.

Phylum Priapulida. Small, wormlike, subtidal.

Phylum Sipuncula. Peanut worms, benthic, exclusively marine.

Phylum Echiura. Spoon worms. Spoon-shaped proboscis. Infaunal or live under rocks.

Phylum Annelida. Segmented worms. Mostly benthic. Includes feather duster worms (Fig. 16.5d), scale worms (Fig. 16.33r), and Christmas tree worms (Fig. 16.9c,d).

Phylum Tardigrada. Meiofaunal. Eight-legged. Can hibernate for long periods.

Phylum Pentastoma. Tongue worms. Parasites of vertebrates.

Phylum Pogonophora. Tube-dwelling worms with no digestive system. Absorb food through body wall. All species are marine. Mostly deep-water (Fig 17.17b).

Phylum Echinodermata. Spiny-skinned animals. Most are benthic epifaunal or infaunal.

Class Asteroidea. Sea stars (Figs. 17.4b, 17.8e).

Class Ophiuroidea. Brittle stars, basket stars (Fig. 16.9b).

Class Echinoidea. Sea urchins (Figs. 16.10a, 16.17d, 16.34b, 17.4c, 17.9b), sand dollars (Fig. 17.4d), sea biscuits.

Class Holothuroidea. Sea cucumbers (Figs. 16.9a, 16.27l, 17.4a).

Class Crinoidea. Sea lilies and feather stars (Fig. 16.15a,b; 16.33g).

Phylum Chaetognatha. Arrowworms, stiff-bodied, mostly planktonic.

Phylum Hemichordata. Unsegmented infauna with primitive nerve cord. Acorn worms. All marine.

Phylum Arthropoda. Jointed, legged animals with segmented body covered by an exoskeleton.

Subphylum Crustacea. Copepods (Figs. 14.18a, 16.31a), krill (Fig. 16.1a), barnacles (Figs. 14.20b; 16.6c,d; 17.12c), amphipods, shrimp (Figs. 16.1b,c; 16.15a,b,j,l; 16.17a; 16.33h–q; 16.34c,e; 17.8d; 17.13d), lobster, crabs (Figs. 14.20a; 16.15f,g,k; 16.17c; 16.33a–e; 16.34a,b; 17.8c; 17.12g), euphausiids (Fig. 14.18b, 16.1a), isopods (Fig. 16.31b,c).

Subphylum Chelicerata. Horseshoe crabs, sea spiders.

Subphylum Uniramia. Insects. Only five species of a single genus are present in the ocean.

Phylum Chordata. Animals with a nerve cord and gills, gill slits, or lungs.

Subphylum Urochordata. Tunicates (Fig. 16.8a–d), sea squirts, salps (Fig. 14.19e, 16.3).

Subphylum Cephalochordata. Lancelets, *Amphioxus*. Found in coarse temperate and tropical sediments.

Subphylum Vertebrata. Spinal column of vertebrae, internal skeleton, brain.

Class Agnatha. Jawless fishes. Cartilaginous skeletons. Lampreys, hagfishes.

Class Chondrichthyes. Cartilaginous skeletons. Rays (Fig. 14.22f,h,i), sharks (Fig. 14.22a–e,g), skates, sawfishes, chimeras.

Class Ostreichthyes. Bony fishes (Figs. 14.21a–d; 16.12a; 16.13a–e; 16.14; 16.15c–e,h,i; 16.16a–g; 16.17b; 16.18a,d–f; 16.19b–f; 16.20; 16.21a–c; 16.22a–g; 16.23a,c; 16.24a–e; 16.25a–g; 16.26a,b; 16.27b,c,k; 16.31a–c; 16.33f,g; 16.34c; 17.6; 17.8g; 17.13c).

Class Amphibia. No marine species, but one species (Asian mud frog) is known to tolerate marine water. Frogs, toads, salamanders.

Class Reptilia. Turtles (Figs. 14.26a,b; 16.11d), sea snakes (Fig. 14.26c,d). One species of iguana (Fig. 14.26e,f) and one species of crocodile are marine.

Class Aves. Birds (Figs. 14.27, 19.13b). Many species live on and feed in the ocean, but all must return to land to breed.

Class Mammalia. Warm-blooded animals that have mammary glands and hair, bear live young.

Order Cetacea. Whales (Fig. 14.24b,c), porpoises, dolphins.

Order Sirenia. Sea cows. Manatee (Fig. 14.24d), dugong.

Order Carnivora. Marine species in two suborders.

Suborder Pinnipedia. Seals (Figs. 14.25a,b), sea lions (Fig. 14.25c, 19.13a), walrus (Fig. 14.25d).

Suborder Fissipedia. Sea otters (Figs. 14.25e, 17.9a).

Order Primates. Primates. Apes, human beings. Only one species is known to tolerate marine water—surfers, swimmers, scuba divers.

The phyla Porifera, Cnidaria, Platyhelminthes, Nematoda, Mollusca, Arthropoda, and Chordata each have more than 10,000 known member species. The phyla Cyanobacteria, Chrysophyta, Pyrrophyta, Chlorophyta, Phaeophyta, Rhodophyta, Zoomastigophora, Sarcodina, Ciliophora, Mycophyta, and Echinodermata each have between approximately 1000 and 10,000 known member species. All other phyla listed have fewer than about 1000 known species. However, it is estimated that there may be between 1 and 10 million species in the oceans, of which only a few hundreds of thousands have yet been identified.

Glossary

absorption spectrum (pl. **spectra**). Wavelengths of electromagnetic energy that substances can absorb and convert to heat. Absorption spectra are plotted as the relative efficiency of absorption versus wavelength. Compare *emission spectrum*.

abyssal fan. Fan-shaped sediment accumulation spreading out and decreasing in thickness from a point of large sediment input. Abyssal fans are located usually off the mouth of a river or at the foot of a submarine canyon.

abyssal hill. Sediment-covered volcanic peak that rises less than 1 km above the deep-ocean floor.

abyssal plain. Flat seafloor extending seaward from the base of the continental slope and continental rise or from the seaward edge of an oceanic trench.

abyssal zone. Benthic environment between 4000 and 6000 m depth.

acid rain. Rain with acidity higher than normal because of dissolved gaseous emissions of industry and automobiles, notably sulfur compounds.

acoustic. Pertaining to sound.

adiabatic expansion. Expansion of a gas without the addition of external heat. Adiabatic expansion causes the temperature of the gas to decrease. Adiabatic compression causes the temperature of the gas to increase.

adsorption (adj. **adsorbed**). Attraction and adhesion of ions to a solid surface.

aerobic. In the presence of oxygen.

algae (sing. **alga**). Simple single-celled or many-celled photosynthetic organisms (plants) that have no root, stem, or leaf systems. See Appendix 3.

algal ridge. Irregular ridge located on the wave-exposed seaward edge of many coral reefs. The ridge is composed largely of encrusting algae.

alkalinity (adj. **alkaline**). The opposite of acidity. Measure of the degree to which the concentration of hydroxyl ions (OH^-) is greater than the concentration of hydrogen ions (H^+) in a solution. Seawater is slightly alkaline.

amino acids. Members of a group of compounds that combine in varied proportions to form proteins.

amoebas. Microscopic single-celled animals that move by making continuous protrusions of the body and that feed by engulfing bits of food. See Appendix 3.

amphidromic system. Tide wave that rotates around an ocean basin during one tidal period. An amphidromic system rotates around an amphidromic point (a node) at which the tidal range is zero.

amphipods. Members of an order of the subphylum Crustacea, phylum Arthropoda, that includes laterally compressed species such as the "sand hoppers." See Appendix 3.

amplitude. Vertical distance from the mean level of a wave (the still sea surface for an ocean wave) to the crest or to the trough. A wave's amplitude equals one-half of the wave height.

anadromous. Pertaining to species of fishes that are spawned in freshwater, migrate to the ocean to live until they reach maturity, then return to freshwater to spawn. Compare *catadromous*.

anemones. Multicelled animals of the class Anthozoa, phylum Cnidaria. There are both single and colonial polypoid forms. See Appendix 3.

angle of incidence. Angle between a wave front (usually an electromagnetic wave) or a wave ray and the plane of an interface that it meets. Angle of approach. The angle of incidence is important in refraction.

anion. Negatively charged ion. Compare *cation*.

annelids. Members of the phylum Annelida. Elongated segmented worms. See Appendix 3.

anoxia (adj. **anoxic**). Total absence of dissolved or free molecular oxygen. Compare *hypoxia*.

anthropogenic. Produced by people and their activities.

antinode. The part of a standing wave where the vertical motion is at a maximum. Compare *node*.

aphotic zone. The part of the ocean in which light is insufficient for photosynthesis. The aphotic zone comprises all the oceans below about 1000 m in areas where ocean water is clear and extends to shallower depths in more turbid waters. Compare *photic zone*.

aragonite. A mineral form of calcium carbonate. Aragonite is less common than calcite, but it comprises shells of many pteropod species.

archaea (sing. **archaean;** adj. **archaeal**). Microorganisms of the domain Archaea (previously called Archaeobacteria) that belong to an ancient group of organisms that are separate from bacteria and from which eukaryotes (multicelled organisms) may have evolved. Many archaeal species are chemosynthetic and live in extreme environments. See Appendix 3.

aspect ratio. Index of the propulsive efficiency of a fish species that is obtained by dividing the square of the height of the caudal fin by the area of the caudal fin.

assimilative capacity. Maximum rate at which a particular segment of the environment or ocean can accommodate the input of a substance. The term usually applies to anthropogenic waste materials.

asteroid. Rocky object orbiting the sun. An asteroid is smaller than a planet. Most asteroids are found in a belt between the orbits of Mars and Jupiter, but some have orbits that take them close to the sun and across the orbits of the planets.

asthenosphere. Plastic or partially molten layer of the Earth's upper mantle upon which the continents "float." The asthenosphere varies in thickness, with its upper boundary at depths of 10 km or more and its lower boundary at depths as great as 800 km.

atoll. Ring-shaped coral reef that grows upward from a submerged volcanic peak and encloses a lagoon. Atolls may support low-lying islands composed of coral debris. Compare *barrier reef* and *fringing reef*.

autotrophs (adj. **autotrophic**). Plants, bacteria, or archaea that can synthesize organic compounds from inorganic nutrients by photosynthesis or chemosynthesis. Compare *heterotrophs*.

back-arc basin. Generally shallow sea created on the nonsubducting plate at an oceanic convergent plate boundary where the subduction is fast enough to stretch the edge of the nonsubducting plate. The back-arc basin lies behind the volcanic island arc. Also called "back-island basin."

backscatter (adj. **backscattered**). The portion of the light or sound randomly reflected off particles suspended in a fluid that is scattered back in the general direction of the light or sound source. Compare *scatter*.

backshore. Inner portion of the shore that is landward of the mean spring-tide high-water line. Waves act on the backshore only during exceptionally high tides and severe storms. Compare *foreshore*.

backwash. Water flowing down a beach after the swash of one wave has stopped and before the swash of the next wave arrives.

bacteria (sing. **bacterium;** adj. **bacterial**). Microscopic single-celled organisms that comprise one of the three recognized biological domains. Bacteria are predominantly autotrophs and decomposers, and most are parasitic. They reproduce by cell division. See Appendix 3.

baleen. Horny material consisting of numerous plates with fringed edges that grows down from the upper jaw of plankton-feeding whales. Baleen is used to filter (strain) the plankton food from the water.

bar-built estuary. Shallow estuary (lagoon) separated from the open ocean by a bar such as a barrier island. Water in bar-built estuaries is usually well mixed vertically.

barnacles. Members of an order of the subphylum Crustacea, phylum Arthropoda, that attach to the substrate, secrete a protective covering of calcareous plates, and strain food particles from the water with a weblike structure. See Appendix 3.

barrier island. Long, narrow island built of wave-transported sand, separated from the mainland by a usually shallow lagoon.

barrier reef. Coral reef that parallels the shore but is separated from the landmass by open water. Compare *fringing reef* and *atoll*.

basalt (adj. **basaltic**). Dark-colored volcanic rock, rich in iron, magnesium, and calcium, characteristic of the oceanic crust.

bathyal zone. Benthic environment from 200 to 2000 m depth.

bathymetry (adj. **bathymetric**). Seafloor mapping; the study of landform features beneath the water surface. A bathymetric map gives the depth contours of the seafloor.

baymouth bar. Shallow bar, usually of sand, that extends partially or completely across the mouth of a bay.

beach. Sand- or sediment-covered zone between the seaward limit of permanent vegetation and the mean low-water line. Beaches sometimes include the seafloor from the mean low-water line to the surf zone.

benthic. Pertaining to the seafloor, or to organisms that live on or in the seafloor.

benthos. Organisms that live on or in the ocean bottom.

berm. Nearly horizontal portion of a beach backshore at the seaward edge of which the beach slopes abruptly seaward. Compare *scarp*.

bioaccumulation (v. **bioaccumulate**). Process by which dissolved chemicals are taken up by organisms until the concentrations in the tissues are at equilibrium with those in the solution. See **CC18**.

bioassay. Test in which organisms are exposed to different concentrations of a substance to determine the concentration at which the substance is toxic to that species. See **CC18**.

bioavailable. Existing in a chemical form that is suitable for uptake by organisms.

biochemical. Pertaining to the synthesis, conversion, use, and decomposition of organic chemicals in the life processes of organisms.

biodegradable. Capable of being decomposed by organisms in the environment. The term generally refers to decomposition to harmless substances.

biodiversity. Poorly defined term that refers to a combination of genetic diversity, species diversity, ecosystem diversity, and physiological diversity within a community or ecosystem. See **CC17**.

biogenous. Pertaining to material of a biological origin. Usually applied to sediments in which the hard-part remains of organisms (e.g., diatom tests, radiolarian shells) constitute a high proportion of the grains. Compare *cosmogenous*, *hydrogenous*, and *lithogenous*.

biogeochemical cycle. Transfers of compounds among the living and nonliving components of an ecosystem.

bioluminescent. Light-producing. Referring to organisms that use chemical reactions to produce light.

biomagnification (v. **biomagnify**). Process by which dissolved chemicals are taken up by organisms and continuously accumulated to higher concentrations throughout the life cycle. Tissue concentrations are not in equilibrium with concentrations in the surrounding environment. See **CC18**.

biomass. The total amount of living matter, expressed in units of weight in the entire water column per unit area of water surface, or weight per unit of water volume.

biota. All living organisms in a given ecosystem, including archaea, bacteria, protists, fungi, plants, and animals.

bioturbation. Reworking (churning and mixing) of sediments by organisms that burrow in them.

bloom. Very dense aggregation of phytoplankton, resulting from a rapid rate of reproduction.

blubber. Outer fat layer of whales.

blue-green algae. See *cyanobacteria*.

brackish. Pertaining to water that is a mixture of freshwater and seawater and has a low salinity.

breakwater. Artificial structure constructed in the ocean to protect a shore from the action of ocean waves.

brine. Water that has a higher salinity (generally much higher) than normal seawater.
brittle stars. Organisms of the class Ophiuroidea, phylum Echinodermata, that have long, slender arms covered in bristles attached to a small central disk. See Appendix 3.
bulkhead. Structure constructed to separate land and water areas at the shoreline and designed to reduce earth slides and slumps or to lessen wave erosion at the base of a cliff.
buoyancy (adj. **buoyant**). Ability of an object to float (or rise through a fluid) that results from a density difference between the object and the fluid in which it is immersed.
bycatch. Fishes and other marine animals caught in fishers' nets that are not the target of the fishing. Bycatch is generally thrown overboard as waste.
calcareous. Composed of or containing calcium carbonate.
calcite. The most common mineral form of calcium carbonate. Compare *aragonite*.
caldera. Crater in a volcano, usually at the center, created by an explosion or collapse of the volcanic cone.
calorie. Obsolete unit of heat energy defined as the amount of heat needed to raise the temperature of 1 g of water by 1°C.
capillary wave. Small ocean wave that has a wavelength less than 1.5 cm. The primary restoring force for capillary waves is surface tension.
carbonate compensation depth (CCD). Depth below which all calcium carbonate particles falling from above are dissolved before they can be incorporated in the sediments.
carcinogens (adj. **carcinogenic**). Compounds that cause cancer in animals. Compare *mutagens* and *teratogens*.
carnivores (adj. **carnivorous**). Animals that depend solely on other animals for their food supply. Compare *herbivores* and *omnivores*.
cartilaginous. Pertaining to vertebrate animals that have skeletons made of cartilage, a tough elastic tissue.
catadromous. Pertaining to species of fishes that are spawned at sea, migrate to a freshwater stream or lake where they live until they reach maturity, then return to sea to spawn. Compare *anadromous*.
cation. Positively charged ion. Compare *anion*.
CCD. See *carbonate compensation depth*.
central rift valley. Steep-sided linear valley that runs along the crest of many parts of oceanic ridges. Central rift valleys are created by pulling apart of the diverging plates.
centripetal force. Force that acts toward the center of rotation of an orbiting body and that is needed to maintain the object in its circular path and prevent it from moving off in a straight line. See **CC12**.
cephalopods. Animals belonging to the class Cephalopoda, phylum Mollusca, that have a well-developed pair of eyes and a ring of tentacles surrounding the mouth. Most have no, or an internal, shell. Cephalopods include squid, octopi, and *Nautilus*. See Appendix 3.
cetaceans. Marine mammals belonging to an order that includes whales, porpoises, and dolphins. See Appendix 3.
CFC. See *chlorofluorocarbons*.
chaos (adj. **chaotic**). Apparently random behavior of systems that involve nonlinear relationships. See **CC11**.
chemosynthesis (adj. **chemosynthetic**). Production of organic compounds from inorganic substances by use of energy obtained from the oxidation of substances such as hydrogen sulfide, ammonia, hydrogen, and methane. Compare *photosynthesis*.
chlorofluorocarbons (CFCs). Class of compounds used as refrigerants and for many other purposes. When released to the atmosphere, CFCs migrate upward to the ozone layer, where they react with ozone, reducing its concentration.
chlorophylls. Group of green pigments that are essential to plants and are active in the capture of light energy for photosynthesis. See **CC14**.
chordates. Members of the animal phylum Chordata, to which mammals, reptiles, birds, fishes, and tunicates belong. See Appendix 3.
chronic toxicity. Sublethal or lethal effects due to long-term exposure of an organism to a toxic substance at concentrations below that at which the organism suffers immediate (within days or weeks) sublethal or lethal effects.
chronometer. Clock that has sufficient long-term accuracy and tolerance of motion that it can be used as a portable time standard.
climate (adj. **climatic**). Long-term averages of weather conditions such as temperature, rainfall, and percentage cloud cover, and the distribution of these averages among the seasons at a given location.
clones. Members of a species that are genetically identical because they are asexually reproduced.
cnidarians. Members of the phylum Cnidaria, consisting of predominantly marine animals with a saclike body and stinging cells on tentacles that surround a single opening to the gut cavity. The two basic body forms are the medusa and the polyp. Medusal forms are pelagic and include jellyfish. Polypoid forms are predominantly benthic and include sea anemones and corals. Also called "coelenterates." See Appendix 3.
coast. Strip of land that extends from the shore inland to the seaward limit of terrain that is unaffected by marine processes.
coastal plain. Low-lying land next to the ocean extending inland until the first major change in the features of the terrain.
coastal-plain estuary. Estuary formed by flooding of a coastal river valley as sea level rises.
coastal upwelling. Vertical transport and mixing of deep, nutrient-rich water into the surface water mass as a result of offshore Ekman (windblown) transport of surface water.
coastal zone. Zone between the coastline and the point offshore to which the influence of freshwater runoff extends (often the shelf break). The term can also mean this area of coastal ocean plus the adjacent coast.
coastline. Landward limit reached by the highest storm waves on the shore. Compare *shoreline*.

coccolithophores. Microscopic planktonic algae surrounded by a cell wall embedded with calcareous disks called "coccoliths." See Appendix 3.
coelenterates. See *cnidarians*.
cohesive. Pertaining to the molecular force between particles within a substance that acts to hold the particles together. Cohesive particles behave as though they were sticky. See **CC4**.
colloidal. Pertaining to substances that have a particle size smaller than clay.
community. All of the organisms that live within a definable area or volume of ocean or land.
compensation depth. Depth at which algae consume oxygen for respiration at the same rate that they produce oxygen by photosynthesis.
complexed. Generally pertaining to a dissolved ion that has an association (a weak electrostatic bond) with the molecules of a polar organic substance. The association can alter the ion's solubility, bioavailability, and toxicity.
conduction (adj. **conducted**). Transfer of heat (or electricity) through a material or between two materials in contact with each other by passing directly from one molecule to another adjacent molecule.
contaminant (v. **contaminate**). Substance added to an ecosystem (or to a sample, an organism, or other object) in an amount that causes the concentration of the substance to exceed its natural level. Contaminants may or may not have harmful effects. Compare *pollution*.
continental collision plate boundary. Region where two lithospheric plates converge, each of which has continental crust at its margin.
continental divergent plate boundary. Region where a continent is splitting apart to form the edges of two new lithospheric plates.
continental drift. Movement of the continents over the Earth's surface. Continental drift is the theory that preceded plate tectonics.
continental margin. Zone of transition from a continent to the adjacent ocean basin. The continental margin generally includes a continental shelf, continental slope, and continental rise and often the land adjacent to the coast.
continental rise. Gently sloping seafloor between the base of a continental slope and the abyssal plain. A part of the deep-ocean floor.
continental shelf. Zone bordering a continent that extends from the low-water line to the depth at which there is a marked increase in the downward slope of the seafloor that continues to the deep-ocean floor.
continental slope. Relatively steeply sloping seafloor seaward of the continental shelf that ends where the downward slope markedly decreases at the edge of the deep-ocean floor.
contour (adj. **contoured**). Line on a chart or graph that connects points of equal value of the parameter charted (e.g., ocean depth, temperature, salinity).
convection (v. **convect,** adj. **convective**). In a fluid being warmed at its bottom and/or cooled at its upper surface, process by which warmer fluid rises and cooler fluid sinks in a density-driven circulation. See **CC1**, **CC3**.
convection cell. Circulatory system established by convection in which water (or other fluid) is warmed, rises, is displaced by newly upwelled warm water, cools, and sinks to be warmed again. See **CC3**.
convergence (v. **converge,** adj. **convergent**). Location at which fluids of different origins come together, usually horizontally (often in a convection cell). Convergence results in sinking (downwelling) or rising (upwelling) when it is at the top or bottom surface of a fluid layer, respectively. See **CC3**. Compare *divergence*.
convergent plate boundary. Lithospheric plate boundary where the relative motion of adjacent plates is toward each other, producing ocean trench/island arc systems or ocean trench/continental mountain and volcano complexes. Compare *divergent plate boundary*.
copepods. Small shrimplike animals of the subphylum Crustacea that are zooplankton. See Appendix 3.
corals. Group of benthic cnidarians that exist as individuals or in colonies and may secrete external skeletons of calcium carbonate. See Appendix 3.
coral reef. Mainly calcareous reef composed substantially of coral, coralline algae, and sand. Coral reefs are present only in waters where the minimum average monthly temperature is 18°C or higher.
Coriolis effect. Apparent deflection of a freely moving object caused by the Earth's rotation. The deflection is *cum sole*. See **CC12**.
cosmogenous. Pertaining to material that originated in outer space (e.g., meteorite fragments). The term is generally applied to particles of such origin within sediments. Compare *biogenous*, *hydrogenous*, and *lithogenous*.
countershading. Coloration of pelagic fishes, in which the upper half is dark and nonreflective and the lower half is light-colored and more reflective. Countershading reduces visibility to predators from above and below.
covalent bond. Chemical bond in which atoms are combined to form compounds by sharing one or more pairs of electrons.
crest. For a wave, the portion that is displaced above the still-water line. The term is often used to refer to the highest point of the wave (or other topographic feature) only. Compare *trough*.
crinoids. Animals of the class Crinoidea, phylum Echinodermata. Also called "feather stars." See Appendix 3.
crust. Outer shell of the solid Earth. The lower limit of the crust is usually considered to be the Mohorovičić discontinuity (top of the asthenosphere). The thickness of the crust ranges from about 6 km beneath the oceans to 30 to 40 km beneath the continents. Compare *mantle*.
crustaceans. Animals of a subphylum (Crustacea) of the phylum Arthropoda that have paired, jointed appendages and hard outer skeletons. Crustaceans include barnacles, copepods, lobsters, crabs, and shrimp. See Appendix 3.
CTD. Abbreviation for conductivity, temperature, and depth. A CTD is an instrument package that measures these parameters continuously while being lowered and raised through the water column.

ctenophores. Animals of the phylum Ctenophora of transparent planktonic animals that are spherical or cylindrical in shape and have rows of cilia. Ctenophores include comb jellies. See Appendix 3.

cum sole. Literally translated from Latin as "with the sun." To the right in the Northern Hemisphere and to the left in the Southern Hemisphere. If you are facing toward the sun's arc in the sky (toward the equator), the sun appears to move from left to right in the Northern Hemisphere and from right to left in the Southern Hemisphere.

current. Horizontal movement of water.

cyanobacteria (sing. **cyanobacterium;** adj. **cyanobacterial)**. Phylum of organisms within the domain Bacteria. Cyanobacteria were originally called "blue-green algae" because they have the ability to photosynthesize. See Appendix 3.

cyclonic. Pertaining to counterclockwise circulation in the Northern Hemisphere and clockwise circulation in the Southern Hemisphere.

cyst. Saclike structure. Sometimes, a protective covering of a microorganism during a metabolic resting phase.

DDT. Dichloro-diphenyl-trichloroethane. Insecticide widely used in the United States in the 1950s and 1960s. DDT caused major damage to birds and marine mammal populations. It is now banned in many countries, but it is still widely used in the tropics.

declination. Measure of the sun's (or moon's) apparent north–south seasonal movement. The angle between a line from the Earth's center to the sun (or moon) and the plane of the equator.

decomposers. Heterotrophic microorganisms (mostly bacteria and fungi) that break down nonliving organic matter to obtain energy. During decomposition, nutrients are released to solution and become available for reuse by autotrophs.

deep-water wave. Ocean wave that is traveling in water depth greater than one-half its wavelength. Compare *intermediate wave* and *shallow-water wave*.

delta. Low-lying deposit of river-borne sediment at the mouth of a river. Deltas are usually triangular in shape.

density, absolute. The mass per unit volume of a substance (usually expressed as grams per cubic centimeter). Compare *density, relative*.

density, relative. The ratio of the mass per unit volume of the substance divided by the mass of the same volume of a standard substance. The standard substance is usually pure water at 4°C. Relative density is a dimensionless number, but since the mass of 1 cm^3 of pure water at 4°C is almost precisely 1 g, relative density and absolute density are numerically almost identical. Compare *density, absolute*.

deposit feeders (adj. **deposit-feeding)**. Organisms that feed by ingesting particles of sediment and metabolizing organic matter in or on the particles.

deposition (adj. **depositional)**. Accumulation of sediment on the seafloor (or of solid particles from the atmosphere onto the land or ocean surface).

detritivores (adj. **detritivorous)**. Organisms that eat organic detritus.

detritus. Any loose material, but generally decomposed, broken, and dead organisms.

diagenesis (adj. **diagenetic)**. Chemical or mineralogical changes that take place in a sediment or sedimentary rock after its formation.

diatoms. Microscopic unicellular algae of the phylum Chrysophyta that have an external skeleton of silica. See Appendix 3.

diffusion (v. **diffuse)**. Movement of a substance (or property such as heat) by random molecular motions from a region of higher concentration to a region of lower concentration (along a concentration gradient).

dinoflagellates. Single-celled microscopic organisms in the kingdom Protista. Dinoflagellates may have chlorophyll, be autotrophic, and belong to the phylum Pyrrophyta; or they may be heterotrophic and belong to the phylum Zoomastigophora. See Appendix 3.

dip angle. Angle to the horizontal of the direction of the magnetic field in rock. The dip angle is approximately equal to the latitude at which the rock formed.

diurnal. Having a cycle that occurs during a 24-h day and that generally recurs daily.

diurnal tide. Tide with one high water and one low water during a tidal day (approximately 24 h 49 min). Compare *semidiurnal tide* and *mixed tide*.

divergence (v. **diverge,** adj. **divergent)**. Horizontal flow of a fluid away from a common center. The fluid is replaced by upwelling if the divergence is at the surface of the fluid and by downwelling if it is at the bottom. Compare *convergence*.

divergent plate boundary. Lithospheric plate boundary where adjacent plates are pulling apart from each other. Compare *convergent plate boundary*.

diversity (adj. **diverse)**. Presence of variety or variation within an ecosystem. The term may be applied to variety of species, variety of genes within a species, variety of habitat, etc. See **CC17**.

DNA. Deoxyribonucleic acid. Complex organic molecule that contains the genetic code.

downwelling (adj. **downwelled)**. Vertical movement of a fluid downward due to density differences or where two fluid masses converge, displacing fluid downward. In the ocean, the term often refers to coastal downwelling, where Ekman transport causes surface waters to converge or impinge on the coast, displacing surface water downward and thickening the surface layer. Compare *upwelling*.

drift net. Net hung vertically like a drape that may extend for many kilometers. Drift nets indiscriminately catch anything that swims into them.

dynamic height. Height of the water column above a depth below which no currents are assumed to be present.

eastern boundary current. Surface layer current that flows toward the equator on the eastern side of a subtropical gyre. Eastern boundary currents are slow, wide, and shallow. Compare *western boundary current*.

ebb. Period of a tidal cycle when the tidal current is flowing seaward or when the tide level is falling. Compare *flood*.

echinoderms. Animals of the phylum Echinodermata, which have bilateral symmetry in larval forms and usually a five-sided radial symmetry as adults. Echinoderms are benthic organisms with rigid or articulating external skeletons of calcium carbonate that have spines. Echinoderms include sea stars, brittle stars, sea urchins, sand dollars, sea cucumbers, and sea lilies. See Appendix 3.

echolocation. Use of sound by some marine animals to locate and identify underwater objects from their echoes.

ecology (adj. **ecological**). Study of relationships between species and between the species and their environment.

ecosystem. Organisms in a community and the nonliving environment with which they interact.

eddy. Circular movement of water.

EEZ. See *exclusive economic zone*.

effluent. Outflow of water from a system. The term is often applied to wastewater discharges that flow out of a treatment plant, sewer, industrial outfall, or storm drain.

Ekman transport. Net wind-driven transport of surface water at an angle *cum sole* to the wind direction as a result of the Coriolis effect.

El Niño. Episodic movement of warm surface water south along the coast of Peru that is associated with the cessation of upwelling in this region. The term is often used to refer to a complex episodic sequence of events in the oceans and atmosphere called "El Niño/Southern Oscillation" (ENSO).

electrical conductivity. Measure of a substance's ability to conduct an electrical current. In seawater, electrical conductivity is related to (and used to measure) salinity.

electromagnetic radiation. Energy that travels at the speed of light as waves. Electromagnetic waves range in wavelength from very long (up to 10 km) radio waves to very short (10^{-12} m) cosmic rays.

electrostatic. Pertaining to the attractive or repulsive force between two electrically charged bodies that does not involve electrical current flow between the bodies.

embryo. Early or undeveloped life stage of an animal.

emission spectrum (pl. **spectra**). Wavelengths at which warm bodies emit electromagnetic energy. Emission spectra are plotted as the relative intensity of emission versus wavelength. Emission spectra vary with an object's temperature. Compare *absorption spectrum*.

encrust. To grow on the surface of and form a crust on a solid substrate.

ENSO. El Niño/Southern Oscillation. See *El Niño*.

environment. Physical and chemical characteristics of a location or area.

enzymes. Organic substances that are synthesized by organisms and behave as catalysts for biochemical reactions.

epifauna. Animals that live on the surface of the seafloor or other substrate, either moving freely or attached. Compare *infauna*.

epipelagic zone. Upper region of the oceanic province. The water column from the surface to a depth of 200 m.

equinox. Time of year when the sun is directly overhead at the equator. Equinoxes occur on March 20 or 21 and September 22 or 23 each year. Compare *solstice*.

erosion (v. **erode,** adj. **erosional**). Process of being gradually worn away. Erosion is usually caused by the action of winds and currents on rocks and sediments.

estuary (adj. **estuarine**). Any region where freshwater and seawater mix.

eukaryotes. Organisms (single-celled or multi-celled) whose cells are surrounded by a membrane and that have a structurally discrete nucleus and other well-developed subcellular compartments. Eukaryotes include all organisms except archaea, and bacteria. Compare *prokaryotes*.

euphausiids. Members of an order of shrimplike planktonic animals of the subphylum Crustacea, phylum Arthropoda. Euphausiids include several species commonly called "krill." See Appendix 3.

eustasy (adj. **eustatic**). The equilibrium level of the ocean surface. Eustatic changes of sea level take place in response to changes in ocean volume and take place worldwide, as distinct from locally. See **CC2**.

eutrophication. Physical and biological changes that occur when excessive nutrients are released into an aquatic environment. Eutrophication may lead to blooms and anoxia.

evaporite. Mineral deposit formed by the evaporation of seawater.

exclusive economic zone (EEZ). Zone in which the coastal state (nation) has ownership of resources including fishes and seafloor minerals. EEZs are generally 200 nautical miles (370 km) wide.

excretion (v. **excrete**). Substances (generally waste products) released to the external environment from the tissues of a living organism, or the process of releasing such substances.

excurrent. In animals that feed by pumping or passing water through the body (e.g., tunicates), pertaining to the opening through which the ingested water is expelled from the organism. Compare *incurrent*.

exotic terrane. Fragments of continental crust or sometimes oceanic crust and sediment that have been accreted (attached) to other continents.

extratropical cyclone. Cyclonic storm formed in high latitudes at the polar fronts. Extratropical cyclones resemble and can be as strong as hurricanes.

family. Level of taxonomic classification of species that is between the levels of order and genus. See Appendix 3.

fault. Fracture in the Earth's crust in which one side has been displaced in relation to the other.

fauna. Animal population of an ecosystem or region. Compare *flora*.

fecal pellet. Solid waste product excreted by animals.

feedback. Reaction to a process of change that either reinforces or moderates that change. See **CC9**.

fetch. Uninterrupted distance over which the wind blows (measured in the direction of the wind) without a significant change of direction.

filter feeders (adj. **filter-feeding**). Animals that feed by sifting or straining small particles suspended in the water. Compare *suspension feeders*.

fjord. Long, narrow, deep inlet. Usually the seaward end of a valley flooded by rising sea level after it was cut by a glacier that has since retreated.

flagellates. Protozoa that have flagella. See Appendix 3.

flagellum (pl. **flagella**). Whiplike appendage used by some microscopic organisms to provide propulsion for locomotion.

flood. Period of a tidal cycle when the tidal current is flowing landward or when the tide level is rising. Compare *ebb*.

flora. Plant population of an ecosystem or region. Compare *fauna*.

focus. Location within the Earth's crust where an earthquake occurs.

food chain. Sequence of organisms in which each is the food source for the next in sequence. Compare *food web*.

food chain efficiency. Percentage of food ingested by organisms at a particular trophic level that is converted into biomass at that trophic level. Also called "trophic efficiency." See **CC14**.

food web. Series of food chains that are interconnected in a complex way to create a mosaic of feeding relationships.

foraminifera (sing. **foraminiferan;** adj. **foraminiferal**). Planktonic and benthic protozoa of the phylum Sarcodina, protected by shells, and usually calcareous. See Appendix 3.

foreshore. Zone between the low- and high-tide lines. Compare *backshore*.

fossil. Remains of an organism or its imprint that has been preserved in rocks.

fossil fuel. Fuel that is derived directly from fossilized organic matter. Fossil fuels include oil, natural gas, coal, and peat.

foul (n. **fouling**). To attach to or lie on the surface of an underwater object. The term applies especially to barnacles and other marine organisms that grow on vessel hulls and other human-made structures.

fracture zone. Linear zone of steep-sided irregular seafloor topography. Most fracture zones are inactive remnants of transform faults.

frequency. Number of periodic events that occur within a specified interval. For waves, the number of crests (or troughs) that pass a given point per unit time; the inverse of wave period.

friction. Retarding force that resists the motion of two objects (or an object and a fluid, or two fluids; see *shear stress*) that are moving in relation to each other and whose surfaces are in contact.

fringing reef. Reef that is attached to the shore of an island or continent with no open water lagoon between the reef and shore. Compare *barrier reef* and *atoll*.

front. Well-defined boundary between two air masses or two water masses of different density.

frustule. Siliceous covering of a diatom.

fundamental niche. Range of environmental variables within which a species can both survive and successfully reproduce. Compare *survival niche*.

fungi. Members of the kingdom Fungi, organisms that reproduce by means of spores. Most marine fungi are microscopic benthic decomposers. See Appendix 3.

galaxy (adj. **galactic**). Assemblage of stars (millions to hundreds of billions) held together by the gravitational attraction of the member stars on one another. Most galaxies are either a flattened, spiral form like the Milky Way, the galaxy in which our sun is located, or elliptical without a spiral pattern.

genetic. Pertaining to the genes of an organism. The term refers to the information (genetic code) encoded in DNA and other substances that describes the characteristics of an organism and that can be passed on to the offspring during reproduction.

genus (pl. **genera**). Level of taxonomic classification of species that is between the levels of family and species. See Appendix 3.

geostrophic. Pertaining to cyclonic fluid motions that are maintained as a result of a near balance between a gravity-induced horizontal pressure gradient and the Coriolis effect. See **CC13**.

gill. Thin-walled organ of marine animals used for respiration.

glaciation. Extent to which glaciers are developed in a given area or at a given time.

glacier (adj. **glacial**). Large mass of ice that forms on land by the recrystallization of old compacted snow. Glaciers flow from the area where they are formed downhill to an area where ice is removed by melting or calving (breaking off) into a water body.

gobies. Members of a family of small bony fishes (class Ostreichthyes, subphylum Vertebrata, phylum Chordata). Many goby species have pelvic fins modified to form a suction disk. See Appendix 3.

graded bed. Vertical sequence of sediments or sedimentary rock in which each layer comprises particles of smaller grain size from bottom to top.

grain size. Diameter of grains that compose a sediment.

gravity. Attractive force between any two bodies in the universe.

grazers (v. **graze**). Strictly, animals that eat plants. More generally, the term includes animals that eat detritus (or other animals) that covers the surface of a substrate.

greenhouse effect. Tendency of the atmosphere or greenhouse glass to be transparent to incoming solar radiation while absorbing (or reflecting) longer-wavelength heat radiation from the Earth. See **CC9**.

groin. Artificial structure that projects into the ocean from the shore. Groins block longshore transportation of sediment and usually are intended to trap sand and prevent its loss from a beach.

groundwater. Water beneath the ground surface that has seeped through the soil and rock from above.

guyot. Conical, volcano-shaped feature on the ocean floor whose top has been eroded to form a flat top.

gyre. Circular motion. The term generally refers to a circular current system centered in the subtropical high-pressure region of a major ocean basin.

habitat. Place where a particular plant or animal lives.

hadal zone. Deepest environment of the oceans. Restricted to ocean trenches deeper than 6 km.

half-life. Amount of time required for half the atoms of a radioactive isotope sample to decay to atoms of another element. See **CC7**.

halocline. Depth range in the water column in which there is a gradient of salinity in the vertical dimension. Compare *pycnocline* and *thermocline*.

hard parts. Rigid structural material of plants and algae and the shells and skeletons of animals. Hard parts are usually siliceous or calcareous.

harmonics. Component simple waves that are added together to make up the complex waveform observed as a result of the interference of waves of different frequencies and/or from different directions.

heat capacity. Amount of heat required to raise the temperature of 1 g of a substance by 1°C.

herbicides. Chemicals that are used to kill or inhibit the growth of plants.

herbivores (adj. **herbivorous**). Animals that rely primarily or solely on plants for their food. Compare *carnivores* and *omnivores*.

hermatypic corals. Reef-building corals that have symbiotic algae in their tissues and that cannot grow successfully below the photic zone.

hermit crabs. Any of a number of species of crabs of the suborder Reptantia, order Decapoda, subphylum Crustacea, phylum Arthropoda that have no shell of their own. Hermit crabs live in shells of dead gastropod mollusks and move to a larger shell as they grow. See Appendix 3.

herrings. Members of a family of small plankton-eating bony fishes (class Ostreichthyes, subphylum Vertebrata, phylum Chordata) that have oily tissues and are extremely abundant in some areas. Herrings include sardines and anchovies. See Appendix 3.

heterotrophs (adj. **heterotrophic**). Animals or bacteria that do not photosynthesize or chemosynthesize and therefore depend for food and energy on organic compounds produced by other species. Compare *autotrophs*.

high seas. Area of the world oceans that is outside the territorial control of any nation.

high-tide line. Highest point on the shore that is covered by water at high tide. Compare *low-tide line*.

high-tide zone. Zone of the shore that is mostly exposed and lies between the lowest (neap) high-tide line and highest (spring) high-tide line. Compare *low-tide zone* and *middle-tide zone*.

holoplankton. Organisms that spend their entire life as members of the plankton. Compare *meroplankton*.

hot spot. Surface expression of a persistent convection plume of molten mantle material rising to the Earth's surface.

humidity. Amount of water vapor in the air. It is measured as relative humidity, the ratio of the water vapor concentration to the saturation concentration at the same temperature and pressure, expressed as a percentage.

hurricane. Tropical cyclonic storm with winds that have velocity greater than 120 $km \cdot h^{-1}$. The term applies to such storms in the North Atlantic Ocean, eastern North Pacific Ocean, Caribbean Sea, and Gulf of Mexico. Such storms in the western Pacific Ocean are known as "typhoons," "cyclones," or "willy willys."

hydrated (n. **hydration**). Chemically combined with water or, for ions, surrounded by water molecules in a weak electrostatic association.

hydrocarbons. Large group of chemicals containing carbon and hydrogen atoms. The carbon atoms may be arranged in chains of varying length or in one or more six-atom rings. These compounds are the predominant components of petroleum, and some are also produced by plankton.

hydrogen bond. Bond between molecules that forms because of the dipolar nature of the molecules. Hydrogen bonds are present in water and a few other compounds.

hydrogenous. Pertaining to solid material formed by chemical precipitation from solution. The term usually applies to the component particles or coatings of sediments that are precipitated from seawater. Compare *biogenous*, *cosmogenous*, and *lithogenous*.

hydrographic. Pertaining to mapping of the oceans and their depth.

hydroids. Group of animal species of the class Hydrozoa, phylum Cnidaria. Polypoid forms attach to the substrate. Most are colonial, many with a branching form resembling a feather or fern on which individual polyps are arranged. Hydroids reproduce by budding and have a pelagic medusal stage. See Appendix 3.

hydrophone. Device that senses underwater sound.

hydrosphere. Gaseous, liquid, and solid water of the Earth's upper crust, ocean, and atmosphere. The hydrosphere includes lakes, groundwater, snow, ice, and water vapor.

hydrostatic pressure. Pressure that results from the weight of the water column that overlies the depth at which the pressure is measured.

hydrothermal minerals. Predominantly fine-grained particles precipitated from the water that is discharged at hydrothermal vents. Hydrothermal minerals are rich in iron and manganese.

hydrothermal vent. Location where heated water is vented through the seafloor. This water is seawater that has percolated down through fractures in recently formed ocean floor and has been heated by underlying magma. Most known vents are near the central axis of oceanic ridges and rises.

hypoxia (adj. **hypoxic**). Presence of dissolved oxygen in the aquatic environment at concentrations low enough to be detrimental to organisms. Hypoxia is usually considered to exist when dissolved oxygen concentrations are at or below 2 $mg \cdot l^{-1}$. Compare *anoxia*.

ice age. Period during which the Earth's average climatic temperature was colder, glaciers were more extensive, and sea level was lower. Several such periods have occurred in the past million years, each lasting for several thousand years.

ice exclusion. Process whereby salts are excluded from the ice that forms as seawater freezes, resulting in a higher salinity in the remaining liquid water.

in situ. Literally translated from Latin as "in place"—that is, not removed from its natural environment.

incubation. Maintenance of eggs or embryos in a favorable environment for hatching and development.

incurrent. In animals that feed by pumping or passing water through the body (e.g., tunicates), pertaining to the opening through which the ingested water is taken into the organism. Compare *excurrent*.

infauna. Animals that live buried in soft sediments (sand or mud). Compare *epifauna*.

intermediate wave. Surface water wave that, at a given water depth, has a wavelength between those of deep-water waves and shallow-water waves. Wave in water depth between one-half and one-twentieth of the wavelength. Compare *deep-water wave* and *shallow-water wave*.

internal wave. Wave that develops below the surface of a fluid at a pycnocline and travels along this boundary.

intertidal zone. Zone covered by the highest normal tides and exposed by the lowest normal tides, and any tide pools within this zone. Also called "littoral zone."

intertropical convergence. Zone where northeast trade winds and southeast trade winds converge.

invertebrates. Animals that have no backbone.

ion (adj. **ionic**). Atom (or combined group of atoms) that becomes electrically charged by gaining or losing one or more electrons to produce a negatively charged anion or a positively charged cation, respectively.

ionic bond. Chemical bond that is formed by the electrical attraction between cations and anions.

isobar (adj. **isobaric**). Line that connects values of equal pressure on a map or graph. Compare *isopycnal* and *isotherm*.

isopods. Animals with flattened bodies belonging to an order of the phylum Crustacea. Most isopods are scavengers or parasites on other crustaceans or fishes. See Appendix 3.

isopycnal. Line that connects values of equal density on a map or graph. Compare *isobar* and *isotherm*.

isostasy (adj. **isostatic**). Equilibrium, comparable to buoyancy, in which the rigid lithospheric plates float on the underlying mantle. See **CC2**.

isostatic leveling. Tendency of lithospheric plates to rise or fall to an equilibrium level with respect to the level at which they float on the asthenosphere after their density or mass has changed—for example, after cooling of the crust or changes in the extent of glaciation. See **CC2**.

isotherm. Line that connects points of equal temperature on a graph or map. Compare *isobar* and *isopycnal*.

isothermal. Of the same or uniform temperature.

isotope (adj. **isotopic**). Atoms of an element that have different numbers of neutrons, and therefore different atomic masses than the atoms of other isotopes of the same element. See **CC7**.

jet stream. Easterly-moving air mass at an altitude of about 10 km that can have speeds exceeding $300\ km{\cdot}h^{-1}$. Jet streams follow a meandering path in the mid latitudes and influence how far polar air masses extend into lower latitudes.

jetty. Elongated structure built outward from the shore into a body of water to protect a harbor or a navigable passage from accumulation of sand transported by longshore drift.

joule. Unit of energy equal to the energy expended in 1 s by an electrical current of 1 ampere with a potential difference of 1 volt.

kelp. Various species of large brown algae (Phaeophyta). See Appendix 3.

kinetic energy. Energy of an object in motion. Kinetic energy increases as the mass of the object or the speed of the object in motion increases. Compare *potential energy*.

krill. Common name applied to euphausiids, members of an order of the subphylum Crustacea, phylum Arthropoda. See Appendix 3.

lagoon. Shallow estuary or area of ocean adjacent to the shore but partly or completely separated from the open ocean by an elongated, narrow strip of land such as a reef or barrier island.

Langmuir circulation. Cellular water circulation set up by strong winds that blow consistently in one direction. The cells are arranged in alternating clockwise and counterclockwise helical spirals aligned parallel to the wind direction.

larva (pl. **larvae**). Animal embryo that lives free from its parents before assuming the adult form.

laser. Abbreviation for "light amplification by stimulated emission of radiation." Instrument that generates a very intense, extremely narrow beam of light of a single wavelength.

latent heat of fusion. For 1 g of a substance at its melting point temperature, the quantity of heat energy that must be added to convert it from solid to liquid (or that must be removed to convert the liquid to solid) without changing the temperature.

latent heat of vaporization. For 1 g of a substance at its boiling point temperature, the quantity of heat energy that must be added to convert it from liquid to gas (or that must be removed to convert the gas into liquid) without changing the temperature or pressure.

latitude (adj. **latitudinal**). Partial designation of location on the Earth's surface. Latitude is expressed as the angular distance north or south of the equator. The equator has a latitude of 0°, the North Pole 90°N, and the South Pole 90°S. Compare *longitude*.

lava. Fluid magma that emerges from an opening in the Earth's surface, or the same material after it cools and solidifies.

leach (v. **leaching**). To dissolve constituents of solids by passing (filtering) a fluid (usually water) through cracks or pores in the solid or sediment.

leeward. The direction toward which the wind is blowing or the waves are moving. The term usually applies to the sheltered downwind or downcurrent side of a barrier or landmass. Compare *windward*.

levee. Low ridge that forms the sides of a river channel. Levees may be human-made or natural (created by sediment deposition during flooding).

lichens. Organisms that are a mutualistic relationship of fungi with algae or cyanobacteria. The algae (or cyanobacteria) are protected by the fungi, which depend on the algae to produce food by photosynthesis. See Appendix 3.

light-limited. Pertaining to the condition in which the rate of production of organic matter by a photosynthetic organism or population of organisms is inhibited by the absence of sufficient light energy when all other requirements for an increase in the rate of production are met.

limestone. Sedimentary rock composed of at least 50% calcium or magnesium carbonate. Limestone may be either biogenous or hydrogenous.

limiting nutrient. Nutrient present at such a low concentration that its lack of availability reduces the rate of growth or prevents the growth of phytoplankton (or other primary producers).

limpets. Mollusks of the class Gastropoda, phylum Arthropoda, that have a low conical shell and adhere to a substrate where they are covered by the shell. See Appendix 3.

lipids. Fats. Lipids are among the principal structural components of living cells.

lithogenous. Pertaining to material derived from the rock of continents and islands and transported to the ocean by wind or running water. The term is usually applied to sediments that have a high proportion of mineral grains of terrestrial origin. Compare *biogenous*, *cosmogenous*, and *hydrogenous*.

lithosphere (adj. **lithospheric**). Outer layer of the Earth. The lithosphere includes the crust and the part of the upper mantle that is fused to the crust. It is the layer that is broken into the lithospheric plates.

lithospheric plates. Sections of the Earth's lithosphere that are separated from each other by boundaries at which the adjacent sections move in relation to each other.

littoral drift. See *longshore drift*.

littoral zone. To biological oceanographers, the benthic zone between the highest and lowest normal water marks reached by the tide. To geological oceanographers, the zone between the seaward boundary of land vegetation (or the base of a cliff if present) and the point where the seafloor reaches a depth at which sediment is no longer disturbed by waves.

longitude (adj. **longitudinal**). Partial designation of location on the Earth's surface. Longitude is expressed as the angular distance east or west of the Greenwich meridian (0° longitude). 180° longitude is the international date line. Compare *latitude*.

longshore bar. Sand mound that extends generally parallel to the shoreline a short distance offshore. The bar may be submerged or exposed, especially at low tide, and is created by sand accumulated by wave action.

longshore current. Current that flows in the surf zone and parallel to the shore. Longshore currents are created by breaking waves.

longshore drift. Sediment transport along the beach within the region from the breaker zone to the top of the swash line. Longshore drift is associated with the longshore current. Also called "littoral drift."

low-tide line. Lowest point on the shore that is not covered by water at low tide. Compare *high-tide line*.

low-tide terrace. Flat section of the foreshore that lies seaward of any scarp and on which most wave energy is dissipated.

low-tide zone. Zone of the shore that is mostly covered with water and lies between the highest (neap) low-tide line and the lowest (spring) low-tide line. Compare *high-tide zone* and *middle-tide zone*.

lunar month. Interval between successive times when the moon and sun are both directly overhead at a specific line of longitude at a specific time of day. The time interval between two successive full moons (or new moons), approximately 29½ days.

macroalgae (sing. **macroalga**). Algae that have massive forms, easily seen by the naked eye. Macroalgae are generally attached to the substrate. Compare *microalgae*.

magma. Molten rock.

magmatic arc. Line of volcanic islands formed on the nonsubducting plate parallel to and near an oceanic convergent plate boundary.

magnetic anomaly. Local variation of the Earth's magnetic field caused by variable magnetization of minerals in the Earth's crust.

manganese nodule. Lump of hydrogenous mineral consisting primarily of oxides of iron and manganese. Manganese nodules are scattered in groups over some parts of the ocean floor.

mangrove. Group of tropical plant species that grow in low marshy areas at latitudes below about 30°. Mangroves have extensive root systems and produce much organic detritus to create a unique coastal environment for marine life.

mantle. 1. The layer of the Earth between the core and crust. 2. In certain mollusks, including clams, mussels, and oysters, the part of the animal's body that secretes shell materials.

marginal sea. Semi-enclosed body of water adjacent to a continent.

marine mammals. Members of the class Mammalia, subphylum Vertebrata, phylum Chordata, that live some or all of their life in the ocean. Warm-blooded animals that have mammary glands and hair, and that bear live young. See Appendix 3.

maximum sustainable yield. Maximum quantity of fish that can be harvested annually while still allowing the population to be sustained by reproduction. See **CC16**.

meander (adj. **meandering**). Sinuous curve or turn in a current (of any fluid such as air, river water, ocean water, or magma).

meroplankton. Planktonic larval forms of organisms that become members of the benthos or nekton when they become adult. Compare *holoplankton*.

mesoplates. Hypothesized segments of the Earth's mid-depth mantle that may move relative to each other below the crustal tectonic plates. Three major mesoplates have been proposed: Hawaiian (mostly beneath the oceanic plates of the Pacific), Tristan (beneath most plates of the Atlantic and Indian Oceans), and Icelandic (beneath Eurasia, the northernmost Atlantic Ocean, and the Arctic Ocean).

metamorphosis (v. **metamorphose**). 1. Change of form of an organism, usually as it passes from one life stage to another, similar to the change from caterpillar to butterfly. 2. Change in the mineral composition of rocks after their initial formation.

methane hydrates. Ice-like solids formed in sediments or sedimentary rock layers by the trapping of methane gas in the crystalline lattice of water at high pressures and low temperatures. At atmospheric pressure, they are unstable and release their methane.

microalgae (sing. **microalga**). Algae that are sufficiently small that they cannot be seen easily by the naked eye unless present in high concentrations. Most microalgae are single-celled, and they may be benthic or planktonic. Compare *macroalgae*.

micronutrient. An element or compound, including certain trace metals and vitamins, that is essential to the life processes of some species of organisms and that is present in seawater at concentrations substantially lower than those of nitrogen and phosphorus compounds.

middle-tide zone. Zone between the high-tide zone and the low-tide zone that is usually covered by water but is exposed to the atmosphere during all or most low tides.

mixed layer. Surface layer of ocean water that is mixed by wave and tide motions. As a result of the mixing, this layer has relatively uniform temperature and salinity.

mixed tide. Tide that has two high and two low tides each tidal day, with the two highs and/or the two lows being markedly different in height. Compare *diurnal tide* and *semidiurnal tide*.

mollusks. Members of a phylum (Mollusca) of soft unsegmented animals that usually are protected by a calcareous shell and use a muscular foot for locomotion. Mollusks include snails, mussels, clams, chitons, and octopi. See Appendix 3.

monsoon (adj. **monsoonal**). Seasonally reversing winds, especially those in the Indian Ocean and southern Asia that blow from the southwest during summer and from the northeast during winter. The term is derived from the Arabic word for season, *mausim*.

mutagens (adj. **mutagenic**). Chemical compounds that can cause mutations in organisms that are exposed to them. Compare *carcinogens* and *teratogens*.

neap tide. Tide that has the smallest range within a lunar month. Neap tides occur twice during the month, when the moon is at its first and third quarters. Compare *spring tide*.

nebula (pl. **nebulae**). Cloud of interstellar gas and dust, often illuminated by stars.

nekton. Pelagic animals that are active swimmers and thus can overcome currents and determine their position in the ecosystem. Nekton include fishes, marine mammals, and squid.

niche. Range of environmental characteristics within which a particular species can survive and reproduce. The term is often used to include the function of the organism itself in the ecosystem.

node. Point on a standing wave where there is no (or minimal) vertical motion. Compare *antinode*.

nonindigenous species. Species that is imported, accidentally or deliberately, to an ecosystem in which it is not present naturally.

nonpoint source. Source of pollution other than discharge pipes of industry and sewage treatment plants. Nonpoint sources include storm drains, street runoff, runoff from agricultural land, deposition of air pollutants, acid rain, and many other widely dispersed sources.

nudibranchs. Sea slugs. Members of the class Gastropoda, phylum Mollusca, that have no protective covering as adults. See Appendix 3.

nutrient. Any organic or inorganic compound that is used by plants in primary production. The most important nutrients include nitrogen and phosphorus compounds.

nutrient-limited. Pertaining to the condition in which the rate of growth of an organism or population of organisms, generally the production of organic matter by a photosynthetic organism or populations of organisms, is inhibited by the absence of a sufficient supply of one nutrient element (often nitrogen or phosphorus) when all other requirements for an increase in the rate of growth or production are met.

oceanic convergent plate boundary. Region where two lithospheric plates converge, each of which has oceanic crust at its margin. Compare *oceanic ridge*.

oceanic plateau. Small area where the seafloor is raised a kilometer or more above the surrounding oceanic crust. Oceanic plateaus are extinct volcanoes, old oceanic ridges, or fragments of continents.

oceanic ridge. Linear undersea mountain range that marks a tectonic plate boundary where two lithospheric plates diverge. Oceanic ridges extend through all the major oceans. Compare *oceanic convergent plate boundary*.

omnivores (adj. **omnivorous**). Animals that feed on both plants and other animals. Compare *carnivores* and *herbivores*.

ooze. Sediment that contains at least 30% skeletal remains of marine organisms. Ooze may be siliceous or calcareous, and it may be diatom ooze, foraminiferal ooze, radiolarian ooze, or pteropod ooze, depending on the organisms that are the major contributors to the sediment.

orbital velocity. For an orbiting object (or particle within a wave), the speed of movement (distance covered per unit time) in the orbital path. See **CC12**.

osmoregulation. Ability of an organism to adjust its internal salt concentration independently of the external salinity.

osmosis. Passage of water molecules from a solution of lower solute concentration into a solution of higher solute concentration through a semipermeable membrane that separates the two solutions.

osmotic pressure. Measure of the tendency for osmosis to occur. The pressure that must be applied to a solution to prevent water molecules from passing into it through a semipermeable membrane that has pure water on the other side.

outfall. Pipeline that extends from the shore across the seafloor. An outfall carries liquid waste some distance offshore, where it is discharged from the submerged end of the pipe.

overfishing (adj. **overfished**). Harvesting a fish species at a rate exceeding the maximum harvest that would still allow a stable population to be maintained by reproduction.

oviparous. Pertaining to animals that release eggs that develop and hatch outside the body. Compare *ovoviviparous* and *viviparous*.

ovoviviparous. Pertaining to animals that incubate eggs inside the reproductive tract (usually the mother's) until they hatch. Compare *oviparous* and *viviparous*.

oxygen demand (adj. **oxygen-demanding**). Quantity of oxygen that is needed to decompose organic matter or oxidize chemicals in an ecosystem. Waste materials often increase the natural oxygen demand.

ozone. Form of oxygen in which the molecule has three oxygen atoms.

ozone layer. Region of the atmosphere between about 15 and 30 km altitude in which there is a natural high concentration of ozone. Ozone in this layer absorbs much of the ultraviolet radiation from the sun.

PAH. See *polyaromatic hydrocarbons*.

paleomagnetism (adj. **paleomagnetic**). The record of the past orientation of the Earth's magnetic field incorporated in rocks during their formation.

parasites. Organisms that take their food and nutrients from the tissues of another organism. Parasites benefit from the host, but the host is disadvantaged.

parasitism. Symbiotic relationship in which the parasite harms the host from which it takes its nutrition.

partial tides. Harmonic components comprising the tide at any location. The periods of the partial tides are derived from the various components of the periodic motions of the Earth, sun, and moon in relation to one another.

partially mixed estuary. Estuary in which a distinct low-salinity layer moves seaward over a distinct higher-salinity layer that moves landward, but with a vertical gradation of salinity between the layers due to substantial vertical mixing. Compare *salt wedge estuary* and *well-mixed estuary*.

passive margin. A continental margin that is not significantly deformed by tectonic processes, because the margin is located away from the edge of a lithospheric plate.

pathogen (adj. **pathogenic**). Any microscopic organism that causes disease.

PCBs. Polychlorinated biphenyls, a group of industrial chemicals with a variety of uses. PCBS are toxic and mutagenic, they are not readily biodegraded, and they may be biomagnified.

pelagic. Pertaining to the open-ocean water environment.

petroleum hydrocarbons. Organic compounds that are present in petroleum and that consist predominantly of carbon and hydrogen. Extremely large numbers of different compounds are present in any petroleum sample.

pH. Measure of acidity or alkalinity. pH is measured on a logarithmic scale of 1 to 14 in which lower values indicate higher hydrogen ion concentration and therefore higher acidity.

phosphorite nodule. Lump composed primarily of phosphate minerals. Phosphorite nodules are scattered throughout certain parts of the ocean floor.

photic zone. Upper part of the ocean in which solar radiation is of sufficient intensity to enable photosynthesis to occur. Compare *aphotic zone*.

photosynthesis (v. **photosynthesize,** adj. **photosynthetic**). Production of carbohydrate from carbon dioxide and water (with the release of oxygen) in the presence of chlorophyll by the use of light energy (see **CC14**), as is done by plants. Compare *chemosynthesis*.

phytoplankton. Plankton that photosynthesize. Compare *zooplankton*.

pinnipeds. Members of a suborder (Pinnipedia) of marine mammals that includes sea lions, seals, and walrus. See Appendix 3.

planetary vorticity. Rate of rotation of a fluid (ocean water or atmospheric air) due to the rotation of the Earth.

plankton (adj. **planktonic**). Organisms that drift passively or swim weakly and are dependent on currents to determine their location. Most plankton are microscopic forms.

plate tectonics (adj. **plate tectonic**). Initially a theory that the lithosphere is divided into plates that are moving relative to each other across the Earth's surface. Now the term refers to the processes affecting plate motions and the effects of these motions.

pollution (v. **pollute**). Addition of substances to or alteration of the ocean ecosystem in a way that is deleterious to the ocean ecosystem or its resources. Compare *contaminant*.

polyaromatic hydrocarbons (PAHs). Organic compounds present in organisms and petroleum, and created by the burning of fossil fuels that have carbon atoms arranged in ring structures. PAHs include the more toxic and carcinogenic compounds in petroleum.

polyp. Single individual of a cnidarian colony or a solitary attached cnidarian that has a central mouth fringed with many small tentacles.

pore water. Solution present between the mineral grains of a sediment or rock.

potential energy. Energy that is the result of the relative position of an object, such as the energy of a compressed coil spring or of an object that is placed at the top of a slope. In each case, the potential energy can be released and converted to kinetic energy by releasing the object.

pressure gradient. Pressure (P) variation on a horizontal surface. A gradient can be straight or curved. The steepness of the gradient is measured as difference in pressure per unit distance within the gradient (e.g., $\delta P \cdot km^{-1}$).

primary production. The process of synthesis of organic material by photosynthetic or chemosynthetic autotrophs. The term is sometimes used to mean primary productivity.

primary productivity. Rate of production of organic matter by autotrophs, measured as the quantity (usually mass) of organic matter synthesized by organisms from inorganic substances within a given volume of water or other habitat in a unit of time.

progressive wave. Wave in which the waveform progressively moves. Compare *standing wave*.

prokaryotes. Organisms (single-celled or multicelled) whose cells are not surrounded by a membrane and that do not have a discrete nucleus and other subcellular compartments. Archaea and bacteria are prokaryotes. Compare *eukaryotes*.

protists. Members of the kingdom Protista. Protists are those eukaryotes that are not animals, plants, or fungi.

protozoa (s. **protozoan**). Single-celled animals that have a nucleus confined within a membrane. See Appendix 3.

pteropods. Members of an order of pelagic animals of the class Gastropoda, phylum Mollusca, in which the foot is modified for swimming and the shell may be absent. See Appendix 3.

pycnocline. Depth range in the water column in which density changes rapidly in the vertical dimension. Compare *halocline* and *thermocline*.

radioactive (n. **radioactivity**). Pertaining to a property of certain elements, or isotopes of an element, whose atomic nuclei are unstable and can spontaneously disintegrate and emit ionizing radiation.

radioisotope. Radioactive isotope of an element that may also have nonradioactive isotopes.

radiolaria (sing. **radiolarian;** adj. **radiolarian**). Planktonic and benthic protozoa of the phylum Sarcodina that are protected by shells that are usually siliceous. See Appendix 3.

radionuclides. Nuclei of radioactive atoms.

reef. Rocky elevation of the seafloor whose upper part is at depths of less than about 20 m. The term is often restricted to such areas that pose a hazard to navigation.

reef flat. Platform of coral fragments and sand that is relatively exposed at low tide. Also called "reef terrace."

reef terrace. See *reef flat*.

refraction (v. **refract;** adj. **refracted**). Process by which part of a wave is slowed, causing the wave to bend as it passes from one zone to another in which it travels at a different speed. Refraction occurs as a water wave enters shallow water, or as a sound wave or light wave crosses an interface between two fluids or crosses a thermocline or halocline.

relict sediment. Sediment that was deposited under a set of environmental conditions that have since changed, but has not been buried by more recent sediment.

reptiles. Species of the class Reptilia, subphylum Vertebrata, phylum Chordata. Reptiles breathe air and are cold-blooded. The few marine species include turtles and sea snakes. See Appendix 3.

residence time. Average length of time that a particle of any substance spends in a defined part of the ocean. See **CC8**.

residual current. Current that remains after the reversing tidal current components have been subtracted. Residual current is indicative of the mean drift after multiple tidal cycles.

respiration (v. **respire**). Process by which organisms use organic materials (food) as a source of energy. Respiration normally uses oxygen and produces carbon dioxide.

restoring force. Force that tends to restore a disturbed ocean surface to a flat configuration.

resuspend (adj. **resuspended**). To lift particles off the sediment surface (by currents) or the ground (by winds) on which they have been temporarily deposited. Particles become suspended sediment and airborne dust in water and air, respectively.

rift zone. Zone where the Earth's crust is being torn apart. Rift zones are often elongated and may be locations where tectonic plates are separating.

rip current. Fast, narrow surface or near-surface current that flows seaward through the breaker zone at nearly right angles to the shore. Rip currents are the seaward return flow of the water piled up on the shore by incoming waves.

rocky intertidal zone. Zone of a rocky coastline between the high- and low-tide lines.

runoff. Freshwater that is returned to the ocean or to a river after falling on the land as rain or snow.

salinity. Measure of the quantity of dissolved salts in ocean water. Salinity is defined in terms of the conductivity of the water relative to the conductivity of a defined salt solution. Salinity has no units but is approximately equal to the weight in grams of dissolved salts per kilogram of seawater.

salps. Members of a genus of pelagic tunicates (subphylum Vertebrata) that are cylindrical and transparent. See Appendix 3.

salt marsh. Relatively flat area of the shore where fine sediment is deposited and salt-tolerant grasses grow.

salt wedge estuary. Estuary, normally deep, that has a large volume of freshwater flow separated by a sharp halocline from a lower wedge-shaped layer of seawater that moves landward. Compare *partially mixed estuary* and *well-mixed estuary*.

sand dune. Rounded mound or hill of sand on the backshore formed by accumulations of windblown sand. Sand dunes may have rooted vegetation.

saturation pressure. Maximum amount of water that can remain in the vapor phase in air at a particular pressure and temperature expressed as a partial pressure of water vapor (the pressure if only the water vapor were present).

saturation solubility. Concentration of a dissolved substance when no more of the substance can be dissolved in the solvent.

scarp. Linear, steep or nearly vertical topographic feature that separates areas of gently sloping or flat surfaces. Compare *berm*.

scatter (adj. **scattered**). Reflection of light or sound in random directions, generally by particles suspended in a fluid. Compare *backscatter*.

scavengers (v. **scavenge**). Animals that feed on dead organisms.

school (adj. **schooling**). Aggregation of fishes, squid, or crustaceans that is organized to remain together as the organisms move.

scuba. Abbreviation for "self-contained underwater breathing apparatus," the means by which humans descend beneath the sea surface carrying their own source of air to breathe.

sea cucumbers. Members of the class Holothuroidea, phylum Echinodermata. See Appendix 3.

sea fans. Corals whose colonies grow out from a point of attachment to the substrate in a form resembling an intricate fan. Sea fans belong to the order Alcyonacea. See Appendix 3.

sea grasses. Any of several species of rooted plants of the class Angiosperma that grow predominantly in marshes and shallow-water lagoons. See Appendix 3.

sea stars. Animals of the class Anthozoa, phylum Cnidaria, that have a number of radial arms, on the underside of which are numerous tube feet to provide locomotion. The mouth is on the underside at the center. Many sea stars prey on mollusks. See Appendix 3.

sea urchins. Animals belonging to the class Echinoidea, phylum Echinodermata. Sea urchins have a fused test (external calcified covering) and well-developed spines. See Appendix 3.

seafloor spreading. Process producing oceanic crust by upwelling of magma along the axis of the oceanic ridges.

seamount. Individual peak of seafloor topography that rises more than 1000 m above the ocean floor.

seawall. Wall built parallel to the shore to protect the shore from erosion by waves.

sediment. Particles of organic or inorganic origin that accumulate in loose form.

sedimentary arc. Chain of low sedimentary islands formed at some oceanic convergent plate boundaries between the boundary and the magmatic arc.

sedimentation. Accumulation of particles on the seafloor by progressive deposition of particles from the suspended sediment to form sediments.

sedimentation rate. Rate of accumulation of sediments, normally measured in millimeters of sediment thickness per thousand years in the open ocean.

seismograph. Instrument that detects and records earthquake waves that have traveled through the Earth from an earthquake focus.

semidiurnal tide. Tide with two high and two low tides each tidal day, with the two highs and the two lows being equal or almost equal to each other in height. Compare *diurnal tide* and *mixed tide*.

sensible heat. Heat that, when added to or removed from a substance, changes the temperature of that substance.

sewage sludge. Solids or slurry remaining after sewage wastewater has been treated. Sewage sludge contains pathogens, trace metals, nutrients, and other contaminants.

shallow-water wave. Wave whose wavelength is at least 20 times the depth of water beneath it. Compare *deep-water wave* and *intermediate wave*.

shear stress. Resistance (friction) that develops at the interface between two fluids that are moving in relation to each other.

shelf break. Depth at which the gradual seaward slope of the continental shelf steepens appreciably, defining the boundary between the continental shelf and the continental rise.

shelf valley. Valley in the seafloor topography that cuts across the continental shelf. Shelf valleys are usually "drowned river valleys."

shellfish. Animals of the phylum Crustacea or Mollusca that have hard outer shells. The term generally is applied to species that are valuable for human consumption.

shoal. Shallow place in a body of water that presents a hazard to navigation. Shoals are often in the form of a sandbank or sandbar whose surface may be exposed when the tide is low.

shore. Zone between the highest level of wave action during storms and the lowest low-tide line.

shoreline. Line that marks the intersection of the water surface with the shore. The shoreline migrates up and down as the tide rises and falls. Compare *coastline*.

siliceous. Pertaining to material containing abundant silica.

sill. Submarine ridge that separates the deeper parts of two adjacent ocean basins. Sills are often present at the mouths of fjords and other coastal embayments or at the entrances to marginal seas.

slack water. The time when the current speed is zero as a reversing tidal current changes direction.

slump. Collapse of a mass of earth or sediments from the sides of sloping topography, and movement of this mass downward on the slope.

soft corals. Species of corals that do not secrete massive calcareous skeletons and do not have zooxanthellae.

solar wind. A stream of charged subatomic particles (mainly protons and electrons) that flows outward from the sun and other stars.

solstice. Time of year when the sun is directly over one of the tropics. Solstices occur on June 20 or 21 when the sun is over the Tropic of Cancer in the Northern Hemisphere and on December 21 or 22 when the sun is over the Tropic of Capricorn in the Southern Hemisphere. Compare *equinox*.

solubility (adj. **soluble**). Ability of a substance to be dissolved in a liquid. The term is generally used to mean saturation solubility.

sonar. Abbreviation of "sound navigation and ranging," a method by which sound pulses can be used to measure the distance to objects in the ocean.

sorted. Describing the range of grain sizes in a sediment. Well-sorted sediment consists of grains that have a restricted range of grain sizes.

sounding. Measuring the depth of water beneath a ship.

spatial. Pertaining to the location of points in three-dimensional space. The term is used in the same way that *temporal* is used with respect to points in time.

spawn. To produce eggs, which may be laid in one location or dispersed in the water.

species. Population of organisms whose members interbreed under natural conditions and produce fertile offspring and that is reproductively isolated from other such groups. See Appendix 3.

species succession. Sequence of dominance by different species of phytoplankton during seasonal changes, especially in mid-latitude marine ecosystems.

spit. Narrow strip of land, commonly consisting of sand deposited by longshore currents, that has one end attached to the mainland and the other terminating in open water.

sponges. Animals of the phylum Porifera, most of which are microscopic but build massive colonies. Sponges are filter feeders. See Appendix 3.

spore. Small reproductive body that is highly resistant to decomposition but is capable of growing and metamorphosing to produce an adult form either immediately or after a prolonged interval of dormancy.

spreading cycle. Period during which the continental crust on the Earth's surface is broken into a number of pieces that move apart on their lithospheric plates. Spreading cycles are preceded and followed by a period when the continents are brought together.

spring tide. Tide that has the greatest range within a lunar month. Spring tides occur twice during the month, when the moon is new and when it is full. Compare *neap tide*.

standing stock. Biomass of a population present at any given time.

standing wave. Waveform that oscillates vertically without progressive movement. Compare *progressive wave*.

steady state. Condition of equilibrium in a system in which the inputs of a substance or energy are equal to the outputs and the distribution of the substance or energy within the system does not change with time.

storm surge. Temporary rise of sea level above its normal height as a result of wind stress and reduced atmospheric pressure during storms.

stratification (adj. **stratified)**. Layering of fluids according to density. Stratification is stable when density decreases continuously (but not necessarily uniformly) with distance from the Earth's center.

subduction (adj. **subducted)**. Process by which one lithospheric plate descends beneath another.

subduction zone. Area in which a lithospheric plate is descending into the asthenosphere.

sublittoral zone. Benthic environment extending from the low-tide line to a depth of 200 m.

submarine canyon. Steep, *V*-shaped canyon cut into the continental shelf or slope.

submersible. Undersea vehicle with an enclosure that has an atmosphere in which human passengers can be transported.

subtropical gyre. Circular current system centered in the subtropical high-pressure region of each major ocean basin, driven by the trade winds and westerly winds. The subtropical gyres rotate clockwise in the Northern Hemisphere and counterclockwise in the Southern Hemisphere.

supernova (pl. **supernovae)**. Violent explosion of a star that is many times more massive than our sun. The exploding star may become temporarily extremely bright. Matter is thrown off into space at high velocity and high energy during the explosion, and the star collapses to become either a neutron star or a black hole.

superplume. Massive upwelling of warmer mantle material that arises from extended areas near the Earth's core. There are thought to be two such superplumes on the present-day Earth. They may each feed numerous hot spots.

supersaturated. Describing an unstable condition in which the concentration of a dissolved substance (or of water vapor in air) is greater than the saturation solubility. The substance (or water vapor) may not precipitate unless provided with nuclei (such as suspended particles or dust) on which to deposit.

supralittoral zone. Splash or spray zone above the spring high-tide shoreline.

surf. Turbulent foam produced by breaking waves. The term is also used to mean "surf zone."

surf zone. Region between the shoreline and offshore in which most wave energy is released by breaking waves.

surface microlayer. Thin (about 0.1 mm thick) layer that covers the entire ocean surface and has different properties than the underlying water has. Often called a "slick."

surface tension. Tendency for the surface of a liquid to contract because of the attractive forces between its molecules.

survival niche. Range of environmental variables within which the individuals of a species can survive. Compare *fundamental niche*.

suspended sediment. Small solid particles that sink slowly or are maintained in suspension and distributed in the water column by turbulence.

suspension feeders (adj. **suspension-feeding)**. Animals that feed by capturing or filtering suspended particles from the water column. Compare *filter feeders*.

swash. Water that washes up over exposed beach as waves break at the shore.

swell. Smoothly undulating ocean wave that is the result of wave dispersion and that is transported with little energy loss across great stretches of ocean.

swim bladder. Gas-containing flexible organ in many fish species that aids in attaining neutral buoyancy.

symbiosis (adj. **symbiotic)**. Association between two species in which one or both benefit. A species in such an association that does not benefit may be harmed or may be unaffected by the association.

synoptic. Describing measurements made simultaneously at many different locations.

tectonic estuary. Estuary whose origin is related to tectonic deformation of the coastal region.

teratogens (adj. **teratogenic)**. Chemicals that cause mutations in the offspring of organisms that are exposed to them. Compare *carcinogens* and *mutagens*.

terrigenous. Pertaining to material derived from the land.

territorial sea. Strip of ocean 12 nautical miles wide adjacent to land within which the coastal nation has control over the passage of ships.

thermocline. Depth range in the water column in which temperature changes rapidly in the vertical dimension. Compare *halocline* and *pycnocline*.

thermohaline circulation. Vertical movements of ocean water masses caused by density differences that are due to variations in temperature and salinity.

tidal range. Height difference between high and low tides. The term can apply to the tides of a single day or to the highest high and lowest low of a specified period, such as a month.

tide. Periodic rise and fall of the ocean surface caused by differences in the gravitational attraction due to the moon and sun on different parts of the Earth.

tide pool. Depression in the intertidal zone that remains wet or filled with water when the tide recedes below its level on the shore.

tomography (adj. **tomographic)**. Method of mapping the internal parts of the Earth, ocean, or other body by observing waves transmitted through the body as they arrive at various points on its surface. The resulting data are used to produce a three-dimensional view of the internal variations of wave transmission velocity.

topography (adj. **topographic)**. Shapes, patterns, and physical configuration of the surface of the land or seafloor, including its relief (local differences in elevation or depth) and the positions of natural and human-made features.

tracer. Chemical constituent that has variable concentrations in seawater that can be used to identify and follow the movement of water masses.

trade wind. Air mass that moves from subtropical high pressure belts toward the equator. Trade winds are northeasterly in the Northern Hemisphere and southeasterly in the Southern Hemisphere. Compare *westerly*.

transform fault. Fault that offsets the boundary between the edges of two lithospheric plates.

transform plate boundary. Boundary between two tectonic plates where the motion is such that two plate edges are sliding past each other.

trench. Long, narrow, and deep depression on the ocean floor that has relatively steep sides.

trophic efficiency. See *food chain efficiency*.

trophic level. With primary producers as the first level, the number of steps in a food chain to a particular organism. Organisms that eat primary producers are at the second level, organisms that eat second-level organisms are at the third level, and so on.

troposphere. Zone of the atmosphere between the Earth's surface and an altitude of about 12 km.

trough. The part of an ocean wave that is displaced below the still-water line. Compare *crest*.

tsunami. Long-period gravity wave generated by a submarine earthquake or volcanic event.

tube worms. Many species of worms that live within a rigid tube that they secrete, drill into the substrate, or construct with shell fragments and sand. Tube worms are usually filter feeders that extend feeding tentacles from the tube into the water column to capture suspended particles.

tuned. Pertaining to a wave oscillation whose wavelength is such that successive wave crests enter a basin or enclosure at the same time that the crest of a preceding wave returns.

tunicates. Any of various species of the subphylum Urochordata, phylum Chordata. Tunicates include both benthic (sea squirts) and pelagic (salps) species. See Appendix 3.

turbidite. Sediment deposit formed by turbidity currents. Turbidites have vertically graded bedding.

turbidity (adj. **turbid**). Reduction of the clarity of a fluid caused by the presence of suspended matter.

turbidity current. Episodic fast current that resuspends sediments and carries them down a submarine slope. Turbidity currents may be initiated by a sudden force such as an earthquake.

turbulence (adj. **turbulent**). Flow of a fluid in which random velocity fluctuations distort and confuse the flow lines of individual molecules.

upwelling (adj. **upwelled**). Vertical upward movement of a fluid due to density differences; or where two fluid masses converge, displacing fluid upward; or where an upper layer diverges, causing fluid from below to rise into the divergence. In the ocean, the term refers to coastal upwelling, where Ekman transport causes surface waters to move away from the coast and deeper (often cold and nutrient-rich) water to be brought to the surface; or to upwelling caused by winds that move surface water masses away from each other, creating a divergence. Compare *downwelling*.

van der Waals force. Weak force between atoms or molecules caused by the slight polarities of atoms and molecules that are the result of small variations in the configurations of the electron clouds surrounding the atom or molecule.

vertebrates. Animals of the subphylum Vertebrata, phylum Chordata. Vertebrates include species that have a well-developed brain and a skeleton of bone or cartilage. Examples are fishes, amphibians, reptiles, birds, and mammals (including human beings). See Appendix 3.

virus (adj. **viral**). Infective agent that can cause disease and can multiply when associated with living cells. Viruses are complex proteins that are not regarded as living organisms.

viscosity (adj. **viscous**). Property of a substance that causes it to offer resistance to flow. Internal friction.

viviparous. Pertaining to animals that give birth to living young. Compare *oviparous* and *ovoviviparous*.

water mass. Body of water identifiable and distinguishable from other water bodies by its characteristic temperature, salinity, or chemical content.

wave dispersion. Separation of ocean waves by wavelength as they travel away from their point of origin. Longer waves travel faster.

wave height. Vertical distance between a crest and the preceding trough.

wave interference. Combination of two or more simple waves of different periods or traveling in different directions to produce complex waveforms.

wave period. Time that elapses between the passage of two successive wave crests past a fixed point.

wave ray. Path across the sea surface that is followed by a point on a wave front as the wave travels and is refracted and reflected.

wave speed. Speed at which the waveform of a progressive wave travels.

wave steepness. Ratio of wave height to wavelength.

wavelength. Horizontal distance between corresponding points on successive waves, such as from crest to crest.

weather. Temperature, humidity, cloud cover, wind velocity, and other atmospheric conditions, and their variations at a specific location on a given day.

weathering (adj. **weathered**). Process by which rocks are broken down by chemical and mechanical (winds, ice formation, etc.) means.

well-mixed estuary. Estuary in which vertical mixing by tides or wind is such that there is no vertical stratification. Salinity in well-mixed estuaries increases progressively toward the ocean. Compare *partially mixed estuary* and *salt wedge estuary*.

westerly. Air mass that moves away from a subtropical high-pressure belt toward higher latitudes. Westerlies blow from the southwest in the North Hemisphere and from the northwest in the Southern Hemisphere. Compare *trade wind*.

western boundary current. Poleward-flowing warm surface layer current that flows on the western side of subtropical gyres. Western boundary currents are fast, narrow, and deep. Compare *eastern boundary current*.

wetlands. Low-lying flat areas that are covered by water or have water-saturated soils for at least part of the year.

windward. The direction from which the wind is blowing. The term usually applies to the exposed side of a landmass or barrier facing the oncoming wind. Compare *leeward*.

zooplankton. Animal plankton. Compare *phytoplankton*.

zooxanthellae (sing. **zooxanthella**). Forms of algae that live symbiotically in the tissue of corals and other animals and provide some of the host animal's food supply by photosynthesis.

Credits

Front End Paper: U.S. Naval Oceanographic Office/ National Imagery and Mapping Agency, U.S. Department of Defense; Digital Image by Dr. Peter W. Sloss, NOAA/NGDC.

Table of Contents: p. i: SeaWiFS/NASA; **p. ii:** *"HMS"Challenger made fast to St. Paul's Rocks,* oil on canvas by Adrianna Cantillo; **p. iii:** © D. Segar and E. Stamman Segar; **p. iv:** © D. Segar and E. Stamman Segar; **p. v:** Jim Sugar/Corbis; **p. vi:** © D. Segar and E. Stamman Segar; **p. vii:** Guy Motil/Corbis; **p. viii:** © D. Segar and E. Stamman Segar; **p. viiii:** World Perspectives/Getty Images; **p. x:** JSC/NASA; **p. ix:** © D. Segar and E. Stamman Segar; **p. xi:** © D. Segar and E. Stamman Segar; **p. xii:** © D. Segar and E. Stamman Segar; **p. xiii:** © D. Segar and E. Stamman Segar; **p. xiiii:** © D. Segar and E. Stamman Segar; **p. xv:** © D. Segar and E. Stamman Segar; **p. xvi:** Scanpix/AP Photo; **p. xvii:** © D. Segar and E. Stamman Segar.

Chapter Opener 1: SeaWiFS/NASA; **fig. 1.2:** Anglo-Austrailan Observatory; **fig. 1.4:** J. Hester and P. Scowen/ NASA.

Chapter Opener 2: *HMS Challenger made fast to St. Paul's Rocks,* oil on canvas by Adrianna Cantillo; **p. 22–23:** © D. Segar and E. Stamman Segar; **fig. 2.1:** © The Granger Collection; **fig. 2.2:** © D. Segar and E. Stamman Segar; **fig. 2.3:** © Douglas Peebles Photography/Alamy; **fig. 2.5:** Central Library, National Oceanic and Atmospheric Administration/ Department of Commerce; **fig. 2.7a:** Bettmann/Corbis.

Chapter Opener 3: (top left) © D. Segar and E. Stamman Segar; (bottom left) © D. Segar and E. Stamman Segar; (right) © D. Segar and E. Stamman Segar; **p. 38–39:** © D. Segar and E. Stamman Segar; **fig. 3.3c:** Courtesy of NOAA; **fig. 3.4:** Courtesy of NOAA, National Ocean Service; **fig. 3.5:** NASA; **fig. 3.7c:** Meridata Finland Ltd. (T. Reiner, 2005; Courtesy of EnCana and M. Enachescu); **fig. 3.7d:** (T. Reiner, 2005; Courtesy of EnCana and M. Enachescu); **fig. 3.10:** Courtesy of General Oceanics; **fig. 3.11:** Adapted from A.C. Duxbury and A. Duxbury, *The World's Oceans,* Addison-Wesley, Reading, MA; **fig. 3.12:** National Oceanography Centre UK; **fig. 3.19:** D. Cavagnaro/Visuals Unlimited; **fig. 3.20:** © 1994 NOAA, NURC, University of North Carolina Wilmington; **fig. 3.21:** © Gregory S. Boland, GB Photographic Services; **fig. 3.22b:** Woods Hole Oceanoraphic Intitution; **fig. 3.23:** Courtesy of Juergen Fischer; IFM-GEOMAR, Kiel.

Critical Concepts Opener: (top center) John Space Center/NASA; (bottom left) provided by Gene Carl Feldman, NASA, Goddard Space Flight Center; (bottom middle) NOAA. **fig. CC4.1:** Adapted from A. Sundorg, 1956, *Geografiska Annaler,* 38:135–316; **fig. CC9.1:** Adapted from R.C. Scott, *Introduction to Physical Geography,* West Publishing, St Paul/Minneapolis, 1996; **fig. CC10.3:** Adapted from James Gleick, 1987, *Chaos: Making a New Science,* Viking Penguin, New York, NY; **fig. CC10.4:** From James Gleick, 1987, *Chaos: Making a New Science,* Viking Penguin, New York, NY.

Chapter Opener 4: (all four images) © D. Segar and E. Stamman Segar; **p. 68–69:** © D. Segar and E. Stamman Segar; **fig. 4.3:** Adapted from E. J. Tarbuck and F.K. Lugens, 1991, *Earth Science,* 6th Edition, Macmillan Publishing Company, Columbus, OH; **fig. 4.7:** Adapted from B. Isacks, J. Oliver and L.R. Sykes, 1968, *Journal of Geophysical Research,* 73:5855–900; M. Barazangi and J. Dorman. 1969. *Seismological Society of America Bulletin* 59:369–380; **fig. 4.8:** Adapted from B. W. Pipkin, 1994, *Geology and the Environment,* West Publishing Company, St. Paul, MN; **fig. 4.9:** From T.H. Van Andel, 1985, *New Views of an Old Planet,* permission of Cambridge University Press; **fig. 4.12a:** David Weintraub/Photo Researchers, Inc.; **fig. 4.12b:** Philip James Corwin/Corbis; **fig. 4.18:** Jakobsson, M., N. Cherkis, J. Woodward, R. Macnab, and B. Coakley, *New grid of Arctic bathymetry aids scientists and mapmakers,* EOS Transactions of the American Geophysical Union, v. 81, p. 89, 93, 96, 2000; **fig. 4.23:** © D. Segar and E. Stamman Segar; **fig. 4.25:** Adapted from J.T. Wilson, 1968, *American Physical Society Proceedings* 112:309–20; **fig. 4.26:** © Douglas Faulkner, Photo Researchers, Inc.; **fig. 4.27:** Adapted from P.R. Pinet, 1992, *Oceanography,* West Publishing Company, St. Paul, MN; **fig. 4.28:** Jean-Pierre Pieuchot/Getty Images; **fig. 4.29:** Adapted from T.H. Van Andel, 1985, *New Views of an Old Planet,* Cambridge University Press; **fig. 4.30:** Data from various sources including World Meteorological Organization, GARP Publication Series No 16, Geneva, and R.G. Fairbanks, 1990, *Paleooceanography* 5(6):937–48.

Chapter Opener 5: Jim Sugar/Corbis; **p. 104–105:** © D. Segar and E. Stamman Segar; **fig. 5.1:** © D. Segar and E. Stamman Segar; **fig. 5.2:** Adapted from J.R. Heirtzler, X.L. Le Pichon and J.C. Bason,1966, *Deep Sea Research,* 427–43; **fig. 5.3:** Adapted from E. Bullard, J.E. Everett, and J.G. Smith, 1965, *Philosophical Transactions of the Royal Society of London* 258; **fig. 5.4a:** Adapted from E.H. Colbert, 1973, *Wandering Lands and Animals,* 72; **fig. 5.4b:** From American Association of Petroleum Geologists, 1982, Theory of Continental Drift, Symposium, Tulsa, Oklahoma; R. Foster, 1983, General Geology, 4:251, Merrill Publishing Company; **fig. 5.4c,d:** From J.S. Monroe and R. Wicander, 1995, *Physical Geology,* West Publishing Company, St. Paul, MN; **fig. 5.5:** From T.H. Van Andel, 1985, *New Views of an Old Planet,* Cambridge University Press; **fig. 5.6b:** Courtesy of U.S. Geological Survey; **fig. 5.7:** Adapted from B. Isacks, J. Oliver and L.R. Sykes, 1968, *Journal of Geophysical Research,* 73:5855–900; M. Barazangi and J. Dorman, 1969, Seismological Society of America Bulletin 59:369–380;

fig. 5.8: Adapted from P.J. Wylie, 1976, The Way the Earth Works, John Wiley & Sons, NY; **fig. 5.10:** Gung, YC and B. Romanowicz (2004) *Q tomography of the upper mantle using three component long period waveforms, Geophys. J. Int.*, 157, 813–830; **fig. 5.14:** Reproduced with modifications from P.J. Wylie, The Way the Earth Works, 1976, with permission of John Wiley & Sons, NY; **fig. 5.15:** from A. Cox "Geomagnetic Reversals," *Science* 163:237 (1969), American Association for the Advancement of Science; **fig. 5.16:** from *Bedrock Geology of the World* by Larson and Pitman, copyright 1984 by W.H. Freeman and Company; **fig. 5.17:** NASA Goddard Space Flight Center.

Chapter Opener 6: © D. Segar and E. Stamman Segar; **p. 130–131:** © D. Segar and E. Stamman Segar; **fig. 6.6a:** Marco Garcia/Getty Images; **fig. 6.6b:** Provided by the SeaWiFS Project, NASA/GSFC, and Orbimage.

Chapter Opener 7: (top left) Craig Tuttle/Corbis; (bottom left) Frans Lanting/Corbis; (center) JSC/NASA; (bottom right) Guy Motil/Corbis; **p. 148–149:** © D. Segar and E. Stamman Segar; **fig. 7.6:** Adapted from R.B. Montgomery, Deep Sea Research 5 (1958): 134–48; **fig. 7.11:** Provided by Gene Carl Feldman, NASA/GSFC; **fig. 7.13:** Partially adapted from L.E. Kinsler and A.R. Frey, 1962, *Fundamentals of Acoustics,* John Wiley & Sons, NY; **fig. 7.15:** Data from U.S. Office of Naval Research.

Chapter Opener 8: © D. Segar and E. Stamman Segar; (bottom right) © D. Segar and E. Stamman Segar; **p. 168–169:** © D. Segar and E. Stamman Segar; **fig. 8.1:** Adapted from M.G. Gross and E. Gross, 1996, Oceanography, 7th Edition, Prentice Hall, NJ; **fig. 8.2:** Adapted from J.D. Milliman and R.H. Meade, 1983, Journal of Geology, 91:1–21; **fig. 8.3:** © Beth Davidow/Visuals Unlimited; **fig. 8.4:** NASA; **fig. 8.5:** Science Photo Library/Photo Researchers, Inc.; **fig. 8.6:** Steve Gschmeissner/Science Photo Library/Photo Researchers, Inc.; **fig. 8.7a:** © D. Segar and E. Stamman Segar; **fig. 8.7b:** Sinclair Stammers/Science Photo Library/Photo Researchers, Inc.; **fig. 8.8:** Eric Grave/Science Source/Photo Researchers, Inc.; **fig. 8.9:** Adapted from W.H. Berger, A.W.H. Be, and W. Sliter, 1975, "Dissolution of Deep-Sea carbonates: An Introduction," Cushman Foundation for Foraminiferal Research, Special Publication 13, Lawrence, KS; **fig. 8.10:** Adapted from H.V. Thurman, 1994, Introductory Oceanography, 7th Edition, Macmillan Publishing Company, NY; **fig. 8.11a:** Institute of Oceanographic Sciences/NERC/Science Photo Library/Photo Researchers, Inc.; **fig. 8.11b:** Charles D. Winters/Photo Researchers, Inc.; **fig. 8.11c:** Adapted from D.S. Cronan, 1977, "Deep Sea Nodules: Distribution and Geochemistry," in Marine Manganese Deposits, G.P. Glasby, ed., Elsevier Science Publishing, New York, NY; **fig. 8.12:** Adapted from R.M. Pratt, 1968, U.S. Geological Survey Professional Paper 529-B, 1–44, U.S. Government Printing Office, Washington, D.C.; **fig. 8.13:** IODP/TAMU-William Crawford; **fig. 8.14:** Adapted from B.C. Heezen and M. Ewing, 1952, American Journal of Science 250:949–73; **fig. 8.16:** Adapted from J. Kennet, 1982, Marine Geology, Prentice Hall Inc., Englewood Cliffs, NJ; **fig. 8.17e,f:** Adapted from K.O. Emery, 1969, Scientific American 221(3):106–22; **fig. 8.18a:** Adapted from T.A. Davies and D.S. Gorsline, 1976, "Oceanic Sediments and Sedimentary Processes, in Chemical Oceanography," J.P. Riley and R. Chester (eds.), vol.5, Academic Press: Orlando, FL; **fig. 8.18b:** Adapted from B.C. Heezen and C.D. Hollister, 1971, The Face of the Deep, Oxford University Press, New York, NY; **fig. 8.19:** IODP/TAMU-William Crawford.

Chapter Opener 9: (top left) © D. Segar and E. Stamman Segar; (right) NASA/Jet Propulsion Laboratory/TheUniversity of California at Los Angeles/The Canadian Atmospheric Environment Service; (bottom left) World Perspectives/Getty Images; **p. 198–199:** © D. Segar and E. Stamman Segar; **pp. 199–200:** *The Sea Around Us* by Rachel Carson, Copyright ©1950 by Rachel L. Carson, Copyright ©1972 by Paul Brooks, Used by permission of Frances Collin, Trustee u-w-o Rachel L. Carson; **fig. 9.2:** Courtesy Paul Newman/NASA; **fig. 9.13:** Adapted from H.U. Sverdrup, M.W. Johnson and R.H. Fleming, 1942, The Oceans, Prentice Hall, Englewood Cliffs, NJ; **fig. 9.14 a,b&c:** NOAA/PMEL; **fig. 9.15:** Adapted from H.U. Sverdrup, M.W. Johnson and R.H. Fleming, 1942, The Oceans, Prentice Hall, Englewood Cliffs, NJ; **fig. 9.16:** Data from H.U. Sverdrup, M.W. Johnson and R.H. Fleming, 1942, The Oceans, Prentice Hall, Englewood Cliffs, NJ and others; **fig. 9.18:** Adapted from C. Colin et al., 1971, Cah. Orstrom. Ser. Oceanogr. 9:167–86; **fig. 9.20:** Data from National Ocean and Atmospheric Administration; **fig. 9.21:** From C.C. Ebbesmeyer, et al., 1989, "Steps in the Pacific Climate: Forty Environmental Changes Between 1968–75," Proceeds 7th Annual Climate Workshop; **fig. 9.22:** Adapted from D.V. Bogdanov, 1963, Deep Sea Research, 10:520–23 and E.E. Enger and B.F. Smith, 1992, *Environmental Science, A Study of Interrelationships,* 4th Edition. Copyright 1992 Times Mirror Higher Education Group, Inc., Dubuque, IA. All rights reserved; **fig. 9.23:** Data from Smithsonian Physical Tables, 1994, and G.Wust et al. 1954; **fig. 9.24:** Data from National Oceanic and Atmospheric Administration, National Climate Data Center; **fig. 9.27:** NOAA; **fig. 9.29:** Adapted from R.C. Scott, 1996, Introduction to Physical Geography, West Publishing Company, St. Paul, MN.

Chapter Opener 10: (all four images) JSC/NASA; **p. 238–239:** © D. Segar and E. Stamman Segar; **pp. 239–40:** *The Sea Around Us* by Rachel Carson, Copyright ©1950 by Rachel L. Carson, Copyright ©1972 by Paul Brooks, Used by permission of Frances Collin, Trustee u-w-o Rachel L. Carson; **fig. 10.13d:** Adapted from J.A. Knauss, 1961, Scientific American, 214:105–19; **fig. 10.16:** Data from National Oceanic and Atmospheric Administration; **fig. 10.17:** Data from National Oceanic and Atmospheric Administration; **fig. 10.18:** Provided by Gene Carl Feldman/NASA/GSFC; **fig. 10.20:** Data from G. Neumann, 1968, Ocean Currents, Elsevier Science Publishing, New York, NY; **fig. 10.21a:** Courtesy of Mobil Oil Corp.; **fig. 10.23:** Adapted from P. Tchernia, 1980, Descriptive Regional Oceanography, Pergamon Press Ltd., Oxford; and G. Dietrich et al., 1980, "General Oceanography: An Introduction," John Wiley & Sons, NY; **fig. 10.24:** Adapted from P. Tchernia, 1980, Descriptive Regional Oceanography, Pergamon Press Ltd.,

Oxford; and G. Dietrich et al., 1980, "General Oceanography: An Introduction," John Wiley & Sons, NY; **fig. 10.25a:** Adapted from G. Wurst, 1961, Journal of Geophysical Research, 66:3261–71; **fig. 10.25b:** Adapted from G.L. Pickard and W.J. Emery, 1982, Descriptive Physical Oceanography, Pergamon Press Ltd., Oxford; **fig. 10.26:** Adapted from H. Stommel, 1958, Deep Sea Research, vol 5; **fig. 10.29:** Adapted from National Oceanic Administration/ University Consortium for Atmospheric Research data and other sources; **fig. 10.31:** With permission from H.G. Ostlund and C.G.H. Rooth, 1990, Journal of Geophysical Research, 95(C11):20,147–20,165; **fig. 10.32:** Adapted from Molinari, Robert L., Rana A. Fine, W. Douglas Wilson, Ruth G. Curry, Jeff Abell, and Michael S. McCartney, "The arrival of recently formed Labrador Sea Water in the Deep Western Boundary Current at 26.5°N" *Geophysical Research Letters,* Vol. 25, No. 13, pp. 2249–52, July 1, 1998.

Chapter Opener 11: AP Photo/Suzanne Plunkett; **p. 278–279:** © D. Segar and E. Stamman Segar; **fig. 11.4:** Adapted from B. Kinsman 1965, Wind Waves: Their Generation and Propagation on the Ocean Surface, Prentice Hall, Englewood Cliffs, NJ; **fig. 11.6d:** Adapted from H.U. Sverdrup, M.W. Johnson and R.H. Fleming, 1942, The Oceans, Prentice Hall, Englewood Cliffs, NJ; **fig. 11.7c:** Courtesy of Eugene L Griessel; Official SAAF Photograph; **fig. 11.11:** Mediacolor's/Alamy; **fig. 11.16:** Adapted from H.V. Thurman, 1988, Introductory Oceanography, Charles E. Merrill, Columbus, OH; **fig. 11.19a:** William Ervin/Science Photo Library/Photo Researchers, Inc.; **fig. 11.19b:** Douglas Faulkner/Photo Researchers, Inc.; **fig. 11.19c:** Scientifica/Visuals Unlimited; **fig. 11.19d:** John Cleare/ Mountain Camera/Alamy; **fig. 11.20:** © D. Segar and E. Stamman Segar; **fig. 11.22b:** NOAA.

Chapter Opener 12: (all four images) © D. Segar and E. Stamman Segar; **p. 310–11:** © D. Segar and E. Stamman Segar; **fig. 12.5:** Data from National Oceanic and Atmospheric Administration; **fig. 12.6a:** Adapted from The Times Atlas of the Oceans, 1983, Van Nostrand Reinhold, NY; **fig. 12.6b:** Adapted from The Oceanographic Atlas of the North Atlantic Ocean, Section 1: Tides and Currents, 1968, H.M. Publication No 700, Naval Oceanographic Office, Washington D.C.; **fig. 12.7:** Data from American Practical Navigator, 1958, H.M. Publication No 9, U.S. Naval Oceanographic Office, Washington D.C.; **fig. 12.12:** Adapted from D.E. Cartwright, 1969, Science Journal, 5:60–67; **fig. 12.13:** Adapted from A. Defant, 1958, Ebb and Flow, University of Michigan Press, Ann Arbor, MI; **fig. 12.14:** Data from H.A. Marmer, 1930, The Sea, Appleton and Company, New York; **fig. 12.16:** Data from National Oceanic and Atmospheric Administration; **fig. 12.17:** Glenn Oliver/Visuals Unlimited; **fig. 12.18:** Data from National Oceanic and Atmospheric Administration; **fig. 12.19:** Data from National Oceanic and Atmospheric Administration; **fig. 12.21:** Attar Maher/Corbis Sygma.

Chapter Opener 13: © D. Segar and E. Stamman Segar; **p. 336–337:** © D. Segar and E. Stamman Segar; **fig. 13.2a:** © Michael Collier/DRK Photo; **fig. 13.2b:** © D. Segar and E. Stamman Segar; **fig. 13.2c:** © D. Segar and E. Stamman Segar; **fig. 13.2d:** © D. Segar and E. Stamman Segar; **fig. 13.2e:** © D. Segar and E. Stamman Segar; **fig. 13.2f:** © Gregory S. Boland, GB Photographic Services; **fig. 13.2g:** © D. Segar and E. Stamman Segar; **fig. 13.2h:** Martin G. Miller/Visuals Unlimited; **fig. 13.2i:** Larry Cameron/Photo Researchers, Inc.; **fig. 13.2j:** © D. Segar and E. Stamman Segar; **fig. 13.2k:** © D. Segar and E. Stamman Segar; **fig. 13.2l:** © D. Segar and E. Stamman Segar; **fig. 13.2m:** Abram Schoenfeld/Photo Researchers, Inc.; **fig. 13.2n:** © Gregory S. Boland, GB Photographic Services; **fig. 13.2o:** © D. Segar and E. Stamman Segar; **fig. 13.3:** adapted from Fig. 2 in Moore J.G., W.R. Normack and R.T. Holcomb, 1994, "Giant Hawaiian Underwater Landslides," *Science* 264:46 (1994)/ U.S. Geological Survey; **fig. 13.4a:** © D. Segar and E. Stamman Segar; **fig. 13.5a:** NASA; **fig. 13.5b:** Adapted from R.C. Scott, 1996, Introduction to Physical Geography, West Publishing Company, St. Paul, MN; **fig. 13.6a:** © D. Segar and E. Stamman Segar; **fig. 13.6b:** © D. Segar and E. Stamman Segar; **fig. 13.7:** Base image courtesy of JSC/NASA; **fig. 13.8b:** © D. Segar and E. Stamman Segar; **fig. 13.9:** © D. Segar and E. Stamman Segar; **fig. 13.10c:** © D. Segar and E. Stamman Segar; **fig. 13.10d:** © D. Segar and E. Stamman Segar; **fig. 13.12a:** © D. Segar and E. Stamman Segar; **fig. 13.12b:** © D. Segar and E. Stamman Segar; **fig. 13.13a:** © D. Segar and E. Stamman Segar; **fig. 13.13b:** © D. Segar and E. Stamman Segar; **fig. 13.13c:** © D. Segar and E. Stamman Segar; **fig. 13.13d:** © D. Segar and E. Stamman Segar; **fig. 13.14a:** © D. Segar and E. Stamman Segar; **fig. 13.14b**:© D. Segar and E. Stamman Segar; **fig 13.16:** Adapted from D.L. Inman and J.D. Frautschy, 1965, *Littoral Processes and the Development of Shorelines*, Coastal Engineering, Santa Barbara Specialty Conference; base image courtesy of NASA; **fig. 13.19a:** © D. Segar and E. Stamman Segar; **fig. 13.19b**:© D. Segar and E. Stamman Segar; **fig. 13.19c:** Marc Epstein/Visuals Unlimited; **fig. 13.21a:** © D. Segar and E. Stamman Segar; **fig. 13.21b:** © D. Segar and E. Stamman Segar; **fig. 13.22b:** © John S. Shelton; **fig. 13.23b:** Berno Wittich/Visuals Unlimited; **fig. 13.24b:** Frank Hanna/Visuals Unlimited; **fig. 13.25b:** © John S. Shelton.

Chapter Opener 14: © D. Segar and E. Stamman Segar; **p. 370–371:** © D. Segar and E. Stamman Segar; **fig. 14.3a:** Andre Seale/Alamy; **fig. 14.3b:** © D. Segar and E. Stamman Segar; **fig. 14.6:** Data from H.U. Sverdrup, M.W. Johnson and R.H. Fleming, 1942, The Oceans, Prentice Hall, Englewood Cliffs, NJ; **fig. 14.7:** Data from H.U. Sverdrup, M.W. Johnson and R.H. Fleming, 1942, The Oceans, Prentice Hall, Englewood Cliffs, NJ; **fig. 14.10:** Adapted from I. Everson, 1977, "The Living Resources of the Southern Ocean," U.N. Development Program, Rome; **fig. 14.12a:** Gene Carl Feldman, NASA/GSFC; **fig. 14.12b:** Gene Carl Feldman/NASA, GSFC; **fig. 14.12c:** Gene Carl Feldman/NASA, GSFC; fig. **14.12d:** Gene Carl Feldman/NASA, GSFC; **fig. 14.14:** D. P. Wilson/Science Source/Photo Researchers, Inc.; **fig. 14.15 a&b:** D. P. Wilson/Science Source/Photo Researchers, Inc.; **fig. 14.15c:** Natural History Museum, London; **fig. 14.15d:** M. Abbey/Photo Researchers, Inc.; **fig. 14.17a:** Adapted from I. Valiela, 1984, Marine Ecological Processes, Springer-Verlag, NY; **fig. 14.17b:**

Adapted from E.M. Hulburt, 1962, Limnology and Oceanography 7:307–15; **fig. 14.18a:** D. P. Wilson, Science Source/Photo Researchers, Inc.; **fig. 14.18b:** Flip Nicklin/Minden Pictures; **fig. 14.18c:** M. I. Walker/Photo Researchers, Inc.; **fig. 14.18d:** © Robert Brons, Biological Photo Service; **fig. 14.19a:** © D. Segar and E. Stamman Segar; **fig. 14.19b:** © D. Segar and E. Stamman Segar; **fig. 14.19c:** Dave B. Fleetham/Visuals Unlimited; **fig. 14.19d:** Bob DeGoursey/Visuals Unlimited; **fig. 14.19e:** © D. Segar and E. Stamman Segar; **fig. 14.20a:** Kjell B. Sandved/Visuals Unlimited;**fig. 14.20b:** Triarch/Visuals Unlimited; **fig. 14.21a:** © D. Segar and E. Stamman Segar; **fig. 14.21b:** Jeffrey L. Rotman/Peter Arnold, Inc.; **fig. 14.21c:** Labat/Lanceau/Awuarium De La Rochelle Jacana/Photo Researchers, Inc.; **fig. 14.21d:** © D. Segar and E. Stamman Segar; **fig. 14.22a:** © 1995 Carl Roessler, Sea Images, Inc.; **fig. 14.22b:** © D. Segar and E. Stamman Segar; **fig. 14.22c:** © D. Segar and E. Stamman Segar; **fig. 14.22d:** © D. Segar and E. Stamman Segar; **fig. 14.22e:** © D. Segar and E. Stamman Segar; **fig. 14.22f:** © D. Segar and E. Stamman Segar; **fig. 14.22g:** © D. Segar and E. Stamman Segar; **fig. 14.22h:** © D. Segar and E. Stamman Segar; fig. **14.22i:** © D. Segar and E. Stamman Segar; **fig. 14.23a:** © D. Segar and E. Stamman Segar; **fig. 14.23b:** © D. Segar and E. Stamman Segar; **fig. 14.23c:** © D. Segar and E. Stamman Segar; **fig. 14.23d:** © D. Segar and E. Stamman Segar; **fig. 14.23e:** © D. Segar and E. Stamman Segar; **fig. 14.23f:** © D. Segar and E. Stamman Segar; **fig. 14.24a:** Adapted from *The Whale Manual,* Friends of the Earth, San Francisco, CA; **fig. 14.24b:** Stephen Frink Collection/Alamy; **fig. 14.24c:** ImageState/Alamy; **fig. 14.24d:** Bruce Coleman Inc./Alamy; **fig. 14.25a:** © D. Segar and E. Stamman Segar; **fig. 14.25b:** © D. Segar and E. Stamman Segar; **fig. 14.25c:** © D. Segar and E. Stamman Segar; **fig. 14.25d:** Dan Guravich/Photo Researchers, Inc.; **fig. 14.25e:** © Phil Degginger; **fig. 14.26a:** © D. Segar and E. Stamman Segar; **fig. 14.26b:** © D. Segar and E. Stamman Segar; **fig. 14.26c:** © D. Segar and E. Stamman Segar; **fig. 14.26d:** © D. Segar and E. Stamman Segar; **fig. 14.26e:** Courtesy of Dr. Carol S. Hemminger; **fig. 14.27a:** Rod Planck/Photo Researchers, Inc.; **fig. 14.27b:** © D. Segar and E. Stamman Segar; fig. **14.27c:** © D. Segar and E. Stamman Segar; **fig. 14.27d:** © D. Segar and E. Stamman Segar.

Chapter Opener 15: © D. Segar and E. Stamman Segar; **p. 410–411:** © D. Segar and E. Stamman Segar; **p. 411:** *The Sea Around Us* by Rachel Carson, Copyright ©1950 by Rachel L. Carson, Copyright ©1972 by Paul Brooks, Used by permission of Frances Collin, Trustee u-w-o Rachel L. Carson; **fig. 15.3a:** © D. Segar and E. Stamman Segar; **fig. 15.3b:** © D. Segar and E. Stamman Segar; **fig. 15.3c:** © D. Segar and E. Stamman Segar; **fig. 15.3d:** © D. Segar and E. Stamman Segar; **fig. 15.3e:** © D. Segar and E. Stamman Segar; **fig. 15.4:** Ring Generation and Evolution Observed with SeaWiFS, Journal of Physical Oceanography, 32,1058–1074; **fig. 15.5:** Provided by Gene Feldman, NASA/GSFC; **fig. 15.6a:** Provided by Gene Carl Feldman, NASA/GSFC; **fig. 15.6b:** Courtesy of NASA/GSFC; **fig. 15.7:** Adapted from M.G. Gross and E. Gross, 1996, Oceanography, 7th Edition, Prentice Hall, NJ; **fig. 15.9:** Provided by Gene Carl Feldman, NASA/GSFC; **fig. 15.10:** Adapted from H.G. Bigelow, 1927, Bulletin of U.S. Bureau of Fisheries 40/2; **fig. 15.11a:** © D. Segar and E. Stamman Segar**; fig. 15.11b:** © D. Segar and E. Stamman Segar; **fig. 15.11c:** © D. Segar and E. Stamman Segar; **fig. 15.11d:** © D. Segar and E. Stamman Segar; **fig. 15.14:** National Oceanic and Atmospheric Administration; **fig. 15.15:** Data from J.H. Ryther, 1969, Science 166:72–76; **fig. 15.16:** © D. Segar and E. Stamman Segar; **fig. 15.18:** JSC/NASA; **fig. 15.19:** from P. R. Pinet, *Invitation to Oceanography*, 2006: Jones and Bartlett Publishers, Sudbury, MA, www.jbpub.com. Reprinted with permission**; fig. 15.20:** Partially adapted from J. Pethick, 1984, An Introduction to Coastal Geomorphology, Edward Arnold, Baltimore, MD; **fig. 15.22:** Courtesy of Landsat MSS, U.S. Geological Survey; **fig. 15.23:** Adapted from A. Remane, 1971, in Biology of Brackish Waters, A. Remane and C. Schlieper, eds., E. Schweizerbartsche Verlagsbuchhandlung, Stuttgart.

Chapter Opener 16: (all six images) © D. Segar and E. Stamman Segar **p. 444–445:** © D. Segar and E. Stamman Segar; **fig. 16.1a:** Peter Parks/Oxford Scientific Films, *Animals Animals*; **fig. 16.1b:** © D. Segar and E. Stamman Segar; **fig. 16.1c:** © D. Segar and E. Stamman Segar; **fig. 16.2a:** © D. Segar and E. Stamman Segar; **fig. 16.2b:** © D. Segar and E. Stamman Segar; **fig. 16.3 (inset):** © D. Segar and E. Stamman Segar; **fig. 16.5a:** © D. Segar and E. Stamman Segar; **fig. 16.5b:** © D. Segar and E. Stamman Segar; **fig. 16.5c:** © D. Segar and E. Stamman Segar; **fig. 16.5d:** © D. Segar and E. Stamman Segar; **fig. 16.6a:** © D. Segar and E. Stamman Segar; **fig. 16.6b:** © D. Segar and E. Stamman Segar; **fig. 16.6c:** © D. Segar and E. Stamman Segar; **fig. 16.6d:** © D. Segar and E. Stamman Segar; **fig. 16.7a:** © D. Segar and E. Stamman Segar; **fig. 16.7b:** © D. Segar and E. Stamman Segar; **fig. 16.7c:** © D. Segar and E. Stamman Segar; **fig. 16.7d:** © D. Segar and E. Stamman Segar; **fig. 16.7e:** © D. Segar and E. Stamman Segar; **fig. 16.7f:** © D. Segar and E. Stamman Segar; **fig. 16.7g:** © D. Segar and E. Stamman Segar; **fig. 16.7h:** © D. Segar and E. Stamman Segar; **fig. 16.7i:** © D. Segar and E. Stamman Segar; **fig. 16.8a:** © D. Segar and E. Stamman Segar; **fig. 16.8b:** © D. Segar and E. Stamman Segar; **fig. 16.8c:** © D. Segar and E. Stamman Segar; **fig. 16.8d:** © D. Segar and E. Stamman Segar; **fig. 16.9a:** © D. Segar and E. Stamman Segar; **fig. 16.9b:** © D. Segar and E. Stamman Segar; **fig. 16.9c:** © D. Segar and E. Stamman Segar; **fig. 16.9d:** © D. Segar and E. Stamman Segar; **fig. 16.10a:** © D. Segar and E. Stamman Segar; **fig. 16.10b:** © D. Segar and E. Stamman Segar; **fig. 16.10c:** © D. Segar and E. Stamman Segar; fig. **16.10d:** © D. Segar and E. Stamman Segar; **fig. 16.11b:** © D. Segar and E. Stamman Segar; **fig. 16.11c:** © D. Segar and E. Stamman Segar; **fig. 16.11d:** © D. Segar and E. Stamman Segar; **fig. 16.12a:** © D. Segar and E. Stamman Segar; **fig. 16.13a:** © D. Segar and E. Stamman Segar; **fig. 16.13b:** © D. Segar and E. Stamman Segar; **fig. 16.13c:** © D. Segar and E. Stamman Segar; **fig. 16.13d:** © D. Segar and E. Stamman Segar; **fig. 16.13e:** © D. Segar and E. Stamman Segar; **fig. 16.14:** © Gregory S. Boland, GB Photographic Services; **fig. 16.15a:** © D. Segar and E. Stamman Segar; **fig. 16.15b:** © D. Segar and E. Stamman Segar; **fig. 16.15c:** © D. Segar and E.

Stamman Segar; **fig. 16.15d:** © D. Segar and E. Stamman Segar; **fig. 16.15e:** © D. Segar and E. Stamman Segar; **fig. 16.15f:** © D. Segar and E. Stamman Segar; **fig. 16.15g:** © D. Segar and E. Stamman Segar; **fig. 16.15h:** © D. Segar and E. Stamman Segar; **fig. 16.15i:** © D. Segar and E. Stamman Segar; **fig. 16.15j:** © D. Segar and E. Stamman Segar; **fig. 16.15k:** © D. Segar and E. Stamman Segar; **fig. 16.15l:** © D. Segar and E. Stamman Segar; **fig. 16.16a:** © D. Segar and E. Stamman Segar; **fig. 16.16b:** © D. Segar and E. Stamman Segar; **fig. 16.16c:** © D. Segar and E. Stamman Segar; **fig. 16.16c** (inset): © D. Segar and E. Stamman Segar; **fig. 16.16d:** © D. Segar and E. Stamman Segar; **fig. 16.16e:** © D. Segar and E. Stamman Segar; **fig. 16.16f:** © D. Segar and E. Stamman Segar; **fig. 16.16f** (inset): © D. Segar and E. Stamman Segar; **fig. 16.16g:** © D. Segar and E. Stamman Segar; **fig. 16.17a:** © D. Segar and E. Stamman Segar; **fig. 16.17b:** © D. Segar and E. Stamman Segar; **fig. 16.17c:** © D. Segar and E. Stamman Segar; **fig. 16.17d:** © D. Segar and E. Stamman Segar; **fig. 16.17e:** © D. Segar and E. Stamman Segar; **fig. 16.18a:** © D. Segar and E. Stamman Segar; **fig. 16.18b:** © D. Segar and E. Stamman Segar; **fig. 16.18c:** © D. Segar and E. Stamman Segar; **fig. 16.18d:** © D. Segar and E. Stamman Segar; **fig. 16.18e:** © D. Segar and E. Stamman Segar; **fig. 16.18f:** © D. Segar and E. Stamman Segar; **fig. 16.19a:** © D. Segar and E. Stamman Segar; **fig. 16.19b:** © D. Segar and E. Stamman Segar; **fig. 16.19c:** © D. Segar and E. Stamman Segar; **fig. 16.19d:** © D. Segar and E. Stamman Segar; **fig. 16.19e:** © D. Segar and E. Stamman Segar; **fig. 16.19f:** © D. Segar and E. Stamman Segar; **fig. 16.20:** © D. Segar and E. Stamman Segar; **fig. 16.21a:** © D. Segar and E. Stamman Segar; **fig. 16.21b:** © D. Segar and E. Stamman Segar; **fig. 16.21c:** © D. Segar and E. Stamman Segar; **fig. 16.22:** (all seven images) © D. Segar and E. Stamman Segar; Concept adapted from P.W. Webb, 1984, Scientific American 251/1:72–82; **fig. 16.23a:** © D. Segar and E. Stamman Segar; **fig. 16.23c:** © D. Segar and E. Stamman Segar; **fig. 16.24:** (all five images) © D. Segar and E. Stamman Segar; Concept adapted from H.V. Thurman, 1987, Essentials of Oceanography, Charles E. Merrill, Columbus, OH; **fig. 16.25a:** © D. Segar and E. Stamman Segar; **fig. 16.25b:** © D. Segar and E. Stamman Segar; **fig. 16.25c:** © D. Segar and E. Stamman Segar; **fig. 16.25d:** © D. Segar and E. Stamman Segar; **fig. 16.25e:** © D. Segar and E. Stamman Segar; **fig. 16.25f:** © D. Segar and E. Stamman Segar; **fig. 16.25g:** © D. Segar and E. Stamman Segar; **fig. 16.26a:** © D. Segar and E. Stamman Segar; **fig. 16.26b:** © Greg Boland, GB Photographic Services; **fig. 16.27a:** © D. Segar and E. Stamman Segar; **fig. 16.27b:** © D. Segar and E. Stamman Segar; **fig. 16.27c:** © D. Segar and E. Stamman Segar; **fig. 16.27d:** © D. Segar and E. Stamman Segar; **fig. 16.27e:** © D. Segar and E. Stamman Segar; **fig. 16.27f:** © D. Segar and E. Stamman Segar; **fig. 16.27g:** © D. Segar and E. Stamman Segar; **fig. 16.27h:** © D. Segar and E. Stamman Segar; **fig. 16.27i:** © D. Segar and E. Stamman Segar; **fig. 16.27j:** © D. Segar and E. Stamman Segar; **fig. 16.27k:** © D. Segar and E. Stamman Segar; **fig. 16.27l:** © D. Segar and E. Stamman Segar; **fig. 16.27l** (inset): © D. Segar and E. Stamman Segar; **fig. 16.27m:** © D. Segar and E. Stamman Segar; **fig. 16.28:** Adapted from H. Jones, 1968, Fish Migration, Edward Arnold Ltd., Baltimore, MD; **fig. 16.29a:** Adapted from C.E. Bond, 1979, Biology of Fishes, Saunders Publishing, Philadelphia, PA; **fig. 16.29b:** Adapted from P.A. Larkin, 1977, in Fish Population Dynamics, J.A. Gulland ed., John Wiley, London; and D.H. Cushing, 1982, Climate and Fisheries, Academic Press, New York, NY; **fig. 16.30:** Adapted from A.C. Duxbury and A.B. Duxbury, 1994, An Introduction to the World's Oceans, 4th Edition, William C. Brown, Dubuque, IA; **fig. 16.31a:** © D. Segar and E. Stamman Segar; **fig. 16.31b:** © D. Segar and E. Stamman Segar; **fig. 16.31c:** © D. Segar and E. Stamman Segar; **fig. 16.31d:** © D. Segar and E. Stamman Segar; **fig. 16.32:** Adapted from Dogiel, Petrushevski and Polyanski, 1970, Parasitology, THF Publications, NJ; **fig. 16.33a:** © D. Segar and E. Stamman Segar; **fig. 16.33b:** © D. Segar and E. Stamman Segar; **fig. 16.33c:** © D. Segar and E. Stamman Segar; **fig. 16.33d:** © D. Segar and E. Stamman Segar; **fig. 16.33e:** © D. Segar and E. Stamman Segar; **fig. 16.33f:** © D. Segar and E. Stamman Segar; **fig. 16.33g:** © D. Segar and E. Stamman Segar; **fig. 16.33h:** © D. Segar and E. Stamman Segar; **fig. 16.33i:** © D. Segar and E. Stamman Segar; **fig. 16.33j:** © D. Segar and E. Stamman Segar; **fig. 16.33k:** © D. Segar and E. Stamman Segar; **fig. 16.33l:** © D. Segar and E. Stamman Segar; **fig. 16.33m:** © D. Segar and E. Stamman Segar; **fig. 16.33n:** © D. Segar and E. Stamman Segar; **fig. 16.33o:** © D. Segar and E. Stamman Segar; **fig. 16.33p:** © D. Segar and E. Stamman Segar; fig. **16.33q:** © D. Segar and E. Stamman Segar; fig. **16.33r:** © D. Segar and E. Stamman Segar; **fig. 16.33s:** © D. Segar and E. Stamman Segar; **fig. 16.34a:** © D. Segar and E. Stamman Segar; **fig. 16.34b:** © D. Segar and E. Stamman Segar; **fig. 16.34c:** © D. Segar and E. Stamman Segar; **fig. 16.34d:** © D. Segar and E. Stamman Segar; **fig. 16.34e:** © D. Segar and E. Stamman Segar.

Chapter Opener 17: © D. Segar and E. Stamman Segar; **p. 496–497:** © D. Segar and E. Stamman Segar; **fig. 17.2:** Adapted from M. Lerman, 1986, Marine Biology: Environment, Diversity, and Ecology, Benjamin-Cummings, Reading, MA; and V.J. Chapmen, ed. 1977, Wet Coastal Systems, Elsevier Science Publishing, New York, NY; **fig. 17.4a:** © D. Segar and E. Stamman Segar; **fig. 17.4b:** © D. Segar and E. Stamman Segar; **fig. 17.4c:** © D. Segar and E. Stamman Segar; **fig. 17.4d:** © D. Segar and E. Stamman Segar; **fig. 17.4e:** © D. Segar and E. Stamman Segar; **fig. 17.4f:** © 1980 John Lidington/Photo Researchers, Inc.; **fig. 17.5:** © D. Segar and E. Stamman Segar; **fig. 17.6:** © D. Segar and E. Stamman Segar; **fig. 17.7a,b:** Adapted from J.A. Estes and J.F. Palmisano, 1974 Science 185:1058–60; **fig. 17.7c:** © D. Segar and E. Stamman Segar; **fig. 17.8a:** © D. Segar and E. Stamman Segar; **fig. 17.8b:** © D. Segar and E. Stamman Segar; **fig. 17.8c:** © D. Segar and E. Stamman Segar; **fig. 17.8d:** © D. Segar and E. Stamman Segar; **fig. 17.8e:** © D. Segar and E. Stamman Segar; **fig. 17.8f:** © D. Segar and E. Stamman Segar; **fig. 17.8g:** © D. Segar and E. Stamman Segar; **fig. 17.9a:** © John Gerlach, Visuals Unlimited; **fig. 17.9b:** © D. Segar and E. Stamman Segar; **fig. 17.9c:** Adapted from J.A. Estes and J.F. Palmisano, 1974 Science 185:1058–60; **fig. 17.12a:** © D. Segar and E. Stamman Segar; **fig. 17.12b:** © D. Segar and E. Stamman Segar; **fig. 17.12c:** © D. Segar and E. Stamman Segar; **fig. 17.12d:** © D. Segar and E. Stamman Segar; **fig. 17.12e:** © D. Segar and E. Stamman Segar; **fig. 17.12f:** © D. Segar and E. Stamman Segar; **fig. 17.12g:** © D. Segar and E. Stamman Segar; **fig. 17.13a:** Bigelow Laboratory for Ocean Sciences; **fig. 17.13b:** Peter Parks/imagequestmarine.com; **fig. 17.13c:** Daniel L. Geiger/SNAP/Alamy; **fig. 17.13d:** Peter Parks/imagequestmarine.com; **fig. 17.16a:** Adapted from H.W. Jannasch and M.J. Mottle, 1985, Science 229:717–25; **fig. 17.16b:** Adapted from R.M. Haymon and K.C. Macdonald, 1985, American Scientist 73/5:441–49; **fig. 17.17a:** Woods Hole Oceanographic Institution; **fig. 17.17b:** Al Giddings Stock Photography.

Chapter Opener 18: (top left) AFP/Getty Images; (middle left) AFP/Getty Images; (bottom left) US Navy/Getty Images; (top right) Jeffrey L. Rotman/Corbis; (middle right) Scanpix/AP Photo; (bottom right) AFP/Getty Images; **p. 526–527:** © D. Segar and E. Stamman Segar; **fig. 18.3a:** Adapted from P.E. Smith, 1978, Rapp. P. V. Cons. Int. Explor. Mer., 173:117–27; **fig. 18.3b,c:** Data from National Oceanic and Atmospheric Administration; **fig. 18.4a:** ©1995 Jon Bertsch/Visual Unlimited; **fig. 18.4b:** © D. Segar and E. Stamman Segar; **fig. 18.4c:** © D. Segar and E. Stamman Segar; **fig. 18.5:** Data from U.S. Department of Energy; **fig. 18.6a:** © D. Segar and E. Stamman Segar; **fig. 18.6b:** © D. Segar and E. Stamman Segar; **fig. 18.6c:** Photographer David Predeger; Photo courtesy of BP Exploration (Alaska) Inc.; **fig. 18.7a:** From U.S. Department of Commerce Public Bulletin 188717; **fig. 18.7b:** Adapted from H.V. Thurman, 1994, Introductory Oceanography, 7th Edition, Macmillan Publishing Company, New York, NY; **fig. 18.7c:** Fred Lyon/Photo Researchers, Inc.; **fig. 18.8a:** Adapted from Sea Frontiers, 1987, 33(4):260–261; produced by the National Climatic Data Center with support from the U.S. Department of Energy; **fig. 18.8b:** Adapted from A.C. Duxbury and A.B. Duxbury, 1994, An Introduction to the World's Oceans, 4th Edition, William C. Brown, Dubuque, IA; **fig. 18.9c,e:** Courtesy, Lockheed Martin Missiles and Space.

Chapter Opener 19: © D. Segar and E. Stamman Segar; **p. 548–549:** © D. Segar and E. Stamman Segar; **fig. 19.1:** from F.H. Nichols et al., "The Modification of an Estuary," *Science,* 231:567 (1986) U.S. Geological Survey; **fig. 19.2a:** © Jeffrey Sylvester, FPG International; **fig. 19.2b:** © D. Segar and E. Stamman Segar; **fig. 19.3a:** © D. Segar and E. Stamman Segar; **fig. 19.3b:** Calvin Larsen/Photo Researchers, Inc.; **fig. 19.4a:** © D. Segar and E. Stamman Segar; **fig. 19.4b:** AFP/Getty Images; **fig. 19.4c:** Simon Fraser/Photo Researchers, Inc.; **fig. 19.4d:** John Gaps III/AP Photo; **fig. 19.4e:** Rob Stapleton/AP Photo; **fig. 19.4f:** AFP/Getty Images; **fig. 19.5a:** Tom Myers/Photo Researchers, Inc.; **fig. 19.5b:** Courtesy of U.S. Coast Guard; **fig. 19.6b:** Adapted from U.S. Commission on Ocean Policy. An Ocean Blueprint for the 21st Century. Final Report.Washington, D.C., 2004; **fig. 19.7b,c:** Data from National Oceanic and Atmospheric Administration; **fig. 19.8a:** Lawrence Migdale/Photo Researchers, Inc.; **fig. 19.10a:** Adapted from D. Segar, 1988, Romberg Tiburon Centers, Technical Report No. 10, San Francisco State University, Tiburon, California; **fig. 19.10b:** Data from U.S. Environmental Protection Agency; **fig. 19.11a:** Adapted from W.E. Smith and A.M. Smith, 1975, Minamata, Holt Rinehart and Winston, New York, NY; **fig. 19.11b,c:** Adapted from S.A.Gerlach, 1981, Marine Pollution, Diagnosis and Therapy, Springer-Verlag, New York, NY; **fig. 19.12b:** © D. Segar and E. Stamman Segar; **fig. 19.13a:** National Marine Mammal Laboratory/NOAA; **fig. 19.13b:** © Fred Bavendam, Peter Arnold, Inc.; **fig. 19.14a:** Data from Y. Yablokov et al., 1993, "Facts and Problems Related to Radioactive Waste Disposal in Seas Adjacent to the Territory of the Russian Federation," Office of the President of the Russian Federation, Moscow; **fig. 19.14b,c:** Adapted from New York Times April 27, C8, 1993.

Every effort has been made to contact the copyright holders of the material used in *Introduction to Ocean Science*. Please contact us with any update information.

Index

Bold entries with a letter followed by a number are grid references to the cited locations on the inside back cover map. Bold italic entries with a number followed by a **G** indicate the page number of the term's definition in the glossary. Page numbers in *italics* refer to figures and illustrations.

180
A
150
B
120
C
90
D
60
E
30
F
1
60
2
30
3
0
4
30
5
60
6
Arctic Ocean
Chukchi Sea
Bering Strait
Beaufort Sea
Mackenzie River
Baffin Bay
Greenland
Nansen
Greenland Sea
Iceland
Bering Sea
AK
Fairbanks
Turnagain Arm
Anchorage
Homer
Valdez
Prince William Sound
Dutch Harbor
Pribilof islands
Cook Inlet
Gulf of Alaska
Aleutian islands
Scotch Cap
Amchitka island
Ketchikan
Vancouver Island
BC
Fraser River
Puget Sound
Columbia River
North America
Hudson Bay
Labrador
Labrador Sea
Canada
New Brunswick
Prince Edward Is.
St. Lawrence River
Newfoundland
Juan de Fuca Ridge
Seattle
WA
Sheridan
Portland
OR
Mt. St.Helens
Great Lakes
Hudson River
ME
MA
Nova Scotia
Bay of Fundy
Gulf of Maine
New York City
Cape Mendocino
Sacramento River
CA
Sierra Nevada Mts.
Rocky Mtns.
CO
NV
MO
PA
NY
NJ
Raritan Bay
Delaware Bay (DL)
Chesapeake Bay
San Francisco
Monterey
San Joaquin River
Los Angeles
San Diego
United States
Appalachians
MD
VA
NC
SC
Cape Hatteras
Pamlico Sound
New Orleans
MS
LA
GA
Savannah River
Bermuda
Sargasso Sea
Scot
Great Br
Ireland
Thames River
English Channel
Portugal
Sp
Gibraltar
Azores
More
Tropic of Cancer
TX
Mississippi Delta
FL
Bahamas
Miami
Florida Strait
Baja
Gulf of California
Galveston
Gulf of Mexico
Jamaica
Canary Islands
Kauai
Oahu
Molokai
Maui
Kona
Hilo
Hawaii
Mexico
Yucatan
Cayman Is.
Puerto Rico
U.S. Virgin Islands
Pacific Ocean
Belize
Caribbean Sea
Atlantic Ocean
Mid-Atlantic Ridge
Nicaragua
Aruba
Bonaire
Netherlands Antilles
Costa Rica
Panama
Orinoco River
East Pacific Rise
Galapagos Islands
Ecuador
Amazon River
St. Paul's Rocks
Peru
South America
Brazil
Andes Mountains
Polynesia
Tonga
Samoa
Cook Islands
Marquesas Islands
Tahiti
Rio de Janeiro
Tropic of Capricorn
Tonga Trench
Macdonald Seamount
Easter Islands
Rio Grande Rise
Chile
Argentina
Tristan da Cunha
Patagonia
Falkland Islands
Tierra del Fuego
South Sandwich Trench
Southern Ocean
Antarctic peninsula
Weddel Sea